Thin Layer Chromatography in Phytochemistry

CHROMATOGRAPHIC SCIENCE SERIES

A Series of Textbooks and Reference Books

Editor: JACK CAZES

Thin Layer Chromatography in Phytochemistry

Monika Waksmundzka-Hajnos
Medical University of Lublin
Lublin, Poland

Joseph Sherma
Lafayette College
Easton, Pennsylvania, U.S.A.

Teresa Kowalska
University of Silesia
Katowice, Poland

CRC Press
Taylor & Francis Group
Boca Raton London New York

CRC Press is an imprint of the
Taylor & Francis Group, an **informa** business

CRC Press
Taylor & Francis Group
6000 Broken Sound Parkway NW, Suite 300
Boca Raton, FL 33487-2742

First issued in paperback 2020

© 2008 by Taylor & Francis Group, LLC
CRC Press is an imprint of Taylor & Francis Group, an Informa business

No claim to original U.S. Government works

ISBN-13: 978-0-367-57754-4 (pbk)
ISBN-13: 978-1-4200-4677-9 (hbk)

Library of Congress Cataloging-in-Publication Data

Thin layer chromatography in phytochemistry / editors, Monika
Waksmundzka-Hajnos, Joseph Sherma, Teresa Kowalska.
      p. cm. -- (Chromatographic science series)
   Includes bibliographical references and index.
   ISBN 978-1-4200-4677-9 (hardback : alk. paper) 1. Plants--Analysis. 2. Thin
layer chromatography. I. Waksmundzka-Hajnos, Monika. II. Sherma, Joseph.
III. Kowalska, Teresa. IV. Title. V. Series.

QK865.T45 2008
572'.362--dc22   2007040781

Visit the Taylor & Francis Web site at
http://www.taylorandfrancis.com

and the CRC Press Web site at
http://www.crcpress.com

Contents

Part I

Part II

Primary Metabolites

Secondary Metabolites—Shickimic Acid Derivatives

Secondary Metabolites — Isoprenoids

Secondary Metabolites — Amino Acid Derivatives

Secondary Metabolites—Compounds Derived from Acetogenine (Acetylocoenzyme A)

Preface

Thin layer chromatography (TLC), including its counterpart high performance TLC, is one of the most important methods in the field of phytochemical analysis because of the great variety of mobile and stationary phase combinations, capillary and forced flow development methods, and detection possibilities; minimum sample preparation required because plates are not reused; high sample throughput because multiple samples can be developed on the same plate; and other advantages outlined in Chapter 1. This is the first book to be published that gives a complete description of the techniques, materials, and instrumentation of TLC and its applications to essentially all primary and secondary plant metabolites in a logically organized and comprehensive manner. We are honored that the book is part of the Chromatographic Science Series, edited by Dr. Jack Cazes and published by CRC Press/Taylor & Francis, the initial book of which was the classic and bestselling landmark *Dynamics of Chromatography: Principles and Theory* by the world renowned pioneer in the field, J. Calvin Giddings, published in 1965; this series, which now has more than 95 volumes, undoubtedly represents the most prestigious collection of reference books and textbooks on the subject of chromatography, the premier analytical method in use today.

The book is organized in two parts comprising a total of 29 chapters. Part I (10 chapters) starts with a chapter introducing the book, followed by chapters on plant materials in modern pharmacy, herbal medicines and nutritional supplements, biological activity of primary and secondary metabolites, chemosystematics, layers, capillary flow development methods, physical and chemical detection methods, biodetection methods, and forced flow development. Part II (19 chapters) reports methods for the TLC separation, identification, and quantification of particular classes of metabolites in a great variety of plant samples, starting with chapters on primary metabolites (carbohydrates; lipids; and amino acids, peptides, and proteins) and followed by those on secondary metabolites (phenols, coumarins, flavonoids, lignans, mono- and sesquiterpenes, diterpenes, triterpines, tetra- and polyterpenes, steroids, iridoids, indole alkaloids, isoquinoline alkaloids, tropane alkaloids, other biosynthetic alkaloids, polyacetylenes, and quinones).

All of the chapters were written by widely recognized experts in the fields of phytochemical analysis and TLC working in laboratories in Europe, Asia, South America, or the United States, which guarantees an authoritative, international view of the field. The authors have included in their chapters tables and figures along with the text, and a selective list of the most important up-to-date and older references, in order to present their topics in a way that would be most practical and useful for an audience of both experienced and inexperienced scientists in diverse disciplines working in all types of laboratories, e.g., academic, industrial, and government regulatory.

The book can serve as a laboratory manual, reference book, or teaching source in university undergraduate or graduate analytical chemistry and separation science courses, and we believe that in these roles it will lead to greater and more effective use of TLC in phytochemistry. We would be pleased to hear from readers of our book about how it may be helpful in their work, as well as reports of any errors they find and suggestions for changes, additions, and/or deletions if a second edition of the book is produced.

The editors thank Dr. Jack Cazes, Senior Editor Barbara Glunn, and the production staff at CRC Press/Taylor & Francis for their unrestricted support of our book project from its proposal through completion.

Monika Waksmundzka-Hajnos
Joseph Sherma
Teresa Kowalska

Editors

Monika Waksmundzka-Hajnos is professor and head of the Department of Inorganic Chemistry at Faculty of Pharmacy, Medical University of Lublin (Lublin, Poland). Her interests involve the theory and application of liquid chromatography, taking into consideration the optimization of chromatographic systems for separation of natural mixtures and plant extracts for analytical and preparative purposes—with a focus on planar chromatography. Another scientific interest of Dr. Waksmundzka-Hajnos involves the optimization of liquid–solid extraction processes of biologically active secondary metabolites from plant material and optimization of purification processes of crude plant extracts from ballast substances before HPLC or TLC. She is author and coauthor of more than 100 papers and approximately 200 conference papers. She has published review articles in journals such as *Journal of Chromatography B*, *Journal of Liquid Chromatography*, and *Research Trends* (India). She has also authored a chapter on preparative planar chromatography of plant extracts in the textbook *Preparative Layer Chromatography*, edited by Teresa Kowalska and Joseph Sherma (published as Volume 95 in the Chromatographic Science Series by CRC Press/Taylor & Francis). Professor Waksmundzka-Hajnos has received five awards from the Ministry of Health in Poland and two awards from the Polish Pharmaceutical Society for her scientific achievements.

For more than 30 years, Professor Waksmundzka-Hajnos has taught courses in inorganic chemistry for pharmacy and medical chemistry students. She has also taught courses in instrumental analysis for students of pharmacy. Over the past 15 years, Dr. Waksmundzka-Hajnos has directed programs for over 30 MSc pharmacy students involved in research on the theory and practice of different liquid chromatographic techniques. She has also supervised three PhD students researching on separation science.

Since 2005, Dr. Monika Waksmundzka-Hajnos has been a member of the editorial board of *Acta Chromatographica*, the annual periodical published by the University of Silesia (Katowice, Poland), and has also devoted her time to the development of many chromatographic and hyphenated techniques.

Joseph Sherma is the John D. and Frances H. Larkin Professor Emeritus of Chemistry at Lafayette College, Easton, Pennsylvania. Professor Sherma has taught courses in analytical chemistry for more than 40 years, was head of the chemistry department for 12 years, and continues to supervise research students at Lafayette College with support from the Camille and Henry Dreyfus Foundation Senior Scientist Mentor Program. He is the author and coauthor of over 600 papers and the author, coauthor, editor, and coeditor of over 60 books and manuals in the areas of analytical chemistry and chromatography. Professor Sherma coauthored with Bernard Fried (Kreider Professor Emeritus of Biology at Lafayette College) *Thin Layer Chromatography* (editions 1–4) and coedited with Professor Fried

The Handbook of Thin Layer Chromatography (editions 1–3), both published by
Marcel Dekker. Professor Sherma edited or coedited with Dr. Gunter Zweig 29
volumes of the *Handbook of Chromatography series* for CRC Press, coedited with
Professor Fried *Practical Thin Layer Chromatography* for CRC Press, and coedited
with Professor Teresa Kowalska *Preparative Layer Chromatography and Thin
Layer Chromatography in Chiral Separations and Analysis* for CRC Press/Taylor
& Francis. For 23 years, he served as editor for the residues and trace elements
section of the *Journal of AOAC International* and is currently a member of the
editorial advisory boards of the *Journal of Liquid Chromatography & Related
Technologies*; *Journal of Planar Chromatography—Modern TLC*; *Acta Chromatographica*; *Acta Universitatis Cibiniensis, Seria F, Chemia*; and *Current Pharmaceutical Analysis*. Since 1999, Professor Sherma has guest edited with Professor
Fried annual special issues of the *Journal of Liquid Chromatography & Related
Technologies* on all aspects of thin layer chromatography, and periodically he also
guest edits special sections of the *Journal of AOAC International* on specific subjects
of TLC. Professor Sherma also writes articles on different aspects of modern
analytical instrumentation for each issue of the *Journal of AOAC International*.
Since 1970, Professor Sherma has written the biennial review of planar chromatography for the American Chemical Society (ACS) journal *Analytical Chemistry*. He
was recipient of the 1995 ACS Award for Research at an undergraduate institution
sponsored by Research Corporation.

Teresa Kowalska is professor and head of the Department of General Chemistry
and Chromatography at the University of Silesia (Katowice, Poland). Her interests
include the physicochemical foundations of liquid chromatography and gas chromatography, with a focus on modeling of planar chromatography both in its analytical
and its preparative mode. Over the past 39 years, Dr. Kowalska has directed
programs for over 70 MSc students involved in research on the theory and practice
of different chromatographic and hyphenated techniques. She has also supervised
eight PhD students researching on separation science. Dr. Kowalska is the author of
more than 200 papers, 300 conference papers, and a vast number of book chapters
and encyclopedia entries in the field of chromatography. It is perhaps noteworthy
that Dr. Kowalska has authored (and then updated) the chapter on "Theory and
Mechanism of Thin-Layer Chromatography" for all three editions of the
Handbook of Thin-Layer Chromatography, edited by Professor J. Sherma and
Professor B. Fried, published by Marcel Dekker. She has also coedited (with
Professor Joseph Sherma) *Preparative Layer Chromatography* (published as Vol-
umes 95 and 98, respectively, in the Chromatographic Science Series by CRC
Press/Taylor & Francis).

Dr. Kowalska has been editor of *Acta Chromatographica*, the annual periodical
published by the University of Silesia (Katowice, Poland), and devoted herself to the
development of many chromatographic and hyphenated techniques, right from its
establishment in 1992.

Over the past 30 years, Dr. Kowalska has participated as an organizer (and
in recent years as a co-chairperson) for the annual all-Polish chromatographic

symposium with international participation, uninterruptedly held each year (since 1977) in the small mountain resort town of Szczyrk in South Poland. Integration of an international community of chromatographers through these meetings has been regarded by Dr. Kowalska as a specific, yet important contribution to chromatography.

Contributors

Anita Ankli
CAMAG Laboratory
Muttenz, Switzerland

Simla Basar
Institut für Experimentelle und Klinische
   Toxikologie
Universitätskrankenhaus Eppendorf
Hamburg, Germany

Ravi Bhushan
Department of Chemistry
Indian Institute of Technology Roorkee
Roorkee, Uttaranchal, India

George Britton
School of Biological Sciences
University of Liverpool
Liverpool, UK

Ioanna Chinou
Division of Pharmacognosy and
   Chemistry of Natural Products
School of Pharmacy
University of Athens
Athens, Greece

Lars P. Christensen
Department of Food Science
University of Aarhus
Research Centre Aarslev
Aarslev, Denmark

Thales R. Cipriani
Departamento de Bioquímica e Biologia
   Molecular
Universidade Federal do Paraná
Curitiba, Paraná, Brasil

Laurie Dinan
Laboratoire de Protéines: Biochimie
   Structurale et Fonctionnelle
Université Pierre et Marie Curie
Paris, France

Lauro M. de Souza
Departamento de Bioquímica e Biologia
   Molecular
Universidade Federal do Paraná
Curitiba, Paraná, Brasil

Tadeusz H. Dzido
Department of Physical Chemistry
Medical University of Lublin
Lublin, Poland

Jolanta Flieger
Department of Analytical Chemistry
Medical University of Lublin
Lublin, Poland

Kazimierz Głowniak
Department of Pharmacognosy
Medical University of Lublin
Lublin, Poland

Michał Ł. Hajnos
Department of Pharmacognosy
Medical University of Lublin
Lublin, Poland

Juraj Harmatha
Academy of Sciences of the Czech
   Republic
Institute of Organic Chemistry and
   Biochemistry
Prague, Czech Republic

Mirosław A. Hawryl
Department of Inorganic Chemistry
Medical University of Lublin
Lublin, Poland

Peter John Houghton
Department of Pharmacy
King's College London
London, UK

Marcello Iacomini
Departamento de Bioquímica
  e Biologia Molecular
Universidade Federal do Paraná
Curitiba, Paraná, Brasil

Henrik B. Jakobsen
BYRIAL A/S
Lejre, Denmark

Ivona Jasprica
Department of Medicinal Chemistry
Faculty of Pharmacy and
  Biochemistry
University of Zagreb
Zagreb, Croatia

Ireneusz Kapusta
Institute of Soil Science and Plant
  Cultivation
State Research Institute
Pulawy, Poland

Angelika Koch
Frohme Apotheke
Hamburg, Germany

Teresa Kowalska
Institute of Chemistry
Silesian University of Silesia
Katowice, Poland

Rene Lafont
Laboratoire de Protéines: Biochimie
  Structurale et Fonctionnelle
Université Pierre et Marie Curie
Paris, France

Željan Maleš
Department of Pharmaceutical Botany
Faculty of Pharmacy and Biochemistry
University of Zagreb
Zagreb, Croatia

Grażyna Matysik
Department of Chemistry
Medical University of Lublin
Lublin, Poland

Anna Matysik-Woźniak
II Clinic of Ophthalmology
Medical University of Lublin
Lublin, Poland

Marica Medić-Šarić
Department of Medicinal Chemistry
Faculty of Pharmacy and Biochemistry
University of Zagreb
Zagreb, Croatia

Emil Mincsovics
OPLC-NIT Ltd.
Budapest, Hungary

Svetlana Momchilova
Institute of Organic Chemistry
Bulgarian Academy of Sciences
Sofia, Bulgaria

Ágnes M. Móricz
Department of Chemical Technology
  and Environmental Chemistry
L. Eotvos University
Budapest, Hungary

Ana Mornar
Department of Medicinal Chemistry
Faculty of Pharmacy and Biochemistry
University of Zagreb
Zagreb, Croatia

Tomasz Mroczek
Department of Pharmacognosy
Medical University of Lublin
Lublin, Poland

Boryana Nikolova-Damyanova
Institute of Organic Chemistry
Bulgarian Academy of Sciences
Sofia, Bulgaria

Wieslaw Oleszek
Institute of Soil Science and Plant
  Cultivation
State Research Institute
Pulawy, Poland

Anna Oniszczuk
Department of Inorganic Chemistry
Medical University of Lublin
Lublin, Poland

Lubomír Opletal
Charles University of Prague
Faculty of Pharmacy in Hradec Králové
Hradec Králové, Czech Republic

Péter G. Ott
Department of Plant Pathophysiology
Plant Protection Institute
Hungarian Academy of Science
Budapest, Hungary

Anna Petruczynik
Department of Inorganic Chemistry
Medical University of Lublin
Lublin, Poland

Eike Reich
CAMAG Laboratory
Muttenz, Switzerland

Rita Richter
Institute of Organic Chemistry
University of Hamburg
Hamburg, Germany

Guilherme L. Sassaki
Departamento de Bioquímica e Biologia
  Molecular
Universidade Federal do Paraná
Curitiba, Paraná, Brasil

Joseph Sherma
Department of Chemistry
Lafayette College
Easton, Pennsylvania

Agnieszka Skalska-Kamińska
Department of Chemistry
Medical University of Lublin
Lublin, Poland

Krystyna Skalicka-Woźniak
Department of Pharmacognosy
Medical University of Lublin
Lublin, Poland

Helena Sovová
Institute of Chemical Process
  Fundamentals
Academy of Sciences of the Czech
  Republic
Prague, Czech Republic

Bernd Spangenberg
University of Applied Sciences
  Offenburg
Offenburg, Germany

Anna Stochmal
Institute of Soil Science and Plant
  Cultivation
State Research Institute
Pulawy, Poland

Tomasz Tuzimski
Department of Physical Chemistry
Medical University of Lublin
Lublin, Poland

Ernő Tyihák
Department of Plant Pathophysiology
Biochemical Group
Plant Protection Institute
Hungarian Academy of Science
Budapest, Hungary

Monika Waksmundzka-Hajnos
Department of Inorganic Chemistry
Medical University of Lublin
Lublin, Poland

Jarosław Widelski
Department of Pharmacognosy
Medical University of Lublin
Lublin, Poland

Valeria Widmer
CAMAG Laboratory
Muttenz, Switzerland

Magdalena Wójciak-Kosior
Department of Chemistry
Medical University of Lublin
Lublin, Poland

Grażyna Zgórka
Department of Pharmacognosy
Medical University of Lublin
Lublin, Poland

Christian Zidorn
Institute of Pharmacy
University of Innsbruck
Innsbruck, Austria

Part I

1 Overview of the Field of TLC in Phytochemistry and the Structure of the Book

Monika Waksmundzka-Hajnos, Joseph Sherma, and Teresa Kowalska

CONTENTS

1.1 SURVEY OF PHYTOCHEMISTRY

Phytochemistry is a broad area, generally termed "plant chemistry." Investigations in the field of phytochemistry are important for numerous research disciplines, such as plant physiology, plant biochemistry, chemosystematics (which is often referred to as chemotaxonomy), plant biotechnology, and pharmacognosy.

Plant physiology focuses on the life processes occurring in plants. Especially important are the investigations on the influence of various external factors, such as ultraviolet–visible (UV–Vis) radiation, temperature, the nature of soil, the climate, etc., on the composition of active compounds contained in plants. One part of this discipline is known as allelopathy. Within the framework of allelopathy, the responses of the plant organisms to external pathological factors (e.g., environmental pollution, the presence of pathogens, insects, etc.) are investigated.

Plant biochemistry focuses on biochemical transformations that play a fundamental role in the biosynthesis of active compounds contained in plants, which are referred to as primary and secondary metabolites.

Chemosystematics involves the classification of plants on the basis of their biochemistry and chemistry. It proves to be of special importance when searching for and collecting floral specimens. Within the framework of chemosystematics, the relations are investigated between the classes of plants and the occurrence of the specific substances or substance groups in the plant tissues.

The most important application of phytochemical investigation methods is to the field of pharmacognosy. Pharmacognosy is a part of the pharmaceutical sciences and is focused on natural products (mainly on plant materials) and the components thereof that show biological activity and are, therefore, used in therapy.

The history of phytotherapy is almost as long as the history of civilization. The term "pharmacognosy" has been in use for little more than a century, but its foundations were laid out by early civilizations. The Assyrian, Egyptian, Chinese, and Greek records of great antiquity make reference to the nature and use of herbs and herbal drugs. Knowledge of medicinal plants spread in West Europe and then in the whole Western World, to a large extent through the monasteries and their schools of medicine. In 16th century, early botanists published herbals—usually illustrated with the woodcut pictures—describing the nature and use of an increasing number of plants. In modern science, phytotherapy appeared in the 19th century, when the first biologically active compounds (basically alkaloids) were isolated from the plant material (e.g., morphine, strychnine, narcotine, caffeine, etc.) The golden age of phytotherapy lasted until 1935, when the first sulfonamides and then antibiotics were synthesized and used in therapy. Then the age of chemotherapy began. However, it is a widely recognized fact that numerous synthetic drugs exert—along with a positive therapeutic effect—also harmful and often irreversible side effects. To the contrary, in the plant world, one very often encounters strongly active substances coexisting with the other compounds that mitigate their negative side effects. Because of this, in recent years a return to phytotherapy has been observed. This return has further been spurred by an appeal of the World Health Organization to screen plant material for the presence of biologically active compounds contained therein and exerting, e.g., a well pronounced anticancer activity. It is firmly believed that a great, yet still not fully revealed, therapeutic potential exists in plants, because so far only a few percent out of 250,000 plant species have been investigated with regard to their usefulness in medicine.

Nowadays, medicines of natural origin are appreciated for their high effectiveness and low toxicity, and they are the widely used commercial products. The market value of herbal preparations selling in United States alone is estimated at several dozen million dollars per year. Plant materials are often obtained from natural sources, although many of the medicinal plants are also cultivated. From these facts, it is clear that there is a high and increasing need for efficient purity control of plant material, and further for the assessment of their identity and chemical composition, in order to obtain the expected therapeutic effect.

The paramount goal of pharmacognosy is comprehensive investigations of plant materials by use of physical, chemical, and biological methods, and also the search for a possibility to use these materials as natural medicines. Modern pharmacognosy focuses on the chemical components of the plant materials, including the structure and pharmacological properties that are responsible for their use in therapy. Thus, it can be concluded that the main area of interest is in the chemistry of biogenic compounds (i.e., the chemistry of natural compounds of plant origin). This new approach to the subject of pharmacognosy is based on the dynamic treatment of the natural sources of drugs that takes into account their biochemical transformations and consequently allows synthesis of the new biologically active substances. In that way, links are being established between pharmacognosy and plant biotechnology, which involves breeding tissue cultures as a source of technological amounts of the biologically active substances.

An interest of modern pharmacognosy, in particular compounds that occur in the plant materials, is due to their already recognized significance in therapy and also to the importance of a steady search for new natural substances with a curing potential. In this sense, plant material has to be treated as a source of suitable medicines. The therapeutic effect can be obtained by direct use of plant materials, by use of the plant confections, or by use of substances or substance groups isolated from the plant tissues. The latter case occurs only when a given plant contains highly active substances, e.g., the alkaloids in *Secale cornutum, Tuber Aconiti,* and *Rhizoma Veratri,* or cardiac glycosides in *Folium Digitalis purpureae* and *Folium Digitalis lanatae.* These materials are an important source of selected alkaloids or cardiac glycosides.

Plant materials, galenic preparations, and isolated compounds proposed for therapy have to meet certain strictly determined standards. With the most important materials, these standards simply are the pharmacopoeial requirements, although a vast number of herbs used in formal and popular medicine are not included in any pharmacopoeia. Standardization of the plant material and of herbal preparations is meant to guarantee their therapeutic value, and it is a result of the investigations on biologically active components. There are a wide number of methods to investigate plant material, namely macroscopic (focused on botanical identity and purity of the plant material); microscopic (mostly histochemical investigations, which provide the basis for identification of the material); biological (microbiological and biomolecular investigations and investigations of biological activity); and chemical methods. Chemical investigations of the plant material have a variety of goals, such as determination of the substance groups, quantitative analysis of active compounds, isolation of substances from the plant tissues for their further identification, or physicochemical characterization, and, finally, structural analysis of the isolated unknown compounds.

1.2   PROCEDURES OF THIN LAYER CHROMATOGRAPHY

Among the chemical methods of plant examination, chromatographic analysis plays a very important role, and it has been introduced to all the modern pharmacopoeias. Because of numerous advantages of the chromatographic methods (such as their specificity and a possibility to use them for qualitative and quantitative analysis), they comprise an integral part of the medicinal plant analysis.

The following chromatographic methods are most frequently applied in phytochemical analysis: one- and two-dimensional paper chromatography, one- and two-dimensional thin layer chromatography (TLC; also called planar chromatography), high-performance column liquid chromatography (HPLC), gas chromatography (GC), and counter current chromatography (CCC). These methods can also be used for the isolation of the individual components from the component mixtures on a preparative and micropreparative scale.

TLC is a chromatographic technique widely used for qualitative analysis of organic compounds, isolation of the individual compounds from multicomopnent mixtures, quantitative analysis, and preparative-scale isolation. In many cases, it outperforms the other chromatographic techniques. Firstly, there is a multitude of chromatographic systems that can be applied in TLC. Many kinds of TLC and high-performance TLC (HPTLC) precoated plates are commercially available, e.g., those with the inorganic adsorbent layers (silica or silica gel and alumina); organic layers (polyamide, cellulose); organic, polar covalently bonded modifications of the silica gel matrix (diol, cyanopropyl, and aminopropyl); and organic, nonpolar bonded stationary phases (RP2, RP8, RP18) with different densities of coverage of the silica matrix (starting from that denoted as W, for the lowest density of coverage and thus wettable with water). Sorbents applied in TLC have different surface characteristics and, hence, different physicochemical properties. Moreover, there is a wide choice of mobile phases that can be used to separate mixture components; these belong to various selectivity groups and, thus, have different properties as proton donors, proton acceptors, and dipoles. In TLC, ultraviolet (UV) absorption of the mobile phase solvents does not play a significant negative role in detection and quantification of the analytes, because the mobile phase is evaporated from the plate prior to the detection. High viscosity of a solvent can be viewed as a sole property limiting its choice as a mobile phase component. These plate and mobile phase characteristics allow a choice from among an unparalleled abundance of TLC systems that offer a broad spectrum of separation selectivities, which is particularly important when complex mixtures of the plant extracts have to be separated.

Another advantage of TLC is that each plate is used only once, thereby allowing simpler sample preparation methods when compared with techniques such as GC and HPLC, in which multiple samples and standards must be applied to the column in sequence. Highly sorbed materials in plant extract samples can be left behind in a column and interfere in the analysis of subsequent samples. Multiple samples can be analyzed at the same time on a single TLC or HPTLC plate, reducing the time and solvent volume used per sample; the processing of standards and samples on the same plate leads to advantages in the accuracy and precision of quantification by densitometry.

Last, but not least, TLC enables usage of numerous special development techniques. Most separations are carried out by a capillary flow development with a single mobile phase (isocratic) in the ascending or horizontal configuration. Gradient elution with stepwise variations in mobile phase composition, which is widely applied in HPLC, is also used in TLC. Besides, there are the following special modes of developing a chromatogram: unidimensional multiple development (UMD), incremental multiple development (IMD), gradient multiple development (GMD), and bivariant multiple development (BMD). Moreover, the circular and anticircular development methods can also be applied. UMD consists of repeated development of the chromatogram over the same development distance, with a given mobile phase of constant composition and with drying the plate between the individual development runs. IMD is performed by the stepwise increase in the development distance (the increment in the development distance is kept constant), using a steady mobile phase composition and drying the plate between the development runs. It results in narrowing of the spots or zones and improved resolution. In GMD, each step of the repeated chromatogram development is performed with a mobile phase of different composition, thus enabling gradient development. The development distance of the consecutive development runs is kept steady and it is only the mobile phase composition that changes, thus enabling the analysis of complex mixtures spanning a wide polarity range. When a low strength mobile phase is used, the separation of the low polarity components is achieved on a silica layer. When a medium polarity mobile phase is used, then the medium polarity components are separated (the first group is then eluted to the upper edge of the plate). With the high polarity mobile phases, separation of the high polar components of plant extracts can be obtained. BMD involves a stepwise change both of the development distance and the mobile phase composition. With use of a special chamber and computer program, an improved version, known as Automated Multiple Development (AMD), can be applied, with the distance of the development increasing and the mobile phase strength decreasing at each step. AMD enables the analysis of complex samples over a wide polarity range and provides focusing (tightening) of the zones. In the circular and anticircular development modes, the mobile phase migrates radially from the center to the periphery or from the periphery to the center, respectively. Analytes with lower R_F values are better resolved by means of circular chromatography than by means of linear chromatography, and the advantage of the anticircular mode is that it allows better resolution of compounds with higher R_F values.

TLC is also the easiest technique with which to perform multidimensional (i.e., two-dimensional) separations. A single sample is applied in the corner of a plate, and the layer is developed in the first direction with mobile phase 1. The mobile phase is dried by evaporation, and the plate is then developed with mobile phase 2 at a right angle (perpendicular or orthogonal direction); mobile phase 2 has different selectivity characteristics when compared with mobile phase 1. In this way, complete separation can be achieved of very complex mixtures (e.g., of the components of a plant extract) over the entire layer surface.

Particularly valuable separation results can be achieved when using various mobile phase systems to benefit from different separation mechanisms. For example, with cellulose one can apply a nonaqueous mobile phase to achieve the adsorption mechanism of retention and an aqueous mobile phase to achieve the partition mechanism. In a similar way, with the polar chemically bonded stationary phases one can use nonaqueous mobile phases to achieve the adsorption mechanism of retention and the aqueous mobile phases to achieve the reversed-phase mechanism. Shifting from the adsorption to the partition mode causes marked differences in the separation selectivity.

After performing the separation with the optimum layer, mobile phase, and development technique combination, the zones must be detected. If the zones are not naturally colored or fluorescent, or do not absorb 254 nm UV light so they can be viewed as fluorescence-quenched zones on special F-plates containing a fluorescent indicator, a detection reagent must be applied by spraying or dipping, usually followed by heating. This derivatization is mainly used in the postchromatographic mode for localization of the separated component zones on the layer. Very often universal reagents are used, such as iodine vapors or sulfuric acid. These reagents can locate almost all of the existing organic compound classes. Selective reagents can be used as derivatizing reagents for individual or group identification of the analytes. For example, the Dragendorff's reagent ($KBiI_4$) is used for identification of hetero-cyclic bases (e.g., of alkaloids), ninhydrin for identification of compounds containing an amino group in their structure (e.g., of the amines and amino acids), and 2-(diphenylboryloxy)-ethylamine + polyethylene glycol (PEG) for identification of polyphenols.

TLC is coupled with densitometry to enable detection of colored, UV-absorbing, or fluorescent zones through scanning of the chromatograms with visible or UV light in transmission or reflectance modes. By comparison of a signal obtained with that for the standards processed with comparable chromatographic conditions, densito-metric measurements can be used for quantitative analysis of the components con-tained in the mixtures. With multiwavelength scanning of the chromatograms, spectral data of the analytes can be directly acquired from the TLC plates and can further be compared with the spectra of the analytes from a software library or from standards developed on the same plate. Thus, a densitometer with a diode array detector enables direct (in situ) identification of the analytes. Other possibilities to identify analytes are offered by off-line or on-line coupling of TLC with Fourier transform infrared spectrometry, mass spectrometry, etc.

Further, it is worth noting that TLC coupled with bioautographic detection of microbiologically active compounds can be successfully applied in the analysis of plant extracts. Especially suitable for this purpose is direct bioautography, which uses microorganisms (e.g., bacteria or fungi) growing directly on a TLC plate with the previously separated mixtures of the plant extracts. In this procedure, antibacter-ial or antifungal compounds appear as clear spots (i.e., without microorganisms growing on them) on an intensely colored background. This approach can be used as an additional analytical option in screening of biological samples, as a standard-ization method for medicinal plant extracts, and as a selective detection method.

Additionally, special instruments enable the use of the forced-flow migration of the mobile phase. Overpressured-layer chromatography (OPLC), also called opti-mum performance laminar chromatography, makes use of a pump that feeds the sorbent bed with mobile phase at a selected flow rate. Rotation planar chromatog-raphy (RPC) uses centrifugal force in order to obtain an analogous effect. Electro-osmotically driven TLC makes use of electroosmotic flow to force mobile phase across a layer. All of these forced-flow methods provide a constant flow rate of the mobile phase; the linear profile of the flow and elimination of vapor phase from the system may improve system efficiency and peak resolution.

The advantages of TLC are particularly important with plant extracts, which are very complex mixtures of the structurally differentiated chemical compounds. Such extracts very often contain polar (e.g., tannins and phenols) and nonpolar (lipids, chlorophylls, and waxes) ballasts, apart from a fraction of active substances that is of main importance for phytochemistry and pharmacognosy. This latter fraction con-tains closely related compounds of a similar structure and physicochemical proper-ties. Isolation of a fraction of interest from such a mixture requires a complicated procedure, usually liquid–liquid or solid-phase extraction. TLC enables separation of a crude plant extract without an earlier purification. For example, in a normal phase system a nonpolar fraction moves with the mobile phase front (or it can be prewashed with a nonpolar mobile phase prior to the development of a chromatogram), and the polar fraction remains strongly retained near to the origin; then the fraction of interest is separated in the central part of the chromatogram.

Summing up, TLC is a principal separation technique in plant chemistry research. It can be used in a search for the optimum extraction solvents, for identification of known and unknown compounds, and—what is at least equally important—for selection of biologically active compounds. TLC also plays a key role in preparative isolation of compounds, purification of the crude extracts, and control of the separation efficiency of the different chromatographic techniques and systems. TLC has many advantages in plant chemistry research and development. These include single use of stationary phase (no memory effect), wide optimization possibilities with the chromatographic systems, special development modes and detection methods, storage function of chromatographic plates (all zones can be detected in every chromatogram by multiple methods), low cost in routine analysis, and availability of purification and isolation procedures.

1.3   ORGANIZATION OF THE BOOK

The book comprises 29 chapters, divided into two parts. Part I consists of 10 chapters and provides general information on those areas of science that are related to phytochemistry and can benefit from the use of TLC. Moreover, it also contains chapters devoted to the technical aspects of TLC, such as the instrumentation and chromatographic systems involved.

Following this chapter, Chapter 2 focuses on medicinal plants as a source of natural drugs and on their role in modern pharmacy. It also provides a brief overview of the methods used for the investigation of the plant material and of the techniques used for the extraction, purification, and final assessment of the drugs having a plant origin.

Chapter 3 is devoted to the medicines and the diet supplements produced from plants. Firstly, the authors introduce definitions of the plant medicines and plant diet supplements, and then they present the history of herbal drugs in the traditional medicines of various cultures throughout the world. The botanical supplements are then discussed, and the chapter ends by discussing the tasks of TLC in the field of botanical drugs and dietary supplements.

Chapter 4 focuses on the primary and the secondary metabolites and their biological activity. Because classification of the metabolites as primary and secondary is not straightforward and can be viewed differently by the different authors, we editors will explain in the next paragraph our own ideas on this very important issue, which shape the structure of the entire volume.

Primary metabolites are those that occur in each plant and fulfill its basic physiological functions (i.e., appear as the building, energetic, or the reserve material). In other words, primary metabolites are indispensable for the life of a plant. Secondary metabolites are the products of metabolism and play no crucial role in the plant's life. This classification can be regarded as a rough and provisional only, as it often happens that the secondary metabolites have a well recognized physiological function in the plants as well. In practice, all metabolites can be classified in different ways. Most often classification is based on their chemical structure, which generally remains in a good agreement with the biogenetic origin. Sometimes, however, problems can arise with a straightforward classification of certain groups of compounds that belong to the same biogenetic group and yet completely differ in terms of chemical structure. For example, steroidal alkaloids are traditionally included in the alkaloid group. However, their biogenetic origin is not from amino acids but from steroids, and for this particular reason in this book they are classified as steroids. In a similar way, taxoids are sometimes classified as pseudoalkaloids due to the presence of the tertiary nitrogen atom in the molecules of certain taxoid representatives. At the same time, all taxoids biogenetically belong to the class of diterpenes. It is noteworthy that classification of metabolites based on their biogenetic origin is sometimes impractical, and then it is recommended to refer to their chemical structure or physicochemical properties. For example, naphthoquinones and anthraquinones may originate both from shickimic acid and acetogenine. In fact, quinones can have a different biogenetic origin, but are joined in one group based on their similar physicochemical properties. Iridoids could formally be included in the class of monoterpenes, but this is not done because of their differentiated physicochemical and pharmacological properties (classical monoterpenes are the volatile compounds present in essential oils, whereas iridoids usually are the nonvolatile species). For the above reasons, we decided to classify the questionable groups of compounds according to their chemical structure. Consequently, all of the metabolites—both primary and secondary—are additionally divided according to their biogenesis.

Chapter 5 focuses on chemosystematics, also known as chemotaxonomy. This chapter starts from definition of this particular branch of phytochemical science, which involves classification of the plant organisms based on the differences at the biochemical level, especially in the amino acid sequences of common proteins. Then the author highlights the areas of the main interest for the chemosystematic studies and discusses applicability of the main chromatographic modes to this area of research.

In Chapter 6, the sorbents and precoated layers that are particularly useful in the analysis and preparative isolation of the primary and secondary metabolites from plant extracts are described. This chapter covers virtually all of the TLC and HPTLC analytical and preparative layers used for separation, determination, and isolation within the field of phytochemistry, including silica gel, reversed phase and hydrophilic bonded phases, nonsilica sorbents (alumina, cellulose, polyamides, modified celluloses, and kieselguhr), and miscellaneous layers (resin, impregnated, mixed, and dual layers).

Chapter 7 starts with description of the chromatographic chambers and mobile phase compositions that can be utilized in phytochemical research. Then the authors discuss the development of the chromatograms in the different thin layer chromatographic modes. This chapter covers the methods of sample application to the adsorbent layer as well.

Then two chapters following deal with the detection of the analytes after their thin layer chromatographic separation. The main part of Chapter 8 is devoted to derivatization of the plant extract components by use of universal and selective reagents. The specificity of TLC—not shared with any other chromatographic mode—results from the possibility of applying several different detection methods in sequence in order to identify groups or individual analytes. In this chapter, physical methods of detection are also discussed (such as UV–Vis light absorption, fluorescence, and mass spectrometry), as well as the methods of quantitative analysis with use of TLC combined with densitometry.

Chapter 9 deals with the methods of biodetection in TLC that enable rapid and selective determination of the biological activity (antibacterial, antifungal, and other) of plant metabolites. In this chapter the mechanisms of bioactivity of the individual compounds are explained.

Part I ends with the description of the forced-flow planar chromatographic development techniques (Chapter 10), including their influence exerted on separability of the plant metabolites.

Part II of the book is divided into the chapters that reflect the types of the metabolites that occur in plants. Chapters 11 through 13 refer to primary metabolites.

Chapter 11 deals with the chemistry of carbohydrates, with their occurrence in the plants as mono-, oligo-, and polysaccharides, and also as glycoconjugates. It provides an overview of the recommended analytical methods, including sample preparation, derivatization, and the most suitable TLC systems.

In Chapter 12, different classes of plant lipids are presented, and the TLC systems applied to their separation (including normal- and reversed-phase and argentation) are discussed. Class separation of lipids, their isolation, and quantification are taken into the account.

Chapter 13 focuses on free amino acids, peptides, and proteins, including their occurrence in plants and the use of TLC to separate the individual groups of these compounds.

The next part of the book deals with the secondary metabolites occurring in plant tissues, and it is divided into sections according to the metabolic pathways in which individual substances are synthesized.

Chapter 14 starts with the phenolic compounds that belong to the metabolic pathway of shickimic acid, i.e., phenols, phenolic acids, and tannins. It describes the structure and classification of these compounds, their biological importance, sample preparation methods, and the various TLC systems and special techniques that are used for their separation and analysis.

Chapter 15 deals with coumarins that belong to the phenol class and are also derived from shickimic acid. Details are provided on sample preparation and isolation of coumarins with aid of classical TLC and HPTLC and the forced-flow planar chromatographic techniques. Application of TLC to the measurement of biological activity of coumarins is also described. The chapter ends with tables of the plant families in which coumarins occur.

Chapter 16 is dedicated to the phenolic compounds originating from a similar pathway as coumarins, i.e., flavonoids. After a short introduction on chemistry, biochemistry, and medical significance of flavonoids, the methods for their analysis using various TLC systems are presented, including forced-flow development techniques. In this chapter, sample preparation methods and quantification of flavonoids by means of TLC combined with slit-scanning and video densitometry are discussed.

The section of the book on secondary metabolites ends with lignans, also originating from shickimic acid. Chapter 17 is focused on the chemistry, occurrence in plant material, and pharmacological activity of the representatives of this group, followed by the sample preparation techniques and the TLC analysis of these compounds. Details about quantification of lignans in herbal extracts and preparations are also reported.

The next section of the book is focused on isoprenoid derivatives, which include several groups of compounds. It starts with Chapter 18 on the volatile compounds (mono- and sesquiterpenes), including their definition, classification, occurrence, and importance. Then the following applications of planar chromatography are discussed: identification of the volatile fractions in pharmaceutical drugs, taxonomic investigations, tracing of various adulterations, and analysis of cosmetics.

Chapter 19 covers diterpenoids and presents their structure, physicochemical properties, natural occurrence, and pharmacological activity. The details of sample preparation and the analytical and preparative TLC separations of this group of compounds with the aid of different chromatographic systems are described, including derivatization and quantification methods. The chapter ends with a comparison of the performance of TLC with that of the other chromatographic and related techniques used in diterpenoid analysis.

The next group of compounds that belong to the isoprenoid methabolic pathway are triterpenes, and they are described in Chapter 20. After a short introduction on structure and properties of this group, chromatographic systems and detection methods applied in the analysis of triterpenes (saponins included) are presented. The chapter emphasizes the role of planar chromatography as a technique supporting column chromatography in identification and determination of the biological activity of triterpenes.

Chapter 21 focuses on tetra- and polyterpenes, and among them carotenoids represent the most important group of compounds. First, structure, occurrence, and properties are presented. Then the special aspects of the TLC analysis (such as detection and instability of carotenoids) are emphasized. The use of silica and alumina, and also of the basic normal phase adsorbents, is discussed. Usefulness of TLC in screening of the plant material, in preparative separations, and in isolation of individual carotenoids is also described.

The next large group of compounds that belong to the isoprenoid pathway are steroids, and they are presented in Chapter 22. In the introductory part of this chapter, the chromatographic systems and techniques useful for planar separation of steroids are described. Then an overview of the literature is provided, taking into the account the classes of phytosterols, steroids (brassinosteroids, bufadienolides, cardenolides, ecdysteroids, steroidal saponins, steroidal alkaloids, vertebrate-type steroids, and withanolides), and of the related triterpenoids (cucurbitacins). Structural diversity, the separation systems, and the detection and quantification for each class of compounds are presented.

Iridoides are the last group of compounds that belong to the isoprenoid pathway, and they are described in Chapter 23. After the introductory part on the structure and physicochemical properties of iridoides, the issues are emphasized related to isolation of this group of compounds from the plant material and to sample preparation. Then the TLC systems and techniques applied to the analysis of iridoids are described, taking into account the detection methods and the forced-flow techniques. The preparative layer chromatography of iridoids is also discussed.

The next four consecutive chapters deal with alkaloids synthesized in the plant organisms from amino acids. There are several groups of alkaloids differing in their structure, properties, and biological activity.

Chapter 24 focuses on indole alkaloids. Firstly, the chemical structure, occurrence, and pharmacological, ecological, and chemosystematic importance of this group are discussed. This preliminary information is followed by a detailed description of the TLC separations of indole alkaloids, including chromatographic systems, techniques, and detection methods. Details on the separations of the particular types of indole alkaloids are also presented.

Chapter 25 is devoted to the structure, properties, and biological activity of isoquinoline alkaloids. Information concerning problems with chromatographic separation of basic compounds is also provided, and the normal phase, reversed phase, and pseudoreversed phase systems are described in detail. The use of TLC plates and grafted plates for two-dimensional separations and forced-flow techniques applied to the separation of the isoquinoline alkaloids are presented. Examples of TLC applications to quantitative analysis are shown, along with the preparative separations.

Tropane alkaloids are handled in Chapter 26. Chemistry and stereochemistry of tropane and the related alkaloids, and their natural occurrence, are presented first. Various methods of extraction of this entire group of compounds from plant material are described, followed by the pretreatment of the extracts by liquid–liquid partitioning (LLP), solid assisted liquid–liquid partitioning (Extrelut), and solid-phase extraction (SPE). Then the information on TLC of tropane alkaloids including their quantification is provided. The chapter gives detailed information on the OPLC analysis of tropane alkaloids, and a comparison is made with the results originating from the other separation techniques in use.

Chapter 27 focuses on the remaining groups of alkaloids, including phenylethylamine derivatives, quinoline derivatives (Cinchona alkaloids), and pyrrolidine, pyrrolizidine, piridine, and piperidine derivatives (Tobacco, Lobelia, Pepper, Pelletierine, Sedum, Senecio alkaloids), quinolizidine alkaloids (Lupine alkaloids), xanthine, imidazole derivatives, and diterpene alkaloids. Preparation of extracts, the most frequently employed TLC systems, and the detection methods applicable to each individual group are presented.

The last two chapters are devoted to the secondary metabolites derived from acetogenine (acetylocoenzym A). Chapter 28 deals with the distribution of polyacetylenes in plants and pharmacological activity of polyacetylenes. Separation, detection, and isolation by means of TLC in various different systems are described. The results are compared with those originating from HPLC.

Chapter 29 is focused on quinones (antraquinones and naphthoquinones), their occurrence in plants, and pharmacological activity. Applicability of the conventional TLC techniques applied to the separation of quinines, and also of the special modes (e.g., gradient or two-dimensional TLC), is discussed.

The authors who agreed to contribute chapters to the book are all recognized international experts in their respective fields. The book will serve as a comprehensive source of information and training on the state-of-the-art phytochemistry methods performed with aid of TLC. It will help to considerably popularize these methods for practical separations and analyses in a field that will undoubtedly grow in importance for many years to come. A computer assisted search has found no previous book on TLC in phytochemistry. Three editions of the book "Phytochemical Methods" (1973, 1984, and 1998) by J.B. Harborne (Chapman and Hall, London, UK) had chapters organized by compound type, most of which contained some information on TLC analysis. A chapter on "Thin Layer Chromatography in Plant Sciences" by J. Pothier was contained in the book "Practical Thin Layer Chromatography" edited by B. Fried and J. Sherma (CRC Press, 1996), a chapter on planar chromatography in medicinal plant research in "Planar Chromatography" edited by Sz. Nyiredy (Springer, 2001) and a chapter on natural mixtures by M. Waksmundzka-Hajnos et al. in "Preparative Layer Chromatography" edited by T. Kowalska and J. Sherma (CRC/Taylor & Francis, 2006) included information on plant extracts. A book by E. Reich and A. Schibli (Thieme Medical Publishers, Inc., 2007) covers the theoretical concepts and practical aspects of modern HPTLC as related to the analysis of herbal drugs.

However, these information sources are not comprehensive, and the first two are now out of date. Our proposed book will solve this void in information in the critical field of phytochemical analysis.

2 Plant Materials in Modern Pharmacy and Methods of Their Investigations

Krystyna Skalicka-Woźniak, Jarosław Widelski, and Kazimierz Głowniak

CONTENTS

Pharmacognosy is the science which treats of the history, production, commerce, collection, selection, identification, valuation, preservation and use of drugs and other economic materials of plant and animal origin.

The term "pharmacognosy" derived from the ancient Greek words *pharmakon*, drug or medicine, and *gnosis*, knowledge, and literally means the knowledge of drugs.

2.1   PLANT MATERIAL AND MARINE PRODUCTS AS SOURCES OF ACTIVE SECONDARY METABOLITES

Drugs are derived from the mineral, vegetable, and animal kingdoms. They may occur in the crude or raw form, as dried or fresh unground or ground organs or organisms or natural exudations of these (juice or gum), when they are termed "crude drugs."

These are known as herbal medicinal products (HMPs), herbal remedies, or phytomedicines and include, for example:

* Herb of St. John's wort (*Hypericum perforatum*), used in the treatment of mild to moderate depression
* Leaves of *Gingko biloba*, used for cognitive deficiencies (often in the elderly), including impairment of memory and affective symptoms such as anxiety

There are also derived substances, such as alkaloids (e.g., caffeine, from the coffee shrub—*Coffea arabica*—used as a stimulan), glycosides (e.g., digoxin and other digitalis glycosides, from foxglove—*Digitalis* spp.—used to treat heart failure), alcohols, esters, aldehydes, or other constituents or mixtures of constituents isolated from the plant or animal.

Finally, also pure chemical entities exist, which are produced synthetically and referred to as "nature identical", but originally discovered from plant drugs. Examples include:

* Morphine, from opium poppy (*Papaver somniferum*), used as an analgesic
* Quinine, from *Cinchona* bark (*Cincoina* spp.), used in the treatment of malaria

Also many foods are known to have beneficial effects on health:

- Garlic, ginger, and many other herbs and spices
- Anthocyanin- or flavonoid-containing plants such as bilberries, cocoa, and red wine
- Carotenoid-containing plants such as tomatoes, carrot, and many other vegetables [1]

2.2 THE DISTRIBUTION AND CONCENTRATION OF NATURAL COMPOUNDS WITH BIOLOGICAL ACTIVITY IN DIFFERENT ORGANS OF MEDICINAL PLANTS

Isolated pure natural products such as numerous pharmaceuticals used in pharmacy are thus not "botanical drugs," but rather chemically defined from nature. Botanical drugs are generally derived from specific plant organs of plant species. The following plant organs are the most important:

- Herba or aerial parts (*herba*)
- Leaf (*folium*)
- Flower (*flos*)
- Fruit (*fructus*)
- Bark (*cortex*)
- Root (*radix*)
- Rhizome (*rhizoma*)
- Bulb (*bulbus*)

Fruits and seeds have yielded important phytotherapeutic products, e.g., caraway (*Carum carvi*), fennel (*Foeniculum vulgare*), saw palmetto (*Serenoa repens*), horse chestnut seeds (*Aesculus hippocastanum*), or ispaghula (*Plantago ovata*), which are used often in phytotherapy.

Numerous drugs contain also leaf material as the main component. Some widely used ones include balm (*Melissa officinalis*), deadly nightshade (*Atropa belladonna*), ginkgo (*Ginkgo biloba*) peppermint (*Mentha × piperita*), bearberry (*Arctostaphylos uva-ursi*), and many others.

Although the flowers are of great botanical importance, they are only a minor source of drugs used in phytotherapy. One of the most important example is chamomile (*Chamomilla recutita* (Matricariase flos)). Other examples include calendula (*Calendula officinalis*) and arnica (*Arnica montana*).

Stem material which is often a part of those drugs is derived from all above-ground parts, e.g., ephedra (*Ephedra sinica*), hawthorn (*Crataegus monogyna* and *Crataegus oxyacantha*), passion flower (*Passiflora incarnata*), or wormwood (*Arthemisia absynthium*). Also parts of the stem are used in phytotherapy like bark of *Rhamnus frangula* (frangula) or bark of *Salix alba* (willow).

Finally, underground organs (rhizome and root) of many species have yielded pharmaceutically important drugs. Examples include: Devil's claw (*Harpagophytum procubens*), tormentill (*Potentilla erecta*), rhubarb (*Rheum palmatum*), and kava-kava (*Piper methysticum*) [1].

2.3   METHODS OF INVESTIGATIONS OF PLANT MATERIAL

The analytical side of pharmacognosy is embraced in the expression "the evaluation of the drug," for this includes the identification of a drug and determination of its quality and purity.

The identification of the drug is of first importance, for little consideration can be given to an unknown drug as regards its quality and purity. The identification of a drug can be established by actual collection of the drug from plant or animal (among them marine organism) which can be positively identified from the botanical or zoological standpoint. This method is rarely followed except by an investigator of the drug, who must be absolutely sure of the origin of his samples. For this reason "drug gardens" are frequently established in the connection with institution of pharmacognostical research.

Quality of a drug refers to intrinsic value of the drug, that is, to the amount of medicinal principles or active constituents present in the drug. A high grade of quality in the drug is such importance that effort should be made to obtain and maintain this high quality. The most important factors to accomplish this include: collection of the drug from the correct natural source at proper time and in the proper manner, the preparation of the collected drug by proper cleaning, drying and garbling and proper preservation of the clean, dry, pure drug against contamination with dirt, moisture, fungi, filth, and insects.

The evaluation of the drug involves a number of methods, which may be classified as follows: organoleptic, microscopic, biological, chemical, and physical [2].

2.3.1   MACROSCOPIC INVESTIGATIONS

Organoleptic (lit. "impression on the organs") refers to evaluation by means of the organs of sense, and includes the macroscopic appearance of the drug, its odor and taste, occasionally the sound or "snap" of its facture, and the "feel" of the drug to the touch.

For convenience of description the macroscopic characteristic of a drug may be divided into four headings, viz.: shape and size, color and external markings, fracture and internal color, and finally odor and taste.

For example, description of linseed (*Linum usitatissimum*) is as follows: The seed is exalbuminous, of compressed ovate or oblonglanceolate outline, pointed at one end, rounded at the other and from 4 to 6 mm in length; externally glabrous and shiny, brown to dusky red with a pale-yellow, linear raphe along one edge; the hilum and microphyle in a slight depression near the pointed end; odor slight, becoming very characteristic in the ground or crushed drug; taste mucilaginous and oily [3].

2.3.2   MICROSCOPIC AND MICROCHEMICAL METHODS OF INVESTIGATIONS

Microscopical methods of valuing drugs are indispensable in the identification of small fragments of crude drugs and of the powdered drugs as well as in the detection of their adulterans, for these possess few features other than color, odor, and taste whereby clues toward their identity may be afforded. Moreover, owing to the similarity of some plant organs of allied species, definite identification even of certain entire, cellular vegetable drugs cannot be made without the examination of mounts of thin sections of them under a microscope. Every plant possess a characteristic histology in respect to its organs and diagnostic features of these are ascertained though the study of the tissues and their arrangement, cell walls and cell contents, when properly mounted in suitable stains, reagents or mounting media [3].

Some characteristic features can be easily used to establish botanical identity and quality of the drugs. The typical example is the various types of crystals formed by calcium oxylate. Several species of the family Solanaceae are used for obtaining atropine, alkaloid used as spasmolytic in cases of gastrointestinal cramps and asthma. Species containing high amount of atropine like *Atropa belladonna* (deadly nightshade), *Datura stramonium* (thorn apple), or *Hyoscyamus niger* (henbane) are characterized by typical crystal structures of oxalate: sand, cluster crystals and microspheroidal crystals, respectively. These are subcellular crystal structures, which can be easily detected using polarized light and are thus a very useful diagnostic means.

Second typical example are the glandular hairs, which are characteristic for two families (Lamiaceae and Asteraceae) containing many species with essential oils. Figure 2.1 shows diagnostic features of botanical drugs—microscopic examination

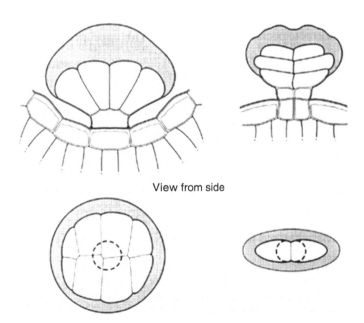

View from side

FIGURE 2.1 Diagnostic features of botanical drugs, that are revealed upon microscopic examination include typical glandular hair as found in the Lamiaceae (a) and Asteraceae (b). *Top*: lateral view; *bottom*: view from above. (From Heinrich, M. et al., *Fundamentals of Pharmacognosy and Phytotherapy*, Elsevier Science, Churchill Livingstone, 2004.)

including typical glandular hair (Lamiaceae and Asteraceae family)—lateral view and view from above.

In many instances, a good idea of the quality of a drug can be ascertained by using *microchemical methods*. These may consist of examining mounts of sections or powdered drug in various reagents which either form salts of contained active principles, that have constant characters (microcrystallization) or show definite color reactions, or of *isolation of constituents* of the powdered drug with a suitable solvent, filtering 2 or 3 drops of the extract on to a slide, permitting the solvent to evaporate and examining the residue. There is also possible isolation of a constituents by *microsublimation*.

It is often possible to detrmine whether a powdered drug has been exhausted by examining the crystals found in its sublimate. These have been found to be characteristic for many drugs. *Microsublimation* upon a slide is a superior technique in comparison with test tube sublimation. The sublimates may be directly examined under the microscope without mechanical alteration.

2.3.3 CHEMICAL METHODS OF INVESTIGATIONS

2.3.3.1 Approximate Group Identification

Identification of the characteristic group or groups of active constituents is one of the basic methods of the evaluation of the drug and the first step in isolation procedure.

For example, Borntrager's test is commonly applied to all anthraquinone drugs. As effect of the reaction a deep rose color is produced.

Another example is a reaction that give acids and flavonoids with Arnov's reagent phenolic (the products of this reaction give purple color). All of these reactions are also used both for qualitative and quantitative analysis (colometric reactions).

Characteristic reaction for flavonoids, like 1% methanolic solution of $AlCl_3$, 5% methanolic solution of KOH, and 1% methanolic solution of Naturstoffreagenz A, are used for derivatization of TLC plates. It enables general evaluation of different groups of active compounds, in this case flavonoids.

2.3.3.2 Quantitative Analysis of Active Compounds in Plant Material by Various Methods (Titration, Spectrophotometric Methods)

Evaluation of plant drugs uses all of the methods known in chemical analysis. Among them we can single out the titration. Titration is a common laboratory method of quantitative chemical analysis which can be used to determine the concentration of known reactant. Because volume measurements play a key role in titration, it is also known as volumetric analysis. A reagent, called titrant, of known concentration (a standard solution) is used to react with a measured quantity of reactant (the analyte). Titration is used in quantitative analysis of tropan alkaloids, where KOH is used as a titrant and methyl red as the indicator.

Spectrophotometric techniques are used to measure the concentration of solutes in solution by measuring the amount of light that is absorbed by the solution in a cuvette placed in the spectrophotometer. According the Beer–Lambert Law there is the linear realationship between absorbance and concentration of an absorbing species. It enables a quantitative determination of compounds in which solutions absorb light. For example total concentration of pyrrolizidine alkaloids in *Symohytum officinale* root were investigated using UV-VIS spectra of adducts of 3,4-dehydroPAs and Erlich's reagent [4].

2.3.3.3 Isolation of Active Compounds

When a crude extract obtained by a suitable extraction procedure shows interesting activity (e.g., an antibacterial activity), demonstrated in bioassay, the next and one of the most difficult steps is to fractionate the extract using a different (sometimes combined) separation method(s) so that a purified biologically active component can be isolated.

Figure 2.2 gives a general isolation protocol starting with selection of biomass (e.g., plant, microbe or tissue culture), which is then extracted using different extraction methods. Hydrophilic (polar) extracts will then usually undergo ion exchange chromatography with bioassay of generated fractions. A further ion exchange method of bioactive fraction would yield pure compounds, which could next be submitted for structure elucidation (MS, NMR).

2.3.4 Biological Methods of Investigations

The biological evaluation of the plant drugs is one of the most important issues of pharmacognosy. For drugs obtained from natural sources, all active compounds present in the plant are responsible for therapeutic effect.

There are plenty of methods for evaluation of biological properties of plant drug. For example, bacteria, such as *Staphylococcus aureus* are used to determine antiseptic value of the drugs. For standardization of *Digitalis* spp. (Foxglove) and other "heart tonic" drugs, pigeons and cats are used. Bioassay is the use of biological system to detect properties of a mixture or a pure compound.

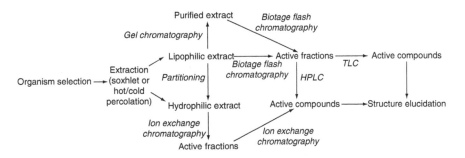

FIGURE 2.2 General isolation strategy for purification of bioactive natural products. (From Heinrich, M. et al., *Fundamentals of Pharmacognosy and Phytotherapy*, Elsevier Science, Churchill Livingstone, 2004.)

Bioassays could involve the use of in vivo systems (clinical trials, whole animal experiments), ex vivo systems (isolated tissues and organs), or in vitro systems (e.g., cultured cells). Collection of materials for testing in bioassays could either be random collection of samples or directed collection, i.e., from plants known to be used traditionally. Bioassays were often linked with the processes of fractionation and isolation, known as bioassay-guided fractionation.

2.4   MODERN EXTRACTION METHODS OF ACTIVE COMPOUNDS FROM PLANT MATERIAL AND MARINE PRODUCTS

Sample pretreatment is one of the most time-consuming steps of the analytical process—and also one of the most important. Plants contain many of bioactive compounds and it is important to extract all of them in the best, short and effective way with minimal solvent usage. Proper extraction technique should also be cheap and simple [5,6]. Important is high recoveries, reproducibility, low detection limits, and automation [7]. Solid–liquid extraction is one of the oldest ways of solid sample pretreatment [5]. Extraction and fractionation of extract is also an important method in isolation of compound groups or individual substances from plant material.

2.4.1   CLASSIC EXTRACTION METHODS IN SOXHLET APPARATUS

Extraction in Soxhlet apparatus has been the leading technique mostly used for a long time and still is considered to be a standard technique and the main reference to which the other new leaching methods are compared [5].

The sample is placed in a thimble holder and during operation is continuously filled with fresh portion of solvent from distillation flask. When the liquid reaches the overflow level, a siphon aspirates the solute of the thimble holder and unloads it back into the distillation flask, carrying the extracted analytes into the bulk liquid. The operation is repeated until complete extraction is achieved [5,6].

Different solvents can be used in extraction process. Addition of co-solvents as a modifier to increase the polarity of liquid phase is possible. Soxhlet extraction mainly depends on characteristic of matrix and of size of particles as the internal diffusion may be a limiting step during extraction [6].

The most important advantage of conventional Soxhlet extraction technique is its continuous character. The sample has a contact with the fresh portion of the solvent. After finishing the process of extraction no filtration is required. It is also very cheap and simple method where small experience is required. The method has the possibility to extract more sample mass than the other methods [5]. Also wide industrial application and better efficiency are advantages of Soxhlet extraction over novel extraction methods [6].

The most significant disadvantages compared with other techniques are the long time of extraction, poor penetration of the matrix by the solvent, and the large amount of solvent required, which is very expensive and causes environmental problems. Some solvents have recently been questioned because of their toxicity. Also we cannot forget that extraction occurs for a long time at the boiling point of the solvent so the thermal decomposition of compounds and creating artifacts is possible [5,6]. Due to all advantages and disadvantages of Soxhlet extraction method it is the most popular one and many scientist have tried to improve it.

First, the changes focused on the new design of basic units (thimble holder, siphons, condenser) what improve application and obtained results [5]. There is also some modification shortening the time of extraction by using auxiliary energies. Thus led to create high-pressure and focused microwave-assisted Soxhlet extraction (FMASE). High pressure was achieved by placing the extractor in autoclave or using supercritical fluid Soxhlet extractor [8].

FMASE shows some differences comparing with other microwave-assisted extraction techniques: extraction is under normal pressure, irradiation is focused on the sample, filtration is not required. The advantages of that technique are shorter extraction time, capability for automation, and quantitative extraction. Extraction efficiencies and precision are better than in conventional method. Still it keeps the advantages of conventional method [8]. Also there are some applications of ultrasound-assisted Soxhlet extraction [9].

2.4.2 SUPERCRITICAL FLUID EXTRACTION

Supercritical fluid extraction (SFE) is one of the most successful techniques. Supercritical state is achieved when the temperature and the pressure of a substance is over its critical value. Under supercritical conditions the fluid has the characteristic of both liquid and gases what makes extraction faster and more effective. Many supercritical fluids have been used, such as freons, ammonia, organic solvents, but the most common is CO_2 because it has lowest toxicity and inflammability. The low supercritical temperature of carbon dioxide makes it attractive for the extraction of thermolabile compounds [5,6]. Application of some of them is limited because of their unfavorable properties with respect to safety and environmental consideration. Water in supercritical state has higher extraction ability for polar compounds, but is not suitable for thermally labile compounds [10].

Many compounds such as phenols, alkaloids, glycosidic compounds are poorly soluble in CO_2 because of its low polarity—difficulties in extracting polar analytes are the main drawbacks of the method. To improve efficiency the polarity of the extractant can be increased (by addition methanol, ethanol, pentane, acetone). The most common solvent is methanol because it is an effective modifier and is up to 20% miscible with CO_2. Sometimes ethanol is a better choice in nutraceuticals because of its lower toxicity The polarity of the analytes can also be reduced (by forming the complex or ion pair) [1,10–14]. Because SFE with CO_2 also extracts lipids from the matrix, further cleanup may be necessary to remove lipids before the analysis.

SFE is frequently used for extraction of fresh plant material. The problem is in high level of moisture what can cause mechanical difficulties. To retain the moisture some chemicals such as Na_2SO_4 or silica gel are mixed with the plant material [10].

Also very important is plant particle size. Large particles can prolong extraction process because the process is controlled by internal diffusion and fine powder can speedup extraction. However, powdered material may cause difficulties in keeping a proper flow rate causing decrease in the total yield of the extracted substances [15].

The most significant advantages of the SFE technique when compared with classical Soxhlet extraction are time reduction, its cleanness and safety, possibility for coupling on-line with detectors and chromatographs, quantitative determination, the preconcentration effect, high mass transference, completeness of extraction (supercritical fluids have a higher diffusion coefficient and lower viscosity than liquid solvent). Very important is that SFE offers possibility for selective extraction and fractionation substances from the plant material by manipulating with pressure and temperature. Choi et al. [16] confirmed it in comparison of supercritical carbon dioxide extraction with solvent extraction of squalene and stigmasterol from *Spirodela polyrhiza*. The SFE of squalene (10 Mpa, 50°C–60°C) was comparable to *n*-hexane Soxhlet extraction but the stigmasterol was not detected under these condition. The method can be easily used in the laboratory for a large scale [17].

Extracts with fewer unwanted analytes may be obtained by careful manipulation of the SFE conditions (pressure, temperature, and use of modifiers). However, the small volume of the extractor, which contains only a few grams of the material, is a disadvantage when a higher sample mass is required.

Supercritical carbon dioxide is a promising solvent for the extraction of natural compounds, especially thermolabile ones. Prevention of degradation could be achieved by eliminating oxygen from the CO_2. Apparently, the addition of antioxidants would be a reasonable solution if there were no mechanical items to adsorb the oxygen [18]. Also SFE eliminates time-consuming process of concentration and uses no or minimal organic solvent what makes method environmental friendly [5,6,12,14,19].

It is worth to notice that sometimes the efficiency is higher than 100% referred to conventional Soxhlet method (some analytes are strongly bound with matrix and not enough energy is involved in the Soxhlet process for their separation) [5].

SFE method with modifier was effective for extraction of coumarins from the peel of *Citrus maxima* [12], furanocoumarins and pentacyclic terpenoids in rhizome of *Dorstenia bryoniifolia* Mart. ex Miq. and bark roots of *Brosimum gaudichaudii* Trécul (*Moraceae*) [20].

Maceration under sonication gave better than SFE extraction of coumarins in *Mikania glomerata* leaves. SF extracts contain a high level of chlorophylls. Also addition of polar modifier (EtOH) did not present significant advantages [21].

SFE with modifier, and pressurized fluid extraction (PFE) with dichloromethane shows that flavanones and xanthones are removed from plant material at similar or slightly higher yields than obtained by solid–liquid extraction, in a much shorter period of time and with decreased amount of solvent [22].

Biologically active substances of rose hip seeds like unsaturated fatty acids and carotene was extracted by SFE with carbon dioxide and propane. Oil yield was higher in comparison with traditional Soxhlet extraction [23].

Effect of low, medium, and high polarity under very high pressure and with polar modifiers has been investigated by Hamburger et al. [24]. *Calendula officinalis*,

Crataegus sp, and *Matricaria recutita* were used as models in this study. Extraction yields under different conditions depended to the large extent on the profiles of secondary metabolites present in the plant material. Extractability of lipophilic compounds increased substantially at pressure higher than 300 bar, the yields of polyphenolics and glycosides remained low even at pressure about 700 bar with 20% of modifier in the extraction fluid. A wide range of applications for the extraction of biologically active substances are described in review article [6,7,25,26].

2.4.3 PRESSURIZED LIQUID EXTRACTION

Pressurized liquid extraction (PLE) also known as an accelerated solvent extraction (ASE), high-pressure solvent extraction (HPSE), or pressurized fluid extraction (PFE) is a technique which uses small volume of conventional solvents at elevated temperatures (>200°C) in a very short time to extract solid samples. The pressure is higher than 200 bar in order to keep the solvent in liquid state (the solvent is still below its critical point), increasing temperature accelerates the extraction kinetics (process of the desorption of analytes from the matrix are faster compared to the condition when solvent at room temperature are used), which gives safe and rapid method [6,7,27,28]. ASE allows the universal use of solvents or solvent mixtures with different polarities.

In PLE, sample with sand, sodium sulfate or Hydromatrix as a dispersant are placed in a cell. Extraction cell is filled with the solvent and the cell is set to certain values, then is heated in an oven to the set values. During the heating cycle, solvent is pumped in and out of the cell to maintain the pressure and to perform the cycles indicated. The fluid coming out of the extraction cell is collected in the collection vial. Before loading the plant materials into the extraction cell, the samples are often pretreated. Proper size of sample enables right diffusion of analytes from the sample to the solvent extract. Drying the sample removes any moisture which may diminish extraction efficiency. The chosen extraction solvent must be able to solubilize the analytes of interest, minimizing the coextraction of other matrix components. Its polarity should be close to that of the target compound [27,29].

PLE can be accomplished in the static (sample and solvent are maintained for a defined time at constant pressure and temperature) or dynamic mode (the solvent flows through the sample in a continuous manner). Because in most cases the dynamic mode uses water as extractant, several authors have preferred to use the term pressurized hot water extraction (PHWE) to refer it. Water is nonflammable, nontoxic, readily available, and an environmentally acceptable solvent. PHWE with and without the addition of a small percentage of organic solvent such as ethanol is highly suited for the chemical standardization and quality control of medicinal plants. At the same time, it can be applied at the pilot scale as a manufacturing process for medicinal plants. Further information about application of PHWE in extraction of active compounds can be found in review papers [27,29,30].

Very important is complete automation of the whole analytical process and highly selective extractions of compounds of different polarities [27].

Compared with classic extraction in Soxhlet apparatus, complete extraction can be achieved in shorter time with a small volume of organic solvent and much better penetration of sample by the solvent in PLE. Keeping the high temperature reduces solvent viscosity and helps in breaking down analyte–matrix bonds. Also no additional filtration is required [27].

The suitability of pressurized liquid extraction (PLE) in medicinal plant analysis was investigated. PLE extracts from representative herbs containing structurally diverse metabolites of varying polarity and solubility were compared with extracts obtained according to Pharmacopoeia monographs with respect to yield of relevant plant constituents, extraction time, and solvent consumption. Experiment shows that one to three extraction cycles of 5 to 6 min at high temperatures afforded exhaustive or almost exhaustive extraction (instead of many hours of Soxhlet extraction). It markedly reduces not only time but also solvent consumption and protect against artifacts of extracted compounds at high temperatures. Reproducibility of results was generally better [31].

Among extraction in Soxhlet apparatus, ultrasonification, microwave-assisted solvent extraction in open and close system and pressurized solvent extraction ASE gives higher yield of furanocoumarins (especially for hydrophobic) from *Pastinac sativa* and *Archangelica officinalis* fruits [32,33]. Dawidowicz et al. have optimized the extraction condition for the analysis of rutin and isoquercitrin in *Sambucus nigra* flowers, leaves, and berries [34].

Ong and Len [35] have developed a method for the analysis of glycosides in medicinal plants using PHWE. The results obtained with this technique were even better than those obtained with Soxhlet extraction. The study showed that hydrophobic and thermally labile components in medicinal plants could be extracted using the combination of surfactant and pressurized hot water at a temperature below its boiling point and lower applied pressure.

The possibility of selectivity changing for extractions of the most typical rosemary antioxidant compounds by means of a small change of temperature has been demonstrated. The experiment shows that it is possible to obtain extracts enriched with different types of polyphenols [36].

PLE was examined as an alternative technology for the extraction of carotenoids in the marine green algae *Haematococcus pluvialis* and *Dunaliella salina* and kavalactones in *Piper methysticum*. The results of this study showed that PLE had comparable extraction efficiency to traditional extraction techniques, however required half the amount of extracting solvent as traditional extraction, and is less time-consuming [37].

PLE with ethyl alcohol of antidiabetic compound, charantin, from fruits of *Momordica charantia* (bitter melon) was proposed as an alternative for conventional Soxhlet extraction with toxic solvent such as chloroform or dichloromethane [38].

PLE was more effective for the extractions of terpenes (terpenic alcohols and phytosterols), fatty acids, and vitamin E from leaves of *Piper gaudichaudianum* Kunth and decreased significantly the total time of extraction, the amount of solvent, and the manipulation of sample and solvents in comparison with Soxhlet [39].

ASE has been used in analysis of polyphenols [40], taxanoids [41], alkaloids [42]. Various aspects of application of PLE of natural products were discussed in literature [6,7,27].

2.4.4   MEDIUM-PRESSURE SOLID–LIQUID EXTRACTION TECHNIQUE

Medium-pressure solid–liquid extraction (MPSLE), introduced by Nyiredy et al. [43], is an extraction technique based on the diffusion-dissolving processes. MPSLE can be performed in a medium-pressure liquid chromatography (MPLC) column filled with the fine-powdered solid phase to be extracted and extractant is pumped through the stationary bed. The extraction distance is relatively short even though the separation is efficient due to the fine particle size of the solid phase. Extraction can be performed with continuous solvent flow and constant applied pressure. Various solvents or their mixtures can be applied.

This extraction process is exhaustive and rapid. Efficiency is comparable to the exhaustive extraction method like Soxhlet extraction, but here the process of concentration is not required. Also all experiment is performed in close and automated system with small amount of organic solvent so it is environment friendly [44]. Choice of solvent systems can be very efficiently performed by analytical HPLC. Transposition to MPLC is straightforward and direct [45].

2.5   PURIFICATION OF CRUDE EXTRACTS AND SAMPLE PREPARATION

For optimization of all the extraction and analytical processes, some interfering compounds having high molecular size such as lipids, pigments, waxes presented in the crude extracts should be eliminated. The next purpose is preconcentration and isolation of analytes.

For removal of coextracted substances different cleanup procedures have been developed such as liquid–liquid partitioning, column or adsorption chromatography on polar adsorbents (Florisil or silica), and gel permeation chromatography (GPC) [27,46].

2.5.1   LIQUID–LIQUID PARTITION

Liquid–liquid partition (LLP) in the past played the major role in the pretreatment the sample (cleanup, concentration, and isolation). LLE is based on the rule that when a third substance is added to a mixture of two immiscible liquids being in equilibrium, the added component will divide itself between the two liquid phases until the ratio of its concentrations in each phase attain a certain value [47].

Method is slow, required the long time, large amount of organic solvents and sample. An important difficulty is formation of emulsion which breaks up very slowly and incompletely [48,49].

Multiple partition steps provide possibility for the preliminary purification of samples, which are to be separated by various chromatographic processes. This is generally achieved by countercurrent methods [45].

2.5.2   SOLID PHASE EXTRACTION

Solid phase extraction (SPE) is an alternative method to liquid–liquid extraction for the separation, purification, and concentration. While LLE relies on partitioning of compounds between two liquid phases in SPE, instead of two immiscible liquid phases, analytes are divided between solid and liquid (sorption step—analytes have greater affinity to the solid phase than for the sample matrix). Retained compounds are removed from solid phase by eluting with an appropriate solvent or mixture of solvents with a greater affinity to the analytes (desorption step). Extracted analytes may also be eluted from SPE disks by a supercritical fluid. Use of carbon dioxide as an SF have some advantages over liquid organic solvents because it is nontoxic, chemically inert, and easy to discard [46,48,49]. The most common retention mechanisms are based on van der Waals forces, hydrogen bonding, dipole–dipole interactions, and cation–anion interactions [50,51].

SPE is usually used to clean up a sample before chromatographic analysis to quantify analytes in the sample and also to remove the interfering components of the complex matrices in order to obtain a cleaner extract containing pure fraction of interesting substances [50].

The applicability of SPE is mainly determined by using different sorbents such as alumina, silica gel, chemically bonded silica phases, ion exchangers, and polymers. Some of them are modified and can be very selective. The polarity of the mobile phase depends on the chemical composition of stationary phase and can change the separation selectivity [46,49,50].

SPE has many advantages. First preconcentration effect is important especially in trace analysis. Usually the volume of solvent needed for complete elution of the analytes is much smaller than the original sample volume. A concentration of the analytes is thus achieved [48,49].

Very important is fractionation of the sample into different compounds or groups of compounds by eluting each fraction with a different liquid phase. Zgórka [52,53] developed a new method for simultaneous determination of phenolic acids and furanocoumarins by use of octadecyl and quaternary amine SPE-microcolumns with changing of mobile phase polarity and eluent strength. Also she fulfills dividing of flavonoids and phenolic acids using the same microcolumns. SPE procedures with modified RP-18 adsorbents were used for elution of proanthocyanidins [54,55]. The combination of molecularly imprinted polymers (MIPs) and SPE is a promising technique that allows selective extraction of specific analytes from complex matrices (LIV). SPE process is easy, fast, requires small amount of samples and small volume of solvents and can be performed either on-line or off-line and can be easily automated. It is also more efficient than LLE and reveals higher recoveries of the analyte [46,48–50,56].

One of the important parameters to control SPE method is the breakthrough volume, which represents the maximum sample volume that can be applied with a theoretical 100% recovery, while recovery is defined as the ratio between the amount extracted and the amount applied [49,57]. An overview of SPE was also published by Żwir-Ferenc and Biziuk [50], Camel [51], Hennion [57], and Poole [58].

Similar chromatographic systems as are used in SPE can be applied in classic column chromatography for purification and fractionation of extracts in preparative scale.

2.5.3 Gel Permeation Chromatography

GPC, also called size-exclusion chromatography, has been used to quantitative separation of molecules in aqueous or buffered solvents on the basis of molecular size, partition, and absorption.

The process is fully automated. It is a highly effective postextraction cleanup method for removing high molecular-weight interferences such as lipids, proteins, and pigments from sample extracts prior to analysis. The procedure can be automated and is suitable for the purification of complicated matrices with a high liquid content [27,46].

The cleaning-up of fatty samples is time-consuming, and often required more than one step. In most cases after GPC the extract can be separated on various adsorbents using different chromatographic methods. The combination of these two techniques results in powerful two-dimensional cleanup by molecular size and polarity. Because of large dimension of GPC column, the concentration of obtained fraction before second step is required [27,59].

To remove lipids from sample in situ cleanup method was developed. Elimination of ballasts can be achieved by adding sorbents such as Florisil, alumina, and silica gel to the PLE cells, which retain co-extractable materials from extract [27].

2.6 CHROMATOGRAPHIC METHODS AND THEIR ROLE IN INVESTIGATIONS OF PLANT MATERIAL

2.6.1 Gas Chromatography

Gas chromatography (GC) is used mostly for analysis of volatile, unpolar (hydrophobe) compounds such as the components of essential oils and other unpolar constituents itself (different kind of terpenes).

Now it is usually coupled with mass spectrometry (GC–MS). This technique allows the measurement of the molecular weight of a compound and once a molecular ion has been identified, it is possible to measure this ion accurately to ascertain the exact number of hydrogens, carbons, oxygens, and the other atoms that may be present in the molecule. This will give a molecular formula. Several techniques are available in MS, of which electron impact is widely used. This techniques gives good fragmentation of the molecule and is useful for structure elucidation as the fragments can be assigned to functional groups present in the molecule. The disadvantage of this technique is that molecular ions are sometimes absent. Softer techniques such as chemical ionization (CI), electrospray ionization (ESI), and fast atom bombardment (FAB) mass spectrometry ionize the molecule with less energy; consequently, molecular ions are generally present, but with less fragmentation information for structure elucidation purposes. For example GC–MS technique was used to analyze volatile compounds present in different types of *Mentha piperita* like menthol, menthone, isomenthone, 1,8-cineole, limonene, and so on [60].

2.6.2 High-Performance Liquid Chromatography

High-performance liquid chromatography (HPLC) is currently popular and is widely used for the analysis and isolation of bioactive natural products. The analytical sensitivity of the technique, particularly when is coupled with UV detection such as photodiode array (DAD), enables the acquisition of UV spectra of eluting peaks from 190 to 800 nm. The flow rates of this system are typically 0.5–2.0 mL/min and sample loading in the analytical mode allows the detection and separation of plant extract components. With DAD UV detection, even compounds with poor UV characteristic can be detected.

HPLC is a highly sensitive technique when is coupled to electronic library searching of compounds with known UV spectra. Modern software enables the UV spectra of eluting peaks to be compared with spectra stored electronically, thereby enabling early identification of known compounds or, usefully, the comparison of novel compounds with a similar UV spectrum, which may indicate structural similarity. It is also possible to increase the size of these electronic libraries and improve the searching power of the technique. HPLC is a powerful technique for fingerprinting of biologically active extracts and for comparisons that can be drawn with chromatograms and UV spectra stored in an electronic library. This is currently very important for the quality control of herbal medicines.

With HPLC, which can be run in fully automated mode, and carousel autosamplers, it is possible to analyze tens to hundreds of samples. HPLC apparatus are computer-driven and chromatographic run may be programmed. Computers are also used for the storage of the chromatographic data. Modern technology can be applied for column sorbents, including standard sorbents such as silica and alumina, reversed phase C18 C8, phenyl stationary phases, and modern sorbents such as polar bonded stationary phases (CN-, Diol-, aminopropyl-silica), chiral stationary phases, gel size–exclusion media, and ion exchangers. This versatility of stationary phase has made HPLC a highly popular method for bioassay-guided isolation.

HPLC is a high-resolution technique, with efficient, fast separations. The most widely used stationary phase is C_{18} (reverse-phase) chromatography, generally employing water/acetonitrile or water/methanol mixtures as mobile phase. These mobile phases may run in gradient elution mode, in which the concentration of a particular solvent is increased over a period of time, starting, for example, with 100% water and increasing to 100% acetonitrile over 30 min, or in isocratic elution mode, in which a constant composition (e.g., 70% acetonitrile in water) is maintained for a set period of time.

2.6.3 ELECTROPHORESIS AND ELECTROCHROMATOGRAPHY

Electrophoresis has been known for many years but gain widespread attention when Jorgenson and Lukacs demonstrated small quartz tubes. Since that time capillary electrophoresis (CE) has been very popular method for separation of active compounds. The separation is obtained by differential migration of solutes in an electric field, driven by two forces, the electrophoretic migration and the electro-osmotic flow. Migration into discrete zones is due to differences into electrophoretic mobilities which is related to the mass-to-charge ratio. In a typical CE instrument, analytes are introduced at the anode and are detected at the cathode [61,62].

The equipment for CE is very simple and cheap. A typical capillary tube is 25–75 cm long and made of silica. The most widely used detector for monitoring of plant metabolites is an ultra-violet (UV) spectrophotometer [61]. CE has been coupled also to mass spectrometry through an ESI interface [62].

Different classes of compounds were analyzed with CE and results were comparable to HPLC while CE offer advantages in terms of high efficiency, simplicity, low solvent consumption, and short analysis times. CE method permits the use of very small amounts of sample because of small dimensions of the capillary. The important advantage is large surface to volume ratio that allows for application of large electric potential across the capillary [62].

The use of capillaries makes possible to use very strong electric fields what significantly reduces the time required for the effective separation [63]. The number of CE applications is growing rapidly especially in pharmaceutical analysis—where not high sensitivity but efficient separations is often a major issue [64].

Disadvantages include low sensitivity comparing to HPLC and limitation in the preparative scale. CE can be an alternative where analysis requires higher efficiency or resolution than HPLC [61].

CE is a micro-analytical method which provides advantages in terms of speed, high efficiency, low cost, and simplicity. It can be applied for the separation of different classes of compounds using the same equipment, changing only the composition of the buffer. CE can be used on-line with other techniques such as HPLC, NMR, MS. It needs small amount of sample and buffer [65].

Several modes of CE are available—capillary zone electrophoresis (CZE), micellar electrokinetic chromatography (MEKC), capillary gel electrophoresis (CGE), capillary isoelectric focusing (CIEF), capillary isotachophoresis (CITP), capillary electrochromatography (CEC), and nonaqueous CE—but CZE and MEKC are the most popular for phytochemical analysis [61].

In CZE, capillary containing the running electrolyte is suspended between two buffer reservoirs and the sample is introduced at the anodic end. A large potential is applied across the capillary and substances migrate according to its electrophoretic mobility. In MEKC, the buffer solution contains surfactant (e.g., sodium dodecyl sulfate) in a concentration above the critical micelle concentration. Analytes may portion themselves on formed micelles as a stationary phase [62,63]. MEKC is also very helpful for poorly water-soluble samples [62].

The use of gel (CGE) has made a significant impact on peptide separation. Cross-linked gels may work as molecular sieve and distinguish substances on the basis of their molecular weight (Ed). Applications of electrophoresis techniques for the separation of different compounds from natural materials are described in some review papers [62,63,65,66].

CEC is a technique that combines the advantages of CZE (high efficiency) with those of HPLC (high selectivity). The solutes are transported through the column by the electro-osmotic flow (EOF) of the solvent or by their own electrophoretic mobility, and there is less zone broadening than in LC [67,68]. On the basis of differences in column format, three modes of CEC can be distinguished: with packed columns, with monolithic columns (silica and polymer-based material) and the open-tubular format, with a stationary phase as a coating on the wall of a capillary (OT-CEC) [69]. An additional advantage of CEC is the lack of pressure limitation allowing the use of smaller particles than in LC. Solutes are separated according to their partitioning between the mobile and stationary phases and, when charged, their electrophoretic mobility [64].

2.6.4 Coupled Methods (GC–MS, LC–MS, LC–NMR)

Rapid detection of biologically active natural products plays a strategical role in the phytochemical investigation of crude plant extracts. Hyphenated techniques such as HPLC coupled to UV photodiode array detection (LC–DAD–UV) and to mass spectrometry (LC–MS or LC–MS–MS) as well as GC–MS) provide on-line numerous structural information on the metabolites prior to isolation. The use of all these hyphenated techniques allows the rapid structural determination of known plant constituents with only a minute amount of plant material. With such an approach, the time-consuming isolation of common natural products is avoided and an efficient targeted isolation of compounds presenting interesting spectroscopical or biological features can be performed.

Crude plant extracts are very complex mixtures containing sometimes hundreds or thousands of different metabolites. The chemical nature of these constituents differs considerably within a given extract. The variability of the physicochemical parameters of these compounds causes numerous detection problems. Although different types of LC detectors exist, such as UV, IR, fluorescence, electrochemical, evaporative laser light scattering, etc., none permits the detection of all the secondary metabolites encountered in a plant extract within a single analysis. For example, a product having no important chromophore cannot be detected by UV detector. Detection of all these compounds can be performed by mass spectrometry (MS) detector, which can be ideally considered as universal detection technique. At present, MS is the most sensitive method of molecular analysis and has the potential to yield an information on the molecular weight as well as on the structure of the analytes. Furthermore, due to its high power of mass separation, very good selection can be obtained. The main problem of the use of LC–MS in natural product chemistry resides in the ionization of compounds found in a crude plant extract. Indeed, if many LC–MS interfaces exist on the market, none of them allows a real universal ionization of all constituents of a plant extract. Each of these interfaces has its own characteristic and range of applications, and several of them are suitable for analysis of plant secondary metabolites [70].

REFERENCES

1. Heinrich, M. et al., *Fundamentals of Pharmacognosy and Phytotherapy*, Elsevier Science, Churchill Livingstone, Edinburgh, 2004.
2. Gathercoal, E.N. and Wirth, E.H., *Pharmacognosy*, Lea & Febiger, Philadelphia, 1947.
3. Youngken, H.W., *Textbook of Pharmacognosy*, 6th edition, The Blakiston Company, Philadelphia, Toronto, 1950.
4. Mroczek, T., Widelski, J., and Glowniak, K., Optimization of extraction pyrrolizidine alkaloids from plant material. *Chem. Anal.*, 51, 567, 2006.
5. Luque de Castro, M.D. and Garcia-Ayuso, L.E., Soxhlet extraction of solid materials: an outlet technique with promising innovative future, *Anal. Chim. Acta*, 369, 1, 1988.

6. Wang, L. and Welle, C.L., Recent advances in extraction of nutraceuticals from plants, *Trends Food Sci. Tech.*, 17, 300, 2006.

7. Raynie, D.E., Modern extraction techniques, *Anal. Chem.*, 76, 4659, 2004.

8. Luque-Garcia, J.L. and Luque de Castro, M.D., Focused microwave-assisted Soxhlet extraction: devices and applications, *Talanta*, 64, 571, 2004.

9. Luque-Garcia, J.L. and Luque de Castro, M.D., Ultrasound-assisted Soxhlet extraction: an expeditive approach for solid sample treatment. Application to the extraction of total fat from oleaginous seeds, *J. Chromatogr. A*, 1034, 237, 2004.

10. Lang, Q. and Wai, C.M., Supercritical fluid extraction in herbal and natural product studies – a practical review, *Talanta*, 53, 771, 2001.

11. Lancas, F.M.M.C., Queiroy, M.E.C., and da Silva, I.C.E., Seed oil extraction with supercritical carbon dioxide modified with pentane, *Chromatographia*, 39, 687, 1994.

12. Teng, W.Y., Chen, C.C., and Chung, R.S., HPLC comparison of supercritical fluid extraction and solvent extraction of coumarins from the peel of Citrus maxima fruits, *Phytochem. Anal.*, 16, 459, 2005.

13. Ashraf-Khorassani, M., Gidanian, S., and Zamini, Y., Effect of pressure, temperature, modifier, modifier concentration and sample matrix on the supercritical fluid extraction efficiency of different phenolic compounds, *J. Chromatogr. Sci.*, 33, 658, 1995.

14. Luque de Castro, M.D. and Jimeènez-Carmona, M.M., Where is supercritical fuid extraction going? *Trends Anal. Chem.*, 19, 223, 2000.

15. Chemat, S. et al., Comparison of conventional and ultrasound-assisted extraction of carvone and limonene from caraway seeds, *Flavour Frag. J.*, 19, 188, 2004.

16. Choi, Y.H. et al., Comparison of supercritical carbon dioxide extraction with solvent extraction of nonacosan-10-ol, α-amyrin acetate, squalene and stigmasterol from medicinal plants, *Phytochem. Anal.*, 8, 233, 1997.

17. King, J.W., Analytical-process supercritical fluid extraction: a synergestic combination for solving analytical and laboratory scale problems, *Trends Anal. Chem.*, 14, 474, 1995.

18. Cocero, M.J. et al., Supercritical extraction of unsaturated products. Degradation of b-carotene in supercritical extraction processes, *J. Supercrit. Fluid*, 19, 39, 2000.

19. Brogle, H., CO_2 as a solvent: its properties and application, *Chem. Ind.*, 11, 385, 1982.

20. Vilegas, J.H.Y. et al., Off-line supercritical fluid extraction—high resolution gas chromatography applied to the study of *Moraceae* species, *Phytochem. Anal.*, 4, 230, 1993.

21. Celeghini, R.M.S., Vilegas, J.H.Y., and Lancas, F.M., Extraction and quantitative HPLC analysis of coumarin in hydroalkoholic extracts of Mikania glomerata Spreng. (guaco) leaves, *J. Braz. Chem. Soc.*, 12, 706, 2001.

22. da Costa, C.T. et al., Comparison of methods for extraction of flavanones and xanthones from the root bark of the osage orange tree using liquid chromatography, *J. Chromatogr. A*, 831, 167, 1999.

23. Szentmichályi, K. et al., Rosa hip (*Rosa canina* L.) oil obtained from waste hip seeds by different extraction methods, *Bioresour. Technol.*, 82, 195, 2002.

24. Hamburger, M., Baumann, D., and Adler, S. Supercritical carbon dioxide extraction of selected medicinal plants—effects of high pressure and added ethanol on yield of extracted substances, *Phytochem. Anal.*, 15, 46, 2004.

25. Angela, M. and Meireles, A., Supercritical extraction from solid: process design data (2001–2003), *Curr. Opin. Solid State Mater. Sci.*, 7, 321, 2003.

26. Modey, W.K., Mulholland, D.A., and Raynor, M.W. Analytical supercritical fluid extraction of natural products, *Phytochem. Anal.*, 7, 1, 1996.

27. Carabias-Martinez, R. et al., Pressurized liquid extraction in the analysis of food and biological samples, *J. Chromatogr. A*, 1089, 1, 2005.

28. Richter, B.E. et al., Accelerated solvent extraction: a technique for sample preparation, *Anal. Chem.*, 68, 1033, 1996.
29. Ong, E.S., Cheong, J.S.C., and Goh, D., Pressurized hot water extraction of bioactive or marker compounds in botanicals and medicinal plant materials, *J. Chromatogr. A*, 1112, 92, 2006.
30. Smith, R.M., Extractions with superheated water, *J. Chromatogr. A*, 975, 31, 2002.
31. Benthin, B., Danz, H., and Hamburger, M., Pressurized liquid extraction of medicinal plants, *J. Chromatogr. A*, 837, 211, 1999.
32. Waksmundzka-Hajnos, M. et al., Influence of the extraction mode on the yield of some furanocoumarins from *Pastinaca sativa* fruits, *J. Chromatogr. B*, 800, 181, 2004.
33. Waksmundzka-Hajnos, M. et al., Effect of extraction method on the yield of furanocoumarins from fruits of *Archangelica officinalis* Hoffm, *Phytochem. Anal.*, 15, 1, 2004.
34. Dawidowicz, A.L. et al., Optimization of ASE conditions for the HPLC determination of rutin and isoquercitrin in *Sambucus nigra* L., *J. Liq. Chromatogr. Relat. Technol.*, 26, 2381, 2003.
35. Ong, E.S and Len, S.M., Evaluation of surfactant-assisted pressurized hot water extraction for marker compounds in *Radix Codonopsis pilosula* using liquid chromatography and liquid chromatography/electrospray ionization mass spectrometry, *J. Sep. Sci.*, 26,1533, 2003.
36. Ibañez, E. et al., Subcritical water extraction of antioxidant compounds from rosemary plants, *J. Agric. Food Chem.*, 51, 375, 2003.
37. Denery, J.R. et al., Pressurized fluid extraction of carotenoids from *Haematococcus pluvialis* and *Dunaliella salina* and kavalactones from *Piper methysticum*, *Anal.Chim. Acta*, 501, 175, 2004.
38. Pitipanapong, J. et al., New approach for extraction of charantin from *Momordica charantia* with pressurized liquid extraction, *Sep. Purif. Technol.*, 52, 416, 2007.
39. Péres, V.F. et al., Comparison of soxhlet, ultrasound-assisted and pressurized liquid extraction of terpenes, fatty acids and Vitamin E from *Piper gaudichaudianum* Kunth, *J. Chromatogr. A*, 1105, 115, 2006.
40. Papagiannopoulos, M. et al., Online coupling of pressurized liquid extraction, solid-phase extraction and high-performance liquid chromatography for automated analysis of proanthocyanidins in malt, *J. Chromatogr. A*, 958, 9, 2002.
41. Hajnos, M.Ł. et al., Influence of the extraction mode on the yield of taxoids from yew tissues – preliminary experiments, *Chem. Anal.*, 46, 831, 2001.
42. Ong, E.S., Woo, S.O., and Yong, Y.L., Pressurized liquid extraction of berberine and aristolochic acid in medicinal plants, *J. Chromatogr. A*, 904, 57, 2000.
43. Nyiredy, Sz., Botz, L., and Sticher, O., Swiss Pat. CH 674 314, 1990.
44. Nyiredy, Sz. and Botz, L., Medium-pressure solid–liquid extraction: a new preparative method based on the principle of counter-current, *Chromatographia*, Suppl. 57, S-291, 2003.
45. Hostettmann, K., Marston, A., and Hostettmann, M., *Preparative Chromatography Techniques: Applications in Natural Product Isolation*, 2nd edition, Springer, Berlin, 1997.
46. Ahmed, F.E., Analytes of pesticides and their metabolites in foods and drinks, *Trends Anal. Chem.*, 20, 649, 2001.
47. Berthod, A. and Carda-Broch, S., Determination of liquid–liquid partition coefficients by separation methods, *J. Chromatogr. A*, 1037, 3, 2004.
48. Fritz, J.F., *Analytical solid-phase extraction*, John Wiley & Sons, New York, 1999.
49. Berrueta, L.A., Gallo, B., and Vicente, F., A review of solid phase extraction: basic principles and new developments, *Chromatographia*, 40, 474, 1995.

50. Żwir-Ferenc, A. and Biziuk, M., Solid Phase extraction technique – trends, opportunities and application, *Polish J. Environ. Stud.*, 15, 677, 2006.
51. Camel, V., Solid phase extraction of trace elements, *Spectrochim. Acta B*, 58, 1177, 2003.
52. Zgórka, G. and Głowniak, K., Simultaneous determination of phenolic acids and linear furanocoumarins in fruits of *Libanotis dolichostyla* by solid-phase extraction and high-performance liquid chromatography, *Phytochem. Anal.*, 10, 268, 1999.
53. Zgórka, G. and Hajnos, A., The application of solid-phase extraction and reversed phase high-performance liquid chromatography for simultaneous isolation and determination of plant flavonoids and phenolic acids, *Chromatographia*, Suppl. 57, S-77, 2003.
54. Hammerstone, J.F. et al., Identification of procyanidins in cocoa (*Theobroma cacao*) and chocolate using high-performance liquid chromatography/mass spectrometry, *J. Agric. Food Chem.*, 47, 490, 1999.
55. Lazarus, S.A. et al., High-performance liquid chromatography/mass spectrometry analysis of proanthocyanidins in foods and beverages, *J. Agric. Food Chem.*, 47, 3693, 1999.
56. Qiao, F. et al., Molecularly imprinted polymers for solid phase extraction, *Chromatographia*, 64, 625, 2006.
57. Hennion, M.C., Solid-phase extraction: method development, sorbents, and coupling with liquid chromatography, *J. Chromatogr. A*, 856, 3, 1999.
58. Poole, C.F., New trends in solid-phase extraction, *Trends Anal. Chem.*, 22, 362, 2003.
59. Rimkus, G.G., Rummler, M., and Nausch, I., Gel permeation chromatography–high performance liquid chromatography combination as an automated clean-up technique for the multiresidue analysis of fats, *J. Chromatogr. A*, 737, 9, 1996.
60. Gherman, C., Culea, M., and Cozar, O., Comparative analysis of some active principles of herb plants by GC/MS, *Talanta*, 53, 253, 2000.
61. Suntornsuk, L., Capillary electrophoresis of photochemical substances, *J. Pharm. Biomed. Anal.*, 27, 679, 2002.
62. Rabel, S.R. and Stobaugh, J.F., Applications of capillary electrophoresis in pharmaceutical analysis, *Pharm. Res.*, 10, 2, 1993.
63. Morzunova, T.G., Structure of chemical compounds, methods of analysis and process control. Capillary electrophoresis in pharmaceutical analysis, *Pharm. Chem. J.*, 40, 3, 2006.
64. Hilhorst, M.J., Somsen, G.W., and Jong, G.J., Capillary electrokinetic separation techniques for profiling of drugs and related products, *Electrophoresis*, 22, 2542, 2001.
65. Issaq, H.J., A decade of capillary electrophoresis, *Electrophoresis*, 21, 1921, 2000.
66. Issaq, H.J., Capillary electrophoresis of natural products, *Electrophoresis*, 20, 3190, 1999.
67. Smith, N.W. and Carter-Finch, A.F., Electrochromatography, *J. Chromatogr. A*, 892, 219, 2000.
68. Vanhoenacker, G. et al., Recent applications of capillary electrochromatography, *Electrophoresis*, 22, 4064, 2001.
69. Eeltink, S. and Kok, W.T., Recent applications in capillary electrochromatography, *Electrophoresis*, 27, 84, 2006.
70. Wolfender, J.L., Rodiguez, S., and Hostettmann, K., Liquid chromatography coupled to mass spectrometry and nuclear magnetic resonanse spectroscopy for the screening of plant constituents, *J. Chromatogr. A*, 794, 299, 1998.

3 Medicines and Dietary Supplements Produced from Plants

Anita Ankli, Valeria Widmer, and Eike Reich

CONTENTS

3.1 INTRODUCTION

This chapter is intended to illustrate the broad range of applications of TLC for the analysis of medicinal plants and derived products. While phytochemical research, in general, focuses on plant constituents and their pharmacological activities we want to describe analytical aspects of quality, efficacy, and safety of raw materials and products. Those aspects are usually related to regulatory issues and typically monitored by authorities. There are different approaches to medicinal plants and their use in Traditional Western, Traditional Chinese, and Ayurvedic Medicine on one hand side and botanical dietary supplements on the other. Correspondingly the analytical challenges can be quite different. Although one technique alone is not able to answer all questions, TLC is most versatile and therefore widely applicable. This chapter raises common analytical questions of plant analysis and works out possible solutions. The available literature is reviewed and key applications of TLC, novel approaches as well as official guidelines and methods are summarized.

3.1.1 HISTORICAL DEVELOPMENT OF HERBAL MEDICINES

Among the earliest documents mentioning the use of ancient plants as medicine are Assyrian and Babylonian clay tablets of 1700–1000 years B.C. The Ebers papyrus of Egypt (1550 B.C.) describes different medicines, treatment of numerous ailments, and incantations to turn away disease-causing demons and other superstitions. Herbal drugs were the principal therapeutic resources. Greek philosophers had a crucial influence on medicine. They did not just observe nature but looked for an explanation and reason behind things. Hippocrates (460–377 B.C.) developed a humoral medicine system, which was used to restore the balance of humors within the body [1].

A similar system is seen in the Indian practice of Ayurveda. It is uncertain whether this represents an independent origin of a parallel idea or is possibly related to the Greek concept. The Unani medicinal system in India is based on the learning of Hippocrates. It was brought to India by Arabic physicians. With the help of natural forces, such as plant medicines, it helps the patient to regain the power of self-preservation to retain health. The Siddha system of Medicine is similar to the ayurvedic system [2].

In China, a system based on the concept of ch'i (breath, air, spirit) including Yin and Yang was created during the era of the Yellow Emperor. The first use of hot, cold, wet, and dry to classify medicine and food was mentioned later. It is known that Greek medicinal doctrine reached China via India by about 500 A.D. In Japanese Kampo Medicine the humoral concept is not present; however, the medicine system is based on TCM and Ayurveda [3].

Hippocrates is called the first official natural healer of Europe. At about 1000 A.D. the progress of the Arabic science influenced the knowledge of the ancient times. When the Renaissance broke with the dogmas of these concepts, a new science of medicine was based on empirical and practical knowledge. Causal analytical thinking of the modern natural science was born [4]. With the advances of chemistry and growth of the pharmaceutical industry the Western World saw a shift to the primary use of synthetic drugs as medicines, while China, India, and other Asian countries still preserved the use of traditional mostly plant derived medicines.

In the last decades and particularly, with the advent of globalization, there is again a growing popularity of medicinal plants and plant derived products in Europe and North America. Natural alternatives to synthetic drugs are in demand. In Europe herbal medicines were developed, in the USA Botanical Dietary Supplements appeared on the market, and in Canada Natural Health Products became available. At the same time Traditional Chinese and Ayurvedic Medicines have entered the world market. These products and their raw materials may belong to very different systems of knowledge or beliefs and their use may be based on diverse traditions, but in today's world they all must meet certain criteria with respect to safety and minimum quality to be in compliance with national and international regulations. There is also an economic aspect: consumers have more choices and they compare value against cost. Marketing strategies based on quality and quantity becomes driving forces for the development of analytical techniques, which can help comparing different products. Plants originating from foreign medicinal systems are emerging on the market and many "unknown" plant drugs come from other parts of the world. There is growing concern about adulteration, when shortages of the most popular plants develop.

3.1.2 REGULATORY ISSUES AND QUALITY CONTROL

Depending on the kind of product derived from a plant and the purpose for which and where it is sold, different regulations may apply, but one of the central elements is always the necessity to properly define the identity of the starting material as well as its consistency with respect to specifications. For some of the most widely used plants, pharmacopoeial monographs are available, which serve as the basis for quality. For other plants, such monographs still have to be elaborated. Aside from botanical and organoleptic characteristics of the plant, monographs usually describe tests for chemical identification and assays.

As analytical techniques thin layer chromatography (TLC) and high-performance thin layer chromatography (HPTLC) are included in the European Pharmacopoeia (PhEur), the United States Pharmacopoeia (USP), the Pharmacopoeia of the Peoples Republic of China (PhPRCh) as well as in the American Herbal Pharmacopoeia (AHP) [5–8]. Also in the Quality Standards of Indian Medicinal Plants (QS-IMP) and the Indian Herbal Pharmacopoeia (IHP), TLC and HPTLC are recommended for qualitative and quantitative evaluation of phytochemical constituents of herbal drugs [9,10]. A general trend towards harmonization and consolidation of monograph contents and the use of comparable analytical tools can be observed in the pharmacopoeias.

The WHO guidelines for the worldwide use of herbal drugs recommend TLC/HPTLC and the use of chromatographic fingerprints for identification and qualitative determination of impurities of herbal medicines [11,12].

If a plant derived product is sold in Europe as Herbal Medicinal Product (HMP) it requires proof of quality, safety, and efficacy for approval. The European Medicines Agency (EMEA) has issued regulations and directives for the quality control of herbal drugs [13–16]. An herbal drug preparation in its entirety is regarded as the active substance. It is recommended to use chromatographic fingerprint techniques such as TLC/HPTLC for identity test, tests for presence of adulterants, and to determine the stability of the HMP. Quality testing must meet International Conference on Harmonization (ICH) standards, regarding specification, validation, and stability testing [17]. The licensing procedure requires full compliance with current Good Manufacturing Practice (cGMP).

If a plant derived product is sold as a dietary supplement the primary difference is that no efficacy has to be demonstrated and that the product from a regulatory perspective is treated as food rather than a drug. cGMP related requirements must still be met.

3.1.3 TLC in the Pharmacopoeias

TLC as a method is described in general chapters of the pharmacopoeias. PhEur, USP, and PhPRCh in their current editions include revised and harmonized description of TLC methodology, specifying the most important parameters of each step. TLC and HPTLC are clearly distinguished and details pertaining to either one are given. This harmonization is an important achievement because it allows modernization of monographs and takes advantage of state of the art HPTLC. Most pharmacopoeial monographs of medicinal plants include TLC as method for identification based on a chromatographic fingerprint, which consists of a sequence of characteristic substance zones. Table 3.1 gives an overview.

The European Pharmacopoeia contains 199 monographs on herbal drugs, including essential oils, gums, and resins. The TLC result is either described or represented as a table (Figure 3.3, Section 3.3.2). The dietary supplement section of the USP 30 includes 26 herbal monographs and a number of monographs on powdered extracts, tablets, and capsules. Chromatograms are described in comparison to the chromatogram of chemical reference substances or reference extracts. Of the 3214 monographs in the Pharmacopoeia of the People's Republic of China, 1078 monographs concern herbal drugs and preparations. The vast majority of those monographs include identification by TLC. In addition to a verbal description of the chromatographic result, an atlas containing images of HPTLC plates has also been published [18] as a supplement. All 18 AHP monographs base the chemical identification on HPTLC fingerprints. Images of HPTLC plates are shown. They compare the fingerprints of botanical reference material and common adulterants to chemical reference substances. A verbal description of the chromatogram focuses on characteristic elements of the fingerprint. In monographs of the IHP and Quality Standards of QS-IMP, about 90% of the plants are identified by TLC. Most monographs feature densitometric evaluation of the TLC fingerprint in addition to images of the plates.

As part of quality control, and in addition of identity, also purity of the herbal drugs must be ensured. This includes the proof of absence of contaminants and adulterants, chemical residues (e.g., pesticides), and fungal and microbial contaminations. Most pharmacopoeias take the same approach by requiring tests for foreign matter, loss on drying, total ash, water, soluble extractive etc. In addition several PhEur monographs include a test for the presence of other species, based on TLC/HPTLC. Such tests are also included in some monographs of USP 30 (e.g., feverfew).

TABLE 3.1

TLC/HPTLC in the Pharmacopoeias

	PhEur	USP 30/ NF 25	PhPRCh 2005 Vol. I	IHP and QS-IMP	AHP
Monographs of herbal drugs	199	26	1078	149	18
Monographs with TLC identification	>90%	23	1523 incl. formulas	132	18
Monographs with TLC assay	None	None	45	87	None

TABLE 3.2

Books Covering TLC for the Analysis of Medicinal Plants

Authors	Title	Publisher	Year
Xie, P. ed.	*TLC Atlas for Chinese Pharmacopoeia of Traditional Chinese Medicine 1992*	Chinese Pharmacopoeia	1993
Wagner, H., Bladt, S.	*Plant Drug Analysis*, 2nd ed.	Springer	1996
Wichtl, M.	*Teedrogen und Phytopharmaka* (German)	WVG	1997
Pachaly, P.	*DC-Atlas: Dünnschichtchromatographie in der Apotheke*, 2nd ed. (German)	WVG	1999
Reich, E., Schibli, A.	*High-Performance Thin-Layer Chromatography for the Analysis of Medicinal Plants*	Thieme, New York	2006
Hänsel, R., Sticher, O. and Xie, P., ed.	*Pharmakognosie Phytopharmazie* (German)	Springer	2007
Xie, P. ed.	*TLC Atlas for Chinese Pharmacopoeia of Traditional Chinese Medicine 2005* (completely revised and extended edition)	Chinese Pharmacopoeia	2007

An increasing number of plant monographs include an assay of active compounds. In 45 monographs of the Chinese Pharmacopoeia TLC is specified as quantitative tool for such assay. Quantitative TLC is also well represented in monographs of India.

3.1.4 TLC Literature Covering the Analysis of Medicinal Plants

Many publications focus on the advantages and utilization of TLC as a technique for quality control of herbal medicines. Some of the important books on the subject are listed in Table 3.2.

The scientific literature covering all aspects of TLC is continuously reviewed and referenced in the CAMAG Bibliography Service (CBS), a database, which is available online [19]. It is the most comprehensive electronic compilation of abstracts of papers on TLC/HPTLC published between 1982 and today. Out of the current 8508 entries in the database 25% are related to medicinal plants. Identification and densitometry (quantitative analysis) are the main topics of publications (Table 3.3). In the Western World identification is the dominant application, whereas in India quantitative densitometric determinations are more frequently reported.

Table 3.4 compares the number of papers on TLC of various dosage forms of TCM. It can be seen that the focus of research is on the first important group, capsules, pills, and tablets, which typically contain dried plant material in a powdered form. The second important group, granules, injections, and powders, includes extracts of medicinal plants.

3.2 ANALYTICAL ASPECTS OF HERBAL MEDICINES AND BOTANICAL DIETARY SUPPLEMENTS

Traditional Western, Traditional Chinese, and Ayurvedic Medicine are based on different philosophies and different approaches to the use of medicinal plants, but all systems commonly prescribe remedies consisting of multiherb preparation. In

TABLE 3.3

Percentages of Papers on Medicinal Plants and Products Thereof Covering Certain Subjects by Origin of Authors

	Western World	China	India
Identification	52	51	40
Densitometry	42	49	73
Preparative TLC	17	16	8
Validation	4	4	23
Stability	3	2	—
Pesticides	2	<1	—
Screening	1	1	—
Total (includes possibility of multiple listings)	121	124	144

Source: From Morlock, G. (ed.), *CAMAG Bibliography Service*, CAMAG Switzerland, Muttenz, 2007. http://www.camag.com/cbs/ccbs.html (accessed March 05, 2007).

many cases little phytochemical information is available about such medicines. Modern herbal medicines and botanical dietary supplements typically include only one or just a few herbal drugs in a product. Particularly for dietary supplements often not much is known about their phytochemistry either. The analysis of single plants already represents an enormous challenge due to natural variability of the individual species. Even more difficult to deal with are multiherb preparations of traditional medicine. The holistic approach to a plant extract in its entirety in comparison to the description of a therapeutically active compound or chemical marker with very specific tools should lead to different analytical approaches, yet modern quality control is asking the same questions about quality and safety when it comes to synthetic drugs and medicinal plants.

TABLE 3.4

Dosage Forms of TCM Analyzed by TLC for Quality Control

Dosage Forms	Publications	Dosage Forms	Publications
Capsules	111	Concentrated decoctions	6
Pills	93	Medicinal wines	6
Tablets	88	Plasters	5
Granules	63	Suppositories	5
Injections	40	Ointments	4
Powders (dry extracts)	22	Ophthalmic preparations	3
Liniment/lotions/smeared films	10	Aerosols and sprays	2
Syrups	7	Medicinal teas	2
Tinctures	7	Cataplasms	1

Source: From Morlock, G. (ed.), *CAMAG Bibliography Service*, CAMAG Switzerland, Muttenz, 2007. http://www.camag.com/cbs/ccbs.html (accessed March 05, 2007).

One technique alone is not able to answer all analytical questions, but with respect to plants TLC/HPTLC as a versatile, reliable, and rapid method is widely applicable.

HPTLC is the method of choice for identification of herbal drugs and derived products as well as for detection of possible adulterations. In-process controls of various production steps and the assessment of stability of final products can conveniently be carried out. TLC/HPTLC convinces with flexibility, specific sensitivity, and simple sample preparation. The method offers the advantage of multiple consecutive detections of separated compounds from many samples analyzed in parallel on the same plate.

Let us now look at some additional aspects of the different herbal medicines before we address common analytical approaches.

3.2.1 TRADITIONAL WESTERN MEDICINES AND EXTRACTS

Traditional Western Medicine is an experience-based folk medicine relying on knowledge handed down from generation to generation. Among the various alternative forms of medicine the use of medicinal plants has always played an important role. Homemade teas of medicinal plants are still popular today. When plants are harvested in the owner's garden for personal use, quality control consists of visual inspection of the plant and the use of organoleptic characteristics such as smell and taste. A current trend favors commercially available rational herbal medicines. Modern concepts of quality control for herbal drugs must be adapted to new circumstances like handling of huge amounts of plant material, dealing with exotic plants coming from foreign medicinal systems, and intentional or unintentional adulteration.

The European Pharmacopoeia defines teas as single or multiherb drugs, which are used for aqueous, oral preparations and are prepared as decoction or maceration. Tinctures are ethanolic extracts made by maceration, percolation, or adequate methods. Both, herbal teas and tinctures are very complex mixtures of compounds. For tinctures, the primary analytical task aside from identification of the plant raw material is to link the chemical profile of the drug with the profile of the product. This can be achieved with an adequate HPTLC fingerprint. On the basis of growing conditions, harvest time, and storage, the pattern of the fingerprint can vary quantitatively and qualitatively. In a screening, raw material can be tested and compared and the plant batch with the highest concentration of desired compounds or classes of compound can be selected for further processing.

From a pharmaceutical point of view the extract as a whole shows the medicinal property and must therefore be seen as the active substance, regardless of whether the herbal tea or tincture is prepared with one or several plant species. Often, neither the therapeutically active constituents nor active markers are known. If they are known they must be analyzed quantitatively. When active ingredients are not yet known, a marker specific for the herbal drug can be chosen for analytical purposes [14].

Complete or primary extracts represent the entirety of the compounds extractable from a plant. Special extracts are made from complete extracts by steps like liquid–liquid extraction or other purification. The aim is to reduce or eliminate ubiquitous or undesirable compounds (e.g., sugars, fats) and to concentrate the therapeutically relevant ingredients. Hence, the special extracts show a qualitatively and quantitatively different spectrum of compounds than the complete extract or the raw material [20]. According to cGMP the process of production must be monitored and documented. This can be done with suitable HPTLC fingerprints focused on the important compounds. The analytical method must be standardized and validated for reliable and repeatable results to be obtained. The documentation step of TLC allows convenient batch-to-batch comparison based on images of chromatograms. For safety reasons, EMEA asks for the determination of the stability of the herbal drug preparation and the expiry date, respectively. Chromatographic fingerprints are a suitable tool also in this respect because they can cover a broad spectrum of compounds.

3.2.2 TRADITIONAL CHINESE MEDICINES

Traditional Chinese Medicines (TCM) is a very old medicinal system with a knowledge that was refined over several thousands of years. The therapeutic basis of TCM lies in supporting the organism in balancing the energy and keeping the body's regulation in good working order. Medicines are usually plants, but also minerals and substances of animal origins. A recipe normally consists of four elements. The principal plant drug (king) is responsible for reinforcement of the weak entity. Associate drugs (minister) help to treat the cause and are used against coexisting symptoms, whereas adjuvant drugs (assistant) can have an opposite effect, weakening the too strong entity in the body. Messenger drugs have a harmonizing effect. It can be assumed that the ingredients of a medicine affect each other as well, which poses a new analytical challenge: a complex medicine might analytically be not the sum of its ingredients [21,22].

Decoctions are the original drug preparations in TCM. They have to be made in a very specific way from mixtures of plants. When preparations from the same mixtures are manufactured on an industrial scale, the resulting medicines may turn out surprisingly different. In [23] Xie demonstrates by HPTLC fingerprints that the formula Sheng Mai Yin consisting of Ginseng root, Ophiopogon root, and Schizandra fruits yields entirely different constituents profiles when prepared in the traditional way as opposed to extraction in an industrial setting. The reason is that ginsenosides in ginseng root are uncontrollably hydrolyzed and then gradually destroyed by the organic acids in Schizandra, when the mixture is excessively heated with water as part of the industrial manufacturing. Traditional gentle boiling over 120 min yields desirably higher content of ginsenoside-Rg3 and -Rh.

The ancient TCM literature acknowledges the great importance of the geographic origin of the herbal drug. Because China covers many climatic regions, each plant species has its ideal region of origin, which stands for a good quality. Fingerprints of plants from different regions are usually different so that the ancient knowledge can be transferred into analytical information.

TCM is older than the current system of botanical nomenclature. Therefore, and because more than just one language was involved over time, it is evident that several plant species may bear the same or similar popular names. In a global trade and exchange of ideas this can easily lead to critical mix-up of species. A well-known example is the mix-up of *Aristolochia* species, containing nephro- and hepatotoxic aristolochic acids, with nontoxic plants of similar name such as "Guan mutong" (*Aristolochia manshuriensis*) and "Chuang mutong" (*Clematis armandii*) or "Guang fangji" (*Aristolochia fangji*), and "Han fangji" (*Stephania tetrandra*). With HPTLC fingerprints all of the mentioned plants and even their mixtures can be easily distinguished [24].

Contaminants such as heavy metals, pesticides, mycotoxins and microbiological impurities are a general issue also for TCM drugs. Most countries have strict regulations concerning the limits of those impurities [25,26].

Because TCM is generally dispensed based on formulas, it is obvious that no single active constituent can be held responsible for the therapeutic effect. The selection of an active substance or a single marker for determination of quality becomes highly questionable. A more comprehensive evaluation is necessary [27].

Another problem for the analyst arises from the fact that according to pharmacopoeial monographs sometimes several species can be regarded as equivalent. The monograph "Cimicifuga radix" of the PhPRCh for example includes three different species: *Cimicifuga foetida, C. heracleifolia,* and *C. dahurica*. The permitted exchangeability of these species in a formula may lead to significant problems, when quality/identity is investigated for a preparation, because the chromatographic fingerprints of the three species are different [28].

TCM drugs are frequently prepared in special ways, such as in vinegar, wine, or honey, or by carbonizing. The resulting medicines have different indications. For example, Rhemanniae radix (Dihuang) and the plant processed with wine, Rhemanniae radix praeparata (Shudihuang), are used as different medicine. Each has a separate monograph in the PhPRCh. From the analytical point of view different chromatographic fingerprints can be expected. Although it can be unclear, what kind of change in the chemical profile takes place during preparation, HPTLC fingerprints can usually reveal the differences [29]. Some drugs can be detoxified by proper processing. This occurs for example when Aconitum is prepared. In case of Aconiti lateralis radix the content of aconitin is reduced by more than 80% [30]. The synergistic effect of the constituents of a formula must also be taken into account for preparations. An interesting example was presented by Tsim et al. [31] who studied one of the simplest yet most widely used TCM preparations (Danggui Buxue Tang), consisting of Astragalus root and roots of *Angelica sinensis* in the specific ratio of 5:1. The study proved that a certain chemical profile of the medicine, the highest amount of active constituents, the lowest amount of unwanted lingustilide, and the highest pharmaceutical activity in various tests were only achieved, when the ancient instructions were exactly followed and material from the specified regions was used.

In the Chinese Pharmacopeia of 2000, *Pueraria lobata* and *P. thomsonii* were combined in the same monograph (Radix Puerariae, "Gegen"). Using HPTLC fingerprints in combination with densitometric evaluation of the chromatograms Chen et al. [32] distinguished the two species clearly as seen in Figure 3.1. The strikingly different chemical profiles of the two plants raised questions about their bioequivalence. Consequently the most recent edition of the PhPRCh (2005) features two individual monographs (Radix Puerariae lobatae, "Ye Ge") and (Radix Puerariae thomsonii, "Gan Ge").

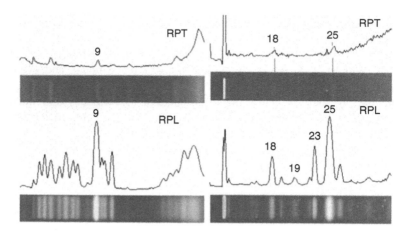

FIGURE 3.1 HPTLC fingerprints of Gan Ge, root of *Pueraria thomsonii* (RPT), and Ye Ge, root of *Pueraria lobata* (RPL) (*bottom*); *left*: Isoflavonoides, *right*: Aglycones. (Courtesy of Chromap Institute, Zhuhai.)

3.2.3 INDIAN SYSTEMS OF MEDICINE

Ayurveda, the Indian ancient traditional health care system, combines medicinal, philosophical, and psychological concepts. The first documented Materia Medica goes back to about 900 B.C., but practice was developed before that time. The fundamental approach of diagnosis and drug development is based on the Tridosha theory of harmony between Vata (energy), Pitta (heat regulation), and Kapha (preservative fluids). A disease passes through five stages with specific symptoms. For the selection of the appropriate medicine, it is important to know the stage of disease. Treatments consist of the use of drugs, diet, and therapies such as meditation or massage. Ayurvedic medicinal preparations are mixtures based mostly on plant products, but can also include minerals, metals, and animal derived products. The drugs are dispensed in a number of powders, solutions, decoctions, pills, and oils. About 1500 plant species are being used in the Indian systems of medicine.

From an analytical perspective very similar problems exist for Ayurvedic Medicines and TCM. The paper by Khatoon et al. [33] illustrates how commonly mixed-up species can be distinguished. Various species of *Phyllanthus* are being sold in India under the trade name "Bhuiamlki." During market surveillance of herbal drugs the authors observed that almost all the commercial samples either comprise *Phyllanthus amarus* or *P. maderaspatensis* or mixtures of *P. amarus*, *P. fraternus*, and *P. maderaspatensis*. HPTLC fingerprints in combination with macroscopic and microscopic characters allow distinguishing the three species.

Typical for Indian medicines are the very complex polyherbal formulas. For example, Chandraprabha vati consists of 37 ingredients of plant and mineral origin and is widely used for various disorders such as anemia, pain, indigestion, and renal calculi. Maha yogaraja guggulu consists of 31 ingredients. The principal constituents of the two medicines were investigated by Bagul et al. [34] using TLC. Analytical work becomes increasingly difficult if such formulas consist of one principal and dozens of minor components.

Quantitative TLC assays of markers are more frequently reported for Ayurvedic formulas than for any other herbal preparation. One of the important issues is the possibility to standardize such products and determine their shelf life [35–37].

3.2.4 Botanical Dietary Supplements

In the USA, herbs and herbal products are regulated under Food Laws as so-called botanical dietary supplements (DSHEA: Dietary Supplement Health and Education Act of 1994 [38]). With a few exceptions such supplements generally do not need approval by the FDA prior to marketing. They are not subjected to the same rigorous testing, manufacturing, labeling standards and regulations as drugs/medicine, but special cGMP [39] have been issued. By law, the manufacturer is responsible for the safety of a dietary supplement product. All ingredients must be stated on the label. In addition, manufacturers may use three types of claims: health claims, structure/function claims, and nutrient content claims. As far as regulation is concerned, all statements on the label must be valid and the manufacturer is responsible for providing the corresponding evidence. For the analyst, this results primarily in the need for identification methods, suitable checks for adulterant as well as assays for the markers with specified content. Under the pending cGMP strong emphasis is placed on proper identification of all materials to be performed before they enter a production process. Certificates of analysis obtained from the vendor of the material are not sufficient.

Products enter the market from all corners and sources. Manufacturers, who take quality control serious, are often confronted with a lack of knowledge concerning the plant constituents and their activity. Quality assurance as well as law enforcement requires analytical methods, which, due to the enormous diversity of products, in many cases are not available. As a highly flexible and rapid method, HPTLC can at least prove the identity and consistency of a material.

3.3 ANALYTICAL TASKS COVERED BY TLC

This section outlines a number of typical analytical tasks and selected examples, which are related to medicinal plants and derived products. It is important to understand that most of the analytical work described here is not carried out for research purpose but as a requirement of regulations enforcing tight quality control. As a consequence this work generally must be in compliance with cGMP. Standard operating procedures (SOPs) describe all activities of the laboratory with the aim to ensure reproducible and traceable results. TLC/HPTLC is usually carried out in a standardized way [40]. Suitable equipment is utilized and all methods are validated [41,42].

3.3.1 Qualitative Description of Herbal Medicines

It has been mentioned earlier that principal difference of herbal medicines and synthetic pharmaceuticals is complexity. Analytical approaches to herbal medicine must therefore include tools to describe this complexity and take it into account when addressing "quality." With an HPTLC fingerprint a plant extract or plant preparation is always shown in its entirety. While some compounds migrate with the solvent front and others remain at the application position, the central portion of the fingerprint can focus on certain groups of constituents. When combined with multiple detections that target general or specific properties of the separated compounds, each fingerprint offers complex information. Generally a fingerprint of a plant should be specific to allow a clear distinction from possible adulterant. The fingerprints of herbal preparations primarily should be comprehensive. Set of fingerprints covering different substance classes of various polarities can be generated. In this respect the lower resolution power of HPTLC in comparison to GC or HPLC can be seen as an advantage. It allows looking at principal components rather than showing the naturally inherent differences of individual samples.

Green tea and green tea extract can be well characterized by a set of four HPTLC fingerprints. Four different developing solvents in combination with four specific derivatization and detection modes visualize characteristic compounds and lead to important quality related information. The flavonoid fingerprint of green tea can be used to obtain information about the geographical origin, whereas the polyphenol pattern allows the discrimination of green tea from, e.g., black tea. Other characteristic constituent profiles including alkaloids and amino acids can help with the identification and determination of adulterants [43]. The example illustrates the flexibility of TLC with respect to mobile phase and specific detection. It also demonstrates the complexity of composition encountered in herbal drugs. For quality control, it is most important to specify acceptance criteria for a material and select appropriate fingerprints accordingly.

3.3.2 IDENTIFICATION OF RAW MATERIAL AND PRODUCTS

The identification of raw material is crucial in order to ensure its authenticity, quality, and safety before it is processed. TLC/HPTLC is the preferred tool for the identification of herbal drugs. Identity is ensured with a chromatographic fingerprint, which visualizes the characteristic chemical constituents of a material. Comparing the fingerprints of a sample with that of a reference drug, the number, sequence, position, and color of the zones must be identical or at least similar. An advantage is that for comparison multiple samples can be analyzed on the same plate [44]. To get meaningful and reliable results, the chemical nature of the separated zone does not have to be known. Not baseline separation of each compound but rather the presence of the most important compound classes in combination with a record of the entire extract is essential for the identity of the sample. For proper identification of a plant, acceptance criteria must be set. The presence or absence of characteristic zones can be a basis for decision. In the test, a sample of the proper species must meet the acceptance criteria and samples of other plants species must fail. HPTLC can give reliable answers rapidly [45,46].

The natural variability of plants is a serious analytical challenge. Even though botanically clearly identified, the different samples of the same species may show different fingerprints. To solve this problem, it is important to work with representative samples, authentic plants, or if possible with botanical reference material (BRM).

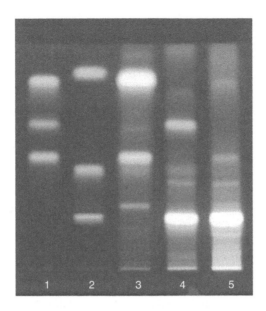

FIGURE 3.2 Fingerprints for differentiation of three *Echinacea* species; Track assignment: 1: caftaric acid, cynarin, cichoric acid (with increasing Rf); 2: echinacoside, chlorogenic acid, caffeic acid (with increasing Rf); 3: *E. purpurea* root; 4: *E. angustifolia* root; 5: *E. pallida* root.

Chemical reference standards are an alternative but they are difficult to obtain and often quite expensive. A screening of several samples from the same species on the same plate gives a visual impression of the natural variety and helps with the selection of a representative sample.

With reference to chemical markers three *Echinaceae* species can be distinguished (Figure 3.2) [47]. *E. angustifolia* contains echinacoside and cynarin in contrast to *E. purpurea*, which is characterized by cichoric acid and caftaric acid. *E. pallida* predominantly contains echinacoside. Images with chromatograms of reference substances or reference drugs are very convenient for comparison, but in the literature description is still dominant. The European Pharmacopoeia adopted a way of describing the chromatogram with a table (Figure 3.3). The chromatogram is

Top of the plate	
Caffeic acid: a strong blue fluorescent zone	
Cynarin: a strong greenish fluorescent zone	A greenish fluorescent zone (cynarin)
Echinacoside: a strong greenish fluorescent zone	A greenish fluorescent zone (echinacoside)
Reference solution	Test solution

FIGURE 3.3 Description of fingerprint of *E. angustifolia* radix according to PhEur.

divided into three Rf sections for reference and test solution. Zones and their colors are noted in the table so that the position in the chromatogram is presented as adequately as possible.

3.3.3 DETECTION OF ADULTERANTS AND IMPURITIES

Fingerprints have become a very important tool not only for identification of herbs but also for distinguishing them from closely related species. Adulterants and mixtures are typically determined by visually comparing fingerprints of the test sample against fingerprints of reference materials with respect to sequence, color, and intensity of the zones prior and after derivatization. On the basis of established acceptance criteria, the identity test can recognize the presence of adulterants instead of the desired species. The presence of specific markers for certain adulterants within the otherwise correct fingerprint of a sample can be used to determine an impurity. This is a simple and fast approach allowing the evaluation of up to 16 samples within a few minutes. If video densitometry or scanning densitometry is employed the reliability of the results can be increased and an impurity in the sample can be quantified.

Let us look at an example. Star anise is the ripe fruit of *Illicium verum*, a tree native to the south of China. In several cases the herbal drug has been mixed up with Shikimi, the fruits of *Illicium anisatum*. Shikimi contains the poisonous anisatin but no or little anethol. The two fruits look morphologically and anatomically very similar. For an identification of the two species by HPTLC, the presence or absence of anethol was proposed as marker. However, the anethol content can vary considerably and for the investigation of mixtures of both species anethol is not adequate. A characteristic zone in the fingerprint, which is specific for the adulterant but not present in *I. verum* is needed. Unfortunately anisatin cannot be detected by TLC. Figure 3.4 shows HPTLC fingerprints of several samples of *I. anisatum* and *I. verum* as well as mixtures of the two species. All samples of *I. anisatum* (tracks 3–6) show a dark zone at Rf ~ 0.22 (arrow). This zone is not seen in *I. verum* and can therefore be used to detect adulteration with *I. anisatum*. The fingerprints of the mixtures show that adulteration is detectable up to 5%.

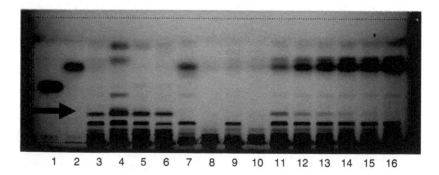

FIGURE 3.4 HPTLC of *Illicium anisatum, Illicium verum*, and mixtures; Track assignment: 1: myristicin; 2: anethol; 3–6: *I. anisatum*; 7–10: *I. verum;* 11–16 mixture of *I. verum* with 50%, 30%, 20%, 10%, 5%, 1% of adulterant.

Another important kind of adulteration found in herbal medicines is that with chemicals such as dyes or synthetic active pharmaceutical ingredients (API). This is a serious problem, especially when the product is marketed as "purely herbal" and the API is not declared. Several herbal anti-impotency medicines, confiscated by customs, were studied concerning content of APIs such as sildenafil (Viagra^R), anti-diabetics, and painkillers [48]. Reddy et al. [49] have published a quantitative HPTLC method for sildenafil citrate in herbal medicines. The rapid and simple densitometric determination is not affected by the matrix and can be applied for the routine screening of preparations marketed as natural herbal medicine.

TLC/HPTLC can be employed for qualitative and quantitative analysis of pesticide residues in herbal medicines. A recent detailed review including methods for insecticides, herbicides, and fungicides was published by Sherma [50].

The presence of aflatoxins is a major risk for some herbal drugs. HPTLC can certainly be employed for safety assurance in this respect [51,52]. Nagler et al. [53] compared the performance characteristics of HPLC and HPTLC for control of aflatoxins in copra and has favored HPTLC.

3.3.4 MONITORING THE PRODUCTION PROCESS AND ENSURING BATCH-TO-BATCH CONFORMITY

For the industrial production of herbal medicines, several cGMP related issues must be considered. The whole production process must be under strict control [54]. The HPTLC fingerprint is an important analytical tool for monitoring of an extraction process. It gives quick and reliable answers about the status of the process in progress or about the finished product. The only requirement is to demonstrate that the fingerprint is selective enough to show any change in the composition of the sample. Because of their comprehensive nature and comparability, HPTLC finger-prints can be used to document how efficiently and completely the starting material is transferred into the product and whether the composition of constituents has changed during production. This information can be combined with semiquantitative and quantitative data regarding one or more markers or active substances.

Another cGMP issue is the batch-to-batch consistency of the production. It involves three elements: consistency of the raw material, full control of the produc-tion process, and proper definition of the finished product. Because the plant raw material comes with a naturally variable composition, it is clear that also the finished product will show a certain variation. Batch-to-batch consistency therefore becomes a matter of definition and specification. In any case final herbal products can be compared to a reference product or to a formerly released batch. A requirement for reliable analytical work is that the HPTLC methodology is standardized and all methods are validated. It is of great advantage to work with a cGMP compliant documentation system, which generates images reproducibly and archives them for comparison with current images.

An example is illustrated in Figure 3.5. On the same plate HPTLC fingerprints of Ginkgo leaves are compared with those of several finished products from a market survey. These include different dosage forms. The comparison targets two sets of constituents, terpene lactones and flavonoids. With respect to the terpene lactone

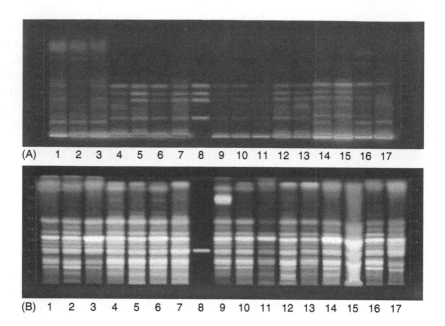

FIGURE 3.5 HPTLC fingerprints of products from Ginkgo leaves in different dosage form; (A) Terpene lactones (Ginkgolides and bilobalide); (B) flavonoids; track assignment: 1–3: Ginkgo leaves; 4–7: dry extracts; 8: reference standards; 9: extract capsule; 10–13: extract tablet; 14–17: tinctures. (From CAMAG Application Note F-16, HPTLC Identification of Ginkgo (Ginkgo biloba), 2003, http://www.camag.com/l/herbal/index.html# (accessed 05 March, 2007).)

profile the three batches of raw material are different from the finished products. As desired constituents the terpene lactones are generally enriched during the extraction process and most of the products show a comparable profile. The sample on track 17 exhibits a different pattern and the zone of ginkgolide C is missing. This sample had already passed its expiration date. The flavonoid patterns of the investigated samples vary much less. Only the liquid extract on track 15 shows considerable matrix interference [55]. A quantitative HPTLC study of several commercial ginkgo products from different countries and Ginkgo leaf extract from China was published by Xie et al. [56].

3.3.5 QUANTITATIVE DETERMINATION

Quantitative determination by HPTLC is generally performed by scanning densitometry. Video densitometry, based on images, is an alternative for the analysis. For the quantitation, some points have to be considered. Most fingerprint methods, used for identification, are suitable for quantification only with some adaptation. Baseline separation for the quantitative evaluation of the selected compound is essential. Hence, robust methods, which are optimized and standardized, are required. The

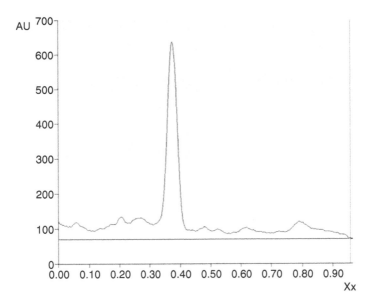

FIGURE 3.6 Densitogram of an Artemisia sample.

sample preparation plays an important role as it does for all quantification techniques [57]. Considering these points, HPTLC quantification leads to reproducible data [58]. Detection limit of absorption measurements is about 10–100 ng. In fluorescence mode, limits of about 0.1–10 ng can be reached. There is a linear relationship between fluorescence signal and substance concentration over a wide concentration range. If the amounts of the nonfluorescing compounds are small enough linear regression can be applied, but generally the calibration function of absorption measurements on a TLC plate is not linear. In this case polynomial regression leads to correct results. For quantitative analysis, suitable methodology, proper instrumentation, and validated methods are a basic requirement.

For example, Artemisinin, an antimalarial substance, is an ingredient of the old Chinese medicinal plant *Artemisia annua*. The determination of artemisinin in the leaves is of great economic importance because synthesis of the compound is not yet feasible. Quantitative HPTLC offers a simple, rapid, cost efficient, and reliable analytical solution [59]. Using a Michaelis–Menten regression it is possible to screen the artemisinin content of 10 samples over a wide concentration range from 20 to 1300 ng. A more accurate assay utilizes a linear working range of 30–100 ng. Artemisinin is baseline separated from all other components of the sample (Figure 3.6). Numerous papers have been published on HPTLC assays of marker compounds of medicinal plants and preparations. Table 3.5 gives an overview of some method parameters.

TABLE 3.5
Overview of Some Quantitative HPTLC Assays of Herbal Drugs

Plant [Reference]	Assay of	Chromatography [V/V]	Derivatization/ Detection
Stephania tetrandra [60]	Tetrandrine L: 50–112 ng,	Toluene, ethyl acetate, methanol, ammonia 28%; 10:10:5:0.3	Iodine reagent, 210 nm
Aristlochia fangji [61]	Aristolochic acid L: 400 pg-8 ng	Toluene, ethyl acetate, water, formic acid; 20:10:1:1	Tin(II)chloride, 366 nm
Acorum calamus [62]	β-Asarone L: 25–300 ng, P: 40–200 ng,	Toluene, ethyl acetate; 93:7	313 nm
Ginkgo biloba [56]	Ginkgo terpenes	Toluene, ethyl acetate, acetone, methanol; 5, 2.5, 2.5, 0.3	Heat, 366 nm
Harpagophytum procumbens [63]	Harpagoside P: 0.4296–4.296 ppm	Ethyl acetate, methanol, water; 77:15:8	Anisaldehyde, 509 nm
Ocimum sanctum [64]	Eugenol, luteolin, ursolic acid	Toluene, ethyl acetate, formic acid; 7:3:0.2	Anisaldehyde-sulfuric acid, 280 nm

Note: L: Linear Regression and P: Polynomial Regression

REFERENCES

1. Mez-Mangold, L., *Aus der Geschichte des Medikaments*, Roche, Basel, 1980, p. 11.
2. Mukherjee, P.K., *Quality Control of Herbal Drugs*, Business Horizons, New Delhi, 2002, p. 2.
3. Anderson, E.N., Why is humoral medicine so popular? *Soc. Sci. Med.*, 25, 331, 1987.
4. Bär, B., Hagels, H., and Langner, E., Phytopharmaka im Wandel der Zeit, *Deutsche Apotheker Zeitung*, 145, 76, 2005.
5. *European Pharmacopoeia*, 5th edn., Council of Europe, European Directorate for the Quality of Medicines (EDQM), Strasbourg, 2005.
6. *The United States Pharmacopeia* 30, *The National Formulary*, 25th edn., Vols. 1–3, The United States Pharmacopeial Convention, Rockville, 2007.
7. *Pharmacopoeia of the People's Republic of China*, Volumes I–III, Chinese Pharmacopoeia Commission, People's Medical Publishing House, Beijing, 2005.
8. *American Herbal Pharmacopoeia*, Santa Cruz, CA, USA.
9. *Quality Standards of Indian Medicinal Plants*, 1–3, Indian Council of Medicinal Research, Indraprastha Press (CBT), New Delhi, 2005.
10. *Indian Herbal Pharmacopoeia*, revised edn., Indian Drug Manufacturers' Association, Mumbai, 2002.
11. World Health Organisation, *Quality Control Methods for Medicinal Plant Materials*, Geneva, 1998, p. 22. http://www.who.int/medicines/services/expertcommittees/pharmprep/QAS05_131Rev1_QCMethods_Med_PlantMaterialsUpdateSept05.pdf (accessed March 05, 2007).

12. World Health Organisation, *Good Manufacturing Practices: Updated Supplementary Guidelines for the Manufacture of Herbal Medicines*, QAS/04.050/Rev.3., Geneva, 2005. http://www.who.int/medicines/services/expertcommittees/pharmprep/QAS04_050Rev3_GMPHerbal_Final_Sept05.pdf (accessed March 05, 2007).
13. Guideline on quality of herbal medicinal products/traditional herbal medicinal products, CPMP/QWP/2819/00 Rev 1 and EMEA/CVMP/814/00 Rev 1. European Medicines Agency (EMEA), London, 2006. http://www.emea.eu.int/pdfs/human/qwp/281900en.pdf (accessed March 05, 2007).
14. Directive 2004/24/EC Traditional herbal medicinal products. European Parliament and the Council, Strasbourg, 2004. http://europa.eu.int/eur-lex/pri/en/oj/dat/2004/l_136/l_13620040430en00850090.pdf (accessed March 05, 2007).
15. Directive 2001/82/EC and Directive 2001/83/EC Veterinary medicinal products and medicinal products for human use. European Parliament and the Council, Strasbourg, 2001. http://ec.europa.eu/enterprise/pharmaceuticals/eudralex/vol-5/dir_2001_82/dir_2001_82_en.pdf and http://europa.eu.int/eur-lex/pri/en/oj/dat/2001/l_311/l_311200 11128en00670128.pdf (accessed March 05, 2007).
16. Annex 7, Manufacture of herbal medicinal products. *Good Manufacturing Practice (GMP) for Medicinal Products*, Vol. 4, European Parliament and the Council. http://ec.europa.eu/enterprise/pharmaceuticals/eudralex/vol-4/pdfs-en/anx07en.pdf (accessed March 05, 2007).
17. Validation of analytical procedures: Text and methodology Q2(R1). *ICH Harmonised Tripartite Guideline*, International conference on harmonisation of technical requirements for registration of pharmaceuticals for human use, 2005. http://www.ich.org/LOB/media/MEDIA417.pdf (accessed March 05, 2007).
18. Xie, P. (ed.), *TLC Atlas for Chinese Pharmacopoeia of Traditional Chinese Medicine*, 2005 ed., People's Medical Publishing House, Beijing, 2007.
19. Morlock, G. (ed.), *CAMAG Bibliography Service*, CAMAG Switzerland, Muttenz, 2007. http://www.camag.com/cbs/ccbs.html (accessed March 05, 2007).
20. Gaedcke, F., Steinhoff, A., and Steinhoff, B., *Phytopharmaka. Wissenschaftliche und rechtliche Grundlagen für die Entwicklung, Standardisierung und Zulassung in Deutschland und Europa*, Wissenschaftliche Verlagsgesellschaft mbH, Stuttgart, 2000.
21. Bensky, D. et al., *Chinese Herbal Medicine: Materia Medica*, 3rd edn., Eastland Press, Seattle, 2004.
22. Stöger, E.A. and Friedel, F., *Arzneibuch der Chinesischen Medizin*, Deutscher Apotheker Verlag, Stuttgart, 2003.
23. Xie, P., The basic requirement for modernization of Chinese herbal medicine, in *Annals of Traditional Chinese Medicine: Vol. 2. Current Review of Traditional Chinese Medicine: Quality Control of Herbs and Herbal Material*, Leung, P.C., ed., World Scientific, Singapore, 2006, chap. 1.
24. Blatter, A. and Reich, E., Analysis of aristolochic acids in Chinese drugs by high-performance thin-layer chromatography (HPTLC), in *Proc. Int. Symp. Planar Sep.*, Visegrád, Nyiredy, Sz., and Kakuk, A., Eds., Research Institute for Medicinal Plants, Budakalász, 2004, p. 45.
25. Arranz, I. et al., Determination of Aflatoxin B1 in medical herbs: Interlaboratory study, *J. AOAC Int.*, 89, 595, 2006.
26. Drasar, P. and Moravcova, J., Recent advances on analysis of Chinese medical plants and traditional medicines, *J. Chromatogr. B*, 812, 3, 2004.
27. Liang, Y.-Z., Xie, P., and Chan, K., Quality control of herbal medicines, *J. Chromatogr. B*, 812, 53, 2004.

28. Reich, E. and Schibli, A., Validation of high-performance thin-layer chromatographic methods for the identification of botanicals in a cGMP environment, *JAOAC* (in press).
29. Meier, R.O., Reich, E., and Berger Büter, K., Quality assurance of Chinese herbal drugs considering as example *Rehmannia glutinosa* and *Stephania tetrandra*, Diploma Thesis, University of Basel, 2004.
30. Hänsel, R. and Sticher, O., *Pharmakognosie – Phytopharmazie*, 8th edn., Springer, 2007. p. 452.
31. Tsim, K.W.K. et al., Danggui Buxue Tang (a Chinese Angelica decoction): A sample trial in TCM standardization, presented at Forum on Int. Standardization of Chinese Medicines and Herbal Products, Shanghai, 2006.
32. Chen, S. et al., High-performance thin-layer chromatographic fingerprints of isoflavonoids for distinguishing between *Radix Puerariae Lobatae* and *Radix Puerariae Thomsonii*, *J. Chromatogr. A*, 1121, 114, 2006.
33. Khatoon, S. et al., Comparative pharmacognostic studies of three Phyllanthus species, *J. Ethnopharmacol.*, 104, 79, 2006.
34. Bagul, M.S. et al., Anti-inflammatory activity of two Ayurvedic formulations containing guggul, *Indian J. Pharmacol*, 37, 399, 2005
35. Elamthuruthy, A.T. et al., Standardization of marketed Kumariasava—an Ayurvedic *Aloe vera* product, *J. Pharm. Biomed. Anal.*, 37, 937, 2005.
36. Agrawal, H. et al., HPTLC method for guggulsterone I. Quantitative determination of E- and Z-guggulsterone in herbal extract and pharmaceutical dosage form, *J. Pharm. Biomed. Anal.*, 36, 33, 2004.
37. Chauhan, B.L., Development of HPTLC fingerprint technique and bioassay to establish the shelf-life of a PHF, *Indian Drugs*, 31,333, 1994.
38. Dietary Supplement Health and Education Act of 1994, 94, 2006. http://www.cfsan.fda.gov/~dms/dietsupp.html (accessed 05 March, 2007).
39. Current Good Manufacturing practice in manufacturing, packing, labelling, or holding operations for dietary supplements (cGMPs), US FDA/CFSAN, Rockville, MD, 2007. http://www.cfsan.fda.gov/~lrd/fr07625a.html (accessed August 21, 2007).
40. Reich, E. and Schibli, A., A standardized approach to modern high-performance thin-layer chromatography (HPTLC), *J. Planar Chromatogr.*, 17, 438, 2004.
41. Koll, K. et al., Validation of standardized high-performance thin-layer chromatographic methods for quality control and stability testing of herbals, *J. AOAC Int.*, 86, 909, 2003.
42. Ferenczi-Fodor, K. et al., Validation and quality assurance of planar chromatographic procedures in pharmaceutical analysis, *J. AOAC Int.*, 84, 1265, 2001.
43. Reich, E. et al., HPTLC methods for identification of green tea extract, *J. Liq. Chromatogr. Relat. Technol.*, 29, 2141, 2006.
44. Reich, E. and Schibli, A., *High-Performance Thin-Layer Chromatography of Medicinal Plants*, Thieme, New York, Stuttgart, 2007.
45. Ravishankara, M.N. et al., Evaluation of antioxidant properties of root bark of Hemidesmus indicus R. Br. (Anantmul), *Phytomedicine*, 9, 153, 2002.
46. Narasimhan, S. et al., Free radical scavenging potential of *Chlorophytum tuberosum* baker, *J. Ethnopharmacol.*, 104, 423, 2006.
47. Reich, E. et al., An AOAC peer-verified method for identification of echinacea species by HPTLC, *J. Planar Chromatogr.*, 15, 244, 2002.
48. Caprez, S., Development of methods for analysis of synthetic adulterants in herbal medicines by HPTLC, Diploma Thesis, University of Basel, 2005.
49. Reddy, T.S., Reddy, A.S., and Devi, P.S., Quantitative determination of sildenafil citrate in herbal medicinal formulations by high-performance thin-layer chromatography, *J. Planar Chromatogr.*, 19, 427, 2006.

50. Sherma, J., Thin-layer chromatography of pesticides—a review of applications for 2002–2004, *Acta Chromatogr.*, 15, 5, 2005.
51. Coker, R.D. et al., Evaluation of instrumentation used for high performance thin-layer chromatography of aflatoxins, *Chromatographia*, 25, 875, 1988.
52. Tomlins, K.I. et al., A bi-directional HPTLC development method for the detection of low levels of aflatoxin in maize extracts, *Chromatographia*, 27, 49, 1989.
53. Nagler, M.J., The application of HPTLC to the control of aflatoxin in Philippine Copra, presented at the TLC Forum Symp., University of Surrey, Guildford, June 3–5, 1996.
54. *EG-Leitfaden einer Guten Herstellungspraxis für Arzneimittel und Wirkstoffe*; 7th edn., Editio Cantor Verlag, Aulendorf, 2003, p. 61.
55. CAMAG Application note F-16, HPTLC Identification of Ginkgo (Ginkgo biloba), 2003. http://www.camag.com/l/herbal/index.html# (accessed 05 March, 2007).
56. Xie, P. et al., Fluorophotometric thin-layer chromatography of Ginkgo terpenes by postchromatographic thermochemical derivatization and quality survey of commercial Ginkgo products, *J. AOAC Int.*, 84, 1232, 2001.
57. Ong, E.S., Extraction methods and chemical standardization of botanicals and herbal preparations, *J. Chromatogr. B*, 812, 23, 2004.
58. Ebel, S., Quantitative analysis in TLC and HPTLC, *J. Planar Chromatogr.*, 9, 4, 1996.
59. Widmer, V., Handloser, D., and Reich, E., Quantitative HPTLC analysis of artemisinin in dried *Artemisia annua* L.—A practical approach, *J. Liq. Chromatogr. Relat. Technol.*, 30, 2209, 2007.
60. Blatter, A. and Reich, E., Qualitative and quantitative HPTLC methods for quality control of *Stephania tetrandra*, *J. Liq. Chromatogr. Relat. Technol.*, 27, 1, 2004.
61. Blatter, A. and Reich, E., High performance thin-layer chromatographic analysis of aristolochic acids in Chinese drugs, *J. Planar Chromatogr.*, 17, 355, 2004.
62. Widmer, V., Schibli, A., and Reich, E., Quantitative determination of β-asarone in Calamus by high-performance thin-layer chromatography, *J. AOAC Int.*, 88, 1562, 2005.
63. Günther, M. and Schmidt, P.C., Comparison between HPLC and HPTLC-densitometry for the determination of harpagoside from *Harpagophytum procumbens* CO_2-extracts, *J. Pharm. Biomed.*, 37, 817, 2005.
64. Anandjiwala, S., Kalola, J., and Rajani, M., Quantification of eugenol, luteolin, ursolic acid, and oleanolic acid in black (Krishna Tulasi) and green (Sri Tulasi) varieties of Ocimum sanctum Linn. using high-performance thin-layer chromatography, *J. AOAC Int.*, 89, 1467, 2006.

4 Primary and Secondary Metabolites and Their Biological Activity

Ioanna Chinou

CONTENTS

4.1 INTRODUCTION

All pathways necessary for the survival of the plant cells are called basic or primary metabolism, while in the secondary metabolism, compounds are produced and broken down that are essential for the whole plant organism. Both metabolisms are overlapping, since it is often not answered or understood why certain chemical constituents have been produced. All plants produce chemical constituents, part of their normal metabolic activities [1–3]. These, can be divided into primary metabolites, such as sugars, amino acids, nucleotides and fats, found in all plants,

59

and secondary metabolites. Secondary metabolites have no obvious function in a plant's primary metabolism as well as in growth, photosynthesis, reproduction, or other "primary" functions of the plant cell. They may possess an ecological role, as pollinator attractants, represent chemical adaptations to environmental stresses, or to be responsible for the chemical defence of the plant against microorganisms, insects and higher predators, or even other plants (allelochemics) [4–7].

Manhood use many of these compounds as high value food (nutraceuticals), spices, flavors and fragrances, vegetable oils, resins, insecticides and other industrial, agricultural raw materials as well as medicinal products and in many cases as drugs. As a result, secondary metabolites that are used commercially as biologically active compounds (pharmaceuticals, flavors, fragrances, etc.) are generally higher value-lower volume products than the primary metabolites.

4.2   PRIMARY METABOLITES

The plants synthesize sugars, amino acids, nucleotides and fats, known as primary plant metabolites as they are produced by them as part of their normal metabolic activities. Primary metabolites can be found in all plants (species, genera, and families) and they are of vital importance to their life. These simple molecules are used to produce polymers essential for the plants. This aspect, of the plants' biochemistry, can be considered as distinct from the production of more complex molecules, produced by more diverse pathways, as the secondary metabolites, which will be discussed in Section 4.3 [6,7].

4.3   SECONDARY METABOLITES

4.3.1   Chemical Classification

Secondary metabolites, in contrast to primary ones, are known to be synthesized in specialized cell types and at distinct developmental stages, making their extraction and purification more difficult, in comparison with primary ones. These chemical constituents are extremely diverse. Each plant family, genus, and species produces a characteristic chemical category or a mix of them, and they can sometimes be used as taxonomic characters in the classification of the plants [8].

Secondary metabolites can be classified on the basis of their chemical structure (e.g., having rings, with or without sugar moieties), composition (containing nitrogen or not), the pathway by which they are biosynthesized or their solubility in various solvents [6–8].

A simple classification includes three main groups:

1. *Terpenoids* (made through mevalonate pathway, composed almost entirely of carbon and hydrogen)
2. *Phenolics* (made from simple sugars, containing benzene rings, hydrogen, and oxygen)
3. *Nitrogen-containing compounds* (or also containing sulphuric parts)

The origin of the main categories of secondary metabolites in relation to the basic metabolic pathways can be shown in Figure 4.1.

Analytically Terpenoids are formed from five-carbon building units, resulting in compounds with C5, C10, C15, and C20 up to C40 skeletons. So, terpenoids is the widespread numerous class of natural products that are derived from a common biosynthetic pathway based on mevalonate as parent, including also the subgroups of

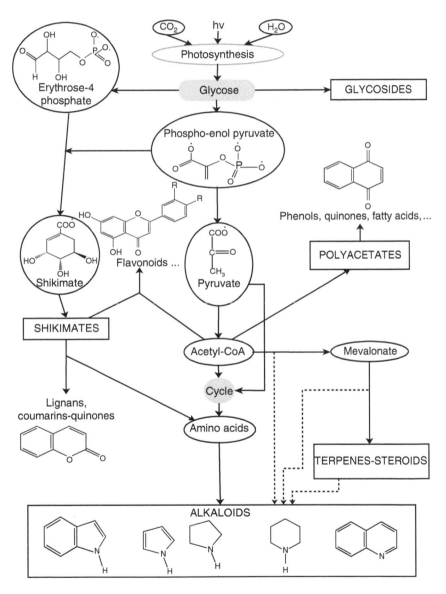

FIGURE 4.1 Origins of the main secondary metabolites in relation to the basic metabolic pathways.

TABLE 4.1
Classification-Occurrence of Terpenoids

C_n	Name	Subclass	Occurrence
10	Monoterpenoids	Iridoids	Oils
15	Sesquiterpenoids	Abscisic acid, C15-lactones	Oils, resins
20	Diterpenoids	Gibberellins	Resins
25	Sesterterpenoids	—	Resins, bitters
30	Triterpenoids	Phytosterols, cardenolides, saponins	Resins, bitters, latex
40	Carotenoids		Green tissue, roots, fruits
103–104	Rubbers		Latex, roots

isoprenoids and steroids among them. Mevalonic acid, a derivative of acetyl-CoA, is thought as precursor for all of them. Terpenoids can be found in higher plants, mosses, liverworts, algae and lichens, as well as in insects or microbes, while steroids can be found in animal and plant kingdoms and in microorganisms [6–8]. The terpenoids and their main classes are listed in Table 4.1. Of all the families of natural products, the diterpenoids have one of the widest ranges of biological activities. The plant growth hormones gibberellins are included among them, the clerodane bitter principles and insect antifeedants, the labdane type sweetener stevioside, tumour inhibitors, cocarcinogens as well as many compounds possessing strong antibiotic activities. Recent interest has centered on the abortifacient isocupressic acid and the antihypertensive agent forskolin. Through centuries, many of these compounds have been used as ingredients of perfumes, drugs, narcotics, or pigments [9,10].

Phenolics are aromatic compounds with hydroxyl substitutions, having as parent compound phenol, but most are more complex polyphenolic compounds, classified by the number of carbon atoms in the basic skeleton [4,6,7]. Their derived classes have one, two, or three side-chains e.g., salicylic acid, p-hydroxyphenylacetic acid, hydroxycinnamic acid and caffeic acids, or substituted phenolic terpenoids, e.g., Δ1-tetrahydrocannabinol. Only plants and microooorganisms are capable of biologically synthesizing the aromatic nucleus. Plant phenolics arise from two main biosynthetic pathways: (1) via shikimic acid (benzoic acid derivatives, lignans, coumarins, etc.) and (2) through acetate, leading to polyketides, which afford by cyclization to products such as xanthones and quinines. The basic structure of flavonoids is derived from the C15 body of flavone. They differ from other phenolic substances in the degree of oxidation of their central pyran ring as well as in their biological properties. The starting product of the biosynthesis of most phenolic compounds is shikimate. The variability of the flavonoids is largely based on the hydroxylation or methylation pattern of the three ring systems. The glycosylation of flavonoids has an additional, ecologically important function [4,6,7,11]. On the basis of their biological functions, phenolic compounds can be classified as shown in Table 4.2.

Nitrogen-containing compounds (extremely diverse, may also contain sulphuric parts). In plants, amino acids are broken down into two groups, protein and nonprotein. There are 20 amino acids, derived from the acid hydrolysates of plant proteins

TABLE 4.2

Important Classes of Phenolics in Plants' Kingdom

C_n Basic Skeleton		Class	Function
6	C_6	Simple phenols, benzoquinones, quinones	Allelopathic substances, fungicide
7	C_6-C_1	Phenolic acids	Fungicide
9	C_6-C_3	Hydroxycinnamic acid, coumarins	Allelopathic substances, phytoalexines
10	C_6-C_4	Naphtoquinone	Protection against pests
13	$C_6-C_1-C_6$	Xanthone	Phytoalexines
15	$C_6-C_3-C_6$	Flavonoids, isoflavonoids, anthocyanes, chalcons, aurones	Flower/fruit pigments, fungicide, phytoalexines
18	$(C_6-C_3)_2$	Lignans	Antioxidants, phytoestrogens
30	$(C_6-C_3-C_6)_2$	Biflavonoids	Antioxidants, antimicrobial activity
N	$(C_6-C_3)_n (C_6)_n$ $(C_6-C_3-C_6)_n$	Condensed tannins	Protection against pests, antimicrobial activity

(as with animal proteins). Plant proteins are essential for carrying out specific cellular functions both internally and externally. Plant proteins are seed-based storehouses for nitrogen and guard against any possible predators. Some are toxic to humans, while some others are necessary in the human diet. Some, furthermore, have been developed into specific drugs as L-Dopa, from Fabaceae families, used in the treatment of Parkinson's disease [6,7]. Alkaloids are a group of nitrogen-containing bases, with their nitrogen atom, as a part of a heterocyclic system, most of which possessing significant pharmacological properties. Structurally, alkaloids are the most diverse class of secondary metabolites, ranging from simple structures to very complex ones. They are classified by the amino acid or derivatives (Table 4.3) from which they are biosynthesized: ornithine and lysine (ornithine is a precursor of the cyclic pyrrolidines such as nicotine alkaloids of tobacco and other Solanaceae), phenylalanine and tyrosine, tryptophan or anthranilic acid, nicotinic acid, polyketides or terpenoids [6,7,12]. Their presence appears to be most prevalent in the Solanaceae, Papaveraceae, Ranunculaceae, Fabaceae, Rubiaceae, and Berberidaceae families. Plant genera providing the highest yield of alkaloids are *Nicotiana*, *Vinca*, *Strychnos*, *Papaver*, and *Rauwolfia* [6,7]. They could be classified into the following five groups:

(1) *Pyridine and piperidine*—This group represents a class which affects mostly the central nervous system, reduces appetite, and contains other properties such as diuretic ones. Nicotine, lobeline, piperine, pilocarpine, sparteine, and coniine are examples [6,7,13].

(2) *Tropine group*—Alkaloids characterized by containing the tropine nucleus. Atropine, cocaine, hygrine, ecgonine, and pelletierine could be served as main examples [6,14].

TABLE 4.3

Important Classes of Alkaloids in Plants' Kingdom

Class	Main Chemical Compounds	Structure
Pyridine group	Piperine; conine; trigonelline, pilocarpine, cytisine; nicotine, sparteine, etc.	Nicotine
Tropine group	Atropine; cocaine; pelletierine	Cocaine
Quinoline group	Quinine, Strychnos bases as strychnine and brucine and veratrum alkaloids: veratrine, cevadine and also dopamine	Dopamine
Isoquinoline group	The opium alkaloids: morphine, codeine, thebaine, papaverine	Morphine
Indole-alkaloids	Serotonin, reserpine, tryptamine	Tryptamine

(3) *Quinoline group*—Quinoline alkaloids developed in the nucleus from tryptophan a well as the strychnos bases: strychnine, brucin, and the veratrum alkaloids: veratrine, cevadine, etc. Quinine, the old antimalarial drug, and quinidine, used both in heart tachycardia and arrhythmia, are included in this group [6,7,13].

(4) *Isoquinoline group*—Alkaloids derived from tyrosine and phenylalanine. The opium alkaloids: morphine, codeine, thebaine, papaverine, narcotine and the more complicated substances, while hydrastine and berberine are included also in this group [7,13,14].

(5) *Indole alkaloids* are derived from tryptophan, and apart from the few with hallucinogenic effects, indoles such as serotonin and reserpine have a sedative effect on the central nervous system. Other constituents in this category are cytostatic, antileukemic, or are able to act on the ratio of oxygen and glucose to the cell [6,7,13–15].

4.3.2 BIOLOGICAL IMPORTANCE FOR THE PLANTS

Secondary plant metabolites have traditionally been regarded as toxic and protective against predators, or acting as insect attractants. As it has been already referred in Introduction part, the production of many secondary metabolites is not, even in our days, absolutely understood. These chemicals have a role as protection in the fight with the animal world [3–7,16]. They are, in most cases, chemically diverse natural products not synthesized outside the plant kingdom. Among them are plant hormones which influence the activities of other cells, control their metabolic activities, and coordinate the development of the whole plant defence mechanisms; possessing toxicological, behavioural, and attractant effects on many different species, including interesting properties in the mammalian central nervous system as stimulants or sedatives or by acting on the cardiovascular system. Many secondary compounds have signalling functions like the flower colors phenolic constituents, anthocyanes, which serve to communicate with pollinators or protect the plants from feeding by animals or infections, known as phytoalexines, inhibiting fungi infections within the plant [15,17].

4.4 NATURAL PLANT CHEMICALS AS SOURCES OF INDUSTRIAL AND MEDICINAL MATERIALS

4.4.1 SECONDARY METABOLITES IN FOODS

Plant secondary metabolites are important sources of many food ingredients and phytochemicals, including flavors, colorants, essential oils, sweeteners, antioxidants, and nutraceuticals. Food ingredients are frequently extracted from source plants without going through strict purification steps and in all cases contain a mixture of many components. The use and search for dietary supplements derived from plants have accelerated in recent years. It is very well known through extended pharmacological (in vitro and in vivo) studies that dietary phytochemicals found in vegetables, fruits, herbs and spices as well as in cereals and nuts play an important role in the well being of the human organism, as a diet that is rich in plant foods contains a variety of secondary metabolites and contributes to protecting the organism mostly against cardiovascular illnesses and cancer [18–21]. In all cases the quality of these foods has to be controlled, as among plant food ingredients a variety of naturally occurring toxins, allergens, or antinutritional agents may be present in the final products.

4.4.2 SECONDARY METABOLITES IN BEVERAGES

Most beverages such as herbal teas, coffee, cocoa, green and black teas, fruit juices, and wine owe their individual properties (flavors and aromas) to the pharmacologically active secondary plant metabolites that they contain. According to recent scientific studies, consumption of fruits and vegetables and also their juices is associated with a decreased risk of heart disease and cancer. This has been ascribed in part to antioxidants in these foods inactivating reactive oxygen species involved in initiation or progression of these diseases. As a result of these studies, many of these active substances (such as vanillin, flavonoids, tannins, anthocyanins, caffeine, and resveratrol) are used in daily diet and high prices are paid for compounds extracted from their natural sources, as they are intended for use as beverages, food additives, flavoring agents, or food supplements [18,19,21–24].

4.4.3 SECONDARY METABOLITES IN PHARMACEUTICALS

Since centuries, many plant compounds have an outstanding role in medicine. Their pharmacological and economical value has lost nothing of its importance until today. They are used either directly or after chemical modification [6,7,19,23–27.]. Some biologically active secondary metabolites have found application as drugs or as chemical model for the design and synthesis (or semisynthesis) of new drug molecules such as the opiates (from morphine and codeine models), aspirin from the naturally occurring salicylic acid (from willow-*Salix* spp.), or etoposide (semi-synthetic antineoplastic agent derived from the mayapple-*Podophyllum peltatum*). Important plant-derived constituents used for their pharmacological properties that are still obtained commercially by extraction from their whole plant sources are listed in Table 4.4, while many others are under clinical trials.

4.4.4 SECONDARY METABOLITES AS PESTICIDES

Plant extracts have been used as insecticides by humans since the time of Roman Empire. More than 2000 plant species have been reported to possess insecticidal properties, among which the most well known are the pyrethrins extracted from pyrethrum flowers. Pyrethrum has been used as an insecticide since 18th century in Persia (Iran), while from the same period it has been used commercially for insect control. In 20th century and especially in 1950s the use of pyrethins declined because the new synthetic analogs (mostly petrochemical derivatives), such as allethrins, appeared as much more effective in the field. Among the most important alkaloids used in insect control have been nicotine and its related nornicotine, used through 16th century, as well as physostigmine, isolated from the calabar beans (*Physostigma venenosum*), which has been served as a chemical model for the development of the group of carbamate insecticides. Another example is rotenone and the rotenoids isolated from the roots of the Legumoninosae *Lonchocarpus*, and *Tephrosia* spp which have also been used as insecticides and piscicides [3,7,28,29].

4.4.5 SECONDARY METABOLITES AS ALLELOCHEMICALS

Allelochemicals (secondary metabolites from higher plants) inhibit selectively the existence of competing species in the surrounding of them, such as soil microorganisms or other plants. This mutual influence of plants is called allelopathy. Allelopathic substances may damage the germination, growth, and development of other plants. These phytotoxic compounds play a role in chemical war between plants (so-called allelopathic interactions) and include natural herbicides, phytoalexins as well as inhibitors of seed germination. Many chemical categories of secondary metabolites are known as allelopathic agents, including phenylpropanoids, quinones, coumarins, flavonoids, tannins, cyanogenic glycosides, and volatile terpenoids [4,7,30,31].

TABLE 4.4
Important Phytochemicals with Potent Medicinal Use and Their Plant Sources

Drug/Chemical	Bioactivity/Clinical Use	Plant Source
Aconitine	CNS activity	*Aconitum napellus*
Adoniside	Cardiotonic	*Adonis vernalis*
Aescin	Anti-inflammatory	*Aesculus hippocastanum*
Aesculetin	Antidysenteric	*Fraxinus rhynchophylla*
Agrimophol	Anthelmintic	*Agrimonia supatoria*
Ajmalicine	Circulatory Disorders	*Rauvolfia sepentina*
Allicin, Ajoene	Antithrombotic activity	*Allium* spp.
Allantoin	Vulnerary	*Symphytum* spp.
Allyl isothiocyanate	Rubefacient	*Brassica nigra*
Aloin, aloe-emodin	Laxative activity	*Aloe vera*
Anabasine	Antismoking, myorelaxant agent	*Anabasis aphylla*
Andrographolide, neoandrographolide	Antidysenteric activity	*Andrographis paniculata*
Anise oil	Digestive	*Pimpinella anisum*
Anthocyanosides	Antioxidative activity	*Vaccinium myrtillus*
Anthraquinones	Laxative activity	*Rheum* spp.
Anthraquinones, emodin, dianthrons	Laxative activity	*Rhamnus* spp.
Arbutin	Anticystitic, antimicrobial activities	*Arctostaphylos uva-ursi*
Arecoline	Anthelmintic	*Areca catechu*
Artemisinin	Antimalarial activity	*Artemisia annua*
Ascaricide (santonin)	Anthelminthic activity	*Artemisia maritima*
Asiaticoside	Vulnerary activity	*Centella asiatica*
Atropine	Anticholinergic activity	*Atropa belladonna*
Azulenes	Antiinflammatory activity	*Matricaria recutita*
Benzyl benzoate	Scabicide activity	*Several plants*
Berberine	Bacillary dysentery, antidysenteric activity	*Berberis vulgaris*
Bergenin	Antitussive activity	*Ardisia japonica*
Betulinic acid	Diuretic, antiseptic activity	*Betula alba*
Biflavonoids	Antioxidative activity	*Citrus sinensis*, *Citrus* sp.
Boldine, benzylbenzoate	Scabicide activity	*Peumus boldus*
Borneol	Antipyretic, analgesic activity	*Zingiber officinale*, several plants
Bromelain	Anti-inflammatory activity, proteolytic enzyme	*Ananas comosus*
Bufotenine, L-Dopa	Anticholinesterase, antiparkinson activity	*Mucuna deeringiana*, *Mucuna pruriens*
Caffeine	CNS stimulant	*Camellia sinensis*, *Coffea arabica*, *Paullinia cupana*
Calendula oil	Anti-inflammatory	*Calendula officinalis*
Camphor	Rubefacient	*Cinnamomum camphora*
Camptothecin	Anticancer activity	*Camptotheca acuminata*

(*continued*)

TABLE 4.4 (continued)
Important Phytochemicals with Potent Medicinal Use and Their Plant Sources

Drug/Chemical	Bioactivity/Clinical Use	Plant Source
(+)-Catechin	Haemostatic, anthelmintic activities	*Potentilla fragarioides, Agrimonia eupatoria*
Cathine, cathinine	Stimulant effect	*Catha edulis*
Charantin	Antidiabetic activity	*Momordica charantia*
Chrysorobin	Antipsoriac activity	*Andira araroba*
Chymopapain	Proteolytic activity, mucolytic enzyme	*Carica papaya*
Citrullol, elaterin	Abortifacient activity	*Citrullus colocynthis*
Clove oil	Antiseptic activity	*Syzygium aromaticum*
Cocaine	Local anaesthetic activity	*Erythroxylum coca*
Codeine	Analgesic, antitussive activity	*Papaver somniferum*
Colchicine	Antitumor agent, anti-gout activity	*Colchicum automnale*
Convallatoxin	Cardiotonic activity	*Convallaria majalis*
Cubebin	Anti-inflammatory activity, antimicrobial	*Piper cubeba*
Curcumin	Choleretic, anticoagulant activity	*Curcuma longa*
Cynarin	Choleretic activity	*Cynara scolymus*
Dianthron	Laxative activity	*Cassia* sp.
L-Dopa	Anti-parkinsonism activity	*Mucuna* sp.
Digitalin, Digitoxin, Digoxin	Cardiotonic activity	*Digitalis purpurea*
Echinacein: arabinogalactan	Pesticide	*Echinacea* spp.
Eleutherosides	Adaptogenic activity	*Eleutherococcus senticosus*
Emetine	Amoebicide, emetic activity	*Cephaelis ipecacuanha*
Ephedrine, pseudephedrine, norpseudephedrine	Sympathomimetic, antihistamine activity	*Ephedra sinica*
Epicatechin	Antidiabetic activity	*Pterocarpus marsupium*
Etoposide	Antitumor agent	*Podophyllum peltatum*
Eucalyptol (cineole), eucalyptus oil	Antimicrobia activity	*Eucalyptus* spp.
Forskolin	Cardiovascular activity	*Coleus forskohlii*
Galanthamine	Cholinesterase inhibitor	*Lycoris squamigera*
Gamma-linolenic acid	Anti-PMS	*Ribes* spp., *Oenothera biennis, Borago officinalis*
Gentiamarin, gentisic acid	Digestive	*Gentiana* spp.
Ginkgosides	Antioxidative activity	*Ginkgo biloba*
Ginsenosides	Adaptogenic activity	*Panax ginseng, Panax quinquefolius*
Glaucarubin	Antiamebic activity	*Simarouba glauca*
Glaucine	Antitussive, hypotensive activity	*Glaucium flavum*
Glaziovine	Antidepressant activity	*Ocotea glaziovii*
Glycyrrhizin	Sweetener, Addison's disease	*Glycyrrhiza glabra*
Gossypol	Male contraceptive	*Gossypium species*

TABLE 4.4 (continued)
Important Phytochemicals with Potent Medicinal Use and Their Plant Sources

Drug/Chemical	Bioactivity/Clinical Use	Plant Source
Harringtonine	Antitumour activity	*Cephalotaxus* spp.
Heliotrine	Antitumour, hypotensive activities	*Heliotropium indicum*
Hemsleyadin	Antidysenteric, antipyretic activities	*Hemsleya amabilis*
Hesperidin	Capillary fragility	*Citrus* species
Humulone, gamma-linolenic-acid	Sedative, anti-inflammatory activity	*Humulus lupulus*
Huperzine	Anticholinesterase activity	*Huperzia serrata*
Hydrastine	Hemostatic, astringent activities	*Hydrastis canadensis*
Hyoscyamine	Anticholinergic activity	*Hyoscyamus niger*
Hypericin	Antistress activity	*Hypericum* spp.
Irinotecan	Anticancer, antitumor agent	*Camptotheca acuminata*
Juglone	Anthelminthic agent	*Juglans* spp.
Kawain	Tranquillizer agent	*Piper methysticum*
Khellin	Bronchodilator, antiasthmatic	*Ammi visaga*
Lanatosides	Cardiotonic activity	*Digitalis lanata*
Lapachol	Anticancer, anti-tumour agent	*Tabebuia* sp.
Lavender oil	Oleochemical industry, perfumery	*Lavandula officinalis*
Liquiritic acid	Skin infections	*Glycyrrhiza glabra*
a-Lobeline	Smoking deterrant, respiratory stimulant	*Lobelia inflata*
Menthol	Rubefacient, anesthetic agent	*Mentha arvensis, Mentha* × *piperita, Mentha spicata*
Mescaline	Hallucinogenic activity	*Lophophora williamsii*
Methyl salicylate	Rubefacient	*Gaultheria procumbens*
Morin	Myorelaxant agent	*Morus alba*
Morphine	Analgesic	*Papaver somniferum*
Nicotine	Insecticide	*Nicotiana tabacum*
Nutmeg oil	Oleochemical industry, perfumery	*Myristica fragrans*
Oleanolic acid	C-AMP, antiallergic activity	*Ziziphus jujube*
Olive oil (mono-unsaturates), oleoeuropeine	Antioxidative	*Olea europaea*
Ouabain	Cardiotonic activity	*Strophanthus gratus, Strophanthus kombe*
Quassinoids	Antimalarial activity	*Ailanthus altissima*
Pachycarpine	Oxytocic activity	*Sophora pachycarpa*
Paeoniflorin	Antiinflammatory activity	*Paeonia albiflora*
Papain	Proteolytic, mucolytic activity	*Carica papaya*
Papavarine	Smooth muscle relaxant agent	*Papaver somniferum*

(*continued*)

TABLE 4.4 (continued)
Important Phytochemicals with Potent Medicinal Use and Their Plant Sources

Drug/Chemical	Bioactivity/Clinical Use	Plant Source
Parthenolides	Antimigraine activity	*Chrysanthemum parthenium*
Peru balsam	Scabicide activity	*Myroxylon balsamum* var. *pereirae*
Phyllanthoside	Antitumor agent	*Phyllanthus* spp.
Physostigmine	Cholinesterase inhibitor	*Physostigma venenosum*
Picrotoxin	Analeptic	*Anamirta cocculus*
Pilocarpine	Parasympathomimetic activity	*Pilocarpus jaborandi*
Platycodin	Analgesic, antitussive activities	*Platycodon grandiflorum*
Podophyllotoxin	Antitumor anticancer agent	*Podophyllum peltatum*
Polygodiol	Antifeedant, antiyeast agent	*Warburgia ugandensis*
Protoveratrines A, B	Antihypertensives	*Veratrum album*
Pseudoephredrine, nor-	Sympathomimetic activity	*Ephedra sinica*
Pyrethrins	Insecticide	*Chrysanthemum cinerariaefolium*
Quassin	Antimalarial	*Quassia amara*
Quinine	Antimalarial, antipyretic activity	*Cinchona ledgeriana*
Quisqualic acid	Ascaricide	*Quisqualis indica*
Reserpine	Antihypertensive, tranquillizer	*Rauwolfia serpentina*
Rhomitoxin	Hypotensive	*Rhododendron molle*
Rorifone	Antitussive	*Rorippa indica*
Rose oil	Oleochemical industry, perfumery	*Rosa* sp.
Rosmarinic acid	Antioxidant activity	*Melissa officinalis*
Rutin	Capillary fragility	*Citrus* sp., *Fagopyrum esculentum*, *Ruta graveolens*
Salicin	Analgesic activity	*Salix alba*, *Populus* spp.
methyl Salicylates	Rubefacient	*Gaultheria procumbens*
Sanguinarine	Dental plaque inhibitor, antiseptic activity	*Sanguinaria canadensis*, *Macleaya cordata*, *Bocconia* spp.
Saussurine	Bronchiorelaxant agent	*Saussurea lappa*
Scillarin A	Cardiotonic activity	*Urginea maritima*
Scopolamine	Sedative activity	*Datura species*
Sennosides A, B	Laxative activity	*Cassia angustifolia*, *Cassaia acutifolia*
Scutellarin	Sedative, antispasmodic	*Scutellaria lateriflora*
Scillaren A	Cardiotonic activity	*Urginea maritima*
Shikonin	Antibacterial, antitussive activity	*Belamcamda chinensis*
Silymarin	Hepatoprotective activity	*Silybum marianum*
Sparteine	Oxytocic activity	*Cytisus scoparius*
Stevioside	Sweetner agent	*Stevia rebaudiana*
Strychnine	CNS stimulant	*Strychnos nux-vomica*
Tannic acid	Free radicals scavenging activity	*Quercus infectoria*
Taxol	Antitumour agent	*Taxus brevifolia*
Teniposide	Antitumour agent	*Podophyllum peltatum*
Delta-9-tetrahydrocannabinol (THC)	Antiemetic, decrease occular tension activity	*Cannabis sativa*

TABLE 4.4 (continued)
Important Phytochemicals with Potent Medicinal Use and Their Plant Sources

Drug/Chemical	Bioactivity/Clinical Use	Plant Source
Tetrandrine	Hypotensive activity	*Stephanie tetrandra*
Theobromine	Diuretic, vasodilator, brochodilator, CNS, stimulant	*Theobroma cacao*
Theophylline	Diuretic, vasodilator, CNS, stimulant	*Camellia sinesis*
Thymol	Spasmolytic, topical antifungal activity	*Thymus vulgaris*
Tolu balsam	Scabicide activity	*Myroxylon balsamum*
Topotecan	Antitumor agent	*Camptotheca acuminate*
Trichosanthin	Abortifacient	*Trichosanthes kirilowii*
Tubocurarine	Myorelaxant agent	*Chondodendron tomentosum*
Valepotriates	Sedative, tranquilizer	*Valeriana officinalis*
Vasicine	Oxytocic, expectorant	*Justicia adhatoda*
Vinblastine, Vincristine	Antitumor, Antileukemic agent	*Catharanthus roseus*
Vincamine	Cerebrotonic, hypotensive activity	*Vinca minor*
Viscin	Antiproliferative activity	*Viscum album*
Wilfordine	Antitumour activity	*Tripterygium wilfordii*
Withanolide	Adaptogenic activity	*Withania somniferum*
Xanthotoxin	Antipsoriac agent	*Ammi majus*
Yohimbine	Aphrodisiac agent	*Pausinystalia yohimbe, Aspidosperma* spp.

4.4.6 SECONDARY METABOLITES AS PLANT GROWTH REGULATORS

The growth and development of plants is regulated by a number of chemical constituents such as gibberellins, cytokinins, abscisic acid and derivatives, as well as ethylene. All these compounds are specific to their action in very low concentrations and regulate cell division, differentiation, enlargement, and organogenesis, and have been classified agriculturally as very important chemical molecules [7,19]. In the framework of the research for finding, secondary metabolites with comparable activities, the steroidal lactone brassinolide, isolated from the pollen of rape (*Brassica napus*), has showed that they can promote plant growth at very low concentrations. Brassinolide as well as its semisynthetic derivatives (brassinosteroids) are expected to be used as plant growth regulators by promoting cell expansion and also cell division [7,19,32,33].

4.5 BIOACTIVITIES OF SECONDARY METABOLITES

4.5.1 LEGACY OF THE PAST

The first generally accepted use of plants as healing agents was depicted in the cave paintings discovered in the Lascaux caves in France, dated between 13,000 and 25,000 BC, [34,35] while from that period till almost 18th century AD, there were no synthetic medicines at all and the 250,000 species of higher plants were the main source of drugs for the world's population. So, the healing power of plants is believed as an ancient idea, very well developed, mostly by the inhabitants of Ancient Egypt, Mesopotamia, and Ancient Greece in the Western World [34,36]. The Ayurvedic sources from India and the Chinese Herbal Medicine give also much more examples of the development of drugs from plants [34,37,38]. The majority of the pharmaceuticals available to Western medical therapy have long history of use as herbal remedies, including opium, quinine, aspirin, digitalis, etc. Approximately 80% of Dioscorides' "Materia Medica"—an encyclopedia of substances used in medicine—consists of plant medicines, while the Greek physician Galen (130 AD) and the the Arab Avicenna (900 AD) with his medical encyclopedia called "the Canon of Medicine" made significant contributions to medicine. Both Galen and Avicenna have been described what we know today as malignant tumors, while Galen referred to karkinos-καρκίνος (which means crab or crayfish in greek), related to the word "carcinoma" [39,40]. Historians have identified many substances found in plants that the ancients used as treatments for infectious diseases as well as against tumors [6,7,34,41], such as the deadly nightshade (belladonna), the squirting cucumber (*Ecballium elaterium* L.), the *Narcissus* bulb, the castor bean (*Ricinus communis* L.), etc. Even in our days, among prescripted drugs, almost 25% contain at least one compound of plant origin. This already high percentage might be higher if we include the OTC (over-the-counter) not prescribed drugs.

4.5.2 Classification of Detected Bioactivities

The study of medicinal plants and their chemical constituents can be focused to their specific bioactivities [6,7,27,41–43]. These bioactivities can be classified according to several scientists as follows:

- *Action on the autonomic nervous system*: (1) Acetylocholine-like drugs as pilocarpine (*Pilocarpus* sp.), arecoline (seeds of *Areca catechu*), muscarine (fungi *Amanita* sp.), physostigmine (seeds of *Physostigma venenosum*), etc., (2) antagonists of acetylocholine as tropane esters alkaloids in Solanaceae plants, tubocurarine (*Chondodrendron tomentosum*), etc., (3) adrenaline-like drugs: ephedrine from *Ephedra* spp., and (4) antagonists of adrenaline as ergot alkaloids from *Claviceps purpurea*
- *Action on the central nervous system*: (1) Drugs affecting mental activity (1a) Hallucinogenics as cannabinoids (*Cannabis sativa*), mescaline from peyote cactus, (1b) stimulating mental activity as purine bases as caffeine, theophylline, theobromine present in coffee, tea, kola and cocoa, (1c) depressing mental activity as reserpine from *Rauwolfia* sp., (1d) analeptics as picrotoxin and lobeline from *Anamirta cocculus* and *Lobelia inflate,* respectively, (2) central depressants of motor function as tropane alkaloids, and (3) possessing analgesic avtivity as morphine from *Papaver somniferum*
- *Action on heart muscle*: cardiac glycosides mostly from *Digitalis* spp., and *Strophanthus* sp.
- *Action on blood vessels*: (1) peripheral vasoconstrictors drugs as ephedrine, nicotine, etc., (2) central vasoconstrictors drugs as picrotoxin, (3) vasodilators as papaverine, ergotamine and *Veratrum* alkaloids, etc.

- *Action on the respiratory system*: (1) bronchodilators as ephedrine, (2) cough depressants as codeine, (3) expectorants as ipecacuahna, liquorice and Senega roots, and (4) antiexpectorants as atropine.
- *Action on the gastrointestinal tract*: (1) anticholinergic drugs, (2) emetics as ipecacuahna, (3) bitters such as *Gentian, Quassia, Cinchona*, (4) carminatives as dill, and aniseed oil, (5) laxatives and purgatives as *Psyllium, Ispaghula, Senna, Aloe*, etc., (6) ulcer therapy as liquorice root, and (7) antidiarrhoeal drugs.
- *Action on the liver*: (1) hepatoprotective activity as *Silibum marianum* flavolignans, or *Cynara scolymus* and (2) hepatotoxic activity as pyrrolizidine alkaloids from Boraginaceae, Asteraceae and Fabaceae families, teucrine from *Teucrium* sp., etc.
- *Action on skin and mucous membranes*: (1) astrigents as tannins, (2) emollients and demulcents as olive and theobroma fixed oils, (3) anti-inflammatory agents, and (4) antiseptics as *Eucalyptus* and *Thymus* oils.
- *Treatment of infections*: (1) Antibiotics mostly from moulds and streptomyces, (2) antimalarials from *Cinchoma* sp., sesquiterpene lactone artemisin from *Artemisia annua*, and (3) amoebicides as the alkaloid emetine from ipecacuahna root and anthelminthics as santonin, *Chenopodium* oil.
- *Treatment of malignant diseases*: Anticancer activity with vinca alkaloids from *Catharanthus roseus*, the famous taxol from *Taxus* sp., podophyllotoxin and semi synthetic derivatives as etoposide and teniposide from *Podophyllum peltatum*, etc.

4.5.3 PLANT SECONDARY METABOLITES AS NEW DRUGS

Historically, plants have provided a source of inspiration for novel drug compounds, as plant-derived medicines have made large contributions to human health. Their roles can be divided (1) to their use as phytomedicines for the treatment of an illness in crude form and (2) to become the base for the development of a medicine through bioguided isolations of its active constituents, detailed biological assays, formulation of dosage forms, followed by several phases of clinical studies designed to established safety, efficacy and pharmacokinetic profile of the new drug. This evaluation is finally followed by acute and chronic toxicity studies in animals [41–44]. Very important example of this way of research is the vinca alkaloids obtained from the Madagascan periwinkle (*Catharanthus roseus* syn. *Vinca roseus*), recently taxol from *Taxus* species as well as homoharringtonine and derivatives of camptothecin [6,7]. The same methods of development and evaluation have been used also for the alkaloid dimers, michellamines (A-C) isolated from the endemic plant of Cameroon *Ancistrocladus korupensis* very promising anti-HIV agents [45].

4.6 PROSPECTS FOR DISCOVERING NEW BIOACTIVE COMPOUNDS FROM PLANTS

It has been estimated that only about 10% percent of the 300,000 to 600,000 existing species of higher plants has been searched for biologically active compounds. It is obvious that the plant kingdom, such a treasure for the research of new chemical molecules, has received little attention as a source of potentially important bioactive compounds and many plant constituents possessing biological properties remain unknown, undiscovered, and unused. In this framework, pharmacologists, botanists, and natural products' chemists have a tough job to do, for the research of new molecules or analogs that could be developed for treatment of various diseases [46–49]. Molecules derived from plants have shown great promise in the treatment of diseases such as malignancies like cancer, respiratory infections, HIV/AIDS, tropical diseases and to the increase in antibiotic resistance in community acquired infections.

Besides, in our new millennium, worldwide, there has been a renewed interest in natural products, as a result of consumer's belief that natural products give more benefits, changes in laws allowing structure–function claims and national concerns for health care cost.

Additionally, the prospects for developing new pesticides and herbicides from plant sources are also very promising, because this area of investigation is newer than the area of medicinal plant research. Natural plant chemicals will undoubtedly play a significant role in the future of pest control in both industrialized and developing countries [50,51].

4.7   CONCLUSIONS

Plant natural products have been, through centuries, and will continue to be, important sources of high value food, beverages, spices, flavors and fragrances, vegetable oils, resins, insecticides and other industrial, agricultural materials, as well as medicinal products and in many cases valuable drugs. However, since most plant species have never been described much less surveyed for chemical or biologically active constituents, it is reasonable to expect that new sources of important secondary metabolites remain to be discovered by phytochemists, botanists, pharmacologists, and related scientists. The advances in chromatographic and spectroscopic techniques permit and facilitate the isolation and structural analyses of potent biologically active plant secondary metabolites that are present in quantities. These new chemical and biological technologies will serve to enhance the continued usefulness of plants as the most important renewable resources of chemicals.

REFERENCES

1. Tyler, V.E., Brady, L.R., and Robbers, J.E., *Pharmacognosy*, Lea & Febiger, Philadelphia, 1981, p. 8.
2. Mann, J., *Secondary Metabolism*, Oxford University Press, Oxford, 1978.
3. Rosenthal, G.A. and Janzen, D.H. (Eds.), *Herbivores: Their Interaction with Secondary Plant Metabolites*, Academic Press, New York, 1979.
4. Harborne, J.B., *Introduction to Ecological Biochemistry*, 2nd ed., Academic Press, New York, 1982.
5. Hedin, P.A., *Plant Resistance to Insects*, American Chemical Society, Washington, D.C., 1983.

6. Bruneton, J., *Pharmacognosy, Phytochemistry, Medicinal Plants*, 2nd ed., Lavoisier Publishing, London, Paris, NewYork, 1999.

7. Evans, W.C., *Trease and Evans' Pharmacognosy*, 13th ed., Bailliere Tindall, London, England, 1989.

8. Bell, E.A. and Charlwood, B.V. (Eds.), Secondary plant products, in *Encyclopedia Plant Physiology*, Vol. 8, Springer-Verlag, Berlin, Heidelberg, New York, 1980.

9. Banthorpe, D.V., in *Methods in Plant Biochemistry* (Series Editors: Dey, P.M. and Harborne J.B.), Vol. 7, Terpenoids, Charlwood, B.V. and Banthorpe D.V., Eds., Academic Press, London, 1991, p. 1.

10. Chinou, I., Labdanes of natural origin–biological activities (1981–2004), *Curr. Med. Chem.*, 12, 1295, 2005.

11. Harborne, J.B., Mabry, T.J., and Mabry, H., *The Flavonoids*, Chapman and Hall, London, 1975.

12. Gilchrist, T.L., *Heterocyclic Chemistry*, 2nd ed., Longman Scientific & Technical, UK, 1993.

13. Joule, J.A. and Mills, K., *Heterocyclic Chemistry*, 4th ed., Blackwell Publishing, Oxford, 2000.

14. Hesse, M., *Alkaloide*, Wiley-VCH, Weinheim, 2000.

15. Geissman, T.A. and Crout, D.H.G., *Organic Chemistry of Secondary Plant Metabolism*, Freeman, San Francisco, 1969.

16. Atsatt, P.R. and O'Dowd, D.J., Plant defense guilds. *Science*, 193, 24, 1976.

17. Luckner, M., *Secondary Metabolism in Microorganisms, Plants, and Animals*, Springer-Verlag, Berlin, Heidelberg, New York, Tokyo, 1984.

18. Leung, A.Y. and Foster, S., *Encyclopedia of Common Natural Ingredients Used in Food, Drugs and Cosmetics*, 2nd ed., Wiley, New York, 1996.

19. Balandrin, M.F., Klocke, J.A., Wurtele, E.S., and Bollinger, W.H., Natural plant chemicals: sources of industrial and medicinal materials, *Science*, 228, 1154, 1985.

20. Crozier, A., Burns, J., Aziz, A.A., Stewart, A.J., Rabiasz, H., Jenkins, G., Edwards, C.A., and Lean, M.E.J., Antioxidant flavonols from fruits, vegetables and beverages: measurements and bioavailability, *Biol. Res.*, 33, 79, 2000.

21. Corder, R., Mullen, W., Khan, N.Q., Marks, S.C., Wood, E.G., Carrier, M.J., and Crozier, A., Red wine procyanidins and vascular health, *Nature*, 444, 566, 2006.

22. Duthie, S.J., Jenkinson, A.M., Crozier, A., Mullen, W., Pirie, L., Kyle, J., Yap, L.S., Christen, P., and Duthie, G.G., The effects of cranberry juice consumption on antioxidant status and biomarkers relating to heart disease and cancer in healthy human volunteers, *Eur. J. Nutr.*, 45, 113, 2006.

23. Mann, J., Davidson, R.S., Hobbs, J.B., Banthorpe, D.V., and Harborne, J.B., *Natural Products. Their Chemistry and Biological Significance*, Addison Wesley Longman, Harlow, UK, 1994.

24. Fraenkel, G.S., The raison d'etre of secondary plant substances, *Science*, 129, 1466, 1959.

25. Massy, Z.A., Keane, W.F., Kasiske, B.L., Inhibition of the mevalonate pathway: benefits beyond cholesterol reduction, *Lancet*, 347, 102, 1996.

26. Bidlack, J. and Wayne, R., *Phytochemicals as Bioactive Agents*, Technomic Publishers, Lancaster, PA, 2000.

27. Cassady, J.M. and Douros, J.D., *Anticancer Agents Based on Natural Product Models*, Academic Press, New York, 1980.

28. Schemltz, I., *Naturally Occurring Insecticides*, Jacobson, M. and Crosby, D.G. Eds., Dekker, New York, 1971, p. 99.

29. Crosby, D.G., *Natural Pest Control Agents*, Crosby, D.G., Ed., American Chemical Society, Washington, D.C., 1966, p. 1.

30. Harbone, J.B., *Phytochemical Ecology*, Academic Press, New York, 1972.
31. Rice, E.L., *Allelopathy*, 2nd ed., Academic Press, New York, 1984.
32. Harborne, J.B., Baxter, H., and Moss, G.P., *Dictionary of Plant Toxins*, Wiley, Chichester, UK, 1996.
33. Takahashi, N., *Chemistry of Plant Hormones*, CRC Press, Boca Raton, FL, 1986.
34. Sneader, W., *Drug Discovery: A History*, Wiley, West Sussex, England, 2005.
35. Lietava, J., Medicinal plants in a middle paleoliothic grave Shanidar IV? *J. Ethnopharmacol.*, 35, 263, 1992.
36. Biggs, R., Medicine in ancient Mesopotamia. *Hist. Sci.*, 8, 94, 1969.
37. Morgan, K., *Medicine of the Gods: Basic Principles of Ayurvedic Medicine*, Mandrake, Oxford, 1994.
38. Triestman, J.M., China at 1000 B.C.: a cultural mosaic. *Science*, 160, 853, 1968.
39. Riddle, J.M., *Dioscorides on Pharmacy and Medicine. History of Sciences Series*, No. 3. University of Texas Press, Austin, 1986.
40. Günther, R., *The Greek Herbal of Dioscorides*, Hafner Publishing Co., New York, 1959.
41. Barz, W. and Ellis, B.E., *Natural Products as Medicinal Agents*, Beal, J.L. and Reinhard, E., Eds., Hippokrates, Stuttgart, Germany, 1981.
42. Aarts, T., *The Dietary Supplements Industry: A Market Analysis*, Dietary Supplements Conference, Nutritional Business International, 1998.
43. Murray, E.M., *The Healing Power of Herbs*, Prima Publishing, Rocklin, CA, 1995, p. 162.
44. Johnston, B., One-third of nation's adults use herbal remedies, *HerbalGram*, 40, 49, 1997.
45. Boyd, M., Hallock, Y., Cardellina II, J., Manfredi, K., Blunt, J., McMahon, J., Buckheit, R., Bringmann, G., Schaffer, M., Cragg, G., Thomas, D., and Jato, J., Anti-HIV michellamines from *Ancistrocladus korupensis. Med. Chem.*, 37, 1740, 1994.
46. Duke, J.A., *Handbook of Medicinal Herbs*, CRC Press, Boca Raton, FL, 1985.
47. Duke, J.A. and Atchley, A.A., *Handbook of Proximate Analysis Tables of Higher Plants*, CRC Press, Boca Raton, FL, 1986.
48. Morton, J.F., *Major Medicinal Plants*. Thomas C.C., Ed., Springfield, IL, 1977.
49. Tyler, V.E., *The New Honest Herbal*, G.F. Stickley Co., Philadelphia, 1987.
50. Farnsworth, N.R. and Bingel, A.S., *New Natural Products and Plant Drugs with Pharamacological, Biological or Therapeutical Activity*, Wagner, H. and Wolff, P., Eds., Springer-Verlag, New York, 1977, p. 1.
51. Nakanishi, K., *Natural Products and the Protection of Plants*, Marini-Bettolo, G.B., Ed., Elsevier, New York, 1977, p. 185.

5 Plant Chemosystematics

Christian Zidorn

CONTENTS

5.1 INTRODUCTION

5.1.1 Definition of Plant Chemosystematics

Long before chemosystematics or even biology and chemistry existed as sciences in our current definition the fact that morphologically similar plant species often resemble each other with regard to their (chemical) constituents was well known. According to Hänsel this fact was already described by James Pettiver in the late 17th century.[1] The close association of plant morphology and certain other qualities of a plant, which today would be attributed to its chemical composition, like e.g., smell, color, or taste, was certainly already known to the earliest human societies.

Today, plant chemosystematics (synonyms: biochemical systematics, chemotaxonomy, comparative phytochemistry, and molecular taxonomy) is an interdisciplinary field in which chemical constituents of plants are used as characters to determine inter- and infraspecific relationships of plant taxa. Initially the term was employed for studies encompassing micro- and macromolecules.[2,3] In the last two decades the field of DNA analysis has vastly expanded and studies employing DNA to infer phylogenies are at present termed molecular studies.[4] These studies are usually no longer included into the definition of the terms chemosystematics and chemotaxonomy. Chemosystematics in its current delimitation conclusively encompasses studies of micromolecular constituents of plants, animals, or microorganisms. While chemosystematic studies in higher animals are quite rare,[5] chemosystematic studies in microorganisms and plants are more frequently published. Chemosystematic investigations in microorganisms are performed to find new compounds potentially of use as pharmaceuticals but also for taxonomic reasons because of the limited morphological diversity of these organisms.[6] Chemical studies of marine organisms are presently in a stage where the great variety and complexity of the chemical compounds of marine organisms are being explored. Therefore, studies focusing on the usage of these compounds as markers for systematic purposes are limited in number.[7]

The focus of this chapter will be on chemosystematic studies of plants. Though such studies are still frequently published, their impact for the classification of plant orders and families has significantly decreased since the direct analysis of DNA has become feasible.[8] Nonetheless chemosystematic studies have a merit in their own right, a fact which will be discussed in more detail in Section 5.1.2.

Examples for techniques commonly employed in chemosystematic studies are gas chromatography (GC), high-performance liquid chromatography (HPLC), paper chromatography (PC), and thin layer chromatography (TLC). While early studies were almost exclusively performed with simple techniques such as PC and TLC, more and more sophisticated techniques have been employed and the evolution of chemosystematics as an interdisciplinary field of science is closely linked to the development of new techniques in analytical chemistry.[9] Currently, the most frequently used techniques in the analysis of natural products are GC/FID and GC/MS for volatile compounds and HPLC/DAD and HPLC/MS for nonvolatile compounds. More elaborate instrumentation such as HPLC/NMR is constantly developed and added to the arsenal of methods employed in chemosystematics.[10]

5.1.2 AIMS OF CHEMOSYSTEMATIC STUDIES

From its early days in the 1940s until the 1980s, plant chemosystematics was considered as a means to contribute to a natural system in plant systematics.[9,11] Chemical plant characteristics were generally studied by chemists and pharmacists. Therefore, a strong historic linkage between pharmacognosy (the field of pharmacy dealing with medicinal plants) and the evolving field of chemosystematics exists.[1] Data about the occurrence of plant secondary metabolites were eventually integrated into the classification systems of Cronquist and Takhtajan.[12,13]

However, with the accumulation of further data, it became gradually evident that like in morphology parallel evolution was also occurring in the accumulation of certain plant secondary metabolites. A classic example for the parallel occurrence of a secondary metabolite in unrelated plant taxa is caffeine, which is accumulated not only in the genus *Coffea* (Rubiaceae) but also in the genera *Camellia* (Theaceae), *Paullinia* (Sapindaceae), *Ilex* (Aquifoliaceae), and *Theobroma* (Sterculiaceae), to name the commercially most important sources only.[14] Recently, caffeine was even identified as a secondary metabolite in the sclerotia of the fungus *Claviceps sorghicola* Tsukiboshi, Shimanuki, and Uematsu.[15] Another example for the occurrence of the same or closely related secondary metabolites in only distantly related plant taxa is the natural product class of sesquiterpene lactones. Sesquiterpene lactones have been isolated from liverworts as well as from members of the Magnoliaceae, Apiaceae, and Asteraceae.[16–19]

With the advance of (macro-)molecular techniques, which enabled the direct investigation of the three plant genomes (nuclear, mitochondrial, and chloroplast),[20] the importance of chemosystematics as a tool to elucidate phylogenetic relationships on higher levels of the systematic hierarchy has decreased. Currently, chemosystematic studies contribute to the solution of scientific problems in the following fields:

1. Provide phenetic markers for newly emerging groupings found in molecular studies (chemotaxonomy).[21]
2. Contribute to the understanding of plant defense mechanisms against abiotic and biotic stress factors (chemical ecology).[22]
3. Help in the rational search for new bioactive chemical entities (pharmacy).[23]
4. Yield new chemical structures not or only difficultly accessible by synthetic chemical routes (organic chemistry).[24]
5. Constitute challenges for modern analytical chemistry because plant extracts are very complex matrices (analytical chemistry).[25]
6. Complement or replace molecular methods in the solution of systematic problems below the species level.[26]

Because of advances in analytical chemistry (better chromatographic resolutions, more sensitive detectors) and progress in data evaluation methods (calculatory powers of modern computers, new methods in multivariate data analysis), chemosystematic studies are getting easier and faster to perform and the resulting data sets contain a higher amount of information than the ones from earlier studies.[20]

5.1.3 Interpretation of Chemosystematic Data

An important consideration in all chemosystematic studies is the fact that the nonoccurrence of a particular compound in a species does not necessarily mean that the genes involved in its biosynthesis were never present in the taxon or its predecessors. Possible other reasons for the nonoccurrence of particular secondary metabolites are secondary loss of the corresponding genes, mutation of these genes or silencing of the genes due to differential gene expression.[27] A drawback of chemosystematic studies using chemical data in a qualitative way (0,1-data matrix) is the fact that often the limit of detection is not known and that minute amounts of certain compounds are not detected unless specific highly sensitive methods have been established to specifically detect these compounds in minute amounts.[28] Therefore, the nondetection of a particular compound is to be weighted less than the detection of a compound. The fact that a particular compound is not detected might either be related to (a) the low amount of the compound present in the given sample or it might be related to (b) the fact that the compound is present in the plant but not in the analyzed organ; the plant might be producing the compound but (c) only at a specific ontogenetic stage; the plant might be synthesizing the compound but (d) only during a particular time of the growing season; (e) the plant might be producing the compound only when certain external factors like herbivore attack or high radiation are acting; (f) some populations of the same (morphologically defined) species might contain a secondary metabolite while in other populations this compound is missing (chemical "polymorphism"); (g) the compound might be present in a parasitic species but the compound was actually biosynthesized by the host species (e.g., compounds from *Galium verum* L. detected in the parasite *Euphrasia stricta* J.F. Lehm.).[29] On the other hand, the confirmed detection of a particular compound in a given taxon usually verifies that the entire genetic information needed to synthesize this chemical is encoded in the genome of the analyzed plant. Notable exceptions to this rule are the parasite host example cited above and many more examples where microorganisms associated with macroorganisms are the actual producers of secondary metabolites but these metabolites are erroneously ascribed to the macroorganism the study was focused on. This phenomenon is particularly frequent in marine organisms.[30]

The fact that certain plant metabolites are only synthesized or accumulated under certain environmental conditions poses severe problems for chemical systematics in the classical sense. On the other hand, these correlations of secondary metabolism and environment open a new field of investigation: chemical ecology. In chemical ecology the ecological function of secondary metabolites for the plants producing them is in the center of attention. Many fascinating insights about plant/plant interactions, plant/animal interactions, and the impact of abiotic factors on plant chemistry have been gained.[22] Detailed accounts of studies on chemical ecology though intimately related to chemosystematic studies are beyond the scope of this chapter.

5.2 CHEMOSYSTEMATIC STUDIES EMPLOYING MACROMOLECULES

Macromolecular studies include DNA sequencing of the various plant genomes (chloroplast, mitochondrial, nuclear) as well as anonymous marker techniques based on gel electrophoresis of DNA fragments, such as amplified fragment length polymorphism (AFLP), randomly amplified polymorphic DNA (RAPD), and restricted fragment length polymorphism (RFLP). Molecular studies directly investigating DNA have emerged as a separate field of science and are generally no longer included in the term chemosystematic studies and they are not dealt with in the present communication.

Apart from DNA, various other macromolecules are found in plants. These macromolecules include, e.g., cell wall components such as cellulose and chitin as well as reserve carbohydrates such as dextran, inulin, and starch. These compounds are usually characteristic for large systematic groups such as families, orders, and therefore generally only of interest for chemosystematic investigations at a higher taxonomic level. For example, inulin (1,2-fructosane) is the reserve carbohydrate in the Asterales (sensu APG) and replaces in this order starch as the reserve carbohydrate.[31]

5.3 CHEMOSYSTEMATIC STUDIES EMPLOYING SMALL MOLECULES

5.3.1 PRIMARY METABOLITES

The general pathways for synthesizing and modifying carbohydrates, proteins, fats, and nucleic acids are essentially the same in all organisms, regardless of their taxonomic affiliation. These pathways are summarized by the term primary metabolism and the compounds involved are termed primary metabolites.[23] Because of their ubiquitous occurrence, primary metabolites are of no use in the delimitation of systematic entities and are usually not employed in chemosystematic studies. However, in the field of chemical ecology there are some comparative investigations on the relative composition of sugars in flower nectar. For example, Galetto and coworkers analyzed some 50 species from Argentinean Patagonia for their chemical floral nectar composition.[32] All analyzed samples contained high levels of mixtures of glucose, fructose, and sucrose, with glucose being the most common major compound. The ratio of the different sugars in the nectars was closely related to the kinds of pollinators that visit the species investigated and not primarily to the systematic position of the investigated species. Analytically, nectar sugars are usually silylated and then identified and quantified by GC. Simpler protocols employ TLC to detect sugars and alditiols.[33]

5.3.2 SECONDARY METABOLITES

Secondary metabolites are compounds, which have a limited distribution in the animal, plant, or fungal kingdom. Secondary metabolites are not necessarily produced under all growing conditions. In most cases the benefit of these compounds for the organisms producing them is not known yet. It is generally assumed—and proven for a growing number of compounds—that secondary natural products have a benefit for example as feeding deterrents for the plants, animals, or fungi producing them.[23] Other potential benefits of secondary metabolites for their producers include the inhibition of the growth or germination of neighboring plants, the alleviation of the effects of abiotic stress factors like UV-radiation, and the inhibition of bacterial and fungal microbes. Because of their nonubiquitous occurrence and because the distribution of many secondary metabolites is in-line with current classification systems, secondary metabolites are the preferred substances for chemosystematic studies.

5.4 METHODS COMMONLY USED IN CHEMOSYSTEMATICS

5.4.1 THIN LAYER CHROMATOGRAPHY/PAPER CHROMATOGRAPHY

PC and TLC were the first techniques employed for chemosystematic studies. PC had been introduced in the 1940s and soon the new technique was also employed for the analysis of plant secondary metabolites.[34–36] Indeed the two classical studies of Alston and coworkers on chemotaxonomy and on chemical ecology were based on results obtained with PC.[37,38] From the early 1960s onwards TLC was introduced[39] and widely applied in natural products research, too.[40]

Today PC is still used in the isolation of flavonoids sometimes but not as an analytical technique. Also TLC is only rarely used directly as an analytical tool in comparative phytochemistry.[41] On the other hand, TLC is still one of the most frequently used techniques in the process of natural product isolation; here it is mainly used to monitor the composition of fractions and to preliminarily assess the purity of natural products. The main advantages of TLC are the wide array of compound polarities in which TLC can be employed, the short analysis time, the moderate price per analysis, and the fact that multiple fractions/compounds can be analyzed in parallel on a single TLC plate.

5.4.2 GAS CHROMATOGRAPHY

GC is the most potent chromatographic separation technique. In plant chemosystematics GC is widely applied in a range of compound classes, especially volatiles such as essential oils.[42] After derivatization (acetylation, silylation) some originally nonvolatile compounds can be separated by GC, too. Interesting examples are sugars and sugar alcohols.[43] These compounds are difficult to detect by other standard techniques such as HPLC/DAD, because of the lack of a chromophore, a prerequisite for HPLC/DAD analyses.

Disadvantages of GC compared with TLC in natural product analysis are the necessity of using sophisticated (and expensive) equipment to perform GC analyses and the fact that unlike in TLC no multiple analyses of different extracts are feasible at the same time (unless of course multiple GC systems are used).

5.4.3 HIGH-PERFORMANCE LIQUID CHROMATOGRAPHY

HPLC is probably the most versatile and most widely applied technique in the analysis of natural products. The main advantage of HPLC is the fact that basically all compounds soluble in water or one of the many organic solvents employed for HPLC can be analyzed. In most cases only compounds, which have a chromophore and therefore are detectable with the widely used UV- or DAD-detectors, are analyzed by HPLC. However, recently the availability of a new generation of evaporative light scattering detectors (ELSD) has also widened the applicability of HPLC for natural product analysis.[44] This technique enables HPLC analysis of all nonvolatile compounds regardless of the absence or presence of a chromophore. Disadvantages of HPLC when compared with TLC are the same as those listed under GC: necessity of expensive equipment and impossibility of parallel analyses.

5.4.4 CAPILLARY ELECTROPHORESIS

Capillary electrophoresis (CE) is a comparatively new analytical technique. Theoretically it could also be applied to chemosystematics. Because of the numerous factors affecting reproducibility and the fact that in many methods time-consuming sample workups are needed, the use of CE for chemosystematic studies is limited. There is a notable exception for the field of bacterial chemosystematics[45] and two recent papers deal with the application of CE in chemosystematic studies on higher plants.[46,47]

5.5 CLASSES OF NATURAL PRODUCTS ANALYZED IN CHEMOSYSTEMATIC STUDIES

5.5.1 POLYKETIDES

5.5.1.1 Anthranoids

Anthranoids occur in various genera of higher plants, e.g., *Aloe* (Asphodelaceae), *Cassia* (Caesalpiniaceae), *Rhamnus* (Rhamnaceae), and *Rheum* (Polygonaceae). Especially in the large genus *Aloe* anthranoids have been employed as chemosystematic markers. In an extensive investigation covering 380 species of *Aloe*, 36 were shown to contain a characteristic combination of aloins A and B, aloinosides A and B, and microdontins A and B.[48–51] This group of 36 species, which was informally named the microdontin chemotype, had no obvious morphological features in common and was also never formally combined to an infrageneric group within the genus *Aloe*. Nonetheless, Viljoen et al. speculate about a common North-East African origin of this group.[49] The largest number of species assigned to the microdontin-type occurs in North-East Africa and the authors discuss the possible evolution as well as routes of migration of a common North-East African ancestor to the current species and their distribution areas.

5.5.1.2 Xanthones

Xanthones, which are generally widespread in the plant kingdom, have been applied as chemosystematic markers in the Gentianaceae family.[52–54] In their comprehensive review Peres et al. compiled data on 192 xanthones from 84 species and 17 genera of the Gentianaceae.[53] Applying the system by Gottlieb and coworkers, which assigns point values to compounds due to their degree of "advancement," these authors list the genera of the Gentianaceae in a decreasing "evolutive order." This approach will potentially have its merits. However, the idea of assigning a point value to a species and averaging this point value based on the point values of all investigated species to get a point value for each genus is an oversimplification of evolutionary history.

Thus, the majority of the phylogenetic information contained in the data compiled by Peres et al. is not expressed by these point values.[53]

5.5.2 PHENYLPROPANOIDS

5.5.2.1 Phenolic Acids

Though, or probably because, phenolic acids are nearly ubiquitous in the plant kingdom, phenolic acids have only been rarely employed as chemosystematic markers in higher plants. On the other hand, in lichen taxonomy the distribution of phenolic acids, so called lichen acids, which are unique to this group of symbiotic organisms have been extensively used for determination and classification. In the exemplary work of Feige et al., a total of 331 lichen compounds, mainly phenolics, were characterized.[55]

Recent examples for the applicability of phenolic acids in higher plant systematics include, for example, our studies in the Lactuceae tribe of the Asteraceae family. The results showed that caffeoyltartaric acid derivatives occur in most members of some genera (e.g., *Crepis* and *Leontodon*) but are completely missing in other related genera (*Hieracium*).[56–60] Moreover, a new class of phenolic acids, dihydrocaffeoyl quinic acids, has been found in the genus *Podospermum* (subtribe Scorzonerinae).[61] The distribution of this class of compounds in other taxa and its potential chemosystematic significance has not been assessed yet.

5.5.2.2 Coumarins

Coumarins are $2H$-1-benzopyran-2-one derivatives; coumarins either derive from cyclization of a C-2 oxygenated *cis*-cinnamic acid or via alternative pathways such as the mixed cinnamic/acetate pathway or completely via the acetate pathway.[62] Coumarins are widely distributed in plants as free coumarins as well as glycosides. In some plants, such as sweet clover, coumarins are a product of enzymatic biosynthesis after damage of the plant tissues containing the precursors (E)- and (Z)-2-coumaric acid.[23] Coumarins occur in great structural variety especially in the Apiaceae and Rutaceae[63] and are additionally found in many other plant families like the Asteraceae,[64] Poaceae,[65] and Rubiaceae.[66] The pleasant smell of hay is partly ascribable to coumarins, which are released from grasses (e.g., *Anthoxanthum odoratum*) and legumes (e.g., *Melilotus officinalis*). Moreover, coumarin itself is important in this respect as it is the main aroma compound of woodruff, *Galium odoratum*, the characteristic ingredient of May bowl.[66] Reviews on chemosystematics employing coumarins include the classical work of Gray and Waterman[62] and a recent review by Campos Ribeiro and Coelho Kaplan.[67]

5.5.2.3 Flavonoids

The potential of flavonoids as chemosystematic markers has been realized early.[68] Their usage as chemosystematic markers was promoted by the fact that they can be detected in almost all investigated plant species, they are chemically stable compounds, and they can be easily separated and detected using simple chromatographic equipment.[69]

Recent examples of the analysis of flavonoids in an explicitly chemosystematic context include the study of the genus *Betula* by Lahtinen et al.,[70] the study of members of the genus *Hieracium* s.l. (i.e., including *Pilosella*) by Zidorn et al.,[59] and a study of the Pinaceae family by Slimestad.[71]

Lahtinen et al. analyzed leaf surface flavonoids from uncrushed fresh leaves to assign 15 *Betula* species to 3 informal groups.[70] These flavonoids, which were all aglyca belonging to three different flavonoid classes: flavanones, flavones, and flavonols, were analyzed by RP HPLC. On the basis of the obtained results three different groups were erected, one formed of the tetraploid *Betula pubescens* and its allies, a second one formed of the diploid *B. pendula* and closely related taxa, and a third group of taxa not assignable to any of the other two groups. According to the authors the proposed fast and simple approach to assign birch species to any of the three groups will be particularly interesting for large scale analyses including multiple replicates per species. These investigations are wanted for the assessment of some supposedly intermediary taxa, potentially of hybrid origin, which form a morphological continuum between *B. pendula* and *B. pubescens*.[72]

The genus *Hieracium* is taxonomically one of the most complicated genera of the European flora. Two of the three subgenera of *Hieracium* s.l. are occurring in Europe and the Alps are one of the centers of diversity of the genus. A broad chemosystematic investigation of phenolics from the genus *Hieracium* based on HPLC analyses of extracts obtained from the flowering heads was performed. The investigation encompassed samples from 76 taxa. One result was that *Hieracium* is chemosystematically well delimited from the morphologically similar genus *Crepis* by the lack of caffeoyl tartaric acid and cichoric acid. However, intrageneric variation of phenolics in flowering heads was quite limited. Principal component analysis (PCA) showed only weak separation of the samples assignable to subgenera Hieracium and Pilosella, respectively. The single compound with the largest loading in PCA, and therefore with the highest relevance for cluster formation, was the rare flavonoid isoetin 4'-O-glucuronide. This compound was common and occurred mostly in high yields in subgenus Pilosella but was a trace compound or entirely missing in most species of *Hieracium* s.str.[59]

In another recent example for the use of flavonoids as chemosystematic markers members of the genera *Abies*, *Picea*, and *Pinus* from the Pinaceae family were investigated.[71] In 18 species a total of 34 different flavonoids was detected. The compounds were not quantified but only peak areas at 340 nm were reported. This simple approach also allowed comparisons of the contributions of each particular flavonoid to the total flavonoid content. Kaempferol 3-O-glucoside was the most common flavonoid in the investigated gymnosperms and it was detected in all but two species.[71]

5.5.3 TERPENES

5.5.3.1 Monoterpenes

5.5.3.1.1 Volatile Monoterpenes

Volatile monoterpenes are among the main constituents of essential oils. As essential oils are not only interesting markers for chemosystematic studies but also have a potential as active ingredients and are defining the taste and smell of various spices, there is a vast array of literature on these compounds. Exemplary studies with a chemosystematic focus include the investigation by Jürgens et al., who reported on the flower scent compounds of *Dianthus* and *Saponaria* (both Caryophyllaceae), which included besides other volatiles also monoterpenes.[73] Skoula et al. used mono- and sesquiterpenes as chemosystematic markers in the genus *Origanum* (Lamiaceae)[74]; and Fäldt et al. quantified 23 monoterpenes, enabling the differentiation between six pine (*Pinus*, Pinaceae) species.[75]

Jürgens et al. analyzed floral fragrance compounds of seven *Dianthus* species and *Saponaria officinalis*. On the basis of the quantification data of volatile compounds Sorensen indices were calculated to indicate similarity between the species.[73] Applying these indices for nonmetrical multidimensional scaling, three groups of taxa were characterized. The floral fragrances of group I consisted mainly of fatty acids. This group encompassed diurnally flowering species. Group II was characterized by high amounts of isoprenoids and floral fragrances of group III were dominated by methylbenzoate and related compounds. These groups corresponded to the respective main pollinators of the investigated taxa.

Skoula et al. found four main classes of volatiles in *Origanum*.[74] On the basis of the prevalence of these compound classes, three chemotypes were defined. Chemoype A was rich in aromatic monoterpenes, chemotype B was rich in thujane-type substances, and chemotype C was rich in acyclic monoterpenes or sesquiterpenoids. The morphologically based sections of *Origanum* were usually assignable to one of these chemotypes. However, *O. majorana* and *O. vulgare* subsp. *gracile* were represented by two different chemotypes, A/B and A/C, respectively.

The main focus of the paper by Fäldt et al. was on quantitative correlations of certain volatiles found in six species of *Pinus*.[75] As a by-product the authors also showed that a PCA based on selectively normalized GC-data of 23 monoterpenes resulted in a clear separation of *Pinus armandii*, *P. caribaea*, *P. sylvestris*, and *P. yunnanensis*. In contrast, *P. cubensis* and *P. tropicalis* were neither separable from each other nor from atypical samples of *P. sylvestris*.

5.5.3.1.2 Iridoids

The huge potential of iridoids as chemosystematic markers had been realized in the early days of modern chemosystematics already.[76] Iridoids occur in a number of families, which according to the results from the angiosperm phylogeny group (APG 2003) all belong to the Asterids subclade of the core eudicots.[77] These iridoid-producing families include: Gronoviaceae, Loasaceae,[78] Valerianaceae, Ericaceae, Gentianaceae, Menyanthaceae, Oleaceae, Rubiaceae, Lamiaceae, Plantaginaceae, Scrophulariaceae, and Verbenaceae.[79] A recent review of the biosynthesis of iridoids and their value as chemosystematic markers for the entire plant kingdom was provided by Sampaio-Santos and Kaplan.[80] More detailed chemosystematic studies focusing on lineages within the Plantaginaceae s.l. are the studies by Grayer, Jensen, Taskova, and coworkers.[81–83] Of general interest is also the detailed account of Kaplan and Gottlieb on iridoids as systematic markers in dicotyledons, though the systematic framework used is predating the "molecular era."[84]

Taskova et al. showed that the iridoid mussaenoside is a chemosystematic marker for a group of taxa including *Veronica officinalis*, *V. aphylla*, *V. urticifolia*, *V. alpina*, and *V. bellidioides*.[85] Formerly *V. officinalis*, *V. aphylla*, and *V. urticifolia* were based on morphological characters grouped into section Veronica, while *V. alpina* and *V. bellidioides* were assigned to section Veronicastrum.[86] Recent (macro-)molecular evidence indicated that these species are indeed closely related.[85] The chemical profile, the seed morphology, and the basic chromosome number are the only nonmacromolecular characters to support this grouping.[82] Taskova et al. also employed iridoids as chemosystematic markers on a higher level of the taxonomic hierarchy.[83] Some taxa of the new Planatginaceae/Veronicaceae family were characterized by the occurrence of iridoids featuring an 8,9-double bond (*Erinus*, *Globularia*, *Plantago*, and *Wulfenia*), by 6-*O*-catalpol esters (*Picrorrhiza*), or by the co-occurrence of both types of iridoids (*Veronica*). Other iridoids like aucubin and catalpol were widespread in the Plantaginaceae and were therefore considered characteristic compounds of nearly all members of the family.

5.5.3.2 Sesquiterpenes

Sesquiterpenes are the largest class of terpenoid compounds featuring more than 4000 representatives.[87,88] Within the sesquiterpenes, the class of sesquiterpene lactones displays the largest structural variety with more than 2000 compounds identified. More than 90% of the latter have been isolated from taxa of the Asteraceae family.[52] Other sources include the Apiaceae, the Lauraceae, the Magnoliaceae, liverworts, and marine organisms.[52,88,89] The comprehensive review by Seaman from 1982 covered data on the occurrence of 1350 sesquiterpenoids in the Asteraceae family.[90] This information was extracted from 969 references. Another classic paper on the chemosystematics of the Asteraceae family with a focus on sesquiterpenoids was provided by Zdero and Bohlmann in 1990.[19] A recent review with a more limited scope is that of Zidorn on the sesquiterpenoids of the subtribe Hypochaeridinae (Asteraceae, Lactuceae).[91]

Asteraceae as a family are characterized by the frequent occurrence of sesquiterpene lactones. Moreover, the distribution of different types of sesquiterpene lactones is in good agreement with tribal and subtribal classifications. For example, in nearly all tribes germacranolides have been found but germacranolides with an 8-β-hydroxysubstitution are virtually restricted to the Asteroideae subfamily. On the other hand, 8-α-hydroxysubstituted germacranolides prevail in the Cichorioideae subfamily. Pseudoguaianolides, which differ from guaianolides in the position of C-15 (attached to C-5 instead of C-4), are prevalent in the Heliantheae tribe.[19] However, pseudoguaianolides are also sporadically occurring in other tribes of the Asteraceae and were even isolated from a member of the Convolvulaceae family.[92]

The review on sesquiterpenoids as chemosystematic markers in the Hypochaeridinae (Asteraceae, Lactuceae) highlighted that the genus *Leontodon* s.l., which was identified as a diphyletic entity using DNA data by Samuel et al.,[93] is also phytochemically inhomogeneous.[91] The sesquiterpene lactone pattern of *Leontodon* subgenus Leontodon, which is characterized by the frequent occurrence of

12,5-guaianolides (hypocretenolides), is more similar to the genus *Hedypnois* than to *Leontodon* subgenus Oporinia. On the other hand, subgenus Oporinia is phytochemically more similar to the genera *Helminthotheca*, *Hypochaeris*, and *Picris*. *Helminthotheca* and *Picris*, which were formerly combined in *Picris* s.l., also were phytochemically more similar to *Hypochaeris* and *Leontodon* subgenus Oporinia, respectively, than to each other. Thus, phytochemical data also support the separation of *Helminthotheca* from *Picris*, which was already proposed by Lack in 1975 based on morphological data.[94]

5.5.3.3 Diterpenes

With the notable exception of ubiquitous compounds such as phytol and giberellic acid, diterpenes are a relatively rare class of plant secondary metabolites.[95] The majority of diterpenes were reported from the Asteraceae family.[95] Diterpenes have been widely employed in chemosystematic studies, including a detailed account of their occurrence and structural variety within the Asteraceae family.[96] Especially rare structural modifications, which feature an irregular substitution pattern, are useful as chemosystematic markers at the species level.[97] However, on higher levels such as the family rank, diterpenes are of little chemosystematic value because of their sporadic occurrence and great variability of oxygenation patterns even between closely related families.[95]

In our study on diterpenes in the genus *Anisotome* and related genera endemic to New Zealand, the occurrence of biosynthetically irregular diterpenes—i.e., compounds with a carbon backbone not complying with the Ružička rules—was restricted to the genus *Anisotome*. Even closely related genera *Aciphylla*, *Gingidia*, or *Lignocarpa* contained no anisotomene-type diterpenes. The chemosystematic value of these compounds is hampered by the following facts: (a) these compounds were not detected in all species of *Anisotome*, (b) amounts often varied pronouncedly within the same species, and (c) some populations of two species (*A. brevistylis* and *A. imbricata*) contained anisotomenes while in other populations anisotomenes were not detectable. Thus, the biosynthesis of anisotomenes might have an ecological significance and the nonoccurrence of anisotomenes might not always indicate the inability of the respective taxon to synthesize these compounds.[97]

5.5.3.4 Triterpenes

Triterpenes occur in an immense number of different compounds ubiquitously in the plant kingdom.[98] This ubiquitous occurrence is probably one of the reasons why these compounds are rarely employed as chemosystematic markers. Practical reasons also contributing to this fact are (a) triterpenes are usually analyzable by GC only after derivatization and (b) many triterpenes lack a chromophore and are therefore not detectable with the widely employed HPLC/DAD detectors. However, these technical difficulties can be overcome and there are some studies using triterpenes as chemosystematic markers. Da Paz Lima et al.[99] investigated various taxa of the Burseraceae and discussed the detected triterpenes in a chemosystematic context and Vincken et al. reviewed the literature on saponins, i.e., triterpenes featuring one or more sugar moieties.[98]

Da Paz Lima et al. used chemosystematic evidence to reassess the taxonomic position of the genus *Trattinnickia* within the Burseraceae family.[99] The Burseraceae are divided into three tribes, one considered to be primitive named Protieae, an intermediate one named Boswellieae/Bursereae, and an advanced one named Canarieae. Traditionally the genus *Trattinnickia* was regarded to be a member of the primitive Protieae tribe but was transferred by Daly based on morphological and anatomical evidence to the advanced Canarieae tribe.[100] However, chemosystematic evidence compiled by Da Paz Lima et al. favored an inclusion of *Trattinnickia* in the Protieae, because dammaranes, which were isolated from *Trattinnickia burserifolia*, are a feature common to members of the Protieae but were never found in any member of the Canarieae tribe.[99]

Vincken et al. did an extensive survey of the literature on saponins including a systematic classification of the known saponins into 11 main carbon skeleton classes.[98] Their data on the distribution of these compound classes throughout the plant kingdom—though based on an outdated classification of the higher plants—indicate that these main classes of triterpenoids are not suitable as chemosystematic markers for higher systematic entities within the plant kingdom.

5.5.4 ALKALOIDS

Because of their pronounced physiological effects, alkaloids were among the first classes of natural compounds to be studied more in detail. Alkaloids display an amazing structural diversity with more than 12,000 compounds known today.[27] Moreover, the distribution of certain types of alkaloids coincides with classical systematic groupings. Therefore, alkaloids have been used extensively in botanical classifications.[101]

Recent reviews also or predominantly focused on alkaloids as chemosystematic markers are the ones by Waterman[102] on alkaloid chemosystematics in general, by Wink[27] highlighting distribution patterns of certain alkaloids in the Fabaceae and in the Solanaceae, by van Wyk[103] also on alkaloids from members of the Fabaceae, and the one by Greger[104] on *Stemona* alkaloids. The review on alkaloids from the Fabaceae family is of general interest, because it combines classical chemosystematics with recent results from (macro-)molecular analyses. The accumulation of quinolizidine alkaloids is a characteristic feature of the "genistoid alliance s.l." within the Fabaceae family, i.e., an informal grouping encompassing the tribes Brongniartieae, Crotolarieae, Euchresteae, Genisteae, Podalyrieae/Liparieae, Sophoreae, and Thermopsideae. Notably, the Crotolarieae are not accumulating quinolizidine alkaloids. It is assumed that the ancestral taxon of the Crotolarieae had the ability to synthesize quinolizidine alkaloids and that this feature was lost due to a turning off of the corresponding genes. This secondary loss of the ability to synthesize certain secondary metabolites is a reoccurring problem in chemosystematics and its potential occurrence has to be kept in mind when constructing phylogenies incorporating chemosystematic data.[27]

Van Wyk demonstrated that alkaloids are better suited as chemosystematic markers in the genistoid tribe of the Fabaceae than flavonoids.[103] The presence of quinolizidine alkaloids is considered to be a potential synapomorphy for the whole of the genistoid group within the Fabaceae family. Examples for particular alkaloids as potential synapomorphies for parts of genistoid group of the Fabaceae include *Ormosia*-type alkaloids for the *Ormosia* group of the Sophoreae, the Brongniartieae, and the Thermopsideae, quinolizidine-type alkaloids of the matrine-type for the Euchresteae and *Sophora* species, and 5-*O*-methylgenistein for the Genisteae tribe.

Greger reviewed the literature on alkaloids of the Stemonaceae and also covered chemosystematic aspects.[104] The 82 known *Stemona* alkaloids are all characterized by a pyrrolo[1,2-*a*]azepine nucleus. The compounds are assigned to three groups— stichoneurine-, protostemonine-, and croomine-type—based on the substitution at C-9 of the pyrroloazepine nucleus. The genus *Croomia* only accumulates croomine-type alkaloids and the genus *Stichoneuron* accumulates only stichoneurine-type alkaloids whereas the genus *Stemona* features all three types of alkaloids. Within *Stemona* the distribution of these three types of alkaloids generally correlates with the infrageneric system of the genus.[104]

5.6 CODING AND ANALYSIS OF CHEMOSYSTEMATIC DATA

Statistical analysis of the available data is one of the most important aspects of chemosystematic studies. In most studies only the presence of particular compounds is reported, more complex studies state presence/absence matrices for a number of taxa and compounds. More sophisticated methods such as PCA or cladistic analysis of natural product data are not frequently employed[105] and publications in which aspects of character coding are addressed directly are a rarity.[58,91,105–107]

A recent paper on the discrimination of different olive (*Olea europaea* L.) varieties based on the HPLC quantification of leave phenolics might serve as an example for an easy yet timely approach to utilize chemosystematic data.[108] A total of 180 samples representing 13 varieties and 6 accessions of one variety were analyzed using a standard extraction and HPLC protocol. Eight phenolic compounds, with only four of them chemically identified, were quantified using their peak areas at the wavelength of their particular maximum absorption. Then, these areas were subjected to PCA and hierarchical cluster analysis (HCA). Both data analysis techniques allowed clear separation for both, the 13 investigated varieties as well as the samples of the same variety grown in various regions of Spain.

In the following paragraph, the data analysis approach most frequently used in our lab is described using an example data set of 24 samples assignable to four species of *Leontodon* s.l. (Asteraceae, Lactuceae). The samples—six per species— were analyzed by HPLC-DAD employing the method described in Zidorn and Stuppner.[58] The quantification data are summarized in Table 5.1. Table 5.2 gives a semiquantitative overview about presence and absence of the detected compounds. In Table 5.3 the percentage of each compound in respect to its compound class, flavonoid or caffeic acid derivative, is listed. On the basis of Tables 5.1 and 5.3 principal component analyses (PCAs) and hierarchical cluster analyses (HCAs) were performed. Figures 5.1 and 5.3 display the results from the analyses based on the absolute quantification data matrix (Table 5.1) and Figures 5.2 and 5.4 display the results based on the relative quantification data matrix (Table 5.3).

TABLE 5.1

Absolute Quantification Results of Phenolics in Extracts from *Leontodon* Flowering Heads[a]

Taxon	Sample	CGA	3,5-DCA	4,5-DCA	CTA	CCA	LUT	L7GC	L7GT	L7GU	L4'GC
Leontodon hispidus	LHP1	3.04	3.37	2.42	2.29	7.61	2.00	6.28	0.25	5.29	4.96
	LHP2	2.37	2.19	0.73	2.34	7.27	1.40	4.78	0.27	3.76	3.35
	LHP3	2.91	3.84	1.73	2.21	6.58	2.42	9.82	0.00	4.24	6.61
	LHP4	2.24	2.87	2.18	2.23	5.85	2.65	7.87	0.10	4.15	5.97
	LHP5	2.30	2.96	1.38	2.38	7.33	1.70	5.75	0.01	4.13	4.27
	LHP6	2.90	3.75	2.32	2.28	7.09	2.09	10.97	0.58	4.21	6.99
Leontodon helveticus	LHV1	5.05	3.95	0.20	1.75	10.07	1.79	5.74	2.15	1.90	1.39
	LHV2	3.96	3.09	0.22	1.28	8.41	1.77	7.05	3.11	2.59	0.96
	LHV3	3.99	2.13	0.22	1.36	8.24	0.96	6.19	2.66	2.08	1.52
	LHV4	4.05	2.20	0.11	1.77	9.85	1.49	7.04	3.19	2.17	1.38
	LHV5	4.01	1.76	0.19	1.66	9.05	1.77	5.67	2.41	2.51	1.31
	LHV6	4.65	1.68	0.11	1.36	8.19	1.38	8.25	3.68	3.01	1.64
Leontodon incanus	LIN1	1.65	2.42	1.70	3.23	13.59	1.64	7.93	1.05	4.73	5.06
	LIN2	1.13	1.78	1.20	1.94	9.97	1.87	5.98	0.34	5.19	5.16
	LIN3	2.56	3.32	1.77	3.52	13.63	1.82	9.11	1.07	6.31	6.54
	LIN4	1.99	2.37	1.09	2.57	10.67	1.72	7.58	0.41	4.37	5.55
	LIN5	1.76	2.08	1.76	2.77	12.08	2.11	6.04	0.59	4.65	5.35
	LIN6	2.23	3.18	1.88	2.40	9.89	1.31	6.88	0.11	4.20	4.20
Leontodon montanus s.l.	LMO1	8.26	3.79	0.00	1.94	6.41	1.57	4.75	1.38	0.00	0.00
	LMO2	6.20	2.41	0.50	0.60	4.71	1.35	4.54	4.28	1.23	0.00
	LMO3	5.61	1.86	0.00	1.04	3.51	1.91	5.85	1.04	2.02	0.00
	LMO4	7.90	2.59	0.00	1.98	5.24	3.07	5.29	2.36	4.47	0.00
	LMO5	8.40	2.08	0.00	1.89	5.89	2.94	4.97	2.27	2.68	0.00
	LMO6	6.46	1.85	0.33	1.03	7.26	1.43	2.96	2.41	1.19	0.00

[a] Collection data and standard deviations are available on request. CGA: chlorogenic acid; 3,5-DCA: 3,5-dicaffeoylquinic acid; 4,5-DCA: 4,5-dicaffeoylquinic acid; CTA: caffeoyltartaric acid; CCA: cichoric acid; LUT: luteolin; L7GC: luteolin 7-O-glucoside; L7GT: luteolin 7-O-gentiobioside; L7GU: luteolin 7-O-glucuronide; L4'GC: luteolin 4'-O-glucoside.

TABLE 5.2
Semiquantitative Data of Phenolics in Extracts from *Leontodon* Flowering Heads

Taxon	Sample	CGA	3,5-DCA	4,5-DCA	CTA	CCA	LUT	L7GC	L7GT	L7GU	L4'GC
Leontodon hispidus	LHP1	+	+	+	+	++	+	++	(+)	++	+
	LHP2	+	+	+	+	++	+	+	(+)	+	+
	LHP3	+	+	+	+	++	+	++	n.d.	+	++
	LHP4	+	+	+	+	++	+	++	(+)	+	++
	LHP5	+	+	+	+	++	+	++	(+)	+	+
	LHP6	+	+	+	+	++	+	++	(+)	+	++
Leontodon helveticus	LHV1	++	+	(+)	+	++	+	++	+	+	+
	LHV2	+	+	(+)	+	++	+	++	+	+	+
	LHV3	+	+	(+)	+	++	+	++	+	+	+
	LHV4	+	+	(+)	+	++	+	++	+	+	+
	LHV5	+	+	(+)	+	++	+	++	+	+	+
	LHV6	+	+	(+)	+	++	+	++	+	+	+
Leontodon incanus	LIN1	+	+	+	+	++	+	++	(+)	++	++
	LIN2	+	+	+	+	++	+	++	+	++	++
	LIN3	+	+	+	+	++	+	++	(+)	+	++
	LIN4	+	+	+	+	++	+	++	+	+	++
	LIN5	+	+	+	+	++	+	++	(+)	+	++
	LIN6	+	+	+	+	++	+	+	+	+	+
Leontodon montanus s.l.	LMO1	++	+	n.d.	+	++	+	+	+	n.d.	n.d.
	LMO2	++	+	(+)	+	+	+	+	+	+	n.d.
	LMO3	++	+	n.d.	+	+	+	++	+	+	n.d.
	LMO4	++	+	n.d.	+	++	+	++	+	+	n.d.
	LMO5	++	+	n.d.	+	++	+	+	+	+	n.d.
	LMO6	++	+	(+)	+	++	+	+	+	+	n.d.

n.d.: not detected; (+) < 0.5 mg; + 0.5 < 5.0 mg; ++ ≥ 5.0 mg.

TABLE 5.3

Relative Quantification Data Matrix of Phenolics in Extracts from *Leontodon* Flowering Heads

Taxon	Sample	CGA	3,5-DCA	4,5-DCA	CTA	CCA	LUT	L7GC	L7GT	L7GU	L4'GC
Leontodon hispidus	LHP1	16.2	18.0	12.9	12.2	40.6	10.6	33.4	1.3	28.2	26.4
	LHP2	15.9	14.7	4.9	15.7	48.8	10.3	35.3	2.0	27.7	24.7
	LHP3	16.9	22.2	10.0	12.8	38.1	10.5	42.5	0.0	18.4	28.6
	LHP4	14.6	18.7	14.2	14.5	38.1	12.8	38.0	0.5	20.0	28.8
	LHP5	14.1	18.1	8.5	14.6	44.8	10.7	36.3	0.0	26.0	26.9
	LHP6	15.8	20.4	12.6	12.4	38.7	8.4	44.2	2.3	16.9	28.1
Leontodon helveticus	LHV1	24.0	18.8	1.0	8.3	47.9	13.8	44.2	16.6	14.7	10.7
	LHV2	23.3	18.2	1.3	7.5	49.6	11.4	45.5	20.1	16.8	6.2
	LHV3	25.0	13.4	1.4	8.5	51.7	7.1	46.2	19.8	15.5	11.3
	LHV4	22.5	12.2	0.6	9.8	54.8	9.8	46.1	20.9	14.2	9.0
	LHV5	24.1	10.5	1.1	9.9	54.3	13.0	41.5	17.6	18.4	9.6
	LHV6	29.1	10.5	0.7	8.5	51.2	7.7	45.9	20.5	16.7	9.1
Leontodon incanus	LIN1	7.3	10.7	7.5	14.3	60.2	8.0	38.8	5.2	23.2	24.8
	LIN2	7.0	11.1	7.5	12.1	62.2	10.1	32.3	1.8	28.0	27.8
	LIN3	10.3	13.4	7.1	14.2	54.9	7.3	36.7	4.3	25.4	26.3
	LIN4	10.6	12.7	5.8	13.7	57.1	8.8	38.6	2.1	22.3	28.3
	LIN5	8.6	10.2	8.6	13.6	59.1	11.3	32.2	3.1	24.8	28.6
	LIN6	11.4	16.2	9.6	12.3	50.5	7.9	41.2	0.7	25.1	25.1
Leontodon montanus s.l.	LMO1	40.5	18.6	0.0	9.5	31.4	20.4	61.7	17.9	0.0	0.0
	LMO2	43.0	16.7	3.5	4.1	32.7	11.8	39.9	37.5	10.8	0.0
	LMO3	46.7	15.5	0.0	8.6	29.2	17.7	54.1	9.6	18.6	0.0
	LMO4	44.6	14.6	0.0	11.2	29.6	20.2	34.8	15.5	29.4	0.0
	LMO5	46.0	11.4	0.0	10.3	32.3	22.9	38.7	17.7	20.8	0.0
	LMO6	38.2	10.9	2.0	6.1	42.9	17.9	37.0	30.2	14.9	0.0

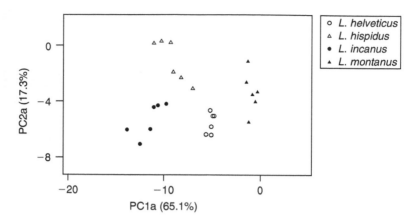

FIGURE 5.1 PCA performed on the absolute quantification data matrix.

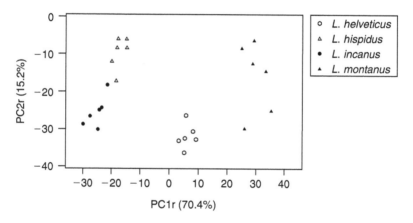

FIGURE 5.2 PCA performed on the relative quantification data matrix.

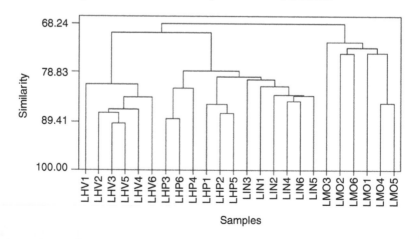

FIGURE 5.3 HCA performed on the absolute quantification data matrix. LHP = *Leontodon hispidus*, LHV = *Leontodon helveticus*, LIN = *Leontodon incanus*, and LMO = *Leontodon montanus* s.l.

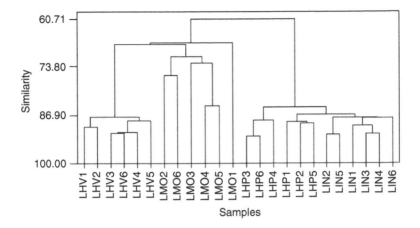

FIGURE 5.4 HCA performed on the relative quantification data matrix. LHP = *Leontodon hispidus*, LHV = *Leontodon helveticus*, LIN = *Leontodon incanus*, LMO = *Leontodon montanus* s.l.

The original data set displayed in Table 5.1 is rather confusing and no clear trends for the four investigated species are visible at first glance. Data reduction resulting in Table 5.2 makes a comparison easier but much of the information contained in Table 5.1 is lost. Table 5.3 retains most of the information of Table 5.1 and, moreover, indicates which percentage the compounds are contributing to the total of their compound class. In this table, *Leontodon montanus* emerges as a species, which is characterized by the absence of luteolin 4'-*O*-glucoside, high relative amounts (>35%) of chlorogenic acid, and rather high amounts (≥10%) of luteolin 7-*O*-gentiobioside. *Leontodon helveticus* is characterized by rather high amounts (≥15%) of luteolin 7-*O*-gentiobioside but lower relative amounts (<30%) of chlorogenic acid. *Leontodon hispidus* and *L. incanus* display qualitatively and quantitatively a similar pattern of phenolics and are both characterized by high relative amounts (≥25%) of luteolin 4'-*O*-glucoside. Both PCA and HCA succeed in separating the four investigated species from each other. This applies for analyses based on the absolute quantification data matrix as well as for those based on the relative quantification data matrix (Figures 5.1 and 5.4). However, the HCA results from the relative quantification data matrix (Figure 5.4) better show the close similarities between *L. hispidus* and *L. incanus*—these species belong to *Leontodon* subgenus Leontodon—on the one hand and between *L. helveticus* and *L. montanus*—these species belong to *Leontodon* subgenus Oporinia—on the other hand.

5.7 SUMMARY/OUTLOOK

As outlined earlier chemosystematics had and still has a number of interesting and important applications. Many studies, which are currently carried out in the search for new drugs or for new pest-resistant crop strains, are in fact chemosystematic studies though their authors usually do not term them as such. On the other hand there is—mainly due to the advent of macromolecular methods and the spectacular new insights into evolution they have yielded—a decrease in interest for chemosystematic studies to solve problems of plant systematics. The current trend towards the overemphasis of molecular techniques is likely to continue in the near future. In this sense Waterman[102] correctly stated that the promised land for chemosystematics is out of reach and that the current situation may be as good as it gets.

However, systematics as the basis and the crown of biology was always able and will always be able to incorporate all disciplines in its quest for a broad based systematic classification. Thus, chemosystematic data like morphological data will continue to contribute to this aim. A perfect phylogenetic system of plant (and/or animal) evolutionary history with entities without any phenetic—morphological or chemical—characters would be no dream come true but a nightmare.

The combination of modern chemosystematic investigations combined with macromolecular studies is therefore an indispensable field of research. In instances where macromolecular studies yield groupings without any morphological differential characters, chemosystematic often contribute new data to support the macromolecular results and help in characterizing the newly emerging systematic entities.

REFERENCES

1. Hänsel, R., Pflanzenchemie und Pflanzenverwandtschaft, *Arch. Pharm.*, 289, 619, 1956.
2. Harborne, J.B. and Turner, B.L., *Plant Chemosystematics*, Academic Press, London, 1984, 562pp.
3. Crawford, D.J., *Plant Molecular Systematics*: *Macromolecular Systematics*, Wiley, New York, 1990, 388pp.
4. Savolainen, V. and Chase, M.W., A decade of progress in plant molecular phylogenetics, *Trends Genet.*, 19, 717, 2003.
5. Jacob, J., Eigener, U., and Hoppe, U., The structure of preen gland waxes from pelecaniform birds containing 3,7-dimethyloctan-1-ol. An active ingredient against dermatophytes, *Z. Naturforsch.*, 52c, 114, 1997.
6. Larsen, T.O. and Frisvad, J.C., Chemosystematics of *Penicillium* based on profiles of volatile compounds, *Mycol. Res.*, 99, 1167, 1995.
7. Valls, R. and Piovetti, L., The chemistry of the Cystoseiraceae (Fucales: Pheophyceae): chemotaxonomic relationships, *Biochem. Syst. Ecol.*, 23, 723, 1995.
8. Graur, D. and Li, W.-H., *Fundamentals of Molecular Evolution*, 2nd ed., Sinauer, Sunderland, 2000, 481pp.
9. Giannasi, D.E. and Crawford, D.J., Biochemical Systematics II. A reprise, *Evol. Biol.*, 20, 25, 1986.
10. Jaroszewski, J.W., Hyphenated NMR methods in natural products research, part 1: direct hyphenation, *Planta Med.*, 71, 691, 2005.
11. Swain, T., *Chemical Plant Taxonomy*, Academic Press, London, 1963, 543pp.
12. Cronquist, A., *The Evolution and Classification of Flowering Plants*, New York Botanical Garden, New York, 1988, 555pp.
13. Takhtajan, A., *Diversity and Classification of Flowering Plants*, Columbia University Press, New York, 1997, 643pp.
14. Hänsel, R. and Sticher, O., *Pharmakognosie—Phytopharmazie*, Springer, Berlin, 2004, 1214pp.
15. Bogo, A. and Mantle, P.G., Caffeine: also a fungal metabolite, *Phytochemistry*, 54, 937, 2000.

16. Mitchell, J.C., Dupuis, G., and Geissman, T.A., Allergic contact-dermatitis from sesqui-terpenoids of plants—additional allergenic sesquiterpene lactones and immunological specificity of compositae, liverworts and lichens, *Brit. J. Dermatol.*, 87, 235, 1972.

17. Schuhly, W., Khan, I., and Fischer, N.H., The ethnomedical uses of Magnoliaceae from the southeastern United States as leads in drug discovery, *Pharm. Biol.*, 39, Suppl. S 63, 2001.

18. Milosvaljevic, S., Bulatovic, V., and Stefanovic, M., Sesquiterpene lactones from the Yugoslavian wild growing plant families Asteraceae and Apiaceae, *J. Serb. Chem. Soc.*, 64, 397, 1999.

19. Zdero, C. and Bohlmann, F., Systematics and evolution within the Compositae, seen with the eyes of a chemist, *Plant Syst. Evol.*, 171, 1, 1990.

20. Waterman, P.G., Chemosystematics—current status, *Phytochemistry*, 49, 1175, 1998.

21. Ronstedt, N., Franzyk, H., Molgaard, P., Jaroszewski, J.W., and Jensen, S.R., Chemo-taxonomy and evolution of *Plantago* L., *Plant Syst. Evol.*, 242, 63, 2003.

22. Dicke, M. and Tacken, W., *Chemical Ecology: From Gene to Ecosystem*, Springer, Berlin, 2006, 189pp.

23. Dewick, P.M., *Medicinal Natural Products*, 2nd ed., Wiley, Chichester, 2002, 507pp.

24. Gutsche, C.D. and Seligman, K.L., Preliminary experiments on the synthesis of colchi-cine: a method for the synthesis of ring B, *J. Am. Chem. Soc.*, 75, 2579, 1953.

25. Urban, S. and Separovic, F., Developments in hyphenated spectroscopic methods in natural product profiling, *Front. Drug Des Disc.*, 1, 113, 2005.

26. Grass, S., Zidorn, C., Stuppner, H., and Blattner, F.R., Comparative molecular and phytochemical investigation of *Leontodon autumnalis* (Asteraceae, Lactuceae) popula-tions from Central Europe, *Phytochemistry*, 67, 122, 2006.

27. Wink, M., Evolution of secondary metabolites from an ecological and molecular phylogenetic perspective, *Phytochemistry*, 64, 3, 2003.

28. Adams, M., Wiedenmann, M., Tittel, G., and Bauer, R., HPLC-MS trace analysis of atropine in *Lycium barbarum* berries, *Phytochem. Anal.*, 17, 279, 2006.

29. Rasmussen, L.K., Rank, C., and Jensen, S.R., Transfer of iridoid glucosides from host plant *Galium verum* to hemiparasitic *Euphrasia stricta*, *Biochem. Syst. Ecol.*, 34, 763, 2006.

30. Proksch, P., Edrada, R.A., and Ebel, R., Drugs from the seas—current status and microbiological implications, *Appl. Microbiol. Biotechnol.*, 59, 125, 2002.

31. Frohne, D. and Jensen, U., *Systematik des Pflanzenreichs*, 5th ed., WVG, Stuttgart, 1998, 371pp.

32. Bernardello, G., Galetto, L., and Forcone, A., Floral nectar chemical composition of some species from Patagonia. II, *Biochem. Syst. Ecol.*, 27, 779, 1999.

33. Han, N.S. and Robyt, J.F., Separation and detection of sugars and alditols on thin layer chromatograms, *Carbohydr. Res.*, 313, 135, 1998.

34. Consden, R., Gordon, A.H., and Martin, A.J.P., Qualitative analysis of proteins: a partition chromatographic method using paper, *Biochem. J.*, 38, 224, 1944.

35. Lugg, J.W.H. and Overell, B.T., Partition chromatography of organic acids on a paper sheet support, *Nature*, 160, 87, 1947.

36. Bate-Smith, E.C., Paper chromatography of anthocyanins and related substances in petal extracts, *Nature*, 161, 835, 1948.

37. Alston, R.E., Mabry, T.J., and Turner, B.L., Perspectives in chemotaxonomy, *Science*, 142, 545, 1963.

38. McClure, J.W. and Alston, R.E., Patterns of selected chemical components of *Spirodela oligorrhiza* formed under various conditions of axenic culture, *Nature*, 201, 311, 1964.

39. van Dam, M.J.D., Kleuver, G.J., and Deheus, J.G., Thin layer chromatography of weakly polar steroids, *J. Chromatogr.*, 4, 26, 1960.

40. Bhandari, P.R., Identification of flavonoids in hops (*Humulus lupulus* Linne) by thin-layer chromatography, *J. Chromatogr.*, 16, 130, 1964.
41. Onyilagha, J., Bala, A., Hallett, R., Gruber, M., Soroka, J., and Westcott, N., Leaf flavonoids of the cruciferous species, *Camelina sativa, Crambe* spp., *Thlaspi arvense* and several other genera of the family Brassicaceae, *Biochem. Syst. Ecol.*, 31, 1309, 2003.
42. Hillig, K.W., A chemotaxonomic analysis of terpenoid variation in *Cannabis, Biochem. Syst. Ecol.*, 32, 875, 2004.
43. Füfzai, Z., Katona, Z.F., Kovacs, E., and Molnar-Perl, I., Simultaneous identification and quantification of the sugar, sugar alcohol, and carboxylic acid contents of sour cherry, apple, and berry fruits, as their trimethylsilyl derivatives, by gas chromatography–mass spectrometry, *J. Agric. Food. Chem.*, 52, 7444, 2004.
44. Ganzera, M. and Stuppner, H., Evaporative light scattering detection (ELSD) for the analysis of natural products, *Curr. Pharm. Anal.*, 1, 135, 2005.
45. Brondz, I., Dahle, U.R., Greibrokk, T., and Olsen, I., Capillary zone electrophoresis as a new tool in the chemotaxonomy of oral treponemes, *J. Chromatogr. B*, 667, 161, 1995.
46. Andersen, K.E., Bjergegaard, C., Moller, P., Sorensen, J.C., and Sorensen, H., Compositional variations for α-galactosides in different species of Leguminosae, Brassicaceae, and barley: a chemotaxonic study based on chemometrics and high performance capillary electrophoresis, *J. Agric. Food Chem.*, 53, 5809, 2005.
47. Phuong, N.T., Lee, K.A., Jeong, S.J., Fu, C.X., Choi, J.K., Kim, Y.H., and Kang, J.S., Capillary electrophoretic method for the determination of diterpenoid isomers in *Acanthopanax* species, *J. Pharm. Biomed. Anal.*, 40, 56, 2006.
48. Bisrat, D., Dagne, E., van Wyk, B.-E., and Viljoen, A., Chromones and anthrones from *Aloe marlothii* and *Aloe rupestris, Phytochemistry*, 55, 949, 2000.
49. Viljoen, A.M., van Wyk, B.-E., and Newton, L.E., The occurrence and taxonomic distribution of the anthrones aloin, aloinoside and microdontin in *Aloe, Biochem. Syst. Ecol.*, 29, 53, 2001.
50. Viljoen, A.M., van Wyk, B.-E., and van Heerden, F.R., The chemotaxonomic value of the diglucoside anthrone homonataloside B in the genus *Aloe, Biochem. Syst. Ecol.*, 30, 35, 2002.
51. Viljoen, A.M. and van Wyk, B.-E., A chemotaxonomic and morphological appraisal of *Aloe* series Purpurascentes, Aloe section Anguialoe and their hybrid, *Aloe broomii, Biochem. Syst. Ecol.*, 29, 621, 2001.
52. Ammon, H.P.T., *Hunnius, Pharmazeutisches Wörterbuch*, 9. Aufl., De Gruyter, Berlin, 2004, 1648pp.
53. Peres, V., Nagem, T.J., Oliveira, F.F., and Oliveira, T.T., The chemosystematic significance of xanthones in Gentianaceae, *Curr. Top. Phytochem.*, 5, 23, 2002.
54. Jankovic, T., Krstic, D., Aljancic, I., Savikin-Fodulovic, K., Menkovic, N., Vajs, V., and Milosavlevic, S., Xanthones and C-glycosides from the aerial parts of four species of *Gentianella* from Serbia and Montenegro, *Biochem. Syst. Ecol.*, 33, 729, 2005.
55. Feige, G.B., Lumbsch, H.T., Huneck, S., and Elix, J.A., Identification of lichen substances by a standardized high-performance liquid chromatograpic method, *J. Chromatogr.*, 646, 417, 1993.
56. Zidorn, C., Zidorn, A., and Stuppner, H., About the occurrence of *Crepis rhaetica* Hegetschw. in Northern Tyrol and its phytochemical characterization, *Veröff. Tiroler Landesmus. Ferdinandeum*, 79, 173, 1999.
57. Zidorn, C. and Stuppner, H., Chemosystematics of taxa from the *Leontodon* section Oporinia, *Biochem. Syst. Ecol.*, 29, 827, 2001.

58. Zidorn, C. and Stuppner, H., Evaluation of chemosystematic characters in the genus *Leontodon* (Asteraceae), *Taxon*, 50, 115, 2001.
59. Zidorn, C., Gottschlich, G., and Stuppner, H., Chemosystematic investigations of phenolics from flowerheads of Central European taxa of *Hieracium* (Asteraceae), *Plant Syst. Evol.*, 231, 39, 2002.
60. Zidorn, C., Pschorr, S., Ellmerer, E.P., and Stuppner, H., Occurrence of equisetumpyrone and other phenolics in *Leontodon crispus*, *Biochem. Syst. Ecol.*, 34, 185, 2006.
61. Zidorn, C., Petersen, B.O., Udovicic, V., Larsen, T.O., Duus, J.Ø., Rollinger, J.M., Ongania, K.-H., Ellmerer, E.P., and Stuppner, H., Podospermic acid, 1,3,5-tri-*O*-(7,8-dihydrocaffeoyl) quinic acid from *Podospermum laciniatum* (Asteraceae), *Tetrahedron Lett.*, 46, 1291, 2005.
62. Gray, A. and Waterman, P.G., Coumarins in the Rutaceae, *Phytochemistry*, 17, 845, 1978.
63. Fernandes da Silva, M.F.D.G., Gottlieb, O.R., and Ehrendorfer, F., Chemosystematics of Rutaceae: suggestions for a more natural taxonomy and evolutionary interpretation of the family, *Plant Syst. Evol.*, 161, 97, 1988.
64. Zidorn, C., Udovicic, V., Spitaler, R., Ellmerer, E.P., and Stuppner, H., Secondary metabolites from *Arnoseris minima*, *Biochem. Syst. Ecol.*, 33, 827, 2005.
65. Tava, A., Coumarin-containing grass: volatiles from sweet vernalgrass (*Anthoxanthum odoratum* L.), *J. Essent. Oil Res.*, 13, 367, 2001.
66. Woerner, M. and Schreier, P., The composition of woodruff volatiles (*Galium odoratum*), *Zeitschr. Lebensm.-Unt. Forsch.*, 193, 317, 1991.
67. Campos Ribeiro, C.V. and Coelho Kaplan, M.A., Evolutionary tendency of coumarin-bearing families in Angiospermae, *Quimica Nova*, 25, 533, 2002.
68. Harborne, J.B., The evolution of flavonoid pigments in plants, in *Comparative Phytochemistry*, Swain, T., Ed., Academic Press, London, 1966, 271.
69. Harborne, J.B., The chromatography of the flavonoid pigments, *J. Chromatogr.*, 2, 581, 1959.
70. Lahtinen, M., Lempa, K., Salminen, J.-P., and Pihlaja, K., HPLC Analysis of leaf surface flavonoids for the preliminary classification of birch species, *Phytochem. Anal.*, 17, 197, 2006.
71. Slimestad, R., Flavonoids in buds and young needles of *Picea*, *Pinus* and *Abies*, *Biochem. Syst. Ecol.*, 31, 1247, 2003.
72. Palme, A.E., Su, Q., Palsson, S., and Lascoux, M., Extensive sharing of chloroplast haplotypes among European birches indicates hybridization among *Betula pendula*, *B. pubescens* and *B. nana*, *Mol. Ecol.*, 13, 167, 2004.
73. Jürgens, A., Witt, T., and Gottsberger, G., Flower scent composition in *Dianthus* and *Saponaria* species (Caryophyllaceae) and its relevance for pollination biology and taxonomy, *Biochem. Syst. Ecol.*, 31, 345, 2003.
74. Skoula, M., Gotsiou, P., Naxakis, G., and Johnson, C.B., A chemosystematic investigation on the mono- and sesquiterpenoids in the genus *Origanum* (Labiatae), *Phytochemistry*, 52, 649, 1999.
75. Fäldt, J., Sjödin, K., Persson, M., Valterova, I., and Bork-Karlson, A.-K., Correlations between selected monoterpene hydrocarbons in the xylem of six *Pinus* (Pinaceae) species, *Chemoecology*, 11, 97, 2001.
76. Bate-Smith, E.C. and Swain, T., The asperulosides and the aucubins, in *Comparative Phytochemistry*, Swain, T., Ed., Academic Press, London, 1966, 159.
77. APG [Angiosperm Phylogeny Group] II, An update of the angiosperm phylogeny group classification for the orders and families of flowering plants: APG II, *Bot. J. Linn. Soc.*, 141, 399, 2003.

78. Weigend, M., Kufer, J., and Müller, A.A., Phytochemistry and the systematics and ecology of Loasaceae and Gronoviaceae (Loasales), *Am. J. Bot.*, 87, 1202, 2000.

79. Sticher, O., Iridoide, in *Pharmakognosie, Phytopharmazie*, Hänsel, R., Sticher, O., Eds., Springer, Berlin, 2004, 390.

80. Sampaio-Santos, M.I. and Kaplan, M.A.C., Biosynthesis significance of iridoids in chemosystematics, *J. Braz. Chem. Soc.*, 12, 144, 2001.

81. Albach, D.C., Jensen, S.R., Özgökce, F., and Grayer, R.J., *Veronica*: chemical characters for the support of phylogenetic relationships based on nuclear ribosomal and plastid DNA sequence data, *Biochem. Syst. Ecol.*, 33, 1087, 2005.

82. Taskova, R.M., Gotfredsen, C.H., and Jensen, S.R., Chemotaxonomic markers in Digitalideae (Plantaginaceae), *Phytochemistry*, 66, 1440, 2005.

83. Taskova, R.M., Gotfredsen, C.H., and Jensen, S.R., Chemotaxonomy of Veroniceae and its allies in Plantaginaceae, *Phytochemistry*, 67, 286, 2006.

84. Kaplan, M.A.C. and Gottlieb, O.R., Iridoids as systematic markers in dicotyledons, *Biochem. Syst. Ecol.*, 10, 329, 1982.

85. Taskova, R.M., Albach, D.C., and Grayer, R.J., Phylogeny of *Veronica*—a combination of molecular and chemical evidence, *Plant Biol.*, 6, 673, 2004.

86. Taskova, R., Peev, D., and Handjieva, N., Iridoid glucosides of the genus *Veronica* s.l. and their systematic significance, *Plant Syst. Evol.*, 231, 1, 2002.

87. Cameron, G.M., Stapleton, B.L., Simonsen, S.M., Brecknell, D.J., and Garson, M.J., New sesquiterpene and brominated metabolites from the tropical marine sponge *Dysidea* sp., *Tetrahedron*, 56, 5247, 2000.

88. Dingermann, T., Hiller, K., Schneider, G., and Zündorf, I., *Schneider Arzneidrogen*, 5. Aufl., Elsevier, München, 2004, 627pp.

89. Asakawa, Y., Toyota, M., Nagashima, F., Hashimoto, T., and El Hassane, L., Sesquiterpene lactones and acetogenin lactones from the Hepaticae and chemosystematics of the liverworts *Frullania, Plagiochila*, and *Porella*, *Heterocycles*, 54, 1057, 2001.

90. Seaman, F.C., Sesquiterpene lactones as taxonomic characters in the Asteraceae, *Bot. Rev.*, 48, 121, 1982.

91. Zidorn, C., Sesquiterpenoids as chemosystematic markers in the subtribe Hypochaeridinae (Lactuceae, Asteraceae), *Biochem. Syst. Ecol.*, 34, 144, 2006.

92. Abdel-Kader, M.S., Two pseudoguaianolides from *Convolvulus oleifolius* growing in Egypt, *Alexandria J. Pharm. Sci.*, 15, 113, 2001.

93. Samuel, R., Gutermann, W., Stuessy, T.F., Ruas, C.F., Lack, H.-W., Tremetsberger, K., Talavera, S., Hermanowski, B., and Ehrendorfer, F., Molecular phylogenetics reveals Leontodon (Asteraceae, Lactuceae) to be diphyletic, *Am. J. Bot.*, 93, 1193, 2006.

94. Lack, H.-W., *Die Gattung* Picris *L., sensu lato, im ostmediterran-westasiatischen Raum*, VWGÖ, Wien, 1975, 184pp.

95. Figueiredo, M.R., Kaplan, M.A.C., and Gottlieb, O.R., Diterpenes, taxonomic markers? *Plant Syst. Evol.*, 195, 149, 1995.

96. Seaman, F., Bohlmann, F., Zdero, C., and Mabry, T.J., *Diterpenes of Flowering Plants. Compositae (Asteraceae)*, Springer, New York, 1990, 638pp.

97. Zidorn, C., Sturm, S., Dawson, J.W., van Kink, J.W., Stuppner, H., and Perry, N.B., Chemosystematic investigations of irregular diterpenes in Anisotome and related New Zealand Apiaceae, *Phytochemistry*, 59, 293, 2002.

98. Vincken, J.-P., Heng, L., de Groot, A., and Gruppen, H., Saponins, classification and occurrence in the plant kingdom, *Phytochemistry*, 68, 275, 2007.

99. Da Paz Lima, M., de Campos Braga, P.A., Lopes Macedo, M., da Silva, M.F. das G.F., Ferreira, G.F., Fernandes, J.B., and Vieria, P.C., Phytochemistry of *Trattinnickia*

burserifolia, *T. rhoifolia*, and *Dacryoides hopkinsii*: chemosystematic implications, *J. Braz. Chem. Soc.*, 15, 385, 2004.

100. Daly, D.C., Studies in neotropical Burseraceae II. Generic limits in New World Protieae and Canarieae, *Brittonia*, 41, 17, 1989.
101. Hegnauer, R., Comparative phytochemistry of alkaloids, in *Comparative Phytochemistry*, Swain, T., Ed., Academic Press, London, 1966, 211.
102. Waterman, P.G., Alkaloid chemosystematics, *Alkaloids*, 50, 537, 1998.
103. van Wyk, B.-E., The value of chemosystematics in clarifying relationships in the genistoid tribes of papilonoid legumes, *Biochem. Syst. Ecol.*, 31, 875, 2003.
104. Greger, H., Structural relationships, distribution and biological activities of *Stemona* alkaloids, *Planta Med.*, 72, 99, 2006.
105. Richardson, P.M. and Young, D.A., The phylogenetic content of flavonoid point scores, *Biochem. Syst. Ecol.*, 10, 251, 1982.
106. Seaman, F.C. and Funk, V.A., Cladistic analysis of complex natural products: developing transformation series from sesquiterpene lactone data, *Taxon*, 32, 1, 1983.
107. Barkman, T.J., Character coding of secondary chemical variation for use in phylogenetic analysis, *Biochem. Syst. Ecol.*, 29, 1, 2001.
108. Japón-Luján, R., Ruiz-Jiménez, J., and Luque de Castro, M.D., Discrimination and classification of olive tree varieties and cultivation zones by biophenol contents, *J. Agric. Food Chem.*, 54, 9706, 2006.

6 Sorbents and Precoated Layers for the Analysis and Isolation of Primary and Secondary Metabolites

Joseph Sherma

CONTENTS

6.1   INTRODUCTION

This chapter will cover mainly commercial precoated layers for thin layer chromatography (TLC) and high performance TLC (HPTLC) that have been used for
phytochemical separations and analyses. Bulk sorbents are available commercially
for those who want to prepare handmade plates, but the variety is much less than
those for precoated plates and the resultant plates are not as reproducible when
compared with precoated plates. As an example of availability, the EMD Chemicals,
Inc. (an affiliate of Merck KGaA, Darmstadt Germany) catalog lists only the
following bulk sorbents for analytical and preparative TLC: silica gel 60, aluminum
oxide 60, kieselguhr G, and silica gel 60 RP-18.

6.1.1   BINDERS

In addition to the chosen sorbent, precoated plates may have a binder to help the
particles adhere to the glass plate or plastic or aluminum sheet that serves as the
backing for preparation of the layer. Plates are designated as "G" if gypsum (calcium
sulfate) binder is used; bulk sorbents for preparing homemade layers by spreading
from a slurry usually contain gypsum. Most commercial TLC and HPTLC plates
contain an organic polymeric binder at a concentration of 1%–2%; these are harder,
smoother, and more durable, and they generally give better separations than G layers.
Plates with no foreign binder are designated with an "H", and high purity silica
"HR". SIL G-25 HR plates (Macherey-Nagel, Dueren, Germany) contain gypsum
binder and a very small quantity of an organic highly polymeric compound; they are
special ready-to-use plates recommended for aflatoxin separations (Figure 6.1).
Layer thickness ranges between 0.1 and 0.25 mm for TLC and HPTLC, with
preparative layers, designated "P", being thicker.

6.1.2   FLUORESCENT INDICATORS

Layers containing an indicator that fluoresces when irradiated with 254 or 366 nm
ultraviolet light are designated as "F" or "UV" layers. These layers are used to
facilitate detection of compounds that absorb at these wavelengths and give dark
zones on a bright background (fluorescence quenching). F_{254} indicators can give
green (zinc silicate) or blue (magnesium tungstate) fluorescence. F_{366} indicators can
be an optical brightener, fluorescein, or a rhodamine dye [1]. Some precoated plates
have both indicators to detect compounds that absorb at both wavelengths ($F_{254+366}$
plates). Plates designated with an "s" have a UV indicator that is acid stable
(e.g., F_{366s} plates). Lux TLC plates (EMD Chemicals, Inc.) and Adamant TLC and
Nano-HPTLC plates (Macherey-Nagel; Figure 6.2) have enhanced brightness F_{254}
indicators.

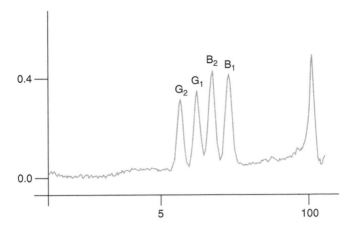

FIGURE 6.1 Separation of aflatoxins on a 0.25 mm SIL G-25 HR plate (0.25 mm layer) developed with chloroform-acetone (90:10) for 30 min. Detection was by densitometric scanning of fluorescence with 366 nm excitation and >400 nm emission wavelengths. (Courtesy of Dr. Detlev Lennartz, Macherey-Nagel. With permission.)

6.1.3 TLC versus HPTLC

High performance plates (10×10 or 10×20 cm) are produced from sorbents having a narrow pore and particle size distribution and an apparent particle size of 5–7 μm instead of 8–10 μm for 20×20 cm TLC plates. Layer thickness is usually 100–200 μm for HPTLC plates when compared with 250 μm for TLC. HP layers are more

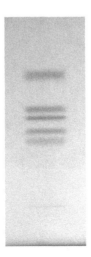

FIGURE 6.2 Steroid chromatogram on an Adamant UV_{254} TLC plate (0.25 mm layer) using a 5 cm development with chloroform–methanol (97:3) mobile phase; from bottom to top the zones are (1) cortisone; (2) corticosterone; (3) testosterone; (4) deoxycorticosterone; and (5) progesterone. (Courtesy of Dr. Detlev Lennartz, Macherey-Nagel. With permission.)

efficient, leading to tighter zones, better resolution, and more sensitive detection. Flow resistance is higher (migration time per cm is slower), but overall development time is shorter because smaller migration distances are used for HPTLC when compared with TLC (i.e., 2–10 versus 5–17 cm for Macherey-Nagel plates, respectively). Sample sizes are generally 0.1–0.2 μL for HPTLC and 1–3 μL for TLC, although the upper levels of these ranges can be exceeded when applying initial zones band-wise with the Camag Linomat instrument or using preadsorbent layers. Silica gel is the most widely used type of HP plate, but other HP layers, including bonded phases, are also commercially available.

Among the newest layers, EMD Chemicals, Inc. offers TLC and HPTLC silica gel 60 plates (60 Å pore size) with imprinted identification codes for use in documentation when performing analyses according to good manufacturing practice (GMP) and good laboratory practice (GLP) standards, e.g., for herbal nutritional supplement manufacturing quality control. TLC and HPTLC plates have particles with irregular shapes (Figure 6.3), but EM Science, Inc. also sells two relatively new HPTLC layers with spherical silica gel, HPTLC plates LiChrospher $Si60F_{254s}$ (0.2 mm layer thickness, 6–8 μm mean particle size) and HPTLC aluminum sheets $Si60F_{254s}$ Raman (0.1 mm thickness and 3–5 μm particle size). The silica gel matrix on the sheets is designed to provide the highest possible signal for direct coupling of TLC with Raman spectrometry. Layers with spherical particles offer improved efficiency, spot capacity, and detection limits compared with irregular particles.

FIGURE 6.3 Electron microscope photograph of a cross section of an irregular-particle silica gel TLC layer through a glass plate, magnification ×500. (Courtesy of Dr. Detlev Lennartz, Macherey-Nagel. With permission.)

Quality control of the traditional herbal medicine Olibanum was carried out by HPTLC on Lichrospher Si60F$_{254}$ and Desaga (Wiesloch, Germany) ProViDoc video documentation [2].

6.1.4 PLATES WITH CHANNELS AND PREADSORBENT ZONE

Kieselguhr (diatomaceous earth) and silica 50,000 are small surface areas, weak adsorbents that are used as the lower 2–4 cm inactive sample application and preadsorbent zone in the manufacture of silica gel and C-18 preadsorbent plates. Samples applied to the preadsorbent zone (also called a concentrating or concentration zone) run with the mobile phase front and form into sharp, narrow, concentrated bands at the preadsorbent-analytical sorbent interface. This leads to efficient separations with minimum time and effort in manual application of samples, and possible sample cleanup by retention of interferences in the preadsorbent. Samples can be applied as either bands or spots to the preadsorbent zone, and they are usually less concentrated than for direct spot application to the analytical layer. The silica 50,000 zone is thinner and has less sample capacity.

Channeled (laned) TLC and HPTLC plates have tracks scribed in the layer that are 9 mm wide with 1 mm between channels (19 useable channels across a 20 cm wide plate). Advantages include no possible cross-contamination of zones during development, and exact location of zones to facilitate lineup of a densitometer light beam for automated scanning. Preadsorbent, channeled plates are optimal for densitometric quantification with manual sample application. Channeled plates also allow more easy removal of separated zones by scraping, prior to recovery by elution, without contamination or damaging adjacent tracks.

6.1.5 PRETREATMENT OF PLATES

To remove extraneous materials that may be present due to manufacture, shipping, or storage conditions, it is recommended to preclean plates before use by development to the top with methanol [3]. The plate is then dried in a clean drying oven or in a clean fumehood on a plate heater for 30 min at 120°C. If the plate is to be used immediately, it will equilibrate with the laboratory relative humidity (which should be controlled to 40%–60% and recorded regularly) during sample application. If necessary, plates are stored in a container offering protection from dust and fumes before use. Plates that are not prewashed do not need activation by heating unless they have been exposed to high humidity. More complete suggestions for initial treatment, prewashing, activation, and conditioning of different types of glass- and foil-backed layers have been published [4].

Plates can be cut to a smaller size using a glass cutter or scissors (plastic or aluminum sheets). Care must be taken to avoid damage or contamination of the layer.

6.2 SILICA GEL

Silica gel (silica) is by far the most frequently used layer material for phytochemical analysis by TLC and HPTLC, e.g., for determination of toxic components such as aristolochic acid in traditional Chinese medicines [5]. It is prepared by dehydration of aqueous silicic acid generated by adding a strong acid to a silicate solution and operates as an adsorbent. Control of temperature and pH during the process is critical. The structure is held together by bonded silicon and oxygen (siloxane groups), and separations take place primarily because of differential migration of sample molecules caused by selective hydrogen bonding, dipole–dipole interaction, and electrostatic interactions with surface silanol groups (Si-OH). The intensity of these adsorption forces depends on the number of effective silanol groups, chemical nature of the sample molecules to be separated, and the elution strength of the mobile phase. Differences in the type and distribution of silanol groups (one, two, or three hydroxyls bonded to a single silicon) for individual sorbents may result in selectivity differences, and separations will not be exactly reproducible for brands of silica gel layers from different manufacturers. Typical properties of silica gels suitable for planar chromatography are as follows [6]: mean hydroxyl (silanol) group density ca. 8 μmol/m^2 (independent of silica gel type); mean pore diameter between 40 and 120 Å (4–12 nm); specific pore volume v_p between of 0.5 and 1.2 mL/g; mean particle size and particle size distribution 5–40 μm; and specific surface area S_{BET} between 400 and 800 m^2/g. Silica gel 60 has 60 Å pore diameter and is the most commonly used type in TLC and HPTLC. Precoated plates used in a laboratory with 40%–60% relative humidity and temperature of 20°C will become equilibrated to a hydration level of 11%–12% water, will not require preactivation unless earlier exposed to high humidity, and will give consistent R_f values within this moisture range.

Silica gel and other precoated layers usually contain a binder to cause the sorbent to adhere to the glass, plastic, or aluminum support backing. Organic binders are most widely used. Plates with no foreign binder are designated as "N".

Silica gel can also act as a support for a liquid stationary phase for separations by liquid–liquid partition TLC. Selective retention of sample molecules results from their differential solubilities in the stationary and mobile liquid phases. In practice, a combination of adsorption and partition mechanisms occurs in many separations on silica gel.

Silica 50,000 is a unique synthetic 100% SiO_2 layer with a mean pore size of 5000 nm that is used as a carrier material for partition TLC, as well as an inert preadsorbent zone placed before the separation layer (see Section 6.1.4).

Silica gel has been used for separations of all classes of compounds important in phytochemistry. Mobile phases are less polar than the silica gel layer (normal phase (NP) or straight phase TLC) and are usually composed of nonpolar and polar constituents with or without an acid or base modifier to improve resolution. The wide applications of silica gel compared with other layers are illustrated in a book chapter on TLC in plant sciences for analysis of alkaloids, flavonoids, anthocyanins, essential oils, cardiac glycosides, and saponins [7].

6.3 SILICA GEL BONDED PHASES

6.3.1 REVERSED PHASE BONDED LAYERS

Hydrophobic reversed phase TLC (RP-TLC), in which the stationary phase is less polar than the mobile phase, was originally carried out on silica gel or kieselguhr layers impregnated with a solution of paraffin or silicone oil. Analtech (Newark, DE) sells RP plates with hydrocarbon liquid phase physically adsorbed onto the silica gel layer surface. Impregnated plates of this kind require the use of aqueous and polar organic mobile phases saturated with the stationary liquid, and they cannot tolerate the use of nonpolar organic solvents that will strip the coating from the support. Bonded phases with functional groups chemically bonded to silica gel eliminate stripping of the stationary liquid from the support by incompatible mobile phases.

Alkylsiloxane-bonded silica gel 60 with dimethyl (RP-2 or C-2), octyl (RP-18 or C-18), octadecyl (RP-18 or C-18), and dodecyl (RP-12 or C-12) functional groups are most widely used for RP-TLC of organic compounds (polar and nonpolar, homologous compounds and aromatics), weak acids and bases after ion suppression with buffered mobile phases, and polar ionic compounds using ion pair reagents. The bonding of the organosilanes containing the hydrophobic group to the accessible silanol groups of silica gel to form new siloxane groups can be done under anhydrous conditions to produce a monolayer bonding, or under hydrous conditions to give a polymer layer; in monolayer formation, mono-, bi-, or tri-functional organosilanes can be used to react with surface silanols. Layers from different companies with the same bonded group can have different percentages of carbon loading and give different results. The hydrophobic nature of the layer increases with both the chain length and the degree of loading of the groups. Alkylsiloxane-bonded layers with a high level of surface modification are incompatible with highly aqueous mobile phases and are used mainly for normal phase separations of low polarity compounds. Problems of wettability and lack migration of mobile phases with high proportions of water have been solved by adding 3% NaCl to the mobile phase to attain better wettability (Whatman C-18 layers; Florham Park, NJ) or preparing "water-wettable" layers (e.g., RP-18W; EMD Chemicals, Inc.) with less exhaustive surface bonding to retain a residual number of silanol groups. The latter layers with a low degree of surface coverage and more residual silanol groups exhibit partially hydrophilic as well as hydrophobic character and can be used for RP-TLC and NP-TLC with purely organic, aqueous-organic, and purely aqueous mobile phases. Chemically bonded phenyl layers are also classified as reversed phase; their separation properties have been reported [1] to be very similar to RP-2 although the aromatic diphenyl bonded group would appear to be very different than ethyl.

RP-2 layers function with an NP mechanism when developed with purely organic mobile phases (Figure 6.4) and an RP mechanism with aqueous-organic or purely aqueous mobile phases. RP-2 layers have been used for determination of coumarins from *Peucedanum tauricum* Bieb. leaves [8] and cytokinin plant hormones [9]. Other examples of RP-TLC in phytochemical analysis are the use of RP-8 layers for polyhydroxylated xanthones and their glycosides obtained from *Chironia krebsii* [10], secoiridoids and antifungal aromatic acids from *Geniana algida* [11], and triterpenoid saponins from the bark of *Quillaja saponaria* [12], and RP-18 layers for determination of peramine in fungal endophyte-infected grasses [13], the flavone phytoalexin suluranetin in rice leaves [14], terpenoids of *Lippia ducis* [15], triterpenoid saponins from *Hedyotis nudicaulis* [16], quinoline alkaloids from *Orixa japonica* [17], and antibacterial phloroglucinols and flavonoids from *Hypericum brasiliense* [18].

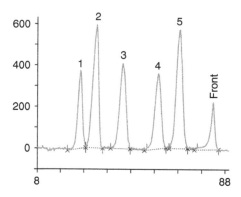

FIGURE 6.4 Steroids densitogram scanned at 254 nm on an RP-2 UV$_{254}$ TLC plate (0.25 mm layer) developed with dichloromethane–methanol (98.5:1.5). Peaks 1–5 are in the same order as for the zones in Figure 6.2 on silica gel, which confirms an NP separation mechanism. (Courtesy of Dr. Detlev Lennartz, Macherey-Nagel. With permission.)

6.3.2 Impregnated Reversed Phase Bonded Layers for Enantiomer Separations

A commercial layer is available under the name Chiralplate (Macherey-Nagel) for separation of enantiomers by the mechanism of ligand exchange. These consist of a glass plate coated with C-18 bonded silica gel and impregnated with the Cu(II) complex of (2S, 4R, 2′RS)-N-(2′-hydroxydodecyl)-4-hydroxyproline as a chiral selector. A book has just been published containing the theory and techniques of chiral TLC, with applications to many compounds of phytochemical interest [19].

6.3.3 Hydrophilic Bonded Layers

Hydrophilic bonded silica gel containing cyano, amino, or diol groups bonded to silica gel through short chain nonpolar spacers (a trimethylene chain [-(CH$_2$)$_3$-] in the case of NH$_2$ and CN plates) are wetted by all solvents, including aqueous mobile phases, and exhibit multimodal mechanisms. Polarity varies as follows: unbonded silica > diol-silica > amino-silica > cyano-silica > reversed phase materials [20].

6.3.3.1 Cyano Bonded Layers

Cyano layers can act as a normal or reversed phase, depending on the characteristics of the mobile phase, with properties similar to a low capacity silica gel and a short chain alkylsiloxane bonded layer, respectively. For example, steroids were separated on a silica gel CN HPTLC plate under NP adsorption conditions using petroleum ether (40°C–60°C)–acetone (80:20) nonpolar mobile phase and RP partition conditions with acetone–water (60:40) polar mobile phase [1]. Two-dimensional separations with different retention mechanisms and selectivity in each direction are possible by development of a sample at right angles in turn with a normal and reversed phase mobile phase. The RP-HPTLC separation of plant ecdysteroids is an example application of CN layers [21].

6.3.3.2 Amino Bonded Layers

Amino layers are used in NP, RP, and weak basic anion exchange modes. In NP-TLC, compounds are retained on amino layers by hydrogen bonding using less polar mobile phases than for silica gel. Activity is less than silica gel, and selectivity is different.

Fewer applications have been reported in the RP-TLC mode where there is limited retention using aqueous-based mobile phases. An example is the separation of oligonucleotides based on differences in hydrophobic properties of the compounds. Charged substances such as nucleotides, purines, pyrimidines, phenols, and sulfonic acids can be separated by anion exchange using neutral mixtures of ethanol–aqueous salt solutions as mobile phases [1].

A special feature of amino precoated layers is that many compounds (e.g., carbohydrates, catecholamines, and fruit acids [22]) can be detected as stable fluorescent zones by simple heating of the plate between 105°C and 220°C (thermochemical activation).

Sugars (Figure 6.5) and sugar alcohols, barbiturates, steroids, carbohydrates, phenols, and xanthin derivatives have been separated on amino layers in various aqueous and nonaqueous mobile phases [23,24].

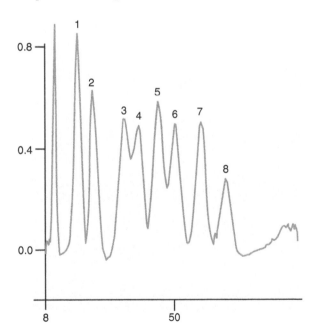

FIGURE 6.5 Densitogram scanned at 254 nm showing the HPTLC separation of sugars on a 0.2 mm Nano-SIL NH_2/UV layer double developed with ethyl acetate–pyridine–water–glacial acetic acid (60:30:10:5) for 8 cm. Peaks: (1) lactose; (2) sucrose; (3) galactose; (4) glucose; (5) fructose; (6) arabinose; (7) xylose; (8) ribose. (Courtesy of Dr. Detlev Lennartz, Macherey-Nagel. With permission.)

6.3.3.3   Diol Bonded Layers

Diol plates have functional groups in the form of alcoholic hydroxyl residues while unmodified silica gel has active silanol groups. The vicinal diol groups are bonded to silica with a quite nonpolar alkyl ether spacer group. Diol layers can operate with NP- or RP-TLC mechanisms, depending on the mobile phase and solutes. Polar compounds show reasonable retention by hydrogen bond and dipole type inter- actions in the former mode, and in the RP mode retention with polar solvent systems is low but higher than with amino layers. A study of mixed mechanisms on cyano, amino, and diol layers was reported [25].

Phenolic constituents [26] and flavonol glycosides [27] in *Monnina sylvatica* root extract were separated on HPTLC diol plates with hexane–isopropanol (4:1) and ethyl acetate–toluene (1:1) mobile phases, respectively. The separation of three urea herbicides on an HPTLC diol plate is demonstrated in Figure 6.6.

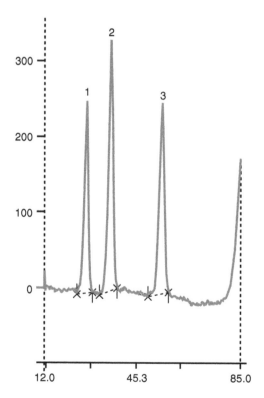

FIGURE 6.6 Densitogram scanned at 238 nm of the HPTLC separation of three pesticides on a 0.2 mm Nano-SIL Diol/UV layer developed with petroleum ether (40°C–60°C)–acetone (80:20) for 7 cm. Peaks: (1) metoxuron; (2) monuron; (3) metobromuron. (Courtesy of Dr. Detlev Lennartz, Macherey-Nagel. With permission.)

6.4 NONSILICA SORBENTS

6.4.1 ALUMINA

Alumina (aluminum oxide) is a polar adsorbent that is similar to silica gel in its general chromatographic properties, but it has an especially high adsorption affinity for carbon–carbon double bonds and better selectivity toward aromatic hydrocarbons and their derivatives. The alumina surface is more complex than silica gel, containing hydroxyl groups, aluminum cations, and oxide anions, and pH and humidity level alter separation properties. It is available in basic (pH 9–10), neutral (7–8), and acid (4–4.5) form. The most used TLC aluminas are alumina 60 (type E), 90, and 150 (type t) with respective pore diameters of 6, 9, and 15 nm; specific surface areas 180–200, 100–130, and 70 m^2/g; and pore volumes 0.3, 0.25, and 0.2 mL/g [24]. The high density of hydroxyl groups (ca. 13 $\mu mol/m^2$) leads to a significant degree of water adsorption, and alumina layers are usually activated by heating for 10 min at 120°C before use [1]. Alumina 150 has been used as a support for partition TLC.

The alumina TLC of plant benzoisoquinoline alkaloids [28] and quaternary ammonium compounds [29] are examples of phytochemical applications.

6.4.2 CELLULOSE

TLC cellulose consists of long chains of polymerized beta-glucopyranose units connected at the 1–4 positions. Precoated TLC and HPTLC crystalline cellulose (AVICEL) layers (400–500 glucose units) and native fibrous cellulose TLC layers (40–200 glucose units) without binder are used for the separation of polar substances such as amino acids and carbohydrates. The mechanism is NP partition with sorbed water as the stationary phase, although adsorption effects cannot be excluded in cellulose separations. Zones are generally less compact and development times longer than on silica gel. Two-dimensional development has been used to obtain "fingerprint" patterns of complex mixtures, such as tannin polymer hydrolyzates [30].

Phytochemical separations have been reported on microcrystalline cellulose for hydroxycinnamic acid esters [31], flavonol glycosides [32], anthocyanins [33], flavone and flavanone aglycones [34], triterpenoid saponins [12,35], and iridoid glucosides [36].

6.4.3 POLYAMIDES

Polyamide 6 (Nylon 6; polymeric caprolactam) and 11 (polymeric undecanamide) are synthetic organic resins that show high affinity and selectivity for polar compounds that can form hydrogen bonds with the surface amide and carbonyl groups. Depending on the type of analyte and mobile phase, three separation mechanisms can operate with polyamide: adsorption, partition (NP and RP), and ion exchange.

Reported applications of polyamide TLC in phytochemistry include flavonoid aglycones [37], flavonol glycosides [38], leaf flavonoids [39], and phenols [40].

6.4.4 Modified Celluloses

Cellulose has been surface modified to produce RP (acetylated cellulose), basic anion exchange (polyethyleneimine (PEI), aminoethyl (AE), diethylaminoethyl (DEAE), and ECTEOLA), or acidic cation exchanger (cellulose phosphate (P) and carboxymethylcellulose (CM)) layers. Acetylated cellulose has been used mostly as a chiral layer for separation of enantiomers [19]. The cellulose ion exchangers have open structures that can be penetrated by large hydrophilic molecules, such as proteins, enzymes, nucleotides, nucleosides, nucleobases, and nucleic acids. Use of these layers has been seldom reported for phytochemistry applications.

6.4.5 Kieselguhr

Kieselguhr is natural diatomaceous earth consisting mostly of SiO_2. It has high porosity and is completely inactive, being used as a support for impregnated reagents (e.g., ethylenediaminetetraacetic acid for separation of tetracycline broad spectrum antibiotics) and as a preadsorbent zone (see Section 6.1.4). Macherey-Nagel manufacturers precoated 0.25 mm kieselguhr layers with and without UV_{254} indicator.

6.5 MISCELLANEOUS LAYERS

6.5.1 Resin Ion Exchange Layers

POLYGRAM IONEX-25 (Macherey-Nagel) are polyester sheets coated with a 0.2 mm mixed layer of silica, a polystyrene-based strong acid cation or strong base anion exchange resin, and a binder. They are suited for separation of organic compounds with ionic groups, such as amino acids, and inorganic ions.

6.5.2 Impregnated Layers

Layers have been impregnated with buffers, chelating agents, metal ions, or other compounds to aid resolution or detection of certain compounds. For example, silica gel buffered to pH 6 was used to separate flavaspidic and filixic derivatives from *Dryopteris fusco-atra* and *D. Hawaiiensis* [41] and silica gel impregnated with 0.2 M sodium phosphate for hydrolyzate sugars of European species of *Hypochooris* [42].

Analtech precoated silica gel plates are available already impregnated with potassium oxalate to facilitate resolution of polyphosphoinositides, magnesium acetate for phospholipids, 0.1 M NaOH for organometallics and acidic compounds, and silver nitrate for argentation TLC. Analtech manufactures plates containing ammonium sulfate for detection of compounds as fluorescent or charred zones after heating (vapor phase fluorescence or self-charring detection), while Macherey-Nagel sells silica gel G plates with ammonium sulfate for separation of surfactants.

Among the many impregnating reagents that have been used, the most common for phytochemistry applications is silver, which serve as pi-complexing metal for separation of organic compounds with unsaturated groups, such as fatty acids [43].

Impregnation is usually carried out in the laboratory by incorporation of the silver salt in the slurry used to prepare a homemade plate, immersing or spraying a commercial plate with methanolic silver nitrate solution, or development or over-development of a commercial plate using the silver solution as the mobile phase. TLC on silver impregnated silica was reported for the determination of acetylated sterols [44], sterols [45], and pentacyclic triterpenoid [46] and sesquiterpenoid [47] phytoalexins (Rhodamine 6G impregnated silica was also used) in plant material.

Preparative as well as analytical silica gel plates have been impregnated with silver ions, e.g., for recovery of alk(en)ylcatechols from *Metopium brownei urushiol* components [48].

Caffeine-impregnated plates are available from Macherey-Nagel and EMD Chemicals, Inc. for charge transfer TLC. An important application is the separation of polycyclic aromatic hydrocarbons (PAHs).

6.5.3 MIXED PHASE LAYERS

Mixed phase layers have been homemade with a particular ratio of sorbents for specific applications. In addition, Macherey-Nagel offers the following 0.25 mm precoated mixed layers: aluminum oxide G/acetylated cellulose for separation of PAHs, cellulose/silica for separation of preservatives, and kieselguhr/silica for carbohydrate, antioxidant, and steroid separations.

6.5.4 DUAL PHASE LAYERS

Combination layers with a C-18 strip adjacent to a silica gel layer (Multi-K CS5) or a silica gel strip adjacent to a C-18 layer (SC5) are available from Whatman for two-dimensional TLC with diverse mechanisms (RP partition and adsorption).

6.6 PREPARATIVE LAYERS

Applications of PLC for the separation of secondary metabolites from plant tissues have been tabulated in a book chapter [49]. The layer was usually silica gel, with occasional reference to the use of alumina, C-18, C-2, and polyamide.

Also listed in this book are commercial precoated plates for PLC that are available from a number of manufacturers [6]. These are mostly silica gel with particle size distributions of 5–40, 5–17, or 10–12 μm; layer thickness 0.5–2 mm; organic, gypsum, or no foreign binder; and with or without fluorescent indicator. Also available as precoated PLC layers are aluminum oxide (5–40 μm particle size distribution, and 1 or 1.5 mm thickness), cellulose (2–20 μm, 0.5 or 1 mm), and RP-18 (5–40 or 10–12 μm, 1 mm). A variety of bulk materials are also available from manufacturers for homemade preparation of preparative layers.

Analtech offers a unique tapered layer with preadsorbent for capillary-flow preparative separations and precast HPTLC silica gel GF rotors with 1000–8000 μm nominal thickness for use with their Cyclograph centrifugal forced-flow PLC instrument.

REFERENCES

1. Wall, P.E. *Thin Layer Chromatography—A Modern Approach*, Royal Society of Chemistry, Cambridge, UK, chap. 2.
2. Hahn-Dienstrop, E., Koch, A., and Mueller, M. *J. Planar Chromatogr.—Mod. TLC* 11, 404–410, 1998.
3. Reich, E. and Schibli, A. *J. Planar Chromatogr.—Mod. TLC* 17, 438–443, 2004.
4. Hahn-Dienstrop, E. *J. Planar Chromatogr.—Mod. TLC* 6, 313–318, 1993.
5. Li, W., Chen, Z., Liao, Y., and Liu, H. *Anal. Sci.* 21, 1019–1029, 2005.
6. Hauck, H.E. and Schulz, M. Sorbents and precoated layers in PLC, in *Preparative Layer Chromatography*, Kowalska, T. and Sherma, J., Eds., CRC/Taylor & Francis, Boca Raton, FL, 2006, chap. 3.
7. Pothier, J., Thin layer chromatography in plant sciences, in *Practical Thin Layer Chromatography*, Fried, B. and Sherma, J., Eds., CRC Press, Boca Raton, FL, 1996, chap. 3.
8. Bartnik, M., Glowniak, K., Maciag, A., and Hajnos, M.L. *J. Planar Chromatogr.—Mod. TLC* 18, 244–248, 2005.
9. Lethan, D.S., Singh, S., and Willcocks, D.A. *Phytochem. Anal.* 3, 218–222, 1992.
10. Wolfender, J.-L., Hamburger, M., Msonthi, J.D., and Hostettmann, K. *Phytochemistry* 30, 3625–3629, 1991.
11. Tan, R.X., Wolfender, J.-L., Ma, W.G., Zhang, L.X., and Hostettmann, K. *Phytochemistry* 41, 111–116, 1996.
12. Higuchi, R., Tokimitsu, Y., Fujioka, T., Komori, T., Kawasaki, T., and Oakenful, D.G. *Phytochemistry* 26, 229–235, 1987.
13. Fannin, F.F., Bush, L.P., Siegel, M.R., and Rowan, D.D. *J. Chromatogr.* 503, 288–292, 1990.
14. Kodama, O., Miyakowa, J., Akatsuka, T., and Kiyosawa, S. *Phytochemistry* 31, 3807–3809, 1992.
15. Souto-Baciller, F.A., Melendez, P.A., and Romero-Ramsey, L. *Phytochemistry* 44, 1077–1086, 1997.
16. Konishi, M., Hano, Y., Takyama, M., Nomura, T., Hamzah, A.S., Ahmad, R.B., and Jasmani, H. *Phytochemistry* 48, 525–528, 1998.
17. Funayama, S., Murata, K., and Nozoe, S. *Phytochemistry* 36, 525–528, 1994.
18. Rocha, L., Marston, A., Potterat, O., Kaplan, M.A.C., Stoecki, H., and Hostettmann, K. *Phytochemistry* 40, 1447–1452, 1995.
19. Sherma, J. Commercial precoated layers for enantiomer separations and analysis, in *Thin Layer Chromatography in Chiral Separations and Analysis*, Kowalska, T. and Sherma, J., Eds., CRC/Taylor & Francis, Boca Raton, FL, 2007, chap. 3.
20. Lepri, L. and Cincinelli, A. TLC sorbents, in *Encyclopedia of Chromatography*, Cazes, J., Ed., Taylor & Francis, Boca Raton, FL, 2005. DOI: 10.1081/E-ECHR-120040171.
21. Bathori, M., Hunyadi, A., Janicsak, G., and Mathe, I. *J. Planar Chromatogr.—Mod. TLC* 17, 335–341, 2004.
22. Klaus, R., Fischer, W., and Hauck, H.E. *LC-GC* 13, 816–823, 1995.
23. Bieganowska, M.L. and Petruczynik, A. *Chem. Anal. (Warsaw)* 42, 345–352, 1997.
24. Hauck, H.E. and Mack, M., Sorbents and precoated layers in thin layer chromatography, in *Handbook of Thin Layer Chromatography*, 2nd ed., Sherma, J. and Fried, B., Eds., Marcel Dekker, New York, NY, 1996, chap. 4.
25. Kowalska, T. and Witkowska-Kita, B. *J. Planar Chromatogr.—Mod. TLC* 9, 92–97, 1996.
26. Bashir, A., Hamburger, M., Gupta, M.P., Solis, P., and Hostettmann, K. *Phytochemistry* 31, 3203–3205, 1992.

27. Bashir, A., Hamburger, M., Gupta, M.P., Solis, P., and Hostettmann, K. *Phytochemistry* 30, 3781–3784, 1991.
28. Castedo, L., Lopez, S., Rodriguez, A., and Villaverde, C. *Phytochemistry* 28, 251–257, 1989.
29. Koheil, M.A.H., Hilal, S.H., El-Alfy, T.S., and Leistner, E. *Phytochemistry* 31, 2003–2008, 1992.
30. Koupai-Abyazani, M.R. and Bohm, B.A. *Phytochemistry* 33, 1485–1487, 1993.
31. Strack, D., Engel, U., Weissenbock, G., Grotjahn, L., and Wray, V. *Phytochemistry* 25, 2605–2608, 1986.
32. Slimestad, R., Andersen, Q.M., Francis, G.W., Marston, A., and Hostettmann, K. *Phytochemistry* 40, 1537–1542, 1995.
33. Odake, K., Terahara, N., Saito, N., Toki, T., and Honda, T. *Phytochemistry* 31, 2127–2130, 1992.
34. Hardorne, J.B. and Williams, C.A. *Phytochemistry* 22, 1520–1521, 1983.
35. Rastogi, S., Pal, R., and Kulshreshtha, D.K. *Phytochemistry* 36, 133–137, 1994.
36. Lahloub, M.F., Zaghloul, N.G., Afifi, M.S., and Sticher, O. *Phytochemistry* 33, 401–405, 1993.
37. Zaidi, F., Voirin, B., Jay, M., and Viricil, M.R. *Phytochemistry* 48, 991–994, 1998.
38. Wollenweber, E., Stueber, A., and Kraut, L. *Phytochemistry* 44, 1399–1400, 1997.
39. Wollenweber, E., Stern, S., Roitman, J.N., and Yatskievych, G. *Phytochemistry* 30, 337–342, 1991.
40. Williams, C.A., Harborne, J.B., and Goldblatt, P. *Phytochemistry* 25, 2135–2154, 1986.
41. Patama, T.T. and Widen, C.-J. *Phytochemistry* 30, 3305–3310, 1991.
42. Gluchoff-Fiasson, K., Favre-Bonvin, J., and Fiasson, J.L. *Phytochemistry* 30, 1670–1675, 1991.
43. Nikolova-Damyanova, B. and Momcholova, Sv. *J. Liq. Chromatogr. Relat. Technol.* 24, 1447–1466, 2001.
44. Huang, L.S. and Grunwald, C. *Phytochemistry* 25, 2779–2781, 1986.
45. Misso, N. and Goad, L. *Phytochemistry* 23, 73–82, 1984.
46. Van der Heijden, R., Threlfall, D.P., Verpoorte, R., and Whitehead, I.M. *Phytochemistry* 28, 2981–2988, 1989.
47. Whitehead, I.M., Ewing, D.F., Threlfall, D.R., Cane, D.E., and Prabhakaran, P.C. *Phytochemistry* 29, 479–482, 1990.
48. Rivero-Cruz, J.F., Chavez, D., Hernandez-Bautista, B., Anaya, A.L., and Mata, R. *Phytochemistry* 45, 1003–1008, 1997.
49. Waksmundzka-Hajnos, M., Wawrzynowicz, T., Hajnos, M.L., and Jozwiak, G. Preparative layer chromatography of natural mixtures, in *Preparative Layer Chromatography*, Kowalska, T. and Sherma, J., Eds., CRC/Taylor & Francis, Boca Raton, FL, 2006, chap. 11.

7 Chambers, Sample Application, and Chromatogram Development

Tadeusz H. Dzido and Tomasz Tuzimski

CONTENTS

7.1 INTRODUCTION

Stages of planar chromatography procedure such as sample application, chromatogram development, registration of chromatogram, and its evaluation cannot be presently performed in one run using commercially available device. It means that planar chromatography analysis is not completely automated at contemporary laboratory practice. Opposite situation is in other liquid chromatography techniques of which the procedures are completely automated and proceed in one run. The present status of thin layer chromatography (TLC) will be probably continued in the near future too and however, there are no signals at present which could indicate that this situation will be changed in a distant future. Chromatographers have to separately optimize each of the mentioned stages using more or less sophisticated devices. There are various equipments for hand- semi- or full-automatic operations at mentioned stages of planar chromatography procedures. At each stage of TLC procedure chromatographer should demonstrate basic skill, which substantially helps to accomplish TLC experiments correctly, to obtain reliable, repeatable, and reproducible results. He/she can meet many pitfalls during work with TLC mode. There are some fundamental books which help to overcome these problems [1–4]. In this chapter we give some information that can draw the reader's attention to the procedures and equipments mentioned earlier, which are more often applied and proven in contemporary planar chromatography practice. We hope that the reader will gain this information to avoid some problems concerned with performing TLC experiments when reading this chapter.

7.2   MODERN CHAMBERS FOR TLC

Chromatogram development can be proceeded using different types of chambers for TLC. The classification of the chromatographic chambers can be performed taking into account volume of vapor atmosphere inside the chamber, direction of solvent front (mobile phase) migration, configuration of the chromatographic plate in the chamber, and degree of automation of chromatogram development.

Regarding volume of vapor atmosphere, two main types of the chamber can be distinguished: normal (conventional) chambers (N-chambers) and sandwich chambers (S-chambers). The volume of vapor atmosphere is determined by distance between chromatographic plate and chamber walls or lid. In the former chamber, this distance is very large and in the last chamber type it does not exceed 3 mm. So in S-chamber volume of vapor atmosphere is very small. Direction of the mobile phase migration through the adsorbent layer can be linear or radial. In the first case solvent migrates through rectangle or square chromatographic plate from one of its edge to opposite edge with constant width of front of the mobile phase, Figure 7.1a. The chamber for radial development can be divided into two groups: chamber for circular and anticircular development. In the former type the mobile phase is delivered at the center of the chromatographic plate and its front migrates towards the periphery of the adsorbent layer, in the form of circle, Figure 7.1b. In the last type the mobile phase migrates in opposite direction and its front again forms circle, Figure 7.1c. The chromatographic plate can be horizontally or vertically positioned in the chamber. Then in the first case the chamber is named as horizontal chamber and in the second as vertical one. There are available chambers that facilitate different degrees of automation during chromatographic plate development including temperature and humidity control, eluent and vapor phase delivery to the chromatographic chamber, and drying the chromatographic plate.

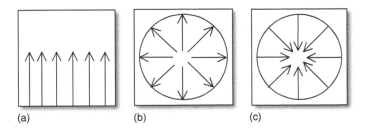

(a)   (b)   (c)

FIGURE 7.1 Modes of development in planar chromatography.

The earlier mentioned classification is not unequivocal because selected chamber can belong to more than one chamber type. Below we describe some of the most popular chambers applied especially in contemporary chromatographic practice.

7.2.1 CONVENTIONAL CHAMBERS (N-CHAMBERS)

The chambers are typically made of glass as a vessel possessing cuboid or cylindrical form. These examples are demonstrated in Figure 7.2. Chromatogram development in such chambers is started by immersing one edge of the chromatographic plate in the mobile phase solution, which was previously poured into the chamber. During development the chamber is covered with a lid. This type of chamber can be very easily applied for conditioning (saturation with vapour phase) of the chromatographic plate, what is very important especially when mixed solution of the mobile phase is used for chromatogram development. Then repeatability of retention values is higher in comparison with development without vapor saturation. Efficiency of this saturation is substantially enhanced by lining the inner walls of the chamber with blotting paper. The chamber is completely saturated after 60 min.

Another type of N-chamber is a cuboid twin-trough chamber that can be conveniently used for chromatography under different conditions of vapor saturation [1]. Schematic view of the chamber is demonstrated in Figure 7.3. The bottom of the chamber is divided by ridge into two parallel troughs. This construction of the chamber enables to perform chromatogram development in three modes: without

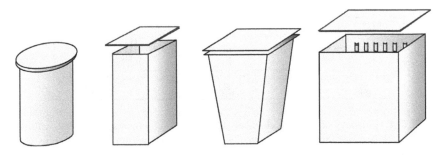

FIGURE 7.2 N-chambers with flat bottom.

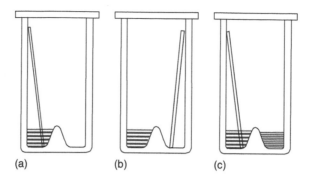

(a)   (b)   (c)

FIGURE 7.3 The twin-trough chamber with various variants of chromatogram development.

chamber saturation, with chamber saturation, and chamber saturation with one solvent followed by development with another one. These three modes of chromatogram development are presented in Figure 7.3.

7.2.2 HORIZONTAL CHAMBERS FOR LINEAR DEVELOPMENT

As it was mentioned earlier the chromatographic plate is positioned horizontally in this chamber type. Horizontal position of the chromatographic plate is advantageous regarding consumption of solvent and velocity of solvent front migration during chromatogram development. One example of horizontal chamber (horizontal developing chamber manufactured by Camag) is presented in Figure 7.4 as a cross section [5,6]. The chromatographic plate (1) is positioned with adsorbent layer face down and is fed with the solvent from the reservoir (3). Chromatogram development is started by tilting the glass strip (4) to the edge of the chromatographic plate. Then a planar capillary is formed between the glass strip and the wall of the solvent reservoir in which the solvent instantaneously rises feeding the chromatographic plate. Maximum distance of development with this chamber is 10 cm. The chambers are offered for 10×10 and 20×10 plates.

Another example of horizontal chamber (Horizontal DS Chamber manufactured by Chromdes) is presented as cross section before and during chromatogram

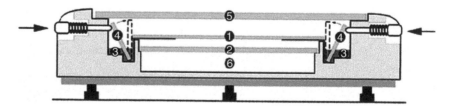

FIGURE 7.4 Horizontal developing chamber (Camag): 1—chromatographic plate with layer face down, 2—counter plate (removable), 3—troughs for solution of the mobile phase, 4—glass strip for transfer of the mobile phase by capillary action to the chromatographic plate, 5—cover glass plate. (Courtesy www.camag.com. With permission.)

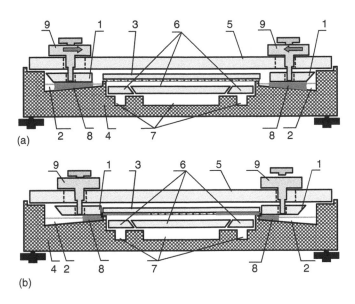

FIGURE 7.5 Horizontal DS-II chamber (Chromdes): (a) before development, (b) during development; 1—cover plate of the mobile phase reservoir, 2—mobile phase reservoir, 3—chromatographic plate with layer face down, 4—body of the chamber, 5—main cover plate, 6—cover plates (removable) of the troughs for vapour saturation, 7—troughs for saturation solvent, 8—mobile phase, 9—mobile phase distributor/injector. (From www.chromdes.com.)

development in Figures 7.5a and 7.5b, respectively [7,8]. The main feature of the chamber is formation of vertical meniscus of the solvent (dark area) between slanted bottom of the mobile phase reservoir (2) and glass strip (1). Chromatogram development is started by shifting the glass strip to the edge of the chromatographic plate (3) with adsorbent layer face down what brings solvent to the contact with the chromatographic plate. During development the meniscus of solvent moves in the direction of the chromatographic plate what makes the chamber very economical (the solvent can be exhausted from the reservoir almost completely). Conditioning of the chamber atmosphere can be performed by pouring some drops of solvent onto the bottom of troughs (lined with blotting paper) (7) after removing glass plates (6). All kinds of plates (foil and glass backed of dimension from 5 × 10 to 20 × 20 cm) can be developed in this chambers dependent on the chamber type and size. Maximum distance of chromatogram development is equal to 20 cm. The two horizontal chamber types described above posses following methodological possibilities:

– Double number of samples in comparison with conventional chambers can be separated on one plate (due to two solvent reservoirs on both sides of the plate which enable simultaneous development of two chromatograms from two opposite edges) [9]
– Saturation of the adsorbent layer with vapors of the mobile phase or another solvent [9,10]

- Two-dimensional (2D) separation of four samples on one plate simultaneously [11],
- Multiple development [12,13]
- Stepwise gradient elution [14,15]
- Zonal sample application for preparative separation (in Horizontal DS Chamber) [16,17]
- Continuous development (in SB/CD and Horizontal DS chambers)* [17]
- Short bed-continuous development (in SB/CD and Horizontal DS chambers)* [17,18]
- Development of six different chromatograms on one plate simultaneously (HPTLC Vario Chamber from Camag or Horizontal DS-M Chamber from Chromdes) [17]

More detailed description of these methodical possibilities can be found by the reader in following references [1,2,17,18].

Another horizontal chamber (H-separating chamber) for TLC is manufactured by Desaga [19]. Its principle of action is based on Brenner–Niedervieser chamber [20]. As it can be seen in Figure 7.6 the chromatographic plate (1) is fed with solvent from the reservoir (2) using wick (3) made of porous glass. The chambers are manufactured for 5×5 cm and 10×10 cm plates.

The elements of the three types of the horizontal chambers described earlier (horizontal developing chamber, horizontal DS chamber, and H-separating chamber) are made of Teflon and glass so they are very resistant to all solvents applied for chromatographic separations.

Another horizontal chambers as Vario-KS-Chamber [21,22], SB/CD-Chamber [23], Sequence-TLC Developing Chamber [24], ES-Chamber [18], ES-Chamber modified by Rumiński [25] and by Wang et al. [26,27] have been described in the literature and applied in some laboratories; however, these are not commercially offered at present.

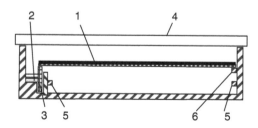

FIGURE 7.6 H-separating chamber (Desaga): 1—chromatographic plate with layer face down, 2—reservoir for the mobile phase, 3—wick of porous glass, 4—cover plate, 5—support for counter plate, 6—support for chromatographic plate. (Adapted from Kraus, L., Koch, A., Hoffstetter-Kuhn, S., *Dünnschichtchromatographie*, Springer-Verlag, Berlin, Heidelberg, 1996.)

---

* SB/CD chamber is not manufactured at present.

7.2.3 HORIZONTAL CHAMBERS FOR RADIAL (CIRCULAR AND ANTICIRCULAR) DEVELOPMENT

The mode of radial development of planar chromatograms is rarely applied in laboratory practice. Radial development using circular mode can be easily performed with Petri dish [28]. This mode, including circular and anticircular development, can be carried out using commercial U-chambers (Camag) [29]. However, at present these chambers are not in the commercial offer of this firm, probably due to low interest of the customers. In spite of some advantages regarding separation efficiency, practitioners prefer to apply linear development than radial development. Two main reasons explain this status: at first too sophisticated chamber construction and its maintenance to develop chromatograms, and the second reason is concerned with shortage of equipment and software for chromatogram evaluation.

7.2.4 AUTOMATIC CHAMBERS

One of the first automatic chambers was described by Omori, Figure 7.7, [30]. The chamber is relatively simple in maintenance and needs only few manual operations. However, this chamber has not been offered on the market. According to the author this chamber can be homemade, which substantially reduces the costs of the device in comparison with the commercially available one.

Some firms (Camag, Desaga, Baron Laborgeräte) offer more sophisticated devices for automatic chromatogram development. In Figure 7.8 Automatic Separating Chamber (TLC-MAT from Desaga [31] or from Lothar Baron Laborgeräte [32]) is demonstrated. All kinds of plates can be developed in this device of 10 cm and

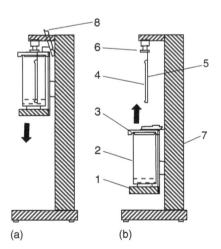

(a)   (b)

FIGURE 7.7 Semiautomatic developing chamber: 1—chamber holder, 2—glass chamber, 3—PTFE main lid, 4—chromatographic late, 5—chromatographic plate holder, 6—sublid, 7—stand, 8—open lid. (Adapted from Omori, T., *Proceedings of the 6th International Symposium on Instrumental Planar Chromatography*, Traitler, H., Voroshilowa, O.I., Kaiser, R.E. (Eds.), Interlaken, Switzerland, 23–26 April, 1991.)

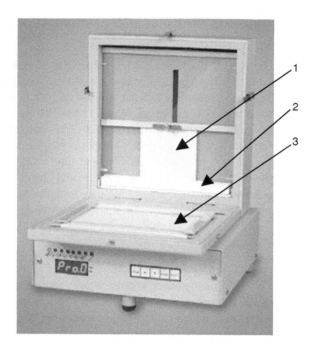

FIGURE 7.8 Automatic separating chamber, TLC-MAT. (From Desaga, Nümbrecht, Germany; Baron Laborgeräte, Insel Reichenau, Germany. With permission.)

20 cm in height. Volume of the chamber space and chromatographic plate can be saturated (for conditioning) for appropriate time preselected by the operator. Volume of the eluent reservoir is very small, which causes the consumption of solvent during development to be very economical. Special sensor is used to control the migration distance of solvent front. This feature enables to break the development of chromatogram when the solvent front is traveled desired migration distance. Vapors of the solvent can be removed from the chamber by special fun, which makes plate dry after development.

Chromatogram development can be performed under conditions of temperature control using device named TLC Thermo Box (Desaga [31] and Lothar Baron Laborgeräte [32]). Separation process can be proceeded at temperature in the range 10°C below to 20°C above room temperature with precision equal to ± 0.5°C.

Few years ago Camag launched new automatic developing chamber (ADC-2), Figure 7.9 [5]. The main features of this chamber according to the manufacturer specification are: fully automatic development of 10×10 cm and 20×10 cm chromatographic plates, twin-trough chamber is applied for chromatogram development (manual methods previously applied with twin-trough chamber can be conveniently adapted for automatic development with ADC 2 chamber). Optional feature of this chamber is development under conditions of controlled humidity. All operations necessary to run the separation process can be introduced from keypad of the chamber or with a computer using manufacturer's software.

FIGURE 7.9 Automatic developing chamber ADC 2. (From Camag, Muttenz, Switzerland. With permission.)

So complete separation process proceeds with no influence of manual operations. All data relevant to separation procedures can be stored in the computer memory. This last feature enables to apply the elaborated procedures in future experiments and is compliant with the requirements of GLP/GMP. Detailed description of the device and its operation is presented in the web site of the manufacturer [5].

Automatic developing chambers described earlier are especially suitable for routine analysis because of repeatable conditions provided by the instrumental control of the chromatographic process—so all chromatograms are repeatable and reproducible. However, it should be noticed that reproducibility of the separation cannot be expected when chromatogram development is performed under the same specified conditions using another automatic chambers. In such case additional optimization procedure for another chamber is required.

The most sophisticated equipment for chromatogram development is AMD 2 (abbreviation AMD means Automated Multiple Development), Figure 7.10 [5]. Operation of this device is based on methodology described by Perry et al. [33] and modified by Burger [34] and involves following stages and features of the chromatographic process:

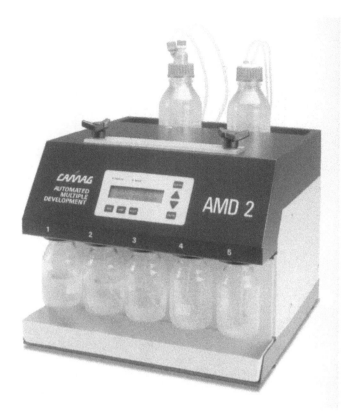

FIGURE 7.10 Device for automated multiple development of chromatograms, AMD 2. (From Camag, Muttenz, Switzerland. With permission.)

- Chromatographic plate (usually HPTLC plate) is developed repeatedly in the same direction.
- Each next step of chromatogram development follows complete evaporation of the mobile phase from the chromatographic plate and is performed over longer migration distance of solvent front than the one before.
- Each next step of chromatogram development uses solvent of lower elution strength than that one used in preceding run; it means that complete separation process proceeds under conditions of gradient elution.
- The focusing effect of the solute bands takes place during separation process, which leads to very narrow sample zones and high efficiency of the chromatographic system comparable with HPLC.

Complete separation procedure is performed automatically and is controlled by the software. Five different solvents (in five bottles) are used for preparation of eluent solutions, so gradient development can be accomplished with similar number of the mobile phase components. Full separation process comprising 20–25 steps seems to take a lot of time. However, this is compensated by simultaneous separation of many samples on one chromatographic plate and using the system outside working hours without inspection. So final analysis is characterized by relatively high throughput. This throughput can be increased by reducing the number of steps of AMD procedure. Application of special software for simulation of planar chromatography process can additionally enhance this procedure [35,36].

7.3 SAMPLE APPLICATION

Resolution of the chromatographic system is dependent on the size of starting zone (spot) of the solute. If this zone is too large then resolution of components of sample mixture, when it is especially more complex and migration of solvent front is short, cannot be satisfactory. Avoiding pitfalls, concerned with sample application, is crucial for final separation in planar chromatography. Conventional application of sample mixture on the chromatographic plate can be performed with calibrated capillary or microsyringe. More advantageous modes of sample application can be performed with semiautomatic applicator or fully automated device. All these modes can be applied for analytical and preparative separations as well. In the following section some information is presented, which will help to introduce the reader in problems of manual and automatic sample application relevant for analytical and preparative separations.

7.3.1 SAMPLE APPLICATION IN ANALYTICAL TLC

The sample spotting can be performed by hand operation using disposal micropipette or calibrated capillary and microsyringe. However, this operation should be performed with care because the adsorbent layer can be damaged by the tip of capillary or syringe needle when pressed too strong against the layer. The adsorbent layer can be prevented from this damage using special device, e.g., Nanomat (Camag) [5] in which the capillary is held by the dispenser. Sample application with microsyringe possess one important advantage—sample volume applied on the chromatographic plate can be conveniently varied depending on requirements of the analysis. Very important variable influencing on the size of the sample spot, when manually applied with capillary or microsyringe, is solvent type of the sample mixture. It is desired that elution strength of this solvent should be as low as possible. When this requirement is fulfilled then it should be expected the sample dimension to be very small. In the other case circular chromatography is realized during sample application leading to widening of starting zone and diminishing final resolution of sample bands on chromatogram. However, for some compounds it is difficult to find an appropriate solvent which fulfill this requirements. Application of the chromatographic plates with preconcentration zone or an aerosol applicator, e.g., manufactured by Camag (Linomat 5 and Automatic TLC Sampler 4) [5] and Desaga (HPTLC-Applicator AS 30) [31] can help to overcome this handicap. In the former case the sample spot is focused in the preconcentration zone when solvent is migrated through it during chromatographic process.

Automatic aerosol applicators have gained higher popularity in laboratory practice in spite of relatively high price due to few following important features:

- Starting sample spot is very small, typical width about 1 mm.
- Dimension of the spot is not dependent on the solvent type of the sample solution.
- High repeatability of sample volume applied on the layer—very important for quantitative analysis.
- Various sample shapes can be obtained—dot (spot), streak, band, or rectangle.
- Various sample volumes can be applied.

Sample shape as the streaks or bands is advantageous with regard to resolution and quantitative analysis. Rectangle shape of the starting zone is advantageous in the case of preparative separations and samples with high matrix content—then large volume of the sample is required for its application [37].

7.3.2 SAMPLE APPLICATION IN PREPARATIVE LAYER CHROMATOGRAPHY

Sample application for preparative separations in planar chromatography usually requires to spot larger volumes of the sample solution on the plate—its solution is usually deposited on almost the whole width of the chromatographic plate in shape of band, streak, or rectangle. Adsorbent layers used for preparative separations are thicker than for analytical separations. This procedure of sample application can be performed manually using capillary or microsyringe. Then the sample solution is spotted side by side on the start line of the chromatographic plate. This mode is tedious and needs a lot of manual operations. Shape of starting band is often not appropriate leading to lower resolution of the zones on final chromatogram. More experience is necessary when sample application is performed by moving tip of pipette or syringe needle over a start line without touching the layer surface [37].

Very good results can be obtained using automatic aerosol applicators what was mentioned above. The starting sample zone can be formed in desired shape. When sample mixture is more complex then starting zone should be formed as very narrow band what leads to higher resolution of bands on the chromatogram. Total volume of the sample applied to the chromatographic plate in one run can be equal to 500 μL (e.g., Linomat 5, Camag) or even more if several repetitions of this procedure is performed.

Convenient mode of sample application in shape of narrow band (1 mm wide) to a plate up to 40 cm wide can be performed using TLC sample streaker from Alltech. In this case volume of sample solution depends on syringe capacity.

Especially large quantities of sample can be applied on the chromatographic plate according to the paper [38]. The sample solution is mixed with specified quantity of bulky adsorbent. The solvent is evaporated and remnant (bulky adsorbent with deposited sample on it) is introduced to the start line of the chromatographic plate. This mode was adapted by Nyiredy and Benkö [39] to extraction and separation of components from plant materials.

Horizontal ES chamber [18,40] or horizontal DS chamber (Chromdes) [17] can be very easily used for band sample application. In the first stage of this procedure the adsorbent layer of the chromatographic plate is fed with sample solution instead of the solvent (the mobile phase). When desired sample volume is introduced then the chromatographic plate is supplied with solvent to proceed chromatographic process. This procedure possess two advantages: no sophisticated equipment is necessary to perform sample application for preparative separation and during sample application frontal chromatography is performed, which leads to preliminary separation of the components of the sample mixture.

7.4   CHROMATOGRAM DEVELOPMENT

As it is was mentioned above chromatogram development in TLC can be performed applying linear or radial modes. Both modes can be performed in simple way using conventional chamber and applying very complicated procedures including sophisticated devices. Involved operations and procedures depend on various variables concerned with properties of sample, adsorbent layer, solvent, mode of detection, and evaluation of chromatogram. Some aspects, especially concerned with mobile phase, will be discussed in this chapter.

7.4.1   MOBILE PHASES APPLIED IN TLC

Mobile phases used for TLC have to fulfill various requirements. It must not chemically affect or dissolve the stationary phase because this leads to modification of properties of the chromatographic system. It must not produce chemical transformations of the separated compounds. The multicomponent mobile phase applied in TLC must be used only once, not repeatedly, because the volatility of solvents produces a continuous modification of quantitative composition of the mobile phase, which negatively affects the chromatographic repeatability. The mobile phase must be easily eliminated from the adsorbent layer and must be compatible with detection methods. The reproducibility can be greatly affected by the conditions and the time of preservation of the mobile phase solution.

Chemical information concerned with mobile phase properties is essential to the initial selection of chromatographic system and detection properties. Choice of the mobile phase (and also the stationary phase) is dependent on many factors concerned with property of the compounds to be separated (Table 7.1) [41].

When properties of the mobile phase and stationary phase of TLC systems are considerably different then separation selectivity is expected to be high. In general, if the stationary phase is polar the mobile phase should be apolar or slightly polar and then such system is named normal-phase system (NP). If stationary phase is nonpolar then the mobile phase should be polar and such chromatographic system is named as reversed-phase one (RP). The choice of the mobile phase is dependent not only on the properties of the adsorbent and its activity but also on the structure and the type of separated analytes. Various solvents can be used as the components of the mobile phase in planar chromatography and their choice to the chromatographic process is based on eluotropic and isoelutropic series. The mobile phase applied in planar chromatography can be composed of one, two, or more solvents.

TABLE 7.1
Hints for Stationary and Mobile Phase Selection with Respect to the Type of Sample Components

	Sample Information		
	1. Polarity of Compound		
	Polarity of		
Compound	Stationary Phase	Mobile Phase	Comments

Normal-phase chromatography (NP) with nonaqueous mobile phases

Compound	Stationary Phase	Mobile Phase	Comments
Low (hydrophobic)	Polar adsorbents (silica or, less often, alumina)	Nonpolar mobile phase (nonaqueous)	It is difficult to separate the compounds of low polarity on silica gel because of their weak retention. Selection of solvents to the mobile phase is limited because most solvents demonstrate too high elution strength (compounds show very high values of R_F or migrate with front of the mobile phase)
High (hydrophilic)			Compounds of high polarity are difficult to separate on silica gel because of strong retention. Selection of solvents to the mobile phase is limited because most solvents are of too lower elution strength for these solutes (compounds have very low values of R_F, stay on the start line of chromatogram)

Normal-phase chromatography (NP) with aqueous mobile phases (pseudo-reversed phase system)

Compound	Stationary Phase	Mobile Phase	Comments
Very high (very hydrophilic)	Polar adsorbents (silica or, less often, alumina)	Polar mobile phase (aqueous)	Compounds of very high polarity (very high hydrophilic, e.g., alkaloids which additionally show strong interactions with silanol groups of silica based stationary phases) are difficult to separate on silica gel with nonpolar mobile phases because of strong retention (stay on the start line of chromatogram even when 100% methanol as the mobile phase is applied) and should be chromatographed with more polar eluents containing water.

Normal-phase chromatography (NP) with nonaqueous mobile phases and reversed-phase chromatography (RP) with aqueous mobile phases

Intermediate polarity	Moderately polar bonded phases, chemically bonded on silica support: cyanopropyl – $(CH_2)_3$-CN, diol – $(CH_2)_3$-O-CH_2-CHOH-CH_2-OH, or aminopropyl – $(CH_2)_3$-NH_2	Nonpolar mobile phase (nonaqueous) and polar mobile phase (aqueous)	Compounds of intermediate polarity are separated on polar chemically bonded stationary phases because their molecules can interact with silanol groups of silica gel. They demonstrate good separation selectivity in both normal-phase (nonaqueous) and reversed-phase (aqueous) systems. The moderately polar stationary phases are compatible with water mobile phase of whole concentration range. Many solvents can be selected to prepare the mobile phase. The only limitation is concerned with miscibility of the mobile phase components. In addition, migration velocity of front of the mobile phase vary less with solvent composition in comparison to typical RP systems.

Reversed-phase chromatography (RP) with aqueous mobile phases

Low (hydrophobic)	Nonpolar adsorbents (chemical modification is based on reactions of the silanol ($\equiv$Si-OH) groups on the silica surface with organosilanes to obtain stationary phases of the type $\equiv$Si-R, where R is aliphatic chain of the type - C_1, C_2, C_8, C_{18})	Polar mobile phase (aqueous)	Compounds of low polarity are difficult to separate in systems with nonpolar adsorbents because of very strong retention. Mobile phase selection is limited because most solvents show too low elution strength for these separations.

Reversed-phase chromatography (RP) with aqueous mobile phases

High (hydrophilic)			Compounds of high polarity are difficult to separate with nonpolar adsorbents because of weak retention. Appropriate mobile phase selection is restricted because most solvents show too strong elution strength for these solutes.

(*continued*)

TABLE 7.1 (continued)
Hints for Stationary and Mobile Phase Selection with Respect to the Type of Sample Components

Sample Information

1. Polarity of Compound

Compound	Polarity of		Comments
	Stationary Phase	Mobile Phase	

Reversed-phase chromatography (RP) with nonaqueous mobile phases

| Low (very hydrophobic) | Nonpolar adsorbents: The stationary phase of the type $\equiv$Si-R (R is aliphatic chain of various length, e.g., $-C_1$, C_2, C_8, C_{18}) is formed after reaction of the silanol ($\equiv$Si-OH) groups of the silica surface with organosilanes. | Polar mobile phase (nonaqueous) | The separation of compounds of low polarity (very hydrophobic samples) is difficult to separate with aqueous mobile phases on nonpolar adsorbents because of very strong retention (stay on the start line of chromatogram even when acetonitrile–water (99:1, v/v) mobile phase is applied. The solutes (e.g., lipids) can be chromatographed using the mobile phase composed of more polar (acetonitrile, methanol) and less polar (tetrahydrofuran, chloroform, methylene chloride, acetone, methyl-t-butyl ether) organic solvents or various mixtures of these solvents. The retention decreases with increasing concentration of the less-polar solvent in the mobile phase (multicomponent eluent may contain even hexane or heptane). |

2. Molecular Mass (MW)

Compound	Solubility	System
MW <1000	Organic soluble	NP systems with silica or with chemically bonded stationary phase (aminopropyl, cyanopropyl, diol)
	Water soluble	*Nonionic*
		(a) RP systems with chemically bonded stationary phase (aminopropyl, cyanopropyl, diol) and alkylsiloxane-bonded stationary phases (C_2, C_6, C_8, C_{18}).
		(b) NP systems on silica and on chemically bonded stationary phases (aminopropyl, cyanopropyl, diol).

Ionic

(a) RP systems on chemically bonded stationary phases (aminopropyl, cyanopropyl, diol) and alkylsiloxane-bonded stationary phases with ligands of the type C_2, C_6, C_8, C_{18}.

(b) IPC (Ion-pair chromatography) in RP systems with chemically bonded stationary phases (aminopropyl, cyanopropyl, diol) and alkylsiloxane-bonded stationary phases with ligands of the type C_2, C_6, C_8, C_{18}.

| MW >1000 | Organic soluble | Precipitation chromatography |
| | Water soluble | Cellulose |

3. pK_a *Value of Compound*

pK_a Value of Compound	Acid–Base Behavior	Stationary Phase	Mobile Phase	Comments
Low values of pK_a High values of pK_a pK_a	Strong acid or weak base Strong base or weak acid	Nonpolar adsorbents: alkylsiloxane-bonded stationary phases (with ligands of the type C_2, C_6, C_8, C_{18}, phenyl) or chemically bonded stationary phases of the type aminopropyl, cyanopropyl, diol).	RP systems: buffered polar mobile phase with controlled pH	When an acidic or basic molecule undergoes ionization (i.e., is converted from an uncharged species into charged one) it becomes much less hydrophobic (more hydrophilic). When pH value of the mobile phase is equal to pK_a of the compounds of interest then values of its ionized and unionized forms are identical (i.e., the values of concentration of B and BH^+ or HA and A^- in the mobile phases are equal). It means that retention changes of these solutes in principle takes place in the pH range from the value $pK_a - 1.5$ to the value $pK_a + 1.5$. The relationship between retention of the solute and mobile phase pH in RP systems is more complicated for compounds with two or more acidic or basic groups.

Source: From Poole, C.F., Dias, N.C., *J. Chromatogr.*, A, 892, 123, 2000; Snyder, L.R., Kirkland, J.J., Glajch J.L., in *Practical HPLC Method Development*, 2nd edn., Wiley, New York, 233–291, 1997.

• Normal-phase planar chromatography

The retention of solutes on inorganic polar adsorbents (silica, alumina) or moderately polar adsorbents (cyanopropyl, diol, or aminopropyl) originates in the interactions of the polar adsorption sites on the surface with polar functional groups of the compounds. This mode was previously called as adsorption or liquid–solid chromatography. Generally, the strength of molecular interactions of the stationary phase with polar molecules of analytes increases in the order: cyanopropyl < diol < aminopropyl ≪ silica ≈ alumina stationary phases. Basic compounds are very strongly retained by silanol groups of silica gel and acidic compounds show increased affinity to aminopropyl stationary phase. Aminopropyl and diol stationary phases show affinity to compounds with proton-acceptor or proton-donor functional groups (e.g., alcohols, esters, ethers, ketones). Other polar compounds are usually more strongly retained on cyanopropyl than aminopropyl chemically modified stationary phases. The alumina surface comprises hydroxyl groups, aluminum cations, and oxide anions and is more complex than silica gel. Alumina favors interactions with π electrons of solute molecules and often yields better separation selectivity than silica for analytes with different number or spacing of double bonds. The stationary phase in NP system is more polar than the mobile phase. The mobile phase in this chromatographic mode is usually a binary (or more component) mixture of organic solvents of different polarity e.g., ethanol + chloroform + heptane. In principle elution strength of solvents applied in NP systems increases according to their polarity, e.g., hexane ≈ heptane ≤ octane < methylene chloride < methyl-t-butyl ether < ethyl acetate < dioxane < acetonitrile ≈ tetrahydrofuran < 1-propanol ≈ 2-propanol < methanol.

The retention of compounds in NP systems generally increases in the order: alkanes < alkenes < aromatic hydrocarbons ≈ chloroalkanes < sulphides < ethers < ketones ≈ aldehydes ≈ esters < alcohols < amides ≪ phenols, amines, and carboxylic acids.

The sample retention is enhanced when the polarity of the stationary phase increases and the polarity of the mobile phase decreases.

• Reversed-phase planar chromatography (RP TLC)

Silica gel chemically modified with various ligands, e.g., C_2, C_8, C_{18} alky chains, or aminopropyl, cyanopropyl, diol, is the most popular stationary phase in RP TLC. The mobile phase used in RP TLC is more polar than adsorbent and usually is composed of two (or more) solvents e.g., water + water-soluble organic solvent (methanol, acetonitrile, tetrahydrofuran, acetone). The organic solvent in the mobile phase solution is often named as modifier. The sample retention increases when its polarity decreases and when polarity of the mobile phase increases. In general the polarity decrease (increase of elution strength) of solvents applied in RP TLC can be presented according to the order: methanol, acetonitrile, dioxane, tetrahydrofuran, 1- and 2-propanol.

Sample containing ionized or ionizable organic analytes are often separated in RP chromatography with buffers as the components of the mobile phase. pH value of the buffer solution should be in the range 2–8 due to lower stability of the stationary phases beside this extent. However, this requirement is often omitted because application of the chromatographic plate in typical experiment is peformed only once. Addition of a buffer to the mobile phase can be applied to suppress the ionization of acidic or basic solutes and to eliminate undesirable chromatographic behavior of ionic species.

Ionic analytes can also be chromatographed in RP systems with additives to the mobile phase. The example is ion-pair chromatography (IPC) performed in RP systems—ionogenic surface-active reagent (containing a strongly acidic or strongly basic group and a hydrophobic moiety in the molecule) is added to the mobile phase. The retention of solutes in IPC systems can be controlled by changing the type or concentration of the ion-pair reagent and of the organic solvent in the mobile phase. Very important parameter of the mobile phase of IPC system is its pH, which should be adjusted to appropriate value. Acid substances can be separated with tetrabutyl-ammonium or cetyl trimethyloammonium salts, whereas basic analytes can be separated by using salts of C_6-C_8-alkanesulphonic acids or their salts in the mobile phase. The retention generally rises with concentration increase of the ion-pair reagent in the mobile phase (higher concentration of this reagent in the mobile phase leads to enhancement of its uptake by the nonpolar stationary phase). However, it should be mentioned that too high concentration of the ion-pair reagent in the mobile phase does not significantly affect the retention. Generally retention of ionogenic solutes also increases with increase of number and size of alkyl substituent in the molecule of ion-pair reagent.

7.4.1.1 Solvent Properties and Classification

The solvents used as components of the mobile phases in TLC should be of appropriate purity and of low viscosity, inexpensive and compatible with the stationary phase and binder being used. The solvent of sample mixture should be of the lowest elution strength as possible (in case of sample application on the chromatographic plate using capillary or microsyringe). Some basic physicochemical parameters (viscosity, dipole moment, refractive index, dielectric constant, etc.) are used for characterization of a solvent's ability to molecular interactions, which are of great importance for chromatographic retention, selectivity, and performance. The physical constants mentioned earlier for some common solvents used in chromatography are collected in few books or articles [1,2,4,42,43].

Solvent strength (eluent strength, elution strength) refers to the ability of the solvent or solvent mixture to elute the solutes from the stationary phase. This strength rises with increase of solvent polarity in NP systems. Reversed order of elution strength takes place for RP systems. Solvent polarity is connected with molecular interactions of solute–solvent, including dispersion (London), dipole–dipole (Keesom), induction (Debye), hydrogen bonding interactions [44].

The first attempts of solvent classifications were performed for characterization of liquid phases applied in gas chromatography (Rohrschneider and McReynolds) [45,46]. Another solvent classification was by Hildebrand [47–49]. In this classification the solubility parameter was derived based on values of cohesion energy of pure solvents.

Snyder's polarity scale has gained very important significance of solvent classification in liquid chromatography practice in which parameter, P', is used for characterization solvent polarity [41]. This parameter was calculated based on distribution constant, K, of test solutes (ethanol, dioxane, nitromethane) in gas–liquid (solvent) systems. Ethanol was chosen for characterization of the solvent with regard to its basic properties (proton-acceptor properties), dioxane—to characterize its acidic properties (proton-donor properties), and nitromethane—to describe dipolar properties of the solvent. The sum of log K values of these three test compounds is equal to parameter, P' of the solvent. In addition each value of log K of the test solutes was divided by parameter, P' then relative values of three types of polar interaction were calculated for each solvent: x_d for dioxane (acidic), x_e for ethanol (basic), and x_n for nitromethane (dipolar). These x_i values were corrected for nonpolar (dispersive) interactions and were demonstrated in three coordinate plot, on equilateral triangle, Figure 7.11. Snyder characterized more than 80 solvents, and obtained eight groups of solvents on the triangle [50,51]. The triangle was named as Snyder–Rohrschneider solvent selectivity triangle (SST). This classification of solvent is useful for selectivity optimization in liquid chromatography. Solvents belonging to various groups should demonstrate different separation selectivity. Especially the solvents located close to various triangle corners should demonstrate the most different separation selectivity. Another advantageous feature of the SST is that the number of solvents applied in optimization procedure can be reduced to the

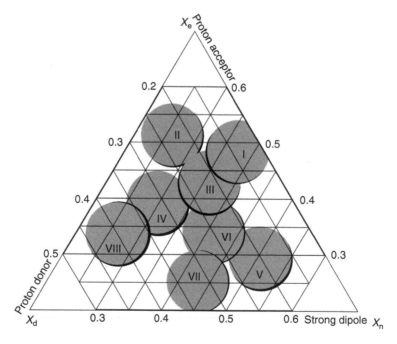

FIGURE 7.11 The solvent selectivity triangle (SST). (Adapted from Snyder, L.R., *J. Chromatogr. Sci.*, 16, 223, 1978.)

members representing each group from the SST. This approach was tested with success for normal [52] and RP systems [53]. However, for some chromatographic systems the prediction of selectivity changes failed [54].

Some empirical scales of solvent polarity based on kinetic or spectroscopic measurements have been described [55] to present their ability to molecular interactions.

There are several solvatochromic classification of solvents, which is based on spectroscopic measurement of their different solvatochromic parameters [55–60]. The ET(30) scale [55] is based on the charge-transfer absorption of 2,6-diphenyl-4-(2,4,6-triphenyl-N-pyridino) phenolate molecule (know as Dimroth and Reichardt's betaine scale). The Z scale [56,57] is based on the charge-transfer absorption of N-ethyl-4-methocycarbonyl) pyridinium iodine molecule (developed by Kosower and Mohammad). The scale based on Kamlet–Taft solvatochromic parameters has gained growing popularity in the literature and laboratory practice [58–61]. Following parameters can be distinguished in this scale: dipolarity/polarizability (π^*), hydrogen-bond acidity (α) and basicity (β), see Table 7.2. The solvatochromic parameters are average values for a number of selected solutes and somewhat independent of solute identity. Some representative values for solvatochromic parameters of common solvents used in TLC are summarized in Table 7.2.

These parameters were normalized in similar way as x_d, x_e, x_n parameters of Snyder. The values of α, β, and π^* for each solvent were summed up and divided by the resulted sum. Then fractional parameters were obtained (fractional interaction coefficients): α/Σ (acidity), β/Σ (basicity), and π^*/Σ (dipolarity). These values were plotted on triangle similarly as in Snyder–Rohrschneider SST. In Figure 7.11 SST based on normalized solvatochromic parameters is plotted for some common solvents applied in liquid chromatography [49].

More comprehensive representation of parameters characterizing solvent properties can be expressed based on Abraham's model in which following equation is used [62–65]:

$$\log K_L = c + l \log L^{16} + rR_2 + s\pi_2^H + a\sum \alpha_2^H + b\sum \beta_2^H \qquad (7.1)$$

where
  $\log K_L$ is the gas–liquid distribution constant
  $\log L^{16}$ is the distribution constant for the solute between a gas and n-hexadecane at 298 K
  R_2 is excess molar refraction (in $cm^3/10$)
  π_2^H is the ability of the solute to stabilize a neighboring dipole by virtue of its capacity for orientation and induction interactions
  $\sum \alpha_2^H$ is effective hydrogen-bond acidity of the solute
  $\sum \beta_2^H$ is hydrogen-bond basicity of the solute

All these parameters with exception of $\log K_L$ are the solute descriptors. As it can be seen the parameters s, a, b represent polar interactions of solvent molecule with solute one as dipole–dipole, hydrogen-bond basicity and hydrogen-bond acidity,

TABLE 7.2
Parameters Applied for Characterization of Solvents for Liquid Chromatography

Solvent	Selectivity Group	Snyder's Classification Based on Selectivity Triangle[a]					Kamlet–Taft and Coworkers Classification					Abraham's Model Classification					
		Solvent Strength		Solvent Selectivity			$E_T(30)$	Solvatochromic Parameters				System Constant for Distribution Between Gas Phase and Solvent (Abraham's Model)					
		(P')	$(S_{S\,RP})$	x_e	x_d	x_n		E_T^N	π_1^*	α_1	β_1	r	s	a	b	l	c
n-Butyl ether	I	2.1		0.44	0.18	0.38	33.0	0.071	0.27	0	0.46						
Diisopropyl ether		2.4		0.48	0.14	0.38	34.1	0.105	0.27	0	0.49						
Methyl tert.-butyl ether		2.7					34.7	0.124									
Diethyl ether		2.8		0.53	0.13	0.34	34.6	0.117	0.27	0	0.47						
n-Butanol	II	3.9		0.59	0.19	0.25	49.7	0.586	0.47	0.79	0.88						
2-Propanol		3.9		0.55	0.19	0.27	48.4	0.546	0.48	0.76	0.95						
1-Propanol		4.0		0.54	0.19	0.27	50.7	0.617	0.52	0.78							
Ethanol		4.3	3.6	0.52	0.19	0.29	51.9	0.654	0.54	0.83	0.77	−0.21	0.79	3.63	1.31	0.85	0.01
Methanol		5.1	3.0	0.48	0.22	0.31	55.4	0.762	0.60	0.93	0.62	−0.22	1.17	3.70	1.43	0.77	0
Tetrahydrofuran	III	4.0	4.4	0.38	0.20	0.42	37.4	0.207	0.58	0	0.55						
Pyridine		5.3		0.41	0.22	0.36	40.5	0.302	0.87	0	0.64						
Methoxyethanol		5.5		0.38	0.24	0.38											
Dimethylformamide		6.4		0.39	0.21	0.40	43.2	0.386	0.88	0	0.69						
Acetic acid	IV	6.0		0.39	0.31	0.30	51.7	0.648	0.64	1.12							
Formamide		9.6		0.38	0.33	0.30	55.8	0.775	0.97	0.71							
Dichloromethane	V	4.3		0.27	0.33	0.40	40.7	0.309	0.82	0.30	0						
1,1-Dichloroethane		4.5		0.30	0.21	0.49	41.3	0.327	0.81	0	0						
Ethyl acetate	VI	4.4		0.34	0.23	0.43	38.1	0.228	0.55	0	0.45						

Methyl ethyl ketone		4.7		0.35	0.22	0.43			0.67	0.06	0.48						
Dioxane		4.8	3.5	0.36	0.24	0.40	36	0.164	0.55	0	0.37						
Acetone		5.1	3.4	0.35	0.23	0.42	42.2	0.355	0.71	0.08	0.48						
Acetonitrile		5.8	3.1	0.31	0.27	0.42	45.6	0.460	0.75	0.19	0.31	−0.22	2.19	2.38	0.41	0.73	0
Toluene	VII	2.4		0.25	0.28	0.47	33.9	0.099	0.54	0	0.11	−0.22	0.94	0.47	0.10	1.01	0.12
Benzene		2.7		0.23	0.32	0.45	34.3	0.111	0.59	0	0.10	−0.31	1.05	0.47	0.17	1.02	0.11
Nitrobenzene		4.4		0.26	0.30	0.44	41.2	0.324	1.01	0	0.39						
Nitromethane		6.0		0.28	0.31	0.40	46.3	0.481									
Chloroform	VIII	4.3		0.31	0.35	0.34	39.1	0.259	0.58	0.44	0	−0.60	1.26	0.28	1.37	0.98	0.17
Dodecafluoroheptanol		8.8		0.33	0.40	0.27											
Water		10.2	0	0.37	0.37	0.25	63.01	1.000	1.09	1.17	0.18	0.82	2.74	3.90	4.80	−2.13	−1.27

Source: From Poole, C.F., Dias, N.C., J. *Chromatogr., A*, 892, 123, 2000; Snyder, L.R., Kirkland, J.J., Glajch J.L., in *Practical HPLC Method Development*, 2nd edn., Wiley, New York, 1997, 233–291; Snyder, L.R., J. *Chromatogr.*, 92, 223, 1974; Snyder, L.R., J. *Chromatogr. Sci.*, 16, 223, 1978; Snyder, L.R., Glajch, J.L., Kirkland, J.J., J. *Chromatogr.*, 218, 299, 1981; Glajch, J.L., Kirkland, J.J., Squire, K.M., Minor, J.M., J. *Chromatogr.*, 199, 57, 1980; Kowalska, T., Klama, B., J. *Planar Chromatogr.*, 10, 353, 1997; Johnson, B.P., Khaledi, M.G., Dorsey, J.G., *Anal. Chem.*, 58, 2354, 1986; Kosower, E.M., J. *Am. Chem. Soc.*, 80, 3253, 1958; Kosower, E.M., Mohammad, M., J. *Am. Chem. Soc.*, 90, 3271, 1968; Kamlet, M.J., Abboud, J.L.M., Abraham, M.H., Taft, R.W., J. *Org. Chem.*, 48, 2877, 1983; Kamlet, M.J., Abboud, J.L.M., Taft, R.W., J. *Am. Chem. Soc.*, 99, 6027, 1977; Laurence, C., Nicolet, P., Dalati, M.T., Abboud, J.L.M., Notario, R., J. *Phys. Chem.*, 98, 5807, 1994; Taft, R.W., Kamlet, M.J., J. *Am. Chem. Soc.*, 98, 2886, 1976; Abraham, M.H., Poole, C.F., Poole, S.K., J. *Chromatogr. A*, 842, 79, 1999; Abraham, M.H., Whiting, G.S., Shuely, W.J., Doherty, R.M., *Can. J. Chem.*, 76, 703, 1998; Abraham, M.H., Whiting, G.S., Carr, P.W., Quyang, H., J. *Chem. Soc., Perkin Trans. 2*, 1385, 1998; Abraham, M.H., Platts, J.A., Hersey, A., Leo, A.J., Taft, R.W., J. *Pharm. Sci.*, 88, 670, 1999; Dallenbach-Tölke, K., Nyiredy, Sz., Meier, B., Sticher, O., J. *Chromatogr.*, 365, 63, 1986; Nyiredy, Sz., Dallenbach-Tölke, K., Sticher, O., J. *Planar Chromatogr.*, 1, 336, 1998; Nyiredy, Sz., Dallenbach-Tölke, K., Sticher, O., J. *Liq. Chromatogr.*, 12, 95, 1989; Nyiredy, Sz., Fatér, Zs., J. *Planar Chromatogr.*, 8, 341, 1995; Nyiredy, Sz., *Chromatographia*, 51, S288, 2000; Reich, E., George, T., J. *Planar Chromatogr.*, 10, 273, 1997.

[a] $S_{S\ RP}$ is an empirical solvent strength parameter used in reversed-phase (RP) system.

respectively. While the parameter r represents ability of solvent molecule to interact with n- or π-electrons of solute molecule. In addition to previous classifications of solvents this model takes into account molecular interactions concerned with cavity formation in solvent for solute molecule and dispersion interactions between solvent and solute. These effects are presented by constants c and l.

The values of discussed parameters are given in Table 7.2. The chromatographer can compare these data and others in this table what can be helpful for optimization of retention and separation selectivity.

One of the first attempts of the solvent systematization with regard to their elution properties was formulated by Trappe as the eluotropic series [66]. The pure solvents were ordered according to their chromatographic elution strength for various polar adsorbents in terms of the solvent strength parameter ε^0 defined according to Snyder [67,68] and expressed by the following equation:

$$\varepsilon^0 = \Delta G_S^0 / 2.3 \, RTA_S \tag{7.2}$$

where

ΔG_S^0 is the adsorption free energy of solute molecules
R is the universal gas constant
T is the absolute temperature
A_S is the area occupied by the solvent molecule on the adsorbent surface

The parameter ε^0 represents adsorption energy of the solvent per unit area on the standard activity surface. Solvent strength is the sum of many types of intermolecular interactions.

Neher [69] proposed an equieluotropic series, which give possibility to replace one solvent mixture by another one: composition scales (approximately logarithmic) for solvent pairs are subordinated to give constant elution strengths for vertical scales. Equieluotropic series of mixtures are approximately characterized by constant retention but these can show often different selectivity. The scales, devised originally for planar chromatography on alumina layers, were later adapted to silica by Saunders [70], who determined accurate retention data by HPLC and subordinated the composition scale to Snyder's elution strength parameter [71].

Snyder [72] proposed calculation of the elution strength of multicomponent mixtures. The solvent strength ε_{AB} of the binary solvent mobile phase is given by the relationship

$$\varepsilon_{AB} = \varepsilon_A^0 + \frac{\log \left(X_B 10^{a \, n_b (\varepsilon_B^0 - \varepsilon_A^0)} + 1 - X_B \right)}{a \, n_b} \tag{7.3}$$

where

ε_A^0 and ε_B^0 are the solvent strength of two pure solvents A and B, respectively
X_B is the mole fraction of the stronger solvent B in the mixture
α is the adsorbent activity parameter
n_b is the adsorbent surface area occupied by a molecule of the solvent B

The solvent strength for a ternary mixture was also derived [72]. These equations were tested for a series of mobile phases on alumina [72–74] and silica [75], demonstrating good agreement with experimental data especially for the last adsorbent. Some discrepancies were observed for alumina when different classes of solutes were investigated [54].

7.4.1.2 Optimization of the Mobile Phase Composition

Identification and quantitation of analytes are objective tasks of each analysis. Reliable results of this analysis can be obtained with TLC mode when resolution, R_S, of sample components is satisfactory, at least greater than 1.0. The resolution can be expressed according to the equation

$$R_S = 0.25 \underbrace{\left(\frac{K_2}{K_1} - 1\right)}_{i} \cdot \underbrace{\sqrt{K_f N}}_{ii} \cdot \underbrace{\left(1 - \overline{R_F}\right)}_{iii} \tag{7.4}$$

where
  N is plate number of the chromatographic system
  K is distribution constant of the solute 1 or 2

Distribution constant is related to retention factor, k, according to the following equation:

$$K = k \, V_s / V_m \tag{7.5}$$

where V_s/V_m is the ratio of the stationary and the mobile phase volumes. Relationships between k and R_F (retardation factor) is as follows:

$$k = \frac{1 - R_F}{R_F} \tag{7.6}$$

As it is seen the resolution in TLC can be optimized by adjusting three earlier mentioned variables: (i) selectivity, (ii) performance, and (iii) retention. If distribution constant of two solutes is the same then separation is impossible. The resolution increases when plate number is higher. In planar chromatography the performance of the chromatographic system is dependent on R_F—higher R_F leads to higher performance. On the other hand retention increase (decrease of R_F) is responsible for resolution increase. It means that both variables, performance, and retention, should be characterized by optimal value of R_F for which resolution reaches maximum value. This value is close to 0.3, compare Figure 7.12 where resolution is plotted vs. retardation factor. Typical mixtures are more complicated (multiple component) and it is not possible to separate all components with R_F values close to 0.3. The optimal R_F range of separated solutes in the chromatogram practically is 0.2–0.8, or thereabouts and if of the correct selectivity, will distribute the sample components evenly throughout this R_F range [71].

Typical selection of the solvent is based on eluotropic series for most popular silica and less often used alumina or ε^0 parameter. Simple choice of mobile phase is

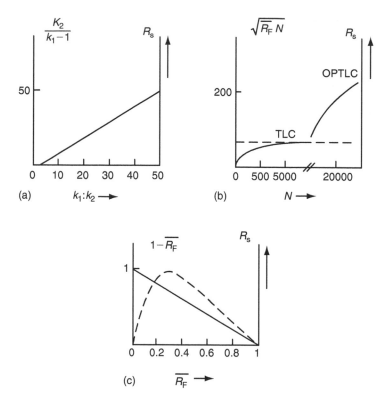

FIGURE 7.12 The influence of (a) selectivity, (b) performance, and (c) retention on resolution. (Adapted from Snyder, L.R., *Principles of Adsorption Chromatography*, Marcel Dekker, New York, 1968.)

possible by microcircular technique on the basis of eluotropic series or for binary or more component mobile phases [76,77]. In the microcircular technique, after spotting the sample mixture in few places on the chromatographic plate the selected solvents are applied in the center of each spot by means of a capillary. Then the sample bands migrate radially, and different chromatographic behavior of spotted mixture can be observed: nonsuitable solvent when spotted mixture give a clench spot (too weak solvent strength) or periphery fringe (too strong solvent strength), and suitable solvent when spotted mixture give the zones which are spread over entire surface of circular development.

A single solvent rarely provides suitable separation selectivity and retention in chromatographic systems. Typical solution of the mobile phase is selected by adjusting an appropriate qualitative and quantitative composition of a two- (binary) or more component mixture. The dependence of retention on the composition of the mobile phase can be predicted using the few most popular approaches reported in the literature and used in laboratory practice.

The semiempirical model of adsorption chromatography (for NP systems) was independently created and published same time ago by Snyder [71] and Soczewiński [78].

This approach has been called as Snyder–Soczewiński model [79,80]. With some simplification, both authors' models lead to identical equations describing the retention as a function of the concentration of the more polar modifier in binary mobile phase comprising less polar diluent (e.g., in NP system of the type: silica-polar modifier (ethyl acetate) + nonpolar diluent (*n*-heptane)).

$$R_M = \log k = \text{const} - m \log C_{mod} \qquad (7.7)$$

where

C_{mod} is mole fraction (or volume fraction) of the polar component (modifier) in the mobile phase

m is constant

k is retention factor

$R_M = \log ((1-R_F)/R_F)$

The value of *m* is interpreted as number of solvent molecules displaced by the solute molecule from the adsorbent surface (or the ratio of the area occupied by solute molecule and by the solvent one).

The typical experimental relationships between R_M and eluent concentration expressed as logarithmic scale are straight lines and usually not parallel. Distance between lines and their slopes give information about variations of selectivity. The slope is dependent on eluent strength and number of polar groups in the solute molecule. For some examples the lines cross (changes in spot sequence on chromatogram). Moreover, for some diluent-modifier pairs, the vertical distances, ΔR_M, between the lines are differentiated showing individual selectivities of the systems relative to various pairs of solutes.

For the RP systems an analogous semilogarithmic equation was reported by Snyder [81] and is presented below:

$$R_M = \log k = \log k_w - S\varphi_{mod} \qquad (7.8)$$

where

$\log k_w$ is the retention factor of the solute for pure water as the mobile phase

φ_{mod} is volume fraction of the modifier (e.g., methanol)

Similar equation was reported for partition systems of paper chromatography by Soczewiński and Wachtmeister much earlier [82]. For $\varphi_{mod} = 1$ (pure modifier), $S = \log k_w - \log k_{mod}$, $S = \log (k_w/k_{mod})$—the logarithm of hypothetical partition coefficient of solute between water and modifier (actually miscible) [79]. The constant *S* increases with decreasing polarity of the organic solvent and is a measure of its elution strength. On the other hand, *S* rises with increase in size in the solute molecule. The earlier mentioned equation can be used for prediction of retention and selectivity for reasonable concentration range. However, for broad concentration range this equation does not predict solute retention with good precision. In cases of broad concentration range of the mobile phase the following equation was reported [83]:

$$\log k = \log k_w + a\varphi + b\varphi^2 \qquad (7.9)$$

where a and b are constants, which are dependent on solute and the mobile phase type. Deviations from this equation occur especially beyond concentration range $0.1 < \varphi < 0.9$, i.e., for high and low concentration of water. These deviations are explained by several reasons. Conformational changes of alkyl chain structure of the stationary phase at high water concentration in the mobile phase can influence this effect. When concentration of water is low then its participation in hydrophobic mechanism of retention is eliminated and additionally molecular interactions of the solute and unreacted silanols can occur.

Important significance of the relationships between retention and composition of the mobile phase for prediction of separation of sample components inspired many authors more deeply to investigate the problem of prediction of retention. One example is finding concerned with dependence of log k vs. $E_{T(30)}$ solvatochromic parameter [84]. This relationships shows very good linearity. Another approach is based on methodology concerned with linear salvation energy relationships. In this mode, solvatochromaic parameters described earlier were applied for formulation of equations which were used for retention prediction in various chromatographic systems. Important advantage of this mode is that sample descriptors were determined from other experiments as chromatographic ones. Disadvantage of the mode is that system constants applied in the equation should be individually determined for each chromatographic system including various qualitative and quantitative composition of the mobile phase.

The dependence of retention on the composition of the mobile phase can be described using different theoretical models too:

- Martin–Synge model of partition chromatography [85,86]
- Scott–Kucera model of adsorption chromatography [87,88]
- Kowalska model of adsorption and partition chromatography [89–91]
- Ościk thermodynamic model [92,93]

It is purposeful to discuss in this moment more in detail about the modes of retention and selectivity optimization, which can be applied to obtain appropriate chromatographic resolution. Various strategies were described in the scientific literature e.g., overlapping resolution mapping scheme (ORM) [53,93–95], window diagram method [96–98], computer-assisted method [99–104], and chemometric methods [105–110]. However, it seems that the strategy of separation optimization based on classification of solvents by Snyder (or solvatochromic parameters) and PRISMA method described by Nyiredy [111–114] is the most suitable in laboratory practice for planar chromatography separations of sample mixtures of phytochemistry origin. This opinion is expressed taking into account simplicity of this procedure and low costs of operations involved (no sophisticated equipment and expensive software are necessary).

As it was mentioned earlier solvents from each group of SST show different selectivity, which can lead to change in separation order. When the average solvent strengths and selectivity values are calculated for each solvent groups of SST, then linear correlations of these quantities were found for solvent groups I, II, III, IV, VIII and for solvent groups I, V and VII [115]. Solvents of group VI do not belong to either correlations due to quite different ability to molecular interactions in comparison with solvents of remaining groups. It was mentioned earlier that the solvents belonging to the groups in corners of SST (groups I, VII, VIII) and from its middle part (group VI) are the most often applied in normal phase systems of planar chromatography. Nyiredy et al. [112] suggested the selection and testing of ten solvents with various strengths from eight selectivity groups of SST (diethyl ether (I); 2-propanol and ethanol (II); tetrahydrofuran (III); acetic acid (IV); dichloromethane (V); ethyl acetate, dioxane (VI); toluene (VII); chloroform (VIII)). All these solvents are miscible with hexane (or heptane) of which solvent strength is about 0. Experiments were performed in unsaturated chambers.

For separation of nonpolar compounds the solvent strength of the mobile phase can be controlled by change of hexane (heptane) concentration. The separation of polar compounds can be regulated (optimized) by adding polar solvent to the mobile phase (e.g., low concentration of water). Thereby, the R_F values of the sample compounds should be brought within the range 0.2–0.8. The next step of mobile phase optimization system is to construct a tripartite PRISMA model, which is used for correlation of the solvent strength (S_T) and selectivity of the mobile phase (Figure 7.13). The upper portion of the frustum serves to optimize polar compounds, the center part does so for nonpolar compounds, while the lower part symbolizes the modifiers. It enables to chose the number of the mobile phase components in the range from two to five. The optimization process is detailed in the literature [111–114]. The PRISMA method represents an useful approach for the optimization of mobile phase, especially in cases of complex samples from plants containing a great number of unknown components [116]. The "PRISMA" model works well also for RP systems [117]. Mentioned procedure was used for selection of the mobile phases to separate synthetic red pigments in cosmetics and medicines [118], cyanobacterial hepatotoxins [119], drugs [120], and pesticides [121].

7.4.2 CLASSIFICATION OF THE MODES OF CHROMATOGRAM DEVELOPMENT

As it was mentioned earlier a chromatogram development can be performed applying linear or radial modes. In this section various methodological possibilities of these modes will be discussed with special attention to linear development.

7.4.2.1 Linear Development

7.4.2.1.1 Isocratic Linear Development
Isocratic linear development is the most popular mode of chromatogram development in analytical and preparative planar chromatography, also in phytochemistry analysis. It can be easily performed in conventional chambers and horizontal chambers of all types. The mobile phase in reservoir is brought to the contact with adsorbent layer and then the movement of eluent front takes place. The chromatogram development is stopped when mobile phase front reaches desired position. In the isocratic mode of chromatogram development plates of different sizes are applied

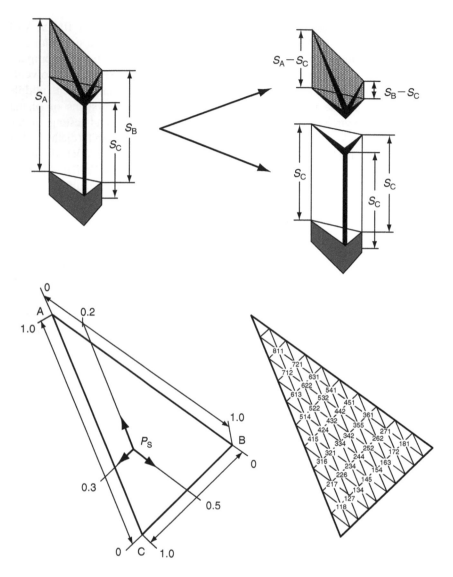

FIGURE 7.13 The "Prisma Model." (From Siouffi, A.M., Abbou, M., Optimization of the mobile phase, in *Planar Chromatography, a Retrospective View for the Third Millennium*, Nyiredy, Sz. (Ed.), Springer Scientific Publisher, Budapest, 2001. With permission.)

for phytochemistry separations (usually 5 × 5 cm, 10 × 10 cm, and 10 × 20 cm and this makes the migration distance equal to about 4, 9, or 18 cm, respectively). Eluent can be supplied to the chromatographic plate simultaneously from its opposite edges (in Horizontal Developing Chamber from Camag or Horizontal DS Chamber from Chromdes) so that the number of separated samples can be doubled in comparison with development in vertical chamber or to development in horizontal chamber when

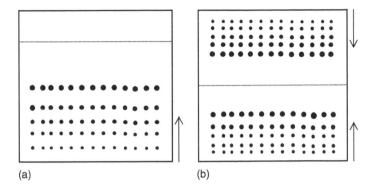

(a)   (b)

FIGURE 7.14 The number of the separated samples can be doubled when development is performed (b) from two opposite sides of the plate in comparison to development (a) from one side. (Adapted from Dzido, T.H., *Planar Chromatography, a Retrospective View for the Third Millennium*, Nyiredy, Sz. (Ed.), Springer Scientific Publisher, Budapest, 2001.)

performed from one edge of the plate. An example of this type of linear development is illustrated in Figure 7.14.

7.4.2.1.2 Continuous Isocratic Development

In the conventional mode of chromatogram development the chromatographic plate is placed in the developing chamber. The development is finished when eluent front reaches the end of the chromatographic plate or desired position on the pate. However, the development can proceed further if some part of the plate extends out the chamber, allowing the mobile phase to evaporate and ensuring that solvent migration is continuous and development is performed over the entire length of the plate with evaporation proceeded with constant efficiency. To enhance the efficiency of evaporation a blower or heating block can be applied to the exposed part of the chromatographic plate. To ensure continuous development the mobile phase can be evaporated at the end of the glass cover plate by use of nitrogen stream also [122]. In Figure 7.15a the cross section of the DS chamber is presented during continuous development, also compare Figure 7.5. Under these conditions the planar chromatogram development is more similar to the column chromatography mode than to the conventional development. In case of not complete separation of the components of lower R_F values some increase of separation can be obtained when applying this mode. In Figure 7.15b and c this procedure is schematically demonstrated. As it is presented the chromatogram development has proceeded to the end (front of the mobile phase reached the end of the chromatographic plate) and mixture components of higher R_F value are well separated as opposed to these of lower values, Figure 7.15b. In this situation the continuous development should be performed. The end part of the chromatographic plate, which comprise the bands of good resolution, needs to be exposed as it is demonstrated in Figure 7.15b. The components of lower R_F values can migrate through a longer distance, which usually leads to improvements of their separation, Figure 7.15c. If necessary a larger part of the chromatographic plate can be exposed in the next stage of continuous development to obtain

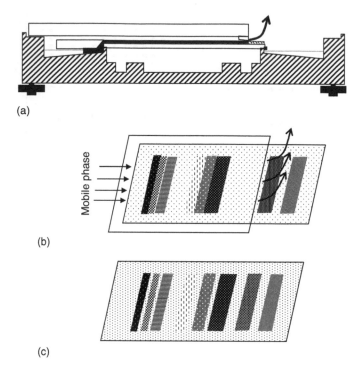

(a)

Mobile phase

(b)

(c)

FIGURE 7.15 Schematic demonstration of horizontal DS chamber applied for continuous development: (a) cross section of DS chamber during continuous development, (b) part of the plate with bands of lower retention exposed but with bands of higher retention covered to enable further development, (c) final chromatogram. (Adapted from Dzido, T.H., Polak, B., in *Preparative Layer Chromatography*, Kowalska, T., Sherma, J. (Eds.), CRC Press, Taylor & Francis, Boca Raton, FL, 2006.)

an improvement of separation of components of even higher retention than those located on the exposed part of the chromatographic plate.

7.4.2.1.3   Short Bed-Continuous Development
The migration distance varies with time according to the equation

$$Z_t = \kappa t^{1/2} \tag{7.10}$$

where Z_t, κ, and t are the distance of the solvent front traveled, constant, and migration time, respectively. The development of planar chromatograms on long distance (e.g., 18 cm) usually takes a lot of time. The development of planar chromatograms is more and more longer with gradual decrease of mobile phase velocity, which takes place in planar chromatography process. Therefore, initially high flow of the mobile phase was used to accelerate the chromatographic analysis in SB/CD. In the SB/CD this path is very short, typically equal to several centimeters [18,123–125]. The eluent strength should then be much weaker than in the

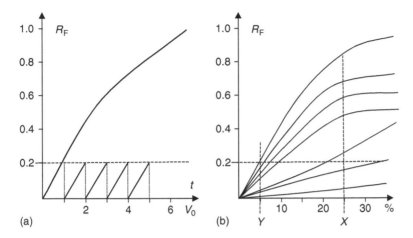

FIGURE 7.16 (a) Principle of SB/CD, elution with five interstitial volumes on 4 cm distance (5 × 4 cm) is faster than single development on 20 cm distance, (b) R_F values of sample components plotted as a function of modifier concentration. Optimal concentration (Y) for SB/CD (5 × 4 cm) is lower than for development on full distance of 20 cm (X). (Adapted from Soczewiński, E., *Chromatographic Methods Planar Chromatography*, Vol. 1, Kaiser, R.E. (Ed.), Dr. Alfred Huetig Verlag, Heidelberg, Basel, New York, 1986.)

conventional development, because several void volumes of eluent migrate through the layer. This is the reason why this mode is preferentially applied for analytical separations. The development of chromatogram on a short distance with simultaneous evaporation of the mobile phase from the exposed part of the chromatographic plate can be very conveniently performed by means of horizontal chambers. The SB/CD mode was introduced by Perry [123] and further popularized by Soczewiński et al. [18,124] using a horizontal equilibrium sandwich chamber.

The principle of the SB/CD technique is demonstrated in Figure 7.16. Instead of chromatogram development over a distance of 18 cm (Figure 7.16a), continuous elution over a short distance, e.g., 5 cm, with simultaneous evaporation of the mobile phase from the exposed part of the chromatographic plate (Figure 7.16b) can be performed. Several void volumes pass throughout the short chromatographic plate bed. However, the flow rate of the eluent depends on the efficiency of solvent evaporation and the flow rate of the mobile phase can be higher if a more volatile solvent is used. It is often necessary to increase the efficiency of evaporation of the mobile phase from the end of the short plate by the application of a heater or blower.

In SB/CD mode, a better resolution relative to the conventional development can be obtained for similar migration distance of solutes. It is well known that the best resolution of the mixture components can be obtained in conventional development if average R_F value is equal to 0.3. However, in the continuous development the applied mobile phase is of lower eluent strength, e.g., eluent strength, which enables to reach the average value $R_F = 0.05$. Under such conditions, several void volumes of the mobile phase should pass through the chromatographic system. If the average

migration distance of component mixture is similar to that of conventional development, then the resolution obtained with continuous development is better. This effect
is explained by higher selectivity of the chromatographic system with mobile phase
of lower eluent strength and by better kinetic properties of the chromatographic
system. At lower eluent strength molecules of the component mixture spend more
time in the stationary phase and flow rate of the mobile phase is higher (closer to
optimal value) under the condition determined by the efficiency of solvent evaporation from the exposed part of the plate.

The SB/CD is especially used in a marked increase of detection sensitivity of
solutes, e.g., to the analysis of trace polyaromatic hydrocarbons in river water
samples. The SB/CD technique can be used to preconcentrate the sample solution
directly on the thin layer. The results of experiments are similar to that when
precoated plates with a narrow weakly adsorbing zone are used. In the first step
the dilute samples are spotted along the layer in series 2–3 cm long. In the second
step the solutes are then eluted with a volatile solvent under a narrow cover plate
forming sharp starting zones and if necessary, evaporation of the eluent can be
accelerated by a stream of nitrogen. Next, the cover plate is removed to completely
evaporate the solvent. After drying the chamber is covered and the plate with starting
zones is developed with suitable eluent. The resolution obtained by the SB/CD mode
is better than by continuous mode and the development time is also shorter.
Additionally, the spot diameter is very small, which leads to better detection level.

7.4.2.1.4 2D Separations

One of the most attractive features of planar chromatography is the ability to operate
in the 2D mode. 2D-TLC is performed by spotting the sample in the corner of a
square chromatographic plate and developing with the solvent in the first direction
with the first eluent. After the development is completed the chromatographic plate is
then removed from the developing chamber and the solvent is allowed to evaporate
from the layer. The plate is rotated through 90° and then developed with the second
solvent in the second direction, which then is perpendicular to the direction of the
first development. In 2D-TLC the layer is usually of continuous composition, but
two different mobile phases must be applied to obtain a better separation of the
components. If these two solvent systems are of approximately the same strength but
of optimally different selectivity, then spots will be distributed over the entire plate
area and in the ideal case the spot capacity of the 2D system will be the product of the
spot capacity of the two constituent 1D systems. If the two constituent solvent
systems are of the same selectivity but of different strength, spot will lie along a
straight line; if both strength and selectivity are identical, spots will lie along the
diagonal.

Computer-aided techniques enable identification and selection of the optimum
mobile phases for separation of different groups of compounds. The first report on
this approach was by Guiochon and coworkers, who evaluated 10 solvents of fixed
composition in 2D separation of 19 dinitrophenyl amino acids chromatographed on
polyamide layers [126]. The authors introduced two equations for calculating the
separation quality—the sum of the squared distances between all the spots, D_A,
and the inverse of the sum of the squared distances between all the spots, D_B.

Streinbrunner et al. [127] proposed other functions for identification of the most appropriate mobile phases—the distance function DF and the inverse distance function IDF, which are the same form as D_A and D_B, respectively, but these use distances rather than the squares of distances. The planar response function *PRF* has been used as optimization criterion by Nurok et al. [128]. Strategies for optimizing the mobile phase in planar chromatography (including 2D separation) [129] and overpressured layer chromatography (including 2D overpressured layer chromatography) [130] has also been described. Another powerful tool is the use of graphical correlation plots of retention data for two chromatographic systems, which differ with regard to modifiers or adsorbents [131].

The largest differences were obtained by the combination of NP systems and revesed-phase system with the same chromatographic layer, e.g., cyanopropyl [132,133]. Nyiredy [2,134] described the technique of joining two different adsorbent layers to form a single plate. Also the largest differences were obtained by a combination of NP systems of the type silica/nonaqueous eluent and revesed-phase system of the type octadecyl silica/water + organic modifier (methanol, acetonitrile, dioxane) on multiphase plates with a narrow zone of SiO_2 and a wide zone of RP-18 (or vice versa) which were commercially available from Whatman (Multi K SC5 or CS5 plates) [135–138].

In 2D development, the mixtures can be simultaneously spotted at each corner of the chromatographic plate so that the number of separated samples can be higher in comparison with "classical 2D development" [8]. An example of this type of 2D development is illustrated in Figure 7.17a–e. Figure 7.17e shows videoscan of the plate which shows separation of three fractions of the mixture of nine pesticides by 2D planar chromatography with NP/RP systems on chemically bonded-cyanopropyl stationary phase.

The multidimensional separation can be performed using different mobile phases in systems with single-layer or bi-layer plates. Graft-thin layer chromatography is a multiple system in which chromatographic plates with similar or different stationary phases are used. Compounds from the first chromatographic plate after chromatogram development can be transferred to the second plate, without the scraping, extraction, or re-spotting the bands by use of a strong mobile phase [2]. Graft-thin layer chromatography, a novel multiplate system with layers of the same or different adsorbents for isolation of the components of natural and synthetic mixtures on preparative scale, was first described by Pandey et al. [139]. Separation of alkaloids by graft-thin layer chromatography on different, connected, adsorbent layers (diol and octadecyl silica) has also been reported [140]. Moreover, Graft-thin layer chromatographic separation (2D planar chromatography on connected layers) of mixture of phenolic acids [141], saponins [142], and three mixtures of pesticides has also been described [143]. The example of this technique is demonstrated in Figure 7.18 [144].

Horizontal chambers can be easily used for 2D separations. The only problem seems to be the size of the sample. In a conventional 2D separation used for analytical purposes the size of the sample is small. The quantity of the sample can be considerably increased when using a spray-on technique with an automatic applicator. Soczewiński and Wawrzynowicz have proposed a simple mode to enhance the size of the sample mixture with the ES horizontal chamber [40].

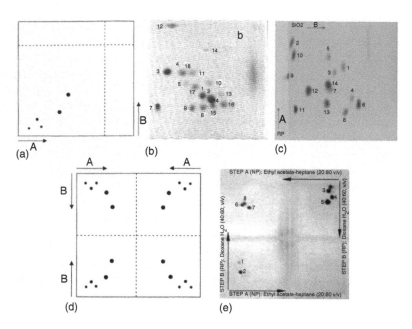

FIGURE 7.17 2D development, (a) schematic presentation of 2D-chromatogram. (Adapted from Dzido, T.H., *Planar Chromatography, a Retrospective View for the Third Millennium*, Nyiredy, Sz. (Ed.), Springer Scientific Publisher, Budapest, 2001.) (b) 2D-chromatogram of 18-component pesticide mixture presented as videoscan of the HPTLC plate (cyanopropyl) in systems with A (first direction): tetrahydrofuran-n-heptane (20:80, v/v) and B (second direction): acetonitrile-water (50:50, v/v). (From Tuzimski, T., *J. Planar Chromatogr.*, 17, 328, 2004. With permission.) (c) 2D-chromatogram of the 14-component mixture of pesticides presented as videoscan of dual-phase Multi-K CS5 plate in systems: A (first direction): methanol-water (60:40, v/v) on octadecyl silica adsorbent, B (second direction): tetrahydrofuran-n-heptane (20:80, v/v) on silica gel. (From Tuzimski, T., Soczewiński, E., *J. Chromatogr. A*, 961, 277, 2002. With permission.) (d) schematic presentation of 2D-chromatogram of four samples simultaneously separated on the plate. (Adapted from Dzido, T.H., *Planar Chromatography, a Retrospective View for the Third Millennium*, Nyiredy, Sz. (Ed.), Springer Scientific Publisher, Budapest, 2001.) (e) 2D-chromatograms of three fractions of the mixture of nine pesticides presented as videoscan of the HPTLC plate (cyanopropyl) in systems with A (first direction): ethyl acetate-n-heptane (20:80, v/v), B (second direction): dioxane – water (40:60, v/v). (From Tuzimski, T., Soczewiński, E., *Chromatographia*, 59, 121, 2004. With permission.)

7.4.2.1.5 Multiple Development

Multiple development is the mode in which the direction development of the mobile phase flow is identical for each development step but the development distance and mobile phase composition can be varied in each step. The chromatogram is developed several times on the same plate and each step of the development follows the complete evaporation of the mobile phase from the chromatographic plate of the previous development. On the basis of the development distance and the composition of the mobile phase used for consecutive development steps, multiple development techniques are classified into four categories [134]:

   – Unidimensional multiple development (UMD), in which each step of chro-
     matogram development is performed with the same mobile phase and the
     same migration distance of eluent front;

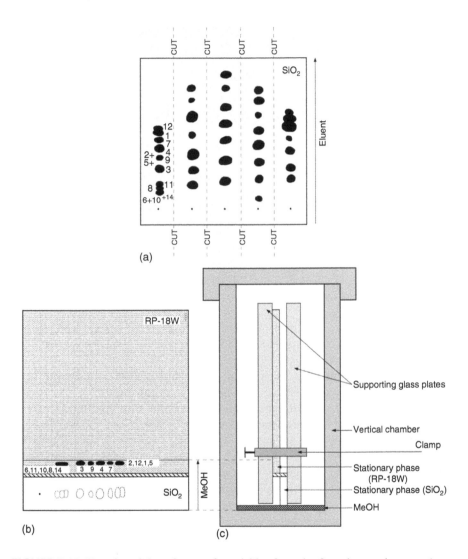

FIGURE 7.18 Transfer of the mixture of pesticides from the first plate to the second one.
(a) First development with partly separated mixtures of pesticides on silica plate. After
development the silica plate was dried and cut along the dashed lines into 2 cm × 10 cm
strips. (b) A narrow strip (2 cm × 10 cm) was connected (2 mm overlap hatched area) to 10
cm × 10 cm HPTLC RP-18W plate along the longer (10 cm) side of the strip. The partly
separated mixture of pesticides was transferred in a vertical chamber to the second plate using
methanol as strong eluent to the distance about of 1 cm. (c) Schematic diagram of cross section
of connected two adsorbents layers.

(continued)

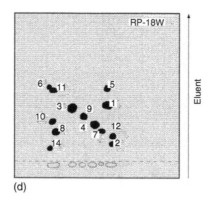

(d)

FIGURE 7.18 (continued) (d) The HPTLC RP-18W plate was developed in the second dimension with organic water eluent in the DS chamber. (From Tuzimski, T., *J. Planar Chromatogr.*, 20, 13, 2007. With permission.)

- Incremental multiple development (IMD), in which the same mobile phase but an increasing development distance in each subsequent step is applied;
- Gradient multiple development (GMD), in which the same development distance but a different composition of the mobile phase in each step is applied;
- Bivariant multiple development (BMD), in which the composition and development distance is varied in each step of chromatogram development.

These modes of chromatogram development are mainly applied for analytical separations because of very good efficiency, which is comparable to HPLC.

The sophisticated device used in this mode is manufactured by Camag and is know as automated multiple development (AMD or AMD 2) system. AMD mode enables both isocratic and gradient multiple development. In a typical isocratic AMD mode the development distance is increased during consecutive development steps while the mobile phase strength is constant. In initial stage of AMD gradient procedure the solvent of the highest strength is used (e.g., methanol, acetonitrile, or acetone), in the next stages—an intermediate or base solvent of medium strength (e.g., chlorinated hydrocarbons, ethers, esters, or ketones), and in the final stage— nonpolar solvent (e.g., heptane, hexane) [145].

Several parameters must be considered to obtain the best separation in AMD mode: choice of solvents, gradient profile of solvents, and number of steps. All modes of multiple development can be easily performed using chambers for automatic development, which are manufactured by some firms. However, these devices are relatively expensive. Typical horizontal chambers for planar chromatography should be considered for application in multiple development in spite of more manual operations in comparison with automatic chromatogram development. Especially Horizontal DS Chamber could be considered for separations with multiple development. This chamber can be easily operated because of its convenient

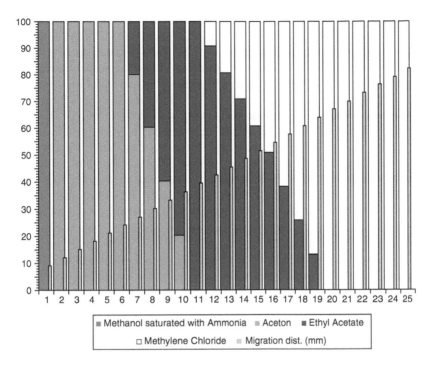

FIGURE 7.19 AMD gradient for antitussivess, opiate derivatives and substitutes. (From Pothier, J., Galand, N., *J. Chromatogr. A,* 1080, 186, 2005. With permission.)

maintenance, including cleaning the eluent reservoir. For the separation of a more complicated sample mixture the computer simulation could be used to enhance the efficiency of optimization procedure [35,36,146–149].

Analysis of three types of opiate alkaloids (the poppy alkaloids: morphine, codeine, thebaine, noscapine, and papaverine; the semisynthetic and synthetic derivatives such as pholcodine, ethylmorphine, and dextromethorphan; the narcotic compounds, diacetylmorphine (heroin) and opiates) employed as substitutes in treatment of addiction (buprenorphine and methadone) by the AMD was described and compared with results obtained by classical TLC method. The AMD system enabled a clean separation of each of three opiate groups (antitussives and substitutes) studied and the best results have been obtained with universal gradient [150]. Two reagents were used for the detection of alkaloids by spraying: Dragendorff and iodoplatinate reagents. In Figure 7.19 AMD gradient for antitussivess, opiate derivatives, and substitutes is shown [150].

7.4.2.1.6 Gradient (Stepwise and Continuous) Development
The separation efficiency is much better than in isocratic development because of the elimination of the "general elution problem" (especially when investigated sample mixtures comprise components of various polarity with a wide range of k values) and presence of compressing effect of the gradient and enhanced mutual displacement of the solutes especially effective for moderate k values [151].

Typical isocratic planar chromatogram of the mixtures containing compounds of various polarity is composed of bands of medium retention (R_F values in the range from 0.1 to 0.8), of lower retention ($0.8 < R_F < 1$), and of higher retention ($0.0 < R_F < 0.1$). For such mixture its bands of lower and higher retention are not well separated and are located close to the mobile phase front and start line, respectively. All the components can be separated only if a suitable continuous or stepwise mobile phase gradient is chosen for chromatogram development.

Gradient elution, both continuous and stepwise, can be performed in sandwich chamber with glass distributor (horizontal DS and ES chambers). Principle of the mode is then based on the introduction of mobile phase fractions of increasing strength following one after another in a series into eluent reservoir in horizontal DS chamber or under the distributor plate in ES chamber [18,35,152–154]. The most important requirement that should be fulfilled during chromatogram development in this case is that each next eluent fraction should be introduced into reservoir when previous one has been completely exhausted.

A more sophisticated equipment is necessary for continuous gradient elution in horizontal chamber. Miniaturized gradient generator for continuous and stepwise gradients with two vessels connected with elastic PTFE tubing and filled up with spontaneously mixing solvents have been proposed by Soczewiński and Matysik [154–156]. This gradient generator was combined to ES chamber. Densitograms of thin layer chromatograms from isocratic and gradient elution (continuous or stepwise) were compared and they showed considerable improvement in separation under gradient elution conditions [155,156]. However, the device is more suitable for narrow plates (e.g., 5 cm wide). Wider plates would not produce an uniform gradient profile across their area.

The device for continuous gradient elution in horizontal chamber described by Nyiredy [39] and presented earlier, Figure 7.20, seems to be a very interesting solution to both analytical and preparative applications.

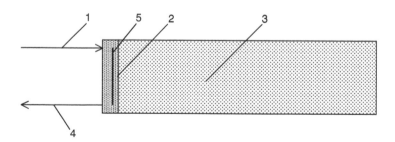

FIGURE 7.20 The cover of prepared chromatoplate for gradient development in a fully on line horizontal chamber: 1—solvent system inlet, 2—Silicoflon cover sheet, 3—chromoplate, 4—solvent system outlet, 5—channel for solvent system. (Adapted from Nyiredy, Sz., Benkö, A., in *Proceedings of the International Symposium on Planar Separations, Planar Chromatography 2004*, Nyiredy, Sz. (Ed.), Research Institute for Medicinal Plants, Budakalász, 2004.)

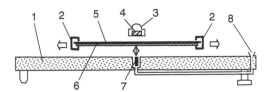

FIGURE 7.21 "Sequence-TLC" developing chamber (Scilab), (1) support with solvent source (reservoir), (2) holding frame, (3) magnet holder, (4) magnet, (5) cover plate, (6) TLC plate, (7) wick with iron core, (8) solvent entry. (Adapted from Bunčak, P., *GIT Fachz. Lab. (Suppl., Chromatographie)*, G-I-T-Verlag, Darmstadt, 3–8, 1982.)

Continuous gradient elution can also be easily created by the saturation of the mobile phase during the development in horizontal DS chamber with another solvent vapors, whose drops are placed on the blotting paper lined on the trough bottom of the chamber [11]. High reproducibility of this mode is difficult to obtain and the gradient range is restricted. But for some separations this mode could be considered for application because of very good selectivity that can be obtained especially for mixture components showing various properties of proton-donor and proton-acceptor interactions [11].

7.4.2.1.7   Sequence Development
Using sequence TLC device by Buncak [24], Figure 7.21, and modified ES chamber by Wang et al. [26,27], Figures 7.22 and 7.23, it is possible to deliver the mobile phase in any position to the chromatographic plate. It means that the solvent entry position on the plate can be changed. This is an advantage for planar chromatography owing to the increased separation efficiency, which can be achieved by the following: multiple development with various eluents being supplied to different positions on the chromatographic plate in each step; changing the solvent entry position during the development leads to increase in the efficiency because of higher flow rate of the mobile phase; cleaning the plate before the development; separating the trace components from the bulk substance; spotting the mixture to be separated in the middle of the chromatoplate; development with mobile phase of lower eluent strength in one direction (the mixture components of higher polarity stay on the start line but the mixture components of lower polarity are separated) and developing after evaporation of the mobile phase with stronger eluent in the opposite direction (then components of higher polarity are separated).

7.4.2.1.8   Temperature Control
Temperature control in planar chromatography is rare. Most planar chromatography analyses are usually performed at room temperature in nonthermostated developing chambers. The optimum chromatographic separation is a compromise between maximum resolution and minimum analysis time. In classical planar chromatography the total analysis time is the same for all solutes and the solutes' mobility is driven by nonforced flow of the mobile phase—capillary action. The efficiency and selectivity of a chromatographic process, and the precision and reproducibility of analysis are temperature dependent. The running time is strongly affected by the developing

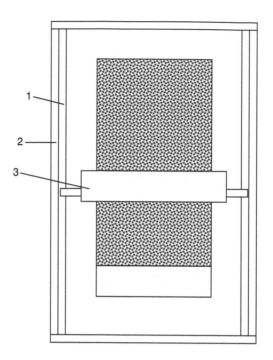

FIGURE 7.22 Top view of ES chamber modified by Wang et al.: 1—supporting plate, 2—spacing plate, 3—distributor. (Adapted from Su, P., Wang, D., Lan, M., *J. Planar Chromatogr.*, 14, 203, 2001.)

distance, the degree of saturation of vapor of the mobile phase, its viscosity and particle size of the stationary phase. The mobile phase viscosity depends on the mobile phase composition and is decreased with temperature increase. The last effect leads to increase of the mobile phase flow rate and eventually to shortening of chromatogram development. The relationship, between the retention parameter of solutes (R_M) and the reciprocal of absolute temperature ($1/T$), is often linear (van't Hoff plot). Zarzycki has described some technical problems associated with temperature-controlled planar chromatography [157]. The author has also described a construction of a simple

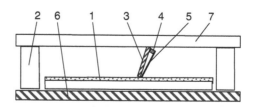

FIGURE 7.23 Cross section of ES chamber with funnel distributor (modified by Wang et al.): 1—spacing plate, 2—base plate, 3—distributor, 4—glue, 5—slide, 6—thin-layer plate, 7—cover plate. (Adapted from Lan, M., Wang, D., Han, J., *J. Planar Chromatogr.*, 16, 402, 2003.)

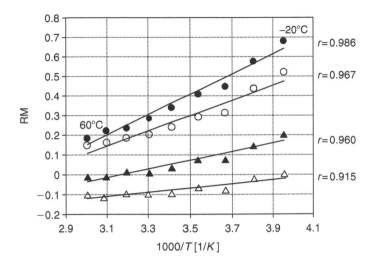

FIGURE 7.24 The Van't Hoff plots on the investigated steroids; estetrol (Δ); estriol (▲); estrone (○); 17-β-estradiol (●). (From Zarzycki, P.K., *J. Chromatogr. A*, 971, 193–197, 2002. With permission.)

developing device designed for temperature control of thin layer chromatographic plates [158]. Figure 7.24 shows the Van't Hoff plots of the steroids [158]. Figure 7.25 illustrates the changes of migration time of eluent front at different temperatures and front distances as the contour map using methanol–water (70:30, v/v) mobile phase and HPTLC RP-18W plates [158].

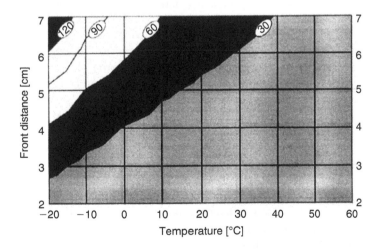

FIGURE 7.25 The contour map of the solvent front migration times at different solvent front distances and temperatures using methanol-water (70:30, v/v) mobile phase and HPTLC RP-18W plates. The spaces between contour lines correspond to 30 min. (From Zarzycki, P.K., *J. Chromatogr. A*, 971, 193–197, 2002. With permission.)

Dzido described also an adaptation of the horizontal DS chamber to planar chromatography with temperature control [159]. The author has also observed the change of development time at RPTLC systems at a temperature of 58°C in comparison with 15°C [159]. This chamber enables precise temperature control of the chromatographic system because the chromatographic plate is located between two heating coils connected to the circulating thermostat.

The influence of temperature and mobile phase composition on retention of different cyclodextrins and two macrocyclic antibiotics has been examined by RP-TLC using wide-range (0%–100%) binary mixtures of methanol–water [160]. Using thermostated chamber for planar chromatography the interactions between cyclodextrins and n-alcohols were investigated [161]. The influence of temperature on retention and separation of cholesterol and bile acids in RP-TLC systems was also reported [162].

7.4.2.2 Radial Development

Radial development of planar chromatograms can be performed as circular and anticircular. Capillary action is the driven force for the mobile phase movement in these modes. Otherwise in rotation planar chromatography, which is another mode of radial development, the centrifugal force is responsible for the mobile phase movement. This mode was described by Hopf [163] at the first time. Different modifications of this technique have also been reported [164–172].

7.4.2.2.1 Circular Development

In circular mode of chromatogram development the samples are applied in a circle close to the center of the plate and the eluent enters the plate at the center. The mobile phase is moved through the stationary phase from the center to periphery of the chromatographic plate and the sample components form zones like rings. In the first report of circular development by Ismailov and Schraiber a chamber was not used [173]. Circular or anticircular mode of chromatogram development in closed system was first carried out in a Petri dish containing eluent and a wick that touches the layer, supported on top of the dish, at its central point. The example of such chromatogram is presented in Figure 7.26 [174]. The chromatogram was obtained with circular U-chamber from Camag, Figure 7.27, which can be used for preparative and analytical separations.

The chamber for circular development described by Botz et al. [175] and modified by Nyiredy [176], is especially suitable for preparative planar chromatogram development in which various sample mixtures (solid or liquid) can be applied on the chromatographic plate.

Obtained separation quality, using circular mode, was considerably higher than that when linear ascending development was performed in twin-trough chamber. Even linear development using plates with a preconcentration zone produced lower separation quality in comparison with circular development, Figure 7.28 [175]. The authors advise that separation using circular development with the chamber described has advantages relative to separation efficiency obtained in linear development.

However, the advantages are referred to chromatogram development in vertical N-chamber (twin-trough chamber). The separation quality was not compared with that using horizontal mode of linear development. The authors have reported one

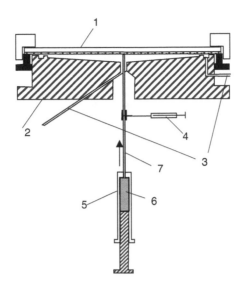

FIGURE 7.26 Cross section view of the U-chamber (Camag, Muttenz); (1) chromatographic plate, (2) body of the chamber, (3) inlet or outlet for parallel or counter gas flow, to remove vaporized mobile phase, to dry or moisten (impregnate) the plate, (4) syringe for sample injection, (5) dosage syringe to maintain the flow of the mobile phase, (6) eluent, (7) capillary. (Adapted from Kaiser, R.E., *HPTLC High Performance Thin-Layer Chromatography*, Zlatkis, A., Kaiser, R.E. (Eds.), Elsevier, Institute of Chromatography, Amsterdam, Bad Dürkheim, 1977.)

disadvantage of the mode: recovering of bands of interest can be performed only by scraping the adsorbent from the plate. This is more complicated than in the case of linear chromatogram development using rectangular plates.

In spite of the advantages mentioned earlier the circular development is rarely used in contemporery practice of planar chromatography.

7.4.2.2.2   Anticircular Development
In anticircular mode of chromatogram development the sample mixture is spotted at the circumference of the plate and mobile phase is moved from the circumference to the center of the plate. The sample application capacity is larger than in circular mode because of the long start line. Especially good separation in this mode of chromatogram development is observed for high R_F values. Anticircular development is very rarely applied in planar chromatographic practice for analytical separation.

This mode of separation was introduced by Kaiser [177]. Studer and Traitler adapted anticircular U-chamber from Camag to preparative separations on 20×20 cm plates [178]. However, the mode has not gained much popularity in laboratory practice probably because a more sophisticated equipment is necessary to perform the separation.

Issaq [179] has proposed the application of conventional chambers to perform anticircular development. In this mode a commercially available chromatographic plate is divided in triangular plates and sample (or samples) is spotted along the base

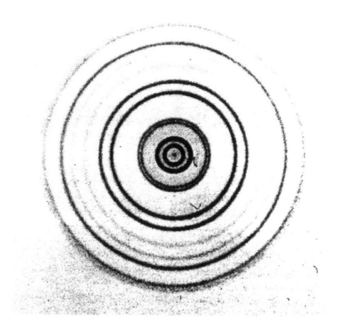

FIGURE 7.27 Circular chromatography of dyes on precoated silica gel high-performance TLC plate; lipophilic dyes, mobile phase: hexane-chloroform-NH₃, 70:30. (From Ripphahn, J., Halpaap, H., *HPTLC High Performance Thin-Layer Chromatography*, Zlatkis, A., Kaiser, R.E. (Eds.), Elsevier, Institute of Chromatography, Amsterdam, Bad Dürkheim, 1977. With permission.)

of triangular plate. Wetting of the mobile phase is started when base of the triangular chromatographic plate is contacted with solvent. It means that all kinds of developing chambers (N-chambers and S-chambers) can be easily used in this mode of chromatogram development. The bands on the plate after preparative chromatogram development are narrower than the original bands on start line of the plate depending on their migration distance, Figure 7.29 [179]. It means that the bands are more concentrated, require less solvent for development and less solvent to elute from the plate as well.

7.5 COMBINATIONS OF DIFFERENT MODES OF CHROMATOGRAM DEVELOPMENT

The application of multidimensional planar chromatography (MD-PC) combined with different separation systems and modes of chromatogram development is often necessary for performing the separation of more complicated multicomponent mixtures. High separation efficiency can be obtained using modern planar chromatographic techniques, which comprise 2D development, chromatographic plates with different properties, a variety of solvent combinations for mobile phase preparation, various forced-flow techniques, and multiple development

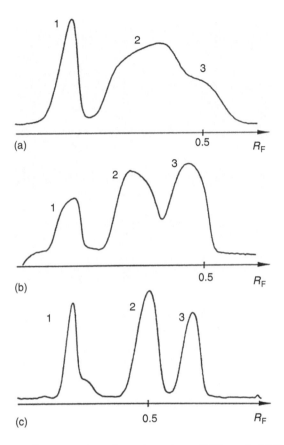

FIGURE 7.28 Preparative separation of test dye mixture I (Camag): (a) linear development using preparative plate without concentrating zone (development time: 83 min), (b) linear development using preparative plate with concentrating zone (development time: 76 min), (c) circular development using preparative plate (development time: 38 min). 1 = oracet blue, 2 = oracet red, 3 = butter yellow; detection: VIS at λ = 500 nm. (From Botz, L., Nyiredy, S., Sticher, O., *J. Planar Chromatogr.*, 3, 401–406, 1990. With permission.)

modes. By combination of these possibilities, MD-PC can be performed in various ways. Multidimensional chromatography definedby Giddings includes two criteria [180]:

- Components of the mixture are subjected to two or more separation steps in which their migration depends on different factors.
- When two components separated in any single steps, they always remain separated until completion of the separation.
  Nyiredy divided MD-PC techniques as follows [2,181,182]:
- Comprehensive 2D planar chromatography (PC × PC)—multidimensional development on the same monolayer stationary phase and two

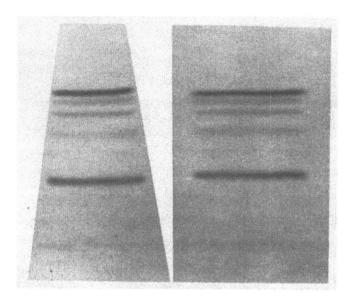

FIGURE 7.29 Comparison of the separation of streaks of dyes on triangular and rectangular 5 × 20 cm plates. (From Issaq, H.J., *J. Liq. Chromatogr.*, 3, 789–796, 1980. With permission.)

developments with different mobile phases or using a bilayer stationary phase and two developments with the same or different mobile phases.

- Targeted or selective 2D planar chromatography (PC + PC), in which following the first development from the stationary phase a heart-cut spot is applied to a second stationary phase for subsequent analysis to separate the compounds of interest.
- Targeted or selective 2D planar chromatography (PC + PC)—second mode-technique, in which following the first development, which is finished and plate dried, two lines must be scraped into the layer perpendicular to the first development and the plate developed with another mobile phase, to separate the compounds that are between the two lines. For the analysis of multicomponent mixtures containing more than one fraction, separation of components of the next fractions should be performed with suitable mobile phases.
- Modulated 2D planar chromatography (^{n}PC), in which on the same stationary phase the mobile phases of decreasing solvent strength and different selectivity are used.
- Coupled-layer planar chromatography (PC–PC) technique, in which two plates with different stationary phases are turned face to face (one stationary phase to second stationary phase) and pressed together so that a narrow zone of the layers overlaps, the compounds from the first stationary phase are transferred to the second plate and separated with a different mobile phase.

– Combination of MD-PC methods, in which the best separation of multi-component mixture is realized by parallel combination of stationary and mobile phases, which are changed simultaneously. By use of this technique, e.g., after separation of compounds in the first dimension with changed mobile phases, the plate is dried and separation process is continued in perpendicular direction by use of grafted technique with changed mobile phase (based on the idea of coupled TLC plates, denoted as graft TLC in 1979 [139]).

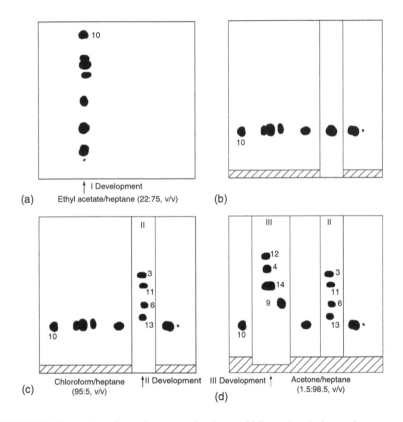

FIGURE 7.30 Illustration of step by step selective multidimensional planar chromatography separation. (a) The dried plate after first separation (I development). (b) The plate prepared for the separation of the second group of compounds: two lines (about 1 mm thick) are scraped in the stationary phase perpendicular to the first development in such a way that the spot(s) of target compounds are between these lines. In addition the strip of adsorbent layer of 5 mm width is removed from the plate along its lower edge to prevent wetting the layer outside the area fixed by these lines during the second development (hatched lines indicate the removed part of the stationary phase). So the mobile phase wets narrow strip of the layer only during the second run. (c) The dried plate after separation of the second group of pesticides (3, 6, 11, 13) by use double development with chloroform-n-heptane (95: 5, v/v) as the mobile phase at the same distance (UMD). (d) The plate after separation of the components of the fourth group of pesticides with acetone-n-heptane (1.5: 88.5, v/v) as the mobile phase in III development.

(continued)

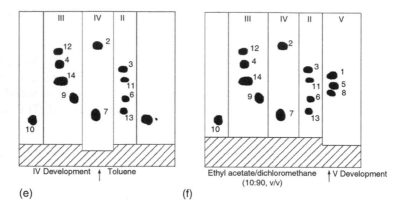

FIGURE 7.30 (continued)

FIGURE 7.30 (continued) (e) The plate after separation of two components of the third group (2, 7) of pesticides with toluene as the mobile phase in the IV development. (f) The plate after separation of three components of the first group (1, 5, 8) of pesticides with ethyl acetate - dichloromethane (10: 90, v/v) as the mobile phase in the V development. (From Tuzimski, T., *J. Separation Sci.*, 30, 964, 2007. With permission.)

A new procedure for separation of complex mixtures by combination of different modes of MD-PC was described [121,183]. By the help of this new procedure 14 or 22 compounds from a complex mixtures were separated on 10 cm × 10 cm TLC and HPTLC plates [121,183]. In Figure 7.30 the example of this procedure is illustrated step by step for separation of 14 compounds from complex mixtures on TLC plate [121].

7.6   CONCLUSIONS

Modes of chromatogram development, sample application, and application of appropriate stationary and mobile phases are the key elements that influence the resolution of the mixture components and efficiency of quantitative and qualitative analysis. Optimization of these elements can be effectively performed on the basis of good understanding of the theoretical fundamentals and practical knowledge of planar chromatography. Sophisticated equipments, methods, and software are inherent elements of today's planar chromatography and can effectively facilitate optimization of chromatographic separation. Thanks to these features, this method is a powerful analytical and separation mode in contemporary analysis, which has gained growing popularity in laboratory practice. The literature on planar chromatography regarding the problems discussed in this chapter is very broad. Only part of this literature is cited in the references of this chapter, which is the additional evidence of the meaning and interest in this mode especially for separation and analysis of compounds of biological origin. The authors hope that the readers will find this chapter to be helpful in their laboratory practice concerned with planar chromatography separations.

REFERENCES

1. Geiss, F., *Fundamentals of Thin-Layer Chromatography*, Huethig, Heidelberg, 1987.
2. Nyiredy, Sz. (Ed.), *Planar Chromatography—A Retrospective View for the Third Millennium*, Springer Scientific Publisher, Budapest, 2001.
3. Hahn-Deinstrop, E., *Applied Thin-Layer Chromatography*, Wiley-VCH, Weinheim, 2000.
4. Fried, B., Sherma, J., *Thin-Layer Chromatography*, 4th edn. (revised and expanded), Marcel Dekker, New York–Basel, 1999.
5. www.camag.com.
6. Jaenchen, D.E., Planar chromatography (instrumental thin-layer chromatography), in: *Handbook of Thin Layer Chromatography*, 2nd edn, Sherma, J., Fried, B. (Eds.), Marcel Dekker, New York, 1996, pp. 109–148.
7. www.chromdes.com.
8. Dzido, T.H., Modern TLC chambers, in: *Planar Chromatography—A Retrospective View for the Third Millennium*, Nyiredy, Sz. (Ed.), Springer, Budapest, 2001, pp. 68–87.
9. Fenimore, D.C., Davis, C.M., *Anal. Chem.*, 53, 253A–266A, 1981.
10. Dzido, T., Polak, B., *J. Planar Chromatogr.*, 6, 378–381, 1993.
11. De Brabander, H.F., Smets, F., Pottie, G., *J. Planar Chromatogr.*, 1, 369–371, 1988.
12. Markowski, W., Matysik, G., *J. Chromatogr.*, 646, 434–438, 1993.
13. Matysik, G., *Chromatographia*, 43, 39–43, 1996.
14. Matysik, G., Soczewiński, E., *J. Planar Chromatogr.*, 9, 404–411, 1996.
15. Matysik, G., Markowski, W., Soczewiński, E., Polak, B., *Chromatographia*, 34, 303–307, 1992.
16. Matysik, G., Soczewiński, E., Polak, B., *Chromatographia*, 39, 497–504, 1994.
17. Dzido, T.H., Polak, B., On methodical possibilities of the horizontal chambers in PLC, in: *Preparative Layer Chromatography*, Kowalska, T., Sherma, J. (Eds.), CRC Press, Taylor & Francis Group, Boca Raton, 2006.
18. Soczewiński, E., Equilibrium sandwich TLC chamber for continuous development with a glass distributor, in: *Planar Chromatography*, Vol. 1, Kaiser, R.E. (Ed.), Dr. Alfred Huetig Verlag, Heidelberg, Basel, New York, 1986, 79–117.
19. Kraus, L., Koch, A., Hoffstetter-Kuhn, S., *Dünnschichtchromatographie*, Springer-Verlag, Berlin, Heidelberg, 1996.
20. Brenner, M., Niederwieser, A., *Experientia*, 17, 237–238, 1961.
21. Geiss, F., Schlitt, H., Klose, A., *Z. Anal. Chem.*, 213, 331–346, 1965.
22. Geiss, F., Schlitt, H., *Chromatographia*, 1, 392–402, 1968.
23. Regis Technologies, Morton Grove, IL., USA.
24. Buncak, P., *Chromatographie Suppl. GIT*, Darmstadt, 1982, 3–8.
25. Rumiński, J.K., *Chem. Anal. (Warsaw)*, 33, 479–481, 1988.
26. Su, P., Wang, D., Lan, M., *J. Planar Chromatogr.*, 14, 203–207, 2001.
27. Lan, M., Wang, D., Han, J., *J. Planar Chromatogr.*, 16, 402–404, 2003.
28. Blome, J., Advantages, limits and disadvantages of the ring ring developing technique, in: *High Performance Thin-Layer Chromatography*, Zlatkis, A., Kaiser, R.E. (Eds.), Elsevier, Institute of Chromatography, Bad Dürkheim, 1977.
29. Kaiser, R.E., The U-Chamber in: *High Performance Thin-Layer Chromatography*, Zlatkis, A., Kaiser, R.E. (Eds.), Elsevier, Institute of Chromatography, Bad Dürkheim, 1977.

30. Omori, T., A simple time-controlled developing chamber for small TLC-plates, in: *Proceedings of the 6th International Symposium on Instrumental Planar Chromatography*, Traitler, H., Voroshilova, O.I., Kaiser, R.E. (Eds.), Interlaken, Switzerland, 23–26 April, 1991.
31. http://www.sarstedt.com/php/main.php?newlanguage = en.
32. www.baron-lab.de.
33. Perry, J.A., Haag, K.W., Glunz, L.J., *J. Chromatogr. Sci.*, 11, 447–453, 1973.
34. Burger, K., *Fresenius Z. Anal. Chem.*, 318, 228–233, 1984.
35. Markowski, W., Gradient development in TLC, in: *Encyclopedia of Chromatography*, Cazes, J. (Ed.), Marcel Dekker, New York, Basel, 2005, pp. 702–717.
36. Markowski, W., *J. Chromatogr. A*, 726, 185–192, 1996.
37. Morlock, G.E., Sample application and chromatogram development, in: *Preparative Layer Chromatography*, Kowalska, T., Sherma, J. (Eds.), CRC Press, Taylor & Francis Group, Boca Raton, 2006.
38. Botz, L., Nyiredy, Sz., Sticher, O., *J. Planar Chromatogr.*, 3, 10–14, 1990.
39. Nyiredy, Sz., Benkö, A., Solvent gradient technique in planar chromatography, in: *Proceedings of the International Symposium on Planar Separations, Planar Chromatography 2004*, Nyiredy, Sz. (Ed.), Research Institute for Medicinal Plants, Budakalász, 2004, pp. 55–60.
40. Soczewiński, E., Wawrzynowicz, T., *J. Chromatogr.*, 218, 729–732, 1981.
41. Poole, C.F., Dias, N.C., Practitioner's guide to method development in thin-layer chromatography, *J. Chromatogr., A*, 892, 123–142, 2000.
42. Snyder, L.R., Kirkland, J.J., Glajch, J.L., Non-ionic samples: Reversed- and normal-phase HPLC, in: *Practical HPLC Method Development*, 2nd edn, Wiley, New York, 1997, pp. 233–291.
43. Lide, D.R., Frederikse, H.P.R., Eds., Fluid properties, in: *CRC Handbook of Chemistry and Physics*, CRC Press, Boca Raton, FL, 1995, Chapter 6.
44. Héron, S., Tchapla, A., *Analusis*, 21, 327–347, 1993.
45. Rohrschneider, L., *Anal. Chem.*, 45, 1241–1247, 1973.
46. McReynolds, W.O., *J. Chromatogr. Sci.*, 8, 685, 1970.
47. Jandera, P., Churáček, *J. Chromatogr.*, 91, 207, 1974.
48. Tijssen, R., Billiet, H.A.H., Schoenmakers, P.J., *J. Chromatogr.*, 128, 65, 1976.
49. Schoenmakers, P.J., Billiet, H.A.H., Tijssen, R., De Galan, L., *J. Chromatogr.*, 149, 519–537, 1978.
50. Snyder, L.R., *J. Chromatogr.*, 92, 223–230, 1974.
51. Snyder, L.R., *J. Chromatogr. Sci.*, 16, 223–234, 1978.
52. Snyder, L.R., Glajch, J.L., Kirkland, J.J., *J. Chromatogr.*, 218, 299–326, 1981.
53. Glajch, J.L., Kirkland, J.J., Squire, K.M., Minor, J.M., *J. Chromatogr.*, 199, 57–79, 1980.
54. Kowalska, T., Klama, B., *J. Planar Chromatogr.*, 10, 353–357, 1997.
55. Johnson, B.P., Khaledi, M.G., Dorsey, J.G., *Anal. Chem.*, 58, 2354–2365, 1986.
56. Kosower, E.M., *J. Am. Chem. Soc.*, 80, 3253–3260, 1958.
57. Kosower, E.M., Mohammad, M., *J. Am. Chem. Soc.*, 90, 3271–3272, 1968.
58. Kamlet, M.J., Abboud, J.L.M., Abraham, M.H., Taft, R.W., *J. Org. Chem.*, 48, 2877–2887, 1983.
59. Kamlet, M.J., Abboud, J.L.M., Taft, R.W., *J. Am. Chem. Soc.*, 99, 6027–6038, 1977.
60. Laurence, C., Nicolet, P., Dalati, M.T., Abboud, J.L.M., Notario, R., *J. Phys. Chem.*, 98, 5807–5816, 1994.
61. Taft, R.W., Kamlet, M.J., *J. Am. Chem. Soc.*, 98, 2886–2894, 1976.
62. Abraham, M.H., Poole, C.F., Poole, S.K., *J. Chromatogr. A*, 842, 79, 1999.

63. Abraham, M.H., Whiting, G.S., Shuely, W.J., Doherty, R.M., *Can. J. Chem.*, 76, 703, 1998.
64. Abraham, M.H., Whiting, G.S., Carr, P.W., Quyang, H., *J. Chem. Soc., Perkin Trans.* 2, 1385, 1998.
65. Abraham, M.H., Platts, J.A., Hersey, A., Leo, A.J., Taft, R.W., *J. Pharm. Sci.*, 88, 670, 1999.
66. Trappe, W., *J. Biochem.*, 305, 150–154, 1940.
67. Snyder, L.R., *J. Chromatogr.*, 63, 15–44, 1971.
68. Gocan, S., Mobile phase in thin-layer chromatography. in: *Modern Thin-Layer Chromatography*, Grinberg, N. (Ed.), Chromatographic Science Series 52, Marcel Dekker, New York, 1990, pp. 427–434.
69. Neher, R., in: *Thin-Layer Chromatography*, Marini-Bettolo B.G. (Ed.), Elsevier, Amsterdam, 1964.
70. Saunders, D.L., Anal. Chem., 46, 470, 1974.
71. Snyder, L.R., *Principles of Adsorption Chromatography*, Marcel Dekker, New York, 1968.
72. Snyder, L.R., *J. Chromatogr.*, 13, 415–434, 1971.
73. Snyder, L.R., *J. Chromatogr.*, 8, 178–200, 1962.
74. Snyder, L.R., *J. Chromatogr.*, 16, 55–88, 1964.
75. Snyder, L.R., *J. Chromatogr.*, 11, 195–227, 1963.
76. *Dünnschicht-Chromatographie*, Stahl, E. (Ed.), Springer Verlag, Berlin, 1st edn. 1962, 2nd edn. 1967.
77. Abboutt, D., Andrews, R.S., *An Introduction to Chromatography*, Longmans, Green, London, 1965, p. 27.
78. Soczewiński, E., *Anal. Chem.*, 41, No. 1, 179, 1969.
79. Soczewiński, E., Quantitative retention-eluent composition relationships in partition and adsorption chromatography, in: *A Century of Separation Science*, Issaq, H. (Ed.), Marcel Dekker, New York, 2002, Chapter 11, pp. 179–195.
80. Soczewiński, E., *J. Chromatogr. A*, 965, 109, 2002.
81. Snyder, L.R., Dolan, J.W., Gant, J.R., *J. Chromatogr.*, 165, 3–30, 1979.
82. Soczewiński, E., Wachtmeister, C.A., *J. Chromatogr.*, 7, 311–320, 1962.
83. Snyder, L.R., Carr, P.W., Rutan, S.C., *J. Chromatogr. A*, 656, 537–547, 1993.
84. Qasimullach, A.A., Andrabi, S.M.A., Qureshi, P.M., *J. Chromatogr. Sci.*, 34, 376–378, 1996.
85. Martin, A.J.P., Synge, R.L.M., *Biochem. J.*, 35, 1358, 1941.
86. Martin, A.J.P., *Biochem. J.*, 50, 679, 1952.
87. Scott, R.P., Kucera, P., *J. Chromatogr.*, 112, 425, 1975.
88. Scott, R.P., *J. Chromatogr.*, 122, 35, 1976.
89. Kowalska, T., *Microchem. J.*, 29, 375, 1984.
90. Kowalska, T., *Monatsh. Chem.*, 116, 1129, 1985.
91. Kowalska, T., *Fat Sci. Technol.*, 90, 259, 1988.
92. Ościk, J., Chojnacka, G., *J. Chromatogr.*, 93, 167–176, 1974.
93. Ościk, J., Chojnacka, G., *Chromatographia*, 11, 731–735, 1978.
94. Glajch, J.L., Kirkland, J.J., *Anal. Chem.*, 55, 319A–327A, 1983.
95. Li, S.F.Y., Lee, H.K., Ong, C.P., *J. Chromatogr.*, 506, 245–252, 1990.
96. Nurok, D., Beker, R.M., Richard, M.J., Cunningham, P.D., Gorman, W.B., Bush, C.L., *J. High Resolut. Chromatogr. Chromatogr. Commun.*, 5, 373–380, 1982.
97. Prus, W., Kowalska, T., *J. Planar Chromatogr.*, 8, 205–215, 1995.
98. Prus, W., Kowalska, T., *J. Planar Chromatogr.*, 8, 288–291, 1995.
99. Coman, V., Măruţoiu, C., Puiu, S., *J. Chromatogr. A*, 779, 321–328, 1997.
100. Kiridena, W., Poole, C.F., *J. Planar Chromatogr.*, 12, 13–25, 1999.

101. Pelander, A., Summanen, J., Yrjönen, T., Haario, H., Ojanperä, I., Vuorela, H., *J. Planar Chromatogr.*, 12, 365–372, 1999.
102. Markowski, W., Soczewiński, E., Matysik, G., *J. Liq. Chromatogr.*, 10, 1261–1275, 1987.
103. Markowski, W., Czapińska, K.L., *J. Liq. Chromatogr.*, 18 (7) 1405–1427, 1995.
104. Matyska, M., Dąbek, M., Soczewiński, E., *J. Planar Chromatogr.*, 3, 317–321, 1990.
105. Perišić-Janjić, N.U., Djaković-Sekulić, T., Jevrić, L.R., Jovanović, B.Ž., *J. Planar Chromatogr.*, 18, 212–216, 2005.
106. Tuzimski, T., Sztanke, K., *J. Planar Chromatogr.*, 18, 274–281, 2005.
107. Djaković-Sekulić, T., Perišić-Janjić, N., *J. Planar Chromatogr.*, 20, 7–11, 2007.
108. Flieger, J., Świeboda, R., Tatarczak, M., *J. Chromatogr. B.*, 846, 334–340, 2007.
109. Tatarczak, M., Flieger, J., Szumiło, H., *Chromatographia*, 61, 307–309, 2005.
110. Tuzimski, T., *Chromatographia*, 56, 379–381, 2002.
111. Dallenbach-Tölke, K., Nyiredy, Sz., Meier, B., Sticher, O., *J. Chromatogr.*, 365, 63–72, 1986.
112. Nyiredy, Sz., Dallenbach-Tölke, K., Sticher, O., *J. Planar Chromatogr.*, 1, 336–342, 1998.
113. Nyiredy, Sz., Dallenbach-Tölke, K., Sticher, O., *J. Liq. Chromatogr.*, 12, 95, 1989.
114. Nyiredy, Sz., Fatér, Zs., *J. Planar Chromatogr.*, 8, 341–345, 1995.
115. Nyiredy, Sz., *Chromatographia*, 51, S288–S296, 2000.
116. Nyiredy, Sz., *J. Chromatogr. B*, 812, 35–51, 2004.
117. Reich, E., George, T., *J. Planar Chromatogr.*, 10, 273–280, 1997.
118. Morita, K., Koike, S., Aishima, T., *J. Planar Chromatogr.*, 11, 94–99, 1998.
119. Pelander, A., Sivonen, K., Ojanperä, I., Vuorela, H., *J. Planar Chromatogr.*, 10, 434–440, 1997.
120. Cimpoiu, C., Hodişan, T., Naşcu, H., *J. Planar Chromatogr.*, 10, 195–199, 1997.
121. Tuzimski, T., *J. Separation Sci.*, 30, 964–970, 2007.
122. Nyiredy, Sz., *J. Planar Chromatogr.*, 15, 454–457, 2002.
123. Perry, J.A., *J. Chromatogr.*, 165, 117–140, 1979.
124. Matysik, G., Soczewiński, E., *J. Chromatogr.*, 446, 275–282, 1988.
125. Lee, Y.K., Zlatkis, K., in: *Advances in Thin Layer Chromatography, Clinical and Environmental Applications*, Touchstone, J.C. (Ed.), Wiley-Interscience, New York, 1982.
126. Gonnord, M.F., Levi, F., Guiochon, G., *J. Chromatogr.*, 264, 1–6, 1983.
127. Steinbrunner, J.E., Johnson, E.K., Habibi-Goudarzi, S., Nurok, D., Computer-aided evaluation of continuous development two-dimensional thin layer chromatography, in: *Planar Chromatography* Vol. 1, Kaiser, R.E. (Ed.), Hüthig Verlag, Heidelberg, 1986.
128. Nurok, D., Habibi-Goudarzi, S., Kleyle, R., *Anal. Chem.*, 59, 2424–2428, 1987.
129. Nurok, D., *Chem. Rev.*, 89, 363–375, 1989.
130. Nurok, D., Kleyle, R., McCain, C.L., Risley, D.S., Ruterbories, K.J., *Anal. Chem.*, 69, 1398–1405, 1997.
131. De Spiegeleer, B., Van den Bossche, W., De Moerlose, P., Massart, D., *Chromatographia*, 23, 407–411, 1987.
132. Hauck, H.E., Mack, M., Sorbents and precoated layers in thin-layer chromatography, in: Sherma, J., Fried, B. (Eds.), *Handbook of Thin Layer Chromatography*, 2nd edn, Marcel Dekker, New York, 1996, Chapter 4, pp. 101–128 (Fig. 5, p. 115).
133. Tuzimski, T., *J. Planar Chromatogr.*, 17, 328–334, 2004.
134. Szabady, B., Nyiredy, Sz., The versatility of multiple development, in: *Dünnschicht-Chromatographie in Memorian Professor Dr. Hellmut Jork*, Kaiser, R.E., Günther, W., Gunz, H., Wulff, G. (Eds.), InCom Sonderband, Düsseldorf, 1996, pp. 212–224.

135. Tuzimski, T., Soczewiński, E., *J. Planar Chromatogr.*, 15, 164–168, 2002.
136. Tuzimski, T., Soczewiński, E., *J. Chromatogr. A*, 961, 277–283, 2002.
137. Tuzimski, T., Soczewiński, E., *Chromatographia*, 56, 219–223, 2002.
138. Tuzimski, T., Bartosiewicz, A., *Chromatographia*, 58, 781–788, 2003.
139. Pandey, R.C., Misra, R., Rinehart, Jr., K.L., *J. Chromatogr.*, 169, 129–139, 1979.
140. Łuczkiewicz, M., Migas, P., Kokotkiewicz, A., Walijewska, M., Cisowski, W., *J. Planar Chromatogr.*, 17, 89–94, 2004.
141. Glensk, M., Sawicka, U., Mażol, I., Cisowski, W., *J. Planar Chromatogr.*, 15, 463–465, 2002.
142. Glensk, M., Cisowski, W., *J. Planar Chromatogr.*, 13, 9–11, 2000.
143. Tuzimski, T., *J. Planar Chromatogr.*, 18, 349–357, 2005.
144. Tuzimski, T., *J. Planar Chromatogr.*, 20, 13–18, 2007.
145. Ebel, S., Völkl, S., *Dtsch. Apoth. Ztg.*, 130, 2162–2169, 1990.
146. Markowski, W., Soczewiński, E., *J. Chromatogr.*, 623, 139–147, 1992.
147. Markowski, W., Soczewiński, E., *Chromatographia*, 36, 330–336, 1993.
148. Markowski, W., *J. Chromatogr.*, 635, 283–289, 1993.
149. Markowski, W., Czapińska, K.L., Błaszczak, M., *J. Liq. Chromatogr.*, 17, 999–1009, 1994.
150. Pothier, J., Galand, N., *J. Chromatogr.*, 1080, 186–191, 2005.
151. Snyder, L.R., Kirkland, J.J., *Introduction to Modern Liquid Chromatography*, 2nd edn., Wiley, New York, 1979.
152. Gołkiewicz, W., Gradient development in thin layer chromatography, in: *Handbook of Thin-Layer Chromatography*, 3rd edn. (revised and expanded), Sherma, J., Fried, B. (Eds.), Marcel Dekker, New York, Basel, 2003, pp. 153–173.
153. Matysik, G., Soczewiński, E., *Anal. Chem. (Warsaw)*, 33, 363–369, 1988.
154. Matysik, G., Soczewiński, E., *J. Chromatogr.*, 446, 275–282, 1988.
155. Soczewiński, E., Matysik, G., *J. Liq. Chromatogr.*, 8, 1225–1238, 1985.
156. Matysik, G., Soczewiński, E., *Chromatographia*, 26, 178–180, 1988.
157. Zarzycki, P.K., *J. Planar Chromatogr.*, 14, 63–65, 2001.
158. Zarzycki, P.K., *J. Chromatogr. A*, 971, 193–197, 2002.
159. Dzido, T.H., *J. Planar Chromatogr.*, 14, 237–245, 2001.
160. Zarzycki, P.K., Nowakowska, J., Chmielewska, A., Wierzbowska, M., Lamparczyk, H., *J. Chromatogr. A*, 787, 227–233, 1997.
161. Zarzycki, P.K., Wierzbowska, M., Nowakowska, J., Chmielewska, A., Lamparczyk, H., *J. Chromatogr. A*, 839, 149–156, 1999.
162. Zarzycki, P.K., Wierzbowska, M., Lamparczyk, H., *J. Chromatogr. A*, 857, 255–262, 1999.
163. Hopf, P.P., *Ind. Eng. Chem.*, 39, 365, 1947.
164. Caronna, G., *Chim. Ind. (Milan)*, 37, 113, 1955.
165. McDonalds, H.J., Bermes, E., Shepherd, *Chromatogr. Methods*, 2, 1, 1957.
166. Herndon, J.F., Appert, H.E., Touchstone, J.C., Davis, C.N., *Anal. Chem.*, 34, 1061, 1962.
167. Heftman, E., Krochta, J.M., Farkas, D.F., Schwimmer, S., *J. Chromatogr.*, 66, 365–369, 1972.
168. Finley, J.W., Krochta, J.M., Heftman, E., *J. Chromatogr.*, 157, 435–439, 1978.
169. Hostettmann, K., Hostettmann-Kaldas, M., Sticher, O., *J. Chromatogr.*, 202, 154–156, 1980.
170. Botz, L., Nyiredy, S., Wehrli, E., Sticher, O., *J. Liq. Chromatogr.*, 13, 2809–2828, 1990.
171. Botz, L., Dallenbach, K., Nyiredy, S., Sticher, O., *J. Planar Chromatogr.*, 3, 80–86, 1992.
172. Nyiredy, S., Planar chromatography, in: Chromatography, Heftmann, E. (Ed.), Elsevier, Amsterdam, Oxford, New York, Basel, 1992, pp. A109–A150.

173. Ismailov, N.A., Schraiber, M.S., *Farmatzija*, 3, 1, 1938.
174. Ripphahn, J., Halpaap, H., Application of a new high-performance layer in quantitative TLC, in: *HPTLC High Performance Thin Layer Chromatography*, Zlatkis, A., Kaiser, R.E. (Eds.), Elsevier, Amsterdam, 1977, pp. 189–221.
175. Botz, L., Nyiredy, S., Sticher, O., *J. Planar Chromatogr.*, 3, 401–406, 1990.
176. Nyiredy, S., in: Sherma, J., Fried, B. (Eds.), *Handbook of Thin-Layer Chromatography*, 3rd edn., Marcel Dekker, New York, Basel, pp. 307–337, 2003.
177. Kaiser, R.E., *J. High Resol. Chromatogr. Chromatogr. Commun.*, 3, 164–168, 1978.
178. Studer, A., Traitler, H., *J. High Resol. Chromatogr. Chromatogr. Commun.*, 9, 218–223, 1986.
179. Issaq, H.J., *J. Liq. Chromatogr.*, 3, 789–796, 1980.
180. Giddings, J.C., Use of multiple dimensions in analytical separations, in: *Multi-Dimensional Chromatography*, Ed. Cortes, H.J., Marcel Dekker, New York, 1990, pp. 251–299.
181. Nyiredy, Sz., *LC-GC Eur.*, 16, 52–59, 2003.
182. Nyiredy, Sz., Multidimensional planar chromatography, in: *Multidimensional Chromatography*, Mondello, L., Lewis, A.C., Bartle, K.D. (Eds.), Chichester, 2002, Chapter 8, pp. 171–196.
183. Tuzimski, T., *J. Planar Chromatogr.*, 21 (2008) (issue in press).

8 Derivatization, Detection (Quantification), and Identification of Compounds Online

Bernd Spangenberg

CONTENTS

8.1   INTRODUCTION

Thin layer chromatography (TLC) and especially high performance thin layer chromatography (HPTLC) are flexible, fast, and inexpensive off-line separation techniques, suitable especially in phytoanalysis [1]. HPTLC, in comparison with TLC, allows a better separation owing to a smaller particle size of the stationary phase. Sample application for separation is done directly on the stationary phase and subsequently a solvent flows through the stationary phase achieved by capillary force forming the mobile and the stationary phase. Special staining detection methods make TLC or HPTLC the method of choice for rapid analysis of medicinal plants and their preparations [2,3].

8.2 PHYSICAL METHODS ONLINE

8.2.1 DIRECT DETECTION IN ABSORPTION

In planar chromatography light is used for detecting separated sample spots by illuminating the TLC plate from the top with light of known intensity (I_0). If the illuminating light shows higher intensity than the reflected light (J), a fraction of light must be absorbed by the sample (the analyte) or the TLC layer. If a sample spot absorbs light, the reflected light intensity of this spot (J) is smaller than the illuminating light. The difference between these light intensities is absorbed by the sample. The definition of the total absorption coefficient a is

$$I_{abs} = I_0 - J = aI_0. \tag{8.1}$$

Increasing sample amounts will induce a decreasing light reflection. Therefore a transformation algorithm is needed which turns decreasing light intensities into increasing signal values. Ideally there should be a linear relationship between the transformed measurement data (TMD) and the analyte amount [4–7].

Theoretical considerations lead to the following equation for transformation purposes that show linearity between the TMD and the absorption coefficient [8]:

$$\text{TMD}(\lambda) = k\left(\frac{I_0}{J} - \frac{J}{I_0}\right) + \left(\frac{J}{I_0} - 1\right) = \frac{a}{1 - a}, \tag{8.2}$$

where
  k is the backscattering factor ($k \geq 0$ and $k \leq 1$)
  I_0 represents the illuminating light intensity at different wavelengths
  J is the intensity of reflected light at different wavelengths
  a represents the absorption coefficient

Equation 8.2 is split into two parts, the absorption share and a fluorescence share. The first term in Equation 8.2 describes the light absorption and is dependent on the backscattering factor, whereas the second term describes the fluorescence of the analyte [8,9]. The value of the backscattering factor k is in the range between 0 and 1. The backscattering factor depends on the scattering quality of the stationary phase.

8.2.1.1 The Reversal Reflectance Formula

For $k = 1$, Expression 8.2 describes a situation where all the light is reflected from the plate surface. No inner parts of the TLC layer are illuminated and light absorption occurs only at layer-top. With $k = 1$, Expression 8.3 can be derived from Equation 8.2.

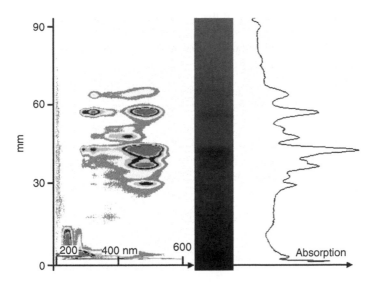

FIGURE 8.1 Contour plot of a ginkgo biloba extract separation calculated according to Expression 8.3. The track was stained by using the Anisaldehyde–sulfuric acid reagent after separation on silica gel with ethyl acetate, acetic acid, formic acid, and water $100 + 11 + 11 + 26$ (v/v).

$$\text{TMD}(\lambda) = \left(\frac{I_0}{J} - 1\right) = \frac{a}{1 - a}. \tag{8.3}$$

Expression 8.3 transforms the light losses caused by absorptions into positive values [8,9]. The measurement result of a single TLC track is best visualized as a contour plot.

A contour plot comprises the measurement data of a single track at different wavelengths. To measure a contour plot, a track of a TLC plate is scanned by use of a diode-array detector. Usually the plate is moved below an interface, which illuminates the plate at different wavelengths and detects the reflected light. For each wavelength the reflected light intensity (J) and the light intensity (I_0) of the illuminating lamp are measured. A contour plot comprises the TMD data at different wavelengths and different track locations. In other words, it summarizes the measured spectra in the y-axis at different separation distances. Mostly ten spectra are measured within 1 mm separation distance.

Figure 8.1 shows a photo of a ginkgo biloba extract separation and the contour plot of the track. In y-axis the different compounds of the ginkgo extract are visualized as peaks at different wavelengths. In the x-axis the positive values of Equation 8.4 are represented as spectra at different separation distances. The wavelength range is 200–600 nm.

8.2.1.2 The *Kubelka/Munk* Theory

The *Kubelka/Munk* theory was first published in the year 1931 and is based on the assumption that half of the scattered flux is directed forwards and half backwards [4–6]. Both fluxes show the same intensity. According to *Kubelka* and *Munk* [5] scattering in each layer will illuminate the next layer above and below with half of its nonabsorbed light intensity. With the abbreviation $R_0 = J/I_0$ and with $k = 1/2$ the following expression results:

$$\text{TMD}(\lambda, k = 1/2) = \frac{(1 - R_0)^2}{2R_0} = \frac{a}{1 - a}. \tag{8.4}$$

The *Kubelka/Munk* equation comprises both the absorption and the fluorescence signals of an analyte [8]. This expression is therefore recommended to get a quick overview of which kind of substances are being separated.

8.2.2 DIRECT DETECTION IN FLUORESCENCE

For $k = 0$ no incident light is reflected to the plate top [8]. Light leaving the TLC plate at the top must therefore be fluorescence light.

$$\text{TMD}(\lambda, k = 0) = \left(\frac{J}{I_0} - 1 \right). \tag{8.5}$$

In general, the fluorescent light is shifted to higher wavelengths (λ_F) in comparison with the absorbed light (λ_A). That means that the fluorescence usually shows lower energy than the absorbed light. A contour plot evaluated by the fluorescence formula instantly reveals compounds at the track showing fluorescence.

Figure 8.2 shows left the fluorescence contour plot of a ginkgo biloba extract evaluated with Equation 8.5 in the wavelength range from 400 to 600 nm. In y-axis the different compounds of the ginkgo extract are visualized as peaks at different wavelengths. In the x-axis the positive values of Equation 8.5 are represented as spectra at different separation distances. The fluorescence densitogram (right) is evaluated according to the fluorescence formula (8.5).

It is well known that fluorescence from an RP-18 phase looks much brighter than from a silica-gel plate, because the coating of RP-18 material blocks a nonradiative deactivation of the activated sample molecules. By spraying a TLC plate with a viscous liquid, e.g., paraffin oil dissolved in hexane (20%–67%), the fluorescence of a sample can be tremendously enhanced. The mechanism behind fluorescence enhancement is to keep molecules at a distance either from the stationary layer or from other sample molecules [10]. Therefore not only paraffin oil but also a number of different molecules show this enhancement effect.

8.2.3 FLUORESCENCE QUENCHING

For the detection of UV absorbing substances simply by eye, TLC plates are very often prepared with a fluorescence indicator. Commonly an inorganic dye (manganese-activated zinc-silicate) is used. This dye absorbs light at 254 nm showing a green fluorescence at ~520 nm. Sample molecules in the layer cover the fluorescent dye and inhibit its light absorption. In comparison with an uncovered

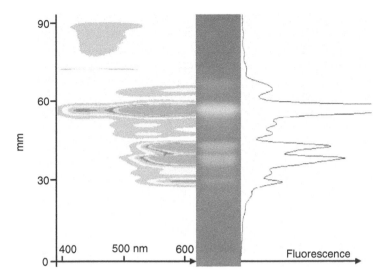

FIGURE 8.2 Fluorescence contour plot of a of a ginkgo biloba extract evaluated with Equation 8.5 (left), the stained track by use of NEU-reagent (middle) and the fluorescence densitogram according to the fluorescence formula (8.5).

area, sample spots or sample bands show lower light intensity in the vis region because the covered fluorescent dye cannot transform absorbed light into a fluorescence emission. Dark zones on a bright fluorescent background will indicate the position of the components. The term "fluorescence quenching" is often used for this decrease of reflected light intensity. However, the sensitivity of this detection method is lower than the sensitivity of reflection measurements.

8.2.4 MASS SPECTROMETRY ONLINE

The combination of TLC and mass spectroscopy has often been described [11–14]. For desorption from the plate and for ionization several techniques are published such as matrix-assisted laser desorption/ionization mass spectrometry (MALDI techniques using an UV/IR-Laser) [15–19], by atmospheric pressure chemical ionization (APCI) [20–22], single ion beam (SIMS) [23], or desorption electro spray ionization (DESI) [24–26]. All these methods are able to measure the molecule mass and characteristic fragments produced during the measurement process. This is very important information to identify—in combination with UV–Vis Spectra—the nature of TLC zones. Nevertheless, quantification is possible only by using an internal standard, because desorption processes are difficult to reproduce in all methods.

Recent publications of three new methods offer a quantitative TLC–MS approach without using an internal standard. Firstly this is the use of a "ChromeXtraktor" device [27], the coupling of TLC- or HPTLC plates with a DART-device (Direct Analysis in Real Time) [28] and the APGD-method (Atmospheric Pressure Glow Discharge) [29]. The TLC-Extractor is a 4×2 mm stamp, which is set on the sample zone. An HPLC-pump pump mobile phase through the part of the TLC plate which is covered by the stamp and extract the analyte which is transported to an MS-device. The spatial resolution of this simple and robust working method is 2 mm. The detection limit is in the pg range [27]. The DART system works with an excited helium gas stream, forming protonated water clusters from the surrounding air. These clusters transfer their energy to the analyt, forming molecule cations. The spatial resolution of this device is better than 7 mm and the detection limit is in the lower ng range.

In the APGD method a simple ion source using plasma under atmospheric pressure desorbs and ionizes the analyt. The spatial resolution of this technique is better than 2.5 mm. All TLC–MS systems offer structural information about the analyt and extend the scope of the TLC technique [32].

8.3   CHEMICAL METHODS ONLINE

A scanner is not necessarily recommended for the identification of fluorescent substances because the fluorescent light is mainly emitted in the visible region (vis region). In this case it is possible to use human eyes as detection system.

8.3.1   NONDESTRUCTIVE UNIVERSAL DETECTION REAGENTS

The commonly used nondestructive location method in TLC or HPTLC is to expose the plate to iodine vapor in a closed chamber, which contains some iodine crystals. Iodine is lipophilic and accumulates in lipophilic sample spots, showing a brown color on a pale yellow-brown background. The same result will occur by spraying with an iodine solution. In nearly any case this iodine accumulation is totally reversible without altering the sample because outside the closed chamber iodine evaporates quickly from the plate. Caution should be taken with this iodine treatment in the case of unsaturated compounds because iodine vapor can react with double bonds [30].

8.3.2   DESTRUCTIVE UNIVERSAL DETECTION REAGENTS

Charring is often the only way of sample staining for molecules showing no reactive groups. Charring breaks down the original components into other visible compounds or in the extreme, into pure carbon. This decomposition process of organic samples mainly results in black-to-brown zones on a white background. A TLC plate with inorganic binder is recommended to avoid a black background. Some of the later mentioned charring reagents produce fluorescent zones at lower temperature (80°C–120°C) before final charring occurs at higher temperatures (above 160°C).

Sulfuric acid in different dilutions is often used as a universal reagent. The decomposition power of H_2SO_4 is enhanced when used in conjunction with $MnCl_2$ or $CuSO_4$. Phosphomolybdic acid oxidizes most organic compounds while forming a blue-grey dye and is often used as a universal reagent for reducing substances.

8.3.3   Group Selective Detection Reagents

In literature a large number of more or less selective detection reagents have been published [1,30,31]. The presented selections of reagents cover reactions, which can easily be performed. The reagents are all stable for days when stored in a refrigerator.

One of the most often-used group selective detection reagent in phytochemistry is *Neu-reagent*. A 1% Biphenyl boric acid β-aminoethylester dissolved in methanol or ethanol readily reacts with compounds showing hydroxyl groups, such as flavonoids, sugars, anthocyanines, or hydroxy acids. Various colors are formed by many natural products, which often show a bright fluorescence. Plate dipping in a viscose liquid such as polyethylene glycol can enhance the fluorescence.

Vanillin in combination with acids reacts with many compounds such as terpenoids, sterols and alkaloids and in general with lipophilic compounds, forming dark-colored zones. The mechanism is a condensation reaction of the carbonyl group with activated CH- or CH_2-groups of the analyte. A reagent with similar properties can be obtained by mixing 50 mg 4-dimethylaminobenzaldehyde with 1 mL concentrated H_2SO_4 dissolved in 100 mL ethanol [1,30,31]. This reagent is known as *van Urks reagent*, whereas mixing a solution of 50 mg 4-dimethylaminobenzaldehyde in 50 mL methanol with 10 mL concentrated HCl is known as *Ehrlich's reagent* [30]. Similar reactions can also be performed using Anisaldehyde (4-methoxybenzaldehyde) in combination with sulfuric acid, which reacts with sugars and glycosides, too. This mixture is known as *Eckert's reagent*.

The mixture of formaldehyde in sulfuric aid is used to detect opiates and is known as *Marquis reagent*.

The *Dragendorff's* reagent has been published in different compositions, which mainly stains nitrogen-containing compounds such as alkaloids. It produces colored zones on a white background. The staining is stable on plate as well as the reagent itself.

Sulfur containing samples show colored spots when sprayed with 2,6-dibromoquinone-4-chlorimide (*Gibbs reagent*). This reagent also creates colored zones when samples contain phenols. For reactions with only phenols, the less reactive 2,6-dichloroquinone-4-chlorimide can be used under the same conditions.

Antimony chloride, such as 4% $SbCl_3$ in 50 mL $CHCl_3$ or 50 mL glacial acid forms various colors, characteristic for compounds showing carbon double bonds (*Carr-Price reagent*).

Ninhydrin (1,2,3,-Indantrione) transforms all samples containing $-NH_2$ groups such as amino acids, peptides, or amines into red or purple products. To perform a reaction at least 5–10 min heating at 120°C is necessary.

Aldehydes and ketones can be located as orange or yellow zones after reacting with 2,4-dinitrophenylhydrazine hydrochloride. In some cases heating to 100°C is necessary.

Potassium hydroxide dissolved in water or ethanol reacts with antraquinones, flavonoides, anthrones, and cumarins, forming colored and fluorescing zones.

In the year 1994, *Takao* published a staining reaction to identify radical scavenging activity [33]. *Takao* used the compound 1,1-diphenyl-2-picrylhydrazyl (DPPH), which changes its color in the presence of antioxidants such as ascorbic acid or rutin from blue-purple to colorless or yellow. The reaction is finished immediately.

Fast blue salt B contains a diazo group, which can readily react with phenolic groups and form red-colored dyes. The same specific staining shows $FeCl_3$ forming red dyes with nearly all phenolic groups. Flavonoids, tannins, and hypericines can be stained.

8.3.4 DETECTING REAGENTS FOR INDIVIDUAL COMPOUNDS

Detecting reagents for individual compounds stain only a single group of compounds specifically. For example acetic anhydride forms fluorescent dyes with ginkgolides, emitting a greenish-blue color when irradiated by UV light.

Aniline–diphenylamine in combination with phosphoric acid specifically stains sugars and glycosides forming grey-blue colors.

Nitric acid forms red colors with ajmaline and brucine.

Hydrochloric acid mixed with acetic acid reacts with valepotriates containing a diene structure to blue colored dyes. Iodine–hydrochloric acid reacts with purine to grey-colored dyes. Iodoplatinate reagent stains all organic compounds containing nitrogen, such as alkaloids forming a blue-violet color. Palladium-II-chloride forms yellow-red colors with thioles. It is the perfect reagent for staining Alium species.

Tin (II) chloride reacts with aristolochic acids forming blue fluorescing colors in UV [3].

8.4 QUANTIFICATION IN PLANAR CHROMATOGRAPHY

8.4.1 QUANTIFICATION BY ABSORPTION

We assume that in TLC two different types of absorption occur simultaneously within the layer. The stationary phase will absorb light as well as sample molecules will do. Besides this loss of incident light, a loss of light can also occur when transmission light leaves the plate at the back or at the plate sides. We can gather all plain plate light-losses without any ($I_{abs,u}$) in the plate absorption coefficient a_u. A sample therefore "sees" the light intensity I_0 of the illumination lamp minus the light absorption of the plain TLC plate. If the sample absorption coefficient is defined as a_m following equation is valid [7,8]:

$$\frac{a}{(1-a)} = m\frac{a_m}{(1-a_m)} + \frac{a_u}{(1-a_u)}. \tag{8.6}$$

There is a linear connection between the mass m of a compound and its corresponding signal. The intercept contains the constant a_u only and therefore describes the plain plate absorption. Equation 8.6 shows linearity between transformed measurement signals and total absorption coefficient a and direct linearity between the transformed data and the sample mass. With transformed intensity data according to Equation 8.6 the part of light absorbed by the sample can be separated from the light absorbed by the layer. The diffuse reflected light intensity of the plain plate J_0 can easily be taken at a surface area free of any other compounds. If this signal is used instead of the lamp intensity I_0 to calculate the relative reflection R, we assume that the TLC plate does not show any loss of light instead of sample absorption. The corrected relative reflection R can be written as

$$R = \frac{J}{I_0 - I_{\text{abs},u}} = \frac{J}{J_0}. \tag{8.7}$$

As a result the plate absorption intensity ($I_{\text{abs},u}$) becomes zero but in fact the light intensity of the lamp I_0 is only replaced by J_0. Mathematically the original light flux of the lamp I_0 is reduced by the whole loss of light in the plate surface to J_0 and hence we pretend $I_{\text{abs},u}$ to be zero and this makes a_u zero as well. From Equations 8.2, 8.6, and 8.7 we obtain the fundamental expression for quantitative TLC.

$$k\left(\frac{1}{R} - R\right) + (R - 1) = m\frac{a_m}{(1 - a_m)}, \tag{8.8}$$

where
  R represents relative reflectance
  k is the backscattering factor ($k \geq 0$ and $k \leq 1$)
  a_m represents mass absorption coefficient
  m is the sample mass

8.4.2 QUANTIFICATION BY FLUORESCENCE

The term fluorescence is used for a transformation of absorbed light into fluorescence (J_F). The extent of this transformation is described by the quantum yield factor q_F. The light intensity absorbed by the sample can be calculated from the light intensity reflected from the clean plate surface (J_0) minus the light intensity (J) reflected from a sample spot.

$$J_F = q_F(I_0 - J) = q_F I_0 \left(1 - \frac{J}{I_0}\right), \tag{8.9}$$

where
  J_F is the emitted fluorescence intensity
  q_F represents fluorescence quantum yield factor
  I_0 is the illuminating light intensity at the wavelengths of absorption
  J represents light intensity reflected from the sample

Taking Equation $J/I_0 = (1 - a)$ into account, which can be extracted from Expression 8.3, a linear relationship results between the absorption coefficient a and the fluorescence signal.

$$J_F = q_F I_0 a. \tag{8.10}$$

If the fluorescence intensity J_F is corrected by the reflected intensity of the clean plate at the fluorescence wavelength (J_{0F}), the result is Expression 8.11.

$$\text{TMD}(\lambda, k = 0) = \left(\frac{J}{J_0} - 1\right)_F = q_F \frac{J_0}{J_{0F}} m a_m, \qquad (8.11)$$

where

J_F represents emitted fluorescence intensity

J_{0F} is the reflected light intensity of a clean plate at the wavelength of fluorescence

q_F represents fluorescence quantum yield factor

J_0 is the reflected light intensity of a clean plate at the wavelength of absorption

a_m mass absorption coefficient

m is the sample mass

The fluorescent light intensity is directly proportional to the sample amount in the layer. In Equations 8.10 and 8.11 a crucial advantage of fluorescence spectrometry can be seen in comparison with measurements in absorption: the fluorescence signal increases with increasing illumination power.

8.4.2.1   Reagents

Acetic anhydride reagent. Spray the TLC plate with or dip in acetic anhydride, heat for about 30 min at 150°C, and then inspect under UV-365 nm. This reagent is used for the detection of ginkgolides [2].

Acetic anhydride sulfuric acid reagent (Liebermann–Burchard reagent). For preparation carefully dilute 1 mL of concentrated H_2SO_4 and 20 mL acetic anhydride to 100 mL with chloroform (or absolute ethanol). Spray or dip the plate and dry. This is used for the detection of sterols and terpenoides [3].

Acetic anhydride sulfuric acid copper-reagent. For preparation carefully mix 10 mL of concentrated H_2SO_4 with 90 mL acetic anhydride. Dissolve 3 g copper-II-acetate in 100 mL of 8% phosphoric acid. After plate dipping or spraying, heat for 15 min at 125°C. This is a universal reagent.

Ammonium vapor (NH_3). Heat the plate at 120°C for 10 min in a closed chamber in the presence of $(NH_4)_2HCO_3$. Carbonyl compounds show a bright fluorescence under UV 365 nm.

Anisaldehyde–acetic acid reagent. Mix 0.5 mL anisaldehyde with 10 mL glacial acetic acid. Spray or dip the plate and then heat at 120°C for 7–10 min. This is used for the detection of petasin/isopetasin [2].

Anisaldehyde–sulfuric acid reagent. Mix 0.5 mL anisaldehyde with 10 mL glacial acetic acid, followed by 85 mL methanol and 5 mL concentrated sulfuric acid, in that order. Spray or dip the TLC plate, heat at 100°C for 5–10 min, then evaluated in vis or under UV-365 nm. The reagent has only limited stability and is no longer useable when the color has turned to red-violet. This is used for the detection of terpenoids, propylpropanoids, pungent and bitter principles, saponins [3].

Antimony-III-chloride reagent (SbCI₃, CarrPrice reagent). Make a 20% solution of antimony-III-chloride in chloroform or ethanol. Spray or dip the TLC plate and heat for 5–6 min at 110°C. Evaluate in vis or UV-365 nm. This is used for the detection of cardiac glycosides, saponins [2].

Barton's reagent

(a) 1 g potassium hexacyanoferrate (III) dissolved in 100 mL water.
(b) 2 g iron-III-chloride in 100 mL water.

Spray or dip the TLC plate in a 1:1 mixture of (a) and (b). Evaluate in vis. This is used for the detection of gingeroles (Zingiberis rhizoma) [2].

Benzidine reagent. Dissolve 0.5 g benzidine in 10 mL glacial acetic acid and fill up to 100 mL with ethanol. Dip the plate, and heat and evaluate in vis (Caution: Benzidine is carcinogenic.). This used for the detection of aucubin (Plantaginis folium) [2].

Chloramine–trichloroacetic acid reagent. Prepare 10 mL 3% aqueous chloramine T solution (sodium tosylchloramide) and mix with 40 mL 25% ethanolic trichloroacetic acid. Spray or dip the plate and heat at 100°C for 5–10 min and evaluate under UV-365 nm. This is used for the detection of cardiac glycosides [2].

Copper (II) sulfate. Dissolve 20 g copper (II) sulfate in 180 mL water and 16 mL *o*-phosphoric acid (85%). Immerse the plate in the reagent for 1 s and then heat at 180°C for 5–10 min [3]. This is a Universal reagent.

2,6-Dichloroquinone chloroimide (Gibb's reagent). Dissolve 50 mg of 2,6-dichloroquinone chloroimide in 200 mL ethyl acetate or methanol. Spray or dip the plate and immediately expose to ammonia vapor. This is used for the detection of arbutin and capsaicin [2,3].

Dinitrophenylhydrazine reagent (DNPH reagent). Dissolve 0.1 g of 2,4-dinitrophenylhydrazine in 100 mL methanol, followed by the addition of 1 mL of 36% hydrochloric acid. Spray or dip the plate and evaluate immediately in vis. Heating may be needed. This is used for the detection of ketones and aldehydes [2,3].

DNPH–acetic acid–hydrochloric acid reagent. Dissolve 0.2 g 2,4-dinitrophenylhydrazine in a solvent mixture of 40 mL glacial acetic acid (98%), 40 mL hydrochloric acid (25%), and 20 mL methanol. Spray or dip the plate and evaluate in vis. The plate is heated at 100°C for 5–10 min and evaluated again in vis. This is used for the detection of valepotriates (Valeriana). Chromogenic dienes react without warming [2].

Dragendorff reagent. Solution (a): Dissolve 0.85 g bismuth nitrate in 10 mL glacial acetic acid and 40 mL water under heating. If necessary, filter.

Solution (b): Dissolve 8 g potassium iodide in 30 mL water.

Just before spraying or dipping, solution (a) and (b) are mixed with 4 mL acetic acid and 20 mL water. This is used for the detection of alkaloids and heterocyclic nitrogen compounds [2,3].

Dragendorff' reagent, followed by sodium nitrite. After treatment with Dragendorff reagent, the plate may be additionally sprayed with 10% aqueous sodium nitrite, thereby intensifying the dark brown colored zones. This is used for the detection of alkaloids and heterocyclic nitrogen compounds [2].

Dragendorff' reagent, followed by H_2SO_4. After treatment with Dragendorff reagent, the plate is additionally sprayed with 10% ethanolic sulfuric acid, thereby intensifying the bright orange colored zones. This is used for the detection of alkaloids, heterocyclic nitrogen compounds [2].

4-dimethylamino benzaldehyde reagent (EP reagent). Dissolve 0.25 g 4-dimethylamino benzaldehyde in a mixture of 45 mL 98% acetic acid, 5 mL 85% *o*-phosphoric acid and 45 mL water, followed by 50 mL concentrated sulfuric acid (under cooling with ice). Evaluate the sprayed plate in vis.

For dipping, 0.25 g 4-dimethylamino benzaldehyde is dissolved in 50 mL acetic acid and 3 mL 85% *o*-phosphoric acid is added [30]. This is used for the detection of proazulene (Matricariae flos) [2].

After heating at 100°C for 5–10 min proazulene gives a blue-green color (vis.) The blue color of azulene is intensified by the reagent [2].

4-dimethylamino benzaldehyde reagent (Ehrlich's reagent). Dissolve 0.3 g of 4-dimethylaminobenzaldehyde in 25 mL methanol. Add 10 mL 32% HCl while cooling. The addition of one drop of a 10% aqueous iron-II-chloride solution mostly gives improved results. The plate can be sprayed or dipped. This is used for the detection of iridoids and proazulenes [3].

Fast blue salt reagent. Dissolve 0.5 g fast blue salt B (3,3'-dimethoxybiphenyl-4, 4'-bis(diazonium)-dichloride) in 100 mL water. Spray or dip the plate, dried and inspected in vis. Spraying may be repeated, using 10% ethanolic NaOH, followed again by inspection in vis. This is used for the detection of phenolic compounds [2,3].

Fast red salt reagent. Dissolve 0.5 g fast red salt B (diazotized 5-nitro-2-aminoanisole) in 100 mL water. Spray or dip the plate, followed immediately by either 10% ethanolic NaOH or exposure to ammonia vapor. This is used for the detection of amarogentin [2].

Hydrochloric acid–glacial acetic acid reagent ($HCl + CH_3COOH$). Carefully mix 8 mL concentrated hydrochloric acid and 2 mL of glacial acetic acid. Spray the plate and heat at 110°C for 10 min. Evaluation in vis or under UV-365 nm. This is used for the detection of valepotriates with diene structure (halazuchrome reaction) [2,3].

Iodine reagent. Place solid iodine in a chromatographic tank and place the developed and dried TLC plate in the tank and expose to iodine vapour.

All lipophilic compounds or compounds containing conjugated double bonds give yellow-brown (vis) zones [2,3].

Iodine–chloroform reagent. Dissolve 0.5 g iodine in 100 mL chloroform. Spray or dip the plate and keep at 60°C for about 5 min. The plate is evaluated after 20 min at room temperature in vis or under UV-365 nm. This is used for the detection of Ipecacuanha alkaloids [2].

Iodine-hydrochloric acid reagent (I + HCl)

(a) Dissolve 1 g potassium iodide followed by 1 g iodine in 100 mL 96% ethanol.
(b) Mix 25 mL 25% HCl with 25 mL 96% ethanol.

The plate is first sprayed with 5 mL of (a) followed by 5 mL of (b). For dipping, solution (a) is mixed with equal volume of (b). This is used for the detection of the purine derivatives (caffeine, theophylline, theobromine) [2,30].

Iodoplatinate reagent. Dissolve 0.3 g hydrogen hexachloroplatinate (IV) hydrate in 100 mL water and mix with 100 mL 6% potassium iodide solution. Spray the plate and evaluate in vis. This is used for the detection of nitrogen-containing compounds, e.g., alkaloids (blue-violet) [2].

For detection of Cinchona alkaloids the plate is first sprayed with 10% ethanolic H_2SO_4 and then with Iodoplatinate reagent [2].

Iron-III-chloride reagent (FeCl₃). Dissolve 1 g Iron-(III)-chloride in 5 mL water and dilute to 100 mL with ethanol. Spray or dip the plate and evaluated in vis. This is used for the detection of phenols, flavonoids, tannins, plant acids, ergot alkaloids, hops bitter principles and hypericines [3].

Kedde reagent (3,5-dinitrobenzoic acid KOH-reagent). Mix equal amounts of freshly prepared 2% methanolic 3,5-dinitrobenzoic acid and 1 M methanolic KOH (5.7 g dissolved in 100 mL methanol). Spray or dip the plate and evaluate in vis. This is used for the detection of cardenolides [2].

Liebermann–Burchard reagent. See Acetic anhydride sulfuric acid reagent.

Manganese-(II)-chloride (MgCl₂). Dissolve 200 mg $MnCl_2 \cdot 4H_2O$ in 30 mL water and subsequently diluted with 30 mL methanol. Finally the amount of 2 mL concentrated H_2SO_4 is carefully added. After spraying, 15 min heating at 120°C is necessary to complete the reaction. All kinds of organic compounds form brown spots on a white background.

Marquis' reagent. Dilute 3 mL formaldehyde to 100 mL with concentrated sulfuric acid. Evaluate the plate in vis, immediately after spraying or dipping. This is used for the detection of morphine, codeine, thebaine [2,3].

Nitric acid (HNO₃). Expose the plate to nitric acid vapor in a chromatographic chamber for 1–10 min [3].

Ninhydrin reagent. Dissolve 30 mg ninhydrin in 10 mL *n*-butanol and mix 0.3 mL 98% acetic acid. Dip or spray and heat for 5–10 min under observation and evaluate in vis. This is used for the detection of amino acids, biogenic amines, and ephedrine [2].

NEU-reagent. Dip or spray the plate with 1% methanolic diphenylborinic acid-β-ethylamino ester (diphenylboryloxyethylamine), followed by a 5% methanolic poly-ethylene glycol-400 solution. Heat the plate at 100°C for 3 min [2,3]. This is used for the detection of flavonoids, aloins. Intense fluorescence is produced under

UV-366 nm. Polyethylene glycol increases the fluorescence intensity. The fluorescence behaviour is structure dependent [2].

Palladium-II-chloride reagent (PdCl₂). Add 1 mL concentrated HCl to a solution of 0.5% palladium-II-chloride in water. Spray or dip the mixture. This is used for the detection of Allium species (yellow-brown zones in vis.) [2].

Phosphomolybdic acid reagent [H₃(P(Mo₃O₁₀)₄)]. Spray solution: Dissolve 5–20 g phosphomolybdic acid in 100 mL ethanol. Dipping solution: Dissolve 250 mg phosphomolybdic acid in 50 mL ethanol. Spray or dip the plate and heat at 100°C for 5 min. The color of zones can be optimized by exposing the plate to ammonia vapor [3]. This is used for the detection of constituents of fatty oils, reducing substances, steroids, essential oils [2].

Phosphomolybdic acid reagent ([H₃(P(Mo₃O₁₀)₄)] and H₂SO₄). Dissolve 4 g phosphomolybdic acid in 40 mL hot water and carefully add 60 mL concentrated sulfuric acid to the cooled solution [2]. Rhaponticoside and deoxyrhaponticoside form strong, blue zones [2].

Potassium hydroxide reagent (KOH, Bornträger reaction). Dissolve 5–10 g KOH in 10 mL water and dissolve in 200 mL ethanol (96%). Spray or dip the plate and evaluate in vis or under UV-365 nm, with or without warming. This is used for the detection of anthraquinones (red), anthrones (yellow, UV-366 nm), and coumarins (blue fluorescence under UV-366 nm) [2].

Potassium permanganate–sulfuric acid reagent. Carefully dissolve 0.5 g potassium permanganate in 15 mL concentrated sulfuric acid, while cooling in ice (warning: explosive manganese heptoxide is formed). This is used for the detection of fenchone the plate is sprayed (or dipped) first with phosphomolybdic acid reagent followed by 10 min heating at 110°C, followed by spraying with permanganate reagent (5 min at 110°C; blue colors are formed) [2].

Sulfuric acid (H₂SO₄). Carefully add 20 mL sulfuric acid to 180 mL ice-cold methanol. Immerse the plate in reagent for 1 s then heat at 100°C for 5 min [3]. This is a General reagent.

Takao's reagent. Dissolve 100 mg of 1,1-diphenyl-2-picrylhydrazyl (DPPH) in 50 mL acetone or methanol [32].

Tin (II) chloride–hydrochloric acid. Dilute 1.5 mL HCl (32%) with 8 mL water. Dissolve 1 g SnCl₂ · 2H₂O in this mixture. Spray the plate heat at 100°C for a maximum of 1 min [3]. This reagent should always be freshly prepared. This is used for the detection of aristolochic acids [3].

Trichloroacetic acid–potassium hexacyanoferrate-iron-III-chloride reagent

(a) 25% trichloroacetic acid in chloroform.
(b) Mix 1% aqueous potassium hexacyanoferrate with an equal volume of 5% aqueous iron-III-chloride.

Spray the plate with solution (a) and heat at 110°C for 10 min. Spray the plate with solution (b) and evaluate in vis. This is used for the detection of sinalbin and sinigrin [2].

Vanillin–glacial acetic acid reagent. Dissolve 0.8 g vanillin (4-hydroxy-3-methoxy-benzaldehyde) in 40 mL glacial acetic acid and add 2 mL concentrated H_2SO_4. Spray the plate, heat for 3–5 min at 110°C and evaluate in vis. The reagent is unstable and should be discarded when coloration appears. This is used for the detection of terpenoids, sterols, salicin, ergot alkaloides, and most lipophilic compounds [3].

Vanillin–phosphoric acid reagent. Dissolve 1 g vanillin in 25 mL ethanol and mix with 25 mL water and 35 mL *o*-phosphoric acid. Spray or dip the plate and heat for 10 min at 100°C and subsequently evaluate in vis or under UV-366 nm [30]. This is used for the detection of terpenoids, lignanes, and cucurbitacins [2].

Vanillin-sulfuric acid reagent

(a) 1% ethanolic vanillin solution.
(b) 10% ethanolic sulfuric acid.

Dip or spray the plate with (a) followed immediately by (b). Evaluate the plate in vis. after heating at 110°C for 5–10 min. For dipping, 0.4 g vanillin are dissolved in 100 mL diethyl ether and finally 0.5 mL concentrated sulfuric acid are added. Heat the plate after dipping for 3–5 min at 110°C and evaluate in vis. This is used for the detection of essential oils (terpenoids, phenylpropanoids) [2].

REFERENCES

1. Nyiredy, Sz., (editor), *Planar Chromatography—A Retrospective View for the Third Millennium*, 11th ed., Springer, Budapest, 2001.
2. Wagner, H. and Bladt, S., *Plant Drug Analysis—A Thin Layer Chromatography Atlas*, Springer, Berlin, 1996.
3. Reich, E. and Schibli, A., *High-Performance Thin-Layer Chromatography for the Analysis of Medicinal Plants*, Thieme, Stuttgart, 2007.
4. Kortüm, G., *Reflextionsspektroskopie*, Springer, Berlin, 1969.
5. Kubelka, P. and Munk, F., Ein beitrag zur optik der farbanstriche, *Z. Tech. Phys.* 11a, 593–601, 1931.
6. Kubelka, P., New contributions to the optics of intensely light-scattering materials. II. Nonhomogeneous layers, *J. Opt. Soc. Am.* 44, 330–335, 1954.
7. Spangenberg, B., Post, P., and Ebel, S., Fibre optical scanning in TLC by use of a diode-array detector—Linearization models for absorption and fluorescence evaluations, *J. Planar Chromatogr.* 15, 88–93, 2002.
8. Spangenberg, B., Does the Kubelka–Munk theory describe TLC evaluations correctly? *J. Planar Chromatogr.* 19, 332–341, 2006.
9. Spangenberg, B. and Klein, K.-F., New evaluation algorithm in diode-array thin-layer chromatography, *J. Planar Chromatogr.* 14, 260–265, 2001.
10. Spangenberg, B., Lorenz, K., and Nasterlack, St., Fluorescence enhancement of pyrene measured by thin-layer chromatography with diode-array detection, *J. Planar Chromatogr.* 16, 331–337, 2003.
11. Ford, M.J., Deibel, M.A., Tomkins, B.A., and Van Berkel, G.J., Quantitative thin-layer chromatography/mass spectrometry analysis of caffeine using a surface sampling probe

electrospray ionization tandem mass spectrometry system, *Anal. Chem.* 77, 4385–4389, 2005.

12. Jautz, U. and Morlock, G., Efficacy of planar chromatography coupled to tandem mass spectrometry for employment in trace analysis, *J. Chromatogr. A*, 1128, 244–250, 2006.
13. Morlock, G. and Schwack, W., Determination of isopropylthioxanthone (ITX) in milk, yoghurt and fat by HPTLC-FLD HPTLC-ESI/MS and HPTLC-DART/MS, *Anal. Bioanal. Chem.* 385, 586–595, 2006.
14. Dreisewerd, K., Kolbl, S., Peter-Katalinić, J., Berkenlcamp, S., and Pohlentz, G., Analysis of native milk oligosaccharides directly from thin-layer chromatography plates by matrix-assisted laser desorption/ionization orthogonal-time-of-flightmass spectrometry with glycerol matrix, *J. Am. Soc. Mass Spectrom.* 17, 139–150, 2006.
15. Wilson, I.D., The state of the art in thin-layer chromatography-mass spectrometry: a critical appraisal, *J. Chromatogr. A* 856, 429–442, 1999.
16. Dreisewerd, K., Müthing, J., Rohlfing, A., Meisen, I., Vukelić, Ž., Peter-Katalinić, J., Hillenkamp, F., and Berkenkamp, St., Analysis of gangliosides directly from thin-layer chromatography plates by infrared matrix-assisted laser desorption/ionization orthogonal time-of-flight mass spectrometry with a glycerol matrix, *Anal. Chem.* 77, 4098–4107, 2005.
17. Salo, P.K., Salomies, H., Harju, K., Yli-Kauhaluoma, J., Ketola, R.A., Kotiaho, T., and Kostiainen, R., Analysis of small molecules by ultra thin-layer chromatography-atmospheric pressure matrix-assisted laser desorption/ionization mass spectrometry, *J. Am. Soc. Mass Spectrom.* 16, 906–915, 2005.
18. Crecelius, A., Clench, M.R., Richards, D.S., and Parr, V., Quantitative determination of Piroxicam by TLC-MALDI TOF MS, *J. Pharm. Biomed. Anal.* 35, 31–39, 2004.
19. Santos, L.S., Haddad, R., Höehr, N.F., Pill, R.A., and Eberlin, M.N., Fast screening of low molecular weight compounds by thin-layer chromatography and "on-spot" MALDI-TOF mass spectrometry, *Anal. Chem.* 76, 2144–2147, 2004.
20. Peng, S., Edler, M., Ahlmann, N., Hoffmann, T., and Franzke, J., A new interface to couple thin-layer chromatography with laser desorption/atmospheric pressure chemical ionization mass spectrometry for plate scanning, *Rapid Commun. Mass Spectrom.* 19, 2789–2793, 2005.
21. Van Berkel, G.J., Llave, J.J., De Apadoca, M.F., and Ford, M.J., Rotation planar chromatography coupled on-line with atmospheric pressure chemical ionization mass spectrometry, *Anal. Chem.* 76, 479–482, 2004.
22. Asano, K.G., Ford, M.J., Tomkins, B.A., and Van Berkel, G.J., Self-aspirating atmospheric pressure chemical ionization source for direct sampling of analytes on surfaces and in liquid solutions, *Rapid Commun. Mass Spectrom.* 19, 2305–2312, 2005.
23. Orinak, A., Vering, G., Arlinghaus, H.F., Andersson, J.T., Halas, L., and Orinakova, R., New approaches to coupling TLC with TOF-SIMS, *J. Planar Chromatogr.* 18, 44–50, 2005.
24. Kertesz, V., Ford, M.J., and Van Berkel, G.J., Automation of a surface sampling probe/electrospray mass spectrometry system, *Anal. Chem.* 77, 7183–7189, 2005.
25. Ford, M.J. and Van Berkel, G.J., An improved thin-layer chromatography/mass spectrometry coupling using a surface sampling probe electrospray ion trap system, *Rapid Commun. Mass Spectrom.* 18, 1303–1309, 2004.
26. Ford, M.J., Deibel, M.A., Tomkins, B.A., and Van Berkel, G.J., Quantitative thin-layer chromatography/mass spectrometry analysis of caffeine using a surface sampling probe electrospray ionization tandem mass spectrometry system, *Anal. Chem.* 77, 4385–4389, 2005.
27. Luftmann, H., A simple device for the extraction of TLC spots: direct coupling with an electrospray mass spectrometer, *Anal. Bioanal. Chem.* 378, 964–968, 2004.

28. Cody, R.B., Laramee, J.A., and Dupont Durst, H., Versatile new ion source for analysis of materials in open air under ambient conditions, *Anal. Chem.* 77, 2297–2302, 2005.
29. Morlock, G., Kopplungen der Planar-Chromatographie mit der Massenspektrometrie, *CLB Chemie Labor und Biotechnik*, 57, 343–348, 2006.
30. Jork, H., Funk, W., Fischer, W., and Wimmer, H., *Thin layer Chromatography*, Vol. 1a and b, VCH Verlagsgesellschaft, Weinheim, 1990 and 1993.
31. Barret, G.C., Nondestructive detection methods in paper and thin layer chromatography, in *Advances in Chromatography*, Vol. 11, Marcel Dekker, New York, 1974, pp. 145–179.
32. Van Berkel, G.J., Ford, M.J., and Deibel, M.A., Thin-layer chromatography and mass spectrometry coupled using desorption electrospray ionization, *Anal. Chem.* 77, 1207–1215, 2005.
33. Takao, T., Kitatani, F., Watanabe, N., Yagi, A., and Sakata, K., A simple screening method for antioxidants and isolation of several antioxidants produced by marine bacteria from fish and shellfish, *Biosci. Biotech. Biochem.* 51, 1780–1783, 1994.

9 Biodetection and Determination of Biological Activity of Natural Compounds

Ernő Tyihák, Ágnes M. Móricz, and Péter G. Ott

CONTENTS

9.1 INTRODUCTION

There are three main mechanisms by which one microorganism may limit the growth of another microorganism: antibiosis, micro-parasitism, and competition for resources [1]. Viullemin in 1899 applied the word antibiosis to situations in which one organism destroyed another to sustain its own life [2]. Among the cell proliferation promoting and retarding factors, antibiotics are the most important molecules originating from the antibiosis processes. The word antibiotic is very broad and includes widely disparate mechanisms. Microbial toxins have recently been discussed in detail [3]; they are usually considered to be poisonous to higher animals and humans. The antibiotic penicillin can be a toxin to a person sensitive to it.

The infectious agents can be the causes or risk factors for various human diseases, which are one of the most changeable and exciting areas of clinical practice [4,5]. It is important to note that 98% of death of children in developing countries has resulted from infectious diseases [6].

Although the "antibiotic era" is barely five decades old, mankind is now faced with the global problem of emerging resistance in virtually all pathogens. Surveys have revealed that almost no group of antibiotics was introduced to which resistance has not been observed [7]. The resistance of microbial strains to antibiotics is a big problem in the field of antimicrobial animal and human therapy [8]. Due to increasing microbial resistance in medicine and agriculture, discovery of new antibiotics is an especially important research objective.

The biological systems as microbes or plants contain thousands of constituents and are a valuable source of new and biologically active molecules, e.g., antibiotics, antineoplastics, herbicides, and insecticides [9,10]. Medicinal plants have been used for centuries as remedies for different human diseases because they contain substances of therapeutic value. For their investigation, it is important to have suitable biological assays and chemical screening methods. Modern aspects of the biological detection mean the discovery of new antimicrobial agents with novel modes of action through mechanism of action investigations to bioactivity-guided fractionation/isolation and development of animal experimental model.

9.2   HISTORY OF BIOAUTOGRAPHY

Goodall and Levi [11] combined for the first time in 1946 the paper chromatography (PC) and the microbial detection of antibiotics to determine the composition and purity of penicillin in a mixture. They used yet a preincubated agar layer which enabled the compounds to diffuse from the paper into the agar. The utility and the first applications of bioautography were reviewed in 1973 by Betina [12] who noted the techniques used so far and the difficulties encountered. Rios et al. [13] summarized in details the bioautographic screening methods for natural products with antimicrobial activity. They emphasized that it is difficult to use these methods as standard procedures, because of the factors (medium composition, microorganism tested, pH, incubation time, solubility of the sample in the culture, staining procedure, etc.) which affect the results. Many other reviews and chapters provide opportunities to study the usefulness of the different fields of bioautography [10,13–15].

9.3   CONVENTIONAL METHODS FOR BIOLOGICAL DETECTION
OF ANTIMICROBIAL ACTIVITY

The methods for detection of antimicrobial activity can be classified into three groups: diffusion, dilution, and bioautographic. The beneficial characters of diffusion methods are the high suitability for screening pure substances. This procedure is the agar-overlay method using a disk, hole, or cylinder, as reservoir. The agar dilution method uses a fixed amount of a substance dissolved or suspended in a suitable solvent; this solution is mixed with nutrient agar which is left to set [15]. The agar dilution method is less suitable than the diffusion technique for qualitative work and can be used for the determination of the minimum inhibitory concentration (MIC) values of a mixture (e.g., extract) or purified as well as pure compounds [15].

Rios et al. [13] proposed classification of the bioautographic technical solutions into three general variants: (1) contact bioautography, (2) the immersion method, and (3) direct bioautography. In the contact bioautography the antibiotic-like compounds are transferred from the adsorbent layer to an inoculated agar plate by direct contact with microbes. After 15–30 min the chromatoplates are removed and the antimicrobial substances which have diffused into the agar inhibit the growth of the microbes. In the immersion bioautography the developed chromatoplate is covered with agar medium. After solidification the chromatoplate is seeded with the microbes and growth or inhibition bands can be visualized.

In direct bioautography the microorganisms grow directly on the adsorbent layer and so all the different steps of method (separation, preconditioning, seeded incubation, visualization) are performed on the adsorbent layer. The principle of direct bioautography is that a suspension of a microorganism growing in a suitable broth is applied to a developed chromatoplate after drying. Incubation of the chromatoplate with the microbes in a humid atmosphere at the optimum temperature enables growth of bacteria, and with a vital dye the alive cells can be visualized, as the dehydrogenases from the living microorganisms convert tetrazolium salt into the corresponding intensively colored formazan. In this procedure the antibacterial compounds

appear as clear spots against a colored background. The basic elements, the optimization, the realization of direct bioautography are well summarized [10,15].

9.4 NEW POSSIBILITIES AND APPROACHES TO BIOLOGICAL DETECTION OF PLANT SUBSTANCES

9.4.1 SEPARATION TECHNIQUES

The adsorbent bed in layer arrangement is suitable for direct bioautographic detection of different biologically active substances using living cells (mainly microbes)
which can proliferate on the adsorbent bed for the direct detection. The column
techniques cannot be used for such investigations because the adsorbent bed in
column arrangement is not suitable—among others—for the application and growth
of microbes; however, the fractions from different gas or liquid column chromatographic or capillary electrophoresis separations can be used for biological evaluation
in a layer adsorbent bed.

9.4.1.1 Conventional Planar Layer Liquid Chromatographic Techniques

In conventional ascending TLC, the capillary forces transport the mobile phase
through the porous adsorbent layer [16,17]. The eluent front velocity in TLC
declines as this distance increases. Thus the rate of migration of the solvent mixture
in the TLC and HPTLC systems caused by capillary forces is under the control of
atmospheric vapor over the adsorbent layer. Furthermore, a part of the solvent
evaporates from the wet surface of the adsorbent layer, because it is usually impossible to fully saturate the chamber—largely due to problems of thermal gradient
[18,19]. This effect can be reduced by the use of small-volume chambers even in this
case. In conventional TLC/HPTLC, the capillary forces during eluent migration
decrease the free energy of the liquid as it enters the porous structure of the adsorbent
layer, so the mechanism of transport is the action of capillary forces [20]. The
possibilities of capillary flow velocities in TLC/HPTLC fundamentally limit separation efficiency.

From the beginning, the TLC was used for the biological detection [16,17] as
well. More recently, the use of two-dimensional TLC (2D-TLC) to bioautographic
detection means a progress and shows the demand for the advancement of the
technique in general [21].

9.4.1.2 Forced Flow Planar Layer Liquid Chromatographic Techniques

In modern technical solutions the eluent system can migrate through the stationary phase in layer arrangement under the additional influence of forced flow,
which can be achieved by the application of a pump system for overpressured
layer chromatography (OPLC) [22], by centrifugal force for centrifugal chromatography [23] (an alternative term is rotation planar chromatography (RPC) [24]), or by
an electric field for high speed TLC (HSTLC) [25]. Although, the forced flow of
eluent is a necessary requirement, it is not enough for efficient separation. The linear
and constant development velocity over the entire separation distance is a further

necessary step but it is yet not enough for optimum mobile-phase velocity. On the basis of these requirements the present forced flow planar layer liquid chromatographic techniques can be divided into two main groups: one group includes techniques complying with these requirements as different linear versions of OPLC [22,26]. The other group includes such forced flow techniques that are only near to these special requirements as two modern versions of planar electro chromatography (PEC): electroosmotically driven TLC [27–29] and dielectrochromatography (DEC) [30,31], RPC [24] as well as circular OPLC [32]. According to more recent publications using pressurized chamber systems the efficiency of these techniques has been further increased [33,34].

All forced flow techniques can be used theoretically and practically for biological detection; however, up-to-date linear OPLC was only used for direct bioautography [35–37].

9.4.2 BIODETECTION

In the bioautographic detection modes, most diverse microbial (e.g., Gram-positive and Gram-negative bacteria) and other biological systems are already used in different laboratories. Direct bioautography makes it possible to locate antimicrobial activity on a chromatogram because the inhibition zones can be observed directly on the chromatoplate.

Weins and Jork [38] proved that mold spores, yeast cells, bacteria, or cell organelles, e.g., chloroplasts, in a suitable nutrient medium can be applied to the chromatoplate. These unique investigations show the future direction in the field of bioautography with the possibility of detecting biochemical interactions. The inhibition or promotion of growth on the same chromatoplate is especially a pertinent observation. Tabanca et al. [39,40] developed a routine method for the measurement of inhibition and promotion activity of (plant) substances. In many cases the small concentrations are stimulative. They compared the first time the potential of TLC and OPLC in the direct bioautography of essential oils [40]. It has been established that OPLC coupled with direct bioautography is a powerful tool especially for separating and determining the number of bioactive compounds in a lipophilic, nonpolar, and even complex mixture. Direct bioautography coupled with OPLC offers numerous advantages for the discovery of plant substances that can be used in medicine and pest management.

Table 9.1 illustrates only a few exemplars for the application of bioautography for the detection and isolation of some natural (mainly plant) substances with diverse biological activity.

9.4.3 SPECTROSCOPIC (MS, FT-IR, NMR, ETC.) EVALUATION OF CHROMATOGRAPHIC SPOTS BEFORE AND AFTER INOCULATION

The near-ideal method generates the advantage of combining information from independent spectroscopic data using complementary hyphenated techniques and integrating biological information at the same time. This solution allows complete or partial identification of the substance group studied and gives, with good confidence,

TABLE 9.1

Application of Conventional Bioautography for the Detection of Some Natural (Mainly Plant) Substances with Diverse Biological Activity

Activity	Origin	Chemical Characters	Reference
Antialgal	*Vallineria spiralis*	2-Ethyl-methyl-maleimide	[41]
Antibacterial	*Cameraria* pupae	Lysozyme	[42]
Anticandidal	Higher plants	Essential oils	[43]
Antifungal	*Haplophyllum* sp	Alkaloids	[44]
	Ruta graveolens	Acridone alkaloids	[45]
Antimalarial	*Newbouldia laevis*	Naphthoquinone–anthraquinones	[46]
Antimyco-bacterial	*Thymbra spicata*	Essential oil	[47]
	Tetradium sp	Geranylated furocoumarins	[48]
Molluscicidal	*Erigeron speciosus*	Steam distillate	[49]
Against multidrug resistance	Indian plants	Extracts	[50]

valuable information on the presence of putative new natural products as well. For the on-line identification of the main constituents of the crude plant extract (LC chemical screening), LC-nuclear magnetic resonance spectrometry (NMR), high-resolution LC-tandem mass spectrometry (MS–MS), high-resolution LC–APCI-Q-TOF-MS–MS, Raman IR as well as Fourier transform (FT)-IR, and LC-UV with postcolumn addition of UV shift reagents can be used [51]. Of course, these different LC hyphenated techniques have been used separately or in partial combination before. In fact, a complete technical hyphenation of all techniques can be solved theoretically by the creation of a "total analysis device" [52], but it is sometimes problematic and not necessary. LC microfractionation can be performed just after LC–NMR detection and all peaks after collection can be submitted to bioautography assays. The coupling of LC–NMR is one of the most powerful methods for the separation and structural elucidation of unknown substances in mixtures [53]. The structures of the new compounds from plants can be determined effectively by 1D- and 2D-NMR analyses. GC–MS is already a conventional technique in phytochemistry as well. The use of high-resolution LC–MS–MS in phytochemical analysis is relatively new and only few papers have demonstrated the potential of this technique for crude plant extract analysis [54]. The superiority of Q-TOF (time of flight)-LC–MS–MS for the analysis of plant substances is demonstrated [55]. Raman IR spectroscopy and FT-IR technique can be used successfully in phytochemical analysis, including bioautographic detection (e.g., measurement of trans-resveratrol) as well (Figure 9.1) [56].

9.4.4 OPTIMIZATION OF PARAMETERS IN DIRECT BIOAUTOGRAPHY AND SOME OF THEIR TECHNICAL SOLUTIONS

Many factors can influence the results in the direct bioautography such as the method of extraction, composition of the culture medium, pH (e.g., acids in eluent systems), incubation time, the age of microbes as well as solvents and additives and adsorbents

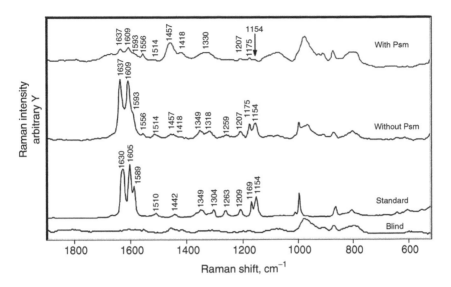

FIGURE 9.1 Raman spectra of trans-resveratrol with and without *Pseudomonas* sp. (Psm) inoculation.

in a TLC/OPLC system [57]. Measurement of ATP content is a sensitive and specific indicator of the viability of bacteria and is related to the activity of many enzymes during their life. Nagy et al. [58] found that the bioluminescence ATP assay [59] was a suitable method for studying the viability of test bacteria by direct bioautography without agar gel medium. For correct interpretation of ATP data the measured values should be referred to a cellular property (protein, cell number, etc.) [15,58]. For the chromatographic conditions the precondition of TLC/HPTLC or OPLC adsorbents (e.g., at 130°C for 3 h) is a determining step [15]. Without this procedure the adsorbent layers become partly detached when it is soaked. A complete review in this chapter is not possible. The occurrence and elimination of the disturbing phenomena in conventional bioautography were summarized by Botz et al. [15,57,60].

9.5 INTERACTION GUIDED SEPARATION AND DETECTION OF PLANT SUBSTANCES IN THE ADSORBENT LAYER—BIOARENA SYSTEM

The biological potential of the basic bioautography is much bigger than it seems from excellent up-to-date results. The exploitation of this potential occurs in the so-called BioArena system [35,37,61], which is the first further development of the original direct bioautography.

9.5.1 BASIC ELEMENTS OF BIOARENA SYSTEM AND ITS IMPORTANCE

Figure 9.2 shows the basic elements of the BioArena [37,61]. It has to be pointed out that BioArena is based on the direct bioautography with exploitation of its all

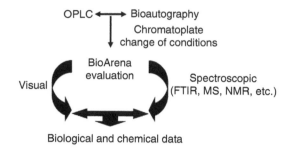

FIGURE 9.2 The main steps of the BioArena. (From Tyihák, E. et al., *J. Planar Chromatogr.*, 18, 67, 2005. With permission.)

advantages. It is a system that integrates the advantages of planar layer liquid chromatography (ideally, versions of linear OPLC [22,26]) and basic direct bioautography [15,60] as well as visual and spectroscopic evaluation of spots before and after inoculation. This integration utilizes the possibilities of interactions of microbes with dye substances and with other small and large cofactor molecules in the planar layer adsorbent bed after a chromatographic separation. These endogenous or exogenous molecules are available for the cells of the indicator organism in the culture medium on the adsorbent layer [37,61].

9.5.2 Occurrence and Role of HCHO in Biological Systems—Formaldehydome System

Study of the interactions between microbes or other living cells and antibiotic-like (more exactly, cell proliferation influencing) compounds, and key endogenous and characteristic exogenous molecules on a planar layer adsorbent bed promises to furnish new information about the function and role of formaldehyde (HCHO) as a small, indispensable, endogenous molecule during the cell proliferation, differentiation, and disease resistance in general [62].

HCHO—similar to hydrogen peroxide (H_2O_2) another small molecule—is an endogenous component, mainly in the form of hydroxymethyl groups, of all biological systems [63]. It is typically formed during enzymatic methylation and demethylation processes, but there are other origins of this simplest aliphatic aldehyde in biological systems [62–64]. It has become increasingly evident that there is a primary HCHO cycle in biological systems in which the S-methyl group of L-methionine can be formed from HCHO originating from natural HCHO generators and that the HCHO-yielding function of S-adenosyl-L-methionine (SAM) is an essential component of this fundamental biological pathway [62,63].

The term "formaldehydome" means the complete set of HCHO-cycle-mediated and nonmediated HCHO pathways of a given biological unit (Figure 9.3). The quality of the distribution of HCHO molecules in methyl or hydroxymethyl/formyl groups (e.g. hypo- or hypermethylated regions in DNA or methyl groups in histones) and at the same time the demethylation of the different methylated/hydroxymethylated compounds (e.g., demethylation of N-methylated L-lysine molecules in histones [65])

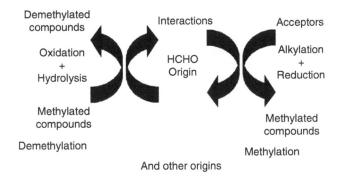

FIGURE 9.3 Basic elements of the formaldehydome. (From Tyihák, E. et al., *J. Planar Chromatogr.*, 18, 67, 2005. With permission.)

fundamentally affect the composition of the formaldehydome and give a characteristic profile for a biological unit.

Although the role of HCHO in the induction of neoplasia and atherosclerosis has been widely discussed [66], the real function of HCHO in these processes is unknown. The unknown reactions of HCHO demand a new approach to analyze parts of the formaldehydome with special emphasis on the detailed mechanisms of different means of endogenous and exogenous HCHO capture, donation, and transfer, etc. Although conventional bioautography [15] can be used for biological detection of the antibiotic effect it is not suitable for studying and understanding all the complicated reactions involved. It is supposed that these HCHO reactions in cells occur among other chemical reactions. Model experiments are necessary with complex separation and detection systems. One solution is the BioArena system. HCHO is practically present in all "-ome" systems: all big biological units as the genome, proteome, metabolome, etc., are interlaced by dynamic methylation–demethylation processes [67].

9.5.3 INTERACTIONS BETWEEN HCHO AND H_2O_2

HCHO—similar to H_2O_2—is an endogenous, indispensable component of all biological systems. These small molecules can be continuously formed by almost all cells, both intracellularly and extracellularly [62,63]. The inhibiting–killing activity of HCHO released can be further increased by means of the interaction with H_2O_2, that is, HCHO and H_2O_2 can meet in different regions of cells and so they can interact endogenously. In this interaction the very reactive singlet oxygen ($^1O_2$) and excited HCHO can be formed [68,69].

It is a recent observation that antibodies can catalyze the generation of very reactive biological oxidants, including dihydrogen trioxide (H_2O_3), with a ring skeleton and ozone (O_3) in the interaction of $^1O_2$ with H_2O [70,71]. It is supposed that HCHO, originating from the formaldehydome system, is also present in the activity process of antibodies and this may be, in general, one basis of the formation of $^1O_2$ in this complicated system. Figure 9.4 illustrates the reaction series

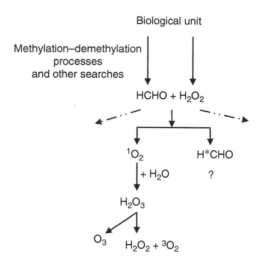

FIGURE 9.4 Possibilities of formation of different reactive oxidants in the interactions of HCHO and H_2O_2.

supposedly, taking into account the earlier and more recent observations [69–73]. It is probable that the mechanism of bacterial killing by antibodies and antibiotics is based on similar or the same principle. These very reactive small molecules—from HCHO to O_3—may be the crucial factors of the innate (natural) resistance and form a common bridge between the innate (natural) and adaptive (induced) resistance in plant and animal organisms, alike [67].

These reactive small molecules can also be formed in the chromatographic spots after inoculation and so they can be studied in a relatively separated form from the biological system.

9.5.4 EFFECT OF DEPRIVATION AND MOBILIZATION OF HCHO ON THE ANTIBACTERIAL ACTIVITY

9.5.4.1 Interaction and Double Effect of Trans-Resveratrol [37,61]

It has been established that trans-resveratrol (TR), a common food component in grapes, wines, etc. can interact with HCHO in model chemical experiments and in biological systems (e.g., pig heart tissue) [63]. The mobilization of HCHO from a biological unit by TR is a totally new approach to the "red wine" topic and so is its double effect: first, the elimination of HCHO by TR may cause a chemopreventive effect; second, the reaction products between TR and HCHO may act as killing/inhibiting factors against pathogens or cancer cells. This double effect of TR may be one basis of its most diverse beneficial activities.

The BioArena system gives a good possibility for studying the microchemical reactions of TR with HCHO using endogenous HCHO capturer and mobilizing molecules in the culture medium. Figure 9.5 illustrates that TR really behaves as

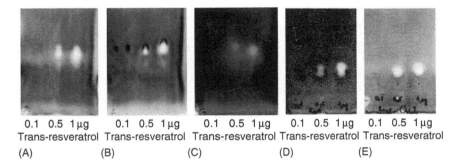

0.1 0.5 1 µg   0.1 0.5 1 µg   0.1 0.5 1 µg   0.1 0.5 1 µg   0.1 0.5 1 µg
Trans-resveratrol Trans-resveratrol Trans-resveratrol Trans-resveratrol Trans-resveratrol
(A)   (B)   (C)   (D)   (E)

FIGURE 9.5 Effect of endogenous and exogenous substances on the antibacterial activity of trans-resveratrol Chromatographic conditions: silica gel 60F$_{254}$ (Merck, after preconditioning at 120°C for 3 h), chloroform–methanol, $80 + 8$ (v/v); automatic OPLC instrument: external pressure 5.0 MPa, flow rate 250 mL min^{-1}, mobile phase volume 9700 µL, eluent flush 450 µL, and separation time 2375 s. Biological conditions: (A) *Pseudomonas savastanoi* bacterial suspension (Psm); (B) Psm + 2 mg L-arginine in 1 mL Psm; (C) Psm + 2 mg glutathione in 1 mL Psm; as well as (E) 2 mg CuSO$_4 \cdot$ 5H$_2$O in 100 mL bacterial suspension. A and D are controls. Incubation at optimum temperature and staining with MTT.

an antibacterial antibiotic against *Pseudomonas savastanoi pv. phaseolicola*. The control (bacterial cell suspension only) sheet shows this clearly. With the addition of endogenous HCHO-capturing molecules such as L-arginine and glutathione to the culture medium with bacterial cells, the antibacterial activity of TR was found to decrease dose-dependently. TR can mobilize the HCHO from the bacterial cells to TR spots as well.

The second part of the figure shows the effect of Cu(II) ions on the antibacterial activity of TR Cu(II) ions mobilize, and coordinate the HCHO molecules in hydroxymethyl groups [72] and so they can transport HCHO molecules to TR spots. The antibacterial activity of TR can be increased dramatically using a higher amount of Cu(II) ions in the culture medium [61].

Using the aqueous suspension of *Saccharomyces cerevisiae* for the biological detection (Figure 9.6) it can be seen that the HCHO capturer molecules in the culture medium can decrease the antimicrobial activity of the TR on the silica gel adsorbent layer while for the effect of Cu(II) ions as HCHO transporting ions the antibacterial activity of TR of the same amount is increased markedly. (The unique big spots of TR in this case generate questions about planar layer liquid chromatography as well. It seems that the distribution of TR molecules in the adsorbent layer is much wider than it was imaginable earlier.)

9.5.4.2 Study of "Classical Medicines" in BioArena System

Figure 9.7 illustrates the antibacterial activity of *Chelidonium* alkaloids and authentic chelidonine. It can be seen that all alkaloids of the extract generate antibacterial activity, which can be increased in the presence of Cu(II) ions in the culture medium. The *Chelidonium* alkaloids have N- and O-methyl groups, that is, they are potential

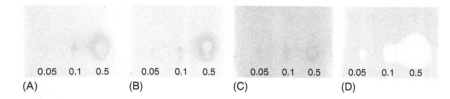

0.05 0.1 0.5   0.05 0.1 0.5   0.05 0.1 0.5   0.05 0.1 0.5
(A)   (B)   (C)   (D)

FIGURE 9.6 Effect of endogenous and exogenous substances on yeast (*Saccharomyces cerevisiae*) influencing biological activity of trans-resveratrol Conditions: as at Figure 9.5, but eluent composition was chloroform–methanol $80 + 4$ (v/v); Biological conditions: (A) yeast suspension (3 g yeast in 100 mL distilled water); (B) $A + 2$ mg L-arginine in 1 mL yeast suspension; (C) $A + 2$ mg glutathione in 1 mL yeast suspension; (D) $A + 4$ mg $CuSO_4 \cdot 5H_2O$ in 100 mL yeast suspension.

HCHO generators [73]. These HCHO molecules participate in the killing–inhibiting process together with the ones of other origin, e.g., through interactions.

Figure 9.8 compares positive and negative densitograms of *Chelidonium* alkaloids: positive densitogram was made before the biological detection procedure, while the so-called negative densitograms show the antibacterial activities. The negative densitograms are suitable in this case and in general also suitable for the illustration and measurement of quantitative changes [73], that is, for the quantitative evaluation of biochemical reactions as well.

Salicylic acid is a well-known "old medicine." The antimicrobial activity of this phenolic acid has also been reported. Figure 9.9 demonstrates the antibacterial activity of salicylic acid and when using HCHO capturer molecules, similar to

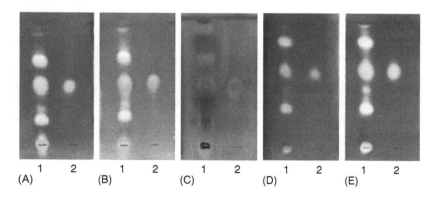

1   2   1   2   1   2   1   2   1   2
(A)   (B)   (C)   (D)   (E)

FIGURE 9.7 Effect of endogenous and exogenous substances on the antibacterial activity of components of (1) *Chelidonium* root extract (2.5 μL) and (2) *Chelidonium* alkaloid test mixture (1.5 μL). Chromatographic conditions: silica gel $60F_{254}$; mobile phase: dichloromethane–methanol, $97 + 3$ (v/v), Desaga glass TLC chamber; Biological conditions: (A) *Pseudomonas* sp. suspension, Psm, control; (B) $A + 5$ mg L-arginine in 1 mL Psm; (C) $A + 5$ mg glutathione in 1 mL Psm, and (E) $A + 4$ mg of $CuSO_4 \cdot 5H_2O$ in 100 mL in Psm. A and D are controls. (From Sárközi, Á. et al., *J. Planar Chromatogr.*, 19, 267, 2006. With permission.)

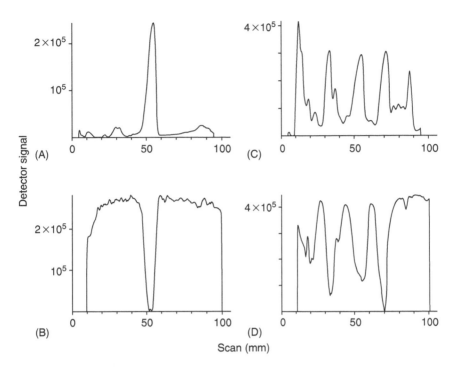

FIGURE 9.8 Densitograms obtained from *Chelidonium* alkaloid test mixture and *Chelidonium* root extract before (A, C) and after (B, D) bioautography ($\lambda = 305$ nm (A, C); $\lambda = 590$ nm (B, D). (Shimadzu CS 930 densitometer). (From Sárközi, Á. et al., *J. Planar Chromatogr.*, 19, 267, 2006. With permission.)

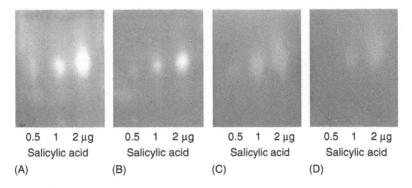

FIGURE 9.9 Effect of endogenous and exogenous substances on the antibacterial activity of salicylic acid. Chromatographic conditions: silica gel 60F$_{254}$ (Merck); mobile phase: chloroform-methanol $80 + 8$ (v/v); Desaga glass TLC chamber. Biological conditions: (A) 100 mL *Pseudomonas* sp., Psm; (B) A + 2 mg L-arginine in 1 mL Psm; (C) 2 mg glutathione in 1 mL Psm; (D) A + 1 mg L-ascorbic acid in 1 mL Psm. Incubation at optimum temperature and staining with MTT.

earlier investigations, this activity is also characteristically decreased. It follows from these results that salicylic acid acts through HCHO and its reaction products. This statement is supported by the last chromatogram sheet where the effect of vitamin C is shown. As a reducing agent, the l-ascorbic acid reduced the HCHO molecules, so the antibacterial activity of the salicylic acid is decreased appreciably.

9.5.5 Elimination of O_3 from the Chromatographic Spots and Its Effect on the Antibacterial Activity of Antibiotics

It follows from the essence of BioArena system [63] that there is an unlimited possibility for the interactions between microbes and biologically active small and big compounds in the adsorbent layer bed of a chromatoplate. We exploited this possibility using cinnamic acid as an antibiotic-like compound. Figure 9.10 shows the antibacterial activity of the cinnamic acid (Figure 9.10A). It is supposed that O_3 can be formed in the chromatographic spots of cinnamic acid, which is responsible mainly for this antibacterial effect. O_3 can originate from the interaction between HCHO and H_2O_2 in which in the first time singlet oxygen can be formed, which can oxidize the water molecules and so H_2O_3 can be formed. From H_2O_3 with dispro-portion—among others—O_3, a super killing factor forms (see Figure 9.4). Using Indigo Carmine for elimination of O_3 molecules from the chromatographic spots of cinnamic acid it has been established that in the presence of Indigo Carmine decreased really dose-dependently the antibacterial activity of cinnamic acid, as it can be seen in Figure 9.10.

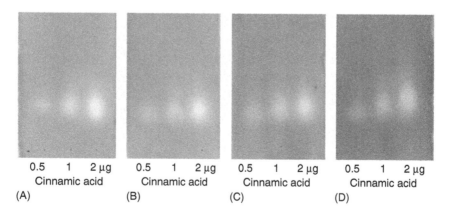

0.5　1　2 µg	0.5　1　2 µg	0.5　1　2 µg	0.5　1　2 µg
Cinnamic acid	Cinnamic acid	Cinnamic acid	Cinnamic acid
(A)	(B)	(C)	(D)

FIGURE 9.10 Effect of Indigo Carmine on the antibacterial activity of cinnamic acid. Chromatographic conditions: silica gel $60F_{254}$ (Merck); mobile phase: chloroform–methanol $80 + 8$ (v/v); Desaga glass TLC chamber. Biological conditions: (A) 100 mL *Pseudomonas savastanoi* suspension, Psm; (B) A + 10 mg Indigo Carmine; (C) A + 20 mg Indigo Carmine; (D) A + 40 mg Indigo Carmine. Incubation and staining with MTT.

9.5.6 TIME-DEPENDENT CHANGE OF SPOT COLOR OF TRANS-RESVERATROL (TR) IN BIOARENA AS A MANIFESTATION OF BIOCHEMICAL CHANGES

It is characteristic for the BioArena system that the cells (e.g., bacterial cells) grow freely on the adsorbent layer because of supportive conditions. There is practically an opportunity to monitor the multiplication of bacterial cells on the adsorbent layer.

Figure 9.11 summarizes and shows schematically the evaluation of a 9 day experiment with TR in a *Pseudomonas savastanoi* system. The figure depicts seven consecutive daily stages of and around the same TR spot. The first day's stage shows the original start detection, which is equal with the conventional bioautographic detection (antibiotic effect): it is always a nice white spot. On the second day around the resveratrol spot a black ring is formed, which increases with time and after 6 days the TR spot becomes totally black. It is supposed that in this stage there is an intensive multiplication of bacterial cells on the adsorbent layer and so the black color is originated from the increased activity of living bacterial cells to reduce the tetrazolium dye. On the following day a white ring appears around TR, which increased with time. Finally the last stage spot becomes again white (a second direct effect in the same spot after 9 days) [37,61]. This last killing effect can be originated from the last growth phase of bacterial cells when killing factors as HCHO, H_2O_2, $^1O_2$, and O_3 can be formed consecutively or in parallel.

This unique phenomenon could be observed with other substances (e.g., Cu(II) ions) as well.

9.5.7 BIOARENA AS AN ALTERNATIVE ANIMAL EXPERIMENTAL MODEL

The animal experiments are very expensive and sometimes raise severe moral issues. Therefore, searching for new solutions is both useful and actual. Figure 9.12 illustrates the applicability of BioArena as an alternative animal experimental model. In this complex system there is a possibility for isolation of potential medicinal products as antibiotics or antineoplastic substances. It is a BioArena-guided isolation procedure

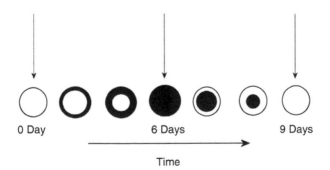

FIGURE 9.11 Schematic drawing of time-dependent change of spot color in BioArena. General phenomenon. (From Tyihák et al., *J. Planar Chromatogr.*, 18, 67, 2005. With permission.)

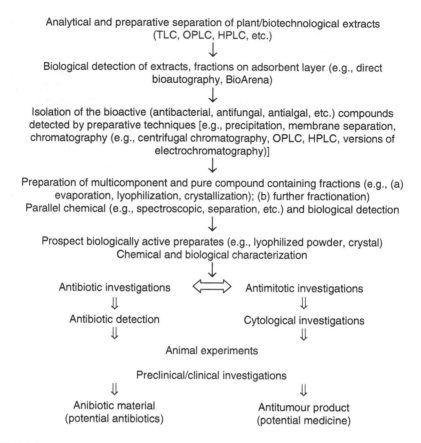

Analytical and preparative separation of plant/biotechnological extracts
(TLC, OPLC, HPLC, etc.)
↓

Biological detection of extracts, fractions on adsorbent layer (e.g., direct
bioautography, BioArena)
↓

Isolation of the bioactive (antibacterial, antifungal, antialgal, etc.) compounds
detected by preparative techniques [e.g., precipitation, membrane separation,
chromatography (e.g., centrifugal chromatography, OPLC, HPLC, versions of
electrochromatography)]
↓

Preparation of multicomponent and pure compound containing fractions (e.g., (a)
evaporation, lyophilization, crystallization); (b) further fractionation)
Parallel chemical (e.g., spectroscopic, separation, etc.) and biological detection
↓

Prospect biologically active preparates (e.g., lyophilized powder, crystal)
Chemical and biological characterization
↓

Antibiotic investigations ⟸⟹ Antimitotic investigations
⇓   ⇓
Antibiotic detection   Cytological investigations
⇓   ⇓

Animal experiments

Preclinical/clinical investigations
⇓   ⇓
Anibiotic material   Antitumour product
(potential antibiotics)   (potential medicine)

FIGURE 9.12 Place of BioArena as a potential alternative animal experimental model in the process of separation, isolation, and biological detection of new biologically active compounds. Schematic model.

and at the same time an evaluating system that leads to production of new pure ingredients.

9.6 BIOASSAY-GUIDED SEPARATION AND ISOLATION OF BIOLOGICALLY ACTIVE PLANT SUBSTANCES

The combination of chemical and biological screening has been proved to be very efficient for the survey of a number of extracts and for the isolation of new plant ingredients and several examples of this conventional approach can be found in the literature [41,44]. Plant active components of biological interest are usually isolated following a bioactivity-guided fractionation procedure which include conventional and modern (BioArena system) biological detection. For the rapid and efficient isolation of active components, the dereplication of crude plant extracts in the first time with LC-hyphenated techniques (e.g., LC/UV-DAD, LC–MS, LC/NMR) and other modern techniques (e.g., MALDI MS) represents a strategic element to avoid finding already known compounds, guiding towards the isolation and identification of new plant bioactive constituents. The parallel development and application of different biological detection methods is also crucial for exact and efficient localization of the active components [74]. However, the development of BioArena system (e.g., coordination of operating steps, using aimed series of endogenous and/or exogenous molecules [61]) opens new horizons in the bioassay-guided fractionation and isolation works.

9.7 CONCLUSIONS

The exploitation of the unique potentialities of the planar layer adsorbent bed for the separation, isolation, and identification of the constituents of plant and other extracts [74] as well as complicated reaction mixtures (e.g., in combinatorial chemistry) [75] is an especially actual and auspicious task. The possibilities of the biological detection also include the application of the conventional versions of bioautography to new antimicrobial agents with novel modes of action. However, these conventional solutions are not suitable for studying and understanding all the complicated biochemical reactions involved. Model experiments are necessary with complex separation and detection systems in micro- and ultramicroassay level. One solution is the BioArena system, which integrates the advantages of the layer chromatographic separation (ideally linear OPLC [22,26]), and direct bioautography [15]. This integration exploits the possibilities of interaction of microbes with dye substances and with other small and large cofactor molecules in the layer adsorbent bed after chromatographic separation, which makes these endogenous or exogenous cofactor molecules available as a culture medium for pathogenic cells. In these interactions HCHO and its reaction products appear as key molecules of the formaldehydome. Thus, in the biological world, they participate in a series of practically unknown interactions in different tissues (cells). Although, HCHO as a key molecule and its reaction products can generate surprising and new reactions in model experiments, further biochemical key molecules can hopefully also be studied in BioArena or in similar new model systems.

REFERENCES

1. Dehority, B.A. and Tirabasso, P.A., Antibiosis between ruminal bacteria and ruminal fungi, *Appl. Environ. Microbiol.*, 66, 2921, 2000.
2. Cook, R.J. and Baker, K.F., *The Nature and Practice of Biological Control of Plant Pathogens*, American Phytopathological Society, St. Paul, MN, 1983, p. 32.
3. Ajl, S.J. et al., *Microbial Toxins*, Vols. I–VIII, Academic Press, New York, 1970–1972, p. 4057.
4. Lorber, B., Update in infectious diseases, *Ann. Intern. Med.*, 137, 974, 2002.
5. Lorber, B., Update in infectious diseases, *Ann. Intern. Med.*, 145, 354, 2006.
6. Murray, C.J.L. and Lopez, A.D., Mortality by cause for eight regions of the world: global burden of disease study, *Lancet*, 349, 1269, 1997.
7. Eloff, J.N., On expressing the antibacterial activity of plant extracts—a small first step in applying scientific knowledge to rural primary health care, *South African J. Sci.*, 96, 16, 2000.

8. Wedge, D.E., Galindo, J.C.G., and Macias, F.A., Fungicidal activity of natural and synthetic sesquiterpene lactone analogs, *Phytochemistry*, 53, 747, 2000.
9. Chomnawang, M.T. et al., Antimicrobial effects of Thai medicinal plants against acne-inducing bacteria, *J. Ethnopharmacol.*, 101, 330, 2005.
10. Wedge, D.E. and Camper, N.D., In: *Biologically Active Natural Products: Agrochemicals and Pharmaceuticals* (Cutler, H.G. and Cutler, S.J., Eds.), CRC Press, Boca Raton, FL, 2000, p. 1.
11. Goodall, R.R. and Levi, A.A., A microchromatographic method for the detection and approximate determination of the different penicillin in the mixture, *Nature*, 158, 675, 1946.
12. Betina, V., Bioautography in paper and thin-layer chromatography and its scope in the antibiotic field, *J. Chromatogr.*, 78, 41, 1973.
13. Rios, J.L., Recio, M.C., and Villar, A., Screening methods for natural products with antimicrobial activity: a review of the literature, *J. Ethnopharmacol.*, 23, 127, 1988.
14. Hostettmann, K. et al. The role of planar chromatography in the rapid screening and isolation of bioactive compounds from medicinal plants, *J. Planar Chromatogr.*, 10, 251, 1997.
15. Botz, L., Nagy, S., and Kocsis, B. In: *Planar Chromatography—A Retrospective View for the Third Millennium* (Nyiredy, Sz., Ed.), Springer, Budapest, Hungary, 2001, p. 175.
16. Stahl, E. (Ed.), *Thin-Layer Chromatography: A Laboratory Handbook*, 2nd edn., Springer, Berlin, 1969.
17. Kirchner, J.G. (Ed.), Thin-layer chromatography, 2nd edn. In: *Techniques of Chemistry*, Vol. XIV, Wiley, St. Louis, MO, 1990.
18. Zlatkis, A. and Kaiser, R.E. (Eds.), *HPTLC—High Performance Thin-Layer Chromatography*, Elsevier, Amsterdam, 1977.
19. Sherma, J. and Fried, B. (Eds.), *Handbook of Thin-Layer Chromatography*, 3rd edn., Marcel Dekker, New York, 2003.
20. Geiss, F., *Fundamentals of Thin-Layer Chromatography: Planar Chromatography*, Hüthig, Heidelberg, 1987.
21. Wedge, D.E. and Nagle, D.G., A new 2D-TLC bioautography method for the discovery of novel antifungal agents to control plant pathogens, *J. Nat. Prod.*, 63, 1050, 2000.
22. Tyihák, E., Mincsovics, E., and Kalász, H., New planar liquid chromatographic technique: overpressured thin-layer chromatography, *J. Chromatogr.*, 174, 75, 1979.
23. Hopf, P.P., Radial chromatography in industry, *Ind. Eng. Chem.*, 39, 938, 1947.
24. Nyiredy, Sz., Planar chromatography. In: *Chromatography* (Heftmann, E., Ed.), Elsevier, Amsterdam, Oxford, New York, 1992, p. A109.
25. Pretorius, V., Hopkins, B.J., and Schieke, J.D., A new concept for high-speed liquid chromatography, *J. Chromatogr.*, 99, 23, 1974.
26. Tyihák, E. and Mincsovics, E., Overpressured layer chromatography (optimum performance laminar chromatography) (OPLC). In: *Planar Chromatography—A Retrospective View for the Third Millennium* (Nyiredy, Sz., Ed.), Springer, Budapest, 2001, p. 137.
27. Dzido, T.H., Mroz, J., and Jozwiak, G.W., Adaptation of a horizontal DS chamber to planar electrochromatography in a closed system, *J. Planar Chromatogr.*, 17, 404, 2004.
28. Malinowska, I., Rózylo, J.K., and Krason, A., The effect of electric fields on solute migration and mixture separation in TLC, *J. Planar Chromatogr.*, 15, 418, 2002.
29. Nurok, D., Planar electrochromatography, *J. Chromatogr. A*, 1044, 83, 2004.
30. Kreibik, S. et al., Planar dielectrochromatography on non-wetted thin-layers, *J. Planar Chromatogr*, 14, 246, 2001.
31. Coman, V. and Kreibik, S., Planar dielectrochromatography—A perspective technique, *J. Planar Chromatogr.*, 16, 338, 2003.

32. Mincsovics, E., Ferenczi-Fodor, K., and Tyihák, E., Overpressured layer chromatography. In: *Handbook of Thin-Layer Chromatography, Third Edition, Revised and Expanded* (Sherma, J. and Fried, B., Eds.), Dekker, New York, 2003, p. 175.

33. Dzido, T.H., Plocharz, P.W., and Slazak, P. Apparatus for pressurized planar electrochromatography in a completely closed system, *Anal. Chem.*, 78, 4713, 2006.

34. Tate, P.A. and Dorsey, J.G., Linear voltage profiles and flow homogeneity in pressurized planar electrochromatography, *J. Chromatogr. A*, 1100, 150, 2006.

35. Tyihák, E. et al., The combination of the overpressured layer chromatography and bioautography and its applications to the analysis of molecules influencing cell proliferation, *Chem. Anal. (Warsaw)*, 48, 543, 2003.

36. Móricz, Á. et al., Separation and detection of aflatoxins using overpressured-layer chromatography and bioautography, *J. Planar Chromatogr.*, 16, 417, 2003.

37. Tyihák, E. et al., Antibiosis, antibiotics and formaldehyde cycle: the unique importance of planar chromatographic techniques to progress in these fields, *J. Planar Chromatogr.*, 17, 84, 2004.

38. Weins, C. and Jork, H., Toxicological evaluation of harmful substances by in situ enzymatic and biological detection in high-performance thin-layer chromatography, *J. Chromatogr. A*, 750, 403, 1996.

39. Tabanca, N. et al., Bioactive constituents from Turkish *Pimpinella* species, *Chem. Biodiver.*, 2, 221, 2005.

40. Tabanca, N. et al., Characterization of volatile constituents of *Scaligeria tripartita* and studies on the antifungal activity against phytopathogenic fungi, *J. Chromatogr. B*, 2006.

41. Xian, Q. et al., Isolation and identification of antialgal compounds from the leaves of *Vallisneria spiralis* L. by activity-guided fractionation, *Environ. Sci. Pollut. Res. Int.*, 13, 233, 2006.

42. Fiolka, M.J., Ptaszynska, A.A., and Czarniawski, Antibacterial and antifungal lysozyme-type activity in *Cameraria ohridella* pupae, *J. Invertebr. Pathol.*, 90, 1, 2005.

43. Pauli, A., Anticandidal low molecular compounds from higher plants with special reference to compounds from essential oils, *Med. Res. Rev.*, 26, 223, 2005.

44. Cantrell, C.L. et al., Isolation and identification of antifungal and antialgal alkaloids from *Haplophyllum sieversii*, *J. Agric. Food Chem.*, 53, 7741, 2005.

45. Meepagala, K.M. et al., Algicidal and antifungal compounds from the roots of *Ruta graveolens* and synthesis of their analogs, *Phytochemistry*, 66, 2689, 2005.

46. Eyong, K.O. et al., Newbouldiaquinone A: a naphthoquinone–anthraquinone ether coupled pigment, as a potential antimicrobial and antimalarial agent from Newbouldia laevis, *Phytochemistry*, 67, 605, 2006.

47. Kilic, T., Analysis of essential oil composition of *Thymbr spicata* var. spicata: antifungal, antibacterial and antimycobacterial activities, *Z. Naturforsch C*, 61, 324, 2006.

48. Adams, M. et al., Antimycobacterial activity of geranylated furocoumarins from *Tetradium daniellii*, *Planta Med.*, 72, 1132, 2006.

49. Meepagala, K.M. et al., Molluscicidal and antifungal activity of *Erigeron speciosus* steam distillata, *Pest. Manag. Sci.*, 58, 1043, 2002.

50. Ahmad, I. and Aqil, F., In vitro efficacy of bioactive extracts of 15 medicinal plants against Esbetal-producing multidrug-resistant enteric bacteria, *Microbiol. Res.*, 162, 264, 2007.

51. Queiroz, E.F. et al., On-line identification of the antifungal constituents of *Erythrina vogelii* by liquid chromatography with tandem mass spectrometry, ultraviolet absorbance detection and nuclear magnetic resonance spectrometry combined with liquid chromatographic micro-fractionation, *J. Chromatogr. A*, 974, 123, 2002.

52. Louden, D. et al., Spectroscopic characterisation and identification of ecdysteroids using high-performance liquid chromatography combined with on-line UV-diode assay,

FT-infrared and H-1-nuclear magnetic resonance spectroscopy and time of flight mass spectrometry, *J. Chromatogr. A*, 910, 237, 2001.

53. Albert, K. et al., On-line coupling of separation techniques to NMR, *J. High Resolut. Chromatogr.*, 22, 135, 1999.

54. McLuckey, S.A. and Wells, J.M., Mass analysis at the advent of the 21st century, *Chem. Rev.*, 101, 571, 2001.

55. Waridel, P. et al., Evaluation of quadrupole time-of-flight tandem mass spectrometry and ion-trap multiple-stage mass spectrometry for the differentiation of C-glycosidic flavonoid isomers, *J. Chromatogr. A*, 926, 29, 2001.

56. Horváth, E. et al., Unpublished data, 2007.

57. Botz, L. et al., Chromatographic aspects of direct bioautography and its use for detecting antimicrobial activity of compounds from higher plants, *Fundam. Clin. Pharmacol.*, 13 (Suppl. 1), 359, 1999.

58. Nagy, S. et al., Optimal life condition of test bacteria for direct bioautographic detection. In: *Proceedings of the International Symposium on Planar Chromatography, Planar Chromatography 2000*, Lillafüred, Hungary (Nyiredy, Sz., Ed.), RIMP, Budapest, p. 77.

59. Turner, G.K. In: *Bioluminescence and Chemiluminescence: Instruments and Applications* (Van Dyke, K., Ed.), CRC Press, Boca Raton, FL., 1985, p. 43.

60. Botz, L., Kocsis, B., and Nagy, S., Bioautography. In: *Encyclopedia of Analytical Science* (Worsfold, P., Townshend, A., and Poole, C. Eds.), Vol. 1, Elsevier, Oxford, 2005, p. 271.

61. Tyihák, E., et al., The potential of BioArena in the study of the formaldehydome, *J. Planar Chromatogr.*, 18, 67, 2005.

62. Tyihák, E., Trézl, L., and Szende, B., Formaldehyde cycle and the phases of stress syndrome, *Ann. N Y Acad. Sci.*, 851, 259, 1998.

63. Tyihák, E. et al., Formaldehyde cycle and the natural formaldehyde generators and capturers, *Acta Biol. Hung.*, 49, 225, 1998.

64. Kalász, H., Biological role of formaldehyde and cycles related to methylation, demethylation, and formaldehyde production, *Mini Rev. Med. Chem.*, 3, 175, 2003.

65. Klose, R.J. et al., The transcriptional repressor JHDM3A demethylates trimethyl histone H3 lysine 9 and lysine 36, *Nature*, 442, 312, 2006.

66. Monticello, T.M. and Morgan, K.T., Chemically-induced nasal carcinogenesis and epithelial cell proliferation: a brief review, *Mutat. Res.*, 380, 33, 1997.

67. Tyihák, E., Double immune response of plants to pathogens and its biochemical basis. In: *Foliculture, Ornamental and Plant Biotechnology, Advances and Topical Issues, Vol. III: Global Science Books* (Teixeira da Silva, J.A., Ed.), London, UK, 2006, p. 380.

68. Trézl, L. and Pipek, J., Formation of excited formaldehyde in model reactions simulating real biological systems, *J. Mol. Struct.—Theochem.*, 170, 213, 1988.

69. Tyihák, E. et al., Possibility of formation of excited formaldehyde and singlet oxygen in biotic and abiotic stress situations, *Acta Biol. Hung.*, 45, 3, 1994.

70. Wentworth, A.D. et al., Antibodies have the intrinsic capacity to destroy antigens, *Proc. Natl. Acad. Sci. U S A.*, 97, 10930, 2000.

71. Wentworth, P. Jr. et al., Evidence for the production of trioxygen species during antibody-catalyzed chemical modification of antigen, *Proc. Natl. Acad. Sci. U S A.*, 100, 1490, 2003.

72. Tyihák, E. et al., Double effect of trace elements. In: *Proceedings of the Trace Elements in the Food Chain*, Budapest, May 25–27, 2006 (Szilágyi, M. and Szentmihályi, K., Eds.), p. 394.

73. Sárközi, Á. et al., Investigation of *Chelidonium* alkaloids by use of a complex bioautographic system, *J. Planar Chromatogr.*, 19, 267, 2006.

74. Hostettmann, K., Marston, A., and Wolfender, H., Strategy in the search for new lead compounds and drugs from plants, *Chimia*, 59, 291, 2005.
75. Gombosuren, N. et al., A multidimensional overpressured layer chromatographic method for the characterization of tetrazine libraries, *J. Biochem. Biophys. Methods*, 69, 239, 2007.

10 Forced-Flow Planar Layer Liquid Chromatographic Techniques for the Separation and Isolation of Natural Substances*

Emil Mincsovics

CONTENTS

---

* In Memoriam of Professor Szabolcs Nyiredy.

10.1 INTRODUCTION

Column and layer liquid chromatographic techniques as supplementary techniques due to their arrangements have always been characteristically developed in constant mutual interaction. Hence it is not surprising that the intensive development of forced-flow column liquid chromatography (high-performance liquid chromatography, HPLC) as originally potential planar layer version of HPLC entailed the need for the fundamental renewal of the most popular planar layer liquid chromatographic technique, TLC. HPTLC is based on the use of fine particle chromatoplates with narrow particle size distribution of adsorbent and is carried out with capillary-driven separation and sophisticated instrumentation [1]. However, the greatly increased developing time on a fine particle adsorbent layer (quadratic law exists concerning the front migration distance against the time [2]) made it necessary to employ forced mobile phase flow generating or at least approaching the optimum linear velocity to yield the highest efficiency of separation allowed by the layer adsorbent bed characteristic.

In the mirror of this it can also be understood that the latest efforts aimed at the further development of layer liquid chromatography are characterized by the desire to introduce sophisticated instrumental techniques similar to HPLC [3–5].

Interactions can be caught out among forced-flow planar layer liquid chromatography techniques, too; e.g., the application of counter plate contacted with the adsorbent surface in centrifugal layer chromatography (CLC) (alternative term is rotation planar chromatography, RPC) [6], and the application of external pressure on the adsorbent surface as well as the sealing of the parallel layer edges in pressurized planar electrochromatography (PPEC) [7], which were used earlier in overpressured layer chromatography (OPLC) [8].

Whilst CLC/RPC and OPLC techniques were applied in phytochemistry up to know, they are discussed here in detail.

10.2 CLASSIFICATION OF FORCED-FLOW PLANAR LAYER LIQUID CHROMATOGRAPHIC TECHNIQUES

To increase the speed and the efficiency of separation in planar layer liquid chromatography, different driving forces can be applied and the forced-flow techniques can also be classified accordingly (Figure 10.1).

Centrifugal force was applied by Hopf [9] to accelerate the separation speed in 1947 and the proposed name for this type of plant was "chromatofuge." Further on, the name has been changed to centrifugal thin layer chromatography (centrifugal TLC, CLC) [10] and Nyiredy et al. introduced a new, alternative term for the variations of this technique, RPC [11].

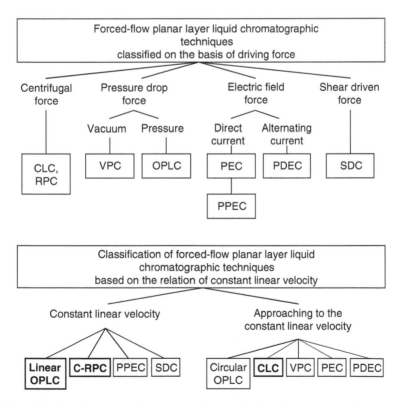

FIGURE 10.1 Classification of forced-flow planar layer liquid chromatographic techniques based on the forcing power and the relation to constant linear velocity (lower figure). Box with thick lines and bold letters, techniques applied up to know in phytochemistry.

The mobile phase can also be forced through the planar adsorbent bed by means of pressure drop. The first pressure drop driven separation system is OPLC, in which a pump system is applied for the eluent admission (generating the pressure force) into a pressurized ultramicro chamber enveloping the adsorbent layer by means of external pressure. This real planar version of HPLC was introduced by Tyihák et al. in 1979 [12–14].

Another possibility of pressure drop driven separations is vacuum planar chromatography (VPC) and was introduced by Delvordre et al. in 1992 where the driving force is the vacuum applied [15]. Regrettably this idea was not followed.

The next group in forced-flow planar layer liquid chromatography applies electric field force. TLC with electroosmotic flow generated by direct current was first described by Pretorius et al. in 1974, who called the technique high-speed thin layer chromatography (HSTLC) [16]. After a suspension of 20 years Pukl et al. reintroduced the technique using initially dry layer and called it firstly as planar electrochromatography (PEC) [17]. Different chamber constructions and also their modifications were developed in which initially dry [18–20] and prewetted [21–27] normal and reversed phase adsorbent layers can be applied for separation; however, constant linear velocity along the chromatoplate was not achieved due to the vaporization generated by Joule heating even if sandwich configuration were used.

To get more reproducible results and to increase the separation distance as well as the speed of separation, the PEC was improved in 2004 by Nurok et al. applying external pressure to the adsorbent layer surface as well as sealed layer edges on parallel sides in the direction of development (both adopted from OPLC) and it was termed as pressurized planar electrochromatography (PPEC) [7]. Experimental pressurized chamber was constructed by Dzido et al. [28,29] and Tate and Dorsay [30] suitable to generate external pressure up to 12 MPa and they showed the importance of the soaking time prior to sample application and separation concerning the separation reproducibility and linear velocity along the chromatoplate using RP-18 phase. Necessary to note that the higher external pressure applied on the bed results in lower linear velocity at constant voltage due to the bed compression. The effectiveness of separation can be characterized by plate height to theoretical plate (HETP) and the lowest value published is 10.6 μm using 9 KV and 10 cm long RP-18 chromatoplate [31].

Alternating current was applied by Kreibik et al. in 2001 as another solution of the electric field force and it was called planar dielectrochromatograhy (PDEC) where transverse electric field of AC to porous media effects a dielectroosmotic flow (DEOF) [32]. Both vertical [32,33] and horizontal chamber [34] were constructed and the separation time in PDEC yields about 30% shortening in comparison with conventional linear TLC development. It should be noted in contrast with planar electrochromatography thermal effects are absent during the development [35].

A new forced-flow technique called shear-driven (liquid) chromatography (SDC) was introduced by Desmet and Baron in 1999 where the mobile phase is between two plates and it is forced to flow above as well as inside the very thin adsorbent layer with shear stress mediation generated by a moving plate [36]. Due to its flat format and the open-channel architecture, the system in fact offers a mixed form of the functionalities of both TLC and HPLC. The detection can be fulfilled during the separation at flowing conditions or at stopped-flow conditions by means of a CCD camera [37]; however, the detection sensitivity is Achilles' heel of this miniaturized LC system. In agreement with the theory 0.5 μm theoretical plate height has been realized using porous silicon layer and this corresponds to more than 2 million plates/m for a retained component ($k = 3$) [38]. The typical linear velocity is 5–15 mm/s, consequently the separation is very quick and it can be fulfilled in few seconds.

When fully developed, this shear-driven method has the potential to be a powerful analytical separation technique.

Classification of forced-flow planar layer liquid chromatographic techniques based on the relation of constant linear velocity is also possible (Figure 10.1). Accordingly, linear OPLC, "column"-RPC (C-RPC), PPEC, and SDC is able to generate constant linear velocity along the adsorbent layer; however, the other solutions, such as circular OPLC, CLC/RPC techniques, VPC, PEC and PDEC, approach only to constant linear velocity. This means that the later techniques are less efficient.

10.3 PROGRESS IN ROTATION PLANAR CHROMATOGRAPHY

10.3.1 INTRODUCTION

In 1947 Hopf [9] developed the chromatofuge, the first forced-flow planar layer liquid chromatography system in which centrifugal force was applied to accelerate the separation speed. After several modifications of centrifugally mediated paper and layer liquid chromatographic techniques, five CLC/RPC instruments have been introduced into the market: centrifugal preparative liquid chromatograph with variable rotation speed (0–1000 rpm) Model CLC-5 (Hitachi, Tokyo, Japan), Chromatotron (Harrison Research, Palo Alto, CA, USA) centrifugal thin layer chromatograph having a constant rotation speed (750 rpm), Rotachrom (Petazon, Zug, Switzerland) Model P rotation planar chromatograph with manageable rotation speed (100–1500 rpm), CycloGraph 100–1400 rpm (Analtech, Newark, DE, USA) and a multifunctional instrument Extrachrom (Research Institute for Medicinal Plants, Budakalász, Hungary) rotation planar separator in which chromatographic separation as well as extraction can be performed and the rotation speed can be regulated up to 2000 rpm [39,40].

The term rotation planar chromatography (RPC) introduced by Nyiredy et al. [6,11] covers all the off-line and online analytical and preparative solutions (CLC as well) using the principle of rotation based separations, where the mobile phase flows from the center to the periphery of the stationary phase by means of capillary action and/or centrifugal force, irrespectively of the quality and type of adsorbent and of the volume of the vapor phase.

10.3.2 CLASSIFICATION AND SEPARATION MODES OF RPC

The rotation planar chromatography can be classified on different bases: size of vapor phase (chamber type), development/separation mode related to the shape of adsorbent bed and to the connections of principal steps of chromatography, number of separation steps, mobile phase composition admitted, number of samples applied, and target of the separation, etc.

Vapor phase (chamber type)—The volume of vapor phase may be varied by the space above the adsorbent layer and this depends on the chamber type applied; accordingly, normal chamber (N-RPC), microchamber (M-RPC), ultramicro chamber (U-RPC), and column (C-RPC) have been defined and in the last two the vapor phase above the adsorbent layer is absent [6].

Development/separation modes—RPC is essentially dedicated for circular (radial) development mode (Figure 10.2); however, linear, anticircular (triangular) separations, are also possible.

On the basis of the connections of principal chromatographic steps—sample application, separation, detection, isolation—RPC can also be classified. Fully off-line mode can be applied for analytical separation, where the principal steps are fulfilled separately; however, it is also applicable for micropreparative and preparative isolation similarly to conventional TLC process (band scrape off and elution). In this separation mode the separation starts with dry adsorbent bed with spotted samples and can be developed to the periphery of the adsorbent disc with or without overrun (Figure 10.2a and b) and detection can be fulfilled in situ on the layer. In the fully online mode (the steps are not performed separately but connected online) one sample is applied onto the dry, prewetted or equilibrated adsorbent layer at the central hole (Figure 10.2c) and the bands can be eluted continuously by centrifugal force. The effluent at the outlet can be detected online with an appropriate flow-cell detector; moreover fractions can also be collected for isolation (Figure 10.2d) [6,11].

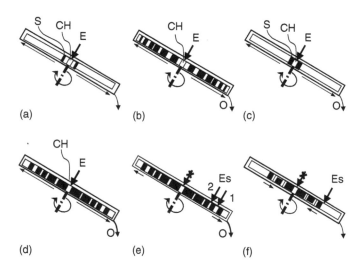

FIGURE 10.2 Schematic drawing of RPC processes. (a) Starting period of the separation with dry layer and off-line sample application, (b) full development or overrun of a, (c) starting period of online sample application onto dry, prewetted or equilibrated layer, (d) full development or overrun of c, (e) sequential RPC with elution by strong eluent at 1st and 2nd position, and (f) sample concentration for the next development/elution. CH, central hole; S, sample; O, outlet of the chamber; E, eluent for development/elution; Es, strong eluent for elution/band concentration.

If the cross-sectional area of the adsorbent bed is constant in the direction of development the development can be carried out with constant linear velocity along the radius. For the analytical layer with constant thickness the adsorbent bed should be modified by scratching the lines into the layer, generating constant cross-sectional area. This process can be applied for anticircular development with the difference that the cross-sectional area is reduced from the center to the periphery. Another solution of linear mobile phase movement has been applied for the column arrangement (C-RPC), in which the thickness of the bed is reduced in the radius direction using a special geometric design.

Number of separation steps—Multiple developments can also be fulfilled in RPC. Depending on the separation problem, the number of re-chromatography steps can freely be used by combining the development distance, solvent strength, and the selectivity and accordingly multiple developments can be differentiated as unidimensional (UMD), incremental (IMD), gradient (GMD), and bivariate (BMD) [41,42].

A special combination of circular and anticircular development modes are applied in the sequential technique (S-RPC), in which circular development with centrifugal force is applied for the separation (Figure 10.2c and d) and the well-separated bands can be eluted with parallel concentration by a strong eluent (Figure 10.2e). The nonseparated components may be pushed back to the center by anticircular development with capillary action at low rotation speed using a strong eluent (Figure 10.2f). Of course, at the start the band will be concentrated during this step (Figure 10.2c) and after a drying period it is ready for the next separation with a new eluent (Figure 10.2). The number of separation/elution/concentration steps is practically unlimited [43,44].

Mobile phase composition admitted—Eluent with constant constituent ratio is admitted for isocratic development; however, mobile phase gradient is also applicable for separation. Starting the separation with dry layer the preferred gradient is stepwise. If the stationary phase is equilibrated prior to separation (U-RPC or C-RPC) continuous gradient can be used similarly to HPLC.

Number of samples applied—For isolation, preparative or micropreparative RPC is used, where one sample is applied as a circle. Starting with dry adsorbent bed sample is applied off-line close to the center (Figure 10.2a). For wetted or equilibrated layer, online sample application is used at the center hole of the rotating adsorbent bed (Figure 10.2c). In contrast to preparative separation for analytical separation, samples can be applied in parallel and up to 72 samples can be spotted to one plate.

Target of the separation—Analytical separation can be applied for qualitative and quantitative determinations; however, the main goal of RPC is the micropreparative and preparative separations for isolation or purification of compounds.

10.3.3 MAIN FACTORS INFLUENCING THE SEPARATION SPEED

10.3.3.1 Mobile Phase Movement

In RPC the movement of mobile phase is accelerated by centrifugal force and in the most commonly used circular development mode with constant layer thickness results in continuously regressive velocity along the radius as well as in time of development. This means, in contrast with conventional TLC development, that in RPC techniques the speed of separation can be influenced by rotation speed; however, the mobile phase velocity may only be approached to the optimum velocity. In general, the higher the speed of rotation the faster is the migration of the mobile phase. As it is obvious from the figure, threefold increase of rotation speed yields more than twice mobile phase velocity.

Another tool of increasing the mobile phase velocity in circular RPC is to increase the diameter of the hole in the center of the stationary phase at constant rotation speed. Approximately fourfold increase of diameter results in twofold mobile phase velocity. The optimum rotation speed depends on the separation problem and also on the mobile phase used [11].

Nonetheless C-RPC generates circular preparative separation and the mobile phase velocity along the radius is constant. The reason is the constant cross-sectional area of the adsorbent bed in the development (radius) direction and this is valid for rotation planar extraction (RPE) as well [40,45]. The geometric design can be described by the following equation:

$$h = K/(a + br + cr^2)$$

where
    h is the actual height of the planar column at radius r
    r is the radius of the planar column
    a, b, c, and K are constants

10.3.3.2 Mobile Phase Flow Rate

In N-RPC, M-RPC, and U-RPC the flow rate at constant rotation speed is limited by the amount of mobile phase, which can be accommodated by the adsorbent layer without flooding over the surface. The greater the amount of solvent applied, the higher the rotation speed must be to keep the mobile phase within the adsorbent bed [6,11,39]. Because of the column arrangement of C-RPC the flooding over the adsorbent surface is essentially excluded; however, flow rate of feeding and take away ratio should be harmonized ($\geq 100\%$).

10.3.4 Adsorbent Bed

Analytical separation can be performed on glass or foil backed precoated adsorbent layers having different phases of bare or chemically modified silica gels (e.g., RP-18, RP-8, amino, cyano). Because precoated adsorbent layers are well bonded to the carrier plate, they tolerate the centrifugal force and can be used without any damages. For preparative work homemade layer is used for isolation. In most cases the preparation is fulfilled by slurry packing and to prevent the damages higher binder content should be used in contrast to the layers used for conventional TLC [39]. For C-RPC the bed can be packed without binder using dry unmodified or chemically modified silica gels at 2000 rpm and in contrast with other RPC fine particle adsorbent can also be used. Ready-to-use preparative layers (called rotors) dedicated for CycloGraph (Analtech, Newark, DE, USA) and Chromatotron (Harrison

Research, Palo Alto, CA, USA) instruments have been introduced by Analtech. Silica gel rotors with 1, 2, 4, 6, and 8 mm thickness and aluminum oxide rotors in 2 and 4 mm sizes are also available [46].

10.3.5 ELUENT SELECTION

TLC pre-assay with unsaturated or saturated chamber can be performed for mobile phase selection using, e.g., "PRISMA" optimization system, taking into account the R_f differences existing between linear and circular developments and also the chamber saturation conditions of RPC systems. For U-RPC and C-RPC the eluent of unsaturated TLC can be transferred without modification. For N-RPC and S-RPC the eluent should be modified by dilution with an apolar solvent for the separation of apolar compounds; however, for separation of polar compounds the change of eluent composition is proposed. The eluent of saturated TLC separation can be applied for M-RPC separation without modification. If required the eluent can be verified with the chosen analytical RPC system for preparative separation [47]. HPLC eluent can be transferred to U-RPC and C-RPC without modification if the phase is equilibrated [6].

10.3.6 TYPICAL PROBLEMS AND THEIR ELIMINATIONS IN RPC

In N-RPC, M-RPC, and U-RPC the mobile phase administration is limited and over this limit a "surface effect" (flooding over the surface) may appear impairing the separation. This unwanted effect can be eliminated by means of higher rotation speed, which can be fulfilled in all RPC instruments except Chromatotron in which the rotation speed is constant. Another solution is the application of lower flow rate [6,11,39].

Owing to the rotation of the adsorbent layer in RPC, two more undesirable effects may occur in terms of temperature. An extra evaporation of mobile phase may yield the change of eluent composition; moreover this extra evaporation may generate the effect of "standing front" in N-RPC and M-RPC (starting with dry layer). In both cases it necessary to take into consideration the eluent selection [48,49].

10.3.7 APPLICATIONS

TLC pre-assays have been used successfully for mobile phase optimization of M-RPC and U-RPC separations of ginsenosides using saturated and unsaturated chambers. The eluent composition is methyl ethyl ketone–methanol–water–hexane. The selected ratio in saturated chamber (80:10:10:15, v/v) is well fitted to the analytical M-RPC separation while another ratio received in unsaturated chamber (70:22:8:30, v/v) is used for U-RPC separation, resulting in the baseline separations of eight compounds in both cases (700 rpm, 5 mL/h) [11].

The isolation of components in pure from a biological system, e.g., plant extract is not an easy task; thus, in most cases more different separation techniques should be applied. As known, the RPC is mainly dedicated to preparative isolation of synthetic and naturally occurring compounds; however, analytical separation is also possible and it is well suited for modeling the preparative separation conditions such as eluent composition, speed of rotation, flow rate, and sample loading as well as detection conditions.

A complex reliable isolation procedure has been described for the purification of ecdysteroids from *Serratula wolffii*. Starting with 209 g crude extract, the previously purified extract with cleanup on polyamide and vacuum reversed-phase column chromatographic separation ajugasteron C, dacryhainansterone, 22-deoxy-integristerone A and two new ecdysteroids have been isolated in 1–10 mg range by means of single or repeated RPC using 1 mm thick silica gel and stepwise gradient separations. The adsorbent bed can be reused after a methanol washing procedure [50].

A scale-up procedure has been performed with furocoumarin isomers using U-RPC, precoated analytical TLC plate (0.2 mm, average particle size 11 μm), and homemade preparative silica gel layers (4 mm, average particle size 15 μm), *n*-hexane–dichloromethane–chloroform–tetrahydrofuran (72.8:10.8:8.3:8.1, v/v) eluent, and 860 rpm for every separations. Because the thickness ratio between preparative and analytical adsorbent bed is 20, this factor can be used for scale-up. Accordingly, on the basis of the analytical off-line separation conditions (8 mg extract, 0.17 mL/min, densitometry at UV 313 nm), the online isolation of components from *Heraclenum spondylium* has been fulfilled with 160 mg loading, 3.5 mL/min flow rate, 860 rpm rotation speed, and 279 nm for UV detection. Because of the larger particle size (130%) generates less and the longer separation distance (125%) more flow resistances of the bed applied for isolation comparing with TLC separation, these effects almost neutralize each other and 20 times higher flow rate can be applied at same rotation speed without any problem for layer having 4 mm thickness [51].

The separation power of C-RPC has been demonstrated by Nyiredy using the eluent of analytical HPLC for the separation of flavonolignans from a purified *Silybum marianum* extract. As can be seen the chromatographic profiles of analytical HPLC and C-RPC are very similar (Figure 10.3) [6].

A method for the isolation of parasorboside and gerberin from the *Gerbera hybrida* has been developed. Two closely related glycosides were extracted and purified using Extrachrom (Research Institute for Medicinal Plants, Budakalász, Hungary) multifunctional separation tool in comparison with medium pressure solid–liquid extraction and medium pressure liquid chromatography. Structures of compounds isolated from the purified extracts by TLC have been confirmed by ^{1}H- and ^{13}C-NMR spectroscopy [52].

Extrachrom has been applied for the extraction and isolation of some components of oak (*Quercus robur* L.) bark. The crude extract has efficiently been separated on silica gel by C-RPC (7.6 mg (+)-catechin/840 mg extract) and the fractions have been checked by analytical TLC and RPC on cellulose layer with pure water [53].

10.4 PROGRESS IN OVERPRESSURED LAYER CHROMATOGRAPHY

10.4.1 INTRODUCTION

The basic ancestor of overpressured layer chromatography (OPLC) is the ultramicro (UM) chamber [54], in which the adsorbent layer is covered by a counterplate

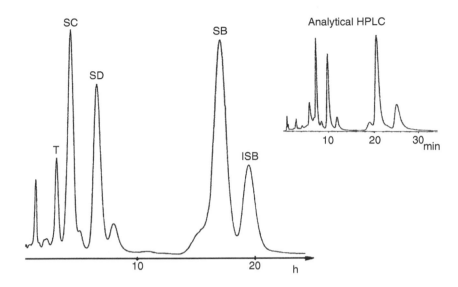

FIGURE 10.3 Column RPC separation of flavonolignans from purified *Silibum marianum* extract by use of the mobile phase for analytical high-performance liquid chromatography. Conditions: stationary phase, Sepralyte C18 (15 μm); mobile phase, water–tetrahydrofuran–methanol–acetonitrile, 75.24 + 17.78 + 3.85 + 3.13; flow rate, 3 mL/min; rotation speed, 1200 rpm; temperature 23.1°C; T, taxifoline; SC, silycristine; SD, silydianine; SB, silybine; ISB, isosilybine. (Reproduced from Nyiredy, Sz., *Planar Chromatography—A Retrospective View for the Third Millennium*, Nyiredy, Sz., ed., Springer, Hungary, 2001, 177. With permission.)

contacting with the adsorbent layer surface eliminating totally the vapor phase of mobile phase above the adsorbent layer and the development is fulfilled in a normal tank with capillary action. Based on that, different experimental pressurized UM chambers has been developed, where an external pressure was applied to the adsorbent layer surface by a cushion system and the eluent is forced to flow by means of overpressure through the adsorbent layer using a pump system [12–14].

Accordingly, the flow rate of eluent admission can be adjusted to the optimum linear velocity results in the higher efficacy; moreover contrary to HPLC, the adsorbent bed can be open. The essence of OPLC is that the adsorbent layer is closed and pressurized during the separation and the development/elution can be carried out at the highest separation efficiency (given by the adsorbent layer) using the optimal linear velocity generated by a pump. In addition, the separation can be checked in the adsorbent layer at any distance or at any time of development/separation (TLC feature). It follows from the principle of OPLC that low (2–5 bar), medium (10–30 bar), and high (50–100 bar or more) external pressures can be used in this forced-flow planar layer liquid chromatographic technique; moreover both off-line and online separations can be used for analysis and isolation as well [55].

It should be noted that OPLC can be found in the literature with different alternative fugitive terms such as overpressured thin layer chromatography (OPTLC), overpressurized thin layer chromatography, optimum performance laminar chromatography, overpressure planar chromatography, high pressure planar chromatography (HPPLC), and forced-flow TLC (FF-TLC).

10.4.2 Separation Modes and Methods of OPLC

10.4.2.1 Transfusion, Infusion, and Infusion–Transfusion OPLC

Similarly to conventional TLC/HPTLC development, three geometric arrangements can be used in OPLC: linear, circular (radial), and triangular (anticircular). In the linear developing mode, one-directional, two-directional, and two-dimensional development is possible; moreover multiple linear developments can also be performed [56,57].

Fully off-line OPLC has two operations the *infusion* and the *transfusion* one. In *infusion* mode the eluent is introduced into the totally closed layer (layer surface is closed by external pressure, the layer edges are sealed on four sides and the chamber outlet is closed). The air originally containing the adsorbent layer is continuously compressed during the process helping on the pore filling of particles. This yields a continuously reduced waviness of front of total wetness and an increased efficiency. The eluent introduction is finished when the inlet pressure reaches pressure limit previously selected. The infusion process is suitable only for off-line developments (one- and two-directional, two-dimensional) without overrun and adsorbent layer sealed on four sides should be used [58].

The *transfusion* operation corresponds to the original OPLC technique permitting the pass through for both the air and eluent [8,55,59]. In this system both off-line and online operation as well as their combination is possible. In the online separation–detection, the eluted air disturbs the first part of online detection. The *infusion–transfusion* operation is suitable for continuous development and serves better conditions for online detection, flushing out the compressed air [58,60].

10.4.2.2 Connection Possibilities of Operating Steps of OPLC

OPLC is an instrumentalized version of planar layer liquid chromatography, and can be classified on the basis of principal operating steps of the chromatographic process (sample application, separation, detection, and isolation). Steps can be connected off-line or online modes; moreover their combinations (partial off-line/online) also applicable, yielding a high flexibility for OPLC (Figure 10.4) [61].

In the *fully off-line* mode, all the principal steps of the chromatographic process are performed off-line as separate operations. For this both infusion and transfusion operating modes are applicable and it is dedicated for parallel analytical separation.

In the online separation–detection mode with off-line sample application (partial online process), the solutes are measured in the drained eluent by connecting a flow-cell detector to the eluent outlet. The process can be fulfilled by transfusion as well as infusion–transfusion operations.

The process of online sample application-separation with off-line detection (*partial off-line* process) is also possible and it can be carried out by transfusion, infusion, and infusion–transfusion operation as well. It was not practiced yet;

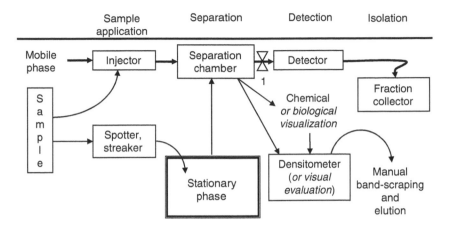

FIGURE 10.4 Scheme of transfusion, infusion, and infusion–transfusion OPLC processes: off-line step (thinner line); online step (thick line); 1 indicates switching valve at outlet.

however, the multichannel injection onto the prewetted/equilibrated adsorbent has a great chance for parallel separation with off-line detection.

The entire chromatographic process can be performed *fully online* by connecting a loop injector to the eluent inlet and a flow through detector (UV, DAD, MS, etc.) to the eluent outlet, in much the same way as in HPLC, applying transfusion or infusion–transfusion operation [62,63].

In analytical fully off-line OPLC, several samples can be processed in parallel. The technique offers further advantages, similarly to TLC, only the spots or bands of analytical interest may be assessed, quantitative evaluation can be repeated with various detection parameters, including the application of specific chemical reagent or biochemical reaction, and chromatogram spots or bands can be evaluated visually as well.

In OPLC systems, the change in the ratio of eluent composition provides good possibilities for special separation modes, i.e., isocratic, gradient, and stepwise gradient and this solution is suitable for both analytical and preparative investigations.

In preparative fully off-line OPLC, after the development a drying, scraping of the adsorbent layer, elution, and crystallization procedures can be performed similarly to conventional preparative TLC methods. However, in preparative off-line OPLC, the resolution is considerably increased and thick, fine-particle adsorbent layers can also be used. The isolation of only the selected component of interest from the adsorbent layer is possible.

In the *combined online/off-line* OPLC system, sample components eluted can be measured online (as in HPLC), although others that remain on the adsorbent layer after the separation can be evaluated off-line by means of a densitometer [61].

Combination of a *multilayer* system with a forced eluent flow complicates really to a certain extent the original simple and flexible TLC technique and also partly conventional OPLC. However, the result is an efficient and promising technique in the field of layer liquid chromatography, which is applicable to analytical and preparative separations in various types of laboratories. The layers can be coupled in parallel as well as in series.

Parallel version of overpressured multilayer chromatography (OPMLC) using two or more chromatoplates is very attractive because a large number of samples (50–100 or more) can be separated during one development [64].

Serial coupled OPMLC (called "long distance" OPLC) can be used for the elevation of the theoretical plate number and resolution alike as elaborated by Botz et al. [65]. Long distance OPLC is a multilayer development technique employing specially prepared chromatoplates. Several plates are placed on top of each other to extend the development distance. The end of the first uppermost chromatoplate has a slit-like perforation to enable the mobile phase to flow to a second layer, in which migration continues until the opposite end of the chromatoplate; here the chroma-tography can be continued on to a subjacent chromatoplate or the eluent be led away. Owing to the special arrangement of the prepared layers and the use of forced eluent flow, the mobile phase can travel through the stationary phases at optimum flow velocity. Of course, in this technical solution the development distance of chroma-toplates can easily be increased to the extent desired. The potential of serial con-nected layers can be increased by use of different (hetero) stationary phases during a single development [66].

It is obvious that the layer can couple in the *parallel/serial* mode as well. This version is well suited for efficient micropreparative isolation [8].

10.4.2.3 Sample Transfer by OPLC to HPLC Separation

Analysis is rather difficult when the sample contains impurities in high concentration together with the components to be measured. This situation is typical for biological samples. The sample has to be purified in one or more steps before chromatographic analysis.

OPLC itself can also be used as a sample cleanup unit of multidimensional systems for separation and identification of components of complex mixtures. The efficiency of the HPLC separation can be improved by direct coupling it with OPLC transferring selected, preliminary separated components to the column [67]. The system can be combined with conventional TLC separation as well and OPLC works as an interface unit transferring the preliminary separated components from the localized area of the adsorbent layer. This combination can also be extended for off-line sample application onto the layer using "spray-on" technique.

10.4.2.4 One- and Multichannel OPLC Separations in the Adsorbent Layer Segmented by Flowing Eluent Wall

A new general concept has been developed for single-channel and multichannel OPLC separations using a nonsegmented adsorbent bed and a flowing eluent wall (FEW) system for operational segmentation [68]. According to this concept, for single online sample application-separation and online or off-line detection, the sample as well as the eluent can be introduced into the same place of the adsorbent layer. For mobile phase administration only one pump can be used. The mobile phase distributor scatters the stream of eluent to the mobile phase line of the FEW and to the mobile phase line of separation, in which sample is applied into the eluent stream by an injector. The distributor space (trough) of the eluent which forms the FEW and the distributor space of sample application transfer separately the applied liquid to the connecting adsorbent bed. These are close to each other and have small volume with a trace of flow resistance perpendicular to the development. The linear velocity of FEW parts as well as the separation parts is the same. The velocity at the sides sealed is distinct from the homogeneous velocity of the bulk. At the outlet side of a nonsegmented adsorbent bed, the sample or mobile phase collector space collects and transfers the effluent stream to the flow-cell detector, which can be connected to a fraction collector for isolation purposes. To eliminate the contaminants originated from the sealing material, the separation line is collected separately from the FEW lines [69].

For parallel fully online separation the FEW can be used for the operating segmentation of nonsegmented adsorbent bed, detaching it into active and nonactive parts regarding separation. Only mobile phase is introduced into the nonactive part, while for the active separation part, eluent and also the sample can be admitted; thus, the unsuitable part of the adsorbent layer is excluded from the separation process. Accordingly, FEW can help in the elimination of edge effect of OPLC for single sample injection; moreover it can also eliminate the sample mixing effect of neighboring lanes for multichannel separation process [68,70].

10.4.3 THEORETICAL BASIS OF OPLC

10.4.3.1 The Rule of Mobile Phase Movement and Factors Influencing Retention

10.4.3.1.1 Formation and Migration of Alpha and Front of Total Wetness in OPLC

It is well known in conventional TLC/HPTLC that the eluent front migrates by means of capillary forces and quadratic equation exits ($z_f^2 = k \cdot t$; z_f, distance of α front; t, time of development; k, velocity constant) [2,71].

In transfusion OPLC starting the separation with dry adsorbent the eluent can be forced through the adsorbent layer by means of a pump system using a chosen flow rate. Feeding the eluent by constant flow rate onto the chromatoplate, the speed of the α front (F_α) depends on the free cross-sectional area of the adsorbent layer in the direction of the development. Only linear development is able to result in constant linear velocity, chromatoplates with circular and triangular shapes are not. In contrast with linear development circular OPLC yields decreasing velocity of F_α along the radius. The basic flow rule of linear transfusion OPLC is

$$z_f = u \cdot t$$

where
    z_f is the migration distance of the eluent front
    u is the linear velocity of the eluent front
    t is the time of the development [12,14]

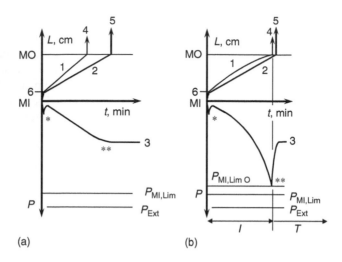

FIGURE 10.5 Variation of front position and eluent pressure during off-line OPLC development using (a) transfusion and (b) infusion–transfusion operating modes. 1, alpha front; 2, front of total wetness; 3, mobile phase inlet pressure; 4, leaving alpha front; 5, leaving front of total wetness; 6, point of off-line sample application; L, distance on the adsorbent layer; P, pressure; t, time; MI, mobile phase inlet position; MO, mobile phase outlet position; P_{Ext}, external pressure; $P_{MI,Lim\ O}$, limit of mobile phase inlet pressure for outlet; $P_{MI,Lim}$, limit of mobile phase inlet pressure; *transition period of rapid mobile phase admission; **transition period of leaving the partially wetted zone; I, infusion period; T, transfusion period. (From Mincsovics, E. et al., *J. Planar Chromatogr.—Mod. TLC*, 15, 280, 2002. With permission.)

Figure 10.5 shows the temporal procession of (a) transfusion and (b) infusion–transfusion operations. To ensure straight front shape starting with dry adsorbent layer in linear OPLC initially, a rapid eluent flow rate (rapid period) is given, exceeding the flow of the capillary action, which is followed by a lower speed of eluent administration (separation period) (line 1). This high velocity step yields straight front line by quick filling of the adsorbent layer at the area of eluent inlet trough and remains straight during the separation. The automated personal OPLC 50 system manages the whole process automatically. In the early generations of OPLC systems a pressurized buffer space has been applied for linear front formation. At a certain distance (position 6) the velocity becomes constant and sample should be applied over this distance (cf. H versus u curve).

So, in the linear off-line transfusion and also in fully online OPLC separations the velocity of F_α as well as mobile phase is constant along the plate (except the rapid period) and it can proportionally be modified by the flow rate (Q):

$$u = \frac{Q}{a \cdot b \cdot \varepsilon}$$

where
  a is the width
  b is the thickness
  ε the total porosity of the adsorbent layer

Starting the separation with dry porous adsorbent layer, two zones can be found that arises due to significant differences in their refractive indexes, even if single eluent and conventional or forced-flow layer chromatographic techniques are used [62,72,73]. In fully off-line transfusion OPLC in the zone under the F_α the external (space between particles) and internal porosity (pores) of the bed is filled partially with air and eluent. This is called the partially wetted zone, which sometimes disturbs the separation, in this narrow range [62,63]. The next zone toward the eluent inlet point is a totally wetted, which is completely filled by the eluent. The border between these zones is the front of total wetness (F_{tw}) (Nyiredy et al. termed "disturbing front" [72]), which is a sharp zigzag line. F_{tw} migrate proportionally with F_α at a constant flow rate (Figure 10.5a, line 2) and during the separation its bandwidth increases with migration distance [62,72,73]. The inlet pressure (curve 3) increases gradually with time until the mobile phase begins to flow from the chamber (line 4) and after a transition period in which as the F_{tw} reaches the outlet (line 5) the inlet pressure becomes constant. Velayundhan et al. [73] found that R_{tw} linearly increases with the flow rate but F_α shows slight nonlinearity at higher flow rates. Within experimental error their incompressible model is in agreement with experiments. It should be noted that this effect is greater on TLC plate than on HPTLC plate. In transfusion OPLC the waviness of F_{tw} can be dramatically reduced applying a higher external pressure on the layer surface [59] or can be totally eliminated by applying a prerun with an eluent in which analyzable components do not migrate using a stepwise gradient [52].

In contrast with the transfusion operation, infusion process (Figure 10.5b), yields a continuously decreasing waviness of front of F_{tw} as well as sample band shape. The speed of F_α decreases continuously whereas the inlet pressure increases and when the inlet pressure reaches the pressure limit the eluent feeding is automatically terminated. The infusion–transfusion operation enables flushing out the compressed air at the first infusion period. Contrary to the transfusion process, the infusion–transfusion process yields a quick stabilization concerning the eluent/adsorbent ratio resulting better conditions for online separation–detection process [58,60].

Owing to the partially filled zone in transfusion off-line operation the adsorbent/eluent ratio is not constant along the plate. Because of this the front distance is always longer than the real one measured at totally filled conditions. It can be concluded that for full development of infusion operation needs about 10% more eluent comparing with transfusion separation [58].

10.4.3.1.2 Formation and Migration of Secondary Fronts
It is a well-known fact in classical TLC that the eluent components adsorbed strongly by the sites of adsorbent can cause secondary fronts (F_β, F_γ, ...) [74]. This effect can be found during adsorption as well as in reversed-phase development when the eluent consists of solvents of different strength. Owing to the total elimination of vapor space, the effect of this chromatographic solvent demixing is stronger in fully off-line OPLC systems than in chambers having small vapor spaces, e.g., in sandwich chambers. These secondary fronts are independent of F_{tw} and divide the adsorbent layer into zones of different eluting strength, within which the solvent strength and polarity are practically the same, while at the fronts themselves there

is a sudden increase in eluent strength that gives rise to "polarity steps." This phenomenon takes place in HPLC [75] and in fully online OPLC as well during equilibration process, when the apparatus is not used for separation [62,63].

If the secondary front collects components to be analyzed from the preceding zone (α zone), shorter development or higher sample origin is needed. When sample components are not sensitive and the mobile phase of α zone can elute the component collected by secondary front, double development with same eluent can be used; moreover the second run can be performed by the apolar constituent of the mixture. If the phenomenon cannot be overcome new eluent should be used. If the polar constituent of the used eluent is replaced by a weaker one of the same volumetric ratio, higher $R_{f,\beta}$ and lower ε value of β zone arise. Replacing the apolar constituent of this new eluent to stronger one yields higher $R_{f,\beta}$ and a higher ε value of the β zone. At a given eluent–adsorbent system, the $R_{f,\beta}$ value is constant irrespective of migration distance and eluent velocity used [55,62]. $R_{f,\beta}$ depends on the eluent composition and it was found that a plot of R_M [$R_M = (1/R_f) - 1$] versus the logarithm of the mole fraction of polar constituent in the mixture did not show linear relationship, unlike the compounds' migration in the β zone. $R_{f,\beta}$ increases with increasing concentration of polar modifier as well as with decreasing specific surface area of the adsorbent [76].

Similar results were found by Wawrzynowicz and Soczewinski in the case of a sandwich chamber and binary eluents [77].

10.4.3.1.3   Retention Transfer among TLC, Off-line and Online OPLC and HPLC

Retention data of fully off-line OPLC can be converted to online separation–detection and also to fully online conditions by using the equation found by Schlitt and Geiss [78]

$$k = \frac{1}{R_f} - 1$$

where
   R_f is the retention factor
   k is the capacity factor of a given component in the off-line and online system, respectively ($k = (t_R/t_0) - 1$, where t_R is the retention time of a component and t_0 is the retention time of the nonretained peak or dead time)

Strong correlation was found on silica gel layers among fully off-line, partially off-line, and fully online OPLC using more component eluent mixture; however, due to the adsorbent bed difference (activity) the slope of the line is not 1. If the β front collects some components, the following equation can be used to convert these data into those of the fully online system:

$$R_{M,i}^w = R_{M,\beta} + R_{M,i}$$

where
   $R_{M,i}^w$ is the R_M value in the wet system
   $R_{M,\beta}$ is the R_M value of the β front
   $R_{M,i}$ is the R_M value of the given collected components in the α zone [63]

If the number of silanols in silica gel is reduced, e.g., by diol modification the chromatoplate is less sensitive to relative humidity and yields less retention than bare silica gel; thus, it is suitable for the separation of nonpolar and polar compounds using simple less polar eluent. The retention data between fully off-line and fully online OPLC yield stronger correlation than it is on silica gel layers [76].

Reversed-phase ion-pair chromatography can be optimized by fully off-line OPLC [79]. Good agreement was found in selectivity between HPLC and OPLC ion-pair systems using the same eluent composition, and this serves a modeling possibility of HPLC ion-pair systems by fully off-line OPLC [80].

Similar mobile phase selectivity was found for coumarins in the case of TLC, off-line OPLC, and HPLC separations. The change of eluent strength had the same effect on retention using TLC and fully off-line OPLC [81].

The retention transfer is influenced by the eluent/adsorbent ratio as well as the adsorbent activity. In infusion operation the eluent/adsorbent ratio is practically constant along the adsorbent layer and the value is equal to the ratio of the totally wetted zone of transfusion operation. Because the off-line processes starts with dry adsorbent bed, during the equilibration the adsorbent bed activity may change yielding an altered retention for online OPLC and this activity change is the main factor in adsorption system and besides this the eluent/adsorbent ratio is less important in general [8]. Using diagonal sample application and single eluent in transfusion OPLC, R_f values were practically independent of spotting location, and their values were higher on HPTLC layers than on TLC layers [14].

10.4.3.2 Efficiency Characteristics of OPLC and Their Influencing Factors

10.4.3.2.1 Theoretical Plate Height
The theoretical plate height (HETP, H) in TLC can be calculated according to Guiochon and Siouffi [82], which can be applied for off-line OPLC as well [14]:

$$H = \frac{\sigma^2}{(L_f - s_0)R_f}$$

where
    σ is the spot variance
    L_f is the front distance
    s_0 is the distance between the spotting location and the eluent inlet trough
    R_f is the retention factor

In transfusion off-line OPLC the HETP is practically constant along the adsorbent layer and it is independent of the development distance if sample deposited has a small diameter in the direction of development (Figure 10.6). The figure shows the essence of OPLC and the basic differences between conventional TLC, HPTLC, and off-line OPLC regarding efficiency [14,59]. The H value depends on the characteristics of the plate (mainly from the particle size) and decreases in the following order: preparative layer, TLC, HPTLC, and Raman plate [59,62,63]. H depends on the R_f and the front distance as well and it is slightly influenced by the thickness [62,63,83].

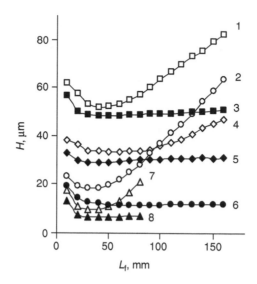

FIGURE 10.6 Correlation between theoretical plate height (H) and the front distance (L_f) using PTH-leucine ($R_f = 0.61$, 0.4 mg/mL), chloroform–dichloethane–ethyl acetate (45:45:10, v/v) eluent and various irregular and spherical silica gel layers developed in conventional unsaturated chamber and personal OPLC 50 system with infusion operation. 1, 3 preparative layer (18–22 µm, thickness 0.49 mm, 8 µL/8 mm sample); 4, 5 TLC layer (10–12 µm, thickness 0.21 mm, 3 µL/8 mm sample); 2, 6 HPTLC layer (5–6 µm, thickness 0.18 mm, 2 µL/8 mm sample); 7, 8 Raman plate (spherical, 3–4 µm, thickness 0.098 mm, 1 µL/8 mm sample); linear velocity (cm/min) of transfusion OPLC development, preparative layer, 1.32, TLC layer, 1.48, HPTLC layer, 2.58, Raman plate, 4.33.

In contrast with transfusion off-line OPLC the infusion one is more efficient owing to the reduced waviness of component band at the area of F_{tw} [58].

H versus migration distance of different compounds yields a gradually decreasing curve which can be linearized by plotting H versus the inverse migration distance [62,76]. It was found that the slope of the line depends on the spot size deposited. Based on the linear relationship between the H and the inverse migration distance of a component (L_i) the following H calculation has been suggested.

$$H_i = \frac{(\sigma_{i0}^*)^2}{L_i} + H_\infty$$

where
σ_{i0}^* is the initial spot variance
H_∞ is the HETP at the point of intersection

It was assumed that $H_\infty = hd_p$, where h is the reduced plate height and d_p is the particle diameter [76].

Similarly to HPLC, the linear velocity also influences the H values in OPLC [83,84]. Different operating modes of transfusion OPLC yield very similar curves

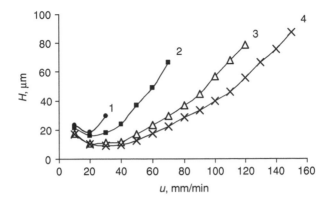

FIGURE 10.7 Effect of linear velocity (u) on theoretical plate height (H) using different external pressure and an experimental OPLC system with fully off-line transfusion operation. (1) 1 MPa; (2) 2.5 MPa; (3) 5 MPa; (4) 7.5 MPa; sample, PTH-valine, 0.4 mg/mL, 5 μL/10 mm band; HPTLC silica gel; dichloromethane–ethyl acetate, 92:8, v/v; front distance, 170 mm.

of H versus linear velocity; however, H values are different. Fully off-line OPLC produces the lowest and owing to the "extra-column" band broadening the fully online OPLC the highest H values. Between them are two curves of partial off-line (or partial online) OPLC [63,76]. In fully online OPLC the total variance of a peak is the sum of all contributions (injector, connecting tubing, troughs, adsorbent layer, detector, and electronic), similarly to HPLC [85]; however, in fully off-line OPLC the peak variance of the adsorbent layer exists only. This is the origin of fundamental differences concerning the H values among different operating modes of OPLC.

The quality of the packing can also influence H value and it can be decreased by compression of the adsorbent layer up to a limited external pressure [86]. Figure 10.7 shows the effect of linear velocity on H value using different external pressures and fully off-line transfusion OPLC. The high external pressure significantly increases the efficiency and the optimum range of linear velocity becomes broader while it does not occur during online development [59,87].

10.4.3.2.2 Theoretical Plate Number
The theoretical plate number (N) can be calculated in off-line OPLC with the equation

$$N = L_f/H$$

where
    L_f is the distance between start and front
    H is the average theoretical plate height

In online OPLC, similar to HPLC the following equation [88] can be used for the N calculation:

$$N = 4 \cdot \left(\frac{t_R}{t_{wi}}\right)^2$$

where

    t_R is the retention time of the component

    t_{wi} is peak width at the points of inflexion

As known, H is practically constant along the plate [14]. This means that the theoretical plate number increases linearly with development distance contrary to conventional TLC/HPTLC.

The theoretical plate number (as well as the spot or peak capacity) can be increased by multilayer system using long distance OPLC and 70×10^3 theoretical plate number was achieved using butter yellow and 70 cm long development [45,89].

10.4.3.2.3 Spot and Peak Capacity

The spot capacity of conventional TLC/HPTLC is limited, as found Guiochon and Siouffi, and about 20 spots can be resolved with a resolution of unity. In fully off-line OPLC the maximum value of spot capacity (n_M) is

$$n_M = \frac{1}{2}\sqrt{\frac{L}{H}}$$

where

    L is the distance of development

    H is the average HETP value of compounds at given conditions [90]

Peak capacity (n) of a column in HPLC is given by the well-known equation [90], and it is applicable for OPLC using online separation/detection systems [61]:

$$n = 1 + \frac{\sqrt{N}}{4}\ln(1 + k)$$

where

    N is the theoretical plate number of the adsorbent bed

    k is the capacity factor of the most retained compound eluted

Because of the adsorbent bed is openable in OPLC after an online separation–detection, compounds separated but remaining on the layer can also be detected in situ by densitometry. In this combined online/off-line OPLC system at given bed length and isocratic conditions, higher spot or peak capacity can be perceived than by single online or single off-line OPLC separation, because the peak and spot capacities are additive [61].

10.4.3.2.4 Resolution

In the case of off-line detection the resolution (R_s) of a neighboring pair of spots or bands can be described by the following equation:

$$R_s = \left[\frac{K_1}{4 \cdot K_2} - 1\right]\sqrt{R_f N}\,(1 - R_f)$$

where

K_1, K_2 are distribution coefficients of two substances

R_f is the average retention factor of pairs and the N is the theoretical plate number

The plot of resolution versus R_f shows high differences between conventional TLC/HPTLC and OPLC. The optimum range of resolution in TLC is R_f 0.3–0.4; contrarily in OPLC this R_s range is R_f 0.3–0.65 [91].

Three factors were studied by Hauck and Jost regarding the resolution using fully off-line OPLC [83]. The resolution increases linearly with increasing front distance and the use of the optimal linear velocity in relation to H produces the highest resolution. The effect of layer thickness on resolution shows an optimum value which is between 80 and 160 μm.

10.4.4 BASIC ELEMENTS AND INSTRUMENTS OF OPLC

10.4.4.1 Adsorbent Layers and Adsorbent Layer Systems for OPLC Separations

Contrary to circular OPLC technique, the linear development requires a special chromatoplate that is sealed at the edges (two, three, or four sides). This prevents the eluent flowing off the chromatoplate in the unwanted direction. Linear migration of the eluent front can be achieved by placing a narrow channel in the adsorbent at the eluent inlet and outlet. The eluent inlet trough directs the eluent to form a linear front while eluent outlet trough collects it at the end of chromatoplate, serving the possibility of overrun and online detection. To generate these functions, troughs can be formed in the PTFE cover plate as well. Ready-to-use OPLC layers sealed on four sides are available commercially (OPLC-NIT Ltd., Budapest, Hungary) which can be used for transfusion, infusion, and infusion–transfusion off-line and online developments alike [8,56].

10.4.4.2 Gas- and Water-Cushion Experimental and Commercial Instruments

On the basis of experience gained with experimental pressurized chambers, the LABOR Instrument Works (Budapest, Hungary) developed CHROMPRES 10 and CHROMPRES 25, the first commercial OPLC systems. In CHROMPRES 10 maximum cushion pressure permitted is 1.0 MPa and it can be used with plastic, aluminum, or glass backed chromatoplates up to 20×40 cm sizes, coated with fine-particle (5–6 μm) or superfine-particle (2–3 μm) adsorbent.

CHROMPRES 25 system allows 2.5 MPa maximum cushion pressure and the maximum size of layer is 20×20 cm. The optimum eluent front velocity is higher and the increase of eluent front velocity means a higher eluent inlet pressure [8,55]. The system gives better conditions for such separations than the CHROMPRES 10.

Kaiser and Rieder developed a high-pressure circular OPLC (3.8–4.0 MPa) called high-pressure planar liquid chromatography (HPPLC) [92].

An OPLC instrument has been developed by Witkiewicz et al. where gas is used to form the external pressure and the eluent is fed to the chromatoplate by a syringe pump [93].

The last solutions [92,93] are not practically in laboratory praxis.

10.4.4.3   OPLC Instruments with Cassette Systems

Automated personal OPLC 50 system (developed by OPLC-NIT Ltd., Budapest, Hungary), in which 5.0 MPa maximum cushion pressure is permitted, includes the separation chamber with tray like layer cassette and the liquid delivery system. A microprocessor-controlled liquid delivery unit is the heart of system. The pump has two heads, one for the eluent delivery, and the other for the hydraulic liquid delivery. All parameters (external pressure, eluent volume of rapid period and of development, eluent flow rate) can be given for development and can also be stored in the software of the delivery system. The separation chamber is equipped with one eluent inlet and one outlet capable for online detection. The automatic developments (single isocratic or stepwise gradient with max. three steps) are absolutely repeatable. Linear, one- and two-directional, two-dimensional, circular cassettes can be used for infusion off-line and for transfusion off-line and online developments using analytical or preparative adsorbent layer and the appropriate cassette [8,59].

The OPLC Separation Unit is a stand alone chamber having a built-in hydraulic pump to form 5 MPa of external pressure and for eluent delivery an independent pump, e.g., an HPLC pump can be connected. The unit has one inlet and one outlet. The cassette enveloping the adsorbent layer can be inserted into the chamber similarly to personal OPLC 50 system [68].

10.4.4.4   One and Multichannel Experimental OPLC Systems with Flowing Eluent Wall Configuration

For flowing eluent wall (FEW) configurations the original hydraulic unit of OPLC Separation Unit has been changed to a new one which is equipped with two eluent inlet connections, one for sample injection and another one for the FEW formations, and the outlet can be connected to a flow-cell detector and/or a fraction collector. The experimental four- and eight-channel FEW versions are suitable for parallel fully online separations [68–70].

10.4.5   APPLICATIONS OF OPLC FOR SEPARATION OF PLANT SUBSTANCES

It should be noted that the mobile phases can be used for analytical and preparative separations independently from the instrument generations of OPLC, and however, the new systems having the possibility to apply higher eternal pressure result in more effective separations.

10.4.5.1   Analytical Applications of OPLC in Phytochemistry

10.4.5.1.1   Increasing the Separation Efficiency and the Sample Throughput
In contrast to HPLC, the planar layer liquid chromatography serves the possibility of parallel sample separation and it is valid for OPLC as well. In linear fully

off-line one-directional OPLC 16 samples whilst in two-directional 72 samples can be separated on a 20×20 cm layer during a run. Resolution in HPTLC is limited by the development distance, because it cannot be increased beyond 8–9 cm. OPLC permits longer development distances, due to the forced flow, and the theoretical plate number as well as the resolution can be significantly increased.

Biogen amine-content was determined in cauliflower and broccoli. The dansylated derivatives were separated on HPTLC silica by a stepwise gradient separation. Approximately 26 min were needed for the separation of 14 samples simultaneously with 2.5–15 ng/spot detection limit [94]. The effect of cold-hardening of wheat [95] and cadmium stress of wheat seedlings [96] on the polyamine content were also studied by this method.

Bis-indol alkaloids extracted from *Catharanthus roseus* were separated and determined using a homemade amino-bonded HPTLC silica gel and OPLC. PRISMA model followed by factorial experimental design was used for mobile phase optimization [97].

Camptotheca alkaloids were separated on normal particle and fine particle silica gel layers with THF-methylene chloride (1:3, v/v) eluent using conventional TLC and OPLC separations [98].

Tertiary and quaternary alkaloids from *Chelidonium majus* L. were separated at optimized mobile phase velocity using silica gel layer and personal OPLC 50 system [99].

Two phenol isomers thymol and carvacrol with a classical binary eluent (hexane–ethyl acetate 95:5 v/v) were separated by a single run within 15 min from different chemotypes of *Thymus vulgaris* L. [100].

The essential oils obtained from stems and leaves, fruits and roots of *Scaligeria tripartita* were separated by two steps. The first elution was conducted with *n*-hexane-Et_2O (95:5, v/v) and due to the secondary front formation some components were collected by F_β. Using a second run with pure hexane, these collected components can also be separated [101].

Glycosidically bound volatile components of fruit were separated by the second generation of OPLC system [102]. Flavonoid glycosides from *Betula folium* were separated by optimized eluent made by PRISMA model [103]. Ginsenosides were also separated in parallel using the first and last generations of OPLC instruments and the eluent were optimized by PRISMA model (Figure 10.8) [103–105]. Cannabinoids from *Cannabis sativa* L. were separated on silica gel layer by TLC, AMD, and OPLC and these separations techniques were also compared [106–108]. The main coumarins from *Peucedanum palustre* were also separated in parallel by the first generation of OPLC system [109]. Resveratrol and other stilbene isomers from white and blue grape and also from red wine were separated by OPLC using RP-18 adsorbent layer [110,111]. Separation of bound vitamin C, ascorbigen and its methylated form were separated by transfusion OPLC [112]. This was used for the separation of cabbage extract and ascorbigen content was determined in *Brassica* vegetables [112,113].

Chrompres 10 OPLC chamber was used for the closely related furocoumarins [114].

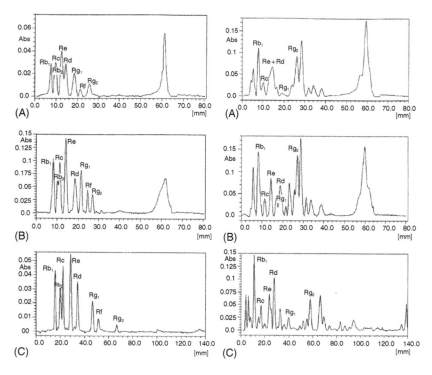

FIGURE 10.8 Fully off-line OPLC separation of ginsenoside standards (left) and purified extract from *Panax quinquefolium* L. using different silica gels and chloroform–methanol–ethyl acetate–water–hexane (20:22:60:8:4, v/v) eluent. A, TLC; B, HPTLC; C, OPLC. (Reproduced from Ludwiczuk, A., Nyiredy, S., and Wolski, T., *J. Planar Chromatogr.—Mod. TLC*, 18, 104, 2005. With permission.)

As a consequence of the forced mobile phase flow in OPLC, more viscous eluent can also be applied for real time separation even if fine particle adsorbent and longer development are used. The separation time is significantly shorter than that in conventional TLC/HPTLC. Polar quaternary alkaloids in plant extract were separated in 10 min on silica gel at a distance of 14 cm using ethyl acetate–tetrahydrofurane–acetic acid (60:20:20, v/v) eluent [115].

Rapid separation of xanthines from tea leaf extracts was performed on HPTLC silica with chloroform-acetic acid (6:4, v/v) eluent and the two-directional separation was finished within 5 min (4 s per sample) [59].

Because of the short separation time, the OPLC system is useful for rapid high throughput screening and it is especially valid for the parallel multilayer OPLC. Circular parallel OPMLC was used for screening the plant species of poppy with high alkaloid content with a separation speed of 5 s per sample [116].

The main protein amino acids were separated by *n*-butanol–acetonitrile–0.005 M KH_2PO_4 aq.–acetic acid (10:5:30:10, v/v) eluent in a serial connected double-layer system, where the total bed length of the connected HPTLC silica gel layers was

34 cm and after visualization by ninhydrin reagent densitometry was used for evaluation at 490 nm [117].

The spot and peak capacity can be increased in OPLC by the combination of online and off-line operation mode, where the eluted components are measured online first and after stopping the elution the components remained on the adsorbent layer are measured off-line by densitometry. Because the spot and peak capacity are additive, the total capacity is increased as demonstrated by PTH-amino acid separations, and its number approximately is two times higher than that available by single online or single off-line separations alone [61].

With two-dimensional off-line OPLC separation the spot capacity can dramatically increase as demonstrated by the perfect separation of 16 closely related coumarins from the genus *Angelica*. The optimization process of eluent systems was also discussed [118].

10.4.5.1.2   Sample Cleanup

If the sample contains impurities in high concentration the analysis is rather difficult. In the case of biological samples this situation is typical and the sample has to be simplified by purification separately before chromatographic analysis using one or more steps. OPLC itself can also be used for in situ sample cleanup if the disturbing components migrate with the eluent front or remain at the origin on the chromatoplate.

Sample cleanup for OPLC separation of complex plant-extract can be solved by two-step development. The separation of fumonisins of B series from inoculated rice culture was performed on reversed phase RP-18 adsorbent layer. The first eluent for sample cleanup was acetonitrile–1% KCl aq. (1:9, v/v), and the second eluent for the real separation of the components was acetonitrile–4% KCl aq. (2.5:1, v/v). Detection was performed by fluorescent measurement at 365 nm after the derivatization with fluorescamin reagent [119].

Aflatoxins in wheat were separated by OPLC using silica gel adsorbent layer. The prepared samples were prewashed in situ on the adsorbent layer by predevelopment in the reverse direction, from the outlet-side of the OPLC-equipment with diethyl ether–hexane (1:1, v/v). The aflatoxins were then separated in normal direction with chloroform–toluene–tetrahydrofuran (15:15:1, v/v) [120–123].

10.4.5.1.3   Direct and Indirect Connection Possibility of OPLC
with Other Techniques

Because of the low solvent consumption and short development time of OPLC, and the linear correlation between OPLC relative retention values and logarithms of the capacity ratios obtained by HPLC it can be used as pilot technique for HPLC [79,80,124].

Indirectly coupled OPLC–GC–MS system was used for the investigation of acetylenic thiophene derivatives in extracts of *Tagetes patula*. The fractions of effluent of online OPLC separation were collected and injected to a GC–MS system [125]. Similar solutions are possible for FTIR and OPLC [126].

Direct coupling of OPLC to HPLC can be used for separation and identification of components in a complex mixture, where OPLC is the cleanup unit of the multidimensional system. Only analyzable components can be transferred to the other chromatographic system, whilst the disturbing compounds remain on the OPLC layer at the front and at the start. The efficiency of HPLC separation can be improved by this system as was demonstrated by the separation of tea leaf extract [67].

The direct coupling of online OPLC to a sophisticated mass spectrometer is the most powerful method for the study of a complex biological mixture. This combined system was used for separation and electrospray mass spectrometric detection of glycolipids [127].

10.4.5.1.4   Application of Special Detection

Because the adsorbent layer in OPLC can be open, it is suitable for biological detection using bioautography as well as BioArena system. These special detections can be found with more details in Chapter 9.

10.4.5.2   Preparative Applications

As with analytical OPLC, off-line and online methods can be distinguished in pre-parative OPLC applications. The off-line OPLC method is similar to conventional TLC method: development, drying, scraping of the adsorbent layer, elution, and crystalliza-tion. Phorbol diester constituents of croton oil were identified by off-line OPLC separation followed by extraction and chemical ionization mass spectrometry [128].

The online OPLC method is more effective for preparative applications because of the time-consuming scraping and elution can be eliminated. It was used for the isolation of hemp constituents and the cannabinoid acid fractions were analyzed by different spectroscopic methods [106].

Biologically active compounds of plants (*Rhamnus frangula, Heracleum sphon-dylium, Gentiana purpurea*) were separated by online OPLC using eluents optimized by PRISMA model and the loading of the 2 mm thick layer was 50–500 mg [129].

Alkaloids and glycosides were isolated from different plants (*Simaba multiflora, Crossopteryx febrifuga, Steblus asper*) by online OPLC [130].

Because long distance OPLC (serial multilayer) allows longer running distance, compounds from extremely complex biological matrices can be separated and isolated by this technique. The effectiveness of the system was demonstrated by fully online separation of a raw extract from roots of *Pseucedanum palustre* [89].

Micropreparative OPLC separation–isolation can be modeled by fully off-line OPLC using different sample volume of an extract and 10 mm band sizes. The selected volume of optimum loading capacity can be converted to the analytical as well as preparative adsorbent layer for isolation. This process was used for the isolation of ascorbigen from cabbage extract using analytical silica gel layer [60] and also for the isolation of xanthines from tea leaf extract using 0.5 mm preparative layer [59]. Typical loading of an analytical layer is 20–25 mg per run, whilst the 0.5 mm thick layer can be loaded with 50–60 mg sample; however, the loading depends on the separation problem to be solved.

Sample cleanup can be performed in situ on the chromatoplate in the case of off-line sample application and online separation/detection using a partial elution first in which impurities can be eluted in the reverse direction from the outlet-side of the system. The second step of isolation is a normal elution/detection/fraction

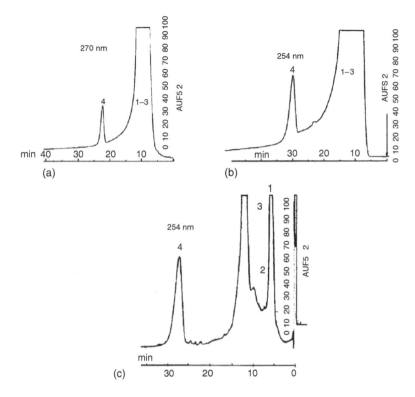

FIGURE 10.9 Isolation of trigonelline from a *Leuzea* extract by different transfusion OPLC processes using normal particle analytical silica gel adsorbent layer. (a) Off-line sample application, partial prewetting from outlet and online separation–detection, sample, 300 μL/180 mm; (b) fully online OPLC, sample injected, 1000 μL; (c) off-line sample application, prewetting and partial elution from outlet to inlet side, as a cleanup step, separation–detection step for isolation using, sample, 1000 μL/180 mm; 1, 2, 3 refer to plant pigments, 4 indicates trigonelline. (Reproduced from Mincsovics, E. et al., *J. Planar Chromatogr.—Mod. TLC*, 15, 280, 2002. With permission.)

collection. Trigonelline isolation from Leuzea extract was performed accordingly, yielding pure component (Figure 10.9) [60].

Fully online OPLC system with one-channel FEW configuration was used for micropreparative isolation of components of *Matricaria chamomilla* L. oil using analytical silica gel layer and 22.5 mg loading (Figure 10.10) [70]. This system was also used for the activity-guide isolation of red wine compounds [131].

10.5 THE ROLE OF FORCED-FLOW PLANAR LIQUID CHROMATOGRAPHIC TECHNIQUES FOR THE SEARCH OF NATURAL PRODUCTS

The efficient forced-flow techniques are well suited for high throughput screening as well as for analysis of naturally occurring compounds because the separation of

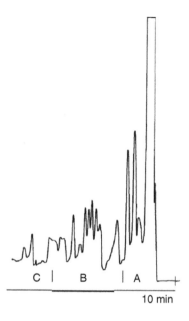

10 min

FIGURE 10.10 Fully online stepwise gradient chamomile oil isolation using OPLC Purification Unit with one-channel FEW configuration and fine particle silica gel. Eluent, hexane–ethyl acetate; (A) 98:2; (B) 92.5:7.5; (C) 85:15; flow rate 2 mL/min; injection (eluent A), 22.5 mg/250 μL. (Reproduced from Mincsovics, E., *J. Planar Chromatogr.—Mod. TLC*, 17, 411, 2004. With permission.)

samples can be performed in parallel at optimum velocity which is, e.g., in OPLC, generally 5–10 times higher than in conventional TLC. It is especially valid for parallel coupled multilayer OPLC using circular, one- or two-directional separations in which 2–4 layers can be developed in parallel. Characteristically, in linear OPLC for one-directional separation 16 while for two-directional 72 samples can be applied onto a 20 × 20 cm layer [59]. Because the adsorbent layer can be open, after the separation it is capable for in situ evaluation by densitometry with or without special chemical visualization reactions; moreover biochemical interactions with microbes are also possible for bioassays. It can be performed by direct bioautography or with the recently developed BioArena which integrates the advantages of OPLC and bioautography [132]. On the basis of the chromatographic, chemical, and biochemical information, an activity-guided quick isolation can be fulfilled using online preparative OPLC with a loading of 50–200 mg or CLC/RPC/C-RPC with loading capacity 100–1000 mg. The structure elucidation of pure isolated biologically active substances can be performed by the usual spectroscopic methods (e.g., UV–visible, MS, MS–MS, ^{1}H NMR, ^{13}C NMR) [131,133].

ACKNOWLEDGMENT

The author thanks Professor Ernő Tyihák for helpful discussion.

REFERENCES

1. Kaiser, R.E., *Einführung in die Hochleistungs-Dünnschicht-Chromatographie*, Institute for Chromatography, Bad Dürkheim, 1976.
2. Guiochon, G. and Siouffi, A.M., Study of the performances of thin-layer chromatography. III. Flow velocity of the mobile phase, *J. Chromatogr. Sci.*, 16, 598, 1978.
3. Brown, P.R., *High Pressure Liquid Chromatography*, Academic Press, New York, NY, 1973.
4. Engelhardt, H., *Hochdruck-Flüssigkeitschromatographie*, 2nd ed., Springer, Heidelberg, 1977.
5. Horváth, Cs., *High-performance Liquid Chromatography. Advances and Perspectives*, Vol. 1, Academic Press, New York, NY, 1980.
6. Nyiredy, Sz., Rotation planar chromatography, in *Planar Chromatography—A Retrospective View for the Third Millennium*, Nyiredy, Sz., ed., Springer, Hungary, 2001, 177.
7. Nurok, D. et al., Apparatus and initial results for pressurized planar electrochromatography, *Anal. Chem.*, 76, 1690, 2004.
8. Mincsovics, E., Ferenczi-Fodor, K., and Tyihák, E., Overpressured layer chromatography, in *Handbook of Thin Layer Chromatography*, 3rd ed., Sherma, J. and Fried, B., eds., Marcel Dekker, New York, NY, 2003, 175.
9. Hopf, P.P., Radial chromatography in industry, *Ind. Eng. Chem.*, 39, 938, 1947.
10. Hostettmann, K., Hostettmann-Kaldas, M., and Sticher, O., Rapid preparative separation of natural products by centrifugal thin-layer chromatography, *J. Chromatogr. A*, 202, 154, 1980.
11. Nyiredy, Sz., Dallenbach-Toelke, K., and Sticher, O., Analytical rotation planar chromatography, in *Recent Advances in Thin-Layer Chromatography*, Dallas, F.A.A., Read, H., Ruane, R.J., and Wilson, I.D., eds., Plenum Press, New York, NY and London, 1988, 45.
12. Tyihák, E., Mincsovics, E., and Kalász, H., New planar liquid chromatographic technique: overpressured thin-layer chromatography, *J. Chromatogr.*, 174, 75, 1979.
13. Kalász, H. et al., Circular development with overpressured thin-layer chromatography, *J. Liquid Chromatogr.*, 3, 845, 1980.
14. Mincsovics, E., Tyihák, E., and Kalász, H., Resolution and retention behavior of some dyes in overpressured thin-layer chromatography, *J. Chromatogr.*, 191, 293, 1980.
15. Delvordre, P., Regnault, C., and Postaire, E., Vacuum planar chromatography (VPC): A new versatile technique of forced flow planar chromatography, *J. Liq. Chromatogr. & RT.*, 15, 1673, 1992.
16. Pretorius, V., Hopkins, B.J., and Schieke, J.D., Electro-osmosis—A new concept for high-speed liquid chromatography, *J. Chromatogr. A*, 99, 23, 1974.
17. Pukl, M., Prosek, M., and Kaiser, R.E., Planar electrochromatography, Part. 1: Planar electrochromatography on non-wetted thin-layers, *Chromatographia*, 38, 83, 1994.
18. Malinowska, I. and Rózyło, J.K., Planar electrochromatography on silica and alumina, *J. Planar Chromatogr.—Mod. TLC*, 11, 411,1998.
19. Malinowska, I., Planar electrochromatography on nonwetted layers with binary mobile phases, *J. Planar Chromatogr.—Mod. TLC*, 13, 307, 2000.
20. Malinowska, I., Rózyło, J.K., and Krasoń, A., The effect of electric fields on solute migration and mixture separation in TLC, *J. Planar Chromatogr.—Mod. TLC*, 15, 418, 2002.
21. Nurok, D. et al., The performance of planar chromatography using electroosmotic flow, *J. Planar Chromatogr.—Mod. TLC*, 11, 244, 1998.

22. Howard, A.G. et al., Electroosmotically driven thin-layer electrochromatography on silica media, *J. Chromatogr. A*, 844, 333, 1999.
23. Nurok, D., Frost, M.C., and Chenoweth, D.M., Separation using planar chromatography with electroosmotic flow, *J. Chromatogr. A*, 903, 211, 2000.
24. Nurok, D. et al., Variables that affect performance in planar chromatography with electroosmotic flow, *J. Planar Chromatogr.—Mod. TLC*, 14, 409, 2001.
25. Nurok, D. et al., The performance of planar electrochromatography in a horizontal chamber, *J. Planar Chromatogr.—Mod. TLC*, 15, 320, 2002.
26. Dzido, T.H. et al., Application of a horizontal DS chamber to planar electrochromatography, *J. Planar Chromatogr.—Mod. TLC*, 16, 176, 2003.
27. Nurok, D., Koers, J.M., and Carmichael, M.A., Role of buffer concentration and applied voltage in obtaining a good separation in planar electrochromatography, *J. Chromatogr. A*, 983, 247, 2003.
28. Dzido, T.H., Mróz, J., and Jóźwiak, G.W., Adaptation of a horizontal DS chamber to planar electrochromatography in a closed system, *J. Planar Chromatogr.—Mod. TLC*, 17, 404, 2004.
29. Dzido, T.H., Płocharz, P.W., and Ślązak, P., Apparatus for pressurized planar electrochromatography in a completely closed system, *Anal. Chem.*, 78, 4713, 2006.
30. Tate, P.A. and Dorsey, J.G., Linear voltage profiles and flow homogeneity in pressurized planar electrochromatography, *J. Chromatogr. A*, 1103, 150, 2006.
31. Novotny, A.L. et al., Results with an apparatus for pressurized planar electrochromatography, *Anal. Chem.*, 78, 2823, 2006.
32. Kreibik, S. et al., Planar dielectrochromatography on non-wetted thin layers, *J. Planar Chromatogr.—Mod. TLC*, 14, 246, 2001.
33. Kreibik, S. et al., Enhancement of mobile phase velocity in TLC by means o fan external alternating electric field, *J. Planar Chromatogr.—Mod. TLC*, 4, 355, 2001.
34. Kreibik, S. et al., Horizontal planar dielectrochromatography. I. Preliminary results, *J. Planar Chromatogr.—Mod. TLC*, 15, 425, 2002.
35. Coman, V. and Kreibik, S., Planar dielectrochromatography—A perspective technique, *J. Planar Chromatogr.—Mod. TLC*, 16, 338, 2003.
36. Desmet, G. and Baron, G.V., On the possibility of shear-driven chromatography: A theoretical performance analysis, *J. Chromatogr. A*, 855, 57, 1999.
37. Clicq, D. et al., High-resolution liquid chromatographic separations in 400 nm deep micro-machined silicon channels and fluorescence charge-coupled device camera detection under stopped-flow conditions, *Anal. Chim. Acta*, 507, 79, 2004.
38. Clicq, D. et al., Porous silicon as a stationary phase for shear-driven chromatography, *J. Chromatogr. A*, 1032, 185, 2004.
39. Nyiredy, Sz., Preparative layer chromatography, in *Handbook of Thin-layer Chromatography*, 2nd ed., Sherma, J. and Fried, B., eds., Marcel Dekker, New York, NY, 1996, 307.
40. Nyiredy, Sz., Rotation planar extraction (RPE): A new exhaustive, preparative forced-flow technique. Part 1: Description of the method and practical aspects, *J. Planar Chromatogr.—Mod. TLC*, 14, 246, 2001.
41. Nyiredy, Sz., Multidimensional planar chromatography, in *Dünnschicht-Chromatographie in memoriam Prof. Dr. Hellmut Jork*, Kaiser, R.E., Günther, W., Gunz, H., and Wulff, G., eds., Düsseldorf, InCom Sonderband, 1996, 166.
42. Szabady, B. and Nyiredy, Sz., The versatility of multiple development in planar chromatography, in *Dünnschicht-Chromatographie in memoriam Prof. Dr. Hellmut Jork*, Kaiser, R.E., Günther, W., Gunz, H., and Wulff G., eds., Düsseldorf, InCom Sonderband, 1996, 212.

43. Nyiredy, Sz., Erdelmeier, C.A.J., and Sticher, O., Sequential centrifugal layer chromatography (SCLC): A new technique for the isolation of natural compounds. Part 1: Description of the method and practical aspects, *J. High Resolut. Chromatogr. Chromatogr. Commun.*, 8, 73, 1985.

44. Erdelmeier, C.A.J., Nyiredy, Sz., and Sticher, O., Sequential centrifugal layer chromatography (SCLC): a new technique for the isolation of natural compounds. Part 2: Comparative study on centrifugal layer chromatography (CLC) and SCLC for the separation of furocoumarin isomers, *J. High Resolut. Chromatogr. Chromatogr. Commun.*, 8, 132, 1985.

45. Nyiredy, Sz. et al., Centrifugal planar-column chromatography (CPCC): A new preparative technique. Part 1: Description of the method and practical aspects, *J. High Resolut. Chromatogr. Chromatogr. Commun.*, 9, 605, 1986.

46. www.analtech.com

47. Nyiredy, Sz., Dallenbach-Tölke, K., and Sticher, O., The PRISMA optimization system in planar chromatography, *J. Planar Chromatogr.—Mod. TLC*, 1, 336, 1988.

48. Nyiredy, Sz., ROTACHROM—neue Möglichkeiten in der Trenn- und Isolierungstechnik, *GIT Suppl. Chromatogr.*, 3, 51, 1986.

49. Nyiredy, Sz., Botz, L., and Sticher, O., ROTACHROM A new instrument for rotation planar chromatography (RPC), *J. Planar Chromatogr.—Mod. TLC*, 2, 53, 1989.

50. Kalász, H. et al., Role of preparative rotation planar chromatography in the isolation of ecdysteroids, *J. Liq. Chromatogr. & RT.*, 29, 2095, 2006.

51. Nyiredy, Sz. et al., Ultra-micro-chamber rotation planar chromatography (U-RPC): A new analytical and preparative forced-flow technique, *J. Planar Chromatogr.—Mod. TLC*, 1, 54, 1988.

52. Yrjönen, T. et al., Application of centrifugal force to the extraction and separation of parasorboside and gerberin from *Gerbera hybrida*, *Phytochem. Anal.*, 13, 349, 2002.

53. Vovk, I. et al., Rotation planar extraction and rotation planar chromatography of oak (*Quercus robur* L.) bark, *J. Chromatogr. A*, 991, 267, 2003.

54. Tyihak, E. and Held, G., Thin-layer chromatography in pharmacognosy, in *Progress in Thin-Layer Chromatography and Related Methods*, Vol. II, Niederwieser, A. and Pataki, G., eds., Ann Arbor Science Publishers, Ann Arbor, Michigan, MI, 1971, chap. 6.

55. Tyihák, E. and Mincsovics, E., Forced-flow planar liquid chromatographic techniques, *J. Planar Chromatogr.—Mod. TLC*, 1, 6, 1988.

56. Tyihák, E., Mincsovics, E., and Körmendi, F., Overpressured layer chromatography I. Basic principles, instruments and developments, *Hung. Sci. Instr.*, 55, 33, 1983.

57. Tyihák, E., Overpressured layer chromatography and its applicability in pharmaceutical and biomedical analysis, *J. Pharm. Biomed. Anal.*, 5, 191, 1987.

58. Mincsovics, E., Efficiency characteristics of off-line infusion and conventional transfusion OPLC, in *Proceedings of Planar Chromatography 2000*, Nyiredy, Sz., ed., Research Institue for Medicinal Plants, Budakalász, 2000, 109.

59. Mincsovics, E. et al., Personal overpressured layer chromatography (OPLC) basic system 50, flexible tool in analytical and semipreparative work, *J. AOAC Intern.*, 82, 587, 1999.

60. Mincsovics, E. et al., Micro-preparative OPLC—Rapid isolation by transfusion and infusion-transfusion processes, *J. Planar Chromatogr.—Mod. TLC*, 15, 280, 2002.

61. Mincsovics, E. and Tyihák, E., Combination of off-line and on-line operating steps in OPLC, *J. Planar Chromatogr.—Mod. TLC*, 1, 309, 1988.

62. Mincsovics, E., Tyihák, E., and Siouffi, A.M., Characteristics of the one-dimensional on-line separation and detection in a modified CHROMPRES chamber, in *Proceedings*

of *International Symposium TLC with Special Emphasis on Overpressured Layer Chromatography (OPLC)*, Tyihák, E., ed., Labor MIM, Budapest, 1986, 251.

63. Mincsovics, E. and Tyihák, E., in *Recent Advances in Thin-Layer Chromatography*, Dallas, F.A.A., Read, H., Ruane, R.J., and Wilson, I.D., eds., Plenum Press, New York, NY, London, 1988, 57.

64. Tyihák, E., Mincsovics, E., and Székely, T.J., Overpressured multi-layer chromatography, *J. Chromatogr.*, 471, 375, 1989.

65. Botz, L., Nyiredy, Sz., and Sticher, O., The principles of long distance OPLC, a new multi-layer development technique, *J. Planar Chromatogr.—Mod. TLC*, 3, 352, 1990.

66. Nyiredy, Sz., The bridge between TLC and HPLC: overpressured layer chromatography (OPLC), *Trends Anal. Chem.*, 20, 91, 2001.

67. Mincsovics, E., Garami, M., and Tyihák, E., Direct coupling of OPLC with HPLC: Clean-up and separation, *J. Planar Chromatogr.—Mod. TLC*, 4, 299, 1991.

68. Mincsovics, E. et al., Single- and multichannel OPLC separation on non-segmented sorbent bed using flowing eluent wall for operating segmentation, *J. Liq. Chromatogr. & Relat. Technol.*, 26, 2611, 2003.

69. Mincsovics, E., Manach, M., and Papillard, D., Flowing eluent wall as a tool of parallel on-line OPLC separations on non-segmented sorbent bed, in *Proceedings of International Symposium on Planar Separations, Planar Chromatography 2003*, Nyiredy, Sz., ed., Budapest, Hungary, 2003, 163.

70. Mincsovics, E., Flowing eluent wall processes in OPLC: Using segmentation of non-segmented adsorbent layer for single and parallel separations, *J. Planar Chromatogr.—Mod. TLC*, 17, 411, 2004.

71. Ruoff, A.D. and Giddings, J.C., Paper geometry and flow velocity in paper chromatography, *J. Chromatogr.*, 3, 438, 1960.

72. Nyiredy, Sz. et al., The disturbing zone in overpressured layer chromatography (OPLC), *J. High Resolut. Chromatogr. Chromatogr. Commun.*, 10, 352, 1987.

73. Velayudhan, A., Lillig, B., and Horváth, Cs., Analysis of multiple front formation in the wetting of thin-layer plates, *J. Chromatogr.*, 435, 397, 1988.

74. Niederwieser, A. and Brenner, M., Polyzonal thin-layer chromatography. Chromatographic separation of non-solic mixtures and its use in the separation of solid mixtures. I. 2-zone thin-layer chromatography. Theory and practice, *Experientia*, 21, 50, 1965.

75. Snyder, L.R. and Kirkland, J.J., *Introduction to Modern Liquid Chromatography*, 2nd ed., John Wiley, New York, NY, 1979, chap. 9.

76. Mincsovics, E., Tyihák, E., and Siouffi, A.M., Comparison of off-line and on-line overpressured layer chromatography (OPLC), *J. Planar Chromatogr.—Mod. TLC*, 1, 141, 988.

77. Wawrzynowicz, T. and Soczewinski, E., Solvent demixing effects in continuous thin-layer chromatography and their elimination, *J. Chromatogr.*, 169, 191, 1979.

78. Schlitt, H. and Geiss, F., Thin-layer chromatography as a pilot technique for rapid column chromatography, *J. Chromatogr.*, 67, 261, 1972.

79. Szepesi, G. et al., Optimization of reversed-phase ion-pair chromatography by over-pressurized thin-layer chromatography. I. Over-pressurized thin-layer chromatography, *J. Chromatogr.*, 290, 127, 1984.

80. Gazdag, M. et al., II. High-performance liquid chromatographic separation on silica by utilizing pre-investigation data from overpressurized thin-layer chromatographic experiments, *J. Chromatogr.*, 290, 135, 1984.

81. Vuorela, P. et al., Overpressured layer chromatography in comparison with thin-layer and high-performance liquid chromatography for the determination of coumarins with reference to the composition of the mobile phase, *J. Chromatogr.*, 670, 191, 1994.

82. Guiochon, G. and Siouffi, A.M., Study of the performances of thin layer chromatography II. Band broadening and plate height equation, *J. Chromatogr. Sci.*, 16, 470, 1978.

83. Hauck, H.E. and Jost, W., Investigations and results obtained with overpressured thin-layer chromatography, *J. Chromatogr.*, 262, 113, 1983.

84. Kalász, H. and Nagy, J., Ion exchange application of overpressured TLC, *J. Liquid Chromatogr.*, 4, 985, 1981.

85. Rowlen, K.L. et al., Measurement of column efficiency in whole column detection chromatography, *Anal. Chem.*, 63, 575, 1991.

86. Roeraade, J. and Flodberg, G., Potentials and limiting factors in high resolution forced flow TLC, in *Proceedings of the Fourth International Symposium on HPTLC (Planar Chromatography)*, Traitler, H., Studer, A., and Kaiser, R.E., eds., Institute for Chromatography, 1987, 341.

87. Tyihák, E., Mincsovics, E., and Siouffi, A.M., Effect of the external pressure on the efficiency of off-line and on-line OPLC, *J. Planar Chromatogr.—Mod. TLC*, 5, 121, 1990.

88. Snyder, L.R. and Kirkland, J.J., *Introduction to Modern Liquid Chromatography*, 2nd ed., John Wiley, New York, NY, 1979, chap. 2.

89. Botz, L., Nyiredy, Sz., and Sticher, O., Applicability of long distance overpressured layer chromatography, *J. Planar Chromatogr.—Mod. TLC*, 4, 115, 1991.

90. Guiochon, G. and Siouffi, A.M., Study of the performances of thin-layer chromatography. Spot capacity in thin-layer chromatography, *J. Chromatogr.*, 245, 1, 1982.

91. Tyihák, E. et al., Behavior of the dinitrophenylhydrazones of saturated aldehydes and ketones in normal and reversed phase TLC and OPLC, *J. Planar Chromatogr.—Mod. TLC*, 5, 376, 1992.

92. Kaiser, R.E., *Einführung in die HPPLC*, Hüthig, Heidelberg, 1987.

93. Witkiewicz, Z., Mazurek, M., and Bladek, J., A new instrument for overpressured layer chromatography, *J. Planar Chromatogr.—Mod. TLC*, 6, 407, 1993.

94. Kovács, Á., Simon-Sarkadi, L., and Mincsovics, E., Stepwise gradient separation and quantification of dansylated biogenic amines in vegetables using personal OPLC instrument, *J. Planar Chromatogr.—Mod. TLC*, 11, 43, 1998.

95. Simon-Sarkadi, L. et al., OPLC investigation of the effect of cold-hardening on the level of polyamines in wheat, *J. Planar Chromatogr.—Mod. TLC*, 14, 326, 2001.

96. Leskó, K. et al., OPLC analysis of polyamines in wheat seedlings under cadmium stress, *J. Planar Chromatogr.—Mod. TLC*, 17, 435, 2004.

97. Nagy-Turák, A. and Végh, Z., Extraction and in situ densitometric determination of alkaloids from Catharanthus roseus by means of overpressured layer chromatography on amino-bonded silica layers. 1. Optimization and validation of the separation system, *J. Chromatogr. A*, 668, 501, 1994.

98. Erdelmeier, C.A.J. et al., The use of overpressure layer chromatograophy (OPLC) for the separation of natural products with antineoplastic activity, *J. Nat. Prod.*, 49, 1133, 1986.

99. Malinowska, I. et al., Mobile-phase velocity—A tool for separation of alkaloids by OPLC, *J. Planar Chromatogr.—Mod. TLC*, 18, 176, 2005.

100. Pothier, J. et al., Comparison of planar chromatographic methods (TLC, OPLC, AMD) applied to essential oils of wild thyme and seven chemotypes of thyme, *Il Farmaco*, 56, 505, 2001.

101. Tabanca, N. et al., Characterization of volatile constituents of *Scaligeria tripartita* and studies on the antifungal activity against phytopathogenic fungi, *J. Chromatogr. B*, 850, 221, 2007.

102. Salles, C., Jallageas, J.-C., and Crouzet, J., Chromatographic separation and partial identification of glycosidically bound volatile components of fruit, *J. Chromatogr.*, 522, 255, 1990.

103. Dallenbach-Toelke, K. et al., Optimization of overpressured layer chromatography of polar naturally occurring compounds by the PRISMA model, *J. Chromatogr.*, 365, 63, 1986.

104. Dallenbach-Toelke, K. et al., TLC, HPTLC and OPLC separation of ginsenosides, *J. High Resolut. Chromatogr. Chromatogr. Commun.*, 10, 362, 1987.

105. Ludwiczuk, A., Nyiredy, S., and Wolski, T. Separation of the ginsenosides fraction obtained from the roots of Panax quinquefolium L. cultivated in Poland, *J. Planar Chromatogr.—Mod. TLC*, 18, 104, 2005.

106. Oroszlán, P. et al., Separation, quantitation and isolation of cannabinoids from *Cannabis sativa* L. by overpressured layer chromatography, *J. Chromatogr.*, 388, 217, 1987.

107. Szabady, B., Hidvégi, E., and Nyiredy, Sz., Determination of neutral cannabinoids in hemp samples by overpressured-layer chromatography, *Chromatographia (Suppl.)*, 56, 165, 2002.

108. Galand, N., Ernouf, D., Montigny, F., Dollet, J., and Pothier, J., Separation and identification of cannabis components by different planar chromatography techniques (TLC, AMD, OPLC), *J. Chromatogr. Sci.*, 42, 130, 2004.

109. Vuorela, H. et al., Separation of the main coumarins of *Peucedanum palustre* with various planar chromatographic methods, *J. Planar Chromatogr.—Mod. TLC*, 1, 123, 1988.

110. Király-Véghely, Zs. et al., Separation of stilbene isomers from red wine by overpressured-layer chromatography, *J. Planar Chromatogr.—Mod. TLC*, 17, 4, 2004.

111. Király-Véghely, Zs. et al., Identification and measurement of resveratrol and formaldehyde in parts of white and blue grape berries, *Acta Biol. Hung.*, 49, 281, 1998.

112. Kátay, Gy. et al., Comparison of thin-layer chromatography and overpressured layer chromatographic techniques for the separation of ascorbigen and 1'-methylascorbigen, *J. Chromatogr. A*, 764, 103, 1997.

113. Kátay, Gy. et al., Ovarpressured-layer chromatographic determination of ascorbigen (bound vitamin C) in Brassica vegetables, *J. Planar Chromatogr.—Mod. TLC*, 17, 360, 2004.

114. Zogg, G., Nyiredy, Sz., and Sticher, O., Overpressured layer chromatographic (OPLC) separation of closely related furocoumarins, *J. Liq. Chromatogr.*, 10, 3605, 1987

115. Pothier, J. et al., Separation of quaternary alkaloids in plant extracts by overpressured layer chromatography, *J. Planar Chromatogr.—Mod. TLC*, 6, 220, 1993.

116. Szűcs, Z. et al., High-throughput analytical strategy with combined planar and column liquid chromatography for improvement of the poppy (*Papaver somniferum* L.) with a high alkaloid content, *Chromatographia (Suppl.)*, 56, 49, 2002.

117. Tyihák, E. et al., Analysis of amino acids by personal OPLC instrument: 1. Separation of the main protein amino acids in a double-layer system, *J. Planar Chromatogr.—Mod. TLC*, 11, 5, 1998.

118. Härmälä, P. et al., Two-dimensional planar chromatographic separation of a complex mixture of closely related coumarins from the genus Angelica, *J. Planar Chromatogr.—Mod. TLC*, 3, 515, 1990.

119. Kátay, Gy., Szécsi, Á., and Tyihák, E., Separation of fumonisins by OPLC, *J. Planar Chromatogr.—Mod. TLC*, 14, 53, 2001.

120. Papp, E. et al., Validation and robustness testing of an OPLC method for the determination of aflatoxins in wheat, *J. Planar Chromatogr.—Mod. TLC*, 13, 328, 2000.

121. Otta, K.H., Papp, E., and Bagócsi, B., Determination of aflatoxins in food by over-pressured-layer chromatography, *J. Chromatogr. A.*, 882, 11, 2000.
122. Papp, E. et al., Liquid chromatographic determination of aflatoxins, *Microchem. J.*, 73, 39, 2002.
123. Móricz, Á.M. et al., Overpressured layer chromatographic determination of aflatoxin B1, B2, G1 and G2 in red paprika, *Microchem J.*, 85, 140, 2007.
124. Zogg, G.C., Nyiredy, Sz., and Sticher, O., Comparison of fully off-line and fully-on-line linear OPLC with HPLC, illustrated by the separation of furocoumarin isomers, *J. Planar Chromatogr.—Mod. TLC*, 1, 351, 1988.
125. Betti, B. et al., Use of overpressured layer chromatography and coupled OPLC-GC-MS for the analysis of acetylenic thiophene derivatives in extracts of *Tagetes patula*, *J. Planar Chromatogr.—Mod. TLC*, 7, 301, 1994.
126. Horváth, E. et al., Critical evaluation of experimental conditions influencing the surface-enhanced Raman spectroscopic (SERS) detection of substances separated by layer liquid chromatographic techniques, *Chromatographia (Suppl.)*, 51, 297, 2000.
127. Chai, W. et al., On-line overpressure thin-layer chromatographic separation and electrospray mass spectrometric detection of glycolipids, *Anal. Chem.*, 75, 118, 2003.
128. Erdelmeier, C.A., Van Leeuwen, P.A., and Kinghorn, A.D., Phorbol diester constituents of croton oil: separation by two-dimensional TLC and rapid purification utilizing reversed-phase overpressure layer chromatography, *Planta Med.*, 54, 71, 1988.
129. Nyiredy, Sz. et al., Preparative on-line overpressure layer chromatography (OPLC): A new separation technique for natural products, *J. Nat. Prod.*, 49, 885, 1986.
130. Erdelmeier, C.A.J., Kinghorn, A.D., and Farnsworth, N.R., On-plate injection in the preparative separation of alkaloids and glycosides, *J. Chromatogr.*, 389, 345, 1987.
131. Mincsovics, E. et al., Isolation of some antimicrobial compounds of red wine by OPLC flowing eluent wall technique, *Chromatographia* (Suppl.), 62, 51, 2005.
132. Tyihák, E. et al., Combination of overpressured layer chromatography and bioautography in the analysis of molecules influencing cell proliferation, *Chem. Anal. (Warsaw)*, 48, 543, 2003.
133. Nyiredy, Sz. and Glowniak, K. et al., Planar chromatography in medicinal plant research, in *Planar Chromatography—A Retrospective View for the Third Millennium*, Nyiredy, Sz., ed., Springer, Hungary, 2001, 250.

Part II

Primary Metabolites

11 TLC of Carbohydrates

Guilherme L. Sassaki, Lauro M. de Souza,
Thales R. Cipriani, and Marcello Iacomini

CONTENTS

11.1 INTRODUCTION

Carbohydrates are carbon compounds that contain large quantities of hydroxyl groups. The simplest carbohydrates also contain either an aldehyde moiety (these are termed polyhydroxyaldehydes) or a ketone moiety (polyhydroxyketones). They are the most abundant biological molecules, and fill numerous roles in living organisms, such as the storage and transport of energy (starch, glycogen and sucrose) and structural components (cellulose in plants, chitin in animals).[1,2] All carbohydrates can be classified as monosaccharides, oligosaccharides, or polysaccharides. Anywhere from two to ten monosaccharide units, linked by glycosidic linkage, make up an oligosaccharide. Polysaccharides are much larger, containing hundreds of monosaccharide units. The presence of the hydroxyl groups allows carbohydrates to interact with the aqueous environment and to participate in hydrogen bonding, both within and between chains. Additionally, carbohydrates and their derivatives play major roles in the functioning of the immune system, fertilization, pathogenesis, blood clotting, and development. Derivatives of the carbohydrates can contain nitrogens, phosphates, and sulfur groups. Some carbohydrates can attach by *C-, O-* or *N*-glycosidic bonds, giving rise to complex structures namely glycoconjugates. When the sugar moiety is combined with lipid it forms a glycolipid and when it is combined with protein it forms another glycoconjugate molecule, a glycoprotein. The predominant carbohydrates encountered in the body are structurally related to the aldotriose glyceraldehyde and to the ketotriose dihydroxyacetone. All carbohydrates contain at least one asymmetrical (chiral) carbon and are, therefore, optically active. In addition, carbohydrates can exist in either of two conformations D or L, as determined by the orientation of the hydroxyl group about the farthest asymmetric carbon from the carbonyl group. With a few exceptions, those carbohydrates that are of physiological significance exist in the D-conformation. The mirror-image conformations, called enantiomers, are in the L-conformation. The carbohydrates found in plants are monosaccharides, disaccharides, higher oligosaccharides, polysaccharides, and their derivatives: glycosides and glycoproteins. However, plants have many polysaccharides and glycoconjugates structures containing L-arabinose and L-fucose, which have been found as complex heteropolysaccharides, namely arabinogalactans and gums, respectively.[1–3]

11.1.1 SUGAR CHEMISTRY

Monosaccharides are classified according to three different characteristics: the placement and function of its carbonyl group, the number of carbon atoms, and its chiral center. If the carbonyl group is an aldehyde, the monosaccharide is an aldose; if the carbonyl group is a ketone, the monosaccharide is a ketose. The smallest possible monosaccharides, those with three carbon atoms, are called trioses, just as those with four, five, and six are tetroses, pentoses, hexoses, and so on.[1,2] Combining the function and number of carbons there is, for example, an aldohexose, aldopentose, and ketohexose—glucose (six-carbon aldose), arabinose (five-carbon aldose) and fructose (six-carbon ketone), respectively (Figure 11.1). The carbon atoms of sugars linked to a hydroxyl group (-OH), with the exception of the carbonylic groups and

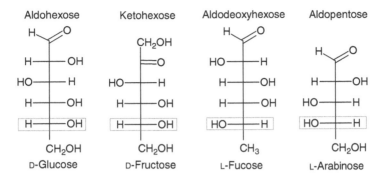

FIGURE 11.1 Examples of aldohexose, ketohexose, aldodeoxyhexose, and aldopentose. Boxes indicate the last asymmetrical carbon, which define the absolute configuration.

the last carbon, are asymmetric, making them stereocenters with two possible configurations. D-glucose is one isomer for an aldohexose (noncyclic form) with the formula $(CH_2O)_6$, four of its six carbons atoms are chiral centers, making D-glucose one of the $2^4 = 16$ possible stereoisomers in straight-chain form. Sugars belonging to the D-series are the most common isomers found in nature.

The aldehyde or ketone group of a straight-chain monosaccharide will react reversibly with a hydroxyl group on a different carbon atom to form a hemiacetal or hemiketal, forming a heterocyclic ring with an oxygen bridge between two carbon atoms. Rings with five and six atoms are called furanose and pyranose forms, respectively, and exist in equilibrium with the straight-chain form. For example, D-glucose can adopt two types of ring conformation, which result in the formation the D-glucopyranose (D-Glc*p*) and D-glucofuranose (D-Glc*f*); however, the pyranosyl conformation is most commonly found in nature. For these reasons the ketone and aldehyde groups of carbohydrates make these structures highly reactive, so most of sugar reactions follow the classical reactions of these chemistry functions, being catalyzed by acidic or basic compounds.

11.1.1.1 Mutarotation

In water and polar solvents, the free reducing carbohydrates can convert the straight-chain form to cyclic form. The carbon atom containing the carbonyl oxygen, called the anomeric carbon, becomes a chiral center with two possible configurations: the oxygen atom may take a position either above or below the plane of the ring. The possible resulting stereoisomers are called anomers. In the α anomer, the -OH substituent on the anomeric carbon rests on the opposite side of the ring from the CH_2OH attached to the asymmetric carbon farthest from the anomeric carbon. The alternative form, in which the CH_2OH and the anomeric hydroxyl are on the same side of the plane of the ring, is called the β anomer. Because the ring and straight-chain forms readily interconvert, both anomers exist in equilibrium. So the glucose monosaccharide could be present in 10 possible structures, straight-chain forms of D- or L-glucose, α-D-Glc*p*, β-D-Glc*p*, α-L-Glc*p*, β-L-Glc*p*, α-D-Glc*f*, β-D-Glc*f*, α-L-Glc*f*, and

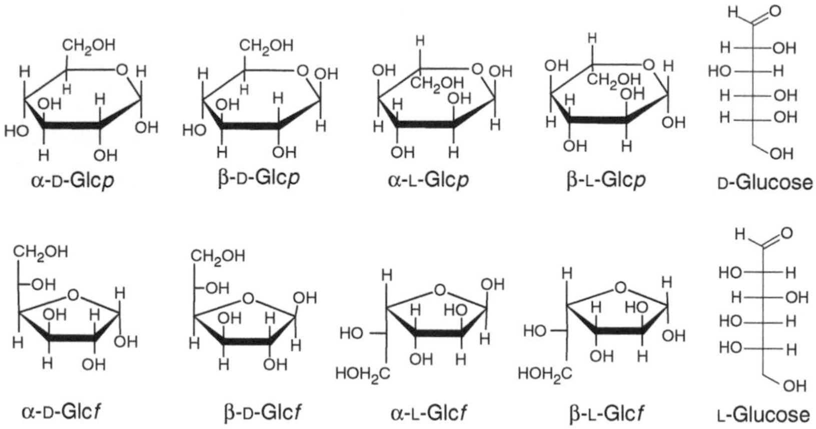

FIGURE 11.2 Possible isomers for glucose.

β-L-Glc$f$ (Figure 11.2). All of these characteristics made the carbohydrates the most complex molecules to study in nature, and the many possibilities of combinations, types of linkage and sequences of the sugar units give rise to many conformations and these structures are used in many functions by the living organisms.

11.1.1.2 The Reducing Sugars

Carbohydrates that contain the free aldehydes or α-hydroxymethyl ketone groups can be oxidized by Cu(II) ion and are classified as reducing sugars. They reduce the Cu(II) ion to Cu(I), the sugar is oxidized and its carbonyl group (i.e., aldehyde or ketone group) is converted to a carboxyl group. This reaction can be performed by Benedicts reagent ($CuSO_4$/citrate) and Fehlings reagent ($CuSO_4$/tartrate) in aqueous alkaline solutions. So the oxidation will always occur when the hemi-acetal form (cyclic) became ring-open, giving the reactive aldehyde group (Figure 11.3A). However, the glycosides (acetals) are nonreducing sugars (Figure 11.3B), unable to become ring-open, therefore, they cannot be oxidized.

11.1.1.3 Bases for Sugar Detection on TLC

The ketone and aldehyde groups of carbohydrates are the central point of reaction for detection in TLC or in paper chromatography (PC). However, the hydroxyl groups are very important in the formation of by-products in the dehydration of the sugars in the presence of strong inorganic acids and heat, as observed by the formation of furfuraldehyde (Figure 11.4). The classical work of Trevelyan, Procter & Harrison (1950) used silver nitrate–sodium hydroxide to detect sugars and polyalcohols in PC.[4] The detection of sugar with silver nitrate would be performed on TLC using aqueous ammonia instead of sodium hydroxide. However, the most common technique use the furfuraldehyde in conjunction with many phenolic reagents, which give rise to different colors, as observed by orcinol, α-naphtol, resorcinol, anisaldehyde,

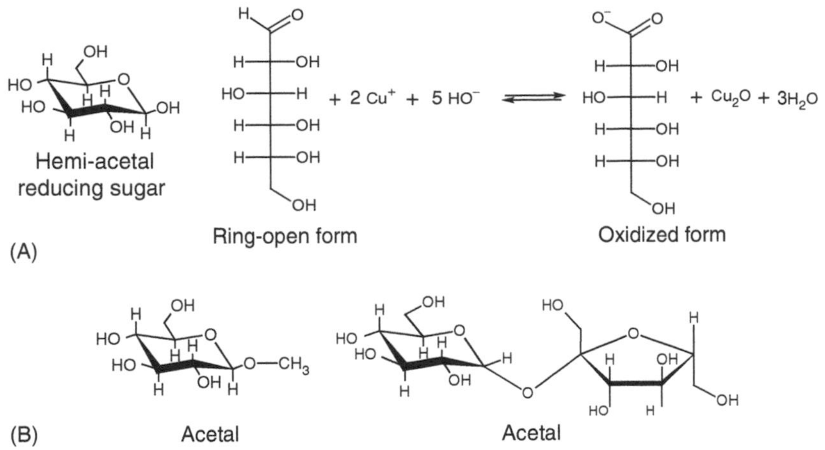

FIGURE 11.3 (A) Example of oxidization reaction. (B) Nonreducing sugars.

and aniline–diphenylamine. Another advantage of this type of detection is the peculiar color obtained for each monosaccharide. The most popular use of TLC of carbohydrates is performed in Silica Gel G-60 plates, followed by detection of the carbohydrates with Orcinol–H_2SO_4 spray, since it is highly sensitive to sugars and each monosaccharide class gives rise to a specific color. However, depending on the sugar concentrations, the color intensities may vary, so some color variations may appear.

11.1.2 PLANT CARBOHYDRATES

In plants, carbohydrates exist as monosaccharides, oligosaccharides, polysaccharides, and their derivatives, such as cyanogenic glycosides, phenolic and flavonoid glycosides, glycolipids, and glycoproteins.

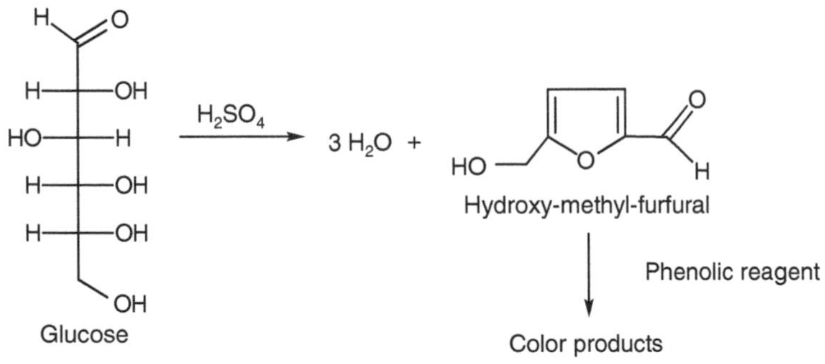

FIGURE 11.4 Furfuraldehyde formations during acidic treatment and reaction with phenolic reagents.

11.1.2.1 Monosaccharides and Polyols

There are several kinds of monosaccharides and they are the basic constituents of all carbohydrate structures. The main monosaccharides found in plants are glucose, fructose, galactose, mannose, rhamnose, arabinose, xylose, glucuronic acid, galacturonic acid, and 4-*O*-methylglucuronic acid. These monosaccharides are usually found as constituents of oligo- and polysaccharides, glycoproteins, glycolipids, etc.

The polyols are constituted by reduced forms of monosaccharides. Both acyclic (alditols) and cyclic (cyclitols) polyols occur in plants, mostly in combined forms as glycosides. Glucitol, mannitol, dulcitol, and glycerol, for example, are alditols that may occur in free form. Inositols are the most widely occurring cyclitols, and *myo*-inositol is the most important of them, playing various roles in plant metabolism.[5] An example of evaluation of polyos in plant extracts is shown in Figure 11.5.

11.1.2.2 Oligosaccharides

As the name suggests, oligosaccharides (Greek *oligos*—few) are carbohydrate structures composed of few monosaccharides units, linked by *O*-glycosidic bonds, usually consisting from two to ten monosaccharide residues. According to the number of monosaccharide contents per mole, the oligosaccharide may be classified as

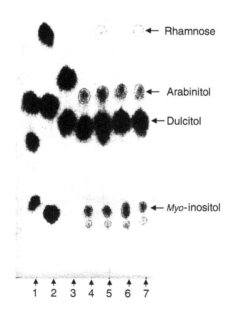

FIGURE 11.5 Paper chromatography (Whatman No. 1–40 × 20 cm) of alcoholic fractions of the plant *Maytenus ilicifolia*. Solvent buthanol:pyridine:H_2O (5:3:3 v/v), for 18 h, and detection with silver nitrate–sodium hydroxide. Standards are listed from origin to front: (1) Lac, Gal, Man; (2) *myo*-inositol, Ara, Rha; (3) Glc, Xyl. Samples: (4) MeOH:H_2O (2:1 v/v) extract; (5) MeOH:H_2O (2:1 v/v) extract, after hydrolysis; (6) ethanolic supernatant of aqueous extraction; (7) ethanolic supernatant of aqueous extraction, after hydrolysis.

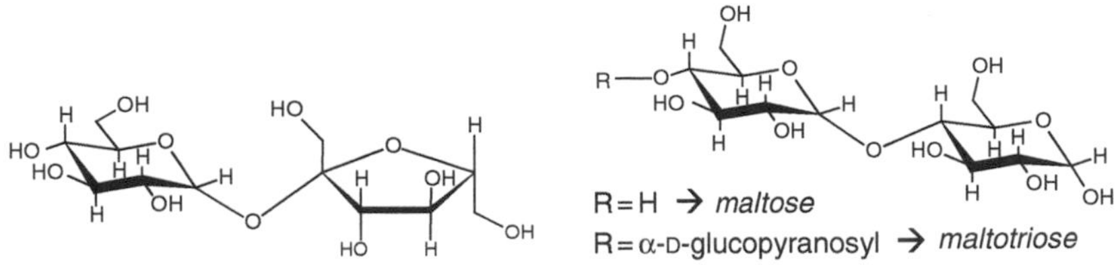

FIGURE 11.6 (A) *Sucrose* – [α-D-glucopyranosyl-(1→2)-β-D-fructofuranoside], (B) *maltose* [α-D-glucopyranosyl-(1→4)-α-D-glucopyranose] and *maltotriose* [α-D-glucopyranosyl-(1→4)-α-D-glucopyranosyl-(1→4)-α-D-glucopyranose]. Higher homologs comprise the chain elongation.

disaccharide (two residues), trisaccharide (three residues), tetrasaccharide (four residues), and so on. Changing the number and the type of monosaccharide contents, as well as, the glycosidic linkages may form several oligosaccharides.

Two classes of oligosaccharides are found in plants: the primary oligosaccharides, which are always synthesized in vivo, and the secondary oligosaccharides, which are produced as a result of polysaccharide, glycoprotein, or glycolipid degradation, in vivo or in vitro.

The most common primary oligosaccharide, and widespread mainly by its sweet taste, is sucrose (Figure 11.6A), a nonreducing disaccharide composed by glucopyranose and fructofuranose. However, some oligosaccharides may be found as primary or secondary products. *Maltose* and its higher homologs (Figure 11.6B) are examples, since they may be produced by enzymatic catalysis, through the action of glucosyltransferases on glucans degradation,[6,7] or in the laboratory by partial hydrolysis. On the other hand, *maltose* may arise from activated glucosyl units during the photosynthesis.[8]

Sucrose is a very important plant disaccharide, since it is the main form in which photosynthetic energy is transported throughout the plant. Moreover, sucrose is the main precursor for synthesis of other oligosccharides. Two enzymes are known to be capable of sucrose synthesis in plants: sucrose synthase and sucrose-6-P synthase.

Although sucrose is the most ubiquitous and abundant disaccharide, in pteridophytas *trehalose* and *selagiose* (Figure 11.7) serve as the main soluble reserve of carbohydrates, replacing sucrose.[9]

Many oligosaccharide families from plants, such as *fructans* or fructo-oligosaccharides and *raffinose* have sucrose as a common precursor.

Fructans are fructose (Fru) polymers (oligo- or polysaccharides) synthesized from sucrose by a combined enzymes action, Sucrose:Sucrose 1-Fructosyl Transferase (1-SST), and Fructan:Fructan 1-Fructosyl Transferase (1-FFT). In the first step, 1-SST synthesizes the 1-*kestose* from two *sucrose* units and, in the second step, the 1-FFT makes the chain elongation by transferring Fru units from one *fructan* to another.[10–13] In some plants, such as grasses, or those from the families Asteraceae, Poaceae, and Liliaceae, fructans are the main energetic storage.[14]

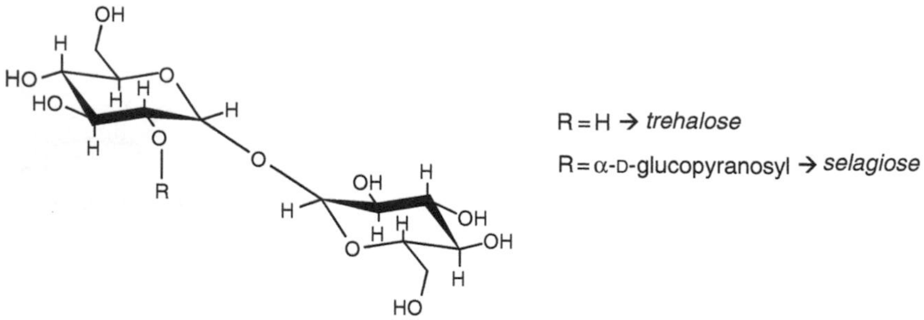

R = H → *trehalose*

R = α-D-glucopyranosyl → *selagiose*

FIGURE 11.7 α,α-*Trehalose* [α-D-Glucopyranosyl-(1→1)-α-D-Glucopyranoside] and *Selagiose* [α-D-Glucopyranosyl-(1→1)-α-D-Glucopyranosyl-(2→1)-α-D-Glucopyranoside].

Some examples of fructo-oligosaccarhides, such as trisaccharides 1-*kestose*, 6-*kestose* and *neokestose*, and the tetrasaccharide *nystose*, are shown in Figure 11.8. *Raffinose* family oligosaccharides are found in many seeds, especially legumes. Oligosaccharides belonging to the *raffinose* series are important reserves of carbohydrates, and are involved in the carbohydrate translocation throughout the plant tissues.[15,16] Some examples of oligosaccharides from *raffinose* series include the trisaccharide *raffinose* [α-D-galactopayranosyl-(1→6)-α-D-glucopyranosyl-(1→2)-β-D-fructofuranoside], which lends its name to the group, and its derivatives, tetrasaccharide *stachyose* and the pentasaccharide *verbascose* (Figure 11.9).

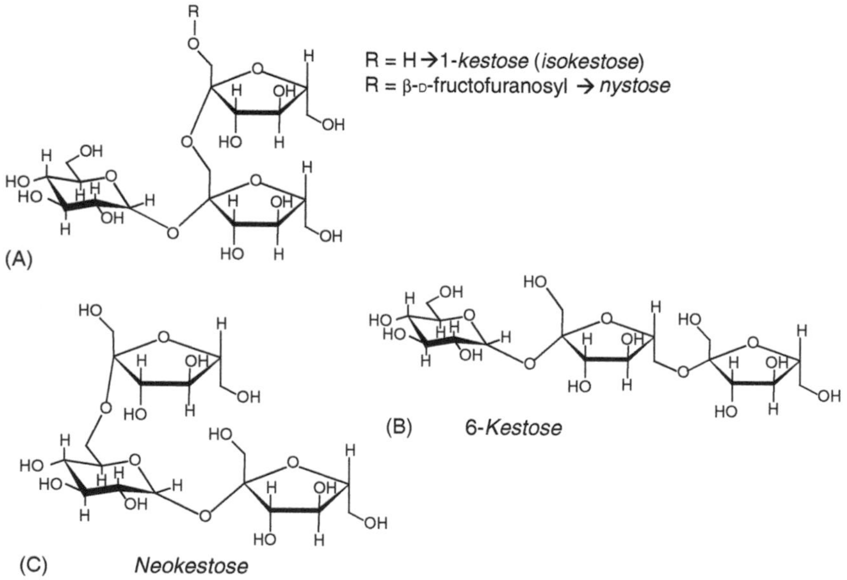

R = H → 1-*kestose* (*isokestose*)
R = β-D-fructofuranosyl → *nystose*

(A)

(B)   6-*Kestose*

(C)   *Neokestose*

FIGURE 11.8 Fructans series (Kestoses). (A) 1-*kestose* [α-D-glucopyranosyl-(1→2)-β-D-fructofuranosyl-(1→2)-β-D-fructofuranoside], *nystose* [α-D-glucopyranosyl-(1→2)-β-D-fructofuranosyl-(1→2)-β-D-fructofuranosyl-(1→2)-β-D-fructofuranoside]. (B) 6-*kestose* [α-D-glucopyranosyl-(1→2)-β-D-fructofuranosyl-(6→2)-β-D-fructofuranoside]. (C) *neokestose* [β-D-fructofuranosyl-(2→6)α-D-glucopyranosyl)-(1→2)-β-D-fructofuranoside].

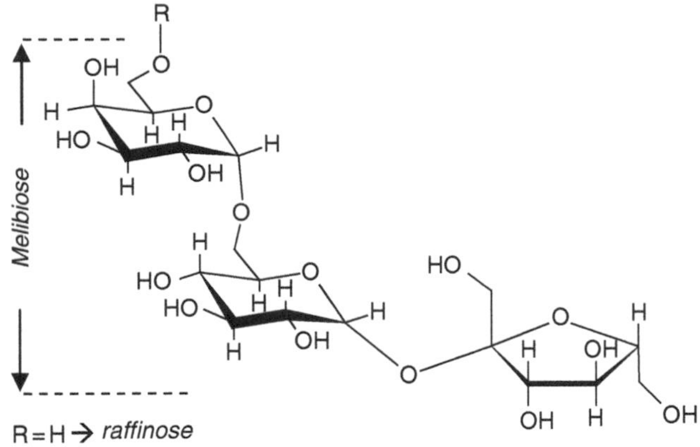

R = H → *raffinose*
R = α-D-Galactopayranosyl → *stachyose*
R = α-D-Galactopayranosyl (1→ 6) α-D-Galactopayranosyl → *verbascose*

FIGURE 11.9 *Raffinose* oligosaccharides.

If the fructosyl end is removed from the *raffinose*, it will give rise to a disaccharide, *melibiose*, α-D-galactopyanosyl-(1→6)-D-glucopyranose. When the fructosyl end is removed from stachyose, it gives rise to the trisaccharide *manninotriose*, α-D-galactopyanosyl-(1→6)-α-D-galactopyanosyl-(1→6)-D-glucopyranose.

There are many oligosaccharide series which also have sucrose as a common precursor. Some examples, including *umbelliferose, planteose, lychnose,* and *gentianose*, are shown in Figure 11.10. Although not represented, these oligosaccharides may present higher homologs (higher DP). An example of HPTLC of mono- and oligosaccharides are shown in Figure 11.11.

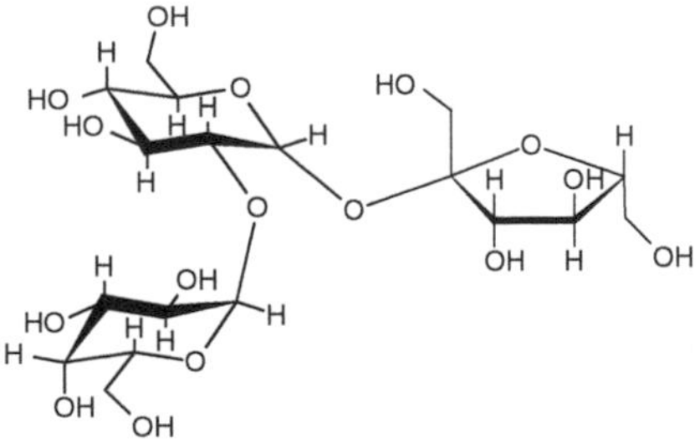

Umbelliferose—α-D-galactopayranosyl-(1→2)-α-D-glucopyranosyl-(1→2)-β–D-fructofuranoside

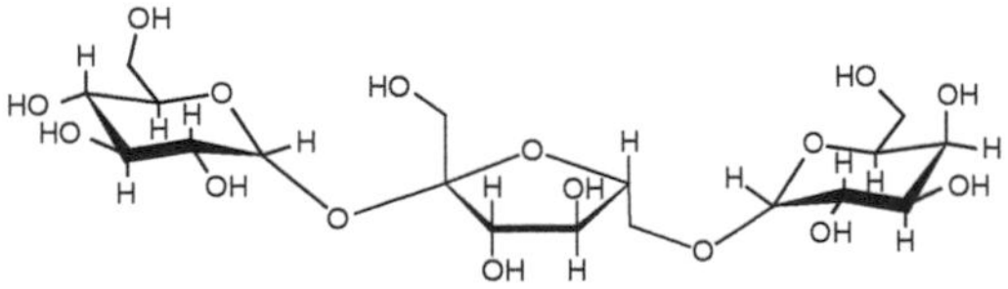

Planteose—α-D-glucopyranosyl-(1→2)-β-D-fructofuranosyl-(6→1)α-D-galactopayranoside

FIGURE 11.10 Fructose-containing oligosaccharides.

(*continued*)

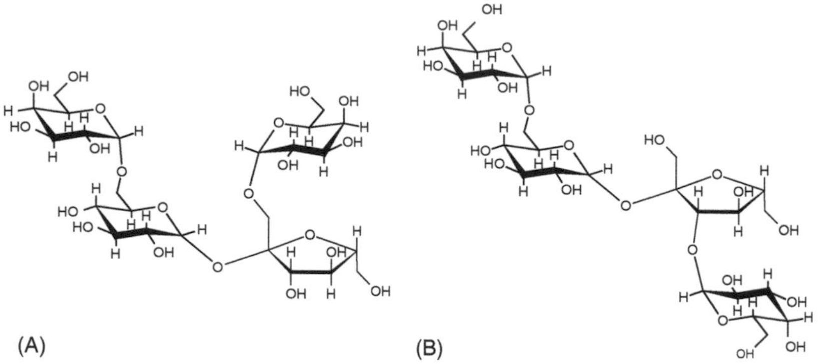

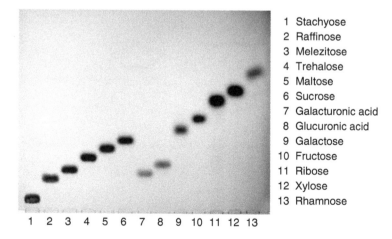

A: *Lychnose*—α-D-galactopayranosyl-(1→6)-α-D-glucopyranosyl-(1→2)-β-D-fructofuranosyl-
(1→1)-α-D-galactopyranoside;

B: *Isolychnose*- α-D-galactopayranosyl-(1→6)-α-D-glucopyranosyl-(1→2)-β-D-fructofuranosyl-
(1→3)-α-D-galactopyranoside

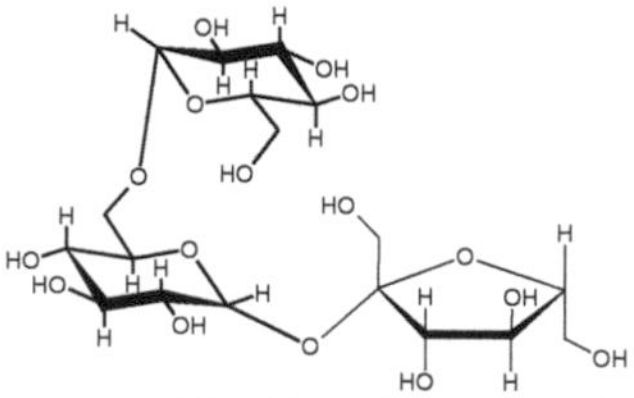

Gentianose—β-D-glucopayranosyl-(1→6)-α-D-glucopyranosyl-(1→2)-β-D-fructofuranoside

FIGURE 11.10 (continued) Fructose-containing oligosaccharides.

1 Stachyose
2 Raffinose
3 Melezitose
4 Trehalose
5 Maltose
6 Sucrose
7 Galacturonic acid
8 Glucuronic acid
9 Galactose
10 Fructose
11 Ribose
12 Xylose
13 Rhamnose

1 2 3 4 5 6 7 8 9 10 11 12 13

FIGURE 11.11 Oligo- and monosaccharides standards developed on HPTLC in ethyl acetate: *n*-propanol:acetic acid:H_2O (4:2:2:1 v/v) and stained with orcinol–H_2SO_4. (1–6) Tetra-, tri-, and disaccharides; (7–13) monosaccharides, including uronic acids.

11.1.2.3   Polysaccharides

The plant polysaccharides are distributed in three main groups: storage, structural, and gum exudate polysaccharides.

Storage polysaccharides are found in tubers and in seeds of plants. Starch, mannans, and xyloglucans are the most common found polysaccharides. Starch is an α-glucan constituted by two polymers, amylase and amylopectin. Amylase consists of long, unbranched chains of (1→4)-linked α-D-glucose units. Amylopectin, unlike amylase, is highly branched. The glycosidic linkages joining successive α-D-glucose residues in amylopectin chains are (1→4), and the branch points (about one per 24–30 residues) are (1→6)-linked.

Mannans are a heterogeneous family of polysaccharides subdivided into four groups: pure mannans, glucomannans, galactomannans, and galactoglucomannans. The (1→4)-linked β-D-mannose units constitute the basic structure of mannans.

Xyloglucans are a group of polysaccharides based on a main chain constituted by (1→4)-linked β-D-glucose units, to which short side chains are attached at C-6 of at least one-half of the glucose residues. α-D-xylopyranose, β-D-galactopyranose, and α-L-fucopyranose units appear in the side chains.[3,5]

Structural polysaccharides are the main chemical constituents of most plant tissues, because they are the main constituents of the cell wall of plants. The polysaccharides components of the cell wall are cellulose, hemicelluloses, and pectins. These groups of polysaccharides can be distinguished on the basis of extractability: pectins (water), hemicelluloses (alkaline solutions), cellulose (residue after extraction of the pectins and hemicelluloses).[17]

The pectins are a complex group of acid polysaccharides, containing galacturonic acid, rhamnose, arabinose, and galactose as main monosaccharide constituents. Homogalacturonans, rhamnogalacturonans, arabinans, galactans, and arabinogalactans are examples of pectic polysaccharides. The homogalacturonans consist of (1→4)-linked α-D-GalAp units in a linear structure, while the rhamnogalacturonans are constituted by repeating (1→4)-α-D-GalpA-(1→2)-α-L-Rhap groups, often having C-4 of the rhamnosyl units substituted by an arabinan, galactan, or arabinogalactan sequence.[18] Arabinans often have (1→5)-linked α-L-Araf main chains substituted at O-3 by α-L-Araf units, while galactans often show a (1→4)-linked β-D-Galp linear chain. Arabinogalactans can be constituted by a (1→4)-linked β-D-Galp main chain (type I) or by a (1→3)-linked β-D-Galp main chain, substituted at O-6 by (1→6)-linked β-D-Galp chains (type II—mainly found in hemicellulosic fractions). Both are substituted by α-L-Araf side chains.[18,19]

Hemicelluloses are a class of polysaccharides that form a network with cellulose. Xylans, xyloglucans, glucomannans, and galactoglucomannans are some examples of hemicelluloses. Plant xylans are characterized by a (1→4)-linked β-Xylp main chain, which can carry side chains with many kinds of monosaccharides, including acid monosaccharides.[3] Xyloglucans and mannans were described earlier.

The gums exudates form a group of polysaccharides, many of which are of considerable economic importance by virtue of their abundance and their special solution properties.[3] Gum exudates are formed by stem injury and show striking resemblances to cell wall polysaccharides. Most of them contain type II arabinoga-lactans, which contain a $(1\rightarrow3)$-linked β-D-Galactopyranose main chain, substituted at O-6 by $(1\rightarrow6)$-linked β-D-Galactopyranose side chains, which are mainly substituted at O-3 by $(1\rightarrow5)$- and $(1\rightarrow3)$-linked α-L-Arabinofuranose chains.[20,21] Gum exudates polysaccharides having a main chain of $(1\rightarrow4)$-linked β-Xylopyranose units have also been reported in various species.[22,23]

11.1.2.4 Glycoconjugates (Glycolipids)

Glycolipids are important membrane components in most living organisms. With other important lipids, such as phospholipids, sphingolipids, and sterols, they comprise the main components of biological membranes. As their structural variability, their biological function is equally diversified.

Plant glycolipids may be divided in two main classes, Glycoglycerolipids and Glycosphingolipids.

Glycoglycerolipids may contain one or two fatty acid chains bound to hydroxyl groups of glycerol, as well as its sugar moiety, whose monosaccharide component may vary, but is usually composed by D-galactopyranosyl units or sulfoquinovosyl, a modified 6-deoxyglucopyranosyl containing a sulfonyl group linked to C-6. The carbohydrate moieties from glycolipids are usually referred as polar head group (PHG), and in most cases define the biological function of the lipid. The most common lipids of this class are: monogalactosyldiacylglycerol (MGDG), digalactosyldiacylglycerol (DGDG), trigalactosyldiacylglycerol (TGDG), and sulfoquinovosyldiacylglycerol (SQDG). These lipids may also occur in monoacylated form, in which only one fatty acid is linked to the primary hydroxyl group of glycerol. These lipids are called *lysolipids*.

Photosynthetic tissues from plants can contain high proportions of galactolipids such as MGDG and DGDG.[24,25] The role of these lipids involves thylakoid aggregation and stacking.[26–28]

Glycosphingolipids consist of a sphingoide base (sphingosine), referred as long chain base (LCB), characterized by an amide linkage with a fatty acid. The hydroxyl group presents in C-1 links the glycosyl moiety to the sphingosine. As well as glycosylglycerides, glycosphingolipids may contain only one (cerebroside) or more sugar units and the glycosidic linkages vary according to the nature of the lipid. Glycosphingolipids di-, tri-, and tetrahexosides containing D-glucosyl and D-mannosyl units were isolated from wheat flour.[29–31]

There is a great variety of plant glycosphingolipids, some of them presenting a phosphate diester group, making the linkage between the sphingosine and the glycosyl moieties were isolated from leaves and seeds.[32,33] These sphingolipids, called phytoglycolipids, isolated from *Phaseolus vulgaris* leaves have a complex glycosyl moiety, presenting a trisaccharide, inositol-hexuronic-hexosamine, attached to mannose, galactose, and arabinose. The main fatty acids found were 2-hydroxy C_{22}, C_{24}, and C_{26} and long chain bases dehydrophytosphingosine and phytosphingosine.[32] The extraction and analytical methods for phytoglycolipids on chromatographic paper and TLC plates were previously described.[33,34] Glycolipids structures are represented in Figure 11.12.

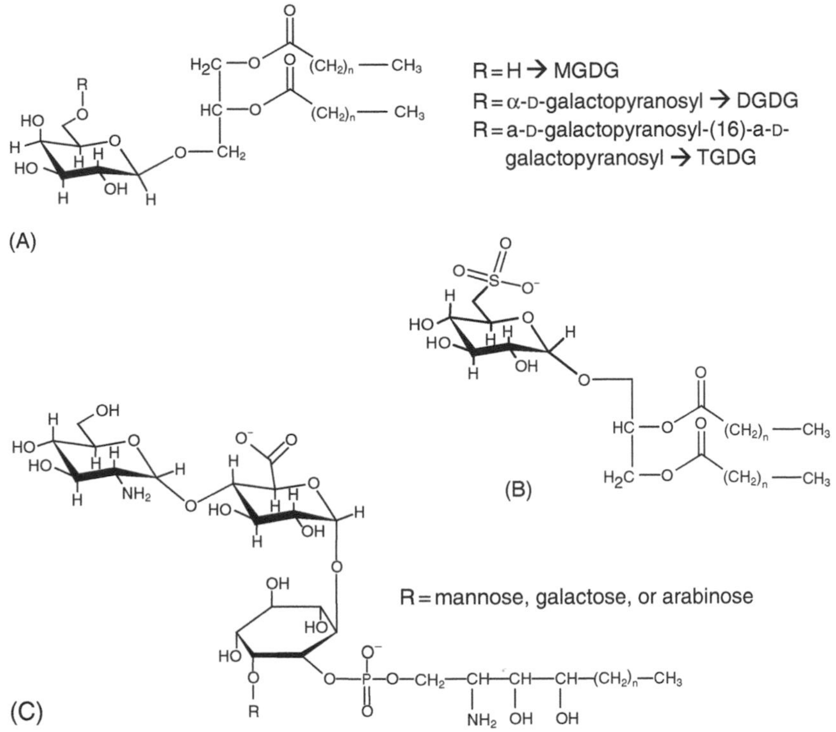

FIGURE 11.12 Glycolipids found in plants: (A) Galactolipids series, mono-, di-, and tri-galactosyldiacylglicerol, (B) Sulfonolipid – sulfoquinovosyldiacylglycerol, and (C) General structure of phytoglycolipids, a sphingolipid.

Other glycoconjugates called heteroglycosides are very common in plant extracts. In these compounds, the carbohydrate moiety presented glycosydic linkage to phenols, steroids, flavones, etc.[35]

11.2 EXPERIMENTAL TLC OF SUGARS

11.2.1 ADSORBENTS AND SOLVENTS

Carbohydrates are very hydrophilic compounds and for this reason they attach strongly to Silica Gel, Alumina, and Cellulose, so highly polar solvents were necessary as mobile phase in TLC development.

To date, the most common adsorbents are obtained from Merck, Sigma, and Carlo Erba. TLC aluminum or Glass plates containing Silica gel G 60 as adsorbent are widespread for rapid monosaccharide identification; the resolution and sensibility can be increased when HPTLC-Silica gel G 60 plates are used. Cellulose aluminum plates or PC (Whatman No. 1) have been used for monosaccharide and oligosac-charide separation; however, the development is slower than Silica Gel G 60.

Since carbohydrates attach strongly to the adsorbents the mobile phase must be very polar. PC or Cellulose Aluminum plates have been developed with variations in the ratio of butanol:pyridine:H_2O. Analytical chromatography for monosaccharide composition uses this solvent system in the ratio of 5:3:3 v/v in Whatman No. 1, and preparative isolation of oligosaccharides have been carried out with butanol:pyridine:H_2O, 1:1:1, v/v, in Whatman No. 3 PC. The same solvent system has been used in Cellulose plates, sometimes small variations in the ratio are necessary. The development of Silica Gel G 60 plates are also made with polar solvents. The literature shows the use of many organic mixtures of solvents systems; however, for plant monosaccharides the solvent system ethyl acetate:n-propanol:acetic acid:H_2O (4:2:2:1 v/v) has provided fast separation and high monosaccharides resolution (pentoses, hexoses, and deoxihexoses), as well as for sucrose and trealose oligosaccharides. Sometimes the plates of silica Gel G 60 were buffered with boric acid, sodium borate, and sodium acetate at ~0.02 M, this procedure enhances the separation of the monosaccharide in TLC. The separation of oligosaccharides, such as aldobiuronic, kestoses, fructooligosaccharides, fructans series, and maltose series can be done with n-propanol:H_2O (70:30 v/v). As the adsorbents and the mobile phase are important in the sugar separation and determination, the use of specific detection is applied, since different reagents would be necessary to evaluate reducing or nonreducing sugars, cyclitols (inositol), and alditols.

11.2.2 UNIVERSAL REAGENTS STAINING

The visualization of sugars by nonspecific reagents has been performed with sulfuric acid in silica gel and alumina, since they are inorganic compounds and are not carbonized by strong inorganic acids. Sugars, cyclitols, and polyalcohols are carbonized by solutions of 10%–40% of H_2SO_4 dissolved in MeOH or in water after heating at 100°C for 5–40 min. The heating time is inversely proportional to the sulfuric acid concentration. Generally, reducing sugars are more sensitive than inositols and polyalcohols. Iodine vapors have been applied to detect carbohydrates; however, it is less sensitive than sulfuric acid. Although it is not destructive for 5–20 min of exposure, it can be used in quantitative and preparative chromatography.

11.2.3 SPECIFIC REAGENTS

The ketone and aldehyde groups of the reducing or nonreducing sugars are the preferential point to react specifically in the sugars, as described earlier. The detection of sugars in TLC uses the furfuraldehyde in conjunction with many phenolic reagents, which give rise to different colors, as observed by orcinol, α-naphtol, resorcinol, anisaldehyde, and aniline–diphenylamine (Table 11.1).

11.2.3.1 Preparation and Application of TLC-Staining Reagents

Orcinol–H_2SO_4 for Sugar: 250 mg dissolved in 95 mL EtOH:5 mL H_2SO_4. The TLC plates must be evenly sprayed or quickly dipped into the solution, dried, and heated at 100°C for 5–15 min. The solution is stable for a month if stored at 4°C.

TABLE 11.1
Color Reagents of Carbohydrates on TLC

Sugar	Anisaldehyde–Sulfuric Acid	Naphtol–Resorcinol–Sulfuric Acid	Aniline–Diphenylamine	Orcinol–Sulfuric Acid	Sulfuric Acid
L-rhamnose	Green	Green	Pale green	Pale brown	Gray-brown
D-ribose	Blue	—	—	Blue	Gray-brown
D-xylose	Gray	Light blue	Bright blue	Blue	Gray-brown
L-arabinose	Yellow-green	Blue-green	Bright blue	Blue	Gray-brown
D-mannose	Green	Light blue	—	Violet	Gray-brown
D-glucose	Light blue	Blue-violet	Gray-green	Violet	Gray-brown
D-galactose	Green-gray	Blue-violet	Gray-green	Violet	Gray-brown
D-glucoronic acid	—	Blue	—	Blue	Gray-brown
D-galacturonic acid	—	Blue	—	Blue	Gray-brown
Sucrose	—	—	—	Red-brown	

- *Silver Nitrate–Sodium Hydroxide for Sugars and Polyols*: *Spray reagent 1*—1 mL saturated aqueous silver nitrate solution is diluted to 200 mL with acetone and 5–10 mL of water are then added until the precipitated has dissolved.
- *Spray reagent 2*—0.5 N aqueous-methanolic sodium hydroxide (20 g sodium hydroxide are dissolved in a minimum of water and the solution diluted to 1 L of methanol).

Spraying is carried out with reagent 1, the plate is dried, then sprayed with reagent 2, and finally heated 1–2 min at 100°C. To clarify the background the plate must be sprayed with a 5% aqueous sodium thiosulphate solution.[36]

Naphtoresorcinol–H_2SO_4 for Sugar—Solution A: 200 mg of naphtoresorcinol dissolved in 100 mL of ethanol. Solution B—20% H_2SO_4 in EtOH (v/v). For staining, mixed solutions A and B (1:1 v/v) are sprayed on the plates, which are then heated at 100°C for 5–10 min.[36]

Ninhydrin for Aminocompounds: 200 mg of ninhydrin dissolved in 100 mL of acetone. The plates may be sprayed or quickly dipped into the solution, dried, and heated at 100°C, until the colors arise (5–10 min).

Sulfuric Acid—General Staining: 10% of concentrate H_2SO_4 dissolved in methanol (v/v). The users must take proper care during the mixture preparation to avoid burns. The plates are sprayed and heated at 120°C–140°C, for 5–15 min.

Anisaldehyde–H_2SO_4 for Sugars: a fresh solution of 0.5 mL of anisaldehyde is dissolved in 9 mL EtOH, 0.5 mL H_2SO_4 and 0.1 mL of acetic acid. The TLC plates must be evenly sprayed or quickly dipped into the solution, dried, and heated at 100°C for 5–15 min.[36]

p-Anisidine Hydrochloride for Reducing Sugars: 4 g of *p*-anisidine hydrochloride are dissolved in 4 mL of H_2O and the volume completed with *n*-butanol until 100 mL. TLC plates must be evenly sprayed or quickly dipped into the solution, dried, and heated at 100°C for 5–15 min.

Aniline–Diphenylamine–Phosphoric Acid for Reducing Sugars: 4 g of diphenyla-mine, 4 mL of aniline, and 20 mL of 85% phosphoric acid (v/v) are dissolved in 200 mL of acetone. After the spray the plates must be heated at 85°C for 10 min.[36]

11.2.4 SAMPLE PREPARATION

11.2.4.1 Water Soluble Sugars (Mono-, Oligo-, and Polysaccharides)

The monosaccharide components of oligomers and polymers can be evaluated by their acidic hydrolysis with 2 M TFA at 100°C for 10 h. The product evaporated at room temperature and then dissolved in water and applied on TLC (Figure 11.13A) or PC, and examined by silver nitrate or orcinol–H_2SO_4 spray to detect carbohydrate spots or bands.[4,37,38] These reagents are more sensitive than a variety of methods that use aromatic amines combined with acids, including *p*-anisidine hydrochloride.[39] However, the latter and orcinol–H_2SO_4 have great advantage of being more specific, distinguishing by the color spots arising from hexose, pentose, and methylpentose.

11.2.4.1.1 Partial Acid Hydrolysis

Partial hydrolysis can be effected by using conditions somewhat milder than those necessary for complete hydrolysis, for example 0.5 M TFA at 70°C–100°C for periods of 1–10 h, instead of 2 M TFA at 100°C for 10 h. The optimum time for maximum production of oligosaccharide can be determined at intervals by PC or TLC (Figure 11.13B). Generally high quantities of polysaccharide are needed, since

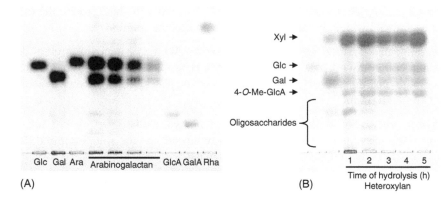

FIGURE 11.13 (A) TLC of total hydrolyzed arabinogalactan at 100°C, for 8 h, with 2M TFA (determination of sugar composition, including the uronic acid). The TLC plates were devel-oped in ethyl acetate:*n*-propanol:acetic acid:H_2O (4:2:2:1 v/v) and stained with orcinol–H_2SO_4. (B) TLC of hetoroxylan (partial acid hydrolysis using 0.5 M TFA at 100°C). At 1 h of hydrolysis, oligosaccharides are formed, in conjunction with more labile monosaccharides (generally nonreducing ends). After 5 h of hydrolysis only monosaccharides were observed.

the predominant product is a monosaccharide, and in order to isolate the minor oligosaccharide component a chromatography can be carried out on a column of charcoal-Celite 545 (1:1 w/w). Monosaccharides are eluted with water and varying concentrations of aqueous EtOH up to 30%–40% have been used to elute oligo-saccharides.[40] As the oligosaccharide product is often mixed with coeluted silicate, a further fractionation is generally carried out on a column of powdered Whatman CF-11 cellulose,[41,42] which has the advantage of being renewable after each run. Elution is recommended with relatively cheap acetone containing increasing propor-tions of water, which successively elutes monosaccharides (acetone–water, 10:1 v/v), disaccharide (7:1 to 4:1), trisaccharide (3:1), tetrasaccharide (5:2), etc.[43] The eluates from the charcoal or cellulose columns, when they contain mixtures of oligosaccharides, can be further fractionated on Whatman No. 3 filter paper.

Analysis of fractions can be carried out by TLC, PC, or HPLC. The latter has a great number of possibilities. Examples of columns are Whatman Partisil-10 PAC for maltodextrins (DP 2–10),[44] and cellodextrins (DP 2–6),[45] Waters Bondapak for sucrose, raffinose, stachyose, and β-cyclodextrin, and Waters Altech Carbohydrate 10 for mannose-containing oligosaccharides.[46] Although preparative HPLC is pos-sible, the simpler and much more economical PC method is sometimes preferred, since the resolution is often better.[47] Also, direct cellulose column chromatography can be employed before preparative PC.

11.2.4.2   Organic Soluble Compounds Containing Sugars

11.2.4.2.1   Glycolipids

TLC for glycolipids analysis has been well developed using mixtures of organic solvents. These mixtures of solvents and H_2O, in appropriate amounts, may play an important role on TLC development, and it is possible to have a considerable gain in separation ratios on the plates. Moreover, it is possible to adapt the solvent according to specific needs.

An important way of improving the results consists in the sample preparation, since many types of lipids may be present in plant extracts. Extraction methods consist in the dissolution of the material of interest in specific solvent or mixtures, usually MeOH and $CHCl_3$, at different concentrations, which have shown good results in total lipid extraction. The extraction process may be performed at room temperature or under high temperature (60°C), and the process must be repeated until the lipids are completely removed the matrix. When the matrix for extraction consists of leaves or young stems, several components may be removed together with the lipids, such as chlorophylls, and other pigments, low molecular mass organic compounds, so crude extracts should be pretreated in order to avoid interference.

Fractionation on chromatographic column using high nonpolar solvents (ethers or hexanes) is employed to eliminate the pigments and other nonpolar components such as triglycerides. The increase of the polarity, by replacing the nonpolar solvent to $CHCl_3$ and gradually add MeOH (i.e., 5% to 50%), makes possible to get lesser complex mixtures of lipids, which can be analyzed on TLC plates. Usually solvents such as $CHCl_3$:MeOH (9:1, v/v) or $CHCl_3$:MeOH:H_2O (65:25:4, v/v) are used as

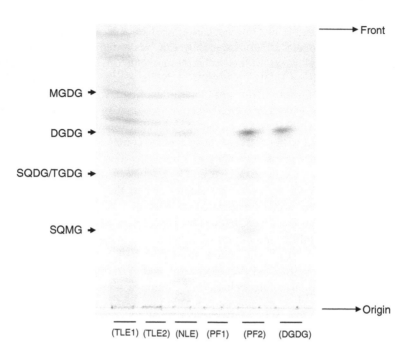

FIGURE 11.14 Total lipid extract from *Maytenus ilicifolia* (TLE1); total lipid extract from *Phyllanthus niruri* (TLE2); neutral lipid extract (NLE); and polar fractions (PF1 and PF2) after fractionation on DEAE-Sepharose; digalactosylglyceride standard (DGDG).

developers. To selectively visualize the glycolipids, the plate may be stained by orcinol/sulfuric acid method,[38] specific for carbohydrate moieties. Free sugars and oligosaccharides may cause interference if present in the samples.

Neutral, polar, and charged lipids may be separated on chromatographic columns as shown in Figure 11.14. In this case the lipids were fractionated on DEAE-Sepharose column. The TLC was developed in $CHCl_3:MeOH:H_2O$ (65:25:4, v/v) and stained by orcinol/H_2SO_4.

A method for glycolipid analysis from invertebrates[48] may be useful if applied in plant glycolipid analysis. The procedure was first carried out by suspension of the thawed sample in cold water followed by sonication, until the nematodes were broken. Into this solution $CHCl_3$ and MeOH were added, giving rise to final $CHCl_3:MeOH:H_2O$ ratio of 4:8:3, v/v. The solution was held for 4 h at 37°C, and then centrifuged to remove the insoluble residues. The supernatant was partitioned in $CHCl_3:MeOH:H_2O$ (4:8:5.6, v/v), and the phases were separated by centrifugation. The lower phase was dried down and resuspended in 1:1 ($CHCl_3:MeOH$) for analysis, while the upper phase was appropriated desalted, dried down, and resuspended in methanol. The two phases obtained were analyzed on HPTLC plates at different conditions. The lipid samples were developed in a TLC chamber containing $CHCl_3:MeOH:H_2O$ (4:4:1, v/v) for upper phase lipids, or $CHCl_3:MeOH:H_2O$ (45:18:3, v/v) for lower phase lipids, which were stained by the orcinol/H_2SO_4

method. Glycolipids containing complex carbohydrate moieties were well resolved by this methodology.

11.2.5 QUANTITATIVE ANALYSIS AND SUGAR COMPOSITION

Analysis on TLC (or HPTLC) provides a rapid and reproducible method for quantitative and qualitative analysis of carbohydrates and glycoconjugates, since the methods are easier and cheaper than other common techniques, such as HPLC and GC.

The qualitative method consists in the TLC development with the appropriate solvents and staining reagents (see section 11.2.1–11.2.3), containing authentic standards. The R_f values of standards are useful, since they are compared to the samples R_f values. Moreover, the color obtained by the staining reagents aid in the identification of sugars. To obtain a quantitative analysis on TLC it is necessary to create a calibration curve (i.e., 0.1–1 µg/µL of known standards). After staining, the TLC is digitalized and the quantification is carried out by densitometry techniques.

To quantify sugars on TLC, it is necessary to examine the image using appropriate software. Sassaki and coworkers[49] got accurate quantification for methylglycosides partially methylated using Scion Image Program (Scion Corporation, Maryland, USA), based on perception of the image of each spot on TLC.

TLC quantification methods have been employed in sucrose and fructo-oligosaccharides determination in order to provide a more rapid analysis than is possible by high-performance liquid chromatography (HPLC), since the time-consuming in samples pre-treatment is not needed.[50] So, TLC seems to have some advantages over HPLC, such as in sample preparation, detection by specific reagents and the low consumption solvents and short time of analysis.[51]

11.3 CONCLUSIONS

The great advantage in TLC analysis is its simplicity. In this review different solvents have been shown, since they are the simplest and easiest to obtain from chemical industries. These solvents present efficiency and high resolution for plant carbohydrates and glycoconjugates analysis. Specific detection of sugars on TLC uses the furfuraldehyde allied to orcinol, since it can resolve the most common monosaccharide classes present in plant by its peculiar colors and R_f values. Since reduced sugars (alditols) are not stained by orcinol, the use of silver nitrate–sodium hydroxide can provide their detection, being useful to observe these components. TLC of polysaccharides is not possible; however, it has been used to determine their monosaccharide composition after hydrolysis. The latter, also, has been monitored by TLC, provides the ideal time for oligosaccharides formation. Glycoconjugates have been isolated and characterized by TLC using standards. The presence of substitution in the sugar moiety, such as phosphate and amino groups, has been observed on TLC using specific staining methods, molybdate reagent[52] and ninhydrin, respectively, in conjunction with orcinol visualization. However, TLC technique does not provide information about O-glycosidic or N-glycosidic-linkages, position of substituents, and sugar configuration. So, the conformational and fine structural characterization has been carried out using complementary techniques, such as methylation, NMR, (nuclear magnetic resonance), GC (gas chromatography), LC (liquid chromatography), and MS (mass spectrometry).

REFERENCES

1. Collins, P. and Ferrier, R., *Monosaccharides, Their Chemistry and Their Roles in Natural Products*, Wiley, Chichester, 1995.
2. Solomons, G. and Fryhle, C., *Organic Chemistry*, 7th ed., Wiley, New York, NY, 2000, chap. 22.
3. Aspinall, G.O., Chemistry of cell wall polysaccharides, in *The Biochemistry of Plants*, vol. 3, Stumpf, P.F. and Conn, E.E., Eds., Academic Press, New York, NY, 1980, chap. 12.
4. Trevelyan, W.E., Procter, D.P., and Harrison, J.S., Detection of sugars on paper chromatograms, *Nature*, 166, 444, 1950.
5. Avigad, G. and Dey, P.M., Carbohydrate metabolism: storage carbohydrates, in *Plant Biochemistry*, Dey, P.M. and Harborne, J.B., Eds., Academic Press, London, 1997, chap. 4.
6. Edelman, J. and Keys, A.J., A maltose–glucose transglucosylase from wheat germ, *Biochem. J.*, 79, 12, 1961.
7. Linden, J.C., Tanner, W., and Kandler, O., Properties of glucosyltransferase and glucan transferase from spinach, *Plant Physiol.*, 54, 752, 1974.
8. Linden, J.C. et al., Asymmetric labeling of maltose during photosynthesis in CO_2, *Z. Pflanzenphysiol.*, 76, 176, 1975.
9. Kandler, O. and Hopf, H., Occurrence, metabolism, and function of oligasaccharides, in *The Biochemistry of Plants*, vol. 3, Stumpf, P.K. and Conn, E.E., Eds., Academic Press, 1980, chap. 7.
10. Edelman, J. and Jefford, T.G., The mechanism of fructosan metabolism in higher plants as exemplified in *Helianthus tuberosus*, *New Phytol.*, 67, 517, 1968.
11. Koops, A.J. and Jonker, H.H., Purification and characterization of the enzymes of fructan biosynthesis in tubers of *Helianthus tuberosus* Colombia. II. Purification of sucrose: sucrose 1-fructosyltransferase and reconstitution of fructan synthesis in vitro with purified sucrose:sucrose 1-fructosyltransferase and fructan:fructan 1-fructosyltransferase, *Plant Physiol.*, 110, 1167, 1996.
12. Lüscher, M. et al., Inulin synthesis by a combination of purified fructosyltransferases from tubers of *Helianthus tuberosus*, *FEBS Lett.*, 385, 39, 1996.
13. Van den Ende, W. and Van Laere, A., *De novo* synthesis of fructans from sucrose in vitro by a combination of two purified enzymes (sucrose:sucrose 1-fructosyltransferase and fructan:fructan 1-fructosyltransferase) from chicory roots (*Cichorium intybus* L.), *Planta*, 200, 335, 1996.
14. Pollock, C.J., Fructans and the metabolism of sucrose in vascular plants, *New Phytol.*, 104, 1, 1986.
15. Trip, P., Nelson, C.D., and Krotkov, G., Selective and preferential translocation of C^{14}-labeled sugars in white ash and lilac, *Plant Physiol.*, 40, 740, 1965.
16. Senser, M. and Kandler, O., Galactinol, ein galactosyldonor fur die biosynthese der zucker der raffinosefamilie in blattern, *Z. Pflanzenphysiol.*, 57, 376, 1967.
17. Reid, J.S.G., Carbohydrate metabolism: structural carbohydrates, in *Plant Biochemistry*, Dey, P.M. and Harborne, J.B., Eds., Academic Press, London, 1997, chap. 5.

18. Carpita, N.C. and Gibeaut, D.M., Structural models of primary cell walls in flowering plants: consistency of molecular structure with the physical properties of the walls during growth, *Plant J.*, 3, 1, 1993.

19. Carpita, N.C. and McCann, M., The cell wall, in *Biochemistry and Molecular Biology of Plants*, Buchanan, B.B., Gruissem, W., and Jones, R.L., Eds., American Society of Plant Physiologists, Rochville, 2000, chap. 2.

20. Fincher, G.B., Stone, B.A., and Clarke, A.E., Arabinogalactan-proteins: structure, biosynthesis, and function, *Annu. Rev. Plant Physiol.*, 34, 47, 1983.

21. Whistler, R.L., Exudate gums, in *Industrial Gums: Polysaccharides and their Derivatives*, Whistler, R.L. and BeMiller, J.N., Eds., Academic Press, New York, NY, 1993, chap. 12.

22. Maurer-Menestrina, J. et al., Structure a highly substituted glycanoxylan from the gum exudate of the palm *Livistona chinensis* (chinese fan), *Carbohydr. Res.*, 338, 1843, 2003.

23. Simas, F.F. et al., Structure of a heteroxylan of the gum exudate of *Scheelea phalerata* (uricuri), *Phytochemistry*, 65, 2347, 2004.

24. Webb, M.S. and Green, B.R., Biochemical and biophysical properties of thylakoid acyl lipids, *Biochim. Biophys. Acta*, 1060, 133, 1991.

25. Lee, A.G., Membrane lipids: it's only a phase, *Curr. Biol.*, 10, 377, 2000.

26. Webb, M.S., Tilcock, S.P.S., and Green, B.R., Salt-mediated interactions between vesicles of the thylakoid lipid digalactosyldiacylglycerol, *Biochim. Biophys. Acta*, 938, 323, 1988.

27. Menikh, A. and Fragata, M., Fourier transform infrared spectroscopic study of ion binding and intramolecular interactions in the polar head of digalactosyldiacylglycerol, *Eur. Biophys. J.*, 22, 249, 1993.

28. Hincha, D.K., Effects of calcium-induced aggregation on the physical stability of liposomes containing plant glycolipids, *Biochim. Biophys. Acta*, 1611, 180, 2003.

29. Carter, H.E. et al., Wheat flour lipids. II. Isolation and characterization of glycolipids of wheat flour and other plant sources, *J. Lipid Res.*, 2, 215, 1961.

30. MacMurray, T.A. and Morrison, W.R., Composition of wheat-flour lipids, *J. Sci. Food Agric.*, 21, 520, 1970.

31. Laine, R.A. and Renkonen, O TI., Ceramide di- and trihexosides of wheat flour, *Biochem. J.*, 13, 2837, 1973.

32. Carter, H.E. and Koob, J.L., Sphingolipids in bean leaves (*Phaseolus vulgaris*), *J. Lipid Res.*, 10, 363, 1969.

33. Carter, H.E. and Kisic, A., Countercurrent distribution of inositol lipids of plant seeds, *J. Lipid Res.*, 10, 356, 1969.

34. Kaul, K. and Lester, R.L., Characterization of inositol-containing phosphosphingolipids from tobacco leaves, *Plant Physiol.*, 55, 120, 1975.

35. Elbein, A.D., Glycolipids, in *The Biochemistry of Plants*, vol. 3, Stumpf, P.K. and Conn, E.E., Eds., Academic Press, New York, NY, 1980, chap. 15.

36. Krebs, K.G., Heusser, D., and Wimmer, H., Spray reagents, in *Thin-Layer Chromatography*, Stall, E., Ed., Spring-Verlag, Berlin, 1969, chap. Z.

37. Hough, L., Application of paper chromatography to the separation and detection of polyhydric alcohols, *Nature*, 165, 400, 1950.

38. Skipski, V.P., Thin layer chromatography of neutral glycosphingolipids, *Meth. Enzymol.*, 35, 396, 1975.

39. Hough, L., Jones, J.K.N., and Wadman, W.H., Quantitative analysis of mixtures of sugars by the method of partition chromatography. V. Improved methods for the separation and detection of the sugars and their methylated derivatives on paper chromatograms, *J. Chem. Soc.*, 1702, 1950.

40. Whistler, R.L. and Durso, D.F., Chromatographic separation of sugars on charcoal, *J. Am. Chem. Soc.*, 72, 677, 1950.

41. Hough, L., Jones, J.K.N., and Wadman, W.H., Quantitative analysis of mixtures of sugars by the method of partition chromatography. V. Improved methods for the separation and detection of the sugars and their methylated derivatives on paper chromatograms, *J. Chem. Soc.*, 1702, 1950.

42. Whistler, R.L. and BeMiller, J.N., Cellulose column chromatography, *Methods Carbo- hydr. Chem.*, 1, 47, 1962.

43. Gorin, P.A.J., Spencer, J.F.T., and Magus, R.J., Comparison of proton magnetic reson- ance spectra of cell wall mannans and galactomannans of selected yeasts and their chemical structures, *Can. J. Chem.*, 47, 3569, 1969.

44. Rabel, F.M., Caputo, A.G., and Butts, E.T., Separation of carbohydrates on a new polar bonded phase material, *J. Chromatogr.* 166, 731, 1976.

45. Gum, E.K. Jr., and Brown, R.D. Jr., Two alternative HPLC separation methods for reduced and normal cellooligosaccharides, *Anal. Biochem.*, 82, 372–375, 1977.

46. Black, L.T. and Bagley, E.B., Determination of oligosaccharides in soybeans by high pressure liquid chromatography using an internal standard, *J. Am. Oil Chem. Soc.*, 55, 228, 1978.

47. Brunt, K., Rapid separation of linear and cyclic oligosaccharides on a cation-exchange resin using a calcium ethylenediaminetetraacetic acid solution as eluant, *J. Chromatogr.*, 246, 145, 1982.

48. Griffitts, J.S. et al., Glycolipids as receptors for *Bacillus thuringiensis* crystal toxin, *Science*, 307, 922, 2005.

49. Sassaki, et al., Rapid synthesis of partially *O*-methylated alditol acetate standards for GC-MS: some relative activities of hydroxyl groups of methyl glycopyranosides on Purdie methylation, *Carbohyd. Res.*, 340, 731, 2005.

50. John, J.A.St., Bonnett, G.D., and Simpson, R.J., A method for rapid quantification of sucrose and fructan oligosaccharides suitable for enzyme and physiological studies, *New Phytol.*, 134, 197, 1996.

51. Günther, M. and Schmidt, P.C., Comparison between HPLC and HPTLC-densitometry for the determination of harpagoside from *Harpagophytum procumbens* CO_2-extracts, *J. Pharm. Biomed. Anal.*, 37, 817, 2005.

52. Dittmer, L.C. and Lester, R.L., A simple, specific spray for the detection of phospholipids on thin-layer chromatograms, *J. Lipid Res.*, 15, 126, 1964.

12 TLC of Lipids

*Svetlana Momchilova and
Boryana Nikolova-Damyanova*

CONTENTS

12.1 INTRODUCTION

12.1.1 LIPIDS: DEFINITION, OCCURRENCE, AND ROLE OF CHROMATOGRAPHY

The term *lipid* in the following text is restricted to the naturally occurring esters of the long-chain aliphatic monocarboxylic acids, the fatty acids (FAs), and to compounds that are closely related biosynthetically or functionally. Most abundant in the living organisms, including plants, are the FA esters with glycerol (1,2,3-trihydroxy propanol), denoted as glycerolipids. In their turn, glycerolipids are usually classified according to the number of hydrolytic products per mole. Simple (often denoted as neutral) lipids release two types of products: glycerol and FA, while complex (denoted usually as polar) lipids give three or more products such as FA, glycerol, phosphoric acid, and aminobase, for example. Triacylglycerols (TAG) comprise the most abundant simple lipid group, and phospholipids (PLs) and glycolipids (GLs) are the most abundant complex lipid groups. Closely related to these according to the earlier-mentioned definition are the sterols and the sterol esters usually accompanying the glycerolipids in plants. Plant lipids, therefore, present a complex mixture of several lipid types, denoted later as groups: simple and complex glycerolipids, sterol esters, and sterols. Free FA, mono-, and diacylglycerols are usually considered as artifacts.

FAs, the key structural element of lipids, are carboxylic compounds with even numbered straight chains of 12–22 carbon atoms and zero to (mostly in plants) three double bonds in, predominantly, *cis* (Z) configuration. To simplify the description of lipids, a shorthand notation of FA is accepted, which presents the configuration and the position of double bonds, followed by the number of carbon atoms and the number of the double bonds. Thus, *c*9-18:1 denotes an FA with 18 carbon atoms, one double bond at the ninth carbon atom in *cis* configuration. Compounds with chains of 14, 16, and 18 carbon atoms are the most abundant saturated FA in plants. Monounsaturated (monoenoic) FA comprise most often compounds with 16 and 18 carbon atoms. The double bond position may vary and there are plant families characterized by the specific position of the double bond in the octadecenoic FA. Thus, the most abundant is undoubtedly the *c*9-18:1 (oleic acid), but *c*6-18:1 is the main FA in Umbelliferae, *c*11-18:1 (*cis*-vaccenic acid) often accompanies oleic acid in many seed oils and erucic acid (*cis*13-22:1) is the main component in Criciferae seed oil.[1] Linoleic (*c*9,*c*12-18:2, sunflower oil) and linolenic (*c*9,*c*12,*c*15-18:3, linseed oil) acids are the main representatives of the polyenoic FA in plant lipids with the so-called *methylene interrupted* system of double bonds. Important isomer of linolenic acid is *c*6,*c*9,*c*12-18:3 (γ-linolenic acid) found in significant amounts in borage and evening primrose oils. These polyenoic FA are often denoted as *essential* since they are not synthesized by humans and animals but are important precursors of other polyunsaturated biologically active compounds in these organisms. Plant lipids are the single source of essential FA. Other rare isomers of unsaturated FA have been found in plants and again they characterize specific plant families. For example, polyunsaturated FA with double bonds separated by more than one methylene group, such as *c*5,*c*9- and *c*5,*c*13-systems, was found in coniferous plants. FAs with *trans* configuration of the double bond have also been identified, and *t*3-16:1, for example, is typical for chloroplast lipids, *t*9,*t*12-18:2 is found in *Chilopsis linearis* seed oil and α-eleostearic acid (*c*9,*t*11,*t*13-18:3) is typical for tung oil. Crepenynic acid, which contains a triple bond (octadeca-*c*9-en-12-ynoic acid), was identified in *Crepis* and *Afzelia* oils.

We are still learning about the role of lipids as primary metabolites in the living organisms. Two of their main functions are clear so far: (1) as an efficient source of energy and (2) as an important part of the cell membrane responsible for its integrity and normal function. There are evidences that lipids may participate in many other processes in the living organism.[2] Plant lipids, on the other hand, are important part of the human diet, beside the other valued features, because of supplying the essential FA necessary for the normal function of the human metabolism.

Thus, we are studying plant lipids for two major reasons: (1) to understand better the plant metabolism and physiology and (2) to estimate the usefulness of a given plant as a nutrition or industrial source.

12.1.2   TLC: PAST, PRESENT, AND FUTURE IN THE ANALYSIS OF LIPIDS

Since the lipids present a complex mixture of different groups comprising in their turn numerous individual components with closely related chemical and physical properties, chromatography appears to be the relevant method for separation and investigation of the lipid components. Between the many chromatographic techniques, thin layer chromatography (TLC) has its special place, being one of the first separation methods introduced in the lipid analysis. Most of the present basic knowledge on the structure and function of lipids was achieved by employing various TLC techniques.

The first description of a technique close to TLC goes back to 1938 but only about 20 years later, Kirchner in USA (1951) and Stahl in Germany (1965) first converted the idea into the present TLC, providing the materials and devises, and developing the basic principles. In the early sixties, Kaufmann and coworkers[3] in Germany and Privett and coworkers[4] in USA introduced TLC in lipid analysis, developing the basis of silica gel-, silver ion-, and reversed-phase TLC (RP-TLC) techniques. Later, Chobanov and coworkers[5,6] and Nikolova-Damyanova and Amidzhin[7] converted analytical silver ion TLC (Ag-TLC) of FA and TAG and RP-TLC of TAG into methods for separation and quantification (see also the reviews by Nikolova-Damyanova[8] and Nikolova-Damyanova and Momchilova[9]).

The rapid development and the broad application of high-performance liquid chromatography (HPLC) in lipid analysis stopped the further development of TLC and in many lipid laboratories TLC was practically swept being considered too laborious and time consuming, and requiring certain experimental skills and attention. This is true indeed in some extent as scientists using TLC do not push buttons only but are fully engaged in the analysis, but at the same time TLC is easy to perform, versatile, and does not require expensive instrumentation. TLC also allows direct quantification by using scanning densitometry with the requirement to calibrate the measurements against a reference sample, a requirement that is valid also for all other chromatographic techniques. TLC is the single chromatographic technique that gives a full picture of the sample composition, allowing for a rapid and less ambiguous identification of the separated compounds since the sample and the reference material are applied on the same thin layer and are chromatographed under identical conditions. Separation and identification of lipid groups on unmodified silica gel TLC is the easiest and the most rapid way for screening the lipid content of a large number of samples when studying plants. Also as will be shown later, combining the TLC modes in an efficient sequence provides results that are not inferior to those obtained by HPLC.

12.2 TLC TECHNIQUES

The application of TLC in the analysis of plant lipids will be discussed later. Inevitably, substantial part of the attention will be paid to the TLC techniques in use in our laboratory, where silica gel TLC, analytical and preparative Ag-TLC, and analytical RP-TLC are almost in everyday use. The Chromarod modification of TLC is not considered here and the comprehensive review by Shanta[10] is recommended for understanding this technique. TLC of lipids was discussed in many essential books and reviews, such as Thin Layer Chromatography by Fried and Sherma;[11] Lipid Analysis by Christie,[2] Silver Ion Chromatography of Lipids by Nikolova-Damyanova,[12] for example, and these are highly recommended.

12.2.1 GENERAL PRINCIPLES, TLC MODES

TLC is a separation technique, in which the stationary phase is spread as a thin layer on a plate of inert material (glass, aluminum foil, or plastic), whereas the mobile phase is a liquid. During the movement of the mobile phase along the plate, the lipid components are retained with different strengths on the layer, depending on their molecular distinctions, thus forming separated spots or bands. The migration of each spot is presented quantitatively by the R_f value and the lower the R_f value the stronger is the retention.

Three TLC modes are utilized for analysis of plant lipids: (1) silica gel TLC, where the stationary phase is a layer of fine particles of silica gel; (2) silver ion TLC (Ag-TLC), where silver ions are introduced in the stationary phase; and (3) RP-TLC, where layers of silica gel or Kieselguhr are chemically treated with long-chain organic compounds to give a nonpolar stationary phase.

In silica gel, TLC the stationary phase contains a large number of hydroxyl groups that interact with the polar functional groups of the lipid components. As a result, the lipid molecules are separated according to the number of the polar functionalities and, in some cases, according to their spatial location.

In Ag-TLC, silver ions, which are mechanically introduced in the stationary phase, interact with the lipid molecules, forming weak reversible complexes with the double bonds. Thus, Ag-TLC separates lipids according to the number, configuration and, to some extent, the position of the double bonds in the molecule.

In RP-TLC, lipids are distributed between a stationary phase containing nonpolar long hydrocarbon chains and a polar mobile phase, leading to separation of the lipid molecules according to both their chain length and number of double bonds.

These three TLC techniques can be performed in preparative or analytical mode.

12.2.2 PLATES AND PLATE PREPARATION

All TLC modes share the standard TLC facilities: plates, development tanks, micropipettes or microcapillars for sample application, spraying devices and reagents for detection of separated zones, densitometers for quantification, etc. Layers of silica gel G (with calcium sulfate as a binder), silica gel H (no binding substance), Kieselguhr or, very rarely, alumina, are used as stationary phases. The plates are either handmade in laboratory or commercially precoated. The handmade plates are prepared by manual spreading of aqueous slurry of the adsorbent on glass support. Sets of devices needed for such preparation are commercially available. On the other hand, in commercially precoated plates the adsorbent is spread on glass, aluminum foil, or plastic sheet automatically; therefore the layer thickness is always uniform. Nevertheless, in many cases the handmade plates are preferred because of the lower cost and the possibility to modify easily the layer as well as to vary the thickness. The layer thickness ranges between 0.2 and 0.3 mm for analytical plates and 0.5 and 1.0 mm for preparative plates. After spreading the adsorbent slurry, the plates are air-dried for at least 24 h. In silica gel TLC, the laboratory prepared plates are put briefly in an oven at 110°C immediately before their use.

In silver ion TLC, the adsorbent, usually silica gel G, is impregnated with silver ions. The impregnation can be performed by the incorporation of silver salt (normally $AgNO_3$) into the silica gel slurry, by immersing or spraying the plate with methanol, ethanol, acetonitrile, water, or ammonia solution of $AgNO_3$, or by "dynamic impregnation" (development of the plate with $AgNO_3$ solution).[13] The concentrations of impregnating solutions vary in broad limits (from 0.5% to 40%) depending on the analytical purpose. After impregnation, the plates are air-dried in a dark place and activated for 5–30 min in an oven at 110°C–120°C before sample application. We have shown in several papers that impregnation with high silver ion percentages is a disadvantage since the high percentage does not improve substantially the resolution while causing numerous problems with the further handling of the plates.

In reversed-phase TLC, the adsorbent is usually Kieselguhr, which is practically inert and is treated by different manners with chlorosilanes or with long-chain hydrocarbons such as paraffin. Details of the procedures are given in Section 12.5.

12.2.3 GENERAL ON MOBILE PHASES

The mobile phases for TLC of simple lipids contain usually two or three components. Four or more solvents are needed for separation of complex lipids. Silica gel TLC and Ag-TLC, having polar stationary phases, require less polar mobile phases. In most cases hexane, petroleum ether (bp 40°C–60°C), toluene, and chloroform are the major components, while acetone, diethyl ether, methanol, ethanol, acetic acid are added in small proportions. Since the separation of complex lipids requires more polar mobile phases, water or ammonia solutions are often used as mobile phase components. Moreover, two-dimentional development of the plate is often performed. In RP-TLC, polar mobile phase is used, consisting of water, methanol, acetonitrile, or acetone.

In all TLC modes, a common practice to improve the resolution is the two- or even threefold development of the plate with either the same mobile phase or with mobile phase(s) of gradually decreasing polarity. The latter approach is often applied in Ag-TLC.

In silica gel TLC and Ag-TLC, the atmospheric humidity affects directly the reproducibility of separation. Since it is not easy in the practice to control the humidity, its effects are compensated by minimal changes in the mobile phase composition. RP-TLC does not suffer by this problem.

12.2.4 SAMPLE PREPARATION

A special feature of sample preparation in the analysis of plant lipids is the necessity for preliminary extraction of plant tissues. Isopropanol is recommended as extraction solvent in order to deactivate the enzymes. The preliminary treatment is followed by various extraction procedures, and the most often used is the "Folch" procedure.[2] Then the lipid extract is usually fractionated by silica gel TLC into groups of different polarity. Generally, the lipid components of lower polarity (mainly TAG) are subjected successively to Ag-TLC and RP-TLC as intact species.[14–17] The more polar lipid components, such as partial acylglycerols, free FA (if and when present), and complex lipids, are converted to less polar derivatives for further analyses. The choice of appropriate derivative depends on the nature of the lipid component, as well as on the separation needed.

In all TLC modes, the lipid sample is applied on the plate as a solution of appropriate concentration in suitable solvent (hexane, dichloroethane, methanol). The sample size strongly affects the resolution, the overload having considerably negative effect. Generally, TAG sample subjected to preparative TLC should not exceed 80–100 mg, whereas a maximum of 20–30 μg should be applied on an analytical plate. For Ag-TLC and RP-TLC as well as for complex lipids, these limits are considerably lower. Thus, in general, the appropriate sample size depends on the sample composition, the quantitative proportions between the components, and on the resolution required.

12.3 SEPARATION OF LIPIDS BY SILICA GEL TLC

Silica gel TLC is the most widely spread TLC mode. We advise this to be the first step for general view on the sample composition. This technique is employed for lipid groups separation, identification, and isolation, as well as to obtain additional information about unusual components in the sample.[18,19] Beside this, silica gel TLC is widely used for fast and easy verification of the products after chemical transformations of the lipids.

12.3.1 SEPARATION AND IDENTIFICATION OF LIPID GROUPS

As stated earlier (Section 12.1.1) plant lipids comprise a multicomponent mixture of simple and complex lipids. The mixture of simple and complex lipid classes is too complicated to be separated in a single run; therefore the general strategy is to fractionate, first, the simple lipids, and then the complex lipids. This approach was applied, for example, in the analysis of almond[20] and kachnar seed oils[21] as well as pumpkin and sunflower seeds.[22,23] Thus, simple lipids were fractionated by mobile phase of hexane–diethyl ether–formic acid (80:30:1, by volume), then GLs were developed with acetone–acetic acid–water (100:2:1, by volume) and, finally, PL classes were separated with chloroform–methanol–acetic acid–water (170:30:20:7, by volume).[22,23]

Silica gel G is the adsorbent for separation of simple lipids, and silica gel H is used for the resolution of complex lipids. The most often employed mobile phase for separation of simple lipids is hexane-diethyl ether-acetic (or formic) acid in proportions (80:20:2, by volume).[2,21–23] In this case, the lipid groups migrate in the order (in decreasing retention): sterol esters, methyl esters, TAG, free FA, sterols, 1,3-diacylglycerols, 1,2-diacylglycerols, monoacylglycerols, and complex lipids. Another simple and very useful mobile phase for separation of simple lipid groups is hexane–acetone (100:8, by volume), having the advantage to clearly resolve sterol esters, methyl esters, and TAG. The separation of monoacylglycerols and free FA is, however, less satisfactory.[14,16,17,24]

On the other hand, the separation of polar lipid groups is not an easy task. Three-or more component mobile phases are usually employed in one- or two-dimensional TLC systems. For example, two-dimensional development with chloroform–methanol–water (65:25:4, by volume) in the first direction, and with chloroform–methanol–(25%–28%) ammonia solution (13:7:1, by volume) in the second direction were recommended for separation of the polar lipids from plant cells.[25]

The lipid groups, separated by silica gel TLC, are usually identified by comparison with authentic standards applied and separated on the same plate under the same chromatographic conditions. Unfortunately, there is no specific spray reagent for identification of simple glycerolipids, but sterols and sterol esters can be detected successfully by acidic ferric chloride solution.[2] In more complicated cases, when the identification is unreliable, the suspicious zones are isolated from the plate and analyzed in detail by appropriate methods (chemical transformations, spectral techniques, etc.).

12.3.2 ISOLATION OF LIPID GROUPS BY PREPARATIVE SILICA GEL TLC

Silica gel TLC is often used for preparative purposes in the analysis of plant lipids since it is more rapid and sensitive than the column chromatography. As mentioned earlier, the preparative TLC is performed on plates with 1 mm silica gel layer and bigger sample size. Generally, the mobile phases are the same or similar to those used in the analytical mode. After the development, the lipid zones are visualized by appropriate spraying reagent (usually solutions of 2′,7′-dichlorofluorescein or Rhodamin 6G, under UV light at 366 nm) and the corresponding zones are scraped off, transferred to small glass columns, and eluted with either diethyl ether (for simple lipids) or chloroform–methanol–water mixtures (for complex lipids). The isolated lipid groups are then subjected to further analysis.

12.3.3 SEPARATION OF PLs AND OTHER LIPIDS

Despite the great achievements of HPLC, TLC remains to be the cheaper and widely used method for separation of PLs and other complex lipids. Besides its versatility, TLC enables the use of specific spray reagents for identification of the specific functional groups, e.g., choline in phosphatidylcholine.

As mentioned earlier, the resolution of complex lipids is not easily achieved. TLC in a single dimension allows only a rough fractionation, and PLs tend to overlap with GLs in case mobile phases based on chloroform–methanol mixtures are used. Therefore, two-dimensional developments with multicomponent mobile phases are applied for the separation of individual classes of complex lipids. For instance, chloroform–methanol–water (75:25:2.5, by volume) for the first development, and chloroform–methanol–acetic acid–water (80:9:12:2, by volume) in the second direction were used for complete separation of the GLs and PLs: monogalactosyldia-cylglycerol, digalactosyldiacylglycerol, phosphatidylethanolamine, sulfoquinovosyl-diacylglycerol, phosphatidylglycerol, diphosphatidylglycerol, phosphatidylinositol, phosphatidylcholine, and phosphatidylserine in *Arabidopsis thaliana*.[2]

12.3.4 QUANTIFICATION

In spite of the efforts, the quantification of all lipid groups is a problem not perfectly resolved yet. Various approaches were used but none is universal and fully reliable. Probably the most popular method is the charring of separated lipid groups on the plate, followed by determination by scanning densitometry. The charring is per-formed either by spraying the silica gel layer with a suitable reagent (most often, 50% ethanolic sulphuric acid, 20% aqueous ammonium bisulphate, or 3% cupric acetate in 8% orthophosphoric acid) or by treating the layer with vapors of sulphuryl chloride. Then the plate is heated at 180°C and the lipid components appear as dark spots, whose densities are measured. Among the main problematic points in this approach are the reproducibility of charring and the availability of authentic stand-ards for preparation of calibration curves for each lipid group. Instead of scanning densitometer, a liquid scintillation counter can be used, but unfortunately, with similar problems. Since all lipids contain FA, another approach includes transmethy-lation of each lipid group in the presence of suitable internal standard (17:0 is suitable for plant lipids) and quantification of FAME by GC. Unfortunately, this procedure is time consuming and is not convenient for routine analysis.

Different approaches have been proposed for complex lipids, too. For example, GL and PL classes were quantified by spectrophotometric measurement of, respect-ively, the hexose and the phosphorus content.[21]

On the other hand, nondestructive methods for quantification of lipid groups have been developed as well. Among these, fluorometric quantification is the most often used, e.g., the measurement of the fluorescence of an appropriate dye (usually Rhodamine 6G or 1-anilino-8-naphthalene sulphonate) in the presence of lipids. In other cases, the preferred approach was to separate the lipid groups by TLC so as to elute the substances from the adsorbent and determine gravimetrically the respective content.[14,16,17,21,24,26] However, the gravimetric measurement is unreliable when the lipid group is a minor component.

12.4 SEPARATION OF LIPIDS BY SILVER ION TLC

Ag-TLC is a valuable method for the separation of lipid components differing in the number of double bonds, their configuration and, occasionally, the position in the acyl chain.[12,27,28] The resolution is based on the ability of unsaturated FA moieties to form weak reversible charge-transfer complexes with silver ions. As a result, in general, the migration order of any lipid group is determined by the overall number of double bonds in the molecule. Thus, the retention of common FA from plants (chain length of 16–22 carbon atoms, methylene-interrupted double bonds) increases with increasing the number of double bonds from zero to three. Furthermore, *cis*-configuration of the double bond ensures stronger retention than the *trans* configuration. The position of the double bond in the acyl chain influences the retention rather weak. It appeared that the resolution of FA positional isomers could be increased by introducing in the ester moiety appropriate derivatives with higher electron density.[12]

Preparative Ag-TLC is the most often and most widely used technique for rapid, simple, versatile, and unambiguous separation of FA before gas chromatographic (GC) or HPLC determinations. The complementary employment of Ag-TLC and GC combined with mass spectrometry (MS) is, for the present, the most powerful tool for elucidation of FA composition in complex lipid samples. In the analysis of plant lipids, Ag-TLC is used as a single method mainly for the separation and quantification of FA and TAG.

12.4.1 Separation of FA Derivatives

FAs have been subjected to Ag-TLC usually as methyl esters (FAME). Such approach is particularly efficient before subsequent GC analysis. Moreover, there are various methods for methylation or transmethylation, which are simple and easy to perform, providing 100% yield.[2,29]

Most of the FA separations performed by Ag-TLC were based on the number of double bonds. Since the plant FAs contain most often up to three double bonds, their resolution as FAME is considered routine. Mixtures of hexane–diethyl ether in proportions between 90:10 and 80:20 (by volume), or hexane–acetone, 100:3 (by volume) are typical mobile phases for the purpose. The percentage of the silver nitrate solution for impregnation of the layer varies between 0.5%[30] and 30%.[31] For example, preparative Ag-TLC was employed for fractionation of saturated, monoenoic, dienoic, and trienoic FAME in the seed lipids from *Hesperopeuce mertensiana* (Pinaceae), using 5% AgNO$_3$ solution for impregnation and hexane–diethyl ether (80:20, by volume) as mobile phase.[32] Also, preparative plates dynamically impregnated with 10% AgNO$_3$ solution, and mobile phase of hexane–diethyl ether–acetic acid (90:9:1, by volume) were employed for fractionation of FAME with zero to three double bonds in *Saussurea* spp.,[33] *Heteropappus Hispidus*, *Asterothamus centrali-asiaticus*, *Artemisia palustris*,[34] and *Androsace septentrionalis* (Primulaceae)[35] seed oils. Then each of these fractions was subjected to capillary GC for quantitative measurement[33–35] or was further derivatized for GC–MS analysis.[32]

Undoubtedly, one of the most important achievements of Ag-TLC is the clear separation of *trans*- and *cis*-FA isomers. Although the *trans* isomers in plants are very rare and in rather low amounts, their analysis is necessary and is possible in details by GC or GC–MS after preliminary fractionation by Ag-TLC. *Trans*-monoenes are usually completely separated from the saturated and the *cis*-monoenes under the chromatographic conditions established for resolution according to the number of double bonds (Figure 12.1). For example, the *trans*- and *cis*-monoenes in bean leaves were separated on a plate, impregnated with 0.5% AgNO$_3$, using

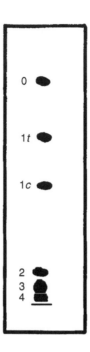

FIGURE 12.1 Schematic presentation of the separation of reference mixture of fatty acid methyl esters by analytical Ag-TLC on silica gel layer impregnated with 0.5% methanolic silver nitrate; mobile phase petroleum ether–acetone, 100:5 (v/v), onefold development in open cylindrical tank; detection by successive treatment for 30 min in each bromine and sulphuryl chloride vapors and heating for 5 min at 180°C. The numbers alongside denote the number of double bonds; *t*—trans, *c*—cis configuration of the double bond.

hexane-acetone (100:2, by volume) as mobile phase.[36] Likewise, *trans*-monoenoic fractionation, namely of *t*3-16:1 and *t*3-18:1, was achieved along with the resolution of FAME with zero to three double bonds in *Heteropappus Hispidus* and *Asterothamus centrali-asiaticus*, using a plate, dynamically impregnated with 10% $AgNO_3$ and developed with hexane–diethyl ether–acetic acid (90:9:1, by volume).[34] Under the same chromatographic conditions was achieved separation also between *t*3,*c*9,*c*12-18:3 and *c*9,*c*12,*c*15-18:3.[34]

Separation of positionally isomeric FAME in plants, such as 6–18:1 and 9–18:1, is also possible. Thus, *c*6-18:1 and *c*9-18:1 in four Umbelliferae seed oils (*Petroselinum sativum*, *Foeniculum vulgare*, *Coriandrum sativum*, and *Pimpinella anisum*) were determined using homemade plates, impregnated by 1% $AgNO_3$ and developed continuously with 5 mL petroleum ether–diethyl ether (100:5, by volume) at −18°C.[30] Substantial improvement in the resolution of these isomers, and what was more important, at room temperature, was achieved, when, prior to analysis, the FA were converted into phenacyl instead of methyl esters.[37] Thus, base-line resolution of 6-, 9-, and 11–18:1 was achieved on both homemade and precoated plates. This approach was applied for the Ag-TLC/densitometric determination simultaneously of saturated, *c*6-, *c*9- and *c*11-18:1, and *c*9,*c*12-18:2 in *Petroselinum* *sativum* and *Pimpinella anisum*, using a homemade plate, impregnated with 1% $AgNO_3$ and developed twice with chloroform–acetone (100:0.25, by volume) at ambient temperature.[38]

12.4.2 Separation of Triacylglycerols

The Ag-TLC separations of TAG are based mainly on the overall degree of unsaturation in the TAG molecule. Out of two species, containing equal number of double bonds, that with a concentration of double bonds in one fatty acyl moiety is held more firmly. A mixture of TAG containing the common range of C16 to C18 FA with up to three double bonds, which are typical for the plants, can be resolved in the following migration order[39,40]:

$$S_3 > S_2M > SM_2 > S_2D > M_3 > SMD > M_2D > SD_2 > S_2T > MD_2 >$$

$$SMT > M_2T > D_3 > SDT > MDT > D_2T > ST_2 > MT_2 > DT_2 > T_3,$$

where S, M, D, and T denote saturated, mono-, di-, and trienoic fatty acyl moieties, respectively. Some changes in the migration of trienoic acid containing TAG may occur depending on the nature of the mobile phase and the quantitative proportions. Generally, separation of TAG with up to six double bonds (D_3) can be performed on a single plate (Figure 12.2). Samples of higher unsaturation require more efforts but distinct resolution of all species up to T_3 has been demonstrated in seed oils from pepper,[41] orange, grapefruit, lemon, and tangerine.[42] Homemade plates impregnated with 0.5% $AgNO_3$ were used and developed continuously in open cylindrical tanks. The mobile phases consisted most often of 4–15 mL petroleum ether–acetone in proportions from 100:4 to 100:8 (by volume), depending on the polarity of the separated TAG classes. Usually, each sample was analyzed on three plates with mobile phases of different composition and volume, in order to achieve complete resolution of, respectively, S_3-SM_2, SM_2-D_3 and the most unsaturated TAG classes. The same technique was employed as a stand-alone method to determine cocoa butter and its replacers,[43] as well as in the development of an approach for precise classification of virgin olive oils.[44] In conjunction with RP-TLC, Ag-TLC was applied to determine the TAG molecular species in numerous plant oils as will be shown later.

Ag-TLC is capable of resolving TAG positional isomers also. This feature is very useful to detect adulteration of natural vegetable oils and fats, since they contain some TAG positional isomers in negligible or zero amounts, while these TAG isomers are usual constituents of interesterified fatty products. For example, the isomeric pairs SMS–SSM, MMS–MSM, SDS-SSD, DSD–DDS, and DMD–DDM (positions *sn*-1 and *sn*-3 are not differentiated; in each pair the firstly denoted isomer migrates first) were separated by continuous development in open tanks with chloroform–methanol mixtures.[45]

12.4.3 Separation of Other Lipids

At present, Ag-TLC is of mostly historical interest in the analysis of lipids other than FA and TAG. It has been almost completely replaced by reversed-phase HPLC.

FIGURE 12.2 Ag-TLC chromatogram of TAG classes in papaya seed oil (right lane) and reference TAG mixture (left lane). Handmade plate with silica gel layer impregnated with 0.5% methanolic silver nitrate; 9 mL mobile phase petroleum ether–acetone, 100:5 (v/v), continuous development in open cylindrical tank; detection by successive treatment for 30 min in each bromine and sulphuryl chloride vapors and heating for 5 min at 180°C. S, M, D, T denote, respectively, saturated, mono-, di-, trienoic fatty acyl residues.

Even at its best days, Ag-TLC was used mainly in preparative mode for biochemical studies of PLs. Some occasional applications for resolution of partial acylglycerols, cholesterol- and wax esters, etc., can be found in the book by Christie[2] and in the review by Nikolova-Damyanova.[27]

12.4.4 QUANTIFICATION

Indirect quantification by Ag-TLC is usually done by extracting the fractions from the adsorbent layer in the presence of internal standard, transmethylating and determination of FAME by GC. Thereby, information is obtained simultaneously on both the composition and the absolute amounts of the lipid fractions.

Undoubtedly, direct, in situ quantification by scanning photodensitometry is more advantageous. Since the lipids do not possess chromogenic groups, they are usually visualized for direct quantification by charring and carbonization on hot plate (180°C–200°C) in the presence of a charring reagent. For this, spraying of the layer with charring reagents (50%–70% aqueous, methanolic, or ethanolic solutions of sulphuric acid) or incorporation of such reagents into the layer have been tested.[12] According to the authors, the best approach to perform uniform carbonization is achieved by treating the developed plate with vapors of sulphuryl chloride in a tightly closed chamber.[5] An important requirement for quantitation is that the charring reagent should react in an equal manner with all components, and especially, if they have different degree of unsaturation. In the first attempts to perform quantitative Ag-TLC, the need for correction coefficients to compensate the different number of double bonds was firmly postulated. Later a simpler solution was suggested by Chobanov et al.[5] Firstly, the silver nitrate content of the adsorbent was significantly lowered by using an immersion procedure with 0.5% methanolic silver nitrate, which made charring easier and simplified the densitometric determination. Secondly, the need for correction factors was eliminated by treating the developed plates initially with bromine vapors, and then with sulphuryl chloride vapors. Although the effect of the bromine is not clear, it ensures high accuracy and precision of the direct densitometric quantifications. More detailed information about the applications of quantitative Ag-TLC can be found in the reviews by Nikolova-Damyanova and Momchilova[9] on FA and by Nikolova-Damyanova on TAG.[8]

12.5 SEPARATION OF LIPIDS BY REVERSED-PHASE TLC

RP-TLC represents a separation technique based on the distribution of lipid compounds between a nonpolar stationary phase and a relatively polar mobile phase (therefore the term "reversed"). Compounds are separated according to their overall polarity expressed by the *Partition Number* (PN), defined as $PN = CN-2n$, where *CN* is the number of carbon atoms in the acyl chain(s) and n is the number of double bonds. The higher the PN, the stronger is the retention of the lipid component on the stationary phase, i.e., the shorter is the migration distance.

In the lipid analysis RP-TLC was introduced by Kaufmann and coworkers when the achievements of paper chromatography were transferred to TLC. To design a nonpolar phase, a thin layer of Kieseguhr (assumed to act as nonactive carrier) was impregnated with tetradecane.[3] In later works by the same group, tetradecane was replaced by impregnation of the Kieselguhr layer by dipping in liquid paraffin in petroleum ether, a change that improved significantly the separation of TAG.[46] Sherma and Bennet demonstrated that commercial precoated C18 silica gel plates are also suitable for separation of lipids.[47] Although the separation was successful, visualization caused problems on these layers and many staining reagents such as iodine/α-cyclodextrin,[48] phosphomolibdic acid,[49] cupric acetate,[47] ethanolic sulphuric acid,[47] and others much less effective[47] were tested. While the technology of modification of the layer was different in laboratory-made and precoated layers, the detection problem persisted and was obviously connected with the long carbon chains attached to the surface. On the other hand, Ord and Bamford demonstrated a modification of the Kaufmann's approach in which a thin layer of silica gel was impregnated by exposure to the vapors of dimethyldichlorosilane (DMCS) for 1 h, thus creating a C1 lipophylic stationary phase. Despite the layer was less lipophylic, the lipid components tested were clearly separated on this phase and were easily visualized by spraying the plate with ethanolic phosphomolibdic acid.[50,51] Finally,

Nikolova-Damyanova and Amidzhin combined and modified the above achievements, carefully examining all elements of the RP-TLC system to offer an approach for RP-TLC of TAG which used Kieselguhr treated with DMCS, a carefully chosen mobile phases and detection by spraying with ethanolic sulphuric acid.[7] The approach combined the efficient resolution achieved on paraffin layers with the easier detection ensured by the C1 layer and allowed correct densitometric quantification of TAG.

12.5.1 SEPARATION OF FA DERIVATIVES

Analytical TLC in all its modes has never been a serious rival to GC in FA analysis. While preparative and analytical Ag-TLC are still of certain interest and often are important complementary techniques to GC of complex FA mixture, especially in the presence of positional and/or configurational isomers, RP-TLC of FA and FAME is of historical interest only and will be very briefly presented here.

Both free FA and FAME were equally good separated on silica gel layer impregnated by treating the plates with DMCS vapors for 1 h.[50] The plates were developed with acetonitrile–acetic acid–water (6:3:1, by volume). The species were separated according to the PN. In order to avoid the formation of critical pairs, reference mixtures of saturated (C12 to C22) and unsaturated FA (C18 with 1 to 3 double bonds) were separated on different plates. The following order of FAME migration can be reconstructed from the results (in order of decreasing retention):

$$22{:}0 > 20{:}0 \sim 22{:}1 > 18{:}0 > 16{:}0 \sim 18{:}1 > 14{:}0 \sim 16{:}1 \sim 18{:}2 > 18{:}3$$

It is evident that components with the same PN values (20:0 and 22:1, PN $= 20$; 14:0, 16:1, and 18:2, PN $= 14$) will remain unresolved when applied as a mixture.

To overcome the formation of unresolved zones, Alvarez et al.[52] employed the idea of Kaufmann and Wessels[48] to combine Ag-TLC and RP-TLC. FA were first tagged with the fluorescence probe 4′-bromomethyl-6,7-dimethoxy coumarin and were then subjected to two-dimensional TLC, the first direction being a RP-TLC on a C18 silica precoated layer, and the second—Ag-TLC. On the first direction the plate was developed with acetonitrile–acetone–methanol–water (60:29:10:10, by volume).

12.5.2 SEPARATION OF TRIACYLGLYCEROLS

RP-TLC of TAG found somehow larger application mainly because of the efforts of Sherma and Fried, and Nikolova-Damyanova and coworkers. If applied as a single analytical method, this technique cannot compete with RP-HPLC because of the much lower resolution power and the formation of numerous critical groups of species with the same PN. It is, however, an efficient aid of Ag-TLC in elucidating the TAG structure of natural lipid mixtures, as will be demonstrated later. The idea was first suggested and performed into complete analytical procedure by Kaufmann and Wessels.[46,48,49,53] Thus, TAG were first separated by preparative

Ag-TLC into fractions with the same unsaturation. Fractions that contained one or two unsaturated moieties were then subjected to RP-TLC on a 20×20 cm Kieslguhr plate impregnated by dipping in 7.5% of liquid paraffin in petroleum ether (bp 35°C–45°C) and the plate was developed with acetone–acetonitrile (8:2 or 7:3, by volume) depending on the overall number of double bonds. Species in each unsaturation class were clearly resolved according to the respective PN which, in the case of TAG, represents the overall number of carbon atoms of the three FA chains and the overall number of double bonds on the one or two unsaturated FA residues. Visualization was performed by spraying the plate with iodine/α-cyclodextrin for the TAG containing at least one unsaturated FA residue and with phosphomolybdic acid for trisaturated TAG. Quantification was performed by a tedious indirect procedure using RP-TLC in preparative mode.[53]

On the other hand, Ord and Bamford demonstrated[51] that impregnation with paraffin is not a must, and that with exception of S_3, S_2U, and SU_2 types TAG (U, unsaturated) can be easily resolved on a layer treated with DMCS, i.e., on a C1 reversed phase. Reference mixtures of TAG were separated with mobile phase water–acetonitrile–ethanol–acetone (8:2:18:72, by volume). Quantification of the separated TAG was not discussed in this or any other paper by the same authors, to the best of our knowledge.

Nikolova-Damyanova and Amidzhin[7] adopted and modified the earlier-mentioned approaches with RP-TLC following an analytical and preparative Ag-TLC separation. The Kieselguhr layer was treated for 6 h with DMCS vapors which allowed the clear separation of TAG containing C20, C22, and C24 FA residues. After this treatment, the plates, when kept in a closed container, could be used for at least 6 months. The mobile phase suggested was acetone–acetonitrile–water by keeping constant the acetone–acetonitrile ratio, 7:3 (by volume), and gradually increasing the water proportion, according to the PN of TAG. Thus, water content in the mobile phase varied in the range from 12 parts by volume for S_2M and S_2D TAG to 20 parts by volume for the SD_2 and ST_2 TAG (see the review of Nikolova-Damyanova[8] for details). Plates were developed twice, each time with fresh 3 mL of the same mobile phase. Detection was performed by spraying the plates with 50% sulphuric acid in ethanol and heating at 220°C to give black spots on nearly white background, suitable for densitometric quantification. Details of the procedure and comments can be found in the paper by Nikolova-Damyanova.[8]

Reference TAG were separated on silica layer chemically modified with C18 chains with a mobile phase of acetonitrile–methyl ethyl ketone–chloroform (50:35:15, by volume). The species were detected by spraying with phosphomolybdic acid.[54]

A limitation of all approaches is the low capacity of the nonpolar phase and, therefore, the sample size is rather small, up to 10 μg. A serious limitation is also that the C1 reversed phase does not allow the separation of trisaturated TAG. Species that differ by the chain length of the unsaturation acyl residues, e.g., 18:1 and 16:1, were not resolved either. No efforts were made so far to examine whether this is disadvantage of the technique or the amount of such TAG in the real sample examined was bellow the detection limit.

12.5.3   SEPARATION OF OTHER LIPIDS

Theoretically all acylglycerols are potential subjects to RP-TLC; however, only few applications have been reported so far. Thus, Sherma and Bennet[47] described the separation and quantification of cholesterol and cholesteryl oleate in a mixture with triolein in blood serum on a C18 precoated plate. The plate was developed with a mobile phase of acetonitrile–methyl ethyl ketone–chloroform–acetic acid (50:30:15:5, by volume) and the separation was acceptable. The migration order was cholesterol, triolein, cholesteryl oleate, i.e., it matched completely those achieved by nonmodified silica gel TLC. Compounds were detected by dipping the plate into 5% solution of phosphomolybdic acid in abs. ethanol and heating at 110°C–120°C for 8–12 min to give blue spots on yellow background.

In another procedure, Kennerly[55] presented the separation of PL classes on two-dimensional RP-TLC and Ag-TLC after conversion to diacylglycerols, phosphorylation, and derivatization of the resulting phosphatidic acids to dimethyl derivatives (see the original paper for more details). Plastic-back silica gel plates were used and part of the layer was impregnated by dipping the plate in 5% octadecylsilane in dry toluene for 5–10 min. In this RP-TLC direction, the plates were developed twice with a mobile phase of acetonitrile–tetrahydrofuran (4:1, by volume). Since separation by Ag-TLC was performed in the first direction, RP-TLC separated species of the same unsaturation and again the resolution was according to the respective PN values. The resolution of species like 18:0/18:2 and 16:0/18:2, and 18:0/18:1, 16:0/18:1, 16:0/16:1 (in order of increasing migration distance) was demonstrated.[55] Since the species were radioactively labeled, detection was performed by radiography for 3 h.

12.5.4   QUANTIFICATION

In part of the cases discussed earlier[46,48,49,55] quantification of separated components was indirect, most often by scraping the spots, methylation or transmethylation and GC.

However, an advantage of TLC is the possibility of direct measurements of the spot area by fulfilling two requirements: (1) tagging the species with either UV-absorbing, fluorescent or radioactive group and (2) finding a stable and easy to use staining reagent and calibrating the measurements. The first approach is suitable for FA or diacylglycerol analysis and successful applications have been reported, i.e., for lipids possessing functional groups that are easily derivatized. Thus, Alvarez et al.[52] used a fluorescent reagent to tag FA and the resulting derivatives were separated and scanned at 352 nm (cut-off filter at 400 nm). The linear detector response was in the interval 300 fmol to 0.13 pmol. Kennerly[55] applied both radiography of radioactively labeled dimethylphosphatidic acids and GC of methyl esters after removing the spots and eluting the substances for identification and quantification.

For other lipids, and mostly for TAG, the second approach is applicable. Sherma and Bennett[47] found phosphomolybdic acid to be the best staining reagent for the commercial C18 precoated TLC plates. To measure the content of cholesterol and cholesteryl oleate, the plate was scanned with fiber optic densitometer in the double beam mode using a wave length of about 500 nm. The calibration graph for cholesteryl oleate was linear in the range of 200–1000 ng. The approach seems suitable to monitor the lipid content in plants but can hardly compete with the much easier and well established procedure of separation lipid groups by nonmodified silica gel TLC followed by charring.

Treating the plate with 50% ethanolic sulphuric acid, heating at 220°C and scanning at 430 nm in reflectance zigzag mode was found the best procedure to quantify TAG species separated on C1 Kieselguhr layer.[56] The measurements were linearly dependant on the concentration in the interval 0.5–3.5 µg per spot. The relative standard deviation did not exceed 10% rel. The accuracy in each studied case was checked by comparing the FA composition once as obtained directly from the GC analysis and as calculated for the TAG composition. An example of species separated as described above in section 12.5.2 is presented on Figure 12.3. The figure demonstrates the clear separation of the TAG species achieved.

Ag-TLC and RP-TLC used in a complimentary way were applied to provide rather full information on the TAG composition of many plant oils as, for example, sunflower,[7] olive,[57] palm kernel,[53] peanut,[58] corn, cotton,[59] soybean, linseed,[60] coffee,[14] sesame,[15] neem,[16] walnut, hazelnut, almond.[17] The data on TAG composition obtained by this combination of TLC techniques is often more detailed than those obtained by RP-HPLC and comparable to that obtainable by combining Ag-HPLC with RP-HPLC.

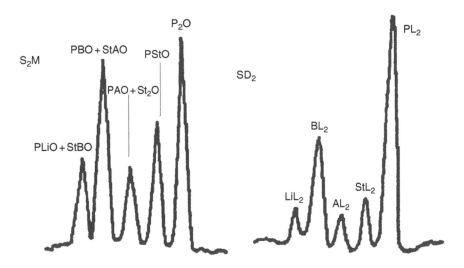

FIGURE 12.3 Densitograms of peanut oil TAG classes isolated by preparative Ag-TLC and separated into TAG species by RP-TLC on Kieselguhr treated with dichloromethylsilane; detection by spraying with 50% ethanolic sulphuric acid and heating at 220°C. Scanning: densitometer Shimadzu CS-930, zigzag reflectance mode at 430 nm, beam slit 1.2 × 1.2 mm. S, M, D—as in Figure 12.2; P—palmitic, St—stearic, A—arachidic, B—behenic, Li—lignoceric, O—oleic, L—linoleic acyl residues.

ACKNOWLEDGMENTS

The partial financial support of NSF of Bulgaria, Contract No. X-1513, is gratefully acknowledged.

REFERENCES

1. Gunstone, F.D., Fatty acid structure, in *The Lipid Handbook*, Gunstone, F.D., Harwood, J.L., and Padley, F.B. (Eds.), Chapman & Hall, London, 1986, chap. 1.
2. Christie, W.W., *Lipid Analysis*, 3rd edn., The Oily Press, Bridgwater, England, 2003.
3. Kaufmann, H.P., Makus, Z., and Das, B., Dunshicht Chromatographie. IV. Trennung der Triglicerides, *Fette-Seifen-Anstrichmittel*, 63, 807, 1961.
4. Privett, O.S. and Blank, M.L., A method for the structural analysis of triglycerides and lecithins, *J. Am. Oil Chem. Soc.*, 40, 70, 1963.
5. Chobanov, D., Tarandjiska, R., and Chobanova, R., Rapid densitometric determination of triglycerides by argentation thin-layer chromatography, *J. Am. Oil Chem. Soc.*, 53, 48, 1976.
6. Chobanov, D., Tarandjiska, R., and Nikolova-Damayanova, B., Quantitation of isomeric fatty acids by argentation TLC, *J. Planar Chromatogr.—Mod. TLC*, 5, 157, 1992.
7. Nikolova-Damyanova, B. and Amidzhin, B., Separation of triglyceride groups by reversed phase thin-layer chromatography on silanized kieselguhr, *J. Chromatogr.*, 446, 283, 1988.
8. Nikolova-Damyanova, B., Quantitative thin-layer chromatography of triacylglycerols. Principles and application, *J. Liq. Chromatogr. & Relat. Technol.*, 22(10), 1513, 1999.
9. Nikolova-Damyanova, B. and Momchilova, S., Silver ion TLC of fatty acids—a survey, *J. Liq. Chromatogr. & Relat. Technol.*, 24, 1447, 2001.
10. Shanta, N.C., Thin layer chromatography—flame ionization detection Iatroscan system, *J. Chromatogr.*, 624, 21, 1992.
11. Fried, B. and Sherma, J., *Thin Layer Chromatography*, Marcel Dekker, New York, 1999.
12. Nikolova-Damyanova, B., Silver ion chromatography of lipids, in *Advances of Lipid Methodology—Five*, Adlof, R.O., Ed., The Oily Press, Bridgwater, England, 2003, chap. 2.
13. Aitzetmuller, K. and Goncalves, L.A.G., Dynamic impregnation of silica stationary phases for the argentation chromatography of lipids, *J. Chromatogr.*, 519, 349, 1990.
14. Nikolova-Damyanova, B., Velikova, R., and Jham, G.N., Lipid classes, fatty acid composition and triacylglycerol molecular species in crude coffee beans harvested in Brazil, *Food Res. Int.*, 31, 479, 1998.
15. Nikolova-Damyanova, B., Velikova, R., and Kuleva, L., Quantitative TLC for determination of triacylglycerol composition of sesame seed oil, *J. Liq. Chromatogr. & Relat. Technol.*, 25(10–11), 1623, 2002.
16. Momchilova, S. et al., Fatty acids, triacylglycerols, and sterols in neem oil (*Azadirachta indica* A. Juss) as determined by a combination of chromatographic and spectral techniques, *J. Liq. Chromatogr. & Relat. Technol.*, 30(1), 11, 2007.
17. Momchilova, S. and Nikolova-Damyanova, B., Quantitative TLC and gas chromatography for determination of the lipid composition of raw and microwaved roasted walnuts, hazelnuts and almonds, *J. Liq. Chromatogr. & Relat. Technol.*, 30, 2267, 2007.
18. Tsevegsuren, N. et al., Occurrence of a novel *cis,cis,cis*-octadeca-3,9,12-trienoic (Z,Z,Z,-octadeca-3,9,12-trienoic) acid in *Chrysanthemum* (*Tanacetum*) *zawadskii* Herb. (Compositae) seed oil, *Lipids*, 38, 573, 2003.

19. Tsevegsuren, N., Aitzetmuller, K., and Vosmann, K., *Geranium sanguineum* (Geraniaceae) seed oil: a new source of petroselinic and vernolic acid, *Lipids*, 39, 571, 2004.

20. Zacheo, G. et al., Changes associated with post-harvest ageing in almond seeds, *Lebensm.-Wiss. u.-Technol.*, 33, 415, 2000.

21. Ramadan, M.F. et al., Characterisation of fatty acids and bioactive compounds of kachnar (*Bauhinia purpurea* L.) seed oil, *Food Chem.*, 98, 359, 2006.

22. Yoshida, H., Hirakawa, Y., and Abe, S., Influence of microwave roasting on positional distribution of fatty acids of triacylglycerols and phospholipids in sunflower seeds (*Halianthus annuus* L.), *Eur. J. Lipid Sci. Technol.*, 103, 201, 2001.

23. Yoshida, H. et al., Microwave roasting effects on the oxidative stability of oils and molecular species of triacylglycerols in the kernels of pumpkin (*Cucurbita* spp.) seeds, *J. Food Comp. Anal.*, 19, 330, 2006.

24. Nguyen, H. and Tarandjiiska, R., Lipid classes, fatty acids and triglycerides in papaya seed oil, *Fat Sci. Technol.*, 97, 20, 1995.

25. Sato, N. and Tsuzuki, M., Isolation and identification of chloroplast lipids, in *Methods in Molecular Biology, Vol. 274: Photosynthesis Research Protocols*, Carpentier, R., Ed., Humana Press, Totowa, NJ, 2004, 149.

26. Yuldasheva, N.K. et al., Lipids from leaves of *Hippophae rhamnoides*, *Chem. Nat. Comp.*, 42, 98, 2006.

27. Nikolova-Damyanova, B., Silver ion chromatography and lipids, in *Advanced in Lipid Methodology—One*, Christie, W.W., Ed., The Oily Press, Ayr, Scotland, 1992, chap. 6.

28. Dobson, G., Christie, W.W., and Nikolova-Damyanova, B., Silver ion chromatography of lipids and fatty acids, *J. Chromatogr. B*, 671, 197, 1995.

29. Christie, W.W., *Gas Chromatography and Lipids*, The Oily Press, Ayr, Scotland, 1989, chap. 4.

30. Nikolova, B.M., Tarandjiska, R.B., and Chobanov, D.G., Determination of petroselinic acid, oleic acid and the major triglyceride groups in some Umbelliferae seed oils, *Compt. rend. Acad. bulg. Sci.*, 38(9), 1231, 1985.

31. Morris, L.J., Separation of higher fatty acids isomers and vinilogues by thin-layer chromatography, *Chem. Ind. (London)*, 1238, 1962.

32. Destaillats, F., Wolff, R.L., and Angers, P., Saturated and unsaturated anteiso-C_{19} acids in the seed lipids from *Hesperopeuce mertensiana* (Pinaceae), *Lipids*, 37, 325, 2002.

33. Tsevegsuren, N., Aitzetmuller, K., and Vosmann, K., Unusual fatty acids in Compositae: γ-linolenic acid in *Saussurea* spp. seed oil, *J. High Resol. Chromatogr.*, 20, 315, 1997.

34. Tsevegsuren, N. et al., Seed oil fatty acids of Mongolian Compositae: the trans-fatty acids of *Heteropappus hispidus*, *Asterothamus centrali-asiaticus* and *Artemisia palustris*, *J. High Resol. Chromatogr.*, 23, 360, 2000.

35. Tsevegsuren, N., Aitzetmuller, K., and Vosmann, K., Isomers of hexadecenoic and hexadecadienoic acids in *Androsace septentrionalis* (Primulaceae) seed oil, *Lipids*, 38, 1173, 2003.

36. Ivanova, A. et al., Gas chromatography and silver ion thin-layer chromatography for the analysis of phospholipid changes in thylakoid membranes from bean leaves, *Compt. rend. Acad. bulg. Sci.*, 50(2), 79, 1997.

37. Nikolova-Damyanova, B., Christie, W.W., and Herslof, B., Improved separation of some positional isomers of monounsaturated fatty acids, as their phenacyl derivatives, by silver-ion thin layer chromatography, *J. Planar Chromatogr.*, 7, 382, 1994.

38. Nikolova-Damyanova, B., Momchilova, S., and Christie, W.W., Determination of petro-selinic, *cis*-vaccenic and oleic acids in some seed oils of the Umbelliferae by silver ion thin layer chromatography of their phenacyl esters, *Phytochem. Anal.*, 7, 136, 1996.

39. Gunstone, F.D. and Padley, F.B., Glyceride studies. III. The component glycerides of five seed oils containing linoleic acid, *J. Am. Oil Chem. Soc.*, 42, 957, 1965.

40. Roehm, J.N. and Privett, O.S., Changes in the structure of soybean triglycerides during maturation, *Lipids*, 5, 353, 1970.

41. Tarandjiiska, R., Nguyen, H., and Lichev, V., Triglyceride composition of tomato, pepper and grape seed oils by argentation thin layer chromatography, *Riv. Ital. Sost. Grasse*, 68, 309, 1991.

42. Tarandjiiska, R. and Nguyen, H., Triglyceride composition of seed oils from Vietnamese citrus fruits, *Riv. Ital. Sost. Grasse*, 66, 99, 1989.

43. Nikolova-Damyanova, B. and Amidzhin, B.S., Comparative analysis of cocoa butter and cocoa butter replacers, *Bulg. Chem. Commun.*, 25, 361, 1992.

44. Tarandjiiska, R.B. and Marekov, I.N., Precise classification of virgin olive oils with various linoleic acid contents based on triacylglycerol analysis, *Anal. Chim. Acta*, 364, 83, 1998.

45. Nikolova-Damyanova, B., Chobanov, D., and Dimov, S., Separation of isomeric triacylglycerols by silver-ion thin-layer chromatography, *J. Liq. Chromatogr.*, 16–18, 3997, 1993.

46. Kaufmann, H.P. and Wessels, H., Die Trennung der Triglyceride durch Kombination der Adsorptions- und der Umkehrphase-Chromatographie, *Fette-Seifen-Anstrichmittel*, 66, 81, 1964.

47. Sherma, J. and Bennet, S., Comparison of reagents for lipid and phospholipid detection and densitometric quantitation on silica gel and C18 reversed phase thin layers, *J. Liq. Chromatogr.*, 6, 1193, 1983.

48. Kaufmann, H.P. and Wessels, H., Die Triglycerid Structuren des Olivenoles und des durch Veresterung von Olivenol-Fettsauren hergestellten Glycerid-Gemiches, *Fette-Seifen-Anstrichmittel*, 68, 249, 1966.

49. Wessels, H. and Rajagopal, N.S., Die DC-Trennung von isomeren und naturlichen Triglycerid-Gemishen, *Fette-Seifen-Anstrichmittel*, 71, 543, 1969.

50. Ord, W.O. and Bamford, P.C., Reversed phase thin layer chromatography of fatty acids and fatty acids methyl esters on silanised kieselgel G, *Chem. Ind. (London)*, 1681, 1966.

51. Ord, W.O. and Bamford, P.C., The thin layer chromatographic separation of fatty acid methyl esters and glycerides according to the chain length and unsaturation, *Chem. Ind. (London)*, 277, 1967.

52. Alvarez, J.G. et al., Determination of free fatty acids by diphasic-two dimensional TLC-fluorescence spectrodensitometry, *J. Liq. Chromatogr.*, 13, 2727, 1990.

53. Wessels, H., Die quantitative DC Analyse von Triglyceriden, *Fette-Seifen-Anstrichmittel*, 75, 478, 1973.

54. Masterson, C., Bernard, F., and Sherma, J., Comparison of mobile phases and detection reagents for the separation of triacylglycerols by silica gel, argentation and reversed phase thin layer chromatography, *J. Liq. Chromatogr.*, 15, 2967, 1992.

55. Kennerly, D.A., Two-dimensional thin-layer chromatographic separation of phospholipids molecular species using plates with both reversed-phase and argentation zones, *J. Chromatogr.*, 454, 425, 1988.

56. Nikolova-Damyanova, B. and Amidzhin, B., Densitometric quantitation of triacylglycerols, *J. Planar Chromatogr.–Mod. TLC*, 4, 397, 1991.

57. Amidzhin, B. and Nikolova-Damyanova, B., Densitometric identification of triglycerides separated by reversed-phase thin-layer chromatography, *J. Chromatogr.*, 446, 259, 1988.

58. Chobanov, D., Amidzhin, B., and Nikolova-Damyanova, B., Direct densitometric determination of triglyceride groups separated by reversed phase thin layer chromatography, *Riv. Ital. Sost. Grasse*, 68, 357, 1991.

59. Tarandjiiska, R. et al., Determination of molecular species of triacylglycerols from highly unsaturated plant oils by successive application of silver ion and reversed phase thin layer chromatography, *J. Liq. Chromatogr.*, 18, 859, 1995.
60. Tarandjiiska, R.B. et al., Determination of triacylglycerol classes and molecular species in seed oils with high content of linoleic and linolenic fatty acids, *J. Sci. Food Agric.*, 72, 403, 1996.

13 Amino Acids

Ravi Bhushan

CONTENTS

13.1   INTRODUCTION

In the process of protein biosynthesis the primary protein amino acids are incorporated one by one as per the instructions contained in the "gene" for the concerned protein. The process is referred to as translation as it entails the translation of the genetic code to the amino acid sequence. Once the protein is synthesized, some modifications, referred to as post-translational modifications, can occur in some amino acid residues of the protein. Thus proteins may contain certain amino acids other than these 20 (coded) amino acids. These are called as secondary protein amino acids. If, however, the modification leads to crosslinking of two amino acids then these are referred to as tertiary protein amino acids.

13.1.1 GENERAL NATURE

In a broad sense, amino acids can be put into two groups—the coded amino acids and the noncoded amino acids.

13.1.1.1 Coded or Primary Protein Amino Acids

There are about 20 α-amino acids (in fact, 19 amino acid and 1 imino acid), which constitute an alphabet for all proteins and differ only in structure of the side chain R. All, except glycine, have L-configuration. These are utilized in the synthesis of peptides and proteins under genetic control, which in turn are vital for life. These α-amino acids are referred to as coded amino acids or primary protein amino acids or proteinogenic amino acids. The amino acids exist as zwitterions at their isoelectric points (pI). The structures, names, abbreviations, and pK_a and pI values for the 20 common amino acids are summarized in Table 13.1.

13.1.1.2 Noncoded or Nonprotein Amino Acids

Nonprotein amino acids (also called nonstandard amino acids) are those amino acids that are neither found in proteins assembled during protein biosynthesis nor generated by post-translational modifications. This may be due to the lack of a specific codon (genetic code) and t-RNA. Hundreds of such amino acids are known and a large number of these are α-amino acids.

13.1.2 OCCURRENCE

Some of the amino acids have also been detected in meteorites, especially in a type known as carbonaceous chondrites. A large number of amino acids are found in nature, most of which are α-amino acids. These occur in free form or as constituents of other biomolecules like, peptides, proteins, coenzymes, hormones, etc.

Microorganisms and plants can produce uncommon amino acids, which can be found in peptidic antibiotics (e.g., nisin or alamethicin). Lanthionine is a sulfide-bridged alanine dimer that is found together with unsaturated amino acids in antibiotics (antibiotic peptides of microbial origin). 1-Aminocyclopropane-1-carboxylic acid (ACC) is a small disubstituted cyclic amino acid and a key intermediate in the production of the plant hormone ethylene.

Amino acids are found in nature in various parts of plants such as stem, leaf, root, and fruits. Many classes of plants containing amino acids are algae, fungi, gymnosperms, angiosperms, etc. Among algae, chlorella is the major source of amino acids, having highest protein content. Chlorella (green algae) is a whole food and is the key to its uniquely varied and concentrated supply of vitamins, minerals, proteins, and other nutrients. Among angiosperms soybean has high protein content. Pteridophytes have been shown to have high amino acid contents and thus are a good nutrient with high protein contents. Foodstuffs that lack essential amino acids are poor sources of protein equivalents, as the body tends to deaminate the amino acids obtained, converting proteins into fat and carbohydrates. The net protein utilization is profoundly affected by the limiting essential amino acid content in the food stuff, e.g., methionine in pulses, lysine in wheat and rice, and lysine and tryptophan in maize. It is therefore a good idea to mix foodstuffs that have different weaknesses in their essential amino acid distributions.

TABLE 13.1
Structures, pka Values, and pI Values of the 20 Common Amino Acids

Structure	Name	Abbreviation	pk_{a_1} (α-Carboxyl)	pk_{a_2} (α-Amino)	pk_{a_3} (α-Side Chain)	pI
Nonpolar side chain						
Glycine structure	Glycine	Gly (G)	2.3	9.6		6.0
Alanine structure	Alanine	Ala (A)	2.3	9.7		6.0
Valine structure	Valine	Val (V)	2.3	9.6		6.0
Leucine structure	Leucine	Leu (L)	2.4	9.6		6.0
Isoleucine structure	Isoleucine	Ilu (I)	2.4	9.7		6.1
Methionine structure	Methionine	Met (M)	2.3	9.2		5.8
Phenylalanine structure	Phenylanine	Phe (F)	1.8	9.1		5.5
Proline structure	Proline	Pro (P)	2.0	10.6		6.3
Neutral polar side chain						
Serine structure	Serine	Ser (S)	2.2	9.2		5.7
Threonine structure	Threonine	Thr (T)	2.6	10.4		6.5

(*continued*)

TABLE 13.1 (continued)
Structures, pka Values, and pl Values of the 20 Common Amino Acids

Structure	Name	Abbreviation	pKa1 (α-Carboxyl)	pKa2 (α-Amino)	pKa3 (α-Side Chain)	pI
Nonpolar side chain						
	Cysteine	Cys (C)	1.7	10.8	8.3 (Sulfhydryl)	5.0
	Asparagine	Asn (N)	2.0	8.8		5.4
	Glutamine	Glu (Q)	2.2	9.1		5.7
	Tyrosine	Tyr (Y)	2.2	9.1	10.1 (Phenolic hydroxyl)	5.7
	Tryptophan	Trp (W)	2.4	9.4		5.9
Charged polar side chain						
	Aspartate	Asp (D)	2.1	9.8	3.9 (β-carboxyl)	3.0
	Glutamate	Glu (E)	2.2	9.7	4.3 (Y-carboxyl)	3.2
	Histidine	His (H)	1.8	9.2	6.0 (Imidazole)	7.6
	Lysine	Lys (K)	2.2	9.0	10.5 (E-amino)	9.8
	Arginine	Arg (R)	2.2	9.0	12.5 (Guanidino)	10.8

The noncoded amino acids are found mostly in plants and microorganisms and arise as intermediates or as the end product of the metabolic pathways. These may also arise in the process of detoxification of the compounds of foreign origin. Some examples are given below:

- N-acetyltyrosine, formed in the metabolism of tyrosine.
- Betaine, involved in glycine biosynthesis and citrulline participates in ornithine cycle.

- Gramicidin-S contains the nonprotein amino acid L-ornithine, which is an important metabolic intermediate in the synthesis of amino acids and other compounds.
- β-alanine, an isomer of alanine, occurs free in nature and as a component of the vitamin pantothenic acid, coenzyme A, and acyl carrier protein.
- Quaternary amine creatine, a derivative of glycine, plays a fundamental role in the energy storage process in vertebrates, where it is phosphorylated and converted to creatine phosphate. The structures of some of the nonprotein amino acids are given in Figure 13.1.

It is difficult to ascribe an obvious direct function to most of these amino acids in an organism. However, most of the function of these amino acids in plants and micro-organisms may be associated with other organisms in the environment.

13.1.2.1 Nonprotein D-Amino Acids

Amino acids having the D-configuration also exist in peptide linkage in nature, but not as components of large protein molecules. Their occurrence appears limited to smaller, cyclic peptides or as components of peptidoglycan (covalent complexes of polypeptides and polysaccharides in which the latter form the bulk of the conjugate) of bacterial cell walls. Some examples are given below:

- Antibiotic gramicidin-S (Figure 13.2) is an example of a peptide that contains two residues of phenylalanine in the D-configuration.
- D-Valine occurs in actinomycin-D, a potent inhibitor of RNA synthesis.
- D-Alanine and D-glutamic acid are found in the peptidoglycan of the cell wall of gram-positive bacteria.

13.1.2.2 Homologs of Protein Amino Acids

In addition to the earlier-mentioned nonprotein amino acids, for which metabolic roles have been described, several hundred other nonprotein amino acids have been detected as natural products. Higher plants are a particularly rich source of these amino acids. In contrast, these do not occur widely, but may be limited to a single species or only a few species within a genus and are usually related to the protein amino acids as homologs or substituted derivatives. The examples are as follows:

- Canavanine—a homolog of arginine, in which the δ-methylene group of arginine is replaced by an oxygen atom, found in alfalfa seeds and acts as a natural defense against insect predators.
- L-azetidine-2-carboxylic acid, a homolog of proline, may account for 50% of the nitrogen present in the rhizome of Solomon's seal (*Polygonatum multiflorum*), and orcylalanine (2,4-dihydroxy-6-methyl phenyl-L-alanine), found in the seed of the corncockle *Agrostemma githago*, may be considered as a substituted phenylalanine or tyrosine (Figure 13.1).

Names and Structures	Names and Structures
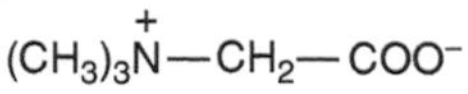 $(CH_3)_3\overset{+}{N}-CH_2-COO^-$ 1. Betaine	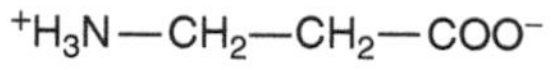 $^+H_3N-CH_2-CH_2-COO^-$ 2. β-alanine
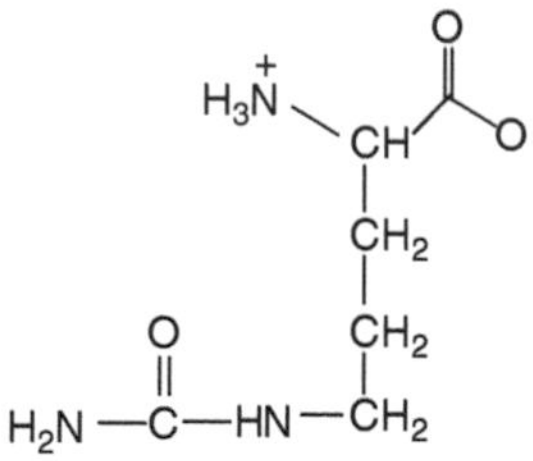 3. Citrulline	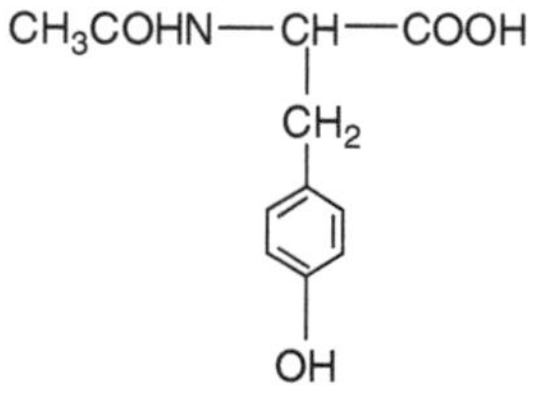 $CH_3COHN-CH-COOH$ with CH_2 and phenol ring with OH 4. N-Acetyltyrosine
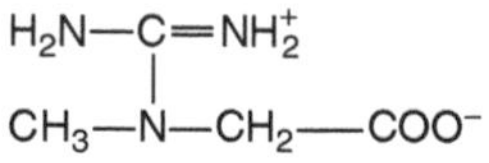 $H_2N-C=NH_2^+$ $CH_3-N-CH_2-COO^-$ 5. Creatine	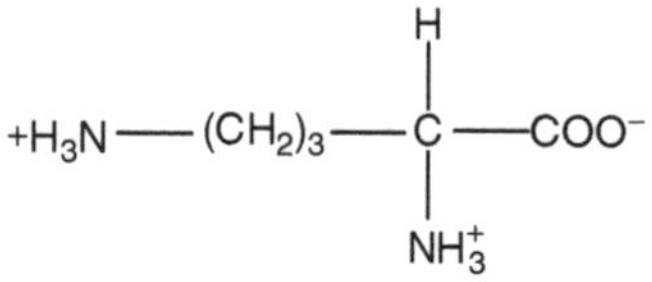 $^+H_3N-(CH_2)_3-\overset{H}{\underset{NH_3^+}{C}}-COO^-$ 6. L-Ornithine
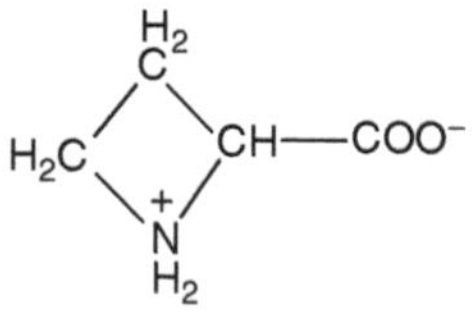 7. Azetidine-2-carboxylic acid	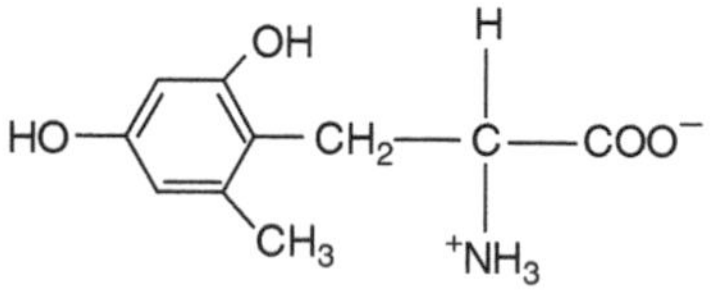 8. Orcyl-L$_8$-alanine
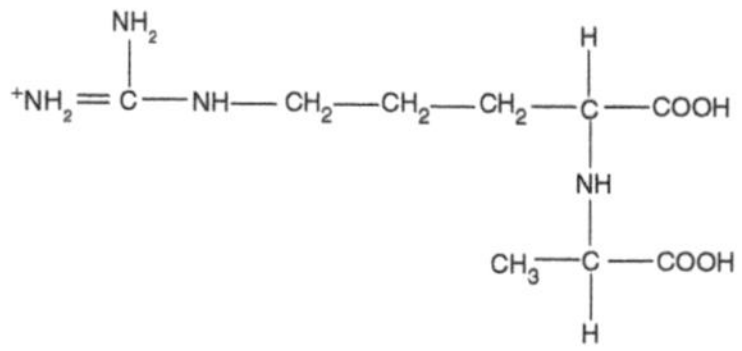 9. Octopine	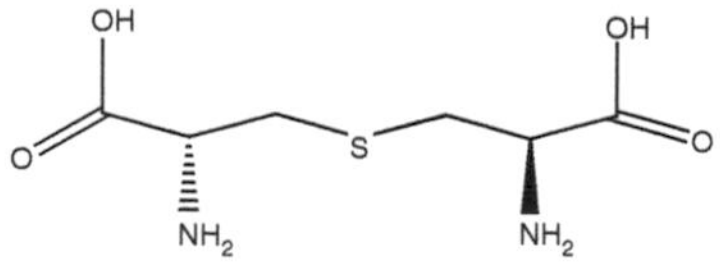 10. Lanthionine

FIGURE 13.1 Names and structures of some nonprotein amino acids.

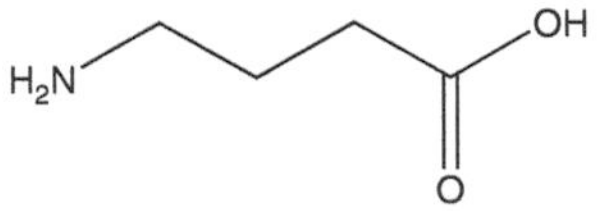

11. γ-Aminobutyric acid

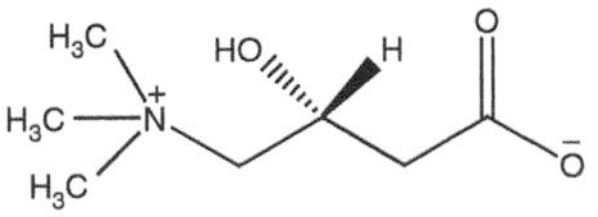

12. Carnitine

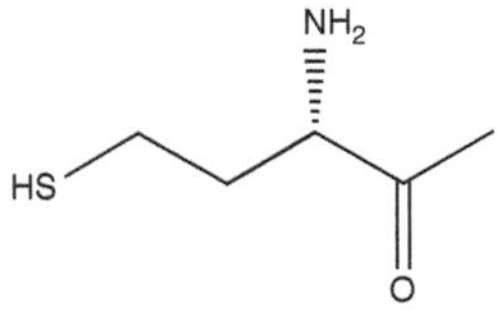

13. Homocysteine

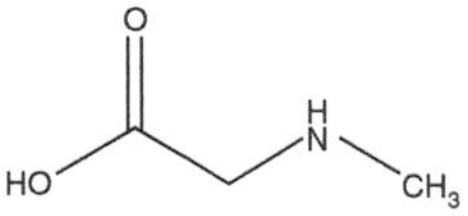

14. Sarcosine

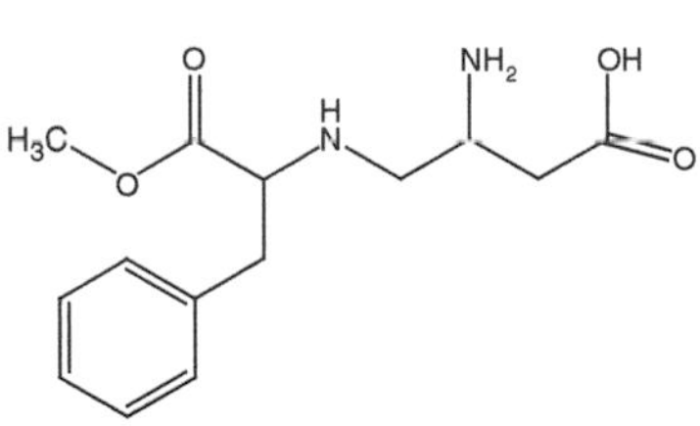

15. Aspartame

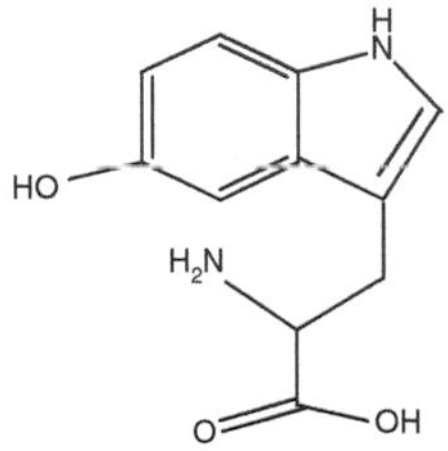

16. Hydroxytryptophan

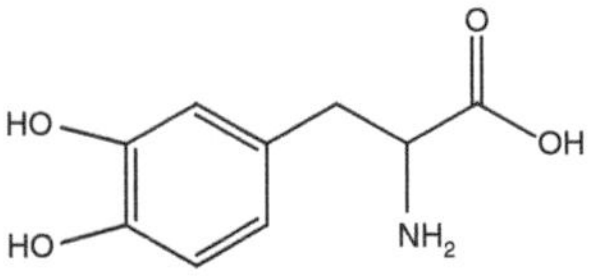

17. Levodopa

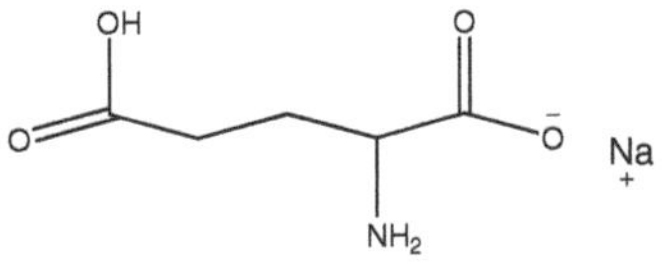

18. Monosodium glutamate

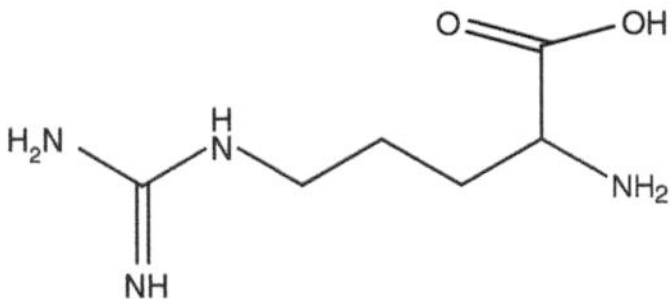

19. Canavanine

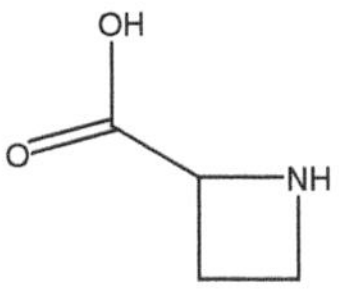

20. Azetidine-2-carboxylic acid

FIGURE 13.1 (continued)

(continued)

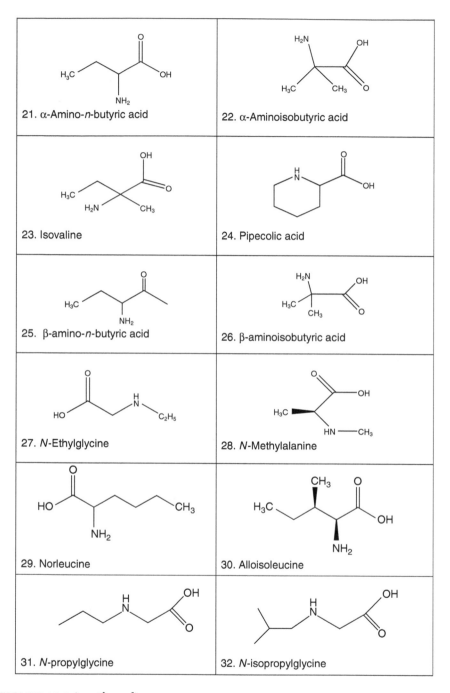

FIGURE 13.1 (continued)

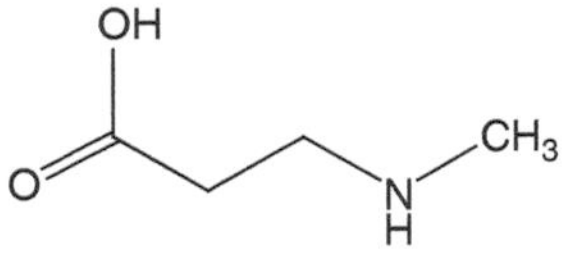

33. *N*-Methyl-β-alanine	34. α,β-Diaminopropionic acid
35. Isoserine	36. α,γ-Diaminobutyric acid
37. α-Hydroxy-γ-aminobutyric acid	

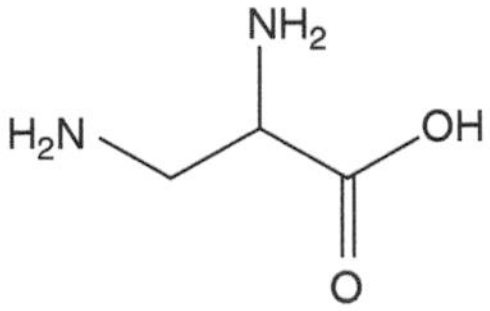

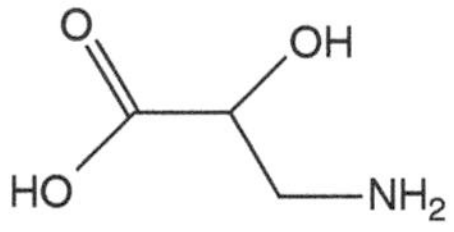

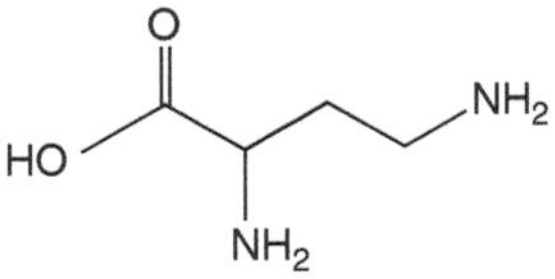

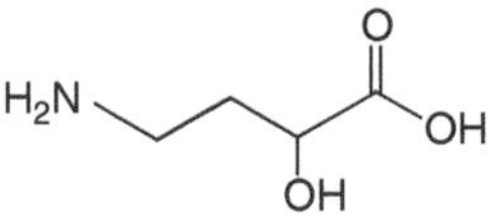

FIGURE 13.1 (continued)

A particularly interesting group of nonprotein amino acids is found among the opines, which are formed in the tumors induced on plants by the crown gall bacterium, *Agrobacterium tumefaciens*. Some opines are conjugates of protein amino acids and α-keto acids. Octopine, for example, is a conjugate of pyruvic acid and arginine (Figure 13.1). It has two chiral centers: one derived from arginine in the L-configuration and the other from pyruvic acid in the D-configuration.

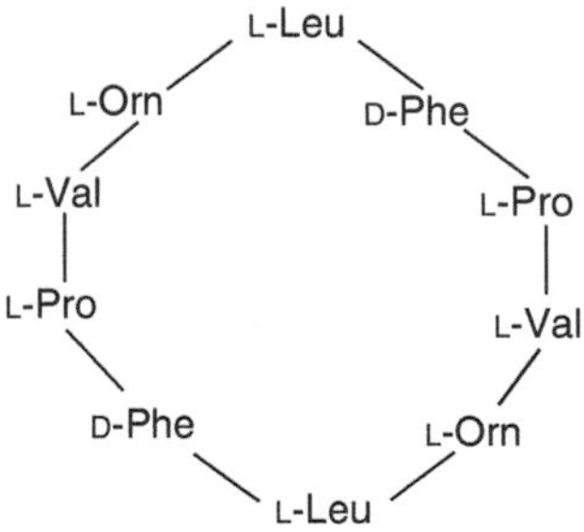

FIGURE 13.2 Structure of gramicidin-S.

13.2 SEPARATION OF AMINO ACIDS

TLC can separate amino acids and their derivatives with high resolution and with many other advantages over other methods. This chapter is not an exhaustive review of the field; however, the methods described include experiences from this laboratory and contributions from other laboratories and can serve as starting points for particular applications.

Modern methods for separation and quantitation of free amino acids either before or after protein hydrolysis include TLC, ion-exchange chromatography, HPLC, GC, and capillary electrophoresis. Chemical derivatization of amino acids may be required to change them into forms amenable to separation by the various chromatographic methods or to create derivatives with properties, such as fluorescence, that improve their detection. LC is currently the most widely used analytical technique, although there is a need for collaborative testing of methods available. Newer developments in chromatographic methodology and detector technology have reduced sample and reagent requirements and improved identification, resolution, and sensitivity of amino acid analyses of food samples.

13.2.1 PREPARATION OF TEST MATERIALS

The analysis of amino acids required either for the determination of the composition of proteins or for the investigation of certain fluids or extracts derived from animals or plants requires the removal of materials such as peptides, proteins, carbohydrates, urea, salts, and lipids from them by specific operations, and proteins and peptides need to be hydrolyzed.

13.2.1.1 Removal of Macromolecules

Various kinds of precipitating agents are used to remove macromolecules. A comparison of deproteinizing methods has shown that in certain cases a considerable loss of amino acids must be taken into account.

13.2.1.2 Enrichment of N-Methylated Amino Acids from Biological Fluids and Protein Hydrolysates

An aliquot of the fluid or hydrolysate is adjusted to 0.1 M in HCl, treated with an equal volume of aqueous Reinecke's salt (2%), and left in the refrigerator overnight. The precipitate is then filtered off and dissolved in acetone. The solution is centrifuged, and the supernatant is mixed with an equal volume of water and extracted several times with ether. The lower layer containing water, acetone, and ether is evaporated to dryness, and the residue is dissolved in aqueous 10% propan-2-ol solution for use in TLC.

13.2.1.3 Hydrolysis of Proteins

Proteins are hydrolyzed to amino acids by treatment with acid, alkali, or enzymes. Each method has certain disadvantages as shown in Table 13.2. The most commonly used methods for total hydrolysis are described below.

TABLE 13.2
Disadvantages of Methods of Hydrolysis of Proteins

Method of Hydrolysis	Disadvantages
Acid hydrolysis	
8 N H_2SO_4 at 110°C for 18 h	1. Tryptophan is destroyed; Ser and Thr are partially destroyed.
	2. Presence of carbohydrates leads to formation of black material, humin.
6 M HCl at 110°C for 18 h	1. Trp, Asn, Gln destroyed; Ser, Thr, Tyr partially lost.
	2. Cys and Met are either partially destroyed or oxidized to cysteic acid and Met-S, S-dioxide, respectively.
Alkali hydrolysis	
Ba(OH)$_2$	1. Partial or complete destruction of Arg, Cys, Ser, Thr.
	2. Causes racemization and some deamination.
NaOH [1] or LiOH [2]	LiOH is reported to be best [2] for tryptophan determination.
Enzymes pepsin, trypsin, papain, chymotrypsin	1. Each enzyme is generaly specific for a particular peptide bond.
	2. May produce hydrolysis of enzymes, which would interfere with amino acid analysis.

13.2.2 METHOD FOR ACID HYDROLYSIS

A sample (50–100 mg) of air-dried or lyophilized protein is weighed into a tube, and 6 M HCl (1 mL for 5 mg of protein) is added. The tube is evacuated using a vacuum desiccator [3], sealed, and placed in a circulating air oven at 110°C with good temperature control. After hydrolysis for the appropriate period of time (24, 48, or 71 h), it is centrifuged gently. Then the tubes are cracked open and the HCl is removed as quickly as possible using a stream of N_2. The HCl can alternatively be neutralized by adding solid Ba(OH)$_2$ (up to pH 7) and removing white $BaSO_4$ by filtration or centrifugation. The clear hydrolysate may be frozen in an acetone–solid CO_2 bath, placed in a vacuum desiccator over NaOH pr KOH, and lyophilized. However, for clear hydrolysis of proteins for amino acid analysis one may consult Light and Smith [4], Moore and Stein, Savoy et al. [5], or Perham [6].

13.2.3 METHOD FOR SULFUR-CONTAINING AMINO ACIDS

Moore et al. [1] determined cysteine and cystine as cysteic acid by performic acid oxidation. However, methionine can also be determined as methionine S,S-dioxide. Performic acid is prepared by adding H_2O_2 (1 mL, 30%) to formic acid (9 mL, 88%) and allowing the mixture to stand at room temperature for 1 h. It is then cooled to 0°C. Performic acid (2 mL) is added to the protein (containing about 0.1 mg cystine) in a Pyrex tube, which is then allowed to stand a 0°C for 4 h for soluble proteins or overnight for proteins that do not dissolve. Then HBr (0.30 mL, 48%) is added with swirling, the mixture is evaporated to dryness at 40°C using a rotary evaporator, and the protein is hydrolyzed in vacuo with HCl (3 mL, 6 M) at 110°C for 18 h. The hydrolysate is treated as mentioned earlier, before analysis. A rapid method of protein hydrolysis by microwave irradiation has been described by Chiou and Wang [2]. They describe a design for a reusable Teflon–Pyrex tube for fast inert gas flushing under microwave irradiation. Results have been compared with those of conventional heating methods in terms of destruction or degradation of certain labile amino acids and their recoveries depending upon hydrolysis time by microwave irradiation.

13.3 CHROMATOGRAPHIC TECHNIQUES

13.3.1 ADSORBENTS AND THIN LAYERS

A variety of adsorbents such as silica gel, alumina, polyamide, and cellulose are available commercially for use in TLC work. Alumina and Silica gel are used with or without suitable binder such as gypsum or starch. Mixtures of two adsorbents or adsorbents impregnated with certain reagents such as 8-hydroxyquinoline and various metal ions have also been used successfully to improve resolution. By far the most thin-layer work has been done on layers prepared from water-based slurries of the adsorbents. Even with the same amount and type of binder, the amount of water that is used for a given slurry varies among the different brands of adsorbents. For example, in the case of cellulose the amount of powder to be mixed with water varies depending on the supplier; Serva, Camag, and Whatman have recommended the use of 60–80, 65, and 25 mL of water, respectively, for 10 g of their cellulose powders. These slurries may be prepared by shaking a stoppered flask or by homogenizing for a few seconds with a mechanical mixer. On the other hand, for the preparation of an aluminum oxide slurry (acid, basic, or neutral) it is recommended that one may use 35 g of aluminum oxide with 30 mL of water with spreading equipment, and 6 g of adsorbent in 15 mL of ethanol–water (9:1) mixture for pouring directly onto the plate without a spreading apparatus. Korzun et al. [7] used slurry of 120 g of alumina G in 110 mL of water to prepare 1 mm thick layers for preparative TLC. Cellulose powders in general contain impurities that are soluble in water or organic solvents, which should be removed by washing the cellulose powder several times with acetic acid (0.1 M), methanol, and acetone and drying before use. The layer is made by "turbo mixing" MN (Macherey-Nagel) cellulose-300 (15 g) for 10 min in distilled water (90 mL) and then spreading it to give a 0.25 mm thick layer. The layers are left overnight to dry.

The cellulose layers have several advantages; they are stable, they can be used with various specific reagents, and they give reproducible data. They are recommended particularly for quantitative evaluation by densitometry. The drawbacks of cellulose layers are that corrosive reagents cannot be used and the sensitivities of detection reactions of certain amino acids are lower than on silica gel layers.

The best known and most widely used adsorbents for TLC purposes are from Merck, but products of other firms can be used satisfactorily. Precoated plates are widely known, and an increasing number of workers use them for the investigation of amino acids and their derivatives. For example, ready-made cellulose layers from Macherey-Nagel (Germany) containing MN cellulose-300 in appropriately bound form are one of the best-known products. Chiralplate from the same firm, for the separation of enantiomers of amino acids and their various derivatives, contains a coating of reversed-phase silica gel impregnated with a chiral selector and copper ions. Use of home made thin-layer plates has been found to be more convenient in our laboratory, and it is recommended that one not change the brand of adsorbent during a particular set of experiments.

13.3.2 PREPARATION OF THIN PLATES

Slurry of silica gel G (50 g) in distilled water (100 mL) is prepared and spread with the help of a Stahl-type applicator on five glass plates of 20×20 cm to obtain 0.5 mm thick layers. The plates are allowed to set properly at room temperature and are then dried (activated) in an oven at an appropriate temperature (60°C–90°C) for 6 h or overnight. The plates are cooled to room temperature before the samples are applied.

The same method has been used successfully to prepare plates with silica gel, silica gel-polyamide, and cellulose and with these adsorbents impregnated with a variety of reagents including HDEHP, TOPO, 8-hydroxyquinoline, dibenzoyl methane, and several metal salts [8–25]. Brucine [26], (−)-quinine [27], tartaric acid and L-histidine [28], and pharmaceutical industrial waste {(1R,3R,5R)-2-azabicyclo-[3,3,0]octan-3-carboxylic acid} [29] were also mixed in slurries of silica gel as impregnating reagents to resolve enantiomers of amino acids and their PTH derivatives. Mixtures of H_2O–EtOH or other organic solvents can also be used depending on the nature of the impregnating reagents. Citrate [30] and phosphate [31] buffers have also been used for preparing the slurry of silica gel in place of water. It is customary to use 0.25 or 0.50 mm thick layers in activated form, but for preparative purposes 1–2 mm thick layers are best [7].

13.3.3 DEVELOPMENT OF CHROMATOGRAMS

Amino acids are generally soluble in water, but some are less soluble than others. Alcoholic 0.5 M or 0.1% HCl should be used to prepare solutions of amino acids that are sparingly soluble in water. Standard solutions of amino acids are prepared in a suitable solvent such as 70% EtOH or 0.1 N HCl in 95% ethanol.

These solutions are applied generally as tight spots, 1–2 cm from the bottom of each layer, by using a glass capillary or Hamilton syringe. In the beginning, a higher concentration, (e.g., 500 ng) or more is applied; however, the detection limits are determined for the system developed by repeating the experiment with lower concentrations.

The chromatograms are generally developed in rectangular glass chambers, which should be paper-lined for good chamber saturation and pre-equilibrated for 20–30 min with solvent prior to placing the plates inside. The time taken depends on several factors such as the nature of the adsorbent, the solvent system, and the temperature. The developed chromatograms are dried in a chromatography oven between 60°C and 100°C and the cooled plates are usually sprayed with ninhydrin reagent. Heating at 90°C–100°C for 5–10 min produces blue to purple zones of all amino acids except proline (yellow spot).

The same method is adopted for both one- and two-dimensional modes. The locating reagent is used after the second run, and a more polar solvent is generally used for developing the chromatogram in the second dimension.

13.3.4 DETECTION OF AMINO ACIDS ON THIN LAYER CHROMATOGRAMS

After drying the chromatogram it may be viewed under ultraviolet (UV) light if the absorbent had a fluorescent indicator or the compounds such as dansyl amino acids-fluoresce. Solvent fronts may be seen, which indicate irregularity of solvent flow. Ninhydrin is the most commonly used reagent for the detection of amino acids, and a very large number of ninhydrin reagent compositions have been reported in the literature. The reagent may be made slightly acidic with a weak acid following the use of an alkaline solvent and vice versa. Constancy of color formed may be attained by the addition of complex-forming cations (Cu^{2+}, Cd^{2+}, or Ca^{2+}), and specific colors may be produced by the addition of bases such as collidine or benzylamine. Some of the ninhydrin compositions and their applications are described as follows:

1. A solution of ninhydrin (0.2% in acetone) is prepared with the addition of a few drops of collidine or glacial acetic acid. The chromatogram is dipped or sprayed with it and dried at 60°C for about 20 min or at 100°C for 5–10 min. Excessive heating causes a dark background. The sensitivity limit is 1 μg. Most amino acids give a violet color, whereas aspartic acid (Asp) gives bluish-red, and praline (Pro) and hydroxyproline (Hyp) give yellow. (See Table 13.1 for abbreviations for 20 common amino acids.)

2. Ninhydrin (0.3 g) in n-butanol (100 mL) containing acetic acid (3 mL) is sprayed on a dried, solvent-free layer, which is then heated for 30 min at 60°C or for 10 min at 110°C [32]. Detection limits range from 0.001 μg for alanine (Ala) to 0.1 μg for proline and aspartic acid [32].

3. Ninhydrin (0.3 g), glacial acetic acid (20 mL), and collidine (5 mL) are made up to 100 mL with ethanol [33] or ninhydrin (0.1% w/v) in acetone–glacial acetic acid–collidine (100:30:4) [34].

4. A solution of cadmium acetate (0.5 g) in water (50 mL) and glacial acetic acid (10 mL) is made up to 500 mL with acetone. Portions of this solution are taken, and solid ninhydrin is added to give a final concentration of 0.2% g. The chromatogram is sprayed and heated at 60°C for 15 min. It is interesting to note the results immediately and again after 24 h, at room temperature [35]. Alternatively, the layer is impregnated thoroughly with the reagent and the colors are allowed to develop in the dark at room temperature for 24 h [36]. This reagent gives permanent colors, mainly red but yellow for praline. Sensitivity is 0.5 nmol.

5. Ninhydrin (1.0 g) in absolute ethanol (700 mL), 2,4,6-collidine (29 mL), and acetic acid (210 mL) has been used for spraying on solvent-free cellulose layers. The chromatogram is then dried for 20 min at 90°C.

6. Development of ion-exchange resin layers in ninhydrin (1%) in acetone containing collidine (10%) at room temperature for 24 h or at 70°C for 10 min has also been recommended.

7. A spray of ninhydrin (0.1% or 0.2%) in acetone on chromatograms followed by heating at 60°C or 90°C for 10–20 min has also been used [8,15,17–20].

8. Polychromatic reagents: Moffat and Lytle [37] developed a polychromatic ninhydrin reagent. It consisted of (a) ninhydrin (0.2%) in ethanol (50 mL) + acetic acid (10 mL) + 2,4,6-collidine (2 mL) and (b) a solution of copper nitrate (1.0%) in absolute ethanol. The two solutions are mixed in a ratio of 50:3 before use. Krauss and Reinbothe [38] replaced ethanol by methanol and also achieved polychromatic amino acid detection by joint application of ninhydrin and primary, secondary, or tertiary amines. The layers were first sprayed with diethylamine, dried for 3 min at 110°C, when the spots of amino acids appeared on a pale blue background. Use of ninhydrin (0.27 g), isatin (0.13 g), and triethylamine (2 mL) in methanol (100 mL) gave spots of amino acids on a yellow background.

Several other reactions have also been used for the detection of specific amino acids (Table 13.2). Oxalic acid (ethanolic 1.25%), dithiooxamide (ethanolic saturated), or dithizone followed by ninhydrin was used to aid identification and detect amino acids with various specific colors [39]. Acetyl acetone-formaldehyde detected amino acids as yellow spots under UV [40]. By using isatin–ninhydrin (5:2) in aqueous butanol or by modifying ninhydrin detection reagent by addition of D-camphor [41] and various acids [42], identification of amino acids was improved. Spraying of layers with 1,3-indanedione or 0-mercaptobenzoic acid prior to ninhydrin improved sensitivity limits and color differentiation in amino acids detection [43]. 3,5-Dinitrobenzoul of timing was achieved by coupling pneumatic nebulization with optical fiber-based detection in a chemiluminescence TLC system to detect dansyl amino acids. A new spray reagent, p-dichlorodicyanobenzoquinone, detected amino acids with 0.1–1 μg detection limits and produced various distinguishable colors that facilitate identification [44]. Chromatograms sprayed with ninhydrin (0.03 g ninhydrin in 100 mL of n-butanol plus 3 mL of glacial acetic acid), air-dried for 5 s, resprayed, and heated in an oven at 110°C for 10 min gave the best sensitivity, stability, and color differentiation in comparison with different recipes of ninhydrin and fluorescamime sprays [45].

13.4 TLC ANALYSIS

A brief account of TLC analysis of amino acids [46–72] extracted from various plants and their parts in either free form or from the hydrolysates of proteins extracted is given in Table 13.3.

13.4.1 PLANT EXTRACTS

Separation and estimation of amino acids in crude plant extracts was carried out by thin-layer electrophoresis and chromatography [73]. Extraction was done with MeOH–CHCl₃–H₂O mixture that enabled subsequent removal of pigments and lipids in the CHCl₃ layer. Components of the aqueous layer were separated by thin-layer high-voltage electrophoresis without further purification. Use of a linear origin permitted higher loadings and minimized tailing. Amino acids were revealed with ninhydrin or by autoradiography. Cellulose acetate was then used to bind the

TABLE 13.3

TLC Analysis of Amino Acids Extracted from Various Plants

S. No.	Name of Plant	Part of the Plant	Amino Acids Analyzed	Remarks	Reference
1	*Crithmum maritimum* L.	Aerial parts	Leucine, phenylalanine, valine, tyrosine, proline, alanine, serine, lysine, and histidine	Qualitative analysis	[46]
2	*Fagus silvatica* (beech)	Whole plant extract	Free amino acids		[47]
3	*Allium sativum* L (garlic)	Aqueous extracts from powdered and fresh bulbs	Free amino acids	• 0.25 mm silica gel G • Detection with ninhydrin • Mobile phase: butanol-*n*-propanol-glacial acetic acid-water (3:1:1, v/v)	[48]
4	Equisetum	Extract	Free amino acids		[49]
5	*Coffea arabica* and *Camellia sinensis*	Leaves of tea seedlings, fruits and mature plants	Free amino acids	2D-TLC	[50]
			Pipecolic acid, hydroxyproline, proline and hydroxyproline	silica gel 60	
6	Medicinal plants		Free amino acids	• Extraction with 1% hydrochloric acid • Cellulose plates	[51]
7	*Phaseolus lunatus*	Leaves collected at pre-infectional, and late infectional stages	Cysteine, arginine, histidine		[52]
8	Sugar cane	Tissue	18 amino acids, 5 amino acids were identified and quantified densitometrically	• Hydrolysis of protein with conc. HCl • Quantitative HPTLC • Derivatization with dansyl chloride • Mobile phase: 5% EDTA–BuOH–Et$_2$O (5:10:35, v/v)	[53]

#					Ref.
9	*Leucaena glauca*	Leaves and immature, unripe fruits	Glutamic acid, aspartic acid, leucine, and isoleucine	Quantitative determination	[54]
10	*Caucalis platycarpos* L.	Dry decoction of the plant	7 amino acids	Ion-exchange chromatography	[55]
11	*Capsella bursa-pastoris*	Aerial parts	22 amino acids (e.g., proline, α- and γ-aminobutyric acid, α-aminoadipic acid)		[56]
12		Rape pollens		• Silica gel G TLC • Detection with ninhydrin • Mobile phase:n-PrOH-H2O (7:3, v/v)	[57]
13	*Crotalaria sessiliflora* L.	Dried whole plant	3.150% aspartic acid, 2.230% glutamic acid, and 1.060% alanine	Quantitative determination	[58]
14	*Vitis vinifera*		Arginine and praline		[59]
15	*Avena sativa*	Plant tissue	3 Amino acid conjugates of chloro-s-triazine		[60]
16	Pine, ash, sycamore, willow, chenopod, ragweed, sorrel, Johnson grass, and rye grass	Pollen	Protein and free amino acid		[61]
17	*Nicotiana tabacum*		L-tyrosine and L-phenylalanine		[62]
18	Vegetable samples		Free amino acids	• Phenylthiohydantoin derivatives Quantitative Determination	[63]
19	Blue-green alga A. *Fertilissima*		Free amino acids		[64]
20	Plant extracts		Methionine and cysteine	• Ion-exchange TLC • Buffer pH 2.5	[65]
21	*Arnica Montana*	Root, rhizomes	Glutamic acid, arginine, proline	Qualitative and quantitative determination	[66]

(continued)

TABLE 13.3 (continued)
TLC Analysis of Amino Acids Extracted from Various Plants

S. No.	Name of Plant	Part of the Plant	Amino Acids Analyzed	Remarks	Reference
22	*Helianthella quinquenervis*	Involucral bracts	Total of 24 amino acids, nonprotein (5), 7–10 nonprotein amino acids		[67]
23	*Hypericum hirsutum, H. montanum*	Flowering tops	16 amino acids	TLC on cellulose	[68]
24	*Viola arvensis* and *Melilotus altissimus*		Aspartic, glutamic acids, threonine, and phenylalanine	Qualitative and quantitative determination	[69]
25	*Parnassia palustris*	Dry plant parts	Glutamic acid, aspartic acid, arginine, tryptophan, histidine, valine, and alanine	Quantitative determination	[70]
26	*Valeriana officinalis* L.	Leaves, stems, flowers, and roots	Glutamine, arginine, alanine, and γ-aminobutyric acid	• Cellulose and silica gel plates • Solvent system: n-butanol-acetone–glacial acetic acid–water (35:35:10:20, v/v)	[71]
27	*Spartium junceum* L.	Stem, leaves, and flowers	Amino acids		[72]

adsorbent particles into a film, so that individual regions of the chromatogram could be cut and lifted out as a coherent piece.

A new method for detecting N-acyl amino acid (substrate of aminoacylase) in vivo was studied [74]. In the acidic portion of nonprotein part of onion seed N-acyl amino acid could be detected. Six kinds of N-acetyl amino acid were developed in a thin silica gel and hydrolyzed by a spray of 6 N HCl at 100°C for 10 min.

TLC was employed for determination of free amino acids of local, air-borne allergenic pollen extracts. A total of 17 local, aqeous, lyophilized pollen extracts were investigated [75] for their free amino acid content, using 2D TLC technique. Quantitative microanalytical TLC of amino acids (along with sugars, oil-soluble pigments and cofactors) from plants of *Arabidopsis thaliana* was carried out [76] using two different plates and two different developers: BuOH–HOAc–H$_2$O (70:15:15) on one plate and PhOH–EtOH–H$_2$O (23:42:35) on the other plate for separation of amino acids; the chromatographic fraction I was with C$_6$H$_6$–cyclohex-ane–acetone–MeOH (25:25:3:1) on a thin layer of alumina, and chromatographic fraction II was on a thin layer of diphenylpropane polycarbonate (K-511-v) buffered at pH 6.8 by dipping in a phosphate buffer.

TLC of amino acids was studied on silica gel, Cellulose MN 300-silica gel (10:4), and cellulose plates and on silica gel-coated chromatogram sheets [77] by developing twice in the first dimension with BuOH–Me$_2$CO–NH$_3$–H$_2$O (10:10:5:2) and twice in the second dimension with *iso*-PrOH–HCO$_2$H–H$_2$O (20:1:5) on cellu-lose plates, when 24 amino acids were separated. The spots were detected by spraying with 0.2% ninhydrin in EtOH. The method gives good separation of free amino acids in purified plant extracts.

13.4.2 SOYBEAN AND KIDNEY BEAN

Changes in free amino acids in pods, seeds, leaves, and stems of the soybean plant were investigated by TLC during formation of soybean seed [78]. Asparagine was widely distributed in a relatively large amount in all the parts of soybean plants, especially in pods and stems. In leaves, acidic amino acids were distributed predominantly in upper leaves and aliphatic neutral amino acids in lower leaves. γ-Amino butyric acid was detected from all the parts analyzed. β-Alanine was found in leaves and stems near leaves but α-aminobutyric acid was found only in leaves. γ-Glutamylphenylalanine and γ-glutamyltyrosine were found only in ripe seeds. Pipecholic acid which has not yet been found in any other part of the soybean plant was found in unripe pods.

Free amino acids, 16–19 in number, were determined in cotyledon, stem, root, leaves, and seed of kidney bean plant at approximately 1 week interval from seeding to harvest by TLC [79]. Asparagine was the most abundant among amino acids determined in every part of the bean plant except around the flowering stage. The amount of this amide showed a characteristic decrease during germination and seed-ripening. γ-Aminobutyric acid was detected in every part of the plant, especially abundant in the leaves (0.2%–3.0%). β-Alanine was detected in trace amounts in every part of the plant. γ-Glutamylleucine and γ-glutamyl-S-methylcysteine were detected only in the germinating cotyledons and ripening seeds.

13.4.3 TLC Combined with Other Techniques

Free amino acids were identified in plant materials [80] by a combination of double electrophoresis (in the first dimension) and TLC (in the second dimension. The first electrophoresis run on cellulose plates was performed with the solvent HCO_2H–HOAc and the second run with pyridine buffer. Ascending chromatography was then performed with the solvent BuOH–MeOH–MeCOEt–(Et)$_2$NH–H$_2$O. The method gave a very good separation of basic and acidic amino acids.

Combined one- and two-dimensional chromatography, gas chromatography, and mass spectrometry, and an improved one-dimensional TLC on a cation exchanger were used to separate and determine 15 amino acids in three varieties of grapes during ripening [81]. Samples were taken from pot- and field-grown grapes and from those receiving N fertilizer. Proline and arginine accumulated in the greatest amounts in the berry during ripening and in response to N nutrition. Amino-N concentrations were greater in the resistant cultivar than in the susceptible cultivar, and greater in shoots than in roots. TLC revealed that, in both cultivars, a greater number of amino acids occurred in shoots than in roots at all stages of plant growth [82]. Large amounts of alanine were detected in roots and shoots of both cultivars. Concentrations of phenylalanine, histidine, aspartic acid, proline, and tyrosine were larger in roots of the resistant cultivar. A measurable concentration of cystine (11.24 μg/g) was detected only in the shoot of the resistant variety. Experiments with *F. udum* culture filtrates suggested that S-containing amino acids, e.g., cystine, can preferentially chelate Fe^{3+}, rendering it unavailable for activation of wilt toxins produced by the pathogen.

Free amino acids in the foliage and phloem exudate of several varieties of palm were detected [83]. The fluorescent label 5-dimethylaminonaphthalene-1-sulfonyl

TABLE 13.4
Free Amino Acid Analysis of Phloem Exudates of Palm Varieties

Palm Species	Ala/Dns-NH$_2$	Asn	Asp	Arg	Glu
Arecastrum romanzoffianum	++	+	++	−	++
Carpentaria acuminate	++	±	++	±	++
Chrysalidorcarpus hutescens	++	+	+	−	+
Cocos nucifera (yellow Malayan dwarf)	++	±	+	−	+
Phoenix dactylifera	++	±	++	+	++
Cocos nucifera (Jamaica tall)	++	±	++	+	++
Veitchia merrillii	++	±	++	+	+

Symbols: (+) observed; (++) bright; (±) faint; (−) not observed.

Source: From Barcelon, Maria de los Angeles, McCoy, R.E., and Donselman, H.M., *Journal of Chromatography*, 260, 147, 1983.

chloride was used for amino acid analysis with two-dimensional TLC on polyamide sheets. A possible correlation between the presence of arginine in the palm varieties studied and susceptibility to lethal yellowing (LY) disease was observed. The results of the TLC analyses are given in Table 13.4. The phloem sap from healthy and diseased *Veitchia merrilfii* and *Cocos nucifera* (Jamaica tall coconut) were analyzed for free amino acids. The results are given in Table 13.5 while Table 13.6 shows the analyses results reported for the 18 different palms, based on the combined results, obtained from an average of nine analyses taken from foliage harvested at three different dates.

TABLE 13.5
Free Amino Acids in Phloem Sap from Coconut

Amino Acids	*Cocos nucifera* (Jamaica Tall)		*Vietchia merrillii*	
	Healthy	Diseased	Healthy	Diseased
Ala/Dns-NH$_2$	++	+	++	++
Asn	++	++	−	−
Asp	+	±	+	±
Arg/m-DA	+	+	±	+
AABA/GABA	++	++	++	+
Cmc	+	++	++	++
Cys	−	±	−	−
Glu	++	+	++	+
Gln	+	++	++	++
Gly	++	+	++	+
His/m-DA	−	+	+	+
OH-Pro	+	+	−	−
Ile	+	+	++	+
Leu/Nle	+	+	++	+
Bis-Dns-Lys	+	±	+	±
Met	+	±	+	+
Meo	−	−	+	+
Phe	+	++	++	++
Pim	+	+	+	+
Pro	++	++	++	++
Ser	++	++	++	++
Thr	+	±	±	±
Trp	+	+	+	+
Bis-Tyr/Trn	±	−	+	+
Val	+	+	++	++

Symbols as in Table 13.4, m-DA: mono dansyl diamino acid, AABA: α-amino Butyric Acid, GABA: γ-amino Butyric Acid, Meo: methionine sulfoxide, Nle: Norleucine, OH-Pro: hydroxy proline, Pim: pimelic acid.

Source: From Barcelon, Maria de los Angeles, McCoy, R.E., and Donselman, H.M., *Journal of Chromatography*, 260, 147, 1983.

TABLE 13.6
Amino Acid Analysis of Different Palms

Species	Ala/Dns-NH$_2$	Asn	Asp	Arg/m-DA	AABA/GABA	m-Cad	Cmc	Cys	DAPA
Arecastrum romanzoffianum	++	−	++	−	+	−	+	−	−
Carpentaria acuminate	++	+	++	−	+	+	±	−	−
Chrysalidocarpus lutescens	++	+	++	±	+	−	+	−	−
Phoenix roebelenii	++	+	++	−	++	−	−	+	+
Ptycosperma elegans	+	±	++	−	++	−	−	+	±
Roystonea regia	++	±+	++	−	+	−	±	+	+
Sabal palmetto	++	+	++	−	++	−	−	±	−
Washingonia robusta	++	+	++	−	++	+	−	−	+
Cocos nucifera (Malayan dwarf)	++	+	++	±	+	−	−	±	±
Cocos nucifera (Maypan)	++	+	++	−	++	−	−	+	+
Caryota mitis	+	+	++	±	++	±	−	±	−
Arikuyroba schizophylla	++	+	++	±	+	−	−	+	−
Phoenix dactylifera (Deglet moor)	++	+	+	±	++	−	−	±	−
Phoenix dactylifera (Halawi)	++	+	+	±	+	−	−	+	−
Cocos Nucifera (Jamaica tall)	++	+	++	+	++	−	−	−	+
Pritchardia thrustonii	++	+	+	+	++	−	−	+	±
Veitchia merrillii	++	+	++	+	+	−	−	+	+
Wallichia disticha	++	+	++	+	++	−	−	+	+

Symbols as in Table 13.4, DAPA: diamino propionic acid, m-Cad: mono dansyl cadaverine.

Source: From Barcelon, Maria de los Angeles, McCoy, R.E., and Donselman, H.M., *Journal of Chromatography*, 260, 147, 1983.

Using thin layer chromatographic technique, poisonous plants from different family members were identified and discriminated [84] by the distribution pattern of leaf amino acids (known and unknown). This study revealed interesting results which were valuable and consistent to differentiate from one taxon to the other by amino acid dispersion pattern. TLC was employed [85] to determine the composition of amino acids of Christ's thorn (*Paliurus spina-christi* Mill, a deciduous thorny shrub).

Thin layer chromatographic analysis [86] of bioactive constituents of chamomile (*Matricaria recutita* L.) on cellulose plates revealed the presence of leucine, phenylalanine, valine, tryptophan, tyrosine, proline, alanine, threonine, glutamic acid, lysine, and histidine. The smallest number of amino acids was detected in white florets while other chamomile parts did not differ significantly with regard to chromatographic profile of these plant constituents. TLC was also applied for quantitative analysis [87] of amino acids (along with sugars, oil-soluble pigments, and cofactors) from plants of *Arabidopsis thaliana*.

13.4.4 AMINO ACIDS FROM WINE

Wine is an alcoholic beverage produced by the fermentation of fruits, typically grapes though a number of other fruits are also quite popular—such as plum, elderberry, and blackcurrant. Nongrape wines are called fruit wine or country wine. Other products are made from starch-based materials, such as barley wine, rice wine, and sake. Beverages made from other fermentable material such as honey (mead), or that is distilled, such as brandy, are not wines. A fortified wine is a wine to which additional alcohol has been added, most commonly in the form of brandy (a spirit distilled fromwine).

The original reason for fortification was to preserve wine; the fortification process survives as consumers have developed tastes for wines preserved this way. Common fortified wines include, Sherry, Port, Marsala, Madeira, Vermouth, and Muscat.

Many papers have been published on wines containing amino acids. Out of these paper chromatography was the oldest technique for determination of amino acids in wines. Table 13.7 shows different techniques (HPLC not included here) for determination of amino acids in different wines [88–92].

13.5 PAPER CHROMATOGRAPHY

Paper chromatography is the oldest technique for identification and determination of amino acids. The solvent systems used for paper chromatographic analysis can serve as a starting point and basic guide line for finding or working out a suitable and successful solvent system for TLC analysis. A brief account of reports [93–103] on amino acid analysis from various plants is given in Table 13.8.

TABLE 13.7
Amino Acids in Different Wines

S. No.	Amino Acids	Source	Technique	References
1	Ala, Asp, Glu, Pro, Gly, Leu, Ser, Val, OH-Pro, Met, nor-val, Thr, Arg, Lys, γ-aminobutyric acid, Try, His, and Cys	Greek wines	Paper chromatography	[88]
2	Ala, Glu, Cys, Lys Thr, Pro, Gly, Ser, His, Arg, Leu, and γ-aminobutyric acid	Barbera red wine	Paper chromatography	[89]
3	Pro, Glu, Gly, and Ala	Mashes	Paper chromatography	[90]
4	Basic and acidic amino acids	Musts and wines	High voltage paper electrophoresis and paper chromatography	[91]
5	Ala, Arg, Asn, Gln, Gly, Lys, Met, Pro, Thr, Try, Tyr, Val	Wine	Thin layer chromatography	[92]

TABLE 13.8

Some Paper Chromatographic Results for Amino Acid Analysis from Plants

S. No.	Plant and its Part	Amino Acids Identified	Solvent System and Detection	Remarks	Reference
1	• *Euphorbia caducifolia,* flowers of different shades	*Free:* Ser, Cys, Glu, Asp *In protein hydrolysate:* Lys, Glu, Ser, Ala, Cys, Pro, Met, Tyr, Leu, Thr, His, Val, Asp	*n*-BuOH–HOAc–H$_2$O (4:1:1,v/v), and Phenol–water (3:1, v/v) for first and second dimension, respectively	—	[93]
2	• *Dactylonium aegyptium* • *D. scindicum,* whole plant	*Free:* Asp, Glu, Cys, Gly, His, Phe, Val, Ser, Tyr	-do-, detection by ninhydrin	• Quantitation by spectrophotometry at 540 nm • Analysis by 2D-TLC also	[94]
3	• *Abutilon indica* • *Grewia tenax* • *Salvadora persica* • *Nicotiana rhombifolia* • *Cadaba indica* • *Tephrosia purpurea*	*Free:* Glu, Pro, Asp, Ser, Thr, Ala, Val, Tyr, Trp, in protein hydrolysate: *Cys, His, Ser, Asp, Glu, Ala, Pro, Val, Met, Leu, Thr, Lys* (*plant wise results not shown*)	-do-	—	[95]

	Source	Amino acids	Method		Reference
4	• *Phaseolus vulgaris,* • *Vigna mungo,* and • *V. radiate* seeds	18 Amino acids except Asn and Glu (in protein hydrolysate)	-do-	Quantitation by spectrophotometry, at 570 nm	[96]
5	• *Euphorbia caducifolia* fruits	*Free:* Ser, Cys, Glu, Asp, γ-aminobutyric acid	-do-	—	[97]
6	• *Citrullus colocynthis* seeds, pulp	Val, Glu, His, Arg	-do-	—	[98]
7	Rice, Wheat, and Maize	Gln, Asn, Orn, Ser	—	—	[99]
8	Jack beans	L-α-amino-δ-hydroxyvaleric acid and γ-L-glutamyl-L-alanine	—	—	[100]
9	Wedgewood iris	Thr, Ala, Pro, Val, Tyr, Phe, Ile, Leu, Cys, His, Gly	—	—	[101]
10	• *Rauwolfia serpentine*	Glu, Ser, Gly, Thr, Ala, β-alanine, Pro, Phe, Leu, Ile, γ-aminobutyric acid	—	—	[102]
11	• *Symphytum officimale* Onion seed	*N*-acyl amino acid	—	—	[103]

Thin Layer Chromatography in Phytochemistry

The readers may find a large number of successful application details for TLC analysis of amino acids, their derivatives (and also the enantiomeric resolution) in literature at one place [104–107]; the methods described therein can be successfully applied for amino acids, from different plant sources, with some experimental skills and manipulations. Therefore, these are not repeated here.

REFERENCES

1. Smyth, D.G., Stein, W.H., and Moore, S., The sequence of amino acid residues in bovine pancreatic ribonuclease: revisions and confirmations. *The Journal of Biological Chemistry*, 238, 227–34, 1963.
2. Chiou, S.H. and Wang, K.T., in (Villafranca, J., Ed.), *Current Research in Protein Chemistry: Techniques, Structure & Function*, Academic Press: San Diego, pp. 3–10, 1990.
3. Phillips, R.D., An apparatus for rapidly preparing protein samples for hydrolysis. *Analytical Biochemistry*, 113(1), 102–7, 1981.
4. Light, A. and Smith, E.L., Amino acid analysis of peptides and proteins, in (Hans Neurath, Ed.), *Proteins, Composition, Structure, and Function*, Vol. 1, Academic Press: New York, pp. 1–44, 1963.
5. Savoy, C.F., Heinis, J.L., and Seals, R.G., Improved methodology for rapid and reproducible acid hydrolysis of food and purified proteins. *Analytical Biochemistry*, 68 (2), 562–71, 1975.
6. Perham, R.N., Techniques for determining the amino-acid composition and sequence of proteins, in *Techniques in the Life Sciences*, B110, pp. 1–39, 1978.
7. Korzun, B.P., Dorfman, L., and Brody, S.M., Separation of some alkaloids, steroids, and synthetic compounds by thin-layer chromatography. *Analytical Chemistry*, 35(8), 950–2, 1963.
8. Srivastava, S.P., Bhushan, R., and Chauhan, R.S., TLC separation of amino acids on silica gel impregnated layers. *Journal of Liquid Chromatography*, 7, 1359–65, 1984.
9. Srivastava, S.P., Bhushan, R., and Chauhan, R.S., TLC separation of some metal ions on impregnated layers. *Journal of Liquid Chromatography*, 7, 1341–4, 1984.
10. Srivastava, S.P., Bhushan, R., and Chauhan, R.S., Use of ligands for improved separation of metal ions on thin silica. *Journal of Liquid Chromatography*, 8, 571–4, 1985.
11. Srivastava, S.P., Bhushan, R., and Chauhan, R.S., TLC separation of some closely related synthetic dyes on impregnated silica gel layers. *Journal of Liquid Chromatography*, 8, 1255–61, 1985.
12. Bhushan, R., Srivastava, S.P., and Chauhan, R.S., Thin layer chromatographic separation of thorium, titanium, zirconium and some rare earths on cellulose impregnated layers. *Analytical Letters*, 18A, 1549–53, 1985.
13. Bhushan, R. and Ali, I., TLC resolution of PTH-amino acid on Zn^{++}, Cd^{++}, Hg^{++}, impregnated silica plates. *Journal of Liquid Chormatography*, 9, 3479–87, 1986.
14. Bhushan, R. and Reddy, G.P., Resolution of PTH-amino acids on impregnated silica plates. *Fresenius Zeitschrift für Analytical Chemistry*, 327, 798–9, 1987.
15. Bhushan, R. and Ali, I., TLC resolution of enantiomeric mixtures of amino acids. *Chromatographia*, 23, 141–2, 1987.
16. Bhushan, R. and Reddy, G.P., Rapid resolution of PTH-amino acids by TLC on silica gel plates impregnated with transition metal ions. *Analytical Biochemistry*, 162, 427–30, 1987.
17. Bhushan, R. and Ali, I., TLC resolution of racemic mixtures of PTH-amino acids on (+)-tartaric acid and impregnated silica gel plates. *Journal of Chromatography*, 392, 460–3, 1987.

18. Bhushan, R. and Ali, I., Effects of halides on TLC resolution of amino acids, below their isoelectric points. *Chromatographia*, 23, 207–8, 1987.
19. Bhushan, R. and Ali, I., TLC resolution of amino acids in a new solvent and effect of alkaline earth metals. *Journal of Liquid Chromatography*, 10, 3647–52, 1987.
20. Bhushan, R., Chauhan, R.S., Reena, and Ali, I., A Comparison of amino acid separation of Zn^{++}, Cd^{++} and Hg^{++} impregnated silica layers. *Journal of Liquid Chromatography*, 10, 3653–7, 1987.
21. Bhushan, R. and Reddy, G.P., TLC of PTH-amino acids on silica gel thin layers impregnated with zinc salts and sulphates of some metals. *Analyst (London)*, 112, 1467–9, 1987.
22. Bhushan, R. and Ali, I., TLC Resolution of constituents of vitamin-B complex. *Archives of Pharmacology*, 320, 1186–7, 1987.
23. Martens, J., Lubben, S., and Bhushan, R., Synthese Eines Neuen Chiralen Selektors Fur Die Dunnschichtchromatographische. *Tetrahedron Letters*, 30, 7181–2, 1989.
24. Bhushan, R. and Reddy, G.P., Some effective solvents for the rapid resolution of DNP-amino acids by TLC. *Chromatographia*, 25, 455–6, 1988.
25. Bhushan, R. and Reddy, G.P., TLC of PTH-amino acids on silica layers impregnated with metal salts. *Analytical Letters*, 21, 1075–84, 1988.
26. Bhushan, R. and Ali, I., TLC Resolution of enantiomeric mixtures of amino acids. *Chromatographia*, 23, 141–2, 1987.
27. Bhushan, R. and Arora, M., Resolution of enantiomers of DL-amino acids on silica gel plates impregnated with optically active (−)-quinine. *Biomedical Chromatography*, 15, 433–6, 2001.
28. Bhushan, R., Martens, J., and Arora, M., Direct resolution of (±)-ephedrine and atropine into their enantiomers by impregnated TLC. *Biomedical Chromatography*, 15, 151–4, 2001.
29. Bhushan, R., Martens, J., Wallbaum, S., Joshi, S., and Parshad, V., TLC resolution of enantiomers of amino acids and dansyl derivatives using (1R,3R,5R)-2-azabicyclo-[3,3,0]octan-3-carboxylic acid as impregnating reagent. *Biomedical Chromatography*, 11, 286–8, 1997.
30. Honegger, C.G., Thin-film ionophoresis and thin-film ionophoresis-chromatography. *Helvetica Chimica Acta*, 44, 173–9, 1961.
31. Borke, M.L. and Kirch, E.R., Separation of some of the opium alkaloids by surface chromatography. *Journal of the American Pharmaceutical Association*, 42, 627–9, 1953.
32. Fahmy, A.R., Niederwieser, A., Pataki, G., and Brenner, M., Thin-layer chromatography of amino acids on Kieselgel G. II. A rapid method for the separation and qualitative determination of 22 amino acids. *Helvetica Chimica Acta*, 44, 2022–6, 1961.
33. Jones, K. and Heathcote, J.G., Rapid resolution of naturally occurring amino acids by thin layer chromatography. *Journal of Chromatography*, 24(1), 106–11, 1966.
34. Brautigan, D.L., Ferguson-Miller, S., Tarr, G.E., and Margoliash, E., Definition of cytochrome c binding domains by chemical modification. II. Identification and properties of singly substituted carboxydinitrophenyl cytochromes c at lysines 8, 13, 22, 27, 39, 60, 72, 87, and 99. *The Journal of Biological Chemistry*, 253(1), 140–8, 1978.
35. Heathcote, J.G. and Washington, R.J., A simple quantitative method for the determination of small amounts of amino-acids. *Analyst*, 92(99), 627–33, 1967.
36. Barrollier, J., Method of documentation of thin-layer chromatograms. *Naturwissenschaften*, 48, 404, 1961.

37. Moffat, E.D. and Lytle, R.I., Polychromatic technique for the identification of amino acids on paper chromatograms. *Analytical Chemistry*, 31, 926–8, 1959.

38. Krauss, G.J. and Reinbothe, H., Polychromatic detection of amino acids with different ninhydrin spray reagents after separation on MN 300 HR-cellulose thin layers. Further occurences of mimosacean amino acids. *Biochemie und Physiologie der Pflanzen*, 161(6), 577–92, 1970.

39. Basak, B. and Laskar, S., Spray reagents for the detection of amino acids on thin-layer plates. *Talanta*, 37(11), 1105–6, 1990.

40. Devani, M.B., Shishoo, C.J., Shah, S.A., Soni, K.P., and Shah, R.S., Localization of amino acids on thin-layer chromatograms with acetylacetone-formaldehyde reagent. *Journal of Chromatography*, 537(1–2), 494–6, 1991.

41. Laskar, S., Bhattacharya, U., and Basak, B., Modified ninhydrin spray reagent for the identification of amino acids on thin-layer chromatography plates. *Analyst*, 116(6), 625–6, 1991.

42. Laskar, S. and Basak, B.K., Identification of amino acids with ninhydrin in acid media. *Journal of Planar Chromatography—Modern TLC*, 3, 535–6, 1990.

43. Laskar, S. and Basak, B.K., Identification of amino acids on thin-layer chromatography plates. *Journal of Planar Chromatography—Modern TLC*, 3, 275–6, 1990.

44. Sinhababu, A., Basak, B., and Laskar, S., Novel spray reagent for the identification of amino acids on thin-layer chromatography plates. *Analytical Proceedings*, 31(2), 65–6, 1994.

45. Norfolk, E., Khan, S.H., Fried, B., and Sherma, J., Comparison of amino acid separations on high performance silica gel, cellulose, and C-18 reversed phase layers and application of HPTLC to the determination of amino acids in *Biomphalaria glabrata* snails. *Journal of Liquid Chromatography*, 17(6), 1317–26, 1994.

46. Males, Z., Plazibat, M., and Petlevski, R., Thin-layer chromatographic analysis of amino acids of *Crithmum maritimum* L. *Farmaceutski Glasnik*, 57(5), 175–80, 2001.

47. Hodisan, T., Culea, M., Cimpoiu, C., and Cot, A., Study of free amino acids from plant extracts. II. Separation, identification, and determination of free amino acids from *Fagus silvatica* by liquid chromatography (LC) and gas chromatography (GC). *Revista de Chimie (Bucharest)*, 49(6), 393–7, 1998.

48. Nikolic, V.D., Stankovic, M.Z., Cvetkovic, D.M., Nikolic, L.B., and Skala, D.U., Separation and identification of amino acids and thiosulfinates in extracts of garlic (*Allium sativum* L.). *Hemijska Industrija*, 52(4), 156–60, 1998.

49. Hodisan, T., Cimpoiu, C., Hodisan, V., and Sarbu, C., Study of free amino acids from equisetum extract, *Chemia*, 40(1–2), 177–81, 1995.

50. Higuchi, K., Suzuki, T., and Ashihara, H., Pipecolic acid from the developing fruits (pericarp and seeds) of *Coffea arabica* and *Camellia sinensis*, in *Colloque Scientifique International sur le Cafe [Comptes Rendus], Vol. 1: Seizieme Colloque Scientifique International sur le Cafe*, 16th edn, pp. 389–95, 1995.

51. Sarbu, C., Cimpoiu, C., and Hodisan, T., Free amino acids in plant extracts. I. Thin-layer chromatographic separation and identification of free amino acids, *Revista de Chimie (Bucharest)*, 46(6), 581–4, 1995.

52. Mandavia, M.K. and Parameswaran, M., Changes in amino acids and phenols in limabean (*Phaseolus lunatus*) varieties resistant and susceptible to stem rot disease (*Macrophomina phaseolina*). *Gujarat Agricultural University Research Journal*, 18 (2), 19–23, 1993.

53. Das, B. and Sawant, S., Quantitative HPTLC analysis of dansyl amino acids. *Journal of Planar Chromatography—Modern TLC*, 6(4), 294–5, 1993.

54. Hilal, S.H., Saber, A.H., Haggag, M.Y., El-Zalabani, S.M., and Ahmed, F.I.F., Protein, common amino acids and mimosine contents of *Leucaena glauca* Benth. *Egyptian Journal of Pharmaceutical Sciences*, 32(1–2), 91–101, 1991.

55. Kujundzic, N., Stanic, G., and Vukusic, I., A study of the chemical composition of the plant *Caucalis platycarpos* L., *Acta Pharmaceutica* (*Zagreb, Croatia*), 42(1), 91–3, 1992.

56. Maillard, C., Barlatier, A., Debrauwer, L., Calaf, R., and Balansard, G., Identification of the amino acids in the aerial parts of *Capsella bursa-pastoris* Moench. Significance for routine assay. *Pharmaceutiques Francaises*, 46(3), 211–16, 1988.

57. Song, C., A method for adding ninhydrin directly to developing agent. *Zhiwu Shenglixue Tongxun*, 3, 59–60, 1988; CAN 109: 207690, 1988.

58. Jing, Y., Li, H., and Chen, R., Composition of amino acids in *Crotalaria sessiliflora* L. *Tongbao*, 12(7), 425–6, 1987.

59. Juhasz, O. and Kozma, P., Comparative analysis of free amino acids in grape berries by chromatographic methods. *Kerteszeti Egyetem Kozlemenyei*, 48, 115–26, 1986.

60. Oluic-Vukovic, V., Babic-Gojmerac, T., and Mihanovic, B., Amino acid conjugates of chloro-s-triazine: model substances for biodegradation study of s-triazine herbicides in plants. *Journal of the Serbian Chemical Society*, 50(3), 121–4, 1985.

61. Ruffin, J., Brown, C.A., and Banerjee, U.C., A physiochemical characterization of allergenic pollen-held proteins. *Environmental and Experimental Botany*, 23(4), 311–19, 1983.

62. Camirand, A., Phipps, J., and Wightman, F., Comparative metabolism of L-tyrosine and L-phenylalanine in tobacco plants in relation to the biosynthesis of phenylacetic acids. *Canadian Journal of Botany*, 61(9), 2302–8, 1983.

63. Bonsignore, L., Fadda, A.M., Loy, G., and Maccioni, A.M., Chromatographic separation and identification of amino acids from vegetable samples. *Rendiconti del Seminario della Facolta di Scienze dell'Universita di Cagliari*, 51(1), 1–6, 1981.

64. Rao, P.S.N. and Talpasayi, E.R.S., Changes in soluble amino acids, cellular carbohydrates and lipids in brine alga *Anabaena fertilissima* under salt stress. *Indian Journal of Experimental Biology*, 20(10), 765–7, 1982.

65. Varadi, A. and Pongor, S., Determination of sulfur-containing amino acids by quantitative ion-exchange thin-layer chromatography. *Journal of Chromatography*, 173(2), 419–24, 1979.

66. Yakovleva, E.V., Kostennikova, Z.P., Baratova, A.A., and Levandovskii, G.S., Composition of amino acids in homeopathic matrix tinctures of *Arnica montana*. *Farmatsiya* (*Moscow*), 5, 11–14, 2002; CAN 139: 235171, 2003.

67. David, W. and Richard, S., The Amino Acids of Extrafloral Nectar from *Helianthella quinquenervis* (Asteraceae). *American Journal of Botany*, 67(9), 1394–6, 1980.

68. Males, Z., Plazibat, M., Vundac, V.B., Zuntar, I., and Pilepic, K.H., Thin-layer chromatographic analysis of flavonoids, phenolic acids and amino acids in some croatian *Hypericum taxa*. *Journal of Planar Chromatography—Modern TLC*, 17(4), 280–5, 2004.

69. Drozdova, I.L. and Bubenchikov, R.A., Amino acids in field violet and tall plaster clover. *Farmatsiya* (*Moscow*) 5, 14–15, 2003; CAN 141: 128952, 2004.

70. Kopyt'ko, Ya.F., Amino acids and fatty acids in homeopathic matrix tinctures of grass of Parnassus (*Parnassia palustris*). *Pharmaceutical Chemistry Journal*, 37(7), 347–9, 2003.

71. Gaspar, R.Z. and Males, Z., Determination of free amino acids in valerian (*Valeriana officinalis* L.) by thin-layer chromatography. *Farmaceutski Glasnik*, 61(12), 691–7, 2005.

72. Males, Z., Plazibat, M., Bilusic, V.V., Kremer, D., and Alar, J., Study of flavonoids and amino acids of Spanish broom (*Spartium junceum* L., Fabaceae). *Farmaceutski Glasnik*, 61(9), 499–509, 2005.

73. Bieleski, R.L. and Turner, N.A., Separation and estimation of amino acids in crude plant extracts by thin-layer electrophoresis and chromatography. *Analytical Biochemistry*, 17(2), 278–93, 1966.

74. Umebayashi, M., Aminoacylase in higher plants. III. New method for detection of N-acyl amino acid by thin-layer chromatography. *Nippon Dojo Hiryogaku Zasshi*, 39 (2), 137–42, 1968.

75. Laserna, G. and Manalo, J.B., Free amino acid determination of local, air-borne, allergenic pollen extracts by thin-layer chromatography. *Philippine Journal of Science*, 97(4), 315–35, 1968.

76. Bounias, M., Quantitative microanalytical thin-layer chromatography of sugars, amino acids, oil-soluble pigments and cofactors from plants of *Arabidopsis thaliana*. *Chimie Analytique (Paris)*, 51(2), 76–82, 1969.

77. Pillay, D.T.N. and Mehdi, R., Separation of amino acids by thin-layer chromatography. *Journal of Chromatography*, 47(1), 119–23, 1970.

78. Watanabe, T., Shima, Y., and Ichihara, T., Changes in free amino acid during ripening of soybean seed. *Kogaku Hen*, 20(2), 76–82, 1970.

79. Watanabe, T., Tsugawa, M., Takayama, N., and Furukawa, Y., Changes of free amino acids of each organ in the development and growth of the kidney bean plant. *Nogaku Hen*, 22(1), 45–54, 1971.

80. Dorer, M., Identification of free amino acids in plant materials by two-dimensional thin-layer chromatography: in the first dimension, double electrophoresis, in the second dimension, thin-layer chromatography. *Farmacevtski Vestnik (Ljubljana, Slovenia)*, 26(4), 252–62, 1975.

81. Juhasz, O., Polyak, D., and Kozma, P., Free amino acid content of grapes during ripening. *Hung. Lippay Janos Tud. Ulesszak Eloadasai*, 8, 985–1005, 1977.

82. Murthy, G.S. and Bagyaraj, D.J., Free amino nitrogen and amino acids in *Cajanus cajan* in relation to fusarium wilt resistance. *Indian Phytopathology*, 31(4), 482–5, 1978.

83. Barcelon, Maria de los Angeles, McCoy, R.E., and Donselman, H.M., New liquid chromatographic approaches for free amino acid analysis in plants and insects. II. Thin-layer chromatographic analysis for eighteen varieties of palm trees. *Journal of Chromatography*, 260(1), 147–55, 1983.

84. Rao, N.R.K. and Rajagopal, T., Thin layer chromatographic identification of poisonous plants and its forensic value. *Indian Journal of Forensic Sciences*, 6(1), 5–16, 1992.

85. Males, Z., Plazibat, M., Hazler-Pilepic, K., and Cetina-Cizmek, B., Investigation of the composition of amino acids in Christ's thorn (*Paliurus spina-christi* Mill.). *Farmaceutski Glasnik*, 57(7–8), 257–65, 2001.

86. Blazekovic, B. and Stanic, G., Thin-layer chromatographic analysis of bioactive constituents in chamomile (*Matricaria recutita* L.). *Farmaceutski Glasnik*, 60(6), 243–54, 2004.

87. Bounias, M., Quantitative microanalytical thin-layer chromatography of sugars, amino acids, oil-soluble pigments and cofactors from plants of *Arabidopsis thaliana*. *Chimie Analytique* (Paris), 51(2), 76–82, 1969; CAN 70: 112242, 1969.

88. Demotakis, P.N., Free amino acids in Greek wines. *Chimica Chronika*, 20, 102–7, 1955.

89. Iwano, S., Separation and determination of amino acids in grape wine by paper chromatography. *Nippon Jozo Kyokai Zasshi*, 52, 569–71, 1957.

90. Burges, L.M., Paper chromatography of Navarre wines (free amino acids). *Anales de Bromatologia (Madrid)*, 12, 19–26, 1960.

91. Van Wyk, C.J. and Venter, P.J., The determination of free amino acids in musts and wines by means of high voltage paper electrophoresis and paper chromatography. *Suid-Afrikaanse Tydskrif vir Landbouwetenskap*, 8(1), 57–67, 69–71, 1965.

92. Washuettl, J., Bancher, E., and Olfath, M.D., Thin-layer chromatographic determination of free amino acids in wine. *Bodenkultur*, 19(4), 353–7, 1968.

93. Garg, S.P. and Bhushan, R., Phytochemical studies of different coloured fruits of *Euphorbia caducifolia*. *Transactions of Indian Society of Desert Technology and University Centre of Desert Studies*, 6, 92–8, 1981.

94. Garg, S.P., Bhushan, R., and Kapoor, R.C., Free amino acids and sugars in Dactylonium species growing on different soil types in arid regions. *Trans. Isdt. & Ucds.*, 5, 125–8, 1980.

95. Bhushan, R., Bhatnagar, R., Garg, S.P., and Kapoor, R.C., Protein and amino acids of certain Indian desert plants II. *Trans. Isdt. & Ucds.*, 2, 306–8, 1977.

96. Bhushan, R. and Pant, N., Composition of globulins from *Phaselus vulgaris, Vigna mungo and V. radiata. Trans. Isdt. & Ucds*, 11, 129–32, 1986.

97. Bhushan, R., Simultaneous identification of free amino acids and sugars in plant tissue by paper chromatography. *Fresenius Zeitschrift für Analytical Chemistry*, 128, 309, 1981.

98. Bhatnagar, R., Bhushan, R., Garg, S.P., and Kapoor, R.C., Protein and amino acids of fruits of *Citrullus colosynthis. Journal of the Indian Chemical Society*, 53, 309, 1979.

99. Palfi, G., Changes in the amino acid content of detached wilting leaves of *Solanum laciniatum* in the light and in the dark. *Acta Agronomica Academiae Scientiarum Hungaricae*, 17(3–4), 381–88, 1968; CAN 69: 65202, 1968.

100. Thompson, J.F., Morris, C.J., and Hodson, R.C., Automatic monitoring of large-scale amino acid chromatography in studies of plant nitrogen metabolism. *Automation in Analytical Chemistry, Technicon Symposium*, 84–93, 1966; CAN 67: 29753, 1967.

101. Madan, C.L., Paper chromatographic study of the free and bound amino acids in *Rauwolfia serpentine. Planta Medica*, 15(1), 118–20, 1967.

102. Iida, T., Hoshino, N., and Murakami, T., Free amino acids isolated from the roots of *Symphytum officimale. Shoyakugaku Zasshi*, 21(2), 131–2, 1967.

103. Umebayashi, M., Aminoacylase in higher plants. III. New method for detection of N-acyl amino acid by thin-layer chromatography. *Nippon Dojo Hiryogaku Zasshi*, 39 (2), 137–42, 1968.

104. Bhushan, R. and Martens, J., Peptides and proteins, thin layer (planar) chromatography, in (Wilson, I.D., Adlard, E.R., Cook, M., and Poole, C.F., Eds.) *Encyclopedia of Separation Science*, Vol. 8, pp. 3626–36, 2000.

105. Bhushan, R. and Martens, J., Amino acids: thin layer (planar) chromatography, in (Wilson, I.D., Adlard, E.R., Cook, M., and Poole, C.F., Eds.) *Encyclopedia of Separation Science*, Vol. 5, pp. 2013–33, 2000.

106. Bhushan, R., and Martens, J., Amino acids and their derivatives, in (Sherma, J. and Fried, B., Eds.), *Hand Book of Thin Layer Chromatography*, 3rd Edition—Revised and expanded, Marcel Dekker, pp. 373–415, 2003.

107. Bhushan, R. and Martens, J., Peptides and proteins, in (Sherma, J. and Fried, B., Eds.), *Hand Book of Thin Layer Chromatography*, 3rd Edition—Revised and expanded, Marcel Dekker, pp. 749–66, 2003.

Secondary Metabolites — Shickimic Acid Derivatives

14 Sample Preparation and TLC Analysis of Phenolic Acids

Magdalena Wójciak-Kosior and Anna Oniszczuk

CONTENTS

14.1 DEFINITION, CHEMICAL CLASSIFICATION BIOSYNTHESIS AND OCCURANCE OF PHENOLIC ACIDS

Phenolic acids are aromatic secondary plant metabolites, widely spread throughout the plant kingdom. These compounds are a subclass of a larger category of metabolites commonly referred as "phenolics." The term "phenolics" encompasses approximately 8000 naturally occurring compounds, all of which possess one common structural feature, a phenol [1].

The name "phenolic acids," in general, describes phenols possessing one carboxylic acid functionality. There are two groups of phenolic acids that are worth of attention: cinnamic and benzoic acids derivatives. In the first group there are, among others, ferulic, caffeic, *p*-coumaric, *m*-coumaric, *o*-coumaric, sinapic, acids, and curcumin (dimer of ferulic acid) (Table 14.1, Figure 14.1). In the second there are gallic, *p*-hydroxybenzoic, vanillic, protocatechuic, salicylic, syringic, veratric, and gentisic acids (Table 14.2, Figure 14.2) [2,3]. The most commonly occurring in

TABLE 14.1
Derivatives of Cinnamic Acid

Cinnamic Acids Derivatives	R1	R2	R3	R4	Source
o-Coumaric acid	OH	—	—	—	*Melilotus* sp
p-Coumaric acid	—	—	OH	—	*Arnica montana* (anthodium)
					Calendula officinalis (flos)
					Plantago sp.
					Polygonum sp.
					Lysimachia sp.
					Aloe sp.
Caffeic acid	—	—	OH	OH	Crataegus sp.
					Artemisia sp.
					Taraxacum sp.
					Calendula sp.
					Achillea sp.
					Cichorium sp.
					Polygonum sp.
Ferulic acid	—	—	OH	OCH$_3$	*Arnica* sp.
					Calendula sp.
					Plantago sp.
					Polygonum sp.
					Veronica sp.
Sinapic acid	—	OCH$_3$	OH	OCH$_3$	*Polygonum aviculare* (herba)
					Polygonum hydropiper (herba)
					Brassica nigra (semen)
					Sinapis alba (semen)
					Lysimachia sp.

Source: From Kohlmunzer, S., *Pharmacognosy*, 5th edn., PZWL, Warsaw, 2003 (in Polish); Borkowski, B., *Herba Pol.*, *Part 1*, 39, 71, 1993 (in polish).

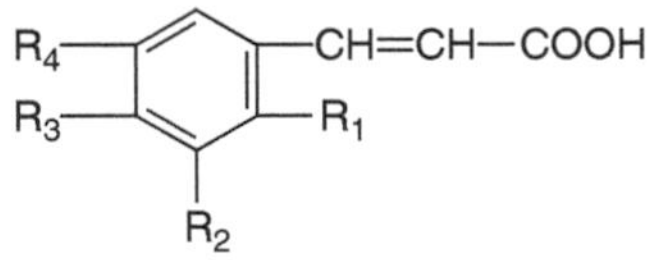

FIGURE 14.1 Cinnamic acid ($R_1 = R_2 = R_3 = R_4 = H$).

TABLE 14.2
Derivatives of Benzoic Acid

Benzoic Acids Derivatives	R_1	R_2	R_3	R_4	R_5	Source
Salicylic acid	OH	—	—	—	—	*Calendula officinalis* (flos)
						Polygonum bistorta (rhizoma)
						Lysimachia nummularia (herba)
						Filipendula ulmaria (flos)
						Polygala senega (radix)
						Viola tricolor (herba)
						Betula alba (cortex)
						Chimaphilla sp.
						Pyrola sp.
Protocatechuic acid	—	OH	OH	—	—	*Erythrea centaurium* (herba)
Gallic acid	—	OH	OH	OH	—	*Rheum* sp.
						Hamamelis sp.
Vanillic acid	—	OCH_3	OH	—	—	*Plantago* sp.
						Polygonum sp.
						Lysimachia sp.
						Erythrea sp.
p-Hydroxybenzoic acid	—	—	OH	—	—	*Arnica montana* (anthodium)
						Calendula officinalis (flos)
						Plantago sp.
						Polygonum sp.
Syringic acid	—	OCH_3	OH	OCH_3	—	*Calendula officinalis* (flos)
						Plantago sp.
Veratric acid	—	OCH_3	OCH_3	—	—	*Veratrum* sp.
						Schoenocaulon sp.
Gentisic acid	OH	—	—	OH	—	*Arnica montana* (anthodium)
						Calendula officinalis (flos)
						Polygonum sp.
						Lysimachia sp.
						Erythrea sp.
						Gentiana sp.

Source: From Kohlmunzer, S., *Pharmacognosy*, 5th edn., PZWL, Warsaw, 2003 (in Polish); Borkowski, B., *Herba Pol.*, *Part 1*, 39, 71, 1993 (in polish).

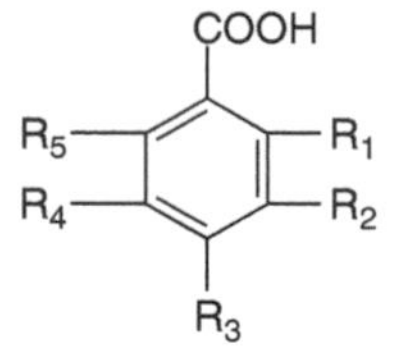

FIGURE 14.2 Benzoic acid ($R_1 = R_2 = R_3 = R_4 = R_5 = H$).

plants are cinnamic acid derivatives, exist as free acids, depsides or glycosides. Caffeic, benzoic, *p*-coumaric, vanillic, ferulic, and protocatechuic are acids present in nearly all plants. Other acids are found in selected plants.

The diversity of these components is not exclusive to their location in plants. Phenolic acids are not homogeneously distributed throughout plant tissues. Insoluble phenolic acids are found in cell walls, while soluble are present within the plant cell vacuoles. In addition, there exists a large variability, found during various stages of maturation. Growing conditions, such as temperature, are known to affect the phenolic acid content. Cinnamic and benzoic acid derivatives are physically dispersed throughout the plant in seeds, leaves, roots, and stems.

Many substituted benzoic and cinnamic acids serve as precursors to more complex compounds, while some of them play regulator and metabolic roles. Only a minor fraction exist as free acids. Instead, the majority are linked through glycoside, ester, ether, or acetal bonds either to structural components of plant (e.g., lignin, cellulose) or to larger polyphenols (e.g., flavonoids, tannins) or smaller molecules (e.g., glucose, tartaric acid) or other natural products (e.g., terpenes). All forms of phenolic acids are well soluble in organic solvents [1,2,4].

Depsides are intermolecular esters that are formed by condensation of two or more molecules of the same or different phenolic acids whereby the carboxyl group of one molecule is esterified with a phenolic hydroxyl group of a second one. These compounds (rosmarinic, chlorogenic, ellagic, *m*-digallic, cetraric, lecanoric acids, and cinarin) are found in lichen and flower plants (Table 14.3, Figure 14.3). The most commonly occurring in plants is chlorogenic acid. Many depsides are precursors to more complex molecules, primary tannins [2,3].

The cinnamic acid derivatives have their biosynthetic origin from the aromatic amino acid L-phenylalanine, synthesized itself from chorismate, the final product in the shikimate pathway. The benzoic acid derivatives have two proposed origins: the main pathway is the side chain degradation of the hydroxycinnamic acids derivatives; the other source is an alternate path stemming from an intermediate in the shikimate pathway, and involves a series of enzymatic reactions converting 3-dehydroshikimate to various benzoic acid derivatives (Figure 14.4) [1,2,5].

14.2 BIOLOGICAL MEANING AND PHARMACOLOGICAL ACTIVITIES OF PHENOLIC ACIDS

Phenolic acids have various biological and pharmacological effects. Although a great deal is still unknown regarding the roles of phenolic acids in plants, they have been

TABLE 14.3
Sources of Depsides

Depside	Source
Rosmarinic acid	*Rosmarinus officinalis* (folium)
	Salvia officinalis (folium)
	Melisa officinalis (folium)
	Symphytum officinale (radix)
Chlorogenic acid	*Salvia officinalis* (folium)
	Nicotina tabacum (folium)
	Thea sinensis (folium)
	Calendula officinalis (flos)
	Ecinacea purpurea (herba)
	Polygonum aviculare (herba)
	Thymus vulgaris (herba)
	Valeriana officinalis (rhizoma)
	Theobroma cacao (semen)
Ellagic acid	*Polygonum* sp.
Cetraric acid	*Cetraria islandica*
Lecanoric acid	*Lecanora* sp. *Parmelia* sp.
Cinarin	*Cynaryna scolymus* (folium)

Source: From Kohlmunzer, S., *Pharmacognosy*, 5th edn., PZWL, Warsaw, 2003 (in Polish); Borkowski, B., *Herba Pol.*, *Part 1*, 39, 71, 1993 (in polish).

connected with diverse functions, including protein synthesis, enzyme activity, photosynthesis of structural components (e.g., integral part of the cell-wall structure), and alleopathy (hydroxybenzoic and hydroxycinnamic acids). They play an important role in the natural host defence mechanism of plants against infectious diseases and inhibit multiplication of plant pathogenic bacteria, viruses, and fungi. Stress conditions such as excessive UV light, wounding, or infection induce biosynthesis of phenolic compounds. Thus environmental factors may have a significant contribution to the content of phenolic acids in plants [1,2,5,6].

Phenolic acids are generally considered as nontoxic and are often found in many traditional herbal medicines. Most of them have showed excellent scavenging activity of active oxygens such as superoxide anion radicals, hydroxyl radicals, and singlet oxygen. The antioxidant activity of phenolic acids and their ethers depends on the number of hydroxyl groups in the molecule that would be strengthened by steric hindrance [7,8].

The monohydroxybenzoic acid shows no antioxidant activity in the ortho and para position, but the *m*-hydroxybenzoic acid has an antioxidant properties. The dihydroxybenzoic acid derivatives show an antioxidant response dependent on the relative positions of the hydroxyl groups in ring. The 3,4,5-trihydroxybenzoic acid (gallic acid) shows strongest antioxidant and antiradical activity. The obtained results show that the antioxidant and antiradical activity of phenolic acids

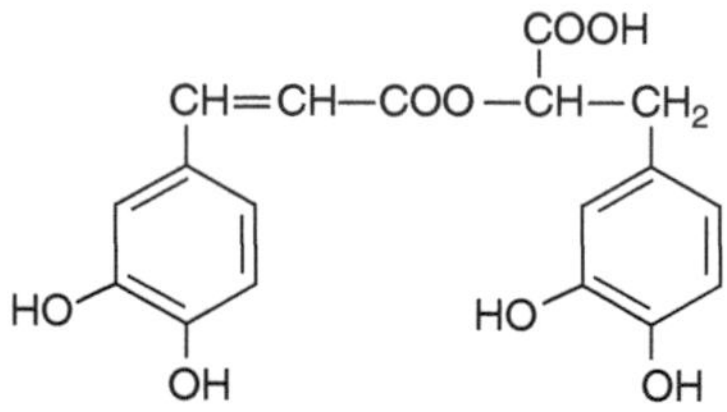

Rosmarinic acid

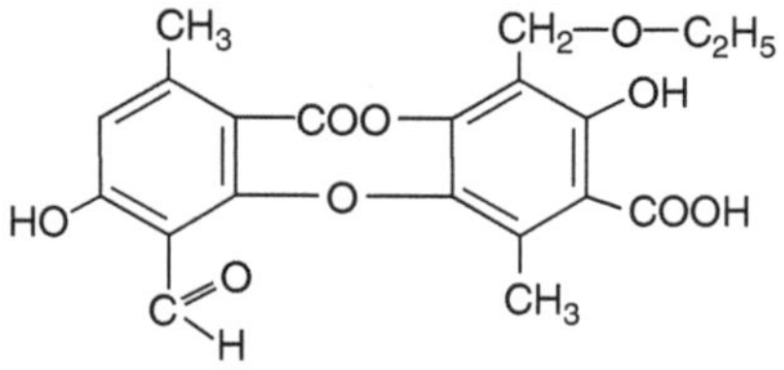

Cetraric acid

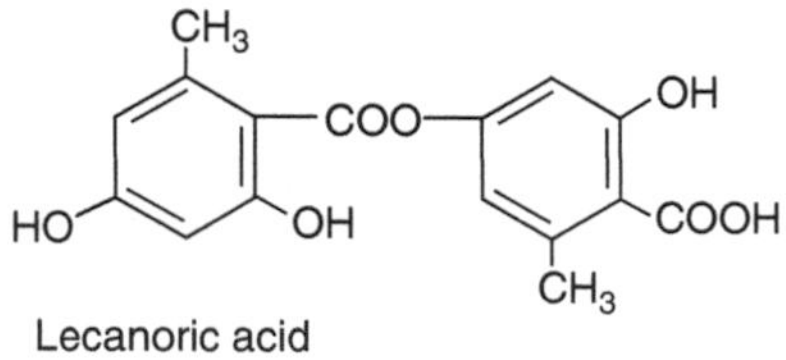

Lecanoric acid

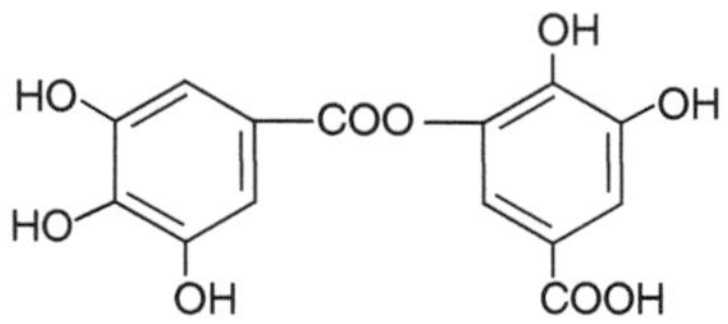

m-Digallic acid

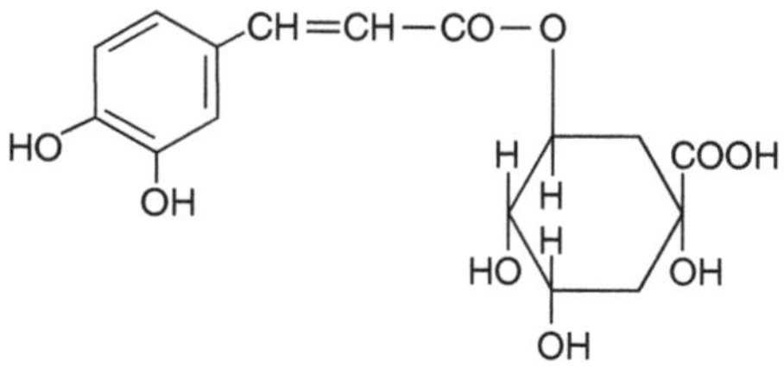

Chlorogenic acid

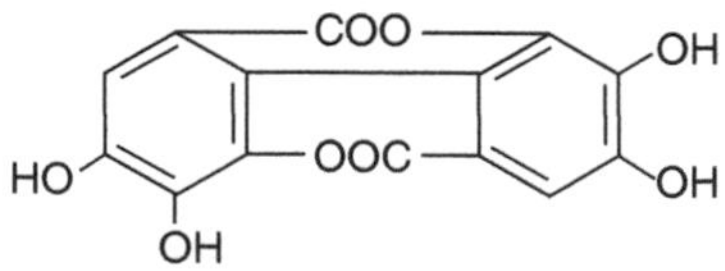

Ellagic acid

FIGURE 14.3 Depsides of phenolic acids.

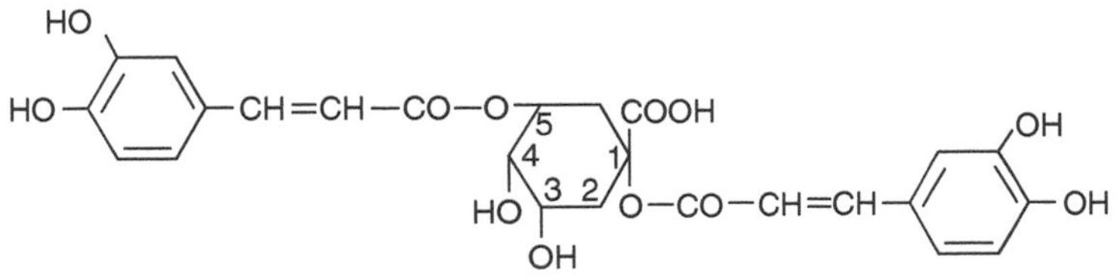

Cinarin

FIGURE 14.3 (continued)

correlated positively with the number of hydroxyl groups bonded to the aromatic ring. Esterification of the carboxylate group of gallic acid decreases antioxidant effectiveness. The presence of the two methoxy groups adjacent to the OH group in p-hydroxybenzoic acid significantly enhances the hydrogen availability and antioxidant properties, while substitution of the 3- and 5-hydroxyl with methoxy groups in syryngic acid (4-hydroxy-3,5-dimethoxybenzoic acid) causes a diminution in antioxidant activity in comparison with the trihydroxy derivative [7,9,10].

The presence of the –CH=CH-COOH group in cinnamic acid ensures greater H-donating ability and subsequent radical stabilization than the carboxylate group in benzoic acid. Thus cinnamic acid derivatives were found to be more active than benzoic ones.

In cinnamic acid derivatives, antioxidant activity also depends on the number of hydroxyl groups bonded to the aromatic ring. Thus diphenolics (e.g., caffeic acid) apparently have a higher radical scavenging ability than monophenolics (e.g., p-coumaric acid) [10–12].

Since free radicals play an important role in the development of cancer, the anticancer effects of some phenolic acids are attributed to their ability in scavenging of free radicals [13,14].

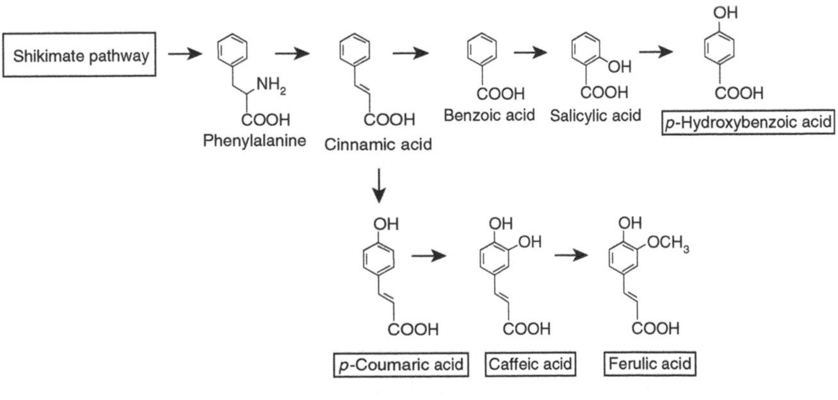

FIGURE 14.4 Biosynthesis of phenolic acids. (From Kohlmunzer, S., *Pharmacognosy*, 5th edition, PZWL, Warsaw, 2003 (in polish); Häkkinen, S., *Flavonols and Phenolic Acids in Berries and Berry Products*, Doctoral dissertation, Department of Clinical Nutrition, Department of Physiology, Department of Biochemistry, University of Kuopio, Kuopio, 2000.)

Recent investigations have linked a series of phenolic acids (ferulic, caffeic, ellagic, tannic, protocatechuic, chlorogenic, rosmarinic acids, and curcumin) with anticancer activity. Caffeic acid and some of its esters might indicate antitumor activity against colon carcinogenesis. Animal studies and in vitro studies suggest that ferulic acid may have direct antitumor activity against breast and liver cancer. Curcumin may suppress the oncogene MDM2, involved in mechanism of malignant tumor formation [1,14–16].

Some of phenolic acids (caffeic, gentisic, rosmarinic, ferulic, salicylic acids, curcumin, and several depsides) have anti-inflammatory and antirheumatic properties. Caffeic acid is known as selective blocker of the biosynthesis of leukotrienes, components involved in inflammatory and immunoregulation diseases, asthma, and allergic reactions. Gentisic acid has been demonstrated to inhibit both the cyclooxygenase and 12-lipoxygenase enzymes and has also been found to decrease the production of leukotrienes. Also, several lichen depsides have been reported to inhibit the formation of leukotrienes. In addition, these compounds have also been described as prostaglandin biosynthesis inhibitors [1,17,18].

In recent years, more and more clinical studies were focused on phenolic acids as a group of potential immunostimulating compounds. Possibly, systemic application of caffeic, gallic, or salicylic acid could result in suppression of leukocytes accumulation in extravascular sites, what may be beneficial in the treatment of chronic inflammatory diseases. Chlorogenic, ellagic, gallic, caffeic, protocatechuic, and salicylic acids have significant stimulatory influence on the production IgG antibodies [6,15,19].

Phenolic acids are currently being investigated for their potential antiviral therapy. Ferulic acid has been shown to possess inhibitory activity on the growth and reproduction of viruses such as influenza virus, respiratory syncytial virus, and AIDS. Caffeic acid derivatives have been shown to be potent and selective inhibitors of HIV-1 integrase.

Antiviral activity of rosmarinic acid is used in the therapy of *Herpes simplex* infections with rosmarinic acid-containing extracts of *Melissa officinalis* [1,14,18,19].

Several phenolic acids have antimicrobial and antifungal properties. Ferulic acid exhibit activity towards Gram-positive bacteria, Gram-negative bacteria, and yeasts. It also shows strong inhibitory effects on the growth of some human gastrointestinal microflora, including *Escherichia coli, Helicobacter pylori, Pseudomonas aeruginosa, Klebsilla pneumoniae, Enterobacter aerogenes, Citrobacter koseri*, and *Shigella sonnei*. Caffeic and protocatechuic acids inhibit the growth of *Escherichia coli* and *Klebsiella pneumoniae*. The same compounds, apart from syringic acid, completely inhibit the growth of *Bacillus cereus*. p-Hydroxy benzoic, vanillic, and p-coumaric completely inhibit the growth of *Escherichia coli, Klebsiella pneumoniae, and Bacillus cereus*. Vanillic and caffeic acids completely inhibit the growth and aflatoxin production by *Aspergillus flavus* and *Aspergillus parasiticus*. Also rosmarinic acid and benzoic acid derivatives substituted by hydroxy group or ether containing oxygen atom have bacteriostatic and fungistatic properties [14,15,18,20,21].

4-Hydroxybenzoic acid shows a hypoglycemic effect and additionally increases serum insulin levels and liver glycogen content in normal rats. The investigations have reported that chlorogenic acid reduces hepatic glycogenolysis and also decreases the absorption of new glucose [2,22].

Ferulic acid has been shown to exhibit cholesterol-lowering activity. From the results of clinical and in vitro studies, the compound has been shown as inhibitor of the activity of tromboxane A_2 synthetase and blood platelet aggregation. Owing to the antioxidant and cholesterol-lowering activities, as well as the preventive effects against trombosis and atherosclerosis, ferulic acid is considered as potential chemo-preventive agent against coronary heart disease. Also red wine polyphenols and gallic and caffeic acids could decrease the hydrolysis of dietary lipid esters and similarly the absorption of free cholesterol [14,23].

There are other biological activities of phenolic acids: adstringent (rosmarinic acid), sedative on the central nervous system (rosmarinic acid, chlorogenic acid), choleretica and cholekinetica (most of phenolic acids), hepatoprotective, analgetic, antipyretic, and keratolytic (salicylic acid) [15,18,24,25].

14.3 ANALYTICAL METHODS OF ISOLATION AND PURIFICATION OF PHENOLIC ACIDS

Extraction of phenolic acids from plant material and traditional preparative tech-niques, including liquid–liquid extraction (LLE) or column chromatography on different sorbents, are well-known procedures applied for isolation and purification of phenolic acids. Solid-phase extraction (SPE) has become a popular procedure, which is used for isolation, purification, and preconcentration of organic compounds present in biological material as an alternative to the methods described earlier [26].

Several papers have been published on the optimal extraction and isolation methods for the analysis of phenolic acids in plants and plant-based foods [5,27–30].

The first step is to mill, macerate, or grind the sample to increase the surface area, which allows a better contact of the extracting solvent with the sample. It also helps in mixing the sample to ensure that the extracted portion is representative of the entire sample. Since many phenolic compounds occur as glycosides or esters, the sample preparation may include alkaline, acid, or enzymatic hydrolysis to unbind bound phenolics [5].

14.3.1 DIFFERENT EXTRACTION METHODS

Extraction yield of phenolic acids from plant materials is influenced by their chem-ical nature (e.g., polarity, acidity, hydrogen-bonding ability of the hydroxyl groups in the aromatic ring). The extraction method employed sample particle size, storage time, and conditions, as well as presence of interfering substances. These compounds may exist as free and estrified phenolic acids or glycosides [1,4].

Most phonetic acid derivatives accumulated in the plant matrix are stored in vacuoles and are commonly extracted with alcoholic or organic solvents. The common solvents for extractions are hot water, ethanol, acetone, diethyl ether, chloroform ethyl acetate, and methanol (or aqueous methanol), the latter being the most common. Escarpa et al. report the successful extraction method of phenolic acids from apples and pears using 100% methanol and from green beans and lentil using 80% aqueous methanol [1]. Recent investigations show that for olive oil, 80% aqueous methanol yields the highest extraction rates [1,6,7]. Another investigations display effectiveness of various extraction methods using methanol and aqueous methanol as extractant. The following liquid solid extraction methods were used: Soxhlet extraction, ultrasonification (USAE), microwave-assisted extraction (MAE), and accelerated solvent extraction (ASE) [26,30–33].

Exhaustive extraction was performed in Soxhlet apparatus for 15 h using the same solvent. Ultrasound assisted extraction was performed for 30 min using pure methanol. Then extract was filtered and the same plant material was soaked with the fresh portion of methanol an again extracted in ultrasonic bath. It was repeated three times. All collected methanolic extracts were combined. The extraction was carried out at ambient temperature (20°C ± 1°C) and at 60°C. MAE was performed with 80% methanol in water using two-step extraction: by 40% generator power during 1 min and by 60% generator power (600 W) during 30 min in open or in closed system. ASE was carried out with 80% methanol in water (for flowers of *Sambucus nigra* L) or with pure methanol (for herb of *Ploygonum aviculare* L) under the following conditions: pressure—60 bars, temperature—100°C, time—10 min [30].

The highest extraction yield of protocatechuic, *p*-hydroxybenzoic, vanillic, and ferulic acids in case of *Sambucus nigra* inflorescence was obtained by Soxhlet extraction. Methods such as ultrasonification, microwave-assisted solvent extraction, and ASE gave similar extraction yield of phenolic acids from *Sambucus nigra* inflorescence.

The highest extraction yield of protocatechuic, *p*-hydroxybenzoic, gallic acids in case of *Polygonum aviculare* herb was obtained by MAE in closed system.

For ferulic acid extracted from *Polygonum aviculare* herb the highest yield was obtained with USAE. Ultrasonification gives for investigated phenolic acids higher (for protocatechuic, *p*-hydroxybenzoic) or similar (for gallic) yield from *Polygonum aviculare* herb as long-lasting exhaustive extraction in Soxhlet apparatus [30].

In the other investigation microwave-assisted extraction has been tested and optimized for the isolation of phenolic acids (gallic, protocatechuic, *p*-hydroxybenzoic, chlorogenic, vanilic, caffeic, syringic, *p*-coumaric, ferulic, sinapic, benzoic, *m*-coumaric, *o*-coumaric, rosmarinic, cinnamic acids) from various plants. The effects of experimental conditions on MAE efficiency, such as solvent composition, temperature, extraction time, have been studied. The extraction efficiencies were compared with those obtained by computer-controlled, two-step Soxhlet-like extractions [33].

In MAE, 1 g of dry plant material was extracted by 10, 20, and 30 mL of water, 0.1 and 2 mol L^{-1} HCl, each containing 0%, 10%, 20%, and 30% of methanol (v/v). The extraction temperature was changed from 40°C to 90°C with 10°C steps and was maintained for 10, 15, 20, 25, and 30 min. Pulsed microwave treatment was used (500 W maximum dose). The best results were obtained with the method proposed for extraction of phenolic compounds from *Dactylis glomerata*. The final analytical procedure employing the MAE with 2 mol L^{-1} HCl for 20 min ($t = 70$°C) was applied for the isolation of phenolic compounds from several plant materials.

Microwave extraction has been found to be more suitable because of shortened extraction time and better RSDs due to more stable conditions (temperature, time) during the extraction. RSDs for MAE were 3.2%–5.8% compared with 4.0%–9.8% in Soxhlet extraction. Recovery of MAE was about 90% for the internal standard (*m*-coumaric acid), i.e., 10% losses appeared during subsequent manipulation with extracts and their preparation for subsequent analysis. Repeatability on a day-to-day and 5-day bases were 3.0%–6.1% and 3.9%–6.8%, respectively [33].

Application of subcritical water extraction (SWE) for isolation of phenolic acids from plant material has recently been reported. The SWE is carried out using hot water, above its boiling point, which is sufficiently pressurized to 15 bar at 200°C and 85 bar at 300°C, to maintain a liquid state. The efficiency of SWE is affected by temperature, extraction time, as well as the presence of small quantities of organic solvents and surfactants. At lower temperatures the water is a very polar system but at higher temperatures ($\geq$250°C) the polarity of pressurized water is similar to that displayed by polar organic solvents. Thus, the polar rosmarinic acid was preferentially extracted at a lower temperature (100°C) while less polar carnosic acid was effectively extracted at a higher temperature (200°C) [34].

Smolarz reported the successful extraction of phenolic acids and their methyl esters from *Polygonum* L. genus with chloroform [27]. Krygier et al. extracted free and estrified phenolic acids from oilseeds using a mixture of methanol–acetone–water (7:7:6 v/v/v) [4]. Chlorogenic acid were efficiently extracted from buckwheat herbs by agitated maceration in 30% ethanol at 60°C for 2 h [4,34]. Different solvents were used for the extraction of major polyphenols in *Orthosiphon stamineus* leaves. Leaf powder was extracted with the following solvents: distilled water, 50% aqueous methanol, methanol, 70% aqueous acetone, and chloroform, at 2, 4, and 8 h, respectively, on a water bath at 40°C. The yield of rosmaric acid was the highest in 50% aqueous methanol extracts at 4 h of extraction; however concentrations of this compound were high in water, methanol, and 70% aqueous acetone extracts. Caffeic acid derivatives, including rosmaric acid, have been reported to constitute 67% of total identified phenolics in aqueous methanol extract and about 94.6% in hot water extract [35].

Lege et al. reported the successful extraction of phenolic acid from cotton. Dried plant samples were extracted with 6 mL 2% glacial acetic acid, placed in a boiling water bath for 10 min, and centrifuged for 15 min. The supernatant was filtered and then hydrolyzed [36].

14.3.2 Hydrolysis of the Esters and Glycosides

Hydrolysis of the esters and glycosides to a free phenolic acid have been a strategy employed to simplify the analysis. There are two main procedures to cleave the ester and glycoside bond reported in the literature, acidic hydrolysis (for glycosides), and alkaline hydrolysis (for esters). A third, less prevalent technique is cleavage through the use of enzymes (esterases or glucosidases) [1,4,37–39].

Although reaction time and temperature for the acidic hydrolysis condition vary a great deal, this general method involves treating the plant extract itself with inorganic acid (HCl) at reflux or above reflux temperatures in aqueous or alcoholic solvent (methanol being the most common, acid ranged from 1 to 2 M aqueous HCl, and the reaction times ranged from 30 min to 1 h). Gao et al. examined a series of acid hydrolysis media in an attempt to find a system for maximum preservation of phenolic acids. A mixture of 2 M HCl and methanol (1:1, v/v) at 100°C for 1 h yielded the highest recovery (30%–65%). Both methyl esters and carboxylic acids were monitored. Solvents such as ethanol, *tert*-butyl alcohol, and 2-propanol gave lower results, and aqueous HCl is reported to have destroyed the hydroxycinnamic acids. Hollman and Venema tested extraction and acid hydrolysis of ellagic acid from walnuts and berries (strawberry, blackberry, red respberry) using different HCl and aqueous methanol concentrations together with a variation of hydrolysis period. The hydrolysis characteristics of ellagitannins of walnuts (optimum 5 M HCl in 57% aqueous methanol for 1 h) different from those of berries (optimum 3.5 M HCl in 72% aqueous methanol for 4–8 h). Krygier et al. report that loss under acidic conditions varies with the phenolic acid, ranging from 15% (for *o*-coumaric acid) to 95% for sinapic acid [1,5].

Alkaline hydrolysis (saponification) entails treating the sample with a solution of NaOH with reported concentrations ranging from 1 to 4 M. Most of the reactions are allowed to stand at room temperature for time ranging from 15 min to overnight, although Shahrzad et al. report hydrolyzing grape juice with 1.5 M NaOH for 62 h. Some investigations mention that the reactions are carried out in the dark, as well as under an inert atmosphere such as nitrogen gas. Minor modifications to this typical method include saponification with constantly agitation for 24 h at 37°C. In the recent investigations alkaline hydrolysis was performed using $NaBH_4$ and $Ba(OH)_2$ at 100°C for 15 min. Lam et al. carried out basic hydrolysis under harsh conditions for comparison and conducted the saponification with 4 M NaOH, for 2 h at 170°C. In the other investigations alkaline hydrolysis was obtained under dark, anaerobic conditions by purging the reaction with N_2 [1].

Krygier et al. reported that alkaline hydrolysis may lead to some significant losses of hydroxycinnamic acid derivatives, but this can be prevented by the addition of 1% ascorbic acid and 10 mM ethylenediaminetetraacetic acid (EDTA) [4,34].

Rommel and Wrolsta tested acid (HCl) and base (NaOH) hydrolysis in the analysis of phenolic composition (e.g., ellagic, hydroxybenzoic, and hydroxycinnamic acids) of red raspberry juice. The phenolic acids pattern of the alkaline-hydrolyzed sample was very similar to that of the acid-hydrolyzed sample of the same juice. Only one ellagic acid compound was hydrolyzed more effectively in alkaline than in acidic conditions [5]. Seo and Morr found that hydrolysis with NaOH leads to better recovery of ferulic acid from soybean protein products than hydrolysis with HCl [5].

14.3.3 SAMPLE PURIFICATION AND CONCENTRATION

Phenolic extracts of plant materials are always a mixture of different classes of phenolics that are soluble in the used solvent system. Additional steps may be required for the removal of unwanted phenolics and nonphenolic ballast substances such as waxes, fats, terpenes, chlorophylls (nonpolar) and tannines, and sugars (polar). LLE, solid-phase extraction (SPE) techniques and fractionation based on polarity and acidity are commonly used to remove unwanted phenolics and nonphenolic substances [4].

Different fractions of phenolic compounds (polymeric procyanidins, phenolic carboxylic acids, and flavonoids) of hawthorn (*Crataegus laevigata*) were fractionated prior to HPLC analysis using column chromatography and SPE. The flowers and leaves of hawthorn were milled to a powder, then extracted with aqueous methanol and pure methanol, and fractioned with polyamide and Sephadex LH-20 columns. Fractions A1, A2, A3, B1, and B2 were obtained (Figure 14.5). The polyamide column was eluted after fraction A3 with acetone and water (7:3) to extract the adsorbed polymeric procyanidins in fraction A4. The first fraction A1

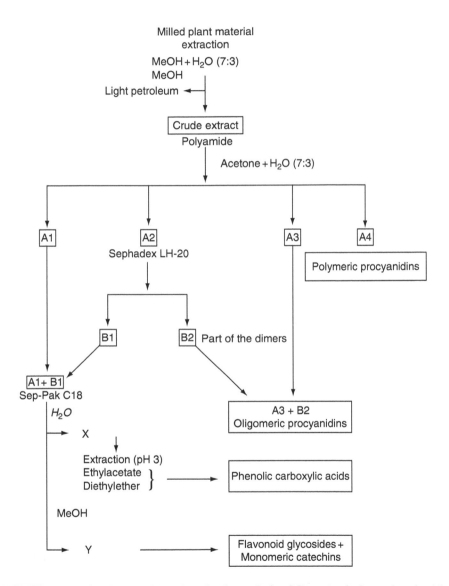

FIGURE 14.5 A fractionation procedure for the analysis of Crataegus lavigata phenols. A1, A2, A3, A4—fractions from the Polyamide column; B1, B2—fractions from the Sephadex LH-20 column; X, Y—fractions from the Sep-Pak C18 plus column. (From Svedröm, U., Vuorela, H., Kostiainen, R., Laakso, I., and Hiltunen, R., *J. Chromatogr. A*, 1112, 103, 2006.)

eluted from the polyamide column and the first fraction B1 eluted from the Sephadex LH-20 column were combined and evaporated to dryness. Five milliliters of water was added to the evaporated residue and the pH was simultaneously adjusted to 7.0 with 1 M NaOH. This neutral solution was loaded onto a Sep-Pak C18 plus cartridge, preconditioned with methanol and next with distilled water (pH 7). After adding the sample, the cartridge was eluted with water (pH 7) to remove the phenolic carboxylic acids (fraction X). The cartridge was then eluted with methanol to extract the adsorbed flavonoids and $(-)$-epicatechin (fraction Y).

Phenolic carboxylic acids were isolated from the aqueous solution (fraction X) by acidifying it to pH 3 with 1 M HCl and extracting sequentially with ethyl acetate and diethyl ether. The solvents were evaporated, the residue was dissolved in methanol, filtered and analyzed by HPLC [40].

14.3.3.1 Liquid–Liquid Extraction

LLE is a well-known procedure applied for purification of phenolic acid fractions.

All extracts (obtained by different extraction methods) were evaporated to dryness under reduced pressure and purified by LLE. Dry residues, soaked with boiling water, were cooled in fridge for 12 or 24 h and then filtered. The resinous precipitate was washed with distilled water and the filtrate was defatted by double extraction with petroleum ether. The aqueous solution (I) was extracted with diethyl ether. Phenolic acids contained in the diethyl ether layers were further treated with 5% sodium bicarbonate to transform the phenolic acids into water-soluble sodium salts. Bicarbonate layers were acidified with 36% hydrochloric acid until pH 3 and, thereafter, extracted against with diethyl ether. The combined ether extracts were dried with anhydrous Na_2SO_4. After evaporation of the solvent the dry fraction of free phenolic acids was obtained [32,37,38].

One part of aqueous solution (I) was acidified with 36% hydrochloric acid (to pH 2), heated for 1 h 60°C to transform the glycosides to a free phenolic acids and then extracted with diethyl ether. The combined ether extracts were dried with anhydrous Na_2SO_4. After evaporation of the solvent the dry fraction of free phenolic acids after acidic hydrolysis was obtained.

To a second part of aqueous solution (I) was added 1 g $NaBH_4$ and 10% $Ba(OH)_2$ (to pH 12) and the mixture was heated for 15 min at 60°C. Next, the solution was acidified with H_2SO_4 to pH 2 and extracted with diethyl ether. The combined ether extracts were dried with anhydrous Na_2SO_4. After evaporation of the solvent the dry fraction of free phenolic acids after basic hydrolysis was obtained [28,32,37,38].

14.3.3.2 Solid-Phase Extraction

SPE makes the sample preparation faster and easier. All extracts (obtained by different method of extraction) were evaporated to dryness under reduced pressure and purified by SPE.

Phenolic compounds from apple musts were fractionated into neutral and acid groups by means of SPE. The experimental procedure was as follows: C_{18} columns were preconditioned for neutral phenols by sequentially passing of methanol and water adjusted to pH 7 and for phenolic acids by sequentially passing methanol and 0.01 M HCl. The sample containing ascorbic acid to avoid oxidation of poly-phenols, was adjusted to pH 7 with NaOH, loaded onto the C_{18} neutral fractionating column, and washed with water. The effluent portion was adjusted to pH 2 with 2 M HCl, passed through the preconditioned acidic column, and washed with 0.01 M HCl. The adsorbed fractions were eluted with methanol and evaporated to dryness. The residue was redissolved in methanol, filtered, and analyzed [41].

In the other investigations extraction of the phenolic acids from the plant extracts was carried out according to the following procedure. The RP-105 SPE cartridge was first conditioned with methanol followed by water. Then the extract was applied onto the cartridge. Co-extracted substances were rinsed from the sorbent with a mixture of 0.1 mol l^{-1} hydrochloric acid: methanol (9:1, v/v). The column was dried by nitrogen gas. All retained compounds of the interest were eluted with diethyl ether. The solvent was evaporated under reduced pressure, the residue was dissolved in the mobile phase, filtered, and analyzed. Methanol was not suitable for the elution of the analytes because of co-elution of many interfering substances, which complicates chromatographic procedure. Comparison of a number of identified and quantified phenolic acids in crude extract and in extracts after purification by LLE and SPE extraction indicated that LLE has no advantage over SPE techniques because of the lower concentration of the individual phenolic acids and the higher variation of the results (RSDs 4.9%–7.5% for LLE and 3.0%–5.1% for SPE). Moreover, less compounds were found in extract of *Dactylis glomerata* after their purification by LLE procedures in comparison with SPE. In the case of SPE (diethylether used as eluent) were also *p*-hydroxybenzoic, *p*-coumaric, sinapic acid, and vanillin were identified and quantified [33].

In an other paper, the aqueous solution of phenolic acids from rice (after acid hydrolysis of ferulic, sinapic, *p*-coumaric, chlorogenic, caffeic, protocatechuic, hydroxybenzoic, vanillic, and syringic acids glycosides) was subjected to SPE using a Sep-Pak Vac C_{18} cartridge previously activated with methanol and conditioned with acidified water. The cartridge was washed first with acidic water–methanol (9:1, v/v) and then the phenolic compounds eluted using acidic water–methanol (3:7, v/v) [42].

Zgórka et al. isolated phenolic acids from *Scutellaria baicalensis*, *Eleutherococcus senticosus*, and *Echinacea* species and other plants. The SPE procedure was carried out using C_{18} and quaternary amine cartridges [26,28,29].

C18 cartridges were conditioned with methanol and with distilled water. Samples of methanolic extracts were evaporated to dryness, diluted with 30% aqueous mathanol, and passed under vacuum through octadecyl cartridges. Eluates, purified from the ballast compounds and containing the complex of phenolics were further adjusted to pH 7.0–7.2 with 5% sodium bicarbonate aqueous solution to obtain anionic form of analytes. Then they were passed under vacuum through quaternary amine SPE-cartridges. These sorbent beds were equilibrated immediately prior to use with bidistilled water followed by 0.1% sodium bicarbonate aqueous solution. The extracts with anionic forms of phenolic acids were introduced to cartridges prepared as described earlier. After introducing the samples, the cartridges were dried under vacuum and phenolic acids were desorbed and eluted from the sorbent beds by use of solvent mixture: 0.2 M phosphoric acid–methanol (1:1; v/v) [26,28,29].

14.4 MATRIX SOLID-PHASE DISPERSION

Matrix solid-phase dispersion (MSPD) is a sample preparation technique that combines both sample homogenization and extraction of the analyzed compounds in one step. This method of extraction is based on the blending of a viscous, solid, or semisolid sample with an abrasive solid support material. The bounded organic phase acts as a solvent or detergent that dissolves and disperses the sample components into the bounded phase. MSPD columns prepared with reversed-phase supports are most frequently eluted with a sequence of solvents beginning with the least polar (hexane) and then with those of increasing polarity (ethyl acetate, acetonitrile, methanol), then water. MSPD has been demonstrated to be a suitable preparation technique, a simple alternative to LLE, SPE, and SFE, for the isolation of phenolic acids from plant material. No homogenization, grinding or milling steps are necessary. It is only necessary to select suitable elution agents giving the highest yields of the analytes and to optimize the volume of the elution medium. Washing steps can be changed according to the amounts of interfering and co-eluting compounds. MSPD was used for sample preparation of *Melissa officinalis* tops prior to liquid chromatography of rosmarinic, caffeic, and protocatechuic acids—phenolic compounds presented in this herb. Different MSPD sorbents (Polygoprep C_{18}, nonendcapped; Bakerbond C_{18} endcapped; Silasorb C_{18}, nonendcapped; Alltech bulk high capacity C_{18}, endcapped; SGX C_{18} slightly endcapped; Florisil) and various elution agents (methanol, methanol acidified with 0.2% formic acid, methanol–acidified water, pH 2.5, and ethyl acetate) were tested. The optimal extraction conditions were determined with the aim to obtain extraction recoveries greater than 90% for all analytes. The experimental procedure was as follows: dried plant tops sample was mixed with previously cleaned sorbent and *n*-hexane. The mixture was homogenized, then the blend was transferred into a syringe with a paper frit on the bottom. The sample was covered with another paper frit and compressed using the syringe plunger. Interfering compounds were washed with *n*-hexane and next with dichloromethane. The syringe was then dried under vacuum. Phenolic compounds were eluted directly with the mixture of methanol–water, pH 2.5 (80:20) and the residue after evaporation to dryness was dissolved in mixture of methanol–water. Eluents were filtered and analyzed [43].

14.5 THE APPLICATION OF TLC IN PHENOLIC ACIDS ANALYSIS

Because of the varied pharmacological activity of phenolic acids plants containing these compounds are widely used in phytotherapy. Therefore, their determination and identification in plant material is very important aspect of phytochemical analysis. Although a lot of methods have been reported for the investigation of phenolics, thin layer chromatography (TLC), owing to its numerous advantages, is still often employed. TLC is fast, inexpensive, and it can be used as a preliminary method in analysis of these compounds preceding the other methods, e.g., HPLC. There is a large variety of solvent systems, adsorbents, and chromatographic techniques employed in analysis of phenolic acids. Silica gel [44–52] or cellulose [53–60] were the most frequently used as the stationary phases but silica modified with octadecyl (RP-18) [61–65], cyanopropyl (CN) [66,67], aminopropyl (NH₂) [65,68] and diol [65,69,70] groups was also reported.

The mixtures of three, four, or even five solvents with various polarity were used as the mobile phases. Typically, nonpolar or weakly polar solvent (e.g., benzene, heptane, toluene), medium polar diluent such as chloroform or ethyl acetate, and strongly polar modifier (e.g., methanol, water) were mixed to obtain the best selectivity of separation. Moreover, acetic or formic acid in concentration range of 0.5%–1% is added to the mobile phase to suppress the ionization of acidic groups and improve the shape of chromatographic bands.

In TLC of phenolic acids, several techniques can be employed depending on analytical task. One-dimensional, isocratic or gradient chromatography is more suitable for quantitative determination whereas two-dimensional technique is usually applied only to preliminary identification of phenolics fraction.

14.5.1 Isocratic TLC

Early investigations used paper chromatography on Whatman filter paper. First attempts using TLC to the phenolic acid analysis have been described by Stahl and Schorn [71], Scrapati and Guiso [72], and van Sumere et al. [73].

Nowadays, many TLC systems have been developed for the separation of phenolic compounds. More polar components which adsorb strongly on silica surface, such as chlorogenic, ellagic, or rosmarinic acid are usually chromatographed in solvent system consisting ethyl acetate/formic acid/water in different ratio (8:1:1, 65:15:20, 88:6:6) [45–47,74]. In some publications formic acid is replaced by methanol, 1-propanol, or 1-butanol [46,47]. The mixture of ethyl acetate/formic acid/acetic acid/water (100:11:11:27) proposed by Wagner is also often employed [74]. The presented chromatographic systems are not adequate for analysis of less polar phenolics because the R_F values are too high, e.g., ferulic and caffeic acids have R_f values above 0.8.

The separation of wider group of popular phenolic acids is obtained by addition of weak polar solvent, e.g., chloroform, toluene, benzene, petroleum ether to the mobile phase. The exemplary R_f values of phenolic compounds in different eluents systems were described by Rastija et al. [48] and Sharma et al. [49]. Table 14.4 presents the examples of different mobile phases suitable for analysis of phenolic acids in plant extracts.

The separation selectivity with use one-dimensional TLC can be improved on silica modified with different chemical groups. The possibility of use aminopropyl silica HPTLC plates has been investigated by Wójciak-Kosior and Skalska [68]. Owing to strong adsorption on this kind of adsorbent, the chromatographic system required mobile phase with addition of acetic acid in concentration above 5%. The satisfactory results were obtained for the mixtures of petroleum ether, heptane, or toluene with diisopropyl ether as a modifier. Figure 14.6 presents the retention of phenolic acids on aminopropyl silica after development with mixture of diisopropyl ether/petroleum ether/acetic acid (6:3:1).

The retention behavior of phenolic acids was also tested on silica modified with diol groups [69]. The mixtures of n-heptane with polar diluents: ethyl acetate, isopropanol, dioxane, and tetrahydrofuran in various concentrations were used as

TABLE 14.4
The Different Chromatographic Systems Applied to Analysis of Phenolic Acids

Adsorbent	Mobile Phase	References
Cellulose	3% NaCl in 0.5% acetic acid–ACN–THF (100:24:1)	[60]
Cellulose	Toluene:ethyl formate:formic acid (5:4:1)	[56,57]
Cellulose	Chloroform:acetic acid:water (4:1:1)	[56,75]
Cellulose	Water–acetic acid (95:5, 85:15) v/v	[53,56,57,76]
Cellulose	Sodium formate:formic acid:water (10:1:100)	[57]
Polyamide	Benzene:methanol:absolute acetic acid (45:8:4)	[77]
Silica gel HPTLC	Ethyl acetate:water:acetic acid:formic acid (100:26:11:11 v/v/v)	[74,78]
TLC silica gel	Ethyl acetate–water–formic acid (8:1:1, 65:20:15, 88:6:6, 85:15:10)	[44,45]
TLC silica gel	Ethyl acetate–methanol–water (100:13.5:10, 77:13:10)	[46,47]
TLC silica gel	Chloroform-methanol-water (65:35:10)	[50]
TLC silica gel	Chloroform–methanol–formic acid (15:3:2, 147:30:23, 37:8:5, 36:9:5)	[48]
TLC silica gel	Chloroform–ethyl acetate–formic acid (21:24:5, 21:23:6)	[48]
TLC silica gel	Chloroform–ethyl acetate–acetic acid (50:50:1)	[49]
TLC silica gel	Chloroform–methanol–acetic acid (90:10:1)	[49]
HPTLC silica gel	Dichlorometan:acetonitrile:formic acid 90% (9.5:0.5:0.1)	[79]
TLC silica gel	Toluene–chloroform–acetone–formic acid (8:4:3:3)	[48]
TLC silica gel	Toluene–acetone–formic acid (7:6:1)	[48]
TLC silica gel	Toluene–ethyl acetate–formic acid (30:10:10, 55:20:25)	[48]
TLC silica gel	Toluene–acetonitrile–formic acid (70:30:1)	[49]
TLC silica gel	Benzene–ethyl acetate–formic acid (30:15:5)	[48]
TLC silica gel	Benzene:methanol:absolute acetic acid (45:8:4)	[77]
TLC silica gel	Benzene–dioxane–acetic acid (85:15;1)	[49]
TLC silica gel	Benzene–chloroform–ethyl acetate–formic acid (5:12:10:3)	[52]
TLC silica gel	Benzene–chloroform–ethyl acetate–formic acid–methanol (10:15:20:10:1)	[52]
TLC silica gel	Petroleum ether–ethyl acetate–formic acid (40:60:1)	[49]
TLC silica gel	Petroleum ether–methanol–acetic acid (90:10:1)	[49]
HPTLC silica gel	Hexane–ethyl acetate–formic acid (20:19:1, v/v/v)	[51]

mobile phases. The plots of hR_f values against the concentration of modifier were determined for eleven derivatives of cinnamic and benzoic acids.

The R_f values for three phenolic acids: ferulic, caffeic, and chlorogenic on cyano-bonded silica in normal phase (NP) and reversed-phase (RP) systems are described by Soczewiński et al. [66]. The differences in selectivities of NP systems were investigated for binary eluents comprising n-hexane and polar solvent: acetone, propan-2-ol, ethyl acetate, tetrahydrofuran, 1,4-dioxane, ethyl–methyl ketone, or diisopropyl ether. Retention vs. eluent composition relationships for RP system were determined for mobile phases comprising water and methanol, tetrahydrofuran, or 1,4-dioxane.

Although the polar stationary phases are the most frequently applied in analysis of phenolic compounds, the nonpolar adsorbents such as RP-2, RP-18 are also sometimes exploited [61–65]. The aqueous solvent systems consisting mixture of methanol with phosphate buffer [61], acetic acid [62], or water [63] are usually

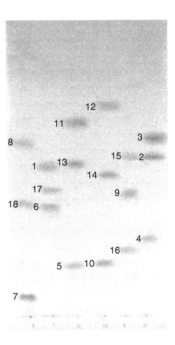

FIGURE 14.6 The separation of phenolic acids on aminopropyl silica with mixture of diisopropyl ether/petroleum ether/acetic acid (6:3:1): (1) salicylic acid, (2) m-hydroxybenzoic acid, (3) p-hydroxybenzoic acid, (4) protocatechuic acid, (5) α-resorcylic acid, (6) β-resorcylic acid, (7) gallic acid, (8) vanillic acid, (9) syringic acid, (10) gentisic acid, (11) veratric acid, (12) cinnamic acid, (13) o-coumaric acid, (14) m-coumaric acid, (15) p-coumaric acid, (16) caffeic acid, (17) ferulic acid, and (18) sinapic acid.

applied for these kinds of adsorbents. Fecka and Cisowski used eluent consist of water/acetonitrile/acetic acid (89:10:1) for the separation of the extract from *Lycopus europaeus* L on RP 18 layer [65].

Figure 14.7 presents the separation of phenolic acids on silica modified with octadecyl groups with use of methanol/water mixture as a mobile phase. The use of mobile phase or adsorbent additives to improve the selectivity of separation was also reported. The effect of addition of β-cyclodextrin on the retention of phenolic acids on HPTLC RP-2 plates were investigated by Bieganowska et al. The mixtures of 0.01 M β-cyclodextrin and 0.04 M orthophosphoric buffer (pH = 3) with acetonitryle, methanol, or tetrahydrofuran were used as a mobile phase. Phenolic acids formed inclusion complexes with β-cyclodextrin and the separation factors were significantly improved by increasing the concentration of β-cyclodextrin [80]. The influence of impregnation of silica layers with $Fe(NO_3)_3$ on behavior of some phenolic acids was tested by Hadzija et al. [81].

14.5.2 ONE-DIMENSIONAL MULTIPLE DEVELOPMENT

Isocratic TLC is often inadequate for resolution of the components present in complex samples. A marked improvement of the separation can be achieved by

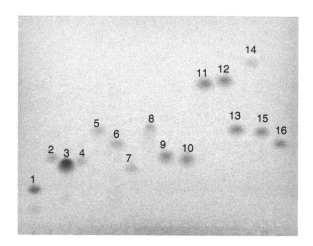

FIGURE 14.7 The separation of phenolic acids on RP-18 layer with use the mixture of methanol/water (50:50) as a mobile phase: (1) cinnamic acid, (2) *p*-coumaric acid, (3) *o*-coumaric acid, (4) *m*-coumaric acid, (5) caffeic acid, (6) hydrocaffeic acid, (7) ferulic acid, (8) chloroggenic acid, (9) *p*-hydroxybenzoic acid, (10) salicylic acid, (11) protocatechuic acid, (12) α-resorcylic acid, (13) β-resorcylic acid, (14) gallic acid, (15) vanillic acid, (16) and syringic acid.

multiple development (MD). In this technique the plate is chromatographed consecutively several times with solvents of the same or varying eluent strength and carefully dried after each step of development. The MD technique enables separation of compounds with similar chemical structures, for example, phenolic acids differing only by the –CH = CH-grouping which has minor effect on the polarity. Figure 14.8 presents MD of the extract from *Polygoni avicularis herba* after alkaline hydrolysis with mixture of diisopropyl ether–heptane–formic acid $(5 + 4 + 1)$. The distance of development was 9 cm [82].

More frequent variant of MD technique is MGD-multiple gradient development in which the mobile phases with decreasing or increasing elution strength are used. MGD enables the analysis of mixtures spanning a wide polarity range.

Program of two-step gradient development to analysis of caffeic acid in some *Dipsacaceae* family plants was proposed by Kowalczyk et al. [83]. The good separation of phenolic acids fraction in extracts of *Lamii albi flos* was obtained by Wójciak et al. after four-step gradient development with mixture of heptane/diisopropyl ether/dichloromethane/formic acid [32]. MGD on diol-silica was also employed by Pobłocka et al. in determination of *p*-coumaric acid in *Atropa belladonna* L., *Datura stramonium* L., and *Hyoscyamus niger* L. [70].

Some gradient programs to the separation of the extract of *Lycopus europaeus* L. for different adsorbents, e.g., diol, aminopropyl, Si60 HPTLC, and LiChrospher plates are described by Fecka and Cisowski [65]. Figure 14.9 shows the results obtained on LiChrospher Si 60 after four step gradient program with the use of cyclohexane/diisopropyl ether/formic acid mixtures in various proportions.

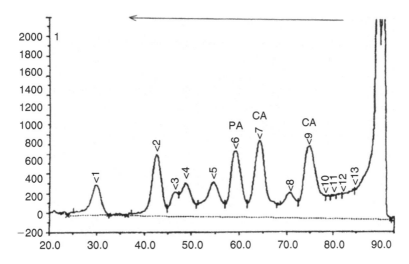

FIGURE 14.8 Densitogram of multiple development of the extract from *Polygoni avicularis herba* after alkaline hydrolysis obtained at $\lambda = 254$ nm. Chromatographic system: HPTLC Si 60 plate/diisopropyl ether–heptane–formic acid $(5 + 4 + 1)$; PA—protocatechuic acid, CA—caffeic acid, GA—gallic acid.

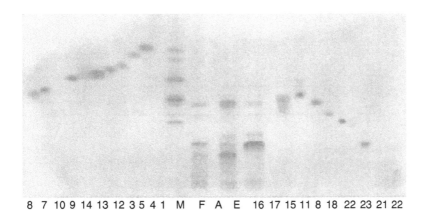

8 7 10 9 14 13 12 3 5 4 1   M   F A E   16 17 15 11 8 18 22 23 21 22

FIGURE 14.9 Chromatogram of extracts from *Lycopus europaeus* L. obtained after gradient program: step 1—diisopropyl ether/formic acid (80:20 v/v), distance 2 cm; steps 2,3—cyclohexane/diisopropyl ether/formic acid (20:78:2 v/v/v), distance 9 cm; step 4—cyclohexane/diisopropyl ether/formic acid (20:79:1 v/v/v), distance 9 cm. M-mixture of standards, F—free phenolic acids, A—phenolic acids after alkaline hydrolysis, E—phenolic acids after enzyme hydrolysis: (1) salicylic acid, (2) *m*-hydroxybenzoic acid, (3) *p*-hydroxybenzoic acid, (4) 2,4-dihydroxybenzoic acid, (5) gentisic acid, (7) 3,5-dihydroxybenzoic acid, (8) gallic acid, (9) vanillic acid, (10) veratric acid, (11) syringic acid, (12) *o*-coumaric acid, (13) *m*-coumaric acid, (14) *p*-coumaric acid, (15) caffeic acid, (16) ferulic acid, (17) isoferulic acid, (18) sinapic acid, (21) isochlorogenic acid, (22) rosmarinic acid, and (23) ellagic acid.

14.5.3 TWO-DIMENSIONAL TLC

The two-dimensional (2D) paper chromatography has been widely applied in phytochemical studies but nowadays the most frequently used adsorbent in 2D-TLC of phenolic acids is cellulose layer. 2D-technique enables the separation of a number of phenolic acids and it is an excellent tool to preliminary analysis. The most investigations on cellulose are based on solvent system proposed by Smolarz and Waksmundzka-Hajnos. The authors successfully separated 14 phenolic acids and their isomeric forms with use the mixture of methanol, acetonitrile, benzene, and acetic acid $(10 + 5 + 80 + 5 \text{ v/v})$ for development in the first direction and eluent consisting sodium formate–formic acid–water $(10 + 1 + 200 \text{ w/v/v})$ for development in the second direction [54]. The described solvent system was used for identification of phenolic acids in the leaves of *Polygonum amphibium* [54], in the petioles of *Rheum* sp. [55], in the flowers of *Lavatera trimestris* [56] and in the leaves, roots and fruits of *Peucedanum verticillare* [57]. Some modifications were also made, e.g., Cisowski et al. replaced methanol and acetonitrile with water [58], Bartnik et al. employed toluene in the place of benzene [59], Kowalski and Wolski used for the first direction the mixture of benzene/methanol/absolute acetic acid (45:8:4) [77]. Figure 14.10 presents the 2D-TLC separation of phenolic acids from the foliage of *Peucedanum tauricum* Bieb. obtained after acid hydrolysis.

2D-TLC technique on cellulose layer was also applied by Ellnain-Wojtaszek and Zgórka to TLC of phenolic acids from *Ginkgo biloba* L. The suitable separation was

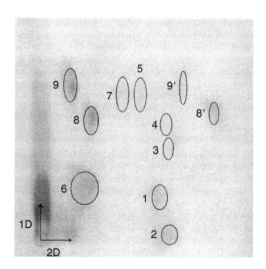

FIGURE 14.10 2D TLC on cellulose of phenolic acids from the foliage of *Peucedanum tauricum* Bieb after obtained after acid hydrolysis. The plate was developed with toluene/methanol/acetic acid/acetonitrile (75.5:10:5:7.5) in the first direction and with sodium formate/formic acid/water (10:1:200) in the second direction: (1) protocatechuic acid, (2) chlorogenic acid, (3) gentisic acid, (4) *p*-hydroxybenzoic acid, (5) vanillic acid, (6) caffeic acid, (7) syringic acid, (8) *p*-coumaric acid, 8′-cis isomer of *p*-coumaric acid, and (9) ferulic acid, 9′-cis isomer of ferulic acid.

achieved with mixtures of benzene/acetic acid/water (6:7:3) and acetic acid/water (3:17) as mobile phases [84].

Silica gel can also be used as a stationary phase in two-dimensional TLC for such purposes [85,86].

2D-TLC on cyanopropyl-bonded stationary phase was performed by Hawryl et al. [67] to investigation of phenolic compounds in the extract from the flowers of *Sambucus nigra* L. The plates were developed with mixture of hexane with acetone or 2-propanol in the first direction and after drying with mixture of water/methanol, water/tetrahydrofuran or water/dioxane in the second one.

The exemplary mobile phases used in 2D-TLC technique are presented in Table 14.5.

2D technique is usually employed only to qualitative determination because comparison of substance with standard is difficult. The interesting resolution of this problem was presented by Medić et al. to quantitative analysis of caffeic acid in propolis [86]. The standard at different concentration was spotted on three places of chromatographic plate and the plate was chromatographed with mixture of *n*-hexane/ethyl acetate/glacial acetic acid (31:14:5) in one direction and after drying with chloroform/methanol/formic acid (44:3.5:2.5) in the second direction. The obtained results are presented in Figure 14.11.

The 2D development can also be coupled with the MGD technique. Glensk et al. obtained good separation of mixture of phenolic acids on silica with use of the

TABLE 14.5
The Examples of Solvent Systems Used in 2D-TLC

No	Mobile Phase		Adsorbent	References
	First Direction	Second Direction		
1	Methanol/acetonitrile/benzene/ acetic acid (10 + 5 + 80 + 5 v/v)	Sodium formate/formic acid/water (10 + 1 + 200)	Cellulose	[54–57]
2	Toluene/methanol/acetonitrile/ acetic acid (75.5 + 10 + 7.5 + 5 v/v)	Sodium formate/formic acid/water (10 + 1 + 20)	Cellulose	[59]
3	Benzene/methanol/absolute acetic acid (45:8:4)	Sodium formate: formic acid: water (10 + 1 + 200)	Cellulose	[77]
4	Benzene/acetic acid/water (6:7:3)	Acetic acid/water (3:17)	Cellulose	[84]
5	Water/acetic acid (85:15)	Benzene/absolute acetic acid/water (6:7:3)	Cellulose	[77]
6	Toluene/ethyl acetate/methanol (85:10:5)	Methanol/ethyl acetate (1:9)	Silica gel	[85]
7	Isopropanol/*n*-hexane (30:70)	Ethyl acetate	Silica gel	[85]
8	*N*-hexane/ethyl acetate/glacial acetic acid (31:14:5)	Chloroform/methanol/formic acid (44:3.5:2.5)	Silica gel	[86]
9	Hexane/acetone (40:60)	Methanol/water (1:1)	Silica-CN	[67]
10	Hexane/propan-2-ol (60:40)	Tetrahydrofuran/water (1:1)	silica-CN	[67]
11	Hexane/propan-2-ol (60:40)	Dioxane/water (1:1)	silica-CN	[67]

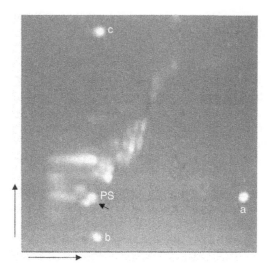

FIGURE 14.11 Two-dimensional chromatogram of caffeic acid in propolis sample. The plate was chromatographed with mixture of *n*-hexane/ethyl acetate/glacial acetic acid (31:14:5) in one direction and with chloroform/methanol/formic acid (44:3.5:2.5) in the second direction.

mobile phase consisting toluene/dioxan/formic acid (7:2:1) in the first direction and the gradient program on increasing distance with decreasing elution strength of the mobile phase in the second direction [87]. The results of separation of eleven phenolic acids are presented in Figure 14.12.

One of possible form of multidimensional separation is procedure in which the different stationary phases are employed. This variant of 2D-TLC was tested by Glensk et al. for the mixture of standards [87]. The first development was performed on an RP-18W HPTLC plates with eluent consisting methanol–water (40:60) and subsequently phenolic acids were transferred on silica gel layer and chromatographed according to three step gradient program in the second direction (see Figure 14.13).

2D-TLC with adsorbent gradient was applied by Hawryl and Waksmundzka in analysis of some phenolic compound in extracts from *Polygonum* sp. and *Verbascum* sp. [88]. The herbal extracts were spotted to RP18W plates and developed with the use of acetone and water mixture (RP mode). After drying, the plates were connected to silica gel or diol and chromatographed with use of a nonaqueous mobile phase comprising ethyl acetate with methanol or propan-2-ol (NP mode).

The variant of 2D-TLC with adsorbent gradient can be also realized on commercially available multiphase plates comprising a silica strip adjacent to an RP layer or RP strip adjacent to a silica layer.

14.5.4 QUANTITATIVE DETERMINATION OF PHENOLIC ACIDS

TLC combined with densitometry is often used for quantitative analysis of phenolic compounds in different parts of plants and herbs. Modern densitometers afford an opportunity of fast, cheap, and sensitive determination for routine analysis involving many samples. Additionally, they enable confirming of identity of investigated

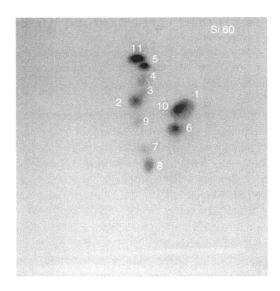

FIGURE 14.12 Chromatogram of separation of phenolic acids on silica after use 2D-TLC coupled with multiple development. The mobile phase used in the first direction: toluene/formic acid (7:2:1), in the second direction: step 1—diisopropyl ether/formic acid (80:20, v/v) on distance 2 cm, step 2 and 3—diisopropyl ether/formic acid/cyclohexane (78:2:20, v/v/v) on distance 9 cm: (1) 3,5-dihydroxybenzoic acid, (2) vanillic acid, (3) p-coumaric acid, (4) p-hydroxybenzoic acid, (5) gentisic acid, (6) caffeic acid, (7) syringic acid, (8) sinapic acid, (9) ferulic, (10) protocatechuic acid, and (11) 2,4-dihydroxybenzoic acid.

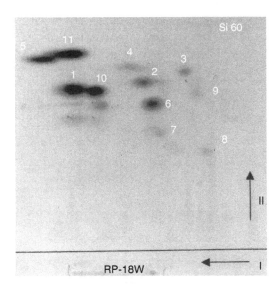

FIGURE 14.13 The separation of phenolic acids with use adsorbent gradient 2D-TLC. The first development was performed on an RP-18W HPTLC plates with mobile phase consisting of methanol–water (40:60) and the second development on Si 60 was performed according to gradient program as in Figure 14.12. Numbering of phenolic acids as in Figure 14.12.

compound on the basis of comparison of UV–VIS spectra of standard and sample directly on the chromatographic plate.

Phenolic acids absorb light at the UV region of the spectrum and derivatization agent is not required to quantification. The measurement of absorbance (absorbance/reflectance mode) at wavelength of maximum absorptions is usually employed [32,70,79–93].

The advantage of TLC is possibility of parallel determination of a few compounds on the same chromatographic plate. An HPTLC-densitometric method has been developed by Hubicka et al. for parallel analysis of caffeic, *p*-coumaric, and ferulic acids in propolis concentrates. The compounds were successfully separated on silica plates with dichloromethane–acetonitrile–90% formic acid, 9.5 + 0.5 + 0.1 v/v as a mobile phase (see Figure 14.14). Densitometry was performed in absorbance mode at $\lambda = 320$ nm. The limits of detection were 29.50 ng for caffeic acid, 8.33 ng for *p*-coumaric acid and 17.20 ng for ferulic acid. The limits of quantification were 47.2, 16.6, and 43 ng, respectively [79].

Three phenolic acids (gallic, caffeic, and protocatechuic) were analyzed simultaneously in *Bistortae* rhizome, *Polygoni avicularis herba* and *Rei* radix [82]. Vundać et al. determined parallel two phenolic acids: chlorogenic and caffeic acid in seven Croatian *Stachys* taxa [78]. The chromatographic conditions for analysis of *p*-coumaric acid in *Atropa belladonna* L., *Datura stramonium* L., and *Hyoscyamus niger* L. on diol-silica were proposed by Pobłocka et al. [70]. Swaroop et al. applied RP-18 layer to analysis of *p*-coumaric acid in flowers of *Rhododendron arboreum* [89]. The content of caffeic acid in different plants of *Dipsacaceae* was densitometrically determined by Kowalczyk et al. [83]. TLC-densitometry was also used for quantification of gallic acid in clove [90] and in tea [91]. Gallic and ellagic acid were also determined in *Phyllanthus amarus* [92], in seeds of *Abrus precatorius*, flowers of *Nymphaea alba*, and in whole plant of *Phyllanthus maderaspatensis* [93].

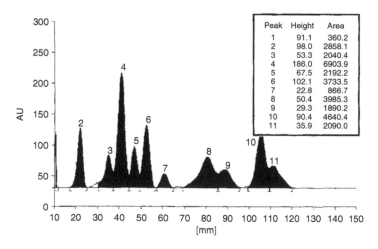

FIGURE 14.14 The densitogram of extract from propolis obtained on silica gel after development of mixture of dichloromethane/acetonitrile/90% formic acid (9.5:0.5:0.1, v/v); 2-caffeic acid, 4-*p*-coumaric acid, 6-ferulic acid.

Although, the measurement of fluorescence usually demonstrates better sensitivity, selectivity, and linearity, it is rarely employed in analysis of phenolic acids. The fluorescence/reflectance mode requires of selection of the proper excitation wavelength and the proper filter should be cut-off to measure only excited irradiation. A TLC-densitometry in fluorescence/reflectance mode for the parallel determination of rosmarinic and caffeic acids in *Salvia* sp. [94] and in 96 *Lamiaceae* taxa [95] was developed by Janicsák et al. The obtained values of limit of detection were very low: 1 ng for rosmarinic and 0.1 ng for caffeic acid [94]. The fluorescence/reflectance mode was also employed by Barthomeuf et al. for the analysis of gallic acid on RP-18 adsorbent [62] and by Petersen et al. to determination of rosmarinic acid on cellulose HPTLC plates [60].

Generally, the obtained validation parameters such as high correlation coefficient ($r^2 > 0.99$), acceptable relative standard deviation, recovery values, and limits of detection and quantification allow the use of TLC with densitometry in routine practice.

14.6 THE METHODS OF DETECTION OF PHENOLIC ACIDS

Phenolic acids are substances which absorb in UV region and their detection on layers containing a fluorescent indicator is relatively easy. Most of phenolics can be detected by quenching the adsorbent fluorescence. After ultraviolet irradiation (usually at $\lambda = 254$ nm), the emission of adsorbent is reduced in regions where UV-active compounds absorb the UV light. They appear as dark zones on fluorescent background. Many phenolic acids possess the ability of fluorescence. They appear as light-bright zones on a dark background. Excitation is usually performed using radiation at $\lambda = 366$ nm. These methods of detection are the most sensitive and they are employed for quantification.

The advantage of separation on thin layer is possibility of additional detection after treatment by various reagents. The simple way of detection is treatment the plates with ammonia vapor. Phenolic compounds should be observed in UV light at $\lambda = 254$ or 366 nm. The effect of reaction of some phenolic acids with ammonia vapor and limits of detection according to Grodzińska-Zachwieja et al. are presented in Table 14.6 [75].

Exposure of the developed plates to iodine vapor in an iodine-saturated chamber has been tested by Sharma et al. [49]. It yielded yellow spots; however the limit of detection after treatment by iodine was not too high, e.g., 80 μg for cinnamic acid, 10 μg for *p*-coumaric acids, 2 μg for ferulic and gallic acid [49].

Derivatization can be also performed by spraying or dipping with various reagents. The methanol solutions of ferric (III) chloride or aluminum chloride in concentration of 2%–3% are often employed. The limit of detection depends on the phenolic acids and ranges from 10 μg for cinnamic acid to 2 μg for gallic, ferulic, *p*-coumaric [49].

The most frequently derivatization method is treatment with diazotized sulfanilic acid and subsequent spraying with 10%–20% aqueous sodium carbonate solution (DSA). Phenolic acids form color spots and are observed in daylight. The limits of detection range from 2 to 10 μg. In some publications the use of diazotiazed *p*-nitroaniline in 10% sodium carbonate solution or diazotiazed benzidine (Linskens

TABLE 14.6

The Effect of Reaction of Some Phenolic Acids with Ammonia Vapor and Limits of Detection

Compound	NH$_3$/UV $\lambda = 366$ nm	LOD (μg)
Chlorogenic	yellow green	0.2
Isochlorogenic	pale green	0.7
Caffeic	Blue	0.3
Ferulic	Blue	0.08
o-Coumaric	Yellow	0.3
m-Coumaric	Grey-yellow	1.0
p-Coumaric	Blue-violet	0.9
Synapic	Blue-green	0.6

reagent) was also described [76]. In Table 14.7 the effects after treatment of popular derivatization agents are presented.

There are also a lot of less popular methods of visualization. Sometimes, visualization can be achieved by spraying with 1% methanol diphenylboryloxyethylamine and then 5% ethanol polyethylene glycol 4000. Phenolic acids appear as blue fluorescent bands [74].

After treatment with 5% ethanol solution of phosphomolybdic acid and heating for 5 min in 80°C phenolic acids color brown [76].

Equal volumes of 1% aqueous solutions of ferric (III) chloride and potassium ferricyanide freshly mixed can be also employed. After treatment of this reagent phenolic acids produce a blue color [96].

TABLE 14.7

The Effect of Treatment of Phenolic Acids with Various Derivatization Agents

Compound	UV $\lambda = 366$ nm	FeCl$_3$	DSA	DNA	DB
p-Coumaric	Deep blue	Orange	Red	Brown-blue	Brown
Chlorogenic	Blue	Brown-green	Light brown	Yellow	Light brown
p-Hydroxybenzoic	Deep blue	Yellow	Light violet	Red	Brown
Caffeic	Blue	Brown-green	Brown	Brown	Brown
Syringic	Violet	–	Red	Blue	Brown
Vanillic	Deep blue	Brown-blue	Orange	Violet	Light brown
Protocatechuic	Violet	Brown-blue	Brown	Red-brown	Brown
Ferulic	Blue	Orange	Violet	Blue	Brown
Gentisic	Blue	Navy blue	Grey	Grey-green	Yellow-brown
Gallic	Violet	Grey	Light brown	Green red	Brown
Ellagic	–	–	Light brown	Red-brown	Light brown
Synapic	Blue	Violet	Blue	Grey	Light brown

DSA—diazotized sulfanilic acid, DNA—diazotiazed p-nitroaniline, DB—diazotiazed benzidine.

Compound with antioxidant activity was visualized using β-carotene-linoleate spray method. The plates were sprayed with solution of β-carotene with addition of linoleic acid and exposure to fluorescence light for 3 h. The color intensity of the bands was related to their antioxidant strength [97]. Some phenolic acids react with vanillin–sulfuric acid reagent forming light purple or purple spots [95]. Several other methods of detection, e.g., treatment with Cartwright–Roberts reagent or Folin–Denis reagent, were described by Hiller [98].

REFERENCES

1. Robbins, R.J., Phenolic acids in foods: An overview of analytical methodology, *J. Agric. Food Chem.*, 51, 2866–2887, 2003.
2. Kohlmunzer, S., *Pharmacognosy*, 5th edn., PZWL, Warsaw, 2003 (in Polish).
3. Borkowski, B., Phenolic acids and their esters, *Herba Pol.*, *Part 1*, 39, 71–83, 1993 (in polish).
4. Krygier, K., Sosulski, F., and Hogge, L., Free, esterified and insoluble-bound phenolic acids. 1. Extraction and purification procedure, *J. Agric. Food Chem.*, 30, 334–336, 1982.
5. Häkkinen, S., *Flavonols and Phenolic Acids in Berries and Berry Products*, Doctoral dissertation, Department of Clinical Nutrition, Department of Physiology, Department of Biochemistry, University of Kuopio, Kuopio, 2000.
6. Barcz, E., Rogala, E., Glinkowska, G., Strzelecka, H., Sikorska, E., Sokolnicka, I., and Skopińska-Różewska, E., Immunotropic activity of plant extracts. IV. The effect of phenolic compounds of poplar leaves (*Populus nigra* L.) extract on human leukocytes movement, *Herba Pol.*, 44, 45–51, 1998.
7. Santos-Gomes, P.C., Seabra, R.M., Andrade, P.B., and Fernandes-Ferreira, M., Phenolic antioxidant compounds produced by in vitro shoots of sage (*Salvia officinalis* L.), *Plant Sci.*, 162, 981–987, 2002.
8. Bakalbassis, E.G., Chatzopoulou, A., Melissas, V.S., Tsimidou, M., Tsolaki, M., and Vafiadis, A., Ab initio and density functional theory studies for the explanation of the antioxidant activity of certain phenolic acids, *Lipids*, 36, 181–191, 2001.
9. Sroka, Z. and Cisowski, W., Hydrogen peroxide scavenging, antioxidant and anti-radical activity of some phenolic acids, *Food Chem. Toxicol.*, 41, 753–758, 2003.
10. Rice-Evans, C.A., Miller, N.J., and Paganga, G., Structure-antioxidant relationships of flavonoids and phenolic acids, *Free Radical Biol. Med.*, 20, 933–956, 1996.
11. Mansouri, A., Makris, D.P., and Kefalas, P., Determination of hydrogen peroxide scavenging activity of cinnamic and benzoic acids employing a highly sensitive peroxyoxalate chemiluminescence-based assay: Structure-activity relationships, *J. Pharm. Biomed. Anal.*, 39, 22–26, 2005.
12. Marinova, E.M. and Yanishlieva, N.V., Antioxidant activity and mechanism of action of some phenolic acids at ambient and high temperatures, *Food Chem.*, 81, 189–197, 2003.
13. Lee, G.-S., Widjaja, A., and Ju, Y.-H., Enzymatic synthesis of cinnamic acid derivatives, *Biotechnol. Lett.*, 28, 581–585, 2006.
14. Ou, S. and Kwok, K.-C., Ferulic acid: Pharmaceutical functions, preparation and application in foods, *J. Sci. Food Agric.*, 84, 1261–1269, 2004.
15. Borkowski, B., Phenolic acids and their esters, *Herba Polon.*, *Part II*, 39, 139–145, 1993 (in polish).
16. Hanif, R., Qiao, L., Shiff, S.J., and Rigas, B., Curcumin, a natural plant phenolic food additive, inhibits cell proliferation and induces cell cycle changes in colon

adenocarcinoma cell lines by a prostaglandin-independent pathway, *J. Lab. Clin. Med.*, 130, 576–584, 1977.

17. Kumar, K.C.S. and Müller, K., Depsides as non-redox inhibitors of leukotriene B_4 biosynthesis and HaCaT cell growth. 2. Novel analogues of obtusatic acid, *Eur. J. Med. Chem.*, 35, 405–411, 2000.

18. Petersen, M. and Simmonds, M.S.J., Rosmarinic acid, *Phytochemistry*, 62, 121–125, 2003.

19. Borkowski, B., Biesiadecka, A., and Litwińska, B., Comparison of virusostatic activity of caffeic, chlorogenic and rosmarinic acids, *Herba Pol.*, 42, 317–321, 1996 (in polish).

20. Obied, H.K., Bedgood, D.R., Jr., Prenzler, P.D., and Robards, K., Bioscreening of Australian olive mill waste extracts: Biophenol content, antioxidant, antimicrobial and molluscicidial activities, *Food Chem. Toxicol.*, 45, 1238–1248, 2007.

21. Fernández, M.A., Garcia, M.D., and Sáenz, M.T., Antibacterial activity of phenolic acids fractions of *Scrophularia frutescens* and *Scrophularia sambucifolia*, *J. Ethnopharmacol.*, 53, 11–14, 1996.

22. Peungvicha, P., Temsiririrkkul, R., Prasain, J.K., Tezuka, Y., Kadota, S., Thirawarapan, S.S., and Watanabe, H., 4-Hydroxybenzoic acid: A hypoglycemic constituent of aqueous extract of *Pandanus odorus* root, *J. Ethnopharmacol.*, 62, 79–84, 1998.

23. Sbarra, V., Ristorcelli, E., Le Petit-Thévenin, J., Teissedre, P.L., Lombardo, D., and Vérine, A., In vitro polyphenol effects on activity, expression and secretion of pancreatic bile salt-dependent lipase, *Biochem. Biophys. Acta*, 1736, 67–76, 2005.

24. Gómez-Caravaca, A.M., Gómez-Romero, M., Arráez-Román, D., Segura-Carretero, A., and Fernández-Gutiérrez, A., Advances in the analysis of phenolic compounds in products derived from bees, *J. Pharm. Biomed. Anal.*, 41, 1220–1234, 2006.

25. Borkowski, B., Skuza, G., and Rogóż, Z., Comparative investigations of action of rosmarinic and chlorogenic acids on central nervous system, *Herba Pol.*, 45, 192–198, 1999 (in polish).

26. Głowniak, K., Zgórka, G., and Kozyra, M., Solid-phase extraction and reversed-phase high-performance liquid chromatography of free phenolic acids in some *Echinacea* species, *J. Chromatogr. A*, 730, 25–29, 1996.

27. Smolarz, H.D., Application of GC-MS method for analysis of phenolic acids and their esters in chloroformic extracts from some taxons of *Polygonum* L. genus, *Chem. Anal.*, 46, 439–444, 2001.

28. Zgórka, G. and Kawka, S., Application of conventional UV, photodiode array (PDA) and fluorescence (FL) detection to analysis of phenolic acids in plant material and pharmaceutical preparations, *J. Pharm. Biomed. Anal.*, 24, 1065–1072, 2001.

29. Zgórka, G. and Hajnos, A., The application of solid-phase extraction and reversed phase high-performance liquid chromatography for simultaneous isolation and determination of plant flavonoids and phenolic acids, *Chromatographia suppl.*, 57, 77–80, 2003.

30. Waksmundzka-Hajnos, M., Oniszczuk, A., Szewczyk, K., and Wianowska, D., Influence of sample preparation methods on the quantitation of some phenolic acids in plant materials by HPLC, *Acta Chromatogr.*, 19, 2007 (in press).

31. Proestos, C., Chorianopoulos, N., Nychas, G.-J.E., and Komaitis, M., RP-HPLC analysis of the phenolic compounds of plant extracts. Investigation of their antioxidant capacity and antimicrobial activity, *J. Agric. Food Chem.*, 53, 1190–1195, 2005.

32. Wójciak-Kosior, M., Matysik, G., and Soczewiński, E., High-performance thin-layer chromatography combined with densitometry for quantitative analysis of phenolic acids in complex mixtures, *J. Planar Chromatogr.*, 19, 21–26, 2006.

33. Štěrbová, D., Matějíček, D., Vlček, J., and Kubáň, V., Combined microwave-assisted isolation and solid-phase purification procedures prior to the chromatographic

determination of phenolic compounds in plant materials, *Anal. Chim. Acta*, 513, 435–444, 2004.

34. Naczk, M. and Shahidi, F., Phenolics in cereals, fruits and vegetables: Occurrence, extraction and analysis, *J. Pharm. Biomed. Anal.*, 41, 1523–1542, 2006.

35. Akowuah, G.A., Ismail, Z., Norhayati, I., and Sadikun, A., The effects of different extraction solvents of varying polarities on polyphenols of *Orthosiphon stamineus* and evaluation of the free radical-scavenging activity, *Food Chem.*, 93, 311–317, 2005.

36. Lege, k.E., Cothren, J.T., and Smith, C.W., Phenolic acid and condensed tannin concentrations of six cotton genotypes, *Environ. Exp. Bot.*, 35, 241–249, 1995.

37. Świątek, L., Phenolic acids and iridoid glycosides in selected polish medicinal plants from *Plantago* genus, *Herba Pol.*, 23, 201–210, 1977 (in polish).

38. Światek, L. and Dombrowicz, E., Phenolic acids in plant materials. I. Investigations of wormwood herb and gentian root, *Farm. Pol.*, 40, 729–732, 1984 (in polish).

39. Ryan, D., Lawrence, H., Prenzler, P.D., Antolovich, M., and Robards, K., Recovery of phenolic compounds from *Olea europaea*, *Anal. Chim. Acta*, 445, 67–77, 2001.

40. Svedröm, U., Vuorela, H., Kostiainen, R., Laakso, I., and Hiltunen R., Fractionation of polyphenols in hawthorn into polymeric procyanidins, phenolic acids and flavonoids prior to high-performance liquid chromatographic analysis, *J. Chromatogr. A*, 1112, 103–111, 2006.

41. Suárez, B., Picinelli, A., and Mangas, J.J., Solid-phase extraction and high-performance liquid chromatographic determination of polyphenols in apple musts and ciders, *J. Chromatogr. A*, 727, 203–209, 1996.

42. Tian, S., Nakamura, K., Cui, T., and Kayahara, H., High-performance liquid chromatographic determination of phenolic compounds in rice, *J. Chromatogr. A*, 1063, 121–128, 2005.

43. Žiaková, A., Brandšteterová, E., and Blahová, E., Matrix solid-phase dispersion for the liquid chromatographic determination of phenolic acids in *Meliss officinalis*, *J. Chromatogr. A*, 983, 271–275, 2003.

44. Vundać, V.B., Maleš, Z., Plazibat, M., Golja, P., and Cetina-Čižmek, B., HPTLC determination of flavonoids and phenolic acids in some Croatian *Stachys Taxa*, *J. Planar Chromatogr.*, 18, 269–273, 2005.

45. Jork, H., Funk, W., Fischer, W., and Wimmer, H., Dunnschicht-Chromatographie, Band 1a, VCH, Weinheim, 1989, p. 149.

46. Maleš, Ž. and Medić-Šarić, M., Optimization of TLC analysis of flavonoids and phenolic acids of *Helleborus atrorubens* Waldst. et Kit., *J. Pharm. Biomed. Anal.*, 24, 353–359, 2001.

47. Maleš, Ž. and Medić-Šarić, M., Investigation of the flavonoids and phenolic acids of *Sambuci flos* by thin-layer chromatography, *J. Planar Chromatogr.*, 12, 345–349, 1999.

48. Rastija, V., Mornar, A., Jasprica, J., Srečnik, G., and Medić-Šarić, M., Analysis of phenolic components in Croatian red wines by thin layer chromatography, *J. Planar Chromatogr.*, 17, 26–31, 2004.

49. Sharma, O.P., Bhat, T.K., and Singh, B., Thin-layer chromatography of gallic acid, methyl gallate, pyrogallol, phloroglucinol, catechol, resorcinol, hydroquinone, catechin, epicatechin, cinnamic acid, *p*-coumaric acid, ferulic acid, and tannic acid, *J. Chromatogr. A*, 822, 167–171, 1998.

50. Wettasinghe, M., Shahidi, F., Amarowicz, R., and Abou-Zadi, M.M., Phenolic acids in defatted seeds of borage (*Borago officinalis* L.), *Food Chem.*, 75, 49–56, 2001.

51. Simonovska, B., Vovk, I., Andrenšek, S., Valentová, K., and Ulrichová, J., Investigation of phenolic acids in yacon (*Smallanthus sonchifolius*) leaves and tubers, *J. Chromatogr. A*, 1016, 89–98, 2003.

52. Hao, S. and Li, J., Determination of phenolic acids in *Salvia miltiorrhiza* B. by thin-layer chromatography, *J. Chin. Herb. Med.* (*Zhongcoayao*), 24, 527–528, 1993.
53. Kader, F., Rovel, B., Girardi, M., and Metche, M., Fractionation and identification of the phenolic compounds of highbush blueberries (*Vaccinium corymbosum*, L.), *Food Chem.*, 55, 35–40, 1996.
54. Smolarz, H.D. and Waksmundzka-Hajnos, M., Two-dimensional TLC of phenolic acids on cellulose, *J. Planar Chromatogr.*, 6, 278–281, 1993.
55. Smolarz, H.D., Medyńska, E., and Matysik, G., Determination of emodin and phenolic acids in the petrioles of *Rheum undulatum* and *Rheum rhaponticum*, *J. Planar Chromatogr.*, 18, 319–322, 2005.
56. Głowniak, K., Skalicka, K., Ludwiczak, A., and Jop, K., Phenolic compounds in the flowers of *Lavatera trimestris* L. (*Malvaceae*), *J. Planar Chromatogr.*, 18, 264–268, 2005.
57. Kozyra, M., Glowniak, K., and Zadubiec, A., Phenolic acids in *Peucedanum verticillare* L. Koch ex Dc, *J. Planar Chromatogr.*, 16, 421–424, 2003.
58. Cisowski, W., Dembińska-Migas, W., Krauze-Baranowska, M., Łuczkiewicz, M., Migas, P., Matysik, G., and Soczewiński, E., Application of planar chromatography to the analysis of secondary metabolites in Callus cultures of different plant species, *J. Planar Chromatogr.*, 11, 441–446, 1998.
59. Bartnik, M., Głowniak, K., and Dul, R., Use of two-dimensional TLC to identify phenolic acids in the foliage and fruit of *Peucedanum tauricum* Bieb., *J. Planar Chromatogr.*, 16, 206–210, 2003.
60. Petersen, H.W., Petersen, L.M., Plet, H., and Ravn, H., A new fluorescence densitometric method for quantitative analysis of rosmarinic acid in extracts of *Zostera marina* L., *J. Planar Chromatogr.*, 4, 235–239, 1991.
61. Mornar, A., Medić-Šarić, M., and Jasprica, J., AMDE data for polyphenols by reversed-phase thin layer chromatography, *J. Planar Chromatogr.*, 19, 409–417, 2006.
62. Barthomeuf, C., Regerat, F., and Combe-Chevaleyre, S., Densitometric assay of gallic acid in fermentation liquor, *J. Planar Chromatogr.*, 6, 245–247, 1993.
63. Fecka, I. and Cisowski, W., TLC determination of tannins and flavonoids in extracts from some *Erodium* species using chemically modified stationary phases, *J. Planar Chromatogr.*, 6, 429–432, 2002.
64. Grodzinska-Zachwieja, Z., Optimization of the separation of cis- and trans-hydroxycinnamic acids by reversed-phase thin-layer chromatography, *J. Chromatogr.*, 241, 217–222, 1982.
65. Fecka, I. and Cisowski, W., Multiple gradient development TLC in analysis of complex phenolic acid from *Lycopus europaeus* L., *Chromatographia*, 49, 256–260, 1999.
66. Soczewiński, E., Hawrył, M.A., and Hawrył, A., Retention behavior of some flavonoids in 2D-TLC systems on cyano bonded polar phases, *Chromatographia*, 54, 789–794, 2001.
67. Hawrył, M.A., Hawrył, A., and Soczewiński, E., Application of normal and reversed-phase 2D TLC on a cyanopropyl-bonded polar stationary phase for separation of phenolic compounds from the flowers of *Sambucus nigra* L., *J. Planar Chromatogr.*, 15, 4–10, 2002.
68. Wójciak-Kosior, M. and Skalska, A., Thin-layer chromatography of phenolic acids on aminopropylsilica, *J. Planar Chromatogr.*, 19, 200–203, 2006.
69. Soczewiński, E., Wójciak, M., and Pachowicz, K., Thin-layer chromatography of phenolic acids on diol silica, *J. Planar Chromatogr.*, 11, 12–15, 1998.
70. Pobłocka, L., Matysik, G., and Cisowski, W., Multiple gradient TLC and densitometric analysis of p-coumaric acid in some medicinal plants from the family *Solanaceae*, *J. Planar Chromatogr.*, 16, 76–79, 2003.

71. Stahl, E. and Schorn, P.J., Dünschitchromatographie hydrophiler Arzneipflanzenauszüge (TLC of hydropfilic extracts from medicinal plants), *Hoppe-Seyler's Z. Physiol. Chem.*, 325, 263, 1961.
72. Scarpati, M.L. and Guiso, M., Acidi caffeil-chinici del caffé e del maté. Nota I. (Caffeoylquinic acids of coffee infusions. Note I.), *Ann. Chim.*, 53, 1315, 1963.
73. van Sumere, C.F., Wolf, G., Teuchy, H., and Kint, J., A new thin-layer method for phenolics substances and coumarins, *J. Chromatogr.*, 20, 48–60, 1965.
74. Wagner, H. and Bladt, S., *Plant Drug Analysis: A Thin Layer Chromatography Atlas*, 2nd edition, Springer-Verlag, Berlin, Heidelberg, New York, 1996.
75. Grodzińska-Zachwieja, Z., Kahl, W., and Klimczak, M., *Dissert. Pharm. Pharmacol.*, 23, 1971.
76. Jończyk, J., Kwasy fenolowe, in *Chromatografia Cienkowarstwowa w Analizie Farmaceutycznej*, Borkowski, B. (Ed.), PZWL, Warszawa, 1973.
77. Kowalski, R. and Wolski, T., TLC and HPTLC analysis of the phenolic acids in *Silphium perfoliatum* L. leaves, inflorescences and rhizomes, *J. Planar Chromatogr.*, 16, 230–234, 2003.
78. Vundać, V.B., Maleš, Z., Plazibat, M., Golja, P., and Cetina-Čižmek, B., HPTLC determination of flavonoids and phenolic acids in some Croatian *Stachys* Taxa, *J. Planar Chromatogr.*, 18, 269–273, 2005.
79. Hubicka, U., Krzek, J., Kaleta, J., and Niedźwiedź, A., Evaluation of densitometric TLC for quantitative analysis of selected phenolic acids for standardization of propolis concentrates, *J. Planar Chromatogr.*, 19, 449–453, 2006.
80. Bieganowska, M.L., Petruczynik, A., and Głowniak, K., Reversed phase retention behavior of β-cyclodextrin complexes of phenolic and cinnamic acids, *J. Planar Chromatogr.*, 8, 63–68, 1995.
81. Hadzija, O., Tonkovic, M., and Isrkic, S., The behaviour of some phenolic acids and aldehydes on thin layers of silica gel impregnated with Fe(III), *J. Liq. Chromatogr.*, 9, 3473–3478, 1986.
82. Sawicka, U., Cisowski, W., Matysik, G., and Kowalczyk, A., HPTLC of phenolic acids in *Bistortae rhizoma*, *Polygoni avicularis herba*, *Rei radix* with densitometric determination, *J. Planar Chromatogr.*, 15, 442–448, 2002.
83. Kowalczyk, A., Matysik, G., Rządkowska-Bodalska, H., and Cisowski, W., Thin layer chromatography and densytometry of caffeic acid in some *Dipsacaceae* family plants, *J. Planar Chromatogr.*, 14, 175–177, 2001.
84. Ellnain-Wojtaszek, M. and Zgórka, G., High-performance liquid chromatography and thin-layer chromatography of phenolic acids from *Ginkgo biloba* L., *J. Liq. Chromatogr. Relat. Technol.*, 22, 1457–1471, 1999.
85. Hawryl, M.A. and Soczewiński, E., Normal phase 2D TLC separation of flavonoids and phenolic acids from *Betula* sp. leaves, *J. Planar Chromatogr.*, 14, 415–422, 2001.
86. Medić-Šarić, M., Jasprica, I., Monar, A., Smolić-Bubalo, A., and Golja, P., Quantitative analysis of flavonoids and phenolic acids in propolis by two-dimensional thin layer chromatography, *J. Planar Chromatogr.*, 17, 459–462, 2004.
87. Glensk, M., Sawicka, U., Mażol, I., and Cisowski, W., 2D TLC-graft planar chromatography in the analysis of a mixture of phenolic acids, *J. Planar Chromatogr.*, 15, 463–465, 2002.
88. Hawryl, M.A. and Waksmundzka-Hajnos, M., Separation of phenolic compounds by NP and RP two-dimensional thin layer chromatography on connected plates, *J. Planar Chromatogr.*, 19, 92–97, 2006.
89. Swaroop, A., Prakash Gupta, A., and Kumar Sinha, A., Simultaneous determination of quercetin, rutin and coumaric acid in flowers of *Rhododendron arboreum* by HPTLC, *Chromatographia*, 649–652, 2005.

90. Pathak, S.B., Niranjan, K., Padh, H., and Rajani, M., TLC densitometric method for the quanification of eugenol and gallic acid in clove, *Chromatographia*, 60, 241–244, 2004.
91. Hachula, U., Anikiel, S., and Sajewicz, M., Application of densitometry and spectrophotometry for determination of gallic acid in tea after chromatographic separation, *J. Planar Chromatogr.*, 18, 290–293, 2005.
92. Dhalwal, K., Biradar, Y.S., and Rajani, M., High-performance thin-layer chromatography densitometric method for simultaneous quantitation of phyllanthin, hypophyllanthin, gallic acid, and ellagic acid in *Phyllanthus amarus*, *J. AOAC Int.*, 89, 619–623, 2006.
93. Bagul, M., Srinivasa, H., Padh, H., and Rajani, M., A rapid densitometric method for simultaneous quantification of gallic acid and ellagic acid in herbal raw materials using HPTLC, *J. Sep. Sci.*, 28, 581–584, 2005.
94. Janicsák, G. and Máthé, I., Parallel determination of rosmarinic and caffeic acids by TLC-densitometry, *Chromatographia*, 46, 322–324, 1997.
95. Janicsák, G., Máthé, I., Miklóssy-Vári, V., and Blunden, G., Comparative studies of the rosmarinic and caffeic acid contents of *Lamiaceae* species, *Biochem. Syst. Ecol.*, 27, 733–738, 1999.
96. Barton, G.M., Evans, R.S., and Gardner, J.A.F., Paper chromatography of phenolic substances, *Nature*, 170, 249–250, 1952.
97. Pratt, D.E. and Miller, E.E., A flavonoid antioxidant in Spanish peanuts, *J. Am. Oil Chem. Soc.*, 61, 1064–1067, 1984.
98. Hiller, K., On the chromatography of cyclic acids, with special reference to isomeric phenol carboxylic acids, *Pharmazie*, 20, 353–356, 1965.

15 Application of TLC in the Isolation and Analysis of Coumarins

Monika Waksmundzka-Hajnos
and Mirosław A. Hawryl

CONTENTS

15.1 CHEMICAL CONSTRUCTION AND PHYSICOCHEMICAL PROPERTIES OF COUMARINS

All coumarins are benzo-α-pyrone derivatives. They are also classified under phenylopropane derivatives because of biogenetic origin and basic skeleton $C_6–C_3$, which represents lactone system. Biologically, coumarins are derived from phenylopropane and through shickimig acid cycle they lead to *trans*-cinnamic acid. *Trans*-cinnamic acid passes to *o*-coumaric acid, its glycoside, and coumarin. In the similar cycle, *p*-coumaric acid passes to umbelliferone. The active isoprene is attached to the simple coumarin molecule and forms furano- and pyranocoumarins by the cyclization [1].

According to the structure, four basic coumarin groups can be distinguished:

- Coumarins of basic benzo-α-pyrone structure
- Isocoumarins
- Furanocoumarins: psoralene type with furane ring condensed with coumarin structure in 6,7 positions, angelicin type with furane ring condensed with coumarin structure in 7,8 positions
- Pyranocoumarins: xanthyletine type with pyrane ring condensed with coumarin structure in 6,7 positions, seseline type with pyrane ring condensed with coumarin structure in 7,8 positions and alloxanthyletine type with pyrane ring condensed with coumarin structure in 5,6 positions [1,2]

More detailed information and structures are presented in Table 15.1.

Coumarins appear in plants as free molecules and glycosides. Most of coumarin aglycones poses substituents (hydroxyl, methoxy, alkyl) in 7 position, sometimes also in 5, 6, 8 positions. Coumarin glycosides undergo enzymolysis in the route of drying and get converted to aglycones.

Solubility of coumarins depends on phenolic hydroxyls and glycoside bond. Coumarin glycosides are soluble in water and in polar solvents such as methanol and ethanol. Furano- and pyranocoumarins are soluble in apolar and medium polar solvents such as dichloromethane, petrol, and chloroform. Coumarins sublimate under reduced pressure and are volatile with water vapors.

Coumarins reveal strong fluorescence in UV light, which makes their identification easy [1,2].

15.2 PHARMACOLOGICAL ACTIVITY OF COUMARINS AND THEIR APPLICATION IN THERAPY

Coumarins, especially furano- and pyranocoumarins, partake in plant protective role and are frequently biosynthesized "de novo" in stress conditions as phytoalexins. Phytoalexins are synthesized by plants as a response to various biotic (bacteria, viruses, fungi, larves) and abiotic (metal ions, herbicides, detergents, etc.) releases [3–6]. Coumarins also have influence on plant cells. In low concentrations, they show synergic activity with plant growth promoting substances and induce plant growth. In high concentrations, they inhibit plant growth, acting as

TABLE 15.1

Structure of Coumarins

Simple coumarins

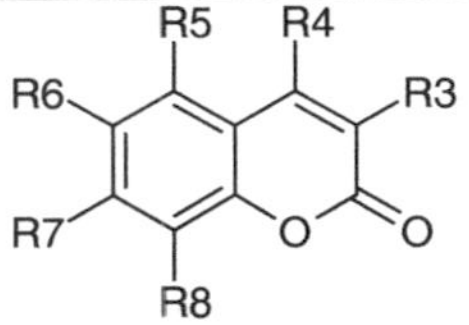

Name	R3	R4	R5	R6	R7	R8
Coumarin	H	H	H	H	H	H
Umbelliferone	H	H	H	H	OH	H
Herniarin	H	H	H	H	OCH_3	H
Aesculetin	H	H	H	OH	OH	H
Scopoletin	H	H	H	OCH_3	OH	H
Scopolin	H	H				
Fraxetin	H	H	H	OCH_3	OH	OH
Isofraxidin	H	H	H	OCH_3	OH	OCH_3
Daphnetin	H	H	H	H	OH	OH
Leptodactylon	H	H	OCH_3	H	OH	OCH_3
Scoparone	H	H	H	OCH_3	OCH_3	H
Osthenol	H	H	H	H	OH	$CH_2-CH=C(CH_3)_2$
Osthol	H	H	H	H	OCH_3	$CH_2-CH=C(CH_3)_2$

(continued)

TABLE 15.1 (continued)
Structure of Coumarins

Name	R3	R4	R5	R6	R7	R8
Ostruthin	H	H	H	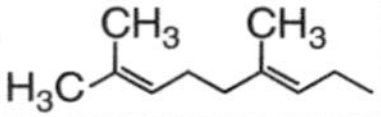	OH	H
Ammoresinol	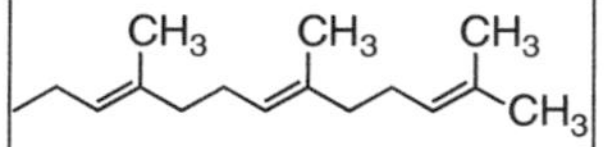	OH	H	H	OH	H
Cichoriin	H	H	H	OH	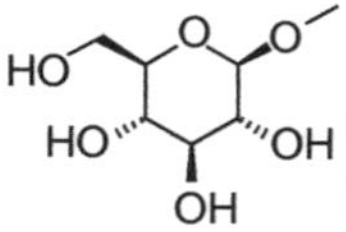	H
Daphnoretin	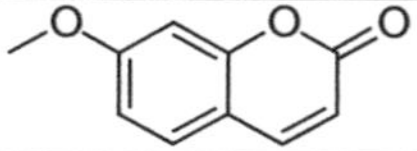	H	H	OCH_3	OH	H
Leptodactylone	H	H	OCH_3	H	OCH_3	OH
Mammeisin	H	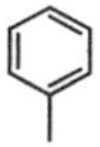	OH	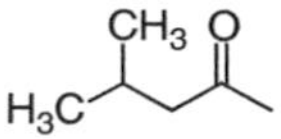	OH	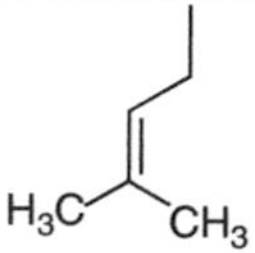
Micromelin	H	H	H	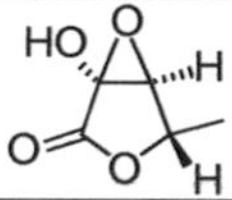	OCH_3	H

Furanocoumarins–psoralen type

Name	R3	R4	R5	R6	R7	R8
Psoralen	H	H	H	H	H	H
Bergapten	H	H	OCH$_3$	H	H	H
Xanthotoxin	H	H	H	H	H	OCH$_3$
Xanthotoxol	H	H	H	H	H	OH
Imperatorin	H	H	CH$_2$–C=C(CH$_3$)CH$_3$	H	H	H
Alloimperatorin	H	H	CH$_2$–C=C(CH$_3$)CH$_3$	H	H	OH
Isopimpinellin	H	H	OCH$_3$	H	H	OCH$_3$
Byakangelicin	H	H	OCH$_3$	H	H	–O–CH$_2$–CH(OH)–C(CH$_3$)(OH)CH$_3$
Chalepensin	H$_3$C–C(CH$_3$)=CH$_2$ prenyl	H	H	H	H	H

(continued)

TABLE 15.1 (continued)
Structure of Coumarins

Name	R3	R4	R5	R6	R7	R8
Oxypeucedanin	H	H	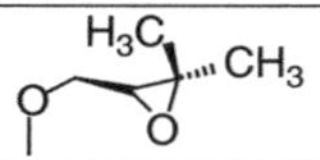	H	H	H
Peucedanin	H	H	H	OCH_3	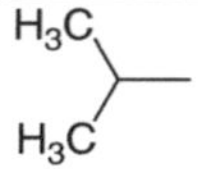	H
Trioxsalen	H	CH_3	H	H	CH_3	H

Furanocoumarins—angelicin type

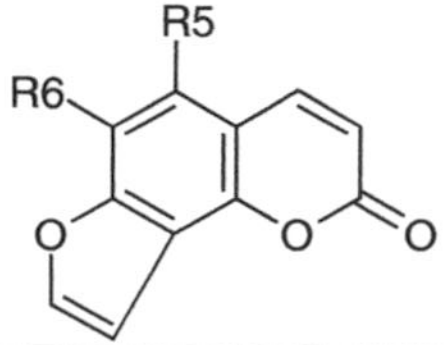

Name	R5	R6
Angelicin	H	H
Isobergapten	OCH_3	H
Sphondin	H	OCH_3
Pimpinellin	OCH_3	OCH_3

Dihydrofuranocoumarins

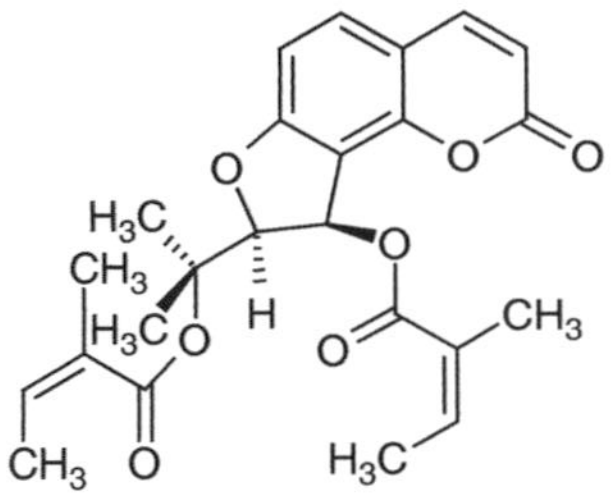

Archangelicin

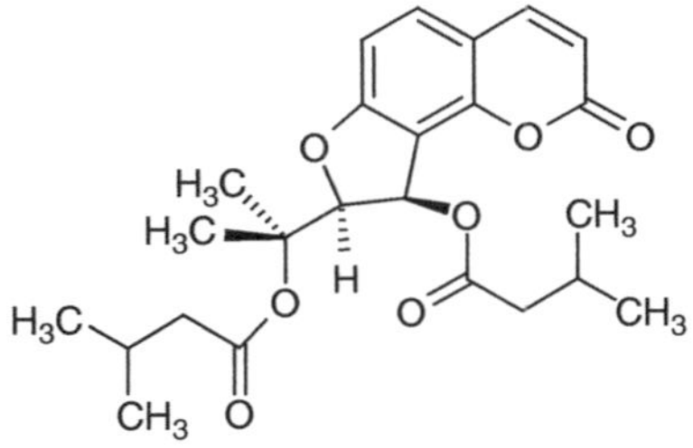

Athamantin

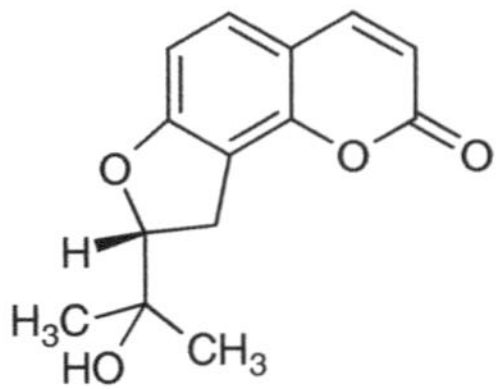

Columbianetin

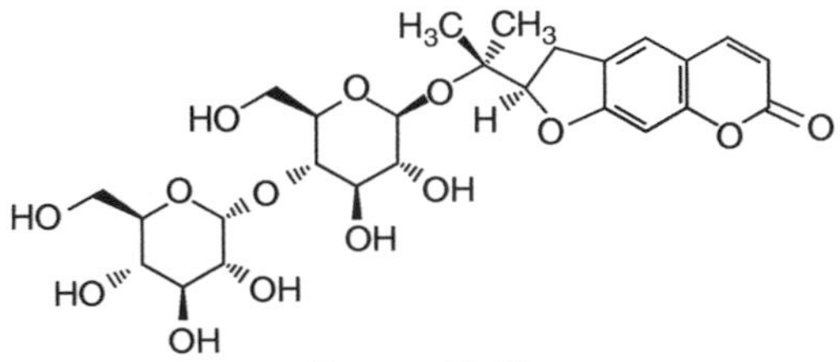

Decuroside III

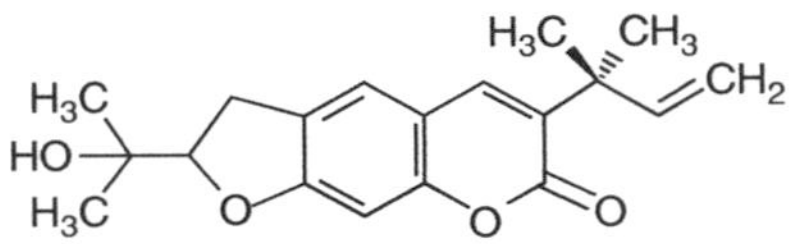

Heliettin

(continued)

**TABLE 15.1 (continued)
Structure of Coumarins**

Pyranocoumarins

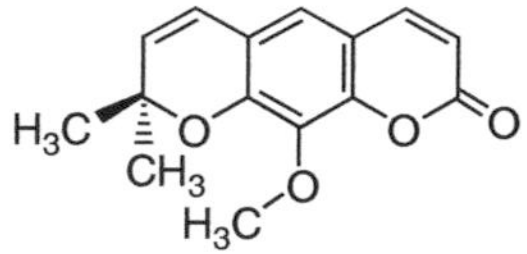

Luvangetin

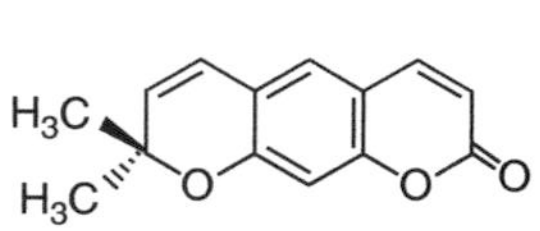

Xanthyletin

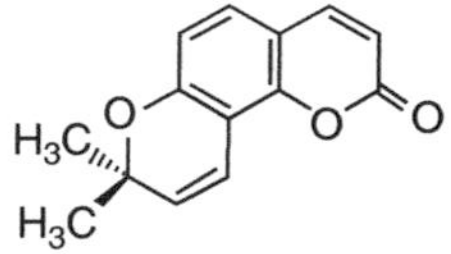

Seselin

Dihydropyranocoumarins

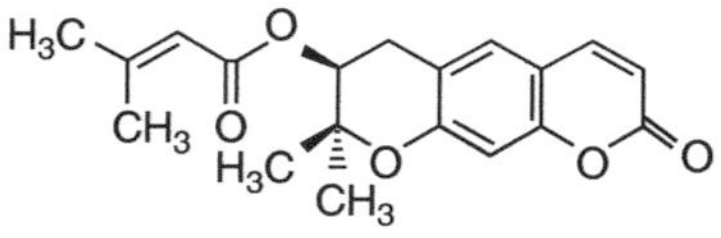

Decursin

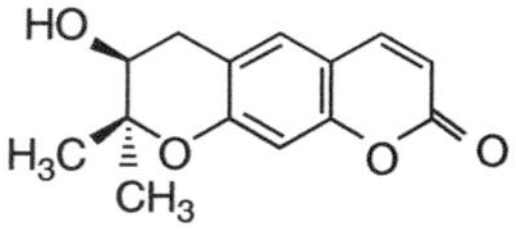

Decursinol

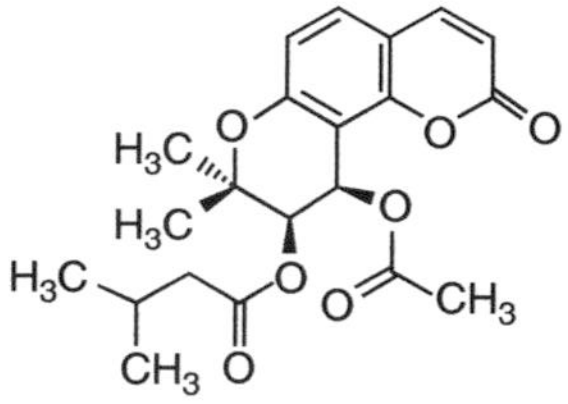

Dihydrosamidin

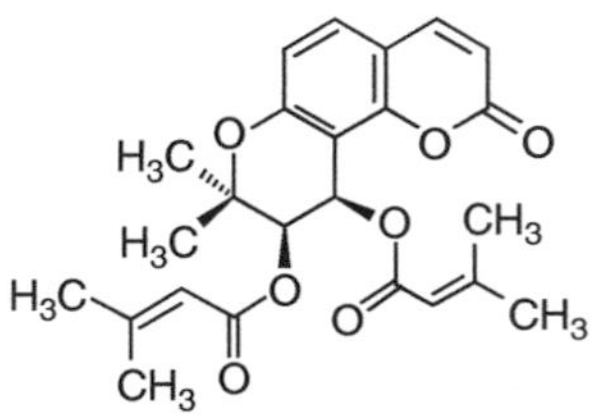

Pteryxin

Isosamidin

Disenecionyl cis-khellactone

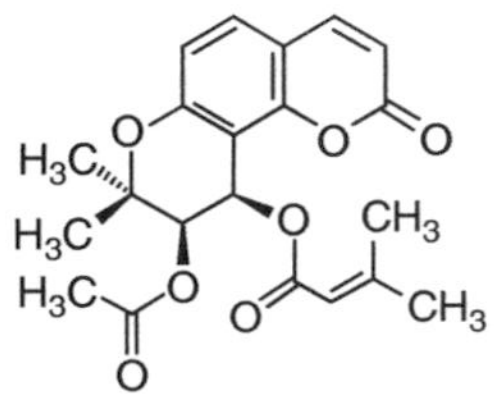

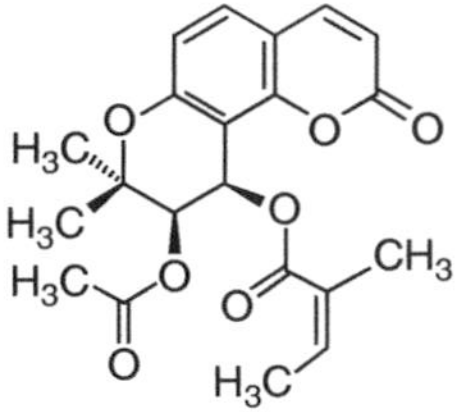

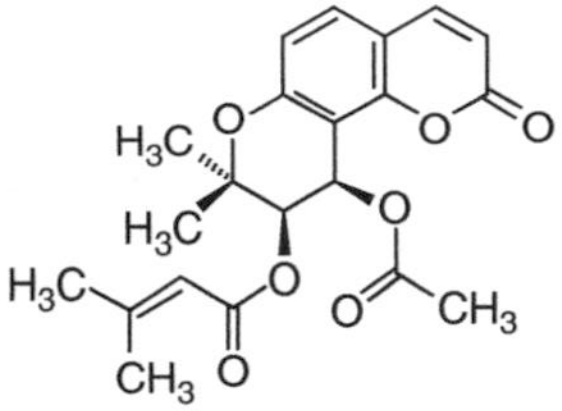

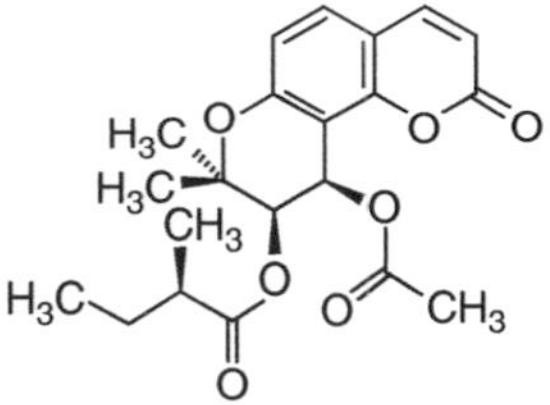

Visnadin

Samidin

(continued)

TABLE 15.1 (continued)
Structure of Coumarins

Other coumarin derivatives

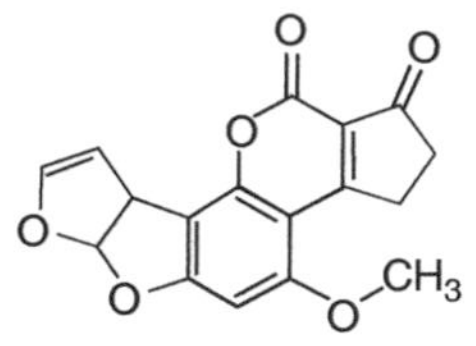

Aflatoxin B1

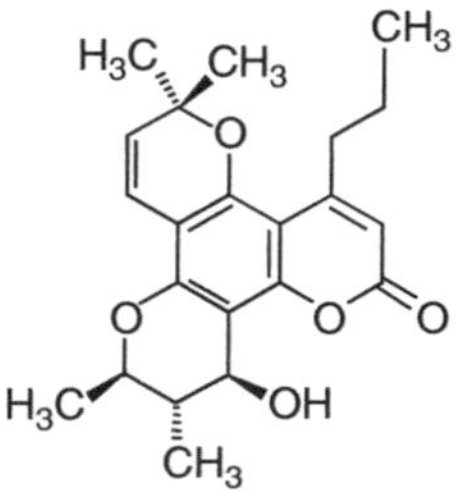

Calanolide A

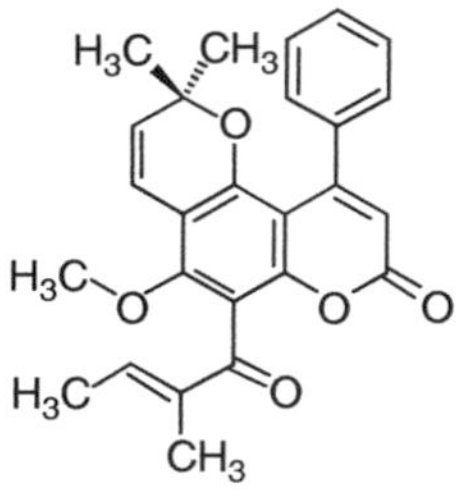

Calophyllolide

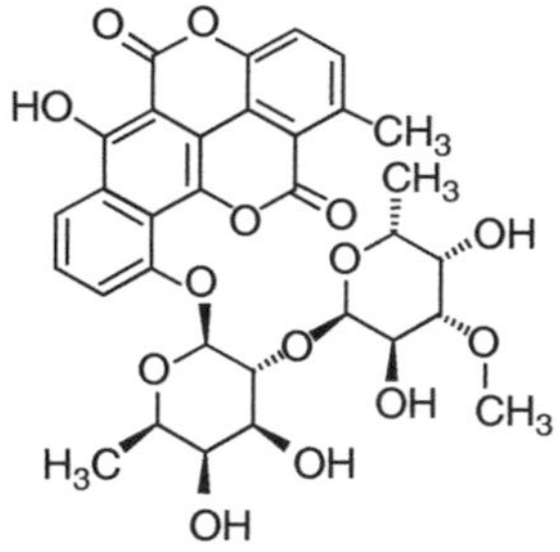

Chartreusin

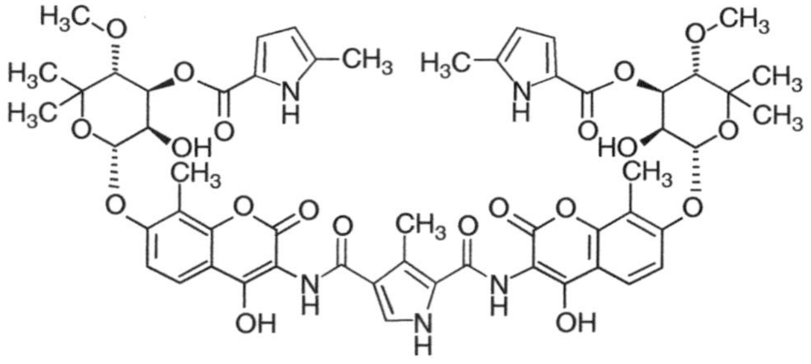

Coumermycin A1

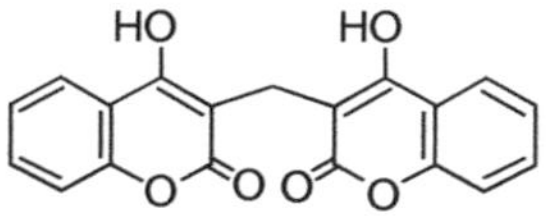

Dicumarol

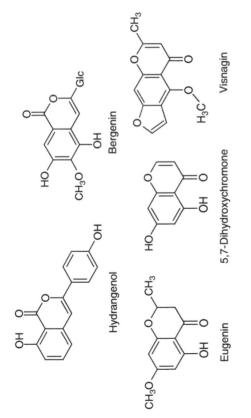

Isocoumarins and chromones

Hydrangenol

Bergenin

Eugenin

5,7-Dihydroxychromone

Visnagin

antagonists of plant growth hormones. In high concentrations, coumarins, such as psoralen and 8-methoxypsoralen, are mutagenic substances, inhibit kernel germination [7].

Biological activity of coumarins is diverse and depends on their chemical structure. The oldest medicine from this group is dicoumarol, isolated from *Melilotus alba* L., which has antitrombotic and anticoagulative properties. Because of that dicoumarol is applied against venous and arterial embolism as well as against hart infarct. Similar properties are possessed by other coumarins, recently isolated from plant material such as psoralen, xanthotoxin, bergapten, imperatorin isolated from *Peucedanum japonicum* and braylin isolated from tree *Toddalia asiatica*. In therapy of circulatory system hypotensic, relaxating coronary vessels and tonic properties of coumarins are also availed. Such properties are also seen in coumarins such as scoparon isolated from *Artemisia scoparia*, angelicin isolated from *Archangelica officinalis*, as well as aesculin isolated from *Aesculus hippocastanum* bark, flowers, and fruits. Aesculin is widely used as the component of gels and ointments against varicose veins of lower limbs and as oral drugs used in therapy of hemorrhoids. It was also ascertained that pyranocoumarins present in *Ammi visnaga* L. fruits such as visnadin, dihydrosamidin, and samidin have strong diastolic properties. Some coumarins have properties of Ca-channel blocking in heart muscle and circular vessels [1,8,9].

Spasmolytic properties of this group of compounds are used in therapy of bile ducts. Besides, fraxidin, isofraxidin, and other coumarins isolated from *Fraxinus exelsior* have cholagogic and choleretic properties [8].

Furanocoumarins are also widely used in therapy because of their photosensibilized properties. It is applied in therapy of leucodermy and psoriasis [8,10,11]. Some hydroxy- and methoxycoumarins have the ability of absorbing UV light and are applied as additives to sunblock creams with anti-UV filters.

Selected coumarins are endowed with bacteriostatic and antifungal properties [8,12,13]. Numerous compounds from this group exhibit analgesic, antiphlogistic, and antipyretic activity. It was also found that selected coumarins (xanthotoxin, imperatorin, scopoletin, qesculin, aesculetin, scoparon) have cytostatic or cytotoxic properties.

Activity of selected coumarins on central nervous system is also known. For example, angelicin has sedative and hypnotic properties, whereas 4-methyl-7-hydroxy-8-piperidinecoumarin acts as a stimulant [1,2,8]. There are also toxic coumarins—aflatoxines produced by fungus *Aspergillus flavus* have carcinogenic activity [14].

15.3 SAMPLE PREPARATION TECHNIQUES OF COUMARINS' FRACTIONS

Coumarins occur in plants as aglycones or glycosides. Aglycones are soluble in petrol, benzene, oil ether, chloroform, diethyl ether, alcohols (methanol, ethanol), but they are not soluble or weakly soluble in water. Glycosides are soluble in water and alcohols.

15.3.1 CLASSICAL METHODS OF ISOLATION OF COUMARINS FROM PLANT MATERIALS

It is difficult to determine optimal criterions of the selection of extrahents with a view to their physicochemical properties (density, surface tension, viscosity), which are directly connected with the penetration of solvent particles to the inside of the plant material. Because coumarins have differentiated chemical structures and various hydrophilic and hydrophobic properties, in classical extraction methods (Soxhlet extraction, maceration, percolation, ultrasonic-aided extraction) many solvents with increasing polarity and elution strength are used [15,16]. Oil ether, petrol, *n*-pentan, *n*-hexan, diethyl ether, acetone, methanol, or their mixtures are mostly applied [17–20].

Głowniak et al. [21] proved that application of mixed binary or ternary solvents (especially N and A class according to Pimentel and McClellan [22] or N, A, or B class) gives significant increase of the extraction efficiency. As A class solvents dichloromethane, chloroform, trichloroethylene; as N class solvents oil ether, petrol, tetrachloromethane; and as B class solvent diisopropyl ether were used.

Härmälä et al. [23] investigated 20 organic solvents to apply them in the extraction of coumarins from root of *Angelica archangelica*. Physicochemical parameters such as density, viscosity, surface tension, and number of carbon atoms in molecule were compared in relation to concentrations of polar and nonpolar substances obtained in extraction processes. Chloroform was chosen as the most efficient solvent in the extraction of coumarins from plant material.

15.3.2 OTHER EXTRACTION METHODS USED IN THE ISOLATION OF COUMARINS FROM PLANTS

The isolation of coumarins from plant materials can also be realized by several other ways. One of these methods is based on lactone type of coumarin structure. Alcohol or water–alcohol solution of KOH crashes the lactone ring in coumarins (on boiling water bath) and coumaric acids arise. Then after acidation these acids cyclize to coumarins again, and then coumarins can be extracted using oil ether.

There are many disadvantages of the described method because coumarins are labile substances and are susceptible to acids and bases, which can destroy their epoxide structure and ester bonds in side chains.

The other two methods of isolation of coumarins from plants are (1) sublimation and fractionating distillation in high vacuum and (2) crystallization from organic solvents [24]. More safe is distillation with water vapor, but this method leads to isolation of coumarins and ballast substances such as ether oils, which can make the separation of single coumarins difficult.

15.3.3 MODERN TECHNIQUES FOR THE EXTRACTION OF COUMARINS

The most modern technique used to isolate coumarins from plant materials is supercritical fluid extraction (SFE). In this method, the supercritical fluid is applied as an extrahent. The fluid is the matter state that is characterized by high diffusion coefficient, low viscosity and density, and lack of surface tension as gases. On the other side, the fluid has an ability to dissolve some substances as the liquid. The connection of these properties causes very good penetration of the matrix (e.g., plant material or other biological sample) and the ability of dissolving the substances in this matrix.

The basic advantages of SFE are as follows:

• The high process speed
• The lack of organic solvents. (Carbon dioxide is most often applied in SFE method. This gas is chemically neutral, nontoxic for environment, cheap and easy to eliminate from the extraction system.)
• The possibility of simultaneous separation of substances present in the matrix (the ability of fluid dissolving is the function of its density—changes of pressure and temperature of the extraction system allow to extract the compounds with various polarities)
• Low risk of oxidation or thermal degradation of extracted compounds
• The possibility of "online" or "off-line" extraction and coupling with other qualitative and quantitative chromatographic methods such as GC, HPLC, and SFC [25,26]

The described method was applied for the separation of furanocoumarins from various genres of *Angelica archangelica* [11,27].

Nyiredy et al. [28] elaborated another modern method of isolation, which is named as medium pressure solid–liquid extraction (MPSLE). In this method, various physical parameters are programmed in extraction system (e.g., mobile phase flow, temperature, pressure, and electrical field) to obtain various changes of distribution of matrix components between solid and liquid phases. Medium pressure chromatographic system (MPLC) is often applied in this technique.

The basic advantages of this method are given below:

• High speed
• Complete material extraction
• The possibility of control of extraction process
• The possibility of control of extraction efficiency
• Obtaining of concentrated, pollution-free extracts

This method was applied by Härmälä et al. [23].

The following extraction methods from plant material were also compared: ultrasonification (USAE), microwave-assisted solvent extraction (MASE) in open and in pressurized mode, accelerated solvent extraction (ASE), also called pressurized liquid extraction (PLE) with exhaustive extraction in Soxhlet apparatus [29,30]. It was found that PSE gives, in most cases, higher yield of extraction of furanocoumarins from *Pastinaca sativa* and *Archangelica officinalis* fruits as exhaustive extraction in Soxhlet apparatus. For more hydrophobic furanocoumarins (bergapten, imperatorin, and phellopterin), it is the most capacitive extraction method of all extraction methods examined. Ultrasound-assisted solvent extraction (USAE) at 60°C (simple and widely used method) also gives, in most cases, higher yield of extraction of furanocoumarins as exhaustive Soxhlet extraction. For more hydrophilic furanocoumarins (santhotoxin and isopimpinellin), it is the most capacitive extraction method of all extraction methods examined. Pressurized MASE gives high yield of extraction of more polar furanocoumarins, but causes change of analytes (hydrophobic ones) during the leaching process. Optimal conditions for the extraction of furanocoumarin fraction from *Archangelica officinalis* fruits [30]—extracting solvent, temperature, and extraction time in PLE—were optimized. It was found that the highest yield by the method was obtained at 100°C–130°C, for 10 min by use of methanol.

15.4 TLC OF COUMARINS IN ANALYTICAL SCALE

From the literature dealing with the analysis of coumarins, it can be mentioned that silica is the first choice for this purpose (see Table 15.2). However, the use of other layers such as Florisil [47,51], polyamide [49], and alumina [49] was also reported. Eluents applied in normal phase systems for separation of pyrano- and furanocoumarins consist of weak polar solvents such as petrol + diethyl ether [41], toluene + ethyl acetate [46], and *n*-hexane + ethyl acetate [52]. For simple coumarins, the use of medium polar eluents can be noticed with dichloromethane as diluent [33,38–40]. Bieganowska et al. [47] report linear relationships of R_M (log k) vs. log C, which proves about displacement model of adsorption of coumarins in NP systems. Linear relationships of R_M (log k) vs. log C for selected coumarins were obtained on aminopropyl silica layers [64]. Similar investigations were performed for alumina and polyamide when aqueous mobile phases were optimized for the separation of coumarins and linear relationships of R_M (log k) vs. C were plotted [49]. From such relationships eluent strength of applied solvent mixture can be fitted and most selective systems for the separation of particular group of coumarins can be chosen.

The use of alkyl-bonded silica layers for the separation of coumarins was also reported—silanized silica [32,55,65] and C18 layers [35,45] eluted with aqueous organic modifiers. In this case, linear relationships of R_M (log k) vs. C were plotted for the optimization of separation of analytes [32,45,65]. From such relationships, lipophilicity—log k_W values—of some coumarins were calculated presuming RP-TLC as well as RP-HPLC data [32].

It should be noticed that special modes of development as multiple development with the decreasing gradient in manual mode [66] or with help of special AMD chamber [39] were used for the improvement of resolution of closely related coumarins.

From the presented examples in Table 15.2, it is seen that thin layer chromatography (TLC) is mainly applied in qualitative analysis for the identification of compounds presented in plant extracts by retention parameters as well as UV spectra taken directly from the layer by use of densitometry.

TLC is also applied for fraction monitoring from preparative column in process of isolation of coumarins from plant material [57,67–75] (see also examples in Table 15.2).

TABLE 15.2
TLC Systems for the Analysis of Coumarins

Compounds	Source	Adsorbent	Eluent	Remarks	References
Coumarins Furanocoumarins Pyranocoumarins	Standards	SiO_2	Binary and ternary eluents	Retention behavior	[31]
Coumarins Furanocoumarins Pyranocoumarins	Standards	RP2	Water + organic modifier	Determination of hydrophobicity	[32]
Coumarins	*Polygala paniculata*	SiO_2	1. T + AcOEt 2. CH_2Cl_2 + MeOH	Optimization of separation	[33]
Coumarins	Essential oil Radix *Pimpinellae*	SiO_2	T + Et_2O + AcOH	Qualitative identification	[34]
Coumarins Furanocoumarins	Standards	SiO_2 RP18	Various solvents and solvent mixtures	Optimization of separation	[35]
Furanocoumarins	Rhizomes *Peucedanum off.*	SiO_2	B + AcOEt	Qualitative identification	[36]
Coumarins	Standards	SiO_2	—	Spray reagent: Chlorosulphonic acid + chloroform	[37]
Coumarins	Standards	SiO_2	CH_2Cl_2 + MeOH	Detection—I_2 vapor, phosphomolibdate reagent	[38]
Coumarins	Standards	SiO_2	CH_2Cl_2 + MeOH + water + HCOOH	AMD	[39]
Coumarins	Genus *Angelica*	SiO_2	1. $CHCl_3$ 2. AcOEt + CH_2Cl_2	2D-TLC 2D-OPLC	[40]
Furanocoumarins	Radix	SiO_2	Petrol + Et_2O	Quantification Densitometry at 300 nm	[41]

Compound	Source/Sample	Stationary phase	Mobile phase	Application	Ref.
Coumarins	Angelicae dahuricae Citrus fruit essential oil	SiO$_2$	1. AcOBu + Hx 2. AcOBu + CHCl$_3$ + Hx	OPLC	[42]
Coumarin isomers	Peucedanum palustre	SiO$_2$	Various mobile phases	OPLC, PRISMA system	[43,44]
Coumarins	Standards	C18	Buff. aqueous sol. of MeOH, MeCN, Dx, THF	Retention behavior	[45]
Furanocoumarins	Cnidium species	SiO$_2$	T + AcOEt + MeOH	Identification, finger print techniques	[46]
Coumarins	Standards	SiO$_2$ Florisil	CH$_2$Cl$_2$ + MeOH, MeCN, Dx or iPrOH	Retention behavior	[47]
Plant phenolics	Phyllanthus emblica leaves	SiO$_2$	T + EtOH + Hx + HCOOH	Identification	[48]
Coumarins	Standards	Alumina Polyamide	Water + MeOH, MeCN, Dx or iPrOH	Retention behavior	[49]
27 Coumarins	Standards	SiO$_2$	Hx + Dx + Et$_2$O + EtOH	Quantitative analysis PRISMA system	[50]
Coumarins	Standards	SiO$_2$ Florisil	CH$_2$Cl$_2$ + Hx + polar modifier (11)	Retention behavior	[51]
Furanocoumarins	Ruta graveolens	SiO$_2$	Hx + AcOEt	Identification	[52]
Coumarins	Decatropis bicolor	SiO$_2$	Hx + AcOEt	Identification, biological activity	[53]
Coumarins	Peucedanum verticillare	SiO$_2$	1. H + CH$_2$Cl$_2$ + AcOEt 2. H + iPr$_2$O + PrOH	Identification	[54]

(continued)

TABLE 15.2 (continued)
TLC Systems for the Analysis of Coumarins

Compounds	Source	Adsorbent	Eluent	Remarks	References
Coumarins	*Peucedanum tauricum*	SiO$_2$ RP2	Cx + AcOEt; CH$_2$Cl$_2$ + MeCN; MeOH + H$_2$O	Identification	[55]
Coumarins + other phenolics	Standards	7 Layers	7 Mobile phases	Identification	[56]
Coumarins	*Ptercocaulon virgatum*	SiO$_2$	T + AcOEt	Fraction monitoring from preparative column	[57]
Sesquiterpene coumarins	*Ferula szowitsiana*	SiO$_2$	Petrol + AcOEt	Fraction monitoring from preparative column	[58]
Coumarins furanocoumarins	*Metrodorea nigra*	SiO$_2$	Hx + CH$_2$Cl$_2$ + AcOEt + MeOH	Fraction monitoring from preparative column	[59]
Coumarins pyranocoumarins	*Cedrelopsis grevei*	SiO$_2$	Hx + AcOEt + MeOH	Fraction monitoring from preparative column	[60]
Coumarins	*Pterocaulon virgatum* species	SiO$_2$	CHCl$_3$ + MeOH	Fraction monitoring from preparative column	[61]
Oxypeucedanin, i-imperatorin, oxypeucedanin hydrate	*Peucedanum ostruthium*	SiO$_2$	CHCl$_3$ + MeOH + H$_2$O	Fraction monitoring from preparative column	[62]
Xanthotoxin, alloxanthotoxin, umbelliferone, pimpinellin	*Diplolaena mollis*	SiO$_2$	1. Hx + AcOEt 2. CH$_2$Cl$_2$ + MeOH	Fraction monitoring from preparative column	[63]

15.5 MULTIDIMENSIONAL SEPARATIONS OF COUMARINS

15.5.1 TWO-DIMENSIONAL TLC IN SEPARATION OF COUMARINS

Polar bonded stationary phases (CN-silica and Diol-silica) can be used with non-aqueous eluents in adsorption mode or with aqueous eluents in partition mode. It enables the application of these systems for two-dimensional separations because of different selectivity and application to the separation of closely related compounds of similar physicochemical properties and retention behavior (see Figure 15.1). On Diol layers 10% methanol in water (containing 1% formic acid) as the first direction eluent and 100% diisopropyl ether as the second direction eluent were used [76]. In this 2D-TLC system, the mixture containing the following coumarins was separated: coumarin, 6-methylcoumarin, 3,4-dihydrocoumarin, aesculetin, aesculin, dimethyl-fraxetin, bergapten, and phellopterin. On cyanopropyl-silica layers 30% acetonitrile in water as the first direction eluent and 35% ethyl acetate in n-heptane as the second direction eluent were used [76]. In this 2D-TLC system, the mixture containing the following coumarins was separated: coumarin, scopoletin, isoscopoletin, fraxidin, aesculetin, umbelliferone, bergapten, xanthotoxin, xanthotoxol, isopimpinellin, heraclenin, angelicin, imperatorin, byacangelicol, and byacangelicin (see Figure 15.2). Using this 2D-TLC system byacangelicol, imperatorin, bergapten, heraclenin, and

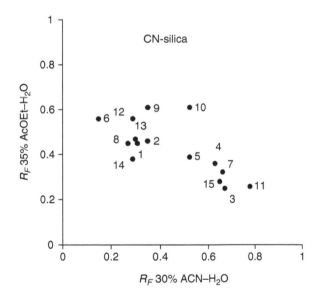

FIGURE 15.1 Correlation of R_F parameters for coumarin standards chromatographed on CN-silica in systems with 30%ACN–water, triple developed and 35% AcOEt–n-heptane double developed. Solutes: 1—isopimpinellin, 2—byacanelicol, 3—fraxidin, 4—umbelliferone, 5—xanthotoxin, 6—imperatorin, 7—scopoletin, 8—byacangelicin, 9—angelicin, 10—coumarin, 11—aesculetin, 12—bergapten, 13—heraclenin, 14—xanthotoxol, 15—isoscopoletin. (From Waksmundzka-Hajnos, M., Petruczynik, A., Hajnos, M.L., Tuzimski, T., Hawryl, A., Bogucka-Kocka, A., *J. Chromatogr. Sci.*, 44, 510, 2006.)

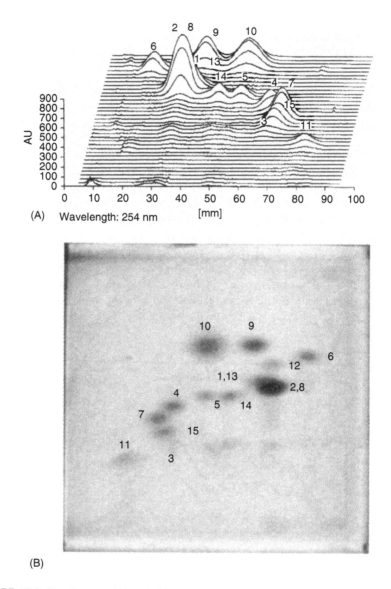

FIGURE 15.2 Densitogram (A) and videoscan (B) of two-dimensional separation of coumarin standards on CN-silica plate. Direction eluent: 30% ACN–water, triple developed, I;
direction eluent: 35% AcOEt–*n*-heptane double developed. Solutes as in Figure 15.1. (From
Waksmundzka-Hajnos, M., Petruczynik, A., Hajnos, M.L., Tuzimski, T., Hawryɫ, A.,
Bogucka-Kocka, A., *J. Chromatogr. Sci.*, 44, 510, 2006.)

xanthotoxol in *Heracleum sphondylium* fruit extract and imperatorin, byacangelicol,
and xanthotoxol in *Heracleum sphondylium* root extract were identified. The extract
components were identified by comparing the retardation factors in both directions
and by UV spectra from DAD densitometric scanner determined for components and

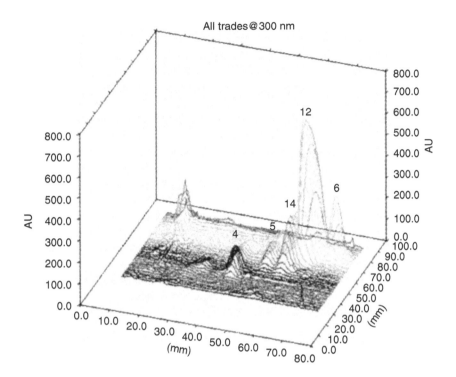

FIGURE 15.3 Densitogram of two-dimensional separation of *Archangelica officinalis* fruit extract on CN-silica plate. Conditions and solutes as in Figure 15.1. (From Waksmundzka-Hajnos, M., Petruczynik, A., Hajnos, M.L., Tuzimski, T., Hawryl, A., Bogucka-Kocka, A., *J. Chromatogr. Sci.*, 44, 510, 2006.)

for standards. Moreover in the same manner byacangelicol, bergapten, isopimpinellin, and xanthotoxol were identified in *Heracleum sibiricum* fruit extract and xanthotoxol, xanthotoxin, bergapten, and imperatorin were identified in *Archangelica officinalis* fruit extract (see Figure 15.3) [76].

Similarly multiphase plates, connected with C18 strips and silica layers, have been used with aqueous (55% methanol + water) and nonaqueous eluents (35% ethyl acetate + *n*-heptane). Such layers were applied for the separation of selected coumarins—mixture of coumarin, scopoletin, isocopoletin, umbelliferone, aesculetin, fraxidin, bergapten, byacangelicol, byacangelicin, angelicin, xanthotoxin, imperatorin, heraclenin, and *Heracleum sphondylium* fruit extract were in the 2D-TLC system separated [76].

Thus, the differences in separation selectivity have been applied for the separation of coumarins' fraction from plant extracts of *Apiaceae* family by 2D-TLC.

15.5.2 COUPLING OF HPLC AND TLC IN THE ANALYSIS OF COUMARINS

The possibility to separate 12 coumarins from *Archangelica officinalis* was also studied by combining an HPLC and a TLC system [77]. HPLC optimized by the use of DryLab for Windows software was performed on RP-18 column by use of 60% methanol in water. Simulation of separation and separation of extract in optimized system are presented in Figure 15.4A and B. This chromatogram shows that column chromatography in RP system does not allow for complete separation of substances: umbelliferone + xanthotoxol $(1+2)$, archangelicin + xanthotoxin $(3+4)$, and pimpinellin + phelopterin + imperatorin $(7+8+9)$ are eluted together in groups. Therefore, fractions from HPLC were collected (fractions I–IX) according to the consecutive peaks and their nonseparated groups on the HPLC chromatogram. The solvents from all fractions were evaporated and then the consecutive fractions analyzed in NP TLC system. The optimization of conditions in TLC system was performed on the basis of retention—composition of nonaqueous solvent relationship: ethyl acetate in dichloromethane–heptane [77]. Optimal selectivity for investigated compounds in NP systems was obtained on silica layers by using 40% (v:v) ethyl acetate in dichloromethane–n-heptane (1:1) as the mobile phase. It makes complete separation of coumarins contained in *Archangelica officinalis* root extract. Videoscan of fractions separated on silica layer is presented in Figure 15.5. In such a way by use of RP-HPLC and NP-TLC, the following coumarins were identified in the extract: umbelliferone, xanthotoxol, archangelicin, xanthotoxin, isopimpinellin, bergapten, pimpinellin, phelopterin, imperatorin, osthol, umbelliprenin, and isoimperatorin.

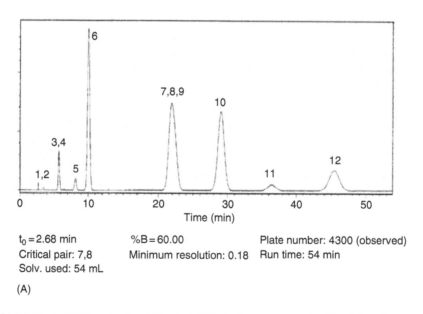

$t_0 = 2.68$ min   %B = 60.00   Plate number: 4300 (observed)
Critical pair: 7,8   Minimum resolution: 0.18   Run time: 54 min
Solv. used: 54 mL

(A)

FIGURE 15.4 (A) The simulated Dry Lab HPLC chromatogram for 60% (v/v) of methanol in water. 1—Xanthotoxol, 2—umbelliferone, 3—archangelicin, 4—xanthotoxin, 5—isopimpinellin, 6—bergapten, 7—pimpinellin, 8—imperatorin, 9—phellopterin, 10—osthol, 11—umbelliprenin, 12—isoimperatorin. (From Hawrył, M.A., Soczewiński, E., Dzido, T.H., *J. Chromatogr. A*, 366, 75, 2000.)

(*continued*)

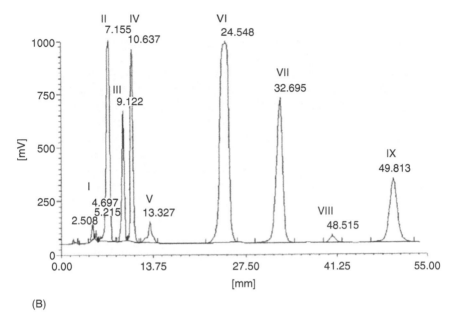

(B)

FIGURE 15.4 (continued) (B) The experimental chromatogram obtained by using 60% (v/v) of methanol in water. Roman numbers are numbers of fractions collected during HPLC separation. (From Hawryl, M.A., Soczewiński, E., Dzido, T.H., *J. Chromatogr. A*, 366, 75, 2000.)

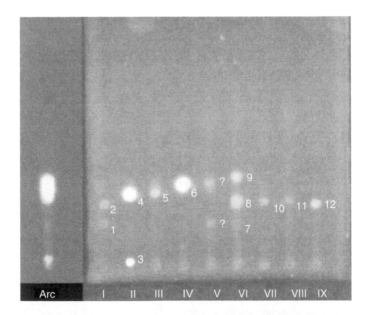

FIGURE 15.5 The Camag Reprostar 3 photography of TLC plate with applied fractions collected during HPLC separation and *Archangelica off.* Extract scanned in black and white mode. Solutes: as in Figure 15.4A, ?: unidentified substances. Arc: *Archangelica officinalis* extract. (From Hawryl, M.A., Soczewiński, E., Dzido, T.H., *J. Chromatogr. A*, 366, 75, 2000.)

15.6 PREPARATIVE LAYER CHROMATOGRAPHY OF COUMARINS

As mentioned earlier coumarins are divided into several groups according to the main skeleton of molecules. It makes possible easy group fractionation of coumarins even in extraction step. However, in particular group of furanocoumarins, pyranocoumarins, etc., exist structural analogues—closely related compounds having similar physico-chemical properties and are difficult for separation. This makes isolation of particular components difficult. For this reason, isolation of individual compounds is usually multistep procedure. Often one chromatographic system is insufficient for the isolation of pure compounds and rechromatography of partly separated fractions, containing several coumarins, is required. In such situation, TLC is a consecutive step of complex procedure also including other chromatographic systems and techniques.

As an example furanocoumarin fraction from fruits of *Apiaceae* genus can be separated for individual compounds in the following manner [78]:

1. Silanized silica layer eluted with methanol–water (6:4) enables fraction-ation of *Heracleum sosnowskyj* fruit extract. Fractions can be separated in system of various selectivities by column chromatography by use of Florisil (magnesium silicate) eluted with 5% acetonitrile in dichloromethane–*n*-heptane (7:3). In such a way, pure xanthotoxin, bergapten, and imperatorin were obtained (see Figure 15.6).

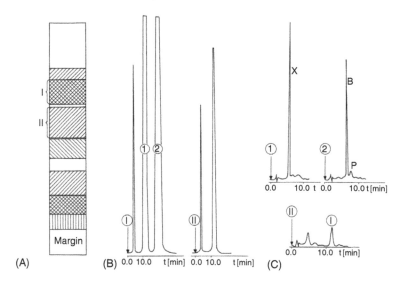

FIGURE 15.6 (A) Micropreparative RP-TLC on silanized silica of crude extract from *Heracleum sosnowskyj* fruit. Sample, 1 mL of eluate from starting band purified on silica layer (with benzene). Eluent: methanol–water (6 + 4). (B) Rechromatography on Lobar-type, 200×15 mm i.d. column (Florisil $d_p = 20$–$50\ \mu$m) of fractions I and II, isolated from RP-TLC. Eluent: 5% acetonitrile in dichloromethane–heptane (7 + 3). (C) Analytical RP-HPLC (RP-18) of column eluates. Eluent: methanol–water (6 + 4). Solutes: X—xanthotoxin, B—bergapten, P—pimpinellin, I—imperatorin. (From Waksmundzka-Hajnos, M., Wawrzynowicz, T., *J. Planar Chromatogr.*, 3, 439, 1990.)

2. Florisil layer eluted with 10% solution of 2-propanol in dichloromethane + *n*-heptane (7:3) and/or with 15% ethyl acetate in benzene enables fractionation of furanocoumarins from *Archangelica officinalis* fruit extract (imperatorin + phelopterin, imperatorin + umbelliprenin, bergapten + pimpinellin, and xanthotoxin + isopimpinellin). For isolation of pure imperatorin and phellopterin, Lobar column filled with Si60 and eluted with 1.5% acetonitrile in dichloromethane + *n*-heptane (7:3) should be used [78].

3. Thin layer chromatography on Florisil (or other polar adsorbent) layer enables separation of crude fruit extracts contaminated with plant oils, which can be purified by pre-elution of starting band with nonpolar solvent (*n*-heptane) and then separated on mixed fractions in selected eluent system (15% ethyl acetate in benzene). Partly separated fractions can be rechromatographed on Florisil layer by use of eluent of different selectivity (5% diisopropyl ether in dichloromethane + *n*-heptane (7:3). In such a way, pure pimpinellin, bergapten, imperatorin, and xanthotoxin from *Heracleum species* were isolated [79].

Extensive chromatography of coumarin fractions by combining two column chromatographic separations on Florisil and silica followed by preparative TLC on silica layer eluted with *n*-heptane-dichloromethane–ethyl acetate mixtures or *n*-heptane diisopropyl ether led to the isolation of seven coumarins from the fruits and nine compounds from the roots of *Peucedanum verticillare*. Among other unknown compounds there were following identified coumarins: umbelliferone, isoimperatorin, xanthotoxin, bergapten, imperatorin, psoralen, *cis*-kellactone, and aesculin [55].

The examples of preparative TLC systems used as a step of isolation of coumarins from plant extracts are presented in Table 15.3. Combinations of RP-TLC (RP-2 layers eluted with aqueous methanol) and NP-TLC (silica layers eluted with dichloromethane + acetonitrile) enable isolation and identification of bergapten, peucedanin, and scopoletin in *Peucedanum tauricum* leaves [80]. However, combination of column chromatography (silica eluted ethyl acetate in dichloromethane in gradient mode), preparative RP-TLC (silanized silica eluted with aqueous methanol), and preparative NP-TLC (silica eluted with cyclohexane + ethyl acetate or other solvent mixtures) enables isolation of isoimperatorin, bergapten, and four unidentified coumarins from *Peucedanum tauricum* fruits [54]. Figure 15.7 presents densitogram from preparative TLC of fraction obtained by CC method in which five isolated compounds are marked. The examples of preparative thin layer chromatography systems used as a step of isolation of coumarins from plant extracts are presented in Table 15.3.

15.7 FORCED-FLOW THIN-LAYER TECHNIQUES IN THE SEPARATION OF COUMARINS

On the basis of TLC experiments and PRISMA system optimization of separation of 16 coumarins present in genus *Angelica* was performed [40]. On the basis of the preassays silica gel layers, methyl-*tert*-butyl ether, dichloromethane, ethyl acetate,

TABLE 15.3
TLC Systems for the Isolation of Coumarins from Plant Extracts by Preparative TLC

Compounds	Source	Adsorbent	Eluent	References
Simple coumarins	*Zanthoxylum schinifolium*	SiO_2	1. $CHCl_3$ + MeOH 2. $CHCl_3$ + AcOEt	[81]
Furocoumarins	*Ticorea longiflora*	SiO_2	Hx + CH_2Cl_2 + MeOH	[82]
Oxypeucedanin, i-imperatorin, oxypeucedanin hydrate	*Peucedanum ostruthium*	SiO_2	$CHCl_3$ + MeOH + H_2O	[62]
Coumarins	*Milletia thonningii*	SiO_2	Petrol + Me_2CO	[83]
Xanthotoxin, alloxanthotoxin, umbelliferone, pimpinellin	*Diplolaena mollis*	SiO_2	1. Hx + AcOEt 2. CH_2Cl_2 + MeOH	[63]
Schinicoumarin, schiniallylol, schinilenol	*Zanthoxylum schinifolium*	SiO_2	Hx + AcOEt	[84]
7-Methoxy-coumarins	*Murraya paniculata*	SiO_2	Hx + T + Me_2CO	[85]
Highly polar oxygenated coumarins	*Pelargonium sidoides*	SiO_2	T + Me_2CO	[86]
5-Methyl coumarins	*Nassauvia pyramidalis, Nassauvia digitata*	SiO_2	Hx + AcOEt	[87]
Oxypeucedanin, umbelliferone, suberosin, saxalin + 2 new furanocoumarins:	*Harbouria trachypleura*	SiO_2	1. Hx + AcOEt 2. CH_2Cl_2 + Me_2CO	[88]
New pyranocoumarin	*Geijera paniculata*	SiO_2	Hx + AcOEt	[89]
Scopoletin ayapin + their glycosides	Crude extract from sunflower	SiO_2	Et_2O	[90]
Scopoletin, fraxetin	*Protium heptaphyllum*	SiO_2	B + AcOEt	[91]
Furanocoumarins: bergapten, xanthotoxin, heraclenol	*Metrodorea flavida*	SiO_2	Hx + AcOEt + MeOH	[92]

Trimethoxy coumarins: marmesin and 2 novel	Chorilaena quercifolia	SiO$_2$	Hx + AcOEt	[93]
Bergapten, xanthotoxin, isopimpinellin, isoimperatorin	Metrodorea nigra	SiO$_2$	1. Hx + CH$_2$Cl$_2$ 2. AcOEt + MeOH	[59]
Simple coumarins and pyranocoumarins	Cedrelopsis grevei	SiO$_2$	Hx + AcOEt + MeOH	[60]
6,7-Dioxygenated coumarins	Pterocaulon virgatum	SiO$_2$	1. CHCl$_3$ + MeOH 2. CHCl$_3$ + Me$_2$CO	[61]
Pyranocoumarins	Paramignya monophylla	SiO$_2$	CH$_2$Cl$_2$ + MeOH	[94]
Dimeric coumarin	Toddalia asiatica	SiO$_2$	B + AcOEt	[95]
Furanocoumarins Pyranocoumarins	Peucedanum japonicum	SiO$_2$	Hx + AcOEt	[96]
Cytotoxic coumarins	Calophyllum dispar	SiO$_2$	1. Pentane + Me$_2$CO 2. CHCl$_3$ + AcOEt	[97]
Umbelliprenin, i-imperatorin, bergapten, byacangelicin	Angelica sylvestris	SiO$_2$	Hx + AcOEt	[98]
Prenyl-coumarins	Boronia inornata, Boronia gracilipes	SiO$_2$	Petrol + AcOEt + MeOH	[99]
Simple coumarins and furanocoumarins	Ferula sumbul	SiO$_2$	CHCl$_3$ + MeOH	[100]
Simple coumarins Furanocoumarins	Ferulago capillaris, Ferulago brachyloba	SiO$_2$	B + AcOEt	[101]
Sesquiterpene coumarins	Ferula szowitsiana	SiO$_2$	Petrol + AcOEt	[102]
Coumarins	Peucedanum officinale Peucedanum ostruthicum	SiO$_2$	1. CHCl$_3$ + AcOEt 2. CHCl$_3$ + AcOEt + MeOH	[103]
Coumarins	Peucedanum tauricum fruits	SiO$_2$ RP2	1. 1Cx + AcOEt 2. H + CH$_2$Cl$_2$ + AcOEt 3. CH$_2$Cl$_2$ + MeCN MeOH + water	[54]

(continued)

TABLE 15.3 (continued)

TLC Systems for the Isolation of Coumarins from Plant Extracts by Preparative TLC

Compounds	Source	Adsorbent	Eluent	References
Coumarins	*Peucedanum verticillare*	SiO_2	1. H + CH_2Cl_2 + AcOEt	[55]
		Florisil	2. H + iPr_2O + iPrOH	
Coumarins	*Peucedanum tauricum* leaves	RP 2	MeOH + water	[80]
		SiO_2	CH_2Cl_2 + MeCN	
Furanocoumarins	*Archangelica officinalis*	Florisil	1. H + CH_2Cl_2 + MeCN	[78]
	Heracleum sosnowskyj fruits	SiO_2	2. H + CH_2Cl_2 + iPr_2OMeOH + water	
		RP 2		
Furanocoumarins	*Heracleum moellendorfi*	Florisil	H + CH_2Cl_2 + iPr_2O	[79]
	Heracleum mantegazzianum fruits		B + AcOEt	
Coumarins	*Polygala paniculata*	SiO_2	Centrifugal TLC petrol + $CHCl_3$	[33]
Coumarins	*Ticorea longiflora*	SiO_2	Hx + CH_2Cl_2 + MeOH	[65]
Coumarins	*Zanthoxylum schinifolium*	SiO_2	Hx + Me_2CO + B + AcOEt	[104]
Coumarins	*Helietta longiflora*	SiO_2	$CHCl_3$ + MeOH	[105]
Coumarins	*Clausena harmandiana*	SiO_2	Hx + AcOEt	[106]
Coumarins	*Citrus hystrix*	SiO_2	Hx + AcOEt	[107]

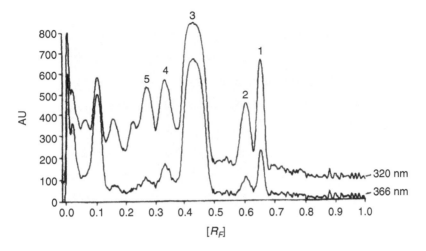

FIGURE 15.7 Densitogram obtained from preparative chromatography of *Peucedanum* taur-icum fraction. The TLC silica plate was developed twice with cyclohexane–EtOAc 80 + 20 (v/v) as mobile phase. 1–5 are the zones isolated. (From Głowniak, K., Bartnik, M., Mroczek, T., Żabża, A., Wierzejska, A., *J. Planar Chromatogr.*, 15, 94, 2002.)

ethyl methyl ketone, and chloroform were chosen as the most effective solvents. However, it was found that on TLC and HPTLC plates none of binary mixtures could separate all the 16 compounds. To improve the separation of investigated compounds the more powerful forced-flow technique, OPLC, was chosen with solvents of lower eluent strength. It is presented in the paper, that one-dimensional OPLC does not cause complete separation of all compounds—some of them are eluted together or spots are adjoined. Thus, the use of 2D-OPLC with solvent of various selectivities was decided and full separation of coumarins was achieved [40]. By use of chloroform in the first direction and 30% ethyl acetate in chloroform as the second direction eluent total baseline separation of all 16 coumarins was achieved [40].

Similar investigations were performed with 15 coumarin standards in normal-phase systems in comparison with TLC and HPLC experiments [44]. It was also ascertained that fully online OPLC can be run as a fully automated method and represents a genuine alternative to HPLC. Figure 15.8 shows full analogy of OPLC and HPLC separations of eight furanocoumarin isomers.

OPLC in system silica/*n*-butyl acetate–hexane or chloroform–butyl acetate–hexane was used as a rapid method to evaluate the composition of citrus essential oils containing coumarins [42]. OPLC was also used for separation of six main coumarins in *Peucedanum palustrae* (see Figure 15.9) [43].

Centrifugal TLC on silica layer eluted with petrol ether–chloroform (1:9) was used for the isolation of *Polygala paniculata* in preparative scale [33].

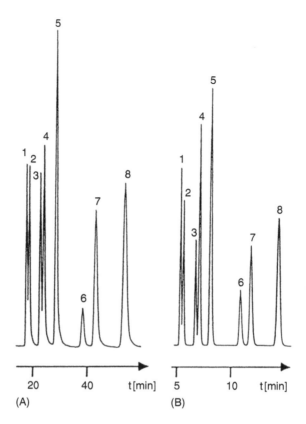

FIGURE 15.8 (A) Fully online OPLC separation of furocoumarins isomers with the modified mobile phase. Flow rate 1.2 mL/min; counterpressure 22 bar. (B) HPLC separation of furocoumarins isomers with the modified mobile phase. Flow rate 2 mL/min. Solutes: 1—isobergapten, 2—angelicin, 3—psoralen, 4—bergapten, 5—pimpinellin, 6—sphondin, 7—xanthotoxin, 8—isopimpinellin. (From Zogg, G.C., Nyiredy, Sz., Sticher, O., *J. Planar Chromatogr.*, 1, 351, 1988.)

15.8 PLANTS WITH NATURALLY OCCURRING COUMARINS

There are many species in which coumarins occur. Some of them are listed below:

Medicago sativa L. (Fabaceae/Leguminosae)—coumarins: medicagol.

Angelica archangelica L. (Apiaceae/Umbelliferae)—coumarins: over 20 furanocoumarins, including angelicin, archangelicin, bergapten, isoimperatorin, and xanthotoxin. Also the coumarins osthol (major constituent in rhizome/root, 0.2%) and umbelliferone. The root also contains the furanocoumarins 2'-angeloyl–3'-isovaleryl vaginate, heraclenol–2'-O-senecioate, and heraclenol–2'-O-isovalerate.

Pimpinella anisum L. (Apiaceae/Umbelliferae)—coumarins: scopoletin, umbelliferone, umbelliprenin; bergapten (furanocoumarin).

Arnica montana L. (Asteraceae/Compositae), Arnica chamissonis *Less. ssp.* foliosa—coumarins: scopoletin and umbelliferone.

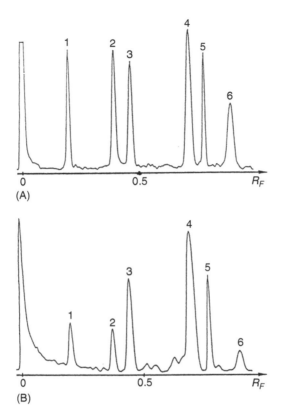

FIGURE 15.9 Circular OPLC separation. (A) Coumarins investigated and (B) extract of *P. palustre*. Stationary phase: Silica Gel 60F$_{254}$ HPTLC plate. Sample application 6 mm from solvent inlet. Mobile phase: tetrahydrofuran–isopropanol $S_T = 0.45$; $P_S = 93$–7. Solutes: 1—oxypeucedanin hydrate, 2—isobyakangelicin, 3—ostruthol, 4—oxypeucedanin, 5—columbianadin, 6—isoimperatorin. (From Vuorela, H., Dallenbach-Tölke, K., Sticher, O., Hiltunen, R., *J. Planar Chromatogr.*, 1, 123, 1988.)

Ferula assafoetida L. (*Ferula rubricaulis* Boiss), *Ferula foetida* (*Bunge*) *Regel* (Apiaceae/Umbelliferae): coumarin derivatives (e.g., umbelliferone), coumarin–sesquiterpene complexes (e.g., asacoumarin A and asacoumarin B).

Menyanthes trifoliata L. (Menyanthaceae)—coumarins: scopoletin.

Agathosma betulina (Berg.) Pillans (*Rutaceae*)—coumarins have been reported for many other *Agathosma* species [108].

Apium graveolens L. (Apiaceae/Umbelliferae)—furanocoumarins: apigravin, apiumetin, apiumoside, bergapten, celerin, celereoside, isoimperatorin, isopimpinellin, osthenol, rutaretin, seselin, umbelliferone, and 8-hydroxy-5-methoxypsoralen [109–117].

Matricaria recutita L. (Asteraceae/Compositae)—coumarins: umbelliferone and its methyl ether, heniarin.

Chamaemelum nobile (L.) All. (Asteraceae/Compositae)—coumarins: scopoletin-7-glucoside.

Taraxacum officinale Weber (Asteraceae/Compositae)—coumarins: cichoriin and aesculin [118].

Aesculus hippocastanum L. (Hippocastanaceae)—coumarins: aesculetin, fraxin (fraxetin glucoside), scopolin (scopoletin glucoside).

Radicula armoracia (L.) Robinson (Brassicaceae/Cruciferae)—coumarins: aesculetin, scopoletin [119].

Glycyrrhiza glabra L. (Leguminosae)—coumarins: glycyrin, heniarin, liqcoumarin, umbelliferone, GU-7 (3-arylcoumarin derivative) [120].

Petroselinum crispum (Mill.) Nyman (Apiaceae/Umbelliferae)—furanocoumarins: bergapten and oxypeucedanin as major constituents (up to 0.02% and 0.01%, respectively); also 8-methoxypsoralen, imperatorin, isoimperatorin, isopimpinellin, psoralen, xanthotoxin (up to 0.003%) [121].

Zanthoxylum americanum Miller (Rutaceae)—coumarins: xanthyletin, xanthoxyletin, alloxanth oxyletin, and 8-(3,3-dimethylallyl) alloxanthoxyletin.

Lactuca virosa L. (Asteraceae/Compositae)—coumarins: aesculin, cichoriin [122].

The distribution profile of naturally occurring furanocoumarins shows large variations according to the plant species. These chemical patterns have been summarized below, respectively for Rutaceae, Moraceae, Leguminosae, and Apiaceae.

15.8.1 Rutaceae

For *Ruta* genus (*Ruta graveolens* and *Ruta pinnata*), four main linear furanocoumarins were described: psoralen, 5-methoxy-psoralen (5-MOP, also known as bergapten), 8-methoxy-psoralen (8-MOP, xanthotoxin), and 5–8-methoxy-psoralen (5–8-MOP, isopimpinellin). Some other minor linear furanocoumarins were reported in Ruta graveolens, such as oxypeucedanin. Citrus plants usually contain 5-MOP as well as other linear furanocoumarins (bergamotin and bergaptol in bergamot tree, phellopterin in *Citrus limonum*, 5–8-MOP in *Citrus acida*). Bergamot oil is the most well-known resource for bergapten (5-MOP), which is the furanocoumarin of major therapeutic interest. The bergamot tree is mainly found in Calabria (Italy) and Ivory Cost. Six other *Rutaceae* species were investigated for their furanocoumarin content (*Thamnosma montana, Fagara shinofolia*). A general rule is that 5-MOP is very common in all this plant family. Indeed, from all the *Rutaceae* plants that were investigated, *Thamnosma montana* and *Z. americanum* were the only species for which 5-MOP was not reported.

15.8.2 Moraceae

Bergapten (5-MOP) was reported in every *Moraceae* species that were investigated, and psoralen was found in 7 out of 11 species. Xanthotoxin (8-MOP) was discovered only in *F. carica*. As a consequence, 5-MOP and psoralen are the most widespread furanocoumarins in *Moraceae*.

15.8.3 LEGUMINOSAE

Among the large *Leguminosae* family, only two plant genres are known to contain furanocoumarins. These genres are *Psoralea* and *Coronilla*. These plants contain the most simple linear and angular furanocoumarins, which are respectively psoralen and angelicin. These two furanocoumarins have been found in most of the *Psoralea* species that was analyzed. On the contrary, some *Coronilla* species, such as *C. juncea* and *C. scorpioides*, lack angelicin. *Psoralea* plants need a special mention as this group gave its name to the psoralen molecule.

15.8.4 APIACEAE

Furanocoumarins are the most widespread in the *Apiaceae* family. So far these molecules have been identified in more than 35 species. The furanocoumarin distribution pattern in *Apiaceae* is quite complex. Most of these plants contain four linear furanocoumarins, which are 5-MOP, 8-MOP, 5–8-MOP, and imperatorin. Psoralen, a precursor of more substituted linear furanocoumarins, was reported at low levels in some species, probably because of a more accurate and sensitive detection method used by the corresponding authors. Other linear furanocoumarins were also found depending on the species. Angelicin is present in *Angelica archangelica*, which gave its name to the molecule. It has been hypothesized by Berenbaum and Feeny [123] that the occurrence of angular furanocoumarins in plants could be related to plant coevolution with insects. Some insects were shown to be able to detoxify linear and angular furanocoumarins with cytochrome P450 enzymes, but the occurrence of chemical mixtures of both angular and linear furanocoumarins is more effective against insects than a single component at equivalent concentrations.

REFERENCES

1. Estevez-Braun, A., Gonzales, A.G., Coumarins, *Nat. Prod. Rep.*, 14, 465–475, 1997.
2. Królikowska, M., *Phytochemical Analysis of Medicinal Plant Material*, Medical University, Łódź, Poland, 1988 (in Polish).
3. Beier, T.M., Oertelin, E.H., Psoralen and other linear furanocoumarins as phytoalexins in celary, *Phytochemistry*, 22, 2595–2597, 1983.
4. Harborne, J.B., *Biochemical Ecology*, PWN, Warsaw, Poland, 1997.
5. Osbourn, A.E., Antimicrobial phytoprotectants and fungal pathogens: a commentary, *Fungal, Genet, Biol.*, 26, 163–168, 1999.
6. Zobel, M.A., Brown, A., Coumarins in the interactions between the plant and its environment, *Allelopathy J.*, 2, 9–20, 1995.
7. Baskin, J.M., Ludlow, C.J., Harris, T.M., Wolf, F.T., Psoralen, an inhibitor in the seeds of *Psoralea subacaulis* (Leguninosae), *Phytochemistry*, 6, 1209–1213, 1967.
8. Cisowski, W., Biological properties of coumarins. Part I. Action on plants and pharmacological and antibacterial properties, *Herba Pol.*, 24, 301–314, 1983 (in Polish).
9. Zgórka, G., Dragan, T., Głowniak, K., Basiura, E., Determination of furanochromones and pyranocoumarins in drugs and Ammi visnage fruits by combined solid-phase extraction–high performance liquid chromatography and thin-layer chromatography–high performance liquid chromatography, *J. Chromatogr. A*, 797, 305–309, 1998.

10. Ceska, O., Chaudhary, S.K., Warrington, P.J., Ashwood-Smith, M.J., Photoactive furanocoumarins in fruits of some umbellifers, *Phytochemistry*, 26, 165–169, 1987.
11. Gawdzik, J., Kawka, S., Mardarowicz, M., Suprynowicz, Z., Wolski, T., Carbon dioxide fractionated supercritical fluid extraction of furanocoumarins from the fruit of *Archangelica officinalis* Hoffm., *Herba Pol.*, 42, 26–35, 1996.
12. Hadaček, F., Müller, C., Werner, A., Greger, H., Proksh, P., Analysis, isolation and insecticidal activity of linear furanocoumarins and other coumarin derivatives from *Peucedanum* (Apiaceae, Apiodeae). *J. Chem. Ecol.*, 20, 2035–2054, 1994.
13. Wolski, T., Gliński, J., Buczek, K., Wolska, A., Receiving and characteristic of furanocoumarin plant extracts of antifungal activity, *Herba Pol.*, 42, 168–173, 1996 (in Polish).
14. Murray, R.D.H., Mendez, J., Brown, S.A., *The Natural Coumarins Occurrence: Chemistry and Biochemistry*, Wiley, Chichester, New York, Brisbane, Toronto, Singapore, 1982.
15. Zgorka, G., Investigation and Isolation of Coumarin Compounds and Phenolic Acids in the Species *Libanotis dolichostyla* Schischk, Doctoral dissertation, Deptarment of Pharmacognosy, Medical Academy, Lublin, 1996 (in Polish).
16. Jerzmanowska, Z., Plant substances. *Methods of Isolation*, PWN, Warszawa, 1967 (in Polish).
17. Balbaa, S.I., Halim, A.F., Halaweish, F.T., Bohlman, F., 6-Hydroxy-4-methoxy-5-methylcoumarin from *Gerbera jamsonii*, *Phytochemistry*, 19, 1519–1522, 1980.
18. Sarker, S.D., Gray, A.I., Waterman, P.G., Coumarins from two *Asterolasia* species, *J. Nat. Prod.*, 57, 1549–1551, 1994.
19. Sione, O.S., Naturally occurring coumarins and related physiological activities, *J. Pharm. Sci.*, 53, 231–264, 1964.
20. Antonious, M.S., Sabry, D.Y., Solvent polarity indicators: extraction of some coumarin indicator dyes, *Microchim. Acta*, 118, 69–79, 1995.
21. Głowniak, K., Wolski, T., Dragan, T., Method of isolation of coumarins especially coumarins glycosides from plant material (in Polish), *Pol. Pat. Appl.*, 158396, 1–5, 1993.
22. Pimentel, G.C., McClellan, A.L., *The Hydrogen Bond*, Freeman, San Francisco, CA, 1960.
23. Härmälä, P., Vuorela, H., Nyiredy, Sz., Tönquist, K., Kaltia, S., Sticher, O., Hiltunen, R., Strategy for the isolation and identification of coumarins with calcium antagonistic properties from the roots of *Angelica archangelica.*, *Phytochem. Anal.*, 3, 42–48, 1992.
24. Beyrich, T., Isolierung von Phellopterin aus Früchten von Heracleum Mantegazzianum (Sommer et Levier) und Angelica archangelica L. 4. Mitt. Über Furanocoumarine., *Pharmazie*, 19, 753–756, 1964.
25. Pipkin, W., Fundamental considerations for SFE method development, *LC–GC Int.*, 5, 8–10, 1992.
26. Poole, C.F., Poole, S.K., *Chromatography Today*, Elsevier Science, Amsterdam, 1991.
27. Nykänen, I., Nykänen, L., Alkio, M., Composition of angelica root oils obtained by supercritical CO_2 extraction and stream distillation, *J. Ess. Oil. Res.*, 3, 229–236, 1991.
28. Nyiredy, Sz., Botz, L., Sticher, O., *Swiss Patent* CH674314, 1990.
29. Waksmundzka-Hajnos, M., Petruczynik, A., Dragan, A., Wianowska, D., Dawidowicz, A.L., Sowa, I., Influence of the extraction mode on the yield of some furanocoumarins from *Pastinaca sativa* fruits, *J. Chromatogr., B*, 800, 181–187, 2004.
30. Waksmundzka-Hajnos, M., Petruczynik, A., Dragan, A., Wianowska, D., Dawidowicz, A.L., Effect of extraction method on the yield of furanocoumarins from fruits of *Archangelica officinalis* Hoffm., *Phytochem. Anal.*, 15, 313–319, 2004.

31. Bieganowska, M.L., Głowniak, K., Retention behaviour of some coumarins in normal-phase high-performance thin-layer and column chromatography, *Chromatographia*, 25, 111–116, 1988.
32. Głowniak, K., Bieganowska, M.L., Effect of modifier and molecular structure of some coumarins on retention in reversed-phase high-performance thin-layer and column chromatography, *J. Chromatogr.*, 370, 281–292, 1986.
33. Hamburger, M., Upta, M., Hostettmann, K., Coumarins from *Polygala paniculata*, *Planta med.*, 50, 215–217, 1985.
34. Kubeczka, K.H., Bohn, I., Radix pimpinellae and its adulterations. Identification by TLC and GC. Revision of the structure formula of the main constituent in the essential oil, *Dtsch. Apoth. Ztg.*, 125, 399–402, 1985.
35. Głowniak, K., Bieganowska, M.L., Reversed-phase systems for the separation of coumarins and furanocoumarins by thin-layer and high-performance liquid chromatography, *J. Liq. Chromatogr.*, 8, 2927–2947, 1985.
36. Dabi-Lengytel, E., Tomps-Farkas, I., Zambo, I., Tetenyi, P., Danos, B., Hetelyi, E., Beleznai, Zs., Analysis of coumarins in the rhizomes of *Peucedanum officinale* L., *Herb. Hung.*, 25, 95–104, 1986.
37. Ghosh, P., Sil, P., Thakur, S., Spray reagent for the detection of coumarins and flavonoids on thin-layer plates, *J. Chromatogr.*, 403, 285–287, 1987.
38. Wall, M.E., Wani, M.C., Manikumar, G., Hughes, T.J., Taylor, H., Mc. Givney, R., Warner, J., Plant antimutagenic agents. 3. Coumarins, *J. Nat. Prod.*, 51, 1148–1152, 1988.
39. Menziani, E., Tosi, B., Bonora, A., Reschiglian, P., Lodi, G., Automated multiple development high-performance thin-layer chromatography analysis of natural phenolic compounds, *J. Chromatogr.*, 511, 396–401, 1990.
40. Harmala, P., Botz, L., Sticher, O., Hiltunen, R., Two-dimensional planar chromatographic separation of a complex mixture of closely related coumarins from the genus *Angelica*, *J. Planar Chromatogr.*, 3, 515–520, 1990.
41. Zogg, G.C., Nyiredy, Sz., Sticher, O., Comparison of fully off-line linear OPLC with HPLC illustrated by the separation of furanocoumarin isomers, *J. Planar Chromatography*, 1, 351–354, 1988.
42. Dugo, P., Mondello, L., Lamonica, G., Dugo, G., OPLC analysis of heterocyclic oxygen compounds from citrus fruit essential oils, *J. Planar Chromatogr.*, 9, 120–125, 1996.
43. Vuorela, H., Dallenbach-Tölke, K., Sticher, O., Hiltunen, R., Separation of the main coumarins of *Peucedanum palustre* with various planar chromatographic methods, *J. Planar Chromatogr.*, 1, 123–127, 1988.
44. Vuorela, P., Rahko, E.-L., Hiltunen, R., Vuorela, H., Overpressured layer chromatography in comparison with thin-layer and high-performance liquid chromatography for the determination of coumarins with reference to the composition of the mobile phase, *J. Chromatogr.*, 670, 191–198, 1994.
45. Bieganowska, M.L., Petruczynik, A., Zobel, M.A., Retention parameters of coumarins and flavonoids in binary reversed-phase HPTLC systems, *J. Planar Chromatogr.*, 9, 273–279, 1996.
46. Qin, L., Zhang, W., Zhang, H., Viang, R., Su, Zh., Study of the interspecific difference and intraspecific variation of coumarins in *Cnidium* species by TLC, *J. Chin. Trad. Med.*, 21, 525–526, 1996.
47. Petruczynik, A., Bieganowska, M.L., Retention parameters of coumarins and flavonoids on Florisil and silica layers. Part II, *J. Planar Chromatogr.*, 11, 267–271, 1998.

48. Summanen, J., Yrj-Nen, T., Hiktunen, R., Vuorela, H., Influence of the densitometer and video-documentation settings in the detection of plant phenolics by TLC, *J. Planar Chromatogr.*, 11, 421–427, 1998.

49. Bieganowska, M.L., Petruczynik, A., Retention behaviour of coumarins and flavonoids on polyamide and alumina layers. Part I., *J. Planar Chromatogr.*, 12, 135–139, 1999.

50. Mousa, O., Vuorela, P., Riekkola, M.-L., Vuorela, H., Hiltunen, R., Evaluation of pure coumarins using TLC-densitometer, spectrophotometer and HPLC with photodiode array detector, *J. Liq. Chromatogr.*, 20, 1887–1901, 1997.

51. Zgorka, G., Retention behaviour of coumarin compounds on silica and Florisil TLC layers, *J. Liq. Chromatogr.*, 24, 1397–1410, 2001.

52. Oliva, A., Meepagala, K.M., Wedge, D.E., Harris, D., Hale, A.L., Aliotta, G., Duke, S.O., Natural fungicides from *Ruta graveolens* L. leaves, *J. Agric Food Chem.*, 51, 890–896, 2003.

53. Garcia-Argaez, A.N., Apan, T.O.R., Delgado, H.P., Velazquez, G., Martinez-Vazques, M., Antiinflammatory activity of coumarins from *Decatropis bicolor* on TPA ear mice model, *Planta med.*, 66, 279–281, 2000.

54. Głowniak, K., Bartnik, M., Mroczek, T., Zabża, A., Wierzejska, A., Application of column chromatography and preparative TLC for isolation and purification of coumarins from *Peucedanum tauricum* Bieb. fruits, *J. Planar Chromatogr.*, 15, 94–100, 2002.

55. Kozyra, M., Głowniak, K., Zabża, A., Zgórka, G., Mroczek, T., Cierpicki, T., Kulesza, J., Mudo, I., Column chromatography and preparative TLC for isolation and purification of coumarins from *Peucedanum verticillare* L. Koch., *J. Planar Chromatogr.*, 18, 224–227, 2005.

56. Betti, A., Lodi, G., Fuzzati, N., Analysis of flavonoid compounds by planar chromatography. A chemometric approach, *J. Planar Chromatogr.*, 6, 232–237, 1993.

57. Maes, D., Debebedetti, S., De Kimpe, N., New coumarins from *Pterocaulon virgatum* (L.) DC, *Biochem. Syst. Ecol.*, 34, 165–169, 2006.

58. Iranshahi, M., Arfa, P., Ramezani, M., Jaafari, M.R., Sadeghian, H., Sesquiterpene coumarins from *Ferula szowitsiana* and *in vitro* antileishmanial activity of 7-prenyloxycoumarins against promastigotes, *Phytochemistry*, 68, 554–561, 2007.

59. Müller, A.H., Vieira, P.C., da Silva, M.F. das G.F., Fernandes, J.B., Dihydrochalcones, coumarins and alkaloids from *Metrodorea nigra*, *Phytochemistry*, 40, 1797–1800, 1995.

60. Um, B.H., Lobstein, A., Weniger, B., Spiegel, C., Yice, F., Rakotoarison, O., New coumarins from *Cedrelopsis grevei*, *Fitoterapia.*, 74, 638–642, 2003.

61. Debenedetti, S.L., Nadinic, E.L., Coussio, J.D., De Kimpe, N., Boeykens, M., Two 6,7-dioxygenated coumarins from *Pterocaulon virgatum*, *Phytochemistry*, 48, 707–710, 1998.

62. Hiermann, A., Schantl, D., Schubert-Zsilavecz, M., Reiner, J., Coumarins from *Peucedanum ostruthium*, *Phytochemistry*, 43, 881–883, 1996.

63. Chlouchi, A., Muyard, F., Girard, C., Waterman, P.G., Bévalot, F., Coumarins from the twigs of *Diplolaena mollis* P.G. Wilson (Rutaceae), *Bioc. Syst. Ecol.*, 33, 967–969, 2005.

64. Bieganowska, M.L., Petruczynik, A., Retention behaviour of some coumarins and flavonoids on aminopropyl silica layers. Part III, *Chem. Anal. (Warsaw)*, 40, 869–878, 1995.

65. Wawrzynowicz, T., Waksmundzka-Hajnos, M., The application of systems with different selectivity for the separation and isolation of some furocoumarins, *J. Liq. Chromatogr.*, 13, 3925–3940, 1990.

66. Wawrzynowicz, T., Czapińska, K., Markowski, W., The use of special development techniques for micropreparative TLC separation of closely related compounds, *J. Planar Chromatogr.*, 11, 388–393, 1998.

67. Barros, M.C.P., Lima, M.A.S., Silveria, E.R., Coumarins and other constituents from *Acritopappus confertus* roots, *Biochem. Syst. Ecol.*, 33, 215–218, 2005.

68. Stein, A.C., Alvarez, S., Avancini, C., Zacchino, S., von Poser, G., Antifungal activity of some coumarins obtained from species of *Pterocaulon (Asteraceae)*, *J. Ethnopharmacol.*, 107, 95–98, 2006.

69. Vilegas, W., Boralle, N., Cabrera, A., Bernardi, A.C., Pozetti, G.L., Coumarins and flavonoid from *Pterocaoulon alopecuroides*, *Phytochemistry*, 38, 1017–1019, 1995.

70. Lee, K.H., Chai, H.B., Tamez, P.A., Pezzuto, J.M., Cordell, G.A., Biologically active alkylated coumarins from *Kayea assamica*, *Phytochemistry*, 64, 535–541, 2003.

71. Chlouchi, A., Girard, C., Tillequin, F., Bevalot, F., Coumarins and furoquinoline alkaloids from *Philotheca deserti* var. deserti (Rutaceae), *Biochem. Syst. Ecol.*, 34, 71–74, 2006.

72. Simpson, D.S., Jacobs, H., Alkaloids and coumarins from *Esenbeckia pentaphylla* (Rutaceae), *Biochem. Syst. Ecol.*, 33, 841–844, 2005.

73. Chlouchi, A., Girard, C., Bevalot, F., Waterman, P., Muvard, F., Taxonomically significant coumarins from three *Philotheca* species (Rutaceae), *Biochem. Syst. Ecol.*, 35, 251–254, 2007.

74. Debenedetti, S.L., Tehrani, K.A., van Puyvelde, L., de Kimpe, N., Isopurpurasol, a coumarin from *Pterocaulon virgatum*, *Phytochemistry*, 51, 701–703, 1999.

75. Torres, R., Faini, F., Modak, B., Urbina, F., Labbe, C., Antioxidant activity of coumarins and flavonoids from resinous exudate of *Haplopappus multifolius*, *Phytochemistry*, 67, 984–987, 2006.

76. Waksmundzka-Hajnos, M., Petruczynik, A., Hajnos, M.L., Tuzimski, T., Hawryƚ, A., Bogucka-Kocka, A., Two-dimensional thin-layer chromatography of selected coumarins, *J. Chromatogr. Sci.*, 44, 510–517, 2006.

77. Hawryƚ, M.A., Soczewiński, E., Dzido, T.H., Separation of coumarins from *Archangelica officinalis* in high-performance liquid chromatography and thin-layer chromatography system, *J. Chromatogr. A*, 366, 75–81, 2000.

78. Waksmundzka-Hajnos, M., Wawrzynowicz, T., Some methodological aspects of the separation of closely related compounds, *J. Planar Chromatogr.*, 3, 439–441, 1990.

79. Waksmundzka-Hajnos, M., Wawrzynowicz, T., On-line purification and micropreparative separation of contaminated fruit extracts by TLC in equilibrium sandwich chambers, *J. Planar Chromatogr.*, 5, 169–174, 1992.

80. Bartnik, M., Gƚowniak, K., Maciąg, A., Hajnos, M., Use of reversed-phase and normal-phase preparative thin-layer chromatography for isolation and purification of coumarins from *Peucedanum tauricum* Bieb. leaves, *J. Planar Chromatogr.*, 18, 244–248, 2005.

81. Chang, C.-T., Doong, S.-L., Tsai, I.-L, Chen, I.-S., Coumarins and anti-HBV constituents from *Zanthoxylum schinifolium*, *Phytochemistry*, 45, 1419–1422, 1997.

82. Toro, M.J.U., Müller, A.H., Arruda, M.S.P., Arruda, A.C., Alkaloids and coumarins from *Ticorea longiflora*, *Phytochemistry*, 45, 851–853, 1997.

83. Asomaning, W.A., Otoo, E., Akoto, O., Oppong, I.V., Addae-Mensah, I., Waibel, R., Isoflavones and coumarins from *Milletia thonningii*, *Phytochemistry*, 51, 937–941, 1999.

84. Chen, I.-S., Lin, Y.-C., Tsai, I.-L., Teng, C.-M., Ko, F.-N., Ishikawa, T., Coumarins and anti-platelet aggregation constituents from *Zanthoxylum schinifolium*, *Phytochemistry*, 39, 1091–1097, 1995.

85. Rahman, A.U., Shabbir, M., Sultani, S.Z., Jabbar, A., Choudhary, M.I., Cinnamates and coumarins from the leaves of *Murraya paniculata*, *Phytochemistry*, 44, 683–685, 1997.
86. Kayser, O., Kolodziej, H., Highly oxygenated coumarins from *Pelargonium sidoides*, *Phytochemistry*, 39, 1181–1185, 1995.
87. Hoeneisen, M., Hernandez, V., Becerra, J., Silva, M., Bittner, M., Jakupovic, J., 5-Methyl coumarins and a new phenol from *Nassauvia pyramidalis* and *N. digitata*, *Phytochemistry*, 52, 1667–1669, 1999.
88. Guz, N.R., Lorenz, P., Stermitz, F.R., New coumarins from *Harbouria trachypleura*: isolation and synthesis, *Tetrahedron Lett.*, 42, 6491–6494, 2001.
89. Bissoué, A.N., Muyard, F., A coumarin from aerial parts of *Geijera paniculata*, *Phytochemistry*, 46, 383–384, 1997.
90. Gutierrez, M.-C., Tena, M., Jorrin, J., Parry, A., Edwards, R., Abiotic elicitation of coumarin phytoalexins in sunflower, *Phytochemistry*, 38, 1185–1191, 1995.
91. Almeida, E.X., Conserva, L.M., Lyra Lemos, R.P., Coumarins, coumarinolignoids and terpenes from *Protium heptaphyllum*, *Biochem. Syst. Ecol.*, 30, 685–687, 2002.
92. Baetas, A.C.S., Arruda, M.S.P., Müller, A.H., Arruda, A.C., Coumarins from *Metrodorea flavida*, *Phytochemistry*, 43, 491–493, 1996.
93. Bissoué, A.N., Muyard, F., Bevalot, F., Coumarins from the aerial parts of *Chorilaena quercifolia*, *Phytochemistry*, 43, 877–879, 1996.
94. Kumar, V., Niyaz, N.M.M., Saminathan, S., Wickramaratne, D.B.M., Coumarins from *Paramignya monophylla* root bark, *Phytochemistry*, 49, 215–218, 1998.
95. Tsai, I.-L., Fang, S.-C., Ishikawa, T., *N*-cyclohexyl amides and a dimeric coumarin from formosan *Toddalia asiatica*, *Phytochemistry*, 44, 1383–1386, 1997.
96. Chen, I.-S., Chang, C.-T., Sheen, W.-S., Teng, C.-M., Tsai, I.-L., Duh, C.-Y., Ko, F.-N., Coumarins and antiplatelet aggregation constituents from formosan *Peucedanum japonicum*, *Phytochemistry*, 41, 525–530, 1996.
97. Guilet, D., Séraphin, D., Rondeau, D., Richomme, P., Bruneton, J., Cytotoxic coumarins from *Calophyllum dispar*, *Phytochemistry*, 58, 571–575, 2001.
98. Murphy, E.M., Nahar, L., Byres, M., Shoeb, M., Siakalima, M., Coumarins from the seeds of *Angelica sylvestris* (Apiaceae) and their distribution within the genus *Angelica*, *Biochem. Syst. Ecol.*, 32, 203–207, 2004.
99. Ahsan, M., Gray, A.I., Waterman, P.G., Armstrong, J.A., Terpenoids, alkaloids and coumarins from *Boronia inornata* and *Boronia gracilipes*, *Phytochemistry*, 38, 1275–1278, 1995.
100. Zhou, P., Takaishi, Y., Duan, H., Chen, B., Honda, G., Itoh, M., Takeda, Y., Coumarins and bicoumarin from *Ferula sumbul*: anti-HIV activity and inhibition of cytokine release, *Phytochemistry*, 53, 689–697, 2000.
101. Jiménez, B., Grande, M.C., Anaya, J., Torres, P., Coumarins from *Ferulago capillaris* and *F. Brachyloba*, *Phytochemistry*, 53, 1025–1031, 2000.
102. Iranshahi, M., Arfa, P., Ramezani, M., Jaafari, M.R., Sadeghian, H., Sesquiterpene coumarins from *Ferula szowitsiana* and *in vitro* antileishmanial activity of 7-prenyloxycoumarins against promastigotes, *Phytochemistry*, 68, 554–561, 2007.
103. Cisowski, W., Pałka-Gudyka, E., Krauze-Baranowska, M., Królicki, Z., Preparative TLC of coumarins with gradient elution, *J. Planar Chromatogr.*, 6, 471–474, 1991.
104. Tsai, I.L., Lin, W.Y., Teng, C.M., Ishikawa, T., Doong, S.L., Huang, M.W., Chen, Y.C., Chen, I.S., Coumarins and antiplatelet constituents from the root bark of *Zanthoxylum schinifolium*, *Planat med.*, 66, 618–623, 2000.
105. de Moura, N.F., Simionatto, E., Porto, C., Hoelzel, S.C.S., Dessoy, E.C.S., Zanatta, N., Morel, A.F., Quinone alkaloids, coumarins and volatile constituents of *Helietta longifoliata*, *Planta med.*, 68, 631–634, 2002.

106. Yenjai, C., Sripontan, S., Sriprajun, P., Kittakoop, P., Jintasirikul, A., Tanticharoen, M., Thebtaranonth, Y., Coumarins and carbazoles with antiplasmodial activity from *Clausena harmandiana*, *Planta med.*, 66, 277–279, 2000.
107. Murakami, A., Gao, G., Kyung Kim, O.E., Omura, A., Yano, M., Ito, C., Furukawa, H., Jiwajinda, S., Identification of coumarins from the fruit of citrus hysterix DC as inhibitors of nitric oxide generation in mose macrophage RVV 264.7 cells, *J. Agric. Food Chem.*, 47, 333–339, 1999.
108. Campbell, W.E. et al., Coumarins of the Rutoideae: tribe Diosmeae, *Phytochemistry*, 25, 655–657, 1986.
109. Garg, S.K. et al., Glucosides of *Apium graveolens*, *Planta Med*, 38, 363–365, 1980.
110. Garg, S.K. et al., Apiumetin—a new furanocoumarin from the seeds of *Apium graveolens*, *Phytochemistry*, 17, 2135–2136, 1978.
111. Garg, S.K. et al., Celerin, a new coumarin from *Apium graveolens*, *Planta Med.*, 38, 186–188, 1980.
112. Garg, S.K. et al., Minor phenolics of *Apium graveolens* seeds, *Phytochemistry*, 18, 352, 1979.
113. Dall'Acqua, F. et al., Biosynthesis of *O*-alkylfuro-coumarins, *Planta Med.*, 27, 343–348, 1975.
114. Garg, S.K. et al., Apiumoside, a new furanocoumarin glucoside from the seeds of *Apium graveolens*, *Phytochemistry*, 18, 1764–1765, 1979.
115. Garg, S.K. et al., Coumarins from *Apium graveolens* seeds, *Phytochemistry*, 18, 1580–1581, 1979.
116. Innocenti, G. et al., Investigations of the content of furocoumarins in *Apium graveolens* and in *Petroselinum sativum*, *Planta Med.*, 29, 165–170, 1976.
117. Kar, A., Jain, S.R., Investigations on the antibacterial activity of some Indian indigenous aromatic plants, *Flavour Ind.*, 2, 111–113, 1971.
118. Williams, C.A. et al., Flavonoids, cinnamic acids and coumarins from the different tissues and medicinal preparations of *Taraxacum officinale*, *Phytochemistry*, 42, 121–127, 1996.
119. Stoehr, H., Herrman, K., Phenolic acids of vegetables. III. Hydroxycinnamic acids and hydroxy benzoic acids of root vegetables, *Z. Lebensm-Unters Forsch.*, 159, 219–224, 1975.
120. Tawata, M. et al., Anti-platelet action of GU-7, a 3-arylcoumarin derivative, purified from *Glycyrrhizae radix*, *Planta Med.*, 56, 259–263, 1990.
121. Chaudhary, S.K. et al., Oxypeucedanin, a major furocoumarin in parsley, *Petroselinum crispum*, *Planta Med.*, 52, 462–464, 1986.
122. Rees, S., Harborne, J.B., Flavonoids and other phenolics of *Cichorium* and related members of the Lactuceae (Compositae), *Bot. J. Linn. Soc.*, 89, 313–319, 1984.
123. Berenbaum, M., Fenny, P., Toxicity of angular furanocoumarins to swallowtail butterflies: escalation in a coevolutionary arms race? *Science*, 212, 927–929, 1981.

16 Application of TLC in the Isolation and Analysis of Flavonoids

Marica Medić-Šarić, Ivona Jasprica, Ana Mornar, and Željan Maleš

CONTENTS

16.1   INTRODUCTION

16.1.1   CHEMISTRY, BIOCHEMISTRY, AND MEDICINAL SIGNIFICANCE OF THE FLAVONOIDS

Flavonoids constitute one of the largest and recently very popular group of phyto-chemicals. They are virtually ubiquitous in green plants and as such are likely to be

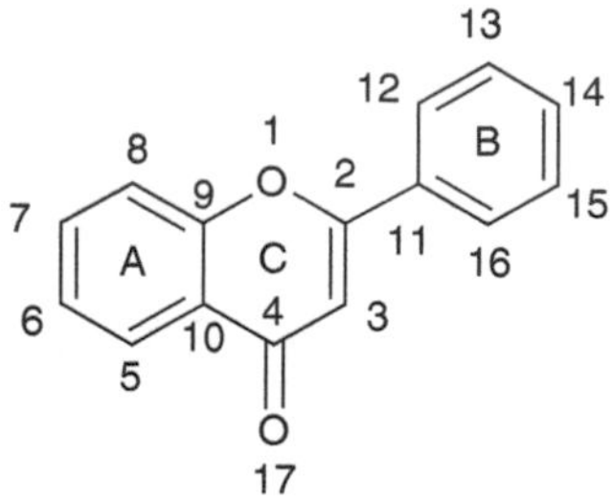

FIGURE 16.1 Basic flavonoid nucleus. Subdivision to different groups is primarily based on the presence (or absence) of a carbonyl substituent on the position C4, the presence (or absence) of a double bond between carbon atoms 2 and 3, and a phenyl substitution at the positions 2 or 3 of the pyrone ring.

encountered in any work involving plant extracts. The term flavonoid is a collective term for plant pigments, mostly derived from benzo-γ-pyrone, which is synonymous with chromone.[1] In plants, flavonoid aglycones (flavonoids without attached sugars) occur in a variety of structures, all containing 15 carbon atoms arranged in a $C_6–C_3–C_6$ configuration (flavonoid nucleus is depicted in Figure 16.1).

Until now, more than 4000 flavonoids have been identified and this number is constantly growing because of the great structural diversity arising from the various hydroxylation, metoxylation, glycosylation, and acylation patterns. Most frequently encountered groups of flavonoid aglycones include flavones, flavonols, anthocyanidins, isoflavones, flavanones, dihydroflavonols, biflavonoids, calchones, and aurones. Flavonoid aglycones possess the chemical properties of phenolics, and thus they are slightly acidic. Those possessing a number of unsubstituted hydroxyl groups, or sugar moieties, are polar substances and soluble in polar organic solvents. The presence of sugar makes flavonoid more water soluble, while less polar aglycones like isoflavones, flavanones, and highly methoxylated flavones and flavonols tend to be more soluble in ether or chloroform.[2] Flavonoids have important roles in plant physiology and are components of the diet of numerous herbivores and omnivores, including humans. This group of compounds exhibit an extraordinary array of biochemical and pharmacological activities in mammalian systems, such as anti-inflammatory, antioxidant, immunomodulatory, hepatoprotective, antimicrobic, and antiviral.[3,4] With the increased popularity and use of herbal medicines containing flavonoids, the question of composition determination and standardization arises. It is almost impossible to compare any kind of biological activity without the chemical characterization of the plant extract. For this reason it is important that scientists working in this field are aware of the importance of the content analysis and for that purpose use the appropriate analytical tools.

16.1.2 Brief Overview of Use of TLC in the Analysis of Flavonoids in Plants

Since the early 1960s, thin layer chromatography (TLC) has been used in flavonoid analysis. Paper chromatography (PC) was the preferred method in the past and valuable descriptions of the use of PC in flavonoid identification and isolation are given in Markham's *Techniques of Flavonoid Identification*.[2] To identify or isolate individual compounds, two-dimensional PC was used. TLC slowly replaced the paper chromatography, as new stationary phases (such as microcrystalline cellulose, polyamide, and silica) were developed. Numerous TLC protocols that can be used in flavonoid separations (including the analysis of very well-known flavonoid drugs such as *Arnicae flos*, *Calendulae flos*, *Crataegi flos et folium*, *Betulae folium*, *Rubi idvaei folium*, *Equiseti herba*, and many others) are listed in the TLC atlas of plant drug analysis.[5] A good review on the use of TLC in this field is given in Sherma's review article as well.[6]

TLC was the method of choice for herbal analysis before instrumental chromatographic techniques such as gas chromatography (GC) or high-performance liquid chromatography (HPLC) were established. Although GC and HPLC are nowadays considered to be the leading techniques (as described in Ref. 7 as well as in the review article of de Rijke et al.[8]), TLC is still the basic tool for the identification of natural compounds given in various pharmacopoeias. It is often used to provide the first characteristic fingerprints of herbs.[9] In addition, TLC technique is constantly improving.[10,11] Modern high-performance TLC (HPTLC) is an efficient instrumental technique and optimized quantitative HPTLC using a densitometric evaluation can produce results that are analogous to those obtained with GC and HPLC.

In this chapter, we will focus on the use of TLC in flavonoid analysis during the past decade, especially on the use of modern instrumental techniques, as well as new methods like rotation planar chromatography (RPC) and overpressured-layer chromatography (OPLC).

16.2   TLC TECHNIQUES

16.2.1   Basic Principles for Flavonoid Separations

Nowadays, TLC analysis of flavonoids is performed mostly on silica gel as stationary phase, using adsorption chromatography as the mode of separation. For normal-phase separations a variety of mobile phases is available, when compared with the reversed-phase chromatography, in which only a few water-miscible organics can be used (in this case methanol is the typical organic phase). As PC was the first form of layer chromatography, identification of flavonoids (prior the silica gel became the most popular stationary phase) was performed on cellulose as well. In Markham's book, recommendations for different solvent and sorbent type (cellulose, polyamide, or silica) combinations can be found, depending on the particular flavonoid group (Table 16.1).[2]

Flavonoid aglycones and glycosides separations listed in the TLC atlas of plant drug analysis were performed on silica gel TLC plates, containing fluorescence indicator. According to the authors, ethyl acetate–formic acid–glacial acetic acid–water (100:11:11:26, v/v) or formic acid–water–ethyl acetate mixed in different proportions (with or without ethylmethyl ketone) are suitable mobile phases for the TLC screening of polar flavonoid glycosides. The good start for investigation of less polar flavonoid aglycones would be a mobile phase composed of toluene–ethyl

TABLE 16.1

Recommended Solvent/Adsorbent Combinations for TLC Identification of Different Flavonoid Types

	Mobile Phase		
Adsorbent Type	Celullose	Polyamide	Silica Gel
Flavonoid group			
Flavonoid glycosides	• *t*-Butanol–acetic acid–water (3:1:1, v/v)[a] • *n*-Butanol–acetic acid–water (4:1:5, v/v)[a]	• water–methanol– ethylmethyl ketone– acetylacetone (13:3:3:1, v/v)	• Ethyl acetate–pyridine– water–methanol (80:20:10:5, v/v), especially flavone C-glycosides
Polar flavonoid aglycones (e.g., flavones, flavonols)	• *t*-Butanol–acetic acid–water (3:1:1, v/v)[a] • Chloroform–acetic acid–water (30:15:2, v/v)[b]	• Methanol–acetic acid–water (18:1:1, v/v)	• Toluene–pyridine–formic acid (36:9:5, v/v)
Nonpolar flavonoid aglycones (e.g., dihydroflavonoids, isoflavones, and methylated flavones)	• 10%–30% acetic acid		• Chloroform–methanol (15:1 to 3:1, v/v)

[a] Mobile phase is mixed thoroughly in a separating funnel and upper phase is used.
[b] Mobile phase is mixed thoroughly in a separating funnel and excess of water is discarded.

formiate–formic acid (50:40:10, v/v) or toluene–dioxan–glacial acetic acid (90:25:4, v/v).[5] Combination of hydrophobic organic solvents such as *n*-hexane or chloroform with more polar ones (ethyl acetate or methanol), with addition of acetic or formic acid (e.g., *n*-hexane–ethyl acetate–acetic acid, 31:14:5, v/v, or chloroform–methanol–formic acid, 44:3.5:2.5, v/v), might be also used.[12]

Detection of flavonoids is usually performed under the UV light at 254 nm (all flavonoids cause fluorescence quenching) or at 366 nm (depending on the structural type, flavonoids show dark yellow, green, or blue fluorescence). Fluorescence can be enhanced using various spray reagents, which will lower the detection limit. The most popular ones are natural product—polyethylene glycol reagent (NP/PEG; the plate is first sprayed with 1% methanolic diphenylboric acid-β-ethylamino ester, followed by 5% ethanolic polyethylene glycol-4000) and 1% ethanolic or methanolic aluminum-chloride (AlCl$_3$) solution. Other derivatization reagents such as diazotized sulphanilic acid (compounds with free phenolic hydroxyl groups will show as yellow, orange or red spots), vanilin-5% HCl (red or purplish-red spots will be produced by catechins and proanthocyanidins immediately after spraying and warming the plates, and by flavanones and dihydroflavonols more slowly), and fast blue salt B (phenolic compounds will form blue or blue-violet azo-dyes with this reagent) can be used as well. Detailed descriptions for preparation of spraying reagents are given in Markham's *Techniques of Flavonoid Identification* and the TLC atlas of plant drug analysis.[2,5]

16.2.2 SAMPLE PREPARATION

Medicinal plant extracts are known to be very complicated samples, which contain a variety of different compounds. Therefore, matrix effect cannot be ignored when the analysis of such sample is needed. Isolation of flavonoids from the sample matrix is thus the most important step prior to any analysis scheme.

The ultimate goal of a sample preparation is the preparation of a sample extract enriched in all compounds of interest and free of interfering matrix impurities. Usually it includes a number of steps from extraction of crude sample, purification, and preconcentration of obtained extract. Sample handling strategies for determination of biophenols (including flavonoids) in food and plants are reviewed in the paper of Tura and Robards,[13] and later in Robard's article.[14] Contaminating non-flavonoid substances such as carbohydrates, proteins, lipids, pigments, etc., do interfere with later chromatographic analysis so the development of modern purification techniques significantly facilitates identification of phenolics in different samples.[11,15] Sample preparation depends on the nature of both analyte and sample, particularly on analyte's physicochemical properties such as solubility and lipophilicity (partitioning between hydrophobic and hydrophilic phases). Liquid–liquid extraction (LLE), which is based on the partition of analyte usually between two immiscible phases (solvents), is a simple method that can be used for purifying herbal extracts. An illustrative example of LLE is removal of chlorophylls, xanthophylls, and other hydrophobic contaminants (various fats and terpenes) from aqueous extracts by extraction with hydrophobic solvents (e.g., chloroform or hexane).[2] In such a way LLE was used in the work of Mbing et al.[16] First described LLE step was purification of aqueous extract of *Ouratea nigroviolacea* Gilg ex Engl. with hexane (to remove the hydrophobic components) and such extract was later partitioned with ethyl acetate. This method was used to purify the extract and facilitate the isolation of two biflavonoids—ouratine A and B. Solid-phase extraction (SPE) appeared as an alternative to LLE and it is now certainly the most widely used cleanup method, which is also applied for fractionation and concentration of analytes from herbal extracts. Nevertheless, its use in flavonoid analysis is relatively new. A brief overview of use of SPE in polyphenols analysis is given in Ref. 7 and in various review articles.[8,13,14] Considering the physicochemical properties of flavonoids, in most cases reversed-phase C18-bonded silica is the sorbent of preference and the sample solution and used elution solvents are usually slightly acidified to prevent ionization of the flavonoids, which would reduce their retention. Under the aqueous conditions, the flavonoids are retained on the column, while the most of the water soluble contaminants are not. After rinsing the cartridge with water to remove the hydrophilic impurities, flavonoids can be eluted with an appropriate volatile solvent such as methanol. Although TLC is known to be an analytical technique that does not require a complicated sample pretreatment, there are more and more examples in the literature in which SPE is used to diminish the matrix effect and in such a way facilitate quantification. In the work of Nowak and Tuzimski[17] SPE was used for preliminary cleaning of rose leaves extracts to remove the interfering matrix components that affected the baseline resolution and caused difficulties during identification and quantification of flavonoids of interest (quercetin and kaempferol). Pretreated aqueous extracts were passed through octadecyl SPE microcolumns. After application of the samples, the microcolumn was washed with water and adsorbed aglycones were desorbed with pure methanol and subjected to further TLC analysis. Chromatography was performed on silica gel and cellulose HPTLC plates and developed in a horizontal chamber. For quantitative determination silica plates were used with 1,4-dioxane–toluene–85% acetic acid $(6 + 24 + 1,$ v/v) as mobile phase. Densitograms were recorded at 373 and 364.8 nm, respectively. Direct application of wine samples to the chromatographic plate in the work of Lambri et al.[18] led to the spot broadening during TLC migration and to nonuniform migration of polar compounds. Such results implied that SPE would be a necessary step in analysis of anthocyanins. Consequently, samples were submitted to purification on a reversed-phase SPE cartridge previously activated by elution of methanol and water. Applied wine samples were first washed with water and then eluted with methanol–formic acid (90:10, v/v). Collected purified samples were applied to the RP-18 WF_{254s} HPTLC plates. The best separation of anthocyanins was achieved by methanol–water–trifluoroacetic acid (55:45:1, v/v). Qualitative and quantitative analysis of obtained chromatograms was performed at 520 nm using scanning densitometry. To improve the separation of flavonoid glycosides (eriocitrin, hesperidin, luteolin-7-O-rutinoside, and diosmin) and rosmarinic acid from *Mentha x piperita* L., aqueous extracts were applied to an SPE octadecyl microcolumn. Obtained methanolic eluates were subsequently subjected to the TLC analysis on unmodified silica gel layers and on silica gel chemically modified with polar (HPTLC NH_2, HPTLC CN) and nonpolar groups (HPTLC RP-18W). The mobile phase acetone–acetic acid $(85 + 15,$ v/v) enabled successful separation on the aminopropyl adsorbent, while optimal mobile phase for octadecyl layer was water–methanol $(60 + 40,$ v/v).[19] SPE was also used to separate the glycerine from the glycerinic extracts of *Ribes nigrum* L., *Sorbus domestica* L., and *Tilia tomentosa* Moench, so rutin could be quantitatively determined by using silica gel 60 F_{254} plates scanned at 260 and 440 nm, respectively, after spraying the plates with 2-aminoethyldiphenylborinate.[20]

16.2.3 Separation of Flavonoid Aglycones and Glycosides on Silica Gel and Reversed-Phase TLC

Analysis of medicinal plants is traditionally one of the oldest fields of application for TLC. In the last decade, regulation of the herbal industry and the demands for analytical methods that can ensure the quality and safety of herbal medicines have been constantly growing. Identification can be considered as the main application of TLC technique. Additionally, HPTLC fingerprints are often used to establish the optimal conditions of extraction, to standardize extracts, and to detect changes or degradation of the plant material during the extract formulation. Methods can also be optimized to identify and quantify a specific target (marker) substance, or a class of substances.[21] In modern approaches in the field of medicinal plants, TLC analysis (including the analysis of flavonoids) are reviewed in the recent article of Gocan and Cimpan,[22] and in two comprehensive papers published in 2006.[8,23]

During the last decade, the attention has been headed to the development of chromatographic fingerprint profiling methods to determine the quality of complex herbal extracts. The HPTLC fingerprint on silica plates combining digital scanning profiling was developed to identify flavonoids puerarin, daidzin, genistin, and daidzein in order to distinguish *Pueraria lobata* (Willd.) Ohwi and *P. thomsonii* Benth. samples (chromatograms showing aglycone fingerprints of *P. lobata* and *P. thomsonii* roots are shown in Figure 16.2).

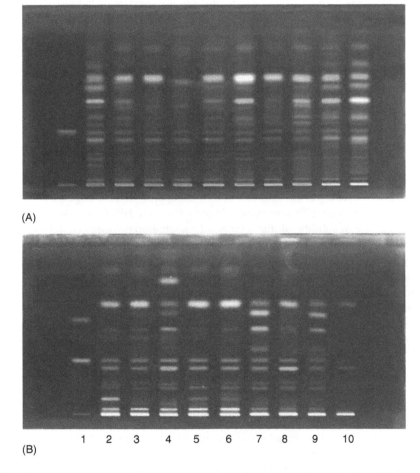

FIGURE 16.2 (A) HPTLC fluorescence images under the excitation wavelength 366 nm of aglycone fraction of root of *Pueraria lobatae* (RPL) and chemical reference substances (track 1: daidzein; tracks 2–11: RPL samples). (B) HPTLC fluorescence images under the excitation wavelength 366 nm of aglycone fraction of root of *Pueraria thomsonii* (RPT) and chemical reference substances (track 1: daidzein; tracks 2–10: RPT samples). (From Chen, S.B. et al., *J. Chromatogr. A*, 1121, 114, 2006. With permission.)

Flavonoid aglycones were identified using toluene–butyl acetate–formic acid (60:30:5, v/v) as the mobile phase, while the separation of glycosides was performed with chloroform–methanol–ethyl acetate–water (20:40:22:10, v/v, lower phase was used).[24] The quality of two samples (Indian and German mother tincture) of *Calendula officinalis* L. prepared from marigold flowers was evaluated in the work of Lalla et al.[25] Two biomarkers, rutin and quercitin, were used for microfingerprinting analysis. Chromatography was performed on silica gel HPTLC plates and for rutin and quercitin identification different mobile phases were used—*n*-butanol–glacial acetic acid–water (3.6 + 0.5 + 0.5, v/v) and toluene–chloroform–acetone–glacial acetic acid (2 + 2 + 2 + 0.5, v/v), respectively. Quantification was performed by scanning densitometry and developed method was validated for both standards. The results enabled clear differentiation between the samples of Indian and German mother tincture. Fingerprint analysis was also performed by Matysik et al.[26] in order to optimize the extraction procedure of *Calendulae flos*. Different solvents were tested and the highest amount of marker flavonoids (rutin, isoquercitrin, narcissin, and isorhamnetin-3-glucoside) was found in the methanolic extract. HPTLC separation of listed flavonoids was performed on silica gel plates containing fluorescent indicator with the mobile phase consisting of toluene–ethyl acetate–formic acid–water (1:9:2.5:2, v/v). NP/PEG reagent was used for the location of the spots under the UV light (366 nm). The spectrum of constituents of different Hawthorn extracts was compared with three different TLC fingerprint chromatograms in different polarity ranges and by means of quantitative determinations of different groups of constituents (procyanidins, flavonoids, total vitexin, total polyphenol) in the paper of Vierling et al.[27] Lederer et al.[28] compared HPTLC fingerprint analysis of Chinese star anise (*Illicium verum* Hook. F.) with HPLC/MS/MS technique to evaluate the reliability of the planar chromatographic method. HPTLC was shown to be an useful tool for the rapid identification of Chinese star anise in the quality control of commercial batches using rutin, hyperoside, chlorogenic acid, and caffeic acid as standards. Furthermore, TLC identification of pinostrobin, pinocembrin, galangin, chrysin, kaempferol, benzyl ferulate, and phenethyl caffeate was used to establish the identity of propolis samples from different regions of Turkey.[29] A combination of planar chromatographic and electrophoretic procedures allowed Adelmann et al.[30] to establish the chemical correlation between an almost homogeneous flora source, poplar buds, and proplis sample based on the content of a particular bioactive flavanone (pinocembrin). In the paper of Prashanth Kumar et al.[31] a simple, specific, and sensitive method for estimation of the rutin content of three therapeutically important Indian plants (*Tephrosia purpurea* (Linn.) Pers., *Leptadenia reticulata* W. & A., and *Ruta graveolens* L.) was reported. Chromatographic analysis was performed on silica gel 60 F_{254} plates with ethyl acetate–*n*–butanol–formic acid–water (5 + 3 + 1 + 1, v/v) as mobile phase and the plates were scanned at 366 nm. The method was validated by determination of precision, accuracy, and repeatability. A new, simple and selective HPTLC method for the separation and determination of kaempferol and quercetin in an extract of *Ginkgo biloba* leaves was reported by Jamshidi et al.[32] Glass-backed silica gel 60 F_{254} HPTLC plates were used and developed in toluene–acetone–methanol–formic acid (46 + 8 + 5 + 1, v/v). Evaluation was performed densitometrically at 254 nm.

Selectivity, repeatability, and accuracy of the method were determined and method was shown to be suitable for standardization of Ginkgo preparations. A rapid and sensitive HPTLC method was developed for determination of apigenin in liquid products of *Matricaria chamomilla* L. Samples were applied directly to silica gel 60 F_{254} HPTLC plates which were developed with toluene–methanol $(10 + 2, v/v)$ as mobile phase. Densitometric evaluation of apigenin was performed at 343 nm.[33] HPTLC and HPLC methods were established and compared for separation and quantitative determination of puerarin, 3′-methoxypuerarin, daidzin, and daidzein, the main isoflavonoid components of several kudzu (*Pueraria lobata* (Willd.) Ohwi) samples. Fluorescence-quenching zones were quantitatively evaluated by UV absorption scanning densitometry at 254 nm after developing HPTLC silica gel F_{254} plates with chloroform–methanol–ethyl acetate–water $(16.2 + 18.8 + 52 + 3, v/v)$ as mobile phase. The results obtained from the use of validated HPTLC and HPLC methods showed that either can be successfully applied for determination of the amounts of different isoflavone components in kudzu root samples.[34] Likewise, Janeczko et al.[35] used HPTLC silica gel F_{254} plates developed with chloroform–methanol–water $(23 + 8 + 1, v/v)$ to determine isoflavone glycosides (genistin and daidzin) in different cultivars of soy (*Glycine max* L.). Another comparison of HPTLC and HPLC methods is given in the paper of Pereira et al.[36] According to the authors, HPTLC method was shown to be suitable for routine fingerprint analysis for quality and authenticity validations of plant material, extracts, and preparations containing *Passiflora* L. species. The objective of the work of Bilušić Vundać et al.[37] was to determine the presence and quantity of flavonoids (hyperoside, isoquercitrin, luteolin, luteolin-7-*O*-glucoside, rutin, vitexin, quercetin, and quercitrin), and phenolic acids (caffeic and chlorogenic acid) in seven plants belonging to Croatian *Stachys* taxa by the use of HPTLC method performed on silica gel 60 F_{254} plates with ethyl acetate–acetic acid–formic acid–water $(100 + 11 + 11 + 26, v/v)$, as mobile phase. After preliminary screening, the amount of each compound was determined using image (chromatogram) analysis technique, following the spraying of the plates with the natural products reagent—polyethylene glycol. A representative chromatogram of rutin analysis in some *Stachys* L. taxa samples is shown in Figure 16.3.

Similar investigation of certain phenolics was performed by Maleš et al.[38] in some Croatian *Hypericum* L. taxa.

Reversed-phase thin layer chromatography is a seldom used technique for the analysis of flavonoids. Examples of such separations are given in previously described papers by Lambri et al. and Fecka et al.[18,19]

16.2.4 USE OF MATHEMATICAL METHODS FOR THE OPTIMIZATION OF CHROMATOGRAPHIC CONDITIONS FOR FLAVONOID SEPARATIONS

Method development is the central element of any applied analytical technique and TLC is not an exception in this respect. Finding a suitable developing solvent for a certain separation problem is one of the most important steps in TLC analysis. When it comes to the separation of flavonoids, two approaches can be found in the

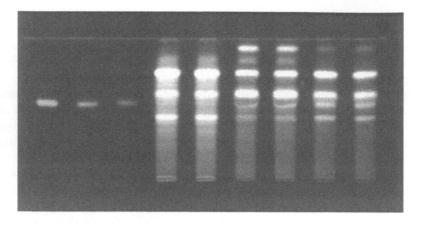

FIGURE 16.3 Analysis of rutin in some *Stachys* taxa (tracks 1–3: rutin in different concentrations; samples 4L – *Stachys salviifolia* leaves, 4C – *Stachys salviifolia* flowers, 6L – *Stachys recta* subsp. *subcrenata* leaves, 6C – *Stachys recta* subsp. *subcrenata* flowers, 7L – *Stachys palustris* leaves, 7C – *Stachys palustris* flowers). (From Bilušić Vundać, V., unpublished data, 2006. With permission.)

literature—PRISMA model (more empirical version of solvent triangle approach proposed by Snyder) developed by Nyiredy and the method of numerical taxonomy and information theory developed by Massart (optimization methods are briefly described in the book *Planar Chromatography—A Retrospective View for the Third Millennium*).[39] The example given in the paper of Reich and George illustrates the application of the PRISMA model to find conditions that enable comparison of essential oils by HPTLC.[40] Guided by the PRISMA model, ten oils (among which *Citronella* oil, rosemary oil, and peppermint oil that all contain flavonoids) were simultaneously developed on silica gel HPTLC plates in the neat solvents that have different solvent strength and according to the Snyder's classification, belong to the different groups: diethyl ether and diisopropyl ether, 2-propanol, tetrahydrofurane–hexane (1 + 5, v/v), dichloromethane, 2-chloropropane, acetone–hexane (1 + 5, v/v), toluene, and chloroform. It should be noted that the optimization of separation conditions was accomplished in approximately two working days. In the work of Medić-Šarić and other authors, the efficiency of different TLC systems is compared using numerical methods.[12,41–43] Parameters such as information content can be used to compare the quality of chromatographic systems and discriminating power is used as a measure of their effectiveness. Flavonoids and phenolic acids in various extracts (different plant, wine, and propolis extracts) were analyzed on silica plates developed with various mobile phases and the methods were optimized using the given procedure of numerical taxonomy. An interesting example of computer-aided optimization of stepwise gradient TLC of plant extracts performed by Matysik and Soczewiński[44] showed that this technique may noticeably increase separation efficiency.

16.2.5 Multidimensional Planar Chromatography (Unidimensional Multiple Development and Two-Dimensional Development)

Multidimensional methods used to improve the separation capacity in planar chromatography include unidimensional multiple development and two-dimensional development. For difficult separation problems the application of multidimensional (MD) TLC is necessary, because the power of one-dimensional chromatography is often inadequate for complete resolution of the components present in complex samples.[39]

The automated multiple development (AMD) system is considered to be a "focusing planar chromatography" which is characterized by the successive and progressive development of a chromatographic plate, with a drying step in between each development. The most important characteristic that allows the successful application of AMD with high-performance thin layer chromatography in complex mixtures analysis is the possibility of carrying out the separation process using a gradient development mode. Multidevelopment produces a band reconcentration effect, which means that bands could migrate over a certain distance without appreciable band broadening. Gradient development and band reconcentration effect allow homogenous spreading over a single chromatogram of many compounds which might belong to very different polarity ranges.[39] In the paper of Menziani et al. a basic approach to an optimized AMD-HPTLC experimental procedure and its ability to separate several classes of natural phenolic compounds is presented.[45] To test the proposed method, *Chamomilla recutita* (L.) Rauschert extracts were chosen as well-known natural sources of phenolics from a wide chemical polarity range. Among other phenolic substances, six flavonoid aglycones and four glycosides were used as standard compounds. Chromatography was performed on silica gel 60 HPTLC plates without fluorescence indicator. The AMD gradient optimization was carried out on mixtures of selected standards. General stepwise gradient conditions were gradually modified and the final optimized gradient composition started from methanol–dichloromethane–water–formic acid (70.5:25:4.5:1, v/v) and proceeded through 15 steps ending with pure dichloromethane. The flavonoids were detected after derivatization with 4% aqueous aluminium sulphate solution. After 10 min exposure to UV light, they exhibited yellow fluorescence that was stable for weeks. Fluorescence was detected densitometrically, using an excitation wavelength of 360 nm and by measuring the emission through a cutoff filter (400 nm). Ten plant extracts (among which were three flavonoid plants—*Carduus marianus* L., *Baptisia tinctoria* L., and *Lycopus europaeus* L.) were analyzed using AMD-TLC. The separation was performed on silica plates with fluorescence indicator and four AMD gradients were tested. Optimal separation was achieved in 25 steps with mobile phase composed of methanol, ethyl acetate, toluene, 1,2-dichloroethane, and 25% ammonia solution and anhydrous formic acid as modifiers. The chromatograms were scanned at 254 nm and quantitatively evaluated. The optimal AMD separation was also compared with the isocratic chromatography described for listed plant extracts in Homeopathic Pharmacopeia and it was shown that AMD-TLC technique can be used to achieve better simultaneous separations of plant extracts than it is in the case of official TLC methods.[46] The separating power of two chromatographic methods—isocratic

TLC and two-dimensional AMD—for the separation of extracts of four related plants—*Artemisia abrotanum* L., *A. absinthium* L., *A. vulgaris* L., and *A. cina* Berg., that contain flavonoids is discussed in the paper of Olah et al.[47] Silica gel 60 F_{254} HPTLC plates were used for isocratic and AMD separations. Isocratic TLC was performed with ethyl acetate–formic acid–water (80:10:10, v/v) as mobile phase, while AMD was performed in two steps (two-dimensionally) by the use of 16-step two gradient systems (gradient I was composed of methanol, dichloromethane, hexane, and 25% ammonia; and gradient II of methanol, diethyl ether, hexane, and 25% ammonia). After development, spots were visualized under UV light at 254 nm in reflection mode. The number of compounds separated was greater by use of AMD than by isocratic TLC. Two- and three-step gradient elution for separation of flavonoid glycosides and multiple development for separation of aglycones from *Vaccinium myrtillus*, L. and *V. vitis-idaea*, L. was performed by Smolarz et al.[48] Silica gel HPTLC plates were developed with mixtures of toluene, hexane, ethyl acetate, and methanol. Stepwise gradient was used for hyperin, isoquercitrin, and avicularin, while multiple gradient elution for quercetin. Qualitative and quantitative evaluation of *Passiflora coerulea* L. flavonoids (isoorientin, orientin, and vitexin) was published in the paper of Pastene et al.[49] Fingerprint analysis of plant extracts was obtained by using the optimized multiple development with a mixture of ethyl acetate–formic acid–water as mobile phase. Soczewinski et al.[50] used two-step gradient development to separate nine flavonoid glucosides and seven aglycones. Chromatography was performed on silica gel HPTLC plates. In the first step, more polar glycosides were separated using an eluent with high solvent strength (ethyl acetate–formic acid–water, $85 + 15 + 0.5$, v/v). After solvent evaporation, the aglycones were separated in a second step in the same direction with another eluent (dichloromethane–ethyl acetate–formic acid, $85 + 15 + 0.5$, v/v).

Two-dimensional TLC (2D-TLC) when compared with the AMD is still seldom used as a qualitative or quantitative technique in spite of its enormous potential. Still, examples in literature prove the usefulness of this method. In the work of Hawryl et al.[51] 2D-TLC identification of 12 flavonoids and 3 phenolic acids from *Betula* sp. leaves was performed using normal-phase systems. Optimization of mobile phases was conducted and complete separation of the investigated compounds was achieved by use of two optimum nonaqueous mobile phases on silica plate. Similarly, 2D-TLC on cyanopropyl-bonded silica was performed by the same author in order to separate eight flavonoids and three phenolic acids in *Sambuci flos*. As the first dimension a normal-phase separation was used for which seven binary eluents were tested, and the second one (a reversed-phase separation) was studied by using three binary eluents.[52] From among the 21 combinations, the three best ones that all contained *n*-hexane in the first, and water in the second dimension were used. Two-dimensional TLC was shown to be a suitable method for quantitative determination of flavonoids and phenolic acids in Croatian propolis samples.[53] Analysis was performed on silica gel 60 F_{254} plates according to the scheme depicted in the Figure 16.4A. The mobile phases used were *n*-hexane–ethyl acetate–glacial acetic acid $(31 + 14 + 5, \text{v/v})$ and chloroform–methanol–formic acid $(44 + 3.5 + 2.5, \text{v/v})$. Quantitative evaluation was based on the image (chromatogram) analysis after spraying the plates with 1% ethanolic solution of aluminium

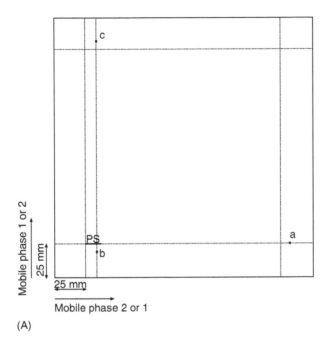

(A)

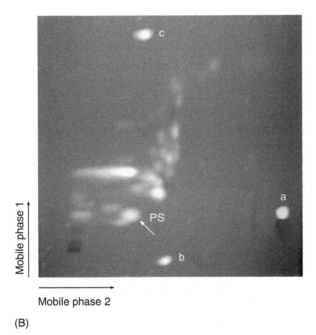

(B)

FIGURE 16.4 (A) Two-dimensional chromatography; scheme for application of standard and sample: PS, propolis sample; a, standard solution ($m_{maximum}$); b, standard solution ($m_{minimun}$); c, standard solution (m_{medium}). (B) Two-dimensional chromatogram of kaempferol in propolis sample from Slavonia ($m_{minimun} = 0.83$ μg, $m_{medium} = 1.67$ μg, $m_{maximum} = 2.50$ μg, and $m_{PS} = 2.82$ μg). (From Medić-Šarić et al., *J. Planar Chromatogr.*, 17, 459, 2004. With permission.)

chloride (a representative chromatogram of kaempferol analysis in one propolis sample is shown in Figure 16.4B).

16.2.6 QUANTITATIVE EVALUATION

Quantitative analysis is carried out in situ by measurement of herbal extract sample and standard zones on layers with a slit-scanning densitometer or video or CCD camera (image processing). Slit-scanning instruments have a number of advantages and were used in most of the representative densitometry papers cited in this chapter.[36,48] However, in some articles quantification based on image analysis is used.[37,53] Densitometry is usually performed at the wavelength of maximum absorption for UV/VIS absorption scanning and the maximum excitation/emission wavelengths for fluorescence scanning.[17,20] Notable examples of quantification and validation of the method are described in the papers of Jamshidi et al.,[32] Prashanth Kumar et al.,[31] Lalla et al.,[25] and Fang et al.[34]

16.3 MODERN TLC TECHNIQUES IN THE SEPARATION OF FLAVONOIDS

Forced-flow planar chromatography (FFPC) is considered to be a modern (improved) planar chromatography technique, in which the solvent system (except the capillary force) moves across the layer under the influence of additional force. Forced flow can be achieved either by the application of external pressure (e.g., vacuum, hydrostatic pressure) like in case of OPLC, while centrifugal force is used in rotation planar chomatography (RPC) and electric field in electroplanar chromatography. This state of the art of planar chromatography is summarized in the book *Planar Chromatography—A Retrospective View for the Third Millennium*[39] and latter paper of Nyiredy.[10] Forced-flow methods are also proposed by Nyiredy[11] as one of the possible planar techniques for the separation of plant constituents in complex mixtures.

16.3.1 OVERPRESSURED-LAYER CHROMATOGRAPHY

Forced-flow development by OPLC involves pumping of the mobile phase through a layer that is sandwiched between a rigid plate and a flexible membrane under pressure. OPLC, together with AMD, is used in the work of Galand et al. for the identification of certain flavones, flavonols, flavanones, isoflavones, and anthocyanins in extracts obtained from bitter orange tree (*Citrus aurantium* L. subsp. *amara* L.), common horse chestnut (*Aesculus hippocastanum* L.), and high mallow (*Malva sylvestris* L.).[54] For OPLC separation, silica gel 60 F_{254} plates were used. A gradient formed from ethyl acetate and chloroform mixed in various proportions was used as mobile phase. Separated flavonoids were detected under UV and visible light at 365 and 530 nm, respectively, after spraying the dry plates with NP/PEG reagent. In this paper, OPLC and AMD (together with densitometric evaluation of the plates) were shown to be appropriate and valuable techniques for identification of flavonoids in herbal extracts. OPLC was also used in the work of Ligor and Buszewski in which both TLC and OPLC methods were developed for the determination of anthocyanins in multicolored *Coleus, Prunus cerasifera* Ehrh., and *Rhus hirta* (L.) Sudworth leaves.[55]

16.3.2   ROTATION PLANAR CHROMATOGRAPHY

RPC is the oldest described FFPC method and it is a technique that can be used for both preparative (using the planar column) and analytical work. An example of use of this method for isolation and identification of flavonoids (quercetin, quercitrin, and rutin) from oak bark (*Quercus robur* L.) is given in the paper of Vovk et al.[56] Oak bark is recognized as a rich source of flavonoid and similar phenolic compounds. The main aim of this work was to study RPC as a tool for preparative (extraction of plant material and fractionation of the obtained extract) and also for analytical separation and detection of some characteristic compounds. Fractionation was performed and obtained fractions were screened by conventional TLC, which was performed on silica gel 60, 20×20 cm TLC plates, with *n*-hexane–ethyl acetate–formic acid (20:19:1, v/v) as mobile phase. Analytical RPC was performed on 20×20 cm cellulose TLC plates. The separation was performed in the ultra-microrotation planar chromatographic (U-RPC) chamber with water as the developing solvent. Fluorescence of flavonoids was detected under the UV light at 366 nm. The advantages of described technique in the preparative separations of complex mixtures are easy and rapid filling of the planar column and possibility to use adsorbent material of small particle size. The possibilities to use commercially available TLC plates, the adjustable volume of vapor phase, and the optimization of developing solvent are beneficial characteristics of the method in analytical separations.[56]

16.4   USE OF TLC IN THE ISOLATION OF FLAVONOIDS FROM PLANT EXTRACTS

To isolate individual flavonoids for further study, PC was used in two-dimensional or one-dimensional mode. After achieving satisfactory separation, spots of interest were cut into small strips and flavonoids were extracted with a suitable solvent.[2] Today, paper is replaced with different sorbents and preparative layer chromatography (PLC) is carried out on thicker layers with application of larger weights and volumes of samples in order to separate and recover 10–1000 mg of compound for further structure elucidation. Silica gel is a layer of preference and many examples of flavonoid isolation could be found in recently published papers. A representative example is the paper of Pieroni et al.[57] The presence of flavonoid structures (luteolin-4′-*O*-glucoside, apigenin-7-*O*-glucoside, luteolin-7-*O*-glucoside, apigenin, and luteolin) from the leaves of *Phillyrea latifolia* L. was revealed by the preparative TLC performed on conventional thick silica gel and reversed-phase layers, followed by UV and MS analysis. Preparative TLC was used in the work of Ponce et al.[58] as well, and isolation and full characterization of four flavonoids from *Anthemis triumfetti* (L.) DC. were performed. For further reading, papers published by Pavlović et al.[59] and Kim et al. are recommended.[60]

16.4.1 FUTURE TRENDS

Nowadays the main application field of TLC technique is quick fingerprint analysis of herbal mixtures. The exclusive feature of chromatographic image of HPTLC coupled with digital scanning profile is more and more attractive to the herbal analysts to construct the chromatographic fingerprints of herbal extracts. With the introduction of automated multidevelopment, gradient elutions and overpressured layer chromatography, planar chromatography may play more and more important role in the analysis of complex mixtures such as herbal extracts. With the coupling of TLC with different mass spectrometric methods (TLC-MS), a more reliable identification in extremely complex matrices could be achieved.

Validation of results to meet good laboratory practice (GLP)/good manufacturing practice (GMP) standards is of a great importance today as well. Complete validation of a TLC/HPTLC would be necessary if the method is proposed for the quality assurance of commercially available herbal product.

Special attention should be paid to the development of PLC methods, as a very useful tool for the separation and isolation of flavonoids from different plants as well.

REFERENCES

1. Havsteen, B.H., The biochemistry and medical significance of flavonoids, *Pharmacol. Therapeut.*, 96, 67, 2002.
2. Markham, K.R., *Techniques of Flavonoid Identification*, Academic Press, London, 1982.
3. Middleton, E., Kandaswami, C., and Theoharides, T.C., The effects of plant flavonoids on mammalian cells; implications for inflamation, heart disease and cancer, *Pharmacol. Rev.*, 52, 673, 2000.
4. Rice-Evans, C.A. and Packer, L., Eds., *Flavonoids in Health and Disease*, 2nd ed., Marcel Dekker, New York, NY, 2003.
5. Wagner, H. and Bladt, S., *Plant Drug Analysis—A Thin Layer Chromatography Atlas*, 2nd ed., Springer, Berlin, 1996, chap. 7.
6. Sherma, J., Thin-layer chromatography in food and agricultural analysis, *J. Chromatogr. A*, 880, 129, 2000.
7. Santos-Buelga, C. and Williamson, G., Eds., *Methods in Polyphenol Analysis*, The Royal Society of Chemistry, Cambridge, 2003.
8. de Rijke, E. et al., Analytical separation and detection methods for flavonoids, *J. Chromatogr. A*, 1112, 31, 2006.
9. Liang, Y.Z., Xie, P., and Chan, K., Quality control of herbal medicines, *J. Chromatogr. B*, 812, 53, 2004.
10. Nyiredy, Sz., Progress in forced-flow planar chromatography, *J. Chromatogr. A*, 1000, 985, 2003.
11. Nyiredy, Sz., Separation strategies of plant constituents—current status, *J. Chromatogr. B*, 812, 35, 2004.
12. Jasprica, I. et al., Investigation of the flavonoids in Croatian propolis by thin-layer chromatography, *J. Planar Chromatogr.*, 17, 95, 2004.
13. Tura, D. and Robards, K., Sample handling strategies for the determination of biophenols in food and plants, *J. Chromatogr. A*, 975, 71, 2002.
14. Robards, K., Strategies for the determination of bioactive phenols in plants, fruit and vegatables, *J. Chromatogr. A*, 1000, 675, 2003.

15. Namiesnik, J. and Górecki, T., Sample preparation for chromatographic analysis of plant material, *J. Planar Chromatogr.*, 13, 404, 2000.
16. Mbing, J.N. et al., Two biflavonoids from *Ouratea nigroviolacea, Phytochemistry*, 67, 2666, 2006.
17. Nowak, R. and Tuzimski, T., A solid-phase extraction–thin-layer chromatographic fiber optical scanning densitometric method for determination of flavonol aglycones in extracts of rose leaves, *J. Planar Chromtogr.*, 18, 437, 2005.
18. Lambri, M. et al., High performance thin layer chromatography (HPTLC) analysis of red wine pigments, *J. Planar Chromatogr.*, 16, 88, 2003.
19. Fecka, I., Kowalczyk, A., and Cisowski, W., Optimization of the separation of flavonoid glycosides and rosmarinic acid from *Mentha piperita* on HPTLC plates, *J. Planar Chromatogr.*, 17, 22, 2004.
20. Cobzac, S. et al., The quantitative determination of rutin in different glycerinic plant extracts by solid-phase extraction and thin-layer chromatography with densitometry, *J. Planar Chromatogr.*, 2, 26, 1999.
21. Schibli, A. and Reich, A., Modern TLC: a key technique for identification and quality control of botanicals and dietary supplements, *J. Planar Chromatogr.*, 18, 34, 2005.
22. Gocan, S. and Cîmpan, G., Review of the analysis of medicinal plants by TLV: modern approaches, *J. Liq. Chromatogr. Relat. Technol.*, 27, 1377, 2004.
23. Gómez-Caravaca, A.M. et al., Advances in the analysis of phenolic compounds in products derived from bees, *J. Pharm. Biomed. Anal.*, 41, 1220, 2006.
24. Chen, S.B. et al., High-performance thin-layer chromatographic fingerprints of isoflavonoids for distinguishing between *Radix Puerariae Lobate* and *Radix Puerariae Thomsonoii, J. Chromatogr. A*, 1121, 114, 2006.
25. Lalla, J.K., Hamrapurkar, P.D., and Gauri, K., Standardization of the homeopathic mother tincture of *Calendula officinalis* Linn., *J. Planar Chromatogr.*, 16, 298, 2003.
26. Matysik, G., Wójciak-Kosior, M., and Paduch, R., The influence of *Calendulae officinalis flos* extracts on cell cultures, and the chromatographic analysis of extracts, *J. Pharm. Biomed. Anal.*, 38, 285, 2005.
27. Vierling, W. et al., Investigation of the pharmaceutical and pharmacological equivalence of different Hawthorn extracts, *Phytomedicine*, 10, 8, 2003.
28. Lederer, I. et al., Combination of TLC and HPLC–MS/MS methods. Approach to a rational quality control of Chinese star anise, *J. Agric. Food. Chem.*, 54, 1970, 2006.
29. Popova, M. et al., Antibacterial activity of Turkish propolis and its qualitative and quantitative chemical composition, *Phytomedicine*, 12, 221, 2005.
30. Adelmann, J. et al., Exotic flora dependence of an unusual Brazilian propolis: the pinocembrin biomarker by capillary techniques, *J. Pharm. Biomed. Anal.*, 43, 174, 2007.
31. Prashanth Kumar, V. et al., High-performance thin-layer chromatographic method for estimation of rutin in medicinal plants, *J. Planar Chromatogr.*, 16, 386, 2003.
32. Jamshidi, A., Adjvadi, M., and Husain, S.W., Determination of kaempferol and quercetin in an extract of *Ginkgo biloba* leaves by high-performance thin-layer chromatography (HPTLC), *J. Planar Chromatogr.*, 13, 57, 2000.
33. Nader, N. et al., HPTLC determination of apigenin in some Iranian liquid products of *Matricaria chamomilla* L., *J. Planar Chromatogr.*, 19, 383, 2006.
34. Fang, C. et al., Comparison of HPTLC and HPLC for determination of isoflavonoids in several Kudzu samples, *J. Planar Chromatogr.*, 18, 73, 2005.
35. Janeczko, Z. et al., Densitometric determination of genistin and daidzin in different cultivars of soy (*Glycine max*), *J. Planar Chromatogr.*, 17, 32, 2004.

36. Pereira, C.A.M. et al., A HPTLC densitometric determination of flavonoids from *Passiflora alata*, *P. edulis*, *P. incarnata* and *P. caerulea* and comparison with HPLC method, *Phytochem. Anal.*, 15, 241, 2004.

37. Bilušić Vundać, V. et al., HPTLC determination of flavonoids and phenolic acids in some Croatian *Stachys* taxa, *J.Planar Chromatogr.*, 18, 269, 2005.

38. Maleš, Ž. et al., Thin-layer chromatographic analysis of flavonoids, phenolic acids, and amino acids in some Croatian *Hypericum* taxa, *J. Planar Chromatogr.*, 17, 280, 2004.

39. Nyiredy, Sz., Ed., *Planar Chromatography—A Retrospective View for the Third Millennium*, Springer, Budapest, 2001.

40. Reich, E. and George, T., Method development in HPTLC, *J. Planar Chromatogr.*, 10, 273, 1997.

41. Medić-Šarić, M., Stanić, G., and Bošnjak, I., The use of information theory and numerical taxonomy methods for evaluating the quality of thin-layer chromatographic separations of flavonoids constituents of *Matricariae flos*, *Pharmazie*, 56, 156, 2001.

42. Maleš, Ž. and Medić-Šarić, M., Optimization of TLC analysis of flavonoids and phenolic acids of *Helleborus atrorubens* Waldst. et Kit., *J. Pharm. Biomed. Anal.*, 24, 353, 2001.

43. Rastija, V. et al., Analysis of phenolic components in Croatian red wines by thin-layer chromatography, *J. Planar Chromatogr.*, 17, 26, 2004.

44. Matysik, G. and Soczewiński, E., Computer-aided optimization of stepwise gradient TLC of plant extracts, *J. Planar Chromatogr.*, 9, 404, 1996.

45. Menziani, E. et al., Automated multiple development high-performance thin-layer chromatographic analysis of natural phenolic compounds, *J. Chromatogr.*, 511, 396, 1990.

46. Gocan, S., Cîmpan, G., and Muresan, L., Automated multiple development thin layer chromatography of some plant extracts, *J. Pharm. Biomed. Anal.*, 14, 1221, 1996.

47. Olah, N.K. et al., Normal-phase high-performance thin-layer chromatography and automated multiple development of hydroalcoholic extracts of *Artemisia abrotanum*, *Artemisia absinthium*, *Artemisia vulgaris*, and *Artemisia cina*, *J. Planar Chromatogr.*, 11, 361, 1998.

48. Smolarz, H.D., Matysik, G., and Wójciak-Kosior, M., High-performance thin-layer chromatographic and densitometric determination of flavonoids in *Vaccinium myrtillus* L. and *Vaccinium vitis-idaea* L., *J. Planar Chromatogr.*, 13, 101, 2000.

49. Pastene, E., Montes, M., and Vega, M., New HPTLC method for quantitative analysis of flavonoids of *Passiflora coerulea* L., *J. Planar Chromatogr.*, 10, 362, 1997.

50. Soczewinski, E., Wojciak-Kosior, M., and Matysik, G., Analysis of glycosides and aglycones of flavonoid compounds by double-development thin-layer chromatography, *J. Planar Chromatogr.*, 17, 261, 2004.

51. Hawryl, M.A. and Soczewinski, E., Normal phase 2D TLC separation of flavonoids and phenolic acids from *Betula* sp. leaves, *J. Planar Chromatogr.*, 14, 415, 2001.

52. Hawryl, M.A., Hawryl, A., and Soczewinski, E., Application of normal- and reversed-phase 2D TLC on a cyanopropyl-bonded polar stationary phase for separation of phenolic compounds from the flowers of *Sambucus nigra* L., *J. Planar Chromatogr.*, 15, 4, 2002.

53. Medić-Šarić, M. et al., Quantitative analysis of flavonoids and phenolic acids in propolis by two-dimensional thin layer chromatography, *J. Planar. Chromatogr.*, 17, 459, 2004.

54. Galand, N. et al., OPLC and AMD, recent techniques of planar chromatography: their interest for separation and characterization of extractive and synthetic compounds, *Fitoterapia*, 73, 121, 2002.

55. Ligor, M. and Buszewski, B., Application of TLC and OPLC in the determination of pigments from natural products, *J. Planar Chromatogr.*, 14, 334, 2001.

56. Vovk, I. et al., Rotation planar extraction and rotation planar chromatography of Oak (*Quercus robur* L.) bark, *J. Chromatogr. A*, 991, 267, 2003.

57. Pieroni, A., Heimler, D., and Huang, Y., A TLC method for separation and identification of flavones and flavone glycosides from biflavones in vegetable extract, *J. Planar Chromatogr.*, 11, 230, 1998.
58. Ponce, M.A. et al., Flavonoids from shoots, roots and roots exudates of *Brassica alba*, *Phytochemistry*, 65, 3131, 2004.
59. Pavlović, M. et al., Phenolic constituents of *Anthemis triumfetti* (L.) DC., *Biochem. Syst. Ecol.*, 34, 449, 2006.
60. Kim, M.R. et al., Antioxidative effects of quercetin–glycosides isolated from the flower buds of *Tussilago farfara* L., *Food Chem. Toxicol.*, 44, 1299, 2006.

17 TLC of Lignans

Lubomír Opletal and Helena Sovová

CONTENTS

17.1 INTRODUCTION

Lignans were already found in plants in the 19th century; the term "lignan" was first introduced by Haworth in 1936. As secondary metabolites of plants, lignans play a significant role in mammalian physiology; they not only constitute a part of medicinal preparations (phytopharmaceuticals), but also occur commonly in food (feed) materials, or more precisely in mammalian food. Hence it is desirable to have rapid chromatographic methods available, which, using selective detectors, detect possible presence of these substances in the analyte.

Thin layer chromatography (TLC) of lignans developed significantly in the 70s–80s of the last century when it played an important role in this field of study of natural substances; according to the number and profile of publications over the last 20 years it is obvious that the prevalent number of important lignans has already been discovered, TLC methods for their observation have already been elaborated and presently this analytical method plays a role of a primary rapid analytical method, which is followed by other methods, namely hyphenation, especially in the field of isolation and identification of new compounds. Although classical TLC is not developing as significantly as in the 70s of the last century, it will still remain a basic method for the primary analysis of extracts field due to its rapidity, unassuming appliances and easy performance; it is possible to obtain good results by TLC especially because most lignans are substances of lower or medium polarity and therefore TLC can be successfully used mainly in the form of adsorption chromatography.

This survey is focused on providing information on the development of TLC used for analysis of lignans, its possibilities, practical use (in the case of important plant materials), and linkage to analytical and separation techniques. It is focused practically only on problems of lignans (hybrid lignans and possibly oligolignans). If we report on neolignans, norlignans, and other conjugates in our study, then only marginally because these substances are not yet (except for rare exception) studied as broadly as lignans.

17.2   BASIC STRUCTURE

The term lignans refers to phenolic compounds whose skeleton is a dimer $(C_6-C_3)_2$ formed by oxidative dimerization of two phenylpropanoid units linked by the central carbons of their side propane chains in position C-8 and C-8′ as given in the brief survey[1] (Figure 17.1).

Types (A–C) derived from the kind of links of phenylpropanoid units are as follows:

- Dibenzylbutane type (A)
- Aryltetraline type (B)
- Dibenzocyclooctane type (C)

Forms (I–VIII) derived from the degree of oxidation (I–IV) and the epoxide structure (VI–VIII) are given below:

- Butane form (I)
- Butandiole form (II)
- 9,9′-Epoxy form (III)
- Hemiacetal form (IV)
- Butanolide form (V)
- Bis-epoxy form (VI)
- 7,9′/7′,9-Epoxy form (VII)
- 7,7′-Epoxy form (VIII)

Forms VI–VIII constitute separate types (D–F).

By linking other bonds in the molecule, upon participation of the aliphatic part of the molecule, a broad spectrum of structural types arises[2,3] with very interesting and exploitable effects.

Besides lignans, also other lignan-like substances with modified structure can be found in plant material:

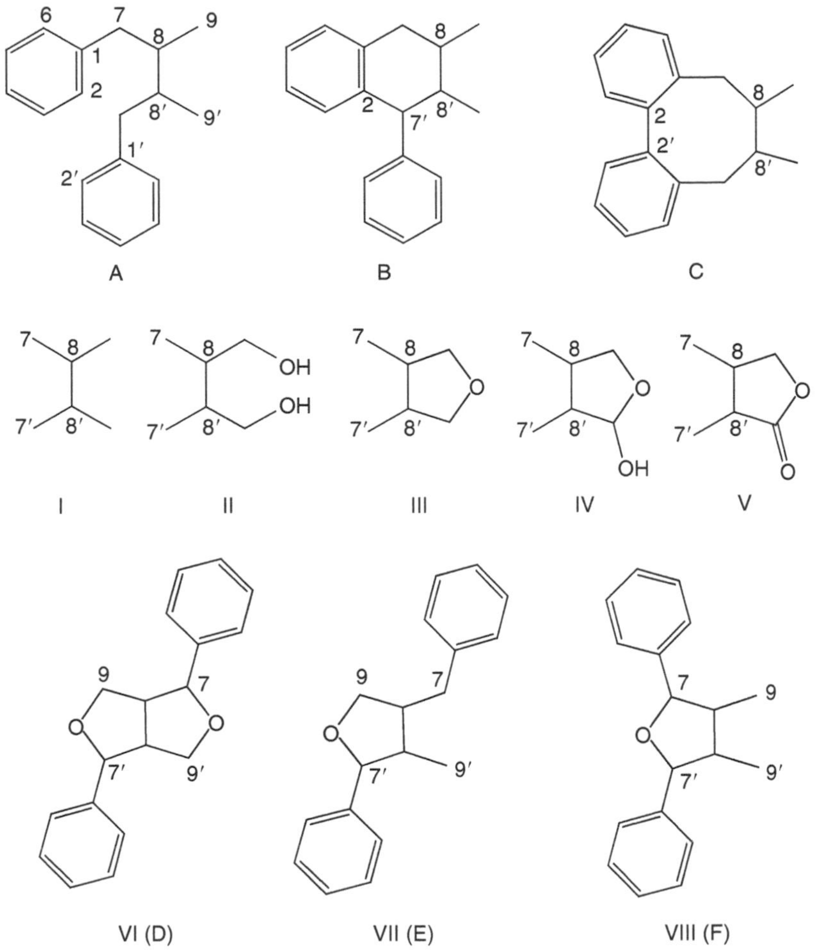

FIGURE 17.1 Basic types of lignans. (Reprinted from Harmatha, J., *Chem. Listy*, 99, 622–632, 2005, Copyright (2005). With permission from Chemické Listy.)

1. *Neolignans*—they are represented by dimers $(C_6–C_3)_2$ but phenylpropanoid units can link in different ways (besides the definition of lignans given above) upon the formation of substances of very different types (e.g., by direct link of phenyls or by means of ether bridges—one or two—possibly by combination of both eventualities; the bond of one unit to an alkyl chain of the other unit is not unusual either). As in this case there is a great variability caused by reduction of some bonds and various types of cyclization, the thought exists that the number of neolignans in nature might be significantly higher than the number of lignans. Practically it is not the case because biosynthesis of these compounds conforms with the greatest probability only to specific strict rules, which enable the formation of only some structural types of neolignans and thus their lower number. This is

substantiated by the fact that this group of substances has not been so far explored as thoroughly (especially from the aspect of biological effect)[4] as lignans.

2. *Norlignans*—these are biogenetically close to lignans (and probably specific for gymnosperm types[5]), though structurally resemble neolignans[6] more; they are *de facto* lignan derivatives with a lower number of carbons (substances of diphenylbutadiene type C_6–C_4–C_6, or conioids C_6–C_5–C_6) and the presence of other, often even conjugate double bonds. It is obvious that they arise during the biosynthesis of lignans (or lignans are their concurrent or additional source) as a result of various reactions as shown in the example of metabolite content in heartwood of *Libocedrus yateensis* (Cupressaceae).[7]

3. *Hybrid lignans*—the term refers to their mixed biogenetical origin: they are flavono-lignans (e.g., so-called silymarin in *Silybum marianum* achenes,[8] hydnocarpin in *Hydnocarpus wightiana* seeds[9]), coumaro-lignans (hyosgerin from *Hyoscyamus niger* seeds[10]), or xantolignans (kielcorin from *Hypericum* spp. roots[11]). Into this group also belong so-called lignoids, e.g., crinasiatin from *Crinum asiaticum* tubers,[12] representing the first lignophenanthridine alkaloid, or a range of macrocyclical spermine alkaloids, e.g., orantine, hordatine, aphelandrine, etc., in whose molecules neolignan substructures can be found.

Lignans can occur in plant material not only in the form of dimer units but also in small amounts also as trimers (sesquilignans) or tetramers (dilignans); the whole group can then be designated as oligomers of Ar-C_3 units, i.e., oligolignans.[13]

Lignan and neolignan type bonds are also important for oligomerization of phenylpropane units during formation of macromolecular substances in which individual basic units are linked during the formation of lignins. Since M_r of these substances is higher than 1 kDa, they are insoluble and thus biologically inactive, hence they are not dealt with.

17.3   SIGNIFICANCE OF LIGNAN-LIKE SUBSTANCES

17.3.1   OCCURRENCE

Occurrence of lignan substances was comprehensively described in the period when the studies of these substances[6,14] developed; they are only identified in systematic units of vascular plants,[15] which is not a surprising fact with respect to their biogenesis. They can be found virtually in all morphological parts (exceptionally in flowers; *Chionanthus virginicus*) in various amounts; noticeable is their content (1.5%–4% w/w) in dry roots and rhizomes of the representants of the *Podophylum* spp. (Berberiolaceae) in which they constitute the prevalent part of secondary metabolites. Significant is their content in core wood of some conifers, e.g., very high content (6%–24% w/w) was found in the core wood of *Picea abies* (Pinaceae).[13] Besides core wood of conifers, resins also represent a rich source (Resina podophylli, Resina guajaci, etc.). Highly instructive information about the occurrence of lignans in plants can be found in Ref. 16).

17.3.2 BIOLOGICAL ACTIVITY

Condensates of phenylpropane units make up, from the aspect of chemical structure, one of the most variable groups of natural substances[17]; it is thus normal to expect a wide range of variation in biological effects. Their role in plants has not been clarified yet. A theory suggests that they represent a defence system against pests (along with the other polyphenols) and can regulate plant growth. Their effects on mammals are also well documented.

So far several hundred lignans have been isolated. However, a significantly narrower range of the compounds have been studied from the biological point of view and only a few of them have been of significant use. Important are their anti-invasive effects (antimicrobial, antifungal, antiviral, antineoplastic); an individual chapter is devoted to their insecticide and antifeedant properties.[3] Antioxidant effects have been described as well (inhibition of superoxide ion formation, increase in the contents of cellular glutathione, inhibition of i-NOS), anti-inflammatory effects (PGE_2, COX inhibition), immunomodulation effects (TNF-α formation increase), effect on apoptosis, and positive effect on cardiovascular system (platelet-antiaggregation activity, influence on diabetes mellitus development, effect on hypertension) and hepatoprotective activity.[3,16,18]

From the aspect of practically usable therapeutic activity, lignans of aryltetraline-type from rhizomes and roots of *Podophyllum peltatum* (Berberidaceae) can be mentioned, namely podophyllotoxin (**1**), 4′-demethylpodophyllotoxin (**2**), peltatines (α-peltatine, **3**; β-peltatine, **4**), semisynthetic derivatives etoposide (**5**), and teniposide (**6**); while the use of podophyllotoxin is limited (external for treatment of condylomas), etoposide and teniposide have become important parts of therapeutic schemes in treatment of some neoplasms.[5,16,19]

| 1 R = CH$_3$ | 3 R = H | 5 R = CH$_3$ |
| 2 R = H | 4 R = CH$_3$ | 6 R = thienyl |

An important part is played by furano-furanoid lignans (more precisely dibenzyl-butanes, Figure 17.1, type A, especially forms II and V), contained in important pharmaceutical and food materials, e.g., seeds of cultivated flax (*Linum usitatissimum*,

Linaceae);[18,20] these substances possess a certain estrogenic activity because they can conformatively induce interaction with estrogen receptors. The main constituent which acts in this manner is secoisolariciresinol (**7**), especially its diglucoside (~0.6%–1.8% according to plant material), (**8**) and matairesinol (**9**); these substances are metabolized in intestines of primates to enterodiol (**10**) and enterolactone (**11**) formerly designated "mammalian lignans" and are present in other tissues of human organism as well (they are subject to enterohepatic circulation); because lignan compounds with —OH groups only in *meta*-position had not been previously known, the notion arose that these are endogenous substances, which was later clarified. Those substances have shown significant biological effect—they can suppress the development of cardiovascular diseases. There is a theory that they are also some effective hormone-dependent neoplasms (breast and prostate carcinomas) and intestinal neoplasms. As the theory is not definitely complete, their observation is still relevant; flaxseed and flax cakes are becoming an important constituent of human food and livestock feedstuff.[21]

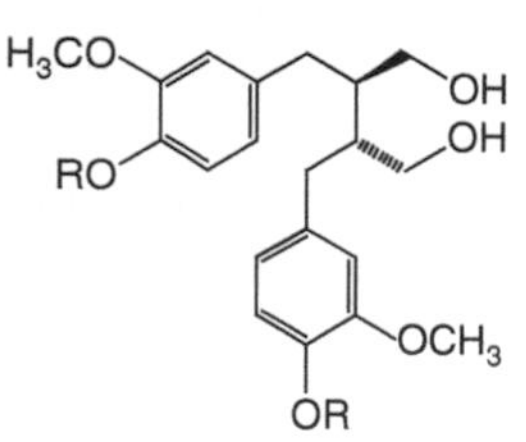

7 R=H
8 R=β-D-glc

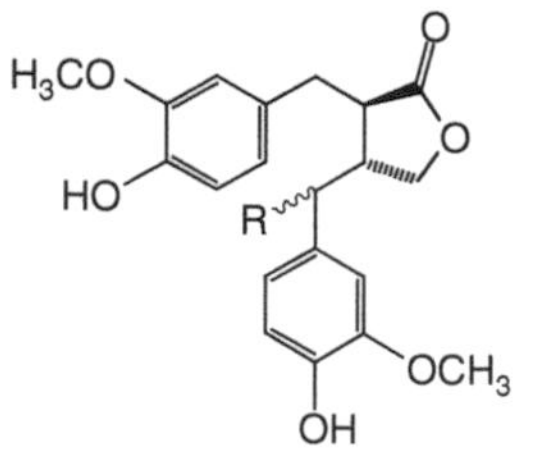

9 R=H
18 R=OH

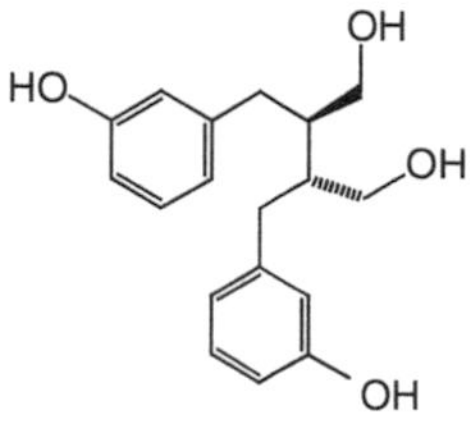

10

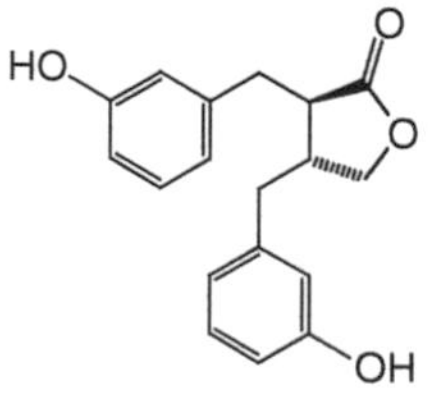

11

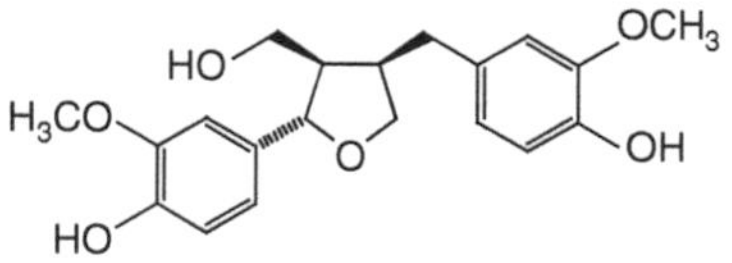

14

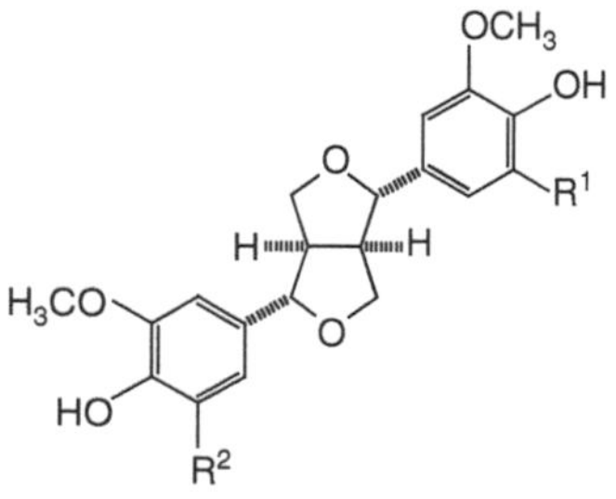

15 R^1=R^2=H
16 R^1=OCH$_3$, R^2=H
17 R^1=R^2=OCH$_3$

19   12   13

Another reason for intensive study of these lignans is the broad range of their occurrence: a great reservoir of these substances is constituted by resins of gymnosperm types, containing e.g. α-conidendrin (12), α-conidendric acid (13), especially abundantly occurring lariciresinol (14), pinoresinol (15) and its derivatives (16,17), hydroxymatairesinol (18) and substances of sesamine type (19), and their occurrence is quite common in everyday plant food (cereals, rice, soya, some nuts, seeds, and fruits Table 17.1) and natural beverages (white and red wine).[22]

TABLE 17.1
Amount of Lignans (mg/kg Dry Weight) in Selected Plants, Food Sources, and in Tree Tissues

Food/Plant Tissue	7	9	14	15	18
Flaxseed	3700	10.9	—	—	—
	2900	5.5	30.4	33.2	—
	12,600	58.6	—	—	—
Sesame seed	293	4.8	95	0.7	—
Rye bran	1.3	1.7	—	—	—
Rye flour	7.2	1.7	—	—	—
Rye bread, dark	0.1	0.1	1.2	1.7	—
Curly kale (*Brassica*)	0.2	0.1	6.0	16.9	—
Broccoli	4.1	0.2	—	—	—
	0.4	0	9.7	3.2	—
Garlic	0.5	0	2.9	2.0	—
Strawberry	15.0	0.8	—	—	—
	0	0	1.2	2.1	—
Olive oil	0	0	0	2.4	—
Black tea	24.2	3.0	—	—	—
Green tea	28.9	2.0	—	—	—
Picea abies (heartwood)	3–370	10–520	17–370	10–20	—
Picea abies (knots)	1400–6800	170–5500	1000–2500	tr	36,000–88,000
Abies alba (heartwood)	140	250	180	20	150
Abies alba (knots)	29,000–36,000	2500–2600	4600–10,000	360–1000	7000–7800

Source: Adapted from Willför, S.M., Smeds, A.I., and Holmbom, B.R., *J. Chromatogr. A*, 1112, 64, 2006. With permission.

tr: trace; "—" indicates not estimated.

Among the lignans of other structural groups, substances of dibenzocyclooctadiene type isolated from *Schisandra chinensis* (Schisandraceae) seeds are beginning to play an important part. Out of the total of ~50 compounds the main lignans schizandrin (**20**), gomisin A (**21**), deoxyschisandrin (**22**), gomisin N (**23**), γ-schizandrin (**24**), and wuweizisu C (**25**) are worth mentioning. They exhibit antioxidative, hepatoprotective and neurotrophic activity, and other effects[23]; the hepatoprotective activity is considered to be the most significant effect. It is based on an increase of glutathione content in tissues and protection against oxidative stress. This effect can nevertheless manifest in other parts too, e.g., in heart or brain tissue.

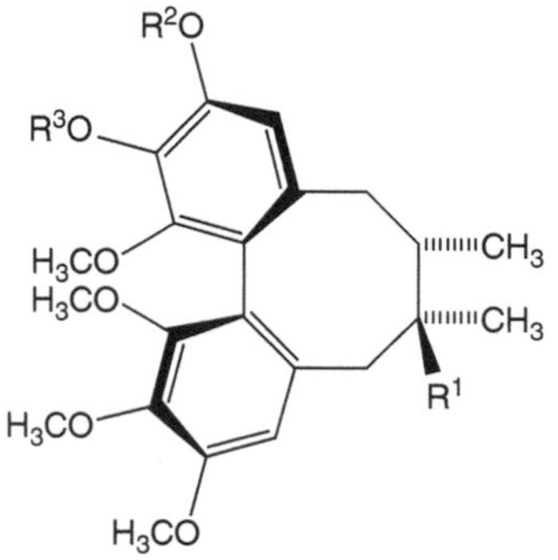

20 R¹=OH, R²=R³=CH₃
21 R¹=OH, R², R³=CH₂
22 R¹=H, R², R³=CH₃

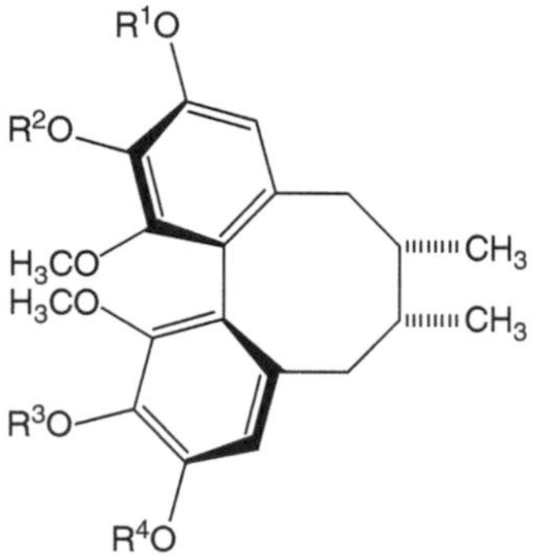

23 R¹, R²=CH₂, R³=R⁴=CH₃
24 R¹=R²=CH₃, R³, R⁴=CH₂
25 R¹, R²=R³, R⁴=CH₂

Much attention is still attracted by other biologically active types of phenylpropane condensates, e.g., flavonolignans from *Silybum marianum* (Asteraceae) achenes represented by so-called silymarin complex, namely the mixture of silybins (silybin A, **26**), isosilybins, silydianin, and silychristin (**27**) with an important antioxidative, but especially hepatoprotective activity, and prospective neolignans magnolol (**28**) and honokiol (**29**), contained in *Magnolia officinalis* (Magnoliaceae) bark is an important medicinal drug in Oriental medicine; both substances possess antioxidative properties, inhibit the flux of Ca²⁺ in thrombocytes, relax smooth muscles by inhibition of calcium influx through calcium channels, possess an antidepressant effect, and have a number of other therapeutically exploitable effects.[16]

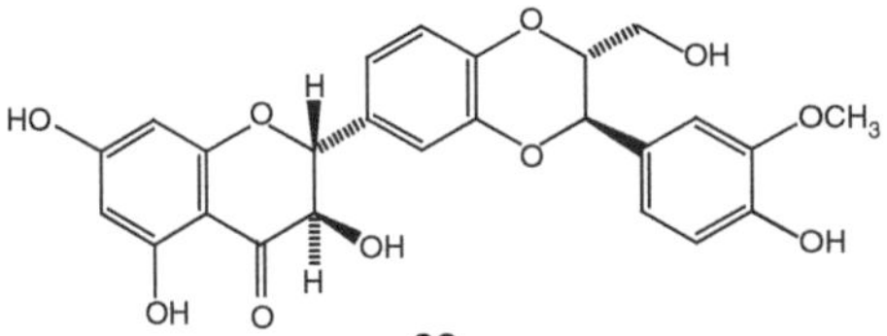

26 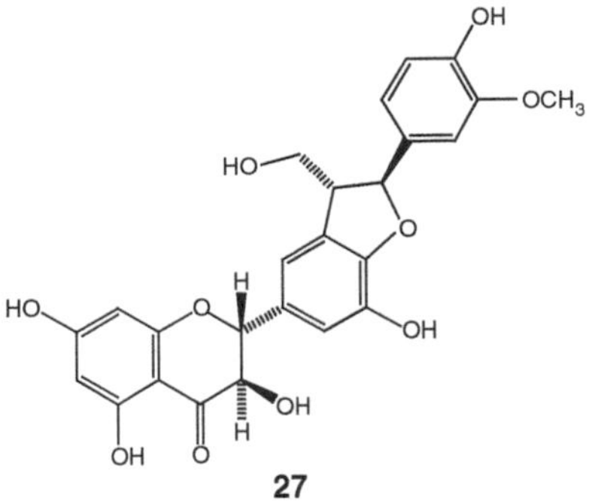   **27**

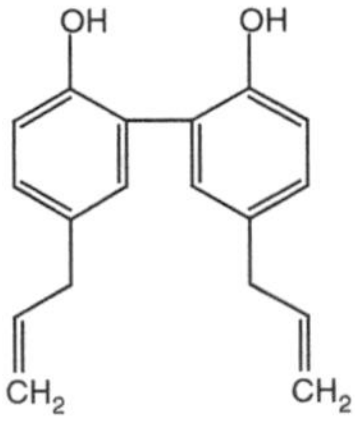

28 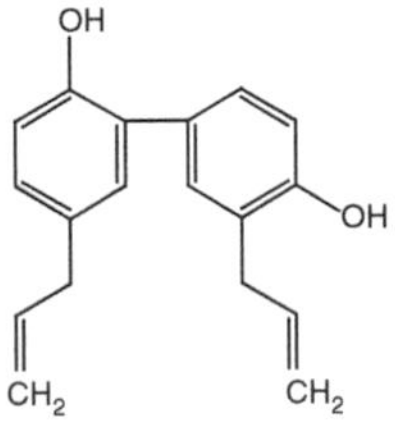   **29**

Also some of neolignans possess important biological activity, e.g., substances of dihydrobenzofurane type are being observed from the aspect of antitumor activity (3′,4-*O*-methylcedrusin **30**; *Croton* spp., Euphorbiaceae). Dihydrobenzofuranic acid **31**, and its methylesters **32**, prepared from **30**, have become the basis for study of substances with antiangiogenic activity. A relatively small group of 8.*O*.4′-neolignans, found only in the representants of Myristicaceae family (e.g., surina-mensin **33** and virolin **34** from *Virola surinamensis*) with antileishmanial activity is also perspective for the preparation of further biologically active derivatives. In some semisynthetic derivatives of 8.*O*.4′-neolignans antifungal activity[4] has been detected.

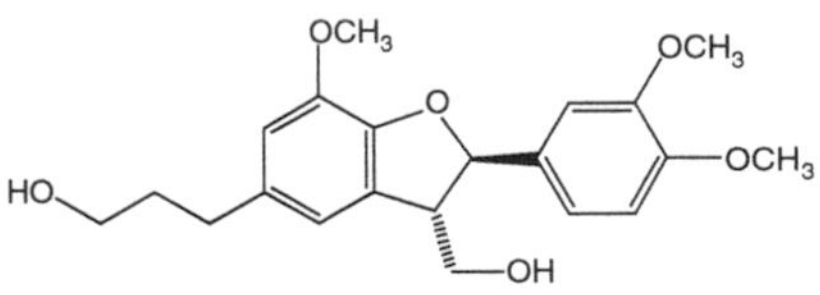

30 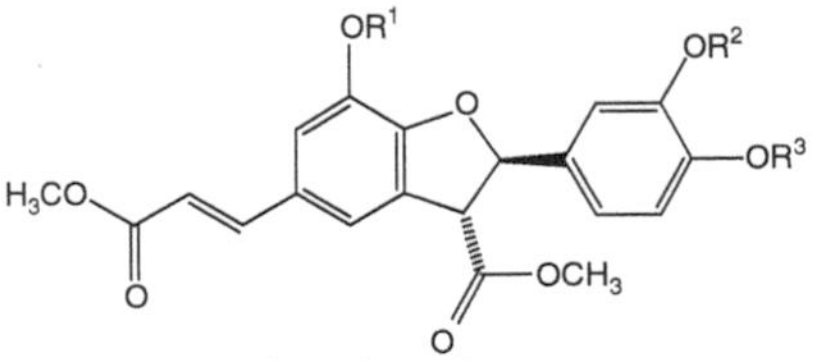   **31** R^1 = R^2 = R^3 = H
              32 R^1 = R^2 = CH$_3$, R^3 = H

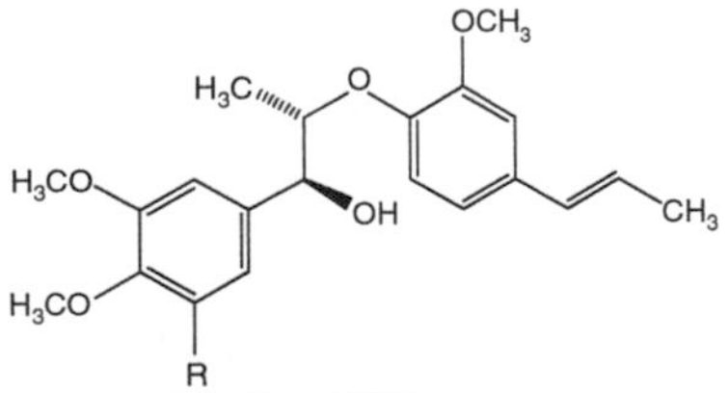

33 R = OCH$_3$
34 R = H

In such a broad structural group of phenolic substances which lignans and their metabolites constitute, a question arises by which mechanism they act; there is a series of explanations but none of them is universal (analogy can be seen in the case of flavonoids as well). That is why explanations must be considered which are applied at the level of individual types of biological effects such as inhibition of aromatase, lowering of 5α-reductase and 17β-hydrogenase activity, the modification of 2- and 16α-hydroxyestrone ratio, inhibition of estrogen bond to α-fetoprotein, stimulation of globulin synthesis, β-glucuronidase activity and other effects which have been comprehensively described.[18]

The prevalent part of these substances is neutral upon ingestion, exceptionally they can be of a bitter savoury taste (cubebine), or their effect can be topically irritative (lignans of *Podophyllum* spp.).

17.4 PHYSICOCHEMICAL PROPERTIES OF LIGNANS

Structural variability of these substances significantly influences their chromatographic separation. If we consider high lipophility of aglycones of this type of substances ($\log P$ ~2–3), which are mainly analyzed, because glycoside-bound forms are not prevalent in nature, then it can be stated that the following phenomena manifest significantly:

1. *The influence of skeletal substitution and partition coefficient* (namely the value of $\log P$): the contribution to hydrophilicity (lower values of $\log P$) is approximately as follows: OH (phenol) > OH (secondary alcohol) > CO (oxo group in lactone) > CH_3; during the interaction with an adsorbent the number of these groups in a molecule and their character plays an important role from the aspect of formation of hydrogen bonds, coordinate bonds, but also dispersive powers. Position of a substituent can also play an important role, e.g., in aryltetraline lignans podophyllotoxine (**1**) and α-peltatine (**3**) the position of OH group is significant: if it is in 7α position, then the substance is less hydrophobic (**1**, $\log P = 2.12$) than in the case of substitution in position 6 on the aromatic ring (**3**, $\log P = 2.51$). If glycosidation of hydroxyl occurs, then hydrophilicity rises significantly (etoposide **5**, $\log P = 1.12$). Nevertheless, prediction of chromatographic behavior of several substances solely on the basis of the number and character of functional groups is practically possible only within series of structurally closely related substances.

   As obvious, the value of $\log P$ plays one of the crucial roles because it is the indicator of lipophilicity of separated substances (it offers a more precise insight into chromatographic behavior regardless of the position and a possible number of functional groups). The modification of a part of a molecule in lignans of dibenzylbutane type (transformation of enterodiol **10** to enterolactone **11**) leads to a quite significant modification of lipophilicity (**10**, $\log P = 3.00$; **11**, $\log P = 3.50$). The introduction of an additional hydroxy group to position 7' (matairesinol **9**, $\log P = 3.25$; hydroxymatairesinol **18**, $\log P = 2.20$) is equally significantly manifested in the increase of polarity. A similar effect occurs during glycosylation (secoisolariciresinol **7**, $\log P = 2.75$; monoglucoside $\log P = 0.79$; diglucoside **8**, $\log P = 0.93$);

in the case of glycosides, though, the chromatographic behavior is influenced by the "envelope" of the molecule.

Opening of the lactone ring upon the formation of α-hydroxyacid also results in the increase of hydrophilicity (α-conidendrin **12**, $\log P = 2.73$; α-conidendric acid **13**, $\log P = 2.19$).

Gradual introduction of methoxy groups on the skeleton does not manifest as significantly as the introduction of a free hydroxy groups, as can be seen in lignans with the furano-furanoid skeleton [type IV (D)] (pinoresinol **15**, $\log P = 1.93$; medioresinol **16**, $\log P = 1.81$; syringaresinol **17**, $\log P = 1.68$).

It seems that formation of methylenedioxy groupings in dibenzocyclooctane type of lignans brings only an unimportant increase in lipophilicity when compared with substances with free methoxy groups (schizandrin **20**, $\log P = 3.65$; gomisin A **21**, $\log P = 3.68$).

If we consider that most lignans contain free phenolic groups (in some cases methoxy groups), then it is logical that a major part of separations on the thin layer is carried out in the mode of adsorption chromatography just on silica gel (even though a range of other adsorbents has been used, such as aluminium oxide,[24] gradient layers of Celite G and silica gel G,[25] magnesol,[26] etc.). On the silica gel presumed results can be unambiguously expected, as has been shown in the instructive paper[27] presenting the separation of aglycones and glycosides from *Podophyllum* spp. by means of gradual development: in grade I (6 cm) the arrangement chloroform–methanol 10:1 has been used, in grade II (12 cm) chloroform–acetone 20:7 has been used. The hRf values found are as follows:[27] aglycones—4′-demethylpodophyllotoxin (**2**), 52; α-peltatine (**3**), 61; podophyllotoxin (**1**), 65; β-peltatine (**4**), 71; 7-desoxypodophyllotoxin (78); glycosides—α-peltatine-β-D-glucopyranoside, 13; podophyllotoxine-β-D-glucopyranoside, 17; β-peltatine-β-D-glucopyranoside, 19.

With respect to structural connection of the substances separated it is obvious that practical results do not differ from the presumption of behavior based on the use of $\log P$ values.

2. *Influence of isomerism and the size of the molecule*: the influence of isomerism has not been profoundly investigated so far. Mathematical prediction of $\log P$ values in dibenzylbutane, aryltetraline as well as in dibenzocyclooctane lignans (in the last case in corresponding pairs with R and S biphenyl configuration) has not virtually shown any differences.

More significant influence can nevertheless be represented by the size of a molecule; the higher the M_r value of an aglycone, the lower its mobility, though in this case as well the chromatographic behavior can only hardly be predicted. It is certain that the hRf values correspond with the total energetic "envelope" of the molecule; this fact can be very clearly observed mainly in glycosides. In the case of separation of phenolic substances on silica gel one can expect the manifestation of behavior stemming from the original Bate-Smith and Westall theory[28] (here completed with a free hydroxy group and its glycosylated form):

$$hR_f(\text{R–CH}_3) > hR_f(\text{R–H}) > hR_f(\text{R–OCH}_3) > hR_f(\text{R–OH})$$
$$> hR_f(\text{R–}O\text{-saccharide})$$

17.5 SAMPLE PREPARATION

17.5.1 COLLECTION AND TREATMENT OF MATERIAL

During the processing of morphological parts of plants with lignan content it must be taken into the account that other phenolic substances susceptible to outer conditions are usually present and thus the formation of condensation products and artefacts is realistic. The method of processing is *de facto* determined by the character of plant part and the accompanying substance.

1. *Soft parts* (leaves, flowers, stems) *and seeds* should be dried at normal temperature and out of direct sunlight.
2. *Hard parts* (roots, rhizomes, wood) *and fleshy fruits* are usually preserved by means of lyophilization; if it is needed to process the exocarp of the fruits with a high content of water, then it is suitable to quickly remove the seeds, which generally contain a significant amount of oil. In other cases it is difficult to detect the quality of lyophilization process. During the processing of wood (knots) with the content of a significant amount of resinoids, it is suitable to split this morphological part into small pieces and put them through lyophilization. Freezing by means of liquid nitrogen and storage in airproof bags at low temperatures is adopted as well. Drying of these plant parts in the air carries the risk of the rise of a greater number of condensation products.

The method of preparation must be chosen so that polymerization, enzyme reactions, and undesirable oxidation are reduced to the minimum.[29] Oxidation can also occur during later processing, namely grinding (oxidative processes and formation of reactive substances are common during grinding of oily seeds or achenes), therefore it is suitable to process this material immediately before the extraction. Sometimes the use of solid carbon dioxide is recommended; since this procedure does not solve the problem unambiguously, then it is suitable to store the samples in aluminium containers under argon after a rapid grinding.

17.5.2 EXTRACTION AND EXTRACTS PROCESSING

Phenylpropane derivatives are present in plant material generally in the form of aglycones; if they occur also as glycosides then their content in tissues is considerably lower. Regarding this it is needed to choose extraction methods (hot or cold):

1. *Extraction in the Soxhlet extractor* is a common, probably the most widely used method; it can be used for sequential extraction (the use of solvents or their mixtures with increasing polarity), which is usually started with petroleum ether, *n*-hexane, or a halogenated hydrocarbon;[30–33] this procedure is especially important in the case of seeds or fruits with a high lipid content. During this "pre-extraction" can nevertheless occur extraction of a part of lignans as well, as was found in the case of highly oily seeds of *Schisandra chinensis*[34] (lipid substances function in this case as certain

solubilization agents and enable to extract partly even the lignans that are insoluble in these solvents under normal conditions). After removing lipidic substances, polar solvents (ethanol, methanol, acetone) are used for preparation of total extracts, often with addition of a certain amount of water. Enhanced penetrating of the solvent into the plant tissue makes the extraction of polar lignans possible. Nevertheless, it is necessary to consider that during the extraction with aqueous polar solvents the distillate does not usually have the composition of the solvent in the distilling vessel, but of certain azeotropes only (the fact is especially observable in the case of aqueous solution of ethanol). For the extraction of polyphenols inclusive lignans of grains 30% acetone can also be successfully used. Most aglycones have low to medium polarity and must be extracted with relatively nonpolar solvents. Although the plant part is usually "pre-extracted," on the basis of physical interactions a significant extraction of undesirable ballast substances occurs too.[33,35] Thus, to analyze lignans and sterols in sesame seeds, the seeds were extracted with n-hexane, the oil was saponified and the unsaponifiables were extracted with diethyl ether.[36,37]

2. *Accelerated solvent extraction* (ASE) is usually carried out at an increased temperature, pressure and in oxygen atmosphere in appropriate devices/apparatus. It is suitable for quick, automated, and sequential extraction upon using relatively small volumes of solvents. Nevertheless, it is not quite usual in working equipment of laboratories. It has been successfully used for extraction of lignans from wood of certain trees (*Picea abies*, Pineceae).[32,33,35,38–40] It is more delicate than the extraction in Soxhlet apparatus.

3. *Percolation* is carried out at room temperature,[41–44] which is very advantageous, a drawback can though be a certain lengthiness of the procedure. If it is performed in the common, simplest way, then it is difficult to use low-boiling and explosive solvents; this drawback can be eliminated by a suitable technical adaptation of apparatus. Samples for TLC are only rarely prepared by percolation, usually only in the case of necessity to pretreat the sample. The use of maceration connected with sonication is very advantageous for qualitative analysis; it is not possible to use this procedure for sequential extraction, the procedure is though quick, very effective, and small amounts of plant material[45] can be utilized without losses. Hairy roots of *Linum leonii* were extracted with 80% methanol and 1 h sonication,[46] air-dried sapwood meals were extracted for 72 h with methanol,[47] and fresh unripe fruits of *Rollinia mucosa* were extracted repeatedly with methanol at room temperature for 7 days.[48] Air-dried leaves of *Phyllanthus amarus* were separately extracted with either n-hexane, chloroform, ethyl acetate, or methanol for 10 h; the maximum yield of two determined lignans was obtained with methanol.[49]

4. *Digestion with hot solvents* (methanol[50] or ethanol[51,52]) is quick and quite delicate; it is suitable for sample preparation mainly for qualitative chromatography.

5. *Supercritical fluid extraction* (SFE) is a relatively new technique for isolation of analytes from natural matrices.[53–55] Low viscosity and very low surface tension of supercritical fluids enables their fast penetration into the plant tissue, and their solvent power is readily controlled by pressure and temperature. Polarity of carbon dioxide (SC-CO_2), the most frequently used supercritical solvent, is low; it dissolves low- and medium-polar substances like lignans but lignan glucosides remain in the plant matrix, unless the solvent is modified with polar solvent (methanol, ethanol). In the case of more polar lignans, extraction with SC-CO_2 is a convenient method for removing lipophillic substances like oil in flaxseed prior to extraction of lignans with a hydrophilic solvent.[22] Tandem on-line coupling of SFE with supercritical fluid chromatography (SFC) was applied to extract and separate two neolignans from *Cortex magnoliae*[56] (CO_2–MeOH 95:5). *Schisandra chinensis* was extracted with pure and modified SC-CO_2; pure solvent extracted 96% lignans from seeds and 26% lignans from leaves but modified solvent (CO_2–MeOH 90:10) extracted 87% lignans from the leaves.[57] Similarly, pure SC-CO_2 extracted 66% biologically active lignan arctigenine from *Forsythia koreana* and with CO_2–MeOH (90:10) the yield was increased to 110% (related to the yield obtained with methanol).[58] Thus, SFE is suitable for lignan extraction prior to analytical TLC, but its application before quantitative analysis requires a validation study.

Preparation of lignan concentrates with a low content of ballast substances by a common extraction is practically impossible; lipophilic solvents extract not only undesirable substances but also lignans which are without OH groups, or possibly with maximum one hydroxy group.[41,43,59] During extraction with polar solvents lignans can be rather successfully separated from other substances (lipids) using SPE on C18 cartridge;[57,60] these methods are used mainly for quantitative analysis, but they are successful also in TLC.

   Purification of total extracts with lignan content is quite time-consuming and laborious, it is though suitable to carry it out because in following TLC it makes analysis significantly easier. Methanol extracts are usually concentrated, diluted with water, this suspension fractioned with *n*-hexane and consequently with chloroform,[48] dichloromethane,[61,62] or ethyl acetate[51] with the aim to obtain a lignan fraction. At our workplace the following procedure proved to be useful for preparative separation of phenolic substances of this polarity, which is relatively laborious, nevertheless it brings good results: a primary alcohol extract is thickened into syrup-like consistence, dissolved in 70%–80% methanol and shaken with petroleum ether (pentane); the prevalent part of impurities (waxes, fats, essential oils) passes into petroleum ether with a small amount of lignans. From the methanol concentrate the solvent is removed, aqueous residue is diluted with water and shaken with solvents of increasing polarity. During the preparation and purification of extracts from vegetable parts the use of acetonitrile proves worth.

   In some cases it is necessary to pre-treat plant material before extraction and chromatographic analysis; the analysis of flaxseeds can be mentioned as an example

(Figure 17.2). Seco-diglucoside (SDG) is present there in the form of ester-bound oligomers, namely oligomers with hydroxymethylglutaric acid;[63,64] these oligomer (polymer) compounds are easily soluble in aqueous methanol or ethanol, their TLC analysis is though problematic, therefore it is necessary to perform alkaline hydrolysis so that SDG is released. If it is needed to obtain free seco-aglycone, it is necessary to carry out a complementary acid-hydrolysis.[63,65–67] Since flaxseed is a very important industrial material, consequential procedures involving various degrees of hydrolysis were developed for SDG and seco-aglycone analysis in food and feedstuff.[21,63]

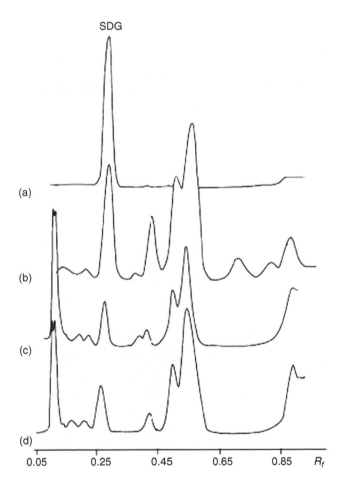

FIGURE 17.2 HPTLC densitograms of secoisolariciresinol diglucoside (SDG) standard (a) and hydrolysis product of ethanol-1,4-dioxane extract of flaxseed (b), defatted flaxseed (c), undefatted flaxseed (d). (Reprinted from Coran, S.A. et al., *J. Chromatogr. A*, 1045, 217, 2004, Copyright (2004). With permission from Elsevier.)

Hydrolysis in the case of lignan glycosides very significantly contributes to simplification of chromatographic analysis of lignans at least in the first stage, when detection of aglycones is concerned. It is important in observation of industrial technologic procedures; interesting is the example of observation of lignans from a water/aqueous phase on preparation of paper pulp;[68] after removing of mechanical particles from aqueous solution, adjustation of the pH to 3.5 (diluted sulphuric acid) lignans are agitated with MTBE (methyl-*tert*-butyl ether) and further analyzed. The method enables the extraction >95% of the content of lignans from aqueous phase, though it is substantially less efficient for extraction/obtaining of sesquilignans and dilignans. For these lignans proved useful also the use of ethyl acetate as the organic phase.

Several synoptic papers[22,69] have been recently devoted to the topic of sample preparation for quantitative analysis, namely the papers of Tura and Robards,[29] stating/mentioning a following procedure utilizable for analysis of phenolic substances from plant material.

It is suitable to store extracts with lignan content in a dry state, covered with argon at decreased temperature. We have found that it is advisable to dry samples in the vacuum over silica gel and not over phosphoric oxide; at higher content of water vapor in the desiccator pollution of samples with phosphoric acid formed by reaction of phosphine with water on insufficiently pure phosphoric oxide can occur, which results in lowering of pH on the surface of a dried sample and formation of condensation products, which complicates later analysis.

17.6   TLC TECHNIQUES

17.6.1   ADSORBENTS AND ELUTION SYSTEMS

Silica gel plates, very often fluorophore impregnated (F_{254}), are commonly used for TLC of lignans. Reversed-phase TLC plates have been rarely applied, e.g., in TLC of lignans and neolignans in *Brassica fruticulosa*.[70]

The composition of elution systems depends on both lignans structure and composition of the sample. Solvents used for TLC, such as lower aliphatic hydrocarbons (petroleum ether,[71] and especially *n*-hexane[48,72]), halogenated hydrocarbons (namely chloroform[27,36,52,73–75] and dichloromethane[22,48]) mostly have low to medium polarity. Benzene[47,76,77] or toluene[46,78,79] suppresses diffusion of zones and can ensure their compactness. Esters of lower carboxylic acids, notably ethyl acetate,[80,81] ketones (butan-2-one,[80] acetone[27,46,78,79]), and especially lower alcohols (methanol[42,48,70] and ethanol[22,76,82]) have been added to the above mentioned solvents. Addition of lower carboxylic acids, namely formic acid,[67] can play an important role because it inhibits dissociation of phenolic hydroxyls and leads to better separation of zones. For the separation of more polar glycosides, mixtures with higher content of polar solvents,[83,84] such as $CHCl_3$–MeOH (1:1) or $CHCl_3$–MeOH–toluene–NH_3 (10:6:3:1) have been used. Addition of water is also suitable.[67]

As analyzed in detail in Part 4, two-stage one-dimensional TLC was used by Stahl and Kaltenbach[27] in order to separate both glycosides and aglycones of *Podophyllum* species. Two-dimensional TLC with *n*-hexane–diethyl ether (7:3) in the first direction and chloroform–diethyl ether (9:1) in the second direction enabled a complete separation of lignans from tocopherols in the extract of *Sesamum* spp.[36]

17.6.2 DETECTION OF LIGNANS ON TLC

As lignans absorb UV light, they are readily detected at 254 nm using the plates with fluorescent indicator. Alternatively, the spots on developed plates can be temporarily visualized by plate exposure to iodine vapor.[36,85] Some lignans, as those in *Linum* spp.,[86] *Phyllanthus* spp.,[79] and *Justicia procumbens*,[87] give characteristic blue fluorescent spots under UV light.

Identification of the predominant lignans is facilitated by specific colors obtained by coloring agents. Spraying with sulphuric acid in ethanol followed by rapid heating gives different colors to lignans: violet, red or red-brown, dark gray, and blue.[22] Sulphuric acid has been frequently applied, either sprayed as anisaldehyde–sulphuric acid reagent,[81] H_2SO_4–AcOH–H_2O (1:20:4) solution,[70] and H_2SO_4–acetic anhydride (1:3),[27] or the TLC plates were immersed[49] in a mixture of vanillin in concentrated H_2SO_4–EtOH (5:95).

Very usable is copulation reaction of phenolic hydroxyl group with diazotized compounds: Barton[76] used diazotized sulphanilic acid reagent and obtained pink-brown to yellow-orange colors of lignan spots on silica gel. Jackson and Dewick[73] detected spots of *Podophyllum* lignans using concentrated HNO_3–AcOH (3:10). As a result of demethylation and oxidation, podophyllotoxin reacts to bright red quinone and other lignans react similarly to give red or brown coloration. The 4′-demethyl derivatives react fast but the trimethoxylignans react best after heating. When the agent was applied on the spots of podophyllotoxin-type and polygamatin-type *Commiphora erlangeriana* lignans; only the lignans possessing a podophyllotoxin-type substitution pattern gave the red coloration.[74]

17.7 ANALYTICAL TLC

TLC as a simple, inexpensive, and rapid method is applied mostly for a first qualitative examination of plant extracts and for monitoring various stages of lignan purification. Qualitative TLC has been applied to elaborate lignan patterns e.g., in *Podophyllum* spp.,[27] *Phyllanthus* spp.,[81] in different trees,[22,76] and in seeds of *Sesamum* spp.[36,37] It provides a good overview of the lignan pattern, although all lignans cannot be separated. In the second case, when extracts from plants were subjected to silica gel column chromatography, the eluted fractions were profiled by TLC and the fractions with corresponding profiles were combined.[48,70,78,79]

17.8 DETERMINATION OF LIGNANS

The progress in instrumentation has led HPTLC densitometry to an improvement of its reliability, making this technique competitive with HPLC-UV detection.[67] Procedures involving a minimum of sample manipulation have been developed recently for routine determination of major lignans in several plant species.

To determine secoisolariciresinol diglucoside (SDG) in flaxseed, methanol solutions of extracts from seed (subjected to alkaline hydrolysis and acidified) were applied along with SDG standard solution on silica gel HPTLC plates and developed with ethyl acetate–methanol–water–formic acid (77:13:l0:5, v/v). The plates were scanned within 2 h in reflectance mode at 282 nm. A good separation of SDG from other components was achieved and the accuracy of SDG was equal for defatted and undefatted flaxseed.[67]

A method for determination of justicidin B in *Linum* spp.[86] is based on fluorescence of justicidin B exposed to UV light at 366 nm. $CHCl_3$–MeOH (99:1) is used as mobile phase and the spot of the lignan is scanned at 366 nm using a K400 filter to cut off UV light.

An RP-HPTLC procedure enables rapid determination of the content of cytotoxic lignan podophyllotoxin in *Podophyllum hexandrum* underground parts[88] (Figure 17.3). RP18 F_{254} TLC plates and elution system acetonitrile–water (50:50, v/v) were used; densitometric scanning was performed in the absorbance–reflection mode at 217 nm.

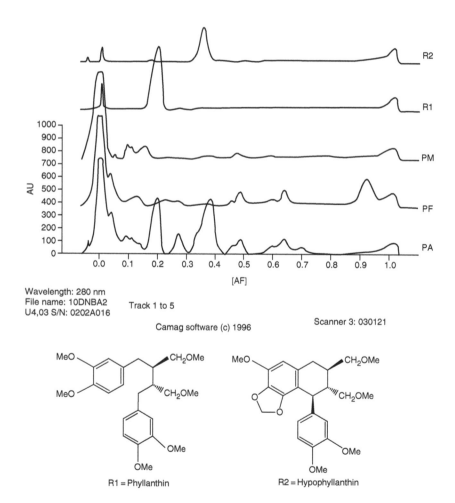

FIGURE 17.3 TLC densitometric scanning profiles of three *Phyllanthus* sp. and their active constituents. (Reprinted from Khatoon, S. et al., *J. Ethnopharmacol.*, 104, 79, 2006, Copyright (2006). With permission from Elsevier.)

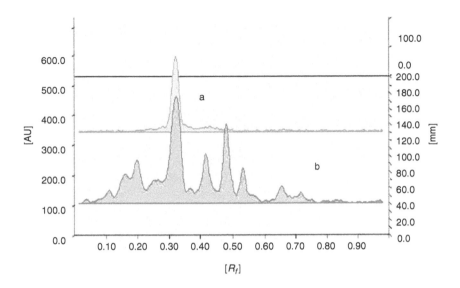

FIGURE 17.4 HPTLC; 3D overlay chromatogram of (a) standard track and (b) resin sample of roots of *Podophyllum hexandrum*. (Reprinted from Mishra, N. et al., *Curr. Sci.*, 88, 1372, 2005, Copyright (2005). With permission from Current Science.)

The amount of phyllanthin and hypophyllanthin in methanol extract of *Phyllanthus amarus* was determined using a similar method.[81] Known quantities of both standards were applied on glass TLC plates along with purified extract and eluted in toluene–ethyl acetate (85:15, v/v); the plate was scanned at 280 nm (Figure 17.4).

According to another method for phyllanthin and hypophyllanthin determination in *Phyllanthus* spp.,[49] purified methanol extract and standards were developed on silica gel plates with *n*-hexane–acetone–ethyl acetate (74:12:8, v/v) mobile phase. The plates were immersed in a mixture of vanillin in concentrated sulphuric acid–ethanol, dried and heated to develop the color; the spots were scanned at 580 nm.

In all studies mentioned earlier, Camag equipment including TLC Scanner was used. In an earlier work, Zhu et al.[89] determined schisandrol A and B (gomisin A and schizandrin) in Sheng Mai San (a Chinese herbal formula with *Panax ginseng*, *Schisandra chinensis* and *Ophiopogon japonicus*) applying its ether extract and standards of both compounds on silica gel plates and developing them with toluene–ethyl acetate (6:4, v/v). Scanning of the spots was carried out using Shimadzu TLC scanner.

17.9   PREPARATIVE TLC

TLC is a useful semipreparative technique possessing a relatively high throughput, depending on sample load and plate thickness. In several studies, individual lignans were isolated from extracts by repeated preparative TLC.[47,52,73] Combination of

TLC with other separation techniques is, however, more frequent, and preparative
TLC serves usually in the final step for purification of lignans.[70,72,75,77,80,90,91]

17.10 AUTOGRAPHIC METHODS

TLC autographic methods have been applied also on lignans. Thus, antioxidative
properties of lignans extracted from *Iryanthera juruensis* fruits were proved when
the developed TLC plates were sprayed with a 0.02% solution of β-carotene in
CH_2Cl_2 and placed under natural light until discoloration of background.[92] All
lignans appeared as yellow spots against white background. A similar antioxidative
assay was performed for lignans in *Orophea enneandra*.[71] These lignans were also
subjected to two other autographic methods: free radical scavenging assay, when
TLC plates were sprayed with 2,2-diphenyl-1-picrylhydrazyl (DPPH) solution in
MeOH and active compounds appeared after 30 min as yellow spots against a
purple background, and antifungal assay with spores of *Cladosporium cucumerinum* sprayed on the developed and dry plates, when inhibition zones indicate the
presence of the original fungitoxic substances after 2–3 days of incubation in a
moist atmosphere.

17.11 CONCLUSION

Research of therapeutical and toxicological properties of lignans has been very
intense in the last years (presently more than 1200 lignans and similar substances
are known). The reason is that the research of new medicines is returning to old
models, even in the field of natural substances in which it searches for (and finds)
new biologically, or, more precisely, therapeutically utilizable properties, is because
it possesses new experimental approaches, enabling the use of isolated cells, cellular
enzymes, and predictive mathematical models.

The second significant reason is because in the last years, markets in Europe
and the US have been supplied with nutritional supplements containing a number of
plants from traditional Oriental medicine (Kampoo, Ayurveda, etc.), about which
little is known in this field. It goes undoubted that they possess a beneficial biological
effect on human organism, they nevertheless must be observed.

Of the equal importance is research in the field of chemical ecology, which could
lead to the use of these substances (in low concentrations) as nontoxic pesticides,
because these substances are commonly present in nature, and there are no significant reports that they should disturb the ecological equilibrium.

Proteomics and genomics are presently quite significantly developed indeed,
therefore it might appear that the study of lignans as chemotaxonomic markers will
not bring any advancement into this field. Presently, it is not the case because even in
these modern fields is a number of empty spaces, and it will take a number of years
before they are filled.

From the aspect of TLC, lignans have been studied for more than 50 years. From
this synoptic paper it is clear that classical chromatographic analysis has already
reached its peak and new, revolutionary knowledge can hardly be expected in this

method. TLC can serve as a quick, routine, qualitative analytical method, which can selectively detect the presence of phenylpropane condensates in a number of cases, and also as a quantitative method for determination of particular lignans in particular plants—see Section 8. More profound analysis must nevertheless be seen in the use of modern instrumental methods, especially hyphenation techniques.

ACKNOWLEDGMENTS

The authors thank the Ministry of Education of the Czech Republic (2B06049) and the National Agency for Agricultural Research (NAZV 1646085) for financial support and to Dr. Veronika Opletalová for calculation of log *P* values (ChemDraw Ultra 7.0, CambridgeSoft Corp., Cambridge 2001) and comments to this review.

REFERENCES

1. Harmatha, J., Structural abundance and biological significance of lignans and related plant phenylpropanoids. *Chem. Listy*, 99, 622, 2005 (in Czech).
2. Umezawa, T., Diversity in lignan biosynthesis. *Phytochem. Rev.*, 2, 371, 2003.
3. Harmatha, J. and Dinan, L., Biological activities of lignans and stilbenoids associated with plant–insect chemical interactions. *Phytochem. Rev.*, 2, 321, 2003.
4. Apers, S., Vlietinck, A., and Pieters, L., Lignans and neolignans as lead compounds. *Phytochem. Rev.*, 2, 201, 2003.
5. Bruneton, J., *Pharmacognosy, Phytochemistry, Medicinal Plants*. Intercept Ltd., Andover, UK, 1995, p. 242.
6. Whiting, D.A., Lignans, neolignans and related compounds. *Nat. Prod. Rep.*, 4, 449, 1987.
7. Erdtman, H. and Harmatha, J., Phenolic and terpenoid heartwood constituents of *Libocedrus yateensis*. *Phytochemistry*, 18, 1495, 1979.
8. Jegorov, A., Flavono-lignans—Modern chemistry of medicinal plants known before Christ. *Chem. Listy*, 90, 859, 1996 (in Czech).
9. Ranganathan, K.R. and Seshadri, T.R., A new flavonolignan from *Hydnocarpus wightiana*. *Tetrahedron Lett.*, 14, 3481, 1973.
10. Sajeli, B., Sahai, M., Suessmuth, R., Asai, T., Hara, N., and Fujimoto, Y., Hyosgerin, a new optically active coumarinolignan, from the seeds of *Hyoscyamus niger*. *Chem. Pharm. Bull.*, 54, 538, 2006.
11. Nielsen, H. and Arends, P., Structure of the xanthonolignoid kielcorin. *Phytochemistry*, 17, 2040, 1978.
12. Ghosal, S., Saini, K.S., Razdan, S., and Kumar, Y., Chemical constituents of amaryllidaceae. XII. Crinasiatine, a novel alkaloid from *Crinum asiaticum*. *J. Chem. Res.*, 3, 100, 1985.
13. Willför, S., Hemming, J., Reunanen, M., Eckerman, C., and Holmbom, B., Lignans and lipophilic extractives in Norway spruce knots and stemwood. *Holzforschung*, 57, 27, 2003.
14. Chatterjee, A., Banerji, A., Banerji, J., Pal, S.C., and Ghosal, T., Recent advances in the chemistry of lignans. *Proc. Indian Acad. Sci.*, 93, 1031, 1984.
15. Hearon, W.M. and MacGregor, W.S., The naturally occurring lignans. *Chem. Rev.*, 55, 957, 1955.

16. Blaschek, W., Ebel, S., Hackenthal, E., Holzgrabe, U., Kellner, K., Reichling, J., Schnetz, V. et al., *Hager ROM 2004: Hager's Handbuch Der Drogen und Arzneistoffe*. Springer & Info II, Universität Würzburg, Würzburg, 2005.

17. Moss, G.P., Nomenclature of lignans and neolignans (IUPAC recommendations 2000). *Pure Appl. Chem.*, 72, 1493, 2000.

18. Westcott, N.D. and Muir, A.D., Flax seed lignan in disease prevention and health promotion. *Phytochem. Rev.*, 2, 401, 2003.

19. Ayres, D.C. and Loyke, J.D., *Lignans: Chemical, Biological and Clinical Properties*. Cambridge University Press, Cambridge, 1990, p. 113.

20. Muir, A.D. and Westcott, N.D., *The Genus Linum*. CRC Press, Boca Raton, FL, 2003, p. 320.

21. Thompson, L.U. and Cunnane, S.C. (eds.), *Flaxseed in Human Nutrition*. 2nd edn., AOCS Press, Champaign, IL, 2003.

22. Willför, S.M., Smeds, A.I., and Holmbom, B.R., Chromatographic analysis of lignans. *J. Chromatogr. A*, 1112, 64, 2006.

23. Opletal, L. and Opletalová, V., *Adaptogens of Natural Origin. Advances in Pharmacy No. 10*, Avicenum, Prague, 1990, p. 133 (in Czech).

24. Kochetkov, N., Khorlin, A.Ya., and Chizhov, O.S., Schizandrin—a lignan of unusual structure. *Tetrahedron Lett.*, 2, 730, 1961.

25. Stahl, E. and Schorn, P.J., II. TLC in the characterisation of animal and plant drugs, in: Stahl, E. (ed.), *Thin-Layer Chromatography. A Laboratory Handbook*. Springer Verlag, Berlin, Heidelberg, New York, 1969, p. 710.

26. Rüttimann, O. and Flück, H., Contributions to the determination of the value of podophyllin. *Helv. Chim. Acta*, 39, 417, 1964.

27. Stahl, E. and Kaltenbach, U., Thin-layer chromatography. IX. Rapid separation of mixtures of digitalis and podophyllum glycosides. *J. Chromatogr.*, 5, 458, 1961.

28. Smith, E.C.B. and Westall, R.G., Chromatographic behavior and chemical structure. I. Some naturally occurring phenolic substances. *Biochim. Biophys. Acta*, 4, 427, 1950.

29. Tura, D. and Robards, K., Sample handling strategies for the determination of biophenols in food and plants. *J. Chromatogr. A*, 975, 71, 2002.

30. Willför, S., Reunanen, M., Eklund, P., Sjöholm, R., Kronberg, L., Fardim, P., Pietarinen, S., and Holmbom, B., Oligolignans in Norway spruce and Scots pine knots and Norway spruce stemwood. *Holzforschung*, 58, 345, 2004.

31. Ekman, R., Analysis of lignans in Norway spruce by combined gas chromatography–mass spectrometry. *Holzforschung*, 30, 79, 1976.

32. Willför, S.M., Ahotupa, M.O., Hemming, J.E., Reunanen, M.H.T., Ecklund, P.C., Sjöholm, R.E., Eckerman, C.S.E., Pohjamo, S.P., and Holmbom, B.R., Antioxidant activity of knotwood extractives and phenolic compounds of selected tree species. *J. Agric. Food Chem.*, 51, 7600, 2003.

33. Willför, S., Hemming, J., Reunanen, M., Eckerman, C., and Holmbom, B., Lignans and lipophilic extractives in Norway spruce knots and stemwood. *Holzforschung*, 57, 27, 2003.

34. Ikeya, Y., Taguchi, H., Yosioka, I., and Kobayashu, H., The constituents of *Schizandra chinensis Baill*. I. Isolation and structure determination of five new lignans, gomisin A, B, C, F and G, and the absolute structure of schizandrin. *Chem. Pharm. Bull.* 27, 1383, 1979.

35. Willför, S., Hemming, J., Reunanen, M., and Holmbom, B., Phenolic and lipophilic extractives in Scots pine knots and stemwood. *Holzforschung*, 57, 359, 2003.

36. Kamal-Eldin, A., Yousif, G., and Appelqvist, L.A., Thin-layer chromatographic separations of seed oil unsaponifiables from four *Sesamum* species. *JAOCS*, 68, 844, 1991.

37. Mohamed, H.M.A. and Awatif, I.I., The use of sesame oil unsaponifiable matter as a natural antioxidant. *Food Chem.*, 62, 269, 1998.
38. Willför, S., Nisula, L., Hemming, J., Reunanen, M., and Holmbom, B., Bioactive phenolic substances in industrially important tree species, Part 1: Knots and stemwood of different spruce species. *Holzforschung*, 58, 335, 2004.
39. Willför, S., Nisula, L., Hemming, J., Reunanen, M., and Holmbom, B., Bioactive phenolic substances in industrially important tree species, Part 2: Knots and stemwood of fir species. *Holzforschung*, 58, 650, 2004.
40. Willför, S., Eklund, P., Sjöholm, R., Reunanen, M., Sillanpää, R., von Schoultz, S., Hemming, J., Nisula, L., and Holmbom, B., Bioactive phenolic substances in industrially important tree species, Part 4: Identification of two new 7-hydroxy divanillyl butyrolactol lignans in some spruce, fir, and pine species. *Holzforschung*, 59, 413, 2005.
41. Koulman, A., Bos, R., Medarde, M., Pras, N., and Quax, W.J., A fast and simple GC–MS method for lignan profiling in *Anthriscus sylvestris* and biosynthetically related plant species. *Planta Med.*, 67, 858, 2001.
42. Kawamura, F., Kawai, S., and Ohashi, H., Sesquilignans and lignans from *Tsuga heterophylla*. *Phytochemistry*, 44, 1351, 1997.
43. Takaku, N., Choi, D.H., Mikame, K., Okunishi, T., Suzuki, S., Ohashi, H., Umezawa, T., and Shimada, M., Lignans of *Chamaecyparis obtusa*. *J. Wood Sci.*, 47, 476, 2001.
44. Hiltunen, E., Pakkanen, T.T., and Alvila, L., Phenolic extractives from wood of birch (*Betula pendula*). *Holzforschung*, 58, 326, 2004.
45. Bártlová, M., Opletal, L., Chobot, V., and Sovová, H., Liquid chromatographic analysis of supercritical carbon dioxide extracts of *Schisandra chinensis*. *J. Chromatogr. B*, 770, 283, 2002.
46. Vasilev, N., Elfahmi, Bos, R., Kayser, O., Momekov, G., Konstantinov, S., and Ionkova, I., Production of justicidin B, a cytotoxic arylnaphthalene lignan from genetically transformed root cultures of *Linum leonii*. *J. Nat. Prod.*, 69, 1014, 2006.
47. Kawamura, F., Kawai, S., and Ohashi, H., Sesquilignans and lignans from *Tsuga heterophylla*. *Phytochemistry*, 44, 1351, 1997.
48. Chen, Y.Y., Chang, F.R., Wu, Y.C., Isoquinoline alkaloids and lignans from *Rollinia mucosa*. *J. Nat. Prod.*, 89, 904, 1996.
49. Tripathi, A.K., Verma, R.K., Gupta, A.K., Gupta, M.M., and Khanuja, S.P.S., Quantitative determination of phyllanthin and hypophyllanthin in *Phyllanthus* species by high-performance thin layer chromatography. *Phytochem. Anal.*, 17, 394, 2006.
50. Halls, S.G. and Lewis, N.G., Reversed-phase HPLC lignan chiral analysis with laser polarimetric detection. *Tetrahedron: Asymmetry*, 14, 649, 2003.
51. Bastos, J.K., Kopycki, W.J., Burandt, C.L., Nanayakkara, N.P.D., and McChesney, J.D., Quantitative determination of podophyllotoxin and related compounds in *Podophyllum* species by reversed-phase high-performance liquid chromatography. *Phytochem. Anal.*, 6, 101, 1995.
52. Jackson, D.E. and Dewick, P.M., Aryltetralin lignans from *Podophyllum hexandrum* and *Podophyllum peltatum*. *Phytochemistry*, 23, 1147, 1984.
53. Bevan, C.D. and Marshall, P.S., The use of supercritical fluids in the isolation of natural products. *Nat. Prod. Rep.*, 11, 451, 1994.
54. Modey, W.K., Mulholand, D.A., and Raynor, M.W., Analytical supercritical fluid extraction of natural products. *Phytochem. Anal.*, 7, 1, 1996.
55. Jarvis, A.P. and Morgan, E.D., Isolation of plant products by supercritical-fluid extraction. *Phytochem. Anal.*, 8, 217, 1997.

56. Suto, K., Ito, Y., Sagara, K., and Itokawa, H., Determination of magnolol and honokiol in Magnoliae Cortex using supercritical fluid chromatography on-line coupled with supercritical fluid extraction by on-column trapping. *J. Chromatogr. A*, 786, 366, 1997.
57. Lojková, L., Slanina, J., Mikešová, M., Táborská, E., and Vejrosta, J., Supercritical fluid extraction of lignans from seeds and leaves of *Schizandra chinensis*. *Phytochem. Anal.*, 8, 261, 1997.
58. Choi, Y.H., Kim, J., Jeon, S.H., Yoo, K.P., and Lee, H.K., High performance liquid chromatography–electrospray Ionization MS–MS analysis of *Forsythia koreana* fruits, leaves, and stems. Enhancement of the efficiency of extraction of arctigenin by use of supercritical-fluid extraction. *Chromatographia*, 57, 73, 2003.
59. Opletal, L., Sovová, H., and Bártlová, M., Dibenzo[a,c]cyclooctadiene lignans of the genus *Schisandra*: Importance, isolation and determination. *J. Chromatogr. B*, 812, 357, 2004.
60. Slanina, J., Taborska, E., and Lojkova, L., Lignans in the seeds and fruits of *Schisandra chinensis* cultured in Europe. *Planta Medica*, 63, 277, 1997.
61. Fernandes, A.M.A.P., Prado, A.L., Barata, L.E.S., Paulo, M.Q., Azevedo, N.R., and Ferri, P.H., A method to separate lignoids from *Virola* leaves. *Phytochem. Anal.*, 8, 18, 1997.
62. Sharma, T.R., Singh, B.M., Sharma, N.R., and Chauhan, R.S., Identification of high podophyllotoxin producing biotypes of *Podophyllum hexandrum Royle* from North-Western Himalaya. *J. Plant Biochem. Biotechnol.*, 9 (Part 1), 49, 2000.
63. Ford, J.D., Huang, K.S., Wang, H.B., Davin, L.B., and Lewis, N.G., Biosynthetic pathway to the cancer chemopreventive secoisolariciresinol diglucoside-hydroxymethyl glutaryl ester-linked lignan oligomers in flax (*Linum usitatissimum*) seed. *J. Nat. Prod.*, 64, 1388, 2001.
64. Kamal-Eldin, A., Peerlkamp, N., Johnsson, P., Andersson, R., Andersson, R.E., Lundgren, L.N., and Åman, P., An oligomer from flaxseed composed of secoisolariciresinoldiglucoside and 3-hydroxy-3-methyl glutaric acid residues. *Phytochemistry*, 58, 587, 2001.
65. Mazur, W. and Adlercreutz, H., Naturally occurring oestrogens in foods. *Pure Appl. Chem.*, 70, 1759, 1998.
66. Johnsson, P., Kamal-Eldin, A., Lundgren, L.N., and Åman, P., HPLC method for analysis of secoisolariciresinol diglucoside in flaxseeds. *J. Agric. Food Chem.*, 48, 5216, 2000.
67. Coran, S.A., Giannellini, V., and Bambagiotti-Alberti, M., High-performance thin-layer chromatographic–densitometric determination of secoisolariciresinol diglucoside in flaxseed. *J. Chromatogr. A*, 1045, 217, 2004.
68. Örså, F. and Holmbom, B., A convenient method for the determination of wood extractives in papermaking process waters and effluents. *J. Pulp Pap. Sci.*, 20, J361, 1994.
69. Slanina, J. and Glatz, Z., Separation procedures applicable to lignan analysis. *J. Chromatogr. B*, 812, 215, 2004.
70. Cutillo, F., D'Abrosca, B., DellaGreca, M., Fiorentino, A., and Zarrelli, A., Lignans and neolignans from *Brassica fruticulosa*: Effects on seed germination and plant growth. *J. Agric. Food Chem.*, 51, 6165, 2003.
71. Cavin, A., Potterat, O., Wolfender, J.L., Hostettmann, K., and Dyatmyko, W., Use of on-flow LC/1H NMR for the study of an antioxidant fraction from *Orophea enneandra* and isolation of polyacetylene, lignans, and tocopherol derivative. *J. Nat. Prod.*, 61, 1497, 1998.
72. Shen, Y.C., Liaw, C.C., Cheng, Y.B., Ahmed, A.F., Lai, M.C., Liou, S.S., Wu, T.S., Kuo, Y.H., and Lin, Y.C., C18 dibenzocyclooctadiene lignans from *Kadsura philippinensis*. *J. Nat. Prod.*, 69, 963, 2006.
73. Jackson, D.E. and Dewick, P.M., Tumour-inhibitory aryltetralin lignans from *Podophyllum pleianthum*. *Phytochemistry*, 24, 2407, 1985.

74. Dekebo, A., Lang, M., Polborn, K., Dagne, E., and Steglich, W., Four lignans from *Commiphora erlangeriana. J. Nat. Prod.*, 65, 1252, 2002.
75. Chang, F.R., Chao, Y.C., Teng, C.M., and Wu, Y.C., Chemical constituents from *Cassytha filiformis* II. *J. Nat. Prod.*, 61, 863, 1998.
76. Barton, G.M., Thin-layer chromatography of guaiacylpropane monomers, selected lignans and phenolic wood extractives. *J. Chromatogr.*, 26, 320, 1967.
77. Chen, I.S., Chen, T.L., Chang, Y.L., Teng, C.M., and Lin, W.Y., Chemical constituents and biological activities of the fruit of *Zanthoxylum integrifoliolum. J. Nat. Prod.*, 62, 833, 1999.
78. Elfahmi, Batterman, S., Koulman, A., Hackl, T., Bos, R., Kayser, O., Woerdenbag, H.J., and Quax, W.J., Lignans from cell suspension cultures of *Phyllanthus nirruri*, an Indonesian medicinal plant. *J. Nat. Prod.*, 69, 55, 2006.
79. Lee, S.S., Lin, M.T., Liu, C.L., Lin, Y.Y., and Liu, K.C.S.C., Six lignans from *Phyllanthus myrtifolius. J. Nat. Prod.* 59, 1061–1065, 1996.
80. Susplugas, S., Hung, N.V., Bignon, J., Thoison, O., Kruczynski, A., Sevenet, T., and Gueritte, F., Cytotoxic arylnaphthalene lignans from a Vietnamese Acanthaceae, *Justicia patentiflora. J. Nat. Prod.*, 68, 734, 2005.
81. Khatoon, S., Rai, V., Rawat, A.K.S., and Mehrotra, S., Comparative pharmacognostic studies of three *Phyllanthus* species. *J. Ethnopharmacol.*, 104, 79, 2006.
82. Ikeya, Y., Taguchi, H., and Yosioka, I., The constituents of *Schizandra chinensis Baill.* XII. Isolation and structure of a new lignan, gomisin R, the absolute structure of wuweizisu C and isolation of achisantherin D. *Chem. Pharm. Bull.*, 30, 3207, 1982.
83. Moazzami, A.A., Andersson, R.E., and Kamal-Eldin, A., HPLC analysis of sesaminol glucosides in sesame seeds. *J. Agric. Food Chem.*, 54, 633, 2006.
84. Schumacher, B., Scholle, S., Hölzl, J., Khudeir, N., Hess, S., and Müller, C.E., Lignans isolated from valerian: Identification and characterization of a new olivil derivative with partial agonistic activity at A_1 adenosine receptors. *J. Nat. Prod.*, 65, 1479, 2002.
85. Kamal-Eldin, A. and Appelqvist, L.A., Variations in the composition of sterols, tocopherols and lignans in seed oils from four *Sesamum* species. *JAOCS*, 71, 149, 1994.
86. Vasilev, N., Nedialkov, P., Ionkova, I., and Ninov, S., HTPLC densitometric determination of justicidin B in *Linum in vitro* cultures. *Pharmazie*, 59, 528, 2004.
87. Chen, C.C., Hsin, W.C., Ko, F.N., Huang, Y.L., Ou, J.C., Teng, C.M., Antiplatelet arylnaphthalide lignans from *Justicia procumbens. J. Nat. Prod.*, 59, 1149, 1996.
88. Mishra, N., Acharya, N., Gupta, A.P., Singh, B., Kaul, V.K., and Ahuja, P.S., A simple microanalytical technique for determination of podophyllotoxin in *Podophyllum hexandrum* roots by quantitative RP-HPLC and RP-HPTLC. *Curr. Sci.*, 88, 1372, 2005.
89. Zhu, Y., Yan, K., and Tu, G., Chemical studies on Shengmai San. I. Quantitative determination of active ingredients of Schizandra in Shengmai San by TLC–densitometry. *Yaowu Fenxi Zazhi*, 8, 71, 1988 (in Chinese).
90. Greger, H., Pacher, T., Vajrodaya, S., Bacher, M., and Hofer, O., Infraspecific variation of sulfur-containing bisamides from *Aglaia leptantha. J. Nat. Prod.*, 63, 616, 2000.
91. Trifunovich, S., Vajs, V., Teshevich, V., Diokovich, D., and Miloslavievich, S., Lignans from the plant species *Achillea lingulata. J. Serb. Chem. Soc.*, 68, 277, 2003.
92. Silva, D.H.S., Pereira, F.C., Zanoni, M.V.B., and Yoshida, M., Lipophyllic atioxidants from *Iryanthera juruensis* fruits. *Phytochemistry*, 57, 437, 2001.

Secondary Metabolites— Isoprenoids

18 TLC of Mono- and Sesquiterpenes

Angelika Koch, Simla Basar, and Rita Richter

CONTENTS

18.1 DEFINITION AND CHEMICAL CLASSIFICATION OF MONO- AND SESQUITERPENES

Terpenes are derived biosynthetically from units of isoprene, which has the molecular formula C_5H_8. The in vivo precursors are isopentenyl pyrophosphate and its isomeric dimethylallyl pyrophosphate. In biosynthesis the isoprene units are linked together to build up the skeletons of the terpenes as a multiple of $(C_5H_8)_n$ [1]. The isoprene units may be linked together head to tail or tail to tail (Figure 18.1).

Although differing in their chemical constitution, mono- and sesquiterpenes have many physical properties in common. The boiling points vary from 140°C–180°C (monoterpenes) to 240°C (sesquiterpenes). Other physical characteristics are densities from 0.759 to 1.096 and high refractive indices. Most of those terpenes are optically active and specific rotation is often a valuable diagnostic property in mixtures of many constituents.

Monoterpenes consist of two isoprene units and have the molecular formula $C_{10}H_{16}$. A further C_5 (isoprene) unit form the C_{15} sesquiterpenes. More of these C_5 units give the diterpenes (C_{20}), sesterpenes (C_{25}), triterpenes (C_{30}), tetraterpenes (C_{40}), and polyterpenes ($C > 40$).

With regard to the monoterpenes, these substances can be further divided into three groups depending on whether they are acyclic, monocyclic, or bicyclic.

Within each group, the monoterpenes may be unsaturated or may have functional groups and be alcohols (e.g., linalool, menthol), aldehydes (e.g., phellandral), or ketones (e.g., carvenone, verbenone, and thujone) (Figure 18.2).

The majority of natural terpenoids have cyclic structures with or without functional groups, sometimes they have aromatic structures (e.g., thymol).

Also included with monoterpenes on biosynthetic reasons are the monoterpene lactones, known as iridoids (Chapter 23 deals with iridoids).

Like the monoterpenes, the sesquiterpenes are grouped according to their basic carbon skeleton. The common ones are either acyclic, monocyclic, or bicyclic (Figure 18.3).

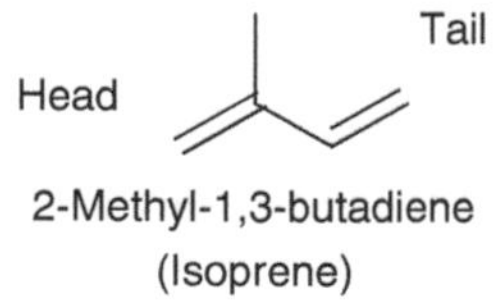

Head   ⟋⟍⟋   Tail

2-Methyl-1,3-butadiene
(Isoprene)

FIGURE 18.1 Structure of the isoprene unit.

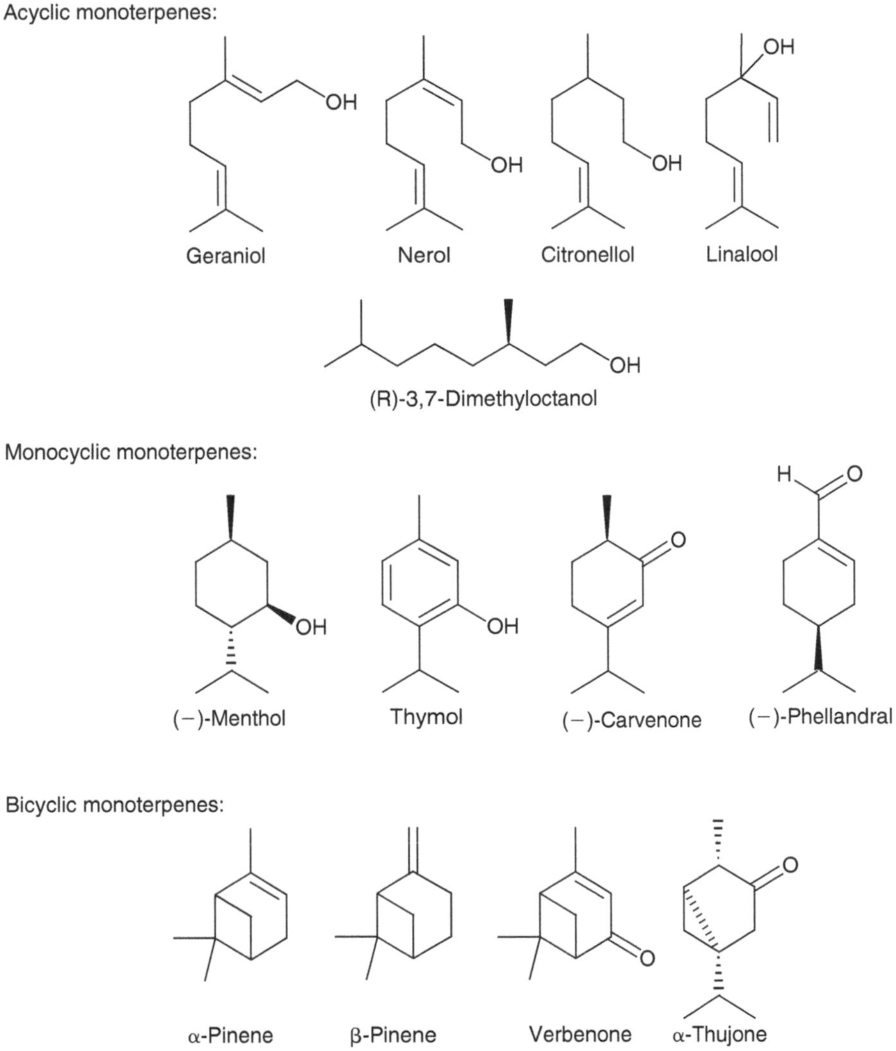

Acyclic monoterpenes:

Geraniol   Nerol   Citronellol   Linalool

(R)-3,7-Dimethyloctanol

Monocyclic monoterpenes:

(−)-Menthol   Thymol   (−)-Carvenone   (−)-Phellandral

Bicyclic monoterpenes:

α-Pinene   β-Pinene   Verbenone   α-Thujone

FIGURE 18.2 Classification of monoterpenes.

Some of the volatile compounds are components of the essential oils and serve as starting material for fragrances and also pharmaceuticals.

Isomerism is common among terpenoids (e.g., geraniol, nerol, see Figure 18.2); the stereochemistry is often difficult to determine. The cyclohexane ring is usually twisted in the chair form, different conformations occur, depending on the substitution of the ring.

Isomerization and structural rearrangement within the molecule occur quite readily and even under relatively mild conditions. That is why artifact formation is always possible during isolation procedures. The mono- and sesquiterpenes occur as glycosides, too (see section 18.6).

Acyclic sesquiterpenes:

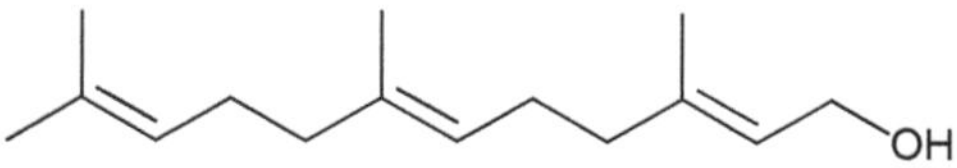

Farnesol

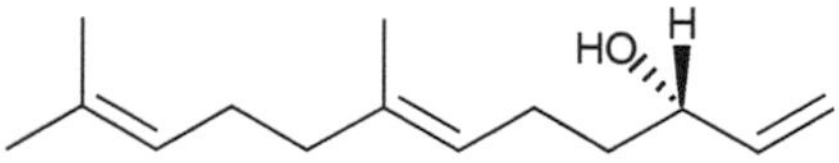

(S)-(+)-Nerolidol

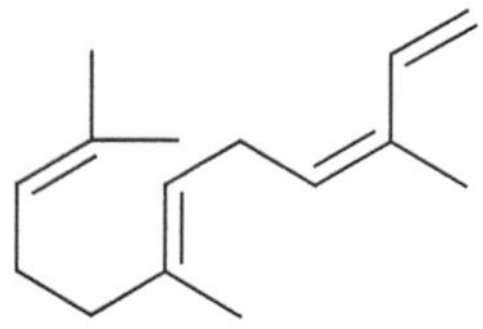

(Z, E)-α-Farnesene

Monocyclic sesquiterpenes:

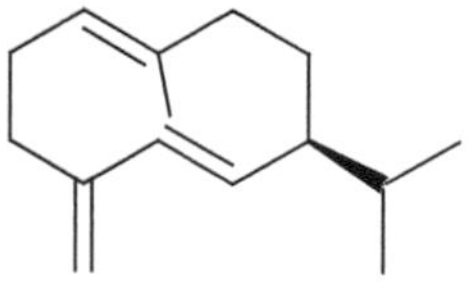

(−)-Germacrene D 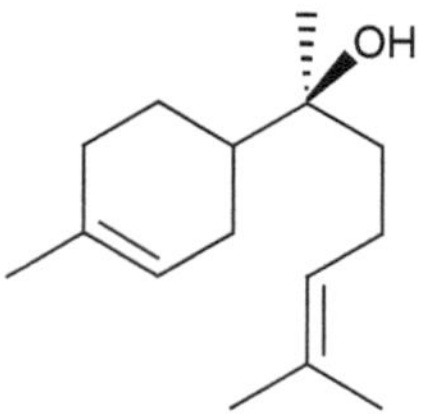   α-Bisabolol

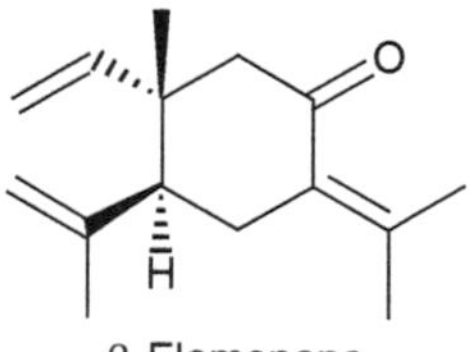

β-Elemenone

Bicyclic sesquiterpenes:

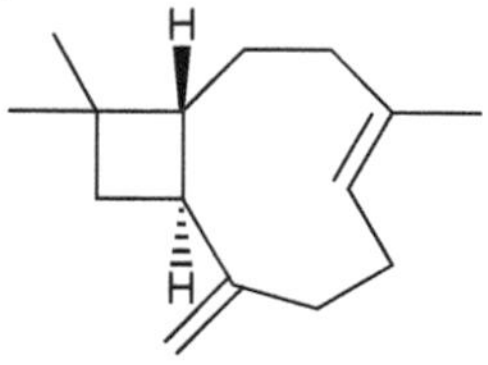

(E)-β-Caryophyllene 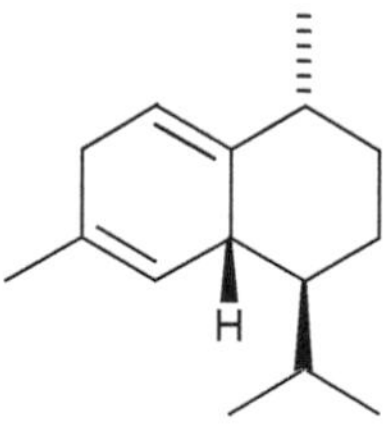   Cadina-1,4-diene

FIGURE 18.3 Classification of sesquiterpenes.

18.2 OCCURRENCE AND IMPORTANCE OF VOLATILE COMPOUNDS IN THE PLANT KINGDOM

Chemically, according to their structure, terpenes are generally lipid-soluble. They are located in the cytoplasm of the plant cell. The volatile essential oils sometimes occur in special glandular cells located at the surface of leaves, flowers, fruits, and seeds. The terpenes of the resins (when dissolved in essential oils they are called balsams) or oils are located in excretion or resin channels of the bark or wood of stems or roots.

The volatile fraction plays a fundamental role in regulating the interactions of a plant with its surrounding environment. More frequently, these fractions are known to have antimicrobial activity. This activity could act as a chemical defence against plant pathogenic diseases.

Terpenes may possess growth-regulating properties (e.g., abscisin), they are agents of protection and defense against phytopathogens and influence interactions between plants and animals (pollination, pheromone).

Volatile terpenes are ubiquitous in the plant kingdom but some families produce a lot in the glandular cells, e.g., Apiaceae, Lamiaceae, and Rutaceae.

18.3 ESSENTIAL OILS

Definition of essential oils in medicine and pharmacy: "volatile, strong smelling essential oils comprise the volatile steam-distillable fraction mainly of terpenoids. Other classes of chemical substances may be present, e.g., aromatic compounds such as phenylpropanoids or coumarins."

The sense of smell and taste are impressions caused by contact of chemical substances with a peripheral receptor system. Flavors dissolved in the saliva stimulate the sensory nerves located in the tongue. Fragrances belong to volatile components and reach the receptors of the olfactory organ by breathing through the nose.

Essential oils are widely used as natural flavor additives for food, as fragrances in perfumery, aromatherapy, and in traditional and alternative medicines. Synthetic variations and derivatives of natural terpenes and terpenoids also greatly expand the variety of aromas used in perfumery and flavors used in food additives.

According to ISO (International Standard Organization) essential oils are of plant origin and are obtained by hydrodistillation or squeezed out of the peels of some citrus fruits.

Terpenes or terpenoids are the primary constituents of the essential oils of many types of plants and flowers. According to their vapor pressure components of essential oils are volatile.

Simple monoterpenes are widespread in the plant kingdom and tend to occur as components of the majority of essential oils. Sometimes some compounds are found in easily detectable amounts, whereas many other terpenoic compounds occur only in trace amounts. There is, however, difficulty in detecting that on microscale.

18.4 SAMPLING OF ESSENTIAL OILS OR VOLATILE COMPOUNDS

The sampling of terpenes depends on their chemical composition. Whereas resins are mostly obtained through incision of the bark (e.g., olibanum, myrrh, and pinus), the classic procedure of sampling the volatile components (essential oils) is the separation from fresh tissue by steam distillation using a Clevenger-like apparatus [2,3].

Artifact formation (dehydration or polymerization) may often occur [4]. For this reason the terpenes (especially the valuable ones, e.g., *Oleum Rosae*, attar of roses) are extracted with pentane, hexane, diethylether, chloroform, or dichloromethane, respectively, or supercritical CO_2.

An ancient sampling and conservation procedure, described by Theophrast 370 BC, is to extract the essential oils with fatty oils. This method is still practiced in Grasse (France), in the center of perfumery, which is called enfleurage when the essential oil is extracted without heat, and maceration when it is extracted with thermal energy. Cold pressed agrumen oils are often obtained from peels of certain citrus fruits.

Headspace is a very mild sampling method for small amounts and is most often connected with gas chromatography/mass spectrometry (GC/MS).

Kubeczka [5,6] introduced a device for special purposes, e.g., sampling small amounts of volatile components for coupling GC with TLC or vice versa.

In the 1960s, a special sample application method was introduced by Stahl [7] for sampling and analyzing small amounts of volatile compounds in drug samples. This method was called TAS, which means Thermomicro Separation and Application of Substances by a special device. This device consists of the TAS furnace (DESAGA Company) and the accessory glass cartridges. By using this thermomicro procedure it is possible to carry out sampling, transfer, and separation as an on-line operation.

The TAS furnace possesses an opening to insert the glass cartridges. These special glass cartridges are sealed with a clamp at one end and at the other end they narrow to a cone point which is directed toward the TLC plate. Small sample amounts of fresh or dried plants are tranferred into this glass cartridge. The vaporization (steam distillation) occurs at temperatures up to 220°C caused by the natural humidity of the samples or by aid of small amounts of starch or moistened silica gel. The vaporized extracts are condensed onto the starting zone of the TLC plate.

In 1977 van Meer and Labadie [8] combined different techniques in the analysis of valepotriates. The authors presented a TAS/TLC separation method of valtrate and isovaltrate on silica gel as stationary phase and dichloromethane–ethyl acetate–acetone $(48 + 1 + 1, \text{v/v})$ as mobile phase. By coupling TAS/TLC it was possible to determine the presence of the isomers valtrate/isovaltrate by their degradated analogs baldrinale and homobaldrinale.

18.5 TLC ANALYSIS OF MONO- AND SESQUITERPENES

18.5.1 SEPARATION CONDITIONS

Fundamentals and recommended TLC techniques of terpene analysis are given by Stahl and Jork [9].

According to small differences in their permitivity ε(2.24–2.76) terpene hydrocarbons are difficult to separate by normal adsorption techniques. But in 1961, Battaile et al. [10] provided a variety of systematic optimization schemes for selecting the mobile phase components and their relative concentrations and the most selective reagents for separation and detection of terpenes. These trials resulted in the most valuable mobile phase composition Skelly-solve B (=hexane)–ethyl acetate $(95 + 5)$

which provided satifactory separation of the peppermint oil terpenes. Spraying with a 4% trichloroacetic acid–chloroform (w/v) solution yielded an initial pink color of menthofuran at room temperature which turned to red. Very diluted sprays of potassium permanganate–sulfuric acid detected the unsaturated terpenes, e.g., piperitenone, limonene, phellandrene, as yellow or white spots on a purple background of unreduced permanganate. An aqueous solution of 0.05% Rhodamine B detected menthol, menthone, menthyl acetate, cineole as red spots against a light pink background. The contrast increased with further applications of the spray.

In 1965, another group of scientists, Attaway et al. [11] considered TLC as a very sensitive technique for identifying volatile organic compounds separated from natural products. Using specific color producing sprays they had been able to make some conclusions regarding their structure. They also considered TLC as a powerful separatory tool, and compounds have been assigned to particular chemical classes on the basis of their R_f values using solvent systems of varying polarity. Attaway concluded that the TLC analysis of terpene hydrocarbons required the use of nonpolar solvents, e.g., volatile hydrocarbon solvents (pentane, hexane). They observed that the corresponding perfluorinated hydrocarbons, in which all hydrogens have been replaced by fluorine, also had nonpolar characteristics.

Using the fluorocarbon solvents, 19 monoterpene hydrocarbons tested had R_f values of zero, while all 11 sesquiterpenes tested had significant difference in R_f values. These values showed sufficient variation to be combined with color reaction data and used to identify individual compounds. In contrast to their behavior with the fluorocarbon solvents, the monoterpenes and sesquiterpenes showed considerable overlap in their R_f values when hexane was used, and no distinction could be made between the two terpene classes.

The two adsorbents, silica gel and alumina, gave significantly different results. When using hexane, the adsorbent activity of alumina was too low to effect good separation of individual terpenes and, consequently, the use of silica gel was compulsory. However, when using the fluorocarbon solvents, the high adsorbent activity of the silica gel prevented the sesquiterpenes from ascending a sufficient distance for separation, and the less active alumina was preferred. They concluded that the fluorocarbon solvents and the alumina adsorbent were extremely specific for the analysis of sesquiterpene hydrocarbons.

Some individual separation problems of mono- and sesquiterpenes could be solved taking advantage of the different values of steam pressure as it is proposed by Demole [12]. A distinction of the separated mono- and sesquiterpenes can be achieved by covering the first plate (with the separated constituents of the essential oil) with a second (new) plate like a sandwich. There must be a space of about 1 mm between both layers. The first plate is slightly warmed up—preferably under reduced pressure—until the monoterpenes are vaporized and adsorbed by the second plate. This procedure seems like a tricky fractionation of mono- and sesquiterpenes.

A technical problem during TLC separation of volatile terpenes is the evaporation of the volatile solvent at room temperature. In 1964, Mathis and Ourisson [13] studied different conditions for separating mono- and sesquiterpenes by TLC with silica gel as adsorbent and hexane as mobile phase. When they developed hydrocarbons at room temperature they received only weak separations. Operating at low

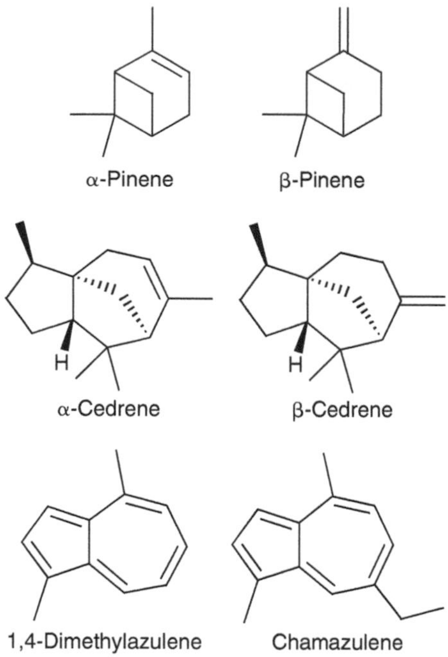

α-Pinene   β-Pinene

α-Cedrene   β-Cedrene

1,4-Dimethylazulene   Chamazulene

FIGURE 18.4 Isomers of pinene and cedrene, examples of azulene sesquiterpenes.

temperatures (−15°C) they achieved separations of saturated hydrocarbons ($R_f > 90$), mono- and sesquiterpenes with one double bond ($R_f = 0.85\text{--}0.65$), mono- and sesquiterpenes with two double bonds ($R_f = 0.65\text{--}0.45$) and mono- and sesquiterpenes with three double bonds ($R_f = 0.45\text{--}0.25$). Even when testing pairs of isomers (Figure 18.4) they received good results: α- and β-pinene ($R_f = 0.78$ and 0.72), α- and β-cedrene ($R_f = 0.72$ and 0.67). The isomers with the exocyclic methylene group reveal the lower R_f value (Table 18.1).

TABLE 18.1

R_f Values of Mono- and Sesquiterpenes on a Silica Gel Plate at −15°C Using Hexane as Mobile Phase

Type of Compound	R_f Value
Saturated hydrocarbons	>0.90
Mono- and sesquiterpenes with one double bond	0.85–0.65
Mono- and sesquiterpenes with two double bonds	0.45–0.25
Constitutional isomers like α-pinene	0.78
β-Pinene	0.72
α-Cedrene	0.72
β-Cedrene	0.67

Source: From Mathis, C. and Ourisson, G., *Phytochemistry*, 3, 115, 1964.

TABLE 18.2

R_f Values of Oxygenated Terpenes on a Silica Gel Plate at Room Temperature Using Hexane–Ethyl Acetate (9 + 1) as Mobile Phase

Type of Compound	R_f Value
Hydrocarbons	>0.90
Epoxides, esters	0.80–0.60
Aldehydes, ketones	0.60–0.40
Alcohols	0.40–0.10
Acids	No migration

Source: From Mathis, C. and Ourisson, G., *Phytochemistry*, 3, 115, 1964.

Testing oxygenated terpenes Mathis and Ourisson changed the mobile phase into hexane–ethyl acetate (9 + 1). When developing at room temperature they received the following separations: hydrocarbons: $R_f > 0.90$, epoxides and esters of terpenes: $R_f = 0.80–0.60$, aldehydes and ketones: $R_f = 0.60–0.40$, alcohols: $R_f = 0.40–0.10$, and acids which actually showed no migration (Table 18.2).

The detection of the separated compounds was carried out by antimony chloride. The initial red color reaction which turns to a dirty green color is specific for monoterpenes and their oxygenated derivatives, whereas their reactive aldehydes give different colors. Sometimes, these colors allow a differentiation of terpenes with a parent skeleton. Sesquiterpenes with an azulene skeleton (see Figure 18.4) appear as blue spots when treated with bromic or nitrogen vapor. 2,4,-Dinitro-phenylhydrazine serves as a reagent for terpenes with an aldehyde or ketone group.

18.5.2 AgNO₃-Impregnated TLC Layers

The impregnation of TLC layers with silver cations has gained considerable interest [14]. The separation takes place according to the degree of unsaturation of the compound, in the separation of *cis-trans*-isomers and in the separation of positional isomers. *Cis*-olefinic structures, complex more tightly with silver ions than do the *trans*-isomers. As a consequence, *cis*-isomers show lower R_f values than do the *trans*-isomers. The R_f data of unsaturated compounds on silver-impregnated layers point to linear relationship with the amount of silver ions present in the layer.

Sonwa and König [15] made profit from the separation power of AgNO₃ when they studied minor constituents of the essential oil of *Cyperus rotundus*. They isolated three new sesquiterpene hydrocarbons. However, the essential oil was first separated at low temperature (–20°C) by column chromatography into a hydrocarbon and an oxygenated fraction. The hydrocarbon fraction was then separated by TLC using petroleum ether as mobile phase in order to eliminate the major hydrocarbon compounds, cyperene and α-copaene, from the interesting unknown compounds of the third band. Further fractionation of these compounds by preparative GC and analysis by GC/MS yielded three major products: α-selinene, valencene, and an

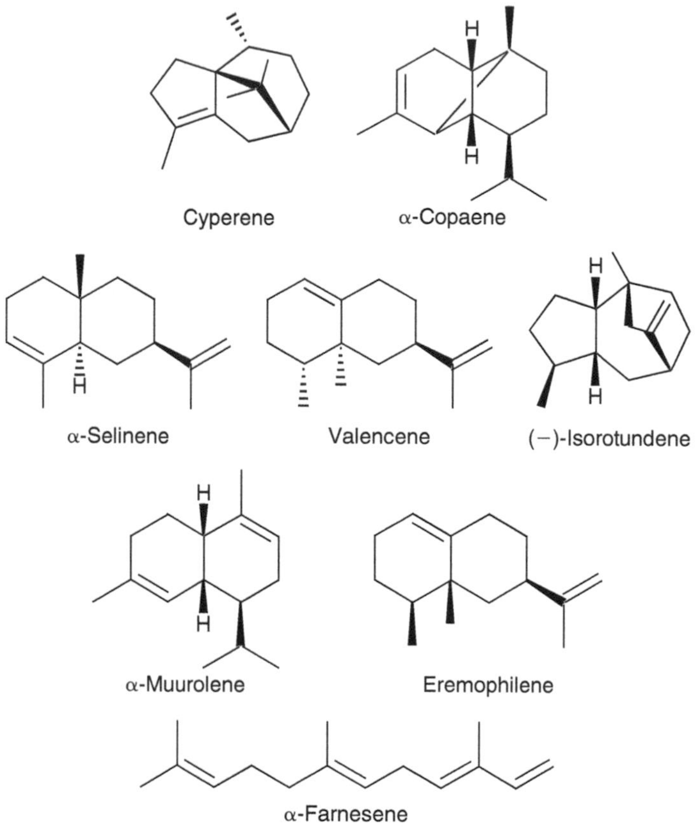

Cyperene   α-Copaene

α-Selinene   Valencene   (−)-Isorotundene

α-Muurolene   Eremophilene

α-Farnesene

FIGURE 18.5 Structures of sesquiterpenes which are separated on silica gel impregnated with AgNO$_3$.

unknown product. Further separation of the unknown product was achieved by preparative TLC on AgNO$_3$ precoated plates using petroleum ether as mobile phase resulting in two bands. One of these bands was purified by preparative GC and identified by spectroscopic means as (−)-isorotundene (Figure 18.5).

The above mentioned AgNO$_3$ precoated plates were prepared by immersing silica 60 F$_{254}$ (Merck) plates into an ethanol–water (4 + 1) solution of AgNO$_3$ (5%). After 30 min the plates were removed from the solution and dried (in an oven). Sulfuric acid (10% in ethanol) was used as a spray reagent.

The hydrocarbon fractions of 30 virgin olive oils were analyzed by Bortolomeazzi et al. [16], focusing in particular on the sesquiterpenes. The hydrocarbon fraction was isolated by silica gel column chromatography of the unsaponifiable fraction of the oils. Accordingly, the sesquiterpene hydrocarbons were then fractionated by AgNO$_3$ TLC and silica gel AgNO$_3$ column on the basis of their degree of unsaturation. Among the sesquiterpenes the more abundant were α-farnesene, α-copaene, eremophilene, and α-muurolene (see Figure 18.5).

TABLE 18.3
Chromatography of Unsaturated Volatile Compounds
on Silver-Impregnated Layers

Composition of Layer	Mobile Phase	Separated Compounds
Silica gel + 5% AgNO$_3$ [15]	Petroleum ether	(−)-Isorotundene
Silica gel + 10% AgNO$_3$ [16]	n-Hexane	α-Farnesene, α-copaene, eremophilane, α-muurolene
Silica gel + 3% AgNO$_3$ [17]	Methanol–water (93 + 7)	Monoterpene alcohols

The development was carried out with n-hexane, and the bands were visualized under UV$_{254}$ light after spraying with a 0.2% (w/v) ethanolic solution of 2,7-dichlorofluorescein. The bands were scrapped off and extracted with dichloromethane. The separated and extracted bands were purified for further analysis and identification by GC/MS.

The earlier mentioned AgNO$_3$ plates were prepared by dipping the plates for 1 min into a mixture of 10% (w/v) AgNO$_3$ solution/water and ethanol in 1:1 (v/v) ratio. The plates were dried at 70°C for 20 min.

Separation of monoterpene alcohols (e.g., menthol, geraniol, citronellol, and linalool, see Figure 18.2) were successfully performed on small AgNO$_3$-impregnated TLC (5 × 5 cm) plates [17]. The plates were prepared by dipping for 3 min into a solution of 3% AgNO$_3$ in methanol–water (93 + 7). The plates were dried at 100°C and protected from light (Table 18.3).

18.6 TERPENE GLYCOSIDES

Phenomena like the increase in essential oil content during storage of essential oil drugs gave rise to the suspicion that terpenes glycosides did occur. Early workers have postulated that glycosidically bound terpenes are involved in essential oil metabolism and possibly serve as a transport form of these essential oil compounds.

A number of plant species have been found to contain monoterpene β-D-glycosides. Francis and Allcock [18] studied in 1969 on the accumulation of free and bound monoterpenes during maturation of the flowers of the Hybrid Tea rose Lady Seton and described the detailed isolation and chemical characterization but neither succeeded in giving indications of their metabolic interrelationship nor did they find a correlation of the concentration of free monoterpenes relative to monoterpene β-D-glycosides. The bound monoterpenes separated from the nonsteam volatile residues of rose petals were liberated by adding 11 N HCl. Another more gentle method of estimating bound monoterpenes was dissolving the residues in sodium acetate buffer and incubating them with β-glucosidase. At least the free monoterpenes were extracted with petroleum ether and estimated by GC and TLC.

Acid hydrolysis of the mixed β-D-glucosides yielded a monoterpene fraction which contained several components not observed in the corresponding fraction obtained by enzymatic hydrolysis. These differences were undoubtedly due to the well-known unstability (or lability) of monoterpenes toward acids.

A correlation of bound and free terpenes was (partly) realized by preparative thin-layer electrophoresis on cellulose plates (CC-41) with 0.005 M borate buffer for about 2 h at 500 V and 200 mA. The developed bands of bound monoterpenes were scraped off, extracted, and purified. The identity of the purified bound monoterpenes was confirmed by TLC analysis compared with authentic material. The major component appeared to be geraniol-β-D-glycoside with smaller concentrations of nerol- and citronellol glycosides. The chromatographic systems to detect terpene glycosides comprised silica gel layers and different mobile phases, e.g., ethanol–ethyl acetate $(15 + 85)$, n-propanol–ethyl acetate–water $(6 + 3 + 1)$ and chloroform–acetic acid–methanol $(85 + 2 + 13)$. To analyze the terpene aglycons they referred to the methods of Battaile et al. [10] (see section 18.5.1).

Zhao et al. [19] reported the isolation and structural determination of five novel sesquiterpene glycosides in the roots of *Dictamnus dasycarpus* used in traditional Chinese medicine. The MeOH extract of the root bark containing the glycosides was subjected to CC on silica gel for fractionation and isolation. TLC was simply used for monitoring the completion of the hydrolysis of the glycosides named dictamnosides. The sugar components were identified by TLC on silica gel with ethyl acetate–methanol–water–acetic acid $(13 + 3 + 3 + 4)$. Again, glucose was found as the only sugar component.

Stahl-Biskup and coworkers [20,21] turned to the topic again. Apart from the usual methods like GC and GC/MS they tried to allocate the released aglycons to the genuine glycosides by introducing two-dimensional TLC development [22,23].

The samples of fresh plant material of thyme were ground in liquid nitrogen and extracted with ethyl acetate. Free volatile terpene aglycons and other nonpolar components were removed by washing with hexane. The refined sample extract was developed with the mobile phase ethyl acetate–chloroform–water–formic acid $(82 + 8 + 5 + 5, v/v)$. After drying, the sample track was exposed to HCl vapor for approximately 15–30 min at 60°C–80°C and after cooling and removal of the HCl vapor the development in the second dimension with the mobile phase hexane–ethyl acetate $(80 + 20, v/v)$ was performed. Detection of the free aglycons was carried out with anisaldehyde reagent. By this, the aglycons could be assigned to the corresponding glycosides.

18.7 APPLICATION OF TLC FOR IDENTIFICATION OF PHARMACEUTICAL DRUGS

18.7.1 VOLATILE COMPOUNDS OF ESSENTIAL OILS ACCORDING TO MONOGRAPHS

Nowadays, to analyze volatile compounds, GC/MS is the method of choice. Today the pharmacopoeias still provide TLC methods for identification of herbal drugs and essential oils. Unfortunately, the majority of these methods do not display recent developments in TLC techniques.

Table 18.4 gives an overview of the herbal monographs listed in the current European Pharmacopoeia [2] which admit TLC methods as proof of identity. It must be mentioned that attention is often directed not to the mono- and sesquiterpenes but

TABLE 18.4
TLC Conditions of Officially Monographed Herbal Drugs with Essential Oil Content. Silica Gel Always Serves as Sorbent Layer

No	Drug Sample, Latin Binomials Synonyms	Mobile Phase/ Development Distance	Detection	Reference Compounds	Essential Oil Content/Identity/Remarks
1	Anisi fructus *Pimpinella anisum* L. Apiaceae Anise	Ethyl acetate–toluene (7 + 93, v/v) 10 cm toluene 10 cm	1. UV$_{254}$ nm 2. PMA-R	RS: anethole	2%–6% Essential oil anethole
2	Anisi aetheroleum	Ethyl acetate–toluene (7 + 93, v/v) 15 cm/6 cm	1: UV$_{254}$ nm 2. MAB-R 3. PMA-R	RS: linalool, anisaldehyde, anethole	Purity/adulteration (Japanese star anise oil, Shikimi oil) See section 18.9.1
3	Anisi stellati fructus *Illicium verum* Hook. fil Illiciaceae Star anise	Toluene 10 cm	AS	RS: anethole	7% Essential oil
4	Anisi stellati aetheroleum	Ethyl acetate–toluene (7 + 93, v/v) 15 cm/6 cm	1. UV$_{254}$ nm 2. MAB-R	RS: linalool, anisaldehyde, anethole	
5	Aurantii amari floris aetheroleum Nerolii aetheroleum *Citrus aurantium* L. subsp. *aurantium* subsp. *amara* Engl. Rutaceae Oil of neroli	Ethyl acetate–toluene (15 + 85, v/v) 15 cm/8 cm	1. UV$_{365}$ nm 2. AS-R 3. UV$_{365}$ nm	RS: anthranile methylate, linalool, linalyl acetate, bergapten	0.2%–0.5% Essential oil exclusion of bergapten (adulteration with fruit peel oil)

(continued)

TABLE 18.4 (continued)

TLC Conditions of Officially Monographed Herbal Drugs with Essential Oil Content. Silica Gel Always Serves as Sorbent Layer

No	Drug Sample, Latin Binomials Synonyms	Mobile Phase/ Development Distance	Detection	Reference Compounds	Essential Oil Content/Identity/Remarks
6	Aurantii dulcis aetheroleum *Citrus sinensis* (L.) Osbeck syn. *C. aurantium* L. var. *dulcis* L.	Ethyl acetate–toluene (15 + 85, v/v) 15 cm	1. UV$_{365}$ nm 2. AS–R 3. UV$_{365}$ nm	RS: bergapten, linalool, linalyl acetate	Fruit peel oil
7	Carvi fructus *Carum carvi* L. Apiaceae Caraway fruits	Ethyl acetate–toluene (5 + 95, v/v) 10 cm	1. UV$_{254}$ nm 2. AS–R	RS: (+)-carvone	3% Essential oil carvone
8	Carvi aetheroleum	Ethyl acetate–toluene (5 + 95, v/v) 10 cm/5 cm	1. UV$_{254}$ nm 2. AS–R	RS: (+)-carvone, carveol	
9	Caryophylli flos *Syzygium aromaticum* (L.) Merrill et. L.M. Perry, syn. *Eugenia caryophyllus* (C. Spreng) Bull. et. Harry Myrtaceae Cloves	Toluene 2 × 10 cm	1. UV$_{254}$ nm 2. AS–R	RS: eugenol	14%–15% Essential oil phenylpropanes: eugenol, eugenol acetate sesquiterpene: β-caryophyllene
10	Caryophylli floris aetheroleum	Toluene 2 × 10 cm	1. UV$_{254}$ nm 2. AS–R	RS: eugenol, eugenol acetate	Eugenol, eugenol acetate, caryophyllene
11	Cinnamomi zeylanici cortex *Cinnamomum zeylanicum* Nees syn. *C. verum* J. Presl Lauraceae Ceylon cinnamon	Dichloromethane	1. UV$_{254}$ nm 2. UV$_{365}$ nm 3. PH–R	RS: cinnamic aldehyde, eugenol	1.2% essential oil hydroxymethoxy- and methoxy cinnamic aldehyde, eugenol

No.	Drug	Solvent system	Detection	RS/LS	Remarks
12	Cinnamomi zeylanici corticis aetheroleum Ceylon cinnamon oil	Methanol–toluene (10 + 90, v/v) 15 cm	AS–R	RS: cinnamic aldehyde, eugenol, linalool, and caryophyllene	
13	Cinnamom cassia aetheroleum Cinnamomum cassia Blume syn. C. aromaticum Nees Chinese or cassia cinnamon	Methanol–toluene (10 + 90, v/v) 15 cm	1. UV$_{365}$ nm 2. AS–R	RS: trans-cinnamic aldehyde, eugenol, coumarin	
14	Citronellae aetheroleum Cymbopogon winterianus Jowitt Rutaceae	Ethyl acetate–toluene (10 + 90, v/v) 15 cm	1. AS–R 2. UV$_{365}$ nm	RS: citronellal	Citronellal (30%–45%), citronellol-geraniol (20%–25%)
15	Colophonium Pinus species resins Pinaceae	Dichloromethane 15 cm	AS–R	LS: thymol, linaloole	Compare the position of red colored bands to the position of the lead compounds
16	Coriandri fructus Coriandrum sativum L. Apiaceae Coriander fruits	Ethyl acetate–toluene (5 + 95, v/v) 10 cm	AS–R	RS: linalool	0.3% Essential oil (incl. 65%–78% linalool) fingerprint of monoterpenes
17	Coriandri aetheroleum	Ethyl acetate–toluene (5 + 95, v/v) 10 cm	See: 16	RS: linalool, geranyl acetate	Linalool, geranyl acetate and geraniol
18	Curcumae xanthorrhizae rhizoma Curcuma xanthorrhiza Roxb., D. Dietrich Zingiberaceae Turmeric	Glacial acetic acid–toluene (20 + 80, v/v) 10 cm	1. DCC-R 2. AS–R 3. UV$_{365}$	LS: fluorescein, thymol	6%–11% Essential oil xanthorrhizol, curcumin, demethoxycurcmin Purity: exclusion of bisdemethoxycurcumines
19	Eucalypti folium Eucalyptus globulus Labill. Myrtaceae Tasmanian Blue	Ethyl acetate–toluol (10 + 90, v/v) 15 cm	AS–R	Cineole	2% Essential oil (incl. 70% 1,8-cineole)

(continued)

TABLE 18.4 (continued)

TLC Conditions of Officially Monographed Herbal Drugs with Essential Oil Content. Silica Gel Always Serves as Sorbent Layer

No	Drug Sample, Latin Binomials Synonyms	Mobile Phase/ Development Distance	Detection	Reference Compounds	Essential Oil Content/Identity/Remarks
20	Eucalypti aetheroleum *Eucalyptus polybractea* R.T. Baker, *E. smithii* R.T. Baker	Ethyl acetate–toluol (10 + 90, v/v) 15 cm	AS–R	Cineole	
21	Foeniculi amari fructus Foeniculum vulgare Miller, ssp. *vulgare*, var. *vulgare* Apiaceae Bitter fennel	Hexane–toluene (20 + 80, v/v) 10 cm	1. UV$_{254}$ nm 2. SA: 140°C	RS: anethole, fenchone	4% Essential oil (min. 60% anethole and 15% fenchone) max. 5% estragol (=methyl chavicol) of the oil
22	Foeniculi dulcis fructus *Foeniculum vulgare* Miller, ssp. *vulgare*, var. *dulce* (Miller) Thellung Sweet fennel	Hexane–toluene (20 + 80, v/v) 10 cm	SA: 140°C	RS: anethole	2% Essential oil min 80% anethole max. 5% estragol (=methyl chavicol) of the oil
23	Foeniculi amari fructus aetheroleum	Ethyl acetate–toluol (5 + 95, v/v) 15 cm	PMA–R	RS: anethole, fenchone	55.0%–75% *trans*-anethole and 12%–25% fenchone
24	Iuniperi pseudo-fructus *Iuniperus communis* L. Cupressaceae Juniper berries	Ethyl acetate–toluene (5 + 95, v/v) 15 cm	AS–R	LS: guaiazulene, cineole	1.0% Essential oil terpineol, terpinene-4-ol

No.	Drug/source/family	Mobile phase	Detection	Reference compounds	Remarks
25	Iuniperi aetheroleum	Ethyl acetate–toluene (5 + 95, v/v) 12 cm	AS–R	RS: terpineol, terpinene-4-ol	
26	Lavandulae flos *Lavandula angustifolia* P.Mill. syn. *L. officinalis* Chaix Lamiaceae Lavender	Ethyl acetate–toluene (5 + 95, v/v) 15 cm	AS–R	RS: linalool, linalyl acetate	1.3% Essential oil linalool, linalyl acetate, and epoxydihidrocaryophyllene
27	Lavandulae aetheroleum	Ethyl acetate–toluene (5 + 95, v/v) 2 × 10 cm, dry between: 5 min	AS–R	RS: linalool, linalyl acetate	Linalool, linalyl acetate, and epoxy-dihidrocaryophyllene
28	Limonis aetheroleum *Citrus limon* (L.) Burman fil. Rutaceae Lemom oil	Ethyl acetate–toluene (15 + 85, v/v) 15 cm	UV_{254} and UV_{365} nm	RL: citroptene, citral	Citroptene and citral
29	Mastix (resin) *Pistacia lentiscus* L. var. *latifolius* Coss. Anacardiaceae	Petrol ether–toluene (5 + 95, v/v) 10 cm	VS–R	LS: eugenol, borneol	1% Essential oil, fingerprint of terpenes
30	Matricariae flos *Matricaria recutica* L. syn. *Chamomilla recutica* (L.) Rauschert Asteraceae German chamomile	Ethyl acetate–toluene (5 + 95, v/v) 10 cm	AS–R	RS: chamazulene, (−)-α-bisabolol LS: bornyl acetate	0.4% Essential oil; steam distillation: genuin matrizin (proazulene) converts to chamazulene (blue); bisabolon, bisabolol oxide A and B
31	Matricariae aetheroleum	Ethyl acetate–toluene (5 + 95, v/v) 10 cm	AS–R	RS: chamazulene, (−)-α-bisabolol LS: guaiazulene, bornyl acetate	

(continued)

TABLE 18.4 (continued)
TLC Conditions of Officially Monographed Herbal Drugs with Essential Oil Content. Silica Gel Always Serves as Sorbent Layer

No	Drug Sample, Latin Binomials Synonyms	Mobile Phase/ Development Distance	Detection	Reference Compounds	Essential Oil Content/Identity/Remarks
32	Melaleucae aetheroleum Melaleuca linariifolia *Smith* *M. dissitiflora* F. Mueller *M. alternifolia* (Maiden and Betch) Myrtaceae Tea tree	Ethyl acetate–heptane (20 + 80, v/v) 10 cm	AS–R	RS: cineole, terpinene-4-ol, α-terpineol	Min. 30% terpinene-4-ol cineole, terpinene-4-ol, α-terpineol
33	Melissae folium *Melissa officinalis* L. Lamiaceae Lemon Balm, Balm	Ethyl acetate–hexane (10 + 90, v/v) 15 cm	AS–R	RS: citronellal, citral	0.03% Essential oil citral, citronellal, and epoxy-caryophyllene
34	Menthae piperitae folium Mentha × piperita L. Lamiaceae Peppermint	Ethyl acetate–toluene (5 + 95, v/v) 15 cm	1. UV$_{254}$ nm 2. AS–R	RS: menthol, cineole, menthyl acetate LS: thymol	0.9%–1.2% Essential oil UV$_{254}$ nm: carvone, pulegone menthol, carvone, pulegone, isomenthon, cineole, menthone, and menthyl acetate
35	Menthae arvensis aetheroleum partim mentholi privum and Menthae arvensis aetheroleum partim mentholi depletum	Ethyl acetate–toluene (5 + 95, v/v) 15 cm	1. UV$_{254}$ nm 2. AS–R	Carvone, pulegone, menthyl acetate, cineole, menthol	Menthone 17%–35% Menthol 30%–50% Pulegone, max. 2.5%
36	Menthae piperitae aetheroleum	Ethyl acetate–toluene (5 + 95, v/v) 15 cm	1. UV$_{254}$ nm 2. AS–R	RS: menthol, cineole, menthyl acetate LS: thymol	

No.	Drug/Plant	Solvent system	Detection	Reference	Notes
37	Myristica fragrants aetheroleum *Myristica fragrans* Houtt Myristicaceae Nut meg	Ethyl acetate–toluene (5 + 95, v/v) 15 cm	VS–R	RS: myristicin	Max. 2.5% safrole
38	Myrrha *Commiphora molmol* Engler Burseraceae Myrrh Resin, Gum myrrh	Ethyl acetate–toluene (2 + 98, v/v) 15 cm	1. UV$_{365}$ 2. AS–R	LS: thymol, anethole	2%–10% Essential oil, Purity:UV$_{365}$: no fluorescence in lower part, exclusion of *C. mukul* Id.: violet zone of furanoeudesma-1,3-dien
39	Origanum herba *Origanum onites* L., *O. vulgare* L. ssp. *hirtum* (Link) Ietsw. Lamiaceae	Dichlormethane 15 cm	AS–R	RS: thymol, carvachrol	2.5% Essential oil, min 1.5% thymol, and carvacrol
40	Rosmarini folium *Rosmarinus officinalis* L. Lamiaceae Rosemary	Ethyl acetate–toluene (5 + 95, v/v) 15 cm	AS–R	RS: borneol, bornyl acetate, and cineole	1.2% Essential oil borneol, bornyl acetate, and cineole
41	Rosmarini aetheroleum	Ethyl acetate–toluene (5 + 95, v/v) 15 cm	VS–R	RS: borneol, bornyl acetate, and cineole	Borneol, bornyl acetate, and cineole
42	Salviae trilobae folium *Salvia fruticosa* Mill, syn. *S. triloba* L. fil. Lamiaceae Greek sage	Ethyl acetate–toluene (5 + 95, v/v) 15 cm	PMA–R	RS: α- and β-thujone, cineole	1.8% Essential oil α- and β-thujone, cineole
43	Salviae officinalis folium *Salvia officinalis* L. Lamiaceae Sage, Dalmatian sage	Ethyl acetate–toluene (5 + 95, v/v) 15 cm	PMA–R	RS: α- and β-thujone, cineole	1.5% Essential oil α- and β-thujone, cineole

(continued)

TABLE 18.4 (continued)

TLC Conditions of Officially Monographed Herbal Drugs with Essential Oil Content. Silica Gel Always Serves as Sorbent Layer

No	Drug Sample, Latin Binomials Synonyms	Mobile Phase/ Development Distance	Detection	Reference Compounds	Essential Oil Content/Identity/Remarks
44	Salviae sclareae aetheroleum *Salvia sclarea* L. Lamiaceae	Ethyl acetate–toluene (5 + 95, v/v) 15 cm	VS-R	RS: linalool, linalyl acetate, and α-terpineol	Essential oil, incl. 56%–78% linalyl acetate
45	Serpylli herba *Thymus serpyllum* L. s.l. Lamiaceae Wild thyme	Dichloromethane 15 cm	1. UV$_{254}$ nm 2. AS–R	RS: thymol, carvacrol	0.3% Essential oil 1. thymol 2. thymol, carvacrol
46	Terebinthinae aetheroleum e pino pinastro *Pinus pinaster* Aiton Pinaceae Turpentine oil	Ethyl acetate–toluene (5 + 95, v/v) 15 cm	AS–R	RS: β-pinene LS: linalool	Essential oil from the resin, incl. 70%–85% α-Pinene
47	Thymi herba *Thymus vulgaris* L., *T. zygis* L. Lamiaceae Thyme	Dichloromethane 15 cm	1. UV$_{254}$ nm 2. AS–R	RS: thymol, carvacrol	1.2% Essential oil 1. thymol 2. thymol, carvacrol, cineole, linalool, and borneol

48	Thymi aetheroleum	Ethyl acetate–toluene (5 + 95, v/v) 15 cm	AS–R		RS: thymol, carvacrol, linalool, and α-terpineol	
49	Valerianae radix *Valeriana officinalis* L. s.l. Valerianaceae Valerian root	Glacial acetic acid–ethyl acetate–hexane (0.5 + 35 + 65; v/v) 10 cm	AS–R		SS: 1g/6 mL methanol evaporate to 2 mL, dissolve in 3mL KOH (10%) (hydrolyzation of acetoxy valerenic acid) LS: fluorescein und sudanred G	0.3% Essential oil, 0.17% sesquiterpene acids (valerenic acid)
50	Zingiberis rhizoma *Zingiber offizinale* Roscoe Zingiberaceae Ginger	Hexane–ether (40 + 60, v/v) 15 cm	1. UV$_{254}$ nm 2. VS–R		LS: citral, resocinol	1.5% Essential oil; gingerols (genuin), shogaols (=degradation product of gingerols), *sesquiterpenes:* α-zingiberene, bisabolene, and curcumene

Source: European Pharmacopoeia 5th ed., incl. Supplements, 2005.

Abbreviations: AS, Anisaldehyde reagent; DCC, Dichloroquinone reagent; LS, Lead substances; MAB, Methyl (4-acetylbenzoat) reagent; PH, Phloroglucinol–HCl reagent; PMA, Phosphomolybdic acid reagent; RS, Reference solutions; SA, Sulfuric acid; SS, Sample solutions; VS, Vanillin–H$_2$SO$_4$ reagent.

to the likewise occuring phenylpropanes. The systems toluene–ethyl acetate (95 + 5, v/v) [2] and toluene–ethyl acetate (93 + 7, v/v) provided by Wagner and Bladt [24] have been proven to be suitable for the analysis of most important essential oils (Table 18.4).

General methods of detection include spraying with 0.2% aqueous $KMnO_4$, 5% antimony chloride in chloroform, concentrated H_2SO_4, or vanillin– or anisaldehyde–H_2SO_4. The plates are heated after spraying at 100°C–105°C until full development of colors has occurred.

More selective agents for detecting terpenes are mentioned in the appropriate sections.

18.7.2 MONO- AND SESQUITERPENES AS PHARMACOLOGICALLY ACTIVE COMPOUNDS

Sesquiterpene lactones often occur in plants of the Asteraceae family in glandular trichomes on the leaf, flower, or seed.

The TLC separation of these compounds is generally conducted on silica gel in more polar mobile phases than used for the separation of the volatile components of essential oils [25], e.g., chloroform–ether (4 + 1), chloroform–methanol (99 + 1), light petroleum–chloroform–ethyl acetate (2 + 2 + 1), benzene–acetone (4 + 1), benzene–methanol (9 + 1), benzene–ether (2 + 3).

The formerly usual solvents chloroform and benzene must be replaced by dichloromethane and toluene, respectively.

The lactones are detected as brown spots by placing the developed plates in a chamber containing iodine crystals. Alternatively, the lactones appear as colored spots on plates sprayed with concentrated H_2SO_4, and heated for about 5 min at 100°C–110°C. The color produced can be diagnostic of certain structural features in the lactone.

18.7.2.1 Tanacetum Parthenium and Chrysanthemum Cinerariaefolium

Valued since the times of Dioscorides (78 AD) as a febrifuge, the leaves of *Tanacetum parthenium* (L.) Schultz Bip. (Syn. *Chrysanthemum parthenium* (L.) Bernh.) with the common name feverfew have now been shown to be useful in reducing the frequency and severity of migraine. The concentration of the main active principal parthenolide, a sesquiterpene lactone, varies widely in different feverfew samples. Standardization of the herbal material on the basis of its parthenolide content is urgently required.

For the proof of the parthenolides, TLC separation on silica gel plates in acetone–toluene (15 + 85) is recommended by the French Pharmacopoeia [26]. Detection is carried out by the standard procedure as, e.g., spraying with vanillin–H_2SO_4 reagent and heating at 100°C for about 5 min to develop a blue color visible at daylight.

Another suggestion [27] comprises a TLC separation on silica gel plates in chloroform–acetone (9 + 1). The R_f value is specified by 0.78 and the spray reagents were solutions of 1% methanolic resorcinol–5% phosphoric acid (1 + 1). After applying the reagent the layer must be heated at 110°C for 2–4 min to show a cherry color for the parthenolides.

Another simple TLC method to separate and identify parthenolides is carried out on silica gel 60 F_{254} in chloroform–methanol $(100 + 1)$ [28]. The sample is extracted by water and purified by taking the parthenolides into chloroform. TLC separation revealed the parthenolides which are detected as dark violet spots after spraying with sulfuric acid–water $(1 + 1)$ and heating for 90 sec at 160°C.

Pyrethrum is closely related to *Tanacetum parthenium* because they belong to the same family (Asteraceae). Formerly, the white *Chrysanthemum* flower, pyrethri flos, of the species *Chrysanthemum cinerariaefolium* Trevir, Schultz, Bip could be found for sale in most European pharmacies as Dalmatian Insect Powder. The primary uses for the pyrethrum flowers were for the control of body lice on humans and animals, and crawling insects in the home [29]. Papers on the chemistry of pyrethrum showed that the insecticidal properties were due to the presence of two monoterpene esters named pyrethrin I and pyrethrin II. Limited supplies of the flowers have encouraged the development and use of synthetic alternatives, especially outside the USA, but none of these have all the unique properties of natural pyrethrum (Figure 18.6).

The main use of pyrethrum at present is in household formulations to kill houseflies, cockroaches, and mosquitoes. There is a general agreement that if supplies of high quality pyrethrum extract were available, the present market would expand in many areas.

Stahl and Pfeifle [30] achieved a TLC separation of pyrethrin I, cinerin I, and jasmolin I on silica gel HF_{254} with the mobile phase hexane–ethyl acetate $(95 + 5)$. Pyrethrin II, cinerin II, and jasmolin II on the other hand are separated by the solvent system hexane–heptane–ethyl acetate $(48 + 40 + 12)$. The pyrethrins were recognized

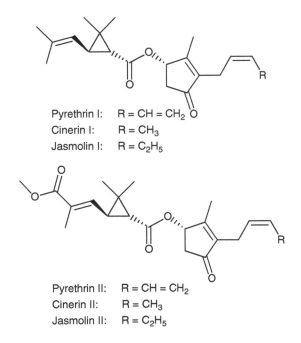

Pyrethrin I: R = CH = CH$_2$
Cinerin I: R = CH$_3$
Jasmolin I: R = C$_2$H$_5$

Pyrethrin II: R = CH = CH$_2$
Cinerin II: R = CH$_3$
Jasmolin II: R = C$_2$H$_5$

FIGURE 18.6 Pyrethrins of *Chrysanthemum cinerariaefolium* with insecticidal properties.

as unstable when exposed to UV light. The detection with anisaldehyde–H$_2$SO$_4$ reagent allowed a differentiation of pyrethrins (grey-black), cinerins, and jasmolins (brown), whereas using antimony(III)-chloride reagent only pyrethrins give a brown color reaction.

A special TAS extraction of pyrethrum flowers was presented in 1981 by Stahl and Schild [31]. For analysis they used silica gel F$_{254}$ plates and as mobile phase hexane–ethyl acetate (90 + 10) without chamber saturation; the development distance was 7 cm. The pyrethrins, cinerins, and jasmolins can be detected in the short UV light. Their R_f values are determined as pyrethrin II: 18–30, pyrethrin I: 65–75, cinerin I: 75–80, and jasmolin I: 80–85.

In 1972, Balbaa et al. [32] presented a TLC separation of pyrethrin I ($R_f = 0.68$) and pyrethrin II ($R_f = 0.40$) on silica gel plates and cyclohexane–ethyl acetate (4 + 1) as mobile phase and detection with Williams reagent (phosphoric acid–acetic acid anhydride, 80 + 20). This color reagent was specific for pyrethrins in giving a pink color after heating at 100°C.

18.7.2.2 Artemisia Annua

Artemisia annua L. (Asteraceae) is a highly aromatic herb and is widely distributed throughout the temperate region and belong to the plants high in demand, as a source for antimalaria medicines. Artemisinin the valuable component of this plant is a sesquiterpene lactone with an endoperoxide bridge and used for the treatment of chloroquine- and piperaquine-resistent malaria (Figure 18.7).

Gabriels and Plaizier-Vercammen [33] analyzed artemisinin (AT) and its lipophilic derivatives artemether (AM) and arteether (AE) using silica gel 60 F$_{254s}$ (Merck) plates and chloroform as mobile phase. After development the plates were

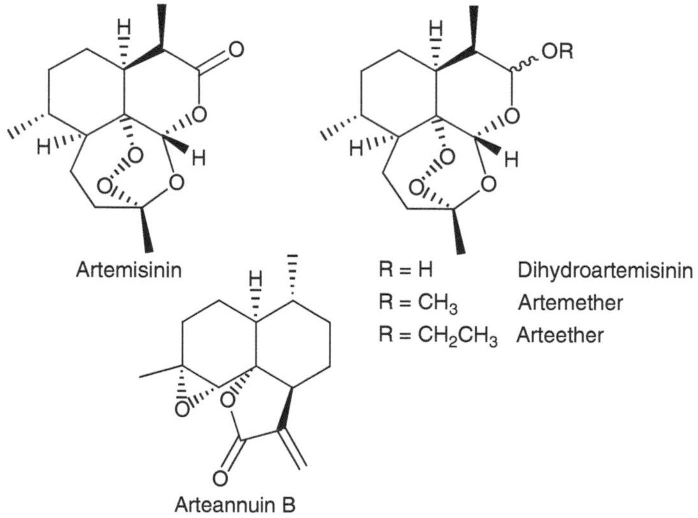

FIGURE 18.7 Active sesquiterpene lactones of *Artemisia annua* for the treatment of malaria.

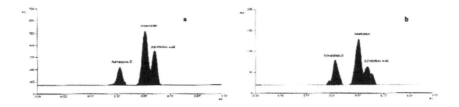

FIGURE 18.8 RP-TLC densitograms of (A) standard artemisinic acid, artemisinin, and arteannuin-B (600,300,150 ng); (B) sample solution (5 mL). (From Bhandari, P. et al., *J. Sep. Sci.*, 28, 2288, 2005. With permission.)

dipped into 4-methoxy benzaldehyde reagent and heated at 110°C for about 8 min. The substances were visualized presenting a purple color (AT) in the daylight and a fluorescent orange color under UV_{366}, whereas AE and AM gave a dark gray color in the daylight and also a fluorescent orange under UV light. The dipping reagent was evaluated in terms of stability of the color, precision, and accuracy. Quantification of the colored substances was carried out by densitometric measurement at 366 nm. Good linearity was obtained in the range of 0,5–8 μg for all analytes.

The same authors presented [34] a further study (2004) on the separation of artemisinin analogs by using reversed-phase TLC. The developed TLC method is carried out on RP-18 plates and acetonitrile–water $(50 + 25, v/v)$ as the mobile phase. Visualization was done by an acidified 4-methoxy benzaldehyde reagent in methanol–water. This method allowed the separation of compounds that have versatile hydrophilic/lipophilic characteristics, namely artemisinin, artesunate, artelinic acid, arteether, both isomers of artemether (α and β), dihydroartemisinin and desoxyartemisinin (see Figure 18.7). The separation of some degradation products and impurities (down to 2%) allows quality and stability control of all pharmacological active compounds in the raw material of *Artemisia annua* and the corresponding pharmaceutical formulations as well. Again, the quantification was carried out by densitometric measurement at 366 nm.

Simultaneous densitometric determination of artemisinin, artemisinic acid, and arteannuin-B was performed by Bhanbdari et al. [35] on RP-18 F_{254s} TLC plates in 0.2% TFA in water–acetonitrile $(35 + 65, v/v)$ as mobile phase (Figure 18.8). The quantitative determination of artemisinin and its derivatives was carried out after derivatization with anisaldehyde reagent at 426 nm in the absorption–reflectance mode by external standards of the above mentioned compounds. The authors recommend this method to be applied for the quantitative analysis of a larger number of samples of *Artemisia annua*.

18.8 APPLICATION OF TLC FOR TAXONOMIC AND CULTIVATION MEANS

18.8.1 *TUSSILAGO FARFARA* AND *PETASITES* SPECIES

Stahl and Schild (1981) [31] describe a method for identifying adulterations of Farfarae folium (*Tussilago farfara* L.) with species of *Petasites* by aid of the marker

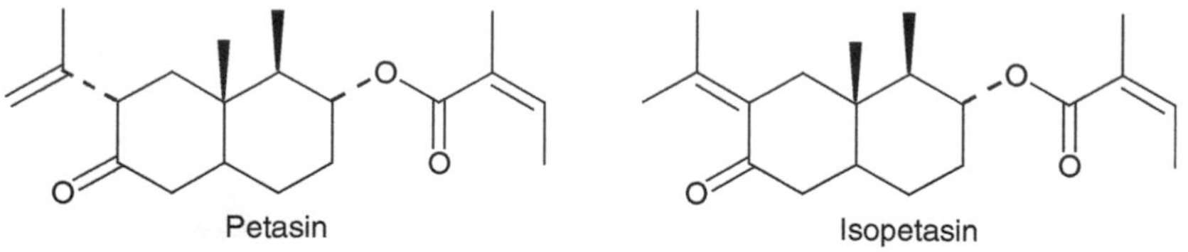

FIGURE 18.9 Active compounds of *Petasites* species to treat migraine attacks.

substance petasin, a sesquiterpene ester, which is not present in *Tussilago*. Petrol ether extracts of the leafs were applied on the sorbent layer (silica gel) and developed in chloroform as mobile phase (development distance 12 cm). For visualization the plate was sprayed with anisaldehyde–H_2SO_4 reagent. Petasin ($R_f = 0.25$) and isopetasin ($R_f = 0.3$) occured as yellow-orange spots or bands after derivatizing in the daylight and under UV_{365} light as light green spots.

Wagner and Bladt [24] suggest another system for differentiation of leaves of *Tussilago farfara* and different *Petasites* species. Petrol ether extracts of leaves are chromatographed in ethyl acetate–formic acid–acetic acid–water ($100 + 11 + 11 + 27$) on silica gel TLC plates. Fluorescence quenching (254 nm) indicates the presence of petasin ($R_f = 0.4$) and isopetasin ($R_f = 0.47$) which is confirmed after spraying and heating with anisaldehyde–H_2SO_4 reagent. Under UV_{365} light petasin and isopetasin (Figure 18.9) show a greenish fluorescence and a faint yellow color in the daylight.

Sesquiterpene esters are the active principles in the medicinal plant *Petasites hybridus* (L.) Gaertn., with the common name butter bur. Two chemotypes, the petasin chemotype and the furanopetasin chemotype are known, but only the petasin chemotype is suitable for pharmaceutical purposes to treat migraine attacks. Chizzola et al. [36] performed experimental crossings within and between plants of both chemotypes to study the genetic basis of the occurrence of these sesquiterpenes. The chemotype was determined by TLC in extracts. For this the dichloromethane extract of the dried plant powder was applied to silica gel 60 F_{254} TLC plates (Merck) and the plate was developed with xylene–acetone ($9 + 1$). The spots were visualized by dipping in anisaldehyde reagent. After heating the petasin chemotype gave a yellow to light purple color.

18.8.2   *Achillea Millefolium* L.

Achillea millefolium L., Asteraceae, is known for its content of proazuelene. A Hungarian taxon of the *Achillea millefolium* group differs in the composition of sesquiterpenes. Five guaianolides of the Hungarian taxon including two pairs of isomers, were characterized analytically [37].

Different TLC systems were developed for the analysis of these compounds. TLC of leucodin, 8α-angeloxy leucodin, achillin, 8α-angeloxy achillin, and desacetylmatricarin was performed on silica gel using two different mobile phases, (I) dichloromethane–acetone ($90 + 10$) and (II) cyclohexane–ethylacetate ($50 + 50$). In mobile phase I the five reference compounds revealed only three spots (UV_{254})

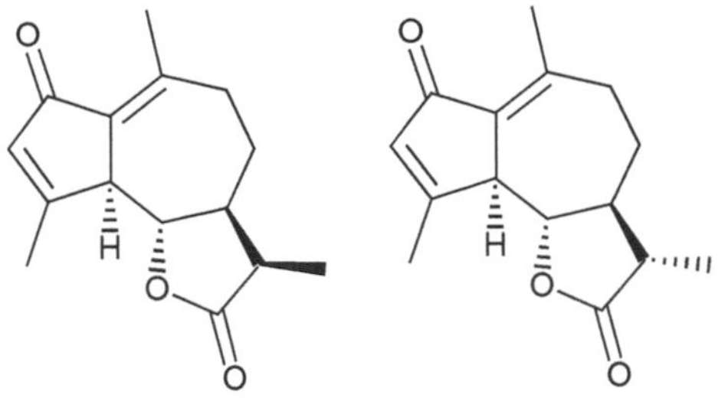

Achillin   Leucodin

FIGURE 18.10 Sesquiterpenes (guaianolides) as lead substances of *Achillea* species.

since the separation of the isomers of leucodins and achillins was not achieved. But separation in mobile phase II resulted in $R_f = 0.51$ for achillin and $R_f = 0.59$ for leucodin. Changing the composition of mobile phase I into dichloromethane–aceton $(95 + 5)$ allowed a separation of 8α-angeloxy achillin ($R_f = 0.45$) and 8α-angeloxy leucodin. These systems provide a good separation of the stereomers achillin/ leucidin and their 8α-angeloxy derivatives, respectively, which are marker substances for certain nonproazulene containing species of *Achillea* (Figure 18.10).

18.9 APPLICATION OF TLC TO DEMONSTRATE ADULTERATIONS OF DISTINCT VOLATILE OILS

18.9.1 ANISE OIL/FENNEL OIL—CHINESE STAR ANISE/JAPANESE STAR ANISE OIL

The European Pharmacopoeia provides a TLC method for identification of anise oil (Anisi aetheroleum, see Table 18.4). TLC plates coated with silica gel F_{254} are developed with toluene–ethyl acetate $(93 + 7)$ and the spots are visualized after spraying or dipping with molybdophosphoric acid reagent and heating (100°C). Contaminations by essential fennel oil may be proven by detecting fenchone. For this purpose the hot plate has to be sprayed again with $KMnO_4$–sulfuric acid reagent to give a light blue color [24].

Pimpinella anisum L. and *Illicium verum* Hook. F (Chinese star anise) are admissible species as a source of the essential oil of anise (Anisi aetheroleum). The essential oil originated from *Pimpinella anisum* seeds contain predominantly *trans*-anethole (80%–95%) with smaller amounts of estragol, anisaldehyde, and pseudoisoeugenyl-2-methylbutyrate [38]. The latter is absent in the essential oil of the Chinese star anise. This substance may serve as a marker to distinguish between both anise oils by GC.

The Chinese star anise oil is often contaminated by *Illicium anisatum* (Shikimi fruits), with its common name Japanese star anise. The Japanese star anise (Shikimi) oil contains the very toxic sesquiterpene lactone anisatine. Pharmaceutical and food authorities are interested in analytical methods to detect those adulterations.

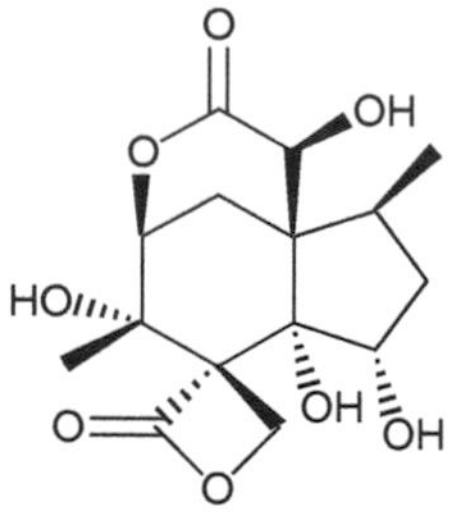

Anisatin

FIGURE 18.11 Sesquiterpene lactone anisatin as the toxic principle of Japanese star anise oil.

In the early 1990s, Schultze et al. [39] presented a TLC method which should be useful to distinguish both oils. They found out myristicin to be a genuin marker substance of Shikimi oil. When developed in hexane–toluene $(50 + 50)$ myristicin appears well separated from the other substances in the lower part of the chromatogram $(R_f \sim 0.3)$ and gives a blue color after spraying with the molybdophosphoric acid reagent.

More recently, in 2006, Lederer and coworkers [40] published a paper on combination of TLC and HPLC/MS methods to allow an effective an reliable control of Chinese star anise. The proposed assay includes a TLC method for comparing the flavonoid spectra of *Illicium* species. *Illicium* sp. showed a characteristic fingerprint compared with the other *Illicium* species. and this method could be regarded as a useful tool for the rapid identification of Chinese star anise in the quality control of commercial batches. The disadvantage of this method was that the toxic anisatine was not detectable. To reach this goal the authors employed the hyphenated HPLC/ESI-MS/MS technique. This system allowed a quantification of the toxic sesquiterpene lactone anisatine (Figure 18.11).

18.10 APPLICATION OF TLC AS A COMPLEMENTARY METHOD

TLC is useful, for example, for monitoring fractions of essential oils separated by preparative LC or GC and on the other hand, vice versa, TLC is useful for large-scale separations with the TLC fractions subsequently being monitored by GC/MS. TLC is also useful for studying antimicrobial activity. In 1997, Hostettmann et al. [41] developed a modification of direct bioautography. A suspension of the appropriate microorganism in a suitable broth was applied to a developed chromatographic plate and incubation in a humid atmosphere promotes the growth of the microorganism. Zones of inhibition were detectable by using special reagents. Though the following article does not present an active terpene it may serve as an example for the suitability of TLC in analyzing essential oils (for bioautography see Chapter 9).

In 1999, Nakamura et al. [42] analyzed the essential oil of *Ocimum gratissimum* L. by TLC on silica gel G plates $(20 \times 10$ cm$)$ using the solvent system hexane–acetyl acetate $(95 + 5)$. TLC plates were run in duplicate and one set was used as the reference chromatogram. Spots and bands were visualized by UV irradiation (254 and 366 nm) and by spraying with vanillin–H_2SO_4 reagent.

The other set was used for bioautography. The chromatogram developed as described earlier was placed in sterile glass square with cover and an inoculum of *Staphylococcus aureus* containing 106 CFU/mL molten Mueller–Hinton agar was distributed over the plates. After solidification of the medium, the TLC plate was incubated overnight at 37°C. Inhibition zones indicated the presence of active compounds. These zones were assigned to the corresponding zone of the reference chromatogram.

REFERENCES

1. Wallach, O., Zur Kenntnis der Terpene und der Ätherischen Öle, *Liebig's Ann. Chem.*, 1, 239, 1887.
2. *European Pharmacopoeia*, 5th ed., incl. Supplements, 2005, German Edition, Deutscher Apothekerverlag, Stuttgart, Govi–Verlag Pharmazeutischer Verlag GmbH, Eschborn.
3. Sprecher, E., Rücklaufapparatur zur erschöpfenden Wasserdampfdestillation ätherischen Öls aus voluminösem Destillatgut, *Dtsch Apoth Ztg*, 103, 213, 1963.
4. Baerheim Svendsen, A., Bestandteile ätherischer Öle, *Dtsch Apoth Ztg*, 127, 2458, 1987.
5. Kubeczka, K.-H., Ein einfaches Verfahren zur Gewinnung kleinster Duftstoffmengen für die DC-Analyse, *Planta Med*, 3, 294, 1969.
6. Kubeczka, K.-H., The combination of chromatographic techniques in natural product analysis, *Planta Med*, 35, 291, 1979.
7. Stahl, E., TAS—ein Thermomikro-Abtrenn- und Applikationsverfahren gekoppelt mit der Dünnschichtchromatographie, *J Chromatogr A*, 37, 99, 1968.
8. van Meer, H. and Labadie, R.P., Combined use of the different techniques in the analysis of Valepotriates, Abstract of the 25th Annual Meeting, *Planta Med*, 32A, 51, 1977.
9. Stahl, E. and Jork, H., Terpenderivate, ätherische Öle, Balsame und Harze, In: Stahl, E. (Ed.), *Dünnschichtchromatographie,* Springer Verlag, Berlin/Heidelberg/New York, NY, 1967.
10. Battaile, J., Dunning, R.L. and Loomis, W.D., Biosynthesis of terpenes. I. Chromatography of peppermint oil terpenes, *Biochim Biophys Acta*, 51, 538, 1961.
11. Attaway, J.A., Barabas, L.J. and Wolfort, R.W., Analysis of terpene hydrocarbons by thin layer chromatography, *Anal Chem*, 37, 1289, 1965.
12. Demole, E. cited in Marini-Bettolo, G.B. In Stahl, E. (Ed.), *Thin-Layer Chromatography*, Elsevier, Amsterdam-London-New York, NY, 1964.
13. Mathis, C. and Ourisson, G., Étude Chimio-Taxonomique du Genre *Hypericum*. II. Identification de constituants de diverses huiles essentielles d'*Hypericum*, *Phytochemistry*, 3, 115, 1964.
14. Maarse, H. and Belz, R., Isolation, *Separation and Identification of Volatile Compounds in Aroma Research*, D. Reidel Publishing company, Dordrecht/Boston/Lancester, 106, 1985.
15. Sonwa, M.M. and König, W.A., Chemical study of the essential oil of *Cyperus rotundus*, *Phytochemistry*, 58, 799, 2001.
16. Bortolomeazzi, R. et al., Sesquiterpene, alkene and alkane hydrocarbons in virgin olive oils of different varieties and geographical origin, *J Agric Food Chem*, 49, 3278, 2001.
17. Kraus, L., *Concise practical book of thin-layer chromatography*, 1st ed., *DESAGA GmbH*, Heidelberg, 1993, 82, and Kleines Praktikumsbuch der Dünnschicht-Chromatographie. 8th ed., *DESAGA GmbH*, Wiesloch, 77, 2005.

18. Francis, M.J.O. and Allcock, C., Geranyl-β-D-Glucoside: Occurrence and synthesis in rose flowers, *Phytochemistry*, 8, 1339, 1969.
19. Zhao, W., Wolfender, J.-L., Hostettmann, K. et al., Sesquiterpene glycosides from *Dictamus dasycarpus*, *Phytochemistry*, 47, 63, 1998.
20. Stahl-Biskup, E., Monoterpene glycosides, state-of-the art, *Flavour Fragr J*, 2, 75, 1982.
21. Stengele, M. and Stahl-Biskup, E., Glycosidally bound volatiles in peppermint (*Mentha piperita* L.), *J Essent Oil Res*, 5, 13, 1993.
22. Schulz, G., Ätherisches Öl und glycosidisch gebundene flüchtige Pflanzeninhaltsstoffe, *Dissertation*, Institut für Pharmazie, Hamburg, 1989.
23. Basar, S. and Koch, A., Test of the stability of olíbanum resins and extracts, *J Planar Chromatogr*, 17, 479, 2004.
24. Wagner, H. and Bladt, S., *Plant Drug Analysis*, 2nd ed., Springer Verlag, Berlin/ Heidelberg, 1996.
25. Harborne, J.B., *Phytochemical Methods*, 3rd ed., Chapman & Hall, London, 1998.
26. American Institute of Homoeopathy, *Pharmacopée Française*, 8th ed., Arlington, VA, 1965.
27. Drozdz, B. and Bloszyk, E., Selective detection of sesquiterpene lactones by TLC, *Planta Med*, 33, 379, 1978.
28. Petasites, in *Hagers Handbuch*, Suppl. 3, Springer Verlag, 622, 1998.
29. Glynne-Jones, A., Pyrethrum, *The Royal Society of Chemistry*, 12, 195, 2001.
30. Stahl, E. and Pfeifle, J., Dünnschicht- und Gas-Chromatographie der insektiziden Pyrethrum—Inhaltsstoffe, *Naturwissenschaften*, 52, 620, 1965.
31. Stahl, E. and Schild, W., *Pharmazeutische Biologie Bd.* 4, Drogenanalyse. II. Inhaltsstoffe und Isolierungen, Gustav Fischer Verlag, 1981.
32. Balbaa, S.I. et al., Comparative study between different methods used for estimation of pyrethrins and a new suggested method, *Planta Med*, 21, 347, 1972.
33. Gabriels, M. and Plaizier-Vercammen, J.A., Densitometric TLC determination of artemisinin and its lipophilic derivatives, artemether and arteether, *J Chromatogr Sci*, 41, 359, 2003.
34. Gabriels, M. and Plaizier-Vercammen, J.A., Development of a reverse-phase thin-layer Chromatographic method for artemisinin and its derviatives, *J Chromatogr Sci*, 42, 341, 2004.
35. Bhandari, P. et al., Simultaneous densitometric determination of artemisinin, artemisinic acid and arteannuin-B in *Artemisia annua* using reversed-phase TLC, *J Sep Sci*, 28, 2288, 2005.
36. Chizzola, R., Langer, T. and Franz, C., An approach to the inheritance of the sesquiterpene chemotypes within *Petasites hybridus*, *Planta Med*, 72, 1254, 2006.
37. Glasl, S., Mucaji, P., Werner, I. and Jurenitsch, J., TLC and HPLC characteristics of desacetylmatricarin, leucodin, achillin and their 8alpha-angeloxy-derivatives, *Pharmazie*, 58, 487, 2003.
38. *Hagers Handbuch, Vol. 6*, Springer Verlag, Berlin-Hamburg-Heidelberg, 135, 1994.
39. Schultze, W., Zänglein, A., Lange, G. and Kubeczka, K.-H., Sternanis und Shikimi, *Dtsch Apoth Ztg*, 130, 1194, 1990.
40. Lederer, I. et al., Combination of TLC and HPLC-MS/MS methods. Approach to a rational quality control of Chinese star anise, *J Agric Food Chem*, 54, 1970, 2006.
41. Hostettmann, K., Strategy for the biological evaluation of plant extracts, *Pure App Chem*, 70, 1109, 1999.
42. Nakamura, C.V. et al., Antibacterial activity of *Ocimum gratissimum* L. essential oil, *Mem Inst Oswaldo Cruz*, Rio de Janeiro, 94, 675, 1999.

19 TLC of Diterpenes

Michał Ł. Hajnos

CONTENTS

19.1   INTRODUCTION

Diterpenes—C_{20} terpenes—are a big group of compounds consisting of four 5-carbon (C_5) units called the active isoprene. These compounds play an important role in plant physiology. They are also known for their pharmacological and toxicological activities. The group of diterpenes consists of about 800 various substances with a known structure, and occurs mainly in higher plants (sometimes in oleoresines and latices), microorganisms, and marine products. In phytochemistry, diterpenes are known as bitter substances, toxic diterpene alkaloids, as well as plant growth hormones (gibberellins) and other secondary metabolites. There is also vitamin A (retinol), which reveals the features of the diterpene structure, and occurs in human and animal organisms. It has also been discovered that pro-vitamin A—the group of tetraterpenoid compounds (carotenoids)—is present in plants.

Diterpene compounds can reveal linear or mono-, di-, tri-, tetra-, polycyclic (macrocyclic), and miscellaneous structures.

From chemotaxonomical point of view, some groups of diterpenes are common in plant kingdom, while others are specific for small systematic groups (families or genera).

19.2   CHEMICAL STRUCTURES OF DITERPENES AND DITERPENOIDS—BASIC CLASSIFICATION OF THE GROUP OF COMPOUNDS

Diterpenoids constitute a vast class of isoprenoid natural products, biosynthesized from mevalonic acid through geranylgeranyl pyrophosphate—GGPP (see Figure 19.1). They are divided into acyclic (phytane), monocyclic (retinol—vitamin A), bicyclic (labdane, clerodane), tricyclic (abietane, pimarane, rosane, podocarpane), tetracyclic (kaurane, trachylobane, beyerane, atisane, gibberellane), macrocyclic (cembrane, taxane, daphnane, tigliane, ingenane, jatrophane), and miscellaneous compounds, also known as prenylsesquiterpenes (meroditerpenes, xenicanes, lobanes, prenylguaianes, dictalanes)—for instance meroditerpenes are polycyclic diterpenes of mixed biogenesis characterized by hydroquinonic methyl nucleus linked to a diterpenic chain.

Ginkgolides and diterpene alkaloids are other groups of diterpenes. The latter is not broadly described in this chapter, because it is included in the chapter, describing different groups of alkaloids.

Structures of main diterpene groups are shown in Figures 19.2–19.10.

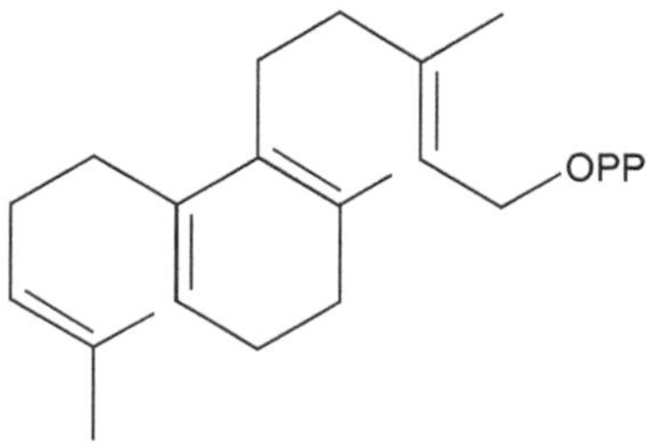

GGPP

FIGURE 19.1   Geranylgeranyl pyrophosphate (GGPP).

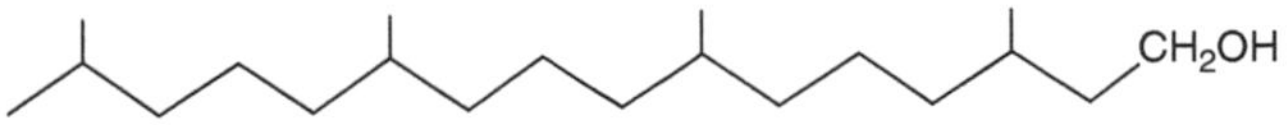

FIGURE 19.2 Phytane diterpene (acyclic) structure.

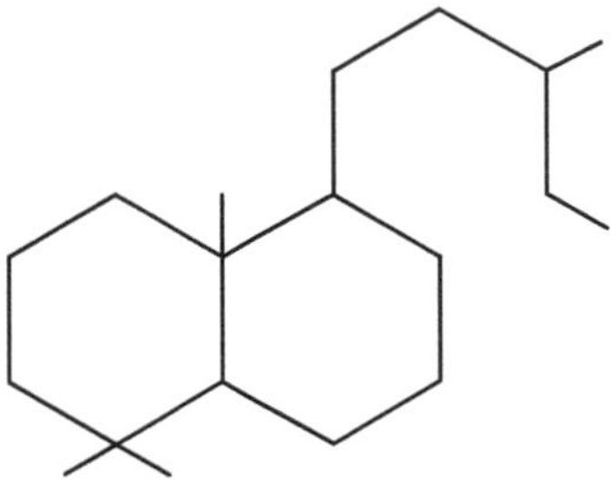

FIGURE 19.3 Labdane structure.

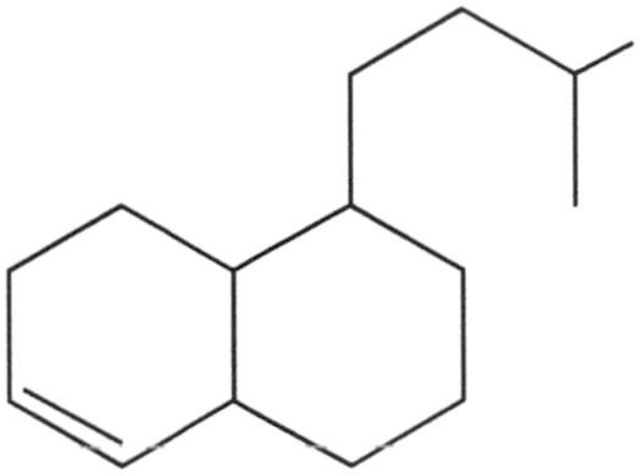

FIGURE 19.4 Clerodane structure.

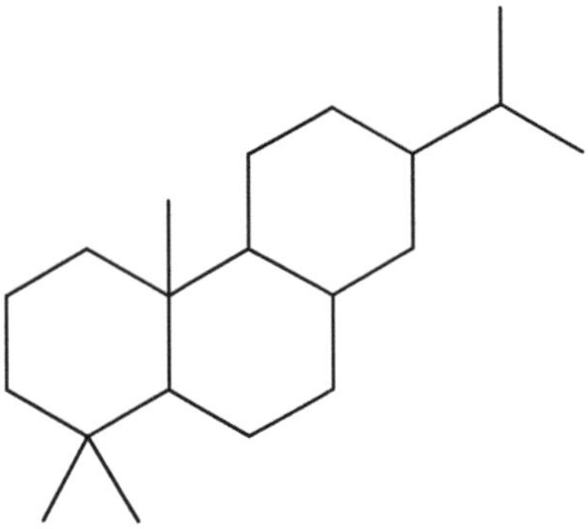

FIGURE 19.5 Abietane structure.

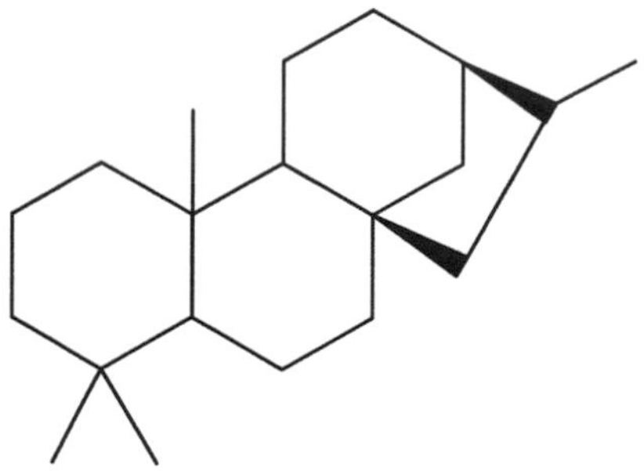

FIGURE 19.6 Kaurene structure.

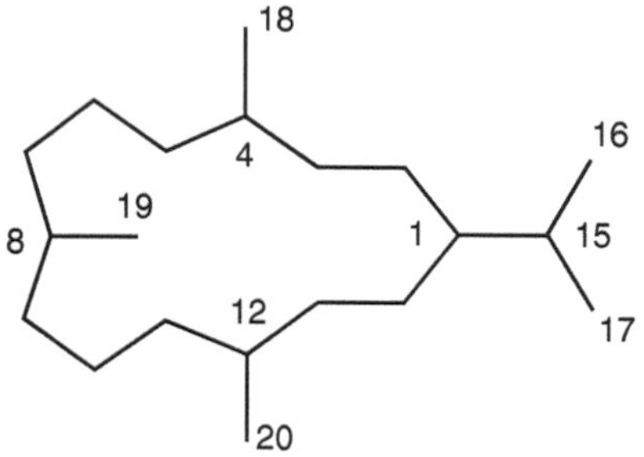

FIGURE 19.7 Cembrane structure.

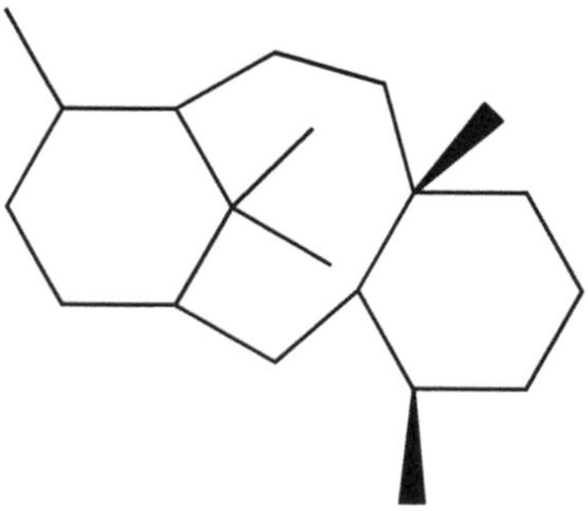

FIGURE 19.8 Taxane structure.

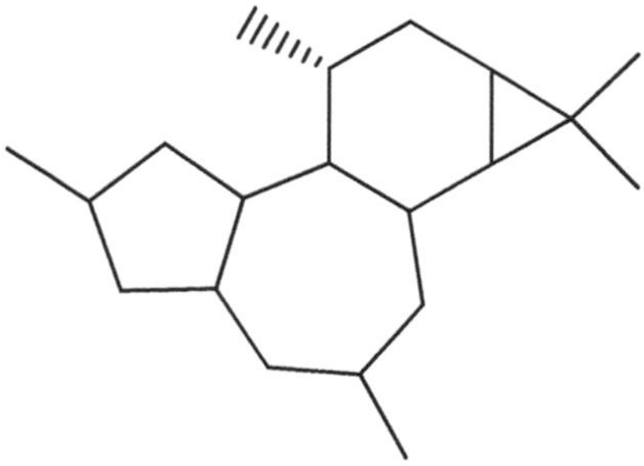

FIGURE 19.9 Tigliane structure.

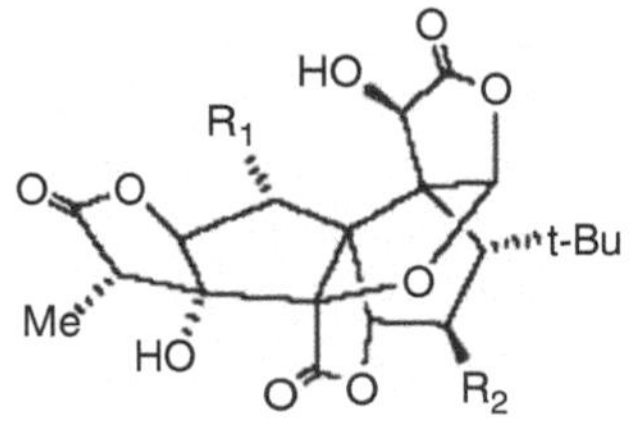

FIGURE 19.10 Ginkgolide structure.

19.3 NATURAL SOURCES OF DITERPENE COMPOUNDS

In some plant families, especially the *Euphorbiaceae* and *Labiatae*, there are a lot of diterpenes of various skeleta. The *Euphorbiaceae* family contains the diterpenes of clerodane, tigliane, jatrophane, lathyrane, or daphnane groups, known for their irritant and toxic properties, whereas clerodane or abietane diterpenes occur in the *Labiatae* family. There are also plant families in which only several genera contain diterpenes, e.g., the *Asteraceae* family. Diterpenes are also a group of compounds characteristic for resins occurring in the *Pinaceae* family (abietanes) [1]. For a change, taxoids and ginkgolides are unique compounds present only in the *Taxus* genus and *Ginkgo biloba*, respectively. Totally different diterpenes, such as the briarane type and meroditerpenes, can be found in marine products (e.g., in corals) [2,3]. Diterpenes such as spongianes are present in sponges [4], whereas dolabellane and dolastane in molluscs [5].

The species of plants, fungi, marine products, and insects containing diterpene compounds are widely described in the chapter and are taxonomically arranged and presented below.

19.3.1 Higher Plants

- *Asteraceae—Aspilia montevidensis* [6], *Stevia* species [7,8], *Mikania banisteriae* [9], *Espeletia* species [10], and *Olearia* species [11]
- *Euphorbiaceae—Croton* species [12,13] (e.g., *Croton lechleri* [14], *Croton oblongifolius* [15], *Croton campestris* [16], *Croton zambesicus* [17,18]), *Euphorbia esula* [19–21], *Euphorbia nubica, Euphorbia helioscopia* [22], *Euphorbia portlandica* [23], *Euphorbia peplus* [24], *Excoecaria bicolor* [25], *Euphorbia ingens* [26], *Euphorbia segetalis* [27], *Euphorbia pubescens* [28,29], and *Euphorbia lagascae* [30]
- *Celastraceae—Tripterygium* Hook species [31]—*Tripterygium wilfordii* [32,33]
- *Capparaceae—Cleome viscosa* [34]
- *Cistaceae—Cistus creticus* ssp. *creticus* [35]
- *Labiatae—Teucrium fruticans* [36], *Teucrium yemense* [37], *Salvia miltiorrhiza* [38–41], *Salvia recognita* [42], *Salvia sclarea* [43], *Salvia divinorum* [44], *Salvia yunnanensis* [45], *Salvia glutinosa* [46], *Salvia eriophora* [47],

Rabdosia longituba [48], *Isodon rubescens* [49], *Isodon japonica* [50,51], *Marrubium* species [52]
* *Leguminosae—Pisum sativum* [53] (gibberellins)
* *Verbenaceae—Vitex agnus castus* [54]
* *Thymelaeaceae—Wikstroemia canescens* [55], *Wikstroemia mekongenia* [55], *Wikstroemia monticola* [56]
* *Orchidaceae—Cymbidium* species [57]
* *Zingiberaceae—Aframomum daniellii* [58], *Alpinia calcarata* [59]
* *Menispermaceae—Penianthus zenkeri* [60]
* *Meliaceae—Azadirachta indica* [61,62]
* *Salicaceae—Salix matsudan* [63]
* *Ginkgoaceae—Ginkgo biloba* [64–66]
* *Pinaceae—Larix kaempferi* [67], *Picea glehnii* [68]
* *Taxaceae—Taxus cuspidata* [69], *Taxus chinensis* [70,71], *Taxus baccata* [72], *Taxus brevifolia* [73], *Taxus wallichiana* [74]

19.3.2 MARINE PRODUCTS

Diterpenes were found in the following marine products:

* brown algae (*Phaeophyceae*): *Stypopodium flabelliforme* [75], *Dictyota dichotoma* [76]
* marine invertebrates: gorgonian octocorals (order: *Gorgonacea*)—*Pseudopterogorgia elisabethae* [77]
* marine sponges (order: *Dictyoceratida*): *Rhopaloeides odorabile* [78], *Hyatella intestinalis* [4]
* molluscs—nudibranch—suborder *Nudibranchia* [78]

19.3.3 FUNGI

* Lower fungi—*Gibberella fujikuroi* [79], Fusarium moniliformae [80]
* Class *Basidiomycetes—Sarcodon glaucopus* [81]

19.3.4 INSECTS

* *Termitidae—Isoptera—Nasutitermes princeps* [82]

All the species mentioned earlier will be widely discussed in the chapter.

19.4 BIOLOGICAL ACTIVITIES OF DITERPENES AND THEIR PHARMACOLOGICAL USE

Diterpenes are supposed to play an important role in the defence system of plants [9]. Insect antifeedant activity is shown by neo-clerodane diterpenes from *Teucrium fruticans* [36], and other species [83]. Trinervitane and kempane compounds are known to be secreted by termite soldiers as their defensive substances [84]. Gibberellins are the group of compounds playing an important role in the physiology of plants as their growth hormones.

Numerous diterpenes exhibit pharmacological activity and are widely used in medicine.

For instance, tanshinones—main pharmacologically active compounds of *Salvia miltiorrhiza*—indicate antiplatelet aggregation, antioxidation, and other activities as well as curative effect in cardiovascular diseases. *Salvia miltiorrhiza* is a traditional Chinese medicinal plant widely used in the treatment of many diseases [38,39,85–87]. Neo-clerodane compounds (mainly salvinorin C) occurring in *Salvia divinorum* are known for their psychotropic (hallucinogenic) properties [44].

Labdane diterpenes from Ladano resin (originating from *Cistus creticus ssp. creticus*) are known for their strong antibacterial properties [35]. It is generally confirmed that labdane diterpenes indicate antimicrobial, antiviral, anti-inflammatory, and cytotoxic activities [88]. Moreover, some of ent-kaurene derivatives indicate antibacterial, antifungal as well as antiinflammatory properties [9]; ent-kaurenes from the *Isodon* genus (*Labiatae*) indicate cytotoxic properties [49]. Other ent-kaurene compounds from the *Stevia* genus (stevioside and related diterpenes) are known for their sweetening purposes [7,8,89].

Trachylobanes—tetracyclic diterpenes occurring in the *Croton* species—are reported to have cytotoxic properties [17]; similar properties are known for bis-labdanic diterpenes from *Alpinia calcarata* [59].

Cembranoids occurring in the *Capparaceae* family (*Cleome viscosa* L.) indicate antibacterial, as well as insecticidal and nematicidal properties [34].

Triptolides—abietane-type diterpenes occurring in the *Tripterygium* genus (*Celastraceae*)—are known as immunosuppressive compounds [90]. The *Tripterygium* species are widely used as antiinflammatory plant materials in Chinese herbal medicine. The main compound from this group—triptolide—has been reported to inhibit autoimmunity, allograft rejection, and graft-versus-host disease [91]. Abietane-type diterpenes from *Larix* sp. (e.g., *Larix kaempferi*) indicate antitumor properties [67], while abietanes from *Salvia officinalis* indicate antiviral properties [92].

Pseudopterosins and seco-pseudopterosins—diterpene glycosides from gorgonian corals—display strong antimalarian, antituberculosis, antiviral, antiinflammatory, and anticancer activities [77]. Meroditerpenoids from brown algae indicate strong cytotoxic properties [75].

Although plants from the *Euphorbiaceae* family are known for their toxic properties, there are some compounds from the clerodane group (occurring in *Croton oblongifolius*) with the confirmed anticancer activities [15]. Anticancer properties (inhibitors of multidrug resistance of tumour cells) are also displayed by lathyrane-type macrocyclic diterpenes [30].

Taxoids (taxanes) are known for their cytotoxic, antimitotic properties and they are used (especially paclitaxel—Taxol) in medicine as antitumor drugs [70,93]. The mechanism of the antitumor activity is based on the promotion of microtubules' creation and the inhibition of their depolimerization.

Ginkgolides are known for antiplatelet properties and have an influence on cerebral circulatory system, and, for these reasons, they are recommended in the treatment of cerebral vascular disease, and in senile dementia [65].

However, some diterpenes exhibit toxic or skin irritant activities. Such properties are reported for diterpenes from the *Thymelaeaceae* and *Euphorbiaceae* families belonging to tigliane, ingenane, jatrophane, and daphnane groups [55,19,25]. It is also reported that ingenane and tigliane diterpenes from the *Euphorbia* species are tumor promoters (experimentally confirmed cancerogenic properties) [22]; other experiments on ingenane diterpenes from *Euphorbia peplus* confirmed their pro-inflammatory activity in animal tests [24].

19.5   CHEMOTAXONOMIC IMPORTANCE OF DITERPENES

Diterpenes are a big and diversified group of compounds, predominantly character-istic for systematic groups of plants and marine products. For instance, cembranes and prenylsesquiterpenes are important chemosystematic distinctive features for different populations of *Dictyota dichotoma*—a cosmopolitan species of marine brown algae [76].

Diterpenes from the kaurene group are characteristic for the *Aspilia* genus of the *Asteraceae* family [6]. Moreover, acyclic diterpenes found in the *Oleria* genus from the *Asteraceae* family can be of chemotaxonomic importance for plants from the *Astereae* tribe [11]. Various groups of diterpenes occur only in some genera of *Asteraceae* family, and for this reason they are chemosystematically important in taxonomy of this family [94].

There are chemotaxonomic differences between various species of the same genus—for example, *Wikstroemia canescens* contains tigliane-type, whereas *Wikstroemia mekongenia* and *Wikstroemia monticola* daphnane-type diterpenes [55,56].

Ginkgolides are a group of compounds occurring only in *Ginkgo biloba* (*Ginkgoaceae*) [65].

Taxoids are a characteristic diterpene group only for the *Taxus* genus [95].

19.6   THIN LAYER CHROMATOGRAPHY—METHODS OF ANALYSIS AND THE ISOLATION OF DITERPENE COMPOUNDS IN NATURAL MATERIALS

19.6.1   SAMPLE PREPARATION—EXTRACTION AND PURIFICATION METHODS

In the chromatography of plant materials using TLC, sample extraction and purifi-cation methods are very important and basic steps of a successful analysis or isolation of natural compounds. Diterpenes are compounds of different physico-chemical and chromatographic properties, but they are generally medium-polar or low-polar compounds. They also occur in plants in a glycoside form, and in such a case they are polar substances (soluble in water, methanol, etc.). It determines the choice of the extrahent, which should be affinitive to the analytes, and thus give exhaustive extraction. Generally, medium-polar extrahents, such as chloroform [14], dichloromethane [36], ethyl acetate, diethyl ether [56], or acetone [25,42] are used, but often strong polar extrahents (e.g., methanol, ethanol, mixtures of alcohols with water, or pure water) are used in the extraction step of diterpenes [7,19,20,55,65,66]. Nonpolar extrahents—*n*-hexane [58]—are sometimes used for the extraction of low-polar diterpenes (with other nonpolar, oily compounds). Mixtures of dichloro-methane and methanol are also used for the extraction of diterpenoids from marine organisms [75,77].

Purification of crude extracts is often necessary for adequate analysis, especially isolation. It consists of removing ballast substances using LLE, SPE, and other techniques. Crude extracts containing diterpenes (in chloroform) can be purified using LLE with water + MeOH/chloroform [96] or methanolic, concentrated extract purification by LLE with water/dichloromethane [19,20] (removing polar ballasts by aqueous phase); the evaporation of the extrahent (dichloromethane) from crude extract and digestion of dry residue with aqueous methanol (30%, 60%, 90%) followed by centrifugation for each portion of aqueous methanol is another separation method. The collected fractions are destined to chromatographic isolation in reversed-phase systems [36].

Sometimes, very complicated procedures of extraction and purification methods are used for the separation of diterpene compounds, especially in isolation procedures. In the isolation of phorbol esters from *Wikstroemia canescens*, the following procedure is used: the extraction with methanol, the evaporation of solvent, LLE of residue in ethyl acetate/water system, organic phase evaporation—the residue dissolved/suspended in dichloromethane, filtration, filtrate concentration—the residue subjected to a Craig distribution in petroleum ether/methanol/water (15/10/0.5); the obtained fraction containing diterpenes is designed for further chromatographic separation [55]. A complicated procedure of purification and separation is also necessary for the isolation of ingenane diterpenes from *Euphorbia peplus*—the procedure consists in several steps of LLE purification followed by three consecutive preparative TLC separations [24].

The precipitation and decantation of ballast substances by freezing of diethyl ether extracts, where ether-soluble fraction is dissolved in isopropyl ether, filtered, and next purified using LLE methods (petrol ether/MeOH + water and benzene: MeOH:water) is another method of purification of crude extracts. Such a prepared extract is separated using CC, and next by TLC [56].

In the preparation of samples for quantitative analysis on TLC, purification procedure is also necessary—in the quantification of neo-clerodane compounds in the *Teucrium* species, the purification of acetone extract on XAD-2 macroporous resin was used to remove interfering compounds (pigments) [97]. In quantitative analysis of ginkgolides, complicated sample preparation (including filtration through polyamide cartridge and normal-phase SPE purification) is necessary [66]. Also in quantitative HPLC analysis of taxanes, multistep procedure of sample preparation, including RP-8 SPE followed by preparative TLC on silica, is necessary. In this case preparative TLC is only one step of purification and preliminary separation process [98].

On the other hand, in qualitative analysis (fingerprint, identification), simple procedures of extraction, without any purification methods, are generally used; for example, diethyl ether extract of *Salvia miltiorrhiza*, evaporated and dissolved in

MeOH, is directly used for HPTLC analysis [38]. In other cases, LLE purification is needed for the identification of more complicated mixtures (water:alcohol extracts of the *Stevia* species leaves partitioned between water/AcOEt; water layer extracted with portions of *n*-butanol and the obtained butanol fraction purified on neutral alumina column [7]) before TLC analysis. The identification of ginkgolides (qualitative TLC) also needs a sample purification method—water extract of *Ginkgo biloba* leaves is purified on celite, next activated charcoal is added to adsorb ginkgolides, supernatant is centrifuged, and compounds are desorbed with acetone. Such a prepared extract is ready for further analysis [65].

19.6.2   QUALITATIVE ANALYSIS—IDENTIFICATION OF DITERPENES IN MATERIALS AND PREPARATIONS

Qualitative analysis using TLC is the method of broad application, used in different analytical cases. As it results from the scientific literature, the main application of qualitative TLC of diterpenes is the monitoring of composition of fractions from preparative chromatography (mainly CC). Figure 19.11 shows TLC monitoring of taxanes' isolation on alumina preparative column.

The second important application is the quality control of known plant materials (e.g., triptolide analysis in the quality control of pharmaceuticals from the *Tripterygium* species [99]), or fingerprint analysis (e.g., in the identification of plants in chemotaxonomic investigations). Figure 19.12. shows fingerprints of tanshinones in *Radix Salviae Miltiorrhizae*.

Qualitative TLC is also a good preliminary method for the optimization of preparative chromatographic methods [72,76]. Sometimes, qualitative TLC analysis not only allows to optimize chromatographic conditions, but also helps to accelerate time-consuming biotechnological processes (e.g., optimization of *Taxus* cell culture process [70]).

19.6.2.1   Chromatographic Systems

The examples of chromatographic systems for the separation of diterpenes are described in Table 19.1.

19.6.2.2   Detection Methods (Including Densitometry), and Derivatization

In the TLC analysis of diterpenes various detection methods, such as UV absorption, fluorescence, as well as postprocess derivatization are used. Table 19.2. presents commonly used reagents and detection methods for such purposes.

19.6.2.3   Qualitative TLC Analysis of Diterpenes as a Preliminary Technique of the Optimization of Separation in PTLC and CC Methods

Analytical TLC is often used in the optimization of separation of compounds by preparative TLC. There are investigations concerning PTLC taxane separation on silica; the optimization covered 27 different eluents and allowed to select the best eluent system: silica/benzene:chloroform:acetone:methanol (8:37:6:3) [72]. The

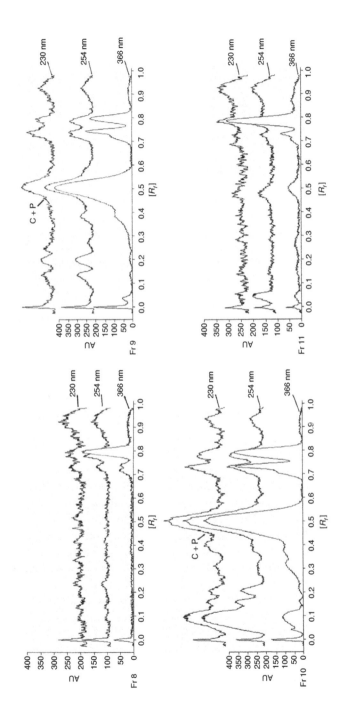

FIGURE 19.11 Monitoring of the effluent taxane—enriched fractions from the alumina column by TLC on silica plate with dichloromethane:dioxane: acetone:methanol (84:10:5:1) as mobile phase. Fraction numbers: A—8, B—9, C—10, D—11; C—cephalomannine, P—paclitaxel. (From Hajnos, M.Ł. et al., *J. Planar Chromatogr.*, 14, 119, 2001. With permission.)

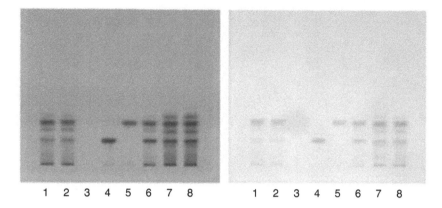

1 2 3 4 5 6 7 8   1 2 3 4 5 6 7 8

FIGURE 19.12 HPTLC fingerprints of tanshinones in *Radix Salviae Miltiorrhizae*. Left, observed under UV 254 nm; right, observed under visible light; R_f value for cryptotanshinone and tanshinone IIA is 0.34 and 0.59, respectively. 1, sample; 2, sample duplicate; 3, solvent blank; 4, cryptotanshinone; 5, tanshinone IIA; 6, spiked sample; 7, reference drug; 8, reference drug duplicate. (From Hu, P. et al., *Chem. Pharm. Bull.*, 53, 481, 2005. With permission.)

similar experiments with taxanes on silica using different eluents gave optimal composition of the quaternary eluent system consisting of: dichloromethane: dioxane:acetone:methanol (84:10:5:1) [93].

TLC experiments guided the semipreparative isolation (using VLC, prep. HPLC methods) of new biologically active meroditerpenoids [75].

19.6.2.4 Analytical TLC of Diterpenes as CC Fraction Composition Control

Analytical TLC is a basic tool of the composition control in preparative liquid column chromatography. There are many examples of TLC application to control the isolation of diterpenes by CC on silica. It is known from simultaneous kaurene-type and steroidal saponine compounds isolation from *Aspilia montevidensis* [6], kaurenes, steroids, and sesquiterpenes from the *Espeletia* species [10], abietanes from *Salvia glutinosa* [46], and also the abietane-type diterpenoids from *Salvia sclarea* [43]. The isolation of clerodane diterpenes from *Croton lechleri* using silica gel column chromatography was monitored by TLC analysis in system: SiO_2/MeOH:CH_2Cl_2 in different proportions [14]. Also, normal-phase CC isolation of furano-clerodanes from the bark of *Croton campestris* was monitored using TLC analysis in system: silica gel/toluene:AcOEt:MeOH (40:9:1)—compounds were visualized by spraying with a 1% solution of 4-dimethylaminobenzaldehyde in EtOH [16]. Neo-clerodanes isolation from *Teucrium fruticans* using RP-18 column was controlled by TLC in system: SiO_2 60 F_{254}/CH_2Cl_2: *tert*-butylmethyl ether (4:1) [36]. The isolation of ginkgolides A, B, C, and J by silica gel column chromatography, benzylation process, and re-chromatography on SiO_2 CC were monitored using TLC on SiO_2 60 F_{254} plates/AcOEt:hexane (1:1) [64]. The isolation of tigliane and ingenane diterpenes from *Euphorbia ingens* using VLC method required TLC method as fraction composition monitoring—in system: silica gel/hexane:Et_2O:AcOEt (1:1:1) [26].

TABLE 19.1
Analytical TLC (Identification)

Compounds	Sample	Adsorbent	Solvent System	Detection	References
Tanshinones	*Salvia miltiorrhiza*	HPTLC silica gel 60 F$_{254}$ plates	Petrol ether: AcOEt: cyclohexane (5:3:2)	UV—254 nm; vis	[38]
Tanshinone IIA	*Radix Salviae Miltiorrhizae*	SiO$_2$	Petrol ether: AcOEt (4:1)	Vis	[100]
Tanshinones	*Salvia miltiorrhiza*	SiO$_2$	• Benzene: AcOEt (14:1) • AcOEt: acetic acid (25:2)	10% H$_2$SO$_4$ in EtOH + heating in 105°C for 10 min	[101]
Tanshinone I, cryptotanshinone, tans-hinone VI	*Salvia miltiorrhiza*	SiO$_2$	• CHCl$_3$ • Hexane:AcOEt (8:2) • CHCl$_3$:MeOH (1:1)	—	[85]
Tanshinone (together with ginsenosides, schisanrols, etc.)	Pharmaceutical granules containing *Salvia miltiorrhiza*	SiO$_2$	• CHCl$_3$: AcOEt: MeOH: water (15:40:22:10) • Toluene:AcOEt (3:2), (19:1) • CHCl$_3$:Me$_2$CO:HCOOH (8:1:1)	10% H$_2$SO$_4$ in EtOH + heating in 105°C for 10 min	[86]
Tanshinone IIA	Pharmaceutical preparation	SiO$_2$	Benzene:AcOEt (19:1)	UV—365 nm; or 10% H$_2$SO$_4$ in EtOH + heating in 105°C for 10 min—vis	[102–104]
Abietanes–nimbiol, nimbionone, nimbosone, nimbonone, nimbonolone	*Azadirachta indica* stem bark	Aluminium oxide SiO$_2$ SiO$_2$	Hexane:AcOEt (17:3) CHCl$_3$:MeOH (39:1) CHCl$_3$:MeOH (97:3) or hexane: AcOEt (9:1)	— —	[61]
Gibberellins	*Pisum sativum*	Silica gel G plates	Benzene: acetic acid: water (8:3:5) (upper phase)	70% sulfuric acid; 30 min heating in 120°C	[53]

(continued)

TABLE 19.1 (continued)
Analytical TLC (Identification)

Compounds	Sample	Adsorbent	Solvent System	Detection	References
Ent-kaurenes—stevioside and related compounds	Stevia sp—leaves	Silica gel GHLF plates	• AcOEt:EtOH:H$_2$O (130:27:20) • CHCl$_3$:MeOH:H$_2$O (6:3:1)—lower layer • AcOEt:HOAc:H$_2$O (8:3:2)	(a) Naphthoresorcinol–H$_2$SO$_4$ (b) Ac$_2$O–H$_2$SO$_4$ (c) 60% H$_2$SO$_4$; in all cases 110°C in 10 min	[7]
		SiO$_2$	• AcOEt:EtOH:H$_2$O (130:27:20) • CHCl$_3$:MeOH (9:1) or (49:1)	UV—366 nm	[8]
Taxanes	Taxus chinensis; T. baccata—cell cult	Aluminium-backed silica gel GF$_{254}$ plates	CHCl$_3$:MeCN (4:1)	UV—254 nm, or vanillin: sulfuric acid	[70]
Taxanes—10-deacetyl-baccatin III, baccatin III, cephalomannine, paclitaxel	Taxus sp—twigs	Silica gel 60 F$_{254}$ plates	Benzene:CHCl$_3$:Me$_2$CO: MeOH (8:37:6:3) CH$_2$Cl$_2$:Dx:Me$_2$CO:MeOH (84:10:5:1)	UV—254 nm, or densitometry at 230 nm UV—254 nm, or densitometry at 230 nm	[72,98] [93]
Taxanes besides taxine alkaloids	Taxus baccata—pollen	Silica gel	CHCl$_3$:MeOH (24:1)	- Dragendorff reagent - Iodoplatinate reagent - 3% H$_2$SO$_4$ in EtOH + heating in 120°C for 5 min	[105]
Taxanes	Taxus brevifolia—bark	Cyano-modified silica HPTLC plates	Two-dimensional TLC: 1. CH$_2$Cl$_2$:hexane: acetic acid (9:10:1) 2. Water:MeCN:MeOH: THF (8:57:0.1)	Anisealdehyde reagent + heating	[73]
		Diphenyl modified silica plates	Two-dimensional TLC: 1. Hexane:iPrOH:Me$_2$CO (15:2:3) 2. MeOH:water (7:3)		

Compound	Source	Stationary phase	Solvent system	Detection	Ref.
Ginkgolides A, B, C	Ginkgo biloba leaves; pharmaceutical preparations	Silica gel K6F plates	AcOEt:toluene:Me$_2$CO:hexane (4:3:2:1)	Heating in 150°C for 35–60 min	[65,66]
Ginkgolides A, B, C, J	Ginkgo biloba leaves; pharmaceutical preparations	Silica gel TLC plates	• Toluene:Me$_2$CO (7:3)	Detection: UV—365 nm Heating; different visualization methods	[66,106]
			• Cyclohexane:AcOEt (1:1) • CHCl$_3$: Me$_2$CO:HCOOH (75:16.5:8.5)	Heating	[66]
GGPP, me-valonic acid	Gibberella fujikuroi incubation mixtures	Silica gel F$_{254}$ plates	n-Propanol:ammonia:water (6:3:1)	Radioscanner LB2722 (Berthold)	[79]
Kaurenes simultaneously with sesquiterpenes and steroids	Espeletia sp—CC-isolated fraction	SiO$_2$	Petrol ether:AcOEt (in different proportions)	Cobalt chloride/H$_2$SO$_4$ + heating	[10]
Ent-kaurenes—xindonguin D,F, melis-soidesin G	Isodon rubescens var. rubescens	Silica gel	• CHCl$_3$:MeOH (7:1),(9:1), (10:1),(20:1) • Cyclohexane:iproh (4:1) • Developed 3x • Cyclohexane:Et$_2$O (1:2)	10% H$_2$SO$_4$ in EtOH + heating	[49]
Spongiane-type diterpenes	Marine sponge: Rhopaloeides odorabile	SiO$_2$	CH$_2$Cl$_2$:AcOEt (4:1)	Vanillin:sulfuric acid reagent + heating	[78]
Diterpenes from different groups—prenylsesquiterpenes, cembranes	Marine brown alga: Dictyota dichotoma	silica gel plates	Two-dimensional TLC: 1. CHCl$_3$:Me$_2$CO:MeOH: HCOOH:water (100:40:20:20:8) 2. Me$_2$CO:benzene:HCOOH: water (200:30:3:10)	10% H$_2$SO$_4$ in MeOH	[76]

(continued)

TABLE 19.1 (continued)
Analytical TLC (Identification)

Compounds	Sample	Adsorbent	Solvent System	Detection	References
Triptolides	*Tripterygium wilfordii*—pharmaceuticals	Silica gel	$CHCl_3$:Et_2O (2:1)	–1% 3,5-dinitrobenzoic acid in ethanol and then 8% NaOH in ethanol	[99]
Triptolides—triptolide, tripchlorolide, wilformine	*Tripterygium wilfordii*	Silica gel	• Et_2O saturated with 1% HCl • $CHCl_3$:Me_2CO (5:3) saturated with 10% acetic acid	Potassium iodobismuthate solution, or densitometry at 218 nm	[33]
Clerodane diterpenes	*Penianthus zenkeri*—roots	SiO_2	• $CHCl_3$:MeOH (in diff. prop.) • AcOEt:MeOH (9:1) • Petrol.ether:Me_2CO (9:1)	Anisaldehyde reagent	[60]
Neo-clerodanes	*Teucrium yemense*	SiO_2	$CHCl_3$:MeCN (3:2)	Anisaldehyde/H_2SO_4 reagent	[37]
Clerodanes—croblongifolin, crovatin, nidorellol	*Croton oblongifolius*	Silica gel	• Hexane:AcOEt (9:1) • $CHCl_3$:MeOH (19:1)	—	[15]
Tiglianes—phorbolesters	*Croton species*—croton oil	Silica gel GF plates; RPS-F plates HPTLC SiO_2 plates	Examination of various chromatographic systems CH_2Cl_2: Me_2CO (3:1)	Vanillin: H_2SO_4:EtOH (3:0.5:100) + heating in 120°C	[12]
				60% H_2SO_4 + heating at 120°C for 20 min	[13]
	Dual-phase plates: SiO_2 + C-18	Two-dimensional TLC: 1. normal phase—$CHCl_3$:Me_2CO (4:1) 2. reversed phase—MeOH:water (9:1)		60% H_2SO_4 in EtOH + heating at 110°C	[107]
Tiglianes, ingenanes	*Euphorbia sp*—latex	Silica gel	Et_2O:petroleum ether (in different proportions)	Vanillin: H_2SO_4 reagent + heating at 100°C	[96]
Tiglianes, ingenanes	*Euphorbia ingens*	Silica gel $60F_{254}$ plates	Hexane:Et_2O:AcOEt (1:1:1)	Vanillin: H_2SO_4 reagent	[26]

Jatrophanes—esulatin A,B,C	Euphorbia esula	Silica gel	• Benzene:CHCl$_3$:Et$_2$O (1:1:3) • CHCl$_3$:Me$_2$CO (19:1) • Cyclohexane:AcOEt:EtOH (20:10:1)	Vanillin: H$_2$SO$_4$ reagent	[21]
Jatrophane polyesters, lathyranes	Euphorbia pubescens— isolated fractions	Silica gel	• CHCl$_3$:MeOH (39:1)- developed 3x • CHCl$_3$:MeOH (19:1)- developed 2x • CHCl$_3$:MeOH (9:1)	H$_2$SO$_4$:acetic acid:water (1:20:4) + heating	[29]
Furano-diterpenes, clerodanes	Croton campestris bark—CC fractions control	Silica gel	Toluene:AcOEt:MeOH (40:9:1)	1% 4-Dimethylamino-benzaldehyde in EtOH	[16]
Ent-trachyloban-3β-ol	Croton zambesicus	Silica gel	Toluene:AcOEt:MeOH (40:9:1)	Anisaldehyde/H$_2$SO$_4$ reagent	[17]
Labdanes—sclareol, dri-mane- and other deriv.	Cistus creticus ssp. creticus— Ladano resin	Silica gel	Hexane:Et$_2$O (49:1)	—	[35]
Bis-labdanic diterpenoids— calcaratin D,E	Alpinia calcarata	Silica gel	Hexane:AcOEt (23:2) or (43:7)— developed 2x	10% Molybdic acid reagent	[59]
Labdanes—rotundifuran	Vitex agnus castus—fruit extract	HPTLC silica gel	• Toluene: AcOEt (9:2)	10% H$_2$SO$_4$ in MeOH + heating in 105°C	[54]
Gibberellins	Secale cereale—immature ears	Silica gel	CHCl$_3$:AcOEt:acetic acid (5:4:1)	—	[108]
Gibberellins	Phaseolus vulgaris—seeds	Silica gel	• Isopropyl ether:acetic acid (95:5) • CHCl$_3$:MeOH:acetic acid:water (40:15:3:2)	—	[109]
Trinervitane diterpenes	Nasutitermes princes— Termitidae	Silica gel	Hexane:AcOEt (7:3)	—	[82]
Cyathane diterpenes— glaucopines A, B	Sarcodon glaucopus— fruiting body	Silica gel	CH$_2$Cl$_2$: MeOH (19:1)	50% H$_2$SO$_4$	[81]

TABLE 19.2

Detection/Derivatization of Diterpenes on Thin Layers

Compounds	Reagent	Detection Method/Colour of Spots	Refs.
Gibberellins	70% Sulfuric acid; 30 min heating in 120°C	Fluorescence	[53]
Stevioside ent-kaurenes	(a) Naphthoresorcinol-H_2SO_4	Various colours of spots; detection in both daylight and long-wave UV	[7]
	(b) Ac_2O-H_2SO_4		
	(c) 60% H_2SO_4; in all cases heating in 110°C during 10 min		
Pseudoptero-sin diterpenes	5% H_2SO_4 in EtOH + heating	Vis	[77]
Prenylsesquiterpenes, cembranes etc.	10% H_2SO_4 in MeOH	Vis	[76]
Taxanes	(a) Plates with fluorescent agent	(a) UV—254 nm (quenching of fluorescence)	[70]
	(b) vanillin:sulfuric acid solution	or	
		(b) Vis	
Taxanes besides taxine alkaloids	(a) Dragendorff reagent	Vis	[105]
	(b) Iodoplatinate reagent		
	(c) 3% H_2SO_4 in EtOH + heating in 120°C for 5 min		
Taxanes	Anisealdehyde reagent + heating	Vis	[73]
Ginkgolides	Heating in 150°C for 35–60 min	UV—365 nm; colours of fluorescence: Ginkgolide A—blue Ginkgolide B—yellow Ginkgolide C—orange	[65]
Ginkgolides	Exposure into ethanolic anhydride for 30 min, heating in 160°C for 30 min,	Densitometry—UV 366 nm	[66,110]
Kaurenes (simultaneously with sesquiterpenes and steroids)	Cobalt chloride/H_2SO_4 + heating	Vis	[10]

Compound	Reagent/treatment	Detection	Ref.
Triptolides	2% 3,5-dinitrobenzoic acid in ethanol and 5% NaOH in ethanol (in proportion 1:3)	densitometry at 535 nm	[31,111,112]
Triptolides—triptolide, tripchlorolide, wilformine	Potassium iodobismuthate solution	Vis; or densitometry at 218 nm	[33]
Tanshinones	—	UV—254 nm—various colours of fluorescence: brown, red, purple, yellow, and others	[40]
Clerodanes (from *Penian-thus zenkeri*)	Anisaldehyde reagent	Vis	[60]
Neo-clerodanes (from *Teucrium yemense*)	Anisaldehyde/H_2SO_4 reagent	Vis	[37]
Furano-clerodanes	1% 4-Dimethylaminobenzaldehyde in EtOH	Vis	[16]
Phorbol	Vanillin: H_2SO_4:EtOH (3:0.5:100) + heating in 120°C	UV (short-wave)	[12]
Phorbolesters	60% H_2SO_4 in EtOH + heating at 120°C (or 110°C) for 20 min	UV—254 nm	[13,107]
Tiglianes, ingenanes	Vanillin: H_2SO_4 reagent + heating at 100°C	Vis brown spots	[96]
Jatrophanes—esulatin A,B,C	Vanillin: H_2SO_4 reagent	Vis	[21]
Jatrophane polyesters, lathyranes	H_2SO_4:acetic acid:water (1:20:4) + heating	Vis	[29]
Spongiane -type diterpenes	Vanillin:sulfuric acid reagent + heating	Vis	[78]
Ent-trachy-loban-3β-ol	Anisaldehyde/H_2SO_4 reagent	—	[17]
Labdanes—marrubiin, premarrubiin	Anisaldehyde/H_2SO_4 reagent + heating to at 105°C for 4 min	Densitometry	[52]
Bis-labdanes—calcaratin D,E	10% Molybdic acid reagent	Vis	[59]
Labdanes—rotundifuran	10% H_2SO_4 in MeOH + heating in 105°C	Densitometry at vis range	[54]
Gibberellins	(a) Prechromatographic derivatization with solution of diazomethane in Et_2O—methylation of gibberellins; and after TLC (b) Postchromatographic derivatization using 50% H_2SO_4 in EtOH + heating at 130°C	Densitometric quantification in fluorescent mode——emission wavelength at 255 nm, excitation filter at 313 nm	[80]

A large scale separation of taxanes from the bark of *Taxus yunnanensis* was also monitored using analytical TLC method in system: SiO_2/AcOEt:hexane (4:3). The visualization of chromatograms was performed under UV-light, or after spraying with 1 M sulfuric acid [113].

19.6.3 QUANTITATIVE TLC ANALYSIS OF DITERPENES

Quantitative TLC analysis, using densitometry technique, of several groups of diterpenes was elaborated and used in practice, giving satisfactory results comparable to those obtained using other analytical techniques. Densitometric analysis is broadly used in the determination of tanshinones, triptolides, or ginkgolides in plant materials (see Figure 19.13) as well as pharmaceuticals. Densitometry in spectrofluorimetric mode is also used in quantitative elucidation of gibberellins in cell culture filtrates or fermentation broths. Figure 19.14 shows exemplary densitograms of gibberellins quantified in *Fusarium moniliforme* cell filtrates.

There are also examples of quantitative TLC determination of diterpenes in medical samples. Figure 19.15 shows exemplary separation of retinol simultaneously with tocopherol in blood plasma sample.

19.6.3.1 Chromatographic Systems

The examples of chromatographic systems used in the separation and quantitiative determination of diterpenes, and also detailed information concerning detection are described in Table 19.3.

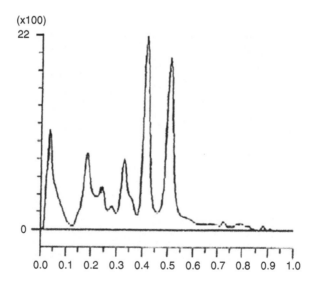

FIGURE 19.13 TLC scanning profile of a Ginkgo leaf extract on a sodium acetate impregnated handmade silica gel plate. Bilobalide, G-A, G-B, and G-C can be observed at R_f values of 0.52, 0.42, 0.33, and 0.18, respectively. (From van Beek, T.A., *J. Chromatogr. A*, 967, 21, 2002. With permission.)

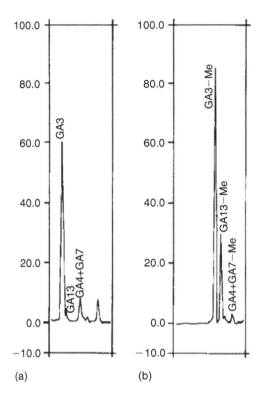

FIGURE 19.14 TLC of gibberellins. (a) Spectrofluorimetric densitogram of GA free acids isolated from the culture filtrate of *Fusarium moniliforme* (spraying with 50% (v/v) sulfuric acid in ethanol). (b) Spectrofluorimetric densitogram of GA methyl (Me) esters isolated from the same sample. (From Rachev, R.C. et al., *J. Chromatogr.*, 437, 287–293, 1988. With permission.)

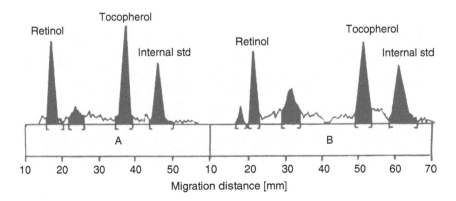

FIGURE 19.15 Densitograms of (A) a standard solution containing retinol, α-tocopherol, and internal standard; and (B) an extract of a volunteer's plasma containing retionl and α-tocopherol. (From Chavan, J.D. and Khatri, J.M., *J. Planar Chromatogr.*, 4, 280, 1992. With permission.)

TABLE 19.3
Quantitative TLC of Diterpenes

Compounds	Sample	Adsorbent	Solvent System	Detection/Quantification	References
Neo-clerodanes	*Teucrium* species	HPTLC silica gel plate	EtOAc–*n*-hexane (6:5)	Ehrlich reagent, and after that densitometric quantification	[97]
Ginkgolides A, B, C, J	*Ginkgo biloba* leaves; pharmaceutical preparations	HPTLC silica gel plate impregnated with NaOAc in methyl acetate	Toluene:AcOEt: Me_2CO: MeOH (10:5:5:0.6)	Exposure into ethanolic anhydride for 30 min, heating in 160°C for 30 min, and after that densitometric quantification in UV 366 nm	[66,110,114]
Ginkgolides A, B, C	*Ginkgo biloba* extract	Silica	AcOEt:Toluene: Me_2CO: hexane (4:4:1:1)	Densitometry at 370 nm	[115]
Triptolides	*Tripterygium* Hook species: *Tripterygium wilfordii*; pharmaceuticals	Silica	$CHCl_3$:Et_2O (2:1)	–2% 3,5-dinitrobenzoic acid in ethanol and 5% NaOH in ethanol (in proportion 1:3) + quantitation by densitometry at 535 nm [31,111,112]; densitometry at 218 nm [32]	[31,32,111,112]
Triptolide	Postsynthetic mixture	Silica	$CHCl_3$:MeOH (4:1)	Densitometry at 217 nm	[116]
Triptolide	Triptolide injection	Silica	Petrol ether:EtOAc (2:3)	Densitometry at 218 nm	[117]
Tanshinone-11; crypto-tanshinone	*Salvia miltiorrhiza*	Silica	Benzene:hexane:MeOH (7:2:1); CH_2Cl_2:hexane: MeOH (50:1:1)	Densitometry at 270 nm	[41]

Compound	Source	Stationary phase	Mobile phase	Detection	References
Tanshinone IIA	Pharmaceuticals, or *Salvia* sp. plant materials	Silica	• Cyclohexane: EtOAc (7:4) • Benzene:CHCl$_3$:Me$_2$CO (10:8:1) • Benzene:EtOAc (19:1)	Densitometry at 485 nm Densitometry at 470 nm [118,122,125]; 275 nm [119]; 280 nm [121]; 490 nm [120]; 270 nm [123]; 465 nm [124];	[87] [118–125]
Labdanes—marrubiin, premarrubiin	*Marrubium* species	Silica	Benzene:Me$_2$CO (17:3)	Anisaldehyde/H$_2$SO$_4$ reagent + heating at 105°C for 4 min, and after that densitometric quantification	[52]
Retinol	Blood serum samples	Silica	Petrol ether:Et$_2$O (1:1)	Densitometry at 330 nm	[126]
Retinol simultaneously with tocopherol	Blood serum samples	HPTLC silica gel 60 F$_{254}$ plate	CHCl$_3$;cyclohexane (55:45)	Densitometry at 290 nm	[127]
Gibberellins (as methylated derivatives)	*Fusarium moniliformae* culture	Silica gel G60 plate	EtOAc:CHCl$_3$:MeOH: acetic acid (20:20:2:0.2)	50% H$_2$SO$_4$ in EtOH and after that densitometric quantification in fluorescent mode	[80]
Gibberellins	Fermentation broths	HPTLC silica gel	Cyclohexane:Me$_2$CO (4:5)	Concentrated hydrochloric acid vapours for 30 min, and after that densitometric quantification in fluorescent mode (excitation at 366 nm, filter—400 nm)	[128]
Taxanes—paclitaxel	*Taxus* species—twigs	Silica plates	Stepwise gradient: heptane:MeOH:CHCl$_3$ 1st stage: (30:19:1), 2nd stage: (140:57:3)	Densitometry at 243 nm	[129]
Taxanes—as dansyl derivatives	*Taxus* species	Silica HPTLC plates	Cyclohexane:EtOAc: toluene:triethylamine: MeOH (9:6:4:3:1)	Densitometry at fluorescent mode (excitation at 254 nm, filter—440 nm)	[130]

19.6.4 PREPARATIVE TLC AS A METHOD OF ISOLATION (OR PURIFICATION) OF DITERPENES ON SEMIPREPARATIVE SCALE

In the literature, PTLC of diterpenes most often is a part of a complicated isolation procedure which also involves preparative column chromatography (CC), flash column chromatography, vacuum liquid chromatography (VLC), low-pressure column chromatography, MPLC, or preparative HPLC [9,19,20,42,46,69]. In some cases, the method of isolation consists in two- (tri-) step re-chromatography in similar chromatographic systems (e.g., CC-TLC-HPLC separation on silica using nonaqueous eluents [14], or CC on silica, CC on Florisil, and prep. TLC on silica [22]), but generally the procedures consist of re-chromatography in different systems (for example, reversed-phase and normal-phase systems) [36]. Sometimes, multistep re-chromatography using different TLC systems is necessary for the isolation of pure compounds [55] (see Table 19.4).

In other cases, a simple procedure of isolation is sufficient—extraction, preparative CC on silica and preparative TLC on silica allow to separate and isolate pure labdane-type diterpenes [58] or cryptotanshinone [39]. The procedure for the isolation of gibberellins from the bulbs of *Cymbidium*, which is based on LLE method (acidified water/ethyl ether) [57] is also simple.

In the isolation of endogenous gibberellins (as methylated derivatives), preparative TLC is used as the method of isolation of purified gibberelline fraction before GC–MS analysis [131].

Preparative TLC of taxanes in normal- and reversed-phase systems (previously optimized on analytical scale [72,93,132]) is frequently used as the step of the purification procedure of taxanes before HPLC analysis, or as one part of multistep isolation procedure of pure taxane diterpenes [72,93,98,133,134]. Figure 19.16 shows densitometric representation of preparative isolation of 10-deacetylbaccatin III from *Taxus baccata* twigs extract after preparative column separation.

19.6.4.1 Chromatographic Systems

TLC is most often one step in complex, multistep isolation procedure of diterpenes. Examples are presented in Table 19.4.

19.6.5 SPECIAL TECHNIQUES IN TLC OF DITERPENES

In the study of biosynthesis of GGPP (geranylgeranyldiphosphate) by fungi strains (*Gibberella fujikuroi*), qualitative TLC analysis with special detection method—radiodetection—was applied [79].

In other experiments, the mixtures of gibberellic, abietic, and kaurenoic acids (in trace amounts) were separated using μ-HPLC (C-18 capillary column) and deposited on various thin layers for ToF-SIMS detection. Thin layers examined are: (1) silica gel Si_{60} aluminium backplate, (2) monolithic silica gel ultra thin layer, and (3) Raman spectroscopy special prepared thin layer. The experiment showed that the first two layers can be used for deposition followed with ToF-SIMS identification in positive ions' scanning [138]. Similar experiments with the use of the same plates

TABLE 19.4
PTLC of Diterpenes

Compounds	Sample	Adsorbent	Solvent System	Detection	References
Phytane-type diterpenes—hanliuine I, II	*Salix matsudan*	SiO$_2$	• Hexane:Me$_2$CO (2:3) • CHCl$_3$:hexane:MeOH (6:1:1)	UV	[63]
Phytane-type diterpenes; clerodanes—bacchotricuneatine	*Olearia* species—fractions from SiO$_2$-CC	Silica gel PF$_{254}$	Et$_2$O:petroleum ether (1:9)	UV	[11]
Clerodanes	*Croton lechleri*—fractions from SiO$_2$-CC	SiO$_2$	MeOH:CH$_2$Cl$_2$ in different proportions	UV 254 nm; 366 nm, different reagents	[14]
Furano-clerodanes	*Croton campestris*—bark; fractions obtained in multistep procedure	SiO$_2$	Hexane:EtOAc (9:1)(7:3) or (1:1)	—	[16]
Neo-clerodanes	*Teucrium fruticans*—fractions from RP-18 prep. HPLC	SiO$_2$ 60 F$_{254}$	CH$_2$Cl$_2$:tert-butylmethyl ether (4:1)	—	[36]
Tigliane-type phorbol esters	*Wikstroemia canescens*—LLE-purified fraction	(1) SiO$_2$ (2) SiO$_2$ (3) RP-18	Petrol ether:acetone (7:3) CH$_2$Cl$_2$: MeOH (95:5) MeOH:water (90:10)	—	[55]
Jatrophane type, esulatin E,F; esulatin C	*Euphorbia esula*— fraction obtained in multistep, CC, VLC isolation in different systems	SiO$_2$ SiO$_2$	Hexane:THF:MeCN (20:5:1) Benzene: EtOAc (7:3)	—	[19] [20]
Jatrophane, ingenane, lathyrane, labdane, and other diterpenes	*Euphorbia segetalis*	SiO$_2$	Different eluents: • CH$_2$Cl$_2$:toluene:methyl-*tert*-butyl ether (9:9:2) (7:7:6) (4:4:1) (2:2:1) • Petrol.ether:methyl-*tert*-butyl ether (3:7)	—	[27]
Lathyrane diterpenes	*Euphorbia lagascae*	SiO$_2$	CHCl$_3$:MeOH (9:1) developed 2x	UV	[30]

(continued)

TABLE 19.4 (continued)
PTLC of Diterpenes

Compounds	Sample	Adsorbent	Solvent System	Detection	References
Pubescenes	*Euphorbia pubescens*	SiO$_2$	CHCl$_3$:Me$_2$CO (9:1)	UV 254 nm,	[28]
Daphnane-type diterpenes	*Wikstroemia monticola* fraction obtained in multistep, LLE, and CC procedure	SiO$_2$ 60 PF$_{254}$	CH$_2$Cl$_2$: EtOAc (7:3)	—	[56]
Daphnane- and Tigliane-type diterpenes	*Excoecaria bicolor* fr. obtained in multistep, LLE, and CC procedure	SiO$_2$	Petrol ether: diethyl ether (1:1); (1:3)	—	[25]
Labdanes	*Aframomum daniellii* (seeds) fraction from SiO$_2$-CC	Kiselgel GF$_{254}$	1% AcOEt in hexane	—	[58]
Ent-kaurenes	*Mikania banisteriae* fraction obtained in multistep procedure (flash column, CC in different systems, MPLC)	SiO$_2$	Cyclohexane: Et$_2$O (2:1)	—	[9]
Neo-clerodanes—salvinorin A,C	*Salvia divinorum*—leaves	SiO$_2$	CHCl$_3$:MeOH:H$_2$O (100:10:1); hexane: EtOAc (1:1)	—	[44]
Neo-clerodanes	*Teucrium yemense*	SiO$_2$ (centrifugal prep. TLC)	Toluene: EtOAc: acetic acid (65:35:1)	—	[37]
Cembranoids	*Cleome viscosa* leaves—fraction after flash chromatography	Silica gel 60F$_{254}$ precoated plates	Petrol.ether:AcOEt (6:100)—developed three times	—	[34]
Cryptotanshinone	*Salvia miltiorrhiza*—callus cell culture—CC-isolated fraction	SiO$_2$	Benzene: AcOEt (8:2)	—	[39]
Tanshinones	*Salvia miltiorrhiza*—rhizomes—CC-isolated fractions	Silica gel 60F$_{254}$ plates (1 mm thickness)	Petrol ether: Et$_2$O: Me$_2$CO (75:20:5)	UV 254	[40]

Tanshinone I, IIA, crypto-, dihydro-tanshinone	$Salvia\ miltiorrhiza$	Silica gel	CH_2Cl_2; or CH_2Cl_2: AcOEt (4:1)	—	[135]
Tanshinones	$Salvia\ miltiorrhiza$ – crude extracts	SiO_2 RP-18	Hexane:EtOAc (10:1) 70% MeOH in water	—	[136]
Abietanes—yunnamin, danshenol C, A, dihydro-tanshinone	$Salvia\ yunnanensis$	Silica gel	Petrol ether:EtOAc (4:1)	UV 254	[45]
Abietanes—danshenol A, 15-epi-danshenol A	$Salvia\ glutinosa$ roots—fraction after VLC and sephadex column	Silica gel	Benzene: AcOEt (6:4)	—	[46]
Abietanes—ferruginol, horminone etc.	$Salvia\ eriophora$ roots	Silica gel	• Petrol ether • Toluene:CH_2Cl_2 (17:3) • CH_2Cl_2:MeOH (49:1)	—	[47]
Tiglianes, ingenanes	$Euphorbia\ nubica$, $E.\ helioscopia$—CC-isolated fraction	SiO_2	Hexane:Me_2CO (in diff. prop.) or benzene: AcOEt (in diff. prop.)	—	[22]
Tiglianes, ingenanes	$Euphorbia$ species—latex	Silica gel	Et_2O:petroleum ether (in different proportions)	Vanillin: H_2SO_4 reagent + heating at 100°C	[107]
Ingenanes	$Euphorbia\ peplus$	• Silica buff. with phosphate buffer (pH 7.0) • Kiselgur coated with 15% dipropylene glycol • SiO_2	• $CHCl_3$:Et_2O:benzene (1:3:3) • Heptane:benzene (17:3) • $CHCl_3$:Et_2O (19:1)	—	[24]

(continued)

TABLE 19.4 (continued)
PTLC of Diterpenes

Compounds	Sample	Adsorbent	Solvent System	Detection	References
Abietanes, quinoid-type diterpenoids	*Euphorbia portlandica*	Silica gel	• CH_2Cl_2: MeOH (49:1) or (19:1)—developed 2x • CH_2Cl_2: Et_2O (19:1) • $CHCl_3$: AcOE (9:1)—developed 4x • CH_2Cl_2: AcOE (in diff. prop.)	UV 254	[23]
Gibberellins	*Cymbidium* sp. pseudobulbs—samples purified using LLE method	Silica gel 60 precoated plates	$CHCl_3$: AcOEt:CH_3COOH (60:40:5)	—	[57]
Gibberellins—as methylated derivatives	Purified fractions of endogenous pre-methylated gibbe-rellins	Silica gel	Toluene: AcOEt:CH_3COOH in different proportions	—	[131]
Seco-ent-kaurenes simultaneously with longi-rabdolides	*Rabdosia longituba*—CC-isolated fraction	SiO_2	Pure Et_2O $CHCl_3$:acetone (7:3) or (8:2)	—	[48]
Ent-kaurenes—shikokianin, rabdoternin A, rabdosichuanin, lasiokaurin	*Isodon japonica* leaves	Silica gel	• $CHCl_3$:MeOH (30:1)—developed two times • $CHCl_3$:acetone (6:1)	UV 254	[50]
Oxetane ent-kauranoid—mayoecrystal I	*Isodon japonica*	Silica gel	Petrol ether:acetone (4:1)—developed three times	UV 254	[51]
Abietane diterpenes	*Larix kaempferi* leaves	Silica gel	$CHCl_3$:MeOH (100:1)	UV 254	[67]

Compound	Source	Stationary phase	Mobile phase	Detection	Ref.
Seco-abietane diterpenes	*Picea glehnii*	Silica gel	• Hexane:EtOAc:MeOH (25:25:1) • CHCl$_3$:MeOH (9:1) or (19:1)	UV 254	[68]
Ent-18-hydroxytrachyloban-3β-ol	*Croton zambesicus*	Silica gel	Toluene: EtOAc:MeCN (5:2:3) or (40:9:1)	Anisaldehyde/H$_2$SO$_4$ reagent	[18]
Bis-labdanic diterpenoids—calcaratin D,E	*Alpinia calcarata*	Silica gel	Hexane:AcOEt (23:2) or (43:7)—developed 2x	UV—254 nm	[59]
Taxanes—10-deacetyl-baccatin III, baccatin III, cephalomannine, paclitaxel	*Taxus* species—twigs—crude extracts or CC-isolated fractions	Silica gel 60 F$_{254}$ plates	Benzene:CHCl$_3$:Me$_2$CO:MeOH (8:37:6:3)	UV—254 nm, or densitometry at 230 nm	[72,98]
			CH$_2$Cl$_2$:Dx:Me$_2$CO:MeOH (84:10:5:1)	UV—254 nm, or densitometry at 230 nm	[93,133,134]
Taxanes	*Taxus* species—twigs—crude extracts or CC-isolated fractions	Silanized silica (RP-2)	MeOH:water (7:3)	UV—254 nm, or densitometry at 230 nm	[132,93]
Taxanes—yunnanxane, etc.	*Taxus wallichiana*—cell suspension cultures	Silica gel	Benzene:Me$_2$CO (8:2)	UV—254	[74]
Taxanes—taxu-yunnanines	*Taxus yunnanensis*	Silica gel	iPrOH:CHCl$_3$ (2:1)	10% H$_2$SO$_4$ in EtOH	[137]
Taxanes	*Taxus chinensis*—needles	Silica gel	Hexane:Me$_2$CO; hexane:AcOEt; CHCl$_3$:MeOH—all eluents in different proportions	10% H$_2$SO$_4$ in EtOH + heating	[71]
Spongiane diterpenes	Marine sponge—*Hyatella intestinalis*	Silica gel	• AcOEt:hexane (1:5) or (1:10) • Me$_2$CO:hexane (1:1) • Benzene:Et$_2$O (2:1)	—	[4]

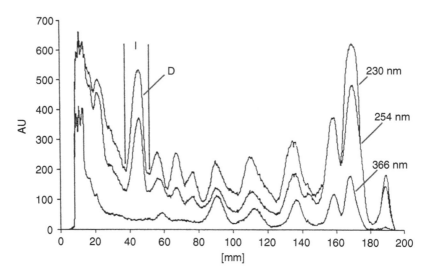

FIGURE 19.16 Densitogram obtained from a preparative thin layer chromatogram of the 10-deacetylbaccatin fraction from the preparative column TLC separation was performed in system: silica/dichloromethane:dioxane:acetone:methanol (84:10:5:1). The chromatogram was scanned at three wavelengths—366, 254, and 230 nm—by means of a Camag 3 scanner. The fraction indicated: "I" contains D—10-deacetylbaccatin III. (From Hajnos, M.Ł. et al., *Chromatographia*, 56, 91, 2002. With permission.)

showed that the separation and ToF-SIMS detection of abietic and gibberellic acid is possible using TLC method (plates used were eluted with EtOH:water, 4:6) [139]. TLC–ToF-SIMS analysis of gibberellins was performed and compared with other identification methods. It was confirmed that the former is more sensitive and specific in comparison with the traditional methods [140].

Special, autobiographic TLC method was applied in the investigation of biological activities of cembrane diterpenoids from *Cleome viscosa (Capparaceae)*. Preliminarily isolated fractions of *Cleome viscosa* leaves extract were chromatographed on TLC silica gel plates, and then they were sprayed separately with different pathogen spore suspensions (bacteria: *Bacillus subtilis, Pseudomonas fluorescens*; fungi: *Cladosporium cucumerinum*). Fractions having a strong inhibitory action on the growth of pathogens appeared as white zones against purple background (after spraying with iodotetrazolium salt solution). Known antibacterial and antifungal compounds were used as the positive controls [34].

It was found that centrifugal preparative TLC on silica (4 mm thickness) eluted with toluene:EtOAc:acetic acid (65:35:1) is an efficient system for the separation of closely related neo-clerodane diterpenoids from *Teucrium yemense* [37].

Overpressure layer chromatography (RP-OPLC) on RP-18 plates eluted with MeOH:water (9:1) (flow-rate 0.1 mL/min) gave a good separation of phorbol diesters from *Croton* oil [107].

REFERENCES

1. Keeling, C.I. and Bohlmann, J., Diterpene resin acids in conifers, *Phytochemistry*, 67, 2415–2423, 2006.
2. Sung, P.J., Sheu, J.H., and Xu, J.P., Survey of briarane-type diterpenoids of marine origin, *Heterocycles*, 57, 535–579, 2002.
3. Sheu, J.H. et al., Novel meroditerpenoid-related metabolites from the Formosan soft coral *Nephtea chabrolii*, *J. Nat. Prod.*, 67, 2048–2052, 2004.
4. Cambie, R.C. et al., Chemistry of sponges. IV. Spongian diterpenes from *Hyatella intestinalis*, *J. Nat. Prod.*, 51, 293–297, 1988.
5. Gavagnin, M. and Fontana, A., Diterpenes from marine opisthobranch molluscs, *Curr. Org. Chem.*, 4, 1201–1248, 2000.
6. Bellini, A.A., de Oliveira, D.C.R., and Vichnewski, W., Steroidal saponin, 7-oxostigmasterol and diterpenes from *Aspilia montevidensis*, *Biochem. Syst. Ecol.*, 27, 317–319, 1999.
7. Kinghorn, A.D. et al., A phytochemical screening procedure for sweet ent-kaurene glycosides in the genus *Stevia*, *J. Nat. Prod.*, 47, 439–444, 1984.
8. Geuns, J.M.C. et al., Metabolism of stevioside by chickens, *J. Agric. Food Chem.*, 51, 1095–1101, 2003.
9. Lobitz, G.O., Tamayo-Castillo, G., and Merfort, I., Diterpenes and sesquiterpenes from *Mikania banisteriae*, *Phytochemistry*, 46, 161–164, 1997.
10. Torrenegra, R.D. and Tellez, A.N., Chemotaxonomic value of melampolides in *Espeletia* species (Asteraceae), *Biochem. Syst. Ecol.*, 23, 449–450, 1995.
11. Warning, U. et al., Diterpenes from *Olearia* species, *J. Nat. Prod.*, 51, 513–516, 1988.
12. Mishra, N.C., Estensen, R.D., and Abdel-Monem, M.M., Isolation and purification of phorbol from croton oil by reversed-phase column chromatography, *J. Chromatogr.*, 369, 435–439, 1986.
13. Bauer, R., Tittel, G., and Wagner, H., Isolation and detection of phorbolesters in croton oil with HPLC. A new method for diterpene ester-screening in *Euphorbiaceae*, *Planta Med.*, 48, 10–16, 1983.
14. Cai, Y., Chen, Z.P., and Philipson, J.D., Diterpenes from *Croton lechleri*, *Phytochemistry*, 32, 755–760, 1993.
15. Roengsumran, S. et al., Croblongifolin, a new anticancer clerodane from *Croton oblongifolius*, *Planta Med.*, 68, 274–277, 2002.
16. El Babili, F. et al., Three furano-diterpenes from the bark of *Croton campestris*, *Phytochemistry*, 48, 165–169, 1998.
17. Block, S. et al., Ent-trachyloban-3β-ol, a new cytotoxic diterpene from *Croton zambesicus*, *Planta Med.*, 68, 647–649, 2002.
18. Baccelli, C. et al., Diterpenes isolated from *Croton zambesicus* inhibit KCl-induced contractions, *Planta Med.*, 71, 1036–1039, 2005.
19. Günther, G. et al., Jatrophane diterpenoids from *Euphorbia esula*, *Phytochemistry*, 47, 1309–1313, 1998.
20. Hohmann, J. et al., Macrocyclic diterpene polyesters of the jatrophane type from *Euphorbia esula*, *J. Nat. Prod.*, 60, 331–335, 1997.
21. Hohmann, J. et al., Macrocyclic diterpene polyesters of the jatrophane type from *Euphorbia esula*, *Acta Pharm. Hung.*, 68, 175–182, 1998.
22. Zayed, S.M.A.D. et al., Dietary cancer risk from conditional cancerogens (tumor promoters) in produce of livestock fed on species of spurge (*Euphorbiaceae*), *J. Cancer Res. Clin. Oncol.*, 127, 40–47, 2001.

23. Madureira, A.M. et al., A new sesquiterpene-coumarin ether and a new abietane diterpene and their effects as inhibitors of P-glycoprotein, *Planta Med.*, 70, 828–833, 2004.

24. Rizk, A. et al., Biologically active diterpene esters from *Euphorbia peplus*, *Phytochemistry*, 24, 1605–1606, 1985.

25. Karalai, C. et al., Medicinal plants of *Euphorbiaceae* occurring and utilized in Thailand. V. Skin irritants of the daphnane and tigliane type in latex of *Excoecaria bicolor* and the uterotonic activity of the leaves of the tree, *Phytother. Res.*, 9, 482–488, 1995.

26. Pieters, L.A.C. and Viletinck, A.J., Vacuum liquid chromatography and quantitative ^{1}H NMR spectroscopy of tumor-promoting diterpene esters, *J. Nat. Prod.*, 52, 186–190, 1989.

27. Jakupovic, J. et al., Diterpenes from *Euphorbia segetalis*, *Phytochemistry*, 47, 1583–1600, 1998.

28. Valente, C. et al., Pubescenes, jatrophane diterpenes, from *Euphorbia pubescens*, with multidrug resistance reversing activity on mouse lymphoma cells, *Planta Med.*, 70, 81–84, 2004.

29. Valente, C. et al., Euphopubescenol and euphopubescene, two new jatrophane polyesters, and lathyrane-type diterpenes from *Euphorbia pubescens*, *Planta Med.*, 70, 244–249, 2004.

30. Duarte, N. et al., New macrocyclic lathyrane diterpenes, from *Euphorbia lagascae*, as inhibitors of multidrug resistance of tumour cells, *Planta Med.*, 72, 162–168, 2006.

31. Nie, K., An, W., and Xia, L., Comparison of the active constituents in three wingnut, *Tripterygium* Hook f., from different regions by thin-layer chromatography, *Chin. J. Herb. Med.* (*Zhongcaoyao*), 25, 539–541, 1994.

32. Yu, L. and Li, J., Determination of triptolide in the common three wingnut, *Tripterygium wilfordii*, by thin-layer chromatography, *Chin. J. Herb. Med.* (*Zhongcaoyao*), 19, 66–68, 1988.

33. Zhou, Y. and Fang, Z., Comparison of the quality of Leigongteng, *Tripterygium wilfordii* Hook. f., wild and cultured, *J. Chin. Trad. Med.* (*Zhongguo Zhongyao Zazhi*), 20, 145–147, 1995.

34. Williams, L.A.D. et al., Biological activities of an extract from *Cleome viscosa* L. (*Capparaceae*), *Naturwissenschaften*, 90, 468–472, 2003.

35. Demetzos, C. et al., Chemical analysis and antimicrobial activity of the resin Ladano, of its essential oils and of the isolated compounds, *Planta Med.*, 65, 76–78, 1999.

36. Coll, J. and Tandrón, Y., Neo-clerodane diterpenes from *Teucrium fruticans*, *Phytochemistry*, 65, 387–392, 2004.

37. Sattar, E.A. et al., Neo-clerodane diterpenoids from *Teucrium yemense*, *Phytochemistry*, 40, 1737–1741, 1995.

38. Hu, P. et al., Quality assessment of Radix Salviae Miltiorrhizae, *Chem. Pharm. Bull.*, 53, 481–486, 2005.

39. Wu, C.-T. et al., Isolation and quantitative analysis of cryptotanshinone, an active quinoid diterpene formed in callus of *Salvia miltiorrhiza* Bunge, *Biol. Pharm. Bull.*, 26, 845–848, 2003.

40. Zhang, K.-Q. et al., Antioxidative components of tanshen (*Salvia miltiorrhiza* Bung), *J. Agric. Food Chem.*, 38, 1194–1197, 1990.

41. Lo, S., Chen, G., and Tong, Y., Determination of tanshinone-11 and cryptotanshinone in plant, *Salvia miltiorrhiza*, by thin-layer chromatography, *Chin. J. Pharm. Anal.* (*Yaowu Fenxi Zazhi*), 8, 154–157, 1988.

42. Tan, N., Topcu, G., and Ulubelen, A., Norabietane diterpenoids and other terpenoids from *Salvia recognita*, *Phytochemistry*, 49, 175–178, 1998.

43. Ulubelen, A., Sönmez, U., and Topcu, G., Diterpenoids from the roots of *Salvia sclarea*, *Phytochemistry*, 44, 1297–1299, 1997.
44. Valdés, L.J. et al., Salvinorin C, a new neoclerodane diterpene from a bioactive fraction of the hallucinogenic Mexican mint *Salvia divinorum*, *Org. Lett.*, 3, 3935–3937, 2001.
45. Xu, G. et al., Two new abietane diterpenoids from *Salvia yunnanensis*, *Planta Med.*, 72, 84–86, 2006.
46. Nagy, G. et al., Danshenol-A and 15-epi-danshenol-A from the roots of *Salvia glutinosa*, *Biochem. Syst. Ecol.*, 26, 797–799, 1998.
47. Ulubelen, A. et al., Cardioactive diterpenes from the roots of *Salvia eriophora*, *Planta Med.*, 68, 818–821, 2002.
48. Takeda, Y. et al., Longirabdolides E,F and G, 6,7-seco-ent-kaurenoids from *Rabdosia longituba*, *Phytochemistry*, 45, 785–789, 1997.
49. Han, Q.-B. et al., Cytotoxic ent-kaurene diterpenoids from *Isodon rubescens* var. *rubescens*, *Planta Med.*, 70, 269–272, 2004.
50. Bai, S.-P. et al., Two novel ent-kauranoid diterpenoids from *Isodon japonica* leaves, *Planta Med.*, 71, 764–769, 2005.
51. Han, Q. et al., A novel cytotoxic oxetane ent-kauranoid from *Isodon japonicus*, *Planta Med.*, 70, 581–584, 2004.
52. Telek, E. et al., Chemical test with *Marrubium* species, *Acta Pharm. Hung.*, 67, 31–37, 1997.
53. Kimura, Y., Identity of the *Pisum* giberellin, *J. Agric. Food Chem.*, 18, 182, 1970.
54. Wahli, F., Stability tests of *Vitex agnus-castus* (chaste tree) extracts, *CBS*, 91, 12–13, 2003.
55. Dagang, W. et al., Oligo- and macrocyclic diterpenes in *Thymelaeaceae* and *Euphorbiaceae* occurring and utilized in Yunnan (southwest China). 4. Tigliane type diterpene esters (phorbol-12,13-diesters) from *Wikstroemia canescens*, *Phytother. Res.*, 7, 194–196, 1993.
56. Jolad, S.D., et al., Daphnane diterpenes from *Wikstroemia monticola*: wikstrotoxins A-D, huratoxin, and ecoecariatoxin, *J. Nat. Prod.*, 46, 675–680, 1983.
57. Zheng, C., et al., Involvement of ethylene and gibberellin in the development of rhizomes and rhizome-like shoots in oriental *Cymbidium* hybrids, *J. Jpn. Soc. Hortic. Sci.*, 74, 306–310, 2005.
58. Kimbu, S., Ngadjui, B., and Sondengam, L.B., A new labdane diterpenoid from the seeds of *Aframomum daniellii*, *J. Nat. Prod.*, 50, 230–231, 1987.
59. Kong, L.-Y., Qin, M.-J., and Niwa, M., New cytotoxic bis-labdanic diterpenoids from *Alpina calcarata*, *Planta Med.*, 68, 813–817, 2002.
60. Achenbach, H. and Hemrich, H., Clerodane-type diterpenes and other constituents of *Penianthus zenkeri*, *Phytochemistry*, 30, 1957–1962, 1991.
61. Ara, I. et al., Tricyclic diterpenoids from the stem bark of *Azadirachta indica*, *J. Nat. Prod.*, 51, 1054–1061, 1988.
62. Ara, I. et al., Diterpenoids from the stem bark of *Azadirachta indica*, *Phytochemistry*, 28, 1177–1180, 1989.
63. Zheng, S. et al., Two new acyclic diterpene-γ-lactones from *Salix matsudan*, *Planta Med.*, 66, 487–489, 2000.
64. Jaracz, S., Maklik, S., and Nakanishi, K., Isolation of ginkgolides A,B,C,J and bilobalide from *G.biloba* extracts, *Phytochemistry*, 65, 2897–2902, 2004.
65. Tallevi, S.G., Detection of ginkgolides by thin-layer chromatography, *J. Nat. Prod.*, 54, 624–625, 1991.
66. van Beek, T.A., Chemical analysis of *Ginkgo biloba* leaves and extracts, *J. Chromatogr. A*, 967, 21–55, 2002.

67. Ohtsu, H. et al., Anti-tumor-promoting rearranged abietane diterpenes from the leaves of *Larix kaempferi*, *Planta Med.*, 65, 664, 1999.
68. Tanaka, R. et al., A new seco-abietane-type diterpene from the stem bark of *Picea glehni*, *Planta Med.*, 70, 877–880, 2004.
69. Cao, C.M. et al., Studies on chemical constituents in heartwood of *Taxus cuspidata*, *Chin. Trad. Med.* (*Zhongguo Zhongyao Zazhi*), 31, 1510–1513, 2006.
70. Srinivasan, V., Roberts, S.C., and Shuler, M.L., Combined use of six-well polystyrene plates and thin layer chromatography for rapid development of optimal plant cell culture processes: application to taxane production by *Taxus* sp., *Plant Cell Rep.*, 16, 600–604, 1997.
71. Shi, Q.-W., Oritani, T., and Sugiyama, T., Three novel bicyclic taxane diterpenoids with verticillene skeleton from the needles of Chinese Yew, *Taxus chinensis* var. *mairei*, *Planta Med.*, 65, 356–359, 1999.
72. Głowniak, K. and Mroczek, T., Investigations on preparative thin-layer chromatographic separation of taxoids from *Taxus baccata* L., *J. Liq. Chromatogr. Relat. Technol.*, 22, 2483–2502, 1999.
73. Stasko, M.W. et al., Multimodal thin-layer chromatographic separation of taxol and related compounds from *Taxus brevifolia*, *J. Liq. Chromatogr.*, 12, 2133–2143, 1989.
74. Agrawal, S. et al., Isolation of taxoids from cell suspension cultures of *Taxus wallichiana*, *Planta Med.*, 66, 773–775, 2000.
75. Sabry, O.M.M. et al., Neurotoxic meroditerpenoids from the tropical marine brown alga *Stypopodium flabelliforme*, *J. Nat. Prod.*, 68, 1022–1030, 2005.
76. Teixeira, V.L., Almeida, S.A., and Kelecom, A., Chemosystematic and biogeographic studies of the diterpenes from the marine brown alga *Dictyota dichotoma*, *Biochem. Syst. Ecol.*, 18, 87–92, 1990.
77. Rodriguez, I.I. et al., New pseudopterosin and seco-pseudopterosin diterpene glycosides from two Colombian isolates of *Pseudopterogorgia elisabethae* and their diverse biological activities, *J. Nat. Prod.*, 67, 1672–1680, 2004.
78. Thompson, J.E. et al., Environmentally induced variation in diterpene composition of the marine sponge *Rhopaloeides odorabile*, *Biochem. Syst. Ecol.*, 15, 595–606, 1987.
79. Knöss, W. and Reuter, B., Biosynthesis of [^{14}C] geranylgeranyl-diphosphate by a prenyl transferase system from a mutant strain of *Gibberella fujikuroi*, *Anal. Biochem.*, 239, 208–212, 1996.
80. Rachev, R.C. et al., Separation of gibberellins by thin-layer chromatography and gas chromatography, and possibilities for their quantitative analysis, *J. Chromatogr.*, 437, 287–293, 1988.
81. Curini, M. et al., Glaucopines A and B, new cyathane diterpenes from the fruiting body of *Sarcodon glaucopus*, *Planta Med.*, 71, 194–196, 2005.
82. Roisin, T., Pasteels, J.M., and Braekman, J.C., Soldier diterpene patterns in relation with aggressive behaviour, spatial distribution and reproduction of colonies in *Nasutitermes princeps*, *Biochem. Syst. Ecol.*, 15, 253–261, 1987.
83. Gebbinck, E.A.K., Jansen, B.J.M., and de Groot, A., Insect antifeedant activity of clerodane diterpenes and related model compounds, *Phytochemistry*, 61, 737–770, 2002.
84. Kato, T. et al., Construction of trinervitane and kempane skeletons based on biogenetical routes, *Helv. Chim. Acta*, 84, 47–68, 2001.
85. Yagi, A. et al., Possible active components of Tan-Shen (*Salvia miltiorrhiza*) for protection of the myocardium against ischemin-induced derangements, *Planta Med.*, 55, 51–54, 1989.

86. Liu, Y. et al., Thin layer chromatographic studies on radix Ginseng, fructus Schisandrae and radix Salviae miltiorrhizae in Yixinfumai granules, *J. Chin. Trad. Med.* (*Zhongguo Zhongyao Zazhi*), 17, 217–219, 1992.

87. Wang, X., Guo, H., and Wang, J., Determination of schizandrin A and tanshinone IIA in Wulingwan by thin-layer chromatography, *Chin. J. Herb. Med.* (*Zhongcaoyao*), 26, 131–132, 1995.

88. Chinou, I., Labdanes of natural origin—biological activities (1981–2004), *Curr. Med. Chem.*, 12, 1295–1317, 2005.

89. Geuns, J.M.C., Molecules of interest. Stevioside, *Phytochemistry*, 64, 913–921, 2003.

90. Qiu, D. and Kao, P.N., Immunosupressive and anti-inflammatiory mechanisms of triptolide, principal active diterpenoid from the chinese medicinal herb *Tripterygium willfordii* Hook. f., *Drug R&D*, 4, 1–18, 2003.

91. Chen, X. et al., Triptolide, a constituent of immunosuppressive Chinese herbal medicine, is a potent suppressor of dendritic-cell maturation and trafficking, *Blood*, 106, 2409–2416, 2005.

92. Tada, M. et al., Antiviral diterpenes from *Salvia officinalis*, *Phytochemistry*, 35, 539–541, 1994.

93. Hajnos, M.Ł. et al., Optimisation of the isolation of some taxoids from yew tissues, *J. Planar Chromatogr.*, 14, 119–125, 2001.

94. Alvarenga, S.A.V. et al., General survey and some taxonomic implications of diterpenes in the *Asteraceae*, *Bot. J. Linn. Soc.*, 147, 291–308, 2005.

95. Parmar, V.S. et al., Constituents of the yew trees, *Phytochemistry*, 50, 1267–1304, 1999.

96. Sosath, S., Ott, H.H., and Hecker, E., Irritant principles of the spurge family (*Euphorbiaceae*). XIII. Oligocyclic and macrocyclic diterpene esters from latices of some *Euphorbia* species utilized as source plants of honey, *J. Nat. Prod.*, 51, 1062–1074, 1988.

97. Lei, W., Shurong, L., and Guangyi, L., Determination of four neo-clerodane diterpenoids of *Teucrium* species by HPTLC scanning, *Yaoxue Xuebao*, 35, 296, 2000.

98. Głowniak, K. et al., The application of zonal thin-layer chromatography to the determination of paclitaxel and 10-deacetylbaccatin III in some *Taxus* species, *J. Planar Chromatogr.*, 12, 328–335, 1999.

99. Zhou, C., Peng, S., and Liao, W., A method for the quality control of the penetrating liquid of Leigongteng, *Tripterygium wilfordii*, *Chin. J. Herb. Med.* (*Zhongcaoyao*), 24, 191–192, 1993.

100. Lin, M., Wang, C., and Wang, Y., An improved procedure for the separation of tanshinone IIA in Radix Salviae Miltiorrhizae, *J. Chin. Trad. Patent Med.* (*Zhongchengyao*), 23, 527–528, 2001.

101. Chen, X., Li, W., and Xia, W., Comparative study of the chemical components of *Salvia miltiorrhiza* Bge. collected from different regions, *J. Chin. Trad. Med.* (*Zhongguo Zhongyao Zazhi*), 22, 522–524, 1997.

102. Liang, H. and Li, J., Identification of chrysophanol and tanshinone IIA in Qingnao Fushen oral liquid by thin-layer chromatography, *Chin. J. Pharm. Anal.* (*Yaowu Fenxi Zazhi*), 18, 207–208, 1998.

103. Li, D. and Peng, D., An improved method for identification of tanshinone IIA in compound Danshen pills by thin-layer chromatography, *J. Chin. Trad. Patent Med.* (*Zhongchengyao*), 21, 51, 1999.

104. Lin, W., Peng, J., and Huang, N., Study of the quality standard for Jiedaiting tablets, *J. Chin. Trad. Patent Med.* (*Zhongchengyao*), 22, 766–768, 2000.

105. Vanhaelen, M. et al., Taxanes in *Taxus baccata* pollen: cardiotoxicity and/or allergenicity? *Planta Med.*, 68, 36–40, 2002.

106. Wagner, H. et al., *Ginkgo biloba*. DC- und HPLC-Analyse von ginkgo-extrakten und ginkgo-extrakte enthaltenden phytopräparaten. (*Ginkgo biloba*. TLC and HPLC analysis of ginkgo extracts and phytopreparation containing ginkgo extracts), *Deutsch. Apoth. Ztg.*, 129, 2421–2429, 1989.
107. Erdelmeier, C.A.J., Van Leeuwen, P.A.S., and Kinghorn, A.D., Phorbol diester constituents of croton oil: separation by two-dimensional TLC and rapid purification utilizing reversed-phase overpressure layer chromatography (RP-OPLC), *Planta Med.*, 54, 71–75, 1988.
108. Datke, W. et al., Reinvestigation of endogenous gibberellins in immature rye ears, *Biochem. Physiol. Pflanz.*, 184, 249–258, 1989.
109. Nagy, M. and Hodur, C., Effect of abscisic acid (ABA) on endogenous gibberellin levels in *Phaseolus vulgaris* L. seeds during their swelling, *Acta Biol. Acad. Sci. Hung.*, 33, 77–86, 1982.
110. Yan, Y. et al., Study on assay of terpenes in the leaves of *Ginkgo biloba* extract and its preparations by insitu fluorometric TLC, *J. Chin. Trad. Med.* (*Zhongguo Zhongyao Zazhi*), 22, 159–162, 1997.
111. Lo, M., Li, J., and Dong, K., Determination of triptolide in Leigongteng tablets by thin-layer chromatography, *Chin. J. Hosp. Pharm.* (*Zhongguo Yaoxue Zazhi*), 16, 76–78, 1996.
112. Yang, C., Zhou, T., and Qin, W., The content variation of triptolide in *Tripterygium wilfordii* in the different seasons, *Chin. J. Hosp. Pharm.* (*Zhongguo Yaoxue Zazhi*), 21, 25–27, 2001.
113. Xue, J. et al., A large scale separation of taxanes from the bark extract of *Taxus yunnanensis* and ^{1}H and ^{13}C-NMR assignments for 7-epi-10-deacetyltaxol, *Chin. J. Chem*, 19, 82–90, 2001.
114. Peishan, X. et al., Fluorophotometric thin-layer chromatography of *Ginkgo* terpenes by postchromatographic thermochemical derivatization and quality survey of commercial *Ginkgo* products, *JAOAC Int.*, 84, 1232–1241, 2001.
115. Tang, Y., Lou, F., and Zheng, W., Determination of terpene lactones in the extract of *Ginkgo biloba* L. by thin-layer chromatography, *Chin. J. Pharm. Anal.* (*Yaowu Fenxi Zazhi*), 18, 305–307, 1998.
116. Chen, D. et al., Determination of semisynthetic triptolide by thin-layer chromatography, *Chin. J. Pharm. Ind.* (*Zhonguo Yiyao Gongye Zazhi*), 21, 211–212, 1990.
117. Xia, Z. et al, Qualitative analysis of triptolide injection by thin-layer chromatography, *Chin. J. Hosp. Pharm.* (*Zhongguo Yiyuan Yaoxue Zazhi*), 11, 118–119, 1991.
118. Li, X. and Tu, W., Determination of tanshinone IIA in"Prostate Healthcare Bag" by thin-layer chromatography, *Chin. J. Herb. Med.* (*Zhongcaoyao*), 26, 520–524, 1995.
119. Liu, J., Determination of tanshinone IIA in Jiannao Yizhi chongji by thin-layer chromatography, *J. Chin. Trad. Med.* (*Zhongguo Zhongyao Zazhi*), 20, 101, 1995.
120. Qi, B., Determination of tanshinone IIA in Jiangzhikang granules by thin-layer chromatography, *J. Chin. Trad. Patent Med.* (*Zhongchengyao*), 18, 13–14, 1996.
121. Zhang, L., Yu, Z., and Xu, X., Study of water–alcohol extraction process of Radix Salviae Miltiorrhizae, *J. Chin. Trad. Patent Med.* (*Zhongchengyao*), 17, 1–3, 1995.
122. Su, Z. et al., Study of degradation dynamics of tanshinone IIA in alcohol-extract of Radix Salviae Miltiorrhizae during extracting process, *J. Chin. Trad. Patent Med.* (*Zhongchengyao*), 19, 1–3, 1997.
123. Hu, J. et al., Determination of tanshinone IIA in three kinds of plants belonging to *Salvia* by thin-layer chromatography, *J. Chin. Trad. Patent Med.* (*Zhongchengyao*), 22, 723–724, 2000.

124. Che, X., Determination of tanshinone IIA in Jinbuhuan pills by thin-layer chromatography, *J. Chin. Trad. Herb. Drug* (*Zhongcaoyao*), 32, 802–803, 2001.
125. Shu, X. et al., Comparison of methods for determination of tanshinone IIA in Huoxue Huayu granules, *J. Chinese Trad. Patent Med.* (*Zhongchengyao*), 27, 483–485, 2005.
126. Tataruch, F., Eine Methode zur quantitativen Bestimmung von Retinol in Blutserum mittels HPTLC (A method for quantitative determination of retinol in blood serum by HPTLC), *Microchim. Acta* (*Wien*), 1, 235–238, 1984.
127. Chavan, J.D. and Khatri, J.M., Simultaneous determination of retinol and α-tocopherol in human plasma by HPTLC, *J. Planar Chromatogr.*, 4, 280–282, 1992.
128. Sackett, P., High-performance thin-layer chromatography of gibberellins in fermentation broths, *Anal. Chem.*, 56, 1600–1603, 1984.
129. Matysik, G. et al., Stepwise gradient thin-layer chromatography and densitometric determination of taxol in extracts from various species of *Taxus*, *Chromatographia*, 41, 485–487, 1995.
130. He, Y.B. and He, L.Y., Separation and determination of taxol and related compounds in *Taxus* by HPTLC and fluorescent derivatisation on the plate, in: *Proceedings of the 9th International Symposium on. Instrumental Chromatography*, Interlaken, April 9–11, 1997, pp. 171–177.
131. Pearce, D.W., Koshioka, M., and Pharis, R.P., Chromatography of gibberellins, *J. Chromatogr.*, 658, 91–122, 1994.
132. Wawrzynowicz, T. et al., Chromatographic separation of taxoids using zonal TLC, *Acta Pol. Pharm.*, 56, 48–52, 1999.
133. Hajnos, M.Ł. et al., Application of pseudo-reversed-phase systems to the purification and isolation of biologically active taxoids from plant material, *Chromatographia*, 56, 91–94, 2002.
134. Hajnos, M.Ł. et al., Influence of the extraction mode on the yield of taxoids from yew tissues—preliminary experiments, *Chem. Anal.* (*Warsaw*), 46, 831–838, 2001.
135. Ren, Y. et al., Novel diterpenoid acetylcholine-esterase inhibitors from *Salvia miltiorrhiza*, *Planta Med.*, 70, 201–204, 2004.
136. Ryu, S.Y., Lee, C.O., and Choi, S.U., In vitro cytotoxicity of tanshinones from *Salvia miltiorrhiza*, *Planta Med.*, 63, 339–342, 1997.
137. Li, S.-H. et al., Taxuyunnanines S.-V., new taxoids from *Taxus yunnanensis*, *Planta Med.*, 68, 253–257, 2002.
138. Oriňák, A. et al., ToF-SIMS characterisation of diterpenoic acids after chromatographic separation, *Appl. Surf. Sci.*, 252, 6668–6671, 2006.
139. Oriňák, A. et al., Post-chromatographic ToF-identification of diterpenes, *Surf. Interf. Anal.*, 38, 599–603, 2006.
140. Oriňák, A. et al., New approaches to coupling TLC with ToF-SIMS, *J. Planar Chromatogr.*, 18, 44–50, 2005.

20 TLC of Triterpenes (Including Saponins)

Wieslaw Oleszek, Ireneusz Kapusta, and Anna Stochmal

CONTENTS

20.1 INTRODUCTION

Triterpenoids are a large group of naturally occurring substances of relatively complex cyclic structures with a carbon skeleton based on six isoprene units. Their immediate biological precursor is the acyclic C_{30} hydrocarbon, squalene (Figure 20.1). Because of multiple substitutions of functional groups they may occur either as alcohols, aldehydes, or carboxylic acids and their appropriate esters, ethers, or glycosides. This makes that multiple triterpenoid structures can be identified in many plant families and also in animal tissues. In general, they can be divided into at least four major classes: true triterpenes,[1] steroids,[2] saponins,[3,4] and cardiac glycosides.[5]

The most widely distributed true terpenes (pentacyclic triterpenoids, C_{30}) belong to several structural groups, which are based on hopane, lupane, or gammacerane

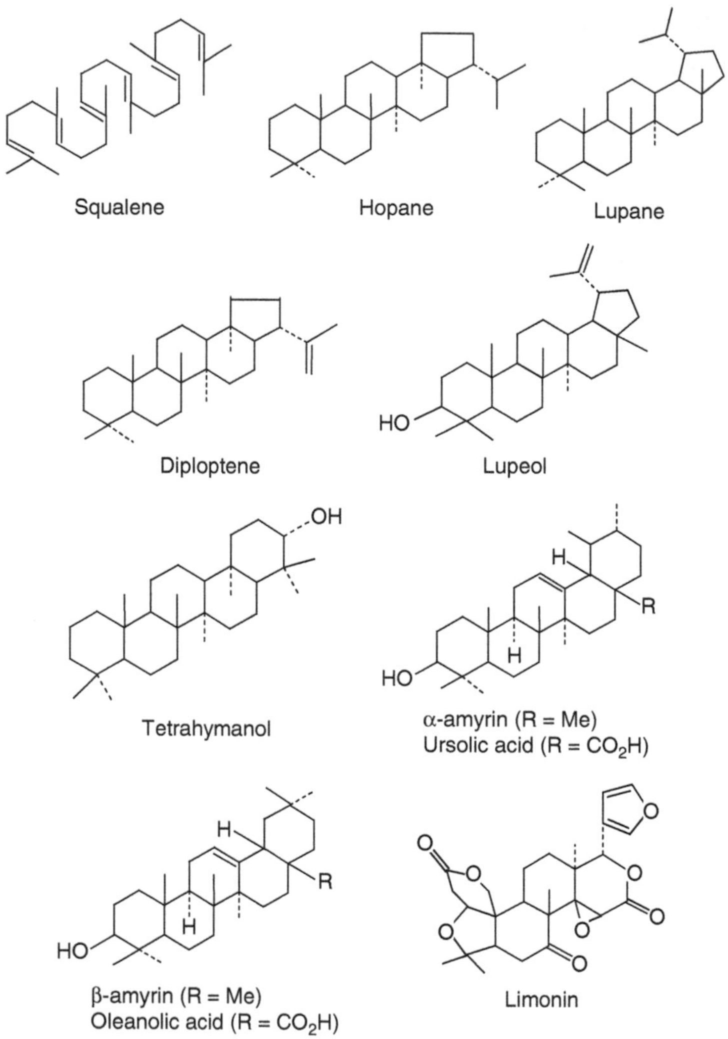

FIGURE 20.1 Chemical structures of common triterpenes.

skeletons. The first two groups are five-ring compounds, in which ring E is five-membered and in gammacerane it is characterized by a six-membered ring. The hopane-derived triterpenes are the most abundant natural product on the earth and widely occur in prokaryotes as membrane rigidifiers. The simplest compound of this series is diploptene and the most abundant in prokaryotes is C_{35} bacterio-hopanepolyols, containing side chain with variable number of vicinal hydroxyl groups. Lupane-based lupeol is most ubiquitous compound found in many fruits and vegetables.

The typical representative of gammacerane-derived compounds is tetrahymanol (gammaceran-21α-ol), widely occurring in living organisms. Commonly distributed

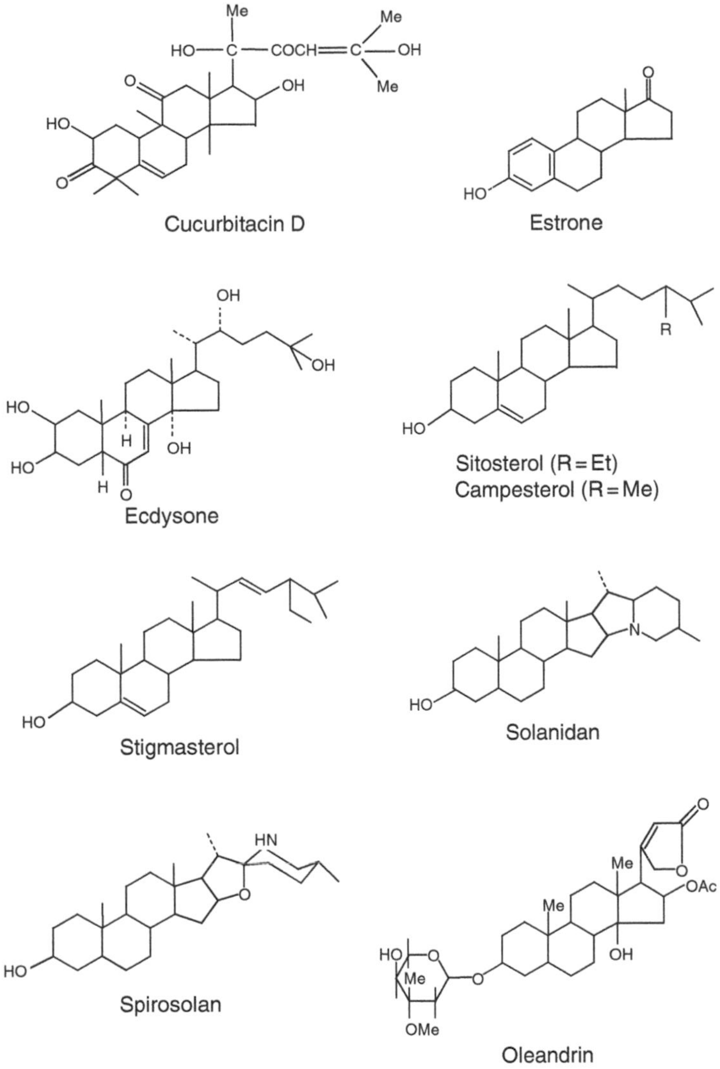

Cucurbitacin D

Estrone

Ecdysone

Sitosterol (R = Et)
Campesterol (R = Me)

Stigmasterol

Solanidan

Spirosolan

Oleandrin

FIGURE 20.1 (continued)

are also α- and β-amyrin and their derived ursolic and oleanolic acids. They can be found in waxy coatings of leaves and fruits, in resins and barks, and also in latex. They play microbial and insect protecting role. Some of them are bitter principles, e.g., limonin occurring in citrus fruits or cucurbitacins found in cucumbers.

Steroids, including sterols, are present in animals and plants (phytosterols) but can also be found in some bacteria, fungi, and yeasts. Certain animal sterols, which are structurally different from phytosterols, also occur in plant tissues, e.g., animal estrogen, estrone, or insect moulting hormones, the ecdysones. The basic nucleus of this group is the tetracyclic C_{17} hydrocarbon 1,2-cyclopentane perhydrophenan-threne. This can be substituted by the methyl groups at C_{10} and C_{13} or by an alkyl side chain at C_{17}. The common sterols such as sitosterol, stigmasterol, and campes-terol occur free and as glucosides.

Saponins are widely distributed in plant species, being reported in nearly 100 families. They are naturally occurring glycosides found mainly, but not exclusively (lower marine animals), in the plant kingdom. They consist of nonsugar aglycone coupled to sugar chain units. These sugars can be attached as one, two, or three sugar chains and the terms monodesmoside, bidesmoside, and tridesmoside have been given to these saponins, respectively (Greek desmos = chain). According to the nature of the aglycone they can be classified into steroidal or triterpene groups. Some authors also include within the saponins steroidal glycoalkaloids of solanidans and spirosolan classes. Over 100 steroidal and probably even larger numbers of triterpene saponins have been identified. This structural diversity and resulting wide range of polarities makes determination of individual saponins very difficult.

Cardiac glycosides (cardenolides) form similarly to saponins very complex group of plant occurring triterpenoids. Most of these glycosides are toxic and influence the heart function. The characteristic feature of cardenolides is the presence of sugar constituents that are not present in other plant natural products. The typical represen-tative of this group of compounds is oleandrin from the leaves of the oleander.

Many triterpenes are known to have biological activities. Some play important role in protection of plants against insect and fungi predation. At the same time they find an application in medicine because of their pharmacological activities. These features create necessity to monitor the occurrence, composition, and concentration of these compounds in plants and plant-derived products. The thin layer chromatog-raphy (TLC) is one of the optional techniques in their analyses.

20.2   ADSORBENTS AND CHROMATOGRAPHY SOLVENTS

Most of triterpenoids occur in plant material as multicomponent mixtures of glyco-sidic forms. Sugars can be substituted in one, two, or three sugar chains. The number of sugars, which can be as high as 8–10, and the multiplicity of the combinations of the sugar connections make big diversity of forms and thus, one plant may contain even 30–40 structurally different triterpenoids. The number of sugars also deter-mines the polarity of these compounds, which can widely range. Thus, the analysis of these compounds is still a challenge and a selection of the TLC adsorbents and solvents is crucial.

There is rather limited choice of ready to use TLC adsorbents for triterpene analysis. The TLC plates covered with different forms of silica gel are most often used as TLC adsorbents (Tables 20.1 through 20.4). In some cases, reversed phases C18 and C8 are also of use and are complementary to silica gel plates. Combination of silica gel and RP18 adsorbent for 2D TLC with sorbent gradient can be the technique of choice to improve separation of complex mixtures. Such a technique was successfully used for the separation of 18 alfalfa saponins.[79] The use of silica gel and RP18 bilayers is complicated because of the possibility of modification of stationary phase by the first solvent system, hence two single layer TLC plates were used instead. First development was performed on RP18 plate and in the perpendicular direction on silica gel plates.

The most frequently used solvents for TLC performed on silica gel include different proportions of chloroform–methanol–water (Tables 20.1 through 20.4).

TABLE 20.1
Analysis of Saponins by Thin Layer Chromatography (TLC)

Sample	Adsorbent	Solvent System	References
Albizia sap.	Silica gel 60 F	CHCl$_3$–MeOH–AcOH–H$_2$O (15:8:3:2)	[6]
		CHCl$_3$–MeOH–H$_2$O (65:40:1)	
Albiziasaponins A–E	Silica gel 60 F	CHCl$_3$–MeOH–H$_2$O (15:3:1)	[7]
	RP18 WF	MeOH–H$_2$O (7:3)	
Alium sap.	Silica gel	n-BuOH–AcOH–H$_2$O (60:15:25)	[8]
Argania sap.	Silica gel	n-BuOH–AcOH–H$_2$O (12:3:5)	[9]
Arnica sap.	Silica gel 60	CH$_2$Cl$_2$–EtOAc (9:1)	[10]
	RP18	MeOH	
Astragalus sap.	RP18	MeOH–H$_2$O (8.5:1:5)	[11]
Azadirachta sap.	DC-cards	Petrol–EtOAc (7:3)	[12]
	Silica gel GF60	Petroleum ether–EtOAc (6.5:3.5)	
Capsicoside E–G	Silica gel	n-BuOH–AcOH–H$_2$O (12:3:5)	[13]
Carpolobia sap.	Silica gel 60 F	CHCl$_3$–MeOH–AcOH–H$_2$O (15:8:3:2)	[14]
		CH$_2$Cl$_2$–MeOH (19:1)	
Chelioclinium sap.	Silica gel	Hexane–EtOAc (4:1) with 0.05% AcOH	[15]
Chenopodium sap.	Silica gel	n-BuOH–HOAc–H$_2$O (60:15:25)	[16]
Cussonia sap.	Silica gel 60 F	MeOH–CH$_2$Cl$_2$–AcOH–H$_2$O (40:55:3:2)	[17]
Cussosaponins A–E	Silica gel	CH$_2$Cl$_2$–MeOH–H$_2$O (17:6:1)	[18]
Dammarane sap.	Silica gel 60 F	CHCl$_3$–MeOH–H$_2$O (65:35:6)	[19]
Dioscorea sap.	Silica gel	CHCl$_3$–MeOH–H$_2$O (13:7:1)	[20]
		CHCl$_3$–MeOH (9:1)	
Draconins A–C	Silica gel 60 F	n-Hexane–AcOEt (20:1, 7:3)	[21]
		Toluene–PrOH (20:1)	
		CHCl$_3$–MeOH (10:1)	
Eranthisaponins A, B	Silica gel 60 F	CHCl$_3$–MeOH–H$_2$O (14:8:1)	[22]
	RP18	MeCN–H$_2$O (2:5, 4:1, 1:3)	
Eucalyptus sap.	Silica gel 60 F	n-HexaneCH$_2$Cl$_2$–EtOAc (16:16:1)	[23]
Fomefficinic acid A–E	Silica gel	CHCl$_3$–MeOH (95:5)	[24]
Gambeya sap.	Silica gel	CHCl$_3$–MeOH–H$_2$O (8:5:1)	[25]
Glinus sap.	Silica gel 60 F	MeOH–H$_2$O (4:1), CHCl$_3$–MeOH–H$_2$O (70:50:4)	[26]
	RP18	MeOH–H$_2$O (4:1)	
	Silica gel Sil G-100	CHCl$_3$–MeOH–H$_2$O (70:50:4)	
Harpulia sap.	Silica gel 60 F	CHCl$_3$–MeOH–H$_2$O (12:8:1)	[27]
		CHCl$_3$–MeOH–HCOOH (65:35:1)	

(*continued*)

TABLE 20.1 (continued)
Analysis of Saponins by Thin Layer Chromatography (TLC)

Sample	Adsorbent	Solvent System	References
Ilex sap.	Silica gel	n-BuOH–AcOH–H$_2$O (65:15:25)	[28]
		CHCl$_3$–MeOH–H$_2$O (70:30:3)	
		HCl$_3$–MeOH-n-PrOH–H$_2$O (5:6:5:1:4, upper phase)	
Lotoidesides A–F	Silica gel 60 F	CHCl$_3$–MeOH–H$_2$O (13:7:1, 28:12:1)	[29]
Lupinus sap.	Silica gel	Hexane–Me$_2$CO (75:25, 60:40, 40:60, 10:90)	[30]
Lyciantosides A–C	Silica gel	n-BuOH–AcOH–H$_2$O (60:15:25)	[31]
		CHCl$_3$–MeOH–H$_2$O (7:3:0.3)	
Maytenus sap.	Silica gel	100% CH$_3$CN	[32]
Medicago sap.	Silica gel	EtOAc–AcOH–H$_2$O (7:2:2)	[33]
Meryta sap.	Silica gel 60 F	CHCl$_3$–MeOH–H$_2$O–EtOAc (28:35:5:32)	[34]
Soyasaponin	Silica gel	CHCl$_3$–MeOH–H$_2$O (65:35:10, lower phase)	[35]
Panax sap.	Silica gel 60	CHCl$_3$–MeOH–H$_2$O (13:7:1)	[36]
Panax sap.	Silica gel 60 F	CHCl$_3$–MeOH–H$_2$O (10:3:1)	[37]
	RP18	MeOH–H$_2$O (7:3)	
Pastuchoside A–E	Silica gel	CHCl$_3$–MeOH–H$_2$O (26:14:3)	[38]
		n-BuOH–AcOH–H$_2$O (4:1:5)	
		CHCl$_3$–MeOH (20:1)	
		CH$_2$Cl$_2$–MeOH–H$_2$O (50:25:5)	
Pittoviridoside	Silica gel	n-BuOH–EtOH–H$_2$O (5:1:4)	[39]
Pittoviridoside	C18	70% MeOH	[40]
	Silica gel	EtOAc–EtOH–H$_2$O (7:2:1, 7:10:1)	
Silene sap.	Silica gel	CHCl$_3$–MeOH–AcOH–H$_2$O (15:8:3:2)	[41]
Solanum sap.	C18	MeCN–H$_2$O (7:3)	[42]
Symplocos sap.	Silica gel 60 F	CHCl$_3$–MeOH (9:1)	[43]
Tribulus sap.	Silica gel 60 F	n-BuOH–AcOH–H$_2$O (60:15:25)	[44]
		CHCl$_3$–MeOH–H$_2$O (40:9:1)	
Tripterygium sap.	Silica gel 60 F	Hexane–EtOH (2:1)	[45]
		CHCl$_3$–hexane (8:2)	
Tuberosides N–U	Silica gel	n-BuOH–AcOH–H$_2$O (4:5:1)	[46]
		CHCl$_3$–MeOH–H$_2$O (7:3:0.5)	

Source: From Oleszek, W., Bialy, Z., *J. Chromatogr. A*, 1112, 78, 2006. With Permission from Elsevier.

Sometimes separation can be substantially improved by mixing in the other alcohols such as ethanol, isopropanol, or their mixtures. For more polar compounds, solvent systems can be acidified with acetic acid. Highly polar triterpenoids are better separated with the mixtures of n-butanol–acetic acid–water (BAW). For glycosides of acidic aglycones or possessing uronic acid in a sugar chain the ethyl acetate–acetic acid–water (7:2:2) is a solvent of choice. Up to 25 medicagenic acid, hederagenin and zanhic acid glycosides of alfalfa were successfully separated with this system.[80]

TABLE 20.2

Thin Layer Chromatography of Glycoalkaloids

Sample	Adsorbent	Solvent System	References
Solanum platanifolium	Silica gel G	CHCl₃–EtOH–1% NH₄OH (2:2:1)	[47]
Solanum arundo	Silica gel	CHCl₃–MeOH–1% NH₄OH (2:4:1)	[48]
Solanum sycophanta	Nanoplates Sill 20 UV	CHCl₃–MeOH–1% NH₄OH (14:6:1)	[49]
Solanum incanum	Silica gel	EtOAc–BuOH–H₂O (2:1:1, upper phase)	[50]
Solanum robustum	Silica gel WF	CHCl₃–MeOH–conc NH₃ (3:3:1)	
	Silica gel RP8	MeOH–buffer soln (2:1)	[51]
Solanum spp.	Silica gel F	MeOH or EtOH–CHCl₃ (2:1)	
		CHCl₃–MeOH–2% NH₄OH (70:30:5)	[52,53]
Solanum lycopersicum	Silica gel	HOAc–EtOAc–MeOH–H₂O (10:30:20:1)	[54]
Archaesolanum sp.	Silica gel G	n-BtOH–diethylamine–MeOH (85:10:2)	
		EtOAc–pyridine–H₂O (3:1:3, upper phase)	[55]
Solanum tuberosum	Silica gel	CHCl₃–95% EtOH–1% NH₄OH (2:2:1, lower phase)	[56]
		n-BuOH–Me₂CO–H₂O (4:5:1)	[57]
		CHCl₃–OHAc–MeOH (50:5:45)	
		Me₂CO–MeOH (3:5)	
		EtOAc–HOAc–H₂O (11:2:2)	
		EtOH–HOAc–H₂O (19:1:1)	
		n-BuOH–HOAc:H₂O (4:1:1)	
		CHCl₃–EtOH aq.–1% NH₄OH (2:2:1, lower phase)	
		Pyridine–EtOAc–H₂O (1:3:3, upper phase)	

For the glycoalkaloids separation, the solvent system should be rather ammonified (NH₄OH) for best separation. In some cases, systems containing ethyl acetate–pyridine–water or diethylamine substantially improve separation.

The aglycones obtained from the hydrolysis of glycosides separate effectively in diisopropyl ether–acetone (75:30), benzene–methanol (9:1→98:2), or petroleum ether–chloroform–acetic acid (7:2:1).[81] The aglycones (steroidal or triterpene) can be separated from the sterols using hexane–acetone (1:4). In this system sterols usually occur as one spot with R_f higher than 0.5, while the other aglycones appear at R_f lower than 0.5. Care has to be taken when products of acid hydrolysis are analyzed. During hydrolysis diene derivatives can be formed and they appear on TLC at R_f about 0.9, and can be by mistake recognized as sterols. The separation of the sterols, e.g., sitosterol, cholesterol, and stigmasterol, is rather difficult to achieve. Like in the other cases in which separation of structurally similar compounds is difficult, the derivatization of the mixture to acetates, followed by TLC of acetyl derivatives is best solution. In this way sterols separate in hexane–ether (97:3) and triterpene soyasapogenols separate in hexane–benzene–acetone (50:45:5), hexane–diethyl ether (65:35), and cyclohexane–acetone (9:1).[82]

TABLE 20.3
Thin Layer Chromatography of Cucurbitacins

Plant	Adsorbent	Solvent System	References
Cucurbita spp.	Silica gel F	Ether–hexane–MeOH (70:30:5)	[58]
		$CHCl_3$–MeOH (95:5)	
Iberis amara	Silica gel	Trichloroethylene–iso-PrOH (4:1)	[59]
		C_6H_6–MeCOEt (2:1)	
		$CHCl_3$–EtOAc–HCOOH (12:12:1)	
		Diisopropylether–$MeCO_2$ (5:2)	
Picrorhiza kurrooa	Silica gel	$CHCl_3$–MeOH (19:1, 9:1, 4:1)	[60–62]
Acanthosicyos horridus	Silica gel GF	Cyclohexane–EtOAc (3:2)	[63]
Cucurbita foetidissima	Silica gel F	$CHCl_3$–MeOH–H_2O (64:40:8)	[64]
		$CHCl_3$–MeOH–AcOH–H_2O	
		(60:32:12:8)	
		EtOAc–AcOH–HCOOH–H_2O	
		(100:11:11:26)	
		Toulene–$MeCO_2$ (4:1)	
Wilbrabdia ebracteata	Silica gel	Et_2O–$MeCO_2$–H_2O (20:1:20,	[65]
		upper phase)	
Ibervillea sonorae	Nanoplates	$CHCl_3$–MeOH–H_2O (75:25:2)	[66]
	sill-20 UV	$CHCl_3$–MeOH (19:1, 9:1, 4:1)	
		Cyclohexane–EtOAc (1:3, 2:3)	
Luffa echinata	Silica gel F	C_6H_6–Et_2O (1:1)	[67]

For separating triterpenes according to the number of the isolated double bonds present in the molecule, argentative TLC on a plate treated with $AgNO_3$ can be employed.[83] For this separation, silica gel plates need to be sprayed with the concentrated aqueous methanolic $AgNO_3$ and then activated for 30 min at 120°C.

20.3   DETECTION

20.3.1   Without Chemical Treatment

Most of triterpenes are colorless substances, crystalline, and optically active but because of the lack of chemical reactivity they are rather difficult to characterize. They are not seen on the TLC plate either in natural light or under UV exposure. The exceptions are glycyrrizic acid and its glycoside conjugates from liquorice, and cucurbitacines present in *Cucurbita* species. These two groups can be detected by exposure to UV-254 nm or UV-365 nm. The other triterpenes, when developed on silica gel 60 F_{254} precoated TLC, can be observed under UV as black quenching spots. These spots, however, are not characteristic for any structural features.

TABLE 20.4
Thin Layer Chromatography of Cardiac Glycosides

Plant	Adsorbent	Solvent System	References
Coronilla varia	Silica gel F	$CHCl_3$–EtOH (2:1)	[68]
Adonis aleppica	Silica gel	$CHCl_3$–MeOH–H_2O (80:19:1)	[69]
		$CHCl_3$–MeOH–H_2O (70:29:1)	
		EtOH–MeOH–H_2O (72:25:8)	
Anodendron affine	Silica gel F	*n*-Hexane–EtOAc (2:1)	[70]
	Silica gel RP8	MeOH–H_2O (9:1)	
Asclepias spp.	Silica gel G	$CHCl_3$–MeOH–formamide (93:6:1)	[71,72]
		EtOAc–MeOH (97:3)	
Crossopetalum gaumeri	Silica gel F	EtOAc–MeOH–H_2O (81:11:8)	[73]
		$CHCl_3$–MeOH–H_2O (64:36:8)	
Strophantus divaricatus	Silica gel	$CHCl_3$–EtOH (10:1)	[74,75]
		C_6H_6–Me_2CO (2:1)	
		n-BuOH–HOAc–H_2O (4:1:5, upper phase)	
		n-BuOH–EtOH–H_2O (5:1:4, upper phase)	
Convallaria majalis	Silica gel	$CHCl_3$–MeOH–H_2O (7:3:1, lower phase)	[76]
		MeCOEt–toulene–H_2O–MeOH–HOAc (80:10:6:5:2)	
		EtOAc–MeOH–H_3BO_3–4.6% HOAc (55:20:15:10)	
Periploca sepium	Silica gel I	EtOAc	[77]
	ODS	$CHCl_3$–MeOH (10:1)	
		MeOH–H_2O (9:1)	
		$CHCl_3$–MeOH–H_2O (25:17:3)	
Digitalis purpurea	Silica gel G	$CHCl_3$–Me_2CO (49:1)	[78]

20.3.2 WITH SPRAY REAGENTS

Detection and preliminary characterization of triterpenes on TLC plates can be performed using different types of spray reagents. Over 50 spray reagents have been used.[84,85] Some of them, which are most often used for different classes of triterpenes, are listed in Table 20.5.

20.3.2.1 Protocols for Preparation of Spray Reagents

Anisaldehyde–sulfuric acid reagent: Mix 0.5 mL anisaldehyde with 10 mL of glacial acetic acid and add 85 mL of methanol and 5 mL of sulfuric acid. Spray with the reagent and heat the plate for 5–10 min at 100°C. The reagent has very limited stability time.

Antimony(III) chloride reagent (SbCl₃): Prepare 20% solution of antimony(III) chloride in chloroform. After spaying heat at 100°C for 5–10 min.

Blood reagent: Ten milliliters of 3.6% sodium citrate are mixed with 90 mL of the fresh bovine blood. Two milliliters of this solution are mixed with 30 mL phosphate buffer pH 7.4. Spray over thoroughly dried TLC plate.

TABLE 20.5
Spray Reagents Most Often Used for Triterpene TLC Detection

Reagent	Color of Spots
Triterpenes and steroids	
Liebermann-Burchard	Blue, green, pink, brown, yellow in visible light also under UV
Vanillin-sulfuric acid	Blue, blue-violet, yellowish
Vaniline-phosphoric acid	Red-violet in visible light and reddish or blue in UV-365
Anisaldehyde-sulfuric acid	Blue red-violet in visible light and reddish or blue in UV-365
Komarowsky reagent	Blue, yellow, red
Antimony (III) chloride (Carr-Price reagent)	Pink, purple in visible light, red-violet, green, blue in UV-365
Ehrlich reagent	Red coloration of furostanol derivatives, spirostanols do not react
Water	Sterols give white, not wetted spots
Blood reagent	White zones on the reddish background
Cucurbitacines	
Liebermann-Burchard	Blue, green, pink, brown, yellow in visible light also under UV
Vanillin–sulfuric acid	Blue, blue-violet, yellowish
Cardiac glycosides	
Kedde reagent	Blue to red-violet in visible light
Chloramine–trichloroacetic acid	Blue, blue-green, yellow-green in UV-365
Vanillin–sulfuric acid	Blue, blue-violet, yellowish

Chloramine–trichloroacetic acid reagent: Mix 10 mL of freshly prepared 3% aqueous chloramine T solution (synonymies sodium sulfamide chloride or sodium tosylchloramide) with 40 mL of 25% ethanolic trichloroacetic acid. Heat for 5–10 min at 100°C.

Ehrlich reagent: Mix 1 g of *p*-dimethylaminobenzaldehyde with 50 mL of 36% HCl and 50 mL ethanol. Spray and heat for 5–10 min at 100°C.

Kedde reagent: Mix 5 mL of the freshly prepared 3% ethanolic 3,5-dinitrobenzoic acid and 5 mL of 2N NaOH.

Komarowsky reagent: Mix 1 mL of 50% ethanolic sulfuric acid with 10 mL of 2% methanolic 4-hydroxybenzaldehyde. Reagent should be prepared shortly before use. Spray and heat the plate at 100°C for 5–10 min.

Liebermann–Burchard reagent: Add carefully 5 mL of acetic anhydride and 5 mL of concentrated sulfuric acid into 50 mL of absolute ethanol, while cooling in ice. Heat the plate at 100°C for 5–10 min.

Vanillin–phosphoric acid reagent: One gram of vanillin dissolves in 100 mL of 50% phosphoric acid. Spray and heat for 10 min at 100°C.

Vanillin–sulfuric acid reagent: Solution I—5% ethanolic sulfuric acid and solution II—1% ethanolic vaniline. Spray the plate with 10 mL of solution I, followed immediately by 10 mL of the solution II. Heat at 110°C for 5–10 min.

Most of the listed sprayers require heating of the plates for several minutes at $105°C–110°C$ for the development of color spots. This requires some attention, as plates overheated produce only dark spots, regardless of the structure of the triterpene. In most cases the color of spots are not very stable, which might be important if documentation of samples or when densitometry is being applied. Most stable are vanillin-based sprayers.

The Ehrlich reagent is highly specific for furostanol steroidal saponins and produces pink-red spots, while spirostanol glycosides are not visualized. This reagent effectively distinguishes these two groups of compounds.

The spray reagents containing aromatic aldehydes, e.g., anisaldehyde and vanillin, in strong mineral acids give the colored products with aglycones. They can be equally used for spraying TLC plates, as well as the reagents for triterpene determination by colorimetric procedures. The absorption maxima of these entities lie between 510 and 620 nm. But the reaction, which is not precisely understood, is not very specific and a number of other classes of substance can react. Certain compounds present in the sample matrix may cause interference. Thus, the crucial issue for colorimetric determination is sample cleanup procedure performed prior to analysis.

20.4 APPLICATION OF TLC FOR THE DETERMINATION OF TRITERPENES

20.4.1 Preparation of Plant Extracts for TLC Analysis

The extraction solvents need to be selected according to expected structural features of analyzed triterpenes. Simple triterpenes, the aglycones or even monoglucosides, can be effectively extracted with nonpolar solvents, including chloroform and ethyl acetate. The extracts after condensation can be applied to TLC plates. For more complex structures such as saponins, aqueous methanol or ethanol is recommended.

In the early work on triterpene glycosides, a hot extraction (under reflux or boiling) was used, followed by the evaporation of alcohol and liquid–liquid extraction of triterpenes or steroids into n-butanol. By this protocol, highly polar compounds, such as saccharides, stay in water solution, providing some purification of the extract. This protocol, however, has some serious disadvantages. First of all, the high temperature degrades some labile structures, e.g., acylated glycosides to nonacylated forms. The extraction with the hot aqueous methanol may also lead to the formation of $–OCH_3$ derivatives, lactones, in both triterpene and steroidal saponins. This may also result in the degradation of genuine compounds into their artefacts, as, e.g., happens with soybean saponins. Such transformations occur on the aglycone skeleton or on the sugar part, or consist in the loss of labile constituents.

The liquid–liquid extraction into butanol can also not be complete and some more polar glycosides remain in the water. This was observed for zanhic acid tridesmoside present in alfalfa aerial parts.[86]

Recent alternative procedure for sample cleanup has been based on solid-phase extraction (SPE) on reversed-phase (C8, C18) sorbents.[87] The saponin extract in aqueous 10%–20% methanol is loaded on a small column, preconditioned with methanol and water. Washing the column with the water removes very polar extract components, such as saccharides. The compounds, remaining on the column, can be selectively washed out with methanol–water mixtures. For most of the extracts, the solvents with a concentration of methanol up to 40% remove most of phenolics, including flavonoids (some isoflavones at these concentrations may retain on the column). Saponins are washed out with methanol concentration higher than 40%, and depending on the structure of a saponin the concentration of the methanol should be optimized. Practically, this optimization is performed by the washing of ready to use reversed-phase cartridge ($1\ cm^3$) with solvents containing an increasing concentration of methanol. The fractions obtained from such an exercise are analyzed with TLC for the presence of saponins.

Selectivity of the C18 sorbent to retain saponins strongly depends on the pH of loaded sample, which is especially helpful in a cleanup of the saponins having COOH groups. In general, properly optimized SPE purification of saponin produces high purity mixtures ready for TLC analysis, colorimetry, or for biological tests.

20.4.2 TLC-Densitometry

Thin layer chromatography in one (1D) and two (2D) dimensional modes is a powerful technique that has been used successfully in the separation and determination of a large number of saponins in plant extracts. A major problem with these techniques is first of all the parallel running of the appropriate standards, minimizing the variation between different plates and color reactions with spraying reagents. The second difficulty is spot detection by means of sophisticated instrumentation for data acquisition and handling to scan the whole plate surface at high speed. This can be achieved, however, by online coupling of a computer with a dual-wavelength flying-spot scanner and 2D analytical software. A great number of saponins are being determined by these techniques and some of them are listed in Table 20.6.

The developing systems and spray reagents are the same as listed in Table 20.5. Linear relationships between the peak area and the amount of standard saponins can be found in the range of 1–5 µg per spot with recovery being at the level of 98% and standard deviation of around 3%–5%. Comparison of the data with those obtained by HPLC shows that the method is sufficiently accurate for quality control monitoring and is particularly suitable for assays in series. With 2D TLC it was possible to analyze 35 saponins extracted from stems and leaves of ginseng, and *Calendula officinalis* saponins. Fifteen ecotypes of quinoa (*Chenopodium quinoa*) used in breeding programs in the United Kingdom were successfully analyzed for the concentrations of three groups of saponins, including oleanolic acid, hederagenin, and phytolaccagenic acid glycosides. The glycyrrizic acid in liquorice extracts, panaxadiol and panaxatriol, and saponins in ginseng and *Avena sativa* saponins were determined with this technique. The triterpene saponin, escin, in horsechestnut extracts and phytopharmaceutical preparations was measured with high-performance TLC (HPTLC) with post-chromatographic derivatization and in situ reflectance measurements. This method enabled determination of large numbers of samples

TABLE 20.6

Determination of Some Saponins by TLC-Densitometry

Saponin	Plates	Solvent System	Spray Reagent
Cucurbitacin B,D,E,I		MeOH–H$_2$O (45:55; 70:30)	254
Cucurbitacin C		EtOAc–C$_6$H$_6$ (75:25)	Anililin-orthophosphate in EtOH
Ginsenosides	Silica gel G,H	CHCl$_3$–MeOH–H$_2$O (65:35:10) BuOH–EtOAc–H$_2$O (4:1:2)	NH$_4$HSO$_4$ in 15% H$_2$SO$_4$
	Silica gel	1,2–dichloroethanol–BuOH– MeOH–H$_2$O (30:40:15:25)	NH$_4$HSO$_4$ in EtOH
	Silica gel H	CHCl$_3$–MeOH–H$_2$O (65:35:10)	10% H$_2$SO$_4$
	Silica gel H	CHCl$_3$–MeOH–H$_2$O (21:11:4) I dir. BuOH–EtOAc–H$_2$O (4:1:1) II dir.	Vaniline in H$_2$SO$_4$
Glycyrrhizin	Silica gel 60	BuOH–OHAc–H$_2$O (5:1:4)	H$_2$SO$_4$
Gypsosides	Silica gel LS	BuOH–OHAc–H$_2$O (4:1:5)	Phosphortungstic acid
Oleanolic acid	Silica gel G	CHCl$_3$–Et$_2$O–MeOH (30:10:1)	Anisaldehyde–H$_2$SO$_4$
	Silica gel G	C$_6$H$_6$–Me$_2$CO (36:13)	10% H$_2$SO$_4$
Triterpene	Silufol	C$_6$H$_6$–Me$_2$CO (8:2)	Phosphortungstic acid

Source: From Oleszek, W., *J. Chromatogr. A*, 967, 147, 2002. With permission from Elsevier.

and did not require any tedious cleanup steps prior to analysis, and was highly recommended in pharmaceutical quality control practice.

20.4.3 TLC-COLORIMETRY

The triterpenes can be analyzed qualitatively/quantitatively with TLC-colorimetry. They are determined colorimetrically in a crude extract and TLC is just a means of confirmation of their presence in the sample, or the TLC-separated bands are scraped, extracted with alcohol, and the extract is treated with a specific reagent. Most frequently used colorants include Earlich or vanillin reagents and measurements are made at $\lambda = 515$–560 nm. The major disadvantage of this procedure is that some other components of the extract such as sterols and bile acids with hydroxyl group at C3 may give a color reaction with the reagent, providing misleading information. An anisaldehyde–sulfuric acid–ethyl acetate reagent gives with steroidal sapogenins a color reaction, which is in general free of any influence from interfering compounds. A good improvement of the colorimetric procedure can be the application of a reversed phase cleanup stage (SPE) prior to the determination. Depending on the nature of saponins, their retention on reversed phase support is different and appropriate selection of the solvents allows their considerable purification.

When silica gel TLC is used prior to colorimetric determination, a considerable portion of the compound can be adsorbed irreversibly in the stationary phase and the quantities present may be underestimated. To use this procedure for routine analysis,

TABLE 20.7
The TLC-Colorimetric Determination of Saponins

Sample	Reagent	Wavelength (nm)
Cyclamiretin A	10% Vaniline–OHAc, $HClO_4$	560
Ginsenosides	8% Vaniline–EtOH, H_2SO_4	544
	8% Vaniline–EtOH, H_2SO_4	544
	Vaniline—OHAc	560
	Vaniline–$HClO_4$	560
	HCl–H_2SO_4	520
Protodioscin (I)	Ehrlich	515
Soyasapogenols	OHAc–H_2SO_4	530
Steroidal aglycones	Anisaldehyde–H_2SO_4–EtOAc	430

Source: From Oleszek, W., *J. Chromatogr. A*, 967, 147, 2002. With permission
from Elsevier.

first it has to be calibrated against a more sophisticated technique, e.g., HPLC.
Examples of the TLC-colorimetric determinations are presented in Table 20.7.

20.5 APPLICATION OF TLC FOR THE DETERMINATION OF BIOLOGICAL ACTIVITY

20.5.1 HEMOLYTIC ACTIVITY OF TRITERPENES

Some saponins show hemolytic activity, which has been used for the development of
semiquantitative tests for their determination. In the simplest hemolytic method,
saponin-containing material or its water extract is mixed with blood or with washed
erythrocytes in isotonic buffered solution (0.9% NaCl). After 20–24 h, samples are
centrifuged and hemolysis is indicated by the presence of hemoglobin in the supernatant. For the evaluation of hemolytic activity, European Pharmacopoeia uses as a
unit the hemolytic index (HI), which is defined as the number of millilitres of ox
blood (2%, v/v) that can be hemolyzed by 1 g of crude saponins or plant material.
The saponin mixture of *Gypsophila paniculata* L. (HI = 30,000) or Saponin white
(Merck, HI = 15,000) is usually used as reference. Hemolytic indices of saponins are
calculated according to the following equation:

$$HI = HI_{std} \times a/b,$$

where
    HI_{std} is the hemolytic index of standard saponin
    a and b are the lowest concentrations of test and standard saponin, respectively,
    at which full hemolysis occurred

Mackie and coworkers[88] measured the absorbance at 545 nm of the supernatant after hemolysis and defined one unit of activity as the quantity of hemolytic material that caused 50% hemolysis.

Another modification of the hemolytic method is the hemolytic micromethod,[89] in which cow's blood stabilized with sodium citrate (3.65% w/v) is mixed with gelatine solution. For this, gelatine (4.5 g) is dissolved in 100 mL of isotonic buffered solution and 75 mL of this is mixed with 20 mL of blood. Gelatine/blood mixture is spread on a glass plate (10 × 20 cm) to a thickness of 0.5 mm and after coagulation, the plates are used for tests. Saponin samples (10 μL) or mashed plant material is placed in localized areas on the gelatine/blood-covered plates and after 20 h the widths of the resulting hemolytic rings are measured. A ring of standard saponin is measured in parallel on each plate (Figure 20.2). The detailed description of HI and microhemolytic methods and their limitations in relation to alfalfa (*Medicago sativa*) saponins can be found in a separate publication.[90] It has been clearly documented that saponins differ in their hemolytic activities depending on their structures and also on the hemolytic method used. Monodesmosidic saponins are usually more active than their bidesmosidic analog. The HI method has been used successfully for the determination of oleane-type saponins in *Aerva lanata* and *A. javanica* and in *Phytolacca dodecandra*.

Hemolytic effects can also be used successfully for spotting hemolytic saponins on TLC plates.[91] For this, developed plates need to be carefully dried from the solvent residue and then covered with a layer of gelatine/blood solution. After few hours, whitening spots can be seen on plates, indicating the presence of saponins. Care has to be taken, as some hydrophobic compounds prevent plates from wetting properly and in fact these areas can be mistakenly identified as hemolytic spots.

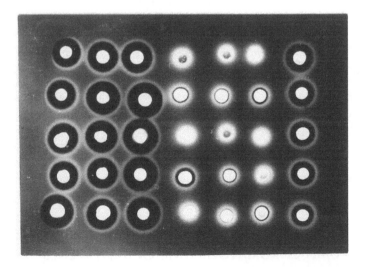

FIGURE 20.2 Hemolytic test performed on blood–gelatine covered plates (negative mode). White spots—mashed plant material; black rings—hemolytic zones; background in original—red.

20.5.2 Antifungal and Antimicrobial Activity Screening

Most of plant extracts searched for a bioactive natural compound are complex mixtures of different classes of phytochemicals. Determination of activity of the extract is just a first step in a long procedure of defining the active principle. Because of the tedious isolation and identification procedures there is a need to introduce fast screening preliminary tests. Most of these tests involve TLC bioautography, based on localization of antibacterial or antifungal activity on chromatogram. Early methods of bioautography were based on the so-called agar-diffusion techniques, in which active compounds were transferred from TLC layer to an inoculated agar plate through a diffusion process during the contact of these two layers. The advantages and disadvantages of this technique in the analysis of antibacterial, antifungal, antiprotozoal, antiphage, phage-inducing, antiviral, and cytotoxic activities of different plant phytochemicals were critically reviewed by Betina.[92] The major disadvantage of agar-diffusion technique was differential diffusion of compounds from chromatogram to the agar plate due to their structure-dependent water solubility.

Further improvement of this technique was the development of direct bioautographic detection on the TLC plate of inhibition of *Erwinia carotovora* and *E. herbicola* by coumarin, streptomycin sulfate, and the extracts of garlic.[93] The TLC plates (20 × 20 cm) were developed with cyclohexane–ethyl acetate (1:1), dried and sprayed (18 ml per plate) with cultures of bacteria suspended in TSB (trypticase-soy-broth). Plates were incubated in a humid environment at 25°C overnight, dried and spayed again with the following solutions: (1) aesculin spray (aesculin 0.2% w/v; ammonium ferric citrate 0.1% w/v; yeast extract 0.5% w/v in distilled water) or (2) tetrazolium salt (2,3,5-triphenyltetrazolium chloride 20 mg/mL). On the plates sprayed with the bacteria and subsequently with an aesculin spray, the hydrolysis of aesculin resulted in the development of a brown color, while the zones of inhibition of the bacteria remained colorless. On the plates sprayed with tetrazolium salt the metabolitically active bacteria convert the tetrazolium salt into an intensively colored formazan and the antibacterial compounds appear as clear spots against a colored background.

The method can be further simplified for testing antibacterial activity against, e.g., *Bacillus subtilis* and *Escherichia coli*. In these methods the TLC plates have been developed in solvent systems, thoroughly dried and a suspension of microorganisms in a suitable broth was applied to the TLC surface. Incubation in a humid environment permits growth of the bacteria. The activity of the compound is determined by measuring the zones of inhibition visualized by appropriate reagents. In the case of *B. subtilis* and *E. coli* visualization can be performed by spraying with tetrazolium salt.

20.6 TLC AS SUPPORTING TECHNIQUE FOR LOW PRESSURE COLUMN CHROMATOGRAPHY OF TRITERPENES

For the elucidation of the structural features of triterpenes their separation into individual compounds is essential. The separation of any crude mixtures containing a number of compounds of similar polarity is still quite a challenge. In the case of

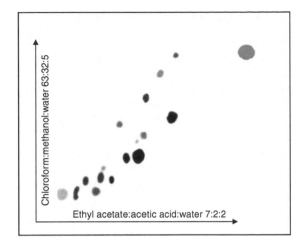

FIGURE 20.3 2D TLC of *Medicago truncatula* saponin mixture (upper) and 1D TLC of individual saponins (lower) obtained from the mixture by the low-pressure liquid chromatography.

triterpenes this is even more difficult, as these compounds do not possess chromophores and the separation cannot be monitored with the UV detector. For the glycosides sometimes a detection with the refractive index (RI) detector is a good solution. However, in many cases, this does not work satisfactorily. In this situation, TLC can be effectively applied for the detection of the separation. The typical example of such TLC application can be the separation of saponins from aerial parts of *Medicago truncatula*.[94] As presented in Figure 20.3 (upper) the extract of *Medicago truncatula* contained a number of saponins that needed to be separated. The first step of separation included low-pressure column chromatography (3×30 cm, C18) using water–methanol linear gradient as a solvent system. Ten millilitre fractions were collected with fraction collector and analyzed with TLC. Fractions showing a similar TLC profiles were combined and concentrated. In this way a number of subfractions containing 2–5 components were obtained. They were further separated on the C18 column (1×40 cm, C18) with an isocratic system consisting of the acetonitrile–water. Again, 2 mL fractions were collected and analyzed with the TLC. Those fractions showing one spot were combined and condensed *in vacuo*. This led to the isolation of individual saponins (Figure 20.3, lower) and the TLC was used as a supporting technique for their detection.

20.7   TLC AS A TOOL IN TRITERPENE IDENTIFICATION

Problems with the separation of the complex triterpene mixtures result in very small quantities, usually a few milligrams of individual compounds being isolated. Thus, for the structure elucidation nondegradative techniques are preferred. These include the mass spectrometry (MS) and the nuclear magnetic resonance spectroscopy (NMR). These sophisticated routine techniques, however, are still not so widely available because of the high costs of the equipment. They are absolutely necessary when the structure of a genuine aglycone, a sequence of the monosaccharides in the sugar chain, the linkage of sugars, the anomeric configuration of each linked sugar, and the location of the sugars on the aglycone need to be established. This takes place when the structure of a new compound is being elucidated. However, in a routine phytochemical work, simpler inexpensive techniques are needed to preliminarily establish the isolated compound—the compound being a new structure—or to confirm its similarity to the known standards. In such a case the TLC analysis can be of a great help.

When the number of standards is available, the 1D TLC performed in a several solvent systems and different sprayers may provide the preliminary information on the chemical nature of compounds present in the extract. The acid hydrolysis of the extract releases the aglycones and sugars, which can be compared by the TLC with reference standards. Figure 20.4 shows the TLC analysis of the aglycones and the sugars obtained from the hydrolysis of *Trifolium repens* saponins. The TLC analysis

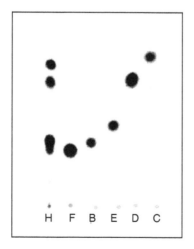
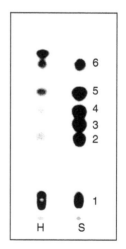

FIGURE 20.4 TLC of the hydrolysis products of *Trifolium repens* saponins. *Left*: soyasapogenol B, C, D, E, and F standards, and H saponin hydrolysate; Silica gel plates, solvent system: petroleum ether–chloroform–acetic acid (7:2:1), sprayer Liebermann-Burchard. *Right*: Cellulose plates, sugar standards S (1—glucuronic acid, 2—galactose, 3—glucose, 4—arabinose, 5—xylose, 6—rhamnose); solvent system benzene–*n*-butanol–pyridine–water (1:5:3:3), sprayer: solution I—20 mL acetone + 0.1 mL of saturated silver nitrate, after drying solution II—0.5N NaOH in EtOH and then solution III—10% aqueous $Na_2S_2O_3$.

of aglycones indicates the presence of the soyasapogenols B, C, D, E, and F and in sugar components the glucuronic acid, galactose, glucose, arabinose, and rhamnose can be confirmed.

For the preliminary information of a sugar sequence of the individual glycosides, the hydrolysis of the compound can be performed directly on the TLC plate. The compound is spotted at the start of the plate and this is placed in the container saturated with hydrochloric acid vapors. The time of the exposure determines the number of the sugars cleaved. After evaporation of acid, the TLC plate can be developed with a solvent system used for the identification of monosaccharides.

20.8 CONCLUSION

None of the actually existing chromatographic techniques provide full fingerprinting of triterpenes and a quality assurance required for the registration of the herbal medicinal products. The TLC on normal and reversed phases (TLC, HPTLC, 2D TLC) provides an excellent qualitative information and in the combination with online coupling of a computer with a dual-wavelength flying-spot scanner, a 2D analytical software can be used for the routine determination of triterpenes in a plant material. The densitometry of triterpenes has been very sensitive, however, to the plate quality; the spraying technique and the heating time strongly influence the results. Therefore, appropriate standards have to be run in parallel with the sample. The TLC methods are still recommended by U.S. Pharmacopeial Forum if confirmation of a peak identity is required.

In the phytochemical work aimed at the isolation and identification of the natural products, the TLC is becoming rather a supporting technique in the analysis of triterpene fractions, obtained from the column chromatography. In this respect this is a fantastic technique, which cannot be replaced with any other procedure.

REFERENCES

1. Charlwood, B.V., Banthorpe, D.V. (Eds). *Methods in Plant Biochemistry*, Vol 7: Terpenoids. Academic Press, London, San Diego, CA, 1991, p. 565.
2. Dence, J.B. *Steroids and Peptides: Selected Chemical Aspects for Biology, Biochemistry, and Medicine*. Wiley, New York, NY, 1980, p. 418.
3. Hostettmann, K., Marston, A. *Saponins*. Cambridge University Press, 1995, p. 544.
4. Oleszek, W., Marston, A. (Eds). *Saponins in Food, Feedstuffs and Medicinal Plants*. Kluwer Academic Publishers, Dordrecht, 2000, p. 291.
5. Greeff, K. *Handbook of Experimental Pharmacology*, Vol 56: Cardiac glycosides. Springer-Verlag, Berlin, New York, NY, 1981.
6. Haddad, M. et al. Two biologically active triterpenoidal saponins acylated with salicylic acid from *Albizia adianthifolia*. *J. Nat. Prod.* 66, 376, 2002.
7. Yoshikawa, M. et al. Characterization of new sweet triterpene saponins from *Albizia myriophylla*. *J. Nat. Prod.* 65, 1641, 2002.
8. Dini, I. et al. Furostanol saponins in *Allium caepa* L. Var. *tropeana* seeds. *Food Chem.* 93, 205, 2005.

9. Alaoui, A. et al. Triterpenoid saponins from the shells of *Argania spinosa* seeds. *J. Agric. Food Chem.* 50, 4601, 2002.

10. Schmidt, T.J., von Raison, J., Willuhn, G. New triterpene esters from flowerheads of *Arnica lonchophylla*. *Planta Med.* 70, 968, 2004.

11. Radwan, M.M. et al. Acetals of three new cycloartane-type saponins from Egyptian collections of *Astragalus tomentosus*. *J. Nat. Prod.* 67, 489, 2004.

12. Siddiqui, B.S. et al. Tetracyclic triterpenoids from the leaves of *Azadirachta indica*. *Phytochemistry*, 65, 2363, 2004.

13. Iorizzi, M. et al. Antimicrobial furostanol saponins from the seeds of *Capsicum annum* L. Var. *acuminatum*. *J. Agric. Food Chem.* 50, 4312, 2002.

14. Mitaine-Offer, A.C. et al. Three new triterpene saponins from two species of *Carpolobia*. *J. Nat. Prod.* 65, 556, 2002.

15. Jeller, A.H. et al. Antioxidant phenolic and quinonemethide triterpenes from *Cheiloclinium cognatum*. *Phytochemistry*, 65, 1977, 2004.

16. Dini, I., Tenore, G.C., Dini, A. Oleanane saponins in "Kancolla," a sweet variety of *Chenopodium quinoa*. *J. Nat. Prod.* 65, 1025, 2002.

17. Tapondjou, L.A. et al. Saponins from *Cussonia bancoensis* and their inhibitory effects on nitric oxide production. *J. Nat. Prod.* 66, 1268, 2003.

18. Harinantenaina, L., Kasai, R., Yamasaki, K. Cussosaponins A–E, triterpene saponins from the leaves of *Cussonia racemosa*, a Malagasy endemic plant. *Chem. Pharm. Bull.* 50, 1292, 2002.

19. Maciuk, A. et al. Four new dammarane saponins from *Zizyphus lotus*. *J. Nat. Prod.* 67, 1642, 2004.

20. Sautour, M. et al. A new steroidal saponins from *Dioscorea cayenensis*. *Chem. Pharm. Bull.* 52, 1355, 2004.

21. Gonzalez, A.G. et al. Steroidal saponins from the bark of *Dracaena draco* and their cytotoxic activities. *J. Nat. Prod.* 66, 796, 2003.

22. Watanabe, K. et al. Eranthisaponins A and B, two new bidesmosidic triterpene saponins from the tubers of *Eranthis cilicia*. *J. Nat. Prod.* 66, 881, 2003.

23. Benyahia, S. et al. Cladocalol, a pentacyclic 28-nor-triterpene from *Eucalyptus cladocalyx* with cytotoxic activity. *Phytochemistry*, 66, 630, 2005.

24. Wu, X. et al. New lanostane-type triterpenes from *Fomes officinalis*. *Chem. Pharm. Bull.* 52, 1377, 2004.

25. Wandji, J. et al. Pentacyclic triterpenoid and saponins from *Gambeya boukokoensis*. *Phytochemistry*, 64, 848, 2003.

26. Endale, A. et al. Hopane-type saponins from the seeds of *Glinus lotoides*. *J. Nat. Prod.* 68, 446, 2005.

27. Voutquenne, V. et al. Haemolytic acylated triterpenoid saponins from *Harpullia austrocaledonica*. *Phytochemistry*, 66, 825, 2005.

28. Andrade, F.D.P. et al. Studies on the constituents of a Brasilian folk infusion. Isolation and structure elucidation of new triterpene saponins from *Ilex amara* leaves. *J. Agric. Food Chem.* 50, 256, 2002.

29. Biswas, T.M. et al. Hopane-type saponins from *Glinus lotoides* Linn. *Phytochemistry*, 66, 621, 2005.

30. Woldemichael, G.M., Montenegro, G., Timmermann, B.N. Triterpenoidal lupin saponins from the Chilean legume *Lupins oreophilus* Phil. *Phytochemistry*, 63, 853, 2003.

31. Piccinelli, A.L. et al. Three new furostanol saponins from the leaves of *Lycianhes synanthera* ("Chomte"), an edible Mesoamerican plant. *J. Agric. Food Chem.* 53, 290, 2005.

32. Shirota, O. et al. Two cangorosin A type triterpene dimmers from *Maytenus chuchuhuasca*. *Chem. Pharm. Bull.* 52, 115, 2004.

33. Tava, A. et al. Stability of saponins in alcoholic solutions: ester formation as artifacts. *J. Agric. Food Chem.* 51, 1797, 2003.
34. Gurfinkel, D.M., Rao, A.V. Determination of saponins in legumes by direct densitometry. *J. Agric. Food Chem.* 50, 426, 2002.
35. Du, Q. et al. Isolation of dammarane saponins from *Panax notoginseng* by high-speed counter-current chromatography. *J. Chromatogr. A*, 1003, 174, 2003.
36. Yoshikawa, M. et al. Structures of new dammarane-type triterpene saponins from the flower buds of *Panax notoginseng* and hepatotoprotective effects of principal ginseng saponins. *J. Nat. Prod.* 66, 926, 2003.
37. Mshvildadze, V. et al. Triterpenoid saponins from leaves of *Hedera pastuchowii. Chem. Pharm. Bull.* 52, 1414, 2004.
38. Seo, Y. et al. A new triterpene saponin from *Pittosporum viridiflorum* from the madagascar rainforest. *J. Nat. Prod.* 65, 65, 2002.
39. Seo, Y. et al. Bioactive saponins from *Acacia tenuifolia* from the Suriname rainforest. *J. Nat. Prod.* 65, 170, 2002.
40. Melek, F.R., Miyase, T., Ghaly, N.S. Triterpenoid saponins from *Meryta lanceolata. Phytochemistry*, 62, 557, 2003.
41. Gaidi, G. et al. New acylated triterpene saponins from *Silene fortunei* that modulate lymphocyte proliferation. *J. Nat. Prod.* 65, 1571, 2002.
42. Gonzalez, M. et al. Antimycotic spirostanol saponins from *Solanum hispidum* leaves and their structure–activity relationship. *J. Nat. Prod.* 67, 941, 2004.
43. Waffo-Teguo, P. et al. Acetylated glucuronide triterpene bidesmosidic saponins from *Symplocos glomerata. Phytochemistry*, 65, 741, 2004.
44. Braca, A. et al. Furostanol saponins and quercetin glycosides from the leaves of *Helleborus viridis. Phytochemistry*, 65, 2921, 2004.
45. Tanaka, N. et al. Terpenoids from *Tripterigium doianum* (Celastraceae). *Phytochemistry*, 61, 93, 2002.
46. Sang, S. et al. New steroid saponins from the seeds of *Allium tuberosum* L. *Food Chem.*, 83, 500, 2003.
47. Puri, R.K., Bhatnagar, J.K. Glycoalkaloids of *Solanum platanifolium. Phytochemistry*, 14, 2096, 1975.
48. Fukuhara, K., Shimizu, K., Kubo, I. Arudonine, an allelopathic steroidal glycoalkaloid from the root bark of *Solanum arundo. Phytochemistry*, 65, 1283, 2004.
49. Usubillaga, A. et al. Steroidal alkaloids from *Solanum sycophanta. Phytochemistry*, 44, 537, 1997.
50. Fukuhara, K., Kubo, I. Isolation of steroidal glycoalkaloids from *Solanum incanum* by two countercurrent chromatographic methods. *Phytochemistry*, 30, 685, 1990.
51. Ripperger, H. Steroid alkaloid glycosides from *Solanum robustum. Phytochemistry*, 39, 1475, 1995.
52. Kuronen, P., Vaananen, T., Pehu, E. Reversed-phase chromatographic separation and simultaneous profiling of steroidal glycoalkaloids and their aglycones. *J. Chromatogr.* 863, 25, 1999.
53. Simanovska, B., Vovk, I. High-performance thin-layer chromatographic determination of potato glycoalkaloids. *J. Chromatogr.* 903, 219, 2000.
54. Lairini, K., Ruiz-Rubio, M. Detection of tomatinase from *Fusarium oxysporum* F. sp. *lycopersici* in infected tomato plants. *Phytochemistry*, 45, 1371, 1997.
55. Lewis, D.C., Liljegren, D.R. Glycoalkaloids from *Archesolanum species. Phytochemistry*, 9, 2193, 1970.
56. Caddle, L.S. et al. Thin-layer chromatographic system for identification and quantitation of potato tuber glycoalkaloids. *Phytochemistry*, 26, 1453, 1978.

57. Shih, M.J., Kuć, J. α- and β-solamarine in Kennebec *Solanum tuberosum* leaves and aged tuber slices. *Phytochemistry*, 13, 997, 1974.
58. Metcalf, R.L., Metcalf, R.A., Rhodes, A.M. Cucurbitacins as kairomones for Diabroticite beetles. *Proc. Natl. Acad. Sci. USA*, 77, 3769, 1980.
59. Nielsen, J.K., Melchior, L., Sorensen, H. Cucurbitacin E and I in *Iberis amara*: feeding inhibitors for *Phyllotreta nemorum*. *Phytochemistry*, 16, 1519, 1977.
60. Laurie, W.A., McHale, D., Sheridan, J.B. A cucurbitacin glycoside from *Picrorhiza kurrooa*. *Phytochemistry*, 24, 2695, 1985.
61. Stuppner, H., Muller, E.P., Wagner, H. Cucurbitacins from *Picrorhiza kurrooa*. *Phytochemistry*, 30, 305, 1991.
62. Stuppner, H., Muller, E.P. Cucurbitacins with unusual side chains from *Picrorhiza kurrooa*. *Phytochemistry*, 33, 1139, 1993.
63. Hylands, P., Magd, M.S. Cucurbitacins from *Acanthosicyos horridus*. *Phytochemistry*, 25, 1681, 1986.
64. Dubois, M.-A. et al. Foetidissimoside A, a new 3,28-bidesmosidic triterpenoid saponin, and cucurbitacins from *Cucurbita foetidissima*. *Phytochemistry*, 27, 881, 1988.
65. Schenkel, E.P. et al. Cucurbitacins from *Wilbrandia ebracteata*. *Phytochemistry*, 31, 1329, 1992.
66. Achenbach, H. et al. Cucurbitanes and cucurbitane-type glycosides from *Ibervillea sonorae*. *Phytochemistry*, 33, 437, 1993.
67. Ahmad, M.U., Huq, M.E., Sutrathard, R.K. Bitter principles of *Luffa echinata*. *Phytochemistry*, 36, 421, 1994.
68. Hembre, J.A. et al. Potential antitumor agents: a cytotoxic cardenolide from *Cornilla varia*. *J. Nat. Prod.* 42, 293, 1979.
69. Pauli, G.F., Matthiesen, U., Fronczek, F.R. Sulfates as novel steroid metabolites in higher plants. *Phytochemistry*, 52, 1075, 1999.
70. Fukuyama, Y. et al. A cardenolide from *Anodendron affine*. *Phytochemistry*, 35, 1077, 1994.
71. Sady, M.B., Seiber, J.N. Chemical differences between species of *Asclepias* from the intermountain region of North America. *Phytochemistry*, 30, 3001, 1991.
72. Seiber, J.N., Roeske, C.N., Benson, J. Three new cardenolides from the milkweeds *Asclepias eriocarpa* and *A. labriformis*. *Phytochemistry*, 17, 967, 1978.
73. Ankli, A. et al. Cytotoxic cardenolides and antibacterial terpenoids from *Crossopetalum gaumeri*. *Phytochemistry*, 54, 531, 2000.
74. Kawaguchi, K., Hirotani, M., Furuya, T. Biotransformation of digitoxigenin by suspension cultures of *Strophantus divaricatus*. *Phytochemistry*, 30, 1503, 1991.
75. Chen, R. et al. Cardenolide glycosides of *Strophantus divaricatus*. *Phytochemistry*, 26, 2351, 1987.
76. Rechtenstamm, R., Kopp, B., Loffelhardt, W. Bioconversion leading to minor cardiac glycosides in *Convalaria majalis*. *Phytochemistry*, 25, 1107, 1986.
77. Xu, J., Tekaya, K., Itokawa, H. Pregnanes and cardenolides from *Periploca sepium*. *Phytochemistry*, 29, 344, 1990.
78. Evans, F.J. Variation in cardenolides and sapogenins in *Digitalis purpurea* during germination. *Phytochemistry*, 11, 2729, 1972.
79. Glensk, M. et al. Two-dimentional TLC with a sorbent gradient for the analysis of alfalfa root saponins. *Chromatographia*, 54, 669, 2001.
80. Biały, Z. et al. Saponins in alfalfa (*Medicago sativa* L.) root and their structural elucidation. *J. Agric. Food Chem.* 47, 3185, 1999.
81. Oleszek, W., Jurzysta, M. Isolation, chemical characterization and biological activity of alfalfa (*Medicago media* Pers.) root saponins. *Acta Soc. Bot. Pol.* 55, 23, 1986.

82. Jurzysta, M., Jurzysta, A. Thin-layer chromatography of acetyl derivatives of soyasapogenols A, B, C, D and E. *J. Chromatogr.* 179, 233, 1979.

83. Ikan, R., Cudzinovski, M. Separation of sterols and corresponding stanols on thin layers of silica impregnated with silver nitrate. *J. Chromatogr.* 18, 422, 1965.

84. Lisboa, B.P. Thin layer chromatography of steroids. *Meth. Enzymol.* 15, 3, 1969.

85. Neher, R. TLC of steroids and related compounds, in *Thin Layer Chromatography*, Stahl, E. (Ed). George Allen and Unwin, London, 1969, p. 311.

86. Oleszek, W. Zanhic acid tridesmoside and other dominant saponins from alfalfa (*Medicago sativa* L.) aerial parts. *J. Agric. Food Chem.* 40, 191, 1992.

87. Oleszek, W. Solid-phase extraction-fractionation of alfalfa saponins. *J. Sci. Food Agric.* 44, 43, 1988.

88. Mackie, A.M., Singh, H.T., Owen, J.M. Studies on the distribution, biosynthesis and function of steroidal saponins in echinoderms. *Comp. Biochem. Physiol.* 56, 9, 1977.

89. Jurzysta, M. Haemolytic micromethod for rapid estimation of toxic alfalfa saponins. *Acta Agrobot.* 32, 5, 1979.

90. Oleszek, W., Structural specificity of alfalfa (*Medicago sativa*) saponin haemolysis and its impact on two haemolysis-based quantification methods. *J. Sci. Food Agric.* 19, 1063, 1990.

91. Wagner, H., Bladt, S., Zgainski, E.M. *Plant Drug Analysis. A Thin Layer Chromatography Atlas.* Springer-Verlag, Berlin, 1984, p. 320.

92. Betina, V. Bioautography in paper and thin-layer chromatography and its scope in the antibiotic field. *J. Chromatogr.* 78, 41, 1973.

93. Lund, B.M., Lyon, G.D. Detection of inhibitions of *Erwinia carotovora* and *E. herbicola* on thin-layer chromatogram. *J. Chromatogr. A*, 110, 193, 1975.

94. Kapusta, I. et al. Triterpene saponins from barrel medic (*Medicago truncatula*) aerial parts. *J. Agric. Food Chem.* 53, 2164, 2005.

21 TLC of Carotenoids

George Britton

CONTENTS

21.1   INTRODUCTION

The carotenoids are very widespread natural pigments, not only in plants but also in bacteria, fungi, yeasts, algae, and nonmammalian animals. Familiar examples of coloration by carotenoids include the yellow of sunflowers, the orange of carrots and oranges, the red of tomatoes, the gold of goldfish, the pink of salmon flesh, and red and yellow feathers of birds such as goldfinches. The recently published *Carotenoids Handbook* [1] lists some 750 different carotenoids that have been isolated from natural sources. Of these, about 500 have been characterized satisfactorily.

Almost all carotenoids are tetraterpenes, with a symmetrical C_{40} skeleton bio-synthesized by tail-to-tail linkage of two molecules of the C_{20} geranylgeranyl diphosphate [2]. This gives the colorless hydrocarbon phytoene (7,8,11,12, 7′,8′,11′,12′-octahydro-ψ,ψ-carotene). A series of desaturation reactions then intro-duces further double bonds and extends the conjugated double-bond (cdb) chromo-phore that gives the carotenoids their distinctive light absorption and color properties. In most cases, intermediates in the desaturation series are phyto-fluene (7,8,11,12,7′,8′-hexahydro-ψ,ψ-carotene), ζ-carotene (7,8,7′,8′-tetrahydro-ψ, ψ-carotene) and neurosporene (7,8-dihydro-ψ,ψ-carotene), and the final product is lycopene (ψ,ψ-carotene), which has a chromophore of 11 cdbs. Subsequent modifications such as cyclization and the introduction of oxygen functions give rise to the diversity of carotenoid structures found in nature.

Some carotenoids have fewer than 40 carbon atoms. The examples found in plants are apocarotenoids, formed by enzymic cleavage of a fragment from one end or both ends of the C_{40} skeleton, for example, 8′-apo-β-caroten-8′-al and bixin (6-methyl 6′-hydrogen (9Z)-6,6′-diapocarotene-6,6′-dioate).

A small number of bacteria, for example, *Staphylococcus aureus,* have been found to contain a series of C_{30} carotenoids (e.g., diaponeurosporene (7,8-dihydro-4,4′-diapo-ψ,ψ-carotene)) which are analogs of the C_{40} series but are biosynthesized as triterpenes with a C_{15}–C_{15} skeleton. Other bacteria and Archaea, for example, *Halobacterium* species and some plant pathogens, contain C_{45} and C_{50} carotenoids. These are biosynthesized by the conventional C_{40} pathway, but one or two additional C_5 units are added as substituents to the end groups, for example, bacterioruberin (2,2′-bis-(3-hydroxy-3-methylbutyl)-3,4,3′,4′-tetradehydro-1,2,1′,2′-tetrahydro-ψ,ψ-carotene-1,1′-diol).

21.1.1 STRUCTURES AND NOMENCLATURE

Hydrocarbon carotenoids collectively are called carotenes; derivatives containing oxygen functions are known as xanthophylls. No natural carotenoids containing directly bonded nitrogen, sulphur, or halogen substituent groups have been discovered. Carotenoids may be acyclic, as lycopene; most examples, however, are monocyclic or dicyclic, having a six-membered (occasionally five-membered) ring at one end or both ends of the molecule. Extensive cyclization, seen in diterpenes and triterpenes, does not occur in the tetraterpene carotenoid series.

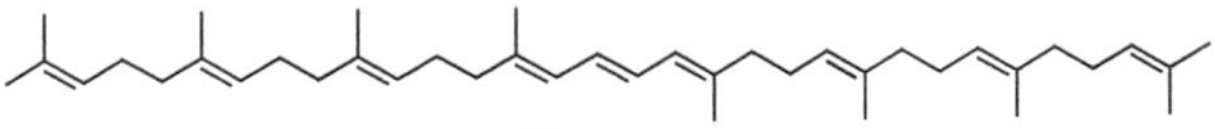

Phytoene

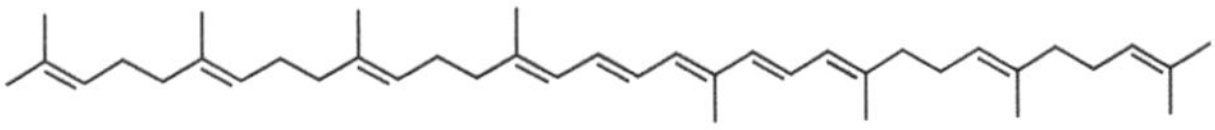

Phytofluene

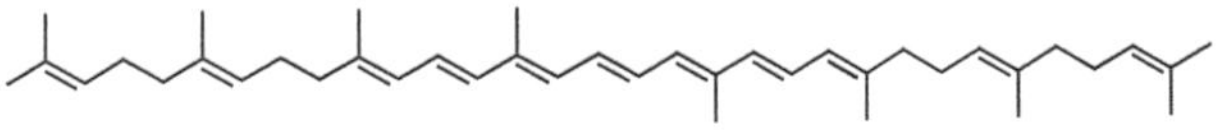

ζ-Carotene

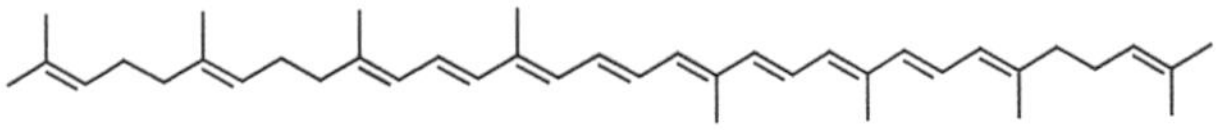

Neurosporene

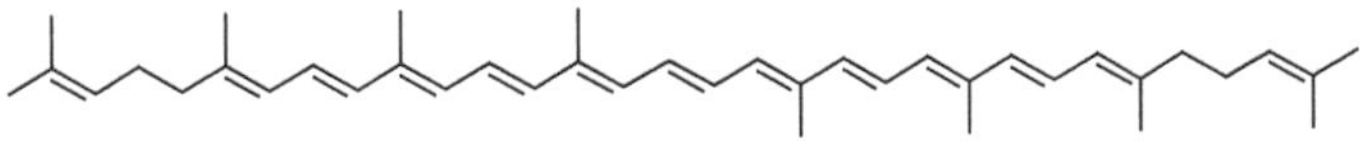

Lycopene

The numbering system for carotenoids is illustrated in Figure 21.1.

It has long been the tradition to give any newly isolated natural carotenoid a trivial name, usually derived from the name of the biological source from which it was obtained. These trivial names, however, generally convey little or no information about structure, so a semisystematic nomenclature has been introduced that

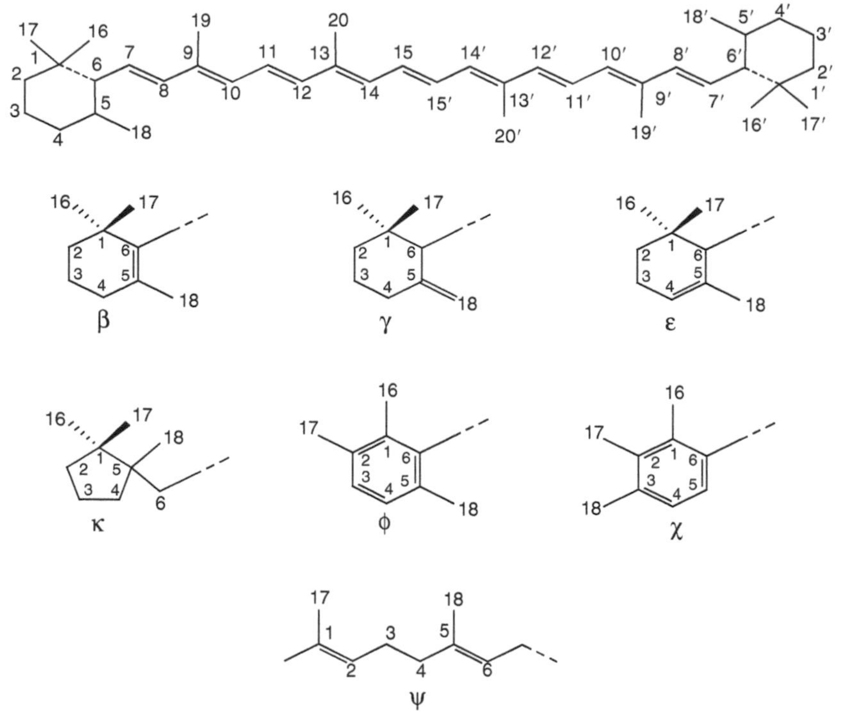

FIGURE 21.1 The numbering scheme for carotenoids, and the seven types of carotenoid end group, with their Greek letter designations.

unambiguously defines the structure of the carotenoid. In this, the carotenoid molecule is considered in two halves. The individual compound is then named as a derivative of the parent carotene, specified by the Greek letters that designate the two end groups. Seven different end groups are recognized and are also shown, in order of priority (Greek alphabetical order) in Figure 21.1. Conventional prefixes and suffixes are used to indicate changes in hydrogenation level and the presence of substituent groups. For example, lycopene becomes ψ,ψ-carotene; β-carotene becomes β,β-carotene; γ-carotene becomes β,ψ-carotene; and lutein becomes β, ε-carotene-3,3′-diol. The rules for the semi-systematic nomenclature are given in full in Refs. [3,4].

Most carotenoids occur naturally mainly as the all-*trans* (usually all-*E*) form, but *cis* (*Z*) isomers, usually in small amounts, are also commonly present. Their presence may be important in relation to the physiological functions of the carotenoids. In some plants, a carotenoid may occur specifically as a *cis* or even a poly-*cis* isomer. Neoxanthin (5′,6′-epoxy-6,7-didehydro-5,6,5′,6′-tetrahydro-β,β-carotene-3,5,3′-triol) in the chloroplasts of green leaves, and bixin, the coloring principle of annatto, are (9′Z) and (9Z) isomers, respectively. An interesting and extreme case is prolycopene which, in the tangerine strain of tomato, replaces (all-*trans*)-lycopene as the main pigment. Prolycopene has the tetra-*cis* structure (7Z,9Z,7′Z,9′Z)-lycopene.

21.1.2 PROPERTIES

21.1.2.1 Solubility

In their solubility properties, carotenoids are like other groups of higher isoprenoids. The carotenes are typical apolar hydrocarbons, insoluble in water. The xanthophylls contain oxygen functions and are, therefore, more polar yet still generally are insoluble in water. Only very polar compounds such as carboxylic acids, diosphenols, glycosides, and glycosyl esters have any appreciable solubility in water. Carotenoids generally have good solubility in a variety of organic solvents. Acetone, tetrahydrofuran, diethyl ether, and ethyl acetate are good solvents for most carotenoids. Carotenes are soluble in hydrocarbon solvents: hexane and petroleum ether. Xanthophylls are soluble in polar solvents such as ethanol. When in crystalline form, however, carotenoids can be difficult to dissolve, even in solvents in which their solubility is good. Benzene, toluene, and dichloromethane will dissolve the crystals efficiently. Some carotenoids, lycopene being a prime example, are very susceptible to crystallizing from solution, and this can cause problems during chromatography.

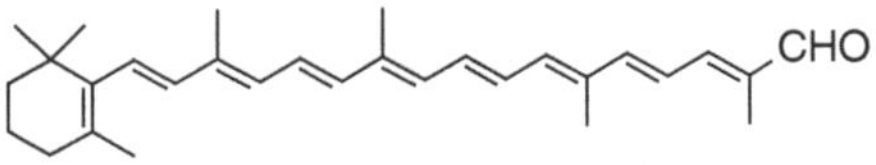

8'-Apo-β-caroten-8'-al

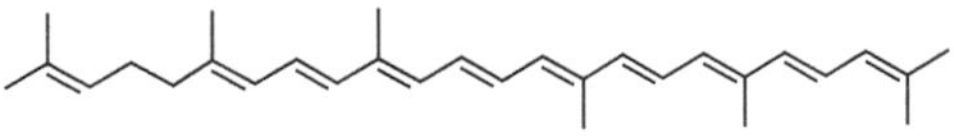

4,4'-Diaponeurosporene

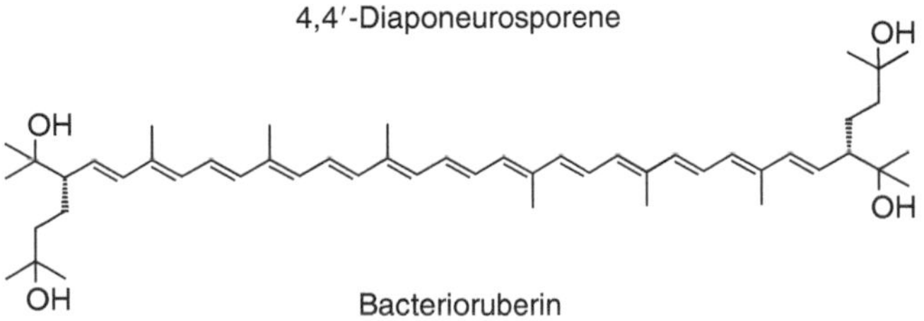

Bacterioruberin

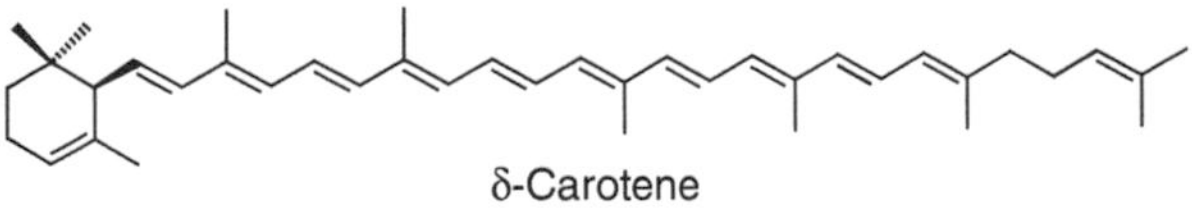

δ-Carotene

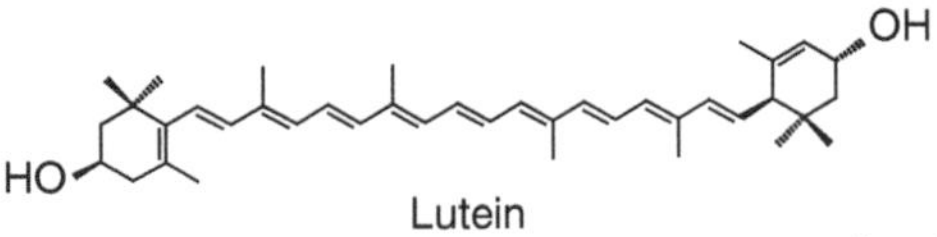

Lutein

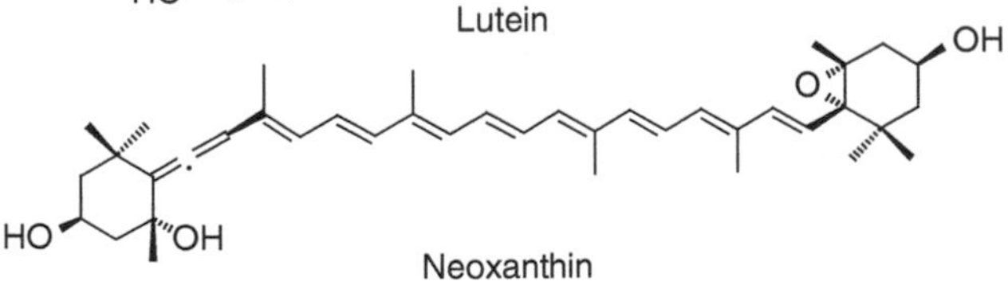

Neoxanthin

21.1.2.2 Light Absorption

The structural feature that gives the carotenoids their distinctive properties and functions is the long cdb system [5]. This is the chromophore that is responsible for the absorption of light in the visible region of the spectrum (usually between 400 and 550 nm) and hence the color of the carotenoids. The longer the chromophore, the longer is the wavelength of maximum absorption. Thus, phytoene and phytofluene, with only 3 and 5 cdbs, respectively, absorb maximally in the UV region at 285 and 348 nm, and are colorless. Neurosporene (9 cdb) is yellow (440 nm) and lycopene (11 cdb) is orange-red (470 nm). The relationship between chromophore and light absorption properties is discussed in detail in Ref. [6].

21.1.2.3 Instability

The electron-rich, delocalized π-electron system of the cdb chromophore is also very susceptible to attack by electrophilic oxidizing agents and oxidizing free radicals, so the carotenoids are unstable and easily degraded under oxidizing conditions.

21.1.2.4 Shape

The cdb system is also responsible for the overall shape of the carotenoid molecule. The all-*trans* isomers are long, rigid, linear molecules, quite distinct from the bent shape of the *cis* isomers.

21.1.3 OCCURRENCE

21.1.3.1 Higher Plants: Photosynthetic Tissues

Carotenoids are known first and foremost as plant pigments and there is, naturally, particular interest in those plants (fruit and vegetables) that are widely consumed in the human diet.

In plants, carotenoids are found universally in the chloroplasts of leaves and other green tissues (i.e., in all green vegetables), though their color is masked by the chlorophylls. The same main carotenoids are found, in similar proportions, in the leaves of virtually all species, namely β-carotene (usually 25%–30% of the total carotenoid), lutein (around 45%), violaxanthin (5,6:5′,6′-diepoxy-5,6,5′,6′-tetrahydro-β,β-carotene-3,3′-diol) (15%), and neoxanthin (15%). Small amounts of α-carotene (β,ε-carotene), β-cryptoxanthin (β,β-caroten-3-ol), zeaxanthin (β,β-carotene-3,3′-diol), antheraxanthin (5,6-epoxy-5,6-dihydro-β,β-carotene-3,3′-diol), and lutein-5,6-epoxide, (5,6-epoxy-5,6-dihydro-β,ε-carotene-3,3′-diol) may also be present [7]. In a very small number of species, notably lettuce (*Lactuca sativa*), lactucaxanthin (ε,ε-carotene-3,3′-diol) is also a major xanthophyll (around 10% of the total) [8]. Most leaves have a similar quantitative carotenoid composition, though changes may be seen in plants under stress (e.g., high light, drought), when β-carotene-5,6-epoxide (5,6-epoxy-5,6-dihydro-β,β-carotene) and increased amounts of zeaxanthin may be found, and some xanthophylls may be esterified, as they are in the autumn leaves of many trees [9]. Storage and processing may also cause changes and degradation [10]. The carotenoid content and concentration are higher in darker green leaves, which contain a greater density of chloroplasts.

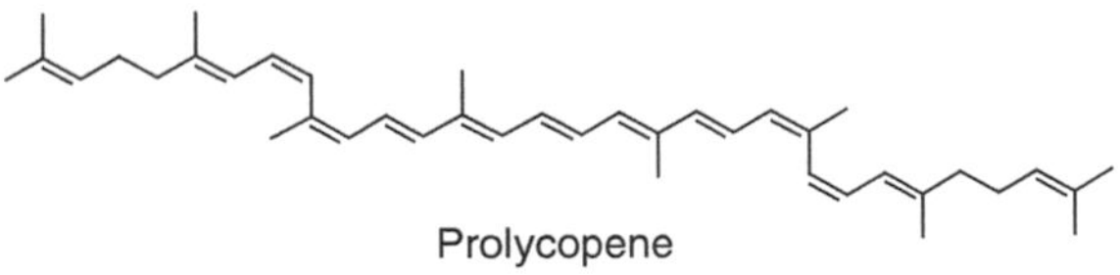

Prolycopene

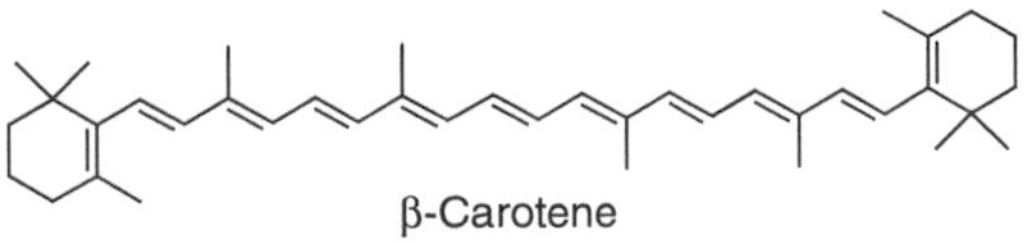

β-Carotene

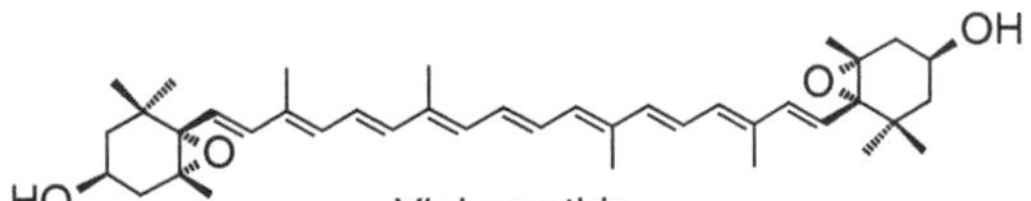

Violaxanthin

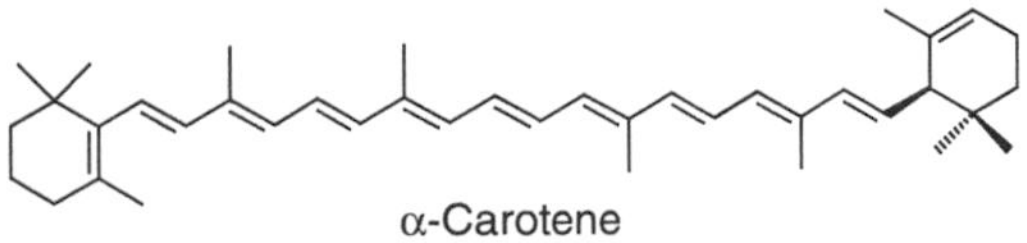

α-Carotene

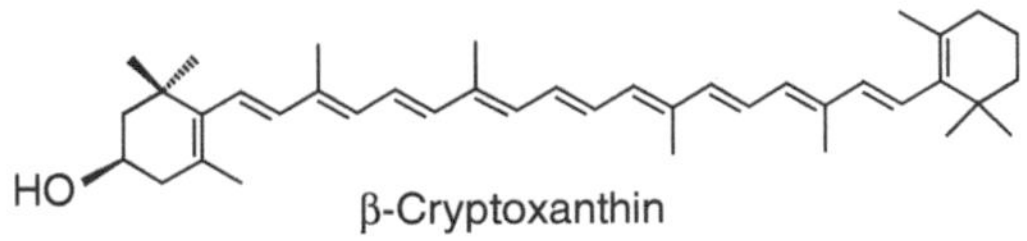

β-Cryptoxanthin

Within the chloroplast, the carotenoids are located, together with chlorophylls, in the pigment–protein complexes of photosystems 1 and 2 in the thylakoid membranes, where they serve as light-harvesting pigments and in protection against photooxidation by singlet oxygen, sensitized by triplet-state chlorophyll [11].

In some coniferous trees and cycads, unusual ketocarotenoids or secocarotenoids, for example, semi-β-carotenone (5,6-seco-β,β-carotene-5,6-dione) may accumulate in leaves or needles at some times of the year or in some stages of development [12,13]. These carotenoids are located not in the chloroplast pigment–protein complexes but in extraplastidic oil droplets.

21.1.3.2 Higher Plants: Nonphotosynthetic Tissues

Carotenoids are also widely distributed in nonphotosynthetic tissues of plants and are responsible for the yellow, orange, and red colors of many flowers and fruits and of some roots [14]. They are normally located in chromoplasts, either in membranes, in oil droplets or, in some notable cases such as carrot (β-carotene) and tomato (lycopene), in crystalline form [15]. The carotenoids are synthesized in the plastids as flowers open or fruit ripen.

Xanthophylls are frequently present as complex mixtures of fatty acyl esters. Modern sensitive analytical methods show that extracts of nonphotosynthetic plant tissues, especially flowers and fruit, may contain complex mixtures of carotenoids, including esters and geometrical isomers and biosynthetic intermediates. More than 120 carotenoids have been detected in *Capsicum annuum*, for example.

21.1.3.2.1 Flowers
Several distinctive carotenoid patterns have been recognized in flowers, where carotenoids may be present in all anatomical parts, for example, sepals, stamens, pollen, and especially petals [14]:

1. Substantial amounts of common carotenes, particularly β-carotene, for example, in *Narcissus*
2. High concentration of lutein (usually as esters) often with some zeaxanthin and lutein-5,6-epoxide, for example, petals of sunflower (*Helianthus annuus*) and marigold (*Tagetes erecta*)
3. Complex mixtures of 5,6- and 5,8-epoxycarotenoids
4. Specific or unusual carotenoids of restricted distribution, for example, eschscholtzxanthin (4',5'-didehydro-4,5'-*retro*-β,β-carotene-3,3'-diol) in *Eschscholtzia californica*)

21.1.3.2.2 Fruit

In ripe fruit, carotenoids may be present in the skin or peel, the internal soft tissues, or the seed and/or seed coat. Different carotenoids may be present in different parts of the fruit. Again, several distinctive carotenoid patterns have been recognized [14]:

1. Usual collection of chloroplast carotenoids (in unripe and green fruit)
2. Large amounts of lycopene and its hydroxy derivatives (e.g., tomato)
3. Large amounts of β-carotene and its hydroxy derivatives β-cryptoxanthin and zeaxanthin (e.g., mango, persimmon)
4. Collections of 5,6- and 5,8-epoxycarotenoids (e.g., star fruit, *Carambola*)
5. Large amounts of apocarotenoids (e.g., *Citrus* species and hybrids)
6. Specific or unusual carotenoids (e.g., capsanthin (3,3'-dihydroxy-β, κ-caroten-6'-one) in red pepper, *Capsicum annuum*)

21.1.3.2.3 Roots

Carotenoids do not occur extensively in roots, but there are some well-known examples, for example, carrot (*Daucus carota*) and sweet potato (*Ipomoea batata*), where large amounts of carotenoid, mainly β-carotene, are present [14].

21.1.3.2.4 Seeds

Some seeds, especially the seed coat, are colored by carotenoids [14], for example, zeaxanthin in sweetcorn (*Zea mays*). The seed coat of *Bixa orellana* contains a very high concentration (up to 10% dry weight) of the apocarotenoid bixin (annatto).

21.1.3.3   Algae

In both microalgae and seaweeds, carotenoids are also present universally as chloroplast pigments with essential roles in photosynthesis. Green algae (Chlorophyta) have a carotenoid complement similar to that of green leaves, though frequently with another xanthophyll, loroxanthin (19-hydroxylutein, β,ε-carotene-3,19,3'-triol) also present. The variety of carotenoid structures found in algae is great and includes compounds containing acetylenic or allenic groups, for example, alloxanthin (7,8,7',8'-tetradehydro-β,β-carotene-3,3'-diol) and fucoxanthin (5,6-epoxy-3'-ethanoyloxy-3,5'-dihydroxy-6',7'-didehydro-5,6,7,8,5',6'-hexahydro-β,β-caroten-8-one), and C_{37} nor-carotenoids, notably peridinin (5,6-epoxy-3'-ethanoyloxy-3,5'-dihydroxy-6',7'-didehydro-5,6,5',6'-tetrahydro-12',13',20'-trinor-β,β-caroten-19,

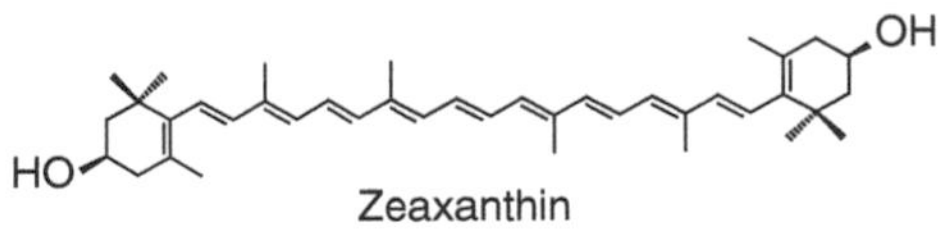

Zeaxanthin

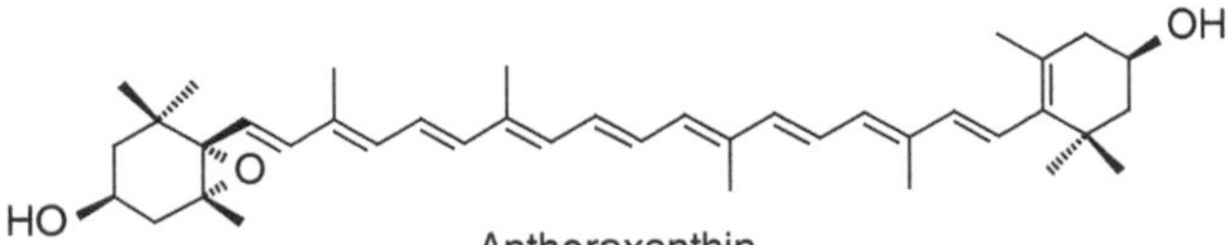

Antheraxanthin

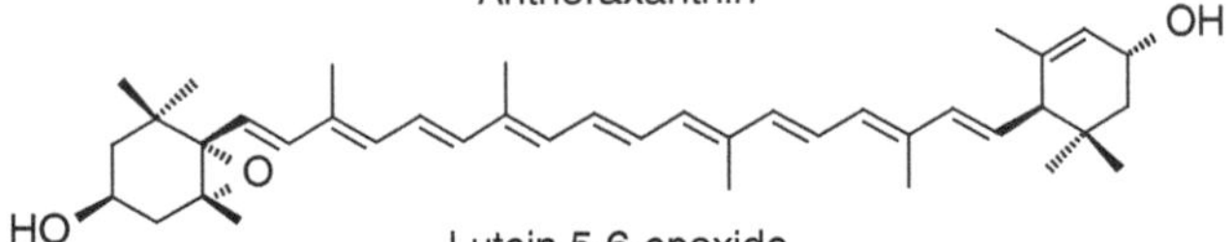

Lutein 5,6-epoxide

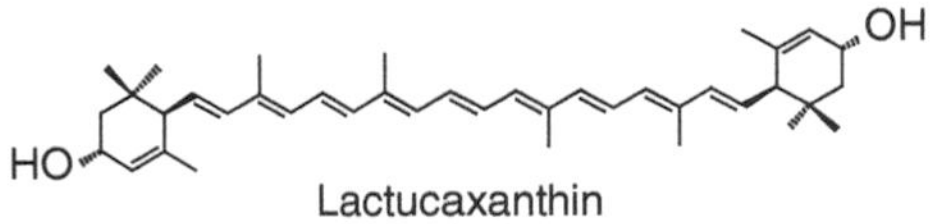

Lactucaxanthin

11-olide), the characteristic xanthophyll of dinoflagellates [7,14]. The presence of certain carotenoids and carotenoid profiles has been used as a systematic marker for algal classification [16].

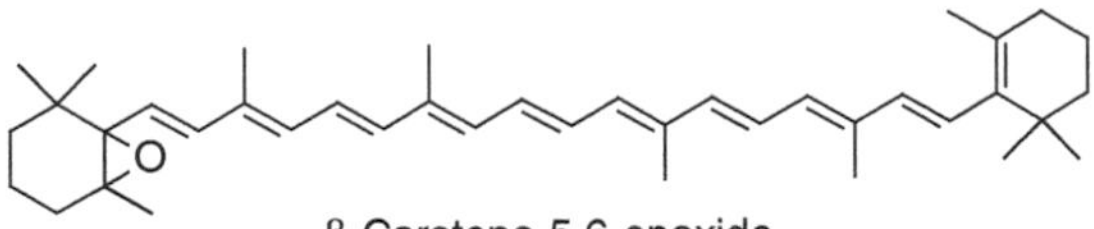

β-Carotene 5,6-epoxide

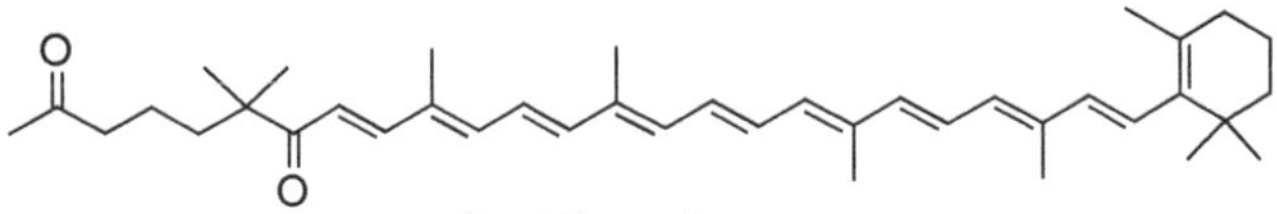

Semi-β-carotenone

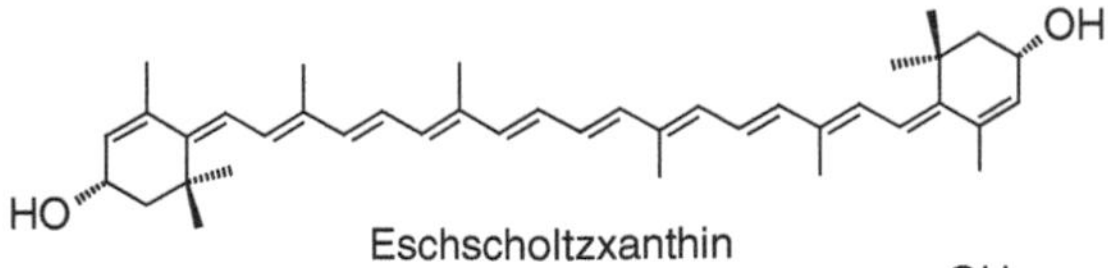

Eschscholtzxanthin

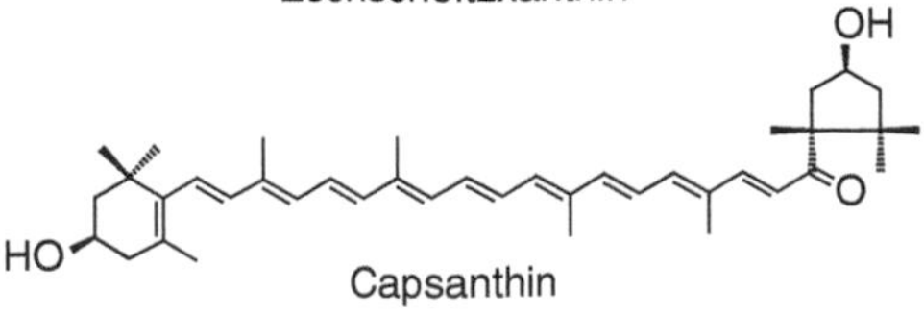

Capsanthin

Some green algae, when subjected to nutritional or environmental stress, respond by synthesizing large amounts of extraplastidic carotenoids. Some examples are exploited commercially, especially *Dunaliella salina* for β-carotene [17] and *Haematococcus pluvialis* for astaxanthin (3,3′-dihydroxy-β,β-carotene-4,4′-dione) [18]. Carotenoid yields of up to a few percent of dry weight can be achieved.

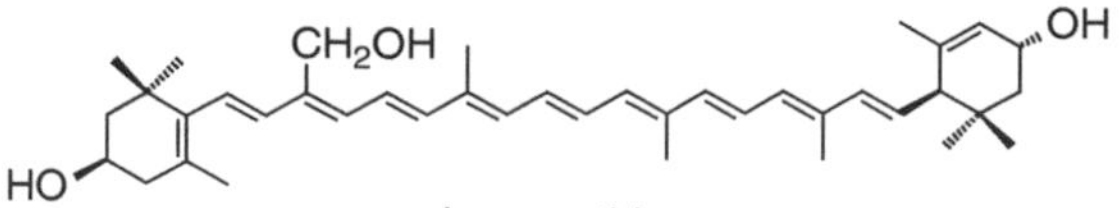

Loroxanthin

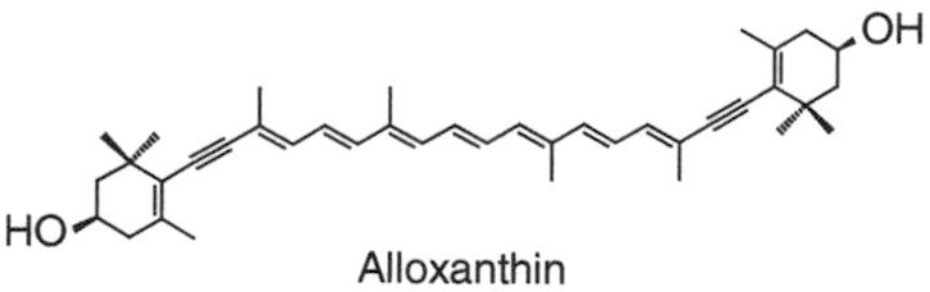

Alloxanthin

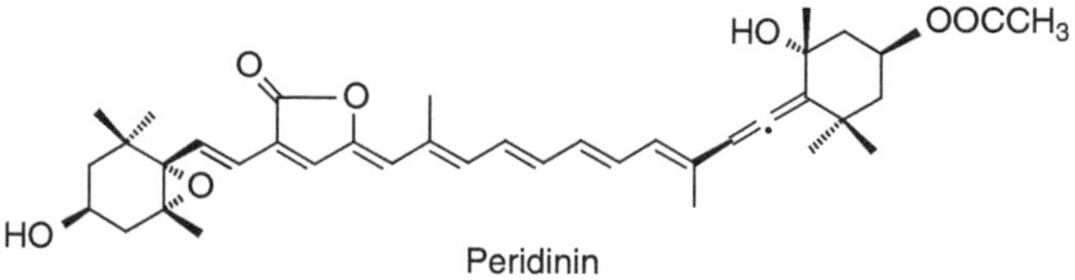

Peridinin

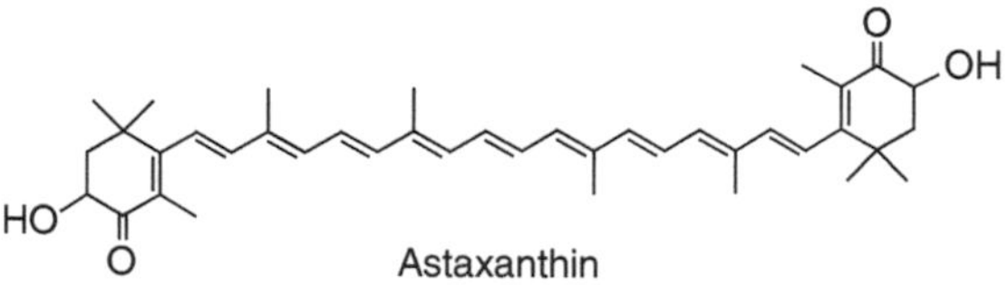

Astaxanthin

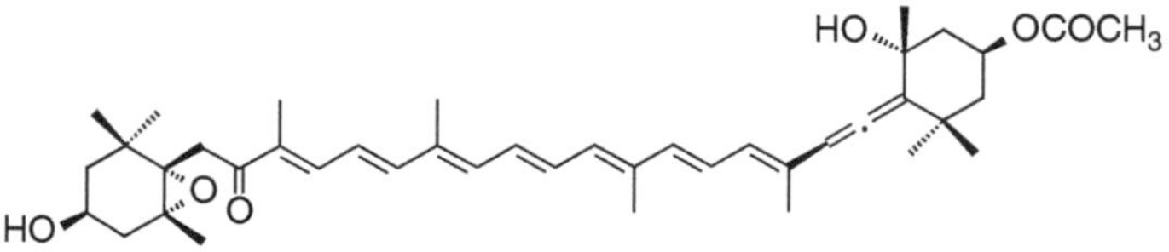

Fucoxanthin

21.1.3.4 Bacteria, Yeasts, and Fungi

Many bacteria, yeasts, and fungi do not contain carotenoids, but there are examples that have a high carotenoid content. The mould *Blakeslea trispora* and the yeast *Xanthophyllomyces rhodorhous* (formerly *Phaffia rhodozyma*) are used commercially as sources of β-carotene and astaxanthin, respectively. Phototrophic bacteria contain carotenoids in their photosynthetic pigment–protein complexes. Acyclic compounds such as spirilloxanthin (1,1′-dimethoxy-3,4,3′,4′-tetradehydro-1,2,1′,2′-tetrahydro-ψ,ψ-carotene) and carotenoids with aromatic ϕ and χ end groups (e.g., okenone (1′-methoxy-1′,2′-dihydro-χ,ψ-caroten-4′-one)) are characteristic [19]. The carotenoids of nonphototrophic bacteria are diverse and include C_{30} and C_{50} compounds as well as familiar C_{40} examples such as zeaxanthin. Carotenoid glycosides are common [14]. There have been no extensive systematic studies of the carotenoids of these bacteria.

21.1.4 Value of TLC Today

In most laboratories today, the method of choice for carotenoid analysis, both qualitative and quantitative, is HPLC, especially when a photodiode array detector is used, which gives the absorption spectrum of each component in the chromatogram on line.

However, because of their strong color, carotenoids are easily detected on TLC, without the need to use possibly damaging spray reagents, so TLC thus remains indispensable, even in laboratories with good HPLC facilities. TLC is widely used for isolating and purifying carotenoids for further study, for rapid preliminary screening of extracts, for comparison of samples with standards, and for monitoring the course of reactions [20]. Although silica is the most commonly used adsorbent for TLC of carotenoids, as it is for other classes of compounds, several other stationary phases, especially basic materials such as magnesium oxide, are extremely useful for carotenoids. These adsorbents are not available in a form suitable for HPLC, so they must be used in open column chromatography (CC) and TLC.

21.1.5 CHROMATOGRAPHIC STRATEGIES

With an unfamiliar extract, preliminary investigation by TLC on silica, with a succession of solvents of increasing polarity (Section 21.6.1), provides an idea of what types of carotenoids are present in the extract and what strategy to use for the preparative separation.

A combination of CC and TLC (and sometimes solvent partitioning) can then generally be used to separate all the carotenoids in the extract. Except in very large-scale or very small-scale work, the usual strategy is first to use CC on alumina or silicic acid to separate the extract into fractions containing groups of carotenoids of approximately similar polarity, for example, a hydrocarbon (carotene) fraction or a dihydroxycarotenoid fraction. TLC is then used to separate and purify the individual carotenoids in each fraction. Successive TLC on silica and then MgO will separate most carotenoid mixtures. HPLC may then be used for final purification or to resolve the most difficult mixtures. For large-scale work it may be more practical to use only CC and crystallization, for very small-scale work only TLC or HPLC [20,21].

21.2 TLC OF CAROTENOIDS—SPECIAL ASPECTS

The key structural feature of the carotenoids, the long cdb chromophore, is a major factor that determines a number of special practical aspects of TLC of carotenoids, i.e., the light absorption properties and hence color that enables easy detection, the susceptibility to oxidative breakdown, and the affinity for some basic inorganic adsorbents. It also determines the linear and bent shape of the all-*trans* and *cis* isomers, respectively.

21.2.1 DETECTION BY COLOR

The carotenoids have high molar absorption coefficients ε and are hence strongly colored and easily detected by eye in submicrogram amounts against the white background of TLC plates. The eye can also detect the subtle differences in color hue between compounds that differ in λ_{max} by only about 4–5 nm. The differences in hue are enhanced when the carotenoids are adsorbed on MgO and other basic inorganic substances; there is, for example, a clear distinction between lutein (λ_{max} 446 nm) which appears yellow and zeaxanthin (λ_{max} 450 nm) which appears orange. The color intensity fades considerably when the solvent evaporates. Because the color is so readily seen, the presence of invisible colorless contaminants is easily missed.

The biosynthetic intermediates phytoene and phytofluene are colorless, absorbing only in the visible region with λ_{max} 285 and 348 nm, respectively, so they cannot be seen in normal visible light. Both are fluorescent, however, when viewed under UV light. Phytofluene exhibits characteristic intense greenish-white fluorescence and is detectable at very low concentration. The fluorescence of phytoene is more difficult to see, being largely in the UV region; dark violet fluorescence may be discernible at high concentration. Phytoene is much more easily detected if fluorescent silica gel GF_{254} plates are used; it absorbs the fluorescence and is seen as a dark area against the bright fluorescent background. On TLC plates that do not contain a fluorescent indicator, a fluorescent spray reagent such as Rhodamine 6G can be used; again phytoene quenches the strong fluorescence.

After TLC, carotenoid zones fade rapidly due to oxidative breakdown (Section 21.2.2.4). Digital cameras allow a photographic record of the chromatogram to be made. This should be done as rapidly as possible after removal from the developing chamber. If photography is not possible, colored spots or zones may be covered with transparent adhesive tape, or a clean glass plate placed on the TLC surface and taped in place. When covered in this way and stored in the dark, carotenoids are protected from oxygen and should remain colored for several days. Plates from which samples are to be recovered should not be stored.

21.2.2 MINIMIZING PROBLEMS OF INSTABILITY

Because of the long cdb system, carotenoids generally are less stable than most other isoprenoids, and are sensitive to oxygen, light, heat, acids and, in some cases, alkali, and especially to combinations of these factors, for example, light and oxygen. Stringent precautions must be observed in all work with carotenoids, including TLC, if losses of material or structural changes are to be minimized [20–22]. Small amounts of *cis* isomers, oxidation products, etc., can readily be detected, so their production as artefacts must be avoided. All procedures that carry risk of oxidation, isomerization, etc., must be carried out as rapidly as possible. Speed of manipulation is important, especially in TLC. The risk of degradation is enhanced when the carotenoids are adsorbed on an active surface.

21.2.2.1 Light

Light is a major promoter of geometrical isomerization of carotenoids, especially in the presence of a photosensitizer such as chlorophyll. So, in all carotenoid work, exposure to direct sunlight or UV light must be avoided. Unless there is particular interest in the *cis/trans* isomer composition in a sample, low intensity diffuse daylight or subdued artificial light is acceptable for most manipulations. If the geometrical isomer composition is important, working under a yellow safelight is recommended.

During TLC, the developing chamber should be covered with black cloth or kept in a dark cupboard. If UV light is used to aid detection, exposure should be kept to a minimum.

21.2.2.2 Heat

Carotenoids also undergo isomerization and sometimes structural modification or degradation, if exposed to heat. They should not be subjected to heat unless this is unavoidable. Solvents should be evaporated below 40°C on a rotary evaporator or, for small volumes, by blowing a gentle stream of nitrogen onto the solution in a tube or vial held in a small water bath at no more than 40°C.

21.2.2.3 Acid and Alkali

Almost all carotenoids are at risk of oxidative decomposition, *cis/trans* isomerization or, in some cases, dehydration, if subjected to acid conditions. The well-known acid-catalyzed isomerization of 5,6-epoxycarotenoids to the 5,8-epoxides can occur during inadvertent exposure to acid during extraction and also during chromatography on acidic adsorbents such as silicic acid [22]. Silica gel plates may need to be treated with base to prevent this isomerization (Section 21.3.4.1). Solvents such as chloroform, which may contain traces of HCl, should not be used. Carotenoids should not be handled in rooms where acidic reagents and strong acids are being used.

Most carotenoids are stable to alkali and not destroyed by the common practice of saponification. Some examples, however, notably α-ketols (such as astaxanthin), fucoxanthin, peridinin, and related compounds, can be altered by treatment with even weak base. Saponification must not be used if it is suspected that such compounds may be present, or if carotenoid esters are of interest. Alumina catalyzes similar changes to astaxanthin and related compounds, and should not be used for CC or TLC if these compounds may be present (Section 21.6.2.4).

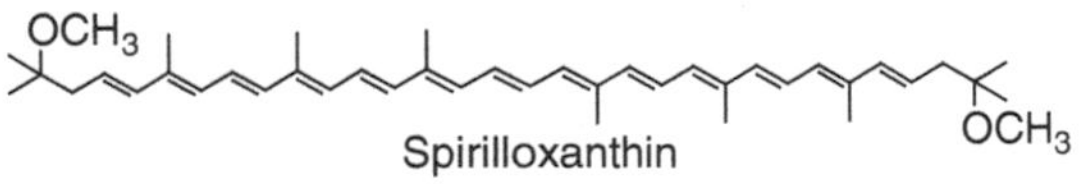

Spirilloxanthin

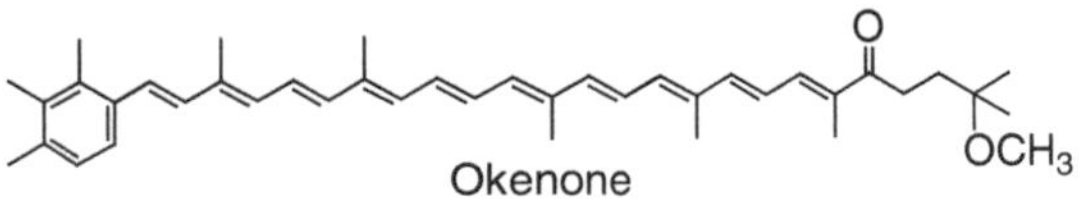

Okenone

Under alkaline conditions, carotenoid aldehydes can undergo aldol condensation with even traces of acetone to produce artefacts. Some inorganic adsorbents, including MgO, can be sufficiently basic and active to catalyze this reaction with acetone present in the developing solvent, unless an inert carrier such as kieselguhr is included in the stationary phase. Similar reactions can also lead to serious contamination by condensation products and polymers of acetone itself.

21.2.2.4   Oxygen

The main cause of destruction of carotenoids is oxidative breakdown, which can lead to the formation of oxidized artefacts or the total destruction and disappearance of the carotenoid. The acyclic carotenoids, especially phytofluene and ζ-carotene, seem to be the most susceptible to oxidative decomposition. The addition of an antioxidant such as BHT (2,6-di-t-butyl-p-cresol, butylated hydroxytoluene) at the time of tissue extraction is a useful way to minimize oxidation, provided the antioxidant can be removed easily and efficiently when pure samples are needed for further study.

Carotenoid-containing extracts and samples, including ones recovered from TLC, must always be stored in the absence of oxygen, either under vacuum or in an inert atmosphere (Ar or N_2). Traces of residual oxygen are sufficient to cause decomposition even at freezer temperature. Solutions should be flushed with N_2 for a few minutes before storage, though it is preferable to store samples dry. Saponification and other reactions should be carried out under an inert atmosphere of N_2; pyrogallol may be added to saponification mixtures as an antioxidant.

For TLC, exposure of carotenoids to air during sample application and when they are adsorbed on thin layers, before and after chromatography, must be as short as possible, and zones to be recovered should be removed and eluted rapidly. TLC of the more unstable carotenoids should be carried out in an inert atmosphere, for example, by flushing the developing chamber with N_2. The final evaporation of solvent from carotenoid samples, including fractions recovered from TLC, is done by blowing a gentle stream of N_2 onto the surface of the solution.

Peroxides, which may accumulate in diethyl ether and related solvents on storage, can also lead to destruction or the formation of artefacts. Freshly distilled, peroxide-free solvents must be used.

21.2.3   USE OF BASIC INORGANIC ADSORBENTS (E.G., MAGNESIUM OXIDE)

The choice of adsorbents for chromatography of carotenoids is extensive. In addition to conventional generally used stationary phases (silica, alumina, kieselguhr, starch) various inorganic materials have been used successfully, including $CaCO_3$, $Ca_3(PO_4)_2$, $Ca(OH)_2$, $MgCO_3$, MgO, and $ZnCO_3$. Some of these, especially MgO, $Ca(OH)_2$, and $CaCO_3$ are used extensively as stationary phases in CC and TLC [21,23]. The value of these materials lies in the fact that separation is based not simply on polarity but on the number and type of double bonds in the carotenoid molecule (Section 21.4.4).

21.3   TLC ON SILICA AND ALUMINA

21.3.1   STRUCTURAL BASIS OF SEPARATION

Silica (silica gel, kieselgel) is the most widely used stationary phase for carotenoids, as it is for other groups of compounds. Alumina is not recommended for routine work because of the instability of some carotenoids on this adsorbent.

TABLE 21.1
TLC of Carotenoids on Silica Gel G: Order
of Increasing Adsorption Affinity
and Decreasing R_f Value for the Main
Carotenoid Groups

Carotenes (hydrocarbons)
Carotene monoepoxides
Carotene diepoxides
Monohydroxycarotenoid acyl esters
Dihydroxycarotenoid diacyl esters
Monomethoxycarotenoids
Dimethoxycarotenoids
Monoketocarotenoids
Carotenoid aldehydes
Monohydroxycarotenoids
Dihydroxycarotenoid monoacyl esters
Diketocarotenoids
Monohydroxymonoketocarotenoids
Dihydroxycarotenoids
Dihydroxyketocarotenoids
Dihydroxycarotenoid monoepoxides
Dihydroxycarotenoid diepoxides
Trihydroxycarotenoids
Tetrahydroxycarotenoids
Carotenoid glycosides and glycosyl esters

Adsorption on silica (and alumina) is primarily determined by the presence and nature of polar substituents in the substances to be separated. For most carotenoids, the strongest influence is that of hydroxy groups; the more hydroxy groups present, the greater the adsorption affinity. Other substituents, for example, carbonyl, ester, ether, and epoxy groups, have a smaller effect. The relationship between increasing polarity, increasing adsorption affinity, and decreasing R_f value is summarized in Table 21.1. There are some carotenoids, however, that show anomalous behavior. The commonly encountered astaxanthin, with two hydroxy groups and two keto groups, is less strongly adsorbed than is zeaxanthin, which has only the two hydroxy groups, because of internal hydrogen bonding. Most striking and surprising, however, is the behavior of carotenoids containing the 5,6-dihydroxy-5,6-dihydro-β end group (e.g. karpoxanthin). The tertiary hydroxy groups are shielded so that they contribute little if anything to the adsorption affinity. The chromatographic behavior of a 5,6-diol is similar to that of the corresponding hydrocarbon.

21.3.2 RECOMMENDED MOBILE PHASES

The most widely used mobile phases for separation of carotenoids by TLC on silica are mixtures of hexane or petroleum ether (petrol) with a solvent of higher polarity,

TABLE 21.2
Mobile Phases Suitable for TLC of Carotenoids on Silica Gel G

Mobile Phase	Carotenoid Group
P^a or 1% E^b/P	Carotenes
5% E/P or 1% A/P	Carotene epoxides
10% E/P or 2%–3% A/P	Carotenol esters, carotenediol diesters
20% E/P or 4%–5% A/P	Monoketocarotenoids
50% E/P or 10% A/P	Monohydroxycarotenoids
50% E/P or 10% A/P	Carotenediol monoesters
50% E/P or 10% A/P	Diketocarotenoids
E or 25% A/P	Dihydroxycarotenoids and their epoxides
E or 30% A/P	Trihydroxy-, tetrahydroxycarotenoids
5% EtOH/E or A	Glycosides, glycosyl esters

P, light petroleum or petroleum ether; E, diethyl ether; A, acetone; EtOH, ethanol.

[a] Hexane may be used instead of petroleum ether.
[b] A higher boiling point ether, for example, t-butyl methyl ether, may be used instead of diethyl ether if the ambient laboratory temperature is high.

usually diethyl ether (ether) or acetone. Table 21.2 lists the solvent compositions that give good results for groups of carotenoids with different functional groups.

Lycopene and some other long-chromophore acyclic compounds have a strong tendency to crystallize when applied to TLC plates, resulting in tailing and poor resolution. The inclusion of 10% benzene, toluene, or dichloromethane in the mobile phase is recommended to keep such compounds in solution without greatly affecting the chromatography.

Mixtures of petroleum ether with tertiary alcohols, particularly t-butanol and t-pentanol, have also been used successfully for the more polar xanthophylls [24]. R_f values for a range of carotenoids on silica gel G plates with acetone/petroleum ether and t-butanol/petroleum ether mixtures are compared in Table 21.3.

For carotenoid glycosides and glycosyl esters, more polar solvent mixtures such as 5%–10% ethanol in ether, 50%–80% acetone in hexane, acetone alone, or ethyl acetate:propan-2-ol:water (60:40:2) have been used successfully.

If a second TLC on silica is needed, different solvent combinations should be used. Mixtures of toluene and methanol, hexane and propan-2-ol, or carbon tetrachloride and ethyl acetate have been found to be suitable. The polarity of the mobile phase should be optimized by adjusting the relative proportions of the two components so that the compound of interest runs with an R_f of about 0.5.

21.3.3 ELUTION

After chromatography, the separated carotenoids are eluted efficiently with ether or ether–acetone mixtures, with the addition of a little ethanol if required. Filtration

TABLE 21.3

Comparison of R_f Values[a] for Some Carotenoids in TLC on Silica Gel G with Two Different Mobile Phases: 25% Acetone in Hexane (25% A/H) and 20% t-Butanol in Hexane (20% Bu/H)

Carotenoids	R_f Values	
	25% A/H	20% Bu/H
β-Carotene	1.0	1.0
Lycopene	1.0	1.0
β-Cryptoxanthin	0.6	0.8
Canthaxanthin	0.55	0.7
Astaxanthin	0.35	0.6
Lutein	0.3	0.55
Zeaxanthin	0.3	0.55
Lutein epoxide	0.25	0.45
Antheraxanthin	0.25	0.45
Capsanthin	0.2	0.4
Violaxanthin	0.15	0.3
Neoxanthin	0.1	0.2

[a] Chromatography is affected by conditions, for example, laboratory temperature and humidity, activation of plates, so selected standards should be used to check accurate R_f values.

through a sintered glass funnel under suction is commonly used to remove the adsorbent, especially in preparative work. If this is not available, a simple but effective alternative is to use a small glass funnel with a small plug of absorbent (defatted) cotton wool in the stem. The eluting solvent (ether, acetone, or ethanol) is added dropwise by Pasteur pipette onto the surface of the adsorbent until this surface is white, and elution is then continued with ether until all the colored carotenoid has been recovered. This usually takes only 2–5 mL of solvent.

The solvent is then removed under a gentle stream of N_2.

21.3.4 MODIFICATION OF PLATES

In order to prevent unwanted structural changes or to improve the chromatography, it may be advantageous to modify the stationary phase either during preparation of the plates or as a pretreatment before chromatography.

21.3.4.1 Neutralization of Acidity

As stated earlier (Section 21.2.2.3), carotenoids containing a 5,6-epoxy-β ring are converted into the corresponding 5,8-epoxides (furanoid oxides) in the presence of traces of acid. Silica gel can be sufficiently acidic to catalyze this isomerization. To prevent this, plates should be neutralized or made slightly basic. This can be achieved by including a small amount of KOH, $KHCO_3$ or pH 7–7.5 buffer in the water used for preparing the plates or, for ready-made ones, by inclusion of a small amount (0.1%–0.5%) of an amine such as triethylamine in the mobile phase.

21.3.4.2 Incorporation of Citric Acid

Compounds with acidic character, such as carboxylic acids, α-ketols, for example, astaxanthin, or those with diosphenol end groups, for example, astacene (3,3'-dihydroxy-2,3,2',3'-tetradehydro-β,β-carotene-4,4'-dione), exhibit tailing on untreated silica plates because of the strong adsorption of their partially ionized carboxy or enolic hydroxy groups.

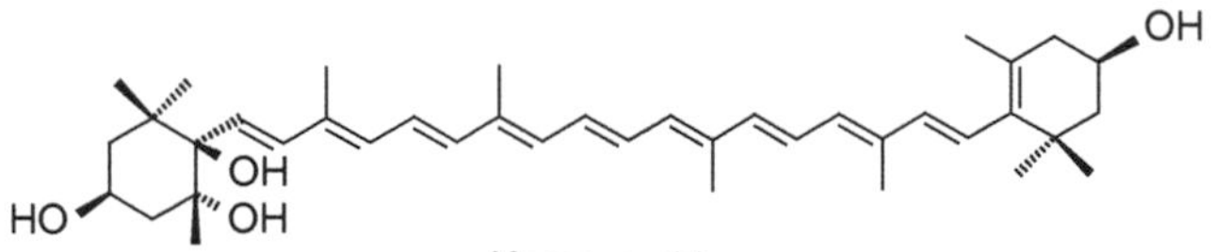

Karpoxanthin

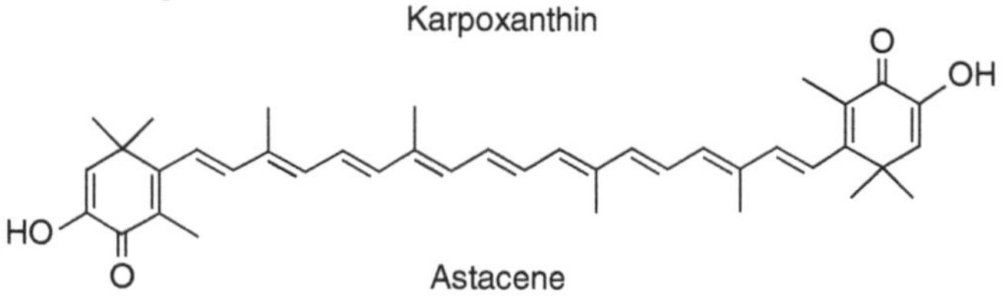

Astacene

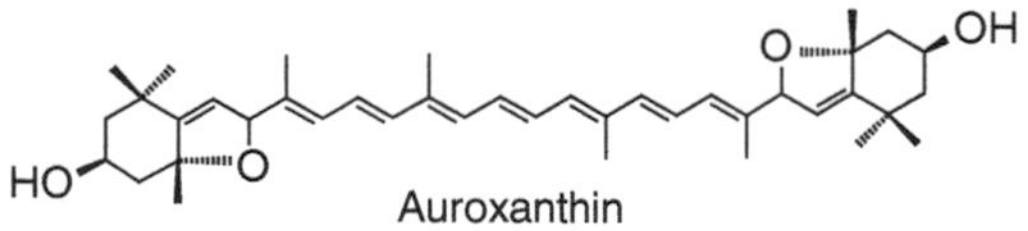

Auroxanthin

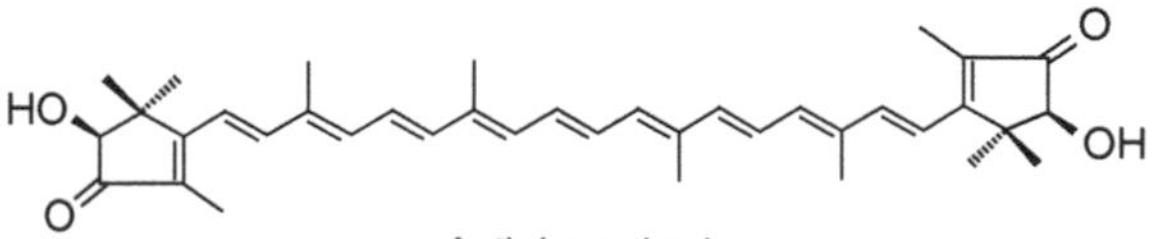

Actinioerythrol

This tailing can be avoided by impregnating the layer with citric acid [20]. The plates are submerged in a solution of 2.5% citric acid in methanol, and dried at room temperature and then for 1 h in a vacuum oven at 80°C under slight vacuum, so that the original activity of the layer is restored. This treatment has no effect on the chromatographic behavior of carotenes or most xanthophylls, but yields an improved resolution of acidic compounds. When these coated plates are used for preparative separations, the citric acid is eluted with the carotenoid and has to be washed out with water. As an alternative, a mobile phase containing a low concentration (0.02 M) of citric acid or acetic acid may be employed, with normal untreated plates. Coating the adsorbent layer with citric acid prior to chromatography has the advantage that the plates can then be used with a solvent of low polarity, such as hexane.

21.3.5 INABILITY TO RESOLVE SOME MIXTURES

Some important pairs of compounds, for example, α-carotene and β-carotene, or lutein and zeaxanthin, are not resolved by TLC on silica and are recovered as a mixture. These mixtures can, however, be separated by subsequent TLC on MgO (Section 21.4.4).

Other examples, in which resolution is not complete, can be separated by repeated development (two or three times) of the plate in the same solvent, which should be selected so that the compounds of interest run with an R_f of around 0.2 in the first development. Development of a 20 cm plate two or three times gives an effective plate length of 40 or 60 cm, respectively.

With some mixtures, for example, of a conventional xanthophyll such as zeaxanthin or astaxanthin and its 7,8-didehydro and 7,8,7′,8′-tetradehydro (acetylenic) analogs, the adsorptive effect of the hydroxy groups is so strong that the very weak effect of the acetylene group(s) is negligible. In such a case, blocking the hydroxy groups by acetylation renders the small effect of the difference in saturation significant and allows separation, especially if the plate is developed two or three times.

21.4 SEPARATION OF CAROTENOIDS BY TLC ON BASIC ADSORBENTS

21.4.1 SUITABLE ADSORBENTS AND MIXTURES

In the carotenoid field, basic inorganic substances such as MgO, $Ca(OH)_2$, $CaCO_3$, and $ZnCO_3$ are used extensively as stationary phases in CC and TLC. These materials have an affinity for double bonds, and this provides the basis for separation, though the mechanism of the interactions is not understood. Separation on these materials is therefore complementary to that on silica. Of these materials, the most commonly used and versatile is MgO. Others, especially $Ca(OH)_2$ and $ZnCO_3$, have more specialized applications but are rarely used since the advent of HPLC.

Thin layers of mixed materials, for example, $MgO:CaCO_3:Ca(OH)_2$ (60:30:4) or $MgO:Ca(OH)_2$:silica gel G: kieselguhr G (9:9:16:14) are prepared in a similar way and have given good separation of a number of carotenoids [25], and of E/Z isomers of peridinin [26], respectively.

21.4.2 PREPARATION OF PLATES

These adsorbents, however, do not adhere well, if at all, to glass plates to form stable thin layers, so a binder has to be included during preparation of the plates: calcium sulphate, kieselguhr G, or silica gel G are usually used for this purpose. TLC plates made from MgO, etc. are not available commercially and have to be prepared in the laboratory with an automatic plate coater/spreader. Simple MgO plates can be prepared with $CaSO_4$ as a binder. For four plates, 20×20 cm, 0.25 mm layer, a mixture of MgO (45 g), $CaSO_4 \cdot 2H_2O$ (4.5 g), and water (90 mL) is mixed for 1 min, preferably with a mechanical homogenizer, then immediately coated onto the glass plates (which should be prewashed with ethanol to ensure good adhesion).

MgO–kieselguhr G plates have also proved very useful. A mixture of MgO (30 g), kieselguhr G (30 g), and water (approximately 90 mL, depending on which MgO product is used) is sufficient for five plates, 20 × 20 cm, 0.25 mm layer. Plates of other adsorbents are prepared in a similar way. All plates are dried at room temperature and then activated in an oven at 120°C for 1.5–2 h. They may be stored for some days in a cabinet in the presence of a drying agent such as $CaCl_2$.

21.4.3 SOME PRACTICAL DIFFICULTIES

TLC on MgO, etc. does have some shortcomings. Adsorbents such as MgO are sold in a form suitable for chromatography by a number of suppliers, but there is considerable variation in properties and efficiencies of separation, sometimes even between different batches from the same supplier. It may be necessary to test products from several suppliers and use the one that gives the best performance for the desired separation.

Carotenoids tend to run as broad, tailing bands, especially when the chromatogram is overloaded (which happens very easily). Because of this and the lack of uniformity between different products, R_f values should not be regarded as reproducible. Also, the carotenoids may be very difficult to recover completely from MgO unless they are eluted as rapidly as possible, preferably before the solvent has fully evaporated. This is a problem especially with long-chromophore acyclic carotenoids, which are adsorbed extremely strongly.

21.4.4 STRUCTURAL BASIS FOR SEPARATION

The decisive factors for separation of carotenoids on MgO are the number and type of double bonds in the carotenoid molecule. Carotenoids with the most extensive cdb system are most strongly adsorbed, for example,

lycopene (11 cdb) > neurosporene (9 cdb) > ζ-carotene (7 cdb)

Acyclic carotenoids are much more strongly held than are ones with the same number of cdbs but with cyclic end groups, for example,

lycopene (ψ,ψ-carotene, acyclic) > γ-carotene (β,ψ-carotene, monocyclic) >
    β-carotene (β,β-carotene, dicyclic)

The position of the double bond in the ring is also important. β-Ring compounds are more strongly adsorbed that the corresponding ε-ring isomers. Hence

β-carotene (β,β-carotene) > α-carotene (β,ε-carotene)
zeaxanthin (β,β-carotene-3,3′-diol) > lutein (β,ε-carotene-3,3′-diol)

The presence of polar groups is not a major factor and this can lead to confusion. Thus the β,β-carotenediol zeaxanthin is less polar than its 5,6:5′,6′-diepoxide, violaxanthin, and is therefore less strongly adsorbed on silica, but the positions are reversed on MgO because the epoxy groups of violaxanthin remove two double bonds from the structure. Violaxanthin does not separate from the corresponding 5,8:5′,8′-diepoxide auroxanthin (5,8:5′,8′-diepoxy-5,8,5′,8′-tetrahydro-β,β-carotene-3,3′-diol) on silica but is much less strongly adsorbed on MgO. For carotenoids with the same double-bond system, however, the presence of hydroxy groups does become the decisive factor. So zeaxanthin (β,β-carotene-3,3′-diol) > β-cryptoxanthin (β,β-caroten-3-ol) > β, β-carotene

The greater importance of the double-bond system is shown convincingly by the fact that the acyclic hydrocarbon lycopene is much more strongly adsorbed on MgO than is the dicyclic diol zeaxanthin. The known correlations and relationships have been derived from experimental observations, however. Insufficient is known about mechanisms of the interactions to allow relative adsorption affinities to be predicted generally.

Calcium hydroxide and $ZnCO_3$ will separate *cis/trans* isomers of carotenoids, something that is difficult to achieve with other adsorbents. They are not now used extensively, however, HPLC being the method of choice.

21.4.5 RECOMMENDED MOBILE PHASES

21.4.5.1 Chromatography on MgO

For TLC on MgO or MgO-kieselguhr G, mixtures of petrol or hexane with ether can be used, but mixtures with acetone generally give more satisfactory results. It is advantageous to include some toluene in the mobile phase for long-chromophore acyclic carotenoids. The adsorptive strength of MgO is somewhat variable, so the solvent compositions need to be optimized for each batch, even of the same product from the same manufacturer. As a guideline, the solvent mixtures given in Table 21.4 should be approximately correct for particular carotenoids and mixtures.

For specific examples see Section 21.7.

TABLE 21.4
Mobile Phases Suitable for TLC of Carotenoids on MgO–kieselguhr G (1:1)

Solvent[a]	Carotenoids
P[b]	Phytoene
2%–6% A/P or 10% T/P	Phytofluene, α-carotene, β-carotene
6%–8% A/P or 10%–15% T/P	ζ-Carotene
10%–15% A/P or 30% T/P	α- and β-Zeacarotene, δ- and γ-carotene
A/T/P (2:2:1)	Lycopene, spirilloxanthin
20% A/P	Lutein, lutein epoxide, violaxanthin
25% A/P	Zeaxanthin
30% A/P	Neoxanthin
A/T/P (1:1:1)	Spheroidene, spheroidenone, hydroxyspheroidene

P, light petroleum or petroleum ether; A, acetone; T, toluene.

[a] The mobile phases listed are usually suitable, but any new batch of MgO should be checked with standard compounds, and the solvent compositions adjusted as necessary by increasing or decreasing the percentage of A or T.

[b] Hexane may be used instead of petroleum ether.

21.4.5.2   Elution

Elution from MgO is best achieved with acetone, plus some toluene and/or ethanol for strongly adsorbed carotenoids. It can be difficult to recover all the carotenoid, however, especially lycopene and other long-chromophore acyclic compounds.

21.4.5.3   Chromatography on Other Stationary Phases

Acetone/hexane or petrol mixtures are also used for TLC separation of *cis/trans* isomers on $Ca(OH)_2$ or $ZnCO_3$. In general, the proportion of acetone in the mobile phase is less than that used with MgO plates, and the composition must be optimized precisely for a particular batch of adsorbent.

The complex mixed stationary phases have not been used much in recent years. The original literature should be consulted for the mobile phase composition appropriate for particular applications.

21.5   SEPARATION OF CAROTENOIDS BY REVERSED-PHASE TLC

The use of RP-HPLC has become widespread and routine but reversed-phase TLC (RP-TLC) on similar phases has not received much attention, in spite of its potential for semipreparative separation and purification.

In the early days of TLC, the only RP systems available were silica or kieselguhr plates that had been dipped into a solution (ca. 5%) of paraffin or other oils in petrol ether. These nonpolar stationary phases require polar mobile phases, usually mixtures of water with methanol, acetone, or acetonitrile, saturated with the RP material. The resulting partition chromatography had limited capacity, tailing, and poor resolution. The use of these systems cannot now be recommended.

Surface-modified silicas with chemically bonded hydrocarbon chains (most commonly C_8 or C_{18}), similar to those used for HPLC, are now available also in the form of ready-made TLC plates [27]. These layers have high capacity and efficiency, but R_f values cannot be predicted merely from polarity because they may exhibit both hydrophilic and hydrophobic character, depending on the degree of carbon loading and endcapping. These RP-TLC plates will separate efficiently carotenol esters with different fatty acid components, but their potential for general separations of carotenoids, either on an analytical or semipreparative scale, has not been fully explored and no systematic surveys have been reported.

An example of the application of RP-TLC for chromatography of carotenoids describes the separation of fucoxanthin and its 19′-butanoyloxy and 19′-hexanoyloxy derivatives on C_8 plates with methanol:water (9:1) as mobile phase [28].

21.6   PRACTICAL APPLICATIONS

21.6.1   PRELIMINARY SCREENING

Preliminary analysis by TLC is a very valuable way to obtain useful information about the complexity of an extract, and an indication of the kind of carotenoids present. Information from such analysis is also used to help devise a strategy for the preparative separation and purification of the individual carotenoids. Even when sophisticated HPLC facilities are available, such preliminary analysis by TLC remains extremely useful. Silica layers coated onto aluminium foil are recommended, because they can easily be cut into pieces of the desired size with scissors, without damaging the layer. Silica gel GF_{254}, which incorporates a fluorescent dye, is recommended to detect phytoene, if this may be present. For simple analytical screening, a piece of the TLC plate approximately 5×2.5 cm is satisfactory. Suitable small developing chambers are available commercially but, for most purposes, a small jar with a screw-on lid, or even a beaker with a suitable cover, will be satisfactory. The most useful developing solvents are a series of petrol (or hexane) mixtures with increasing concentrations of ether or acetone. A great advantage in terms of economy is that only about 1–2 mL of each solvent is required, and separations can be achieved in as little as 2–3 min.

A recommended procedure is to apply a small volume of the extract as a small spot or line and then to run the plate first in petrol or hexane. Anything seen to move in this solvent is a hydrocarbon (carotene). Dicyclic (β-carotene, upper), monocyclic (γ-carotene, middle), and acyclic (lycopene, lower) carotenes separate well and their characteristic colors can easily be distinguished. The TLC result is rapidly documented, either by photography or drawing, and the plate is run again in either 2% acetone or 5% ether in petrol or hexane. This will reveal the presence of compounds such as carotene epoxides, ethers, or peresters. Then, with 3%–4% acetone or 20%–25% ether in petrol or hexane, monoketones, and aldehydes are revealed. Compounds moving in the next solvent in the series, 10% acetone or 50% ether in petrol or hexane, are likely to be diketo- or monohydroxycarotenoids, including monoesters of diols. Ether alone or 25% acetone in petrol or hexane then reveals the presence of dihydroxy, trihydroxy, and tetrahydroxy carotenoids and their epoxy and keto derivatives. Any carotenoid remaining at the origin is likely to be glycoside or glycosyl ester. These will move in acetone alone or 2%–5% ethanol in ether.

If, after running, a plate is left exposed to air for a few minutes, useful diagnostic color changes may be seen. A green color usually indicates a monoepoxide, whereas a diepoxide such as violaxanthin turns blue.

21.6.2 Preparative/Semipreparative TLC and Special Cases

As it is with other classes of compounds, TLC is particularly valuable for isolating microgram to milligram quantities of carotenoids for characterization, further study, or use as standards. Many TLC systems have been described for separating and purifying carotenoids of different levels of polarity, and mixtures of different degrees of complexity. Some of the main features of the TLC behavior of different structural groups or classes of carotenoids are given below and some important examples, namely the separation of all the carotenes in a tomato extract and xanthophylls in a leaf extract, are described in detail in Section 21.7.

21.6.2.1 Carotenes

The separation of some carotenes can be achieved on silica with a solvent of low polarity such as hexane or petrol. Lycopene, γ-carotene (β,ψ-carotene), and β,β-carotene are clearly resolved. With more complex extracts containing a larger number of hydrocarbons, overlap of bands becomes a problem, for example, phytoene virtually cochromatographs with β-carotene. Regioisomers such as β-carotene and α-carotene (β,ε-carotene) are not resolved, nor are geometrical isomers. Mixtures of carotenes that are not separated or only partially resolved on silica can usually be separated on MgO layers, however (Section 21.4.4). In the acyclic series, carotenoids with the most extensive cdb system are most strongly adsorbed and acyclic compounds are more strongly held than cyclic ones that have the same number of double bonds, β-ring compounds being more strongly adsorbed than the corresponding ε-ring isomers. By a combination of TLC on silica and on MgO, all the carotenes in a natural extract can usually be separated. See Section 21.7.1 for separation of the carotenes present in fruit of tomato (*Lycopersicon esculentum*).

21.6.2.2 Hydroxycarotenoids

On silica and alumina, the hydroxy group exerts the greatest influence on adsorption affinity, which increases with increasing number of hydroxy groups. 3-Hydroxy-β-ring carotenoids are more strongly adsorbed than the corresponding 2-hydroxy and 4-hydroxy isomers. Tertiary hydroxy groups have less influence than secondary ones and carotenoids with a 5,6-dihydroxy-β end group have similar R_f values to the parent compounds that lack the hydroxy groups. Obviously, adsorption affinity is much reduced by methylation, esterification, or silylation of the hydroxy groups, so TLC can be used to monitor the course of such reactions.

The β,β-, β,ε- and ε,ε-carotenediols zeaxanthin, lutein, and lactucaxanthin are virtually inseparable on silica but they are easily resolved on layers of MgO.

21.6.2.3 Epoxides

The epoxide oxygen function exerts only a moderate effect on adsorption affinities on silica. Monoepoxides and diepoxides can easily be differentiated but corresponding 5,6-epoxides and 5,8-epoxides are not resolved. These isomers are separated readily on MgO layers, the 5,8-epoxides being much more strongly adsorbed. Because of the rapid isomerization of 5,6-epoxides to 5,8-epoxides catalyzed by traces of acid, acidic adsorbents and solvents should not be used for chromatography of these compounds.

21.6.2.4 Carotenoids Containing the 3-Hydroxy-4-oxo-β End Group

Alumina cannot be used for CC or TLC of carotenoids such as astaxanthin, which contain the 3-hydroxy-4-keto-β end group or actinioerythrol (3,3′-dihydroxy-2,2′-dinor-β,β-carotene-4,4′-dione), which contains the 3-hydroxy-4-keto-2-nor-β end group. On this adsorbent, these carotenoids and their esters are rapidly oxidized to the corresponding 2,3-didehydro derivatives (e.g., astacene and violerythrin (2,2′-dinor-β,β-carotene-3,4,3′,4′-tetrone)) which have acidic character and are almost impossible to elute. For chromatography of these compounds, silica plates impregnated with citric acid should be used (Section 21.3.4.1). Canthaxanthin and astaxanthin esters run close together on silica and may sometimes be confused in biological samples. They are easily distinguished on alumina, however, because of the conversion of the astaxanthin esters into astacene.

21.6.2.5 Acetylenic Carotenoids

Conventional xanthophylls generally cochromatograph with their mono- or diacetylenic analogs. For example, on ready-coated silica gel plates, astaxanthin, and its acetylenic analogs 7,8-didehydroastaxanthin (3,3'-dihydroxy-7,8-didehydro-β,β-carotene-4,4'-dione) and 7,8,7',8'-tetradehydroastaxanthin (3,3'-dihydroxy-7,8,7',8'-tetradehydro-β,β-carotene-4,4'-dione) are not resolved. They can, however, be separated by TLC of their acetates, by repeated development. It is then difficult to recover the parent compounds because the usual ways to remove the acetate groups, namely saponification or reduction with LiAlH$_4$, result in unwanted structural modifications. Xanthophylls which do not contain the 3-hydroxy-4-keto-β end group, for example, zeaxanthin and its acetylenic analog alloxanthin, are stable to saponification, so this problem of unwanted structural changes does not arise.

21.6.2.6 Apocarotenoids

Silica may be used to separate apocarotenals, apocarotenols, and apocarotenoic acids, but compounds with different chain length within each group are not resolved.

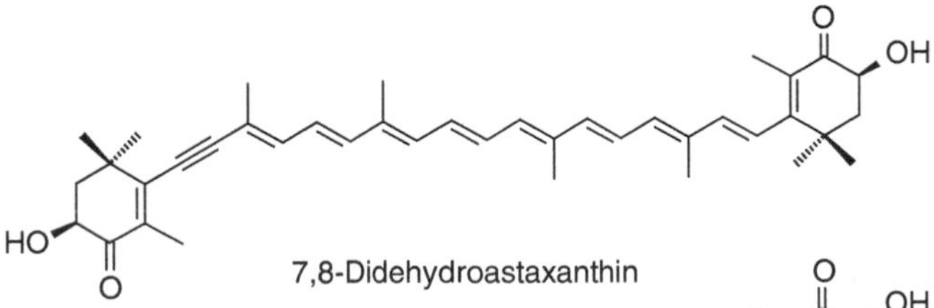

7,8-Didehydroastaxanthin

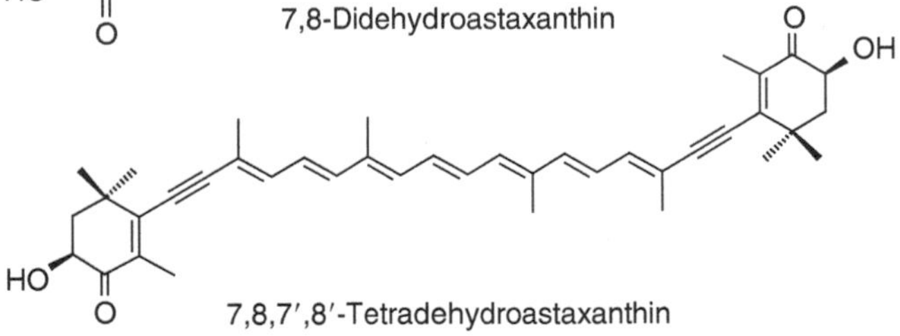

7,8,7',8'-Tetradehydroastaxanthin

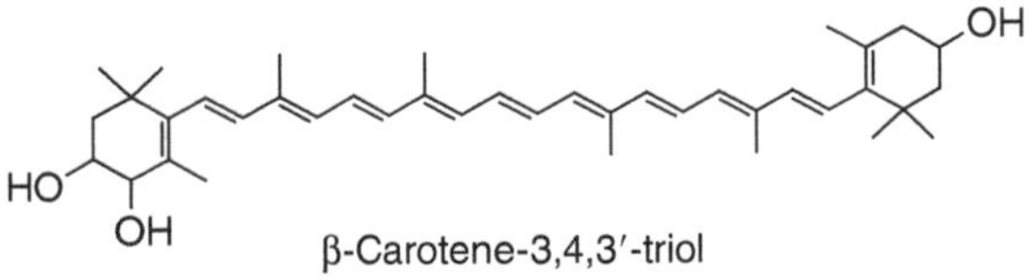

β-Carotene-3,4,3'-triol

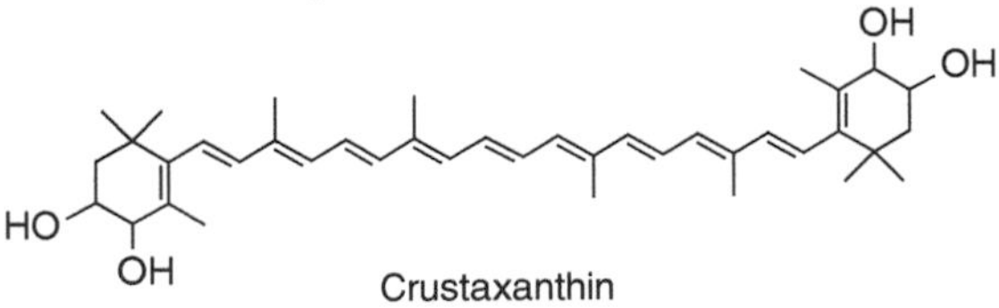

Crustaxanthin

Apocarotenals and apocarotenoic acids of different chain length can be resolved, however, by TLC on MgO or RP-TLC on C_{18} plates [20]. On MgO, the adsorption affinity increases more or less with increasing chain length, except that apocarotenals and apocarotenoic acid methyl esters that have a methyl substituent in the α-position relative to the functional group (i.e., C_{25}, C_{30}, C_{35}) run ahead of the C_{27}, C_{32}, and C_{37} compounds, which have a hydrogen in the α-position and a methyl group in the γ-position. On C_{18} RP plates, however, the order is according to molecular mass and polarity only.

21.6.2.7 Glycosides and Glycosyl Esters

The carotenoid glycosides and glycosyl esters, with their considerably more hydrophilic behavior, require polar chromatographic conditions. Silica layers in combination with a mobile phase containing alcohols (e.g., toluene/ether/methanol), or C_{18} RP plates and water/methanol mixtures, are recommended [20,29]. Surprisingly, esterification of a hydroxy group in the carbohydrate moiety with fatty acid does not greatly alter the chromatographic behavior as long as the remaining free hydroxy groups are not acylated. Peracetylation increases the stability and facilitates separation and further manipulations. The peracetates can be separated on silica gel G plates with ethyl acetate/hexane mixtures as mobile phase.

21.6.2.8 *E/Z* Isomers

On silica TLC of any carotenoid, there are usually two small, poorly resolved zones, sometimes more, which just precede and follow the main band (all-*E*). These minor fractions are probably mixtures of *Z* isomers. It is thought that the preceding band usually contains di-*Z* and poly-*Z* isomers, whilst the mono-*Z* isomers follow the all-*E* isomer. Though rarely used now because HPLC has become the method of choice, $Ca(OH)_2$ and $ZnCO_3$ can give good separation of *E/Z* isomers of carotenes and xanthophylls, respectively, but band or spot shapes are not good and there is considerable tailing. Assignment of bands to any particular isomer is empirical and not precise. When geometrical isomers are to be isolated and identified, it is recommended that CC or TLC is used first to obtain fractions that contain the mixed isomers of one carotenoid only, and these fractions are then subjected to HPLC to separate the *E/Z* isomers.

21.6.2.9 Optical Isomers

In principle, enantiomers cannot be separated on achiral adsorbents. TLC plates with chiral stationary phases are available but these seem not to be suitable for such lipophilic compounds. Some examples of TLC separations of diastereoisomers have been reported, however, for example, of carotenoids with 3,4-dihydroxy-β end groups, particularly β,β-carotene-3,4,3'-triol idoxanthin and β,β-carotene-3,4,3',4'-tetrol (crustaxanthin), which can occur in either the 3,4-*trans* or 3,4-*cis* configuration. The 3,4-*cis* and 3,4-*trans* glycol forms can readily be separated on silica gel 60 plates with 40% acetone or 75% ethyl acetate in hexane.

Some resolution of the three optical isomers of astaxanthin ($(3R,3'R)$, $(3R,3'S)$, $(3S,3'S)$) has been achieved after derivatization to the $(-)$-camphanate diesters [30]. The resulting diastereoisomers can be resolved on silica gel with toluene:ether: propan-2-ol (89:9:4). Multiple development is necessary and separations achieved are not great. HPLC is the preferred method for this separation.

21.6.3 COCHROMATOGRAPHY AND CAROTENOID STANDARDS

Because carotenoids are so easily seen on TLC, comparison by cochromatography is a very useful means of establishing whether an unknown carotenoid may or may not be identical or similar in polarity to a known standard. High purity standards of the most common carotenoids are available commercially, but can be expensive. It is not difficult, however, to use TLC to isolate standards from natural sources with known carotenoid compositions, for example, lycopene from tomato and lutein from green leaves or marigold flowers. Recommendations about suitable sources for particular carotenoids are given in the *Carotenoids Handbook*. Other publications [14] give extensive tables listing what carotenoids have been shown to be present in many flowers, fruits, and other natural sources, though the amounts present are usually not indicated.

If pure standards are not available, it can be advantageous to use chromatographic comparison with a known standard extract as a marker for particular carotenoids or as an indicator of the polarity of an unknown. The extract of tomato, for example, can be used to identify the presence of acyclic (lycopene), monocyclic, and dicyclic carotenes in an unknown sample. Similarly a saponified extract of persimmon fruit contains β-carotene, β-cryptoxanthin, and zeaxanthin as main carotenoids and can be used to indicate the presence of monohydroxy and dihydroxycarotenes in an extract. Saponified extracts of virtually all green leaves contain the same carotenoids—β-carotene, lutein, violaxanthin, and neoxanthin—so are useful indicators of polarity of carotenoids present in an unfamiliar extract.

21.6.4 PURIFICATION OF SAMPLES FOR MS AND NMR

Chromatographic procedures, including TLC, will usually give a carotenoid sample that is free from other contaminating carotenoids and is suitable for UV/Vis absorption spectroscopic and spectrophotometric analysis. During the purification, however, substantial amounts of impurities that originate from the chromatographic adsorbents can be introduced into the sample. They do not interfere with the UV/Vis analysis but can be a serious problem when samples are analyzed by MS or NMR. When samples are to be used for MS or NMR analysis, the solvents used for their preparation must be purified rigorously, and TLC plates must be prewashed by running in a solvent that is at least as polar as that which is to be used for the chromatography and elution. If possible, samples should be eluted from TLC adsorbents with ether; acetone and ethanol are more likely to extract contaminants so, if they are required, only the minimum amounts should be used. Carotenoid samples from TLC should be given an additional purification immediately before being subjected to MS or NMR analysis, for example, by HPLC or by filtering through a small column of neutral alumina (3–4 cm in a Pasteur pipette is usually sufficient).

Two solvents are used, the first to remove less polar contaminants, the second to elute the carotenoid as a concentrated fraction. A small prewashed column of silica may be used in the same way for carotenoids that are not stable on alumina.

21.7 EXAMPLES

To illustrate the use of TLC for the separation of carotenoids from mixtures, two examples are presented here in detail, namely the separation of a series of carotenes from tomato fruit, and the separation of carotenoids, mainly xanthophylls, from a saponified leaf extract. These procedures may be followed for the isolation of carotenoid standards, and may be used or adapted for use for separating carotenoids from other plant sources.

21.7.1 SEPARATION OF TOMATO CAROTENES

Although lycopene is by far the main pigment and is responsible for the red color of tomatoes, a range of other carotenes is also present, in lower concentrations. A tomato extract, or the carotene fraction obtained from this extract by CC on alumina, can be fractionated by successive TLC on silica gel and MgO–kieselguhr. TLC on silica gives three distinct fractions, and these are then resolved into individual carotenes by subsequent TLC on MgO–kieselguhr G. Details of the systems and mobile phases used are given as a flow diagram in Figure 21.2. The procedures used for extraction, saponification, and fractionation by CC are described in detail as a worked example in Ref. [31] together with identification criteria for the individual carotenoids.

21.7.2 SEPARATION OF CAROTENOIDS FROM AN EXTRACT OF GREEN LEAVES

Green leaves of almost all species of plants contain the same set of main carotenoids (Section 21.1.3.1). These may be separated (after saponification to destroy chlorophylls) by CC on alumina to give carotene and xanthophyll fractions.

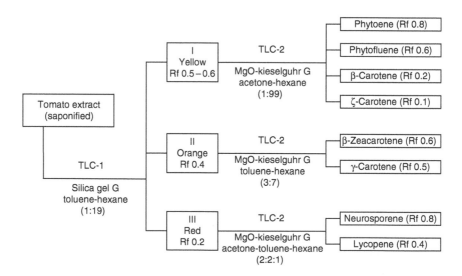

FIGURE 21.2 Flow diagram summarizing the separation by TLC of the carotenes of a saponified extract of ripe tomato fruit (*Lycopersicon esculentum*).

Alternatively, the saponified total extract may be chromatographed directly on silica gel G in a polar solvent that separates the xanthophylls into upto four bands. In this system, the carotenes run with the solvent front are recovered and rechromatographed on silica again with hexane or petroleum ether as solvent. The carotene fraction from this TLC, or from CC, is resolved into α-carotene and β-carotene by subsequent TLC on MgO–kieselguhr. The four bands of xanthophylls are also chromatographed again, on MgO–kieselguhr, to separate the individual

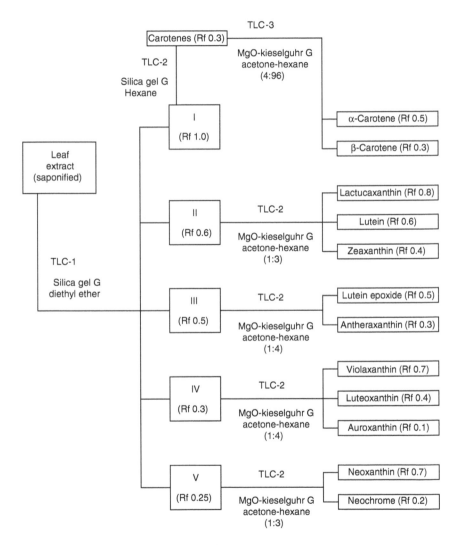

FIGURE 21.3 Flow diagram summarizing the separation by TLC of the carotenoids of a saponified extract of green leaves. Lactucaxanthin is present in very few species, notably lettuce (*Lactuca sativa*). The 5,8-epoxides luteoxanthin, auroxanthin and neochrome may be present as artefacts formed from the natural 5,6-epoxides under acidic conditions.

xanthophylls. Any 5,8-epoxides that may have been formed as artefacts due to acid conditions during extraction and chromatography are separated from the natural 5,6-epoxides during this step.

Details of the systems and mobile phases used for this TLC are given as a flow diagram in Figure 21.3. The procedures used for extraction, saponification, and fractionation by CC are described in detail as a worked example in Ref. [31], together with identification criteria for the individual carotenoids.

REFERENCES

1. Britton, G., Liaaen-Jensen, S., and Pfander, H. (Eds.), *Carotenoids Handbook*, Birkhäuser, Basel, 2004.
2. Britton, G., Overview of carotenoid biosynthesis, in *Carotenoids, Vol. 3: Biosynthesis and Metabolism*, Britton, G., Liaaen-Jensen, S., and Pfander, H., Eds., Birkhäuser, Basel, 1998, chap. 2.
3. IUPAC Commission on the Nomenclature of Organic Chemistry and IUPAC–IUB Commission on Biochemical Nomenclature, Rules for the nomenclature of carotenoids, *Pure Appl. Chem.*, 41, 407, 1975.
4. Weedon, B.C.L. and Moss, G.P., Structure and nomenclature, in *Carotenoids, Vol. 1A: Isolation and Analysis*, Britton, G., Liaaen-Jensen, S., and Pfander, H., Eds., Birkhäuser, Basel, 1995, chap. 3.
5. Britton, G., Structure and properties of carotenoids in relation to function, *FASEB J.*, 9, 1551, 1995.
6. Britton, G., UV/Visible spectroscopy, in *Carotenoids, Vol. 1B: Spectroscopy*, Britton, G., Liaaen-Jensen, S., and Pfander, H., Eds., Birkhäuser, Basel, 1995, chap. 2.
7. Young, A.J., Occurrence and distribution of carotenoids in photosynthetic systems, in *Carotenoids in Photosynthesis*, Young, A. and Britton, G., Eds., Chapman & Hall, London, 1993, chap. 2.
8. Siefermann-Harms, D., Hertzberg, S., Borch, G., and Liaaen-Jensen, S., Lactucaxanthin, an ε,ε-carotene-3,3′-diol from *Lactuca sativa*, *Phytochemistry*, 20, 85, 1981.
9. Barry, P., Evershed, R.P., Young, A.J., Prescott, M.C., and Britton, G., Characterization of carotenoid acyl esters in drought-stressed barley seedlings, *Phytochemistry*, 31, 3163, 1992.
10. Rodriguez-Amaya, D.B., *A Guide to Carotenoid Analysis in Foods*, OMNI, Washington D.C., 1999.
11. Young, A.J., Carotenoids in pigment–protein complexes, in *Carotenoids in Photosynthesis*, Young, A. and Britton, G., Eds., Chapman & Hall, London, 1993, chap. 3.
12. Czeczuga, B., Autumn carotenoids in the leaves of some trees, *Biochem. Syst. Ecol.*, 14, 203, 1986.
13. Cardini, F., Giananneschi, M., Selva, A., and Chelli, M., Semi-β-carotenone from the leaves of two cycads, *Phytochemistry*, 26, 2029, 1987.
14. Goodwin, T.W., *The Biochemistry of the Carotenoids, Vol. 1: Plants*, Chapman & Hall, London, 1980.
15. Britton, G. et al., Physical state of carotenoids in chromoplasts of tomato and carrot: consequences for bioavailability, in *Functionalities of Pigments in Food*, Empis, J.A., Ed., Sociedade Portuguesa de Quimica, Lisbon, 2002, p. 151.
16. Liaaen-Jensen, S., Carotenoids in chemosystematics, in *Carotenoids, Vol. 3: Biosynthesis and Metabolism*, Britton, G., Liaaen-Jensen, S., and Pfander, H., Eds., Birkhäuser, Basel, 1998, chap. 5.

17. Borowitzka, M.A. and Borowitzka, L.J., *Dunaliella*, in *Micro-algal Biotechnology*, Borowitzka, M.A. and Borowitzka, L.J., Eds., Cambridge University Press, Cambridge, 1988, p. 27.

18. Tsavalos, A.T. et al., Secondary carotenoid synthesis in microalgae, in *Research in Photosynthesis*, *Vol. 3*: Murata, N., Ed., Kluwer, Dordrecht, 1992, p. 47.

19. Takaichi, S., Carotenoids and carotenogenesis in anoxygenic photosynthetic bacteria, in *The Photochemistry of Carotenoids*, Frank, H.A. et al., Eds., Kluwer, Dordrecht, 1999, chap. 3.

20. Schiedt, K., Chromatography: Part III. Thin-layer chromatography, in *Carotenoids*, *Vol. 1A*: *Isolation and Analysis*, Britton, G., Liaaen-Jensen, S., and Pfander, H., Eds., Birkhäuser, Basel, 1995, p. 131.

21. Britton, G., General carotenoid methods, *Meth. Enzymol.*, 18c, 654, 1985.

22. Schiedt, K. and Liaaen-Jensen, S., Isolation and analysis, in *Carotenoids*, *Vol. 1A*: *Isolation and Analysis*, Britton, G., Liaaen-Jensen, S., and Pfander, H., Eds., Birkhäuser, Basel, 1995, chap. 5.

23. Pfander, H., Chromatography: Part I. General aspects, in *Carotenoids*, *Vol. 1A*: *Isolation and Analysis*, Britton, G., Liaaen-Jensen, S., and Pfander, H., Eds., Birkhäuser, Basel, 1995, p. 109.

24. Francis, G.W. and Isaksen, M., Thin-layer chromatography of carotenoids with tertiary alcohol—petroleum ether solutions as developing solvents, *J. Food Sci.*, 53, 979, 1988.

25. Hager, A. and Stransky, H., Das carotinoidmuster und die verbreiting des lichtinduzierten xanthophyllcyclus in verschiedenen algenklassen. III. Grünalgen, *Arch. Mikrobiol.*, 72, 68, 1970.

26. Haugan, J.A., Aackermann, T., and Liaaen-Jensen, S., Isolation of fucoxanthin and peridinin, *Meth. Enzymol.*, 213, 231, 1992.

27. Wright, S.W. and Jeffrey, S.W., Fucoxanthin pigment markers of marine phytoplankton analysed by HPLC and HPTLC, *Mar. Ecol. Prog. Ser.*, 38, 259, 1987.

28. Bjørnland, T., Guillard, R.R.L., and Liaaen-Jensen, S., *Phaeocystis* sp. clone 677-3—a tropical marine planktonic prymnesiophyte with fucoxanthin and 19′-acyloxyfucoxanthins as chemosystematic carotenoid markers, *Biochem. Syst. Ecol.*, 16, 445, 1988.

29. Meyer, P., Riesen, R., and Pfander, H., Example 10: Carotenoid glycosides and glycosyl esters, in *Carotenoids*, *Vol. 1A*: *Isolation and Analysis*, Britton, G., Liaaen-Jensen, S., and Pfander, H., Eds., Birkhäuser, Basel, 1995, p. 277.

30. Müller, R.K. et al., Beitrag zur analytik und synthese von 3-hydroxy-4-oxocarotinoiden, *Helv. Chim. Acta*, 63, 1654, 1980.

31. Britton, G., Example 1: Higher plants, in *Carotenoids*, *Vol. 1A*: *Isolation and Analysis*, Britton, G., Liaaen-Jensen, S., and Pfander, H., Eds., Birkhäuser, Basel, 1995, p. 201.

22 TLC of Sterols, Steroids, and Related Triterpenoids

Laurie Dinan, Juraj Harmatha, and Rene Lafont

CONTENTS

22.1 INTRODUCTION AND SCOPE OF THE REVIEW

22.1.1 Usefulness and Validity of TLC for Sterols and Steroids

The general advantages of (HP)TLC as a method for the separation and analysis of low molecular weight, low volatility compounds are well known; it is a simple, convenient, economical, high-throughput method, which in its basic format, at least, needs only portable, low-cost equipment. In addition, the whole polarity spectrum is present between the origin and the solvent front. There are a wide range of detection methods for in situ localization, partial identification, and quantification of compounds, which are not subject to interference from the mobile phase, since this is generally allowed to evaporate off before detection. The detection methods are generally sensitive (nanogram to microgram range) and provide good linearity of response, even if only over a limited range. All these advantages apply to the TLC of sterols and steroids and, in former times, TLC was extensively applied to the analysis of these classes. However, it has now been almost completely supplanted by HPLC as the method of choice, because of the greater possible loading, higher preparative recoveries, greater resolution, and the widespread availability of HPLC equipment nowadays. Consequently, there have been few systematic studies of the chromatographic behavior on TLC of the different classes of phytosteroids, in spite of their major significance in medicine and chemical ecology. Rather, where TLC is used at all today, it is used to solve a specific problem, and not for comparison of a wide range of analogues of a particular phytosteroid class. A notable exception here is the (phyto)ecdysteroids, for which a significant body of TLC data exists, comparing not only many analogues, but also different TLC systems and approaches, including various one- and two-dimensional systems, forced-flow systems, and coupled systems (e.g., TLC–MS).

22.1.2 Scope of the Review

This chapter will update our previous, more general chromatographic review [1], while focusing primarily on the developments in TLC and their applications to phytosteroids and related compounds since then. As pointed out earlier, TLC has several advantages, but it is nowadays rarely the method of first choice for phytochemists, and, as a consequence, few of the recent methodological developments have found much application in the analysis of plant samples. One of the purposes of this chapter is to indicate which of the newer methods could be usefully applied to plant analysis. Representative structures of members of each of the chemical classes considered are shown in Figure 22.1.

22.2 SURVEY OF THE LITERATURE CLASS BY CLASS

22.2.1 Phytosterols

22.2.1.1 Structural Diversity

Phytosterols belong to a large family of structurally related sterol metabolites of plant origin. This structural similarity indicates their common biosynthetic origin, thus evoking also a close biogenetic relation. In a broader sense, they are isoprenoid

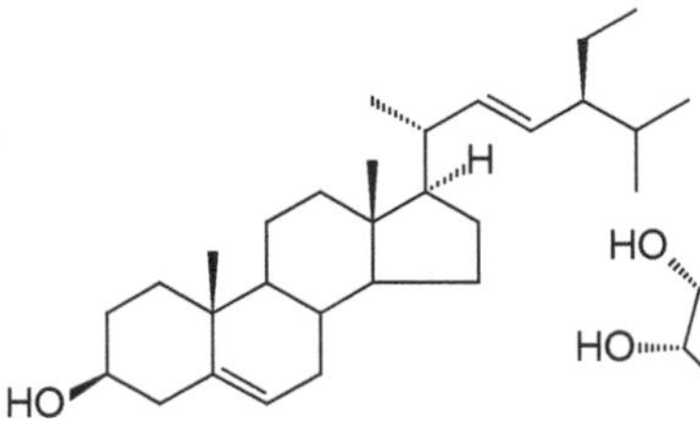

Stigmasterol (**I**)

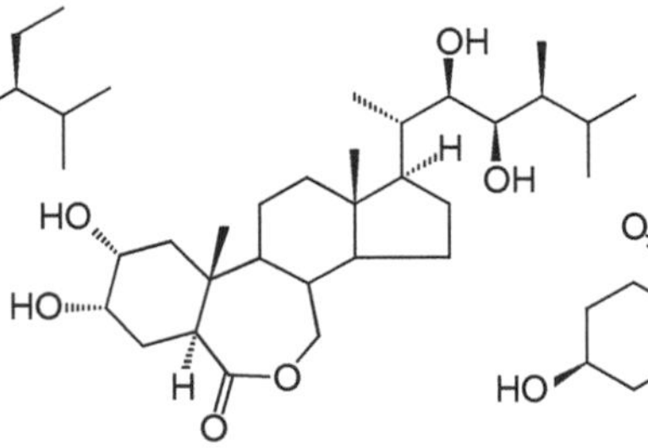

Brassinolide (**II**)

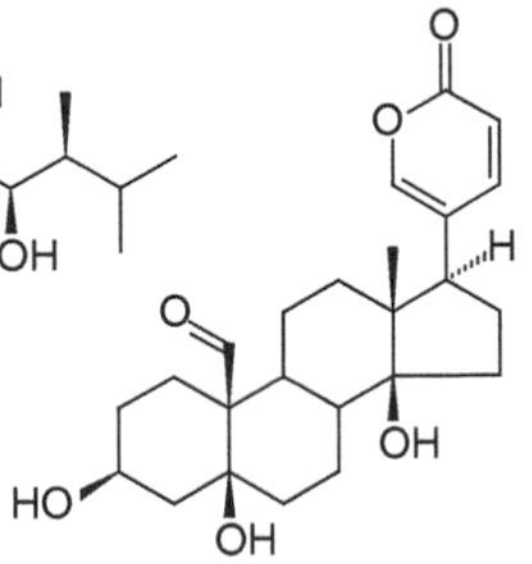

Hellebrigenin (**III**)

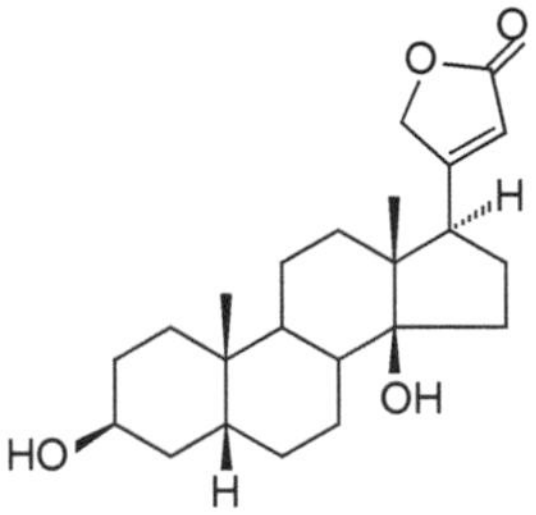

Digitoxigenin (**IV**)

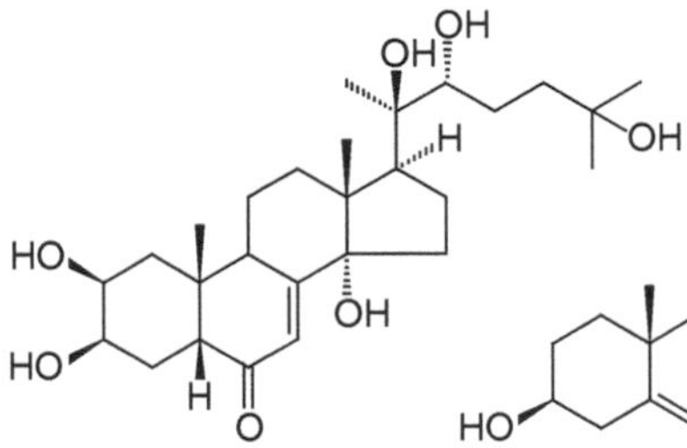

20-Hydroxyecdysone (**V**)

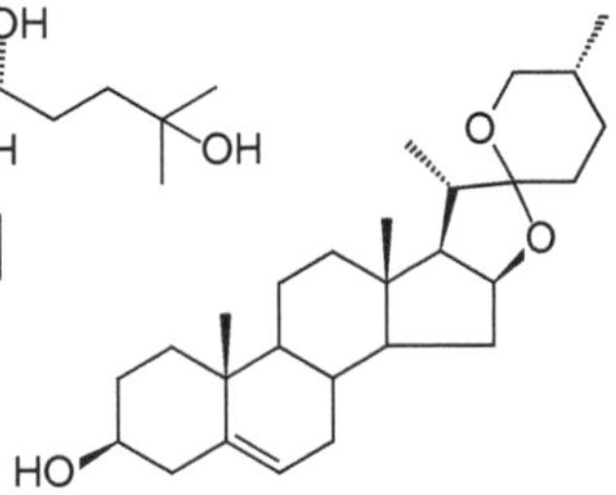

Diosgenin (**VI**)

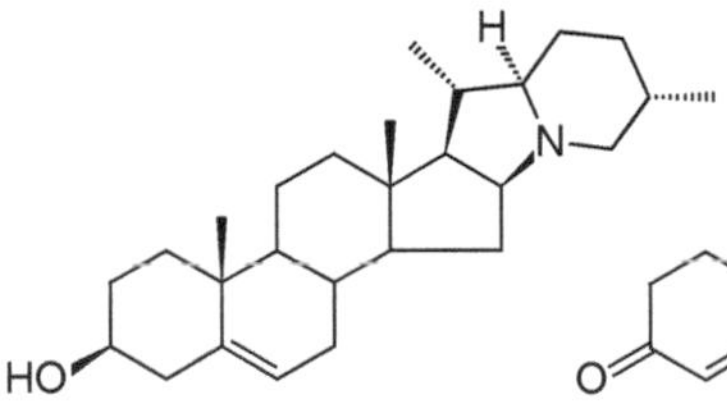

Solanidine (**VII**)

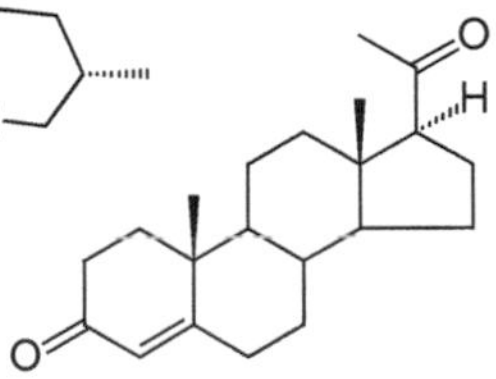

Progesterone (**VIII**)

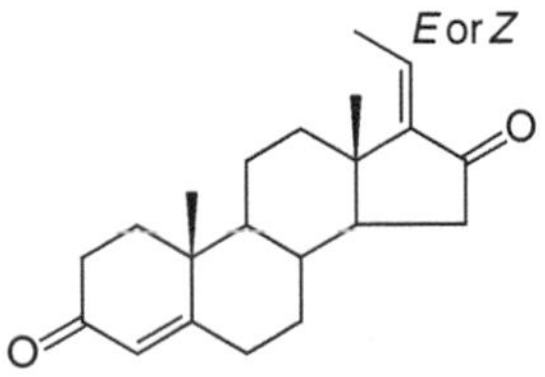

Guggulsterone (**IX**)

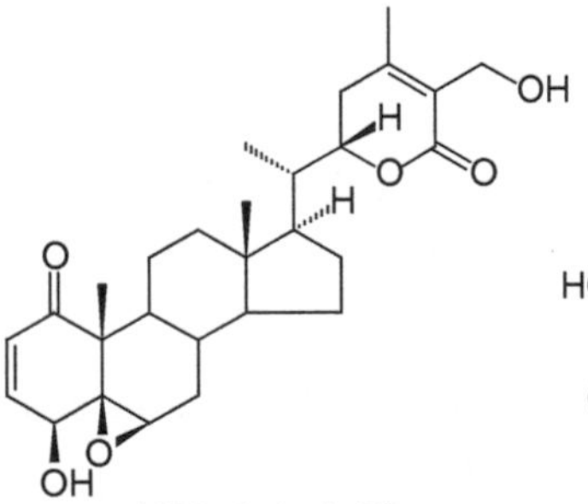

Withaferin A (**X**)

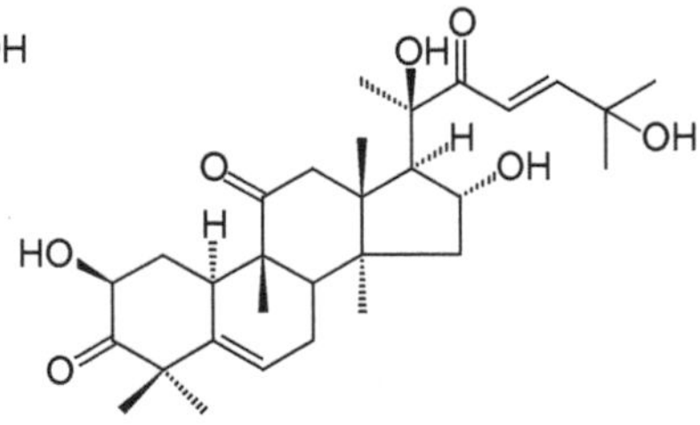

Cucurbitacin D (**XI**)

FIGURE 22.1 Structures of representative members of each of the phytochemical classes considered in this chapter. The configurations are 8β-H, 9α-H, and 14α-H, unless otherwise depicted.

constituents with the nearest biosynthetic relation to triterpenoids [2]. They are, in particular, also biosynthetic precursors to the majority of more oxidized steroid classes in plants discussed in this chapter, as well as the precursors of essential steroids of other coexisting organisms, for example, insect hormones [3]. The sterols comprise several major groups characterized by the presence of a typical steroid (androstane) skeleton and by having a hydroxyl group at C-3, normally in the β-configuration. The phytosterol family possesses, moreover, branching side chains of 8–10 carbon atoms at C-17. Variety in the side chain forms the basis of the structural diversity and thus also the classification of phytosterols. The simplest is the cholestane (C_{27})-type, next is the C-24 methyl homologue ergostane (C_{28})-type and that of its C-24 epimer campestane, the next is the C-24 ethyl homologue stigmastane (C_{29})-type, and again that of its epimer poriferastan. All types can contain 1–3 double bonds in several positions of the sterane skeleton or in the side chain. This increases the number of basic types. Each type can occur also in C-5 epimeric form, which doubles the number of possible analogues. A number of parent hydrocarbon ring and side-chain combinations have been designated in the IUPAC-IUB Nomenclature of Steroids [4] for classification and nomenclature purposes, but, for practical use, the previously introduced, and well-established, trivial names are preferred [5].

Plants often contain phytosterols in various conjugated forms, which enhance the structural diversity even more. The most frequent are steryl fatty acid esters, phenylpropanoate steryl esters, steryl glycosides, and acylated steryl glycosides. Their structural diversity and phylogenetic distribution in plants and plant organs are reviewed by Moreau et al. [6]. Plants regularly contain a free phytosterol fraction with β-sitosterol and stigmasterol (Figure 22.1: **I**) as major components, accompanied by minor sterols, such as campesterol, fucosterol, and other homologues or isomers. These constituents are not easily separable. They readily crystallize together and they do not separate from each other by common chromatography. In former times even chemical transformations have been applied for separation, for example, trifluoroacetylation enhanced plant sterol separation [7].

22.2.1.2 Separation Systems and Detection

For isolation of single sterols or conjugates, generally, solvent extractions with hexane, methylene chloride, or chloroform are used. In the case of conjugates, saponification of samples precedes the chromatography to obtain the total sterol content [8]. Alternatively, the plant samples can be extracted by supercritical fluid extraction. Purification and also partial separation can be achieved by column chromatography or by preparative TLC. Both methods are suitable for sample preparation, purification, qualitative detection, or assays for preliminary estimates of sterol content. However, for separation of single compounds or for quantitative analysis, GC–MS and HPLC–MS techniques are required.

TLC on silica gel is still used for monitoring or preliminary preparation/purification/separation of small amounts of lipid extracts containing sterols. TLC can be used to separate sterol subclasses and their esters [8]. Visualization of TLC spots is largely carried out with UV light (if plates are impregnated with a fluorescence indicator), or by spraying the plates with 50% sulfuric acid in ethanol (or with 80% potassium dichromate in sulfuric acid), followed by destructive carbonization (flame or heat treatment). Separations can be improved by two-dimensional TLC with different solvents or by multiple TLC development. Hexane-diethylether (7:3 v/v) or benzene-diethylether (9:1 v/v) solvent systems are often used for the separation of sterol subclasses and their conjugates in various sample matrices. However, for further separation, GC, HPLC, or, in particular cases, argentation TLC techniques [9] have to be applied. Reports concerning applications of these techniques in lipid analyses are well reviewed by Abidi [8] and more generally by Goad and Akihisa [5]. Commercial HPTLC plates are sometimes used for the lipid analyses [10]. Alternatively, GC and HPLC techniques are combined also with automated multiple development (AMD), utilizing computerized gradient optimization [11]. HPTLC-AMD in conjunction with 25-step mobile phase gradient of hexane, diethyl ether, and methanol was reported for sterol analysis [12]. After visualization with cupric sulfate-phosphoric acid in aqueous methanol, the plates are inspected with a TLC scanning densitometer, and the densitometric chromatogram peaks are quantified. These techniques, however, necessarily require comparison with individual authentic standard sterols, whereas for the most suitable GC–MS analytic method computer data bases (spectral libraries) are now currently available.

22.2.2 STEROIDS

22.2.2.1 Brassinosteroids

22.2.2.1.1 Structural Diversity
Brassinosteroids are derivatives of 5α-cholestane with two vicinal diols (2α,3α and 22R,23R), a 6-keto or a 6-oxalactone group in ring-B and various substituents in position-24 (R/S)-methyl, 24-methylene, etc. Their occurrence seems to be general in the plant kingdom. Over 50 natural brassinosteroid analogues have been identified [13], including brassinolide (Figure 22.1: **II**).

22.2.2.1.2 Separation Systems and Detection
Although preparative TLC was used as a purification step in some early isolations of brassinosteroids from large (kg) amounts of pollen (e.g., [14]), the low levels found in most plant sources have led most workers to use more sensitive methods (e.g., GC–MS or HPLC) for analysis and quantification of brassinosteroids, even if these require prior derivatization to boronates [15] or fluorescent tagging [16]. Consequently, there are no significant studies of the TLC chromatographic behavior of brassinosteroids.

22.2.2.2 Bufadienolides

22.2.2.2.1 Structural Diversity
Bufadienolides are usually C_{24} steroids and their glycosides. They possess a chromophoric six-membered lactone (α-pyrone) ring at C-17β, which imparts a characteristic UV absorbance (λ_{max} at 296–299 nm in methanol [$\varepsilon = 5000$ L/mol/cm] and a second absorption at 200–220 nm). Many possess a 5β-hydroxyl (A/B-*cis* ring junction), a *trans*-B/C ring junction, a 14β-hydroxyl (C/D-*cis* ring junction), and an

aldehyde group at C-19 (e.g., hellebrigenin; Figure 22.1: **III**). Bufadienolides occur in both plants and animals, and over 160 analogues have been identified from members of six plant families (Crassulaceae, Hyacinthaceae, Iridaceae, Melianthiaceae, Ranunculaceae, and Santalaceae) [17]. Helleborin was previously used to treat heart rhythm irregularities, but bufadienolides possess several other biological activities, ranging from antitumor activity to toxicity to mammals and insects.

22.2.2.2.2 Separation Systems and Detection
Dias et al. [18] proposed silica TLC eluted with ethyl acetate/MeOH/H_2O (81:11:8 v/v/v), $CHCl_3$/MeOH/H_2O (70:22:3.5 v/v/v), or methyl ethyl ketone/toluene/H_2O/MeOH/acetic acid (40:5:3:2.5:1 v/v/v/v/v), followed by detection with antimony (III) chloride reagent, 50% ethanolic H_2SO_4, or anisaldehyde/sulfuric acid reagent for screening of cardenolides present in bulbs of *Scilla maderensis*. Luyt et al. [19] used TLC (silica gel, developed with ethyl acetate/MeOH/H_2O [81:11:8 v/v/v], and detected with antimony chloride reagent) for the qualitative comparison of bufadienolides in natural and tissue culture-derived plants of *Drimia robusta*.

22.2.2.3 Cardenolides

22.2.2.3.1 Structural Diversity
The cardenolides are structurally closely related to the bufadienolides, but possess a five-membered lactone (butenolide) ring at C-17β. This imparts a characteristic UV absorption at ca. 220 nm in methanol. The A/B- and C/D-ring junctions are *cis*-fused with a 14β-hydroxyl and a 5β-H (e.g., digitoxigenin; Figure 22.1: **IV**). Cardenolides are widely distributed in the plant world, but are particularly associated with the Asclepiadaceae, Scrophulariaceae, Ranunculaceae, and Convallariaceae. Digoxin and ouabain are significant inhibitors of mammalian Na^+-/K^+-ATPases [20].

22.2.2.3.2 Separation Systems and Detection
No further application of TLC to the analysis of cardenolides from plant sources appears to have been published since our review in 2001 [1].

22.2.2.4 Ecdysteroids

22.2.2.4.1 Structural Diversity
Phytoecdysteroids are predominantly C_{27}, C_{28}, or C_{29} steroids related to 20-hydroxyecdysone (20E: Figure 22.1: **V**). True ecdysteroids possess a 14α-hydroxy-7-en-6-one chromophoric group, which imparts a characteristic UV absorbance at 242 nm in methanol or ethanol solution ($\varepsilon = 12,400$ L/mol/cm). They are typically characterized by the presence of several hydroxyl groups at positions 1, 2, 3, 5, 11, 12, 14, 20, 22, 25, or 26, which gives ecdysteroids considerable water solubility. Polar (glycosides) and nonpolar (acetates) conjugates also occur frequently in plant extracts. To date, over 250 ecdysteroid analogues have been isolated and identified from plant sources [21], ranging from algae, through ferns to angiosperms (monocots and dicots). Ecdysteroid-containing plants generally possess one or a few major ecdysteroid analogues (predominantly 20E and polypodine B [polB]), together with a 'cocktail' of minor ecdysteroids [22].

22.2.2.4.2 Separation Systems and Detection
Many chromatographic procedures have been developed for the separation of phy-toecdysteroids (e.g., [23,24]). Despite the advent of HPLC, the cheapness, simpli-city, and speed of TLC have ensured a continuing role for the technique in ecdysteroid research (e.g., [25]).

22.2.2.4.3 Normal-Phase Systems
Normal-phase (absorption) chromatography on silica gel has been used extensively for the isolation of ecdysteroids and for metabolic work [26–28]. A vast number of solvent systems have been devised for the TLC of ecdysone and related compounds on silica plates, some of which are shown in Table 22.1. A suitable general system consists of 95% EtOH/CHCl$_3$ in a ratio 1:4 (v/v). A wide range of R_f values have been reported by different groups for ecdysteroids on silica gel of different types and with different degrees of activation. For consistent results, plates should be heated at 120°C for 1 h, and then deactivated to constant activity over saturated saline. In addition, TLC tanks should be saturated with the vapor of the solvent used for the chromatography before the plates are developed. Some detection methods, including the use of chromogenic reagents, are described in the following sections.

22.2.2.4.4 Reversed-Phase Systems
Reversed-phase TLC uses either bonded or paraffin-coated plates [29–31]. A variety of bonded silicas with alkyl chain lengths of C$_2$, C$_8$, C$_{12}$, or C$_{18}$ (or, alternatively, various degrees of coating with C$_{18}$) are available and all are suitable for the chromatography of ecdysteroids. Methanol/water, propan-2-ol/water, ethanol/water, acetonitrile/water, and acetone/water solvent systems have been used for chroma-tography, with methanol–water solvents providing a useful general solvent system [29]. A further property of RP-TLC systems is the ease with which they may be devised. The effect of changing the proportion of methanol in the solvent over the range 0%–100% on the R_f values of ecdysone and 20-hydroxyecdysone on either

TABLE 22.1
Solvent Systems for TLC of Ecdysteroids on Silica Gel

Solvent System	Composition (v/v)	R_f Ecdysone	R_f 20-Hydroxyecdysone
CHCl$_3$–95% EtOH	7:3	0.39	0.34
CHCl$_3$–MeOH	9:1	0.10	0.07
CHCl$_3$–Pr-1-OH	9:5	0.21	0.12
CH$_2$Cl$_2$–Me$_2$CO–MeOH	2:1:1	0.69	0.62
CH$_2$Cl$_2$–Me$_2$CO–EtOH	16:4:5	0.32	0.10
CH$_2$Cl$_2$–MeOH–H$_2$O	79:15:1	0.32	0.19
CH$_2$Cl$_2$–MeOH–25% NH$_3$–H$_2$O	77:20:2:1	0.47	0.40
EtOAc EtOH	4:1	0.49	0.46

Source: From the data of Morgan, E.D., and Wilson, I.D. in *Ecdysone: from Chemistry to Mode of Action*, J. Koolman, ed., Georg Thieme Verlag, Stuttgart, 1989.

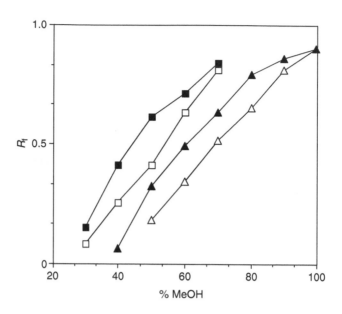

FIGURE 22.2 Influence of solvent composition on the R_f of ecdysone (E) and 20-hydro-xyecdysone (20E); C_{18}: commercial ODS-bonded silica gel plate; PAC: paraffin-coated silica gel plate; $\triangle$ = E on C_{18}; $\square$ = 20E on C_{18}; $\blacktriangle$ = E on PAC; $\blacksquare$ = 20E on PAC. (Drawn from Wilson, I.D., *J. Chromatogr.*, 318, 373, 1985.)

paraffin-coated or C_{18}-bonded silica gel is shown in Figure 22.2. From these results, it can be seen that there is a wide range of solvent compositions over which there is a regular and predictable relationship between the methanol content of the solvent and R_f, thus enabling a rapid fine tuning of separations. Many types of RP-TLC stationary phases are available, either from commercial suppliers or custom-made suppliers [29,30,32,33]. As is the case for HPLC columns, TLC plates from different suppliers may give different results, thus providing evidence for selectivity owing to the stationary phase. This is illustrated in Figure 22.3. The migration of ecdysteroids on RP-TLC is roughly the opposite of that seen using NP-TLC, with nevertheless some variations that can be exploited to achieve some specific separations (Figure 22.4). An advantage of RP-TLC over NP-TLC is the higher recoveries of material from the plate, which can be achieved.

22.2.2.4.5 Two-Dimensional TLC
A plant extract can contain a very complex mixture of phytoecdysteroids. In this case, TLC suffers from severe limitations in efficiency, as usual TLC procedures are unable to resolve such complex mixtures. Two-dimensional techniques may help to overcome this problem, and have proved very satisfactory in this respect [34]. Báthori et al. [34] recommended using NP-plates and a solvent containing a small amount of water for the first dimension, for example, (1) toluene/acetone/96% ethanol/25% aqueous ammonia (100:140:32:9 v/v/v/v) and then (2) chloroform/methanol/benzene (25:5:3 v/v/v). The technique was successfully used with a *Silene italica* subsp. *nemoralis* extract to separate up to 16 different ecdysteroids.

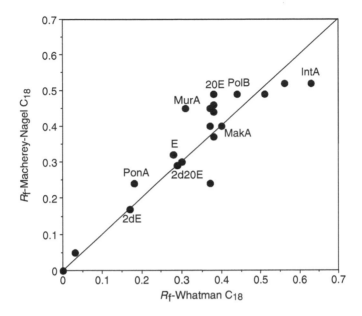

FIGURE 22.3 Retention of ecdysteroids on RP-TLC plates of different origins. Solvent system: MeOH–water 50:50 (v/v). E = ecdysone; 20E = 20-hydroxyecdysone; 2dE = 2-deoxyecdysone; 2d20E = 2-deoxy-20-hydroxyecdysone; IntA = integristerone A; Mak A = makisterone A; MurA = muristerone A; PolB = polypodine B; PonA = ponasterone A; PonC = ponasterone C (see Lafont, R., Harmatha, J., Marion-Poll, F., Dinan, L., and Wilson, I.D., *Ecdybase—The Ecdysone Handbook, 3rd edition*, 2002, online at http://ecdybase.org for structures). (Drawn from Wilson, I.D., *J. Planar Chromatogr.*, 1, 116, 1988.)

More recently, it was proposed to use CN-bonded plates for 2D-TLC [35]. Such plates can be used as either NP or RP systems, thus enabling two very different chromatographic modes to be used to resolve complex samples. Alternatively, there are commercial NP-TLC plates (Whatman Multi-K C-S5) bearing a track of RP-phase, which allow combination of both separation systems, but they have not yet been used for ecdysteroids.

22.2.2.4.6 Programmed Multiple Development (AMD and PMD)
In this system, the plate is repeatedly developed with the same solvent, which is allowed to migrate more and more with each development. This method allows a reconcentration of ecdysteroids at each run, in particular by suppressing tailing, and this finally results in sharper bands and improved resolution [36].

22.2.2.4.7 HPTLC
The use of microparticulate stationary phases results in a significant improvement of separations as compared with conventional TLC. Spot/band broadening is less pronounced, therefore achieving better resolution and higher detection sensitivity. HPTLC can be used for analytical purposes only [32,33,37].

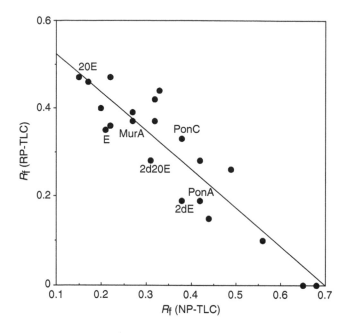

FIGURE 22.4 Relationship between R_f of a set of ecdysteroids on NP- and RP-TLC. E = ecdysone; 20E = 20-hydroxyecdysone; 2dE = 2-deoxyecdysone; 2d20E = 2-deoxy-20-hydroxyecdysone; IntA = integristerone A; Mak A = makisterone A; MurA = muristerone A; PolB = polypodine B; PonA = ponasterone A; PonC = ponasterone C. (Drawn from Wilson, I.D., *J. Chromatogr.*, 318, 373, 1985.)

22.2.2.4.8 *Overpressured Layer Chromatography*

Overpressured layer chromatography (OPLC) is also known as forced-flow TLC (FF-TLC). In OPLC, the plate is held under an inert membrane under hydrostatic pressure (ca. 25 bar) and the solvent is forced through the layer by an HPLC pump. Therefore, the flow of the solvent is controlled, and the plate can be developed within minutes. To be efficient, this technique requires the use of HPTLC plates that can be used with high linear solvent velocities. It can, of course, be used with conventional TLC plates, but in that case flow rates must be reduced to fit with the larger particle size of the silica used. In both cases, this allows development of the plates under optimal conditions which minimize diffusion, while allowing mass transfer to proceed. OPLC has only been used with ecdysteroids in a few instances (e.g., [38]).

With OPLC, the solvent can be run off the end of the plate, and this allows the use of less eluting solvents (which generally result in improved resolution) or, alternatively, to elute compounds from the plate (the solvent outlet can be connected to a fraction collector). In the latter case, there is not much difference from HPLC, except that the TLC plate is disposable and this can be of interest for crude samples that leave strongly adsorbed compounds on the plate.

22.2.2.4.9 Rotation Planar Chromatography

Rotation planar chromatography (RPC) represents a simple and efficient preparative procedure that was used to obtain pure ecdysteroids from *Serratula wolfii* [39]. Like OPLC, this technique allows an accelerated development of the circular TLC plates thanks to centrifugal forces, which are related to the rotation speed [40].

22.2.2.4.10 Use of Boronic Acids for NP- and RP-TLC

Derivatization with (e.g., butyl-, phenyl-) boronic acids can be performed in two different ways, that is, by adding boronic acids (10 mg/mL) to the mobile phase or by in situ derivatization by overspotting before chromatography [41]. Boronic acids react with vicinal diols, especially the 20,22-diol of ecdysteroids, when it is present (the 2,3-diol is poorly reactive). These treatments can be used with both NP- and RP-TLC systems and result in large R_f changes of only those ecdysteroids that bear a 20,22-diol, which may help to separate some closely migrating ecdysteroids.

22.2.2.4.11 Displacement TLC

This technique uses a displacer dissolved in the mobile phase. The role of this displacer (e.g., 3-dimethylaminopropylamine) is to push forward some sample components by displacing them from the stationary phase, and those compounds that move just in front of the displacer can be easily isolated from the TLC plate and are separated from the bulk of other (more polar) molecules. This technique has been successfully applied to purify ecdysteroids from crude plant extracts [42–44].

22.2.2.4.12 Visualization Techniques

Detection of ecdysteroids on the TLC plate can be accomplished using a variety of techniques of varying specificities [1,24]. We shall only describe here the most classical ones. Fluorescence quenching, where a fluor (ZnS) is incorporated into the silica [45], is the easiest technique: under UV light at 254 nm, ecdysteroids appear as violet spots on a bright yellow background. Diluted sulfuric acid can be used to generate various colors after heating ([46], see Table 22.2). More specific reagents such as the vanillin–sulfuric acid spray can be used to give spots of characteristic color ([47], see Table 22.2). The same spots can additionally be examined for fluorescence under UV light at 366 nm [24].

22.2.2.4.13 Scanners

UV scanners can be used at a fixed wavelength (254 nm) for scanning a track of spots, and they can also provide UV spectra of individual spots [48–50] (Figure 22.5).

22.2.2.4.14 TLC–MS and TLC–MS–MS

Offline mass spectrometric analysis of ecdysteroids can be performed directly from silica scraped from the TLC plate or, alternatively, a small piece of the TLC plate can be introduced directly in the mass spectrometer ([51,52], see review by Wilson [53]). Glycerol and DMSO are added to the silica to have a soft ionization procedure, allowing the observation of molecular ions (Figure 22.6). (HP)TLC–MS–MS can also be performed, and the observed fragments provide additional structural information about the different compounds. The technique has been used with crude *Silene otites* samples [37].

TABLE 22.2
Colors of Selected Ecdysteroids after Spraying with 10% H_2SO_4 and Heating at 110°C for 10 min[a] or with the Vanillin–Spray Reagent[b]

Ecdysteroid	Color (H_2SO_4)	Color (Vanillin)
Ajugasterone B	Brown	
Ajugasterone C	Purple	
Cyasterone	Pink	
2-Deoxyecdysone		Blue, then red-brown
2-Deoxy-20-hydroxyecdysone		Olive green, then yellow-brown
20,26-Dihydroxyecdysone		Brown-mauve, then green
Ecdysone	Orange brown	Blue, then red-brown
Ecdysone acetates		Blue
20-Hydroxyecdysone (20E)	Greenish brown	Olive green, then yellow-brown
20E 2/3-Acetate or 2,3-diacetate		Olive green, then yellow
20E 22-Acetate, 2(3),22-diacetate, 2,3,22-triacetate or 2,3,22,25-tetraacetate		Dark green
20E 20,22-Monoacetonide, 20E 2,3; 20,22-diacetonide		Olive green, then brown
Inokosterone	Reddish brown	Yellow-brown
Makisterone A	Brown	Mauve-brown
Makisterone B	Brown	
Makisterone C	Brown	
Makisterone D	Brown	
Polypodine B	Violet	
Ponasterone A	Pink	Mauve, then grey-green
Ponasterone C	Dark blue	
Stachysterone A	Greenish brown	
Stachysterone B	Greenish brown	
Stachysterone C	Brown	
Stachysterone D	Brown	
Taxisterone		Grey-green

[a] From Matsuoka, T., Imai, S., Sakai, M., and Kamada, M., *Ann. Rep. Takeda Res. Labs*, 28, 221, 1969.
[b] From Horn, D.H.S., and Bergamasco, R. in *Comprehensive Insect Physiology, Biochemistry and Pharmacology, Vol. 7*, G.A. Kerkut and L.I. Gilbert, eds, Pergamon Press, Oxford, 1985.

22.2.2.5 Steroidal Saponins

22.2.2.5.1 Structural Diversity
Steroid saponins are glycosidic derivatives of mono- or multihydroxylated steroid compounds possessing one or more oligosaccharide chains (mono-, bi-, or tri-desmosides) linked to a steroid aglycone with specific skeletal features and typically localized substituents. A sugar chain is usually attached to C-3 (less frequently also to C-1, C-2, C-6, or side-chain carbons) and contains predominantly 2–5 linear or branched monosaccharide units. The classical definition of saponins is based on their surfactant or detergent properties, reflected in their stable foam (soap) formation in

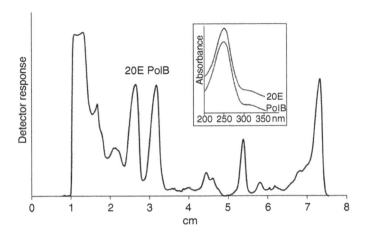

FIGURE 22.5 HP-TLC of a crude extract of *Silene schafta* containing 20-hydroxyecdysone (20E) and polypodine B (PolB) and in situ UV spectroscopy (*inset*) showing typical ecdysteroid spectra. (Redrawn from Wilson, I.D., Lafont, R., Porter, C.J., Kingston, R.G., Longden, K., Fleming, I., and Wall, P., *Chromatography and Isolation of Insect Hormones and Pheromones*, A.R. McCaffery and I.D. Wilson, eds, Plenum Press, London, 1990.)

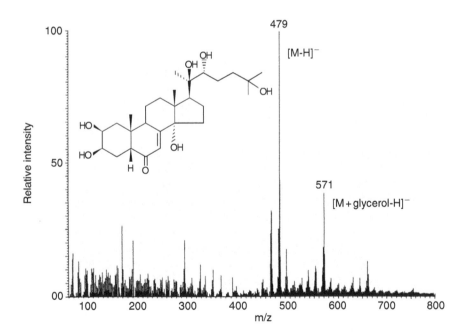

FIGURE 22.6 TLC–mass spectrometry (negative-ion mode) of 20-hydroxyecdysone present in a crude extract of *Silene otites*, obtained directly from the TLC plate after adding glycerol and DMSO. (From Wilson, I.D., Lafont, R., Kingston, R.G., and Porter, C.J., *J. Planar Chromatogr.*, 3, 359, 1990.)

water solutions. This arises because they contain both water- and fat-soluble molecular moieties.

The distribution and biological activities of plant saponins, including the steroidal saponins that occur predominantly in monocotyledonous angiosperms, have been recently reviewed [54]. Occurrence and distribution of saponins in the plant kingdom is extremely frequent and high, in both qualitative and quantitative senses. It is known that over 90 plant families contain saponins [55] and there are still many new occurrences being reported. Saponin content depends on many factors, for example, the cultivar, the age, or the geographic location of the plant. Considerable variation (mainly quantitative) can be observed in organs; high contents are found mostly in reproductive organs (flowers, seeds) or in lateral roots (root hairs), medium levels in stems and leaves, and low levels in roots or bulbs. Steroidal saponins are less widely distributed in nature than the more frequent triterpenoid-type saponins. The main sources of steroid saponins are species in the Liliaceae, Dioscoreaceae, and Agavaceae families, mainly the genera *Agave, Allium, Asparagus, Dioscorea, Lilium*, and *Yucca*.

Steroid saponins of plant origin can be divided into two main groups. The larger group is the spirostanol glycosides, comprising aglycones of the spirostane type (e.g., diosgenin in Figure 22.1: **VI**) with a sugar chain generally situated at position C-3: for spirostanols, it is typical that the spiroketal arrangement is linked at C-22. Structural variations of spirostanols depend mainly on the stereochemistry at position C-5 with a 5α–H or 5β–H configuration (A/B-*trans* or -*cis* ring annelation, respectively) and at position C-25 with the methyl group in the *R*- or *S*-configuration. The presence of a double bond in position 5(6) or 25(27) increases the variability, which can be multiplied by presence of 1–4 hydroxyls at almost any position, but mostly at positions 1, 2, 3, 5, 6, 11, 12, and 15 (and again in α- or β-configuration). Over 45 spirostanol-type aglycones have been structurally identified [55,56] and, when also taking into consideration the oxidized forms (usually carbonyls at positions 6 and 12, or a lactone forming carbonyl at 26), then the number of possible structural analogues amounts to over 100 [57]. The number of structural types rises even more with the variable sugar chains linked to the hydroxyl at position C-3 [58].

The second typical group comprises furostanol glycosides with aglycones of spirostanol-like skeleton, but with an open side chain (transformed F-ring by hydrolysis of C-22 acetal to 22,26-diol). Fewer furostanol aglycone structural forms have been identified than spirostanols, but more glycoside conjugates (saponins), as the sugar chains can be linked not only to position C-3, but often also to C-26. Furostanols can be transformed during enzymatic hydrolysis to spirostanols, and under certain conditions spirostanols also can be transformed by acid hydrolysis to furostanols. Moreover, etherification at position C-22 can also occur (i.e., formation of methyl ethers during methanol extractions), or dehydration of hydroxyl at position C-6 can form 6(7)-ene derivatives. Such side reactions can form many artifacts, which have to be taken into consideration when selecting methods for saponin analysis or separations.

A small and unusual group of steroid saponins are osladin and polypodosaponin [59–61] and polypodosides A, B, and C [62,63] with an open ring E and a preserved six-membered hemiacetal ring F, and with sugar chains attached to both C-3 and

C-26 hydroxyls. These saponins are responsible for the very sweet taste of *Polypodium vulgare* (sweet flag) and *P. glycyrhiza* (licorice fern). The specific 6-keto-7-ene moiety in their structure classifies these compounds also among ecdysteroids [21]. Not only their structural relation, but also the co-occurrence with major phytoecdysteroids (polypodine B and 20-hydroxyecdysone) in *P. vulgare* indicates the biogenetic relationship of these compounds.

Saponins often occur in plants that are used in human and animal nutrition [64], in traditional or modern medicine [65], as herbal remedies in food, pharmaceuticals, cosmetics, etc. [66], or in other commercially important preparations and products [55]. For such purposes, they are produced on a large scale using general or specially modified separation processes (e.g., the formation of water-insoluble complexes with cholesterol). To investigate a wide variety of biological and pharmacological activities [67], chemoecological activities in plant–insect interactions [68] or other antimicrobial, molluscicidal, antifertility activities often requires several specially adapted analytical and separation processes.

22.2.2.5.2   Separation Systems and Detection

TLC on silica is a suitable supporting method for monitoring saponins during fractionation or for final small-scale purification. It is often used also for confirmation of purity and identity of isolated compounds, if suitable standards are available. Such applications were often reported in older papers, currently often substituted by HPLC analyses. For simple TLC separations of saponins, the most frequently used solvent systems are $CHCl_3/MeOH/H_2O$ (14:6:1 v/v/v) or $CHCl_3/MeOH$ (4:1 v/v). For the separation of aglycones, $CHCl_3/MeOH$ (50:1 v/v) was most appropriate (e.g., [69]). They may be detected after chromatography by various visualization reagents as colored spots: (1) spraying with sulfuric acid [70], (2) spraying with vanillin/sulfuric acid [71], (3) spraying with anisaldehyde/acetic acid/sulfuric acid [72], (4) spraying with $SbCl_3$ in chloroform and heating gives characteristic colored spots [73], or (5) spraying with Ehrlich reagent (*p*-dimethylaminobenzaldehyde/concentrated HCl/EtOH) and heating (red colors for furostanol derivatives). Some more systems are indicated in the monograph by Hostettmann and Marston [55] and summarized in detail with references to the original papers by Oleszek [74].

In former times, quantification used spot detection and their intensity evaluation in relation with the parallel running appropriate standards, using specialized spot scanner systems for TLC-densitometry or TLC-colorimetry. Both methods faced a series of problems, but allowed performance of many required quantifications in the pre-HPLC period. A review of such results with comments is provided by Oleszek [74]. The highly polar nature and rather high molecular weight of saponins, as well as their close structural similarities (isomers or epimers of the aglycone or sugar parts) can cause difficulties in TLC, but the greater resolution of HPLC makes this the method of choice. A large number of HPLC analysis reports employing various columns and solvent systems are summarized and commented in the review of Oleszek and Bialy [75]. The main difficulty is the absence of a suitable chromophore for UV detection. Most of the published data are based on recording HPLC profiles at 200–210 nm. This range, however, allows only nonspecific detection, and thus much reduces the use of this technique in quantification of saponins in nonseparated samples. To overcome these problems, the evaporative light scattering detection has been applied (e.g., [76]). Nevertheless, this detection method is not used much for routine quality control of saponin-containing herbal products.

Important progress in detection and quantification of saponins was achieved by combining liquid chromatography with a mass spectrometry detector. Especially, connection of HPLC with electrospray ionization mass spectrometry (HPLC-ESIMS) or with tandem HPLC–ESIMS–MS upgraded facilities for structural analysis of saponins. This technique has, however, still some limitations. For reliable routine use, the availability of appropriate standards obtained by regular separations is still essential. This technique cannot afford sufficient information for particular sugar bounding or sugar branching points without targeted chemical modification, for example, preparation of permethylated derivative [68]. For further consideration of the TLC of saponins, refer to Chapter 20.

22.2.2.6   Steroidal Alkaloids

22.2.2.6.1   Structural Diversity
Steroid alkaloids belong to the very diverse isoprenoid alkaloid family. They represent a large group characterized by the presence of an unaltered or a modified steroid skeleton with nitrogen integrated either into a ring, or as part of the side chain, or within a substituent. According to Hegenauer [77], these nitrogen-containing metabolites are in fact not true alkaloids, but are biogenetically considered as simple nitrogen derivatives of generally occurring steroid types, that is, they are pseudoalkaloids. According to the location of the nitrogen and the skeletal arrangement, several subgroups exist (e.g., [78,79]), representing different structural types: (1) alkaloids with unaltered C_{27} cholestane skeleton, but possessing various heterocyclic rings (six skeletal types from Solanaceae and Liliaceae families), (2) alkaloids with altered C_{27} cholestane (C-nor-D-homo-) ring systems, as well as with 18-nor-cholestane skeleton (five types from Liliaceae, mostly from the genus *Veratrum*), (3) alkaloids with a C_{21} pregnane skeleton and with amino groups at C-3 and C-20 or an imino group between C-18 and C-20 (four types from Apocynaceae and Buxaceae), and (4) alkaloids with C_{21} skeleton and with amino groups at C-3 and C-20, but with 1–3 additional methyl groups at C-4 and C-14 and with a 9,19 cyclopropane or 9(10)-11 *abeo* structure modification (two types from Buxaceae). From a chemical point of view, the nitrogen can be attached as a primary NH_2 group (free or methylated) forming simple steroidal bases, can be ring closed to skeletal or side-chain carbon (as a secondary NH) or annelated into two rings (as a tertiary N, as, for example, in solanidine; Figure 22.1: **VII**). This influences the chemical character and also the chromatographic properties of the compound.

Plants contain steroidal alkaloids often in glycosidic form as glycoalkaloids. Moreover, all the types classified earlier can contain double bonds and hydroxyls in various positions and also sugar chains, as in the case of steroid saponins. However, the number and variety of naturally occurring glycoalkaloids is much lower than is the number of steroid glycosides [80]. Their distribution is also limited to a few plant families (see earlier). Steroidal alkaloids from the family Liliaceae exceed the others in their structural, chemical, and bioactivity diversity [81]. The family

Solanaceae, which includes many important agricultural crop plants, such as potato, tomato, aubergine (eggplant), and capsicum, is similarly rich and complex in steroidal alkaloids [79]. The best known, solasodine (of which the genin is solanidine; Figure 22.1: **VII**), has been found in about 200 *Solanum* species [82]. Glycoalkaloids are usually found in all plant organs, but with the highest concentrations in flowers, unripe berries, young leaves, or shoots (metabolic active parts). They are generally toxic, but in fruits they gradually decompose to nitrogen-free nontoxic constituents during ripening.

22.2.2.6.2   *Separation Systems and Detection*
For isolation of single glycoalkaloids or their aglycones, the same extraction and separation procedures as for steroid saponins or sapogenins are used [1]. However, there is one advantage that can be used in certain cases effectively; alkaloids can form water-soluble salts with acids. Crude alkaloids can be then obtained from the weakly acidic water portions of extracts by neutralization with ammonia. When applying this procedure, it is important to maintain mild conditions and consider the possible formation of artifacts.

For TLC monitoring of steroid glycoalkaloids, generally the same methods and solvent systems are used as for steroid saponins. Silica TLC with $CHCl_3/MeOH/2\%$ NH_4OH (70:30:5 v/v/v) as solvent provides good separation of potato glycoalkaloids and their hydrolysis products [83]. Centrifugal TLC on silica with $CHCl_3/MeOH/H_2O$ (13:7:2 v/v/v) has been used for the separation of steroidal alkaloids from *Fritillaria harelinii* [84]. For detection and color visualization sulfuric acid [70] or $SbCl_3/CHCl_3$ [85] has been used in most cases. Alkaloid-specific spraying reagents can be also applied, for example, Dragendorff's reagent (e.g., [86]). TLC is presently used mainly for monitoring the hydrolysis processes employed in glycoalkaloid analysis (e.g., [87]) or biotransformation processes in detoxification experiments [88], but also to their direct determination by HP-TLC by means of densitometry [89]. Determination by TLC and immunostaining with a specific monoclonal antibody has also been reported [90]. Another specific separation method is the use of derivatized cyclodextrin for TLC separation of enantiomers and diastereomers [91].

For quantification, however, HPLC analysis has mainly been employed in the last decade. A generally applicable method for working with small sample amounts of toxic steroidal glycoalkaloids was developed by Kubo and Fukuhara [92]. For HPLC analyses and preparations, the well-tried steroid saponin systems can be applied with modifications necessitated by the presence of alkaloid nitrogen.

22.2.2.7   Vertebrate-Type Steroids

22.2.2.7.1   *Structural Diversity*
Progesterone and several androgens and estrogens have been detected in plant sources [93], but the levels are low ($\mu g/kg$) relative to the other steroids considered here, and in many cases the compounds have not been conclusively identified (e.g., estrogenic activity could derive from the presence of nonsteroidal phyto-estrogens). However, immunoassay evidence indicates that vertebrate-type steroids are widespread in plants. Progesterone (Figure 22.1: **VIII**) is certainly an intermediate in the

synthesis of cardiac glycosides. Guggulsterone, steroid isomers ([17E/Z]pregna-4,17 (20)-diene-3,16-dione; Figure 22.1: **IX**) isolated from the gum of *Commiphora mukul*, lowers LDL cholesterol levels in humans by acting as an antagonist of the farnesoid X receptor. Exogenous vertebrate steroids bring about a variety of effects on plant growth and reproductive processes [93]. Vitamin D-like activities are present in certain plant species [94].

22.2.2.7.2 Separation Systems and Detection

Agrawal et al. [95] have identified an HPTLC method (silica gel 60F$_{254}$ with toluene/acetone 9:1 v/v for development) for the separation of the guggulsterone isomers with subsequent quantification by densitometric scanning at 250 nm.

22.2.2.8 Withanolides

22.2.2.8.1 Structural Diversity

Over 200 structural analogues of withanolides have been isolated, predominantly as aglycones, although glycosides are known from some sources [96]. They occur predominantly in solanaceous plants, some of which are of great agricultural or medicinal importance. Withanolides are typically C$_{28}$ ergostane-type steroids possessing a 22,26-δ-lactone. The side chain may be linked to the steroid nucleus 17α or 17β. There are also a large number of oxygen-containing functional groups (hydroxyls, ketones, epoxides, and cyclic ethers). Many withanolides possess a 1-oxo-group (e.g., withaferin A; Figure 22.1: **X**). Most withanolides absorb UV light at ca. 220 nm in methanolic or ethanolic solution, owing to the combined contributions of two isolated chromophores (2-en-1-one and side chain α,β-unsaturated-δ-lactone) [97].

22.2.2.8.2 Separation Systems and Detection

There are very few publications dedicated to the chromatography of withanolides. However, preparative TLC has often been used for the isolation of withanolides, using either NP systems (e.g., [98,99]) or RP systems (e.g., [100,101]). Radial chromatography on silica gel eluted sequentially with CHCl$_3$/acetone (3:2 v/v), acetone, and MeOH has been used for the purification of 18-acetoxywithanolides from *Physalis chenopodifolia* [102]. Several methods of detection of withanolides on TLC plates have been employed; fluorescence quenching, exposure to iodine vapor [103], Dragendorff's reagent [104], charring with Ce(SO$_4$)$_2$/10% aqueous H$_2$SO$_4$ [100], MeOH/H$_2$SO$_4$ [99], or vanillin/sulfuric acid [103].

22.2.3 RELATED TRITERPENOIDS—CUCURBITACINS

22.2.3.1 Structural Diversity

The cucurbitacins are a group of highly oxygenated C$_{30}$ triterpenoids possessing a 19 (10→9β)-abeo-10α-lanost-5-ene skeleton (including a gem-dimethyl group at C-4 and further methyls at C-9 and C-14). All cucurbitacins also possess a 5(6)-double bond (e.g., cucurbitacin D; Figure 22.1: **XI**). Many cucurbitacins absorb UV light with a maximum at ca. 230 nm in methanol or ethanol [105], owing to the presence of chromophoric α,β-unsaturated ketones in the A-ring and side chain, although other analogues exist that have a maximum at higher wavelength or no maximum above 210 nm exist. The cucurbitacins are most associated with the Cucurbitaceae, but have also been identified from members of other plant families (Begoniaceae, Cruciferae, Datiscaceae, Desfontainaceae, Elaeocarpaceae, Polemoniaceae, Primulaceae, Rosaceae, Rubiaceae, Scrophulariaceae, and Sterculiaceae). Cucurbitacins B and D occur most commonly, but about 50 other analogues (including glycosides) have been characterized [106].

22.2.3.2 Separation Systems and Detection

Early studies on cucurbitacins used silica gel G plates with a solvent of petroleum ether/n-butanol (4:1 v/v). Several useful chromogenic agents were identified: (1) 0.5% $KMnO_4$ saturated with copper acetate (generating yellow spots), (2) $CHCl_3$/acetic anhydride (1:1 v/v) followed by contact with H_2SO_4 (red spots), (3) 5% vanillin in 70% EtOH containing 21% phosphoric acid and heating (brown spots), and (4) 0.2% triphenyl tetrazolium chloride in ethanolic KOH, exposure to steam, and heating at 100°C (yellow spots, changing to pink) [107]. More recent publications use $CHCl_3$/MeOH (95:5 v/v) (e.g., [108]) for silica gel TLC separation of cucurbitacins. For simultaneous analysis of glycosides of iridoids, cucurbitacins and phenols in methanolic extracts of rhizomes and roots of *Picrorhiza kurroa*, Stuppner and Wagner [109] used silica gel HPTLC plates with $CHCl_3$/MeOH (85:15 v/v) as solvent and RP-18 HPTLC plates with MeOH/H_2O/HOAc (5/5/0.1 v/v/v). Charring with vanillin/sulfuric acid reagent generated different colors for the various glycosides. Centrifugal TLC on silica gel with a solvent of hexane/dichloromethane/ ethyl acetate has been used to separate cucurbitacin glycosides from *Desfontainia spinosa*, and overpressure layer chromatography (OPLC) on silica with ethyl acetate/diethyl ether/chloroform/EtOH (80:10:7:3 v/v/v/v) provided effective separation of closely related cucurbitacins with saturated or unsaturated side chains from *Hemsleya gigantha*) [84].

22.3 SUMMARY OF DETECTION AND QUANTIFICATION METHODS FOR STEROIDS AFTER TLC [110–113]

22.3.1 Nonspecific Methods

22.3.1.1 Fluorescence Quenching

An inorganic fluorescent agent is incorporated into the TLC plate, the distribution or performance of which is not affected by development. Irradiation at 254 or 360 nm (depending on fluorescent indicator present) induces green or blue fluorescence, but where UV-absorbing compounds are located, the fluorescence is quenched and dark areas are seen; nondestructive; useful for all compounds absorbing around 254 or 360 nm; semiquantitative.

22.3.1.2 Charring

Copper Sulfate/Phosphoric Acid: Spray with a solution of 10% $CuSO_4$ in 10% phosphoric acid. Heat for 5–30 min at 110°C. Spots turn colored or fluorescent

(254 or 360 nm) first and then char to brown, grey, or black. Destructive. Suitable for polymer-bound TLC plates.

Ferric Chloride/Sulfuric Acid: Spray with a solution of 2 g $FeCl_3$ in 83 mL *n*-BuOH and 15 mL concentrated H_2SO_4. Heat at 110°C for 5–30 min. Spots turn colored or fluorescent (254 or 360 nm) first and then char to brown, grey, or black. Destructive. Suitable for polymer-bound TLC plates.

Phosphomolybdic Acid: Spray with a solution of 250 mg phosphomoybdic acid in 50 mL EtOH (reagent stable for up to 10 days in dark). Heat at 120°C (with a heat-gun). Spots turn colored or fluorescent (254 or 360 nm) first and then char to brown, grey, or black. Exposure of plate to ammonia vapor can be used to reduce background coloration. Destructive. Suitable for polymer-bound TLC plates. For detection of reducing substances (inc. steroids) and as charring reagent.

Potassium Dichromate/Sulfuric Acid: Spray with a solution of 5 g potassium dichromate in 100 mL concentrated H_2SO_4 and heat to 150°C, if necessary. Gives brown, grey, or black spots with organic compounds. Can only be used with plates with gypsum binder, not newer polymer binder.

Potassium Permanganate/Sulfuric Acid: Spray plate with 1.6% $KMnO_4$ in concentrated H_2SO_4 (prepare with care as there is danger of explosion). Heat plate at 180°C for 15–20 min. Universal reagent for organic compounds. Can only be used with plates with gypsum binder, not newer polymer binder.

Sulfuric Acid: Spray the developed TLC plate with a 5% (v/v) H_2SO_4 in EtOH and then heat the plate at 110°C for 10 min.

Vanillin/Phosphoric Acid: Spray plate with 1 g vanillin in 100 mL 50% aqueous H_3PO_4. Heat at 110°C for 5–30 min. Spots turn colored or fluorescent (254 or 360 nm) first and then char to brown, grey, or black. Exposure of plate to ammonia vapor can be used to reduce background coloration. Destructive. Suitable for polymer-bound TLC plates.

22.3.1.3 Nondestructive General Reagents

Berberine Sulfate: After development, the TLC plate is sprayed with 1% (w/v) berberine sulfate in MeOH/acetone (1:1 v/v). Under UV light (360 nm), organic compounds appear as bright yellow spots on a yellow background.

Iodine Vapor: Charge a chamber with some crystals of iodine. Place the developed, dried chromatogram in the chamber and expose to the iodine vapor. Organic compounds give tan/brown spots.

22.3.2 MORE SPECIFIC METHODS

22.3.2.1 Physicochemical Properties and Chemical Reagents

Where classes of compounds are specified, the reagent may be used to detect these types of compound; however, these may not be the only classes detected by the reagent.

22.3.2.1.1 Absorbance of Visible Light/Fluorescence Activity

View developed plate under visible, short (254 nm), or long-wavelength (360 nm) UV light.

22.3.2.1.2 Chemical Detection by Yielding Colored, Fluorescent, or UV-Active Products

p-Anisaldehyde/Sulfuric Acid: spray with 0.5 mL *p*-anisaldehyde in 50 mL glacial acetic acid and 1 mL 97% sulfuric acid. Heat to 105°C. Steroids turn violet, blue, red, grey, or green.

Antimony (III) Chloride: Spray with a solution of 25 g SbCl$_3$ in 75 mL CHCl$_3$. Heat for 10 min at 100°C. View at 360 nm. Detects steroids, sapogenins, and steroid glycosides.

2,4-Dinitrophenylhydrazine: Spray with 0.4 g 2,4-DNPH in 100 mL 2N HCl + 1 mL EtOH. Steroids containing aldehyde or ketone groups give yellow-red spots.

2,2′-Diphenylpicrylhydrazyl: Spray with a solution of 15 mg 2,2′-DPPH in 25 mL CHCl$_3$. Heat at 110°C for 5–10 min. Steroids containing aldehyde or ketone groups give yellow spots on a purple background.

Dragendorff's Reagent: Mix equal volumes of Solution 1 (1.7 g basic bismuth nitrate and 20 g tartaric acid in 80 mL H$_2$O) and Solution 2 (16 g KI in 40 mL H$_2$O) to give a stock solution (stable for several weeks at 4°C). Spray plate with a solution of 10 g tartaric acid, 50 mL water, and 5 mL of the stock solution. Steroidal alkaloids give orange-brown spots.

Formaldehyde/Sulfuric Acid: Spray with a 1:10 v/v solution of 37% formaldehyde in concentrated H$_2$SO$_4$ immediately after taking the plate from the developing chamber. Alkaloids give variously colored spots.

Formaldehyde/phosphoric Acid: Spray with a solution of 0.03 g formaldehyde in 100 mL phosphoric acid. Detects steroidal alkaloids and steroid sapogenins.

Iodoplatinate: Spray with a freshly prepared mixture of 3 mL 10% (w/v) hexachloroplatinic (IV) acid solution with 97 mL 5% aqueous MeOH or EtOH and 100 mL 6% (w/v) KI in 5% aqueous MeOH or EtOH. Detects alkaloids.

Lead Tetraacetate/2,7-dichlorofluorescein: Mix 5 mL each of Solution 1 (2% [w/v] lead tetraacetate in glacial acetic acid) and Solution 2 (1% [w/v] 2,7-dichlorofluorescein in EtOH) and make up to 200 mL with dry toluene. Mixture stable for only 2 h. Detects vicinal diols.

Nitric Acid/Ethanol: Spray with a solution of 50 drops 65% concentrated HNO$_3$ in 100 mL EtOH. Heat at 120°C, if necessary. Detects alkaloids.

p-Phenylenediamine/Phthalic Acid: Spray with a solution of 0.9 g *p*-phenylenediamine and 1.6 g phthalic acid in 100 mL butan-1-ol saturated with water. Heat at 100°C–110°C. Conjugated 3-ketosteroids give yellow/orange spots.

Phosphoric Acid: Spray the plate heavily with a 1:1 (v/v) mixture of 85% phosphoric acid and H$_2$O. Heat at 120°C for 10–15 min. Sterols, steroids, and bile acids produce various colors under visible and UV light.

Phosphotungstic Acid: Spray with 20% phosphotungstic acid in EtOH. Heat at 110°C for 5–15 min. Detection of cholesterol, cholesterol esters, sterols, and steroids, which produce red spots.

Sulfuric Acid: Spray plate with 10% H_2SO_4. Heat at 110°C for 10 min. Ecdysteroids give a range of colors.

Tetranitrodiphenyl: Spray plate with Solution 1 (saturated 2,3′,4,4′-tetranitrodiphenyl in toluene), dry at room temperature, and then spray with Solution 2 (10% KOH in 50% aqueous MeOH). Blue spots are seen with cardiac glycosides.

Tin (IV) Chloride: Spray with a solution of 10 mL $SnCl_4$ in 160 mL $CHCl_3$/glacial AcOH (1:1 v/v). Heat at 100°C for 5–10 min and view under visible and long-wavelength UV light. Detects sterols and steroids.

p-Toluenesulfonic Acid: Spray with a solution of 20% *p*-toluenesulfonic acid in $CHCl_3$ and heat at 100°C for a few minutes. View under long-wavelength UV light. Used for the detection of steroids.

Trichloroacetic Acid: Spray with either 25% TCA in $CHCl_3$ (for steroids) or 1% TCA in $CHCl_3$ (for vitamin D) or 3.3 g TCA in 10 mL $CHCl_3$ + 1–2 drops H_2O_2 (for *Digitalis* glycosides). Heat at 120°C for 5–10 min. View under visible and long-wavelength UV light.

Trifluoroacetic Acid: Spray with a solution of 1% TFA in $CHCl_3$. Heat at 120°C for 5 min. Detects steroids.

Vanadium (V)/Sulfuric Acid: (1) Spray with a solution of 1.2 g ammonium vanadate in 95 mL H_2O + 5 mL concentrated H_2SO_4 or (2) prepare a solution of 1.82 g vanadium pentoxide in 30 mL 1M Na_2CO_3 (sonicate and allow to cool). Add 46 mL 2.5M H_2SO_4 and make up to 100 mL with acetonitrile. Spray the plate. Detects steroids.

Vanillin/Sulfuric Acid: Spray with a 1% vanillin in concentrated H_2SO_4 or with 0.5 g vanillin in 100 mL concentrated H_2SO_4/EtOH (4:1 v/v). Heat at 120°C until maximum color formation. Detects steroids. This reagent can only be used with plates with a gypsum binder, not the newer polymer binder. A variant of this spray reagent (1.5 g vanillin + 50 mL EtOH + 1 mL concentrated H_2SO_4) and heating the sprayed plate at 110°C is used to detect ecdysteroids, which give different colors depending on the specific structure [48].

Zinc Chloride: Dissolve 30 g $ZnCl_2$ in 100 mL CH_3OH and filter off the insoluble matter. Spray the plate with the solution, heat for 1 h at 105°C and then cover the plate with a glass plate to exclude moisture and view under long-wavelength UV light. Detects steroid sapogenins and steroids.

22.3.2.2   Bioautography

There appear to be no published examples of immunological detection or bioassays used to detect phytosteroids. However, one might envisage that the insect attractant/repellent properties of certain of the compounds (e.g., cucurbitacins) being the basis of a bioassay where the congregation/dispersal of a sensitive insect species on the developed chromatogram is assessed. The ecdysteroid agonist/antagonist properties of some of the classes of compounds (ecdysteroids, cucurbitacins, and withanolides; [114]) could lend itself to a modification of the B_{II} bioassay [115], whereby the developed TLC plate is coated with a thin layer of agar and then medium containing B_{II} cells is plated on top of this in a sterile environment. The active compounds would diffuse through the agar layer and affect the cells in closest proximity. Microscopical examination of the cells would then allow identification of regions of the plate containing agonist activity, or, if 20E, were also included with the cell suspension, antagonist activity.

22.4 CONCLUSIONS AND PROSPECTS

The future directions and potential of TLC in general have been recently considered [116,117], and these could all apply to the separation and analysis of phytosteroids. Yet, over the past decade, TLC has been increasingly less used for the analysis of phytosterols/phytosteroids, even though it would be an appropriate (and economical) method for the solution of many analytical problems. Separations previously performed by TLC now generally use SPE or HPLC. This is probably not a trend which will be reversed in the future. Even the forced-flow techniques (CTLC, OPLC), which can also be used preparatively, have not been used extensively for the separation of phytosteroids. However, the more extensive studies that have been performed on phytoecdysteroids reveal the potential of TLC for the other classes, if similar more systematic studies were to be performed. The advantages of TLC are also demonstrated by the use of parallel TLC analyses of different solvent extracts on silica gel plates with appropriate mobile phases (three of differing polarities) and detection systems (UV light and various spray reagents) for the rapid and economical screening of plant species for pharmaceutically active components, including cardiac glycosides and saponins [118]. Thus, it is to be expected that TLC will continue to be used in important, but niche, applications for the analysis of phytosteroids.

REFERENCES

1. Dinan, L., Harmatha, J., and Lafont, R., Chromatographic procedures for the isolation of plant steroids. *J. Chromatogr. A*, 935, 105–123, 2001.
2. Brown, G.D., The biosynthesis of steroids and triterpenoids. *Nat. Prod. Rep*, 15, 653–696, 1998.
3. Barbier, M., The ecological importance of sterols in invertebrates. In M. Barbier (ed.), *Introduction to Chemical Ecology*, Longman, London and New York, p. 4762, 1976.
4. Moss, G.P., Nomenclature of steroids. *Pure Appl. Chem.*, 61, 1783–1822, 1989.
5. Goad, L.J., and Akihisa, T., *Analysis of Sterols*, Blackie Academic & Professional of Chapman & Hall, London, 1997.
6. Moreau, R.A., Whitaker, B.D., and Hicks, K.B., Phytosterols, phytostanols, and their conjugates in foods: structural diversity, quantitative analysis, and health-promoting uses. *Progr. Lipid Res.*, 41, 457–500, 2002.
7. Harmatha, J., Samek, Z., Novotny, L., Herout, V., and Šorm, F., The structure of adenostylone, isoadenostylone and neoadenostylone—components of *Adenostyles alliariae. Collect. Czech. Chem. Commun.*, 34, 1739–1749, 1969.

8. Abidi, S.L., Chromatographic analysis of plant sterols in foods and vegetable oils. *J. Chromatogr. A*, 935, 173–201, 2001.

9. Nes, W.D., and Heftmann, E., A comparison of triterpenoids with steroids as membrane-components. *J. Nat. Prod.*, 44, 377–400, 1981.

10. Melnik, B.C., Hollmann, J., Erler, E., Verhoeven, B., and Plewig, G., Microanalytical screening of all major stratum corneum lipids by sequential high-performance thin-layer chromatography. *J. Investigat. Dermatol.*, 92, 231–234, 1989.

11. Markowski, M., Computer-aided optimization of gradient multiple development thin-layer chromatography. Part 2. Multistage development. *J. Chromatogr. A*, 635, 283–289, 1993.

12. Zellmer, S., and Lasch, J., Individual variation of human plantar stratum corneum lipids, determined by automated multiple development of high-performance thin-layer chromatography plates. *J. Chromatogr. B*, 691, 321–329, 1997.

13. Zullo, M.A.T., and Kohout, L., Semisystematic nomenclature of brassinosteroids. *Plant Growth Regul.*, 42, 15–28, 2004.

14. Takatsuto, S., Yokota, T., Omote, K., Gamoh, K., and Takahashi, N., Identification of brassinolide, castasterone and norcastasterone (brassinone) in sunflower (*Helianthus annus* L.) pollen. *Agric. Biol. Chem.*, 53, 2177–2180, 1989.

15. Ikekawa, N., The analysis of brassinosteroids-plant growth-promoting substances. *Trends Anal. Chem.*, 9, 337–342, 1990.

16. Winter, J., Schneider, B., Meyenburg, S., Strack, D., and Adam, G., Monitoring brassinosteroid biosynthetic enzymes by fluorescent tagging and HPLC analysis of their substrates and products. *Phytochemistry*, 51, 237–242, 1999.

17. Krenn, L., and Kopp, B., Bufadienolides from animal and plant sources. *Phytochemistry*, 48, 1–29, 1998.

18. Dias, C., Borralho, J.A., and Lurdes Gonçalves, M., *Scilla maderensis*, TLC screening and positive inotropic effect of bulb extracts. *J. Ethnopharmacol.*, 71, 487–492, 2000.

19. Luyt, R.P., Jäger, A.K., and van Staden, J., Bufadienolides in *in vitro* derived *Drimia robusta* plants. *S. Afr. J. Bot.*, 65, 443–445, 1999.

20. Melero, C., Medarde, M., and San Feliciano, A., A short review on cardiotonic steroids and their aminoguanidine analogues. *Molecules*, 5, 51–81, 2000.

21. Lafont, R., Harmatha, J., Marion-Poll, F., Dinan, L., and Wilson, I.D., *Ecdybase—The Ecdysone Handbook, 3rd edition*, 2002 (online at http://ecdybase.org).

22. Dinan, L., Phytoecdysteroids: biological aspects. *Phytochemistry*, 57, 325–339, 2001.

23. Lafont, R., Morgan, E.D., and Wilson, I.D., Chromatographic procedures for phytoecdysteroids. *J. Chromatogr.*, 658, 31–53, 1994.

24. Báthori, M., Kalász, H., Janicsák, G., Pongrácz, Z., and Vámos, J., Thin-layer chromatography of phytoecdysteroids. *J. Liq. Chromatogr. Relat. Technol.*, 26, 2629–2649, 2003.

25. Báthori, M., Hunyadi, A., Janicsak, G., and Mathé, I., TLC of ecdysteroids with four mobile phases and three stationary phases. *J. Planar Chromatogr. Modern TLC*, 17, 335–341, 2004.

26. Horn, D.H.S., The ecdysones. In M. Jacobson and D.G. Crosby (eds), *Naturally Occurring Insecticides*, Marcel Dekker, New York, NY, pp. 333–459, 1971.

27. Morgan, E.D., and Wilson, I.D., Progress in the analysis of ecdysteroids. In J.A. Hoffmann (ed.), *Progress in Ecdysone Research*, Elsevier/North-Holland, Amsterdam, pp. 29–43, 1980.

28. Touchstone, J.C., Ecdysteroids. In *CRC Handbook of Chromatography: Steroids*, CRC Press, Boca Raton, FL, pp. 119–135, 1986.

29. Wilson, I.D., Scalia, S., and Morgan, E.D., Reversed-phase thin-layer chromatography for the separation and analysis of ecdysteroids. *J. Chromatogr.*, 212, 211–219, 1981.

30. Wilson, I.D., Bielby, C.R., and Morgan, E.D., Studies on the reversed-phase thin-layer chromatography of ecdysteroids on C-12 bonded and paraffin-coated silica. *J. Chromatogr.*, 242, 202–206, 1982.

31. Wilson, I.D., Normal-phase thin-layer chromatography on silica gel with simultaneous paraffin impregnation for subsequent reversed-phase thin-layer chromatography in a second dimension. *J. Chromatogr.*, 287, 183–188, 1984.

32. Wilson, I.D., Reversed phase thin-layer chromatography of the ecdysteroids. *J. Planar Chromatogr.*, 1, 116–122, 1988.

33. Wilson, I.D., Comparison of non-hydrophobic C_{18} bonded thin-layer and high-performance thin-layer chromatography plates for reversed-phase thin-layer chromatography of ecdysteroids. *J. Chromatogr. A*, 438, 412–422, 1988.

34. Báthori, M., Blunden, G., and Kalász, H., Two-dimensional thin-layer chromatography of plant ecdysteroids. *Chromatographia*, 52, 815–817, 2000.

35. Kalász, H., Hunyadi, A., and Báthori, M., Novel results of two-dimensional thin-layer chromatography. *J. Liq. Chromatogr. Relat. Technol.*, 28, 2489–2497, 2005.

36. Wilson, I.D., and Lewis, S., Separation of ecdysteroids by high-performance thin-layer chromatography using automated multiple development. *J. Chromatogr.*, 408, 445–448, 1987.

37. Lafont, R., Porter, C.J., Williams, E., Read, H., Morgan, E.D., and Wilson, I.D., The application of off-line HPTLC–MS–MS to the identification of ecdysteroids in plant and arthropod samples. *J. Planar Chromatogr.*, 6, 421–424, 1993.

38. Read, H., Wilson, I.D., and Lafont, R., A note on overpressure thin-layer chromatography of ecdysteroids. In A.R. McCaffery and I.D. Wilson (eds), *Chromatography and Isolation of Insect Hormones and Pheromones*, Plenum Press, London, pp. 127–130, 1990.

39. Kalász, H., Liktor-Busa, E., Janicsak, G., and Báthori, M., Role of preparative rotation planar chromatography in the isolation of ecdysteroids. *J. Liq. Chromatogr. Relat. Technol.*, 29, 2095–2109, 2006.

40. Botz, L., Nyiredy, S., and Sticher, O., A new device for circular preparative planar chromatography. *J. Planar Chromatogr.*, 3, 401–406, 1990.

41. Wilson, I.D., The use of boronic acids for the normal and reversed phase TLC of ecdysteroids. *J. Planar Chromatogr.*, 5, 316–318, 1992.

42. Kalász, H., Báthori, M., Ettre, L.S., and Polyák, B., Displacement thin-layer chromatography of some plant ecdysteroids with forced-flow thin-layer chromatography. *J. Planar Chromatogr.*, 6, 481–486, 1993.

43. Kalász, H., Báthori, M., Kerecsen, L., and Tóth, I., Displacement thin-layer chromatography of some plant ecdysteroids. *J. Planar Chromatogr.*, 6, 38–42, 1993.

44. Kalász, H., Báthori, M., and Máthé, I., Displacement thin-layer chromatography of some ecdysteroids. *J. Liq. Chromatogr.*, 18, 837–848, 1995.

45. Mayer, R.T., and Svoboda, J.A., Thin-layer chromatographic *in situ* analysis of insect ecdysones via fluorescence-quenching. *Steroids*, 31, 139–150, 1978.

46. Matsuoka, T., Imai, S., Sakai, M., and Kamada, M., Studies on phytoecdysones—a review of our work. *Ann. Rep. Takeda Res. Labs*, 28, 221–271, 1969.

47. Horn, D.H.S., and Bergamasco, R., Chemistry of ecdysteroids. In G.A. Kerkut and L.I. Gilbert (eds), *Comprehensive Insect Physiology, Biochemistry and Pharmacology, Vol. 7*, Pergamon Press, Oxford, pp. 185–248, 1985.

48. Báthori, M., Máthé, I., and Guttman, A., Determination of 20-hydroxyecdysone content by thin-layer chromatography and micellar electrokinetic chromatography. *Chromatographia*, 48, 145–148, 1998.

49. Wilson, I.D., Lafont, R., Porter, C.J., Kingston, R.G., Longden, K., Fleming, I., and Wall, P., Thin-layer chromatography of ecdysteroids: detection and identification. In A.R. McCaffery and I.D. Wilson (eds), *Chromatography and Isolation of Insect Hormones and Pheromones*, Plenum Press, London, pp. 117–126, 1990.

50. Báthori, M., Szendrei, K., Kalász, H., Lafont, R., and Girault, J.P., Thin-layer chromatography of ecdysteroids originated from *Silene otites* L. (Wib.). *Chromatographia*, 25, 627–630, 1988.

51. Wilson, I.D., Lafont, R., and Wall, P., TLC of ecdysteroids with off-line identification by fast atom bombardment mass spectrometry directly from the adsorbent. *J. Planar Chromatogr.*, 1, 357–359, 1988.

52. Wilson, I.D., Lafont, R., Kingston, R.G., and Porter, C.J., Thin-layer chromatography–tandem mass spectrometry directly from the adsorbent: application to phytoecdysteroids from *Silene otites*. *J. Planar Chromatogr.*, 3, 359–361, 1990.

53. Wilson, I.D., The state of the art in thin-layer chromatography–mass spectrometry: a critical appraisal. *J. Chromatogr. A*, 856, 429–442, 1999.

54. Sparg, S.G., Light, M.E., and van Staden, J., Biological activities and distribution of plant saponins. *J. Ethnopharmacol.*, 94, 219–243, 2004.

55. Hostettmann, K., and Marston, A., *Saponins*, Cambridge University Press, Cambridge, 1995.

56. Mahato, S.B., Ganguly, A.N., and Sahu, N.P., Steroid saponins. *Phytochemistry*, 21, 959–978, 1982.

57. Patel, A.V., Blunden, G., Crabb, T.A., Sauvaire, Y., and Baccou, Y.C., A review of naturally-occurring steroidal sapogenins. *Fitoterapia*, 58, 67–107, 1987.

58. Agrawal, P.K., Jain, D.C., Gupta, R.K., and Thakur, R.S., Carbon-13 NMR spectroscopy of steroidal sapogenins and saponins. *Phytochemistry*, 24, 2479–2496, 1985.

59. Jizba, J., Dolejš, L., Herout, V., and Šorm, F., Polypodosaponin, ein neuer Saponintyp aus *Polypodium vulgare*. *Chem. Ber.*, 104, 837–846, 1971.

60. Jizba, J., Dolejš, L., Herout, V., and Šorm, F., The structure of osladin—the sweet principle of the rhizomes of *Polypodium vulgare*. *Tetrahedron Lett.*, 18, 1329–1332, 1971.

61. Nishizawa, M. and Yamada, H., Intensely sweet saponin osladin: synthetic and structural study. In G.R. Waller and K. Yamasaki (eds), *Saponins Used in Food and Agriculture*, Plenum Press, New York, NY, pp. 25–36, 1996.

62. Kim, J., Pezzuto, J.M., Soejarto, D.D., Lang, F.A., and Kinghorn, A.D., Polypodoside A, an intensely sweet constituent of the rhizomes of *Polypodium glycyrrhiza*. *J. Nat. Prod.*, 51, 1166–1172, 1988.

63. Kim, J. and Kinghorn, A.D., Further steroid and flavonoid constituents of the sweet plant *Polypodium glycyrrhiza*. *Phytochemistry*, 28, 1225–1228, 1989.

64. Waller, G.R. and Yamasaki, K. (eds), *Saponins used in Food and Agriculture*, Plenum Press, New York, NY, 1996.

65. Waller, G.R. and Yamasaki, K. (eds), *Saponins used in Traditional and Modern Medicine*, Plenum Press, New York, NY, 1996.

66. Oleszek W. and Marston A. (eds), *Saponins in Food, Feedstuffs and Medicinal Plants*, Kluwer Academic Publishers, Dordrecht, 2000.

67. Lacaille-Dubois, M.A., Biologically and pharmacologically active saponins from plants: recent advances. In W. Oleszek and A. Marston (eds), *Saponins in Food, Feedstuffs and Medicinal Plants*, Kluwer Academic Publishers, Dordrecht, pp. 205–218, 2000.

68. Harmatha, J., Chemo-ecological role of spirostanol saponins in the interaction between plants and insects. In W. Oleszek and A. Marston (eds), *Saponins in Food, Feedstuffs and Medicinal Plants*, Kluwer Academic Publishers, Dordrecht, pp. 129–141, 2000.

69. Harmatha, J., Mauchamp, B., Arnault, C., and Sláma, K., Identification of a spirostane-type saponin in the flowers of leek with inhibitory effects on growth of leek-moth larvae. *Biochem. Syst. Ecol.*, 15, 113–116, 1987.

70. Heftmann, E., Ko, S.T., and Bennett, R.D., Response of steroids to sulfuric acid in thin-layer chromatography. *J. Chromatogr.*, 21, 490–494, 1966.

71. Hiai, S., Oura, H., and Nakajima, T., Colour reaction of some sapogenins and saponins with vanillin and sulfuric acid. *Planta Med.*, 29, 116–122, 1976.

72. Dawidar, A.M. and Fayez, M.B.E., Thin-layer chromatographic detection and estimation of steroid sapogenins in fenugreek—steroid sapogenins. *Fresenius J. Anal. Chem.*, 259, 283–285, 1972.

73. Takeda, K., Hara, S., Wada, A., and Matsumoto, N., A systematic simultaneous analysis of steroid sapogenins by thin-layer chromatography. *J. Chromatogr.*, 11, 562–564, 1963.

74. Oleszek, W.A., Chromatographic determination of plant saponins. *J. Chromatogr. A*, 967, 147–162, 2002.

75. Oleszek, W. and Bialy, Z., Chromatographic determination of plant saponins—an update (2002–2005). *J. Chromatogr. A*, 1112, 78–91, 2006.

76. Chai, X.-Y., Li, S.-L., and Li, P., Quality evaluation of *Flos lonicerae* through a simultaneous determination of seven saponins by HPLC with ELSD, *J. Chromatogr. A*, 1070, 43–48, 2005.

77. Hegenauer, R., *Chemotaxonomie der Pflanzen, Vol. 3*, Birkhäuser Verlag, Basel, 1964.

78. Hesse, M., *Alkaloidchemie*, Thieme Verlag, Stuttgart, 1978.

79. Gross, D., Schütte, H.R., and Schreiber, K., Isoprenoid alkaloids. In K. Mothes, H.R. Schütte, and M. Luckner (eds), *Biochemistry of Alkaloids*, VEB Deutscher Verlag der Wissenschaften, Berlin, pp. 354–384, 1985.

80. Heftmann, E., Recent progress in biochemistry of plant steroids other than sterols (saponins, glycoalkaloids, pregnane derivatives, cardiac-glycosides, and sex-hormones). *Lipids*, 9, 626–639, 1974.

81. Li, H.-J., Jiang, Y., and Li, P., Chemistry, bioactivity, and geographical diversity of steroidal alkaloids from the *Liliaceae* family. *Nat. Prod. Rep.*, 23, 735–752, 2006.

82. Schreiber, K., Steroidal alkaloids: the *Solanum* group. In R.H.F. Manske (ed.), *The Alkaloids: Chemistry and Physiology, Vol. X*, Academic Press, New York, NY, p. 1, 1968.

83. Friedman, M., Analysis of biologically active compounds in potatoes (*Solanum tuberosum*), tomatoes (*Lycopersicon esculentum*), and jimson weed (*Datura stramonium*) seeds. *J. Chromatogr. A*, 1054, 143–155, 2004.

84. Marston, A. and Hostettmann, K., Modern separation methods. *Nat. Prod. Rep.*, 391–413, 1991.

85. Wagner, H., Seegert, K., Sonnenbichler, H., Ilyas, M., and Odenthal, K.P., Steroidal alkaloids of *Fontumia africana*. *Planta Med.*, 53, 444–449, 1987.

86. Lábler, L. and Černý, V., Thin-layer chromatography of some steroidal bases and *Holarrhena* alkaloids. *Collect. Czech. Chem. Commun.*, 28, 2932–2940, 1963.

87. Weissenberg, M., Isolation of solasodine and other steroidal alkaloids and sapogenins by direct hydrolysis–extraction of *Solanum* plants or glycosides therefrom. *Phytochemistry*, 58, 501–508, 2001.

88. Quidde, T., Osbourn, A.E., and Tudzynski, P., Detoxification of α-tomatine by *Botrytis cinerea*, *Physiol. Mol. Plant Pathol.*, 52, 151–165, 1998.

89. Bodart, P., Kabengera, C., Noirfalise, A., Hubert, P., and Angenot, L., Determination of alpha-chaconine in potatoes by high-performance thin-layer chromatography/densitometry. *J. AOAC Int.*, 83, 1468–1473, 2000.

90. Tanaka, H., Putalun, W., Tsuzaki, C., and Shoyama, Y., A simple determination of steroidal alkaloid glycosides by thin-layer chromatography immunostaining using monoclonal antibody against solamargine. *FEBS Lett.*, 404, 279–282, 1997.

91. Armstrong, D.W., Faulkner, J.R., and Han, S.M., Use of hydroxypropyl- and hydroxyethyl-derivatized beta-cyclodextrins for the thin-layer chromatographic separation of enantiomers and diastereomers. *J. Chromatogr.*, 452, 323–330, 1988.

92. Kubo, I. and Fukuhara, K., Steroidal glycoalkaloids in Andean potatoes. In G.R. Waller and K. Yamasaki (eds), *Saponins Used in Food and Agriculture*, Plenum Press, New York and London, pp. 405–417, 1996.

93. Janeczko, A. and Skoczowski, A., Mammalian sex hormones in plants. *Folia Histochem. Cytobiol.*, 43, 71–79, 2005.

94. Buchala, A.J. and Pythoud, F., Vitamin D and related compounds as plant growth substances. *Physiol. Plant.*, 74, 391–396, 1988.

95. Agrawal, H., Kaul, N., Paradkar, A.R., and Mahadik, K.R., HPTLC method for guggulsterone. I. Quantitative determination of *E*- and *Z*-guggulsterone in herbal extract and pharmaceutical dosage form. *J. Pharm. Biomed. Anal.*, 36, 33–41, 2004.

96. Anjaneyulu, A.S.R., Rao, D.S., and Lequesne, P.W., Withanolides, biologically active natural steroidal lactones: a review. In A.U. Rahman (eds), *Studies in Natural Product Chemistry, Vol. 20*, Elsevier Science B.V., pp. 135–261, 1998.

97. Ray, A.B. and Gupta, M., Withastereoids, a growing group of naturally occurring steroidal lactones. *Prog. Chem. Org. Nat. Prod.*, 63, 1–106, 1994.

98. Tettamanzi, M.C., Veleiro, A.S., de la Fuente, J.R., and Burton, G., Withanolides from *Salpichroa origanifolia*. *J. Nat. Prod.*, 64, 783–786, 2001.

99. Minguzzi, S., Barata, L.E.S., Shin, Y.G., Jonas, P.F., Chai, H.-B., Park, E.J., Pezzuto, J.M., and Cordell, G.A., Cytotoxic withanolides from *Acnistus arborescens*. *Phytochemistry*, 59, 635–641, 2002.

100. Matsuda, H., Murakami, T., Kishi, A., and Yoshikawa, M., Structures of withanosides I, II, III, IV, V, VI, and VII, new withanolides glycosides, from the roots of Indian *Withania somnifera* Dunal. and inhibitory activity for tachyphylaxis to clonidine in isolated guinea-pig ileum. *Bioorg. Med. Chem.*, 9, 1499–1507, 2001.

101. Nicotra, V.E., Gil, R.R., Vaccarini, C., Oberti, J.C., and Burton, G., 15,21-Cyclowithanolides from *Jaborosa bergii*. *J. Nat. Prod.*, 66, 1471–1475, 2003.

102. Maldonado, E., Torres, F.R., Martinez, M., and Pérez-Castorena, A.L., 118-Acetoxywithanolides from *Physalis chenopodifolia*. *Planta Med.*, 70, 59–64, 2004.

103. Abou-Douh, A.M., New withanolides and other constituents from the fruit of *Withania somnifera*. *Arch. Pharm.*, 6, 267–276, 2002.

104. Atta-ur-Rahman, Dur-e-Shahwar, Naz, A., and Choudhary, I., Withanolides from *Withania coagulans*. *Phytochemistry*, 63, 387–390, 2003.

105. Lavie, D. and Glotter, E., The cucurbitanes, a group of tetracyclic triterpenes. *Fortschr. Chem. Org. Naturst.*, 29, 307–362, 1971.

106. Miró, M., Cucurbitacins and their pharmacological effects. *Phytother. Res.*, 9, 159–168, 1995.

107. Guha, J. and Sen, S.P., The cucurbitacins—a review. *Plant Biochem. J.*, 2, 12–28, 1975.

108. Hussein, H.A., Abdel-Halim, O.B., Marwanm, E-S.M., El-Gamal, A.A., and Mosana, R., Dendrocyin: an isocucurbitacin with novel cyclic side chain from *Dendrosicyos socotrana*. *Phytochemistry*, 65, 2551–2556, 2004.

109. Stuppner, H. and Wagner, H., HPLC- und DC-Untersuchungen von Iridoid-, Cucurbitacin- und Phenol-Glykosiden aus *Picrorhiza kurroa*. *Sci. Pharm.*, 60, 73–85, 1992.

110. Anon, Spray reagents. https://www.macherey-nagel.de/web%5CMN-WEB-DCKatalog.ns (accessed on 09 November 2006).

111. Anon, TLC visualisation reagents. http://www.emdmicro.com/analytics/literature/TLC_ Visualization_Reagents.pdf (accessed on 24 November 2006).
112. Hauck, H.E., Detection in thin-layer chromatography, 2005, http://www.clubdeccm. com/img/10%20novembre%202005/detection%20in%20tlc.pdf (accessed on 10 February 2007).
113. Merck, E., Dyeing reagents for thin-layer and paper chromatography. http://www. clubdeccm.com/img/Dyeing%20Reagents%20for%20TLC.doc (accessed 24 November 2006).
114. Dinan, L. and Hormann, R.E., Ecdysteroid agonists and antagonists. In L.I. Gilbert, K. Iatrou, and S.S. Gill (eds), *Comprehensive Insect Molecular Science, Vol. 3: Endocrinology*, Elsevier Pergamon, Oxford, pp. 197–242, 2005.
115. Clément, C.Y., Bradbrook, D.A., Lafont, R., and Dinan, L., Assessment of a microplate-based bioassay for the detection of ecdysteroi-like orantiecdysteroid activities. *Insect Biochem. Mol. Biol.*, 23, 187–193, 1993.
116. Poole, C.F., Thin-layer chromatography: challenges and opportunities. *J. Chromatogr. A*, 1000, 963–984, 2003.
117. Nyiredy, S., Progress in forced-flow chromatography. *J. Chromatogr. A*, 1000, 985–999, 2003.
118. Pascual, M.E., Carretero, M.E., Slowing, K.V., and Villar, A., Simplified screening by TLC of plant drugs. *Pharm. Biol.*, 40, 139–143, 2002.

23 TLC of Iridoids

Grażyna Zgórka

CONTENTS

23.1 INTRODUCTION

Based on biosynthetic origin, the classical term: iridoids refers to natural monoterpenoid compounds, being secondary plant metabolites with characteristic cyclopenta[c] pyranoid skeleton (Figure 23.1), also described as iridane (cis-2-oxabicyclo[4.3.0]-nonane). In practice, the term is used to cover the broader group of natural products which are considered to arise biogenetically from iridane. Ethymology of the name iridoids indicates that it derives from the volatile monoterpenes, called: irydodial and iridomyrmecin, being the components of the defensive secretion of Australian ants from *Iridomyrmex* genus [1–3]. The older name of iridoids was pseudoindicans as they were known to react with acids giving blue-colored derivatives [4].

23.2 STRUCTURAL DIVERSITY (IRIDOID AND SECOIRIDOID GLYCOSIDES, NONGLYCOSIDIC COMPOUNDS) AND DISTRIBUTION OF IRIDOIDS

Several authors presented the classification of iridoids, based on their biosynthesis and chemical properties. In 1980, El-Naggar and Beal [5] published a list of 258 iridoids (together with their physical data), comprising iridoid and secoiridoid glycosides and some nonglycosidic compounds, however, excluding nitrogen-containing iridoids.

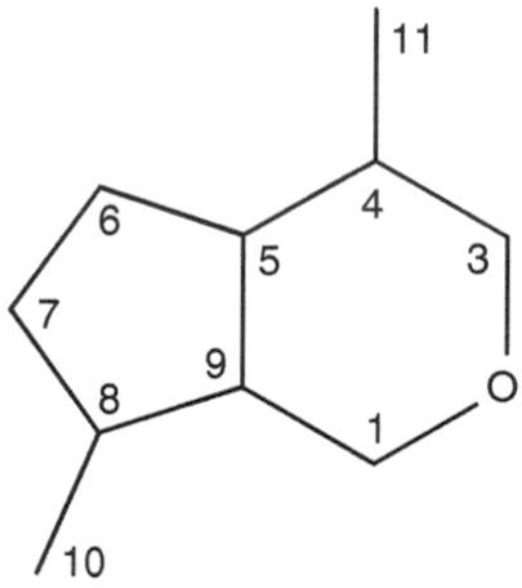

FIGURE 23.1 Chemical structure of cyclopenta[c]pyrane (iridane).

Six years later, Hegnauer [6] elaborated much more broader classification of above 1000 iridoids, dividing these compounds into nine structural groups, among others: cyclopentanoid monoterpenes and secoiridoids, possessing characteristic open 7,8-seco ring (Figure 23.2), as well as pseudoalkaloids, indole, and isochinoline alkaloids, derived from the biosynthetic pathway of a secoiridoid—secologanin. The updated review, published in 1990 by Boros and Stermitz [7], arranged the iridoid structures (346 new compounds) in the similar way that it was earlier presented in the El-Naggar and Beal survey [5]. Five groups of classical iridoids were distinguished. Groups 1–3 comprised iridoids with eight- to ten-carbon skeletons, respectively; Group 4 consisted of iridoid aglycones and Group 5 contained dimeric forms of iridoids, for example, bis-iridoids and bis-iridoid aglycones. Within each group, the structural differentiation of particular compounds is the result of modifications (epoxidation, hydroxylation, and esterification of hydroxyl groups with aromatic acids) of the basic iridoid skeleton. In 1991, Boros and Stermitz [8] published the second part of their review, comprising the next seven groups of iridoid compounds, i.e., valeriana iridoids (Group 6), plumeria-type iridoids (Group 7), miscellaneous iridoid-like compounds (Group 8), simple secoiridoids (Group 9), terpene-conjugated iridoids (Group 10), phenolic-conjugated iridoids (Group 11), bis-secoiridoids (Group 12), and miscellaneous secoiridoids (Group 13).

As currently known, iridoids have been considered as chemotaxonomic markers of numerous botanical families, for example, Bignoniaceae, Cornaceae, Dipsacaceae, Ericaceae, Gentianaceae, Gronoviaceae, Lamiaceae, Loasaceae, Loganiaceae,

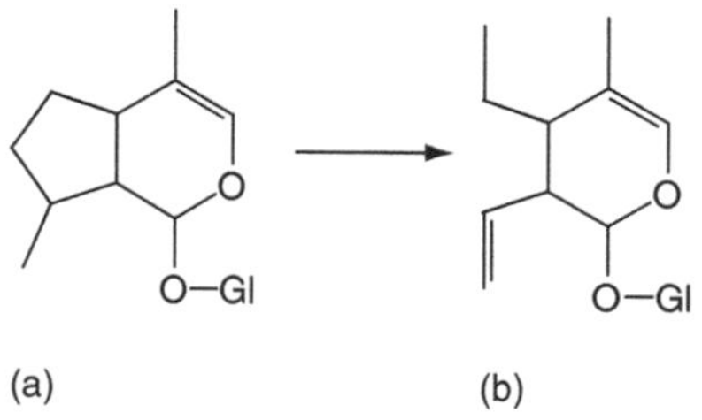

(a)   (b)

FIGURE 23.2 Conversion of iridoid (a) to secoiridoid (b) skeleton.

Oleaceae, Pedaliaceae, Plantaginaceae, Rubiaceae, Scrophulariaceae, Verbenaceae, representing the broader systematic class of Dicotyledones [9–11]. They are often found in a large number of plants in the form of glycosides, much more rarely as nonglycosidic compounds [5]. Aucubin and catalpol belong to the most widespread iridoid glycosides. They occur in abundance in several plant genera (e.g., *Plantago*, *Verbascum*, *Euphrasia,* and *Veronica* spp.), influencing therapeutic properties of individual species [3,12,13]. Well-known herbal medicinal substances obtained from the plants representing the botanical families: Gentianaceae (*Gentiana lutea* L., *Erythraea centaurium* Pers.) and Menyanthaceae (*Menyanthes trifoliata* L.) constitute a rich source of bitter secoiridoids [14]. In the group of therapeutically active iridoid esters, the most important role play valepotriates—characteristic sedative constituents identified in several species from the Valerianaceae family [15–17].

The interesting detail is that some animals, for example, butterflies belonging to the genus *Melitea*, use iridoids in their chemical defence. Their larval forms eat up plants containing bitter tasting iridoid glycosides and gather them in their bodies. Thanks to it, larvae become inattractive to birds—their main predators [18].

23.3 PHYSICAL AND CHEMICAL PROPERTIES

Iridoids are liquid or solid compounds, mostly stable at normal temperature, crystalline or amorphous, possessing high melting points. Polar iridoid glycosides are well soluble in water and alcohols (MeOH, EtOH, *n*-BuOH), whereas solubility of relative aglycones in this media is slightly poorer and increases with the number of hydroxyl groups. Secoiridoids, ester iridoids, and bis-iridoids are usually well dissolved in EtOH or more hydrophobic organic solvents such as EtOAc or CHCl$_3$ [7,8]. Most of iridoids (especially secoiridoids from the Gentianaceae family) reveal bitter taste (*remedia amara*), hence, orally administered, they are traditionally used to stimulate salivary and gastric secretions in the lack of appetite and disturbances in overall metabolism [3].

Iridoids are known to be very sensitive to treatment with acidic reagents, which hydrolyze glycosidic bonds and decompose the structure of aglycones (the disintegration of a pyranoid ring) giving blue-colored products [1,19].

23.4 ISOLATION FROM PLANT MATERIAL COMBINED WITH PRELIMINARY PURIFICATION AND FRACTIONATION OF EXTRACTS

The isolation of iridoids from plant material is still a difficult and complex task because of the instability of these compounds, especially under acidic conditions or at higher temperatures, leading to the degradation of their chemical structures. Hence, even the conditions of sample drying (temperatures not exceeding 40°C) should be rigorously kept [14].

Preliminary solid–liquid extraction techniques used in the isolation of polar iridoid glycosides from plant material are based on a variety of cheap and easily available solvents, such as water, aqueous alcoholic solutions, or alcohols (MeOH,

EtOH) to guarantee the efficient isolation of these constituents from sample matrices [20–25]. An interesting extraction technique was proposed by Iavarone et al. [26] for polar iridoid glucosides. Instead of EtOH, a concentrated solution (25%) of NaCl was used as the solvent, giving higher yields and more rapidness of the extraction process than EtOH as well as the lack of apolar ballasts (chlorophyll) in the final extract. The elaborated method needed only the absorption of the saline solution on charcoal and successive elution of salts with water and iridoid glucosides with aqueous ethanolic mixtures with an increasing content of alcohol.

During the last decade the development of new methods of extraction of solid samples have been observed. For example, in the extraction process of two known iridoid glucosides, catalpol and aucubin from the leaves of *Veronica longifolia* (Scrophulariaceae) pressurized hot water extraction (PHWE) and hot water extraction (HWE) were the most efficient isolation techniques for both compounds in comparison with (also tested) pressurized liquid extraction (PLE) and maceration with various organic solvents [27]. For more effective and safe isolation of iridoid glucosides from aerial parts of *Stachys palustris* (Lamiaceae), ultrasound-assisted extraction (UAE) with MeOH, carried out at room temperature, was used [28]. A large-scale extraction of a secoiridoid—secologanin (from *Symphoricarpos albus*)—comprising UAE with organic solvents and water, microwave-assisted extraction (MAE) and HWE was elaborated. Among the methods tested, UAE with MeOH turned out to be the most efficient technique, i.e., the highest yield of this compound was obtained [29]. As regards to the liquid–liquid extraction of more hydrophobic iridoid compounds (e.g., iridoid esters, nonpolar iridoid aglycones, bis-iridoids, etc.), EtOAc or CH_2Cl_2 are occasionally the solvents of choice [30,31]. Sometimes, an alternative technique—supercritical fluid extraction (SFE)—can be used. The extraction of the roots of a well-known medicinal plant—*Harpagophytum procumbens*—was performed with CO_2 and 25% (w/w) EtOH as a modifier in the supercritical state. Under these conditions the content of the main iridoid glycoside—harpagoside—could be increased to 20%–30% in relation to classical extraction techniques [32].

The important matter, connected with the extraction of solid samples, carried out by means of water or alcohol–water mixtures, is the necessity of precleaning the crude extracts containing polar iridoids from more hydrophobic, ballast compounds using clarifying agents. It is realized by passing extract samples through adsorbent beds of activated charcoal or celite, usually placed in Gooch funnels, and obtaining the irreversible surface adsorption of impurities [22,24,33–35]. Sometimes, this method additionally enables the elution of mono- or oligosaccharides from purified samples using water or 5%–10% solutions of alcohol in water [20,36,37]. Sample cleanup of concentrated alcoholic extracts (rich in iridoid glycosides) from ballast hydrophobic constituents (chlorophyll, triglycerides, waxes, etc.) is also performed in separatory funnels using classical solvent extraction of dry extract residues, earlier redissolved in water, with nonpolar solvents, for example, *n*-hexane, petroleum ether, CH_2Cl_2, or $CHCl_3$ [22,33]. Similar results are often obtained by replacing, earlier mentioned, time-consuming and tedious procedures by the convenient preliminary extraction of plant material in a Soxhlet apparatus using the same hydrophobic solvents [38–40].

Another problem in purification of aqueous, aqueous alcoholic, and alcoholic extracts is the presence of polar phenolics, together with iridoids, which can interfere further separation of the latter compounds. Hence, multistage liquid–liquid extraction techniques, using the range of solvents with increasing polarity (C_6H_{14}, CH_2Cl_2, $CHCl_3$, Et_2O, EtOAc, n-BuOH) are often performed to obtain fractions of iridoids purified from phenolics and concentrated in adequate organic layer, depending on the iridoid affinity for particular solvent medium [41–43]. Additional solution of this problem is the cleanup of aqueous samples, containing both iridoids and phenolics on neutral alumina, i.e., adsorbent showing strong binding ability in relation to the latter compounds [28,44,45]. However, alumina is not suitable in the purification of iridoids possessing aromatic substituents due to the earlier mentioned, irreversible retention of phenolics on the beds of this adsorbent.

The increasing efficiency of solvent extraction technique in the process of preliminary purification and fractionation of iridoid compounds was observed when a droplet counter-current chromatography (DCCC) or current partition chromatography (CPC) were introduced. Both techniques guarantee high recoveries of isolated compounds from crude matrices because of the lack of irreversible adsorption, typical for chromatographic preparative techniques, however, CPC is much more effective because of the application of centrifugal force in the separation process.

Aqueous methanolic fractions from aerial parts of *Gallium album* ssp. *album*, *G. album* ssp. *pychnotrichum*, and *G. lovcense* (Rubiaceae), after separation by ascending DCCC with $CHCl_3$–MeOH–H_2O (43:37:20) combined with column chromatography over silica gel and RP-18 columns, yielded 11 iridoid compounds [46]. DCCC, using the ascending method with $CHCl_3$–MeOH–H_2O–n-PrOH (9:12:8:2) as a mobile phase, was performed by Japanese scientists to separate 10 acylated derivatives of catalpol and asystatioside E from the n-BuOH-soluble fraction of the MeOH extract of *Premna subscandens* (Verbenaceae) leaves [47]. Main iridoid compounds of *Menyanthes trifoliata* (Menyanthaceae)—dihydrofoliamenthin and loganin—were isolated by DCCC using separating system $CHCl_3$–MeOH–H_2O (5:6:4) [48].

Iridoid glycosides from the EtOAc extract, obtained from the roots of *Stachytarpheta cayanensis*, were isolated by step gradient CPC using the sequential increase of n-BuOH in the organic phase of the solvent system and final wash-off of stationary phase [49]. Fractionation of the methanolic extract by CPC ($CHCl_3$–MeOH–H_2O, 9:12:8—using the lower phase as mobile phase) together with gel filtration on Sephadex and chromatography on silica gel afforded three iridoids and two secoiridoids from the roots of *Gentiana linearis* (Gentianaceae) [50].

Traditional and novel partition extraction techniques, described earlier, have been almost always combined with chromatographic separations, for example, open column chromatography (CC), carried out using various classes of adsorbents. During the last two decades, the most popular stationary phases in CC of iridoid compounds comprised: silica gel, polymeric resins (XAD), or reversed-phases (RP-18, RP-8), especially widespread in the form of flash (FC) or vacuum liquid chromatography (VLC) [21,31,40,45,51–54]. Iridoid glucosides and ballast phenolics contained in crude alcoholic extracts have been effectively separated from each other using polyamide [13,53–55] or Sephadex LH-20 [33,34,54,56] columns as well as aqueous alcoholic eluents with decreasing polarity. Apart from traditional CC, other preparative chromatographic methods, including low pressure liquid chromatography (LPLC), medium pressure liquid chromatography (MPLC), and high-performance liquid chromatography (HPLC) have also been used. In these techniques, the application of RP-8 and RP-18 columns in combination with MeOH–H$_2$O mixtures in isocratic and gradient mode has enabled the simplest and most effective separation of polar and medium polar iridoid compounds [21,25,34,55,57,58]. Considering the group of hydrophobic iridoids, the same chromatographic techniques have been performed using silica gel columns and nonpolar eluting solvents [24,59].

23.5 THIN LAYER CHROMATOGRAPHY

23.5.1 Short Overview of Adsorbent and Solvent Systems Used in TLC Qualitative and Quantitative Analysis of Iridoids

As regards to iridoids, analytical TLC has been predominantly used for the control of iridoid fractions obtained from the preparative column chromatography (CC, LPLC, MPLC, and HPLC) as well as for preliminary investigations referring to the proper choice of a mobile phase for further preparative separations using chromatographic and partition (DCCC, CPC) techniques. TLC is also a method of choice in rapid qualitative analysis of iridoids, giving the possibility of fingerprinting of herbal substances used for medical purposes. It has also been employed in chemotaxonomical studies. During last three decades, a series of TLC methods concerning iridoid compounds have been described in the literature—the detailed overview is presented in Table 23.1. It is easy to observe, that silica gel is the most efficient adsorbent in TLC separation of iridoid compounds, only occasionally octadecylsilane (C$_{18}$)-precoated plates have been used. The most popular solvent systems comprise mixtures of CHCl$_3$ or CH$_2$Cl$_2$ with MeOH and H$_2$O as polar modifiers. Significant improvement in the separation of complex samples containing iridoids by the application of gradient development has been obtained [44].

23.5.2 Detection Methods of Various Classes of Iridoid Compounds

Some iridoids, absorbing in the short UV (254 nm), give fluorescence-quenching zones on silica gel plates containing a fluorescent indicator. However, because of the lack of typical chromophores in the majority of these compounds, hindering their direct UV detection, a lot of reagents (Table 23.2) used in postchromatographic derivatization, has been applied. Most of diagnostic sprays contain acidic components causing the degradation of iridoid aglycones, followed by the generation of blue, violet-blue, or grey zones on TLC chromatograms in visible or UV (365 nm) light. As it has been shown in Table 23.2, vanillin-sulfuric acid reagent belongs to the most popular destructive derivatizing agents. Some special diagnostic sprays for the detection of valeriana ester iridoids (valepotriates) have also been elaborated and used. Additionally, as described below, chemical derivatization has been employed in densitometry giving noticeable improvement in qualitative and quantitative analysis

TABLE 23.1
Chromatographic (TLC) Systems Used in Analytical Separation of Iridoids

Iridoids	Adsorbent	Solvent System	References
Iridoid glycosides in *Clerodendrum* sp.	Silica gel 60	CH_2Cl_2–MeOH–H_2O (90:10:1) and (70:30:3) n-BuOH–MeOH–H_2O (4:1:5) or (9:1:10) Toluene–Me_2CO (8:2) or (7:3)	[51]
Mollugoside—an iridoid glucoside from *Galium mollugo*	Silica gel	CH_2Cl_2–MeOH–H_2O–AcOH (7:3:0.2:0.2)	[26]
Iridoids of *Plantago* sp.	Silica gel 60 F_{254}	n-BuOH–H_2O (9:1) n-BuOH–MeOH–H_2O (70:5:20) EtOAc–MeOH–H_2O (100:16.5:13.5)	[12]
Iridoids of *Menyanthes trifoliata*	Silica gel 60 F_{254}	EtOAc–MeOH–H_2O (77:15:8) EtOAc–EtOH–H_2O (87:13:2)	[48]
Iridoid glycosides of *Picrorhiza kurrooa*	Silica gel 60 F_{254}	$CHCl_3$–MeOH (9:1) or (8:2)	[38]
Scorodioside—an iridoid diglycoside from *Scrophularia scorodonia*	Silica gel 60 GF_{254}	$CHCl_3$–MeOH–H_2O (80:20:1)	[55]
Iridoids of *Verbenoxylum reitzii*	Silica gel	$CHCl_3$–MeOH (8:2)	[60]
Iridoid glycosides of *Holmskioldia sanguinea* and *Gmelina philippensis*	Silica gel 60 F_{254} RP-C_{18}	CH_2Cl_2–MeOH–H_2O (70:30:3) MeCN–MeOH–H_2O (25:7.7:67.3)	[61,62]
Aucubin, harpagide, and harpagoside from *Harpagophytum procumbens* and *H. zeyheri*	Silica gel	EtOAc–MeOH (4:1)	[63]
Iridoids of Gronoviaceae and Loasaceae	Silica gel	EtOH–MeOH–H_2O (77:15:8) PrOH–toluene–AcOH–H_2O (25:20:10:10)	[10]
Iridoids of *Rothmannia macrophylla*	Silica gel 60 F_{254}	$CHCl_3$–MeOH–H_2O (9:1:0.1) or (8:2:0.1) or (7:3:0.5)	[42]
Iridoids of *Morinda morindoides*	Silica gel 60 F_{254}	EtOAc–HCOOH–AcOH–H_2O (30:0.8:1.2:8)—upper layer	[39]
Iridoids of *Leonurus cardiaca*	Silufol UV (254 nm)	$CHCl_3$–MeOH–H_2O (80:20:0.1)	[64]
Iridoids of *Vitex agnus castus*	Silica gel 60 F_{254}	EtOAc–MeOH–H_2O (77:15:8)	[65]
Iridoid glucosides of *Stachys palustris*	Silica gel 60 GF_{254}	$CHCl_3$–MeOH–H_2O (25:10:1) $CHCl_3$–MeOH–H_2O (32:11:1.6) EtOAc–HCOOH (7:4)	[28]

(*continued*)

TABLE 23.1 (continued)
Chromatographic (TLC) Systems Used in Analytical Separation of Iridoids

Iridoids	Adsorbent	Solvent System	References
Loganic acid in gentian root	Silica gel 60 F_{254}	EtOAc–MeOH–H_2O (77:15:8)	[14]
Penstemide and serrulatoloside from tissue culture of *Penstemon serrulatus*	Silica gel	Stepwise gradient using three programs (a–c) with changing concentration of iPrOH in EtOAc: (a) 10%,10%,20%,20%,30%,40%, 50%,60%,70%,80%,80% (b) 10%,10%,20%,30%,40%, 50%,60%,70%,80% (c) 10%,10%,20%,20%,30%,30%,40%, 40%,50%	[44]
Secoiridoid glucosides from *Gentiana rhodantha*	Silica gel 60 F_{254} RP-C_{18}	$CHCl_3$–MeOH–H_2O (14:6:1) MeOH–H_2O (13:7)	[33]
Secoiridoids of *Gentiana lutea*	Silica gel 60 F_{254}	$CHCl_3$–EtOH (75:25)	[66]
Valepotriates (iridoids of *Valeriana* sp.)	Silica gel HF_{254} Silica gel 60 F_{254} (HPTLC) RP-C_{18} (HPTLC) Silica gel $HF_{254/366}$ Silica gel F60	*n*-Hexane–MEK (4:1) Benzene–EtOAc (85:15) Toluene–MEK (4:1) Toluene–*n*-hexane–MEK (35:45:20) CH_2Cl_2–EtOAc–Me_2CO (48:1:1) CH_2Cl_2–MEK (98:2) MeOH–H_2O (8:2)	[15]
	Silica gel 60 F_{254} Silica gel 60 F_{254} (HPTLC)	*n*-Hexane–MEK (4:1) Benzene–EtOAc (83:17) CH_2Cl_2–MEK (98:2)	[16]
	RP-C_{18} (HPTLC) Silica gel F_{254}	Benzene–EtOAc (83:17) CH_2Cl_2–EtOAc–Me_2CO (48:1:1) CH_2Cl_2–MEK (98:2) MeOH–H_2O (8:2)	[17] [67]
		Toluene–EtOAc–MEK (80:15:5)	[30]

MEK, methyl ethyl ketone.

of iridoid compounds. For example, densitometric method for the identification and quantification of a known iridoid glucoside—aucubin—in some syrups, containing various plant extracts, was used. Chromatographic separation was performed on silica gel 60 F_{254} HPTLC plates with 1,4-dioxane–xylene–isopropanol–25% ammonia (1:3:5:1) as a mobile phase. Chromatograms were visualized by spraying with a 2:1 mixture of 1% ethanolic 4-dimethylbenzaldehyde and 36% hydrochloric acid, then heating at 100°C for 2 min. Iridoid spots were scanned densitometrically in visible light at $\lambda = 580$ nm using a TLC scanner 3, with CATS 4 software (Camag, Switzerland) [69]. Háznagy-Radnai et al. [28] elaborated rapid and simple densitometric method of quantitative analysis of aucubin, other aucubin-like iridoids

TABLE 23.2
General Spray Reagents Used for the Detection of Various Classes of Iridoid Compounds on Silica Gel Thin Layers

Reagent	References
1% Vanillin/1%–5% H_2SO_4 (EtOH or MeOH solutions)/heating (110°C, 5–10 min)	[10,14,26,34,38,39, 48,51,55,61,62]
Vanillin/conc. HCl in MeOH	[20,36,37]
Vanillin/1M H_2SO_4/conc. HCl in EtOH/heating (110°C, 5 min)	[60]
Anisaldehyde/AcOH/MeOH/H_2SO_4/heating (110°C, 5–10 min)	[63]
1M H_2SO_4	[20,26,36]
10% H_2SO_4 in EtOH/heating (110°C, 3 min)	[12,42]
Resorcin/conc. H_2SO_4 in EtOH	[36]
1% $Ce(SO_4)_2$/10% H_2SO_4/heating	[22,68]
2% phloroglucinol in EtOH/conc. HCl/heating (80°C, 3 min)	[44]
Godin's reagent (1% vanillin in EtOH and 3% aq. $HClO_4$; 1:1, v/v)/heating (80°C, 10 min)	[12,50]
Stahl's reagent (4-dimethylaminobenzaldehyde/AcOH/H_3PO_4/H_2O)	[64]
1% 4-dimethylaminobenzaldehyde in 1M methanolic HCl	[65]
0.2% Fast Red Salt B (diazotized 5-nitro-2-aminoanisole) in H_2O/20% K_2CO_3 in H_2O[a]	[66]
Benzidine in EtOH/conc. HCl/AcOH or CCl_3COOH[b]	[15]
AcOH/HCl (8 + 2)[b] or Hydroxylamine/HCl/NaOH/EtOH[b]	[16]
NH_3 vapor[b]	[67]
DNPH (2,4-dinitrophenylhydrazine) in MeOH/AcOH/HCl[b]	[17]
DNPH in methanol/AcOH + 25% HCl (1 + 1)/heating (110°C; 10 min)[b] or NBP [4-(4-nitrobenzyl)-pyridine] in acetone/heating (110°C, 10 min)[b]	[30]

[a] Spray reagent for gentian secoiridoids (amarogentin).
[b] Diagnostic reagent for valepotriates (valeriana iridoids).

(harpagide, acetylharpagide, ajugoside) in aqueous extracts obtained from the aerial parts of *Stachys palustris* (Lamiaceae). Samples of purified extracts were applied onto silica gel 60 GF_{254} plates and developed over a distance of 10 cm using a mobile phase $CHCl_3$–MeOH–H_2O (25:10:1). Aucubin spots were visualized using Erlich's reagent (a solution of 1% 4-dimethylaminobenzaldehyde in 36% HCl containing acetic anhydride), heated at 105°C for 5 min and then scanned in a Shimadzu CS-9301PC densitometer at $\lambda = 540$ nm (Figure 23.3). The same TLC-densitometric method was performed to evaluate the iridoid composition and content in various organs of 10 *Stachys* species native to or cultivated in Hungary. Aucubin-like iridoids (harpagide, acetylharpagide, ajugoside) were identified and determined quantitatively [70]. The most recent densitometric method used for the determination of aucubin in the leaves of *Plantago palmata* (Plantaginaceae) was published in 2006 by Biringanine and coworkers [71]. Methanolic extracts of the plant material were applied on HPTLC silica plates with a TLC sampler III (Camag, Switzerland) and developed in EtOAc–H_2O–AcOH–HCOOH (67:18:7.5:7.5). The chromatograms

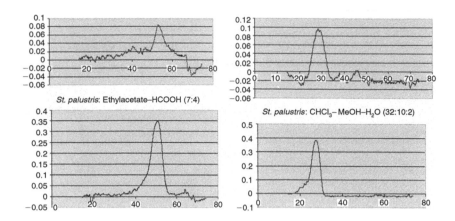

FIGURE 23.3 Densitograms obtained from *Stachys palustris* extracts (top) and from an authentic samples of aucubin as reference compound (bottom). (From Háznagy-Radnai, E., Léber, P., Tóth, E., Janicsák, G., and Máthé, I., *J. Planar Chromatogr.—Mod. TLC*, 18, 314, 2005. With permission.)

obtained were scanned at 330 nm in reflectance mode with a Camag TLC scanner III. Then they were sprayed with a vanillin sulfuric acid reagent, heated at 105°C for 20 min, stabilized at room temperature for 20 min and finally measured at 520 nm.

A TLC-densitometric method, without chemical derivatization, for the determination of therapeutically active secoiridoid—amarogentin in gentian root and tincture has been developed. Chromatographic separation was performed on precoated HPTLC silica gel F_{254} plates. The mixture of $CHCl_3$ and MeOH (4:1) was used as a mobile phase. The spots of amarogentin were scanned and determined densitometrically by measurement of fluorescence quenching at 254 nm on a Vitatron TLD 100 densitometer [72]. A validated HPTLC method was also elaborated to establish the content of other two secoiridoids: swertiamarin and amarogentin in samples of two Himalayan plants *Swertia chirata* and *S. cordata* (Gentianaceae). Chromatographic separation was performed on silica 60 F_{254} HPTLC plates developed with EtOAc–MeOH–H$_2$O (77:8:8), as a mobile phase. Quantitative evaluation of secoiridoids on TLC-chromatograms under UV light ($\lambda = 235$ nm) in reflectance/absorbance mode, using a Camag TLC scanner 3, was carried out (Figure 23.4). High average recoveries of amarogentin and swertiamarin (94.5% and 96.5%, respectively) were obtained [73]. A similar method, combining TLC on silica gel with densitometry (at 240 nm), for simultaneous quantitative determination of two secoiridoids: swertiamarin and gentiopicroside in *Centaurium erythrea* and *C. turcicum* (Gentianaceae) was used [74].

One of the latest densitometric techniques—photodiode-array-(PDA)-densitometry—for the quantification of harpagoside in *Harpagophytum procumbens* CO_2 extracts on Nucleosil 60 F_{254} HPTLC plates has been employed. As a mobile phase the solvent mixture: EtOAc–MeOH–H$_2$O (77:15:8, v/v/v) was used. Visualization of the spots was performed by immersing the plates in anisaldehyde reagent (anisaldehyde–glacial AcOH–MeOH–96% H_2SO_4, 0.5:10:85:5, v/v/v/v). Quantification of harpagoside was done at 509 nm by linear scanning in the reflectance mode

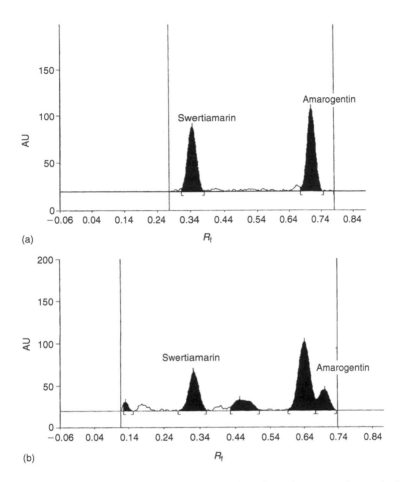

FIGURE 23.4 TLC densitograms showing (a) swertiamarin and amarogentin standards and (b) resolution of swertiamarin and amarogentin in a sample. (From Bhandari, P., Gupta, A.P., Singh, B., and Kaul, V.K., *J. Planar Chromatogr.—Mod. TLC*, 19, 212, 2006. With permission.)

using a PDA scanner. The HPTLC densitometry provided similar reproducibility, accuracy, and selectivity in the quantitative determination of harpagoside to HPLC method simultaneously used [32].

23.5.3 Preparative TLC Including Forced-Flow Planar Techniques

Preparative thin layer chromatography (PTLC), often combined with open column liquid chromatography, still belongs to the most exploited and simple chromatographic techniques involving mobile phase migration through a planar stationary phase by capillary forces. Thicker (0.5–2 mm) adsorbent beds and the bigger format (20 × 20 cm or 20 × 40 cm) of TLC plates have been generally used [75].

Based on literature, silica gel is the most common adsorbent, employed for PTLC of various classes of iridoid compounds. Iridoid fractions obtained from CC of n-BuOH extract from the leaves of *Vitex cymosa* (Verbenaceae) were further purified on silica gel preparative plates, eluted with CHCl$_3$–MeOH (9:1) affording 20 mg of a new iridoid compound—viteoid II [41]. Final purification of the EtOAc extract from epigeal parts of *Allamanda cathartica* performed on silica gel preparative plates with 10% Me$_2$CO in CHCl$_3$, as a mobile phase, gave two bioactive iridoids: plumericin and isoplumericin [31]. Several rare type iridoids, possessing two α,β-unsaturated acid units were isolated from the whole plant of *Tarenna attenuata* using PTLC on silica gel and two mobile phase systems: EtOAc–MeOH (50:1) and CHCl$_3$–MeOH (5:1) [58]. Silica gel CC and PTLC, performed on silica gel 60 PF$_{254}$ plates (CHCl$_3$–Me$_2$CO, 10:1) led to the isolation of six iridoid glycosides (tecomoside, cachineside I, campenoside, 5-hydroxycampenoside, campsiside, and 5-hydroxycampsiside) from the leaves of *Campsis chinensis* (Bignoniaceae) [22]. Damtoft and coworkers [57] separated iridoid and secoiridoid glucosides of *Picconia excelsa* (Oleaceae) by PTLC using 20×40 cm plates coated with 1 mm layers of silica gel PF$_{254}$ and double development in EtOAc–EtOH–toluene (4:1:1). The same two iridoid groups, contained in the leaves of *Syringa afghanica*, were isolated by PTLC on silica gel using CHCl$_3$–MeOH (7:3), Me$_2$CO–CHCl$_3$–H$_2$O (8:2:1), or n-BuOH–AcOH–H$_2$O (8:2:1) [35]. The separation of the crude methanolic extract from the aerial parts of *Villarsia exaltata* (Menyanthaceae), by DCCC and subsequent PTLC on silica plates in EtOAc–EtOH–H$_2$O (87:13:2) and CHCl$_3$–MeOH–H$_2$O (80:19:1), yielded two secoiridoids: exaltoside and 7-epiexaltoside [76]. The n-BuOH-soluble fraction from the dried branches of *Strychnos spinosa* (Loganiaceae) was separated by a combination of preparative HPLC and TLC on silica affording 26 compounds, among them three new secoiridoid glucosides: stryspinoside and strychosides A and B [77]. For the isolation of valepotriates (valeriana iridoids) PTLC on silica gel G using petrol ether–Me$_2$CO–EtOAc (100:8:8) or CH$_2$Cl$_2$–Me$_2$CO–EtOAc (50:1:1) was employed [30].

Because of higher costs, reversed-phase systems are much more seldom used in PTLC.

Ethanolic extracts of two *Linaria* species, subjected to CC on silica gel, were further separated preparatively on RP-8 silica gel F$_{254}$ plates in H$_2$O–MeOH (7:3) affording six iridoid compounds, including a new bis-iridoid—arcusangeloside [37]. Jaslanceoside E—a new secoiridoid glucoside from the aerial parts of *Jasminum lanceoloarium* (Oleaceae)—was purified using preparative RP-18 plates developed with H$_2$O–MeOH (1:1) [56].

The technique of unidimensional multiple development (UMD) was also employed in PTLC of iridoids. Dry methanolic extract, obtained from aboveground parts of *Castilleja minor*, was subjected to the separation process on silica gel plates (mobile phase: CHCl$_3$–MeOH, 65:35; double developed), yielding a mixture of iridoid glycosides: mussaenoside, 8-*epi*-loganic acid, and shanzhiside methyl ester [52]. UMD–PTLC, applied to one of SPE-C18 fractions containing iridoids of *Scrophularia deserti* (Scrophulariaceae), was performed in the normal phase mode on silica using a 95:10 EtOAc–MeOH + AcOH (two drops) solvent system (double development) affording a pure iridoid glucoside—scropolioside B [40].

The employment of forced-flow conditions of eluent in planar chromatography results in much more rapid sample development and higher efficiency of chromatographic process. Hence, such techniques have also been noticed and used in TLC preparative isolation of iridoid compounds. In separation of eight iridoid glucosides from *Veronica officinalis* overpressured-layer chromatography (OPLC), in combination with ultra-microchamber rotation planar chromatography (UCRPC)—belonging to centrifugally accelerated layer chromatographic techniques—have been employed [78]. Final isolation steps of iridoid glucosides from the aerial parts of *Kickxia abhaica* (Scrophulariaceae) comprised preparative rotation planar chromatography (PRPC) using 1–4 mm silica gel P_{254} discs and the following mobile phases: 8% and 10% MeOH–EtOAc, 20% MeOH–CHCl$_3$, and 20% MeOH–CHCl$_3$–NH$_3$. The isolated iridoid compounds were visualized under short- and long-wave UV light, followed by spraying with *p*-anisaldehyde [79]. Fractionation of the acetone extract of the aerial parts of *Gentiana verna* (Gentianaceae) on silica gel CC afforded a glycoside mixture which was further separated over silica gel by PRPC in CHCl$_3$–MeOH (9:1) giving three secoiridoids: sweroside, secologanin, and an acetal derivative of secologanin [80]. The methanolic extract from the aerial parts of *Penstemon ovatus* (Scrophulariaceae) was subjected to PRPC on a chromatotron with a stepwise gradient EtOAc–cyclohexane–MeOH (60:20:2, 5, 10, 20) followed by PTLC with CHCl$_3$–MeOH (8:2) and (7:3), yielding four iridoid glucosides [81]. Furthermore, loganin and related iridoid glycosides were obtained from *Gentiana depressa* by PRPC on silica gel using CHCl$_3$–MeOH (4:1) solvent system [82].

23.6 CONCLUSIONS

Iridoids are a broad group of low-stable natural compounds needing the proper use of isolation and purification techniques in order to prevent the degradation of their chemical structures. Hence, a good selection of extraction or other preparative-scale separation methods is very essential. As it was shown in this review, TLC is one of chromatographic techniques which have been frequently employed both for analytical purposes as well as for the preparative separation and purification of iridoid compounds. In the last decade, the clear progress in the development of new PTLC techniques (including forced flow) as well as densitometry has been observed.

REFERENCES

1. Junior, P., Recent developments in the isolation and structure elucidation of naturally occurring iridoid compounds, *Planta Med.*, 56, 1–13, 1990.
2. Ghisalberti, E.L., Biological and pharmacological activity of naturally occurring iridoids and secoiridoids, *Phytomedicine*, 5, 147–163, 1998.
3. Bruneton, J., *Pharmacognosy, Phytochemistry, Medicinal Plants*, 2nd ed., Lavoisier Publishing, Londres, Paris, New York, 1999, p. 589.
4. Steglich, W., Fugmann, B., and Lang-Fugmann, S., *Römpp Encyclopedia of Natural Products*, Thieme, Stuttgart, New York, 2000, p. 323.

5. El-Naggar, L.J. and Beal, J.L., Iridoids. A review, *J. Nat. Prod.*, 43, 649–707, 1980.
6. Hegnauer, R., *Chemotaxonomie der Pflanzen, Vol. VII*, Birkhäuser Verlag, Basel, 1986, p. 325.
7. Boros, C.A. and Stermitz, F.R., Iridoids. An updated review. I., *J. Nat. Prod.*, 53, 1055–1147, 1990.
8. Boros, C.A. and Stermitz, F.R., Iridoids. An updated review. II., *J. Nat. Prod.*, 54, 1173–1246, 1991.
9. Hegnauer, R. and Kooiman, P., The taxonomic significance of iridoids of Tubiflorae sensu Wettstein, *Planta Med.*, 33, 1–33, 1978.
10. Weigend, M., Kufer, J., and Müller, A.A., Phytochemistry and the systematics and ecology of Loasaceae and Gronoviaceae (Loasales), *Am. J. Bot.*, 87, 1202–1210, 2000.
11. Jensen, S.R., Franzyk, H., and Wallander, E., Chemotaxonomy of the Oleaceae: iridoids as taxonomic markers, *Phytochemistry*, 60, 213–231, 2002.
12. Andrzejewska-Golec, E. and Świątek, L., Chemotaxonomic investigations on the genus *Plantago*. I. Analysis of iridoid fraction, *Herba Pol.*, 30, 9–16, 1984.
13. Andrzejewska-Golec, E., Ofterdinger-Daegel, S., Calis, I., and Światek, Ł., Chemotaxonomic aspects of iridoids occurring in *Plantago* subg. *Psyllium* (Plantaginaceae), *Plant Syst. Evol.*, 185, 85–89, 1993.
14. Carnat, A., Fraisse, D., Carnat, A.P., Felgines, C., Chaud, D., and Lamaison, J.L., Influence of drying mode on iridoid bitter constituent levels in gentian root, *J. Sci. Food Agric.*, 85, 598–602, 2005.
15. Funke, E.D. and Friedrich, H., Valepotriates in the aerial parts of some more Valerianaceae species, *Planta Med.*, 28, 215–224, 1975.
16. Hölzl, J. and Jurcic, K., Valepotriates in the leaves of *Valeriana jatamansii*, *Planta Med.*, 27, 133–139, 1975.
17. Tittel, G., Chari, V.M., and Wagner, H., HPLC-analysis of *Valeriana mexicana* extracts, *Planta Med.*, 34, 305–310, 1978.
18. Suomi, J., Sirén, H., Wiedmer, S.K., and Riekkola, M.L., Isolation of aucubin and catalpol from *Melitaea cinxia* larvae and quantification by micellar electrokinetic capillary chromatography, *Anal. Chim. Acta*, 429, 91–99, 2001.
19. Bianco, A., Jensen, S.R., Olesen, J., Passacantilli, P., and Ramunno, A., Acid rearrangement of secoiridoids related to oleuropein and secologanin, *Eur. J. Org. Chem.*, 22, 4349–4354, 2003.
20. De Luca, C., Guiso, M., and Martino, C., 6-β-hydroksyipolamide, an irydoid glucoside from *Stachytarpheta mutabilis*, *Phytochemistry*, 22, 1185–1187, 1983.
21. Damtoft, S., Hansen, S.B., Jacobsen, B., Rosendal Jensen, S., and Nielsen, B.J., Iridoid glucosides from *Melampyrum*, *Phytochemistry*, 23, 2387–2389, 1984.
22. Imakura, Y., Kobayashi, S., Kida, K., and Kido, M., Iridoid glucosides from *Campsis chinensis*, *Phytochemistry*, 23, 2263–2269, 1984.
23. Takeda, Y., Matsumoto, T., and Ooiso, Y., Nepetacilicioside, a new iridoid glucoside from *Nepeta cilicia*, *J. Nat. Prod.*, 59, 518–519, 1996.
24. Krull, R.E. and Stermitz, F.R., *Trans*-fused iridoid glycosides from *Penstemon mucronatus*, *Phytochemistry*, 49, 2413–2415, 1998.
25. Albach, D.C., Gotfredsen, C.H., and Jensen, S.R., Iridoid glucosides of *Paederota lutea* and the relationships between *Paederota* and *Veronica*, *Phytochemistry*, 65, 2129–2134, 2004.
26. Iavarone, C., Sen, A., Trogolo, C., and Villa, S., Mollugoside, an iridoid glucoside from *Galium mollugo*, *Phytochemistry*, 22, 175–178, 1983.

27. Suomi, J., Sirén, H., Hartonen, K., and Riekkola, M.L., Extraction of iridoid glycosides and their determination by micellar electrokinetic capillary chromatography, *J. Chromatogr. A*, 868, 73–83, 2000.
28. Háznagy-Radnai, E., Léber, P., Tóth, E., Janicsák, G., and Máthé, I., Determination of *Stachys palustris* iridoids by a combination of chromatographic methods, *J. Planar Chromatogr.—Mod. TLC*, 18, 314–318, 2005.
29. Kim, H.K., Choi, Y.H., Luijendijk, T.J.C., Vera Rocha, R.A., and Verpoorte, R., Comparison of extraction methods for secologanin and the quantitative analysis of secologanin from *Symphoricarpos albus* using ^{1}H-NMR, *Phytochem. Anal.*, 15, 257–261, 2004.
30. Bos, R., Woerdenbag, H.J., and Pras, N., Determination of valepotriates, *J. Chromatogr. A*, 967, 131–146, 2002.
31. Abdel-Kader, M.S., Wisse, J., Evans, R., van der Werff, H., and Kingston, D.G.I., Bioactive iridoids and a new lignan from *Allamanda cathartica* and *Himatanthus fallax* from the Suriname rainforest, *J. Nat. Prod.*, 60, 1294–1297, 1997.
32. Günther, M. and Schmidt, P.C., Comparison between HPLC and HPTLC-densitometry for the determination of harpagoside from *Harpagohytum procumbens* CO_2-extracts, *J. Pharm. Biomed. Anal.*, 37, 817–821, 2005.
33. Ma, W.G., Fuzzati, N., Wolfender, J.L., Yang, C.R., and Hostettmann, K., Further acylated secoiridoid glucosides from *Gentiana rhodantha*, *Phytochemistry*, 43, 805–810, 1996.
34. Calis, I., Hosny, M., and Lahloub, M.F., A secoiridoid glucoside from *Fraxinus angustifolia*, *Phytochemistry*, 41, 1557–1562, 1996.
35. Takenaka, Y., Okazaki, N., Tanahashi, T., Nagakura, N., and Nishi, T., Secoiridoid and iridoid glucosides from *Syringa afghanica*, *Phytochemistry*, 59, 779–787, 2002.
36. Bianco, A., Lamesi, S., and Passacantilli, P., Iridoid glucosides from *Satureja vulgaris*, *Phytochemistry*, 23, 121–123, 1984.
37. Bianco, A., Guiso, M., Martino, M., Nicoletti, M., Serafini, M., Tomassini, L., Mossa, L., and Poli, F., Iridoids from endemic Sardinian *Linaria* species, *Phytochemistry*, 42, 89–91, 1996.
38. Stuppner, H. and Wagner, H., Minor iridoid and phenol glycosides of *Picrorhiza kurrooa*, *Planta Med.*, 55, 467–469, 1989.
39. Cimanga, K., Hermans, N., Apers, S., Van Miert, S., Van den Heuvel, H., Claeys, M., Pieters, L., and Vlietinck, A., Complement-inhibiting iridoids from *Morinda morindoides*, *J. Nat. Prod.*, 66, 97–102, 2003.
40. Stavri, M., Mathew, K.T., and Gibbons, S., Antimicrobial constituents of *Scrophularia deserti*, *Phytochemistry*, 67, 1530–1533, 2006.
41. Dos Santos, T.C., Schripsema, J., Monache, F.D., and Leitão, S.G., Iridoids from *Vitex cymosa*, *J. Braz. Chem. Soc.*, 12, 763–766, 2001.
42. Ling, S.K., Tanaka, T., and Kouno, I., Iridoids from *Rothmannia macrophylla*, *J. Nat. Prod.*, 64, 796–798, 2001.
43. Wang, S.C., Tseng, T.Y., Huang, C.M., and Tsai, T.H., *Gardenia* herbal active constituents: applicable separation procedures, *J. Chromatogr. B*, 812, 193–202, 2004.
44. Dzido, T., Matysik, G., Soczewiński, E., Wysokińska, H., and Adamczyk, U., Chromatographic analysis of penstemide and serrulatoloside by high-performance liquid chromatography and gradient thin-layer chromatography, *Chromatographia*, 27, 24–26, 1989.
45. Franzyk, H., Jensen, S.R., and Stermitz, F.R., Iridoid glucosides from *Penstemon secundiflorus* and *P. grandiflorus*: revised structure of 10-hydroxy-8-epihastatoside, *Phytochemistry*, 49, 2025–2030, 1998.

46. Handjieva, N., Mitova, M., Ancev, M., and Popov, S., Iridoid glucosides from *Galium album* and *G. lovcense*, *Phytochemistry*, 43, 625–628, 1996.
47. Sudo, H., Ide, T., Otsuka, H., Hirata, E., Takushi, A., and Takeda, Y., 10-*O*-acylated iridoid glucosides from leaves of *Premna subscandens*, *Phytochemistry*, 46, 1231–1236, 1997.
48. Junior, P., Further investigations regarding distribution and structure of the bitter principles from *Menyanthes trifoliata*, *Planta Med.*, 55, 83–86, 1989.
49. Leitao, G.G., de Souza, P.A., Moraes, A.A., and Brown, L., Step-gradient CCC separation of phenylpropanoid and iridoid glycosides from roots of *Stachytarpheta cayanensis* (Rich.) Vahl, *J. Chromatogr. Relat. Technol.*, 28, 2053–2060, 2005.
50. Bergeron, C., Marston, A., Gauthier, R., and Hostettmann, K., Iridoids and secoiridoids from *Gentiana linearis*, *Phytochemistry*, 44, 633–637, 1997.
51. Jacke, G. and Rimpler, H., Distribution of iridoid glycosides in *Clerodendrum* species, *Phytochemistry*, 22, 1729–1734, 1983.
52. Stermitz, F.R., Ianiro, T.T., Robinson, R.D., and Gardner, D.R., 6-*O*-Acetylmelittoside and other iridoids from *Castilleja* species, *J. Nat. Prod.*, 54, 626–628, 1991.
53. Hosny, M., Secoiridoid glucosides from *Fraxinus oxycarba*, *Phytochemistry*, 47, 1569–1576, 1998.
54. Peng, J.N., Feng, X.Z., and Liang, X.T., Iridoids from *Hedyotis hedyotidea*, *Phytochemistry*, 47, 1657–1659, 1998.
55. Fernandez, L., Diaz, A.M., Ollivier, E., Faure, R., and Balansard, G., An iridoid diglycoside isolated from *Sccrophularia scorodonia*, *Phytochemistry*, 40, 1569–1571, 1995.
56. Shen, Y.C., Lin, S.L., and Chein, C.C., Three secoiridoid glucosides from *Jasminum lanceolarium*, *Phytochemistry*, 44, 891–895, 1997.
57. Damtoft, S., Franzyk, H., and Jensen, S.R., Iridoid glucosides from *Picconia excelsa*, *Phytochemistry*, 45, 743–750, 1997.
58. Yang, X.W., Ma, Y.L., He, H.P., Wang, Y.H., Di, Y.T., Zhou, H., Li, L., and Hao, X.J., Iridoid constituents of *Tarenna attenuata*, *J. Nat. Prod.*, 69, 971–974, 2006.
59. Drewes, S.E. and Kayonga, L., Iridoid molluscocidal compounds from *Apodytes dimidiata*, *J. Nat. Prod.*, 59, 1169–1170, 1996.
60. Von Poser, G.L., Moulis, C., Sobral, M., and Henriques, A.T., Chemotaxonomic features of iridoids occurring in *Verbenoxylum reitzii*, *Plant Syst. Evol.*, 198, 287–290, 1995.
61. Helfrich, E. and Rimpler, H., Iridoid glycosides and phenolic glycosides from *Holmskioldia sanguinea*, *Phytochemistry*, 50, 619–627, 1999.
62. Helfrich, E. and Rimpler, H., Iridoid glycosides from *Gmelina philippensis*, *Phytochemistry*, 54, 191–199, 2000.
63. Baghdikian, B., Guiraud-Dauriac, H., Ollivier, E., N'guyen, A., Dumueuil, G., and Balmsard, G., Formation of nitrogen containing metabolites from the main iridoids of *Harpagophytum procumbens* and *Harpagophytum zeyheri* by human intestinal bacteria, *Planta Med.*, 65, 164–166, 1999.
64. Khishova, O.M., Dunets, L.N., Alekseev, N.A., Petrov, P.T., Tsvilik, G.L., and Golyak, Y.A., Standardization of medicinal preparations based on valerian roots and rhizomes, common motherwort grass, and hawthorn fruits, *Pharm. Chem. J.*, 38, 96–99, 2004.
65. Wahli, F., Stability tests of *Vitex agnus castus* (chaste tree) extracts, *CBS*, 91, 12–13, 2003.
66. Menković, N., Šavikin-Fodulović, K., and Savin, K., Chemical composition and seasonal variations in the amount of secondary compounds in *Gentiana lutea* leaves and flowers, *Planta Med.*, 66, 178–180, 2000.

67. Rücker, G., Neugebauer, M., and El Din, M.S., Qualitative TLC analysis of valepotriates, *Planta Med.*, 43, 299, 1981.

68. Tuntiwachwuttikul, P., Pancharoen, O., and Taylor, C., Iridoid glucosides of *Barleria lupulina*, *Phytochemistry*, 49, 163–166, 1998.

69. Krzek, J. et al., Densitometric determination of aucubin in syrups in the presence of other iridoids—an approach to standardization, *J. Planar Chromatogr.—Mod. TLC*, 15, 196–199, 2002.

70. Háznagy-Radnai, E., Czigle, S., Janicsák, G., and Máthé, I., Iridoids of *Stachys* species growing in Hungary, *J. Planar Chromatogr.—Mod. TLC*, 19, 187–190, 2006.

71. Biringanine, G., Chiarelli, M.T., Faes, M., and Duez, P., A validation protocol for the HPLC standardization of herbal products: application to the determination of acteoside in leaves of *Plantago palmata* Hook. f.s., *Talanta*, 69, 418–424, 2006.

72. Krupińska, A., Segiet-Kujawa, E., Skrypczak, L., and Ellnain-Wojtaszek, M., Quantitative determination of amarogentin by TLC-densitometry, *Sci. Pharm.*, 59, 135–138, 1991.

73. Bhandari, P., Gupta, A.P., Singh, B., and Kaul, V.K., HPTLC determination of swertiamarin and amarogentin in *Swertia* species from the western Himalayas, *J. Planar Chromatogr.—Mod. TLC*, 19, 212–215, 2006.

74. Nikolova-Damyanova, B. and Handjieva, N., Quantitative determination of swertiamarin and gentiopicroside in *Centaurium erythrea* and *C. turcicum* by densitometry, *Phytochem. Anal.*, 7, 140–142, 1996.

75. Hostettmann, K., Marston, A., and Hostettmann, M., *Preparative Chromatography Techniques. Applications in Natural Product Isolation*, 2nd ed., Springer-Verlag, Berlin, Heidelberg, 1998, p. 6.

76. Junior, P., Exaltoside and 7-Epiexaltoside, two novel acylated secoiridoid glucosides from *Villarsia exaltata*, *Planta Med.*, 57, 181–183, 1991.

77. Itoh, A., Oya, N., Kwaguchi, E., Nishio, S., Tanaka, Y., Kawachi, E., Akita, T., Nishi, T., and Tanahashi, T., Secoiridoid glucosides from *Strychnos spinosa*, *J. Nat. Prod.*, 68, 1434–1436, 2005.

78. Dallenback-Toelke, K., Nyiredy, S., and Sticher, O., Application of various planar chromatographic techniques for the separation of iridoid glycosides from *Veronica offcinalis*, *J. Chromatogr. A*, 404, 365–371, 1987.

79. Al-Rehaily, A.J., Abdel-Kader, M.S., Ahmad, M.S., and Mossa, J.S., Iridoid glucosides from *Kickxia abhaica* D.A. Sutton from Scrophulariaceae, *Phytochemistry*, 67, 429–432, 2006.

80. Mpondo, E.M. and Garcia, J., Secologanin and derivatives from *Gentiana verna*, *Planta Med.*, 56, 125, 1990.

81. König, M. and Seifert, K., Iridoid glucosides from *Penstemon ovatus*, *Planta Med.*, 61, 82–83, 1995.

82. Chulia, A.J., Vercauberen, J., and Mariotte, A.M., Iridoids and flavones from *Gentiana depressa*, *Phytochemistry*, 42, 139–143, 1996.

Secondary Metabolites— Amino Acid Derivatives

24 TLC of Indole Alkaloids

Peter John Houghton

CONTENTS

24.1 INTRODUCTION

24.1.1 DEFINITION OF INDOLE ALKALOIDS

Indole alkaloids are so-called because they incorporate an indole ring system **1** in their molecular structure. All indole alkaloids are biogenetically derived from tryptophan **2** and therefore contain two nitrogen atoms, one of which is contained within the five-membered part of the indole nucleus. In a few instances, notably the pharmaceutically important alkaloids such as quinine from *Cinchona* species and camptothecin from *Camptotheca acuminata*, this five-membered ring is expanded by the inclusion of an extra carbon and the alkaloids chemically are classified as quinolines **3** rather than indoles, although their biogenetic origins are the same. The chemistry and biological activity of indole alkaloids have received much attention over the last 50 years and several monographs[1,2] and detailed reviews* have been published.

Three major types of indole alkaloids are recognized by most natural product chemists although several other indole alkaloids are known which do not fit neatly into these three categories. The three major categories are the "simple" indoles that are formed from tryptophan without the addition of other C-skeletal moieties (e.g., harmine **4**), the tryptamine-amino acid congeners found particularly in fungi such as ergot (e.g., ergotamine **5**), and the largest group, numbering several thousand, where the tryptophan skeleton is linked with a monoterpene unit derived from the iridoid secologanin. Because of different rearrangements of the iridoid moiety after conjugation, three major groups of this type of indole alkaloid are generally recognized, these being the *Corynanthe*-type (e.g., ajmalicine **6**), the *Iboga*-type (e.g., catharanthine **7**), and the *Aspidosperma*-type (e.g., tabersonine **8**). In some cases two types are linked together to form bisindoles (e.g., vincristine **9**), which consist of an *Aspidosperma*-type linked to an *Iboga*-type.

---

* See articles in *Alkaloids* edited by Manske, R.H.F., Brossi, A., and Cordell, G.A., Academic Press, London and regular updates in *Natural Product Reports, Royal Society of Chemistry, London.*

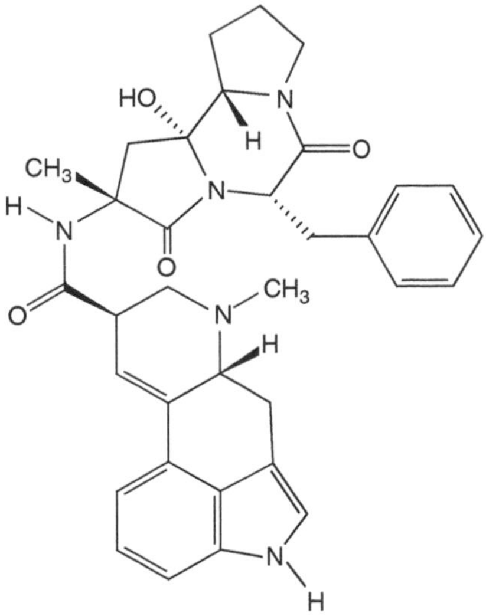

5

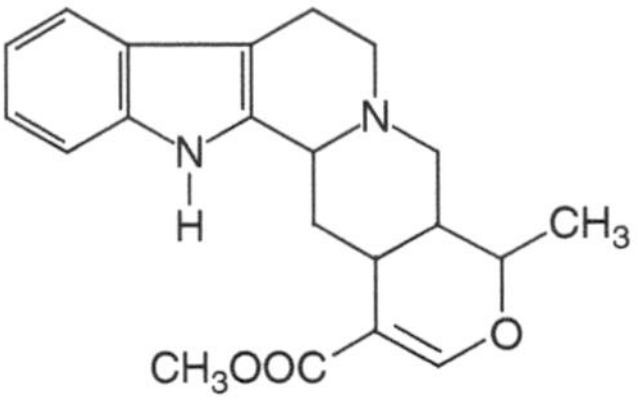

6

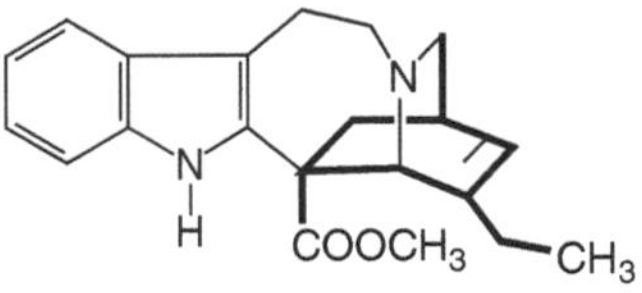

7

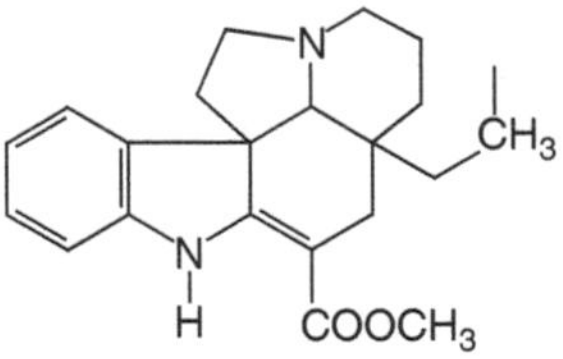

8

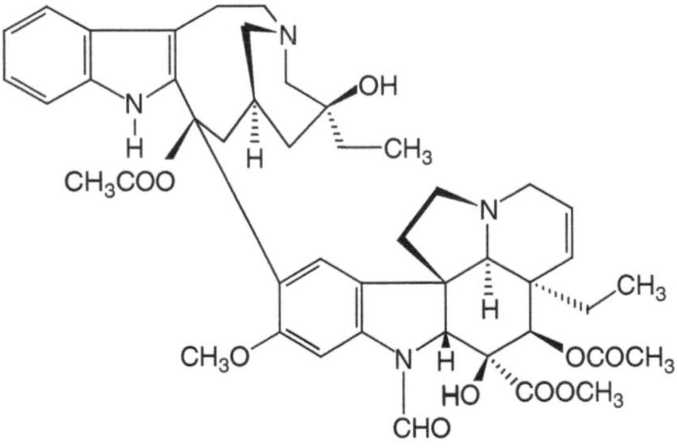

9

24.1.2 Occurrence of Indole Alkaloids

The simple indoles occur sporadically in fungi and flowering plants and also in animals, since it could be argued that the mammalian neurotransmitter serotonin **10** is also a simple indole and others are found in the skin of certain amphibians (e.g., bufotenine **11** found in toads, *Bufo* spp.). The fungal genus *Psilocybe* contains alkaloids such as psilocybin **12**, while harmane alkaloids such as harmine **4** are found in flowering plant species. Fairly simple indole alkaloids are found in the seeds of some members of the Fabaceae, physostigmine **13** probably being the best-known compound of this type. The tryptophan–amino acid alkaloids are often called the ergot alkaloids, since many of them were first isolated from *Claviceps* spp., fungal parasites of grasses that are commonly called ergots. There was considerable surprise when, in the 1960s, they were also found in the seeds of some genera of the Convolvulaceae, considered to be one of the most evolutionary-advanced families of the dicotyledons. The large number of terpenoid–tryptamine alkaloids is largely confined to some genera of three evolutionary-advanced dicotyledonous flowering plant families (i.e., the Apocynaceae, Rubiaceae, and Loganiaceae). An amazing variety of structures belonging to this group have been determined in the last 50 years and they all arise from a Mannich-like condensation between tryptamine and the iridoid secologanin, with subsequent rearrangement of the iridoid components.

10 R=OH, R'=H$_2$
11 R=OH, R'=(CH$_3$)$_2$

12

13

24.1.3 PHARMACOLOGICAL IMPORTANCE OF THE INDOLE ALKALOIDS

The biological activity of some fungi and plants containing indole alkaloids has been known for a very long time, particularly those that have a severe effect on humans or animals, and the elucidation of the structure of the alkaloids responsible is a comparatively recent event. Ergot contamination in rye consumed by humans, and in grasses consumed by domestic grazing animals, has been known to be associated with miscarriage, gangrene, and CNS disturbances for hundreds of years and the effects of the contained alkaloids were exploited in central European traditional medicine for helping in childbirth. Ergot contains several indole alkaloids and two of them have found extensive use in pharmacy. Ergometrine **14** is mainly used in childbirth because of its ability to cause contraction of the uterus and contract blood vessels, thus reducing bleeding, while ergotamine **5** has a stronger dopaminergic effect and is mainly employed in treating migraine because it reduces vasodilation. Several semisynthetic compounds based on ergotamine have been developed for treating Parkinsonism and related neurological disorders. In southern Asia, the toxic nature of *Strychnos nux vomica* was well known and other species of the genus were known as a source of arrow poisons in Africa and South America. Many fungi and plants containing indole alkaloids have been used as hallucinogens or in religious rites because of their effect on consciousness and these include the *Psilocybe* fungi in Oaxaca (Mexico) which were known as "teonanácatl"; the roots of *Tabernanthe iboga* in west central Africa known as "iboga"; and the seeds of *Ipomoea* spp., known in Mexico as "ololiuqui."[3] Physostigmine **13** (also called eserine) is an interesting example of an indole alkaloid that was discovered because of the use of the seeds of a plant *Physostigma venenosum*, in southeastern Nigeria as an ordeal poison. It was the first acetylcholinesterase inhibitor to be discovered and has been used to treat glaucoma and myasthenia gravis. Analogues have been developed as insecticides as well as drugs, the most recent being rivastigmine, which is used for symptomatic relief in Alzheimer's disease.

14

Other plant species containing indole alkaloids were widely used in traditional medical systems (e.g., *Rauvolfia serpentina* roots), known as "sarpagandha" in Ayurvedic medicine have a long tradition of use as a treatment for snakebite and some psychiatric diseases. *Rauvolfia* species contain many indole alkaloids but the one which was most widely used in the mid 20th century was reserpine **15**, which was the first tranquilizing drug but also extensively used at one time as a hypotensive. Ajmaline **16** is another *Rauvolfia* indole alkaloid, which is used to treat cardiac arrhythmia in some countries. Yohimbine **17** from *Pausinystalia yohimbe*, the bark of which has a reputation as a male aphrodisiac, has been used to treat erectile dysfunction. Extracts from barks of *Cinchona* spp. were introduced in Europe for the treatment of malaria not long after the conquest of much of South America by the Spanish in the 16th century and the major alkaloid quinine **18**, which contains a quinoline ring system rather than indole, has been used for many years as an antimalarial drug although now used only where other antimalarials are ineffective.

Possibly the indole alkaloids that are most important clinically are the anticancer bis-indoles vinblastine **19** and vincristine **9**, which are found in *Catharanthus roseus*. These were introduced in the 1960s and have made a marked impact on remission of childhood leukemia and some other cancers. Other anticancer alkaloids derived from tryptophan are the indole ellipticine **20** and the quinoline camptothecin **21**, the latter having greater importance as a template for analogues such as topotecan. In view of the importance of indole alkaloids in modern pharmacy, it is not surprising that considerable attention has been paid to investigating their presence in plants, fungi, and other living organisms, and this has resulted in a considerable amount of knowledge concerning the TLC characteristics of these interesting compounds.

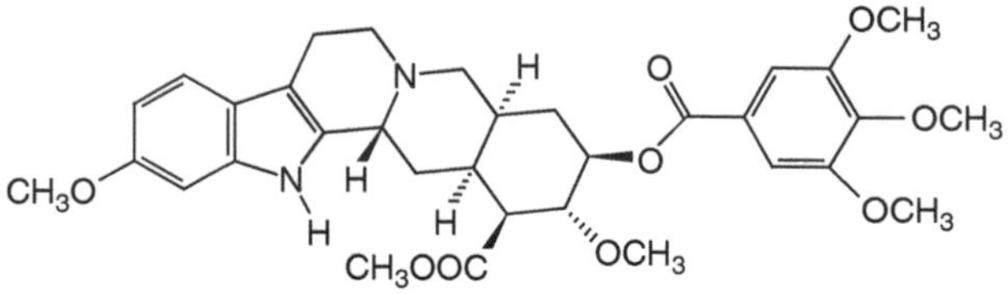

15

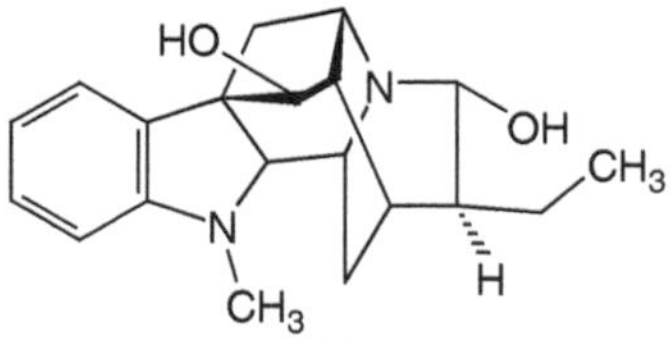

16

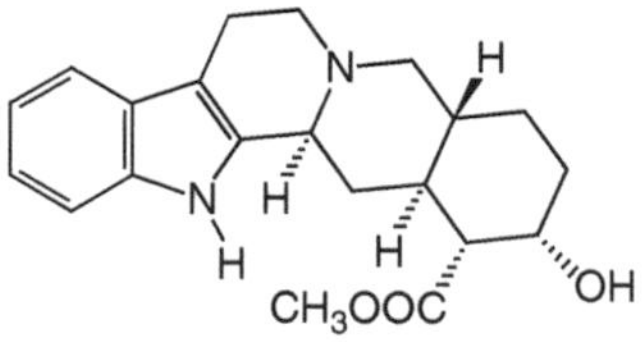

17

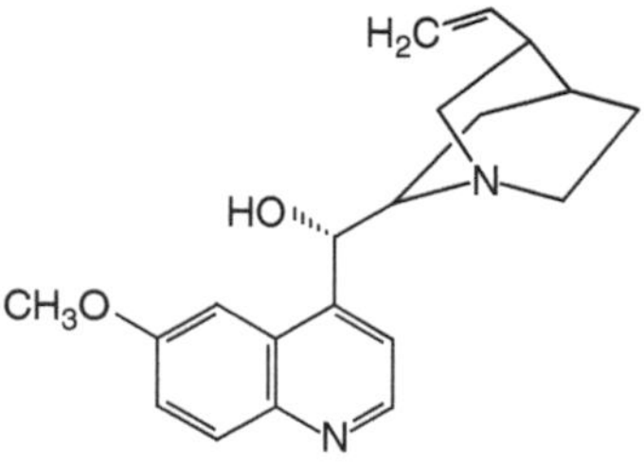

18

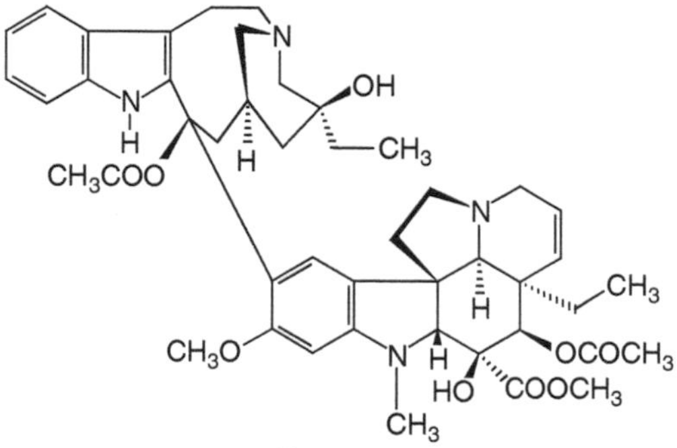

19

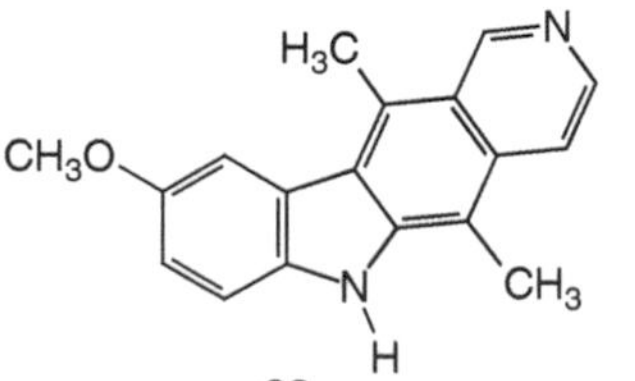

20

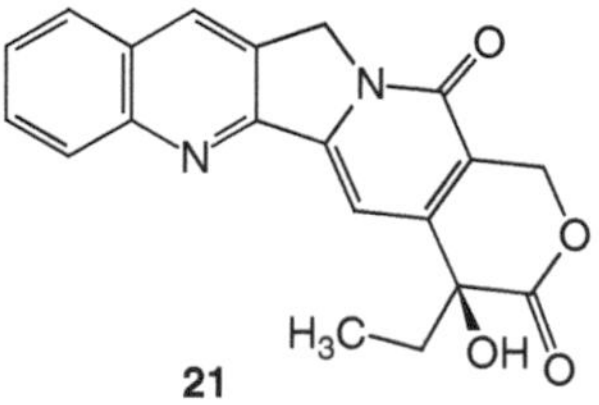

21

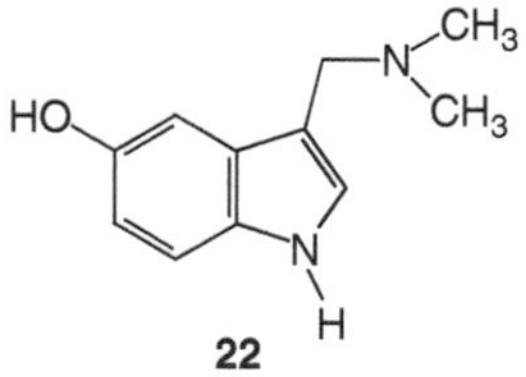

22

24.1.4 ECOLOGICAL ASPECTS OF INDOLE ALKALOIDS

Comparatively little research has been carried out on indole alkaloids, compared
with alkaloids as a group, concerning their ecological significance. In common with

many other types of alkaloids, their bitter taste is thought to act as a feeding deterrent to mammals, as they are often concentrated in parts of the plant vulnerable to attack (e.g., the high percentage of alkaloids in the bark of *Rauvolfia* roots).[4] The deterrent property has been demonstrated for the simple indole gramine as regards sheep.[5] It has also been pointed out that gramine **22** has antifeedant activity against aphids, an example of general antifeedant activity regarding insects, shown by alkaloids.[6]

24.1.5 CHEMOSYSTEMATIC IMPORTANCE OF THE INDOLE ALKALOIDS

Waterman has pointed out that the chemosystematic importance of the alkaloids is not great.[7] The major interest concerning indole alkaloids is that the indole–secologanin indole alkaloids are found in assorted genera of the Apocynaceae, Loganiaceae, and Rubiaceae of the Gentianales, with the Iboga type restricted to the Apocynaceae. These compounds are also found in the lesser families Alangiaceae, Nyssaceae, and Icacinaceae of the Cornales.

24.2 TECHNIQUES

24.2.1 STATIONARY PHASES

Most TLC procedures for separating indole alkaloids use an adsorbent stationary phase such as silica gel G, often with fluorescent agent added, since all indole alkaloids will absorb UV light and can be observed as quenching zones when examined under UV light. Occasionally the silica gel has been treated with 0.1 M sodium hydroxide solution, particularly for the *Rauvolfia* alkaloids which have a low pK_a value.[8] Alumina is less commonly used as a stationary phase but has been employed as a layer treated with 0.5 M lithium hydroxide for the *Catharanthus* alkaloids.[9]

Partition systems are less commonly employed although cellulose impregnated with formamide has been used for the ergot alkaloids.[10]

24.2.2 MOBILE PHASES

A very large number of mobile phases have been used to separate indole alkaloids and can be found in the standard reference sources.[11,12]

Solvents employed reflect the polarity of the alkaloids under investigation. The majority of indole alkaloids are tertiary bases with fairly low polarity, so that mixtures containing a major part of a less polar solvent (e.g., chloroform, toluene) with a small proportion of a more polar solvent (e.g., acetone, ethanol, methanol) are frequently cited. Since tertiary indole alkaloids share the amphiphilic properties of most other alkaloids, resulting in elongated "streaked" zones, sometimes called "tailing," a base is often added to the solvent mixture to ensure that the noncharged form of the alkaloids is predominant and that the zones are more compact. For less polar systems, diethylamine, in proportions up to 10% v/v, is used but in other cases ammonia solution, either concentrated (13.5 M) or in a more dilute form can be used, often in conjunction with ethyl acetate and propan-2-ol. Three different mixtures of ethyl acetate, propan-2-ol, and 13.5 M ammonia, i.e., 100:2:1, 16:3:1, and 9:7:4 have been used to cover a full range of polarities of alkaloids.[13] Quaternary indole alkaloids are sometimes encountered, particularly as constituents of muscle-relaxant arrow and dart poisons. These, together with N-oxides of alkaloids, are very polar in nature and do not move from the baseline with mobile phases unless they are very polar. Methanol:0.2 M ammonium nitrate mixtures have been used to separate such compounds.[14]

24.2.3 Detection Reagents

Most detection of indole alkaloids is carried out by first observing under UV light 254 nm for the presence of fluorescent or quenching zones. If the latter are sought, then a layer with added fluorophor (e.g., silica gel GF$_{254}$) should be used. The *Cinchona* alkaloids in particular give a very strong bright blue fluorescence under UV 365 nm and can be detected at very low concentrations. Many of the ergot alkaloids give a blue fluorescence under UV light at both 254 and 365 nm. It is very common to follow examination under UV light with the use of a chemical chromogenic reagent but there are a few instances of biological detection methods also being applied. Compositions of the chromogenic spray reagents used are given in Table 24.1.

24.2.3.1   Chemical Detection Methods

Dragendorff's reagent (potassium iodobismuthate solution) will give an orange color against a yellow background for all indole alkaloids, based on the precipitate complex of the alkaloid and the bismuth metallic component of the reagent. The colors of zones for some indole alkaloids obtained by spraying with Dragendorff's reagent can be made more intense by subsequent spraying with 10% sodium nitrite solution.[15] Iodoplatinate reagent works on a similar principle and gives a slightly greater range of colors comprising pink, brown, and violet.[16]

4-Dimethylaminobenzaldehyde solution in an acidic medium, such as HCl, is known as Ehrlich's reagent or Van Urk reagent, and gives a blue color with simple indoles such as the harmane series, and particularly with the ergot alkaloids. It has been extensively used to detect the alkaloids of ergot with which it gives blue, violet, or green colors.[17]

Several different reagents have been used to distinguish the alkaloids derived from indole–secologanin conjugation. Cerium(IV) sulfate solution gives a variety of colors with the *Catharanthus* and *Vinca* alkaloids,[18,19] while iron(III) chloride/perchloric acid has been widely utilized for many indole and oxindole compounds, especially those from *Rauvolfia*, *Strychnos*, *Mitragyna*, and *Uncaria* with which a wide variety of colors are seen.[20–22] These latter two reagents are not completely satisfactory since both utilize concentrated acids with strong oxidizing properties which pose a risk to health and of corrosion and fire.

24.2.3.2   Biological Detection Methods

TLC autobiography, in which a developed plate is either overlaid with agar seeded with bacteria or fungal spores or sprayed with a suspension of nonpathogenic bacteria or fungal spores prior to incubation to allow growth to take place, is a standard technique for in situ detection of antibacterial and antifungal

TABLE 24.1

Chromogenic Spray Reagents for Indole Alkaloids

Common Name	Formula	Treatment after Spraying	Notes
Dragendorff's reagent	Solution A: 0.85 g basic bismuth nitrate dissolved in 10 mL glacial acetic acid and 40 mL water with heating Solution B: 8.0 g potassium iodide dissolved in 30 mL water Mix solutions A and B in 1:1 ratio then add to glacial acetic acid and water in 1:2:10 ratio	Observe in daylight	Gives an orange color with many types of alkoloids. Several variations on formula (Stahl)
Iodoplatinate reagent	0.3 g Hydrogen hexachloroplatinate (IV) hydrate dissolved in 100 mL water and mixed with 100 mL 6% w/v potassium iodide solution	Observe in daylight	
Van Urk reagent (Ehrlich's reagent)	0.2 g 4-Dimethylaminobenzaldehyde dissolved in 100 mL 25% HCL and add 1 drop of 10% w/v iron(III) chloride solution	Observe in daylight	
Cerium(IV) sulfate	1 g Ammonium cerium(IV) sulfate dissolved in 100 ms 85% phosphoric acid	Heat at 105°C for 10 min; observe in daylight	Variations of formula exist (Verp & BS)
Iron(III) chloride/ perchloric acid	5.8 g Iron(III) chloride hexahydrate in 100 mL 25% perchloric acid	Heat at 105°C for 10 min; observe in daylight	Highly corrosive and possibility of oxidative damage to materials

substances.[23,24] Although antimicrobial activity has been reported for many indole alkaloids,[25] most work has employed conventional serial dilution methods rather than TLC autobiography.

In recent years another biological activity, inhibition of acetylcholinesterase, has received much attention because of the therapeutic usefulness of compounds displaying this effect. Inhibition of this enzyme results in raised levels of acetylcholine (ACh) and so alleviates symptoms caused by a deficiency of ACh. For many years the muscle weakness observed in myasthenia gravis and the excess intraocular pressure of glaucoma were known to be relieved by acetylcholinesterase (AChE) inhibitors, and the indole alkaloid physostigmine **13** has been extensively employed clinically for many years. In more recent times AChE inhibitors have been used, with some success, to treat symptoms of Alzheimer's disease, since a feature of this condition is low levels of ACh in the brain. A bioautographic method for physostigmine and related compounds had been developed.[26] This technique involved spraying an impregnated cellulose plate with human plasma, leaving it for 30 min before spraying with alkaline bromothymol blue and acetylcholine solution. Any inhibitory zones appeared as blue against a yellow background, the yellow being caused by the change in pH due to liberation of acetic acid by enzyme activity. In more recent years, with the reawakening of interest in AChE inhibitors, two other techniques have been introduced which can be performed on silica gel plates and do not need to use human plasma. Both rely on the detection of products of hydrolysis of a substrate by the enzyme. The first of these was an adaptation of the Ellman colorimetric method which had been widely employed.[27] This relied on hydrolysis of acetylthiocholine by the enzyme and detection of the liberated thiocholine by subsequent spraying with 2,4-dinitrothiobarbituric acid (2,4DTNB), which reacts with the thiocholine to give an orange-yellow color. Inhibitory zones are therefore seen as pale zones against a strong yellow background. It was realized that false positives might be seen if compounds inhibited the reaction between 2,4DTNB and thiocholine, so the need to perform a test on a duplicate plate to detect this was stressed.[28] In this detection of false positives, the plate was sprayed with 2,4DTNB and then with the enzyme incubated with the ACh for 20 min before spraying. A similar approach was taken in the second method, but naphthyl acetate was used as the substrate.[29] The naphthol released by hydrolysis was treated with Fast Blue B salt to give a purple color, inhibitory compounds appearing as white zones.

Apart from physostigmine and related semisynthetic compounds, few indole alkaloids have been shown to be AChE inhibitors but four alkaloids from *Tabernaemontana australis* were recently detected to have this effect using the Ellman TLC method.[30]

24.2.4 QUANTITATIVE TLC

In recent years, HPLC has almost eclipsed TLC as a preferred method of quantification of many natural compounds, including indole alkaloids, although densitometry of these compounds is still sometimes reported, particularly if high performance TLC (HPTLC) is used. All the indole alkaloids absorb UV light at 254 nm and so can be quantified by measuring the decrease in fluorescence of fluorescent silica gel GF$_{254}$ plates.[31] A variety of chromogenic reagents have been used, including Dragendorff's reagent followed by 10% w/v sodium nitrite solution, ceric sulfate, and iron(III) chloride/perchloric acid spray. *Rauvolfia* alkaloids from plant cell cultures were quantified using after spraying with either Ehrlich's reagent or ceric sulfate spray.[32] Another densitometric analysis of the same group of alkaloids was accomplished by fluorimetric analysis of serpentine **23**, irradiated at 365 nm.[33]

23

As well as quantification of particular alkaloids in screening various samples of the same species, or related species, densitometry has been applied to analyzing extracts from tissue cultures and for variation within a particular species. Monthly variation in amounts of indole alkaloids produced by the same tree has been assessed by densitometry, particularly for *Mitragyna*[34] and *Strychnos* species[35] and, using iron(III) chloride/perchloric acid reagent with a filter for 540 nm on the densitometer.

24.3 DETAILS OF SEPARATIONS OF PARTICULAR TYPES OF INDOLE ALKALOIDS

24.3.1 SIMPLE INDOLES

The neurotransmitters serotonin (5-hydroxytryptamine) **10** and the related compounds *N,N*-dimethyltryptamine (DMT) **24**, 5-methoxyDMT **25**, and bufotenine **11**, found in various hallucinogenic snuffs,[36] are included in this group, as are the active constituents of the hallucinogenic fungi of the genus *Psilocybe*. The β-carboline alkaloids (e.g., harmane **4** and harmaline **26**) can also be considered, especially as they too are associated with hallucinogenic brews such as ayahuasca.[36]

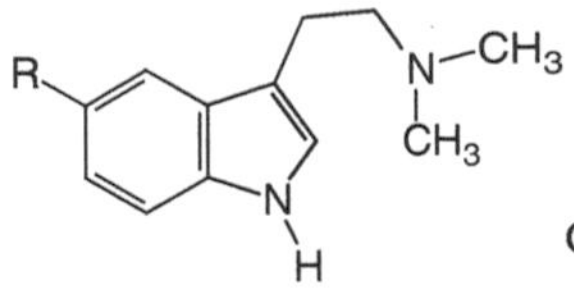

24 R=H
25 R=OH

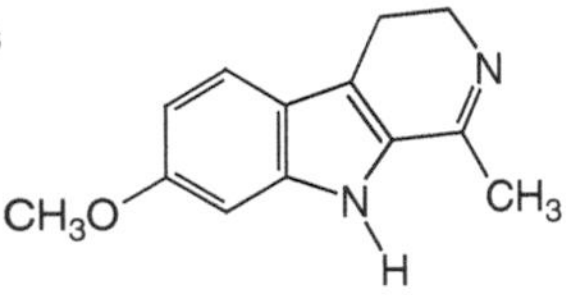

26

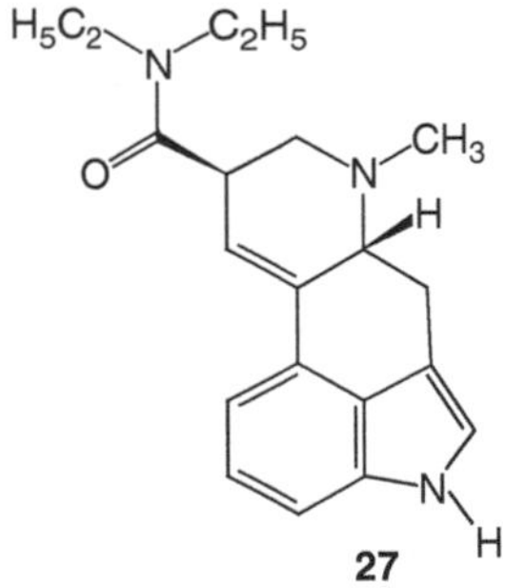

27

Polar systems such as silica gel/methanol:13.5 M ammonia 100:1.5 give good separation of the polar *Psilocybe* alkaloids[37] (e.g., **12**). The substituted tryptamines can be detected with Dragendorff's reagent but a more sensitive method has been developed which utilizes *o*-phthalaldehyde/sulfuric acid and this can detect amounts as low as 20 ng for serotonin and 100 ng for bufotenine.[38] Harmine **4** and harmaline **26**

from *Banistereopsis caapi* stem were separated using silica gel/chloroform:methanol:
water 64:50:10 and a quantitative assessment carried out using Dragendorff's
reagent.[39]

24.3.2 ERGOT ALKALOIDS

The ergot alkaloids comprise both those that are naturally-occurring, being particularly found in species of *Claviceps* (a parasite of cereals), and those that arise from
synthetic modification of the natural compounds, LSD-25 **27** being of particular note
because of its importance as a hallucinatory drug of abuse.

Most of the naturally occurring compounds can be separated on silica with a
mobile phase of medium polarity such as toluene:chloroform:ethanol 28.5:57:14.5.[40]
The ergotoxine group have proved to be more difficult to separate although some
success has been achieved using silica gel impregnated with formamide:ethanol 8:15
with acetone:0.1 M ammonium carbonate:ethanol 32.5:67:5.1 as a mobile phase.[41]
The compounds readily isomerize at the carbon to which the amide group is attached
to the ergoline ring, and so extra zones because of the isomeric artefacts are seen on
TLCs of solutions of the compounds which have been stored for some time.

The ergot alkaloids are detectable as fluorescent zones with UV light 254 and 365
nm, and as blue zones, which change to a purple color on heating, with Van Urk reagent.
A few alkaloids from nonmedicinal species of *Claviceps*, which lack a carbonyl function
on the ergoline ring, give a green color with this reagent. Iodoplatinate reagent gives a
range of colors which enable the common alkaloids to be distinguished from one
another, for example, ergometrine gives a white zone against a pinkish background
while ergotamine gives a strong pink color and ergocristine a pale brown.

Because of the low doses of LSD-25 **27** used illegally for "recreational" use, a
large number of sensitive TLC detection methods have been developed. Impregnation of the silica gel layer with 0.1 M aqueous sodium hydroxide prior to use has
been favored by some and this has given satisfactory separation of **27** with mobile
phases such as chloroform:methanol 4:1.[42] LSD-25 **27** can be detected with the Van
Urk reagent but a method using N,2,6-trichloro-4benzoquinone imine reagent has
also been described which detects as little as 0.25 μg LSD-25.[43]

Densitometric quantification of ergot alkaloids has been carried out either using
fluorescence or after spraying with Van Urk reagent.[44]

24.3.3 TERPENOIDAL INDOLE ALKALOIDS

This group of alkaloids has received the most attention because of the actual
and potential usefulness of these as pharmaceuticals. The ceric sulfate and iron(III)
chloride/perchloric acid spray reagents give a variety of colors with different types of
alkaloids and allow identification of the constituents present in small amounts of
material. The latter spray reagent has been particularly useful in screening herbarium
samples of *Uncaria* for heteroyohimbine indole and oxindole alkaloids (Figure 24.1),[22]
since it not only distinguishes between indole and oxindole alkaloids, but also between
different types of oxindole alkaloids by the colors observed and also the color changes
leading up to the final color. Details of the colors given by different types of alkaloid are
given in Table 24.2. Silica gel was used as the stationary phase in all systems and eight

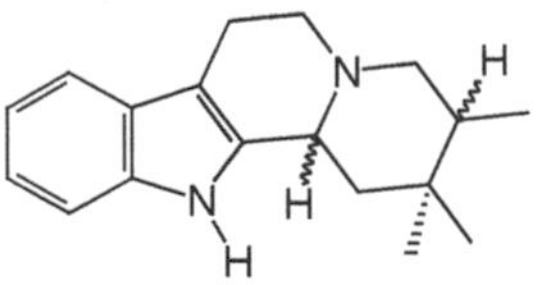

3-H	20-H	Type
α	α	*Allo*
α	β	*Normal*
β	α	*Epiallo*
β	β	*Pseudo*

FIGURE 24.1 Isomeric types of heteroyohimbine alkaloids.

mobile phases were used to account for the range of polarities of alkaloids present, with polar systems being used to detect N-oxides. It was found best to avoid diethylamine being used as a base in the mobile phase, as it was difficult to remove from the plate and interfered with color formation, so 13.5 M ammonia was found to be a suitable substitute. This approach not only enabled differentiation of various ring systems, but also of the stereochemistry of the heteroyohimbine ring system with the *allo*, *normal*, *epiallo* and *pseudo* isomers being seen in order of decreasing R_f value for any one type of alkaloid. In addition, it was found that compounds with 9-OCH₃ had lower R_f values than the corresponding 9-H compounds.

A variety of colors with iron(III) chloride solution/perchloric acid, and also ceric sulfate spray is also observed for the *Rauvolfia* alkaloids, the *Vinca*, and the *Catharanthus* alkaloids.[45] However, little correlation between color and structural features has been elucidated or discussed.

24.3.4 QUINOLINE ALKALOIDS

The major quinoline alkaloids are those found in the bark of *Cinchona* species, especially quinine **18** and quinidine **28**. Separation is achieved using silica gel and a mobile phase consisting of about 10% of a base (e.g., diethylamine in a weakly polar solvent such as chloroform) as used in the European Pharmacopoeia monograph for identification of Cinchona bark. Quinine **18** and quinidine **28** both give a very bright blue fluorescence with UV 365 nm but the corresponding demethoxylated compounds cinchonine **29** and cinchonidine **30** give a much less obvious dark violet fluorescence. With iodoplatinate reagent the four compounds give slightly different colors with quinine giving violet-brown, cinchonidine grey-violet, quinidine red-violet, and cinchonine brown-red.[46] Because of the very low amounts detectable using fluorescence, levels of quinine can be measured in samples using fluorimetric densitometry.

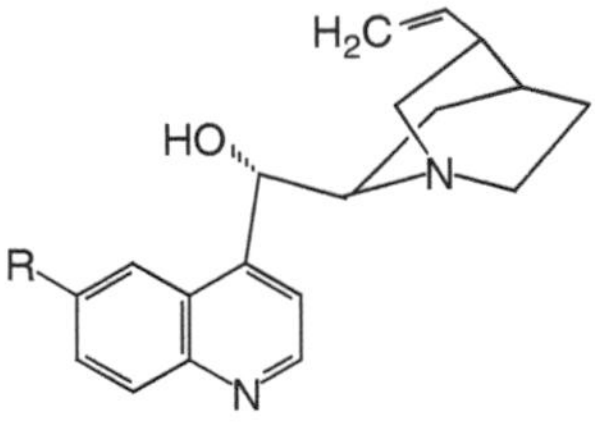

28 R=OCH₃
29 R=H

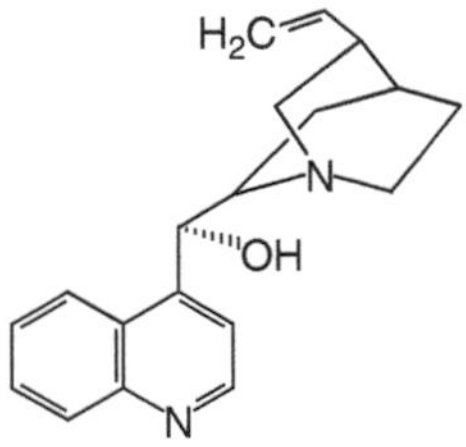

30

TABLE 24.2
Correlation between Color after Spraying with Iron(III) Chloride/Perchloric Acid Spray and Chemical Type for *Mitragyna* and *Uncaria* Heteroyohimbine Alkaloids (H & JDP)

Type of Compound	General Structure	Colors Observed
Indole		
Closed ring E		Grey turning to brown
Open ring E		Grey turning to brown
Oxindole		
Closed ring E *Normal* isomer *Allo, eipallo* isomers		Deep pink Green going to pink
Open ring E 20-ethyl 9-H 9-OH 9-OCH₃ Open ring E 20-vinyl 9-H		Blue going to pink Blue going to purple then brown Blue going to purple Blue to bright pink (rapid change)

24.3.5 INDOLES WITH A QUATERNARY NITROGEN

Not many indole alkaloids exist as quaternary amines, although heteroyohimbines such as alstonine **31** and *Strychnos* alkaloids such as C-mavacurine **32** and the muscle-relaxant bis-alkaloid toxiferine **33** are exceptions. In studies on *Strychnos* spp. alcohol ammonia mixtures were used to separate these compounds[35] as well as another approach, utilizing the ion-pair theory, in adding 0.2 M aqueous ammonium nitrate to methanol (2:3). Detection involved the common chromogenic reagents mentioned earlier. A similar approach, testing different concentrations and types of ion-pair additives to methanol had been applied to alstonine as well as *Strychnos* alkaloids.[47]

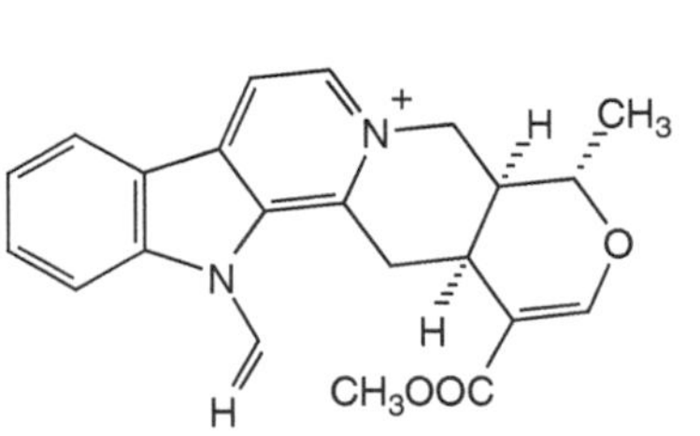

31

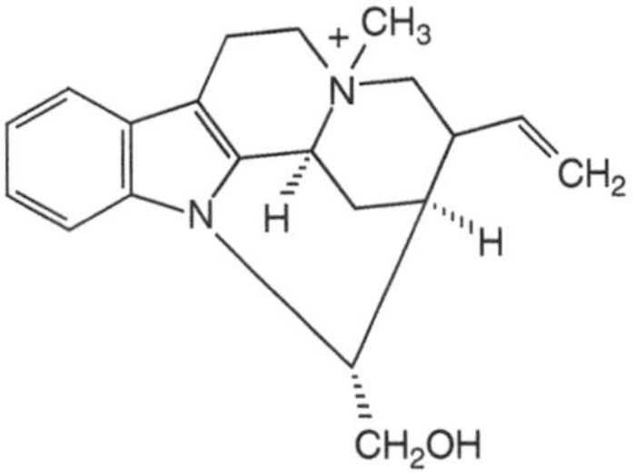

32

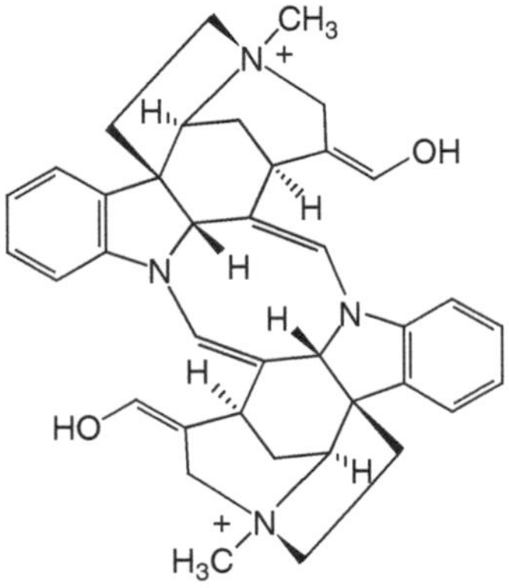

33

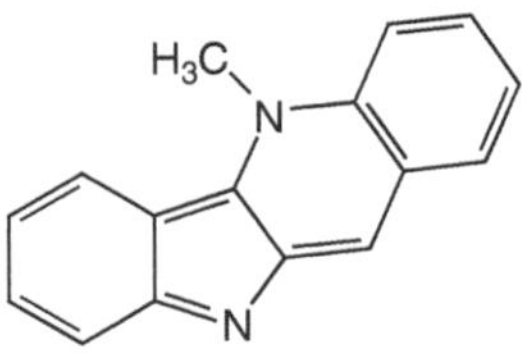

34

24.3.6 OTHER INDOLE ALKALOIDS

The indoloquinolizidine alkaloid cryptolepine **34** is of interest as the major active constituent of *Cryptolepis sanguinolenta*, which has a widespread reputation in West Africa as an effective antimalarial. It can be detected as a bright yellow fluorescent compound under UV light 365 nm after separation on silica gel/ethyl acetate:formic acid:water 6:1:1 and this system was used to screen herbarium samples of several different African species of *Cryptolepis*, of which only *C. sanguinolenta* showed the presence of cryptolepine.[48]

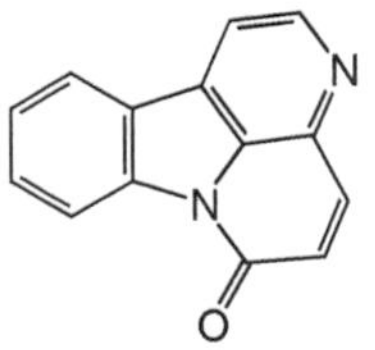

35

Canthinone **35** has been reported from several plant species and has a variety of effects, several of which point towards it playing a biochemical ecological role. It has been analyzed by TLC using silica gel/chloroform:methanol:water 7:3:1 and measuring fluorescence under UV light 365 nm.[49]

Camptothecin **21** has a quinoline ring system but has been shown to arise biogenetically from an indolic precursor. It is of interest because it occurs in several unrelated genera and displays unusual cytotoxic activity by inhibition of topoisomerase I. Since it has been used as a starting material for several anticancer drugs, sources giving high yields were screened fluorometrically at 365 nm with a silica gel/chloroform:methanol 97:3 mobile phase and nanogram quantities could be detected.[50] A similar method was used for quality control of camptothecin formulations used in a hospital.[51]

Ellipticine **20** is another anticancer indole alkaloid with a quinoline moiety which has been used to make derivatives of clinical interest. The 9-hydroxyanalogue is of particular interest but decomposes rapidly in some solvents to give orange and red colors. The breakdown process, and means of preventing it, was monitored by TLC (silica gel/benzene:chloroform:ethanol and noting the zones that were colored in ordinary daylight).[52]

REFERENCES

1. Phillipson, J.D. and Zenk, M.H., *Indole and Biogenetically Related Alkaloids*, Academic Press, London, 1980.
2. Samuelsson, G., Indole alkaloids, in *Drugs of Natural Origin*, 5th ed., Swedish Pharmaceutical Press, Stockholm, 2004, pp. 535–563.
3. Schultes, R.E., Hofmann, A., and Rätsch, C., *Plants of the Gods*, 2nd ed., Healing Arts Press, Rochester, VT, USA, 2001, p. 208.
4. Roberts, M.F. and Wink, M., Introduction, in *Alkaloids: Biochemistry, Ecology and Medicinal Applications*, Roberts, M.F. and Wink, M., Eds., Plenum Press, New York, 1998, pp. 1–7.
5. Arnold, G.W. and Hill, J.L., Chemical factors affecting selection of food plants by ruminants, in *Phytochemical Ecology*, Harborne, J.B., Ed., Academic Press, London, 1972, pp. 72–102.
6. Wink, M., Chemical ecology of alkaloids, in *Alkaloids: Biochemistry, Ecology and Medicinal Applications*, Roberts, M.F. and Wink, M., Eds., Plenum Press, New York, 1998, pp. 265–300.
7. Waterman, P.G., Chemical taxonomy of alkaloids, in *Alkaloids: Biochemistry, Ecology and Medicinal Applications*, Roberts, M.F. and Wink, M., Eds., Plenum Press, New York, 1998, pp. 87–108.
8. Ikram, M., Miana, G.A., and Islam, M., Chromatographic separation of *Rauwolfia serpentina* and opium alkaloids on thin layers of alumina. *J. Chromatogr.* 11, 260, 1963.
9. Jakovljevic, I.M., Seay, L.D., and Shaffer, R.W., Assay methods for some *Vinca rosea* alkaloids. *J. Pharm. Sci.* 53, 553, 1964.
10. McLaughlin, J.L., Goyan, J.E., and Paul, A.G., Thin layer chromatography of ergot alkaloids. *J. Pharm. Sci.* 53, 306, 1964.
11. Stahl, E., *Thin-Layer Chromatography*, 2nd ed., George Allen & Unwin, London, 1969, pp. 446–454.
12. Baerheim Svendsen, A. and Verpoorte R., *Chromatography of Alkaloids, part A: Thin-Layer Chromatography*, Elsevier, Amsterdam, 1983.
13. Bisset, N.G. et al., Muscle-relaxant activity in Asian *Strychnos* species. A re-examination of two western Malaysian dart poisons. *Lloydia* 40, 546, 1977.

14. Verpoorte, R. and Baerheim Svendsen, A., Thin-layer chromatography of some quaternary alkaloids and alkaloid N-oxides. *J. Chromatogr.* 124, 152, 1976.
15. Wagner, H. and Bladt, S., *Plant Drug Analysis: A Thin Layer Chromatography Atlas*, 2nd ed., Springer, Berlin, 1995, p. 30.
16. Baerheim Svendsen, A. and Verpoorte, R., *Chromatography of Alkaloids, part A: Thin-Layer Chromatography*, Elsevier, Amsterdam, 1983, pp. 331, 341.
17. Baerheim Svendsen, A. and Verpoorte, R., *Chromatography of Alkaloids, part A: Thin-Layer Chromatography*, Elsevier, Amsterdam, 1983, p. 392.
18. Šantavý, F., Alkaloids—indole alkaloids, in Stahl, E., Ed., *Thin-Layer Chromatography*, 2nd ed., George Allen & Unwin, London, 1969, pp. 446–454.
19. Baerheim Svendsen, A. and Verpoorte, R., *Chromatography of Alkaloids, part A: Thin-Layer Chromatography*, Elsevier, Amsterdam, 1983, p. 363.
20. Court, W.E. and Timmins, P., Thin-layer chromatographic behavior of some *Rauwolfia* alkaloids on silica-gel layers. *Planta Med.* 27, 319, 1975.
21. Bisset, N.G. and Phillipson, J.D., Alkaloids from seeds of *Strychnos wallichiana* Steud ex DC (*Strychnos-cinnamomifolia* Thwaites var. *wightii* A W Hill). *J. Pharm. Pharmacol.* 25, 563, 1973.
22. Phillipson, J.D. and Hemingway, S.R., Chromatographic and spectroscopic methods for the identification of alkaloids from herbarium samples of the genus *Uncaria. J. Chromatogr.* 105, 163, 1975.
23. Paxton, J.D., Assays for antifungal activity, *in Methods in Plant Biochemistry, Vol. 6: Assays for Bioactivity*, Hostettmann, K., Ed., Academic press, London, pp. 33–46.
24. Vanden Berghe, D.A. and Vlietinck, A.J., Screening methods for antibacterial and antiviral agents from higher plants, in *Methods in Plant Biochemistry, Vol. 6: Assays for Bioactivity*, Hostettmann, K., Ed., Academic press, London, pp. 47–70.
25. Verpoorte, R., Antimicrobially active alkaloids, in *Alkaloids: Biochemistry, Ecology and Medicinal Applications*, Roberts, M.F. and Wink, M., Eds., Plenum Press, New York, 1998, pp. 398–405.
26. Menn, J.J. and Mc Bain, J.B., Detection of cholinesterase-inhibiting insecticide chemicals and pharmaceutical alkaloids on thin-layer chromatograms. *Nature* 209, 361, 1966.
27. Rhee, I.K. et al., Screening for acetylcholinesterase inhibitors from Amaryllidaceae using silica gel thin-layer chromatography in combination with bioactivity staining. *J. Chromatogr. A* 915, 217, 2001.
28. Rhee, I.K., van Rijn, R.H., and Verpoorte, R., Qualitative determination of false-positive effects in the acetylcholinesterase assay using thin layer chromatography. *Phytochem. Anal.* 14, 127, 2003.
29. Marston, A., Kissling, J., and Hostettmann, K., A rapid TLC bioautographic method for the detection of acetylcholinesterase and butyrylcholinesterase inhibitors in plants. *Phytochem. Anal.* 13, 51, 2002.
30. Andrade, M.T. et al., Indole alkaloids from *Tabernaemontana australis* (Muell. Arg) Miers that inhibit acetylcholinesterase enzyme. *Bioorg. Med. Chem.* 13, 4092, 2005.
31. Tam, M.N., Nikolovadamyanova, B., and Pyuskyulev, B., Quantitative thin-layer chromatography of indole alkaloids. II. Catharanthine and vindoline. *J. Liq. Chromatogr.* 18, 849, 1995.
32. Klyushnichenko, V.E. et al., Determination of indole alkaloids from *R. serpentina* and *R. vomitoria* by high-performance liquid chromatography and high-performance thin-layer chromatography. *J. Chromatogr. A* 704, 357, 1995.
33. Monforte-González, M. et al., Quantitative analysis of serpentine and ajmalicine in plant tissues of *Catharanthus roseus* and hyoscymaine and scopolamine in root tissues of

Datura stramonium by thin layer chromatography–densitometry. *Phytochem. Anal.* 3, 117, 1992.

34. Shellard, E.J. and Houghton, P.J., The *Mitragyna* species of Asia part XIX, The alkaloidal pattern in *Mitragyna parvifolia* (Roxb.) Korth. *Planta Med.* 20, 82, 1971.

35. Baser, K.H.C. and Bisset, N.G., Alkaloids of *Strychnos nux-vomica. Phytochemistry* 21, 1423, 1982.

36. Houghton, P.J. and Bisset, N.G., Drugs of ethno-origin, in *Drugs in Central Nervous System Disorders*, Horwell, D.C., Ed., Dekker, New York, 1985, pp. 283–332.

37. Baerheim Svendsen, A. and Verpoorte, R., *Chromatography of Alkaloids, part A: Thin-Layer Chromatography*, Elsevier, Amsterdam, 1983, pp. 409–413.

38. Delima, C.G. et al., Utilization of *o*-phthalaldehyde–sulfuric acid as a spray reagent in thin-layer chromatographic detection of some indolealkylamines and application to cutaneous secretion of toad species. *Talanta* 38, 1303, 991.

39. Schwarz, M.J. et al., Activities of extract and constituents of *Banistereopsis caapi* relevant to parkinsonism. *Pharmacol. Biochem. Behav.* 75, 627, 2003.

40. Wagner, H. and Bladt, S., *Plant Drug Analysis: A Thin Layer Chromatography Atlas*, 2nd ed., Springer, Berlin, 1995, p. 26.

41. Szepesi, G., Molnar, J., and Nyeredy, S., Separation and determination of ergotoxine and dihydroergotoxine alkaloids by a TLC method. *Z. Anal. Chem.* 294, 47, 1979.

42. Fowler, R., Gomm, P.J., and Patterson, D.A., Thin-layer chromatography of lysergide and other ergot alkaloids. *J. Chromatogr.* 72, 139, 1972.

43. Vinson, J.A., Hooyman, J.E., and Ward, C.E., Identification of street drugs by thin layer chromatography and a single visualization reagent. *J. Forensic Sci.* 20, 552, 1975.

44. Baerheim Svendsen, A. and Verpoorte, R., *Chromatography of Alkaloids, part A: Thin-Layer Chromatography*, Elsevier, Amsterdam, 1983, p. 375.

45. Baerheim Svendsen, A. and Verpoorte, R., *Chromatography of Alkaloids, part A: Thin-Layer Chromatography*, Elsevier, Amsterdam, 1983, pp. 363–366.

46. Wagner, H. and Bladt, S., *Plant Drug Analysis: A Thin Layer Chromatography Atlas*, 2nd ed., Springer, Berlin, 1995, p. 32.

47. Baerheim Svendsen, A. and Verpoorte, R., *Chromatography of Alkaloids, part A: Thin-Layer Chromatography*, Elsevier, Amsterdam, 1983, p. 414.

48. Paulo, A. and Houghton, P.J., Chemotaxonomy of the genus *Cryptolepis. Biochem. Syst. Ecol.* 31, 155, 2003.

49. Choo, C.-Y. and Chan, K.-T., High performance liquid chromatography analysis of canthinone alkaloids from *Eurycoma longifolia. Planta Med.* 68, 382, 2002.

50. Nolte, B.A. et al., Rapid micro-assay of camptothecin in *Camptotheca acuminata. Planta Med.* 67, 376, 2001.

51. Gravel, E. et al., A fluorescence detection combined with either HPLC or HPTLC for pharmaceutical quality control in a hospital chemotherapy production unit: application to camptothecin derivatives. *J. Pharm. Biomed. Anal.* 39, 581, 2005.

52. Formisyn, P. et al., Determination of 9-hydroxyellipticine by redox colorimetry. *J. Pharm. Biomed. Anal.* 10, 427, 1992.

25 TLC of Isoquinoline Alkaloids

Monika Waksmundzka-Hajnos
and Anna Petruczynik

CONTENTS

25.1 ISOQUINOLINE ALKALOIDS

25.1.1 DEFINITION, CHEMISTRY, AND CLASSIFICATION OF ISOQUINOLINE ALKALOIDS

To the isoquinoline group belongs numerous alkaloids having as main element molecule isoquinoline or tetrahydroisoquinoline ring. Among isoquinoline alkaloids eight main types (groups) of compounds are distinguished (see Table 25.1). There are following groups of alkaloids: benzylisoquinoline and benzyltetraisoquinoline group, aporphine group, protoberberine group, benzophenantridine group, protopine group, phthalideisoquinoline group, morphinane group, and emetine group. Moreover special group of isoquinoline alkaloids of dimeric structure—bis-benzyloisoquinoline alkaloids—is also included into isoquinoline group.

Biogenetically isoquinoline alkaloids are classified into phenylalanine group or its hydroxyderivative—tyrosine. These precursors take part in synthesis of alkaloid molecules. For example, isoquinoline alkaloids of type benzylisoquinoline are synthesized from phenylalanine and phenylacetaldehyde, tyrosine is the precursor of all opium alkaloids. However, alkaloids of type aporphine, protoberberine, and morphinane arise from the benzylisoquinoline and benzyltetraisoquinoline alkaloids by appropriate condensations.

25.1.2 OCCURRENCE AND MEDICAL SIGNIFICANCE OF ISOQUINOLINE ALKALOIDS

Isoquinoline alkaloids are spread mainly in the following order of plants: *Ranunculales, Papaverales, Geraniales, Rutales, Plumboginales, Myrtiflorae,* and *Rosales.*

Plant materials, which contain isoquinoline alkaloids, can be used in therapy and are included into pharmacopoeias or are used as a source of individual compounds used in therapy.

Opium (*Opium crudum, Laudanum, Meconium*), hardened latex of *Papaver somniferum* L. from Papaveraceae family, is obtained from unripe poppy heads or from poppy straw. Opium is obtained mainly from poppy cultivated in Asia and Europe. Opium should contain about 10% of morphine. Opium is used as alkaloids' material contained 20%–25% of alkaloids. Main alkaloids of opium are morphine, noscapine, codeine, papaverine, tebaine, laudanozine, and retykuline. Opium is used in therapy as *Tinctura Opii, Extractum Opii,* and *Opium pulveratum* with determined morphine content. Opium can also be used for production of morphine as well as other alkaloids. Opium has strong narcotic properties and demonstrates analgesic, spasmolytic, antitussive, and obstructive properties.

Recently *Papaver bracteatum* Lindl is widely cultivated because it does not contain narcotic alkaloids such as morphine. Material can be used for industrial isolation of tebaine, which can be easily transformed into codeine and morphine.

Radix Ipecacuanhae coming from *Cephaëlis ipecacuanha* Rich (*Psychotria ipecacuanha* Stokes, *Uragoga ipecacuanha* Bail, *Cephaëlis acuminata* Karst) from *Rubiaceae* family grow in South America and Asia and are also cultivated. Pharmacopoeial material is *Radix Ipecacuanhae*, which contains isoquinoline alkaloids (2%–6%) such as emetine, psychotrine, α-methylpsychotrine, emetamine, and protoemetine. Ipecacuanhae alkaloids irritate strongly mucous membranes and have expectoratic and secretolitic properties, in high dose have emetic activity. Emetine also has antiprotozoan properties. Radix Ipecacuanhae are used as expectorans in various recipe forms.

Herba Chelidonii from *Chelidonium majus* L. Papaveraceae family, growing in countries from temperate zone, contains about 20 different isoquinoline alkaloids.

TABLE 25.1

Structures of Isoquinoline Alkaloids

Type	Structure	Substituents	Alkaloid	Plant Raw Materials	Activity
Benzoisoquinoline alkaloids	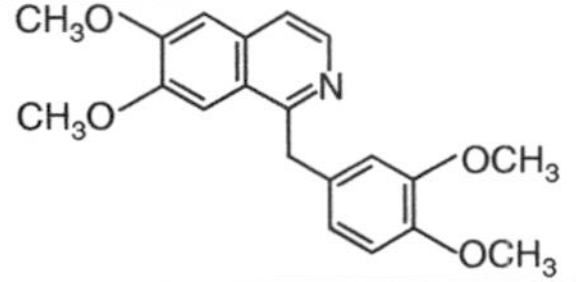	—	Papaverine	Opium	Spazmoliticum
Benzyltetrahydroisoquinoline alkaloids	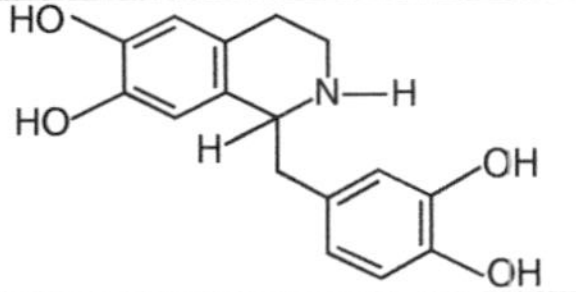	—	Norlaudanosoline		
	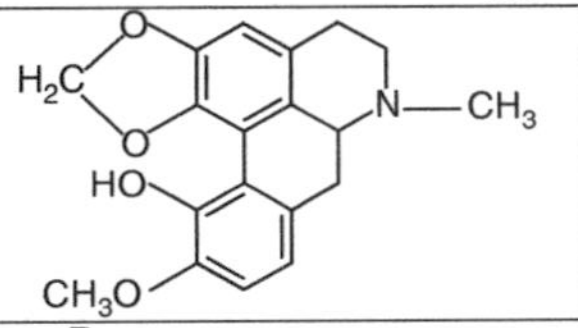	—	Bulbokapine	Species: *Corydalis*	Narcotic
Aporphine alkaloids	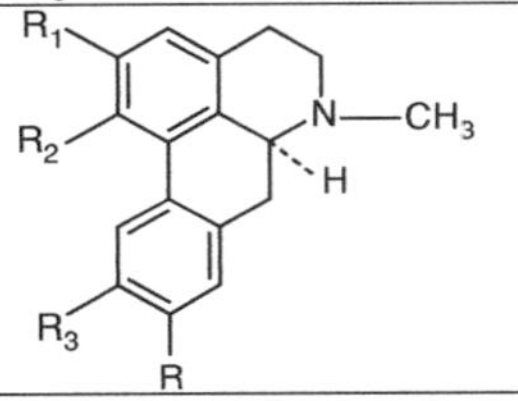	$R_1 = OCH_3$ $R_2 = OCH_3$ $R_3 = OCH_3$ $R_4 = OCH_3$	Glaucine	Species: *Glaucium*	Spasmolytic, expectorant

(continued)

TABLE 25.1 (continued)
Structures of Isoquinoline Alkaloids

Type	Structure	Substituents	Alkaloid	Plant Raw Materials	Activity
Aporphine alkaloids		$R_1 = OH$ $R_2 = OCH_3$ $R_3 = OCH_3$ $R_4 = OH$	Boldine	*Peumus boldus*	Choleretic agent
		—	Magnoflorine	Family: Ranunculaceae	Hypotensive, curare-like drug
Subgroup proaporphine		$R_1 = OH$ $R_2 = OCH_3$ $R_3 = H$	Crotonosine	Species: *Croton*	Psychotropic
		$R_1 = OCH_3$ $R_2 = OH$ $R_3 = CH_3$	Glaziovine	*Ocotea glaziovii*	Sedative
Protoberberine alkaloids		$R_1 + R_2 =$ $-O-CH_2-O-$ $R_3 = OCH_3$ $R_4 = OCH_3$	Berberine	*Chelidonium majus, Hydrastis canadensis, Berberis vulgaris* Family: Berberidaceae, Papaveracae, Ranunculacae	Cholagogic, antibacterial, antidiarrheal

Class	Structure	R substituents	Name	Family/Species	Pharmacological action
Protoberberine alkaloids		$R_1, R_2, R_3, R_4 = OCH_3$	Palmatine	Family: Berberidaceae, Papaveracae, Ranunculacae, Lauraceae	Antiarrhythmic, analgetic, antibacterial
		$R = OCH_3$	Noscapine (narcotine)	Family: Papaveraceae, Species: Hydrastis, Berberis	Stimulator of respiratory center
		$R = H$	Hydrastine	Rhizoma Hydrastidis	Hemostaticum uterinum
			Bicuculine	Species: Papaveraceae	Hypertensive
		$R_1 + R_2 = CH_2$	Chelidonine	Family: Papaveraceae, Ranunculaceae, Rutaceae	Interaction on central nervous system, spasmolytic, mitosis-arresting toin, bacteriostatic, fungistatic
Benzphenanthridine alkaloids		$R_1, R_2 = CH_3$	Homochelidonine	Family: Papaveraceae, Ranunculaceae, Rutaceae	Bacteriostatic, fungistatic
		$R_1 + R_2 = CH_2$	Sanguinarine	Family: Papaveraceae, Ranunculaceae, Rutaceae	Weak narcotic, inhibitor acethylcholinesterase, bacteriostatic, fungistatic

(continued)

TABLE 25.1 (continued)
Structures of Isoquinoline Alkaloids

Type	Structure	Substituents	Alkaloid	Plant Raw Materials	Activity
Benzphenanthridine alkaloids		$R_1, R_2 = CH_3$	Cheleritrine	Family: Papaveraceae, Ranunculaceae, Rutaceae	Irritant on skin, local anaesthetic, strongly toxic, bacteriostatic, fungistatic
		$R_1 + R_2 = CH_2$	Nitidine	Kind *Zanthoxylum*	Antileukaemic
		$R_1, R_2 = CH_3$	Fagaronine	*Fagara zantoxyloides*	Antitumour
Protopine alkaloids		$R_1 + R_2 = CH_3$	Protopine	Family: Papaveraceae	
		$R_1 = R_2 = CH_3$	Cryptopine	Family: Papaveraceae, Rutaceae, Berberidaceae	Antiarrhythmic, interaction on central nervous system
Morphine alakloids		$R_1 + R_2 = H$	Morphine	*Papaver somniferum,* Family: Menispermasceae	Narcotic, analgesic
		$R_1 = CH_3$ $R_2 = H$	Codeine	*Papaver somniferum,* Family: Menispermasceae	Antitussive, analgesic
			Tebaine	Family: Papaveraceae	

Class	Structure	R	Name	Source	Action
Emetine alkaloids		$R = CH_3$	Emetine	*Cephaëlis Ipecacuanha*	Expectorant, emetic
		$R = H$	Cephaline	*Cephaëlis ipecacuanha*	
		$R = H$	Psychotrine	*Cephaëlis ipecacuanha*	
		$R = CH_3$	O-Methylpsychotrine	*Cephaëlis ipecacuanha*	
			Talicarpine	*Thalictrum dasycarpum*	
Bis-benzylisoquinoline alkaloids			Tubocurarine	Family: Menispermaceae	Curare-like
			Tetrandrine	*Stephania tetrandra*	Hypotensive, antipyretic, antiphlogistic

The main *Chelidonium majus* alkaloids are chelidonine, α-homochelidonine, chelerithrine, sanguinarine, berberine, and protopine. The material indicates spasmolytic, cholagogic, and middle sedative properties. Herba Chelidonii is mainly used as spasmolyticum in diseases of alimentary canal, inflammation of bileduct, and cholecystolithiasis. Chelidonium alkaloids also reveal certain cytostatic, fungostatic, and antiviral activity. Herba Chelidonii is part of some herbal mixtures and spasmolytic and antifungal confections.

Rhizoma Hydrastis comes from the plant, *Hydrastis candadensis* L. from Ramunculaceae family, growing in North America forest and are also cultivated. Material contains 2%–6% of isoquinoline alkaloids—hydrastine, berberine, tetrahydroberberine (canadine)—used in the form of *Extractum Hydrastidis canadensis fluidum* as hemostaticum in metrorrhagia.

Cortex Berberidis is material obtained from roots and twigs of *Berberis vulgaris* from Berberidaceae family growing in Europe. It contains about 1% of berberine and with regard to cholagogic and spasmolytic properties is applicable in liver and bile duct diseases.

Flos Rhoeados from *Papaver rhoeas* L. from Papaveraceae family contains as main alkaloid rheadine of structure of papaverubine and other isoquinoline alkaloids, is used rarely in herbal antitussive mixtures.

Herba Fumariae from *Fumaria officinalis* L. belonging to Papaveraceae family contains numerous of isoquinoline alkaloids such as protopine and tetrahydroberberine derivatives—synactine, tetrahydrocoptizine, corydaline, bulbocapine, and fumarocitine. Herba Fumariae used in herbal mixtures of cholagogue properties has also application in preparations against migraines connected with hyperfunction of gall bladder.

Herba Glaucii flavi coming from *Glaucium flavum* Crantz from Papaveraceae family contains numerous of alkaloids such as galaucine, protopine, sanguinarine, chelidonine, chelerythrine, magnoflorine, isoboldine, coridine, izocoridine, norchelidonine, and other. Material serves to obtain glaucine and also some herbal preparations of antitussive activity.

Folium Boldo originates from *Peumus boldus* Bailon from Monimiaceae family growing in South America. Material contains among other compounds about 2% of isoquinoline alkaloids. Many of them are boldine used in therapy as cholereticum.

25.2 SAMPLE PREPARATION FOR TLC ANALYSIS

In the extraction and isolation of alkaloids one has to consider that alkaloids usually occur in plants as salts of organic or inorganic acids, sometimes exist as tannin complexes, and always together with many types of nonalkaloidal compounds [1]. The procedure of extraction depends on ballast substances coexisting with alkaloidal fraction. Plant material with a high lipid or other nonpolar compounds content (chlorophylles, waxes) should preferably be extracted with water containing dilute acids to obtain the alkaloids in aqueous solution as salts (e.g., hydrochlorides). Plant material containing a large number of water soluble compounds (tannines, phenols) should be extracted with organic solvents immiscible with water after addition of alkali to obtain the alkaloids in the organic solvent as free bases [1].

25.2.1 DIFFERENT METHODS OF EXTRACTION

Before extraction procedure the plant tissue should be reduced into small particles in grinder or homogenizer. Various methods of liquid solid extraction are used for extraction of isoquinoline alkaloids from plant material. There are several reports taking the advantage of solubility of bases in aqueous acidic solutions. The following acids were used as extractants: aqueous solutions of H_2SO_4 (0.05M) [2], HCl (3%) [3], (0.1M) [4], and acetic acid [5]. Acidified methanol or ethanol with H_2SO_4 (0.5M) [6], HCl (0.05M) [7] (1M) [8] was also applied.

Most often methanol or 96% ethanol without any additives has been used as extractants and alkaloid fraction is reextracted with water containing various acids. In these cases alcohol is evaporized and alkaloids are dissolved in the following acids: 2.5% aqueous tartaric acid [9], aqueous hydrochloric acid 10% [10], 5% [11], 0.02M [12], sulfuric acid 1M [13], 10% aqueous phosphoric acid [14], or 5% aqueous citric acid [15].

Sometimes a series of organic solvents are used for extraction of isoquinoline alkaloids from plant material. For example, methanol + dichloromethane [16], methanol followed with chloroform [17] or dichloromethane [18].

Usually simple techniques of extraction were applied. It was long lasting maceration or percolation at room temperature with aqueous extractants [19] or alcohols [10,11,14,18,20–22]. Sometimes extraction was performed by use of water bath with hot solvent (usually under reflux): diluted acetic acid [5], methanol, or ethanol [9,12,15,23,24]. The extraction was often assisted by shaking [4,23] or ultrasonification [7,8,25–27]. The use of turbo extraction was also reported [28].

25.2.2 SAMPLE PURIFICATION AND CONCENTRATION

Purification of the alkaloids present in the crude plant extracts can be achieved by reextraction of the alkaloids as free bases with an organic solvent immiscible with water from the aqueous solution after addition of alkali (at pH > 8) [1]. Reextraction of alkaloids as free bases can be performed with an organic solvent from the acidic aqueous solution after addition of alkali. For crude extracts water acidic solutions have been adjusted to high pH (8–11) usually with aqueous ammonia [2–4,10,11,13–15,29] with sodium hydroxide [12] or sodium bicarbonate [9]. Then alkaloid fraction was reextracted with dichloromethane [2,4,14], chloroform [6,8–10,15,29], or diethyl ether [3,12,13].

Sometimes extraction from the aqueous neutral solutions or water suspensions was performed [22,28] by use of organic solvents such as ethyl acetate [28], diethyl ether, or n-butanol [22].

Usually liquid column chromatography is used as the next step of alkaloid isolation. The main adsorbent used in classic column chromatography is silica, which should be introduced to the column as suspension with nonpolar or medium polar solvent. Fraction of alkaloids, dissolved in medium polar solvent (of lower polarity than eluent), after introduction to the column was eluted with various medium polar or polar eluents such as dichloromethane + methanol [8,17], chloroform + methanol [3,9,15], n-butanol [6], or n-hexane + ethyl acetate + ethanol (in gradient mode) [28]. In some cases ion-exchange chromatography is used for the separation of base fraction [3,12,15,29] and more rarely reversed-phase-SPE (RP-SPE) has been used for sample preparation [7,30].

25.3 SYSTEMS AND TLC TECHNIQUES USED IN THE ISOQUINOLINE ALKALOIDS

25.3.1 NORMAL PHASE SYSTEMS

Many TLC systems have been developed for the separation of the isoquinoline alkaloids. The first choice is silica but the use of alumina layers [10,31] and polar bonded stationary phases such as aminopropyl-silica layers [32] and cyanonopropyl-silica layers [33] with nonaqueous eluents was also reported. Usually silica plates and strongly polar eluents are used for these purposes—strongly polar modifier (methanol, ethanol) and medium polar diluent such as chloroform or dichloromethane [4–6,8,34–37] more rarely benzene or toluene [38,39]. Because bisbenzylisoquinoline and most of isoquinoline alkaloids are strong bases, therefore often rather strongly basic TLC systems have been used (see Table 25.2) The use of basic additives to the eluent such as aqueous ammonia or amines is often reported [1,2,9–11,16,28,39,70–73]. Sometimes silica layer was deactivated with ammonia vapors before chromatographic run [6,27]. Figure 25.1 shows that the use of conditioning of the plate with ammonia vapors before the development markedly improved separation of protoberberine alkaloids when eluent, containing ammonia was used for their separation (compare Figure 25.1A and B). The use of silica with eluents containing acetic acid has also been reported [42,74–76].

The retention of alkaloids can be optimized by the change of modifier concentration. When the analysis of retention–eluent composition relationships is performed it can be concluded that R_M versus $\log c$ relationships are in most cases linear, which show accordance with the displacement theory of adsorption model [77]. The correlation coefficient was between 0.97 and 0.99, which indicates the agreement of the results with the adsorption model. Absolute values of the slopes of $R_M = f(\log c)$ are close to 1 ($m < 1$) in most cases. This indicates one-site binding of these multifunctional compounds with the active centers of adsorbent. This may be result of the formation of internal hydrogen bonds or steric hindrance. The example of R_M versus $\log c$ plots for isoquinoline alkaloids is presented in Figure 25.2 [77]. The change of modifier concentration causes marked differences in retention—the optimum values of R_M can be obtained. Also with the change of eluent composition differences in separation selectivity can be received and used for particular purposes.

Addition of diethylamine or aqueous ammonia resulted in marked improvement of spot shapes and symmetry. It is apparent that addition of DEA to the nonaqueous mobile phases usually resulted in increased system efficiency and improvement of peak symmetry [33,67]. It was also observed that increasing the amine concentration usually resulted in reduced retention. Figure 25.3 presents plots of R_M versus ammonia concentration obtained on cyanopropyl plates with eluents consisting of methanol and diisopropyl ether containing 0%–5% ammonia. The marked decrease in retention is observed in small range of ammonia concentration (to 2%) and then

TABLE 25.2

Systems Used for the TLC of Alkaloids in Analytical Scale

Alkaloids	Source	Adsorbent	Eluent	Detection	Remarks	References
Narceine, morphine, codeine, papaverine, thebaine, narcotine	*Papaver rhoeas*	SiO$_2$ 60	T/Me$_2$CO/EtOH NH$_3$ (25%)	UV 365 nm Dragendorff reagent	Identification	[19]
Hydrastine, berberine	*Hydrastis Canadensis* L.	SiO$_2$ G	Cyclohexane/DEA	UV 250 nm Dragendorff reagent Iodoplatine reagent	Qualitative analysis	[25]
Chelidonine, sanguinarine, coptisine, berberine, chelerythrine	*Chelidonium majus* L.	SiO$_2$ 60	CH$_2$Cl$_2$/MeOH CHCl$_3$/MeOH	UV 254 nm	Qualitative analysis, densitometry	[5]
Naphthylisoquinoline alkaloids	*Triphyophyllum peltatum*	SiO$_2$ 60	CHCl$_3$/AcOEt/MeOH/HCl	UV light		[21]
Berbamine, berberine, columbamine, magnoflorine, oxyacanthine, palmatine	*Berberis crataegina*	Al$_2$O$_3$ SiO$_2$	CHCl$_3$/MeOH CHCl$_3$/MeOH/NH$_3$			[10]
Protoberberine alkaloids	*Chelidonium majus* L.	SiO$_2$	CHCl$_3$/MeOH/H$_2$O/ CH$_3$COOH		Monitoring fractions	[42]
Morphine, codeine, thebaine, papaverine, noscapine	Opium	Al$_2$O$_3$ 60 type E	AcOEt/CH$_2$Cl$_2$	UV 365 nm	OPLC Densitometry	[95]
Morphine, codeine, thebaine, papaverine, noscapine, pholcodine, ethylmorphine, diacetylmorphine, buprenorphine		SiO$_2$	(iPr)$_2$O/EtOH/DEA CHCl$_3$/EtOH/DEA T/AcOEt/DEA tert-butanol/NH$_3$/MeOH/H$_2$O	Dragendorff reagent Iodoplatine reagent	OPLC AMD	[4]
Morphine, codeine, thebaine, narcotine, papaverine, narceine	*Papaver somniferum*	SiO$_2$	T/Me$_2$CO/EtOH/NH$_3$	Dragendorff reagent with NaNO$_2$	Qualitative analysis, densitometry	[28]

(continued)

TABLE 25.2 (continued)

Systems Used for the TLC of Alkaloids in Analytical Scale

Alkaloids	Source	Adsorbent	Eluent	Detection	Remarks	References
Isoquinoline alkaloids	*Corydalis cava, Corydalis intermedia, Corydalis solida*	SiO_2 with enzymes	T/AcOEt/MeOH	UV 365 nm, Dragendorff reagent	Bioautography	[26]
Berberine, chelidonine, sanquinarine, palmatine, hydrastine	*Berberis vulgaris, Chelidonium majus, Jatrorrhiza palmate, Hydrastis canadensis*	SiO_2	AcOEt/THF/CH_3COOH	UV 365 nm, Dragendorff reagent	OPLC	[76]
Capnoidine, adlumine, tetrahydropalmatine, bicuculine, protopine, salutarine, scoulerine	*Corydalis sempervirens*	SiO_2	AcOEt/MeOH $CHCl_3$/MeOH	Dragendorff reagent	Qualitative analysis	[13]
7-Hydroxy-6-methoxy-isoquinoline, 7-hydroxy-6-methoxy-1(2H)isoquinoline, 6,7-dimethoxy-N-methl-3,4-dioxo-1(2H)-isoquinoline	*Menispermum dauricum*	SiO_2	$CHCl_3$/MeOH/Me_2CO	NMR	Qualitative analysis	[3]
Jatrorrhizine, columbamine, palmatine	*Jatrorrhiza palmata*	SiO_2	iProH/ammonium carbonate/Dx	UV 254 or 365 nm	Circular TLC	[94]
Morphine, codeine		SiO_2	MeOH/CH_2Cl_2/NH_3 MeOH/CH_2Cl_2/ether/NH_3 MeOH/CH_2Cl_2/Hx MeOH/Hx	UV 210 nm	AMD	[60]

Alkaloids	Plant source	Stationary phase	Mobile phase	Detection	Method	Ref.
Allocryptopine, berberine, chelerythrinhe, chelidonine, chelirubine, chelilutine, chelirubine, coptisine, corysamine, dihydrochelerithrine, dihydrosanquinarine, homochelidonine, magnoflorine, protopine, sanquinarine	Chelidonium majus	SiO_2	T/AcOEt/MeOH EtOH/CHCl$_3$/AcOH	UV 254 or 365 nm Dragendorff reagent	UMD	[31]
Chelidonine, codeine, dionine, emetine, glaucine, narcotine, papaverine, paracodine, protopine		SiO_2	CHCl$_3$/AcOEt CHCl$_3$/AcOEt/2-PrOH CHCl$_3$/AcOEt?MeOH	UV 254 or 365 nm	SiO_2 with addition of NaHCO$_3$ Dragendorff reagent	[77]
Chelirubine, sanquinarine, chelilutine, chelerithrine, allocryptopine, protopine, chelidonine, isocorydine	Chelidonium majus	SiO_2	MeOH/phosphate buffer MeOH/T T/AcOEt/MeOH	UV 365 nm Dragendorff reagent	Pseudo-RP	[87]
Boldine, berberine, emetine, glaucine, codeine, laudanosine, narceine, noscapine, papaverine, protopine, tubocurarine	Fumaria officinalis	SiO_2	MeOH/H$_2$O MeOH/buffer (phosphate or acetate) MeOH/buffer/HDEHP MeOH/buffer/SOS MeOH/buffer/DEA MeOH/H$_2$O/NH$_3$ MeOH/Me$_2$CO/DEA MeOH/CH$_2$Cl$_2$/DEA MeOH/Me$_2$CO/iPr$_2$O/DEA	UV 254	Densitometry 2D TLC	[67]
Dihydrochelerythrine, dihydrosanquinarine	Bocconia arborea	SiO_2	Hx/CHCl$_3$/MeOH	Bioautography	Antimicrobial activity	[97]

(continued)

TABLE 25.2 (continued)
Systems Used for the TLC of Alkaloids in Analytical Scale

Alkaloids	Source	Adsorbent	Eluent	Detection	Remarks	References
Sanquinarine, chelerythrine	*Psoralea corylifolia Sanguinaria canadensis*	SiO_2	$CHCl_3/MeOH$	Dragendorff reagent	Monitoring fraction	[58]
Morphine		SiO_2	$AcOEt/MeOH/NH_3$	UV 254 nm	Determination in urine	[98]
Morphine, heroin, codeine, thebaine, papaverine, narcotine		SiO_2	$CHCl_3/EtOH$	$FeCl_3$ then 1% acidified alcoholic 2,2'-dipyridyl + 100°C for 10 min.	Qualitative analysis	[66]
Heroin, morphine, codeine, thebaine, papaverine, narcotine	*Opium*	SiO_2	$CHCl_3/EtOH$	$HgCl_2$ $K_4Fe(CN)_6$	Qualitative analysis	[59]
Narceine, paracodine, morphine, codeine, thebaine, papaverine, dionine, apomorphine	*Papaver somniferum, Chelidonium majus*	SiO_2	$T/MeOH/Me_2CO$ $B/Me_2CO/iPr_2O/NH_3$ $AcOEt/MeOH/NH_3$ $AcOEt/iPrOH/NH_3$ $MeOH/T$	UV 254, or 365 nm, Dragendorff reagent	Silica gel impregnated with salts of the metals Ni, Zn, Cr, Co	[81]
Heroin	*Opium*	SiO_2	$CHCl_3/AcOEt$	$CuCl_2$ + $K_4Fe(CN)_6$	Qualitative analysis	[40]
Heroin, morphine, papaverine, narcotine		SiO_2	$B/MeCN/MeOH$ Cyclohexane/T/DEA	UV 254 nm	Qualitative analysis	[43]
Heroin		SiO_2	$CHCl_3/MeOH$	UV 254 nm	Study of the stability of heroin in methanol	[46]

Alkaloids	Plant source	Sorbent	Solvent system	Detection	Application	Ref.
Protopine, allocryptopine, sanguinarine, chelerythrine, berberine, chelidonine		Polyamide	EtOH CH$_2$Cl$_2$	Radiodetection	Study of biodistribution	[48]
Naphthylisoquinoline alkaloids	Ancistrocladus heyneanus	SiO$_2$	MeOH/CH$_2$Cl$_2$ saturated with NH$_3$ vapor	Dragendorff reagent	Centrifugal TLC	[16]
Quaternary isoquinoline alkaloids	Romneya coulteri	SiO$_2$	CHCl$_3$/MeOH			[29]
Morphine, codeine	Poppy capsules	SiO$_2$	AcOEt/MeoH/T/NH$_3$	UV 275 nm	Quantitative analysis	[30]
Sanquinarine, homochelidonine, chelerythrine, allocrptopine, protopine, chelidonine	Chelidonium majus	SiO$_2$	T/MeOH T/AcOEt/iPrOH T/AcOEt/MeOH	UV 366 nm, Dragendorff reagent	Qualitative analysis	[60]
6,7-methylenedioxy-2-(6-acetyl-2,3-methylenedioxybenzyl)-1(2H)-isoquinoline alkaloids	Corydalis bungeana	SiO$_2$	CHCl$_3$/MeOH		Fraction cntroll	[49]
Sanquinarine, chelerythrine, chelirubine, allocryptopine, sanquirubine, sanquilutine	Sanguinaria canadensis	SiO$_2$	B/MeOH CHCl$_3$/B/MeOH/formamide	UV light	2D TLC	[50]
Napthylisoquinoline alkaloids	Triphyollum peltatum	SiO$_2$ deactivated with NH$_3$	CH$_2$Cl$_2$/MeOH	X-ray	Isolation	[51]
Canadine, stylopine, homochelidonine, protopine, dihydronitidine, dihydroberberine, allocryptopine, norchelerythrine, norsanguinarine		SiO$_2$	AcOEt-EtOH-0.1M NaOH	UV 365 Dragendorff reagent	Qualitative analysis	[52]

(continued)

TABLE 25.2 (continued)
Systems Used for the TLC of Alkaloids in Analytical Scale

Alkaloids	Source	Adsorbent	Eluent	Detection	Remarks	References
Morphine and its derivatives		SiO_2	$CHCl_3$/MeOH-buffer AcOEt-MeOH-NH_3 BuOH/NH_3/H_2O/MeOH	UV 254 nm	Qualitative analysis	[54]
Boldine, chelidonine, dionine, glaucine, narcotine, papaverine, protopine, sanguinarine		SiO_2, aminopropyl, CN, C18	iPrOH/Hx AcOEt/Hx EtMeCO/Hx THF/Hx DX/Hx MeCN/buffer	UV 254 nm	Qualitative analysis	[32]
Sanguinarine, dihydrosanguinarine		SiO_2	Hx/Me_2CO/MeOH	UV 366 nm	Quantitative analysis	[68]
Sanguinarine	*Argemone mexicana*	SiO_2	BuOH/CH_3COOH/H_2O	UV 366 nm	Qualitative analysis	[74]
Erythrine, erythrinine, norprotosinomenine, coclaurine, norreticuline, reticuline, nororientaline, norisoorientaline, norcoclaurine, protosinomenine, orientaline	*Erythrina crist-galli*	SiO_2	AcOEt/EtOH/NH_3 $CHCl_3$/Me_2CO/DEA	Autoradiography	2D TLC	[89]
Morphine, codeine, dihydrocodeine		SiO_2	AcOEt/MeOH/NH_3 MeOH/NH_3 T/DEA/MeOH/AcOEt Me_2CO/H_2O/NH_3	Dragendorff reagent, Marquis reagent, Fast blue B salt, Fast black K salt, Dansyl chloride derivatization	2D TLC	[96]

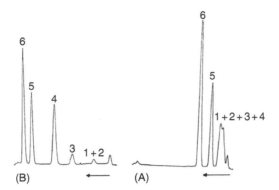

FIGURE 25.1 Influence of ammonia vapor on the chromatographic behavior of protoberberine alkaloids. System: silica/benzene–ethyl acetate–isopropanol–methanol–aqueous ammonia
concentration $(6 + 3 + 1.5 + 0.5, v/v)$. (A) plate conditioned in eluent vapors. (B) Plate
conditioned in ammonia vapors for 15 min. Numbers: 1, columbamine; 2, jatrorrhizine; 3,
palmatine; 4, berberine; 5, epiberberine; 6, coptisine. (From Peishan, X., Yuzhen, Y., and
Qiaoling, L., *J. Planar Chromatogr.*, 5, 302, 1992. With permission.)

retention does not change or slightly increases. Both effects are caused by blocking
of surface silanols and by suppressing of alkaloids' ionization by the basic component of eluent.

NP systems were often used for identification of plant extract components
[31,33,34,36,40,41,44] or for quantitative analysis [5,28,32,34,38,39,45,47,70–73,
78–80]. Figure 25.4 [28] presents separation of isoquinoline alkaloid standards of
morphine group in the system silica HPTLC plates/$T + Me_2O + EtOH + NH_3$ and

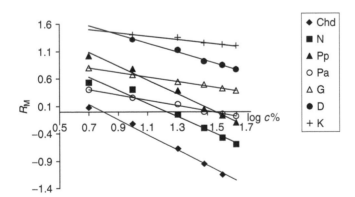

FIGURE 25.2 Plots of R_M against the logarithm of the concentration (% v/v) of modifier in
the mobile phase for isoquinoline alkaloids. Mobile phase: ethyl acetate–chloroform. Solute
identification: Chd, chelidonine; N, narcotine; Pp, protopine; Pa, papaverine; G, glaucine;
D, dionine; K, codeine. (From Soczewiński, E. and Flieger, J., *J. Planar Chromatogr.*, 9, 107,
1996. With permission.)

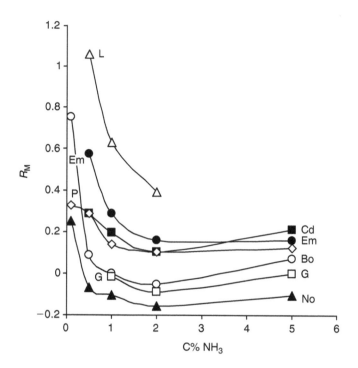

FIGURE 25.3 Plots of R_M against the concentration (% v/v) of ammonia in the mobile phase for isoquinoline alkaloids. System: CN-silica plates/MeOH–iPr$_2$O (20:80). Solute identification: L, ludanosine; Cd, codeine; Bo, boldine; Em, emetine; G, glaucine; P, papaverine; No, noscapine. (From Petruczynik, A., Waksmundzka-Hajnos, M., Michniowski, T., Plech, T., Tuzimski, T., Hajnos, M., Gadzikowska, M., and Jozwiak, G., *J. Chromatogr. Sci.*, 45, 447, 2007. With permission.)

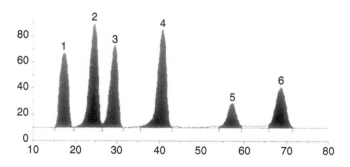

FIGURE 25.4 Densitogram obtained at $\lambda = 600$ nm from the HPTLC separation of morphine alkaloid standards. Numbers: 1, narceine; 2, morphine; 3, codeine; 4, thebaine; 5, papaverine; 6, narcotine. System: Silica 60 HPTLC plates/toluene–acetone–ethanol–aqueous ammonia $(40 + 50 + 6 + 2)$. (From Szabo, B., Lakatos, A., Kászegi, T., and Botz, L., *J. Planar Chromatogr.*, 16, 293, 2003. With permission.)

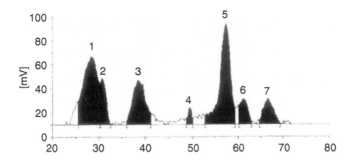

FIGURE 25.5 Densitogram obtained at $\lambda = 600$ nm from the HPTLC separation of the extract of *Papaver somniferum*. Other details as in Figure 25.4. (From Szabo, B., Lakatos, A., Käszegi, T., and Botz, L., *J. Planar Chromatogr.*, 16, 293, 2003. With permission.)

Figure 25.5 shows the separation of the extract of *Papaver somniferum* in the same system for quantitative purposes. Plates were scanned after fluorescence detection.

Figure 25.6 [81] shows densitogram of *Chelidonium majus* alkaloids separated on silica layers by use of $T + MeOH + $ aqueous ammonia. Peaks of berberine, allocryptopine, protopine, homochelidonine, chelidonine, chelerythrine, and sanguinarine are sufficiently well separated and peaks have good symmetry and shape [81].

Figure 25.7 presents video scan and densitogram of separation of individual alkaloids and their mixture on silica layer by use of methanol–acetone–diisopropyl ether eluent containing DEA as mobile phase [67].

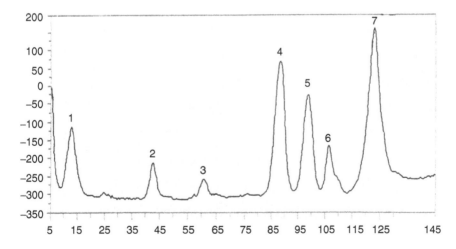

FIGURE 25.6 Densitogram obtained from the alkaloids extracted from *Chelidonium majus* L. on silica $60F_{254}$ with toluene–methanol–aqueous ammonia $(0.5 + 1 + 0.5)$. Compounds: 1, berberine; 2, allocryptopine; 3, protopine; 4, homochelidonine; 5, chelidonine; 6, chelerithrine; 7, sanguinarine. (From Szumiło, H. and Flieger, J., *J. Planar Chromatogr.*, 12, 466, 1999. With permission.)

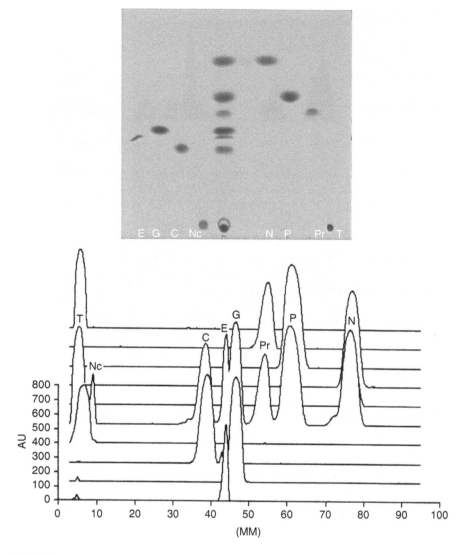

FIGURE 25.7 Densitogram and video scan from chromatogram of the individual isoquinoline alkaloids and their mixture on silica with methanol–acetone–diisopropyl ether–DEA (15 + 15 + 69 + 1). Solutes: E, emetine; G, glaucine; C, codeine; Nc, narceine; N, noscapine; P, papaverine; Pr, protopine; T, tubocurarine. (From Petruczynik, A., Waksmundzka-Hajnos, M., and Hajnos, M.Ł., *J. Planar Chromatogr.*, 18, 78, 2005. With permission.)

25.3.2 RP SYSTEMS

RP-TLC is rarely applied in analysis of alkaloids. However, the use of these systems for the separation of alkaloids is sometimes reported. The RP-TLC plates were used for these purposes with buffered aqueous eluents [53,82,83], with the addition of amines, ion-pair (IP) reagents [53,83], or chiral selectors [84]. Of course for eluents

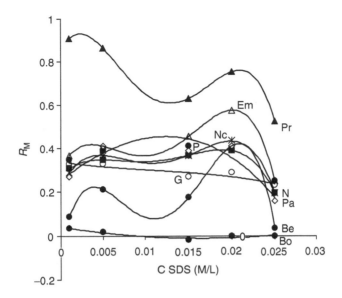

FIGURE 25.8 Relationships between R_M and SDS concentration in mobile phase for selected alkaloids. System C18 W/MeOH–H$_2$O (80:20) buffered with phosphate buffer 0.01 mL at pH 3. Solutes: Em, emetine; Pr, protopine; N, noscapine; Pa, papaverine; Nc, narceine; Be, berberine; Bo, boldine. (From Petruczynik, A., Waksmundzka-Hajnos, M., and Hajnos, M.Ł., *J. Chromatogr. Sci.*, 43, 183, 2005. With permission.)

containing various components of various concentration there are a lot of possibilities for system optimization. The retention parameters and separation selectivity can be controlled by the change of modifier kind and concentration, buffer pH, ion-pair kind and concentration, or amine kind and concentration. By the change of pH in range 2–10 units typical S-curve plots of retention versus pH were obtained. However, in most cases the change of pH does not cause marked differences in spot symmetry and system efficiency [53]. The use of IP reagents or silanol blockers was necessary. Figure 25.8 shows the dependency of the R_M values on the concentration of sodium dodecyl sulfate (SDS) as an IP reagent [53]. The retention of alkaloids initially increases with the increase of SDS concentration in the mobile phase (ranging from 0.001 to 0.02 M SDS) after which a little decrease of retention was observed in most cases. Simultaneously the influence of SDS concentration on the peak symmetry and system efficiency can be noted. The increase of IP reagent concentration causes improvement of system efficiency reaching maximum by 0.025 M of SDS, but peak symmetry was the best by 0.015 M SDS concentration [53]. The retention of alkaloids can also be controlled by the change of amine kind and concentration. Figure 25.9 shows retention versus DEA concentration dependence for isoquinoline alkaloids [53]. The retention of alkaloids decreases with the increase of DEA concentration due to the effect of silanol blocking and then increases due to ion-suppressing effect for alkaloids (see Figure 25.9). The increasing concentration of amine also causes improvement of peak shapes—the highest

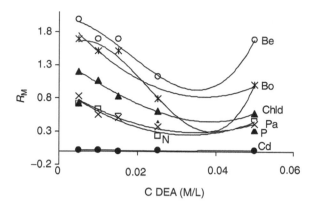

FIGURE 25.9 Relationships between R_M and DEA concentration in mobile phase for selected alkaloids. System C18 W/MeOH–H_2O (90:10). Solutes: Be, berberine; Bo, boldine; Chld, chelidonine; Pa, paracodine; P, papaverine; Cd, codeine. (From Petruczynik, A., Waksmundzka-Hajnos, M., and Hajnos, M.Ł., *J. Chromatogr. Sci.*, 43, 183, 2005. With permission.)

system efficiency and the best peak symmetry. In Figure 25.10 [53], densitograms obtained for protopine in six different chromatographic systems are presented. In a system containing methanol and water alkaloid is strongly retained on RP18 adsorbent—the spot is in chromatogram startline. After adding phosphate buffer to the mobile phase at pH 3 to suppress the silanol ionization, and at pH 8 to suppress the alkaloid ionization, a decrease of retention is observed, but the peaks are still wide and asymmetric. Systems containing amines in eluents were more effective. Thus, the improvement of peak symmetry and of efficiency is noticed with narrow and very symmetric peaks. In the system containing SDS as an IP reagent in a mobile phase, the efficiency gets worse in comparison with systems with amines. However, peak shape and symmetry in IP systems are better in comparison with those of buffered systems [53]. Systems containing DEA in aqueous mobile phases were used for separation of isoquinoline alkaloids' mixture. Figure 25.11 [53] presents video scan and densitogram of plate RP18W triple developed [82] with 80% MeOH in water + 0.05 M DEA for the separation of isoquinoline alkaloid standards and mixture.

Separation of alkaloids in RP systems can be realized by use of CN-silica plates. On CN-silica plates in RP systems pH changes also do not improve the system efficiency [33]. The change of the IP reagent in aqueous eluent causes changes in retention and separation selectivity of the investigated isoquinoline alkaloids, but only for a few alkaloids were system efficiency and spot symmetry satisfactory. In aqueous eluent systems with 2% ammonia, theoretical plates per meter of over 1000 were obtained for most alkaloids, but the spot symmetry was still unsatisfactory. The best results were obtained with 0.1 M DEA as the additive to aqueous methanol. For most of the investigated alkaloids improved system efficiency and symmetric spots were also obtained [33]. Such systems of quite different selectivity than normal phase systems can be applied in multidimensional separations [33,83].

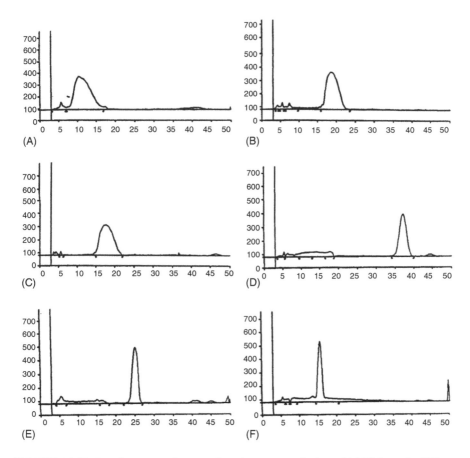

FIGURE 25.10 Densitograms of protpopine chromatographed on C18 W layer in different eluent systems: (A) MeOH–H$_2$O (80:20). (B) MeOH–H$_2$O (80:20) buffered with phosphate buffer at pH 3. (C) MeOH–H$_2$O (80:20) buffered with phosphate buffer at pH 8. (D) MeOH–H$_2$O (80:20) containing 0.1 M of DEA. (E) MeOH–H$_2$O (80:20) containing 0.05 M of TBA–Cl. (F) MeOH–H$_2$O (80:20) buffered with phosphate buffer at pH 3 containing 0.025 M SDS. (From Petruczynik, A., Waksmundzka-Hajnos, M., and Hajnos, M.Ł., *J. Chromatogr. Sci.*, 43, 183, 2005. With permission.)

RP systems are widely used for assessment of lipophilicity of various compounds, where determination by RP-TLC systems is widely applied [86]. RP18 layers and different aqueous eluent systems containing various organic modifiers, buffers at various pH, IP reagents, and amines were applied to obtain retention parameters of selected isoquinoline alkaloids. The systems in which the highest efficiency and the most symmetric peaks were obtained were applied for determination of logk_W (R_{MW}) values. The log$k$ versus modifier concentration relationships were extrapolated to obtain log$k$ values for pure water. The experimental results of lipophilicity have been correlated by different algorithm. The best results in determination of lipophilicity by RP-TLC method for isoquinoline alkaloids were obtained on C18 layers and dioxane–water–ammonia eluent systems [86].

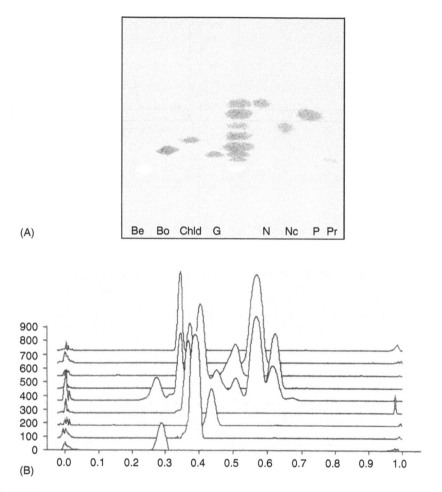

FIGURE 25.11 Video scan (A) and densitogram (B) of RP18W plate with the separated mixture and isoquinoline alkaloid standards. Plate triple developed with the eluent: MeOH–H₂O (80:20) containing 0.05 mL DEA. Solutes: Be, berberine; Bo, boldine; Chld, chelidonine; G, glaucine; N, noscapine; Nc, narceine; P, papaverine; Pr, protopine. (From Petruczynik, A., Waksmundzka-Hajnos, M., and Hajnos, M.Ł., *J. Chromatogr. Sci.*, 43, 183, 2005. With permission.)

25.3.3 PSEUDO-RP SYSTEMS

It was reported [27] that addition of water to the mobile phase improved separation of protoberberine alkaloids and made separation analogous to those obtained by use of aqueous ammonia to the mobile phase (see Figure 25.1). Water as the mobile phase component is often used in separation of strongly adsorbed alkaloids [4,25,79].

The use of typical mobile phases for RP systems in case of silica layers was also reported. Buffered aqueous–organic mobile phases were used to separate isoquinoline alkaloids [44,85–88]. Such systems, called pseudo-RP systems, were used for separation of *Chelidonium majus* alkaloids into two fractions. Large differences between retention coefficients of tertiary and quaternary alkaloids on unmodified silica with aqueous acetate buffer solution + methanol were used to separate *Chelidonium majus* alkaloids into two fractions—tertiary and quaternary alkaloids.

Of course the retention of alkaloids can be controlled by the change of modifier concentration and buffer pH. The use of mobile phase additives, such as those used in RP systems: IP reagents, ammonia, and amines, was also reported [53]. Most of the isoquinoline alkaloids are strongly retained on silica plates developed with the mobile phases comprising of organic modifier and water, and separation selectivity, as well as system efficiency, was poor. Addition of buffer at pH 2 as well as at pH 8 reduces retention, but system efficiency, was still poor. Addition of IP reagents (HDEHP, SOS) as well as short-chain amines (DEA) or ammonia leads to improvement of peak symmetry and efficiency, leading to better selectivity of separation. The most selective separation of isoquinoline alkaloids on silica layers in pseudo-RP systems was obtained by use of mobile phases with ammonia and diethylamine. Figure 25.12 [67] presents densitogram and video scan from chromatogram of the individual isoquinoline alkaloids and their mixture on silica with aqueous eluent containing ammonia.

25.3.4 Special Techniques Used in TLC of Isoquinoline Alkaloids

25.3.4.1 Multidimensional TLC

Multidimensional techniques give possibilities for separation of very complex mixtures, for example, plant extracts by use of systems with various selectivity. This technique can be easily realized with GC but it is a difficult task when liquid chromatography has to be used because of mobile phase components, which play important role in liquid chromatography system. Multidimensional HPLC can be realized when the same or compatible mobile phase flows through each column (IE-RP systems, CN-RP-C18). TLC gives the possibility of two-dimensional separation by use of the same stationary phase with different eluent systems or by use of gradient stationary phase with two layers and two eluent systems. In the last combination we can realize selectivity differences of normal phase systems with adsorption mechanism and RP systems with partition mechanism. It causes wide possibilities of separation of structural analogs.

In case of isoquinoline alkaloids two-dimensional TLC was applied for screening and identification of opiates—morphine, codeine hydrocodone, and dihydrocodeine in biological samples [96]. Silica layers and eluent systems of various selectivities, nonaqueous but also aqueous typical for RP systems with addition of ammonia or diethylamine, were used in experiments. 2D TLC was found to be a suitable and reliable method for targeted detection of low concentrations of opiates. 2D TLC was also applied for investigations of biosynthesis of alkaloids in *Erythrina cristagalli* [89].

Two-dimensional chromatography can be realized on one plate when the use of eluent systems of various properties and selectivity permits full separation of mixture components. When cyanopropyl-silica layer is applied 2D TLC separations can be

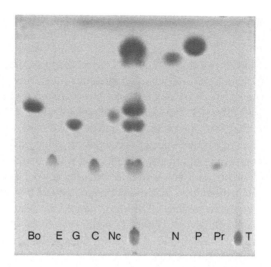

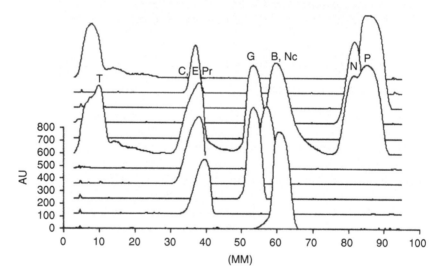

FIGURE 25.12 Densitogram and video scan from chromatography of the individual isoquinoline alkaloids and their mixture on silica with MeOH–water 80 + 20 containing 1% ammonia as mobile phase. Solutes: Bo, boldine; E, emetine; G, glaucine; C, codeine; Nc, narceine; N, noscapine; P, papaverine; Pr, protopine; T, tubocurarine. (From Petruczynik, A., Waksmundzka-Hajnos, M., and Hajnos, M.Ł., *J. Planar Chromatogr.*, 18, 78, 2005. With permission.)

performed by use of nonaqueous eluents in normal phase systems and aqueous ones in RP systems. The CN-silica layer developed with 60% methanol in water + 2% ammonia in first direction and 10% methanol in diisopropyl ether + 2% ammonia in the second direction was used for the separation of the mixture of selected isoquinoline alkaloids (see Figure 25.13) as well as *Chelidonium majus* and *Fumaria*

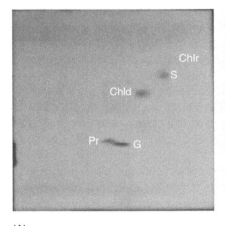
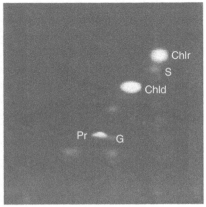

(A)

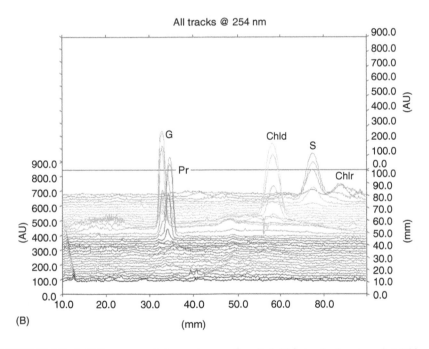

(B)

FIGURE 25.13 A. Video scans of chromatogram for alkaloids' standards scanned at 366 and 254 nm. Eluent systems: I direction: 60% MeOH in water + 2% ammonia; II direction: 10% MeOH in diisopropyl ether + 2% ammonia. B. Densitogram of the plate scanned at 254 nm. Symbols indicate: G—Glaucine, Pr—Protopine, Chld—Chelidonine, S—Sanguinarine, Chlr—Chelerithrine. (From Petruczynik, A., Waksmundzka-Hajnos, M., Michniowski, T., Plech, T., Tuzimski, T., Hajnos, M., Gadzikowska, M., and Jozwiak, G., *J. Chromatogr. Sci.*, 45, 447, 2007. With permission.)

officinalis alkaloid extracts [33]. The same plate with the modified eluent system (I direction: 10% MeOH in diisopropyl ether + 2% ammonia; II direction: 7% MeOH in water + 0.1 M of DEA) was used for the separation of *Glaucium flavum* herb extract (see Figure 25.14) [33]. The method connected with DAD densitometry was applied for identification of alkaloids in plant extracts.

Also silica layer can be applied for two-dimensional separations of isoquinoline alkaloids. When in the first direction aqueous eluent has been used (80% MeOH in water + 1% ammonia) and in the second direction nonaqueous eluent (MeOH + acetone + diisopropyl ether + DEA) nine-component mixture of isoquinoline alkaloids was separated [67].

Two-dimensional TLC can also be realized with connected plates precoated with various adsorbents. It gives wide possibilities for obtaining different separation selectivity and for separation of structural analogs, for example, isoquinoline alkaloids. The use of adsorbent gradient—CN-silica plates developed with nonaqueous eluent and RP18W layers with the aqueous eluent or CN-silica plates developed with nonaqueous eluent and silica layers with the nonaqueous eluent permits full separation of selected isoquinoline alkaloids and gives possibility for their identification on the basis of location of spots on the layer and by UV spectra of components and alkaloid standards by use of DAD scanner in such plant material as herbs of *Chelidonium majus, Fumaria officinalis,* or *Glaucium flavum* [83]. Figure 25.15 shows exemplary chromatogram from the separation of *Chelidonium majus* herb extract obtained in system: CN-silica/MeOH + diisopropyl ether + ammonia–RP18/MeOH + water + DEA, where berberine, protopine, chelidonine, and chelerithrine were identified [83]. Also CN-silica plates connected with silica layers

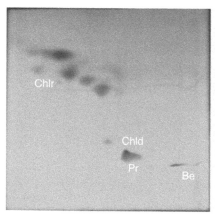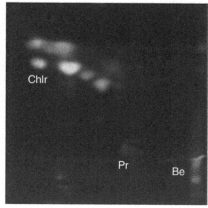

(A)

FIGURE 25.14 (A) Video scans of chromatogram of *Chelidonium majus* herb extract scanned at 366 and 254 nm. Eluent systems: I direction: 60% MeOH in water + 2% ammonia; II direction: 10% MeOH in diisopropyl ether + 2% ammonia;

(*continued*)

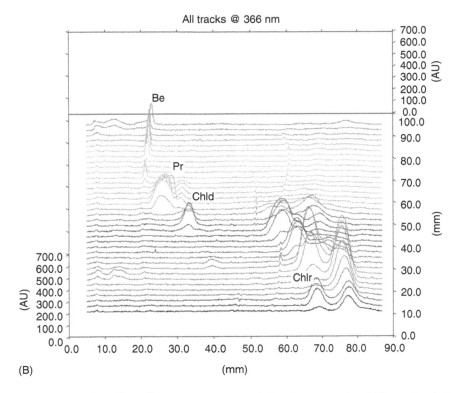

All tracks @ 366 nm

(B)

(mm)

FIGURE 25.14 (continued) (B) Densitogram of the plate scanned at 254 nm. Symbols indicate: Be—Berberine, Pr—Protopine, Chld—Chelidonine, Chlr—Chelerithrine. (From Petruczynik, A., Waksmundzka-Hajnos, M., Michniowski, T., Plech, T., Tuzimski, T., Hajnos, M., Gadzikowska, M., and Jozwiak, G., *J. Chromatogr. Sci.*, 45, 447, 2007. With permission.)

developed with nonaqueous eluents permit the separation of complex plant extracts. Figure 25.16 shows the separation of *Glaucium flavum* alkaloids which permits the identification of glaucine, protopine, chelidonine, sanguinarine, and chelerithrine.

It should be noticed that circular (or anticircular) planar chromatography is also the method in which movement of the mobile phase is two-dimensional, because the mobile phase migrates radially. As a result compounds are better resolved in the lower R_F range. The method was applied for the separation of protoberberine alkaloids in plant material [94].

Multidimensional techniques can also be realized in one direction. The plate can be developed several times in various modes. Unidimensional multiple development (UMD), the repeated development over the same distance with the mobile phase of constant composition, incremental multiple development (IMD), the repeated development over the increasing distance with the mobile phase of constant composition have improved selectivity of separation of isoquinoline alkaloids' extracts from herbs of *Chelidonium majus* [90] as well as of *Fumaria officinalis* [91].

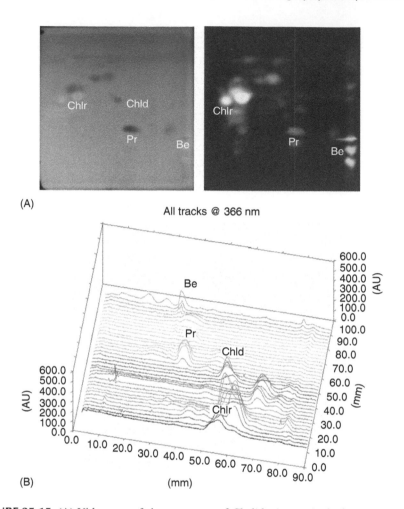

FIGURE 25.15 (A) Video scan of chromatogram of *Chelidonium majus* herb extract scanned at 366 and 254 nm. Systems: I direction: CN/10% methanol diisopropyl ether + 2% ammonia (double developed); II direction: RP18W/80% methanol acetate buffer at pH 3.5 + 0.05 mL^{-1} diethylamine (double developed). (B) Densitogram of the plate scanned at 254 nm. Solutes: Chlr, chelerithrine; Chld, chelidonine; Pr, protopine; Be, berberine. (From Petruczynik, A., Waksmundzka-Hajnos, M., Plech, T., Tuzimski, T., Hajnos, M.Ł., Jozwiak, G., Gadzikowska, M., and Rompała, A., *J. Chromatogr. Sci.*, in press. With permission.)

Gradient development, which can be realized as the stepwise mode, was used for the separation of *Chelidonium majus* alkaloids [69], can also be realized as multiple development technique with change of mobile phase composition at the constant distance or increasing distance. Often the eluent strength of mobile phase decreases. The method can be realized manually with drying after each run. An improved version of this technique is automated multiple development (AMD) in special chambers. The last method was applied for the separation of opiate alkaloids and was found as efficient planar chromatographic technique for these purposes [4].

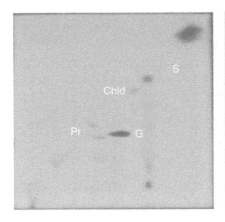
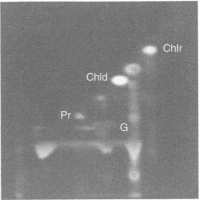

(A)

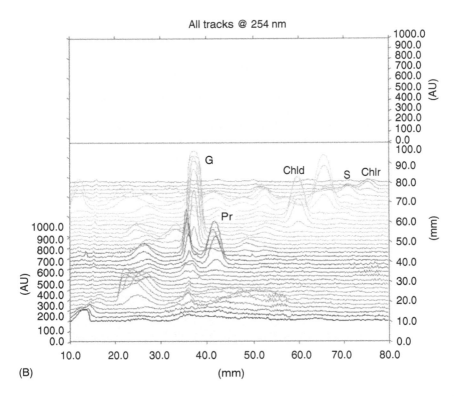

(B)

FIGURE 25.16 (A) Video scan of chromatogram of *Glaucium flavum* herb extract scanned at 254 and 366 nm. Systems: I direction: CN/10% methanol diisopropyl ether + 2% ammonia (double developed); II direction: SiO_2/5% Me_2CO + diisopropyl ether + 0.1 M DEA; 10% Me_2CO + 10% MeOH + diisopropyl ether + 0.1 M DEA. (B) Densitogram of the plate scanned at 254 nm. Solutes: Chlr, chelerithrine; Chld, chelidonine; Pr, protopine; G, glaucine; S, sanguinarine.

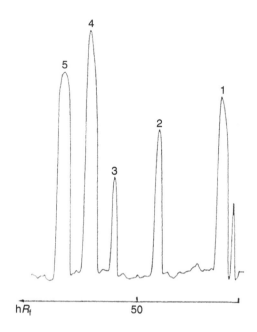

FIGURE 25.17 Densitogram of alkaloids extracted from opium: 1, morphine; 2, codeine; 3, tebaine; 4, papaverine; 5, noscapine. System: alumina 60 F_{254} type E/CH_2Cl_2–ethyl acetate $(80 + 20;$ v/v). Technique: OPLC. (From Pothier, J., Galand, N., and Viel, C., *J. Planar Chromatogr.*, 4, 392–396, 1991. With permission.)

25.3.4.3   Forced-Flow Techniques

Techniques with the forced flow of the eluent make use of different modes of eluent flow than capillary forces. There is electroosmotic flow in electroplanar chromatography, high pressure in overpressured-layer chromatography (OPLC), or centrifugal force in centrifugal TLC. It causes different mobile phase flow profiles in forced-flow techniques, which results in separation efficiency and therefore resolution of neighboring spots in a chromatogram. OPLC was applied for successful separation of isoquinoline alkaloids contained in *Berberis vulgaris* (bark), *Chelidonium majus* herb, and *Hydrastis canadensis* [76] as well as opium alkaloids [23]. Figure 25.18 shows OPLC separation of hydrastine, berberine, and hydrastinine from *Hydrastis canadensis* [76] and Figure 25.17 shows separation of opium alkaloids [23]. Centrifugal TLC was used in separation of naphthylisoquinoline alkaloids from *Ancistrocladus heyneanus* [16].

25.4   QUANTITATIVE DETERMINATION BY TLC AND DENSITOMETRY

As it is seen in Table 25.2 and in cited papers [5,28,34,38,39,47,70–73,78–80] TLC connected with densitometry is often used for quantitative determination of isoquinoline alkaloids. For example, this method was used for quantitative evaluation of

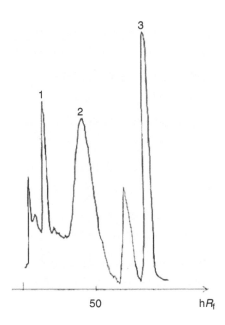

FIGURE 25.18 Densitogram obtained from an extract of *Hydrastis canadensis*: 1, hydrastinine; 2, hydrastinine; 3, berberine. System: silica 60 F_{254}/ethyl acetate–tetrahydrofuran–acetic
acid (60 + 20 + 20 v/v). Technique: OPLC. (From Pothier, J., Galand, N., Tivollier, P., Viel, C.,
J. Planar Chromatogr., 6, 220, 1993. With permission.)

sanguinarine in edible mustard oil [68]. The quantitation was achieved by densitometric scanning of the chromatogram in fluorescence/reflectance mode. Quantitative
analysis was done by external standard method on the basis of peak area in linear
regression mode. The calibration range was set from 5 to 300 ng/band and the high
correlation coefficient ($r^2 > 0.9998$) was obtained. The limit of detection (LOD) and
limit of quantitation (LOQ) were estimated as 1 and 3 ng/band, respectively. It allows
the use of TLC + densitometry method in routine analysis of argemone oil adulteration in edible mustard oil [68].

In order to determine whether densitometric assay is sufficiently reliable, accurate,
and reproducible for routine use, Sarkozi et al. [5] have compared results of quantitation of five *Chelidonium majus* alkaloids obtained by TLC and HPLC. The differences
in quantitative analysis of selected alkaloids by both methods were acceptable for all
determined compounds. According to the results, the TLC–densitometry is proven
reliable and precise enough for routine quantitative analysis of *Chelidonium alkaloids*.

TLC was also used for determination of morphine and codeine in poppy capsules
[30]. The solutions of extracts and standards were applied manually using the Hamilton microliter syringe. TLC silica plates were developed, dried and the densitometric
analysis was performed at 275 nm, which corresponded to the maxima of investigated
compounds. For the quantification, the external standard method was used. Linear
responses were obtained for determined alkaloids in the range of 2–20 μg/spot.
The method was precise and reproducible with low SD values (<0.15). The results

of quantitation by TLC + densitometry were confirmed by GC–MSD method. TLC + densitometry method can be used in routine analysis of morphine and codeine [30].

25.5 TLC OF ISOQUINOLINE ALKALOIDS IN PREPARATIVE SCALE

Preparative layer chromatography is the method of separation and isolation of small quantities of mixture components. It can be applied for isolation of individual compounds before their identification by various physicochemical methods [92]. Preparative layer chromatography was widely applied for the isolation of isoquinoline alkaloids (see Table 25.3). Very important is the choice of system with high selectivity and efficiency in analytical scale, because with the increase of sample retention parameters change and system efficiency decreases. Of course the choice of chromatographic system depends on the chemical character of extracts being separated. The mobile phase should accomplish all requirements for PLC determined as volatility and low viscosity, because nonvolatile components (e.g., ion-association reagent and the most buffers) should be avoided. It means that for PLC of plant extracts normal phase chromatography is much more preferable than RP systems. In the latter situation mixtures, methanol–acetonitrile–water are mostly used. If buffers and acids have to be added to either aqueous or nonaqueous mobile phase, only volatile buffers—acetate, formate, ammonium carbonate, and acids: acetic, formic are preferred. In case of alkaloids when the use of silanol blockers is necessary ammonia or short-chain amines easy to evaporation can be applied.

Investigators, dealing with optimization of chromatography systems for planar preparative separation, selected stationary and mobile phase to obtain a resolution $R_S > 1.5$ for each pair of touching bands. Such a resolution permits to introduce an extract ca.3 mg/g adsorbent.

25.5.1 MODE OF OVERLOADING

Apart from the choice of appropriate system, the optimization of preparative layer chromatography depends on the sample application mode of large volume of sample into the adsorbent layer, layer thickness as well as development mode to ensure high purity and yield of isolation of mixture components. Of course kind of overloading volume or mass changes band profiles of separated compounds. Mass overloading causes unsymmetric bands with back tailing, volume overloading causes rapid widening of bands. In our experiments with kind of overloading in separation of isoquinoline alkaloids, it was found that mass overloading gives much better results than volume one. When the same volumes of *Chelidonium majus* alkaloids were introduced from 0.25 to 5 mg resolution of bands overlaps but the separation of four partly separated zones is still possible. However, when various volumes of sample of the same concentration were applied fast loss of resolution was observed. Similar conclusions result from the experiments with *Fumaria officinalis* extract. Figure 25.19 shows mass overloading in case of introduction of increasing mass of *Fumaria officinalis* extract [65]. It seems that mass overloading gives better results, if sample solubility allows. One should also take into account that introduction of concentrated solutes may be disadvantageous owing to possibility of crystallization of substances

TABLE 25.3

Systems Used for Isolation of Isoquinoline Alkaloids by Preparative Layer Chromatography

Alkaloids	Source	Adsorbent	Eluent	Detection	References
Secondary and tertiary isoquinoline alkaloids	Xylopia parviflora	SiO$_2$	CHCl$_3$/MeOH/NH$_3$		[24]
Benzoyl benzyltetrahydroisoquinoline alkaloids	Aristolochia elegans	SiO$_2$ 60	C$_6$H$_6$/Me$_2$CO		[99]
Licorine, tazettine, pretazettine	Hipeastrum glaucescens	SiO$_2$	CH$_2$Cl$_2$		[20]
Quaternary isoquinoline alkaloids	Xylopia parviflora	SiO$_2$	CHCl$_3$MeOH/NH$_{3aq}$	UV light	[9]
Naphthylisoquinoline alkaloids	Triphyophyllum peltatum	SiO$_2$	MeOH/CH$_3$	Dragendorff reagent	[8]
Capnoidine, adlumine, tetrahydropalmatine, bicuculine, protopine	Corydalis sempervirens	SiO$_2$	AcOEt/MeOH CHCl$_3$/MeOH AcOEt/Hx	Dragendorff reagent	[13]
Roemeridine, proaporphine-tryptamine dimers	Roemeria hybrida	SiO$_2$	C$_6$H$_6$/CHCl$_3$.MeOH		[55]
Morphine, codeine	Papaver somniferum	SiO$_2$	T/Me$_2$CO/MeOH/NH$_3$		[56]
Quaternary isoquinoline alkaloids	Romneya coulteri	SiO$_2$	CHCl$_3$/MeOH		[29]
Bisbenzylisoquinoline alkaloids	Isopyrum thalictroides	SiO$_2$	CHCl$_3$/MeOH CHCl$_3$/MeOH/NH$_3$ CHCl$_3$/Me$_2$CO/MeOH		[57]
Tetrahydropalmatine, stephanine, crebanine, ayuthiamine, corydalmine, sebiferine, stepharine, laudanidine, amurine sanquinarine	Stephania bancroftii, Stephania baculeata, Stephania japonica	SiO$_2$	CH$_2$Cl$_2$/NH$_3$		[61]
	Opium poppy	SiO$_2$	AcOEt/MeOH		[62]
Domesticine, isoboldine, dihydrosanguinarine, noroxyhydrastinine, oxyhydrastinine, coptisine, stylopine, protopine, densiflorine, parfumine	Fumaria sepium, Fumaria agraria	SiO$_2$	AcOEt/Hx B/AcOEt/Et$_3$N		[63]
Chelidonine	Chelidonium majus	SiO$_2$	AcOEt/MeOH/NH$_3$		[64]

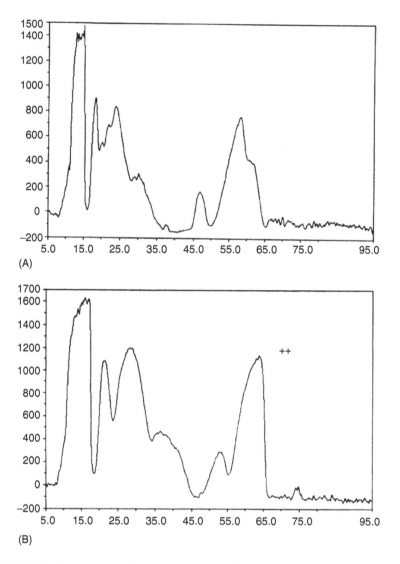

FIGURE 25.19 Densitograms of preparative layer chromatogram of fraction I of *Fumaria officinalis* herb extract introduced by automatic applicator. System: silica/AcOH + PrOH + CH$_2$Cl$_2$ (1:4:5). Plates scanned at 520 nm after derivatization with Dragendorff's reagent. (A) 2 mg of fraction. (B) 5 mg of fraction. (From Józwiak, G. and Waksmundzka-Hajnos, M., *Acta Chromatogr*, 18, 207, 2007. With permission.)

into the adsorbent pores and disturbance in desorption–adsorption process and worsening the separability.

Sampling mode—the way in which sample is introduced to the layer is also very important. We can use capillaries or pipette for application of a sample as a narrow band across the plate. Good results were obtained by the application method—from the eluent container to the edge of the layer, where frontal chromatography stage causes partly separation of starting band components [92].

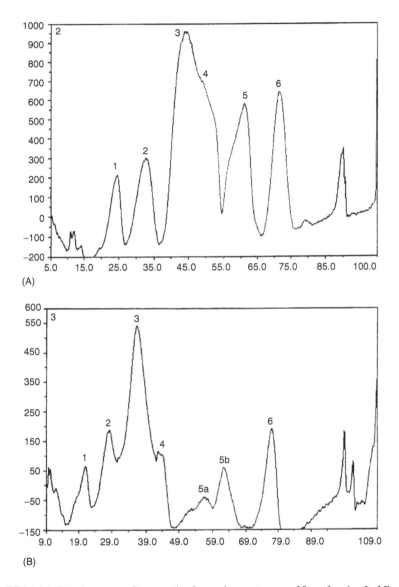

FIGURE 25.20 Densitograms of preparative layer chromatogram of 5 mg fraction I of *Fumaria officinalis* herb extract. System: see Figure 25.19. Plates triple developed and scanned at 520 nm after derivatization with Dragendorff's reagent. (A) Sample introduced by automatic aplicator. (B) Sample dissolved in 100 μL of methanol introduced from the edge of the layer with eluent distributor. Consecutive numbers indicate sequential fractions. (From Józwiak, G. and Waksmundzka-Hajnos, M., *Acta Chromatogr*, 18, 207, 2007. With permission.)

However, the best results were obtained by use of sample applicator with the evaporation of sample solvent. It causes formation of extremely narrow starting band and high resolution of neighboring bands after elution. Figure 25.20 shows densitograms of *Fumaria officinalis* chromatograms applied from the edge of the layer and by automatic applicator [65]. It is clearly seen that components are much better separated when the sample was applied by automatic applicator with the evaporation of solvent.

25.5.2 MODE OF DEVELOPMENT

Second important factor influencing resolution of neighboring bands is mode of development. TLC gives possibilities of the use of special techniques of development, which can be realized without additional technical device. There are gradient elution and multiple development techniques. Especially easy for application are linear multiple development techniques such as UMD and IMD. The techniques consist in repeated development of the plate with the same eluent on the same (UMD) or increasing (IMD) distance with the evaporation of eluent after each run. It causes changes in retention but also increases resolution of neighboring spots or bands, especially in the range of low or medium R_F coefficients. Other techniques such as gradient multiple development or bivariant multiple development [93] are more rarely used for preparative purposes. Figures 25.21 and 25.22 present the effect of UMD and IMD technique on the separation of bands of *Fumaria officinalis* isoquinoline alkaloids [91].

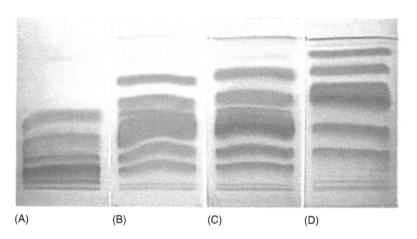

(A)   (B)   (C)   (D)

FIGURE 25.21 Video scans of preparative layer chromatogram of 5 mg fraction I of *Fumaria officinalis* herb extract introduced by automatic applicator system: see Figure 25.19. (A) Plate single developed. (B) Double developed. (C) Triple developed. (D) Fourfold developed. Plates scanned at 520 nm after derivatization with Dragendorff's reagent.

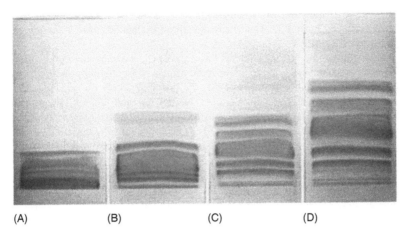

(A)   (B)   (C)   (D)

FIGURE 25.22 Video scans of preparative layer chromatogram of 5 mg fraction I of *Fumaria officinalis* herb extract introduced by automatic applicator. System: see Figure 25.19. Plates developed by IMD technique: (A) After the first step (2 cm distance of development). (B) After the second step (4 cm distance of development). (C) After the third step (6 cm distance of development). (D) After the fourth step (8 cm distance of development). Plates scanned at 520 nm after derivatization with Dragendorff's reagent.

REFERENCES

1. Svendsen, A.B., Thin layer chromatography of alkaloids, *J. Planar Chromatogr.*, 2, 8–18, 1989.
2. de Wet, H., van Heerden, F.R., and van Wyk, B.E., Alkaloids of *Antizoma miersiana* (Menispermaceae), *Biochem. Syst. Ecol.*, 33, 799–807, 2005.
3. Zhang, X., Ye, W., Zhao, S., and Che, C.T., Isoquinoline and isoindole alkaloids from *Menispermum dauricum*, *Phytochemistry*, 65, 929–932, 2004.
4. Pothier, J. and Galand, N., Automated multiple development thin-layer chromatography for separation of opiate alkaloids and derivatives, *J. Chromatogr. A*, 1080, 186–191, 2005.
5. Sarkozi, A., Janicsak, G., Kursinszki, L., and Kery, A., Alkaloid composition of *Chelidonium majus* L. studied by different chromatographic techniques, *Chromatographia Suppl.*, 63, S81–S86, 2006.
6. Bringmann, G., Teltschik, F., Michael, M., Busemann, S., and Rückert, M., Ancistrobertsonines B, C, and D as well as 1,2-didehydroancistrobertsonine D from *Ancistrocladus robertsoniorum*, *Phytochemistry*, 52, 321–332, 1999.
7. Krusinszki, L., Sarkozi, A., Kery, A., and Szoke, E., Improved RP-HPLC method for analysis of isoquinoline alkaloids in extracts of *Chelidonium majus*, *Chromatographia*, 63, S131–S135, 2006.
8. Bringmann, G., Wenzel, M., Rübenacher, M., Schäffer, M., Rückert, M., and Laurent, A. A., Dioncophylline and 8-O-methyldioncophylline D, 7,8'-coupled naphthylisoquinoline alkaloids from *Triphyophyllum peltatum*, *Phytochemistry*, 49, 1151–1155, 1998.

 9. Nishiyama, Y., Moriyasu, M., Ichamaru, M., Iwasa, K., Kat, A., Mathenge, S., Chalo
    Mutiso, P.B., and Juma, F.D., Quaternary isoquinoline alkaloids from *Xylopia parviflora*,
    Phytochemistry, 65, 939–944, 2004.
10. Küpeli, E., Kosar, M., Yesilada, E., and Basser, K.H.C., A comparative study on the anti-
    inflammatory, antinociceptive and antipyretic effects of isoquinoline alkaloids from the
    roots of Turkish *Berberis* species, *Life Sci.*, 72, 645–657, 2002.
11. Suau, R., Rico, R., Lopez-Romero, J.M., Najera, F., and Cuevas, A., Isoquinoline
    alkaloids from *Berberis vulgaris* subsp. *australis*, *Phytochemistry*, 49, 2545–2549, 1998.
12. Sturm, S., Strasser, E.M., and Stuppner, H., Quantification of *Fumaria officinalis* iso-
    quinoline alkaloids by nonaqueous capillary electrophoresis–electrospray ion trap mass
    spectrometry, *J. Chromatogr. A*, 1112, 331–338, 2002.
13. Majak, W., Bai, Y., and Benn, M.B., Phenolic amides and isoquinoline alkaloids from
    Corydalis sempervirens, *Biochem. Syst. Ecol.*, 31, 649–651, 2003.
14. Tempone, A.G., Treiger Borborema, S.E., de Andradem, H.F., de Amorim Gualda, N.C.,
    and Yogi, A., Antiprotozoal activity of Brazilian plant extracts from isoquinoline alkal-
    oid-producing families, *Phytomedicine*, 12, 382–390, 2005.
15. An, T., Huang, R., Yang, Z., Zhang, D., Li, G., Yao, Y., and Gao, J., Alkaloids from
    Cynanchum komarovii with inhibitory activity against the tobacco mosaic virus,
    Phytochemistry, 58, 1267–1269, 2001.
16. Yang, L.K., Glover, R.P., Yoganathan, K., Sarnaik, J.P., Godbole, A.J., Soejarto, D.D.,
    Buss, A.D., and Butler, M.S., Ancisheynine, a novel naphthylisoquinolinium alkaloid
    from *Ancistrocladus heyneanus*, *Tetrahedron Lett.*, 44, 5827–5829, 2003.
17. Bringmann, G., Messer, K., Wolf, K., Mühlbacker, J., and Grüne, M., Dioncophylline E
    from *Dioncophyllum thollonii*, the first 7,3'-coupled dioncophyllaceous naphthylisoqui-
    noline alkaloid, *Phytochemistry*, 60, 389–397, 2002.
18. Montagnac, A., Remy, F., Pais, M., Hadi, A., and Hamid, A., Isoquinoline alkaloids from
    Ancistrocladus tectorius, *Phytochemistry*, 39, 701–704, 1995.
19. Soulimani, R., Younos, C., Jarmouni-Idrissi, S., Bousta, D., Khalouki, F., and Laila, A.,
    Behavioral and pharmaco-toxicological study of *Papaver rhoeas* L. in mice, *J. Ethno-
    pharmacology*, 74, 265–274, 2001.
20. Hofmann, Jr. A.E., Sebben, C., Sobral, M., Dutilh, J.H.A., Enriques, A.T., and Zuanazzi,
    J.A.S., Alkaloids of *Hippeastrum glaucescens*, *Biochem. Syst. Ecol.*, 31, 1455–1456,
    2003.
21. Bringmann, G., Günter, C., Saeb, W., Mies, J., Brun, R., and Assi, L.A., 8-*O*-
    methyldioncophyllionol B and revised structures of other 7,6'-coupled naphthylisoquino-
    line alkaloids from *Triphyophyllum peltatum* (Dioncophyllaceae), *Phytochemistry*, 54,
    337–346, 2000.
22. Zhang, Q., Tu, G., Zhao, Y., and Cheng, T., Isoquinoline and isoindole alkaloids from
    Menispermum dauricum, *Tetrahedron*, 58, 6795–6798, 2002.
23. Pothier, J., Galand, N., and Viel, C., Separation of alkaloids in plant extracts by overpressured
    layer chromatography with ethyl acetate as mobile phase, *J. Planar Chromatogr.*, 4, 392–
    396, 1991.
24. Nishiyama, Y., Moriyasu, M., Ichimaru, M., Iwasa, K., Kato, A., Mathenge, S.G., Chalo
    Mutiso, P.B., and Juma, F.D., Secondary and tertiary isoquinoline alkaloids from *Xylopia
    parviflora*, *Phytochemistry*, 67, 2671–2675, 2006.
25. Govindan, M., and Govindan, G., A convenient method for the determination of the
    quality of goldenseal, *Fitoterapia*, 71, 232–235, 2000.
26. Adsersen, A., Gauguin, B., Gudiksen, L., and Jäger, A.K., Screening of plants used in
    Danish folk medicine to treat memory dysfunction for acetylcholinesterase inhibitory
    activity, *J. Ethnopharmacol.*, 104, 418–422, 2006.

27. Peishan, X., Yuzhen, Y., and Qiaoling, L., Optimization of the TLC of protoberberine alkaloids and fingerprint evaluation of the coptidis rhizome, *J. Planar Chromatogr.*, 5, 302–307, 1992.
28. Szabo, B., Lakatos, A., Käszegi, T., and Botz, L., HPTLC and HPLC determination of alkaloids in poppies subjected to stress, *J. Planar Chromatogr.*, 16, 293–297, 2003.
29. Valpuesta, M., Diaz, A., and Suau, R., Coulteroberbinone, a quaternary isoquinoline alkaloid from *Romneya coulteri*, *Phytochemistry*, 51, 1157–1160, 1999.
30. Popa, D.S., Oprean, R., Curea, E., and Preda, N., TLC-UV densitometric and GC-MSD methods for simultaneous quantification of morphine and codeine in poppy capsules, *J. Pharm. Biomed. Anal.*, 18, 645–650, 1998.
31. Waksmundzka-Hajnos, M., Gadzikowska, M., and Gołkiewicz, W., Special modes of development of *Chelidonium majus* L. alkaloids in systems of the type polar adsorbent-multicomponent mobile phase by TLC, *J. Planar Chromatogr.*, 13, 205–209, 2000.
32. Waksmundzka-Hajnos, M. and Petruczynik, A., Retention behavior of some alkaloids in thin-layer chromatography with bonded stationary phases and binary mobile phase, *J. Planar Chromatogr.*, 14, 364–370, 2001.
33. Petruczynik, A., Waksmundzka-Hajnos, M., Michniowski, T., Plech, T., Tuzimski, T., Hajnos, M., Gadzikowska, M., and Jozwiak, G., Thin-layer chromatography of alkaloids on cyanopropyl bonded stationary phases. Part I, *J. Chromatogr. Sci.*, 45, 447–454, 2007.
34. He, L.Y. and Zhang, Y.B., TLC separation and densitometric determination of six isoquinoline alkaloids in *Corydalis bungeana*, *Yao Xue Xue Bao*, 20, 377–382, 1985.
35. Chia, J.C., Chang, F.R., Li, C.M., and Wu, I.C., Protoberberine alkaloids from Fissistigma balansae, *Phytochemistry*, 48, 367–369, 1998.
36. Chlabicz, J., Paszkiewicz, A., Grochowska, K., and Galasinski, W., Substances of plant origin with anticipated cyto- and oncostatic activity inhibit protein biosyntesis, *Herba Hung.*, 30, 61–71, 1991.
37. Valpuesta, M., Posadas, N., Ruiz, I., Silva, M.V., Gomez, A.I., Suau, R., Perez, B., Pliego, F., and Cabezudo, B., Alkaloids from *Ceratocapnos heterocarpa* plants and in vitro cultures, *Phytochemistry*, 38, 113–118, 1995.
38. Salama, Q.M. and Walash, M.I., Densitometric determination of papaverine in opium, poppy capsules and certain pharmaceutical dosage forms, *Anal. Lett.*, 25, 69–82, 1991.
39. Agyangar, M.R., Biswas, S.S., and Tambe, A.S., Separation of opium alkaloids by thin-layer chromatography combined with flame ionization detection using the peak pyrolysis method, *J. Chromatogr.*, 547, 538–543, 1991.
40. Kamble, V.W., Garad, M.V., and Dongre, V.G., A new chromatographic spray reagent for the detection, *J. Planar Chromatogr.*, 9, 280–281, 1996.
41. Castedo, L., Lopez, S., Rodriguez, A., and Villaverde, C., Alkaloids of *Sarcocapnos crassifolia* subsp. *Speciosa*, *Phytochemistry*, 28, 251–257, 1989.
42. Gołkiewicz, W., Błażewicz, A., and Jóźwiak, G., Isolation of miligram quantities of sanguinarine and chelerithrine from the roots of *Chelidonium majus* L. by zonal micro-preparative TLC, *J. Planar Chromatogr.*, 14, 95–99, 2001.
43. Krishnamurthy, R. and Srivastava, A.K., Simltaneous detection of adulterans and coextractants in illicit heroin by HPTLC with two successive mobile phase, *J. Planar Chromatogr.*, 10, 388–390, 1997.
44. Kanyinda, B., Diallo, B., Fastre, R.V., and Vanhaelen, M., Benzylisoquinoline alkaloids from *Anisocycla cymosa* roots, *Planta Med.*, 55, 394, 1989.
45. Clarke, C. and Raja, R., Improved thin-layer chromatographic system for evaluation of (+)-tubocurarine chloride and comercial curare, *J. Chromatogr.* 244, 174–176, 1982.

46. Varshney, K.M., HPTLC study of the stability of heroin in methanol, *J. Planar Chromatogr.*, 15, 46–49, 2002.

47. Wu, Zh. and Jin, W., Quantitative determination of tetrandrine in Chinese drugs fourstamen stephania, Stephanie tetrandra, and its preparation by TLC, *J. Chin. Herb. Med.*, 19, 349–350, 1988.

48. Wei, Y., Wei, X., Wang, Y., Liu, X., Chu, T., Hu, S., and Wang, X., Iodination and radiolabeling of α-allocryptopine with iodine-125, *Appl. Radiat. Isot.*, 62, 55–62, 2005.

49. Xie, C., Veitch, N.C., Houghton, P.J., and Simmonds, M.S.J., Flavonoid glycosides and isoquinoline alkaloids from *Corydalis bungeana*, *Phytochemistry*, 65, 3041–3047, 2004.

50. Lemre, S.W. and Busch, K.L., Chromatographic separations of the constituents derived from *Sanguinaria Canadensis*. Thin-layer chromatography and capillary zone electrophoresis. *J. Planar Chromatogr.*, 7, 221–228, 1994.

51. Bringmann, G., Saeb, W., God, R., Schäffer, M., Francois, G.K., Peters, E., Peters, M., Proksch, P., Hostettmann, K., and Assi, L. A., 5'-*O*-demethyldioncophylline A, a new antimalarial alkaloiod from *Triphyollum peltatum*, *Phytochemistry*, 49 (6), 1667–1673, 1998.

52. Kopyt'ko, Y.F., Dargaeva, T.D., Sokol'skaya, T.A., Grodnitskaya, E.I., and Kopnin, A.A., New methods for the quality control of a homeopathic matrix tincture of greater celandine, *Pharm. Chem. J.*, 39, 603–609, 2005.

53. Petruczynik, A., Waksmundzka-Hajnos, M., and Hajnos, M.Ł., The effect of chromatographic conditions on the separation of selected alkaloids on silica layers, *J. Chromatogr. Sci.*, 43(4), 183–194, 2005.

54. Kalasz, H., Kerecsen, L., Csermely, T., Gotz, H., Friedmann, T., and Hosztafi, S., TLC investigation of some morphine derivatives, *J. Planar Chromatogr.*, 8, 17–22, 1995.

55. Gunes, H.S. and Golzer, B., Two novel proaporphine-tryptamine dimmers from *Roemeria hybrida*, *Fitoterapia*, 72, 875–887, 2001.

56. Poeknapo, C., Fisinger, U., Zenk, M.H., and Schmidt, J., Evaluation of the mass spectrometric fragmentation of codeine and morphine after ^{13}C-isotope biosynthertic labelling, *Phytochemistry*, 65, 1413–1420, 2004.

57. Istatkova, R.S. and Philipov, S.A., Alkaloids from *Isopyrum thalictroides* L., *Phytochemistry*, 54, 959–964, 2000.

58. Newton, S.M., Lau, C., Gurcha, S.S., Desra, G.S., and Wright, C.W., The evaluation of forty-three plant species for in vitro antimycobacterial activities; isolation of active constituents from *Psoralea corylifolia* and *Sanguinaria Canadensis*, *J. Ethnopharmacol.*, 79, 57–67, 2002.

59. Kamble, V.W. and Dongre, V.G., TLC detection and identification of heroin (diacetylmorphine) in street samples, *J. Planar Chromatogr.*, 10, 384–386, 1997.

60. Essig, S. and Kovar, K.A., The efficiency of thin-layer chromatographic systems: a comparison of separation numbers using addictive substances as an example, *J. Planar Chromatogr.*, 10, 114–117, 1997.

61. Blanchfield, J.T., Sands, D.P.A., Kennard, C.H.L., Byriel, K.A., and Kitching, W., Characterisation of alkaloids from some Australian *Stephania* (menispermaceae) species, *Phytochemistry*, 63, 711–720, 2003.

62. Facchini, P.J., Temporal correlation of tyramine metabolism with alkaloid and amide biosyntesis in elicited opium popy cell cultures, *Phytochemistry*, 49 (2), 481–490, 1998.

63. Suau, R., Cabezudo, B., Rico, R., Lopez-Romero, J.M., and Najera, F., Alkaloids from *Fumaria sepium* and *Fumaria agraria*, *Biochem. Syst. Ecol.*, 30, 263–265, 2002.

64. Jalilian, A.R., Seyfi, P., Afarideh, H., and Shafiee, A., Synthesis of a [18F]labelled chelidonine derivative as a possible antitumor agent, *Appl. Radiat. Isot.*, 54, 407–411, 2001.

65. Jóźwiak, G. and Waksmundzka-Hajnos, M., Preparative-layer chromatography of an extract of alkaloids from *Fumaria officinalis*, *Acta Chromatogr.*, 18, 207–218, 2007.
66. Dongre, V.G. and Kamble, V.W., HPTLC detection and identification of heroin (diacetyl-morphine) in forensic samples. Part III, *J. Planar Chromatogr.*, 16, 458–460, 2003.
67. Petruczynik, A., Waksmundzka-Hajnos, M., and Hajnos, M.Ł., The effect of chromatographic conditions on the separation of selected alkaloids on silica layers, *J. Planar Chromatogr.*, 18, 78–84, 2005.
68. Ghosh, P., Reddy, M.M.K., and Sashidhar, R.B., Quantitative evaluation of sanguinarine as an index of argemone oil adulteration in edible mustard oil by high performance thin layer chromatography, *Food Chem.*, 91, 757–764, 2005.
69. Matysik, G. and Jusiak, L., Stepwise gradient in thin-layer chromatography of *Chelidonium* alkaloids, *J. Chromatogr.*, 446, 273–276, 1990.
70. Niu, Ch. and He, L., Determination of isoquinoline alkaloids in Chelidonine majus L. by thin-layer chromatography, *Acta Pharm. Sinica*, 27, 69–73, 1991.
71. Jiang, Sh., He, Y., Guo, P., and Zhou, Y., Separation of alkaloids in opium by thin-layer chromatography, *Chin. J. Pharm. Anal.*, 15, 46–47, 1995.
72. Li, W., Zhang, H., Chen, J., and Jiang, J., Determination of morphine in pulverized opium by thin-layer chromatography, *Clin. J. Anal. Lab.*, 13, 92–93, 1994.
73. Gu, Y., Li, L., and Lin, K., Determination of morphine in opium by thin-layer chromatography, *Chin. J. Pharm. Anal.*, 13, 236–237, 1993.
74. Devi, S.K. and Sastry, G.S.R., HPTLC method for the determination of sanguinarine in argemone and in argemone-adulterated edible oils, *J. Planar Chromatogr.*, 15, 223–225, 2002.
75. Hahn-Deinstrop, E. and Koch, A., Planar chromatographic determinations of *Chelidonium majus* L., in *Proceedings of the 9th International Symposium on Instrumental Planar Chromatography*, Interlaken, April 9–11, 1997, pp. 115–122.
76. Pothier, J., Galand, N., Tivollier, P., and Viel, C., Separation of quaternary alkaloids in plant extracts by overpressured layer chromatography, *J. Planar Chromatogr.*, 6, 220–222, 1993.
77. Soczewiński, E. and Flieger, J., Thin layer chromatography of alkaloids, *J. Planar Chromatogr.*, 9, 107–112, 1996.
78. Yang, Y., Zhao, S., and Lin, C., Determination of alkaloids in *Stephaniae tetrandrea* radix by thin-layer chromatography, *Bull. Chin. Trad. Med.*, 13, 740–742, 1988.
79. Xie, P.S., Yan, Y.Z., and Lin, Q.L., Optimization of the TLC of protoberberine alkaloids and fingerprint evaluation for the coptidisrhizome, *J. Planar Chromatogr.*, 5, 302–307, 1992.
80. Bebawy, L.I., El Tantawy, M.E., and El Bayoumi, A., Thin layer chromatographic scanner method for the determination of some natural products in raw materials and pharmaceutical preparations, *J. Liq. Chromatogr.*, 21, 741–753, 1998.
81. Szumiło, H. and Flieger, J., Application of differently modified silica gel in the TLC analysis of alkaloids, *J. Planar Chromatogr.*, 12, 466–470, 1999.
82. Nyiredy, Sz., Ed. *Planar Chromatography*. Springer Scientific Publisher, Budapest, Hungary, 2001, p. 103.
83. Petruczynik, A., Waksmundzka-Hajnos, M., Plech, T., Tuzimski, T., Hajnos, M.Ł., Jozwiak, G., Gadzikowska, M., and Rompała, A., TLC of alkaloids on cyanopropyl bonded stationary Phases. II. Connection with RP18 and silica plates, *J. Chromatogr. Sci.*, In press.
84. Duncan, J.D. and Armstrong, D.W., Chiral mobile phase additives in reversed-phase TLC, *J. Planar Chromatogr.*, 3, 65–67, 1990.
85. Gołkiewicz, W., Kuczyński, J., and Jusiak, L., Porównanie selektywności rozdzielania niektórych alkaloidów w układach: wodno-metanolowa faza ruchoma i niemodyfikowany oraz modyfikowany żel krzemionkowy, *Chem. Anal. (Warsaw)*, 36, 209–213, 1991.

86. Petruczynik, A., Matosiuk, D., Waksmundzka-Hajnos, M., and Patel, J., The use different chromatographic systems in theoretical methods of organic bases lipophilicity assessment, *VII Polish Conference on Analytical Chemistry*, Toruń, Poland, July 3–7, 2005, p. 265.

87. Gołkiewicz, W., Gadzikowska, M., Kuczyński, J., and Jusiak, L., Micropreparative chromatography of some quaternary alkaloids from the roots of *Chelidonium majus* L., *J. Planar Chromatogr.*, 6, 382–395, 1993.

88. Gołkiewicz, W. and Gadzikowska, M., Isolation of some quaternary alkaloids from the extract of roots of Chelidonium majus L. by column and thin-layer chromatography, *Chromatographia*, 50, 52–56, 1981.

89. Maier, U.H., Rodl, W., Deus-Neumann, B., and Zenk, M.H., Biosynthesis of *Erythrina* alkaloids in *Erythrina crista-galli*, *Phytochemistry*, 52, 373–382, 1999.

90. Waksmundzka-Hajnos, M., Gadzikowska, M., and Hajnos, M.Ł., Strategy for preparative separation of quaternary alkaloids from *Chelidonium majus* L. by thin-layer chromatography, *J. Planar Chromatogr.*, 15, 289–293, 2002.

91. Waksmundzka-Hajnos, M. and Jóźwiak, G., Special modes of development in preparative layer chromatography of alkaloid extract from *Fumaria officinalis* herb. *J. Planar Chromatogr.*, 2007 (in press).

92. Waksmundzka-Hajnos, M., Wawrzynowicz, T., Hajnos, M.Ł., and Jóźwiak, G., Preparative layer chromatography of natural mixtures, in: *Preparative Layer Chromatography*, Chromatographic Science Series vol 95. (T. Kowalska and J. Sherma eds.), CRC Taylor & Francis, Boca Raton, London, New York, 2006, pp. 251–298.

93. Szabady, B., The different modes of development in planar chromatography (Sz. Nyiredy ed.), Springer Scientific Publisher, Budapest, Hungary, 2001, pp. 88–102.

94. Swendsen, A.B., Van Kepipen-Verleun, A., and Verpoorte, R., Circular thin-layer chromatography of quaternary alkaloids. Quaternary protoberberine alkaloids in plant material, *J. Chromatogr. A.*, 291, 389–391, 1984.

95. Pothier, J., Galand, N., and Viel, C., Determination of some narcotic and toxic alkaloidal compounds by overpressured thin-layer chromatography with ethyl acetate as eluent, *J. Chromatogr. A.*, 634, 356–359, 1993.

96. Novakova, E., The detection of opiates by two-dimensional high-performance thin-layer chromatography, *J. Planar Chromatogr.*, 13, 221–225, 2000.

97. Navarro, V. and Delgado, G., Two antimicrobial alkaloids from *Bocconia arborea*, *J. Ethnopharmacol.*, 66, 223–226, 1999.

98. Krishnamurthy, R., Malve, M.K., and Shinde, B.M., Simultaneous determination of morphine, caffeine and paracetamol in the urine of addicts by HPTLC and HPLC, *J. Planar Chromatogr.*, 13, 171–175, 2000.

99. Shi, L.S., Kuo, P.C., Tsai, Y.L., Damu, A.G., and Wu, T.S., The alkaloids and other constituents from the root and stem of Aristolochia elegans, *Bioorg. Med. Chem.*, 12, 439–446, 2004.

26 TLC of Tropane Alkaloids

Tomasz Mroczek

CONTENTS

26.1 CHEMISTRY AND STEREOCHEMISTRY OF TROPANE AND RELATED ALKALOIDS

This kind of amino acid (mainly ornithine as well as phynylalanine and isoleucine) derivatives is composed of heterocyclic base named tropane, which chemically is 8-azabicyclo[3.2.1]octane. In its hydroxylated form it is esterified with different types of acids usually optically active.[1,2] There are two kinds of hydroxytropane derivatives that differ in the spatial orientation of free –OH group at position C-3. They comprise tropine (tropanol, *endo*-8-azabicyclo[3.2.1]octane-3-ol) and pseudo-tropine (pseudo-tropanol, *exo*-8-azabicyclo[3.2.1]octane-3-ol), which usually are esterified with the following acids: (−)-(*S*)-tropic, apotropic, cinnamic, angelic, tiglic, isovaleric, and α-truxillic ones.[2] Figure 26.1 presents stereochemical structures of typical tropane alkaloids representatives. They include the most often encountered alkaloids such as: (−)-(*S*)-hyoscyamine (named also as L-hyoscyamine),

685

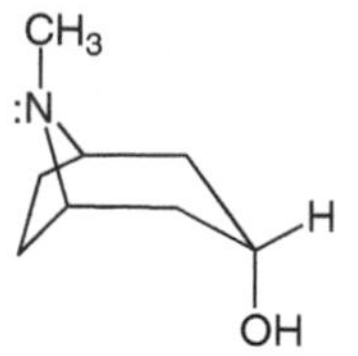
Tropine

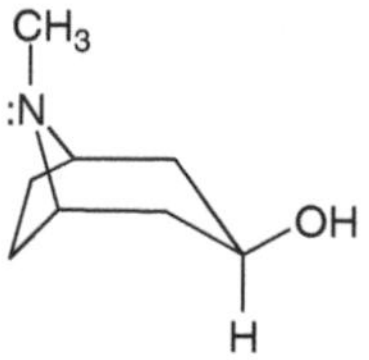
Pseudotropine

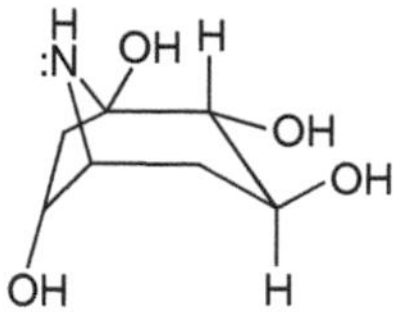
Calystegine B_1

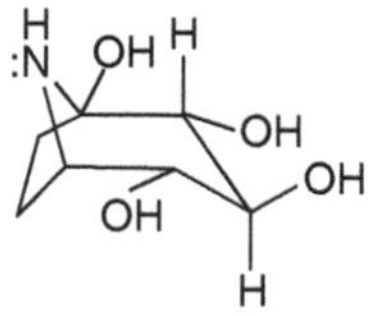
Calystegine B_2

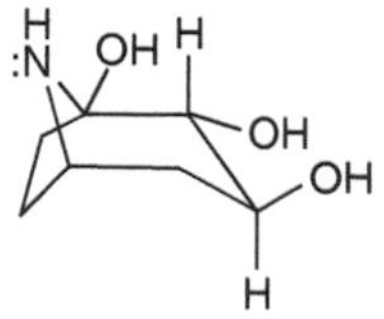
Calystegine A_3

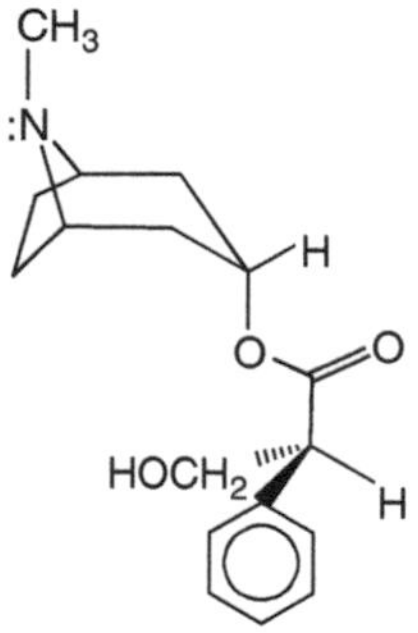
(–)-(*S*)-Hyoscyamine

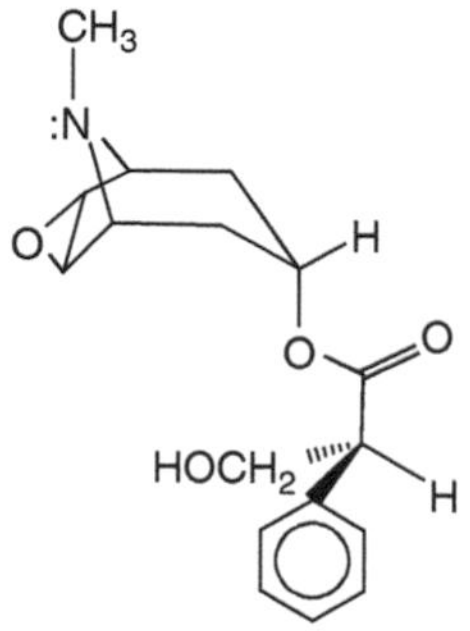
(–)-(*S*)-Scopolamine

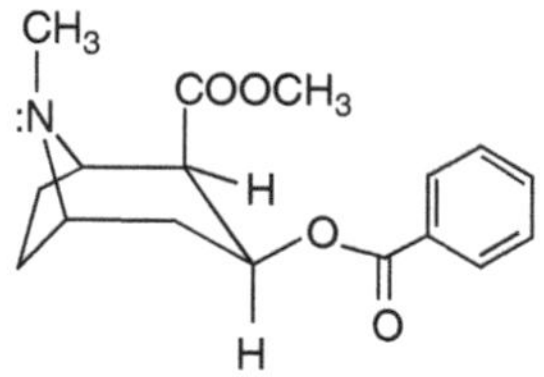
Cocaine

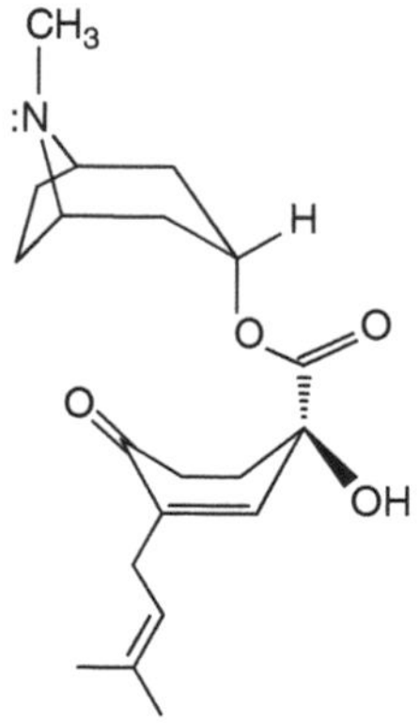
Consabatine

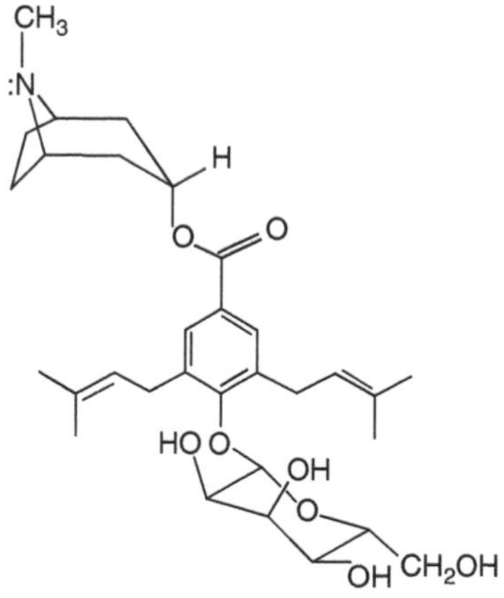
Merresectine B

FIGURE 26.1 Chemistry and stereochemistry of selected tropane alkaloids.

its racemic form called atropine, scopolamine (so-called hyoscyne), some new compounds such as consabatine and merresectine B.[4] Among pseudo-tropine type of alkaloid, cocaine is derivative of ecgonine (pseudotropine-2α-carboxilic acid). Other esterified derivatives of ecgonine such as with cinnamic acid are also encountered.

Several years ago, a new class of *nor*-pseudotropine-polyhydroxylated alkaloids called calystegines from various plants of Convolvulaceae (*Calystegia*, *Convolvulus*) and also Solanaceae plants (undergrounds parts of e.g., *Atropa belladonna*, *Solanum tuberosum*) was isolated.[3,4] Three main representatives named calystegine A_3, B_1, and B_2 are shown in Figure 26.1.

26.2   OCCURRENCE OF TROPANE ALKALOIDS

Tropane alkaloids mainly occur in Solanaceae, Erythroxylaceae, Convolvulaceae plant families.[1] They were found in the following plant genera: e.g., *Atropa*, *Datura*, *Hyoscyamus*, *Brugmansia*, *Duboisia*, *Mandragora*, *Solanum*, *Scopolia*, *Withania*, *Anisodus* from Solanaceae family, *Erythroxylum* from Erythroxylaceae, and *Convolvulus* and *Calystegia* from Convolvulaceae. The alkaloids are localized both in underground (roots) and aerial parts (especially leaves and seeds) of the plants.

26.3   EXTRACTION METHODS OF TROPANE ALKALOIDS FROM PLANT MATERIALS

26.3.1   AMBIENT PRESSURE MODES

Tropane alkaloids are usually termally unstable and sensitive to strong acidic and basic conditions. In the extracts pirrolidines, pyridines, or other heterocyclic bases are present together with emulsifying agents. Therefore, choice of suitable solvent and sample preparation procedure should be carefully selected. Tropane alkaloids can be extracted in their acidic form with diluted hydrochloric (e.g., 5% HCl), sulfuric VI (e.g., 0.01 M H_2SO_4), or acetic acid.[5–7] Jia et al.[8] found the extraction of *Datura* alkaloids the most efficient at pH 2–3.

For the extraction of tropane alkaloids free bases from plant material, alkaline organic phases are also used. Prior to the extraction, the sample is dropped with mainly 10% (or sometimes 25%) ammonia solution followed by organic solvent addition, that is, chloroform, benzene (toxic and cancerogenic, nowadays not recommended), toluene, dichloromethane, or ethanol.[5,9–11]

Fliniaux et al.[12] compared the efficiencies of both acidic and alkaline solutions for the extraction of tropane alkaloids from plant material. They used 0.2 M sulfuric VI acid, methanol–0.1 M HCl (24:1, v/v), methanol–27% ammonia (24:1, v/v), methanol–chloroform–27% ammonia (24:1, v/v). Similar results were obtained in this way.

Calystegines are usually extracted with methanol–water mixtures as they possess hydrophilic properties.[3]

Mroczek et al.[2] analyzed the content of L-hyoscyamine and scopolamine extracted from thorn apple leaves. When 1% tartaric acid in methanol was used at 90°C ± 5°C on heating mantle for 15 min the highest amounts of scopolamine were measured, also in comparison to more sophisticated methods such as UAE (ultrasound assisted extraction) or PLE (pressurized liquid extraction). However, the amounts of L-hyoscyamine were comparable to USE at 60°C and lower than PLE procedures. It shows that sometimes for a given alkaloid suitable extraction procedure should be elaborated.

Among UAE methods when methanol and 1% tartaric acid in methanol were analyzed at room temperature, 40°C or 60°C, it was found that the higher temperature the higher yield of L-hyoscyamine and scopolamine measured.[2] Again 1% tartaric acid in methanol was found to be a better solvent than pure methanol.

For cocaine and benzoylecgonine extraction from coca leaves, focused microwave-assisted extraction (FMAE) was optimized with respect to the nature of the extracting solvent, the particle size distribution, the moisture of the sample, the applied microwave power, and radiation time.[13] FMAE generated extracts similar to those obtained by conventional solid–liquid extraction but in a more efficient manner, that is, 30 s were sufficient to extract cocaine quantitatively from leaves.

26.3.2 PRESSURIZED LIQUID EXTRACTION

Pressurized liquid extraction (PLE) uses an organic solvent at high pressures and temperature above boiling point. It works according to principle of static extraction with superheated liquids. It ensures the higher solubility of analytes in solvent at higher temperatures, higher diffusion rate as a result of higher temperatures, and disruption of the strong solute–matrix interaction caused by van der Waals forces, hydrogen bonding, and dipole–dipole attractions between solute molecules and active sites on the matrix.

The method was used for cocaine and benzoylecgonine extraction from coca leaves.[14]

Mroczek et al.[2] optimized PLE conditions for extraction of L-hyoscyamine and scopolamine from thorn apple leaves. In PLE experiments with methanol, similarly as for ambient pressure UAE method, at the highest temperature conditions (110°C) the highest levels of L-hyoscyamine were measured and increased considerably with increase in temperature. Therefore, methanol PLE procedure at default conditions turned out to be the most efficient method of L-hyoscyamine extraction from plant materials. It was slightly better than 1% tartaric acid in methanol PLE at these conditions. In scopolamine extraction the same tendency as for L-hyoscyamine was observed in PLE method but differently to this compound the yields of scopolamine were lower to measure after rapid extraction (15 min) at 90°C with 1% tartaric acid in methanol solution on heating mantle.

26.3.3 SUPERCRITICAL FLUID EXTRACTION

Supercritical fluid extraction (SFE) method comprises supercritical fluid above its critical point for both temperature and pressure. That fluid has a higher diffusion coefficient and lower viscosity than liquids. Absence of surface tension allows for its rapid penetration into pores of heterogeneous matrices, which helps enhance extraction efficiencies. Selectivity may be manipulated by varying the conditions of temperature and pressure, affecting the solubility of various components in the supercritical fluid.

For extraction of hyoscyamine and scopolamine from plant material SFE method was described.[15] It involved supercritical carbon dioxide at temperature of 60°C and pressure 34 MPa. Some modifiers were also investigated and included: methanol, water, 10% water, and methanolic solution of diethylamine. The best results for 10% methanolic solution of diethylamine were obtained.

26.4 SAMPLE PRETREATMENT OF THE EXTRACTS

26.4.1 LIQUID–LIQUID PARTITIONING

Obtained extracts possess majority of coextractive compounds which may interfere with tropane alkaloids to be analyzed. Therefore, in some cases at first the acidic extract is extracted with chloroform in separation funnel to remove acidic coextractives. Then, remaining extract is alkalized with ammonia solution to pH about 9–9.5 and extracted with nonpolar organic solvents such as chloroform, benzene, toluene, and dichloromethane.[5,9,10] However, liquid–liquid partitioning (LLP) often uses large volumes of costly solvents and is also not amenable to automation because several disjointed steps are usually required. It is also limited to small sample volumes and solutes with large distribution constants.

26.4.2 SOLID ASSISTED LIQUID–LIQUID PARTITIONING

Solid assisted liquid–liquid partitioning (SELLP, Extrelut) replaces the separation funnel and permits routine LLP to be performed simply and efficiently using a single step. It consists of a specially processed, wide-pore kieselguhr with a high pore volume, chemically stable within pH 1–13. In the process an aqueous sample is applied to the dry column filled with granular Extrelut support. It spreads over the chemically inert matrix as a thin layer and functions as a stationary phase. Elution is carried out using organic solvents that are immiscible in water. As the solvent passes through the column, all lipophilic compounds are extracted from the aqueous phase into the organic phase. Main advantages of its method include: total lack of emulsions, saving of solvent, material and time, simple and proper performance, higher recoveries, and cleaner eluates when compared with typical LLP.

El-Shazly et al.[16] applied SELLP procedure for isolation of pure alkaloidal fractions from *Hyoscyamus* sp. The basified aqueous extract was applied into dry Extrelut column, and the liquid was completely absorbed by the kieselguhr. Tropane alkaloid-free bases that are exposed on the surface of the kieselguhr particles are eluted by organic solvents such as chloroform.

26.4.3 SOLID-PHASE EXTRACTION

Solid-phase extraction (SPE) involves selective extraction of the analytes from liquid samples onto solid support of different varieties and types (e.g., silica gel, alumina, florisil, kieselguhr, reversed-phase sorbents such as octyl, octadecyl, diol, cyano, amino, ion-exchange sorbents). The sample is directly passed through previously conditioned (in some new types of sorbents so-called mixed-mode or hydrophilic-lipophilic balance (HLB) this step can be omitted) cartridge filled with a given sorbent and analytes are directly collected, whereas coextractives are retained on the sorbent. It can be performed also in the other way when a sample is applied in solvent of low elution strength. The analytes adsorbed are eluted with solvent of higher elution power.

SPE of tropane alkaloids was usually performed on RP-18 columns. The procedure was applied for plant extracts as well as blood serum, urine, and egg yolk samples.[3,17,18]

Molecularly imprinted polymer (MIP) as selective sorbent in SPE of scopolamine from human urine and serum was also investigated.[19] However, low recoveries (46%–79%) of tropane alkaloids extracted were measured.

Keiner and Dräger[20] applied cation-exchange SPE for isolation of calystegines from plant samples, where they were retained by charge of the secondary amino group.

Mroczek et al.[2] optimized mixed-mode reversed-phase-cation-exchange solid-phase extraction procedure for simultaneous recoveries of L-hyoscyamine, scopolamine, and scopolamine-N-oxide from plant extracts as well as quaternary alkaloid representative scopolamine-N-methyl bromide. Using Oasis MCX cartridges bimodal mechanism of retention (cation-exchange and RP) was considered for efficient purification of different types of alkaloids (tertiary bases, N-oxides) from plant samples as well as quaternary alkaloids potentially used as therapeutic drugs. First three alkaloids (L-hyoscyamine, scopolamine, scopolamine-N-oxide) were efficiently eluted (recoveries 80%–100%) from Oasis MCX cartridge with methanol–10% ammonia (3:1, v/v) solution, whereas for the quaternary salt tetrahydrofurane–methanol–25% ammonia (6:1:3, v/v) was used with recoveries 52%–76%. The authors found this SPE procedure useful for further HPTLC-densitometric and RP-HPLC investigations.

26.5 THIN LAYER CHROMATOGRAPHY OF TROPANE AND RELATED ALKALOIDS

26.5.1 Optimization of the Separation

In TLC (or HPTLC) of tropane alkaloids various stationary phases, including both polar (i.e., silica gel, alumina, cellulose) and nonpolar (e.g., reversed-phase, C_{18}) stationary phases are used. In Table 26.1 literature dates on development of different separation systems in tropane alkaloids separations were listed.

Most papers dealt with TLC separation of major tropane alkaloids, i.e., hyoscyamine (or atropine) and scopolamine. These compounds were separated on TLC aluminium oxide plates, PC Whatman paper No. 1 impregnated, silver halide impregnated silica gel plates, silica gel 60 F_{254} plates, HPTLC silica gel plates, glass sticks coated both with silica gel 0.1 mm layer, and microcristalline cellulose or RP_w18 phases. Because of differences in polarities these compounds are easily separated by single development with mobile phases containing dilution solvents such as chloroform, benzene, toluene, cyclohexane, 1,1,1-trichloroethane, acetone,

TABLE 26.1

TLC Separation Systems of Tropane Alkaloids Investigated

Compounds Separated	Separation Systems	Results Obtained	Ref.
Scopolamine hydrobromide and atropine in pharmaceutical dosage forms	For scopolamine hydrobromide: TLC, aluminium oxide/chloroform—25% ammonia (100:0.15 v/v); for atropine: TLC, silica gel/chloroform–benzene–ethanol (40:10:20 v/v/v)	Scopolamine hydrobromide was well separated from morphine hydrochloride in injection dosage forms and atropine from ergotamine and phenobarbital in tablets	[21]
Atropine, scopolamine	Paper chromatography (PC): Whatman paper No. 1, impregnated with 0.5 M KCl/n-butanol:HCl:water (100:2:to saturation); TLC: silica gel G/acetone:CHCl$_3$: methanol:25%ammonia (20:20:3:1); silica gel G/CHCl$_3$:methanol (1:1)	Efficient separation of atropine from scopolamine and scopoletin in samples from *Scopolia* Jacq. genus	[22]
Cocaine and its metabolites and congeners (benzoyl ecgonine, ecgonine methyl ester, ecgonine, benzoyl *nor*-ecgonine, *nor*-ecgonine)	Glass fiber silica gel impregnated sheets (ITLC): S1, chloroform:acetone:conc. ammonia (5:94:1); S2, benzene:ethyl acetate:methanol:conc. ammonia (80:20:1.2:0.1); S3, chloroform:acetone:diethylamine (5:4:1); S4, ethyl acetate:methanol: conc. ammonia (17:2:1); S5, ethyl acetate:methanol: conc. ammonia (15:4:1); S6, n-butanol–acetic acid–water (35:3:10)	Complete separation of all compounds was observed in system S6. The order of elution was as follows ($R_f \times$ 100): benzoyl *nor*-ecgonine (95), benzoyl ecgonine (88), cocaine (83), *nor*-ecgonine, ecgonine methyl ester (66), and ecgonine (48)	[23]
Hyoscyamine, scopolamine, and others	Silver halide impregnated silica gel for TLC in isocratic or gradient modes	Lower retention but better specificity, good chemical and physical stability of this TLC system in tropane alkaloids separation may make it applicable in routine analysis	[24]

(continued)

TABLE 26.1 (continued)
TLC Separation Systems of Tropane Alkaloids Investigated

Compounds Separated	Separation Systems	Results Obtained	Ref.
Different tropane alkaloids	Silica gel/chloroform:acetone:diethylamine (5:4:1) or chloroform:diethylamine (9:1) or butanone–methanol–7.5 ammonia (6:3:1) or cyclohexane–chloroform (3:7) + 0.05% diethylamine	Good separation of various types of tropane alkaloids	[10]
Hyoscyamine and scopolamine in leaves and fruits of *Datura* sp.	Silica gel/1,1,1-trichloroethane–diethylamine (90:10, v/v)	hR_f values for hyoscyamine and scopolamine were 22 and 36, respectively	[25]
Tropane alkaloids in *Hyoscyamus niger* samples	Silica gel/chloroform–25% ammonia (98:2, v/v)	The alkaloids were well separated from neighboring compounds	[26]
Atropine and scopolamine in *Belladonna* tincture	Silica gel silufol/methanol–chloroform–25% ammonia (30:70:2)	The compounds were separated to be further scraped off the plates for spectrophotometric quantitative analysis	[26]
Atropine and scopolamine in some preparations of pharmacopoeia Japan	Thin layer stick chromatography on glass sticks coated with 0.1 mm layer of silica gel and microcrystalline cellulose (5:2). Developing system: chloroform-acetone-methanol-25% ammonia (73:10:15:2)	The alkaloids were determined as follows: atropine (R_f 0.3–0.35), scopolamine (R_f 0.58–0.64), and papaverine (R_f 0.91)	[26]
Atropine, scopolamine, and tubocurarine in African arrow poison	Silica gel/chloroform–cyclohexane–diethylamine (3:6:1, v/v)	The following R_f values for tropane alkaloids were obtained: atropine (R_f 0.33), scopolamine (R_f 0.6), and d-tubocurarine (R_f 0.71)	[26]
Cocaine among 14 local anesthetics	2D HPTLC on silica gel plates with the following solvent systems: 1D, cyclohexane–benzene–diethylamine (75:15:10) and 2D, chloroform-methanol (80:10)	Efficient separation of cocaine from synthetic anesthetics	[26]
Calystegines in Solanaceae plant family	Silica gel/chloroform–methanol-0.6N ammonia	Good separation of calystegine A and calystegines B zones	[3]
Major tropane alkaloids (hyoscyamine and hyoscyne) in some solanaceous plants	Silica gel/toluene–acetone–methanol–ammonia (20:20:6:1)	Good resolution, symmetrical peaks for both compounds were obtained. The R_f values were 0.35 and 0.57 respectively for hyoscyamine and hyoscyne	[27]

Calystegines in *Atropa belladonna*, *Hyoscyamus muticus*, *Solanum tuberosum*, and *Calystegia sepium*	HPTLC silica gel, 0.2 mm layer. Automated multiple development (AMD). Optimization of type of the mixture of solvents, the developing time, the number of development steps, the drying time between each run and precondition parameters	Excellent separation of calystegines A, B, and C and their putative metabolites: tropinone, tropine, pseudotropine, and *nor*-tropine by gradient development with chloroform-methanol and 9 M ammonia in gas phase. For separation of only individual calystegines the gradient system comprised: methanol-propanol-ethanol and 14 M ammonia for preconditioning	[28]
Hyoscyamine and scopolamine in some of solanaceous plants	Silica gel/chloroform-methanol-25% ammonia (85:15:0.7)	Analyzed compounds were separated from each other and coeluting ones but rather broad peak of hyoscyamine was obtained	[29]
Tropane alkaloids from *Datura inoxia*	NP separation: silica gel/MeOH-acetone-aq. ammonia (50:40:5); silica gel/MeOH-acetone-diethylamine (50:48:2). RP separation: RP_w18/MeOH-water (phosphate buffered at pH 3.4 $(1 + 3)$ containing 0.001 M HDEHP. 2D separation: 1D, silica gel/MeOH-acetone-diethylamine (50:48:2) and 2D, RP_w18/MeOH-water (phosphate buffered at pH 3.4 $(1 + 3)$ containing 0.001 M HDEHP	Good separation of atropine (or L-hyoscyamine), scopolamine, scopolamine-*N*-oxide, tropine, and tropic acid can be achieved by NP systems explored. It was unable to separate L-hyoscyamine from atropine and homatropine in this conditions. 2D system with an ion pair reagent (HDEHP) enabled complete separation of the plant extract alkaloids (excluding pair of L-hyoscyamine and atropine) with considerably improved peak symmetries	[30]
Atropine and scopolamine from *Datura inoxia* herb	Silica gel/MeOH-acetone-aq. ammonia (50:45:5)	Relatively symmetrical peaks for atropine and scopolamine	[31]
L-hyoscyamine and scopolamine from *Datura* sp. together with scopolamine-*N*-oxide and scopolamine-*N*-methyl bromide	Two mobile phases system: acetone-methanol-water-25% ammonia (82:5:5:8, v/v) over a distance of 9.5 cm for scopolamine and L-hyoscyamine determination (after 20-min preconditioning), and then after removing of the solvents at the distance of 5.5 cm with acetonitrile-methanol-85% formic acid (120:5:5, v/v) for scopolamine-*N*-methyl bromide and scopolamine-*N*-oxide analysis	Major *Datura* alkaloids can be easily separated by the former mobile phase but for scopolamine-*N*-methyl bromide and scopolamine-*N*-oxide second development with pseudoreversed-phase system on silica stationary phase is necessary	[2]

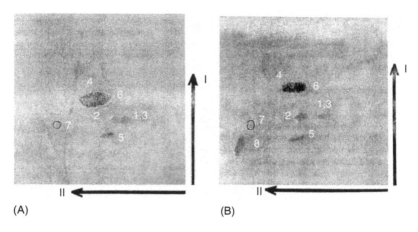

(A)   (B)

FIGURE 26.2 2D TLC of alkaloid mixture (A) and plant extract (B). First direction: SiO$_2$/MeOH–acetone–diethylamine, $50 + 48 + 2$; second direction: RP$_w$ 18/MeOH–water (buffered at pH 3.4) containing 0.01 M HDEHP, $1 + 3$. 1, atropine; 2, homatropine; 3, L-hyoscyamine; 4, scopolamine; 5, scopolamine-N-oxide; 6, tropine; 7, tropic acid; 8, unidentified alkaloid. (From Gadzikowska, M., Petruczynik, A., Waksmundzka-Hajnos, M., Hawryl, M., and Jóźwiak, G., *J. Planar Chromatogr.—Mod. TLC*, 18, 127, 2005. Taken after kind permission of the owner of JPC Akademiai Kiadó.)

and butanone. Polar modifiers were applied: methanol, acetone, acetic acid, and water. To improve peak symmetries pH controlled separation was mostly proposed at high pH conditions operated by aq. ammonia, conc. ammonia, diethylamine or ion pair reagents (e.g., HDEHP at pH 3.4; Figure 26.2).

Quite a few papers described cocaine and its metabolites and congeners determination of what could be done, e.g., by glass fiber silica gel impregnated sheets. In this way several structurally similar compounds could be separated. 2D TLC system was efficiently applied for the separation of cocaine among 14 local anesthetics.

Calystegines were also considered by TLC methods. For separation of only calystegine A from calystegines B, similar systems as for scopolamine were used. However, for determination of various types of calystegines A, B, C and their metabolites, HPTLC automated multiple development (AMD) systems were elaborated.

Sometimes other tropane derived alkaloids or intermediates were analyzed, i.e., scopolamine-N-oxide, scopolamine-N-methyl bromide, tropine, tropic acid, and homatropine. For these reasons pseudoreversed phase silica gel systems or 2D TLC were proposed.

No paper described the completed separation of stereoisomers L-hyoscyamine and atropine. Even in 2D TLC they were coeluted.

26.5.1.1 Quantitative Analysis and Densitometry

Tropane alkaloids possess poor chromophores, i.e., that maximum wavelength of UV absorption is about 205 nm, which is similar to mostly coeluted compounds. Therefore, in majority of proposed detection and quantitation systems derivatization with Dragendorff reagent.[27,30,31] or Dragendorff var. Munier reagent[28] was applied at first. Obtained orange bands of alkaloids could be further scanned by absorption densitometric method at 530 or 520 nm. Usually narrow peaks were obtained in this way with low background absorbance level especially for HPTLC plates. However, stability of its complex was not investigated. Detection limits obtained were typically as follows: 3 μg/spot (scopolamine), 1.7 μg/spot (atropine[30]). In similar approach about 0.5 μg/spot LOD values were measured.[32]

Berkov and Pavlov[29] described densitometry obtained after derivatization with Dragendorff reagent by QuantiScan image analysis software. Reproducibility of the method was rather good; however, it was not compared with typical absorption scanning densitometry. Detection limits were also quite poor for both scopolamine and hyoscyamine (1 μg/spot).

Duez et al.[25] used modified van Urk reagent for tropane alkaloids detection (DMBA in ethanol and 8 M H_2SO_4 1:1). Plates were scanned at 495 nm with good precision.

Mroczek et al.[2] elaborated HPTLC-densitometric assay on silica gel plates at 205 nm without derivatization. All peaks of tropane alkaloids were symmetrical, and suitable separation factor of two the most polar tropane alkaloids was obtained (Figures 26.3 and 26.4). Low background level noticed at merely 205 nm was obtained by effective purification step on Oasis MCX sorbent. The method was

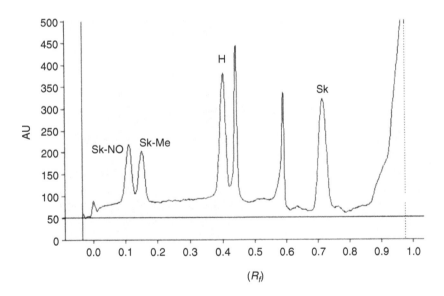

FIGURE 26.3 HPTLC-densitometric assay of the mixture of scopolamine-*N*-oxide (Sk-NO), scopolamine-*N*-methyl bromide (Sk-Me), L-hyoscyamine (H), and scopolamine (Sk). Stationary phase: silica gel 60 F_{254} HPTLC plates, 20 × 10 cm, 0.25 mm thickness; mobile phase: (a) acetone–methanol–water–25% ammonia (82:5:5:8, v/v) over a distance of 9.5 cm, then after evaporation of the solvents: (b) acetonitrile–methanol–85% formic acid (120:5:5, v/v) at the distance of 5.5 cm. Scan was recorded at 205 nm. (From Mroczek, T., Głowniak, K., and Kowalska, J., *J. Chromatogr. A*, 1107, 9, 2006.)

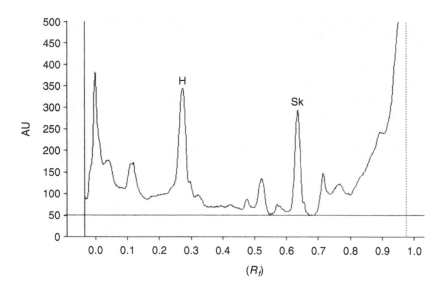

FIGURE 26.4 HPTLC-densitometric assay of L-hyoscyamine (H) and scopolamine (Sk) in the extract from seeds of *Datura fastuosa* purified by SPE procedure. Stationary phase: silica gel 60 F$_{254}$ HPTLC plates, 20×10 cm, 0.25 mm thickness; mobile phase: acetone–methanol–water–25% ammonia (82:5:5:8, v/v) developed over a distance of 9.5 cm. Scan was recorded at 205 nm. (From Mroczek, T., Głowniak, K., and Kowalska, J., *J. Chromatogr. A*, 1107, 9, 2006.)

also considerably more sensitive to those described after derivatization steps. Only about 25–30 ng of the compound could be accurately analyzed (LOD values ranged from 0.13 ng/μL for scopolamine-*N*-oxide to 0.58 ng/μL for scopolamine-*N*-methyl bromide).

For cocaine and its metabolites and congeners, derivatization with iodoplatine and Ludy Tenger's reagent was used.[23]

Calystegines were detected on TLC plates by AgNO$_3$/NaOH reagent.[3,28]

After PC described by Szymańska[22] the bands of alkaloids were eluted from Whatman paper, derivatized with bromothymol and analyzed spectrophotometrically at 620 nm.

26.5.1.2 Over-Pressure TLC

Over-pressure TLC (OPTLC) was a very seldom applied for separation and analysis of tropane alkaloids. Only one paper dealt with this method.[33] The author isolated in this way alkaloids from *Datura stramonium*.

26.5.1.3 Comparison of TLC and Other Chromatographic Methods (HPLC, GC, CE, CZE, MEKC)

Mule et al.[33] compared TLC method of benzoylecgonine (cocaine metabolite in urine) with gas liquid chromatography (GLC) and ^{125}I radioimmunoassay (RIA).

The TLC method, more simple and rapid seemed quite good correlated with RIA method (87% agreement). Between GLC and RIA methods about 95.5% results in the samples were comparable.

Berkov and Pavlov[29] compared quantitative results of hyoscyamine and scopolamine analysis in hairy roots of *Datura stramonium* by TLC and GC methods. The results were also quite well correlated with each other, especially for hyoscyamine. Similar CV% results were also obtained.

Mroczek et al.[2] compared HPTLC-densitometry and RP-HPLC-DAD of L-hyoscyamine and scopolamine determination in broad range of *Datura* sp. samples. Good correlation between HPLC and HPTLC quantitative results was measured although two various quantitation modes were applied (correlation coefficients of mean values in overall analyses were 0.92086 and 0.99995 for L-hyoscyamine and scopolamine, respectively). Each value correlated was in the range of $r \pm 2SD$. In term of precision, HPLC method turned out to be slightly more precise than HPTLC.

Therefore, TLC-densitometry although less sensitive than HPLC or GC method can be regarded also efficient quantitative tool in tropane alkaloids analysis.

Although atropine enantiomers or littorine were not separated so far by TLC (or HPLC) method, these compounds may be efficiently separated by capillary zone electrophoresis method (CZE) using sulfated β-cyclodextrin in less than 5 min.[34]

In similar way structurally related tropane alkaloids, including quaternary salts, can be quickly separated by micellar electrokinetic chromatography[35] (MEKC).

REFERENCES

1. Bruneton, J., Tropane alkaloids, in *Pharmacognosy, Phytochemistry, Medicinal Plants*, 2nd ed., TEC & DOC, Paris, 806, 1999.
2. Mroczek, T., Głowniak, K., and Kowalska, J., Solid-liquid extraction and cation-exchange solid-phase extraction using a mixed-mode polymeric sorbent of *Datura* and related alkaloids, *J. Chromatogr. A*, 1107, 9, 2006.
3. Dräger, B., van Almsick, A., and Mrachatz, G., Distribution of calystegines in several solanaceae, *Planta Med.*, 61, 577, 1995.
4. Jenett-Siems, K., Mann, P., Kaloga, M., Siems, K., Jakupovic, J., Witte, L., and Eich, E., Tropane alkaloids with unique type of acyl moiety from *Convolvulus* species, *Phytochemistry*, 49, 1449, 1998.
5. Strzelecka, H., *Chemiczne metody badania roślinnych surowców leczniczych (Chemical methods of medicinal plants research)*, 3rd ed., PZWL, Warsaw, 1987.
6. Dorer, M. and Lubej, M., Quantitative estimation of total alkaloids in Solanaceae herbs, *Deut. Pharm. Ges.*, 305, 273, 1972.
7. Kuczyński, J., Jusiak, L., and Gołkiewicz, W., Method for separation of scopolamine from vegetable raw material, *Polish Patent*, PL 157.603.
8. Jia, Y., Xie, H., Deng, G., and Sun, M., Optimization for extraction of total alkaloids in *Datura metel* L., *Zhongguo Zhonggao Zazhi*, 198, 480, 1994.
9. *The Polish Pharmacopoeia*, 4th ed., Vol. 2, PZWL, Warsaw, 1970.
10. Cordell, G.A., *The Introduction to Alkaloids (a Biogenetic Approach)*, Wiley, New York, 1981.
11. Plank, K.H. and Wagner, K.G., Determination of hyoscyamine and scopolamine in *Datura inoxia* plants by high performance liquid chromatography, *Z. Naturforsch. C*, 41, 391, 1986.

12. Fliniaux, M.A., Manceau, F., and Dubrenil, A.J., Simultaneous analysis of L-hyoscyamine, L-scopolamine and DL-tropic acid in plant material by reversed-phase high performance liquid chromatography, *J. Chromatogr. A*, 644, 193, 1993.
13. Brachet, A., Christen, P., and Veuthey, J.L., Focused microwave-assisted extraction of cocaine and benzoylecgonine from coca leaves. Optimisation of accelerated solvent extraction of cocaine and benzoylecgonine from coca leaves, *Phytochem. Anal.*, 13, 162, 2002.
14. Brachet, A., Rudaz, S., Mateus, L., Christen, P., and Veuthey, J.L., Optimisation of accelerated solvent extraction of cocaine and benzoylecgonine from coca leaves, *J. Sep. Sci.*, 24, 865, 2001.
15. Choi, Y.H., Chin, Y.W., Kim, J., Jeon, S.H., and Yoo, K.P., Strategies for supercritical fluid extraction of hyoscyamine and scopolamine salts using basified modifiers, *J. Chromatogr. A*, 863, 47, 1999.
16. El-Shazly, A., Tei, A., Witte, L., El-Domiaty, M., and Wink, M., Tropane alkaloids of *Hyoscyamus boveanus*, *H. desertorum*, *H. muticus* and *H. albus*, *Z. Naturforsch. C/J. Biosci.*, 52, 729, 1997.
17. Ylinen, M., Naaranlahti, T., Lapinjoki, S., Huhtikangas, A., Salonen, M.L., Simola, L.K., and Lounasmaa, M., Tropane alkaloids from *Atropa belladonna*; Part I. Capillary gas chromatographic analysis, *Planta Med.*, 52, 85, 1986.
18. Kursinszki, L., Hank, H., Laszlo, I., and Szoke, E., Simultaneous analysis of hyoscyamine, scopolamine, 6beta-hydroxyhyoscyamine and apoatropine in Solanaceous hairy roots by reversed-phase high-performance liquid chromatography, *J. Chromatogr. A*, 1091, 32, 2005.
19. Theodoridis, G., Kantifes, A., Manesiotis, P., Raikos, N., and Tsoukali-Papadopoulou, H., Preparation of a molecularly imprinted polymer for the solid-phase extraction of scopolamine with hyoscyamine as a dummy template molecule, *J. Chromatogr. A*, 987, 103, 2003.
20. Keiner, R. and Dräger, B., Calystegine distribution in potato (*Solanum tuberosum*) tubers and plants, *Plant Sci.*, 150, 171, 2000.
21. Bićan-Fišter, T., Quantitative analysis of tropane alkaloids in pharmaceutical preparations, *J. Chromatogr.*, 55, 417, 1971.
22. Szymańska, M., Pharmacognostic research on some species of *Scopolia* Jacq. genus, *Pol. J. Pharmacol. Pharm.*, 25, 201, 1973.
23. Misra, A.L., Pontani, R.B., and Mule, S.J., Separation of cocaine, some of its metabolites and congers on glass fibre sheets, *J. Chromatogr.*, 81, 167, 1973.
24. Aigner, R., Spitzy, H., and Frei, R.W., Separation of drug substances by modern liquid chromatography on silver impregnated silica gels, *J. Chromatogr. Sci.*, 14, 381, 1976.
25. Duez, P., Chamart, S., Hanocq, M., Molle, L., Vanhaelen, M., and Vanhaelen-Fastre, R., Comparison between thin-layer chromatography-densitometry and high-performance liquid chromatography for determination of hyoscyamine and hyoscyne in leaves, fruits and seeds of *Datura* (*Datura* spp.), *J. Chromatogr.*, 329, 415, 1989.
26. Popl, M., Fahnrich, J., and Tatr, V., *Chromatographic analysis of alkaloids*, Marcel Dekker Inc., New York/Basel, 1990.
27. Gupta, A.P., Gupta, M.M., and Kumar, S., Estimation of tropane alkaloids using high performance thin layer chromatography, *J. Indian Chem. Soc.*, 76, 277, 1999.
28. Scholl, Y., Asano, N., and Dräger, B., Automated multiple development thin layer chromatography for calystegines and their biosynthetic precursors, *J. Chromatogr. A*, 928, 217, 2001.

29. Berkov, S. and Pavlov, A., A rapid densitometric method for the analysis of hyoscyamine and scopolamine in Solanaceous plants and their transformed root cultures, *Phytochem. Anal.*, 15, 141, 2004.
30. Gadzikowska, M., Petruczynik, A., Waksmundzka-Hajnos, M., Hawryl, M., and Jóźwiak, G., Two-dimensional planar chromatography of tropane alkaloids from *Datura inoxia* Mill., *J. Planar Chromatogr.—Mod. TLC*, 18, 127, 2005.
31. Gadzikowska, M., Oniszczuk, M., Waksmundzka-Hajnos, M., and Jóźwiak, G., Comparison of sample pre-treatment techniques in HPTLC analysis of some tropane alkaloids in *Datura inoxia*, Proceedings of the International Symposium on Planar separations, Planar Chromatography 2005, Siofok, Hungary, 29–31 May 2005, pp. 339–346.
32. Montre-Gonzalez, M., Ayora-Talavera, T., Moldonado-Mendoza, E., and Loyola-Vargas, V.M., Quantitative analysis of serpentine and ajmalicyne in plant tissues of *Catharanthus roseus* and hyoscyamine and scopolamine in root tissues of *Datura stramonium* by thin layer chromatography-densitometry, *Phytochem. Anal.*, 3, 117, 1992.
33. Mule, S.J., Jukofsky, D., and Kogan, M., Evaluation of the radioimmunoassay for benzoylecgonine (a cocaine metabolite) in human urine, *Clin. Chem.*, 23, 796, 1977.
34. Mateus, L., Cherkaoui, S., Christen, P., and Vuethey, J.-L., Enantioseparation of atropine by capillary electrophoresis using sulfated β-cyclodextrin: application to a plant extract, *J. Chromatogr. A*, 868, 285, 2000.
35. Wu, H.-L., Huang, C.-H., Chen, S.-H., and Wu, S.-M., Micellar electrokinetic chromatography of scopolamine-related anticholinergics, *J. Chromatogr. A*, 802, 107, 1998.

27 TLC of Alkaloids from the Other Biosynthetic Groups

Jolanta Flieger

CONTENTS

27.1 PHENYLETHYLAMINE DERIVATIVES

This group of alkaloids is derived biosynthetically from the aromatic amino acids tyrosine and phenylalanine. Their chemical structure is based on phenylethyl-amine (1):

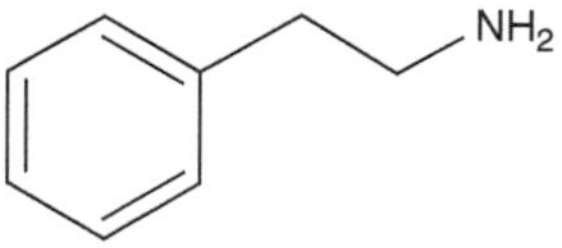

(1) 2-Phenylethylamine

Derivatives of phenethylamine are pharmacologically active compounds due to their structural similarity to the monoamine neurotransmitters. Alkaloids belonging to this group may act as stimulants (ephedrine, cathinone), hallucinogens (mescaline) and bronchodilators (ephedrine).

Substituted phenethylamines have been isolated from the cactuses: *Accacia promi-nens*, *Anhalonium lewinii*, and *Trichocereus candicans*. The abundant research devoted to the isolation of cactus alkaloids concerns determination of the most important one: mescaline (2) and its derivatives—*N*-methylmescaline, *N*-acetylmescaline, candicine as the drugs of abuse.

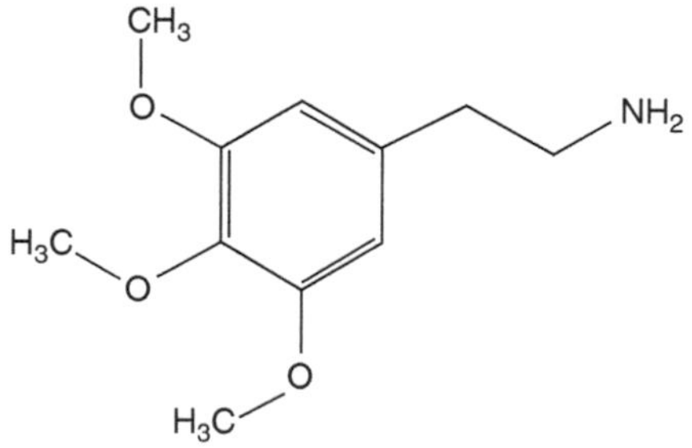

(2) Mescaline

Another subgroup is constituted by ephedrine and its derivatives isolated from various plants in the genus *Ephedra* (e.g., *Ephedra vulgaris*, family Ephedraceae). Ephedrine (3) is optically active owing to the presence of two asymmetric carbon atoms. Therefore ephedrine can form four enantiomers and two racemates. (1R,2R)- and (1S,2S)-enantiomers are designated as pseudoephedrine, while (1R,2S)- and (1S,2R)-enantiomers are designated as ephedrine. In plant derivatives of ephedrine, norephedrine, *N*-methylephedrine, and diastereoisomers (dextrorotarory and levo-rotatory φ-ephedrine) have been found. Ephedrine exerts sympathomimetic effect and is widely used in pharmaceutical preparations as a topical decongestant and as a bronchodilator in the treatment of asthma.

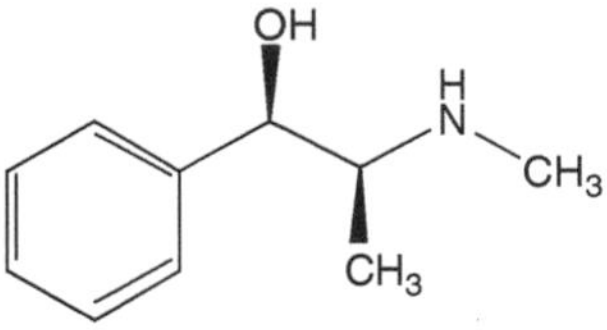

(3) (–)-(1R,2S)-Ephedrine

Colchicine (4) and other alkaloids isolated from plants of genus *Colchicum* (*Autumn crocus*, also known as the "meadow saffron") contains amide group in which nitrogen is not involved in a heterocyclic ring. Colchicine is a highly poisonous alkaloid, therefore its therapeutic value is strictly limited by its toxicity. However, colchicine is now used in medicine mainly in the treatment of gout. It is worth noticing that it is being investigated for its potential use as an anticancer drug, too.

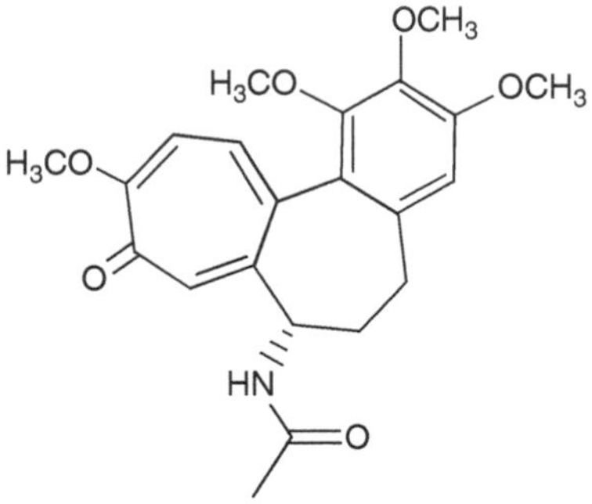

(4) Colchicine

27.1.1 PREPARATION OF EXTRACTS

Thin layer chromatography is usually used for the analysis of plant materials, biological materials, or pharmaceutical preparations. Extraction procedure depends primarily on a sample kind but also on the alkaloid contents.[1]

27.1.1.1 General Method for Extraction of Alkaloids

Powdered drug (1 g) is mixed with 1 mL 10% ammonia solution or 10% Na_2CO_3 solution and then extracted for 10 min with 5 mL methanol under reflux. The filtrate is then concentrated according to the total contents alkaloids of the specific drug.

For low total alkaloid contents (<1%) the presented method could be modified as follows:

Powdered drug (2 g) is ground in a mortar for about 1 min with 2 mL of 10% ammonia solution and then thoroughly mixed with 7 g of basic aluminium oxide. The mixture is then packed loosely into a glass column and 10 mL $CHCl_3$ is added.

Alkaloid bases are eluted with 5 mL $CHCl_3$ and the eluate is collected, evaporated to 1 mL and used for TLC.

Colchicine is exception to this procedure. Because of its neutral character, colchicine can be extracted by the use of neutral solvent systems:

Powdered seeds (1 g) are defatted with 20 mL n-hexane for 30 min under reflux. The defatted seeds are then extracted for 15 min with 10 mL $CHCl_3$. After this, 0.4 mL 10% ammonia is added, then the mixture is shaken vigorously and allowed to stand for about 30 min before filtration. The filtrate is evaporated to dryness and the residue solved in 1 mL ethanol.

27.1.2　THIN LAYER CHROMATOGRAPHIC SYSTEMS

Numerous chromatographic systems have been elaborated for the identification of phenylethylamine alkaloids. TLC data were collected and published in comprehensive reviews.[2–5] Chosen systems found to be the most effective are summarized in Tables 27.1 through 27.4.

For TLC investigation free amines, salts, or dansyl derivatives giving colored or fluorescent compounds can be used. When amine salts are present in the sample and neutral or weak acidic solvents are used as mobile phase multiple spots are formed. This phenomenon could be explained by ion-pair formation mechanism.

In some studies, alkaloids are identified not only in plant material but also in biological samples such as urine or pharmaceutical preparation, wherein they are accompanied by other drugs.

TLC provides a chromatographic drug fingerprint. It is therefore suitable for monitoring the identity and purity of drugs and for detecting adulterations and substitutions. That is why this method was chosen in many pharmacopoeias as a method of choice for fast screening systems suitable for analysis of preparations containing alkaloids. The German Pharmacopoeia[32] recommened TLC systems consisting of silica gel with ethyl acetate–glacial acetic acid–formic acid–water (100:11:11:26) or ethyl acetate–methanol–water (100:13.5:10) as eluent systems for analysis of alkaloids extracted from Colchici semen (*Colchicum autumnale* L., Liliaceae) (Figure 27.1). These systems enable the determination of major alkaloids in this plant drug: colchicine (65%), colchicoside (30%), demecolcine, and lumialkaloids (artifacts). In turn, the Japan Pharmacopoeia[33] and The Chinese one[34] use TLC method for the analysis of Herba Ephedrae (*Ephedra sinica* STAPF, *Ephedra shennungiana* TANG, *E. distachya* L., Gnetaceae (Ephedraceae)). Separation of this group containing L-Ephedrine, Norephedrine (+)-Pseudoephedrine, Norpseudoephedrine is possible in system consisting of silica gel and toluene–chloroform–ethanol (28,5:57:14.5) or ethyl acetate–cyclohexane–methanol–ammonia (70:15:10:5) as eluent systems.

Bodoki et al.[35] described validation of densitometric determination of colchicine extracted from seeds of meadow saffron as well as in pharmaceutical products. Alkaloid extract was separated by TLC using silica gel and a mixture of chloroform–acetone–diethylamine (5:4:1) as mobile phase. Densitometric measurements were carried out at the absorption maximum (350 nm) of colchicine (Figure 27.2).

TABLE 27.1
TLC Systems for Separation of Cactus Alkaloids Collected by Popl et al.

Alkaloids	Stationary Phase	Mobile Phase
Phenolic alkaloids: Anhalamine, anhalidine Anhalonidine, hordenine N-methyltyramine, pellotine	Silica gel	Chloroform–acetone–diethylamine (5:4:1) Cyclohexane–chloroform–diethylamine (5:4:1)
Pilocereine, tyramine	Silica gel impregnated with boric buffer pH 9.2	Methyl ethyl ketone–dimethylformamide–conc. ammonia (13:1.9:0.1)
Nonphenolic alkaloids: Anhalinine, anhalonine, Lophophorine, mescaline N-acetylmescaline N-methylmescaline N-methylanhalomidine O-methylanhalomidine	Silica gel	Chlorofrm: n-butanol–conc. ammonia (50:50:2.5) Chlorofrm–ethanol–conc. ammonia (85:15:0.4) Methanol–ammonia (25%)
	Silica gel impregnated with boric buffer pH 9.2	Methyl ethyl ketone–dimethylformamide–conc. ammonia (13:1.9:0.1)
Dansylated conjugate: β-Phenethylamine, N-methyl-β- phenethylamine, 4-methoxy-β- phenethylamine, N-methyl-4-methoxy- phenethylamine, N-methyl-4-methoxy- β-phenethylamine, 3,4-dimethoxy-β- phenethylamine, mescaline, N-methylmescaline	Silica gel Two-dimensional TLC	Two dimensional I. chloroform–butyl acetate (10:1) II. benzene–triethylamine (10:1)

Source: Popl, M., Fähnrich, J. and Tatar, V., Chromatographic analysis of alkaloids, In Cazes J. (ed.), *Chromatographic Science Series*, Vol. 53, New York, Marcel Dekker, 1990.

TABLE 27.2
TLC Systems Used for Identification of Mescaline Accompanied by Other Drugs of Abuse Collected by Popl et al.

Other Compounds	Adsorbent	Solvent System
Opium alkaloids, cocaine, and nicotine in urine	Silica gel	EtOH–AcOH–H_2O (6:3:1)
	Aluminum oxide	n-BuOH–n-Bu_2O–AcOH (4:5:1)
Amphetamines, opium alkaloids, barbiturates, phenothiazines, antihistamines, caffeine, quinine, nicotine, and cocaine in urine	Silica gel	$CHCl_3$–MeOH–NH_3 (90:10:1)
		EtOAc–MeOH–H_2O–NH_3 (85:10:3:1)
Opium alkaloids, amphetamines, catecholamines, LSD, and marihuana as dansyl derivatives in urine	Polyamide	H_2O–HCOOH (100:1.5)
		Benzene–AcOH (9:1)
		H_2O–AcOH (50:1)
		EtOH–H_2O–n-BuOH–HCOOH (93:150:4:3)
Various amines as dansyl derivatives	Silica gel	Two dimensional
		I. EtOAc–cyclohexane (75:50)
		II. Benzene–MeOH–cyclohexane (85:5:10)
		Two dimensional
		I. $CHCl_3$–BuOAc (10:2)
		II. (A) (isopr)$_2$O– 2×; (B) (isopr)$_2$O–TrEA
		(10:2) – 2×
Opium alkaloids, amphetamines, tryptamines, LSD, psilocin, and psilocybin in drug of abuse seizures	Silica gel	MeOH–ammonia (25%) (100:1.5)
Atropine, phenylethylamine, ephedrine, and various basic compounds in drug of abuse seizures as NDB-Cl derivatives	Silica gel	Et_2O–benzene (1:1)

Source: Popl, M., Fähnrich, J. and Tatar, V., Chromatographic analysis of alkaloids, In Cazes J. (ed.), *Chromatographic Science Series*, Vol. 53, New York, Marcel Dekker, 1990.

TABLE 27.3
TLC Separation of Colchicine Alkaloids

Examined Sample	Stationary Phase	Mobile Phase	Detection and Quantification	Ref.
Colchicine (pure standard)	Silica gel	Chloroform–methanol 95:5	Quantification by densitometry at 243 nm	[6]
Turkish colchicum and merendera species (colchicine, 3-demethylcolchicine, and demecolcine)	Silica gel	Benzene–ethyl acetate–butylamine 5:4:1 or 7:2:1	—	[7]
Colchicines and colchiceinamides	Reversed-phase TLC on silica, impregnated with 5% liquid paraffin	Phosphate buffer (pH 7.4) in various mixtures with acetone	Detection by spraying with an aqueous solution of $FeCl_3$ and exposure to gaseous HCl. After eating at 120°C, green spots appear	[8]
Racemic mixtures of (±)-hyoscyamine atropine and (±)-colchicine	Silica, impregnated with optically active acid, L-spartic acid (0.03%) as chiral selector	Butanol–chloroform–cetic acid–water 3:6:4:1	Detection by exposure to iodine vapor	[9]
Colchicine in different parts of Gloriosa superba	Silica gel	Ethyl acetate–methanol–water 00:27:20	Densitometry at 260 nm	[10]

TABLE 27.4
TLC Separation of Ephedrine Alkaloids

Source of Sample	Stationary Phase	Mobile Phase	Detection and Quantification	Ref.
Ephedrine hydrochloride in Herba Ephedrae and its preparations	Silica gel	Chloroform–ethyl acetate–methanol–water 13:40:22:10	Detection by spraying with 0.5% ninhydrin in ethanol and heating at 105°C for 10 min Identification by comparison with the standard Quantitation by densitometry at 525 nm	[11]
Ephedrine in eyedrops	Silica gel	Ethyl acetate–acetic acid–water 5:5:1	Detection with ninhydrin reagent	[12]
Ephedrine hydrochloride in pharmaceuticals	Silica gel	Butanol–ethanol–water 6:3:1	Detection by spraying with ninhydrin and heating at 900°C for 20 min. Quantification by densitometry at 520 and 700 nm	[13]
Ephedrine in Jiufen powder	Silica gel	Chloroform–ethanol–cyclohexane–NH_3 30:10:10:0.75	Quantification after eluting the spot with ethanol by colorimetry after adding 0.1% bromophenol blue	[14]
Ephedrine-pure standard	$CdSO_4$ or $NiSO_4$ or $NiCl_2$ containing silica gel	Butanol–formic acid 1:1	Location of spots by iodine chamber	[15]
Ephedrine and 2,3,5,6-tetramethyl-pyrazine in Chinese Ephedra, *Ephedra sinica* Stapf, and Mongolian Ephedra, *E. equisitina* Bunge	Silica gel	Ethyl acetate–ethanol–diethylamine 70:30:2	Detection by spraying with Dragendorff's reagent. Quantification by densitometry at 520 and 274 nm, respectively	[16]
Alkaloids in Jiufen San.: strychnine (I), brucine (II) and ephedrine (III)	Silica gel	Chloroform—methanol—NH_3 4:7:2	Detection under UV for I and II, and by spraying with ninhydrin reagent and heating at 110°C for III Quantification by densitometry at 254 nm for I and II, and 510 nm for III	[17]
Ephedrine in Tangxuan Lifi oral	Silica gel	Chloroform—methanol—NH_3 100:20:1	Detection by spraying with 0.3% ninhydrin in butanol—acetic acid 95:5. Quantification by densitometry at 505 nm	[18]

Application	Stationary phase	Mobile phase	Detection	Reference
Ephedrine in rheumatic tablets in presence of brucine, strychnine	Silica gel	Chloroform–methanol–NH$_3$ 55:45:1 and at 510 nm after spraying	Detection by spraying with 0.2% ninhydrin in ethanol and heating at 105°C for 10 min Quantification by densitometry at 254 and 325 nm before spraying	[19]
Ephedrine in Xicoerfeiyan microenema, Herba ephedrae, and Radix scutellarae	Silica gel	(1) Hexane–ethyl acetate 8:2, (2) benzene–ethyl acetate 95:5	Detection by spraying with 5% phosphomolybdic acid in ethanol and 1% vanillin in sulfuric acid Quantification by spectrophotometry	[20]
Ephedrine in Maxingshigan puls	Silica gel	Chloroform–methanol–NH$_3$ 100:8:1	Detection by spraying with 0.5% ninhydrin in ethanol. Quantification of ephedrine by densitometry at 510 nm	[21]
Ephedrine contents in Rhizoma Pinelliae	Silica gel	(1) Butanol–acetic acid–water 4:1:1, 2 chloroform–methanol–isopropanol–NH$_3$ 20:9:20:4 and 100:10:1:1	Detection by spraying with 0.2% ninhydrin in ethanol and heating at 110°C for 10 min Quantification of 1-ephedrine by densitometry at 500 nm	[22]
Ephedrine in psychotropic drugs in presence of phencyclidine, LSD, morphine, and codeine, ethylmorphine, scopolamine, physostigmine, and cocaine	Silica gel and alumina	68 Alkaline mobile phases	TLC of ionic associates of the basic substances with bromoxylenol blue, cresol red, and eriochromecyanine-R	[23]
Ephedrine hydrochloride in its oral compound	Silica gel	Chloroform–butanol–ethanol 2:2:3	Detection by spraying with 0.5% ninhydrin in ethanol and heating at 110°C for 15 min Quantitation by densitometry at 235 nm	[24]
The quality control of Zhenkeling oral	Silica gel	(1) n-Pentanol–methanol–formic acid–water 7:1:1:1, (2) ethyl acetate–methanol–water 4:1:1, (3) chloroform–methanol–NH$_3$ conc. 40:10:1, and (4) ethyl acetate–cyclohexane 9:1	Detection by spraying (1) with 10% sulfuric acid and heating at 110°C for 10 min, (2) with potassium iodobismuthate, (3) with 0.3% ninhydrin in acetic acid–butanone 1:20 and heating at 105°C for 10 min, and (4) under UV 254 nm. Quantitation of ephedrine by densitometry at 505 nm	[25]

(continued)

TABLE 27.4 (continued)
TLC Separation of Ephedrine Alkaloids

Source of Sample	Stationary Phase	Mobile Phase	Detection and Quantification	Ref.
Zhichuan adhesive plasters		(1) Chloroform–ethyl acetate–formic acid 5:4:2, (2) ethyl acetate–acetone–formic acid–water 10:6:2:1	Detection by (1) spraying with 2% $FeCl_3$ and 2% potassium ferricyanide solution, (2) by exposure to iodine vapors. Identification by fingerprint technique	[26]
TLC of diastereomers (including quininequinidine and cinchonine–cinchonidine) and enantiomers (including pseudoephedrine, ephedrine, norephedrine, and epinephrine)	Synthetic polymers imprinted with quinine as chiral stationary phase	Different concentrations of acetic acid (0, 1, 5, 10%) in either methanol or acetonitrile	Detection under UV 366 nm	[27]
Analysis of biogenic amines in fish meal (i.e., putrescine, cadaverine, histamine, tyramine, and ephedrine)	AMD HPTLC on silica gel	Dichloromethane–triethylamine 10:1 as the base solvent with increasing concentrations of n-hexane in order to change the polarity of the gradient	Quantitation by densitometry at 366 nm	[28]
Determination of ephedrine in Xiaoqinglong granules	Micellar TLC on silica gel	3.5% SDS—0.1 mol/L tartaric acid–methanol 1:3	Detection by spraying with 0.2% ninhydrin in ethanol. Quantitation by densitometry at 528 nm	[29]
Ephedrin in Kexinkang puls	Silica gel	Chloroform–methanol–NH_3 20:10:1	Detection by spraying with 1% ninhydrin in ethanol and heating at 105°C for 5 min. Quantitation by densitometry at 520 nm	[30]
Ephedrine in Xichuan pills	Silica gel	Chloroform–methanol–NH_3 100:8:1	Detection by spraying with 5% ninhydrin in ethanol and heating at 105°C. Quantitation by densitometry at 520 nm	[31]

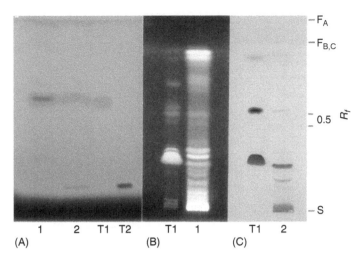

FIGURE 27.1 Colchici semen. Drug sample: 1 Colchici semen (alkaloid extraction method A, 30 μL) 2. Colchici semen (MeOH extract 3 g/10 mL, 10 μL). Reference compound: Tl-colchicine, T2-colchicoside. Solvent systems: A ethyl acetate–glacial acetic acid formic acid–water (100:11:11:26) B ethyl acetate–methanol–water (100:13.5:10). Detection A UV-254nm, B UV-365nm, C Dragendorff's reagent/NaNO$_2$. A: Colchici semen (1, 2). Both extracts are characterized by colchicine, which is seen as a prominent quenching zone at $R_f \sim 0.6$ (Tl), while colchicoside ($R_f \sim 0.15$/T2) is found in the methanolic extract (2) only. B: In the alkaloid fraction (1) a series of seven to nine prominent blue and yellow-white fluorescent zones from the start till $R_f \sim 0.35$, six weaker blue zones at R_f 0.4–0.85 and two zones at the solvent front are detected in UV-365 nm. Besides colchicine at $R_f \sim 0.25$ (Tl), minor alkaloids such as colchiceine, N-acetyl demecolcine and L-ethyl-2-demethylcolchiceine also show a yellow-white fluorescence, while O-benzoyl colchiceine, N-formyl-deacetyl colchiceine, and N-methyl demecolcine fluoresce blue. C: Colchicine and minor alkaloids react as brown zones with DRG reagent. Artifacts of colchicine ($R_f \sim 0.6$) appear as a blue zone at $R_f \sim 0.5$ (vis). (From Wagner, H., Bladt, S., *Plant drug analysis. A thin layer chromatography atlas*, Springer-Verlag, Berlin, Heidelberg, Heidelberg, 1996. With permission.)

27.1.3 DETECTION METHODS

For the detection of colchicine and related alkaloids, Dragendorff's reagent and iodoplatinate reagent can be used. Spraying 10% sodium nitrate solution after the use of Dragendorff's reagent causes the color of alkaloid zones to be intensified and increases the sensitivity to 0.01–0.1 μg.

Special reagent used for detection of alkaloids isolated from *Colchicum autumnale* is 2-methoxy-2,4-diphenyl-3(2H)-furanone (MDPF).[36] As a result, colchicine appears as a yellow fluorescent zone on a dark background in UV light (365 nm). The detection limit is 10 ng in this case. The fluorimetric analysis could be carried out with excitation at $\lambda_{exc} = 313$ nm, and evaluation at $\lambda = 313$ nm.

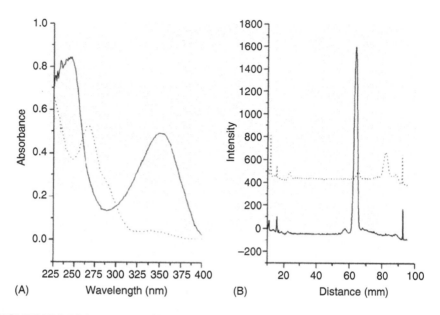

(A) Wavelength (nm)   (B) Distance (mm)

FIGURE 27.2 Light exposure effect over the (A) spectral and (B) chromatographic behavior of colchicines (-) and colchicine (---) degradation products. (From Bodoki, E. et al., *J. Pharm. Biomed. Anal.* 37, 971, 2005. With permission.)

27.2 QUINOLINE DERIVATIVES

27.2.1 *CINCHONA* ALKALOIDS

Cinchona is a genus of about 25 species in the family Rubiaceae, native to tropical South America. The bark of *Cinchona* trees is the source of a variety of alkaloids which are biogenetic group of antranil acid (2-aminobenzoic acid).

(5) Cinchona alkaloids: quinine (R = vinyl, R' = methoxy), cinchonidine
(R = vinyl, R' = hydrogen) and dihydroquinidine & dihydroquinine
(enantiomers with R = ethyl, R' = methoxy)

Typically, two major groups of Cinchona alkaloids have been selected. The first group covers *Erythro* ones: quinine, quinidine, cinchonine, cinchonidine, cupreine, and cupreidine. This group consists of four major alkaloids referred to as the parent alkaloids being two pairs of stereoisomers: quinine–quinidine and cinchonone–cinchonodine.

The second group includes *threo* alkaloids such as epiquinine, epiquinidine, epicinchonine, and epicinchonidine, which are the stereoisomers of the parent alkaloids.

The most important Cinchona alkaloid used as drug is quinine and its stereoisomer quinidine. Quinine was extracted from the bark of the South American cinchona tree, isolated and named in 1817 by French researchers Pierre Joseph Pelletier and Joseph Caventou. The name was derived from the original Quechua (Native American) word for the cinchona tree bark, "Quina" or "Quina–Quina," which roughly means "bark of bark" or "holy bark."

Quinine presents antipyretic, antimalarial, analgesic, and anti-inflammatory properties. Quinine was first used to treat malaria in Rome in 1631. The most recent studies indicate that the (+) stereoisomers are more potent in vitro against *Plasmodium falciparum*. Among *erythro* alkaloids with antimalarial activity quinidine apears to be the most effective drug. Quinine is also used in food and beverages to provide bitter taste.

Quinine is an extremely basic compound and is therefore always presented as a salt. Various preparations that exist include the hydrochloride, dihydrochloride, sulfate, bisulfate, and gluconate.

Cinchona alkaloids have frequently been separated by TLC either in plant material, as pure substances, in complex pharmaceutical preparations or biological materials and food.

27.2.1.1 Thin Layer Chromatographic Systems

Isolation of the alkaloids from the bark of *Cinchona* requires pretreatment with formic acid to release the compounds from their adducts with tannins. After making the medium alkaline, alkaloids are extracted with organic solvent immiscible with water by liquid–liquid extraction procedure.

Cinchona alkaloids are separated in normal phase systems consisting of silica gel plates and organic mobile phases containing solvents such as chloroform, acetone, ethyl acetate to which alcohols such as methanol, propanol, and butanol are added as polarity adjusters with diethylamine or ammonia used to reduce spot tailing. Base impregnated silica gel eluted by neutral solvents was also applied.

The most suitable solvent systems for separation of parent alkaloids, vinyl and dihydro derivatives, phenolic alkaloids, and stereoisomers on silica gel plates are selected by Svendsen and Verpoorte et al.[4]

- Chloroform–diethylamine (9:1)
- Chloroform–methanol–25% ammonia (85:14:1)
- Chloroform–acetone–diethylamine (5:4:1)
- Chloroform–acetone–25% ammnonia–absolute ethanol (5:4:1)
- Chloroform–ethyl acetate–isopropanol–diethylamine (20:70:4:6)
- Chloroform–dichloromethane–diethylamine (20:15:5)

- Dichloromethane–diethyl ether–diethylamine (20:15:5)
- Kerosine–acetone–diethylamine (23:9:9)
- Acetone–25% ammonia (58:2)
- Ethyl acetate–isopropanol–25% ammonia (45:35:5)
- Toluene–ethyl acetate–diethylamine (7:2:1) or (10:10:3)
- Toluene–diethyl ether–diethylamine (20:12:5)
- Toluene–diethyl ether–dichloromethane–diethylamine (20:20:20:8)
- Carbon tertrachloride–*n*-butanol–methanol–10%ammonia (12:9:9:1)
- Methanol–25% ammonia (100:1)
- Cyclohexanol–cyclohexane–*n*-hexane (1:1:1) + 5% diethylamine

Improvement of separation selectivity could be achieved by application of two-dimensional TLC technique on silica gel or 0.1 M sodium hydroxide impregnated silica plates.

Mroczek and Głowniak[37] also elaborated on TLC systems for HPTLC determination of the alkaloids in Chinae tincture, crude extract and for the analysis of quinidine and quinine in pharmaceutical preparations (Figure 27.3). They proposed new quaternary mobile phases consisting of benzene–chloroform–diethyl ether–diethylamine (40 + 15 + 35 + 10) and toluene–chloroform–diethyl ether–diethylamine (40 + 15 + 35 + 10) suitable for simple, rapid and accurate qualitative and quantitative analysis of the most important Cinchona alkaloids occurring at different concentrations.

Chosen TLC systems were adopted in quality control of Cinchona alkaloids: Aluminum oxide or silica gel precoated plates with chloroform–diethylamine (9:1) is thought of as fast screening system suitable for cinchona alkaloids.[1]

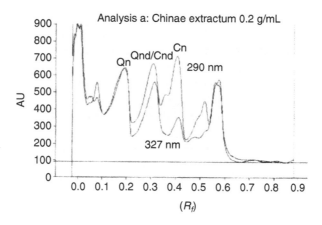

FIGURE 27.3. Assay of alkaloids in Chinae extract. HPTLC–densitometry was performed in absorbance–reflectance mode. Scans were recorded at $\lambda = 327$ and 290 nm. HPTLC conditions: 10 cm × 10 cm silica gel 60 F_{254} plates with toluene–chloroform–diethyl ether–diethyamine 40 + 15 + 35 + 10 (v/v), as mobile phase, development distance 9 cm. Qn, quinine; Qnd, quinidine; Cnd, cinchonidine; Cn, cinchonine. (From Mroczek, T. and Glowniak, K., *J. Planar Chromatogr.*, 13, 457, 2000. With permission.)

British Pharmacopoeia recommended base impregnated silica plates developed with chloroform–methanol–diethylamina whereas *The United States Pharmacopoeia* (USP XVIII) described system consisting of silica gel and chloroform–acetone–diethylamine (5:4:1) as a mobile phase.

To achieve further separation of spots poorly resolved in the first run, other chromatographic techniques like two-dimensional development were proposed. Pound and Sears[38] applied acetone–water–25% ammonia (80:20:1) and benzene–diethylamine (1:1) for the separation of the parent alkaloids and the dihydro bases of quinine and quinidine.

Quantitative determination of Cinchona alkaloids could be performed by off-line method (colorimetry after creation of color products with variety of reagents)[39] or online TLC (densitometry).[40]

Quinine bonded to stationary phase was used as enantioselective chiral selector in TLC as well as in capillary electrophoresis,[41] HPLC[42] and TLC.[27]

Some tertiary furoquinoline and quaternary dihydroquinoline alkaloids were separated by TLC on silica gel, alumina, and octadecyl-bonded silica gel layers. However, the limit of detection obtained by in situ flourometry was substantially lower in comparison with RP-LC determination.[43]

TLC systems suitable for analysis of different samples containing cinchona alkaloids are summarized in Table 27.5.

TABLE 27.5
Detection of Cinchona Alkaloids

Cinchona Alkaloid	Fluorescence Color at 366 nm after Spraying with Sulphuric Acid	Iodoplatinate Reagent
Quinine	Light blue	Violet-brown
Quinidine	Light blue	Violet-brown
Dihydroquinine	Light blue	Violet-brown
Dihydroquinidine	Light blue	Violet-brown
Cinchonine	Dark blue	Blue-violet-brown
Cinchonidine	Dark blue	Blue
Dihydrocinchonine	Dark blue	Blue-violet
Dihydrocinchonidine	Dark blue	Blue-violet
Epiquinine	Light blue	Violet-brown
Epiquinidine	Light blue	Violet-brown
Dihydroepiquinine	Light blue	Violet-brown
Dihydroepiquinidine	Light blue	Violet-brown
Epicinchonine	Dark blue	Blue-violet-brown
Dihydroepicinchonine	Dark blue	Blue-violet
Epicinchonidine	Dark blue	Blue-violet-brown
Dihydroepicinhonidine	Dark blue	Blue-violet
Quinidinone	Yellow-green	Yellow-violet
Cinchoninone	Yellow-green	Yellow-violet
Cupreine	Orange-red	Light blue-violet
Dihydrocupreine	Orange-red	Light blue-violet
Cupreidine	Orange-red	Blue-violet-brown
Dihydrocupreidine	Orange-red	Blue-violet-brown

27.2.1.2 Detection Methods

Detection methods for Cinchona (Figure 27.4) alkaloids applied the fluorescence of these compounds at 366 nm in acidic conditions (dilute sulfuric acid, formic acid spray) or creation of colored spots after spraying of the plate with the iodoplatinate reagent. Colors obtained after spraying with dilute sulphuric acid or iodoplatinate reagent are collected in Table 27.6.[4] Sensitivity of TLC detection of Cinchona alkaloids by these methods is 0.01 μg. Similar level of sensitivity could be achieved by the use of the modification of Dragendorff's reagent according to Munier.

For quantitative determination of alkaloids several direct and indirect methods have been applied. Indirect methods are based on the elution of the spots by different solvents (chloroform, absolute ethanol, methanol–ammonia, and dilute hydrochloric acid) followed by fluorimetric or spectrophotometric analysis. Direct quantification is possible by the use of photodensitometry after spraying with dilute 1% sulfuric acid

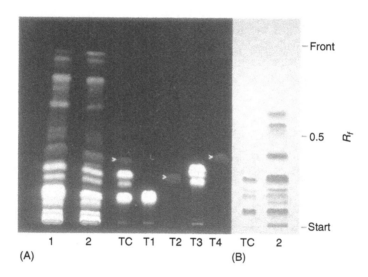

FIGURE 27.4 Chinae cortex. Reference compounds: T1 – quinine, T2 – cinchonidine, T3 – quinidine, T4 – cinchonine, TC – China alkaloid mixture, 1 – Cinchona calisaya, 2 – Cinchona succirubra. Solvent system: chloroform–diethylamine (9:1). Detection A – 10% ethanolic H_2SO_4, UV-365 nm, B – 10% H_2SO_4 followed by iodoplatinate reagent. A: In the R_f range 0.05–0.25 both Cinchona (Chinae Cortex) extracts show six light blue fluorescent alkaloid zones in UV-365 nm. They can be differentiated on the basis of their quinine (T1) content. Quinine (T1) and quinidine (T3) fluoresce bright blue after spraying with 10% ethanolic H_2SO_4, while cinchonidine (T2) and cinchonine (T4) turn dark violet and are hardly visible in UV-365 nm. In the extracts (1) and (2) the zone of cinchonidine (T2) is overlapped by the strong blue fluorescence of quinidine (T1). B: Treatment with iodoplatinate reagent results in eight mostly red-violet zones in the R_s range 0.05–0.65 (vis). The violet-brown zone of quinine is followed by the grey-violet zone of cinchonidine, a weak red-violet zone of quinidine and the more prominent brown-red cinchonine (TC). Three additional red-violet zones are found in the R_f range 0.4–0.6. (From Wagner, H., Bladt, S., *Plant drug analysis. A thin layer chromatography atlas*, Springer-Verlag, Berlin, Heidelberg, Heidelberg, 1996. With permission.)

TABLE 27.6
TLC Analysis of Cinchona Alkaloids

Alkaloid	Aim	Stationary Phase	Solvent System	Detection	Ref.
Anthraquinones, quinine, dihydroquinine, dihydrocinchonine, dihydrocinchonidine, cinchonine, cinchonidine, quinidine, cinchoninone quinamine	Cinchona ledgeriana callus culture	Silica gel	Ethyl acetate–isopropanol–NH₃ 25% 85:10:5	Detection by UV 254 and 366 nm, with ferrichloride solution in perchloric acid, iodoplatinate reagent; 5% KOH in ethanol	[39]
17 Cinchona alkaloids formed during enantioselective hydrogenation of ethyl pyruvate	Identification of hydrogenation products	Silica gel two-dimensional development	Chloroform–acetone–methanol–25% aqueous NH₃ 60:20:20:1 (optimum mobile phase).	Fluorescence detection after post-chromatographic derivatization with 10% sulfuric acid solution. Detection limit 0.5 –0.7 ng/spot	[40]
Quinine	TLC of opiates (morphine, codeine, methadone, and norpropoxyphene)	Silica gel	(a) Ethyl acetate–cyclohexane–methanol–water 70:15:2:8:0.5, (b) ethyl acetate–cyclohexane–methanol–NH₃ 56:40:0.8:0.4, (c) ethyl acetate–cyclohexane–methanol–NH₃ 70:15:2:8:0.5, (d) ethyl acetate–cyclohexane–NH₃ 50:40:0.1,	Detection reagents which were used: (a) silver acetate 1% in water, (b) mercuric sulfate solution, (c) diphenylcarbazone 0.01% in acetone I water, (d) 0.5% sulfuric acid, (e) iodoplatinate, and (f) iodine–potassium iodide reagent	[44]
Quinine	Mixture with atropine, berberine, brucine, ephedrine	Cadmium sulfate, zinc sulfate or nickel chloride containing silica gel	Butanol–formic acid 1:1	Location of spots by iodine chamber	[9]

(continued)

TABLE 27.6 (continued)
TLC Analysis of Cinchona Alkaloids

Alkaloid	Aim	Stationary Phase	Solvent System	Detection	Ref.
Quinine, quinidine, and hydroquinidine	Biomedical analysis	OPLC on silica gel	Methanol–acetonitrile–0.2M NH_3 130:68:2	Quantification by fluorescence densitometry at 348 nm Detection limit 10 ng or 0.1 pg/mL of plasma for quinine and quinidine; 5 ng or 0.05 pg/mL of plasma for hydroquinidine	[45]
Chloroquine, quinine, amodiaquine, mefloquine, pyrimethamine, and sulfadoxine	Human body fluids	Semiquantitative HPTLC analysis on silica gel	Toluene–diethylamine–methanol 8:1:1	Visualization under UV Detection limits between 1 and 50 nm (for amodiaquine)	[46]
Chloroquine, quinine, and their metabolites	In urine	HPTLC on silica gel	Toluene—diethylamine—methanol 8:1:1	Detection under UV 366 nm	[47]
Picrates (atropine, cocaine, novocaine, morphine, codeine, caffeine, theobromine, quinine, euquinine, lupanine, ephedrine, choline, betaine, nicotine, and papaverine)	Identification	Silica gel aluminum oxide	13 Different eluents	Visualization with Dragendorff's reagent or by iodine exposure	[48]
Quinine sulfate	Drug evaluation	Silica gel	Ethanol–NF_3–water 85:4:1	Detection by immersion into a solution of iodine and acidified potassium iodide; quantification by densitometry at 254 nm.	[49]

Diastereomers(quinine–quinidine cinchonine–cinchonidine) and enantiomers (pseudoephedrine, ephedrine, norephedrine, and epinephrine) on as chiral stationary phase	Separation of enantiomers	Synthetic polymers, imprinted with quinine	The mobile phases contained different concentrations of acetic acid (0, 1, 5, 10%) in either methanol or acetonitrile	Detection under UV 366 nm	[27]
Quinine	Mixture with Atropine scopolamine papaverine codeine strychnine	RP-18	Different concentrations of methanol in mixture with citrate buffer of pH 1.5 As counter ions: trifluoroacetic acid, pentafluoropropionic acid and heptafluorbutyric acid were used	Detection under UV 254 nm	[50]
Quinine	Mixture with caffeine	Silica gel	Isopropanol–cyclohexane–25% NH$_3$ 7:2:1	Quinine densitograms were combined from spectral data in the range 400 to 485 nm. Diode-array scanner, densitometry	[51]
Quinine	Qualitative identification in mixture with anthracene	Silica gel	(1) petroleum ether–ethyl acetate–formic acid 30:10:1, (2) petroleum ether–ethyl acetate 25:1, and (3) chloroform–methanol–water 13:7:2	Detection (1) under 365 nm, (2) by spraying with 1% vanillin–H$_2$SO$_4$ solution, (3) by spraying with 10%H$_2$SO$_4$ in ethanol Identification by fingerprint techniques, combined with microscopy with visualization by color reaction of anthracene and quinine with 1% NaOH solution	[52]

(continued)

TABLE 27.6 (continued)
TLC Analysis of Cinchona Alkaloids

Alkaloid	Aim	Stationary Phase	Solvent System	Detection	Ref.
Quinine (Qn), cinchonine (Cn), and cinchonidine (Cnd)	The stem bark of Cinchona officinalis and some herbal and homeopathic formulations	Silica gel	Chloroform/diethylamine (9.6:1.4 v/v)	The plates were scanned and quantified at 226 nm for Qn, Cn, Cnd and for Qnd at 366 nm in fluorescence and reflectance mode	[53]
The separation of ten plant extracts	*Cinchona succirubra, Aesculus hippocastanum, Berberis vulgaris. Artemisia abrotanum, Carduus marianus, Thuja occidentalis, Baptisia tinctoria, Paulinia cupana, Lycopus europaeus and Echinacea angustifolia*	Automated multiple development thin layer chromatography (AMD-TLC)	AMD was achieved in 25 steps using methanol, ethyl acetate, toluene, 1,2-dichloroethane, 25% ammonia solution and anhydrous formic acid as modifiers	The chromatograms were evaluated with a Shimadzu CS-9000 dual-wavelength flying-spot scanner	[54]
Quinidine	Analysis of antiarrhythmia drugs: Lidocaine and diphenylhydantoin Procainamide, propranolol in serum	Silica gel	Two developments with solvents of different polarity	Quinidine is scanned at 290 nm after the second development clozapine—an internal standard	[55]

in ethanol at different wavelengths (330 nm for cichonine and 288 nm for cinchonidine, limit of detection was found to be 100 ng) or fluorimetrically (quinine and quinidine – 365 nm excitation wavelength, measured at 450 nm, cinchonine, cinchonidine – 313 nm excitation wavelength, measured at 410 nm, detection limit – 1 ng).

27.3 PYRROLIDINE, PYRROLIZIDINE, PIRIDINE, AND PIPERIDINE DERIVATIVES

27.3.1 Tobacco Alkaloids

Tobacco alkaloids occur in the nightshade family of plants (Solanaceae), predominantly in tobacco, and in lower quantities in tomato, potato, eggplant (aubergine), and green pepper.

The volatile properties of these alkaloids make gas chromatography a more suitable method for their analysis. Most publications dealing with TLC of tobacco alkaloids concern determination of the well-known one—nicotine (6).

(6) Nicotine   (7) Arecoline

Nicotine contains pyridine and pyrrolidine rings. Dehydrogenation of this molecule leads to creation of optical isomers: nornicotine and nicotyrine. The isomer of nicotine contains a piperidine ring is called anabasine.

Nicotine is a potent neurotoxin but in lower concentrations acts as a stimulant in mammals and is one of the main factors responsible for the dependence-forming properties of tobacco smoking. That is why nicotine and its metabolites are present in urine of smokers.

Dihydropyridine ring contains molecule of ricinine. This alkaloid was found in *Ricinus communis*. However, tetrahydropyridine structure is part of alkaloids isolated from *Areca catechu* such as: arecaidine, its norderivative guvacine, and its methyl ester arecoline (7).

27.3.1.1 Thin Layer Chromatographic Systems

TLC of Tobacco alkaloids has been succesfully realized either in neutral or alkaline (addition of ammonia or diethylamine) and acidic (addition of acetic acid) solvent systems. Fejer-Kossey[56] elaborated suitable eluents for Tobacco alkaloids analysis (Table 27.7).

Recently elaborated on TLC systems have been listed in Table 27.8. Nicotianae folium (Tobacco leaves, *Nicotiana tabacum* L., *N. rustica* L., and other varieties of Solanaceae family) is a plant drug containing 0.06%–10% of alkaloids such as L-nicotine, nornicotine, anabasine, and nicotyrine. Screening system suitable for major alkaloids extracted from Nicotianae folium by methanol consists of silica gel and

TABLE 27.7

Detection and Identification of Tobacco Alkaloids

	hR_F Values Obtained on Silica Gel Plates Developed with Different Solvent Systems					
Tabacco Alkaloid	$CHCl_3$–MeOH–NH_3 (60:10:1)	$CHCl_3$–MeOH–CH_3COOH (60:10:1)	$CHCl_3$–MeOH (100:20)	$CHCl_3$–Et_2O–THF (80:15:5)	Dragendorff's Reagent	König's Reaction (1% Benzidine Spray + BrCN Vapors)
Nicotine	77	8	73	14	Red	Orange
Nicotine N-oxide	8	5	6	2	Red	Raspberry red
Nornicotine	34	5	27	5	Red	Purple
Nicotone	—	—	74	28	Purple-red	Cyclamen violet
Nicotyrine	87	92	81	76	Red	Cherry
Anabasine	50	6	44	7	Red	Carrot
Anatabine	—	—	57	7	Bright red	Pink
Myosmine	—	—	74	23	Red	Pale yellow

TABLE 27.8

TLC Systems for Determination of Tabacco Alkaloids in Materials of Different Origin

Alkaloid	Aim	Adsorbent	Solvent System	Detection and Quantification	Ref.
Cytisine and nicotine	Rapid TLC identification	Silica gel	Dichloromethane–methanol–10% NH_3 83:15:2	Detection by spraying with Dragendorff's reagent Quantification by densitometry at 565 nm	[57]
Nicotine	Determination of nicotine in serum	Silica gel	(A) chloroform–methanol–acetic acid 11:8:1 (B) chloroform–methanol acetic acid 17:2:1	Quantification by densitometry at 262 nm	[58]
Codeine, caffeine, papaverine, methaaqualone, nicotine (as reference compounds), and LSD, MDA (3,4-methylene dioxyamphetamine)	Qualitative identification, LSD, and alkaloids	Silica gel	Methanol–chloroform 1:9	Detection by spraying with Dragendorff's reagent	[59]
Nicotine	Estimation of nicotine from Gutka	Silica gel	chloroform–methanol–NH_3 6:5:1	Densitometry at 255 nm	[60]
Nicotine and its main metabolite—cotinine	Urine from smoking and passively smoking pregnant women	Silica gel bound with C-18 alkyl chains	Acetonitrile–water 88:12	Nicotine was identified after derivatization with Dragendorff's reagent under visible light. Cotinine was quantified by the use of densitometer in reflectance mode at 260 nm	[61]
Main nicotine metabolites (cotinine, trans-3′-hydroxycotinine)	Urine, isolation by means solid–liquid extraction technique using resin Amberlite XAD-2	Silica gel bound with C_{18}	Acetonitrile–water (88:12) with addition of sodium 1-octanesulfonate (50 mg/100 mL)	Visualization under UV illumination at 254 nm. The spots were quantified by scanning in reflection mode	[62]
Caffeine and nicotine and other basic drugs	In human urine	thin layer chromatography–tandem mass spectrometry (TLC–MS–MS)	Ethyl acetate–methanol–ammonium hydroxide (85:10:5)	UV light	[63]
Tobacco alkaloid myosmine	Nuts and nut products	Silica gel	Chloroform–methanol (9:1)	UV light	[64]
Arecoline	Arecae semen	Silica gel	chloroform–methanol–NH_3 240:5:4	Quantification by densitometry at 214 nm	[65]

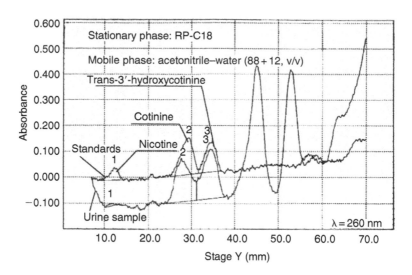

FIGURE 27.5 TLC chromatogram of fraction containing nicotine (1), cotinine (2) and trans-3'-hydroxycotinine (3) isolated from a child's urine sample. (From Tyrpien, K. et al., *Biomed Chromatogr.*, 15, 50, 2001. With permission.)

toluene–ethyl acetate–diethylamine (7:2:1).[1] TLC has been also applied in the analysis of tobacco alkaloids appearing in urine of smokers. Tyrpien et al.[61] applied planar chromatography for the determination of cotinine in urine of active and passive smoking pregnant women. Obtained results are presented in Figure 27.5.

27.3.1.2   Detection of Tobacco Alkaloids

Nicotine and related alkaloids display quenching in UV-254 nm. They can be detected with the usual alkaloid spray reagent: Dragendorff's reagent, iodoplatinate giving black-blue color spots with nicotine. A commonly used method for the detection of tobacco alkaloids is based on König's reaction. Recently Kato and coworker[66] described a new TLC visualization reagent for tertiary amines. Application of citric acid/acetic anhydride reagent (CAR), improves detection. The sensitivity of this method is 2.5 to 15-times greater than that of conventional detection with Dragendorff's reagent. This method appears also suitable for other drugs such as phenethylamine, phenothiazine, xanthine derivative, and narcotics.

27.3.2   *LOBELIA* ALKALOIDS

Lobelia alkaloids are isolated from Lobelia plant species e.g., *Lobelia inflata*. Some botanists place the genus and its relatives in the separate family Lobeliaceae. Herba lobeliae contains 0.2%–0.6% total alkaloids: lobeline (8) containing piperidine ring, isolobinine with dehydropiperidine ring, lobelanine, and DL-lobelidine. Shibano, et al.[67] isolated two new pyrrolidine alkaloids, radicamines A and B from *Lobelia chinensis* LOUR (Campanulaceae). Radicamines A and B were formulated as (2S, 3S,

4S, 5S)-2-hydroxymethyl-3,4-dihydroxy-5-(3-hydroxy-4-methoxyphenyl)-pyrrolidine and (2S, 3S, 4S, 5S)-2-hydroxymethyl-3, 4-dihydroxy-5-(4-hydroxyphenyl)-pyrrolidine determined on the basis of spectroscopic analyses and chemical methods.

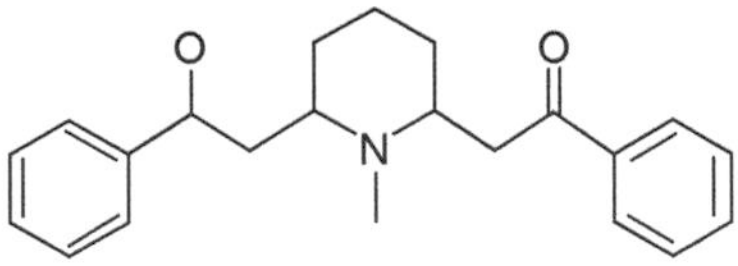

(8) Lobeline

Lobelia's medicinal properties include the following ones: emetic, stimulant, anti-spasmodic, expectorant, diaphoretic, relaxant, nauseant, sedative, diuretic, and nervine. In Chinese folk medicine the plant, *Lobelia chinensis* L. has been used as a diuretic, an antidote, hemostat, and as carcinostatic agent for stomach cancer.

27.3.2.1   Thin Layer Chromatographic Systems

TLC system for fast screening of alkaloids in Lobeliae herba consists of silica gel and mobile phase: toluene–ethyl acetate–diethylamine (70:20:10). Lobeline (R_F 0.65) can be visualized by the use of iodoplatinate reagent.[1]

Bebawy et al.[68] described thin-layer chromatographic scanner method for the determination of lobeline in presence of emetine hydrochloride and khellin in raw material and pharmaceutical preparations. Silica gel developed with pure methanol contains trace amount of ammonia (methanol–ammonia 200:3) appears to be suitable for this purpose. Quantification of lobeline, khellin, and emetine has been performed by densitometry at 285 nm, 250 nm, and 239 nm, respectively.

27.3.3   Pepper Alkaloids

Piperine (9) is the main alkaloid obtained by extraction of *Piper nigrum*. This component is responsible for the hot taste of pepper. Piperine is accompanied by its stereoisomer—chavicine.

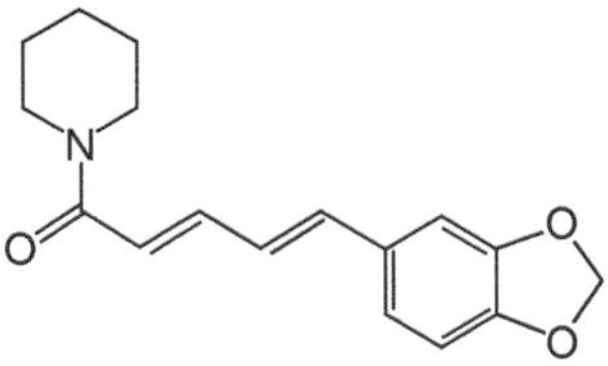

(9) Piperine

Conventionally, pepper alkaloids are extracted by heating under reflux with methanol or stirring with solvents such as ethyl alcohol, chloroform, petroleum ether, diethyl ether, etc. Traditional extraction in a Soxhlet device can be replaced by supercritical fluid extraction method. Separation and identification of pepper alkaloids can be achieved, among others, by TLC.

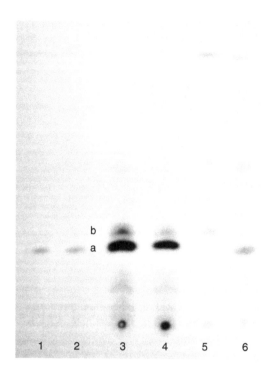

FIGURE 27.6 Thin layer chromatograms obtained from black pepper extracts. Stationary phase: silica gel 60 F_{254}. mobile phase: heptane–ethyl acetate: $60 + 40$ (v/v). Lines 1 and 2, piperine standards; lane 3, chloroform extract; lane 4, ethyl alcohol extract; lane 5, diethyl ether extract; lane 6, petroleum ether extract, a – piperine, b – chavicine. (From Marutoiu, C. et al., *J. Planar Chromatogr.*, 19, 250, 2006. With permission.)

Densitometric determination of piperine on silica gel was reported by Jansz et al.[69] Mobile phases usually consist of nonaqueous solvent systems. Detection of piperine is possible in UV-365 nm. Piperine gives dark blue fluorescence.

Use of TLC and mass spectrometry for the separation and identification of piperine and chavicine in extracts of black pepper was described by Marutoiu et al.[70]

Thin layer chromatograms obtained from black pepper extracts are presented in Figure 27.6.[71] Examples of TLC systems suitable for analysis of tobacco alkaloids are summarized in Table 27.9.

27.3.4 PELLETIERINE ALKALOIDS

Monocyclic tropane metabolites such as hygrine and ring-enlarged tropane homologues such as pelletierine group have been isolated from the bark of tree, tree root, and leaf of *Punica granatum*.

These alkaloids can be divided into the following subgroups:[77]

TABLE 27.9
TLC Systems for Analysis of Tobacco Alkaloids

Examined Material	Adsorbent	Solvent System	Detection	Ref.
Biological tissues	Silica gel	Petrol ether (60–80°C)—acetone 65:35, in the dark at +−24°C	Detection under UV. Quantification by densitometry in fluorescence at 310 and 390 nm	[72]
Plant materials	Silica gel	Cyclohexane–acetone–benzene–ether 7:3:1:1	Detection by spraying with an improved reagent of potassium iodobismuthate first, and then with 25% ethanol. Quantitation by densitometry at 508 nm	[73]
Piper nigrum Linn	Silica gel	Hexane–ethyl acetate 7:3	Quantitation by densitometry at 340 nm	[74]
Piperine and embelin	Silica gel G	Toluene–ethyl acetate 7:3 propanol–butanol–NH₃ 6:1:3	Quantitation by densitometry at 337 nm (piperine) and 333 nm (embelin)	[75]
The fruit of Piper chaba contains: piperine, piperamine, piperlonguminine, and methyl piperate	Analytical HPTLC on silica gel RP-18 and preparative TLC on silica gel	n-hexane–ethyl acetate 1:1	Detection by spraying with 1% cerium sulfate–10% aqueous sulfuric acid followed by heating on a plate heater	[76]

- Pelletierine alkaloids: N-Methylpelletierine, Pseudopelletierine, Norpseu-dopelletierine
- Piperidine alkaloids: Sedridine, 2-(2'-hydroxypropyl) Δ'-piperidine, N-(2'5'-Dihydroxyphenyl) pyridium chloride
- Pyrrolidine alkaloids: Hygrine, Norhygrine

They sometimes occur together with tropanes because they are formed by similar biosynthetic steps.[78] In comparison with other alkaloids that have higher octanol–water partition coefficient, they are fairly water-soluble at least at acidic pH and can be selectively extracted with aqueous acid.

Ferrara et al.[79] described two different extraction methods: extraction by Soxlet and extraction by steam distillation. Then the extracts were compared by TLC chromatography using different solvents and specific chromogen reagents.

TLC systems and detection methods are similar to those for tropane alkaloids.

27.3.5 SEDUM ALKALOIDS

Sedum species also contain pyrrolidine and piperidine alkaloids. Among all *Sedum* species investigated so far *S. acre* L. received the most attention in connection to the major content of piperidine alkaloids. The most important alkaloids of *S. acre* are the two 2-monosubstituted piperidine alkaloids: sedridine and sedamine.

TLC of Sedum alkaloids was performed on silica gel with eluent system consisting of chloroform–methanol–ammonia (85:18:1). Spots were visualized with Dragendorff's reagent and with o-toluidine reagent for specific detection of secondary amines.[80]

27.3.6 SENECIO ALKALOIDS

Senecio alkaloids are derived from the pyrrolizidine ring. This class of alkaloids has been isolated from species such as Senecio *vulgaris*, Senecio *anonymus*, Senecio *alpinus*, Senecio *inaequidens*. Their determination has attracted attention because of their toxicity. Pyrrolizidine alkaloids are accumulated in the form of N-oxides in plants.

TLC has been used for conformation of their presence in plant extracts as well as on preparative scale.

The biosynthesis and accumulation of senecionine and senecionine N-oxide, integerrimine N-oxide, agmatine, arginine, spermine, spermidine, ornithine, putrescine, and citrulline in root cultures of *Senecio Vulgaris* were analyzed according to the results obtained by TLC separation on silica gel with chloroform–methanol–25% NH_3–pentane 82:14:2.6:20 and acetone–methanol–25% NH_3 40:30:20 followed by radioscanning.[81–83]

To identify the biochemical mechanisms of sequestration of pyrrolizidine alkaloids by adults and larvae of chrysomelid leaf beetles, TLC was applied. Crude methanol extracts were separated on silica gel using the solvent consisted of dichloromethane–methanol–25% ammonia (84:14:2) and evaluated quantitatively by means of a multichannel radioactivity detector.[84]

TABLE 27.10

R_F **Values Obtained on Silica Gel and Detection Limits of Pyrrolizidine Alklaoids Isolated from** *Senecio Jacobaea*

Alkaloids	Chloroform–Methanol–Propionic Acid (36:9:5)	Chloroform–Methanol–Ammonium Hydroxide (85:14:1)	1% o-Chloranil in Toluene and Ehrlichs Spray Reagent with Heating at 80°C	Dragendorff's Spray Reagent and 5% Aqueous Sodium Nitrate
Heliotrine	0.47	0.25	6 μg/ml	4 μg/ml
Lasiocarpine	0.67	0.71	10	8
Monocrotaline	—	0.42	8	4
Tansy PA mixture				
Jacobine	0.70	0.59	10	6
Seneciphylline	0.80	0.62	10	6

Metabolism of pyrrolizidine alkaloids by *Peptostreptococcus heliotrinreducens* was studied by GC–MS and TLC method. A mixed culture of ovine ruminal microbes metabolizes the pyrrolizidine alkaloids present in plant Senecio jacabaea including jacobine, seneciphylline, heliotrine, lasiocarpine. Alkaloids were analyzed on silica gel plates developed in either an acidic solvent system or a basic solvent system.[85] R_F values together with detection limits are presented in Table 27.10.

27.4   QUINOLIZIDINE DERIVATIVES

27.4.1   LUPINE ALKALOIDS

Lupinine (10) is a bitter tasting alkaloid present in *Lupinus* spp. of leguminose plants. The fundamental part of its molecule is quinolizidine or more exactly octahydroquinolizidine. Lupinine belongs to lysine biogenetic group of alkaloids. About thirty chemically related compounds have been isolated so far.

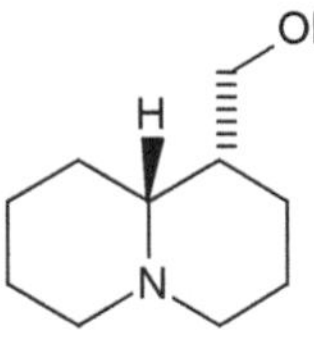

(10) Lupinine 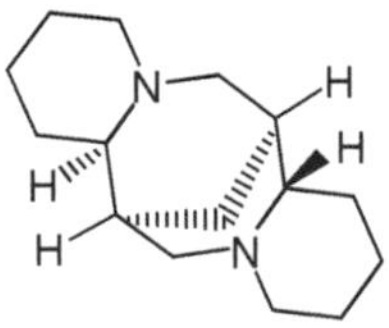   (11) Sparteine

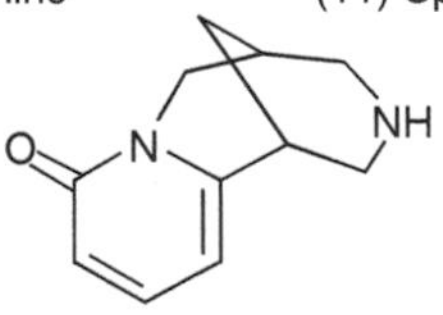

(12) Cytisine

Tetrahydroquinolizidine structure can be found in more complex alkaloids such as cytosine (12), matrine, and sparteine (11).

The properties of quinolizidine alkaloids are favorable for their separation preferably by GC method. Not many papers describing TLC separation of these alkaloids extracted from such species as *Lupinus arbustus* subsp. *calcaratus*, *Genista anatolica* have been published.

TLC was used mainly in preparative scale for the clean up of isolated alkaloids and for separation and characterization by the use of R_F values.

To obtain satisfactory separation selectivity of the lupine alkaloids it is necessary to use two or more TLC systems, for instance two-dimensional TLC with sorbent gradient or overpressured conditions. Other techniques such as paper electrophoresis[86] were also used for this purpose.

Lupine alkaloids can be visualized by the use of Dragendorff's reagent [modification Munier and Macheboef] giving yellow-orange to orange-red spots.

Cytisine exhibits similar properties to nicotine because of structural similarity. Plants that contain cytisine include severals ones from the *Faboideae* subfamily: *laburnum*, *anagyris*, *thermopsis*, *cytisus*, *genista*, and *sophora*.

Cytisine and nicotine are usually separated together by the use of the same TLC system. For rapid identification of nicotine and cytisine, TLC on silica gel with dichloromethan–methanol–19% ammonia (83:15:2) was described. Quantification of cytosine was done by densitometry at 290 nm whereas nicotine after spraying with Dragendorff's reagent was scanned at 565 nm.[57]

Thin layer chromatographic systems for analysis of alkaloids in plant materials require more components. Separation and identification of alkaloids in *Sophora alopecuroides* L. enabling determination of sophorine, matrine, sophocarnine, aloperine, sophocarpine, cytisine, oxymatine, and *N*-methyl catisine was possible by TLC on silica with benzene–acetone–ethyl acetate–methanol 14:6:6:1 before equilibration with NH_3 vapor.[87]

Lupanine and 3β-hydroxylupanine were determined in agriculture products on silica gel plates with cyclohexane–diethylamine (7:3) and visualized by Dragendorff's reagent.[88]

Svendsen[89] collected R_F values of lupine alkaloids (Table 27.11) obtained on commonly used stationary phases silica gel, aluminum oxide and cellulose developed by four chosen solvent systems. Other TLC systems are presented in Table 27.12.

27.5 XANTHINE ALKALOIDS

Naturally occurring purine alkaloids (biosynthetically derived from glycine) are xanthine derivatives isolated from various plants of *Rubiaceae*, *Theace*, and *Sterculiaceae* families.

Methylated xanthine derivatives include caffeine (13), theophylline (15), and theobromine (14).

TABLE 27.11

hR_F Values of Some Lupine Alkaloids

Lupine Alkaloids	hR_F			
	Silica Gel G CHCl$_3$–MeOH– 25% NH$_3$ (85:14:1)	Aluminum Oxide Benzene– Acetone–MeOH (34:3:2)	Celluloze MN 300 Butanol–36% HCl–Water (70:7.5:13.5)	Silica Gel HF254 Chloroform– Ethanol (3:2)
Anagyrine	69	64	7	—
Cytisine	28	19	10	—
N-methylcytisine	61	59	32	—
Δ^{11}-Dehydrolupanine	5		53	—
Δ^5-Dehydrolupanine	83	75	31	—
13-Hydroxylupanine	22	44	—	92
Epi-13-hydroxylupanine	—	—	—	76
13-Hydroxy-α-isolupanine	—	—	—	83
Epi-13-hydroxy-α-isolupanine	—	—	—	54
4-Hydroxylupanine	61	3	40	—
8-Ketosparteine	88	79	27	—
Lupanine	72	74	52	—
Lupinine	23	56	58	—
Matrine	83	73	60	—
17-Oxolupanine	82	74	88	—
17-Oxysparteine	85	76	82	—
Sparteine	8	67	50	—

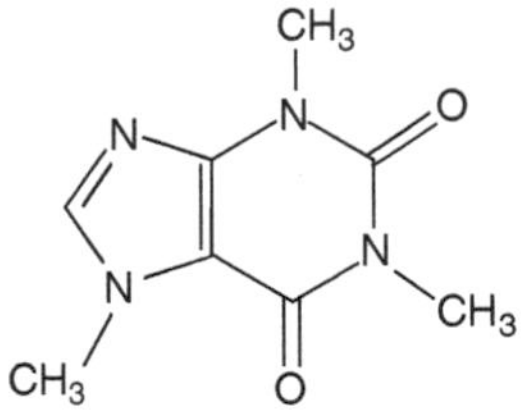

(13) Caffeine

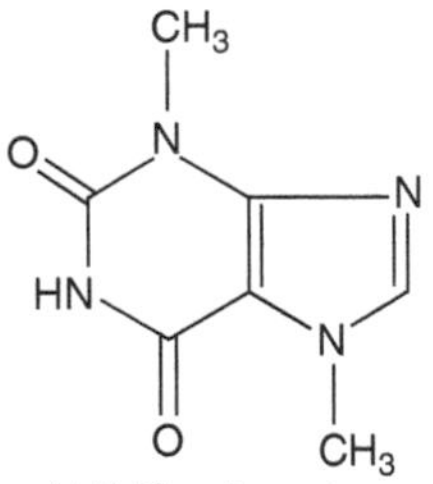

(14) Theobromine

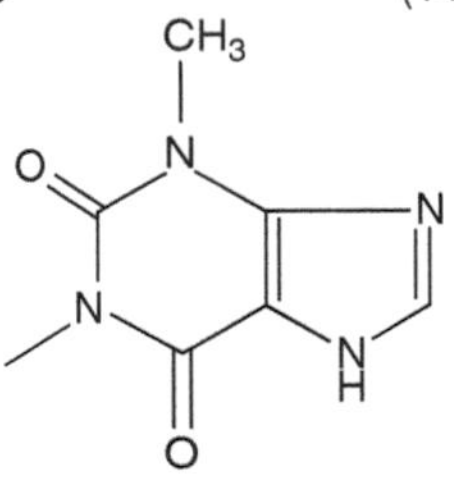

(15) Theophylline

TABLE 27.12
TLC Systems for Analysis of Lupine Alkaloids

Compounds Separated	Sample	Stationary Phase	Mobile Phase	Detection	Ref.
Lupinine, sparteine, virgiboidine, virgiline, oroboidine, virgiline-pyrrolyl-carboxyl ester	Genus *virgilia*	Silica gel	Methanol–chloroform–NH$_3$ 85:14:1	Quantification by GC after elution with methanol	[90]
Lupinine, sparteine, lupanine, 13-hydroxylupanine, and dihydrofuro (2,3-(b) quinolinium salts	Pure patterns	OPLC on alumina Silica gel	Ethyl acetate Ethyl acetate–formic acid–water 10:1:1	Visualization by spraying with Dragendorff's reagent and by fluorodensitometry at 310/420 nm	[91]
Sparteine, a-isosparteine, retamine, hydroxylupanine, cytisine, methylcytisine, sophocarpine, lupanine	The herb and in vitro cultures of several *Genista* species	One-dimensional TLC on silica gel	Hexane–diethyl amine 100:0 to 1:1	Visualization by spraying with Dragendorff's reagent and documentation by photography	[92]
		Silica gel and DIOL	Chloroform–methanol–NH$_3$ 100:0:1, 95:5:1 to 85:15:1 acetonitrile–water–Hydrochloric acid 30:100:5,		
		DIOL and RP-18	30:100:10, 30:100:15 and 30:100:20		
		2D-TLC on DIOL (I) and RP-18 plate with the mobile phase (II)	I. Chloroform–methanol–NH$_3$ 85:15:1 II. Acetonitrile–water–hydrochloric acid 30:100:7		
Quinolizidine, indolizidine, and dipiperidine alkaloids	Species of the *Papilionaceoustribes Sophoreae, Dalbergieae, Loteae, Brongniartieae,* and *Bossiaeeae*	Silica gel	Methanol–28% NH$_3$ 131:2, chloroform–methanol–28% NH$_3$ 85:15:1 or hexane–diethylamine 7:3	Detection with Dragendorff's reagent Also GLC–MS and purification of crude extracts by PLC	[93]

Compound	Source	Layer	Solvent system	Detection	Ref.
Calpaurine and related compounds	*Ethiopian calpurnia aurea* subsp. *aurea*	Silica gel	Chloroform–methanol–28% NH₃ 90:9:1, isopropanol–ethyl acetate–chloroform–28% NH₃ 11:4:4:1, diethylamine–chloroform 1:9 or ether–methanol–28% NH₃ 44:5:1	Detection with Dragendorff's reagent	[94]
Quinolizidine alkaloids	The genus *Pearsonia*	Silica gel, Aluminium oxide	Chloroform–cyclohexane–butylamine 5:4:1, 1.5% Methanol in chloroform	After heating at 100°C for 3 min visualization under UV 254 and 365 nm and by spraying with iodoplatinate reagent	[95]
Lupanine, 13-hydroxysparteine, anagyrine, cytosine, and calycotomine	*Genista anatolica*	Layers: Silica gel GF 254	Chloroform–diethylamine 7:3, 9:1	Visualization UV 254, 366 nm, Dragendorff's reagent	[96]
Gramodendrine, amodendrine, lusitanine, and gramine	*Lupinus arbustus* subsp. *calcaratus*	Silica gel 60 PF 0.5 mm; Preparative TLC Layers Silica gel 60 GF 254, 0.25 mm; Silica gel 60 PF 254 1 mm	Chloroform–methanol–25% ammonia 85:15:1; Chloroform–methanol–18 M ammonia 100:10:1; CH–diethylamine 7:3	Visualization Dragendorff's and Ehrlich's reagents	[97]
Matrine	Meilupijiling	Silica gel	Benzene–acetone–ethyl acetate–NH₃ conc. 10:15:20:1	Detection by spraying with 5% potassium iodobismuthate solution and 5% sodium nitrite solution Identification by finger print technique. Quantation of matrine by densitometry at 480 nm	[98]
Sophocarpine, matrine, sophoridine, oxymatrine, and oxysophocarpine	*Sophara flavescens* and sophora lopecuroides	Silica	Benzene–acetone–methanol 8:3:0.5	Determination by densitometry at 510 nm	[99]

(continued)

TABLE 27.12 (continued)
TLC Systems for Analysis of Lupine Alkaloids

Compounds Separated	Sample	Stationary Phase	Mobile Phase	Detection	Ref.
Sophocarpine, matrine, and sophoramine	Standards	Silica gel	Ether–methanol–28% NH$_3$ 40:4:1	Detection with UV and iodine vapors	[100]
Matrine	Jieshanbao lotion	Silica gel	Benzene–acetone–methanol 16:5:1	Detection by spraying with potassium iodobismuthate reagent. Densitometry at 515 nm	[101]
Matrine, glycyrrhetinic acid, indirubin, indigotin, berberine, and hydroxycholic acid	Vietnamese sophora root, Liquorice root, natural indigo, and golden thread	Silica gel	(1) Chloroform–methanol–NH$_3$ 58:2:1 (2) Hexane–ethyl acetate–acetic acid 25:10:3 (3) Ethyl acetate–methanol–acetic acid–water 15:1:1:2 (4) Chloroform–acetic acid 5:1	Detection by (1) spraying with potassium iodobismuthate reagent, (2) exposing into iodine vapor, (3) spraying with 10% sulfuric acid in ethanol and heating, and (4) under UV-365 nm	[102]
Matrine	The quality standard of Sanjufencilu	Silica gel	(1) Benzene–chloroform–ethyl acetate–formic acid 30:15:20:3 (2) Hexane–ethyl acetate 7:3 (3) Benzene–acetone 9:1 (4) Ethyl acetate–acetone–benzene–NH$_3$ 20:15:10:1	Quantification of matrine by densitometry at 520 nm	[103]
Matrine	Kushen, Sophora flarscens Ait	Silica gel	Chloroform–methanol–NH$_3$ 25:3:1	Detection by exposure to iodine vapor. Quantitation by fluorescence quenching after elution	[104]
Indirubin, matrine	Guilin Xiguashuang spray	Silica gel	(1) Benzene–chloroform–acetone 5:4:1 (2) Toluene–acetone–ethanol–NH$_3$ conc. 20:20:3:1	Detection by spraying with 5% potassium Iodobismuthate in water. Quantification by densitometry at 540 nm and at 510 nm	[105]

Matrine	Xianchen Fencijing	Silica gel	(1) Benzene–ethyl acetate–15:1 (2) Benzene–ethyl acetate–formic acid 20:10:1 (3) Cyclohexane–ethyl acetate–diethylamine 5:5:1 (4) Benzene–diethylamine 5:5:1.4, benzene–ethyl acetate 9:1 (5) Benzene–acetone–methanol 16:6:1	Detection (1) under UV, (2) by spraying with 5% potassium iodobismuthate, (4) by spraying with 10% phosphomolybdic acid in ethanol and heating at 105°C. Quantitation of matrine by densitometry at 355 nm	[106]
Matrine	Quality control of Jiyan oral liquid	Silica gel	(1) Benzene–acetone–ethyl acetate–NH₃ conc. 10:15:20:1, (2) Chloroform–methanol–water 200:40:3, (3) butanol–NH₃ 3 mol/L –ethanol 5:2:1	Detection (1) by spraying with potassium iodobismuthate reagent, (2) by spraying with 10% sulfuric acid in ethanol and heating at 110°C for 5 min. (3) by spraying with 50% sulfuric acid in ethanol and 1% vanillin in ethanol and heating at 120°C for 5 min. Quantitation of matrine by spectrophotometry at 412 nm	[107]
Matrine	Compound Shiwei Tablet	Silica gel	Toluene–acetone–ethanol–NH₃ conc. 100:100:10:1	Detection by spraying with potassium iodobismuthate reagent. Quantitation by densitometry at 515 nm	[108]
Matrine	Shikaugfu granules	Silica gel	With toluene–acetone–ethanol–conc. NH₃ 20:20:3:1	Detection by spraying with potassium iodobismuthate reagent. Identification by comparison of the Rf and in situ spectra with the standard. Quantitation by densitometry at 515 nm	[109]
Matrine and oxymatrine	Different products of Radix sophorae flavescentis	Silica gel	(1) chloroform– methanol–NH₃600:80:3 (2) chloroform–methanol 10:1	Detection by spraying with potassium iodobismuthate reagent Identification by comparison with the standards. Quantitation by densitometry at 510 nm	[110]

(*continued*)

TABLE 27.12 (continued)
TLC Systems for Analysis of Lupine Alkaloids

Compounds Separated	Sample	Stationary Phase	Mobile Phase	Detection	Ref.
Matrine and oxymatrine	Constituents of sophoratonkinensis Gagnep	Silica gel	Toluene–butanone–2-isopropanol–NH₃ 20:20:3:1	Detection by spraying with a solution of potassium iodobismuthate. Identification by finger-print technique. Quantiation of matrine and oxymatrine by densitometry at 510 nm	[111]
Matrine	Huganbadu ointment	Silica gel	(1) Chloroform–methanol–conc. ammonia 80:4:1, (2) benzene–ethyl acetate 20:1, (3) Chloroform–methanol–water 13:6:2, (4) ethyl acetate–formic acid–water 10:1:1	Detection (1) by spraying with 5% solution of potassium iodobismuthate, (2) by spraying with 10% H₂SO₄ in ethanol and heating at 105°C, (3) by spraying with 10% H₂SO₄ in ethanol, heating at 105°C and under UV-365 nm, or under 254 nm. Quantitative determination of matrine by densitometry at 520 nm	[112]
Matrine	Traditional herbal drugs	Silica gel	(1) Benzene–acetone–ethyl acetate–conc. ammonia 10:15:20:2	Detection by spraying with 5% potassium iodobismuthate solution. Quantification by densitometry at 524 nm	[113]

These compounds display pharmacological activity and they are used as drugs inhibiting phosphodiesterase and acting as adenosine antagonists.

Natural sources of xanthine alkaloids usually contain widely varying mixtures of caffeine, theophylline, and theobromine. TLC systems applied for the examination of plant materials should be selective according to all main derivatives.

Additionally, caffeine as a psychoactive substance is the world's most widely consumed ingredient of coffee, tea, soft drinks, and energy drinks.

Caffeine as an analgesic and antipyretic and theophylline used in therapy of respiratory diseases are the components of pharmaceutical preparations. Most of TLC analysis concerns the separation of such mixtures or determination of drugs in blood or urine.

27.5.1 Thin Layer Chromatographic Systems

TLC system suitable for analysis of purine alkaloids extracted from plant drugs: Coffeae semen, Mate folium, Theae folium, Cacao semen consists of silica gel and ethyl acetate–formic acid–glacial acetic acid–water (100:11:11:26) or ethyl acetate–methanol–water (100:13.5:10).[1]

Some TLC systems suitable for separation of naturally occurring purine alkaloids in samples of different origin are given in Table 27.13.

Reich et al.[145] elaborated on HPTLC method for identification of green tea and green tea extract (Figure 27.7). To investigate the alkaloid profile, experiments were performed on silica gel with ethyl acetate–methanol–water (20:2.7:2).

Overpressured-layer chromatography (OPLC) combined with near-infrared (NIR) spectroscopy was used for rapid quantitative determination of multi-component pharmaceutical preparations consisting of caffeine, paracetamol and acetylsalicylic acid. Representative UV densitogram obtained by the use of infusion method with stepwise gradient and mobile phases consist of n-hexane–toluene–diethyl ether–glacial acetic acid–methanol $(50 + 25 + 15 + 9,5 + 0,5)$ and toluene–diethyl ether–glacial acetic acid–methanol $(50 + 30 + 19 + 1)$ is presented on Figure 27.8.[146]

A comparison of overpressured and conventional TLC separation of mixture of doping agents containing caffeine is given in Figure 27.9.[131] Development was performed on HPTLC silica plates with n-butanol–chloroform–methyl ethyl ketone–water–acetic acid (25:17:8:4:6).

Separation of caffeine and theophylline in tea by TLC on silica gel with chloroform–methanol (50:3) gives similar result as high-speed counter current chromatographic technique.[147]

27.5.2 Detection and Quantitative Determination

Xanthine derivatives cannot be detected with Dragendorff's reagent (except millophylline and bamifylline containing amino group in the side-chain). For the colors obtained, the plates should be first sprayed with silver nitrate or 20% nitric acid. Xanthine derivatives do not react with iodoplatinate reagent, either.

TABLE 27.13
Application of TLC Systems for Determination of Xanthine Derivatives (Mainly Caffeine) in Plant Materials, Food, and Pharmaceutical Preparations

Other Compound		Adsorbent	Solvent System	Visualization	Ref.
Paracetamol, chlorphenamine maleate, and calculus bovis	Suxiao-Shangfeng capsules	Silica gel	Acetonitrile–methanol–chloroform (1:1:8)	Detection under UV 254 nm, and by spraying with 10% phosphomolybdic acid in ethanol and heating	[114]
Acetaminophen, salicylamide, and potassium salicylate	Diuretic tablets and capsules	Silica gel with a preadsorbent zone, after prewashed with dichloromethane–methanol (1:1)	Double development for a distance of 6 cm with ethyl acetate–methanol–NH_3 (16:3:1)	UV absorption densitometry	[115]
Aspirin, phenacetin, and caffeine	APC tablets	Micellar TLC of on silica gel	2% Sodium dodecyl sulfate–sodium acetate/disodium hydrogen phosphate (pH 8)–acetonitrile 3:1:1	Detection under UV 254 nm. Quantification by densitometry at 273 nm	[116]
Phenacetin, aminopyrine, caffeine, and phenyl ethyl barbituric acid	Zhitong pili	Silica gel	Chloroform–ether–acetone 9:6:1 with traces of NH_3	Detection under UV 254 nm	[117]
Uric acid, cytosine, guanine, caffeine, adenine, uracil, and thymine		Silica gel	Isopropanol–cyclohexane–25% NH_3 7:2:1 with isopropanol–toluene–25% NH_3 6:3:1.	The measurement of spectra between 198 and 612 nm using a diode-array detector	[118]
Salicylic acid, acetyl salicylic acid, ascorbic acid, paracetamol, caffeine, chloroacetanilide, and dehydroascorbic acid	Complex analgesics and antipyretic pharmaceuticals	Silica gel	n-Hexane–chloroform–acetic acid 4:1:1	Detection under UV, except for dehydroascorbic acid, which was identified by means of a color reaction with potassium iodoplatinate. Quantification by densitometry	[119]

Compounds	Sample	Layer	Mobile phase	Detection	Reference
Nicotine, cotinine, theobromine, caffeine, and theophylline		Silica gel	chloroform–methanol 9:1	UV visualization with a solution of $FeCl_3$ (5 g) and iodine (2 g) in acetone and 20% aqueous tartaric acid	[120]
Nicotine, 7-methylxanthine, 1,7-dimethylxanthine, and 1-methylxanthine		Silica gel	Ethyl acetate–methanol–0.1 M NH_3 80:25:16		
Nicotine, cotinine, caffeine, 3-methylxanthine, and theophylline		NH_2 layers	Ethanol–5% aqu. diethylamine 10:1		
Caffeine, theophylline, and ephedrine	Tablets	Silica gel	Chloroform–methanol–ethyl acetate–NH_3 5:10:10:1	Detection by spraying with 0.3% ninhydrin in butanol–acetic acid 19:1. Quantification by densitometry at 272 nm	[121]
Heroin, opium, procaine hydrochloride, alprazolam, diazepam, lorazepam, nitrazepam, oxazepam, mandrax, phenobarbitone, and caffeine	Street samples	Silica gel	Chloroform–ethanol 9:1	After drying visualization by spraying with 1% mercuric chloride–1% potassium ferricyanide 1:1 followed by heating at 80°C	[122]
Aminopyrine, caffeine, phenacetin, and phenobarbital	Analgesic tablets	Silica gel	Cyclohexane–chloroform–diethylamine 5:4:2	Determination by densitometry	[123]
Acetylsalicylic acid, caffeine, and phenacetin	Analgesic tablets	RP-18	Dichloromethane–methanol 1:1	Detection at 254 nm and densitometric scanning	[124]
Heroin, morphine, methaqualone, papayerine, narcotine, caffeine, and paracetamol	Illicit heroin	Silica gel	Benzene–acetonitrile–methanol 8:1:1, first development to 5 cm, after drying second development with cyclohexane–toluene–diethylamine 75:15:9 to 8 cm	Visualization under UV 254 nm	[125]

(continued)

TABLE 27.13 (continued)
Application of TLC Systems for Determination of Xanthine Derivatives (Mainly Caffeine) in Plant Materials, Food, and Pharmaceutical Preparations

Other Compound		Adsorbent	Solvent System	Visualization	Ref.
Acetaminophen, caffeine, alachlor, and amantadine	Suxiao Ganmao Jiaonang capsules	Silica gel	Chloroform–methanol–acetone 18:3:2 with traces of NH_3	Detection under UV 255 nm	[126]
Paracetamol, caffeine, chlorpheniramine, and Niuhuang (cholic acid, deoxycholic acid, and bilirubin)	Suxiao-Shangfeng capsule	Silica gel	Ethylacetate–acetone–NH_3 20:10:0.1 and petrol ether ethyl acetate–50% acetic acid–methanol 8:32:1.5:1	—	[127]
Caffeine, phenacetin, and aspirin	The main components in acetylsalicylic acid tablets	Silica gel	Benzene–ethyl ether–acetic acid–methanol 100:50:15:1	Quantification by densitometry at 277, 250, and 238 nm, respectively	[128]
Acetaminophen, caffeine, and chlorpheniramine	Fast-acting syrup for cold	Silica gel	Chloroform–methanol–ethyl acetate 5:1:1	Quantification by UV spectrophotometry at 257, 265, and 272 nm	[129]
Caffeine, theophylline, isocaffeine, paraxanthin, and theobromine	Standards mixture	Silica gel	Formic acid–ethanol–dichloromethane 1:5:2	Detection under 254 nm	[130]
Strychnine, ephedrine, amphetamine, coramine, caffeine, and other doping agents	Determination of some doping agents	Overpressured TLC using HPTLC silica plates	Butanol–chloroform–MEK–water–acetic acid 25:17:8:4:6	Detection by UV. Quantitative scanning by absorbance at 210 nm	[131]
Codeine, caffeine, papaverine, methaaqualone, nicotine (as reference compounds), and LSD, MDA (3,4-methylen-dioxyamphetamine)	Standards mixture	Silica gel	Methanol–aqueous NH_3 200:3	Detection with Dragendorff's reagent for reference samples and with van Urk's reagent for LSD	[59]

Compound	Sample	Stationary phase	Mobile phase	Detection	Ref.
Theophylline and structural related substances (i.e., theophyllidine, methylxanthine, theobromine, etophylline, caffeine)	Standards mixture	Silica gel	Toluene–2-propanol–acetic acid 16:2:1	Quantification by densitometry at 274 nm	[132]
Caffeine, theophylline	Standards mixture	Silica, CN, NH2, and RP-8	Methanol–phosphate buffer pH 3 9:1	Detection by UV 254 nm	[133]
Morphine, caffeine, and paracetamol	Biological sample; the urine sample	Silica gel	Ethyl acetate–methanol–ammonia 17:2:1	Visualization under UV 253 nm. Detection limits were 0.5 pg for morphine and 0.2 pg for caffeine and paracetamol	[134]
Theobromine, theophylline, and caffeine	Sample of horse urine after injection of Guarana powder	Silica gel	Ethyl acetate–methanol–NH_3 17:2:1	Detection under UV 254 nm, and by spraying with Dragendorff's iodine reagent	[135]
Theophylline and caffeine	Blood serum	Silica gel	Chloroform–acetone–isopropanol–NH_3 6:4:2:0.5	Quantification by densitometry with internal standard of 7-(2-hydroxyethyl)-theophylline	[136]
Theophylline, caffeine, and acetaminophen	Determination of theophylline in plasma	Silica gel after pre-cleaning of the plates with chloroform–methanol 1:1	Toluene–isopropanol–acetic acid 16:2:1	Densitometry at 278 nm in reflectance/absorbance mode	[137]
Theophylline, theobromine, and caffeine	Sample of serum	Silica gel	Chloroform–isooctane–methanol 1:2:1	Detection by UV 270 nm	[138]
Theophylline, caffeine, theobromine, 7-dimethylxanthine,diphylline, 1-methyluric acid, 3-methylxanthine, xanthine, and hypoxanthine	Biological samples	silica (prewashed)	Ethanol–5% aqueous diethylamine solution 95:5	Quantification by densitometry at 275 nm. Identification can be performed using online spectrophotometric methods (UV/VIS, FTIR, and Raman)	[139]

(continued)

TABLE 27.13 (continued)

Application of TLC Systems for Determination of Xanthine Derivatives (Mainly Caffeine) in Plant Materials, Food, and Pharmaceutical Preparations

Other Compound	Adsorbent	Solvent System	Visualization	Ref.
Caffeine	Pure standard	Silica gel, prewashed with methanol	Dichloromethane–methanol 9:1	Quantitative densitometry at 270 nm. MS after elution of the spots with methanol	[140]
Caffeine	Cola sample	Silica (with concentration zone)	Ethyl acetate–methanol 19:1	Densitometry by absorbance at 254 nm	[141]
Caffeine, theobromine, and theophylline	In food samples	Silica gel	Dichloromethane–ethanol–NH₃ 28% 180:17:3	Quantification by densitometry at 275 nm. Determination limits were 2 pg/mL	[142]
Caffeine	Caffeine in tea polyphenols	Silica gel	Toluene–acetone 3:2	Quantification by densitometry at 276 nm	[143]
Caffeine, theobromine, and theophylline	Purine alkaloids in daily foods	Silica gel	Dichloromethane–ethanol–NH₃ 28% 180:17:3	Quantification by densitometry at 275 nm. Determination limits were 2 pg/mL	[142]
Caffeine, theobromine	Epicuticular wax of *Ilex paraguariensis* A	Silica gel	Chloroform–methanol 95:5	UV fluorescence quenching	[144]
Caffeine, theobromine, and theophylline	Daily foods and health drinks	Silica gel	Dichloromethane–ethanol–28% ammonia 180:17:3	Densitometry at 275 nm, p-hydroxybenzaldehyde as internal standard	[142]

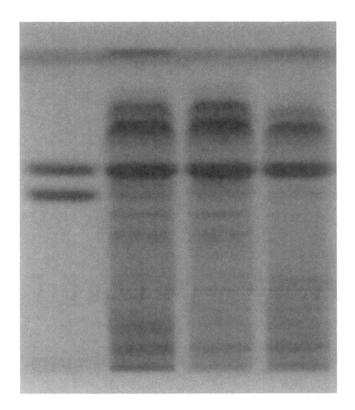

FIGURE 27.7 HPTLC fingerprints of green tea and green tea extract. Track assignment: 1. reference substances: theobromine and caffeine, 2. green tea BRM, 3. green tea extract, 4. green tea commercial product. (From Reich E. et al., *J. Liq. Chromatogr. Relat. Technol.*, 29, 2141, 2006. With permission.)

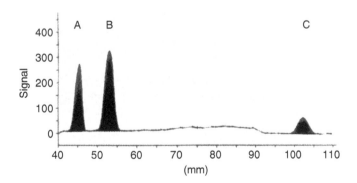

FIGURE 27.8 UV densitogram obtained by infusion method with stepwise gradient: A. caffeine, B. paracetamol, C. acetylsalicylic acid. Mobile phases: (1) *n*-Hexane–toluene–diethyl ether– glacial acetic acid–methanol 50 + 25 + 15 + 9,5 + 0,5 (v/v), (2) Toluene–diethyl ether–glacial acetic acid–methanol 50 + 30 + 19 + 1 (v/v). External pressure 50 bar, rapid period: 300 μL mobile phase 1. Flow rate: 500 μL min⁻¹. Gradient: 1900 μL mobile phase 1, 2500 μL mobile phase 2. (From David, A.Z. et al., *J. Planar Chromatogr.*, 19, 355, 2006. With permission.)

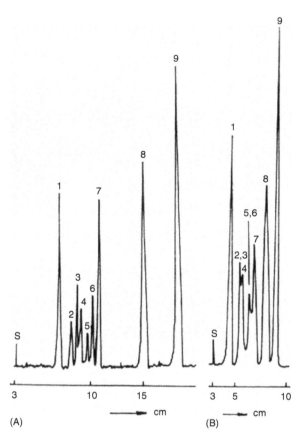

FIGURE 27.9 Densitogram of standard mixture of doping agents. A: overpressured development, B: conventional development. Detection at 210 nm in absorption mode. Reference compounds: 1 – strychnine, 2 – ephedrine, 3 – methamphetamine, 4 – phenmetrazine, 5 – methylphenidate, 6 – amphetamine, 7 – desopimon, 8 – coramin, and 9 – caffeine. (From Gulyos, H. et al., *J. Chromatogr.*, 291, 471, 1984. With permission.)

The most widely applied method for detection is quenching of UV light on fluorescent plates containing a luminophor and iodine–potassium iodide–HCl or iodine–iron(III) chloride–tartaric acid in acetone–water spray reagents giving different colors of xanthine derivatives.

Quantification of zones that quench fluorescence under UV at 254 nm on layers with an indicator has been performed densitometrically. Campbell et al.[148] described a modified flatbed scanner for this purpose and Vovk et al.[149] proved suitability of an image-analyzing system in comparison with a slit-scanning densitometer for quantitative evaluation of TLC plates for the determination of caffeine in Coca Cola.

Quantitative determinations could be performed by direct TLC–FTIR (diffuse reflectance infrared Fourier spectroscopy) coupling.[150] Improvement of the DRIFT spectra and the increase of the signal-to-noise ratio by application of a combination of silica gel 60 and a weak infrared-active, reflection-enhancing

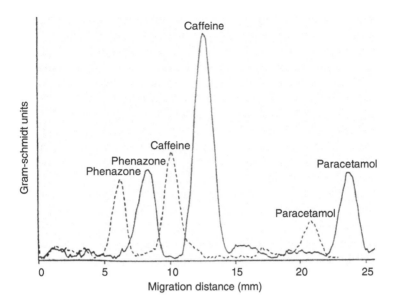

FIGURE 27.10 Gram-Schmidt chromatograms of test substances on 1:1 mixed silica gel 60-magnsium tungstate (-) and on silica gel 60 (- - -). (From Bauer, G.K. et al., *J. Planar Chromatogr.*, 11, 84, 1998. With permission.)

material—magnesium tungstate—was described by Bauer and coworkers (Figure 27.10).[151] For this type of layer the signal-to-noise ratios for caffeine was increased by factors of 3.4 compared with those obtained on pure silica gel 60.

Diode-array scanner has also been used for quantification of purine alkaloids.[51] Simultaneous recording at different wavelengths improves accuracy and reliability of HPTLC analyses, resulting in substantial improvement of in situ quantitative densitometric analyses of caffeine in beverages. Caffeine densitograms were obtained from the spectra in the wavelength range 271–276 nm.

27.6 IMIDAZOLE ALKALOIDS

Imidazole alkaloids of pilocarpine type is derived biosynthetically from histidine. Pilocarpine (16) is a muscarinic alkaloid obtained from the leaves of tropical American shrubs from *Pilocarpus* genus. Pilocarpine acts as a muscarinic receptor agonist in the parasympathetic nervous system.

(16) Pilocarpine

Pilocarpine has been used in ophthalmology to lower the intraocular pressure. Chromatographic methods were used for determination of pilocarpine in eye-drops as well as for the analysis of inactive epimerization product—isopilocarpine, hydrolysis products—pilocarpic and isopilocarpic acids. Separation was conducted on silica gel 60 F_{254} by the use of ethanol–chloroform–28% ammonia (53:30:17). Prior to the detection with Dragendorff's reagent, lactones were regenerated from acids by spraying with a solution p-toluenesulfonic acid in ether.[152]

Quantification of pilocarpine in eyedrops was described also by Rudasits-Kiraly.[12] Analysis was performed on silica plates with chloroform–methanol 3:1, detection by spraying with iodine solution. Simultaneously other components of drops such as atropine, tropine, and ephedrine were determined.

Imidazole alkaloids were isolated from the plant *Pilocarpus Jaborandi*. Roots of *Pilocarpus trachyllophus* yielded together with pilocarpine a new imidazole alkaloid-13-nor-7(11)-dehydropilocarpine.[153] Its structural conformation was determined by spectroscopic analysis. However, TLC analysis showed one spot of their nitrate salts but under HPLC conditions two peaks appeared.

27.7 DITERPENE ALKALOIDS

This group of compounds have diterpene skeletal structure. Representative alkaloids come from *Aconitum napellus* (aconitine (17)), *Aconitum yesoense* (jesaconitine), *Delphinium staphisagria* (delphinine). The European *Aconitum napellus* group comprises three species: *A. napellus*, *A. pentheri*, and *A. angustifolium*.

(17) Aconitine

Aconitine is a highly poisonous alkaloid, which is a neurotoxin. Aconitine is used for creating models of cardiac arrhythmia. For the separation of aconitine and related alkaloids from plant materials and mixtures in combination with other compounds, numerous TLC systems have been applied. Most investigations concern chromatography on silica gel with neutral, acidic, or basic mobile phases. Katz and Rudin[154] used cyclohexane–chloroform–ethanol–diethylamine 4:4:1:1 to separate aconitine and its hydrolysis products. Liu et al.[155] separated alkaloids from *Aconitum Kusnezoffii*, aconitine, mesaconitine, and hypaconitine with ether–chloroform (2.5:1) saturated with 25% NH_3.

Several solvent systems were tested for analysis of aconitine containing plant extracts.[156–158] Li et al.[159] applied TLC on silica with chloroform–ethyl acetate–NH₃ (15:20:1) for determination of aconitine in pharmaceutical preparation.

Determination of aconitine in biological samples was performed on silica gel with (1) cyclohexane–diethylamine 9:1, (2) cyclohexane–ethyl acetate 1:1, (3) benzene–dioxane–diethylamine 7:2:1.[160] Detection by spraying with potassium iodobismuthate solution enabled achieving detection limit 0.3 μg.

Waksmundzka-Hajnos and Petruczynik investigated retention behavior of a number of alkaloids, including aconitine on silica gel and aminopropyl-modified silica gel plates as well as on cyanopropyl-, diol- and RP 18-modified silica gel plates using a variety of binary mobile phases[161] prepared from n-hexane and polar modifiers—2-propanol, ethyl acetate, methyl ethyl ketone, tetrahydrofuran, dioxane—in different concentrations for normal phase chromatography and reversed-phase systems with buffered aqueous-acetonitrile mobile phases (pH 9.15).

Haridutt et al.[162] described a new minor norditerpenoid alkaloid, merconine isolated from *Aconitum napellus* L. The isolation was carried out by vacuum liquid chromatography and purification by centrifugally accelerated radial TLC on an Al₂O₃ rotor of a Chromatotron. The known alkaloids 3-deoxyaconitine, aconitine and mesaconitine were also isolated by this method.

Aconite alkaloids can be visualized by the use of Dragendorff's reagent as well as iodoplatinate reagent or iodine vapor. With iodoplatinate reagent aconitine gives a red-brown color as can be seen on Figure 27.11.[1]

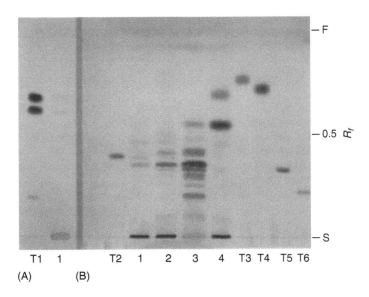

FIGURE 27.11 Chromatograms of Aconiti tuber extracts. Reference compounds: T1 aconitine/mesaconitine, T2 aconitine, T3 deoxyaconitine, T4 hypaconitine, T5 benzoylaconitine, and T6 aconine. Drug sample: 1. trade sample, 2. *A. napellus* L. ssp. *napellus*, 3. trade sample, 4. *A. paniculatum* ssp. *Paniculatum*. TLC system: silica gel/A – toluene–ethyl acetate–diethylamine (7:2:1), B – cyclohexane–ethanol–diethylamine (8:1:1). Detection: Dragendorff's reagent with NaNO₃. (From Wagner, H., Bladt, S., *Plant drug analysis. A thin layer chromatography atlas*, Springer-Verlag, Berlin, Heidelberg, Heidelberg, 1996. With permission.)

Quantitative analysis of aconitine is possible owing to densitometry at 400 nm after spraying with Dragendorff's reagent.

REFERENCES

1. Wagner, H. and Bladt, S., *Plant Drug Analysis: A Thin Layer Chromatography Atlas*, Springer-Verlag, Berlin, Heidelberg, 1996.
2. Stahl, E., *Dünnschichtchromatographie-ein Laboratoriumshandbuch*, Zweite Auflage, Springer, 1967.
3. Macek, K., *Pharmaceutical Application of Thin Layer and Paper Chromatography*, Elsevier, Amsterdam, 1972.
4. Svendsen, A.B. and Verpoorte, R., *Chromatography of Alkaloids, Part A: Thin-layer Chromatography*, Elsevier, Amsterdam, 1983.
5. Popl, M., Fähnrich, J., and Tatar, V., Chromatographic analysis of alkaloids, In Cazes J. (ed.), *Chromatographic Science Series*, Vol. 53, New York, Marcel Dekker, 1990.
6. Sarg, T.M. et al., Thin-layer chromatographic scanner, spectrophotometric and high performance liquid chromatographic methods for the determination of colchicines, *Analyst*, 114, 575, 1989.
7. Stimanek, V. et al., Phytochemical studies of Turkish *Colchicum* and *Merendera* species, *Herb. Hung.*, 29, 64, 1990.
8. Glavic, D., R_M values of some colchicines and colchiceinamides determined by reversed-phase thin-layer chromatography, *J. Chromatogr.*, 591, 367, 1992.
9. Bhushan, R. and Ali, I., Resolution of racemic mixtures of hyoscyamine and colchicine on impregnated silica gel layers, *Chromatographia*, 35, 679, 1993.
10. Chauhan, S.K. et al., Development of HPTLC method for the estimation of colchicine in different parts of *Gloriosa superba*, *Ind. Drug.*, 35(5), 266, 1998.
11. Han, G. et al., Determination of ephedrine hydrochloride in Herba Ephdrae and its preparations by thin-layer chromatography, *J. Chin. Trad. Patent Med.*, 25(3), 203, 2003.
12. Rudasits-Kiraly, E., Data on the cryopreservation of eye-drops. *Gyogyszereszet*, 30, 449, 1986.
13. Zhan, M. and Ma, S.H., Determination of ephedrine hydrochloride in pharmaceuticals by thin-layer chromatography, *Chin. J. Pharm.*, 20, 719, 1985.
14. Luo, G., Lan, G., and Zhang, Q., Determination of ephedrine in Jiufen powder by TLC/acidic dye colorimetry, *J. Chin. Herb. Med.*, 19, 253–255, 1988.
15. Bhushan, R. and Ali, I., TLC resolution of alkaloids on silica layers impregnated with metal ions, *J. Planar Chromatogr.*, 2, 397, 1989.
16. Jia, Y. et al., Determination of ephedrine and 2,3,5,6-tetramethyl-pyrazine in Chinese Ephedra, *Ephedra sinica* Stapf, and Mongolian Ephedra, *E. equisitina* Bunge, by thin-layer chromatography, *J. Chin. Herb. Med.*, 21, 19, 1990.
17. Yang, W. et al., Application of optimization methods to thin-layer chromatographic determination of alkaloids in Jiufen San, *Chin. J. Pharm. Anal.*, 10, 258, 1990.
18. Wang, W. and Wang, Y., Determination of ephedrine in Tangxuan Lifi oral liquid by thin-layer chromatography, *J. Chin. Herb. Med.*, 23, 245, 1992.
19. Ding, T. and Zhang, Y., Determination of brucine, strychnine and ephedrine in rheumatic tablets by thin layer chromatography, *Chin. J. Pharm. Anal.*, 13, 334, 1993.
20. Tao, J. et al., Quality control of a Chinese medicine, Xicoerfeiyan micro enema, Herba ephedrae, Radix scutellarae by thin-layer chromatography, *J. Chin. Trad. Patent Med.*, 17, 2, 1995.

21. He, Q. et al., Comparative study of the extraction process for the preparation of Maxingshigan puls by thin-layer chromatography, *J. Chin. Herb. Med.*, 26, 463, 1995.
22. Wu, H. et al., Study of the effect of ginger-processing on 1-ephedrine contents in Rhizoma Pinelliae, *J. Chin. Trad. Med.*, 21, 157, 1996.
23. Skalican, Z. et al., Study of the potential of thin-layer chromatographic identification of psychotropic drugs in field analysis, *J. Planar Chromatogr.*, 10, 208, 1997.
24. Wu, X. and Su, R., Determination of ephedrine hydrochloride in its compound oral liquid by thin-layer chromatography, *Chin. J. Hosp. Pharm.*, 18, 22, 1998.
25. Jin, B., The quality control of Zhenkeling oral liquid by thin-layer chromatography. *Chin. J. Pharm. Anal.*, 18, 173, 1998.
26. Wang, A., Study of the quality standards of Zhichuan adhesive plasters, *J. Chin. Trad. Patent Med.*, 24(2), 101, 2002.
27. Suedee, R. et al., Thin-layer chromatography using synthetic polymers imprinted with quinine as chiral stationary phase, *J. Planar Chromatogr.*, 11, 272, 1998.
28. Vega, M.H. et al., Use of AMD-HPTLC for analysis of biogenic amines in fish meal, *J. Planar Chromatogr.*, 12, 72, 1999.
29. Wei, H. et al., Determination of ephedrine in Xiaoqinglong granules by micellarthin-layer chromatography, *J. Pharm. Ind.*, 30, 23, 1999.
30. Suo, W. and Zou, L., Determination of ephedrin in Kexinkang puls by thin-layer chromatography, *J. Chin. Trad. Herb. Drug.*, 31(8), 593, 2000.
31. He, Y. and Wang, Y., Determination of ephedrine in Xichuan pills by thin-layer chromatography, *J. Chin. Trad. Herb. Drug.*, 32(4), 324, 2001.
32. Deutscher Arzneimittelcodex, 1986 mit bis zu 3 Ergänzungen 1991. Issued by the Bundesvereinigung Deutscher Apothekerverbände, Deutscher Apotheker Verlag, Stuttgart, or Govi-Verlag GmbH, Frankfurt.
33. *The Pharmacopoeia of Japan*, 11th edn., 1986.
34. Arzneibuch der chinesischen Medizin, Monographien des Arzneibuchs der Volksrepublik China, 1985.
35. Bodoki, E. et al., Fast determination of colchicine by TLC–densitometry from pharmaceuticals and vegetal extracts, *J. Pharm. Biomed. Anal.*, 37, 971, 2005.
36. Jork, H., Funk, W., Fischer, W., and Wimmer, H., *Thin-Layer Chromatography: Reagents and Detection Methods*, Vol. 1a, VCH, Weinheim, 1990.
37. Mroczek, T. and Glowniak, K., TLC and HPTLC assay of quinoline and quinuclidine alkaloids in Cinchonae Cortex and pharmaceutical preparations, *J. Planar Chromatogr.*, 13, 457, 2000.
38. Pound, N.J. and Sears, R.W., HPLC and TLC determination of quinidine and quinine and related alkaloids in pharmaceuticals. *Can. J. Pharm. Sci.*, 10, 122, 1975.
39. Mulder-Krieger, T. et al., Identification of the alkaloids and anthraquinones in *Cinchona ledgeriana* callus culture, *Planta Med.*, 46, 19, 1982.
40. Talas, E. et al., Planar chromatographic separation of cinchona alkaloids formed during enantioselective hydrogenation of ethyl pyruvate, *J. Planar Chromatogr.*, 5, 28, 1992.
41. Piette, V., Lindner,W., and Crommen, J., Enantiomeric separation of N-protected amino acids by non-aqueous capillary electrophoresis with dimeric forms of quinine and quinidine derivatives serving as chiral selectors, *J. Chromatogr. A*, 943, 295, 2002.
42. Park, J.H. et al., Separation of racemic 2,4-dinitrophenyl amino acids on 9-*O*-(phenyloxycarbonyl)quinine-bonded carbon-clad zirconia in reversed-phase liquid chromatography, *J. Chromatogr. A*, 1050, 151, 2004.
43. Montagu, M. et al., Fluorodensitometric assay of 6H-pyrido[4,3-b]-5,11-dimethylcarbazoles (ellipticine and derivatives) biosynthesized by *Ochrosia elliptica* cultures in vitro, *J. Chromatogr.*, 351, 144, 1986.

44. Kaistha, K.K. and Tadrus, R., Improved cost effective thin-layer detection techniques for routine surveillance of commonly abused drugs in drug abuse urine screening and proficiency testing programs with built-in quality assurance, *J. Chromatogr.*, 267, 109, 1983.

45. Detolle, S. et al., Quantitative overpressure layer-chromatography of quinine, quinidine, and hydroquinidine in biomedical analysis, *J. Liq. Chromatogr.*, 13, 1991, 1990.

46. Betschart, B., Determination of antimalarial drugs under laboratory and field conditions using thin-layer chromatography, in *Proceedings of the 6th International Symposium on Instrumental Planar Chromatography*, Interlaken, April 23–26, 1991.

47. Betschart, B. et al., Determination of antimalaria drugs under field conditions using thin-layer chromatography, *J. Planar Chromatogr.*, 4, 111, 1991.

48. Tombesi, O.L. et al., Purification of alkaloids by thin-layer chromatographic decomposition of their picrates, *J. Planar Chromatogr.*, 7, 77, 1994.

49. Kenyon, A.S. et al., Rapid screening of pharmaceuticals by thin-layer chromatography: Analysis of essential drugs by visual methods, *J. Assoc. Off. Anal. Chem.*, 78, 41, 1995.

50. Shalaby, H. and Khalil, H., Reversed-phase ion-pair thin layer chromatography of some alkaloids, *J. Liq. Chromatogr. Relat. Technol.*, 22, 2345, 1999.

51. Spangenberg, B. et al., Fiber optical scanning in TLC by use of a diode-array detector— Linearization models for absorption and fluorescence evaluation, *J. Planar Chromatogr.*, 15, 88, 2002.

52. Yang, D. et al., Study of the identification of Huanhua Shudai Paocha. *J. Chin. Trad. Patent Med.*, 26(1), 74, 2004.

53. Ravishankara, M.N. et al., HPTLC method for the estimation of alkaloids of *Cinchona officinalis* stem bark and its marketed formulations, *Planta Med.*, 67, 294, 2001.

54. Gocan, S., Cimpan, G., and Muresan, L., Automated multiple development thin layer chromatography of some plant extracts, *J. Pharm. Biomed. Anal.*, 14, 1221, 1996.

55. Lee, K.Y., Nurok, D., and Zlatkis, A., Simultaneous determination of antiarrhythmia drugs by high-performance thin-layer chromatography, *J. Chromatogr.*, 1, 403, 1978.

56. Fejer-Kossey, O., The separation of ten tobacco alkaloids by thin-layer chromatography, *J. Chromatogr.*, 31, 592, 1967.

57. Tschirch, C. and Kraus, L.J., Laburnum-alkaloid cytisine. Rapid TLC identification, *Dtsch. Apoth. Zitg.*, 131, 1876, 1991.

58. Liakopouloukyriakides, M. et al., Determination of nicotine in serum by TLC–densitometry, *Anal. Lett.*, 25, 485, 1992.

59. Krishnamurti, R. and Srivastava, A.K., Identification of LSD and MDA by HPTLC in the absence of standard reference samples, *J. Planar Chromatogr.*, 11, 145–148, 1998.

60. Handral, R.D. et al., Estimation of nicotine from gutkha, a chewable tobacco preparation by new HPTLC method, *Ind. J. Pharm. Sci.*, 63(3), 250, 2001.

61. Tyrpien, K. et al., Application of planar chromatography to the determination of cotinine in urine of active and passive smoking pregnant women, *Biomed. Chromatogr.*, 15(1), 50, 2001.

62. Tyrpien, K. et al., Application of liquid separation techniques to the determination of the main urinary nicotine metabolites, *J. Chromatogr. A*, 870(1–2), 29, 2000.

63. Henion, J. et al., Determination of drugs in biological samples by thin-layer chromatography–tandem mass spectrometry, *J. Chromatogr.*, 271, 107, 1983.

64. Zwickenpflug, W. et al., Occurrence of the tobacco alkaloid myosmine in nuts and nut products of *Arachus hypogaea* and *Corylus avellana*, *J. Agric. Food Chem.*, 46, 2703, 1998.

65. Lin, L. et al., Determination of arecoline content in Arecae semen by thin-layer chromatography, *J. Chin. Trad. Med.*, 17, 491, 1992.

66. Kato, N. and Ogamo, A., A TLC visualisation reagent for dimethylamphetamine and other abused tertiary amines, *Sci. Justice*, 41, 239, 2001.

67. Shibano, M. et al., Two new pyrrolidine alkaloids, radicamines A and B, as inhibitors of a-glucosidase from *Lobelia chinensis* LOUR, *Chem. Pharm. Bull.*, 49, 1362, 2001.

68. Bebawy, L.I. et al., Thin layer chromatographic scanner method for the determination of some natural products in raw materials and pharmaceutical preparations, *J. Liq. Chromatogr. Relat. Technol.*, 21, 741, 1998.

69. Jansz, E.R. et al., Determination of piperine in pepper, *J. Natl. Sci. Counc. Sri Lanka*, 103, 140401b, 1985.

70. Mörujtoiu, et al., Separation of some active compounds from plants by TLC coupled with mass spectrometry, *Ann. West Univ. Timisoara—Ser. Chem.*, 12(3), 1629, 2003.

71. Marutoiu, C. et al., Separation and identification of piperine and chavicine in black pepper by TLC and GC–MS, *J. Planar Chromatogr.*, 19, 250, 2006.

72. Bhat, B.G. and Chandrasekhara, N., Determination of piperine in biological tissues by thin-layer chromatography and ultraviolet absorption densitometry, *J. Chromatogr.*, 338, 259, 1985.

73. Lin, S.H. et al., Determination of piperine in Mongolian medicine, Anari 5 by thin-layer chromatography, *J. Chin. Trad. Patent Med.*, 22, 476, 2000.

74. Kulkarni, D. et al., HPTLC method for determination of piperine from *Piper nigrum* Linn, *Ind. Drug.*, 38(6), 323, 2001.

75. Eapen Saumy, M.S. and Grampurohit, N.D., Chemical evaluation of Navayasa churna, *Ind. Drug.*, 39(2), 101, 2002.

76. Morikawa, T. et al., New amides and gastroprotective constituents from the fruit of *Piper chaba*, *Planta Med.*, 70, 152, 2004.

77. Lansky, E.P. and Newman, R.A., *Punica granatum* (pomegranate) and its potential for prevention and treatment of inflammation and cancer, *J. Ethnopharmacol.*, 109, 177, 2007.

78. Hemscheidt, T., Biosynthesis—Aromatic polyketides, isoprenoids, alkaloids, *Top. Curr. Chem.*, 209, 175, 2000.

79. Ferrara, L. et al., Identification of the root of *Punica granatum* in galenic preparations using TLC, *Boll. Soc. Ital. Biol. Sper.*, 65, 385, 1989.

80. Kim, J.H., Hart, H., and Stevens, J.F., Alkaloids of some Asian *Sedum* species, *Phytochemistry*, 41, 1319, 1996.

81. Ehmke, A. et al., Alkaloid N-oxides as transport and vacuolar storage compounds of pyrrolizidine alkaloids in *Senecio vulgaris* L., *Planta*, 176, 83, 1988.

82. Hartmann, T. et al., Sites of synthesis, translocation and accumulation of pyrrolizidine alkaloids *N*-oxides in *Senecio vulgaris* L., *Planta*, 177, 98, 1988.

83. Hartmann, T. et al., Metabolic links between the biosynthesis of pyrrolizidine alkaloids and polyamines in root cultures of *Senecio vulgaris*, *Planta*, 175, 82, 1988.

84. Hartmann, T. et al., Biochemical strategy of sequestration of pyrrolizidine alkaloids by adults and larvae of chrysomelid leaf beetles, *J. Insect Physiol.*, 45, 1085, 1999.

85. Hovermale, J.T. and Craig, A.M., Metabolism of pyrrolizidine alkaloids by *Peptostreptococcus heliotrinreducens* and a mixed culture derived from ovine ruminal fluid, *Biophys. Chem.*, 10, 387, 2002.

86. Nehring, K. and Brandhoff, H., Separation of lupine alkaloids by paper electrophoresis, *Pharmazie*, 16, 402, 1961.

87. Liang, Z. et al., Determination of sophocarnide in Kexieling capsules, alkaloids of *Sophora alopecuroides* L. by thin-layer chromatography, *J. Chin. Trad. Patent Med.*, 19, 8, 1997.

88. Touche, A. et al., A general NMR approach for the structural determination of alkaloids: Application to 3-β-hydroxylupanine, *J. Agric. Food Chem.*, 45, 2148, 1997.

89. Svendsen, A.B., Thin layer chromatography of alkaloids, *J. Planar Chromatogr.*, 2, 8, 1989.
90. Greiswald, R. et al., Distribution taxonomic significance of major alkaloids in the genus *Virgilia, Biochem. Syst. Ecol.*, 17, 231, 1989.
91. Pothier, J. et al., Separation of quinolizidine and dihydrofuro (2,3-b) quinolinium alkaloids by overpressured layer chromatography (OPLC), *J. Planar Chromatogr.*, 3, 356, 1990.
92. Luczkiewicz, M. et al., Two dimensional TLC with sorbent gradient for the analysis of quinolizidine alkaloids in the herb and in vitro cultures of several *Genista* species, *Proceedings of the International Symposium on Planar Separations, Planar. Chromatogr.*, 285–289, 2003.
93. King Horn, A. et al., Alkaloid distribution in some species of the Papilionaceous tribes Sophoreae, Dalbergieae, Loteae, Brongniartieae and Bossiaeeae, *Phytochemistry*, 21, 2269, 1982.
94. Asres, K. et al., Alkaloids of Ethiopian *calpurnia aurea* subsp. *aurea, Phytochemistry*, 25, 1443, 1986.
95. Van Wyk, B.E. and Verdoorn, G.H., Biochemical variation in the genus *Pearsonia, Biochem. Syst. Ecol.*, 19, 685, 1991.
96. Nasution, M.P. et al., A new type of quinolizidine alkaloid from *Genista sessilifolia. Planta Med.* 56, 523, 1990.
97. Hatfield, G.M. et al., Identification of toxic alkaloids from the calcaratus subspecies of *Lupinus arbustus. J. Agric. Food Chem.*, 33, 909, 1985.
98. Zhang, M. and Li, J., Determination of matrine in Meilupijiling by thin-layer chromatography, *J. Chin. Trad. Patent Med.*, 25(6), 518, 2003.
99. Cui, J., Zhang, G., and Wang, M., HPLC and TLC determination of alkaloid constituents in saphora flavescens and saphora lopecuroides. *Chin. Bull. Pharm.*, 20, 59, 1985.
100. Matsuda, K., Yamada, M., Kimura, M., and Hamada, M., Nematicidal activity of matrine and its derivatives against pine wood nematodes, *J. Agric. Food Chem.*, 39, 189, 1991.
101. Li, X. and Tuu, W., Determination of matrine in Jieshanbao lotion by thin-layer chromatography, *J. Chin. Herb. Med.*, 26, 405, 1995.
102. Lu, W. et al., Studies on changes of chemical constituents of radix sophorae flavescentis in complex formulas of traditional Chinese medicine, *J. Chin. Trad. Med.*, 21, 412, 1996.
103. Zhang, L. et al., Study of the quality standard of Sanjufencilu by thin-layer chromatography., *J. Chin. Herb. Med.*, 27, 657, 1996.
104. Wang, B. et al., Determination of matrine in Kushen, Sphora flarscens Ait by thin-layer chromatography—Fluorescence quenching, *Chin. Anal. Chem.*, 25, 693, 1997.
105. Chen, X. et al., Determination of indirubin, matrine in Guilin Xiguashuang spray by thin-layer chromatography, *J. Chin. Trad. Patent Med.*, 20(7), 13, 1998.
106. Shan, Y., The quality control of Xianchen Fencijing by thin-layer chromatography, *J. Chin. Trad. Patent Med.*, 20(8) 11, 1998.
107. Cao, M., Liu, J., and Zhang, J., Quality control of Jiyan oral liquid, *J. Chin. Trad. Patent Med.*, 21(1) 46, 1999.
108. Dong, H. and Yang, Y., Determination of matrine in compound Shiwei Tablet by thin-layer chromatography, *J. Chin. Herb. Med.*, 30, 22, 1999.
109. Li, R. and Su, J., Determination of matrine in Shikaugfu granules II by thin-layer chromatography, *J. Chin. Herb. Med.*, 31(1), 16, 2000.
110. Wang, H. et al., Determination of matrine and oxymatrine in different products of Radix sophorae flavescentis by thin-layer chromatography, *J. Chin. Trad. Patent Med.*, 23(3), 186, 2001.

111. Cui, J. et al., Study of the chemical constituents of sophoratonkinensis Gagnep in its compound formulates, *J. Chin. Trad. Herb. Drug.*, 32(7), 613, 2001.
112. Xie, H. et al., Study of the quality standard of Huganbadu ointment, *J. Chin. Trad. Patent Med.*, 11, 882, 2004.
113. Li, X. et al., An improved procedure for the determination of matrine in Chinese traditional herbal drugs and preparations, *J. Chin. Trad. Patent Med.*, 26(2), 172, 2004.
114. Luo, J. and Liao, L., Identification and isolation of paracetamol, caffeine, chlorphenamine-maleate and calculusbovis in Suxiao-Shangfeng capsules by TLC, *Chin. J. Hosp. Pharm.*, 22, 290, 2000.
115. Ruddy, D. and Sherma, J., Assay of the active ingredient, potassium salicylate, in diuretic tablets and capsules by HPTLC with ultraviolet absorption densitometry, *J. Liq. Chromatogr.*, 24, 321, 2002.
116. Zhao, H. et al., Simultaneous determination of the contents of three components in by micellar thin-layer chromatography, *J. Shenyang Pharm. Univ.*, 18, 338, 2001.
117. Lu, Q., Fu, X., and Wan, T., Simultaneous identification of phenacetin, aminopyrine, caffeine and phenyl ethyl barbituric acid in by thin-layer chromatography, *Chin. J. Pharm. Anal.*, 20, 207, 2000.
118. Spangenberg, B. et al., Thin layer chromatography using diode array detection, *GIT Spez. Sep.*, 112001, 37, 2001.
119. Krzek, J. and Starek, M., Densitometric determination of active constituents and impurities in complex analgesics and antipyretic pharmaceuticals, *J. Planar Chromatogr.*, 12, 356, 1999.
120. Baranowska, I. and Zydron, M., Analysis of biogenic amines, alkaloids and their derivatives by TLC and HPLC, *J. Planar Chromatogr.*, 13, 301, 2000.
121. Qu, A. et al., Determination of three components in compound theophylline tablets by thin-layer chromatography, *Chin. J. Pharm. Ind.*, 26, 366, 1995.
122. Kamble, V.W. and Dongre, V.G., TLC detection and identification of heroin (diacetylmorphine) in street samples. II., *J. Planar Chromatogr.*, 10, 384, 1997.
123. Guo, D., Analysis of analgesic tablets by thin-layer chromatography, *Chin. J. Pharm. Anal.*, 5, 306, 1985.
124. Sherma, J. et al., Analysis of analgesic tablets by quantitative high performance reversed phase TLC, *J. Liq. Chromatogr.*, 8, 2961, 1985.
125. Krishnamurtthy, R. and Srivastava, A.K., Simultaneous detection of adulterants and coextractants in illicit heroin by HPTLC with two successive mobile phases, *J. Planar Chromatogr.*, 10, 388, 1997.
126. Huang, F. et al., Identification of Suxiao Ganmao Jiaonang capsules by thin-layer chromatography, *Chin. J. Pharm. Anal. (Yaowu Fenxi Zazhi)*, 19, 193, 1999.
127. Li, J., Identification of Suxiao Shangfeng capsule by thin-layer chromatography, *Bull. Chin. Trad. Med. (Zhongyao Tongbao)* 13, 355, 1988.
128. Zhang, Q. et al., Determination of the main components in acetylsalicylic acid tablets by thin-layer chromatography, *J. Chin. Chromatogr.*, 8, 193, 1990.
129. Zhang, Y. and Hu, Q., Determination of fast-acting syrup for cold by thin-layer chromatography and UV spectrophotometry, *Chin. J. Pharm. Ind.*, 22, 506, 1991.
130. Ropte, D. and Volkmann, B., Thin-layer chromatography in DAB. Innovation in TLC determinations—In DAB, *Dtsch. Apoth. Ztg.*, 135, 2956, 1995.
131. Gulyos, H. et al., Determination of some doping agents by over pressured thin-layer chromatography, *J. Chromatogr.*, 291, 471, 1984.
132. Renger, B. et al., Validation of analytical procedures in pharmaceutical analytical chemistry, *J. Planar Chromatogr.*, 8, 269, 1995.

133. Omori, T. et al., Cyanoalkyl-bonded HPTLC plates and their chromatographic charac-
     teristics, *J. High Resol. Chromatogr.*, 6, 47, 1983.
134. Krishnamurthy, R. et al., Simultaneous determination of morphine, caffeine, and para-
     cetamol in the urine of addicts by HPTLC and HPLC, *J. Planar Chromatogr.*, 13, 171,
     2000.
135. Salvadori, M.C. et al., Determination of xanthines by high performance liquid chroma-
     tography in horse urine after injection of guarana powder, *Analyst*, 119, 2701, 1994.
136. Kawamoto, H. and Yamane, T., Determination of theophylline and caffeine in blood
     serum by thin-layer chromatography–densitometry, *Jpn. J. Clin. Pathol.*, 33, 217, 1985.
137. Jamshidi, A. et al., A new high performance thin-layer chromatography (HPTLC)
     method for determination of theophylline in plasma, *J. Liq. Chromatogr. Relat. Tech-
     nol.*, 22, 1579, 1999.
138. Heilweil, E. and Touchstone, J., Theophylline analysis by direct application of serum to
     thin-layer chromatograms, *J. Chromatogr. Sci.*, 19, 594, 1981.
139. Wagner, J. et al., HPTLC as a reference method in clinical chemistry: On-line coupling
     with spectroscopic methods, *J. Planar Chromatogr.*, 6, 446, 1993.
140. Prosek, M. et al., Quantification of caffeine by online TLC–MS, *J. Planar Chromatogr.*,
     13, 452, 2000.
141. Sherma, J. and Miller, R., Quantification of caffeine in beverages by densitometry on
     preadsorbent HPTLC layers, *Am. Lab.*, 16, 126, 1984.
142. Kunugi, A. and Tabel, K., Simultaneous determination of purine alkaloids in daily foods
     by high performance thin-layer chromatography, *J. High Resol. Chromatogr.*, 20, 456,
     1997.
143. Liu, F. et al., Determination of caffeine in tea polyphenols by thin-layer chromatog-
     raphy, *J. Chinese Herb. Med.*, 28, 20, 1997.
144. Athayde, M.L. et al., Caffeine and theobromine in epicuticular wax of *Ilex paraguar-
     iensis* A. St.-Hil., *Phytochemistry*, 55, 853, 2000.
145. Reich, E. et al., HPTLC methods for identification of green tea and green tea extract,
     J. Liq. Chromatogr. Relat. Technol., 29, 2141, 2006.
146. David, A.Z. et al., OPLC combined with NIR spectroscopy—A novel technique for
     pharmaceutical analysis, *J. Planar Chromatogr.*, 19, 355, 2006.
147. Yuan, L. et al., Separation of alkaloids in tea by high-speed counter current and thin-
     layer chromatography, *Chin. J. Chromatogr.*, 16, 161, 1998.
148. Campbell, A., Chejlava, M.J., and Sherma, J., Use of a modified flatbed scanner for
     documentation and quantification of thin layer chromatograms detected by fluorescence
     quenching, *J. Planar Chromatogr.*, 16, 244, 2003.
149. Vovk, I. et al., Validation of an HPTLC method for determination of caffeine, *J. Planar
     Chromatogr.*, 10, 416, 1997.
150. Frey, O.R. et al., Possibilities and limitations of assays by on-line coupling of thin layer
     chromatography and FTIR spectroscopy, *J. Planar Chromatogr.*, 6, 93, 1993.
151. Bauer, G.K. et al., Development of an optimized sorbent for direct HPTLC–FTIR on-
     line coupling, *J. Planar Chromatogr.*, 11, 84, 1998.
152. Durif, C. et al., Thin layer chromatography for rapid semiquantitative estimation of
     pilocarpic and isopilocarpic acids in pilocarpine eyedrops. *Pharm. Acta Helv.*, 61, 135,
     1986.
153. Andrade-Neto, M. et al., An imidazole alkaloid and other constituents from *Pilocarpus
     Trachyllophus*, *Phytochemistry*, 42(3), 885, 1996.
154. Katz, A. and Rudin, H., Mild alkaline hydrolysis of aconitine, *Chim. Acta*, 67, 2017,
     1984.

155. Liu, C.H. et al., Separation and determination of the poisonous alkaloids in Aconitum Kusnezoffii and its processed products, *J. Chin. Trad. Med.*, 12, 83, 1987.
156. Zhu, S.H. and Wang, T., Identification of aconitine and lappaconitine in various extracts by thin layer chromatography, *J. Chin. Trad. Patent Med.*, 20(8), 18, 1998.
157. Hau, L. et al., Determination of aconitine in Jianpigao by thin-layer chromatography, *Chin. J. Pharm. Anal.*, 17, 258, 1997.
158. Zhu, Z.H. et al., Determination of mesaconitine, hypaconitine and aconitine in *Xiaohuoluo puls* by thin-layer chromatography, *Chin. J. Pharm. Anal.*, 16, 154, 1996.
159. Li, H. et al., Determination of aconitine in Zhentongning injection by thin-layer chromatography, *Chin. Herb. Med.*, 27, 58, 1996.
160. Liu, X. et al., Determination of aconitine in biological samples, *Chin. J. Chromatogr.*, 20(1), 81, 2002.
161. Waksmundzka-Hajnos, M. and Petruczynik, A., Retention behaviour of some alkaloids in thin layer chromatography with bonded stationary phases and binary model phases, *J. Planar Chromatogr.*, 14, 364, 2001.
162. Haridutt, K. et al., Merckonine, a new aconitine-type norditerpenoid alkaloid with a $-N=C-19H$ functionality, *Heterocycles*, 48, 1107, 1998.

28 Polyacetylenes: Distribution in Higher Plants, Pharmacological Effects and Analysis

Lars P. Christensen and Henrik B. Jakobsen

CONTENTS

28.1   INTRODUCTION

Polyacetylenes belong to a class of molecules containing two or more triple bonds and constitute a distinct group of relatively unstable, reactive, and bioactive natural products. They are found in plants, fungi, microorganisms, and marine invertebrates.[1–8] Polyacetylenes are common among higher plants, wherein they occur regularly in seven families, namely Apiaceae (formerly Umbelliferae), Araliaceae, Asteraceae (formerly Compositae), Campanulaceae, Olacaceae, Pittosporaceae, and Santalaceae, and sporadically in further eight plant families.[5,7–12] Several families of the lower plants (e.g., some fungi families within Basidiomycetes) are also frequent producers of polyacetylenes.[1]

    The biological activities of several polyacetylenes isolated from higher plants are now well-documented. They have been shown to be strong photosensitizers, that is, their toxic activities are dependent on or are augmented by UV-A light, wherein the primary mechanism of action seems to be either a photodynamic disruption of membranes, involving singlet oxygen, or through a predominantly nonphotodynamic (oxygen-independent) mechanism of action.[5,13–15] Aromatic polyacetylenes seem to have access to both a photodynamic and a nonphotodynamic mechanism of action.[5,6,13] Many polyacetylenes from higher plants also possess considerable light-independent toxicity. The bioactivity of aliphatic C_{17}-polyacetylenes occurring in many plants of the Apiaceae and Araliaceae plant families is interesting. These polyacetylenes have shown to possess considerable cytotoxic activity and a potential anticancer effect. Further, it has been demonstrated that aliphatic C_{17}-polyacetylenes also have anti-inflammatory, antiplatelet-aggregatory, antibacterial, antimycobacterial, antifungal, and antiviral properties as well as an immune stimulating effect due to their allergenicity. Finally neuroprotective and neurotoxic effects have also been demonstrated for this class of polyacetylenes.[1,2,4,5] Biologically, active polyacetylenes having unusual structural features have also been reported from cyanobacteria, algae, sponges, and other sources. Naturally occurring aquatic polyacetylenes are also of interest since many of them display important biological activities and possess antitumor, antibacterial, antimicrobial, antifouling, antifungal, phototoxic, and immunosuppressive properties.[1,3] There is therefore no doubt that polyacetylenes are of great interest for the medicinal or pharmaceutical industries. Many of the bioactive polyacetylenes isolated from higher plants occur in well-known medicinal or food plants. This may explain some of the beneficial effects associated with these products. Further, as the beneficial effects of most bioactive polyacetylenes from higher plants seem to occur at nontoxic concentrations they are obvious targets for the development of drugs and healthier foods and food products.[4] This review highlights the present state of knowledge on the distribution of bioactive polyacety- lenes in higher plants with special focus on polyacetylenes occurring in the Apia- ceae, Araliaceae, and Asteraceae as well as their biosynthesis, pharmacological effects, and analysis by TLC and other chromatographic methods.

28.2 DISTRIBUTION OF POLYACETYLENES IN HIGHER PLANTS

More than 1400 different acetylenes and related compounds have been isolated from higher plants and approximately half of these are polyacetylenes. The majority has been isolated from the botanically closely related plant families Apiaceae, Araliaceae, and Asteraceae.[6–12,16] The polyacetylenes isolated from the Apiaceae and Araliaceae are characterized by aliphatic acetylenes, in particular C_{17}-polyacetylenes, whereas the structure diversity among the polyacetylenes from the Asteraceae is considerable and includes thiophenes, dithiacyclohexadienes (thiarubrines), thioethers, sulfoxides, sulfones, alkamides, chlorohydrins, lactones, spiroacetal enol ethers, furans, pyrans, tetrahydropyrans, tetrahydrofurans, aromatic, and aliphatic polyacetylenes.

Polyacetylenes have, besides the earlier-mentioned families, been isolated from approximately 12 other plant families, including Annonaceae,[17–20] Campanula- ceae,[4,8,21–28] Euphorbiaceae,[8] Loranthaceae,[1,29] Meliaceae,[30] Olacaceae,[1,8,31–38] Opiliaceae,[8] Passifloraceae,[39] Pittosporaceae,[8,40] Santalaceae,[8] Solanaceae,[4,8,41–43] and Torricelliaceae.[44]

The Campanulaceae, which has been suggested to be closely related to Aster- aceae, seem to be very uniform in its polyacetylenic chemistry. In the Campanula- ceae polyacetylenic C_{14}-tetrahydropyranyl ethers (Figure 28.1; e.g., compounds 2–5) are the most widespread in this family.[4,8,26,27] Further, lobetyol (6) and lobetyolin (7) and related polyacetylenes are accumulated in hairy roots of various Campanulaceae plant species,[21–25] and related acyclic C_{14}-polyacetylenes have been isolated from the roots of several *Lobelia* and *Siphocampylus* spp.[28]

C_{18}-Polyacetylenic fatty acids bound to triglycerides or as free polyacetylenic fatty acids seem to characterize the polyacetylenic chemistry of the plant families Loranthaceae, Olacaceae, Opiliaceae, and Santalaceae, and a few examples of such polyacetylenic compounds are 10–14 and 16–19 (Figure 28.1). The polyacetylenic chemistry of these families suggests that they are closely related, which is also more or less in accordance with plant systematics. C_{18}-Polyacetylenic fatty acids such as oropheic acid (15) have also been isolated from the Annonaceae.[17,18] The polyace- tylenic chemistry of this family is, however, also characterized by the occurrence of *O*-heterocyclic polyacetylenes such as the lactone sapranthin (20) isolated from the stem bark of *Sapranthus palanga*[19] and polyacetylenes containing a terminal furan ring (21) that have been isolated from the stem bark of *Alphonsea ventricosa*.[20]

The Pittosporaceae is characterized by the occurrence of C_{17}-polyacetylenes such as falcarinol (22) and falcarinone (39) (Figure 28.2) and C_{15}-acetylenes such as the polyacetylenes 8 and 9 (Figure 28.1).[8,40] This clearly indicates a close relationship to

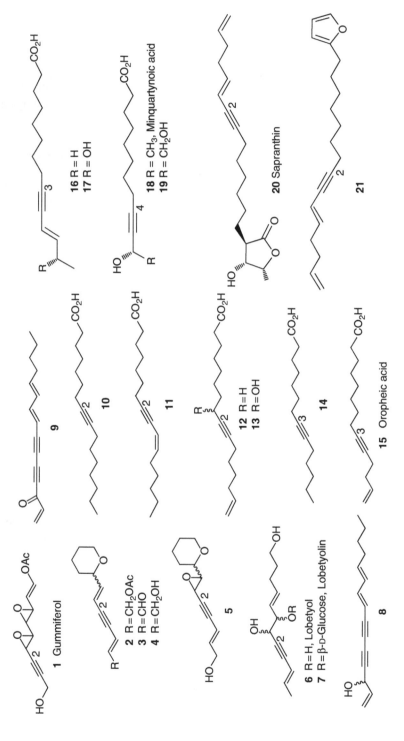

FIGURE 28.1 Examples of polyacetylenes occurring in plant families wherein these compounds are not typical constituents.

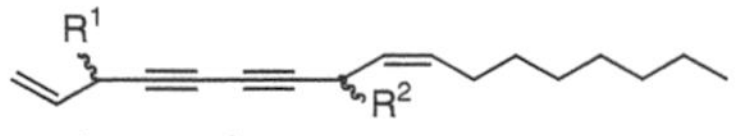

22 $R^1 = OH$, $R^2 = H$, Falcarinol (syn. panaxynol)
23 $R^1 = R^2 = OH$, Falcarindiol
24 $R^1 = OH$, $R^2 = OCH_3$, 8-O-Methylfalcarindiol
25 $R^1 = OCH_3$, $R^2 = OH$, 3-O-Methylfalcarindiol
26 $R^1 = OAc$, $R^2 = OH$, 3-O-Acetylfalcarindiol
27 $R^1 = OH$, $R^2 = OAc$, 8-O-Acetylfalcarindiol

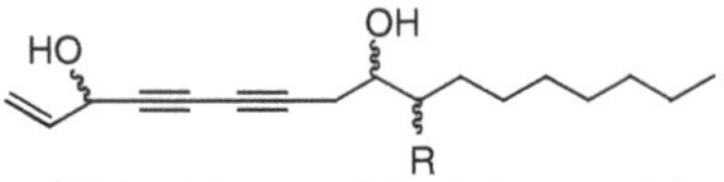

28 $R = OCH_3$, 10-O-Methylpanaxytriol
29 $R = OAc$, 10-O-Acetylpanaxytriol
30 $R = OH$, Panaxytriol
31 $R = Cl$, Panaxydol chlorohydrine

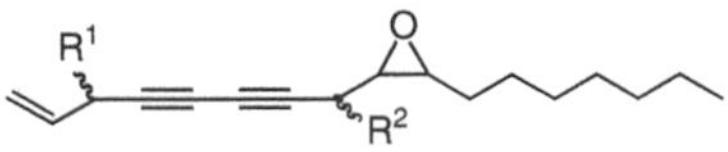

32 $R^1 = OH$, $R^2 = H$, Panaxydol
33 $R^1 = OAc$, $R^2 = H$, Acetylpanaxydol
34 $R^1 = OH$, $R^2 = OCH_3$, 8-Methoxypanaxydol
35 $R^1 = OH$, $R^2 = OAc$, 8-Acetoxypanaxydol
36 $R^1 = R^2 = OH$
37 $R^1 = $ linoleoyl, $R^2 = OH$

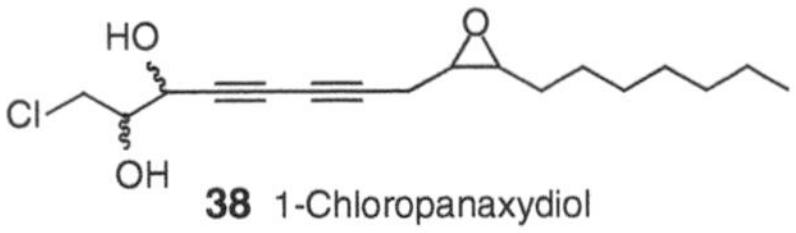

38 1-Chloropanaxydiol

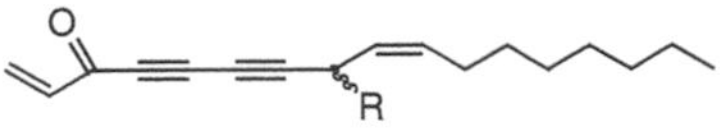

39 $R = H$, Falcarinone
40 $R = OH$, Falcarinolone

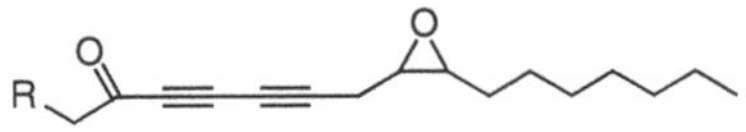

41 $R = CH_3$
42 $R = CH_2OH$
43 $R = CH_2OCH_3$

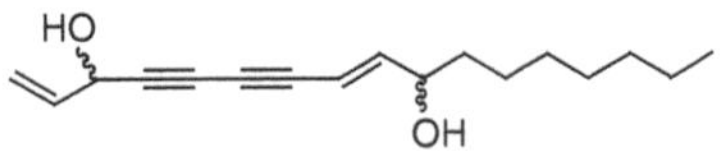

44 Panaxydiol

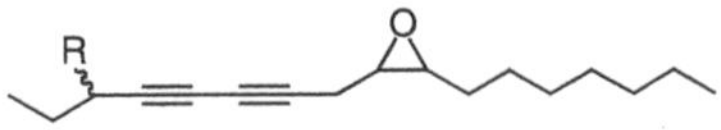

45 $R = OAc$, Ginsenoyne G
46 $R = OH$, Ginsenoyne D

FIGURE 28.2 Examples of aliphatic C_{17}-polyacetylenes of the falcarinol-type that is widespread within the Apiaceae and Araliaceae plant families.

Araliaceae and Apiaceae, wherein especially falcarinone and falcarinol seem to be widespread (see Section 28.2.1). Polyacetylenes of the falcarinol-type have also been isolated from Solanaceous food plants such as tomatoes (*Lycopersicum esculentum*) and aubergines (*Solanum melongena*), wherein they appear to be phytoalexins.[41–43] From the Euphorbiaceae, Meliaceae, Passifloraceae, and Torricelliaceae polyacetylenes have so far only been isolated from a limited number of plant species. Especially the new polyacetylenic diepoxide, gummiferol (**1**), isolated from *Adenia gummifera* (Passifloraceae)[39] is interesting because of its broad cytotoxic activity (see Section 28.4.5).

28.2.1 Polyacetylenes in Apiaceae and Araliaceae

The most common polyacetylenes in the Apiaceae are aliphatic C_{17}-acetylenes in particular falcarinol (**22**) and related polyacetylenes (**22–27, 39, 40** and more) that are widespread in this plant family (Figure 28.2).[4,6,8,16,45–71] However, in some genera other aliphatic C_{17}-polyacetylenes are also typical constituents (Figure 28.3, e.g., **47–68**) of which the majority are unrelated to polyacetylenes of the falcarinol-type. These polyacetylenes predominates in, for example, the genera *Azorella* (**67, 68**),[8] *Cicuta* (e.g., **47–52, 54, 57, 59–62, 64**),[8,51,55,57] *Heracleum* (**58, 66**),[8] *Oenanthe* (e.g., **53–59, 62, 65**),[8] and *Opopanax* (e.g., **53–55, 63, 64, 66**).[8] Also polyacetylenes of shorter chain lengths primarily aliphatic C_{13}- and C_{15}-polyacetylenes occur in

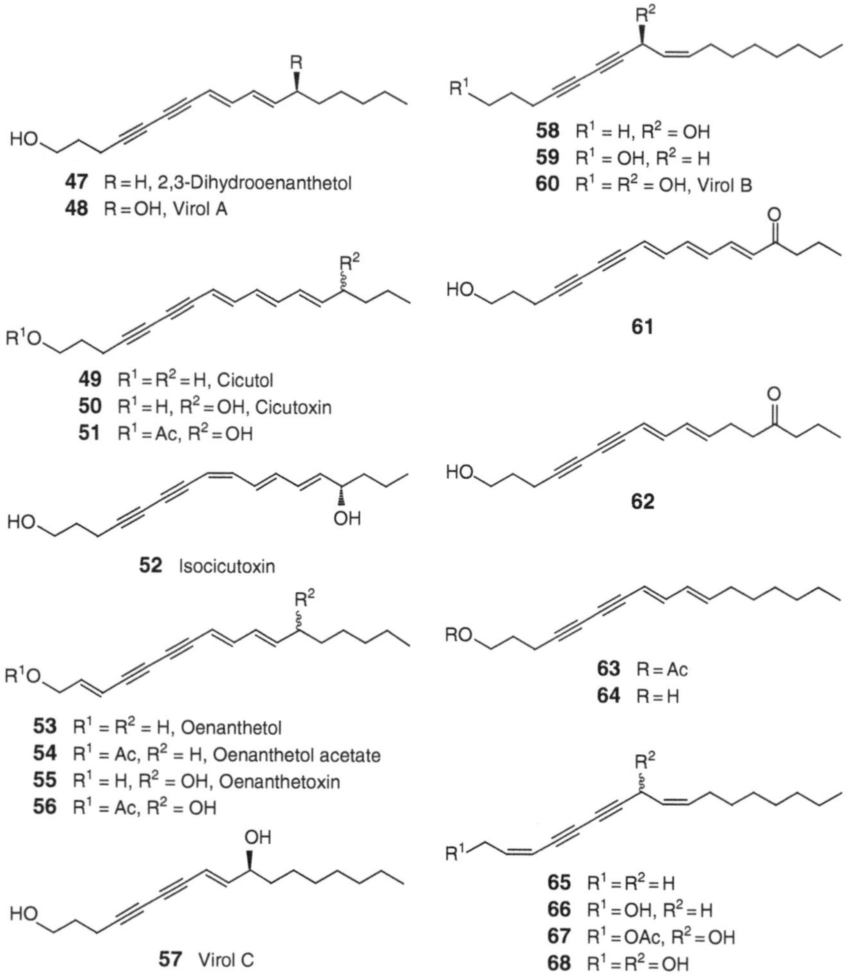

FIGURE 28.3 Examples of aliphatic C_{17}-polyacetylenes that occur frequently in Apiaceae plant species that are unrelated to polyacetylenes of the falcarinol-type.

Apiaceae (Figure 28.4, e.g., **69–74** and **78–86**). These polyacetylenes are clearly derived from their corresponding C_{17}-polyacetylenes by oxidative degradation and oxidative transformations (see Section 28.3). Although these polyacetylenes are not widespread in the family they predominate in genera such as *Bupleurum* (e.g., **82–86**),[8,72,73] *Aethusa* (e.g., **69–74**),[8] and to some extent also in some *Oenanthe* spp. (e.g., **78–81**), where they co-occur with C_{17}-polyacetylenes.[8,74,75]

Only a relatively limited number of plants within the Araliaceae have been investigated for polyacetylenes, in contrast to the numerous investigations of plants within the Apiaceae and Asteraceae. The most common polyacetylenes isolated from Araliaceae are falcarinol and related C_{17}-polyacetylenes (**23**, **27**, **39–46** and many more) of the falcarinol-type, which have been isolated from the genera *Aralia*,[7,8,76]

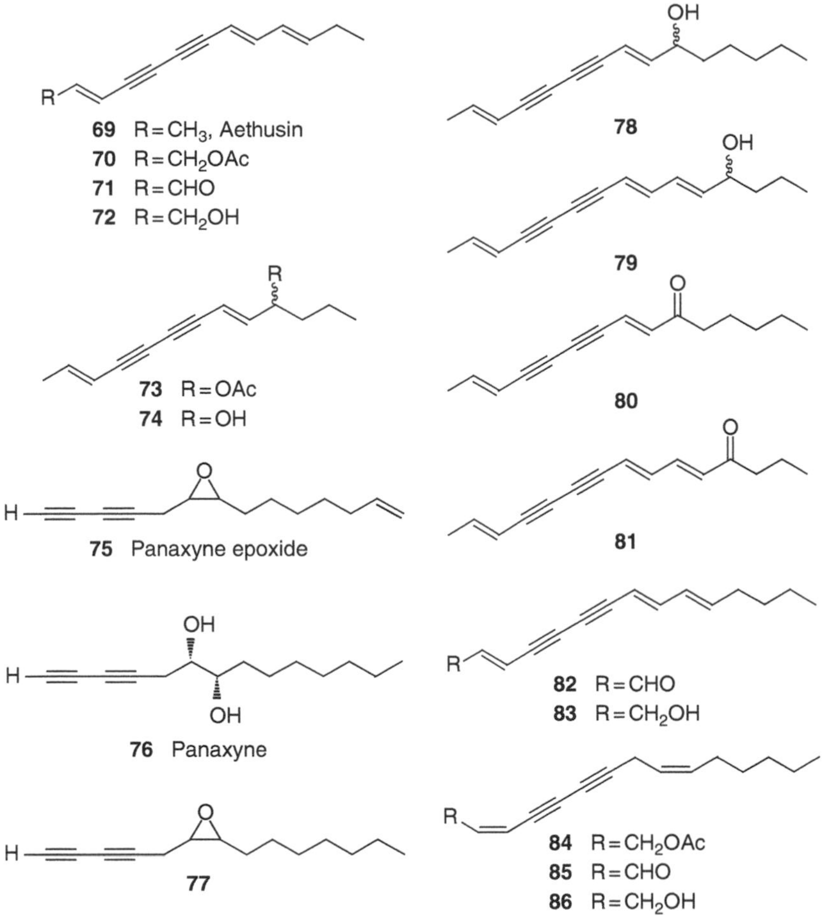

FIGURE 28.4 Examples of aliphatic C_{13}-, C_{14}-, and C_{15}-polyacetylenes isolated from plants of Apiaceae or Aralicaeae.

Dendropanax,[7,77,78] *Didymopanax*,[79] *Evodiopanax*,[80] *Fatshedera*,[81] *Hedera*,[7,8,81–85] *Oplopanax*,[86] *Panax*,[7,8,87–100] *Polyscias*,[7,90] *Schefflera*,[7,8,101] and *Tupidanthus*.[81] However, the isolation of dehydrofalcarinol (**87**) and related polyacetylenes (Figure 28.5, e.g., **88–90**, **96–99**, **102**, **103**) from plant species such as *Aralia cordata*,[76] *Dendropanax arboreus*,[77,78] *Didymopanax*,[79] and *Panax ginseng*[92,93] is interesting because dehydrofalcarinol and polyacetylenes of the dehydrofalcarinol-type so far only have been found in the family Asteraceae, wherein they occur regularly in certain tribes (see Section 28.2.2). Other types of aliphatic polyacetylenes do not seem to be widespread in the Araliaceae, although a few bioactive C_{14}-polyacetylenes (Figure 28.4, **75–77**) that are clearly derived from the corresponding C_{17}-polyacetylenes have been isolated from the roots of *Panax ginseng* (**75**, **76**)[102,103] and *P. quinquefolium* (**75**, **77**).[89,91] The presence of C_{18}-polyacetylenes (Figure 28.6, **104–108**) closely related to C_{17}-polyacetylenes of the falcarinol-type in the genera *Dendropanax*

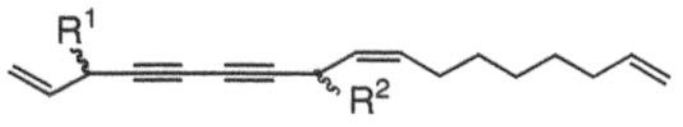

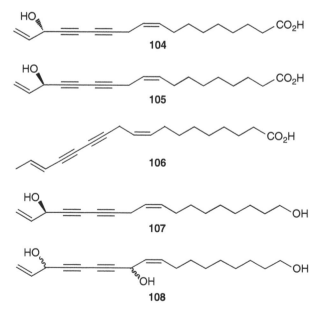

87 R¹=OH, R²=H, Dehydrofalcarinol
88 R¹=R²=OH, Dehydrofalcarindiol
89 R¹=OAc, R²=OH, 3-O-Acetyldehydrofalcarindiol
90 R¹=OH, R²=OAc, 8-O-Acetyldehydrofalcarindiol

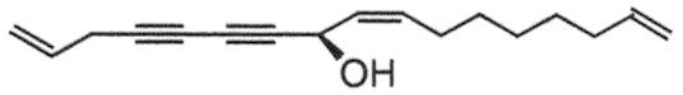

91

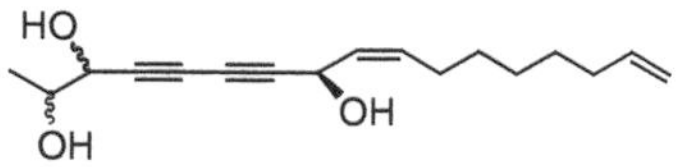

92 Gymnasterkoreayne E

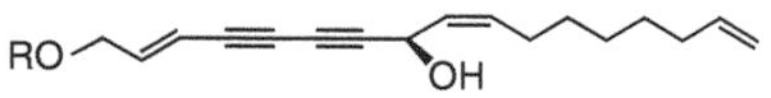

93 R=Ac, Gymnasterkoreayne D
94 R=H, Gymnasterkoreayne F

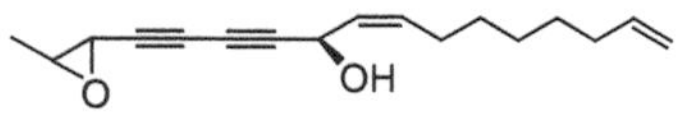

95 Gymnasterkoreayne B

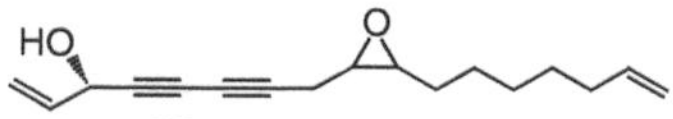

96 Ginsenoyne A

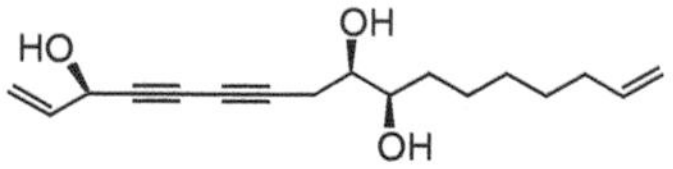

97 Ginsenoyne C

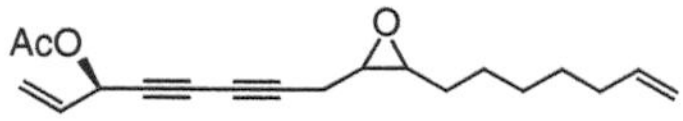

98 Ginsenoyne H

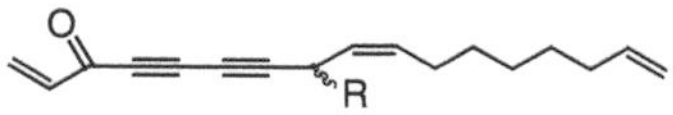

99 R=H, Dehydrofalcarinone
100 R=OH, Dehydrofalcarinonol
101 R=OAc

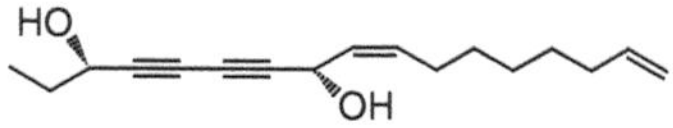

102 Dendroarboreol A

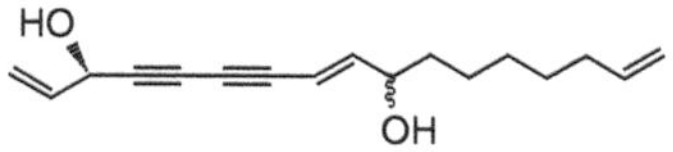

103 Dendroarboreol B

FIGURE 28.5 Examples of aliphatic C₁₇-polyacetylenes of the dehydrofalcarinol-type that is typical for several tribes within Asteraceae but also seem to be characteristic for Araliaceae.

104

105

106

107

108

FIGURE 28.6 Aliphatic C₁₈-polyacetylenes closely related to polyacetylenes of the falcarinol-type isolated from several Araliaceae plant species.

(**104–107**),[7,104–106] *Didymopanax* (**105**),[79] *Cussonia* (**108**),[107] and *Fatsia* (**105**)[105] indicates that this type of polyacetylenes could be characteristic for Araliaceae.

Many polyacetylenes of both the falcarinol-type and dehydrofalcarinol-type have been shown to be highly bioactive. Consequently, these compounds may contribute to at least some of the health effects associated with medicinal plants and food plants from Apiaceae and Araliaceae (see Section 28.4).

28.2.2 POLYACETYLENES IN ASTERACEAE

The Asteraceae consists of 16 tribes according to recent classifications.[108] In the Asteraceae some types of polyacetylenes are widely distributed in the family, whereas others are more or less confined to specific tribes or genera.

A characteristic feature for the majority of the aliphatic polyacetylenes occurring in the Asteraceae is the relatively high number of conjugated triple and double bonds in their structures, resulting in very characteristic UV-spectra.[8,109–111] The most common aliphatic polyacetylenes in Asteraceae are those with an ene-diyne-diene, ene-diyne-ene, diyne-ene, triyne-diene, triyne-ene chromophore but also compounds with four or five triple bonds in conjugation are widespread in most tribes of Asteraceae (Figures 28.7 through 28.9).[5–12,16]

Aliphatic C_{10}-, C_{13}-, C_{14}-, and C_{17}-polyacetylenes are widespread in the Asteraceae, whereas C_{11}-, C_{15}- and C_{16}-polyacetylenes only seem to be confined to certain

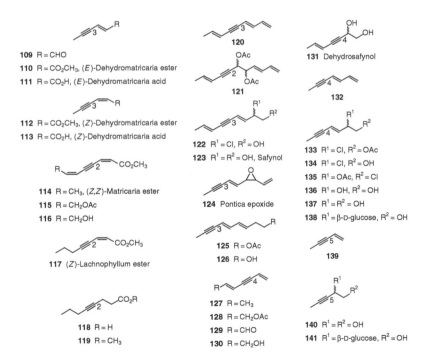

FIGURE 28.7 Examples of aliphatic C_{10}- and C_{13}-polyacetylenes isolated from Asteraceae plant species of which many are widespread within this family.

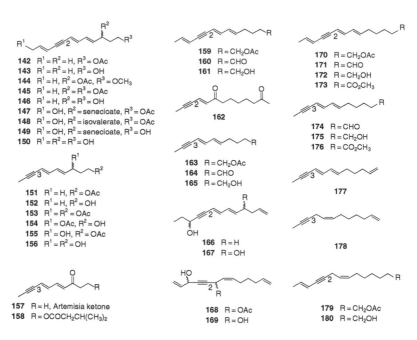

FIGURE 28.8 Examples of aliphatic C_{14}-, C_{15}-, and C_{16}-polyacetylenes isolated from Asteraceae plant species of which many are widespread within this family.

tribes or genera of the Asteraceae. Aliphatic C_{11}-polyacetylenes are characteristic for the genus *Cineraria* (tribe Senecioneae)[8,112–116] but occur otherwise only sporadically in Asteraceae.[8–12] Aliphatic C_{15}- and C_{16}-acetylenes occur mainly in the tribes Cardueae (former Cynareae), Heliantheae and Anthemideae and some typical examples are the compounds **159–180** (Figure 28.8).[8–10,12,117] Aliphatic C_{12}-polyacetylenes are rare in this family as well as aliphatic polyacetylenes with chain lengths less than 10 carbon atoms or above 17 carbon atoms. The aliphatic C_{17}-polyacetylenes that occur in Asteraceae includes, besides C_{17}-polyacetylenes of the dehydrofalcarindiol-type (Figure 28.5, e.g., **87–95** and **99–101**), also a broad range of other C_{17}-polyacetylenes, and examples of the latter type include compounds **181–198** (Figure 28.9).[8–10,12,118–122] Aliphatic C_{13}- and C_{14}-polyacetylenes (e.g., compounds **120–158**) are together with C_{17}-polyacetylenes the most widespread groups of polyacetylenes in this family, whereas aliphatic C_{10}-polyacetylenes such as the Z- and E-isomers of dehydromatricaria ester (**110, 112**) and related compounds (**109, 111, 113–119**) (Figure 28.7) are mainly found in the tribes Anthemideae, Astereae, and Lactuceae.[8,9,11,123–125]

The co-occurrence of C_{10}-polyacetylenic esters and lactones and C_{17}-polyacetylenes in the tribes Astereae and Anthemideae surely is a good indication for a close relationship between these tribes, whereas the connection to the Heliantheae is less pronounced with only the C_{17}-polyacetylenes being common to all three tribes. The regular occurrence of C_{17}-polyacetylenes, such as dehydrofalcarinol and related polyacetylenes as well as other types of C_{17}-polyacetylenes in the tribes Anthemideae, Astereae, and Heliantheae support a close relationship between Asteraceae and the

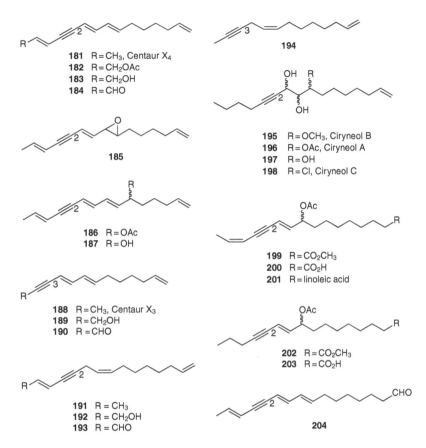

FIGURE 28.9 Examples of aliphatic C_{17}- and C_{18}-polyacetylenes isolated from Asteraceae plant species of which many are widespread within this family.

plant families Araliaceae and Apiaceae and suggests a common pathway in the evolution of these families. The bioactivity of aliphatic polyacetylenes in Asteraceae is largely unknown, although it has been demonstrated that some aliphatic polyacetylenes are phototoxic, possess antibiotic, anti-inflammatory, and anticancer activity (see Section 28.4).

Spiroacetal enol ethers are unique and characteristic for the tribe Anthemideae of which five-membered C_{12}- or six-membered C_{13}-spiroacetal enol ethers are most common (Figure 28.10, e.g., **205–226**).[8,9,109,126–128] The bioactivity of this special type of polyacetylenes is largely unknown, although anti-inflammatory, antitumor, and antibiotic activities have been demonstrated for a few spiroacetal enol ether polyacetylenes (see Section 28.4). Also characteristic for the tribe Anthemideae is the occurrence of C_{10}-polyacetylenic esters (e.g., compounds **109, 110, 112, 114–119**) and their corresponding lactones such as dehydromatricaria lactone (**245**) that otherwise only seem to be present in Astereae.[8,9,11,109] The tribe Anthemideae is also characterized with the presence of polyacetylenic aromatics, alkamides, furans and tetrahydropyrans, and sulfur compounds of which some display interesting bioactivities.[5,8,9,109]

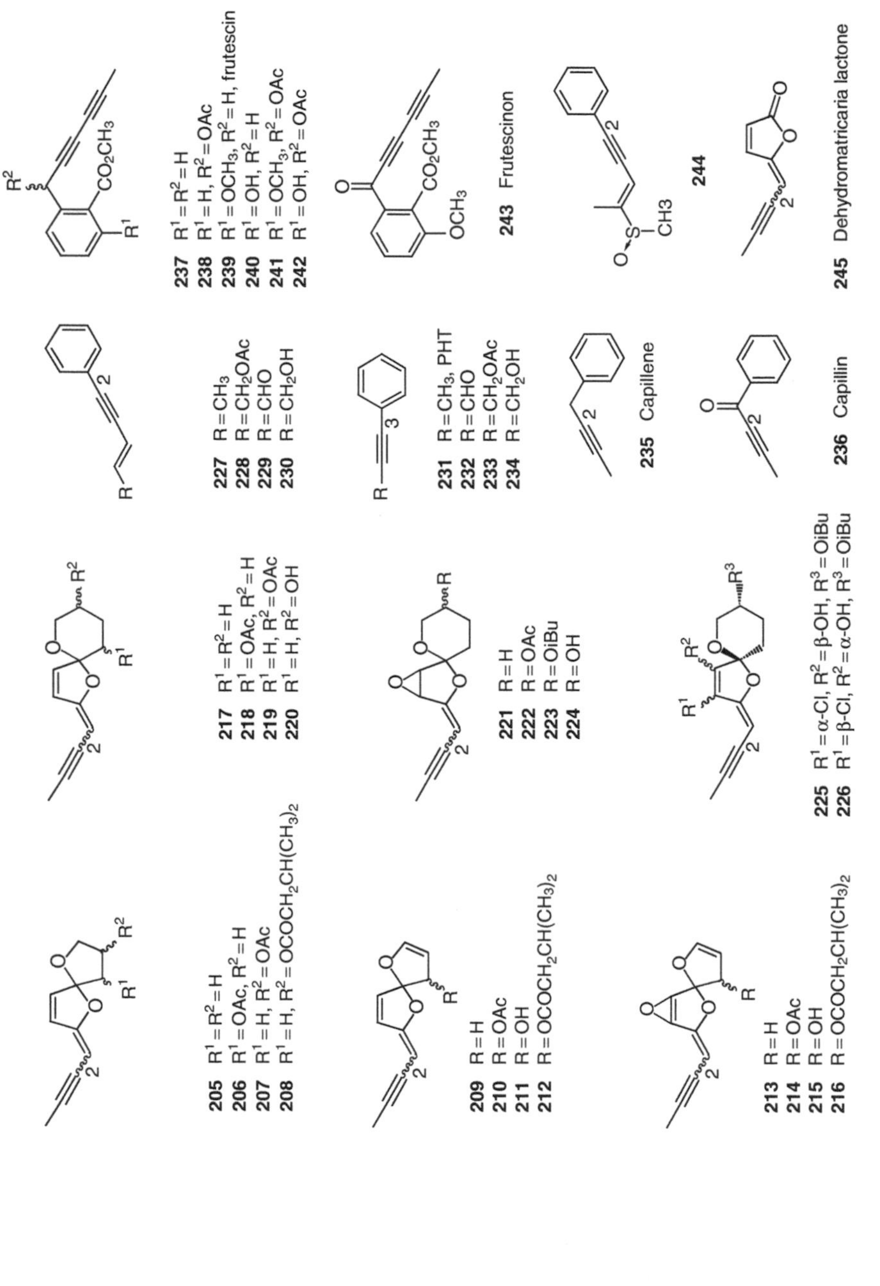

FIGURE 28.10 Examples of polyacetylenic spiroacetal enol ethers, lactones, and aromatics isolated from Asteraceae.

Aromatic polyacetylenes have been found primarily in plants belonging to Anthemideae, Cardueae, and Heliantheae.[8–10,12] Phenylheptatriyne (231, PHT) present in all three tribes has shown to be highly phototoxic (see Christensen[5] and Section 28.4) and capillin (236) and related aromatic polyacetylenes isolated from *Artemisia* spp. (Anthemideae),[8,9,109,129–132] have shown marked antifungal,[5,133] anti-inflammatory[133–135] as well as cytotoxic activities.[136,137] Cytotoxic and antimicrobial activity has also been reported for aromatic polyacetylenes such as 240 and 241 isolated from several *Argyranthemum* spp. (Anthemideae),[138,139] whereas the bioactivity of many other aromatic polyacetylenes occurring in the Cardueae and Heliantheae is almost unknown.[5,8,10,12] However, it appears that aromatic polyacetylenes of the Asteraceae, in particular those occurring in Anthemideae and Heliantheae, seem to constitute a very promising group of bioactive compounds. Special types of aromatic acetylenes that only seem to occur in Anthemideae are the isocoumarins that are derived from polyacetylenic aromatic precursors.[8,9,131,140,141]

Alkamides are mainly found in the plant families Piperaceae, Aristolochiaceae, Rutaceae, and Asteraceae.[8,142,143] However, polyacetylenic alkamides have been found only in Asteraceae, wherein they occur frequently in the tribes Anthemideae and Heliantheae.[8–10,142,143] Within the tribe Anthemideae, the accumulation of olefinic as well as acetylenic alkamides with up to three triple bonds has been shown to be a typical trend of the genus *Achillea*.[6,8,9,143–147] This biogenetic capacity apparently replaces those polyacetylenes, which are otherwise typical for the tribe Anthemideae. The genus *Achillea* comprises approximately 100 species and is especially rich in olefinic and acetylenic pyrrolidides, piperidides, and the corresponding dehydro derivatives (piperideides, pyrrolideides). It seems that the distribution of cyclic amides is mainly restricted to the tribe Anthemideae.[6,8,9,142,143] Most polyacetylenic cyclic amides found in this tribe consist of pyrrolidides (e.g., 258 and 259) and piperidides (e.g., 253 and 261) whereas polyacetylenic piperideides (e.g., 254) are rare (Figure 28.11). Also widespread polyacetylenic isobutyl- and phenethyl amides are found in Anthemideae (e.g., 249, 252, 257, and 262) together with the rare isopentyl amides (e.g., 250 and 260). In Heliantheae polyacetylenic alkamides are primarily found in the genera *Spilanthes* and *Echinacea*, in which polyacetylenic isobutyl- and phenethyl amides (e.g., 248, 251, and 262–264), occur regularly.[6,8,10,142,143,148–158] On the other hand polyacetylenic 2-methylbutyl- and styryl amides such as compounds 246, 247, 255, and 256 appear to be typical for the tribe Heliantheae. Olefinic alkamides possess strong insecticidal activity and appears to have some pharmacological uses,[5,6,143,159] whereas the pharmacological effects of polyacetylenic alkamides are less known, although attention has been drawn to the antibiotic and anti-inflammatory activity of some polyacetylenic alkamides isolated from *Achillea*[160,161] and *Echinacea* spp.[143,148,159,160]

Sulfur-containing polyacetylenes are widely distributed within Asteraceae. Beside Anthemideae they have been found to occur regularly in the tribes Arctoteae, Cardueae, Heliantheae, Inuleae, Mutisieae, and Plucheeae.[6,8–10,12,162,163] Monothiophenes are by far the largest group of these polyacetylenic sulfur compounds. Most dithiophenes and all trithiophenes cannot be classified as polyacetylenes because of the absence of two or more triple bonds in their structures. Nevertheless these compounds are biosynthesized from polyacetylenic precursors.[6,8–10,12,163] Some

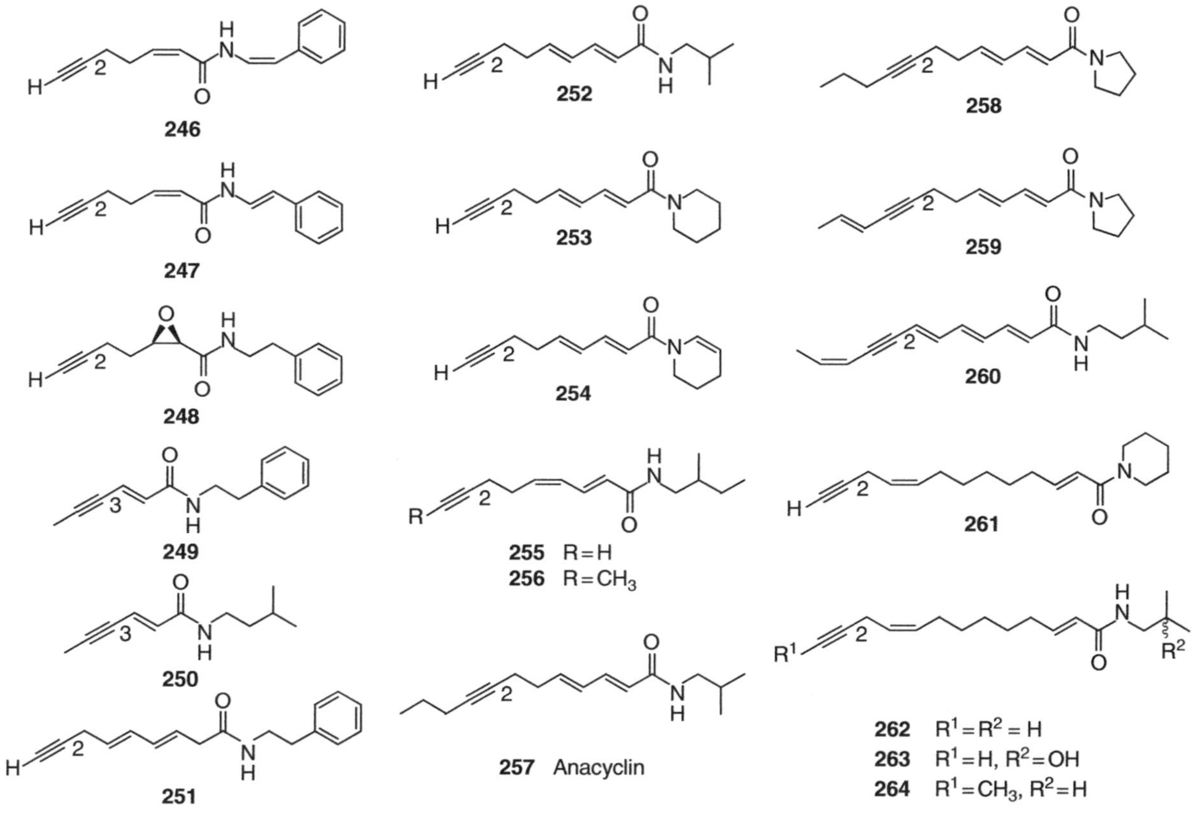

246, 247, 248, 249, 250, 251

252, 253, 254

255 R=H
256 R=CH_3

257 Anacyclin

258, 259, 260, 261

262 $R^1=R^2=H$
263 $R^1=H$, $R^2=OH$
264 $R^1=CH_3$, $R^2=H$

FIGURE 28.11 Examples of polyacetylenic alkamides isolated from Asteraceae that occur frequently in certain genera of the tribes Anthemideae and Heliantheae.

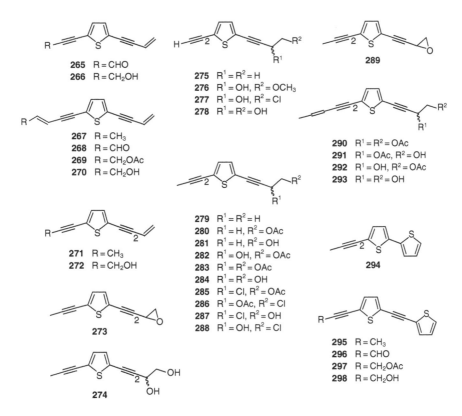

FIGURE 28.12 Examples of polyacetylenic mono- and dithiophenes isolated from Asteraceae of which some are widely distributed in the tribes Anthemideae, Arctoteae, Cynareae, Heliantheae, or Inuleae.

thiophenes are highly phototoxic, whereas their activity in the absence of light seem to be significantly less pronounced or lacking (see Section 28.4). Examples of polyacetylenic mono- and dithiophenes (**265–298**) that occur regularly in Asteraceae are shown in Figure 28.12. Anthemideae is also characterized by the occurrence of C_{10}-polyacetylenic methylthioethers in the genus *Anthemis*[8,9,164,165] as well as methylthioethers and oxidized derivatives in the genus *Chrysanthemum* containing one triple bond, which are biosynthesized from polyacetylenic lactone, furan, cumulene, α-pyrone, aromatic, or spiroacetal enol ether precursors (see Section 28.3).[8,9,166–168] However, polyacetylenic C_{13}-methylthioethers, methylsulfoxides, or methylsulfones (Figure 28.13, e.g., **299–311**) are primarily present in the tribes Heliantheae, Helenieae and Inuleae, wherein they have been found in the genera *Acanthospermum*, *Coreopsis*, *Eclipta*, *Guizotia*, *Ratibida*, and *Wedelia* (Heliantheae),[8,10,169–171] in the genera *Baileya*, *Dugaldia*, *Flaveria*, *Gaillardia*, and *Helenium* (Helenieae),[8,10,172–175] and in the genera *Allagopappus*, *Buphthalmum*, and *Inula* (Inuleae).[8,176,177] From South African *Berkheya* spp. (Arctoteae) some unusual thietanones (Figure 28.13, **312–317**) have been isolated which seem to

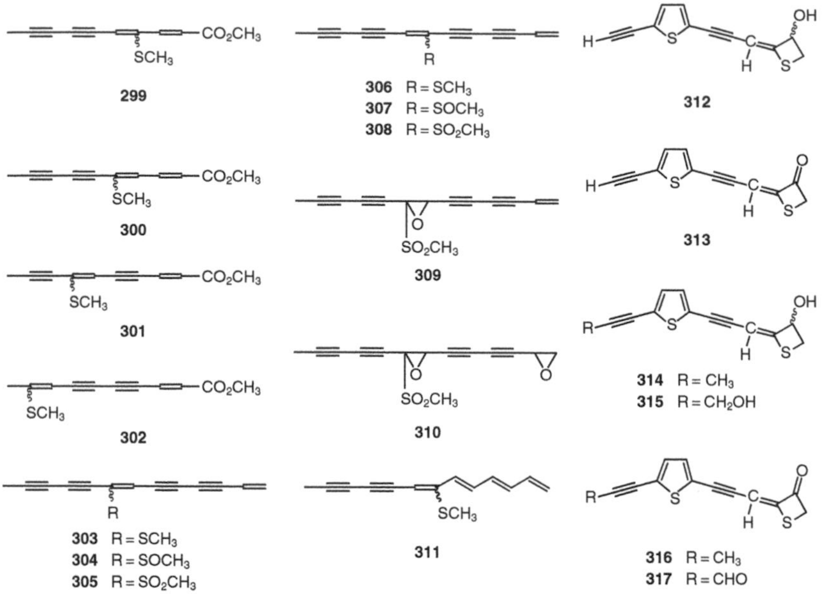
CO2CH3
SCH3
299
CO2CH3
SCH3
300
CO2CH3
SCH3
301
CO2CH3
SCH3
302
R
303 R = SCH3
304 R = SOCH3
305 R = SO2CH3
R
306 R = SCH3
307 R = SOCH3
308 R = SO2CH3
O
SO2CH3
309
O
O
SO2CH3
310
SCH3
311
OH
H
S
C
H
S
312
O
H
S
C
H
S
313
OH
R
S
C
H
S
314 R = CH3
315 R = CH2OH
O
R
S
C
H
S
316 R = CH3
317 R = CHO

FIGURE 28.13 Examples of polyacetylenic methylthioethers, methylsulfoxides, and methylsulfones isolated from Asteraceae and special thietanones that only seem to be characteristic for the tribe Arctoteae.

be characteristic for this tribe only.[178–180] The bioactivity of polyacetylenic methylthioethers, methylsulfoxides, and methylsulfones as well as thietanones has to the best of our knowledge not been investigated.

Another special group of sulfur-containing polyacetylenes is the red dithiacyclohexadienes (1,2-dithiins) trivially known as thiarubrines (Figure 28.14, **318–327**), which are restricted to about 15 genera of the Asteraceae, including the genera *Ambrosia*, *Aspilia*, *Eclipta*, *Melampodium*, *Oyedaea*, *Rudbeckia*, *Schkuhria*, *Steiractnia*, *Verbesina*, *Wedelia* (Heliantheae),[8,10,181–190] *Chaenactis*, *Eriophyllum*, *Lasthenia* and *Palafoxia* (Helenieae),[8,10,191,192] and finally the genus *Geigera* (Inuleae).[193] Plant species belonging to the genera *Aspilia* and *Chaenactis* have been used in traditional medicine of African countries against skin infections, stomach problems, eye infections, and venereal diseases.[194] The thiarubrines **319** (thiarubrine A) and **326** have shown a pronounced antibiotic activity against human pathogens (see Section 28.4). Hence many of the pharmacological effects of plants containing thiarubrines may be due to this type of compounds.[190,194]

Polyacetylenic tetrahydropyrans (Figure 28.15; e.g., **328–333**) are not widespread in Asteraceae, although they occur in the tribes of Anthemideae, Cardueae, Heliantheae, and Inuleae.[8–10,12,176] The C_{14}-polyacetylenic tetrahydropyrans ichthythereol (**328**) and its acetate (**329**) display neurotoxic effects and are known as fish poisons. These compounds were first isolated from *Ichthyothere terminalis* (Heliantheae) a plant species, which have been widely used by native South American

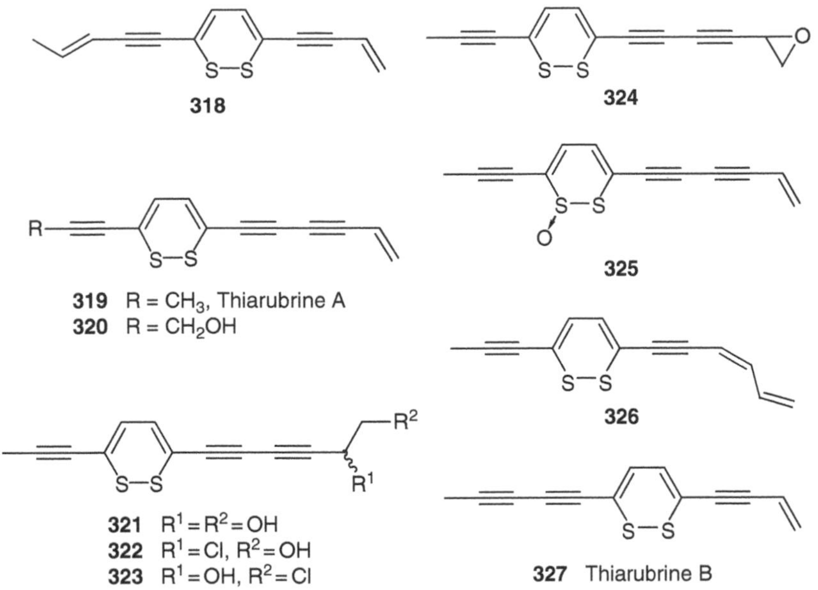

FIGURE 28.14 Polyacetylenic dithiacyclohexadienes (thiarubrines) isolated from tribes Heliantheae, Helenieae, or Inuleae (Asteraceae).

Indians as a fish poison.[195,196] Compounds **328** and **329** was later found in the genus *Dahlia* (Heliantheae) together with closely related tetrahydropyrans (**330**, **331**) that probably also display neurotoxic effects (see Section 28.4.1).[10,197–200] The tetrahydropyrans **328–331** have also been isolated from a few species within the genera *Calea* (**329**) and *Clibadium* (**328–331**) (Heliantheae),[10,201–203] and from the genus *Inula* (**329**) (Inuleae)[176] as well as from several genera within the tribe Anthemideae.[8,9,109] C_{14}-Tetrahydropyrans (**332**, **333**) closely related to ichhyothereol are also known from the genus *Centaurea* (Cardueae), which is characterized by the occurrence of related C_{16}-tetrahydropyrans.[8,12]

Further characteristic polyacetylenes isolated from Asteracee are furanopolyacetylenes (Figure 28.15, e.g., **342–350**). This type of polyacetylenes have been found to occur regularly in the genus *Leucanthemum* (**342**, **343**) (Anthemideae)[8,9,204,205] and in the genus *Atractylodes* (**344–350**) (Cardueae).[8,12,206–208] Other polyacetylenic *O*-heterocycles are the C_{13}-chloroenol ethers and related enol ethers (e.g., **334–338**), which appears to be characteristic for the tribe Gnaphalieae, including the genera *Achyrocline*, *Anaphalis*, *Gnaphalium*, and *Helichrysum*.[8,209–212] C_{13}-Enol ethers such as **339** occur regularly in the genus *Carlina* (Cardueae),[8,12,213,214] while the C_{13}-enol ether **340** is characteristic for the genus *Leuchanthemum*.[8,9,215] Finally polyacetylenic tetrahydrofurans are rare in the Asteraceae and so far they seem to occur in single species within the tribes Anthemideae, Cardueae, and Heliantheae.[8–10,12] For example, the tetrahydrofuran **341** that is closely related to compounds **340** and **342** have to the best of our knowledge only been isolated from *Leucanthemum crassifolium*.[204]

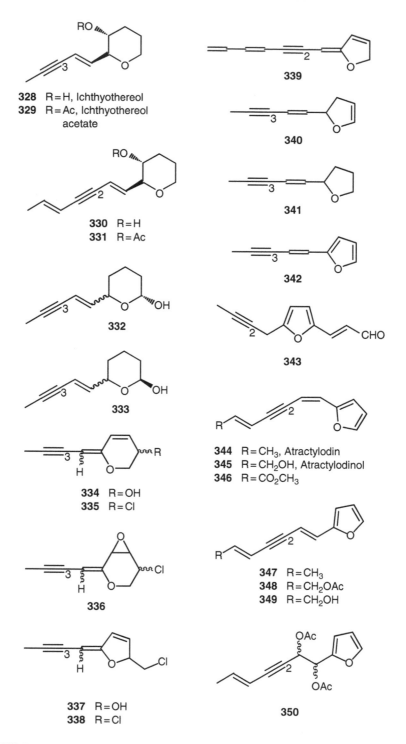

328 R=H, Ichthyothereol
329 R=Ac, Ichthyothereol acetate

330 R=H
331 R=Ac

332

333

334 R=OH
335 R=Cl

336

337 R=OH
338 R=Cl

339

340

341

342

343

344 R=CH₃, Atractylodin
345 R=CH₂OH, Atractylodinol
346 R=CO₂CH₃

347 R=CH₃
348 R=CH₂OAc
349 R=CH₂OH

350

FIGURE 28.15 Examples of different types of polyacetylenic *O*-heterocycles that are frequently detected in Asteraceae of which some are restricted to certain tribes.

28.3 BIOSYNTHESIS

The large structural variation observed among polyacetylenes seems to indicate the involvement of many different precursors in their biosynthesis. A comparison of the polyacetylene structures, with those of oleic, linoleic, crepenynic, and dehydrocrepenynic acids (see Figures 28.16 and 28.17) makes it, however, reasonable to assume that most polyacetylenes are biosynthesized with the latter acids as precursors, although there are some examples of polyacetylenes that are derived from carotenoids and other terpenoids.[8] Many feeding experiments with [14]C- and ^{3}H-labelled precursors have

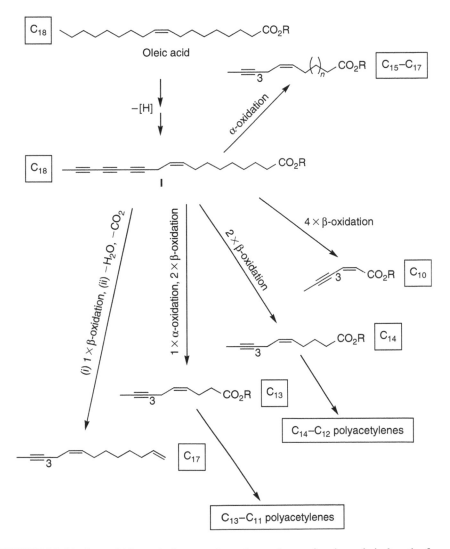

FIGURE 28.16 General biosynthetic route for polyacetylenes of various chain lengths from oleic acid.

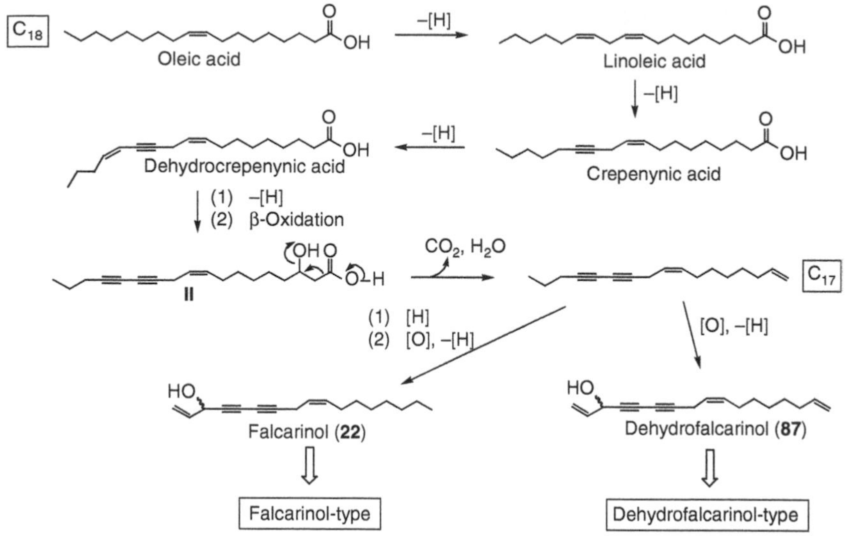

FIGURE 28.17 The possible biosynthesis of aliphatic C_{17}-polyacetylenes of the falcarinol and dehydrofalcarinol-type. [O] = oxidation, [H] = reduction, −[H] = dehydrogenation (oxidation followed by the loss of water).

confirmed both this assumption and further that polyacetylenes are built up from acetate and malonate units.[6–12,216–220] Further evidence for C_{18}-acids as precursors in the biosynthesis of most acetylenes have recently been obtained through identification of a gene coding for a triple bond forming enzyme (fatty acid acetylenase), which occurs in a plant species that produce acetylenes in reponse to fungal infection.[221]

The first step in the biosynthesis of polyacetylenes is the dehydrogenation of oleic acid and linoleic acid to crepenynic acid, a reaction that has become much clearer after the identification of a triple bond forming enzyme as described earlier. Further dehydrogenation of crepenynic acid leads to di- and triyne C_{18}-acids (Figure 28.16, e.g., **I**) that by α- and β-oxidation and other oxidative degradation reactions leads to polyacetylene precursors of various chain lengths, which then are transformed to a large variety of polyacetylenes (Figures 28.16 and 28.17).[6–12] In the following, the biosynthesis of various types of polyacetylenes will be briefly described. Further information about the biosynthesis of polyacetylenes can be found in several reviews on the subject.[6–12]

The biosynthesis of the falcarinol and related polyacetylenes follows the normal biosynthetic pathway for aliphatic C_{17}-acetylenes with dehydrogenation of oleic acid leading to a C_{18}-diyne acid intermediate (Figure 28.17, **II**), which is then transformed to C_{17}-diyne polyacetylenes by β-oxidation followed by the loss of CO_2 and H_2O.[7,8] Further oxidation and dehydrogenation leads then to polyacetylenes of the falcarinol-type or the dehydrofalcarinol-type as outlined in Figure 28.17.

The most widespread polyacetylene in the Asteraceae is trideca-3,5,7,9, 11-pentayn-1-ene (**139**).[7–12] This highly unstable hydrocarbon is the precursor of many polyacetylenes and in particular polyacetylenic thiophenes, methylthioethers,

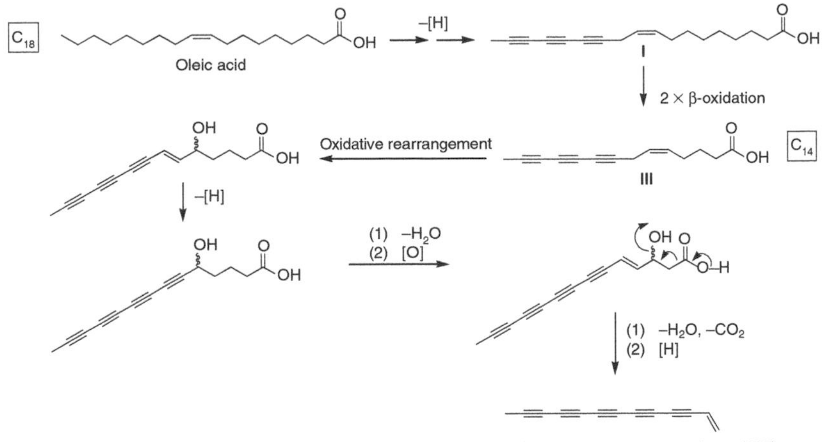

FIGURE 28.18 The possible biosynthesis of trideca-3,5,7,9,11-penayn-1-ene the most widespread polyacetylene within Asteraceae and an important precursor for polyacetylenic sulfur compounds. [O] = oxidation, [H] = reduction, −[H] = dehydrogenation (oxidation followed by the loss of water).

methylsulfoxides, methylsulfones, thiarubrines, and thietanones.[7,8,222–224] The biosynthesis of **139** is initiated with a 2 × β-oxidation of a C_{18}-triyne acid intermediate (Figure 28.18, **I**) leading to the corresponding C_{14}-triyne acid (**III**), which is a very important precursor in the biosynthesis of C_{14}-triyne polyacetylenes.[7–12] An oxidative rearrangement initiated by an allylic oxidation followed by dehydrogenation and decarboxylation leads to **139** as shown in Figure 28.18. Compound **139** is then transformed by addition of H_2S or its biochemical equivalent to polyacetylenic or acetylenic thiophenes, thiarubrines, or thietanonens whereas addition of methanethiol (CH_3SH) or the biochemical equivalent to CH_3SH leads to aliphatic methylthioethers, which may be transformed into their corresponding sulfoxides and sulfones by oxidation as outlined in Figure 28.19. The biosynthesis of thiophene derivatives, methylthioethers, and the corresponding oxidation products from polyacetylene precursors by addition of H_2S or CH_3SH is a general biosynthetic route for these secondary metabolites.[7,8,222–224] Thus, acetylenic thiophenes, methylthioether, and methylsulfoxide spiroacetal enol ethers present in certain tribes of Anthemideae are biosynthesized from the corresponding polyacetylenic spiroacetal enol ethers (Figure 28.20).

The most common spiroacetal enol ethers are biosynthesized from the C_{18}-triyne acid **I** (Figure 28.20). An α-oxidation followed by two β-oxidations of this C_{18}-triyne acid leads to the corresponding C_{13}-triyne acid and further reduction followed by an allylic oxidation leads then to a peroxide intermediate (**IV**) that by further oxidation and ring closure will lead to five-membered C_{13}-spiroacetal enol ethers such as compound **205** as outlined in Figure 28.20. Similarly, two β-oxidations of the C_{18}-triyne acid **I** will lead to the corresponding C_{14}-triyne acid and then finally to six-membered C_{14}-spiroacetal enol ethers.

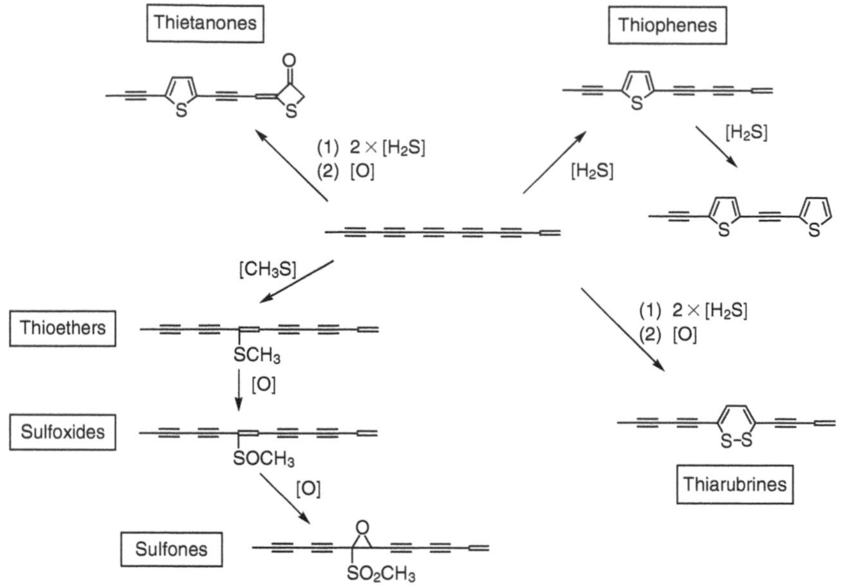

FIGURE 28.19 The possible biosynthesis of various types of polyacetylenic sulfur compounds from trideca-3,5,7,9,11-penayn-1-ene. [O] = oxidation, [H] = reduction, −[H] = dehydrogenation (oxidation followed by the loss of water).

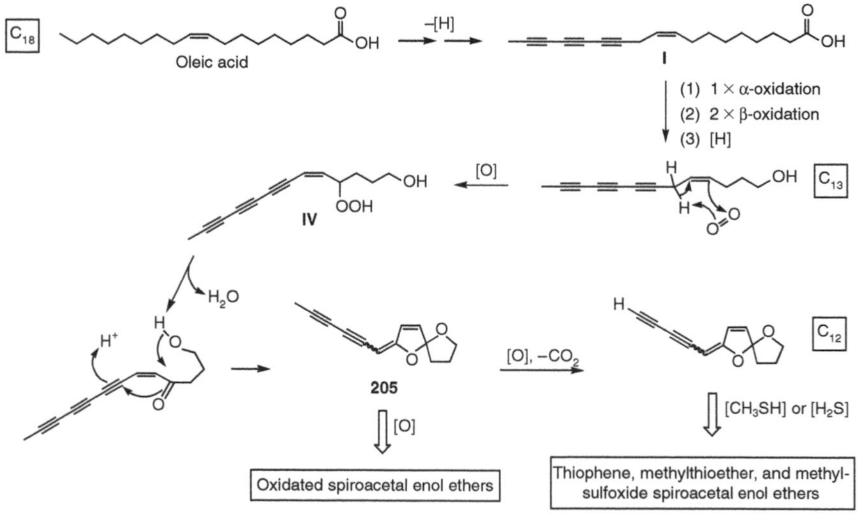

FIGURE 28.20 The possible biosynthesis of C_{13}-polyacetylenic spiroacetal enol ethers and various types of sulfur-containing compounds derived from C_{12}-polyacetylenic spiroacetal enol ethers. [O] = oxidation, [H] = reduction, −[H] = dehydrogenation (oxidation followed by the loss of water).

The majority of aromatic polyacetylenes isolated from higher plants appear to follow almost the same biosynthetic route as the spiroacetal enol ethers starting from the C_{18}-triyne acid **I**. The possible biosynthesis of aromatic polyacetylenes has been described in detail by Bohlmann et al.[8] and Christensen.[9] Also other O-heterocycles such as the tetrahydropyrans and tetrahydrofurans seem to follow a biosynthetic route that in many ways is similar to that of the spiroacetal enol ethers. In Figure 28.21 the likely biosynthesis of the tetrahydropyran ichyothereol and its acetate is shown as an example of the biosynthesis of tetrahydropyrans.

The biosynthesis of the alkamides is still not fully clarified. The amide group must certainly come from the corresponding peptide, thus the isobutyl group from a valine peptide, etc. It has been demonstrated by feeding experiments that C_{14}-polyacetylenic alkamides arise from a C_{18}-polyacetylenic precursor that has lost four carbon atoms by a double β-oxidation.[8,225] It has also been shown that oxidative degradations at the terminal methyl group of the acid moiety of C_{14} alkamides may lead to the loss of two or more carbon atoms.[225] Thus C_{11} and C_{12} polyacetylenic alkamides may be regarded as biosynthetically directly related to the corresponding C_{14} alkamides. Polyacetylenic C_{10}-alkamides is probably biosynthesized from C_{12} alkamides by a β-oxidation

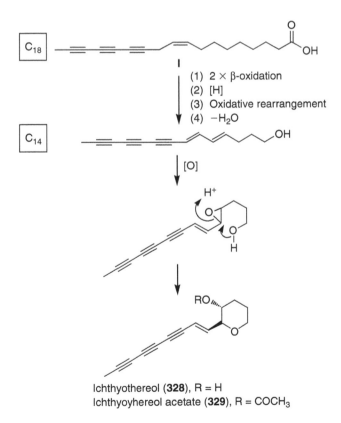

Ichthyothereol (**328**), R = H
Ichthyoyhereol acetate (**329**), R = COCH$_3$

FIGURE 28.21 Possible biosynthesis of the neurotoxic tetrahydropyran ichthyothereol and its acetate.

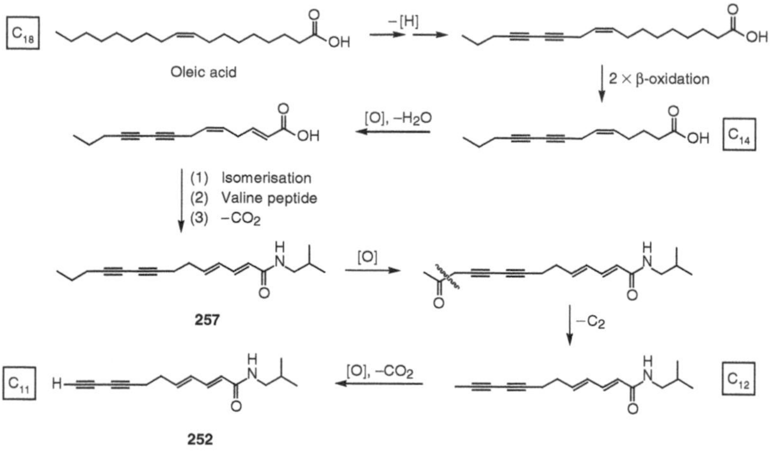

FIGURE 28.22 Possible biosynthesis of aliphatic C_{11}- and C_{12}-alkamides from the C_{14}-polyacetylenic alkamide anacyclin (**257**).

and polyacetylenic C_9 alkamides from the corresponding C_{10} derivatives by oxidative degradation of the methyl end of the molecules. The possible biosynthesis of polyacetylenic C_{14}, C_{12}, and C_{11} isobutylamides is shown in Figure 28.22.

28.4 PHARMACOLOGICAL EFFECTS OF POLYACETYLENES

28.4.1 NEUROTOXIC, NEURITOGENIC, AND SEROTONERGIC EFFECTS

The neurotoxic effects of some polyacetylenes have long been known. Polyacetylenes with this activity include the fish poisons ichthyothereol (**328**) and its acetate (**329**), which have been shown to be quite toxic to goldfish and guppies at a concentration of 0.3 μg/mL. At this concentration the tested fish showed agitated and hyperactive activity followed by loss of coordination, paralysis, and finally death.[133] The presence of relatively large amounts of the tetrahydropyrans in **330** and **331** in *Clibadium grandiflorum*,[10] a plant similar to *Ichthyothere terminalis*, is used by native South American Indians as a fish poison suggests that the tetrahydropyrans **330** and **331** have the same neurotoxic effects as ichthyothereol and its acetate. It has been suggested that the toxicity of the fish poisons **328** and **329** may be attributed to their ability to uncouple oxidative phosphorylation and inhibiting ATP-dependent contractions.[133]

The polyacetylene cicutoxin (**50**), a major toxic principle of water-hemlock (*Cicuta virosa*), a famous poisonous plant of the Apiaceae family, which also occurs in related toxic plants such as spotted water-hemlock (*Cicuta maculata*) and from the hemlock water dropwort (*Oenanthe crocata*),[51,55,57,64,226,227] is well-known to act directly on the central nervous system, causing tonic–clonic convulsions and respiratory paralysis.[57,228] Water-hemlock and related toxic plants have been responsible for the death of numerous human beings and livestock.[227] The toxic effects are probably due to the neurotoxicity of cicutoxin and related polyacetylenic derivatives (Figure 28.3, e.g., **47–49**, **51–62**, **64**, and **65**) present in these plants. The mode of action of cicutoxin and related polyacetylenes is probably related to their ability to interact with the γ-aminobutyric acid type A (GABA$_A$) receptors by inhibiting the specific binding of GABA antagonists to GABA-gated chlorine channels of GABA$_A$ receptors as demonstrated in the rat brain.[51] Binding to these ion channels therefore plays an important role in the acute toxicity of these compounds and hence their pharmacological mode of action. Structure–activity relationship studies in mice with cicutoxin and related derivatives have shown that the length of the π-bond conjugated system in these polyacetylenes and the geometry of the double bonds are critical for the toxicological effects. Moreover, the terminal O-functional group and the allylic alcohol are essential for the toxicity.[51]

Less well-known is the effects of falcarinol (22), which produces pronounced neurotoxic symptoms upon injection into mice with an LD$_{50}$ of approximately 100 mg/kg whereas the related falcarindiol (23) does not seem to have any acute effect with an LD$_{50}$ > 200 mg/kg.[51,229] The type of neurotoxic symptoms produced by falcarinol is similar to those of cicutoxin. Cicutoxin with an LD$_{50}$ < 3 mg/kg is, however, much more toxic than falcarinol.[51]

C$_{17}$-Polyacetylenes of the falcarinol-type also seem to have an effect on neuritogenesis of cultured paraneurons. It has, for example, been demonstrated that the polyacetylenes falcarinol and panaxytriol (30) have a significant neuritogenic effect on paraneurons such as PC12h and Neuro2a cells at concentrations above 2 μM. Further, it has been demonstrated that falcarinol improve scopolamine-induced memory deficit in mice, which is probably due to its ability to promote neuritogenesis of paraneurons.[230]

Several serotogernic agents have been isolated from *Angelica sinensis* a well-known medicinal plant in China.[44] Among the isolated serotonergic agents from the roots of this plant were several polyacetylenes, including falcarindiol and related C$_{17}$- and C$_{18}$-polyacetylenes (Figures 28.2 and 28.23, compounds 36 and 351–353), which indicates that polyacetylenes of the falcarinol-type may also act on serotonin receptors, and thus exhibit pharmacological effects related to improvement of moods, behaviors, and other pharmacological effects mediated by serotonin.[45]

28.4.2 ALLERGENICITY

Many plants containing aliphatic C$_{17}$-acetylenes have been reported to cause allergic contact dermatitis and irritant skin reactions, primarily due to occupational exposure (e.g., nursery workers).[231] The relation between clinical effect and content of polyacetylenes has been investigated in *Dendropanax trifidus*, *Fatsia japonica*, *Hedera helix*, and *Schefflera arboricola* (Araliaceae), and the results showed that falcarinol (22) and the related C$_{18}$-polyacetylenes (105 and 107) are potent contact allergens.[84,85,101,105,106] On the other hand polyacetylenes of the falcarinol-type such as falcarindiol (23) and falcarinone (39) had no effect to the skin.[101] Falcarinol and related compounds have been shown to be responsible for most of the allergic skin reactions caused by plants of the Apiaceae and Araliaceae.[231,232] The allergenic properties of the polyacetylenes 22, 105, and 107 indicate that they are very reactive towards mercapto and amino groups in proteins, forming a hapten–protein complex (antigen) (Figure 28.24). The reactivity of these polyacetylenes towards proteins is probably due to their hydrophobicity and their ability to form an extremely stable

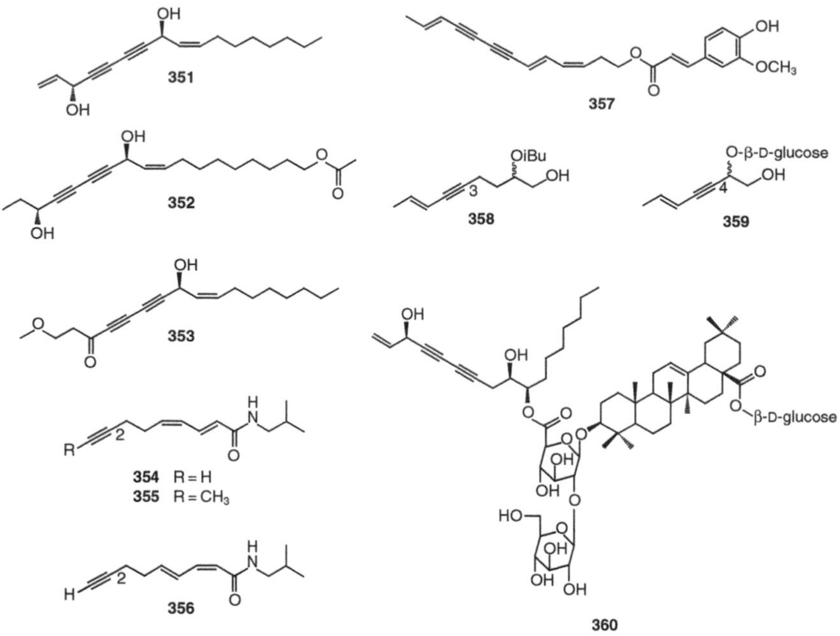

FIGURE 28.23 Examples of bioactive polyacetylenes with serotogernic (**351–353**), anti-inflammatory (**354–359**), and antiviral (**360**) activity.

carbocation with the loss of water, as illustrated in Figure 28.24, thereby acting as very reactive alkylating agent towards various biomolecules. For falcarinol this mechanism may also explain its anti-inflammatory and antibacterial effects (see Sections 28.4.3 and 28.4.4), its cytotoxicity (see Section 28.4.5) and perhaps its bioactivity in general.

28.4.3 ANTI-INFLAMMATORY AND ANTIPLATELET-AGGREGATORY EFFECTS

Herbal remedies based on *Echinacea* spp. (*E. angustifolia*, *E. pallida*, and *E. purpurea*) is one of the best selling herbal medicines in Europe and North America for the treatment of various infections. Numerous bioactive compounds have been isolated from *Echinacea* spp., including caffeic acid derivatives, poly-assaccharides, and alkamides. Of particular interest are the alkamides isolated from *E. angustifolia* and *E. purpurea* and from *Achillea* spp. of which several have shown anti-inflammatory activity in vitro. Inhibition of 5-lipoxygenase (5-LOX) and cyclooxygenase isoforms (COX-1 and COX-2) has been demonstrated for several polyacetylenic isobutyl- and 2-methylbutylamides, such as **252, 255, 256, 354–356**,[151,160] of which some apparently also display other forms of anti-inflammatory activity related to the production of macrophages.[148] Inhibition of 5-LOX and COX-1 activity has also been demonstrated for polyacetylenes isolated from *Atractylodes lancea* of which **357** showed the strongest activity (IC$_{50}$ (5-LOX) = 3.4 μM; IC$_{50}$ (COX-1) = 1.1 μM),[207] and among aliphatic polyacety-lenes isolated from *Bidens campylotheca* of which safynol-2-*O*-isobutyrate (**358**)

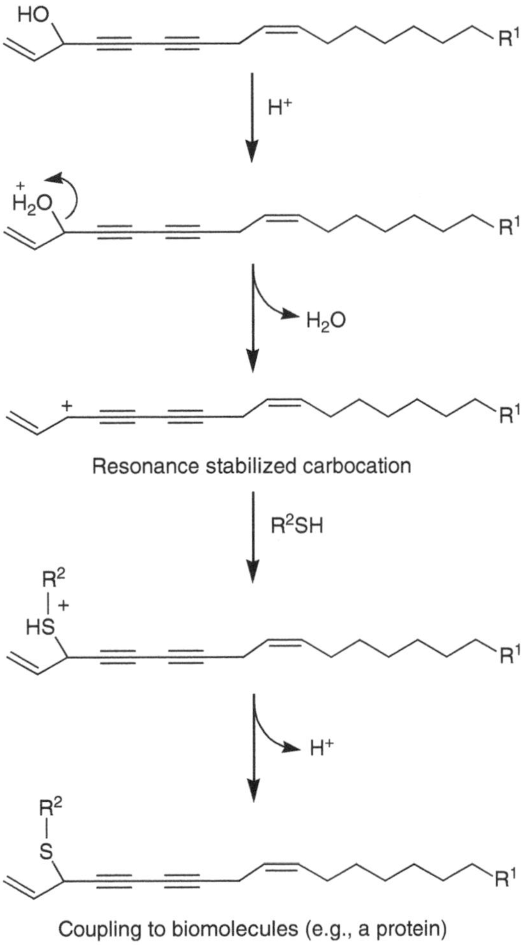

FIGURE 28.24 The possible reaction of falcarinol and related polyacetylenes with biomolecules, which may explain their interaction with the immune system leading to allergenic reactions (Type IV). The bioactivity of polyacetylenes of the falcarinol and dehydrofalcarinol-type, in particular their cytotoxic activity, may be explained by a similar mechanism. $R^1 = CH_3$, CH_2CH_2OH or CH_2CO_2H; R^2SH = thiol residue of a biomolecule.

showed the strongest inhibitory effect (100% 5-LOX inhibition <9.6 µg/mL; IC_{50} (COX-1) = 10.0 µM).[233] Furthermore, anti-inflammatory activity due to inhibition of 5-LOX and COX-1 has also been demonstrated for falcarindiol (**23**) and a C_{18}-polyacetylenic derivative both isolated from *Angelica pubescens* f. *biserata*.[234] In addition falcarindiol have shown antiplatelet-aggregatory effect that is probably related to its anti-inflammatory action.[60]

Anti-inflammatory and antiplatelet-aggregatory effects have also been demonstrated for falcarinol (**22**)[1,235–237] and it has been suggested that this pharmacological action is related to the ability of the compound to modulate prostaglandin catabolism by inhibiting the prostaglandin-catabolizing enzyme PGDH (15-hydroxy-prosta-glandin dehydrogenase).[238] Consequently, polyacetylenes of the falcarinol-type may have a protective effect on the development of cardiovascular diseases such as atherosclerosis. Cholesterol acyltransferase (ACAT) plays a role in the absorption, storage, and production of cholesterol and has been explored as a potential target for pharmacological intervention of diseases such as atherosclerosis. The isolation of aliphatic C_{17}-polyacetylenes from *Panax ginseng* (e.g., **41** and **43**) and from *Gymnaster koraiensis* (**91–95**) with an inhibiting effect on ACAT just demonstrates that falcarinol and related polyacetylenes may be potential bioactive compounds in the prevention of cardiovascular diseases.[1,239]

Furthermore, it has been shown that extracts from *Plagius flosculosus* (Aster-aceae) can inhibit the induction of the transcription factor NF-κB, which plays a key role for the inducible expression of genes mediating proinflammatory effects and thus is an important target for the development of anti-inflammatory agents. The anti-inflammatory agent responsible for this effect was identified as the spiroacetal enol ether **205**,[128] which is widespread in Anthemideae.[8,9] This could indicate that more plants from this tribe as well as more spiroacetal enol ether polyacetylenes in fact may possess anti-inflammatory activity. The polyacetylene glycoside **359** isol-ated from *Bidens pilosa* may explain the immunosuppressive action of this plant, suggesting a promising application as an anti-inflammatory drug.[240] The same polyacetylene has previously been isolated from *B. campylotheca* also known for its anti-inflammatory action as described earlier.[241] Finally, it has been demonstrated that aromatic polyacetylenic ketones such as capillin (**236**) in the oil of *Artemisia capillaris* display marked anti-inflammatory activity.[133,134] and that capillin and capillene (**235**) strongly inhibit induced TGF-β1 apoptosis of primary cultured hepatocytes.[135] The latter seem to indicate that capillin and capillene may be used in the treatment of various inflammatory liver diseases.

28.4.4 ANTIBACTERIAL, ANTIMYCOBACTERIAL, ANTIFUNGAL, AND ANTIVIRAL EFFECTS

Several polyacetylenes from the Asteraceae have been tested for activity towards various microorganisms, and have shown antibacterial effect against gram-positive bacteria (e.g., *Bacillus* spp., *Staphylococcus* spp., *Streptococcus* spp.) and gram-negative bacteria (e.g., *Escherichia* ssp., *Pseudomonas* ssp.) as well as antifungal activity against fungi such as *Candida albicans* and *Microsporum* spp. From these investigations it appears that both aliphatic polyacetylenes (e.g., **22, 23, 87, 88, 109, 110, 112, 114, 118–120, 123, 125–127, 131, 139**, and **141**), polyacetylenic spiroacetal enol ethers (e.g., **205, 210**, and **211**), aromatic polyacetylenes (e.g., **229–231, 235, 236, 238, 241**, and **242**) and polyacetylenic thiophenes/thiarubrines (e.g., **271, 285, 319**, and **326**) possess bacteriostatic or fungistatic activity,[5,13,14,133,134,139,242–251] and that this activity in some cases can be enhanced by UV light (see Section 28.4.6).

The red polyacetylenic dithiacyclohexadienes is another group of highly antibi-otic polyacetylenes. The best studied thiarubrines are **319** and **326** and they have shown a pronounced antibiotic activity against several human pathogens, including *Candida albicans*, *Aspergillus fumigatus*, *Bacillus subtilis*, *Escherichia coli*, and

Mycobacterium phlei, an activity that is slightly augmented by UV-A light.[5,191,245,247,249] The antibiotic activity of the red thiarubrines may explain the use of plants containing these compounds in traditional medicine of African countries against infectional diseases. More recently it has been reported that wild chimpanzees selectively pick and swallow entire leaves of *Aspilia* spp. without chewing them.[182,190] This observation could suggest that the leaves from these plants offer peculiar stimuli to the chimpanzees or have some kind of pharmacological effects.[182,190] The pharmacological effects of these plants are therefore most likely due to the presence of thiarubrines.

Falcarinol and falcarindiol are known as phytoalexins in tomatoes[42,43] but are also well-known antifungal compounds in carrots and related vegetables as well as other plant species.[5,7,66,90,168,251–256] Recent studies have also shown that falcarinol, falcarindiol, and related C_{17}-polyacetylenes have antibacterial effects as well as antimycobacterial effects.[47,66,86] Especially the antimycobacterial effects towards *Mycobacterium* spp. including *M. tuberculosis*,[86] and antibaterial effects towards resistant strains of *Staphyllococus aureus*,[47] is interesting, and seem to indicate pharmacological activities by which falcarinol and related polyacetylenes could have a positive effect on human health.

Polyacetylenes may show high antiviral activity in the presence of light (phototoxic effect, Section 28.4.6) whereas their antiviral activity in the dark does not seem to be pronounced. However, a special antiviral triterpene saponin polyacetylene derived from panaxytriol (Figure 28.23, **360**) have been isolated from *Panax ginseng*. This polyacetylene has been found to inhibit the replication of human immunodeficiency virus type 1 (HIV-1) with an IC_{50} value of 13.4 µg/mL.[257]

28.4.5 Cytotoxicity and Anticancer Effect

Panax ginseng C. A. Meyer (Araliaceae) is one of the most famous and valuable herbal drugs in Asia. Since the anticancer activity of petrol extracts of the roots of *P. ginseng* was discovered in the beginning of 1980s,[97,258] and the lipophilic portion of this plant has been intensively investigated. This had led to the isolation and identification of several cytotoxic polyacetylenes (e.g., **22**, **30–33**, and **38**, **44**, **46**, **75**, **76**, and **96–98**) of which falcarinol (**22**), panaxydol (**32**), and panaxytriol (**30**) are the best studied of the isolated polyacetylenes from this plant (Figure 28.2).[1,96–99,259–262] The polyacetylenes **22**, **30**, and **32** have, for example, been shown to be highly cytotoxic against numerous cancer cell lines, including leukemia (L-1210), human gastric adenocarcinoma (MK-1), mouse melanoma (B-16), and mouse fibroblast-derived tumor cells (L-929),[99,259,260] although being most toxic towards MK-1 cancer cells with ED_{50} values of 0.027 (**22**), 0.016 (**32**), and 0.171 µg/mL (**30**).[259] In addition, compounds **22**, **30**, and **32** have been shown to inhibit the cell growth of normal cell cultures such as human fibroblasts (MRC-5), although the ED_{50} against normal cells was around 20 times higher than for cancer cells. In particular, panaxytriol did not inhibit the growth of MRC-5 cells by 50% even at concentrations higher than 70 µg/mL.[259] The selective in vitro cytotoxicity of the polyacetylenes **22**, **30**, and **32** against cancer cells, when compared with normal cells, indicates that they may be useful in the treatment of cancer. For falcarinol this has recently been demonstrated in a preclinical trial demonstrating inhibitory effects of this compound on the development of azoxymethane (AOM) induced colon preneoplastic lesions in rats. These results clearly indicate that falcarinol may have a protective effect towards development of colon cancer and potentially other types of cancer.[263]

Falcarinol and related cytotoxic polyacetylenes have also been isolated from the Araliaceae species *Acanthopanax senticosus*,[1,264] *Panax quinquefolium*,[89,91,95] *Dendropanax arboreus*,[1,77,78,265] and from several Apiaceae species, which includes *Angelica acutiloba*,[1,266] *Apium graveolens*,[46] *Cicuta maculata*,[64] *Crithmum maritimum*,[59] and *Seseli mairei*.[63]

The cytotoxix polyacetylenes isolated from the roots of *P. quinquefolium* (American ginseng) includes, besides the well-known falcarinol, acetylpanaxydol (**33**), panaxydol, panaxytriol also several other closely related C_{14}- and C_{17}-polyacetylenes (e.g., **28**, **35**, **36**, **44**, **45**, **75**, and **77**; Figures 28.2 and 28.4).[89,91,95] The polyacetylenes isolated from *P. quinquefolium* have been tested against L-1210 cells and the majority of C_{17}-polyacetylenes exhibited strong cytotoxic activities (IC$_{50}$ = 0.5–1.0 μg/mL), whereas the the cytotoxic effect of the related C_{14}-polyacetylenes **75** and **77** were slightly lower (IC$_{50}$ = 10 μg/mL).[89,91]

From the aerial parts of *Dendropanax arboreus*, which is naitive to North America, several aliphatic C_{17}-polyacetylenes have been isolated of which falcarinol, falcarindiol (**23**), dehydrofalcarinol (**87**), and dehydrofalcarindiol (**88**) (Figures 28.2 and 28.5) were found to exhibit in vitro cytotoxicity against human tumor cell lines, with falcarinol showing the strongest activity.[77] Preliminary in vivo evaluation of the cytotoxic activity of **22**, **87**, and **88** using the LOX melanoma mouse xenocraft model demonstrated some potential in vivo antitumor activity of falcarinol and dehydrofalcarinol, with the latter showing the strongest therapeutic effect.[77]

The cytotoxic principles isolated from *C. maculata* consist primarily of cicutoxin (**50**), which has shown high in vitro cytotoxic activity against human nasopharyngeal carcinoma (KB) cells with an ED$_{50}$ value of 2 μg/mL. Furthermore, it appears that this polyacetylene is a strong antileukemic agent.[64] The cytotoxic compounds isolated from *Angelica acutiloba* (**22**, **23**, **40**), *Apium graveolens* (**22**, **23**, **27**, **44**), *Crithmum maritimum* (**22**, **23**), and *Seseli mairei* (Z-isomer of **44** = seselidiol) are more or less well-known cytotoxic principles in Araliaceae or Apiaceae, as described earlier.

Special cytotoxic polyacetylenes derived from falcarindiol have been isolated from *Aciphylla scott-thomsonii* (**361**)[50] and *Angelica japonica* (**362–365**)[52,267] (Figure 28.25) both belonging to Apiaceae. The novel dimeric C_{34}-polyacetylene aciphyllal (**361**) have shown weak cytotoxic activity to P388 cells (IC$_{50}$ > 25 μg/mL), whereas the falcarindiol furanocoumarin ethers **362–365** have shown moderate inhibitory activity towards MK-1, human uterus carcinoma (HeLa) and murine melanoma (B16F10) cells with ED$_{50}$ values in the range 5.0–16.2 (MK-1), 13.9–22.6 (HeLa), and 15.1–28.7 μg/mL (B16F10). The polyacetylenes **362–365** were found to be more toxic towards HeLa and B16F10 cells when compared with falcarinol, falcarindiol, and 8-*O*-acetylfalcarindiol (**27**) (ED$_{50}$ > 19.5 μg/ml), which, however, were found to be very toxic towards MK-1 cells, with ED$_{50}$ values of 0.3 (**22**), 3.9 (**23**), and 3.2 μg/ml (**27**).[52]

The mechanism for the cytotoxic acitivity and potential anticancer effect of falcarinol and related C_{17}-acetylenes is still not known but may be related to their

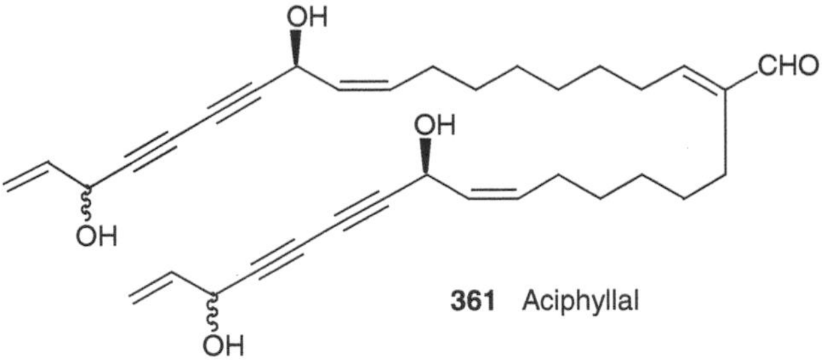

361 Aciphyllal

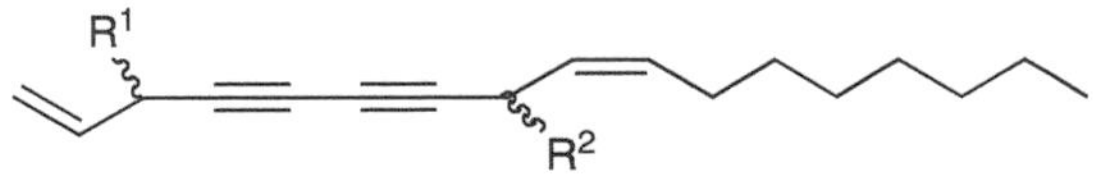

362 R¹ = OH, R² = A, Japoangelol A
363 R¹ = A, R² = OH, Japoangelol B
364 R¹ = OH, R² = B, Japoangelol C
365 R¹ = B, R² = OH, Japoangelol D

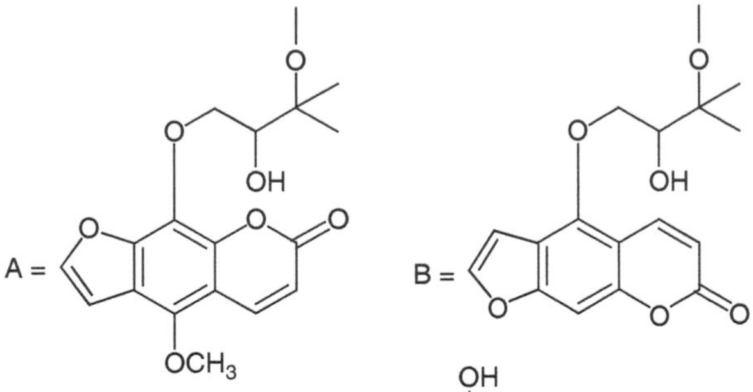

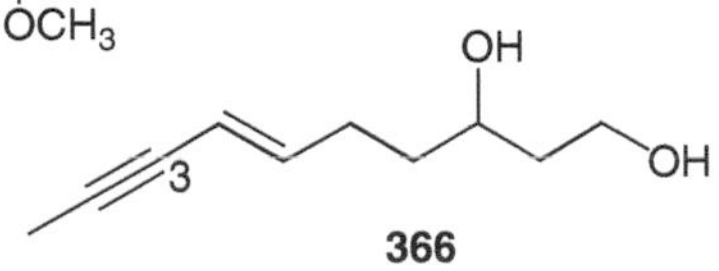

366

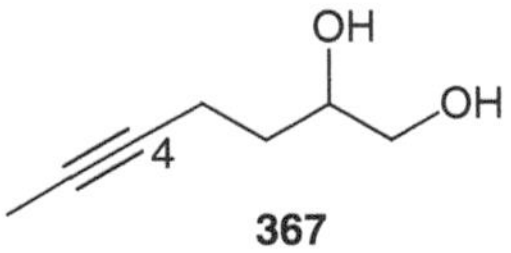

367

FIGURE 28.25 Examples of special polyacetylenes (**361–365**) and highly unsaturated polyacetylenes (**366**, **367**) with cytotoxic or antitumor activity.

reactivity and hence their ability to interact with various biomolecules as mentioned earlier in Section 28.4.2. This is in accordance with an in vitro study, which showed that the suppressive effect of falcarinol on cell proliferation of various tumor cells (e.g., K562, Raji, Wish, HeLa, and Calu-1), probably was due to its ability to arrest the cell cycle progression of the tumor cells into various phases of their cell cycle.[268] Falcarinol have also shown apoptotic characteristics, which could be another mechanism of action of C_{17}-polyacetylenes.[264,269]

From the Asteraceae family several cytotoxic C_{17}-polyacetylenes have been isolated from the roots of *Cirsium* spp. (**195–198**)[118–121] and *Gymnaster koraiensis*

(**88, 90–95**) (Figures 28.5 and 28.9).[1,122] The C_{17}-polyacetylenes from *G. koraiensis* exhibited significant cytotoxicity against L-1210 tumor cells with ED_{50} values of 0.12–3.3 μg/mL, whereas those isolated from *Cirsium japonicum* (**195–198**) were found to be moderately toxic against KB cells with ED_{50} values of 8.6–39.5 μg/mL.[118–120] Also, aliphatic C_{10}-polyacetylenes from the Asteraceae have shown to possess cytotoxic and potential antitumor effects, which includes the well-known (*Z*)- and (*E*)-dehydromatricaria ester (**110, 112**) and their corresponding acids (**111, 113**) (Figure 28.7).[123–125] It has been found that the (*Z*)-dehydromatrica ester is able to inhibit the growth of cancer cell lines such as MK-1 ($ED_{50} = 0.59$ μg/mL), L-929 ($ED_{50} = 0.98$ μg/mL), and B-16 ($ED_{50} = 1.87$ μg/mL), making it almost as cytotoxic as C_{17}-polyacetylenes of the falcarinol-type such as panaxytriol.[125] Other types of aliphatic polyacetylenes from Asteraceae that have been shown to possess cytotoxic and potential antitumor effects are the highly unsaturated aliphatic polyacetylenes **366** and **367** (Figure 28.25) isolated from plant extracts of *Bidens pilosa*. These polyacetylenes have shown a significant antiangiogenic effect and an effect on anticell proliferation and antitube formation against human umbilical vein endothelium cells with IC_{50} values <2.5 μg/mL. Hence these compounds may be potential anticancer agents.[270] The potential anticancer activity demonstrated for aliphatic C_{10}-, C_{13}-, C_{14}-, and C_{17}-polyacetylenes from the Asteraceae is interesting and could imply that more aliphatic polyacetylenes of this plant family may possess important bioactivities yet to be discovered.

The C_{18}-polyacetylenes **16**, **18**, and **19** (Figure 28.1) isolated from the Olacaceae plant species *Coula edulis* (**18**),[34] *Minquartia guianensis* (**18**),[36,37] and *Ochanostachys amentacea* (**16, 18, 19**),[32,33] have been shown to be highly cytotoxic against several human cancer cell lines in vitro of which minquartynoic acid (**18**) is the most toxic.[32] Minquartynoic acid has, for example, been found to be cytotoxic against P388 murine lymphocytic leukemia cells with an ED_{50} value of 0.18 μg/mL.[32] A new polyacetylenic diepoxide compound, gummiferol (**1**), has recently been isolated from *Adenia gummifera* (Passifloraceae) by cytotoxic-guided chromatographic fractionation using carcinoma cells.[39] Besides being toxic to carcinoma cell lines this polyacetylene have also shown a broad cytotoxic spectrum against 10 other human cancer cell lines. *Adenia gummifera* is used to improve animal health in Tanzania and perhaps this plant may also be used to improve human health.

Acetone extracts of various *Argyranthemum* spp. (Asteraceae) have shown cytotoxic activity against HeLa and Hep-2 cell lines. The compounds responsible for this activity were aromatic polyacetylenes (e.g., **240** and **241**) and the spiroacetal enol ether **210** (*E*-isomer). The compounds inhibited growth (IC_{50}) of HeLa and Hep-2 cell lines in concentrations between 18–50 μg/mL. Aromatic polyacetylenes such as capillin (**236**) and capillene (**235**), which are present regularly in *Artemisia* spp., appears to be cytotoxic and to have potential anticarcinogenic activity.[136,137] Concentrations of 1–10 μM of capillin was found to inhibit cell proliferation of human tumor cell lines such as colon carcinoma HT29, pancreatic carcinoma MIA PaCa-2, epidermoid carcinoma of the larynx Hep-2, and lung carcinoma A549 by inducing DNA fragmentation and cell death. These findings suggest that capillin has cytotoxic activity and can induce apoptosis.[137] The spiroacetal enol ethers **221–226** isolated from white mugwort *Artemisia lactiflora* have been shown to be inhibitors of HL-60 cells, and thus potential anticancer agents.[271–273] In addition the spiroacetal enol ether **223** has shown inhibitory effects on a variety of tumor promoter-induced biological responses such as oxidative stress.[274]

28.4.6 PHOTOTOXICITY

Many polyacetylenes and thiophene derivatives require UV-A light (wavelength 320–400 nm) for their toxicity and other biological activities. The phototoxic poly-acetylenes extend their toxicity to a wide range of organisms, including nematodes, fungi, bacteria, viruses, and mammalian cells.[5,13–15,133,162,245–247,250,275–287]

The phototoxicity of polyacetylenic derivatives was discovered by Gommers,[285,286] who showed that the nematicidal activity of polyacetylenic derived thiophene compounds α-terthienyl (α-T) and 5-(buten-1-ynyl)-2,2′-bithienyl (BBT), was tremendously enhanced by light. Since the discovery of the phototoxic action of polyacetylenic derivatives by Gommers many investigations have been performed to elucidate the phototoxicity of polyacetylenes and related compounds.

As described in Section 28.4.4 the thiarubrines, and especially thiarubrine A (**319**) has a strong antibiotic effect on human pathogens and on several bacteria. This activity is light-independent but in some cases can be augmented by UV-A light.[245,247] However, against the gram-positive bacteria *Staphylococcus albus* and *Streptococcus faecalis*, thiarubrine A displays only phototoxic activity.[247] Thiarubrine A is also toxic to nematodes and mammalian cells and the toxicity is considerably enhanced in the presence of UV-A light.[247,279] The activity of thiarubrine A towards nematodes in the dark confirms its traditional use as a stomachic.[247] Thiarubrine A is highly phototoxic to membrane-containing viruses but only slight phototoxic to viruses without membranes.[13,15,287–290] Although UV-A light in some cases enhances the toxicity of thiarubrine A, it also rapidly converts thiarubrine A to the corresponding thiophene (**271**) by extrusion of a sulfur atom.[13,245] The thiophene **271** is inactive in the dark but phototoxic against different microorganisms, thus the observed inhibition of microorganisms by thiarubrine A in UV-A light could be due to the compound itself prior to photoconversion, to the phototoxic thiophene **271** or to a mechanism of toxicity involving the conversion process itself.[245,279]

Phenylheptatriyne (PHT, **231**) is one of the best studied polyacetylenes and is phototoxic to a wide range of organisms, including fungi, bacteria, viruses, cercariae, nematodes, and mammalian cells.[15,246,248,249,279–282,287,288,291–295] PHT also possess activity in the dark towards, for example, nematodes[278,296] and cercariae.[294] However the light-independent effects are in most cases of smaller magnitude when compared with its phototoxicity.

PHT shows potent phototoxic activity against gram-positive bacteria and to lesser extent gram-negative bacteria.[5,284] The phototoxicity of PHT to membrane-containing viruses such as murine cytomegalovirus (MCMV), Sindbis virus (SV), and the lipid-coated phage PM2 seems to indicate that membranes are the target of PHT.[284,292,297] When tested to these membrane-containing viruses, PHT readily inactivated the viruses in concentrations less than 1 μg/mL. In contrast, viruses without membranes, such as bacteriophage T4 and fish infectious pancreatic necrosis virus (IPNV) were resistant.[13,292,297] PHT causes photodamage to erythrocytes, and like other polyacetylenes it does not form cross-links in DNA or other kinds of cytotoxic damage.[248,249,284,293,295,298] Although PHT is very phototoxic it does not induce photodermatitis in man, and there is no evidence to implicate it as a contact allergen.[249]

Many other polyacetylenes have shown phototoxic activities to a wide range of organisms, including mammalian cells. As mentioned in Section 28.4.4 bacteriostatic as well as fungistatic activity of polyacetylenes has been clearly demonstrated. Polyacetylenes that have been shown to be phototoxic to bacteria and fungi includes, **109, 110, 112, 114, 120, 122–127, 132, 139, 153, 156, 181, 188, 209–211, 227–230, 232, 234–236, 267, 271, 285, 288, 326**, and **328**, of which some also have a light-independent antibiotic activity.[5,13,14,241,246–250,281] Falcarinol (**22**) and falcarindiol (**23**) are examples of very strong antibiotics (see Section 28.4.4), whose activity is not enhanced by UV-A.[5,248] Further, it has been demonstrated that the polyacetylenes **156, 267, 326, 328**, and **342** are very toxic towards nematodes in the presence of UV-A light.[133,299]

The antiviral activity of polyacetylenes has been extensively studied.[5,13–15, 162,276,279,282,287–290,292,297] The antiviral activity of these compounds in the dark does not seem to be pronounced, whereas many of these compounds show high antiviral activity in the presence of light. It has been demonstrated that the polyacetylenic thiophene **271** and the furanopolyacetylene **342** are highly phototoxic to membrane-containing viruses (e.g., MCMV and SV) but not to viruses without a membrane,[5,13–15,162,276,289,290] indicating that the main target of these compounds are membranes. Studies of the antiviral and cytotoxic activity of naturally occurring acetylenes have shown that there is some correlation between cytotoxicity and antiviral activity as might be expected considering the similarity of the membranes.[13]

Photoexcitation of polyacetylenes may have a very large effect on their biological activity. However, they do not appear to have a common mechanism of action. Many thiophenes, when exposed to light and air, act as a classical sensitizer and it has been shown that thiophene derivatives, including α-T, are efficient generators of singlet oxygen ($^1O_2$, excited oxygen).[5,291,300] Singlet oxygen is a highly reactive species and can easily oxidize unsaturated fatty acids, sterols, free amino acids, and amino acid residues in enzymes and membrane proteins. The primary mechanism of action of thiophenes therefore seems to be a Type II photodynamic disruption of membranes (an oxygen-dependent Type II mechanism).[5,291] The rates of production by polyacetylenes are considerably less than for thiophenes and in contrast to the thiophenes, the aliphatic polyacetylenes appear to have a predominantly nonphotodynamic (oxygen-independent) mechanism of action, which may involve photoaddition reactions of the photolabile unsaturated structures or the formation of free radicals upon photoexcitation.[5,291,300] However, partly cyclized polyacetylenes, as for example PHT, may be regarded as having an intermediate structure between the aliphatic polyacetylenes and the thiophenes and they seem to have access to both a Type II photodynamic and a nonphotodynamic mechanism of action.[5,300,301]

Studies of the structure–activity relationship of polyacetylenes and thiophenes have shown that at least three conjugated carbon–carbon triple bonds are required for phototoxic activity among the aliphatic polyacetylenes.[5,246] The phototoxicity is enhanced significantly with further unsaturation. Aromatic polyacetylenes are more phototoxic than aliphatic polyacetylenes but in general less phototoxic than thiophenes. These results fit in well with the ability of polyacetylenes and thiophenes to participate in photodynamic (Type II) and nonphotodynamic reactions, the first being the most toxic one.

28.5 DETECTION, SEPARATION, AND ISOLATION OF POLYACETYLENES

Polyacetylenes are, in general, thermally unstable and may undergo photodecomposition if exposed to daylight. Consequently gentle methods should be used for extraction, detection, and isolation of these compounds. Polyacetylenes are usually extracted from fresh or air-dried plant material by an organic solvent such as hexane, diethyl ether, ethyl acetate, methylene chloride, or methanol. Sometimes, the extracts are first subjected to bulk chromatography to obtain fractions/extracts containing polyacetylenes of different polarity. This is, however, not necessary when focusing on lipophilic polyacetylenes that constitute the majority of polyacetylenes isolated from plants (see Section 28.2). For the separation and isolation of polyacetylenes in extracts several methods can be employed of which column chromatography (CC), TLC, and high-performance liquid chromatography (HPLC) are the most used techniques.[4,110,157,302,303] For the detection and quantification of polyacetylenes in extracts/fractions TLC, UV spectroscopy, and HPLC combined by diode array detection (DAD) or simple UV detection are the most frequently used methods. As mentioned in Section 28.2, polyacetylenes are often highly conjugated substances with characteristic chromophores consisting of at least two triple bond(s) in conjugation. Therefore, polyacetylenes usually exhibit UV spectra with characteristic patterns between 200 and 410 nm and with relatively high extinction coefficients (ε up to 350×10^3) at some wavelengths, typically below 380 nm.[8,109–111] Consequently, most polyacetylenes are very easy to detect and quantify in plant extracts by UV spectroscopy even at very low concentrations. Therefore, the application of UV spectroscopy may not only provide useful information about the structure of polyacetylenes but may also be useful for the detection and quantification of these compounds.[4,110,157,302] In the following, a description of the most used techniques and methods for the detection, separation, and isolation of polyacetylenes will be described with focus on the application of TLC.

28.5.1 ANALYTICAL TLC

Analytical TLC is an excellent method for obtaining information on the number of polyacetylenes in extracts or fractions and for the development of isocratic or gradient solvent systems for the separation of these compounds by CC (Section 28.5.5). Normal phase TLC is, in general, the best suited type of thin layer plates for the separation of polyacetylenes, including polyacetylenic glycosides. Separation of polyacetylenes by TLC is usually performed on silica gel, wherein the typical adsorbent used is Merck silica gel 60 such as silica gel 60 F_{254}, 40–63 μm. For silica gel plates of a size of 20×20 cm the maximum number of samples to be analyzed by TLC should not exceed 11 and the amounts applied to the analytical TLC plates is usually 20 μL of extract or fraction.[110,199]

For the analysis of lipophilic polyacetylenes in an extract the following method is a good approach. After separation of the extract by silica gel CC by an organic solvent gradient with increasing polarity the resulting fractions are analyzed by a combination of TLC and UV spectroscopy, which normally will give important information about the purity and the polarity of the polyacetylenes in each single fraction and, in some cases, it will also allow a semiquantitative determination of the polyacetylenes in the fractions. To investigate the fractions resulting from the fractionation by CC, 20 μL of each fraction are applied to thin layer plates, so that 20 μL of the first 11 fractions are placed on the first silica gel plate. Then the silica gel plate is placed in a chamber saturated with a suitable mobile phase usually starting with *n*-hexane or petroleum ether. The second plate is then first spotted with 20 μL of fraction 11, which is also applied to the first plate. On the second plate 20 μL of fractions 12–21 are applied subsequently. The second plate may now be developed with an increase in the polarity of the mobile phase by using diethyl ether (Et$_2$O), ethyl acetate (EtOAc), or CH$_2$Cl$_2$, which are the most frequently used solvents for increasing the polarity of the mobile phase for separation of lipophilic polyacetylenes by TLC.[113,117,127,132,157,215,303–305] The second plate may be developed in 10% Et$_2$O or EtOAc in petroleum ether or *n*-hexane. By using increasing amounts of Et$_2$O or EtOAc for the development of the successive plates, it is possible to get information about the increasing polarity of the polyacetylenes, which combined with information about their UV spectra may form the basis for further analytical separation of the fractions by preparative TLC (Section 28.5.2) or HPLC (Section 28.5.6). However, for the separation of more polar poly-acetylenes, such as polyacetylenic glycosides, on silica gel TLC a mobile phase consisting of different proportions of solvents such as CH$_2$Cl$_2$ and methanol or similar solvents is an advantage.[244] For visualization of the spots on TLC plates a number of spray reagents can be used (see Section 28.5.4).

Analytical TLC has been shown to be useful to identify plant species based on their content of polyacetylenes and related compounds. One-dimensional TLC plates (silica gel 60 F$_{254}$) developed in Et$_2$O–petroleum ether (2:3) has, for example, been used to separate characteristic root acetylenes from closely related *Artemisia* species as illustrated in Figure 28.26.[132] Analytical TLC has also been shown to be useful for the separation and identification of closely related alkamides and related compounds from extracts of *Echinacea* species as described in more detail in Section 28.5.4.[157]

28.5.2 Preparative TLC

Preparative TLC is an excellent method for the separation of polyacetylenes. Fur-thermore, the results from the separation by analytical TLC can be more or less adopted. For preparative TLC an application of a fraction or solution may be performed by placing a great number of spots on a TLC plate approximately 1 cm above the mobile phase. Alternatively, the application can be drawn out slowly with a pipette as a continuous band. After development in a chamber saturated with the same solvent used for the development of the plate, the plate is partially covered, so that only 10% of the developed material at either edge of the plate is kept free of the spraying reagents, as illustrated in Figure 28.27. In this way the unsprayed bands are visualized and the nonsprayed areas of each band are scrapped off into filters and

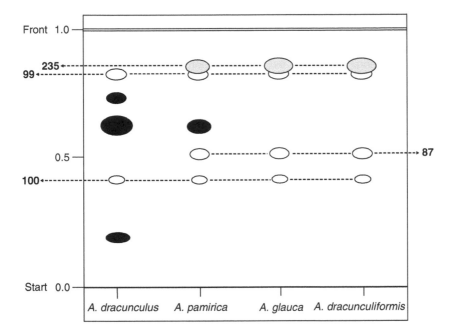

FIGURE 28.26 One-dimensional TLC (silica gel 60 F_{254}) of characteristic root polyacetylenes of *Artemisia* species. Polyacetylenes were extracted from the roots with petroleum ether–Et_2O (2:1) and the extracts revealed the presence of capilene (**235**, gray spots), dehydrofalcarinol-type polyacetylenes (**87**, **99**, **100**; white spots), and butenyl-isocoumarins (artemidins; black spots). Mobile phase Et_2O–petroleum ether (2:3). The spots on the TLC plate were located by UV absorbance. TLC plate redrawn from Greger. (From Greger, H., *Phytochemistry*, 18, 1319, 1979.)

the purified compounds are eluted with a convenient solvent. The fractions obtained are investigated by analytical TLC, UV, or HPLC in order to check their purity. Finally, the purified compounds may be subjected to NMR and other spectroscopic analysis in order to elucidate their chemical structure.

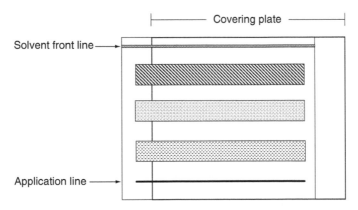

FIGURE 28.27 Preparative TLC. The covering plate is placed above the TLC plate so about 10% of the TLC plate is visualized before spraying with the visualizing agent.

As described in Section 28.5.1, the best separation results for most polyacety-lenes are obtained on silica gel TLC plates. The possibility of using alumina (Al_2O_3) for separation of these compounds has also been investigated.[110,306] However, the high rate of decomposition has led to the abandonment of the use of alumina for the separation of polyacetylenes by preparative TLC as well as for analytical TLC and CC. Reversed phase materials are widely used as stationary phases for the separation of polyacetylenes by HPLC (see Section 28.5.6) and has also been used for the separation of polyacetylenes by TLC.[154] However, reversed phase materials are not widely used for the separation of polyacetylenes by TLC probably because many polyacetylenes has a tendency to bind too strongly to the stationary phase resulting in tailing of the spots or bands. The use of impregnated silica gel plates with caffeine has been shown to be advantageous for separating polyacetylene hydrocarbons and other relatively nonpolar compounds (see Section 28.5.3). Furthermore, it has been shown that caffeine offer protection of polyacetylenes for degradation of these highly unsaturated and unstable compounds on the plates.[306,307]

An alternative separation method to the traditional form of preparative TLC is preparative centrifugal thin layer chromatography (CTLC).[110,308] This technique has been developed in order to ensure a more rapid separation and a more restricted use of solvents. CTLC relies on the action of a centrifugal force to accelerate mobile phase flow across a circular TLC plate and is performed on a commercial available Chro-matotron.[110,308] Most applications of CTLC have employed silica gel as the sorbent (TLC-grade silica gel GF_{254}).[110,308] The plate is rotated at approximately 800 rpm by an electric motor and the sample introduction takes place at the centre and eluent is pumped across the sorbent. Solvent elution produces concentric bands across the plate, which are spun off at the edges and collected for TLC analysis. CTLC has been shown to be quite effective in separating polyacetylenes, such as compounds **227** and **231**, that are difficult to separate by preparative TLC (see Section 28.5.3). Despite some clear advantages of CTLC when compared with preparative TLC this method has been rarely used for the separation of polyacetylenes.[28,309]

28.5.3 IMPREGNATED TLC PLATES

The complexing properties of caffeine can be used to separate polyacetylenic hydrocarbons of comparable structure, which differ slightly in the number or posi-tions of π-electrons. The electronic system of caffeine has the ability to interact with π-electrons in unsaturated compounds. Thus otherwise poorly or nonseparable compounds by TLC may be separated by the addition of caffeine to silica gel on plates or columns.[110,199,215,307,308] For example, the aromatic polyacetylenes **227** and **231** that occur frequently in a mixture in *Bidens* and *Dahlia* spp. (Asteraceae, tribe Heliantheae)[8,10] cannot be separated by TLC but by addition of 10% of caffeine to silica gel plates these compounds are easily separated.[110] Another example is the separation of the polyacetylenic hydrocarbons **177**, **188** and **178**, **194**, respectively, from *Chrysanthemum leucanthemum* (Heliantheae).[215] These compounds are not easily separated on preparative and analytical silica gel plates. However, by using silica gel enriched with 5%–10% of caffeine it is possible to separate both types of polyacetylenes due to differences in the number of π-electrons, whereas it is not possible to separate **177** from **188** and **178** from **194** by means of caffeine as there is no differences in the number of π-electrons in their structures.[110,215]

As described earlier the use of caffeine may be an useful method for separating polyacetylenes, which differ slightly in the number of π-electrons. However, the use of caffeine for separation is restricted to nonpolar and less polar compounds as solvents used for development of TLC plates or columns with added caffeine cannot be very polar.[110,306,307] Caffeine is dissolved by Et_2O, EtOAc, and more polar solvents, and hence the use of such solvents would move the caffeine ahead of the polyacetylenes to be separated by these solvents so that the separating effect would be negligible. As a general rule the eluting solvent should not contain more than 10% of Et_2O in *n*-hexane or petroleum ether. Many polyacetylenic compounds that occur in nature are less polar compounds with slight differences in their π-electrons systems, and hence the use of, for example, caffeine-impregnated TLC plates could be advantageous for the separation of many types of polyacetylenes, including hydrocarbons, esters, aldehydes, ketones, and sulfur compounds. Although the use of caffeine-impregnated TLC plates has been used occasionally for the separation of polyacetylenes the method has not yet received much attention,[110,199,215,310,311] probably because it is limited to the separation of less polar compounds. Procedures for making silica gel plates containing caffeine are available from the literature.[110,215]

Silver nitrate ($AgNO_3$) is another complexing agent reported in the literature for the separation of unsaturated compounds containing different numbers of π-electrons.[312] However, this method has not yet been used for polyacetylene separations, although it may be useful.

28.5.4 VISUALIZATION REAGENTS

For the visualization of the separated compounds on the plates several possibilities exist: One method, which has proven to be an excellent choice for visualization of polyacetylenes, is the use of a neutral solution of potassium permanganate ($KMnO_4$, 0.32%) sprayed onto the plates. This oxidizes most organic compounds, leaving yellow spots.[110,306] For highly unsaturated compounds, such as the polyacetylenes, the oxidation appears to be very fast in contrary to saturated compounds. This can give some important information about the type of compounds present in extracts or fractions. Other methods of visualization that may be used for detection of pure polyacetylenic compounds on TLC plates are application of vanillin, *p*-dimethylaminobenzaldehyde, and anisaldehyde dissolved in a solution consisting of ethanol and sulfuric acid.[110,157,303,313] These visualization agents give rise to spots of characteristic colors that may also give important information about the type or structure of polyacetylenes present in extracts or fractions. For example, the investigation of root extracts and extracts of aerial parts of *Echinacea* species (*E. angustifolia*, *E. pallida*, and *E. purpurea*), commonly used in *Echinacea* drugs, for polyacetylenic alkamides by TLC on silica plates with *n*-hexane–EtOAc (2:1) as mobile phase, revealed the presence of 19 isobutyl- or 2-methylbutylamides, of which the majority were polyacetylenic alkamides, and 7 were ketone polyenes or polyacetylenes (Figures 28.28 and 28.29). Despite the complex mixture of constituents in each of the *Echinacea* extracts, in particular the roots, the plant species could easily be

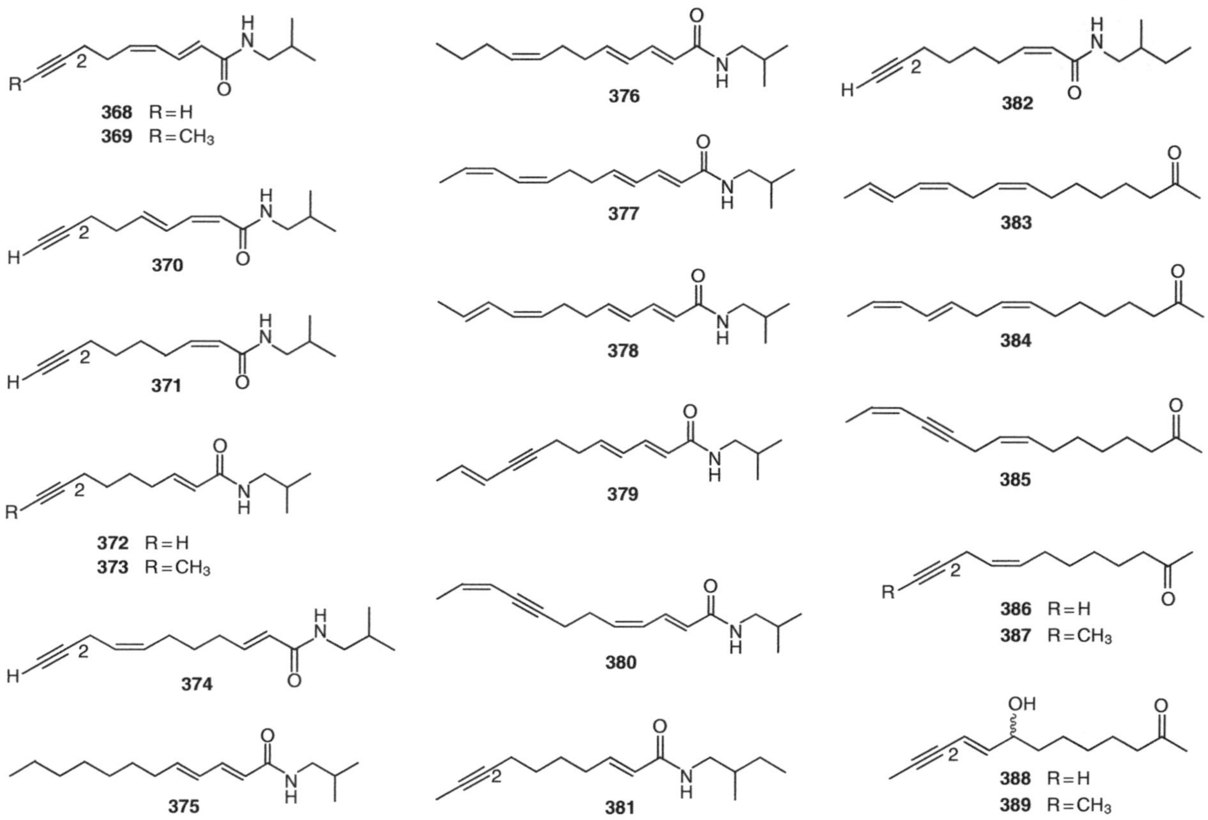

FIGURE 28.28 Alkamides and related compounds detected in extracts of *Echinacea* species by TLC, which also includes the 2-methylbutylamides 255 and 256 and the isobutylamides 262 and 264 (Figure 28.11). (From Bauer, R. and Remiger, P., *Planta Med.*, 55, 367, 1989.)

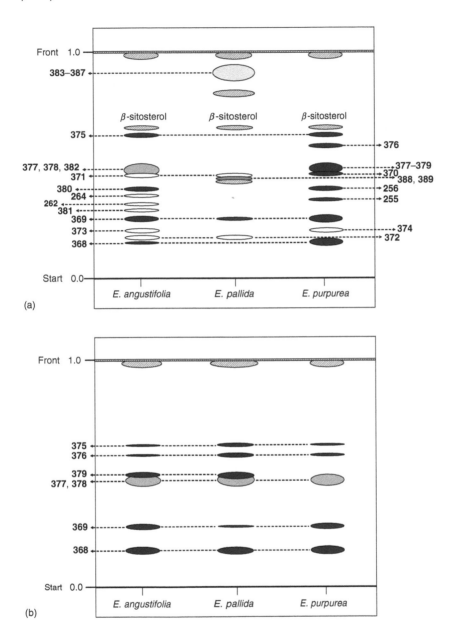

FIGURE 28.29 One-dimensional TLC of chloroform extracts from (a) *Echinacea* roots and (b) *Echinacea* aerial parts. Mobile phase *n*-hexane–EtOAc (2:1). Alkamides and polyenic and polyacetylenic ketones detected by spraying with anisaldehyde/sulfuric acid. White spots (yellow color reaction, see Table 28.1) = 2-monoenamides (**262, 264, 371–374, 381, 382**), black spots (violet color reaction, see Table 28.1) = 2,4-dienamides (**255, 256, 368–370, 375, 376, 379, 380**), grey spots (blue-black color reaction) = compounds **377** and **378**, dark-grey spot (dark-grey color reaction, see Table 28.1) = compounds **388** and **389**, and light-grey spot (yellow-brown color reaction) = compounds **383–387**. TLC plates redrawn from Bauer et al. (From Bauer, R. and Remiger, P., *Planta Med.*, 55, 367, 1989.)

distinguished based on their content of alkamides (Figure 28.29). From the TLC analysis of alkamides it appears that their R_f values are influenced by the stereochemistry of the double bonds, especially in the 2-position. 2-Z-oriented alkamides (e.g., **370** and **371**) show significantly higher R_f values than their corresponding 2-E-isomers (e.g., **368** and **372**). The alkamides were detected on the TLC plates by spraying with anaisaldehyde/sulfuric acid. Alkamides with 2-monoenamide structures (**262**, **264**, **371–374**, **381**, **382**) yielded characteristic yellow spots, while 2,4-dienamides gave violet color reactions (**255**, **256**, **368–370**, **375**, **376**, **379**, **380**) except for the polyenic isobutylamides **377** and **378**, which gave blue-black spots. The hydroxylated polyacetylenic ketones **388** and **389** gave dark-grey spots while the remaining polyenic or polyacetylenic ketones (**383–387**) resulted in yellow-brown color reactions (Figure 28.29).[157] In Table 28.1 characteristic color reactions are listed for selected polyacetylenes.

As described in Section 28.4.4, many polyacetylenes have antifungal and antibacterial properties. Polyacetylenes with such properties may be identified in plant extracts/fractions by thin layer autobiography. Bioautography on TLC plates involves running samples in a suitable solvent and then drying of the plate. A nutrient medium is then inoculated with the microorganism and then sprayed on the plate. Using the plant pathogenic fungus *Cladosporium cucumerinum*, white colored areas of inhibition are visible against a grey background of fungal spores after incubation,[314] which has resulted in the identification of dehydrofalcarinol and dehydrofalcarindiol as the main antifungal compounds in *Artemisia borealis*, inhibiting the growth of *C. cucumerinum* in amounts of 1.25 µg.[243] Bioautography TLC has also been used to identify falcarinol and falcarindiol as the main antibacterial compounds in the dichloromethane extract of *Chaerophyllum aureum*.[251] For the detection and identification of these antibacterial compounds the developed plates were overlaid with nutrition medium, which was inoculated with suspensions of the test organism and then incubated. After incubation, bacterial growth was detected by spraying with a NaCl solution containing a tetrazolium salt. Dehydrogenase activity of the living bacteria converted the tetrazolium salt into violet-colored formazan, whereas spots with antibacterial activity remained as clear zones, identified by comparison with a reference plate sprayed with vanillin/sulfuric acid.[251]

The advantage of TLC bioautographic methods is the potential for determination of bioactivities at specific spots on the plate. This simplifies the location of the bioactive components and helps the design of subsequent isolation strategy by preparative TLC, CC, or HPLC. So far TLC bioautography has only been applied for the isolation of antifungal and antibacterial polyacetylenes but may also be used to identify polyacetylenes with other bioactives which, however, implies the use of other spraying reagents.

28.5.5 COLUMN CHROMATOGRAPHY

Column chromatography (CC) is usually the first step in the separation of polyacetylenes in plant extracts. For the separation of polyacetylenes the preferred method is flash CC described by Stahl, Kahn, and Mitra,[315] which allows separation of grams of extract within an hour or two. Moreover, the method has the advantage over the

TABLE 28.1
Examples of Color Reactions of Polyacetylenic Compounds with Spray Reagents Commonly Used to Visualize Polyacetylenes on TLC Plates

| Polyacetylenes | Spray Reagents[a] | | | |
	Vanillin	p-Dimethylamino-benzaldehyde	Anisaldehyde	KMnO$_4$
Falcarinol (**22**)	Greyish brown	Brown	—[b]	Yellow
Falcarindiol (**23**)	Greyish brown	Dark brown	—	Yellow
(Z)-Lachnopyllum ester (**117**)	Dark blue	Grey	—	Yellow
122	Olive	Greyish brown	—	Yellow
Safynol (**123**)	Greyish brown	Olive	—	Yellow
142	Greyish brown	Yellowish brown	—	Yellow
145	Olive brown and deep magenta	Olive and greyish ruby	—	Yellow
150	Brownish beige	Olive yellow	—	Yellow
156	Dull blue	Olive yellow	—	Yellow
Centaur X$_4$ (**181**)	Grey	Yellowish brown	—	Yellow
191	Reddish grey	Light brown	—	Yellow
210 (Z-isomer)	Olive brown	Olive brown	—	Yellow
210 (E-isomer)	Deep violet	Greyish violet	—	Yellow
229	Dark yellow	Olive yellow	—	Yellow
230	Brownish beige	Greyish yellow	—	Yellow
PHT (**231**)	Dark blue	Greyish yellow	—	Yellow
Ichthyothereol (**328**)	Greyish brown	Brownish orange	—	Yellow
342	Greenish grey	Greenish grey	—	Yellow
388, 389			Dark-grey	Yellow
2-Monoenamides (e.g., **264, 381**)	—	—	Yellow	Yellow
2,4-Dienamides (e.g., **256, 257**)	—	—	Violet	Yellow

[a] Color reactions taken from Picman, A.K. et al., *J. Chromatogr.*, 189, 187, 1980; Bauer, R. and Remiger, P., *Planta Med.*, 55, 367, 1989; and Christensen, L.P., unpublished data.
[b] Not investigated.

conventional open-column chromatography, in that a small pressure of nitrogen is used to protect against oxidation and hence reduce the formation of artifacts during column separation. Also the shorter separation time reduce the possibility for hydrolysis of the polyacetylenes. Separation of polyacetylenes by flash CC or open-column chromatography is usually performed on silica gel 60, 40–63 μm (230–400 mesh) from Merck.[110,215,308,310] The solvents used for development on the column are *n*-hexane or petroleum ether at the beginning with an increasing amount of Et$_2$O, EtOAc, CH$_2$Cl$_2$, or methanol.[28,89,110,113,120,127,130,132,199,215,304,305,310,311] Finally, 100% methanol is often used to elute the most polar compounds from the column. As very unstable and light sensitive polyacetylenes may be present in the individual fractions it is very important that they are stoppered immediately and protected against UV light by use of tin foil around each fraction, which again should be stored in a refrigerator.

28.5.6 ANALYTICAL AND PREPARATIVE HPLC

Numerous publications have described the analysis of polyacetylenes by analytical or preparative HPLC, of which the majority of publications deal with the use of reversed phase (RP) C18 columns such as LiChrospher 100 RP-18,[117,156,157,233,303,316] Zorbax Rx-C18,[46,48] Sperisorb 5S or Develosil ODS,[55,71,155] Varian RP-MCH-10,[184,187–189,191] LiChrosorb RP-18,[158,233,241,317] and Luna 3μ C18(2) 100A.[4,318] In most cases, RP-HPLC separations are performed by gradient elution using different proportions of methanol/water or acetonitrile/-water as mobile phases. Polyacetylenes are normally easy to identify in extracts by diode array detection (DAD) because of their often characteristic UV spectra.[4,8,55,71,110] The detection wavelength(s) depend on the nature of the polyacety-lenes. For example, polyacetylenes of the falcarinol-type are detected at 205 nm, although their characteristic UV maxima occur above 225 nm (Figure 28.30).

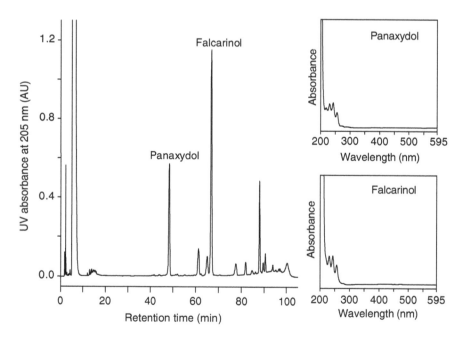

FIGURE 28.30 Analytical RP-HPLC chromatogram of an ethyl acetate extract of an American ginseng root showing a good separation of the two major polyacetylenes falcarinol (**22**) and panaxydol (**32**) and their characteristic UV-spectra. Separations performed on an RP-Luna 3μ C18(2) 100A column (3 μm; 150 × 4.6 mm i.d.) at 40°C using the following solvent gradient: CH₃CN–H₂O [0–5 min (20:80), 10 min (50:50), 30 min (53:47), 45–50 min (65:35), 70–72 min (75:25), 90–95 min (95:5), 100–110 min (20:80)]. Flow rate: 1 mL/min. Injection volume: 20 μL. Detection: DAD from 200–595 nm and acquisition off at 105 min.

Falcarinol and related polyacetylenes have in most cases only two conjugated unsaturated bonds in their structures and therefore the excitation coefficients ($\varepsilon < 6000$ for two triple bonds in conjugation) of these compounds at their characteristic UV-maxima is low.[4,8,55,71] By detecting the compounds at 205 nm instead, the UV sensitivity is improved approximately 10 times and hence the detection of these polyacetylenes. A typical analytical RP-HPLC chromatogram at 205 nm of an EtOAc extract of the roots of American ginseng (*Panax quinquefolium*) is shown in Figure 28.30, with falcarinol and panaxydol being the major polyacetylenes. Quantification of polyacetylenes by analytical RP-HPLC is performed by using an appropriate internal standard[46] or using a calibration curve of authentic polyacetylene standards.[315,317]

After separation of the polyacetylenes from extracts by CC the polyacetylenes can be directly isolated from the crude fractions by preparative HPLC techniques.[4,48,77,94,95,263,315] Preparative or semipreparative RP-columns such as ODS-Hypersil, Ultracarb-ODS, and ODS-HG-5 has, for example, been shown to be very efficient in separating polyacetylenes of the falcarinol-type from crude fractions.[4,48,263,317,319]

28.6 CONCLUSION

Polyacetylenes constitute a class of highly unsaturated secondary metabolites with a relatively limited distribution in higher plants being mainly present in the plant families Apiaceae, Araliaceae, and Asteraceae. The limited distribution of polyacetylenes in the plant kingdom makes them important as chemotaxonomic markers. However, the most interesting feature about the polyacetylenes is their various pharmacological activities, which may be used in the development of effective herbal medicines and drugs for the prevention of, and cures against, various diseases. A particular interesting class of polyacetylenes is the highly bioactive polyacetylenes of the falcarinol-type that occur in many important medicinal plants and are constituents of important food plants, such as carrots, parsnip, parsley, celery, and celeriac.[4] The traditional view of polyacetylenes in food as generally undesirable toxicants may therefore need to be revised and perhaps these compounds may instead be regarded as important neutraceuticals. In any case, it is important to have accurate and reliable analytical methods for qualitative and quantitative measurements in extracts. The present review has clearly demonstrated that TLC methods constitute an important part of these analytical tools. The major challenge with regard to bioactive polyacetylenes is to test their pharmacological activities in vivo, in reclinical as well as clinical studies. This will require large amounts of purified polyacetylenes. The present review has also demonstrated that large scale isolation of these compounds can be performed effectively by simple preparative TLC methods and other isolation techniques such as preparative HPLC.

REFERENCES

1. Dembitsky, V.M., Anticancer activity of natural and synthetic acetylenic lipids, *Lipids*, 41, 883, 2006.
2. Dembitsky, V.M. and Levitsky, D.O., Acetylenic terrestrial anticancer agents, *Nat. Prod. Commun.*, 1, 405, 2006.

3. Dembitsky, V.M. et al., Acetylenic aquatic anticancer agents and related compounds, *Nat. Prod. Commun.*, 1, 773, 2006.

4. Christensen, L.P. and Brandt, K., Bioactive polyacetylenes in food plants of the Apiaceae family: occurrence, bioactivity and analysis, *J. Pharm. Biomed. Anal.*, 41, 683, 2006.

5. Christensen, L.P., Biological activities of naturally occurring acetylenes and related compounds from higher plants, *Rec. Res. Dev. Phytochem.*, 2, 227, 1998.

6. Bohlman, F., Naturally-occurring acetylenes, in Lam, J. et al., Eds., *Chemistry and Biology of Naturally-Occurring Acetylenes and Related Compounds (NOARC), Bioactive Molecules, Vol. 7*, Elsevier, Amsterdam, 1988, p. 1–19.

7. Hansen, L. and Boll, P.M., Polyacetylenes in Araliaceae: their chemistry, biosynthesis and biological significance, *Phytochemistry*, 25, 285, 1986.

8. Bohlmann, F., Burkhardt, T., and Zdero, C., *Naturally Occurring Acetylenes*, Academic Press, London, 1973, p. 547.

9. Christensen, L.P., Acetylenes and related compounds in Anthemideae, *Phytochemistry*, 31, 7, 1992.

10. Christensen, L.P. and Lam, J., Acetylenes and related compounds in Heliantheae, *Phytochemistry*, 30, 11, 1991.

11. Christensen, L.P. and Lam, J., Acetylenes and related compounds in Astereae, *Phytochemistry*, 30, 2453, 1991.

12. Christensen, L.P. and Lam, J., Acetylenes and related compounds in Cynareae, *Phytochemistry*, 29, 2753, 1990.

13. Hudson, J.B. and Towers, G.H.N., Therapeutic potential of plant photosensitizers, *Pharm. Ther.*, 49, 181, 1991.

14. Hudson, J.B. et al., Ultraviolet-dependent biological activities of selected polyines of the Asteraceae, *Planta Med.*, 57, 69, 1991.

15. Hudson, J.B., Plant photosensitizers with antiviral properties, *Antiviral Res.*, 12, 55, 1989.

16. Bohlmann, F., Acetylenic compounds in the Umbelliferae, in Heywood, V.H., Ed., *The Biology and Chemistry of the Umbelliferae*, Academic Press, London, 1971, p. 279.

17. Zgoda, J.R. et al., Polyacetylene carboxylic acids from *Mitrephora celebica*, *J. Nat. Prod.*, 64, 1348, 2001.

18. Cavin, A. et al., Use of on-flow LC/^{1}H-NMR for the study of an antioxidant fraction from *Orophea enneandra* and isolation of a polyacetylene, lignans, and a tocopherol derivative, *J. Nat. Prod.*, 61, 1497, 1998.

19. Etse, J.T. and Waterman, P.G., Alkaloids and an acetylenic lactone from the stem bark of *Sapranthus palanga*, *Phytochemistry*, 25, 1903, 1986.

20. Gopinath, K.W. et al., Polyacetylene compounds. 237. Polyynes from *Alphonsea ventricosa* Hook F. and Th., *Tetrahedron*, 32, 737, 1976.

21. Ishimaru, K. et al., Polyacetylene glycosides from *Pratia nummularia* cultures, *Phytochemistry*, 61, 643, 2003.

22. Tanaka, N. et al., Secondary metabolites in transformed root cultures of *Campanula glomerata*, *J. Plant Physiol.*, 155, 251, 1999.

23. Brandt, K. and Ishimaru, K., Campanula (Bellflower) species: in vitro culture, micropropagation, and the production of anthocyanins, polyacetylenes, and other secondary metabolites, in Bajaj, Y.P.S., Ed., *Biotechnology in Agriculture and Forestry, Vol. 41—Medicinal and Aromatic Plants X*, Springer-Verlag, Berlin, Heidelberg, 1998. Chapter IV.

24. Ando, M. et al., Polyacetylene production in hairy root cultures of *Wahlenbergia marginata*, *J. Plant Physiol.*, 151, 759, 1997.

25. Tada, H., Shimomura, K., and Ishimaru, K., Polyacetylenes in *Platycodon grandiflorum* hairy root and Campanulaceous plants, *J. Plant Physiol.*, 145, 7, 1995.

26. Badanyan, S.O. et al., Natural acetylenes. XXXVII. Polyacetylenes from the campanulaceae plant family. Tetrahydropyranyl and open chain C_{14} polyacetylenic alcohols from *Campanula pyramidalis* L. and *Campanula medium* L., *J. Chem. Soc. Perkin Trans.*, 1, 145, 1973.

27. Lam, J. and Kaufmann, F., Acetylenic constituents of the family Campanulaceae, *Chem. Ind.*, 1430, 1969.

28. Malgalhães, A.F. et al., C_{14} Polyacetylenes from Brazilian Lobelioideae, *Phytochemistry*, 27, 3827, 1988.

29. Ohashi, K. et al., Indonesian medicinal plants. XXV. Cancer cell invasion inhibitory effects of chemical constituents in the parasitic plant *Scurrula atropurpurea* (Loranthaceae), *Chem. Pharm. Bull.*, 51, 343, 2003.

30. Wakabayashi, N., Spencer, S.L., and Waters, R.M., A polyacetylene from Honduras mahogany, *Swietenia mahagoni*, *J. Nat. Prod.*, 54, 1419, 1991.

31. Ohashi, K. et al., Preparation and cancer cell invasion inhibitory effects of C_{16}-alkynic fatty acids, *Chem. Pharm. Bull.*, 51, 463, 2003.

32. Ito, A. et al., Cytotoxic polyacetylenes from the twigs of *Ochanostachys amentacea*, *J. Nat. Prod.*, 64, 246, 2001.

33. Rashid, M.A. et al., Absolute stereochemistry and anti-HIV activity of minquartynoic acid, a polyacetylene from *Ochanostachys amentacea*, *Nat. Prod. Lett.*, 15, 21, 2001.

34. Fort, D.M. et al., Minquartynoic acid from *Coula edulis*, *Biochem. Syst. Ecol.*, 28, 489, 2000.

35. Kraus, C.M. et al., New acetylenes isolated from the bark of *Heisteria auminata*, *J. Nat. Prod.*, 61, 422, 1998.

36. El-Seedi, H.R., Hazell, A.C., and Torssell, K.B.G., Triterpenes, lichexanthone and an acetylenic acid from *Minquartia guianensis*, *Phytochemistry*, 35, 1297, 1994.

37. Marles, R.J., Farnsworth, N.R., and Neill, D.A., Isolation of a novel cytotoxic polyacetylene from a traditional anthelmintic medicinal plant, *Minquartia guianensis*, *J. Nat. Prod.*, 52, 261, 1989.

38. Miller, R.W. et al., Oxygenated fatty acids of isano oil, *Phytochemistry*, 16, 947, 1977.

39. Fullas, F. et al., Gummiferol, a cytotoxic polyacetylene from the leaves of *Adenia gummifera*, *J. Nat. Prod.*, 58, 1625, 1995.

40. Bohlmann, F. and Zdero, C., Polyacetylenic compounds. 236. Further acetylenic compounds from the family Pittosporaceae, *Chem. Ber.*, 108, 2541, 1975.

41. Imoto, S. and Ohta, Y., Elicitation of diacetylenic compounds in suspension cultured cells of eggplant, *Plant Physiol.*, 86, 176, 1988.

42. Elgersma, D.M. et al., Occurrence of falcarinol and falcarindiol in tomato plants after infection with *Verticillium albo-atrum* and charcterization of four phytoalexins by capillary gas chromatography–mass spectrometry, *Phytopathol. Z.*, 109, 237, 1984.

43. De Wit, P.J.G.M. and Kodde, E., Induction of polyacetylenic phytoalexins in *Lycopersicon esculentum* after inoculation with *Cladosporium fulvum* (syn. *Fulvia fulva*), *Physiol. Plant Pathol.*, 18, 143, 1981.

44. Pan, W.D. et al., Two new naturally occurring optical polyacetylene compounds from *Torricellia angulata* var. *intermedia* and the determination of their absolute configurations, *Nat. Prod. Res.*, 20, 1098, 2006.

45. Deng, S.X. et al., Serotonergic activity-guided phytochemical investigation of the roots of *Angelica sinensis*, *J. Nat. Prod.*, 69, 536, 2006.

46. Zidorn, C. et al., Polyacetylenes from the Apiaceae vegetables carrot, celery, fennel, parsley, and parsnip and their cytotoxic activities, *J. Agric. Food Chem.*, 53, 2518, 2005.

47. Lechner, D. et al., The anti-staphylococcal activity of *Angelica dahurica* (Bai Zhi), *Phytochemistry*, 65, 331, 2004.
48. Zidorn, C. et al., Chemosystematic investigations of irregular diterpenes in Anisotone and related New Zealand Apiaceae, *Phytochemistry*, 59, 293, 2002.
49. Hiraoka, N. et al., Furanocoumarin and polyacetylenic compound composition of wild *Glehnia littoralis* in North America, *Biochem. Syst. Ecol.*, 30, 321, 2002.
50. Perry, N.B., Span, E.M., and Zidorn, C., Aciphyllal—a C_{34} from *Aciphylla scott-thomsonii* (Apiaceae), *Tetrahedron Lett.*, 42, 4325, 2001.
51. Uwai, K. et al., Exploring the structural basis of neurotoxicity in C_{17}-polyacetylenes isolated from water hemlock, *J. Med. Chem.*, 43, 4508, 2000.
52. Fujioka, T. et al., Antiproliferative constituents from Umbelliferae plants. V. A new furanocoumarin and falcarindiol furanocoumarin ethers from the root of *Angelica japonica*, *Chem. Pharm. Bull.*, 47, 96, 1999.
53. Nakano, Y. et al., Antiproliferative constituents in Umbelliferae plants. II. Screening for polyacetylenes in some Umbelliferae plants, and isolation of panaxynol and falcarindiol from the root of *Heracleum moellendorffii*, *Biol. Pharm. Bull.*, 21, 257, 1998.
54. Liu, J.H. et al., Inhibitory effects of *Angelica pubescens* f. *biserrata* on 5-lipoxygenase and cyclooxygenase, *Planta Med.*, 64, 525, 1998.
55. Wittstock, U. et al., Polyacetylenes from water hemlock, *Cicuta virosa*, *Planta Med.*, 61, 439, 1995.
56. Fujita, T. et al., A new phenylpropanoid glucoside and other constituents of *Oenanthe javanica*, *Biosci. Biotech. Biochem.*, 59, 526, 1995.
57. Wittstock, U., Lichtnow, K.H., and Teuscher, E., Effects of cicutoxin and related polyacetylenes from *Cicuta virosa* on neuronal action potentials: a comparative study on the mechanism of the convulsive action, *Planta Med.*, 63, 120, 1997.
58. González, A. et al., Biological activity of secondary metabolites from *Bupleurum salicifolium*, *Experientia*, 51, 35, 1995.
59. Cunsolo, F. et al., Bioactive metabolites from sicilian marine fennel, *Crithmum maritimum*, *J. Nat. Prod.*, 56, 1598, 1993.
60. Appendino, G. et al., An anti-platelet acetylene from the leaves of *Ferula communis*, *Fitoterapia*, 64, 179, 1993.
61. Lam, J., Christensen, L.P., and Thomasen, T., Acetylenes from the roots of *Eryngium bourgatii*, *Phytochemistry*, 31, 2881, 1992.
62. Appendino, G. et al., A sesquiterpene ketal from *Bunium paucifolium*, *Phytochemistry*, 30, 3467, 1991.
63. Hu, C.-Q., Chang, J.-J., and Lee, K.-H., Antitumor agents, 115. Seselidiol, a new cytotoxic polyacetylene from *Seseli mairei*, *J. Nat. Prod.*, 53, 932, 1990.
64. Konoshima, T. and Lee, K.-H., Antitumor agents, 85. Cicutoxin, an antileukemic principle from *Cicuta maculata*, and the cytotoxicity of the related derivatives, *J. Nat. Prod.*, 49, 1117, 1986.
65. Kern, J.R. and Cardellina, II, J.H., Native American medicinal plants. Falcarindiol and 3-O-methylfalcarindiol from *Osmorhiza occidentalis*, *J. Nat. Prod.*, 45, 774, 1982.
66. Kemp, M.S., Falcarindiol: an antifungal polyacetylene from *Aegopodium podagraria*, *Phytochemistry*, 17, 1002, 1978.
67. Schulte, K.E. and Pötter, B., Polyacetylene aus *Pituranthus tortuosus* (Desf.) Bnth. u. Hook, *Arch. Pharm.*, 310, 945, 1977.
68. Bohlmann, F., Grenz, M., and Zdero, C., Neue Cumarine aus *Peucedanum*- und *Pteronia*-Arten, *Chem. Ber.*, 108, 2955, 1975.
69. Bohlmann, F., Zdero, C., and Suwita, A., Ein neues Sesquiterpen-Chinon aus *Seseli*-Arten, *Chem. Ber.*, 108, 2818, 1975.

70. Bohlmann, F. and Zdero, C., Ein neues Polyin aus *Centella*-Arten, *Chem. Ber.*, 108, 511, 1975.

71. Hadaček, F., Werner, A., and Greger, H., Computerized HPLC—diode array screening on characteristic acetylene patterns within *Peucedanum* (Umbelliferae–Apioideae), in Lam, J. et al., Eds., *Chemistry and Biology of Naturally-Occurring Acetylenes and Related Compounds (NOARC), Bioactive Molecules*, Vol. 7, Elsevier, Amsterdam, 1988, p. 107–114.

72. Barrero, A.F. et al., Polyacetylenes, terpenoids and flavonoids from *Bupleurum spinosum*, *Phytochemistry*, 48, 1237, 1998.

73. Morita, M. et al., Polyacetylenes from the roots of *Bupleurum falcatum*, *Phytochemistry*, 30, 1543, 1991.

74. Vincieri, F.F. et al., Oxygenated C_{15}-polyacetylenes from *Oenanthe aquatica* fruits hydrocarbons, *Planta Med.*, 51, 107, 1985.

75. Vincieri, F.F. et al., Isolation and structural elucidation of *Oenanthe aquatica* (L.) fruit C_{15} polyacetylene hydrocarbons, *Chem. Ber.*, 114, 468, 1981.

76. Dang, N.H. et al., Inhibitory constituents against cyclooxygenases from *Aralia cordata* Thunb., *Arch. Pharm. Res.*, 28, 28, 2005.

77. Bernart, M.W. et al., Cytotoxic falcarinol oxylipins from *Dendropanax arboreus*, *J. Nat. Prod.*, 59, 748, 1996.

78. Setzer, W.N. et al., A cytotoxic diacetylene from *Dendropanax arboreus*, *Planta Med.*, 61, 470, 1995.

79. Magalhães, A.F. et al., Polyacetylenes of Brazilian species of *Didymopanax* (Araliaceae), *Biochem. Syst. Ecol.*, 20, 783, 1992.

80. Terada, A., Tanoue, Y., and Kishimoto, D., (−)-(9Z)-1,9-Heptadecadiene-4,6-diyne-3-ol as a principal component of the resinous sap of *Evodiopanax innovans* Nakai, Japanese name Takanotsume, and its role in an ancient golden varnish of Japan, *Bull. Chem. Soc. Jpn.*, 62, 2977, 1989.

81. Boll, P.M. and Hansen, L., On the presence of falcarinol in Araliaceae, *Phytochemistry*, 26, 2955, 1987.

82. Christensen, L.P., Lam, J., and Thomasen, T., Polyacetylenes from the fruits of *Hedera helix*, *Phytochemistry*, 30, 4151, 1991.

83. Gafner, F., Reynolds, G.W., and Rodriguez, E., The diacetylene 11,12-dehydrofalcarinol from *Hedera helix*, *Phytochemistry*, 28, 1256, 1989.

84. Bruhn, G. et al., Naürliche Allergene. I. Das Autreten von Falcarinol und didehydrofalcarinol in Efeu (*Hedera helix* L.), *Z. Naturforsch.*, 42b, 1328, 1987.

85. Hausen, B.M. et al., Allergic and irritant contact dermatitis from falcarinol and didehydrofalcarinol in common ivy (*Hedera helix* L.), *Contact Dermatitis*, 17, 1, 1987.

86. Kobaisy, M. et al., Antimycobacterial polyynes of Deveil's Club (*Oplopanax horridus*), a North American naitive medicinal plant, *J. Nat. Prod.*, 60, 1210, 1997.

87. Christensen, L.P., Jensen, M., and Kidmose, U., Simultaneous determination of ginsenosides and polyacetylenes in American ginseng root (*Panax quinquefolium* L.) by high-performance liquid chromatography, *J. Agric. Food Chem.*, 54, 8995, 2006.

88. Hirakura, K. et al., Linoleoylated polyacetylenes from the root of *Panax ginseng*, *Phytochemistry*, 35, 963, 1994.

89. Fujimoto, Y. et al., Polyacetylenes from *Panax quinquefolium*, *Phytochemistry*, 35, 1255, 1994.

90. Lutomski, J., Luan, T.C., and Hoa, T.T., Polyacetylenes in the Araliaceae family. IV. The antibacterial and antifungal activities of two main polyacetylenes from *Panax vietnamensis* Ha et Grushv. and *Polyscias fruticosa* (L.) Harms, *Herba Pol.*, 38, 137, 1992.

91. Fujimoto, Y. et al., Acetylenes from *Panax quinquefolium, Phytochemistry*, 31, 3499, 1992.

92. Hirakura, K. et al., Three acetylenic compounds from the roots of *Panax ginseng, Phytochemistry*, 31, 899, 1992.

93. Hirakura, K. et al., Three acylated polyacetylenes from the roots of *Panax ginseng, Phytochemistry*, 30, 4053, 1991.

94. Hirakura, K. et al., Polyacetylenes from the roots of *Panax ginseng, Phytochemistry*, 30, 3327, 1991.

95. Fujimoto, Y. et al., Cytotoxic acetylenes from *Panax quinquefolium, Chem. Pharm. Bull.*, 39, 521, 1991.

96. Ahn, B.-Z., Kim, S.-I., and Lee, Y.-H., Acetylpanaxydol und Panaxydolchlorhydrin, zwei neue, gegen L1210-Zellen cytotoxische Polyine aus Koreanischen Ginseng, *Arch. Pharm.*, 322, 223, 1989.

97. Fujimoto, Y. and Satoh, M., A new cytotoxic chlorine-containing polyacetylene from the callus of *Panax ginseng, Chem. Pharm. Bull.*, 36, 4206, 1988.

98. Ahn, B.-Z. and Kim, S.-I., Heptadeca-1,8t-dien-4,6-diin-3,10-diol, ein weiteres, gegen L1210-Zellen cytotoxisches Wirkprinzip aus der Koreanischen Ginseng-wurzel, *Planta Med.*, 54, 183, 1988.

99. Ahn, B.-Z. and Kim, S.-I., Beziehung zwischen Struktur und cytotoxischer Aktivität von Panaxydol-Analogen gegen L1210 Zellen, *Arch. Pharm.*, 321, 61, 1988.

100. Shim, S.K. et al., A polyacetylenic compound from *Panax ginseng* roots, *Phytochemistry*, 26, 2849, 1987.

101. Hansen, L., Hammershøy, O., and Boll, P.M., Allergic contact dermatitis from falcarinol isolated from *Schefflera arboricola, Contact Dermatitis*, 14, 91, 1986.

102. Kim, S.I. et al., Panaxyne, a new cytotoxic polyyne from *Panax ginseng* root against L1210 cell, *Saengyak Hakhoechi (S. Korea)*, 20, 71, 1989.

103. Kim, S.I., Kang, K.S., and Lee, Y.H., Panaxyne epoxide, a new cytotoxic polyyne from *Panax ginseng* root against L1210 cell, *Arch. Pharm. Res. (S. Korea)*, 12, 48, 1989.

104. Park, B.-Y. et al., Isolation and anticomplement activity of compounds from *Dendropanax morbifera, J. Ethnopharmacol.*, 90, 403, 2004.

105. Oka, K. et al., The allergens of *Dendropanax trifidus* Makino. and *Fatsia japonica* Decne. et Planch. and evaluation of cross-reactions with other plants of the Araliaceae family, *Contact Dermatitis*, 40, 209, 1999.

106. Oka, K. et al., The major allergen of *Dendropanax trifidus* Makino, *Contact Dermatitis*, 36, 252, 1997.

107. Papajewski, S. et al., Bioassay guided isolation of a new C_{18}-polyacetylenes, (+)-9 (Z),17-Octadecadiene-12,14-diyne-1,11,16-triol, from *Cussonia barteri, Planta Med.*, 64, 479, 1998.

108. Bremer, K., *Asteraceae: Cladistics and Classification*, Timber Press, Portland, Oregon, 1994.

109. Wallnöfer, B., The polyacetylenes within the *Artemisia*-"Vulgares"-group (Anthemideae-Compositae) in Morawtz, W., Ed., *Biosystematics and Ecology Series No. 7*, Austrian Academy of Sciences Press, Wien, Austria, 1994.

110. Lam, J. and Hansen, L., Polyacetylenes and related compounds: analytical and chemical methods, in Bowyer, J.R. and Harwood, J., Eds., *Methods in Plant Biochemistry, Vol. 4*, Academic Press, London, 1990. Chapter 6.

111. Hadaček, F., Vergleichende phytochemische Untersuchungen in der Gattung *Peucedanum* (Apiaceae–Apioideae), *Stapfia*, 18, 190, 1989.

112. Gonser, P. et al., A diels–alder adduct and other constituents from *Cineraria* species, *Phytochemistry*, 29, 3940, 1990.

113. Lehmann, L. et al., Acetylenic compounds and other constituents from *Cineraria* species, *Phytochemistry*, 27, 3307, 1988.
114. Bohlmann, F. and Zdero, C., Polyacetylenic compounds. 267. Biogenetically interesting acetylenic compounds from *Cineraria saxifraga* DC., *Liebigs Ann. Chem.*, 885, 1983.
115. Bohlmann, F., Singh, P., and Jakupovic, J., Sesquiterpenes and a dimeric spiroketone from *Cineraria fruticulorum*, *Phytochemistry*, 21, 2531, 1982.
116. Bohlmann, F. and Abraham, W.-R., Neue sesquiterpene und acetylenverbindungen aus *Cineraria*-arten, *Phytochemistry*, 17, 1629, 1978.
117. Pellati, F. et al., Isolation and structure elucidation of cytotoxic polyacetylenes and polyenes from *Echinacea pallida*, *Phytochemistry*, 67, 1359, 2006.
118. Fei, Z., Kong, L., and Peng, S., Progress in chemical and pharmacological studies on *Cirsium japonicum*, *Zhongcaoyao*, 32, 664, 2001.
119. Takaishi, Y. et al., Absolute configuration of a triolacetylene from *Cirsium japonicum*, *Phytochemistry*, 30, 2321, 1991.
120. Takaishi, Y. et al., Acetylenes from *Cirsium japonicum*, *Phytochemistry*, 29, 3849, 1990.
121. Yim, S.H., Kim, H.J., and Lee, I.S., A polyacetylene and flavonoids from *Cirsium rhinoceros*, *Arch. Pharm. Res.*, 26, 128, 2003.
122. Jung, H.J. et al., Gymnasterkoreaynes A–F, cytotoxic polyacetylenes from *Gymnaster koraiensis*, *J. Nat. Prod.*, 65, 897, 2002.
123. Matsumoto, A. et al., Antitumor polyacetylene extraction from plants, Japanese patent JP 03287532 A2 19911218 Heisei, 1991.
124. Ionkova, I. and Alferman, A., Use of DNA for detection and isolation of potential anticancer agents from plants, *Farmatsiya (Sofia)*, 47, 10, 2000.
125. Matsunaga, H. et al., Inhibitory effect of cis-dehydromatricaria ester from *Solidago altissima* on the growth of mammalian cells, *Chem. Pharm. Bull.*, 38, 3483, 1990.
126. Wurz, G. et al., Absolute configurations of three naturally occurring acetylenic spiroacetal enol ethers, *Liebigs Ann. Chem.*, 99, 1993.
127. Tan, R.X. et al., Sesquiterpenes and acetylenes from *Artemisia feddei*, *Phytochemistry*, 31, 3135, 1992.
128. Calzado, M.A. et al., Inhibition of NF-κB activation and expression of inflammatory mediators by polyacetylenes spiroketals from *Plagius flosculosus*, *Biochim. Biophys. Acta*, 1729, 88, 2005.
129. Wu, T.S. et al., Phenylalkynes from *Artemisia capillaris*, *Phytochemistry*, 47, 1645, 1998.
130. Yano, K. and Ishizu, T., Capillene, a seed germination inhibitor from *Artemisia capillaris* roots, *Phytochemistry*, 37, 689, 1994.
131. Jakupovic, J. et al., Acetylenes and other constituents from *Artemisia dracunculus*, *Planta Med.*, 57, 450, 1991.
132. Greger, H., Aromatic acetylenes and dehydrofalcarinone derivatives within the *Artemisia dracunculus* group, *Phytochemistry*, 18, 1319, 1979.
133. Towers, G.H.N. and Wat, C.-K., Biological activity of polyacetylenes, *Rev. Latinoam. Quim.*, 9, 162, 1978.
134. Fukumaru, T. et al., Synthesis and bioactivity of novel acetylenic compounds, *Agric. Biol. Chem.*, 39, 519, 1975.
135. Yamamoto, M. et al., The herbal medicine inchin-ko-to inhibits liver cell apoptosis induced by transforming growth factor beta 1, *Hepatology*, 23, 552, 1996.
136. Spiridonov, N.A., Konovalov, D.A., and Arkhipov, V.V., Cytotoxicity of some Russian ethnomedicinal plants and plant compounds, *Phytother. Res.*, 19, 428, 2005.
137. Whelan, L.C. and Ryan, M.F., Effects of the polyacetylene capillin on human tumour cell lines, *Anticancer Res.*, 24, 2281, 2004.

138. Badisa, R.B. et al., Pharmacological activities of some *Argyranthemum* species growing in the Canary Islands, *Phytother. Res.*, 18, 763, 2004.
139. Gonzalez, A.G. et al., Biological activities of some *Argyranthemum* species, *Phytochemistry*, 45, 963, 1997.
140. Engelmeier, D. et al., Antifungal 3-butylisocoumarins from Asteraceae–Anthemideae, *J. Nat. Prod.*, 67, 19, 2004.
141. Riggins, C.W. and Clausen, T.P., Root acetylenes from *Artemisia arctica*, *Biochem. Syst. Ecol.*, 31, 211, 2003.
142. Greger, H., Comparative phytochemistry of the alkamides, in Lam, J. et al., Eds., *Chemistry and Biology of Naturally-Occurring Acetylenes and Related Compounds (NOARC), Bioactive Molecules, Vol. 7*, Elsevier, Amsterdam, 1988, p. 159–178.
143. Greger, H., Alkamides: structural relationships, distribution and biological activity, *Planta Med.*, 50, 366, 1984.
144. Greger, H. and Hofer, O., Highly unsaturated isopentyl amides from *Achillea wilhelmsii*, *J. Nat. Prod.*, 50, 1100, 1987.
145. Kuropka, G. and Glombitza, K.-W., Further polyenic and polyynic carboxamides and sesamin from *Achillea ptarmica*, *Planta Med.*, 53, 440, 1987.
146. Greger, H., Hofer, O., and Werner, A., Biosynthetically simple C_{18}-alkamides from *Achillea* species, *Phytochemistry*, 26, 2235, 1987.
147. Greger, H., Zdero, C., and Bohlmann, F., Pyrrolidine and piperidine amides from *Achillea*, *Phytochemistry*, 23, 1503, 1984.
148. Chen, Y. et al., Macrophage activating effects of new alkamides from the roots of *Echinacea* species, *J. Nat. Prod.*, 68, 773, 2005.
149. Wu, L. et al., Diacetylenic isobutylamides of *Echinacea*: synthesis and natural distribution, *Phytochemistry*, 65, 2477, 2004.
150. Binns, S.E., Arnason, J.T., and Baum, B.R., Phytochemical variation within populations of *Echinacea angustifolia* (Asteraceae), *Biochem. Syst. Ecol.*, 30, 837, 2002.
151. Clifford, L.J. et al., Bioactivity of alkamides isolated from *Echinacea purpurea* (L.) Moench., *Phytomedicine*, 9, 249, 2002.
152. Hudaib, M. et al., GC–MS analysis of the lipophilic principles of *Echinacea purpurea* and evaluation of cucumber mosaic cucomovirus (CMV) infection, *J. Pharm. Biomed. Anal.*, 29, 1053, 2002.
153. Ramsewak, R.S., Erickson, A.J., and Nair, M.G., Bioactive *N*-isobutylamides from the flower buds of *Spilanthes acmella*, *Phytochemistry*, 51, 729, 1999.
154. He, X. et al., Analysis of alkamides in roots and achenes of *Echinacea purpurea* by liquid chromatography–electrospray mass spectrometry, *J. Chromatogr. A*, 815, 205, 1998.
155. Nakatani, N. and Nagashima, M., Pungent alkamides from *Spilanthes acmella* L. var. *oleraceae* Clarke, *Biosci. Biotechnol. Biochem.*, 56, 759, 1992.
156. Bauer, R. and Foster, S., Analysis of alkamides and caffeic acid derivatives from *Echinacea simulata* and *E. paradoxa* roots, *Planta Med.*, 57, 447, 1991.
157. Bauer, R. and Remiger, P., TLC and HPLC analysis of alkamides in *Echinacea* drugs, *Planta Med.*, 55, 367, 1989.
158. Bauer, R., Remiger, P., and Wagner, H., Alkamides from the roots of *Echinacea angustifolia*, *Phytochemistry*, 28, 505, 1989.
159. Barnes, J. et al., *Echinacea* species (*Echinacea angustifolia* (DC.) Hell., *Echinacea pallida* (Nutt.) Nutt., *Echinacea purpurea* (L.) Moench): a review of their chemistry, pharmacology and clinical properties, *J. Pharm. Pharmacol.*, 57, 929, 2005.
160. Müller-Jakic, B. et al., In-vitro inhibition of cyclooxygenase and 5-lipoxygenase by alkamides from *Echinacea* and *Achillea* species, *Planta Med.*, 60, 37, 1994.

161. Bohlmann, F. et al., Synthesis of biologically-active unsaturated amides, *Tetrahedron*, 39, 123, 1983.

162. Kagan, J., Naturally occurring di- and trithiophenes, in Kagan, J. and Asselineau, J., Eds., *Progress in the Chemistry of Organic Natural Products, Vol. 56*, Springer-Verlag, New York, 1991, p. 87.

163. Bohlmann, F. and Zdero, C., Naturally occurring thiophenes, in Gronowitz, Ed., *Thiophene and its Derivatives, The Chemistry of Heterocyclic Compounds, Vol. 44(1)*, Wiley, New York, 1985, p. 261.

164. Bohlmann, F. and Zdero, C., Naturally occurring terpene derivatives. 68. New constituents of genus *Anthemis*, *Chem. Ber.*, 108, 1902, 1975.

165. Bohlmann, F. et al., Polyacetylenverbindungen. 78. Neue inhaltstoffe der Gattung *Anthemis*, *Chem. Ber.*, 98, 1616, 1965.

166. Sanz, J.F., Falcó, E., and Marco, J.A., New acetylenes from *Chrysanthemum coronarium* L., *Liebigs Ann. Chem.*, 303, 1990.

167. Bohlmann, F. and Fritz, U., Neue Lyratolester aus *Chrysanthemum coronarium*, *Phytochemistry*, 18, 1888, 1979.

168. Christensen, L.P. and Brandt, K., Acetylenes and psoralens, in Crozier, A., Clifford, M.N., and Ashihara, H., Eds., *Plant Secondary Metabolites: Occurrence, Structure and Role in the Human Diet*, Blackwell, 2006. Chapter 5.

169. Ellmauerer, E. et al., 6β-Lactonized xanthanolides from *Ratibida* species, *Phytochemistry*, 26, 159, 1987.

170. Bohlmann, F. and Zdero, C., Polyacetylenic compounds. 243. Constituents of *Coreopsis parvifolia* Blake, *Chem. Ber.*, 110, 468, 1977.

171. Bohlmann, F. and Zdero, C., Über die Inhaltstoffe aus *Eclipta erecta* L., *Chem. Ber.*, 103, 834, 1970.

172. Jakupovic, J. et al., New pseudoguaianolides from *Gaillardia megapotamica* var. *scabiosoides*, *Planta Med.*, 52, 331, 1986.

173. Bohlmann, F., Misra, L.N., and Jakupovic, J., New sesquiterpene lactone type from *Dugaldia hoopesii*, *J. Nat. Prod.*, 47, 658, 1984.

174. Bohlmann, F. and Zdero, C., Notiz über ein neues sulfoxide aus *Baileya multiradiata* Harv. et Gray, *Chem. Ber.*, 109, 1964, 1976.

175. Bohlmann, F., Rode, K.-M., and Zdero, C., Polyacetylenverbindungen. 121. Polyine aus dem Tribus Helenieae, *Chem. Ber.*, 100, 537, 1967.

176. Bohlmann, F., Singh, P., and Jakupovic, J., Further ineupatorolide-like germacranolides from *Inula cuspidata*, *Phytochemistry*, 21, 157, 1982.

177. Bohlmann, F. and Zdero, C., Polyacetylene compounds. 190. Components of *Buphthalmum salicifolium* L., *Chem. Ber.*, 104, 958, 1971.

178. Bohlmann, F. and Zdero, C., Über die Inhaltstoffe der Tribus Arctotideae, *Chem. Ber.*, 105, 1245, 1972.

179. Bohlmann, F. and Skuballa, W., Neuartige natürliche Schwefelacetylen-Verbindungen, *Chem. Ber.*, 106, 497, 1973.

180. Bohlmann, F. and Suwita, A., Weitere Inhaltsstoffe aus Arten der Tribus Arctotideae, *Chem. Ber.*, 108, 515, 1975.

181. Zid, S.A. and Orihara, Y., Polyacetylenes accumulation in *Ambrosia maritima* hairy root and cell cultures after elicitation with methyl jasmonate, *Plant Cell Tiss. Organ Cult.*, 81, 65, 2005.

182. Page, J.E. et al., Chemical basis for *Aspilia* leaf-swallowing by chimpanzees: a reanalysis, *J. Chem. Ecol.*, 23, 2211, 1997.

183. Guillet, G. et al., Multiple modes of insecticidal action of three classes of polyacetylene derivatives from *Rudbeckia hirta*, *Phytochemistry*, 46, 495, 1997.

184. Ellis, S., Balza, F., and Towers, G.H.N., A dithiacyclohexadiene polyyne alcohol from *Ambrosia chamissonis, Phytochemistry*, 33, 224, 1993.
185. Lu, T. et al., Sesquiterpenes and thiarubrines from *Ambrosia trifida* and its transformed roots, *Phytochemistry*, 33, 113, 1993.
186. Vasquez, M. et al., Sesquiterpene lactones and other constituents from *Rudbeckia mollis*, *Phytochemistry*, 31, 2051, 1992.
187. Balza, F. and Towers, G.H.N., Dithiacyclohexadiene chlorohydrins and related sulphur containing polyynes from *Ambrosia chamissonis, Phytochemistry*, 29, 2901, 1990.
188. Balza, F. et al., Dithiacyclohexadienes and thiophenes from *Ambrosia chamissonis*, *Phytochemistry*, 28, 3523, 1989.
189. Constabel, C.P., Balza, F., and Towers, G.H.N., Dithiacyclohexadienes and thiophenes from *Rudbeckia hirta, Phytochemistry*, 27, 3533, 1988.
190. Rodriguez, E. et al., Thiarubrine A, a bioactive constituent of *Aspilia* (Asteraceae) consumed by wild chimpanzees, *Experientia*, 41, 419, 1985.
191. Norton, R.A., Finlayson, A.J., and Towers, G.H.N., Two dithiacyclohexadiene polyacetylenes from *Chaenactic douglasii* and *Eriophyllum lanatum, Phytochemistry*, 24, 356, 1985.
192. Bohlmann, F. and Zdero, C., Neue guajanolide aus *Lasthenia*-Arten, *Phytochemistry*, 17, 2032, 1978.
193. Bohlmann, F., Zdero, C., and Ahmed, M., New sesquiterpene lactones, geranyllinalol derivatives and other constituents from *Geigeria* species, *Phytochemistry*, 21, 1679, 1982.
194. Towers, G.H.N. and Champagne, D.E., Medicinal phytochemistry of the Compositae: the activities of selected acetylenes and their sulfur derivatives, in Lam, J. et al., Eds., *Chemistry and Biology of Naturally-Occurring Acetylenes and Related Compounds (NOARC), Bioactive Molecules, Vol. 7*, Elsevier, Amsterdam, 1988, p. 139.
195. Chin, C. et al., A toxic C_{14} polyacetylenic tetrahydropyranyl alcohol from the Compositae, *J. Chem. Soc. Chem. Commun.*, 152, 1965.
196. Cascon, S.C. et al., Ichthyothereol and its acetate, the active polyacetylene constituents of *Ichthyothere terminalis* (Spreng.) Malme, a fish poison from the lower Amazon, *J. Am. Chem. Soc.*, 87, 5237, 1965.
197. Lam, J., Polyacetylenes of *Dahlia, Biochem. Syst.*, 1, 83, 1973.
198. Lam, J. and Kaufmann, F., Polyacetylenic C_{14}-epoxide and C_{14}-tetrahydropyranyl compounds from *Dahlia scapigera, Phytochemistry*, 10, 1877, 1971.
199. Lam, J., Kaufmann, F., and Bendixen, O., Chemical constituents of the genus *Dahlia*— III. A chemotaxonomic evaluation of some *Dahlia coccinea* strains, *Phytochemistry*, 7, 269, 1968.
200. Lam, J., Christensen, L.P., and Thomasen, T., Polyacetylenes from *Dahlia* species, *Phytochemistry*, 30, 515, 1991.
201. Bohlmann, F. et al., Germacrene and eudesmanederivatives from *Calea reticulata*, *Phytochemistry*, 21, 1793, 1982.
202. Czerson, H. et al., Sesquiterpenoid and acetylenic constituents of seven *Clibadium* species, *Phytochemistry*, 18, 257, 1979.
203. Jones, G. and Pauling, P.J., Conformations of tetrahydro-2-(non-1-en-3,5,7-triynyl)pyran-3-acetate (cunaniol acetate), Part 1: crystal and molecular structure of cunaniol acetate (violet modification), *J. Chem. Soc. Perkin Trans. II*, 1482, 1979.
204. Bohlmann, F., Zdero, C., and Kocur, J., Polyacetylenic compounds. 223. Constituents of *Chrysanthemum crassifolium* (Lange) P. Fourn., *Chem. Ber.*, 106, 3775, 1973.
205. Bohlmann, F. et al., Polyacetylenverbindungen. 56. Neue acetylenverbindungen aus *Chrysanthemum*-Arten, *Chem. Ber.*, 97, 1179, 1964.

206. Nakai, Y. et al., Effect of the rhizomes of *Atractylodes lancea* and its constituents on the delay of gastric emptying, *J. Ethnopharmacol.*, 84, 51, 2003.
207. Resch, M. et al., Further phenols and polyacetylenes from the rhizomes of *Atractylodes lancea* and their anti-inflammatory activity, *Planta Med.*, 67, 437, 2001.
208. Lehner, M., Steigel, A., and Bauer, R., Diacetoxy-substituted polyacetylenes from *Atractylodes lancea*, *Phytochemistry*, 46, 1023, 1997.
209. Bohlmann, F. et al., Neue Diterpene und neue Dihydrochalkon-derivate sowie weitere Inhaltstoffe aus *Helichrysum*-Arten, *Phytochemistry*, 19, 873, 1980.
210. Hänsel, R. et al., Neue Pyron-derivate aus *Helichrysum*-Arten, *Phytochemistry*, 19, 639, 1980.
211. Bohlmann, F. and Ziesche, J., Neue Diterpene aus *Gnaphalium*-Arten, *Phytochemistry*, 19, 71, 1980.
212. Bohlmann, F. and Abraham, W.-R., Neue Chlorsubstituerte Thiophenacetylen-verbindungen mit ungewöhnlicher Struktur aus *Helichrysum*-Arten, *Phytochemistry*, 18, 839, 1979.
213. Bohlmann, F., Schuster, A., and Meusel, H., A carlina oxide derivative from *Carlina diae*, *Phytochemistry*, 20, 823, 1981.
214. Bohlmann, F. and Rode, K.-M., Polyacetylenverbindungen. 127. Die Polyine der Gattung *Carlina* L., *Chem. Ber.*, 100, 1507, 1967.
215. Wrang, P.A. and Lam, J., Polyacetylenes from *Chrysanthemum leucanthemum*, *Phytochemistry*, 14, 1027, 1975.
216. Jente, R. et al., Experiments on biosynthesis and metabolism of acetylenes and thiophenes, in Lam, J. et al., Eds., *Chemistry and Biology of Naturally-Occurring Acetylenes and Related Compounds (NOARC), Bioactive Molecules, Vol. 7*, Elsevier, Amsterdam, 1988, p. 187–199.
217. Barley, G.C., Jones, E.R.H., and Thaller, V., Crepenynate as a precursor of falcarinol in carrot tissue culture, in Lam, J. et al., Eds., *Chemistry and Biology of Naturally-Occurring Acetylenes and Related Compounds (NOARC), Bioactive Molecules, Vol. 7*, Elsevier, Amsterdam, 1988, p. 85.
218. Jones, E.R.H., Thaller, V., and Turner, J.L., Natural acetylenes. XLVII. Biosynthetic experiments with the fungus *Lepista-diemii* (Singer). Biogenesis of C_8 acetylenic cyano acid diatretyne 2, *J. Chem. Soc. Perkin Trans. I*, 424, 1975.
219. Bu'Lock, J.D. and Smith, G.N., The origin of naturally-occurring acetylenes, *J. Chem. Soc. C*, 332, 1967.
220. Bu'Lock, J.D. and Smalley, H.M., The biosynthesis of polyacetylenes. V. The role of malonate derivatives, and the common origin of fatty acids, polyacetylenes, and "acetate-derived" phenols, *J. Chem. Soc.*, 4662, 1962.
221. Cahoon, E.B. et al., Fungal responsive fatty acid acetylenases occur widely in evolutionarily distant plant families, *Plant J.*, 34, 671, 2003.
222. Arroo, R.R.J. et al., Thiophene interconversions in *Tagetes patula* hairy-root cultures, *Phytochemistry*, 38, 1193, 1995.
223. Croes, A.F. et al., Thiophene biosynthesis in *Tagetes* roots: molecular versus metabolic regulation, *Plant Cell Tiss. Organ Cult.*, 38, 159, 1994.
224. Gomez-Barrios, M.L. et al., Studies on the biosynthesis of hiarubrine A in hairy root cultures of *Ambrosia artemisiifolia* using ^{13}C-labelled acetates, *Phytochemistry*, 31, 2703, 1992.
225. Bohlmann, F. and Dallwitz, E., Biogenesis of polyamides, *Chem. Ber.*, 107, 2120, 1974.
226. Ohta, T. et al., Absolute stereochemistry of Cicutoxin and related toxic polyacetylenic alcohols from *Cicuta virosa*, *Tetrahedron*, 55, 12087, 1999.

227. Anet, E.F.L.J. et al., Oenanthotoxin and cicutoxin. Isolation and structure, *J. Chem. Soc.*, 309, 1953.
228. Starreveld, E. and Hope, E., Cicutoxin poisoning, *Neurology*, 25, 730, 1975.
229. Crosby, D.G. and Aharonson, N., The structure of carotatoxin, a natural toxicant from carrot, *Tetrahedron*, 23, 465, 1967.
230. Yamazaki, M. et al., Effect of polyacetylenes on the neurite outgrowth of neuronal culture cells and scopolamine-induced memory impairment in mice, *Biol. Pharm. Bull.*, 24, 1434, 2001.
231. Hausen, B.M., *Allergiepflanzen—Pflanzenallergene: Handbuch und Atlas der allergie-induzierenden Wild- und Kulturpflanzen*, Ecomed Verlagsgesellschaft mbH, Landsberg/München, 1988.
232. Machado, S., Silva, E., and Massa, A., Occupational allergic contact dermatitis from falcarinol, *Contact Dermatitis*, 47, 113, 2002.
233. Redl, K. et al., Anti-inflammatory active polyacetylenes from *Bidens campylotheca*, *Planta Med.*, 60, 58, 1994.
234. Liu, J.-H., Zschocke, S., and Bauer, R., A polyacetylenic acetate and a coumarin from *Angelica pubescens* f. *biserrata*, *Phytochemistry*, 49, 211, 1998.
235. Alanko, J. et al., Panaxynol, a polyacetylene compound isolated from oriental medicines, inhibits mammalian lipoxygenases, *Biochem. Pharmacol.*, 48, 1979, 1994.
236. Kuo, S.-C. et al., Antiplatelet components in *Panax ginseng*, *Planta Med.*, 56, 164, 1990.
237. Teng, C.-M. et al., Antiplatelet actions of panaxynol and ginsenosides isolated from ginseng. *Biochim. Biophys. Acta*, 990, 315, 1989.
238. Fujimoto, Y. et al., Inhibition of 15-hydroxyprostaglandin dehydrogenase activity in rabbit gastric antral mucosa by panxynol isolated from oriental medicines, *J. Pharm. Pharmacol.*, 50, 1075, 1998.
239. Rho, M.-C. et al., Polyacetylenic compounds, ACAT inhibitors from the roots of *Panax ginseng*, *J. Agric. Food Chem.*, 53, 919, 2005.
240. Pereira, R.L.C. et al., Immunosupressive and anti-inflammatory effects of methanolic extract and the polyacetylenes isolated from *Bidens pilosa* L., *Immunopharmacology*, 43, 31, 1999.
241. Bauer, R., Redl, K., and Davis, B., Four polyacetylenes glucosides from *Bidens camphylotheca*, *Phytochemistry*, 31, 2035, 1992.
242. Avato, P. et al., Antimicrobial activity of polyacetylenes from *Bellis perennis* and their synthetic derivatives, *Planta Med.*, 63, 503, 1997.
243. Wang, Y. et al., Polyacetylenes from *Artemisia borealis* and their biological activities, *Phytochemistry*, 29, 3101, 1990.
244. Rücker, G. et al., Acetylenic glucosides from *Microglossa pyrifolia*, *Planta Med.*, 58, 266, 1992.
245. Constabel, C.P. and Towers, G.H.N., The complex nature of the mechanism of toxicity of antibiotic dithiacyclohexadiene polyines (thiarubrines) from the Asteraceae, *Planta Med.*, 55, 35, 1989.
246. McLachlan, D., Arnason, T., and Lam, J., Structure-function relationships in the photo-toxicity of acetylenes from the Asteraceae, *Biochem. Syst. Ecol.*, 14, 17, 1986.
247. Towers, G.H.N. et al., Antibiotic properties of thiarubrine A, a naturally occurring dithiacyclohexadiene polyine, *Planta Med.*, 51, 225, 1985.
248. Wat, C.-K. et al., Phototoxic effects of naturally occurring polyacetylenes and α-terthienyl on human erythrocytes, *Photochem. Photobiol.*, 32, 167, 1980.
249. Towers, G.H.N. et al., Phototoxic polyacetylenes and thiophene derivatives (effects on human skin), *Contact Dermatitis*, 5, 140, 1979.

250. Towers, G.H.N. et al., Ultraviolet-mediated antibiotic activity of species of the Compositae caused by polyacetylenic compounds, *Lloydia*, 40, 487, 1977.
251. Rollinger, J.M. et al., Lignans, phenylpropanoids and polyacetylenes from *Chaerophyllum aureum* L. (Apiaceae), *Z. Naturforsch.*, 58c, 553, 2003.
252. Olsson, K. and Svensson, R., The influence of polyacetylenes on the susceptibility of carrots to storage diseases, *J. Phytopathol.*, 144, 441, 1996.
253. Nitz, S., Spraul, M.H., and Drawert, F., C_{17} Polyacetylenic alcohols as the major constituents in roots of *Petroselinum crispum* Mill. ssp. *tuberosum*, *J. Agric. Food Chem.*, 38, 1445, 1990.
254. Garrod, B., Lewis, B.G., and Coxon, D.T., *Cis*-heptadeca-1,9-diene-4,6-diyne-3,8-diol, an antifungal polyacetylene from carrot root tissue, *Physiol. Plant Pathol.*, 13, 241, 1978.
255. Chou, S.-C. et al., Antibacterial activity of components from *Lomatium californicum*, *Phytother. Res.*, 20, 153, 2006.
256. Villegas, M. et al., Isolation of the antifungal compounds falcarindiol and sarisan from *Heteromorpha trifoliata*, *Planta Med.*, 54, 36, 1988.
257. Zhang, H. et al., Polyacetyleneginsenoside-Ro, a novel triterpene saponin from *Panax ginseng*, *Tetrahedron Lett.*, 43, 973, 2002.
258. Shim, S.C., Koh, H.Y., and Han, B.H., Polyacetylene compounds from *Panax ginseng* C.A. Meyer, *Bull. Korean Chem. Soc.*, 4, 183, 1983.
259. Matsunaga, H. et al., Cytotoxic activity of polyacetylene compounds in *Panax ginseng* C.A. Meyer, *Chem. Pharm. Bull.*, 38, 3480, 1990.
260. Matsunaga, H. et al., Studies on the panaxytriol of *Panax ginseng* C.A. Meyer. Isolation, determination and antitumor activity. *Chem. Pharm. Bull.*, 37, 1279, 1989.
261. Kim, J.Y. et al., Inhibitory effect of tumour cell proliferation and induction of G2/M cell cycle arrest by panaxytriol, *Planta Med.*, 68, 119, 2002.
262. Hirakura, K. et al., Cytotoxic activity of acetylenic compounds from *Panax ginseng*, *Nat. Med. (Tokyo)*, 54, 342, 2000.
263. Kobæk-Larsen, M. et al., Inhibitory effects of feeding with carrots or (−)-falcarinol on development of azoxymethane-induced preneoplastic lesions in the rat colon, *J. Agric. Food Chem.* 53, 1823, 2005.
264. Bae, H.O. et al., Acetylene compounds separated from *Acanthopanax senticocus* and composition containing the same to induce apoptosis, Korean patent: KP 2003059643 A 20030710, 2003.
265. Pino, J.A. et al., Leaf oil of *Dendropanax arboreus* L. from Cuba, *J. Essent. Oil Res.*, 17, 547, 2005.
266. Tanaka, S. et al., Anti-noiceptive substances from the roots of *Angelica acutiloba*, *Arzneim. Forsch.*, 27, 2039, 1977.
267. Furumi, K. et al., Novel antiproliferative falcarindiol furanocoumarin ethers from the root of *Angelica japonica*, *Bioorg. Med. Chem. Lett.*, 8, 93, 1998.
268. Kuo, Y.-C. et al., A tumor cell growth inhibitor from *Saposhnikovae divaricata*, *Cancer Invest.*, 20, 955, 2002.
269. Young, J.F. et al., Biphasic effect of falcarinol on Caco-2 cell proliferation, DNA damage and apoptosis, *J. Agric. Food Chem.* 55, 618, 2007.
270. Wu, L.-W. et al., Polyacetylenes function as anti-angiogenic agents, *Pharm. Res.*, 21, 2112, 2004.
271. Murakami, A. and Ohigashi, H., Cancer preventive potentials of edible plants from subtropical countries, *Food Style*, 3, 35, 1999.
272. Nakamura, Y. et al., Isolation and identification of acetylenic spiroketal enol ethers from *Artemisia lactiflora* as inhibitors of superoxide generation induced by a tumor promoter in differentiated HL-60 cells, *J. Agric. Food Chem.*, 46, 5031, 1998.

273. Ohigashi, H., Nakamura, Y., and Murakami, A., Active components and antitumor effects of tropical Asian foods, *Food Style*, 2, 31, 1998.

274. Nakamura, Y. et al., A diacetylenic spiroketal enol ether epoxide, AL-1, from *Artemisia lactiflora* inhibits 12-*O*-tetradecanoylphorbol-13-acetate-induced tumor promotion possibly by suppression of oxidative stress, *Cancer Lett.*, 140, 37, 1999.

275. Kagan, J. et al., Photosensitization by 2-chloro-3,11-tridecadiene-5,7,9-triyn-1-ol: damage to erythrocyte membranes, *Escherichia coli*, and DNA, *Photochem. Photobiol.*, 55, 63, 1992.

276. Hudson, J.B. and Towers, G.H.N., Antiviral properties of photosensitizers, *Photochem. Photobiol.*, 48, 289, 1988.

277. Gong, H.-H. et al., The phototoxicity of phenylheptatriyne: oxygen-dependent hemolysis of human erythrocytes and inactivation of *Escherichia coli*, *Photochem. Photobiol.*, 47, 55, 1988.

278. Kagan, J., Phenylheptatriyne: occurrence, synthesis, biological properties, and environmental concerns, *Chemosphere*, 16, 2405, 1987.

279. Towers, G.H.N. and Hudson, J.N., Potentially useful antimicrobial and antiviral phototoxins from plants, *Photochem. Photobiol.*, 46, 61, 1987.

280. Arnason, J.T. et al., Disruption of membrane functions in *Fusarium culmorum* by an acetylenic allelochemical, *Biochem. Syst. Ecol.*, 14, 569, 1986.

281. Marchant, Y.Y. and Towers, G.H.N., Phototoxicity of polyacetylenes to *Cryptococcus laurentii*, *Biochem. Syst. Ecol.*, 14, 565, 1986.

282. Hudson, J.B., Graham, E.A., and Towers, G.H.N., Investigation of the antiviral action of the photoactive compound phenylheptatriyne, *Photochem. Photobiol.*, 43, 27, 1986.

283. DiCosmo, F., Towers, G.H.N., and Lam, J., Photo-induced fungicidal activity elicited by naturally occurring thiophene derivatives, *Pest. Sci.*, 13, 589, 1982.

284. Wat, C.-K. et al., Ultraviolet-mediated cytotoxic activity of phenylheptatriyne from *Bidens pilosa* L., *J. Nat. Prod.*, 42, 103, 1979.

285. Gommers, F.J. and Geerligs, J.W.G., Lethal effect of near ultraviolet light on *Pratylenchus penetrans* from roots of *Tagetes*, *Nematologica*, 19, 389, 1973.

286. Gommers, F.J., Increase of the nematicidal activity of α-terthienyl and related compounds by light, *Nematologica*, 18, 458, 1972.

287. Hudson, J.B. et al., Antiviral properties of thiarubrine-A, a naturally occurring polyine, *Planta Med.*, 52, 51, 1986.

288. Hudson, J.B. et al., Comparison of the antiviral effects of naturally occurring thiophenes and polyacetylenes, *Planta Med.*, 52, 453, 1986.

289. Hudson, J.B. et al., Comparative anti-bacteriophage activity of naturally-occurring photosensitizers, *Planta Med.*, 53, 536, 1987.

290. Hudson, J.B. and Towers, G.H.N., Antiviral properties of acetylenes and thiophenes, in Lam, J. et al., Eds., *Chemistry and Biology of Naturally-Occurring Acetylenes and Related Compounds (NOARC)*, *Bioactive Molecules*, Vol. 7, Elsevier, Amsterdam, 1988, p. 315.

291. McLachlan, D., Arnason, T., and Lam, J., The role of oxygen in photosensitizations with polyacetylenes and thiophene derivatives, *Photochem. Photobiol.*, 39, 177, 1984.

292. Warren, R.A.J. et al., Bacteriophages as indicators of the mechanism of action of photosensitizing agents, *Photobiochem. Photobiophys.*, 1, 385, 1980.

293. MacRae, W.D. et al., Membrane lesions in human erythrocytes induced by the naturally occurring compounds α-terthienyl and phenylheptatriyne, *Photobiochem. Photobiophys.*, 1, 309, 1980.

294. Graham, K., Graham, E.A., and Towers, G.H.N., Cercaricidal activity of phenylheptatriyne and α-terthienyl, naturally occurring compounds in species of Asteraceae (Compositae), *Can. J. Zool.*, 58, 1955, 1980.

295. Yamamoto, E. et al., Photoinactivation of human erythrocyte enzymes by α-terthienyl and phenylheptatriyne, naturally occurring compounds in the Asteraceae, *FEBS Lett.*, 107, 134, 1979.
296. Kawazu, K., Nishi, Y., and Nakajima, S., Two nematicidal substances from roots of *Cirsium japonicum*, *Agric. Biol. Chem.*, 44, 903, 1980.
297. Hudson, J.B., Graham, E.A., and Towers, G.H.N., Nature of the interaction between the photoactive compound phenylheptatriyne and animal viruses, *Photochem. Photobiol.*, 36, 181, 1982.
298. MacRae, W.D. et al., Examination of naturally occurring polyacetylenes and α-terthienyl for their ability to induce cytogenic damage, *Experientia*, 36, 1096, 1980.
299. Wat, C.-K. et al., Photosensitization of invertebrates by natural polyacetylenes, *Biochem. Syst. Ecol.*, 9, 59, 1981.
300. McLachlan, D. et al., Excited-states of photoxic polyacetylenes elucidated by magnetic circular-dichroism, *Photobiochem. Photobiophys.*, 9, 233, 1985.
301. Kagan, J. and Tuveson, R.W., Are there any photocytotoxic reactions of phenylheptatriyne that are not oxygen dependent? in Lam, J. et al., Eds., *Chemistry and Biology of Naturally-Occurring Acetylenes and Related Compounds (NOARC), Bioactive Molecules, Vol. 7*, Elsevier, Amsterdam, 1988, p. 71.
302. Zschocke, S., Liu, J.-H., Stuppner, H., and Bauer, R., Comparative study of roots of *Angelica sinensis* and related Umbelliferous drugs by thin layer chromatography, high-performance liquid chromatography, and liquid chromatography–mass spectrometry, *Phytochem. Anal.*, 9, 283, 1998.
303. Bauer, R., Khan, I.A., and Wagner, H., TLC and HPLC analysis of *Echinacea pallida* and *E. angustifolia* roots, *Planta Med.*, 54, 426, 1988.
304. Bendixen, O., Lam, J., and Kaufmann, F., Polyacetylenes of *Dahlia pinnata*, *Phytochemistry*, 8, 1021, 1969.
305. Chen, Z.-L., The acetylenes from *Atractylodes macrocephala*, *Planta Med.*, 53, 493, 1987.
306. Lam, J., Separation and protection of π-electron-rich compounds on caffeine-impregnated silica gel layers, *Planta Med.*, 24, 107, 1973.
307. Lam, J. and Thomasen, T., Complexing agents for protection of highly-conjugated compounds against photodegradation, in Lam, J. et al., Eds., *Chemistry and Biology of Naturally-Occurring Acetylenes and Related Compounds (NOARC), Bioactive Molecules, Vol. 7*, Elsevier, Amsterdam, 1988, p. 47.
308. Marston, A. and Hostettmann, K., Modern separation method, *Nat. Prod. Rep.*, 8, 391, 1991.
309. Magalhães, A.F. et al., Polyacetylenes from *Pterocaulon* species, *Phytochemistry*, 28, 2497, 1989.
310. Christensen, L.P. and Lam, J., Polyacetylenes and other constituents of *Leuzea centauroides*, *Phytochemistry*, 28, 2697, 1989.
311. Andersen, A.B., Lam, J., and Wrang, P., Polyunsaturated compounds from *Centaurea scabiosa*, *Phytochemistry*, 16, 1829, 1977.
312. Stahl, E., *Thin Layer Chromatography*, Springer-Verlag, Berlin/Heidelberg/New York, 1969.
313. Picman, A.K. et al., Visualization reagents for sesquiterpene lactones and polyacetylenes on thin-layer chromatograms, *J. Chromatogr.*, 189, 187, 1980.
314. Hostettmann, K. and Marston, A., Twenty years of research into medicinal plants: results and perspectives, *Phytochem. Rev.*, 1, 275, 2002.
315. Still, W.C., Kahn, M., and Mitra, A., Rapid chromatographic technique for preparative separations with moderate resolution, *J. Org. Chem.*, 43, 2923, 1978.

316. Brandão, M.G.L. et al., Antimalarial activity of extracts and fractions from *Bidens pilosa* and other *Bidens* species (Asteraceae) correlated with the presence of acetylene and flavonoid compounds, *J. Ethnopharmacol.*, 57, 131, 1997.
317. Washida, D. and Kitanaka, S., Determination of polyacetylenes and ginsenosides in *Panax* species using high performance liquid chromatography, *Chem. Pharm. Bull.* 51, 1314, 2003.
318. Christensen, L.P. and Kreutzmann, S., Determination of polyacetylenes in carrot roots (*Daucus carota* L.) by high-performance liquid chromatography coupled with diode array detection. *J. Sep. Sci.*, 30, 483, 2007.
319. Czepa, A. and Hofmann, T., Structural and sensory characterization of compounds contributing to the bitter off-taste of carrots (*Daucus carota* L.) and carrot puree, *J. Agric. Food Chem.*, 51, 3865, 2003.

29 Quinone Derivatives in Plant Extracts

Grażyna Matysik, Agnieszka Skalska-Kamińska, and Anna Matysik-Woźniak

CONTENTS

29.1 INTRODUCTION

Thin layer chromatography (TLC) still finds wide use in investigation of quinone derivatives as a convenient means of analytical monitoring. Determination of quinones in TLC analysis is simple due to their chromophore groups which give characteristic color of each compound (yellow to red) [1,2].

Quinone derivatives have a cholagogic action and their bitter taste stimulates gastric mucosa to hypersecretion that improves digestion. Long-term use of medications containing quinones may cause serious side effects and diseases such as neoplasms. Raw materials containing described substances were used in prevention of urolithiasis. Such activity comes out of quinones ability to form complex compounds. Due to their toxicity they are not used in treatment anymore. Quinone derivatives irritate the walls of large intestine accelerating the displacement of the intestinal contents. Some of them have antidepressant activity.

Many papers, in which the use of TLC in traditional [3–5] preparative [6], two-dimensional [7], gradient [8], and circural [9–11] techniques were described, referring to an important group of compounds.

The molecular structures similar to quinones can be found in many naturally occurring substances, for example, vitamin K, antibiotics (e.g., Ciprofloxacinum).

Analytical chemists have tried to maximize the information that can be obtained from TLC separations of quinone derivatives, in the same way that was shown for

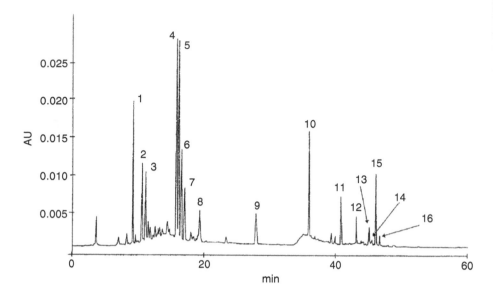

FIGURE 29.1 Profile of *Hypericum perforatum* extract with the HPLC-MS attributions of the components detected: 1—chlorogenic acid isomer, 2—3-*O-p*-coumaroylquinic acid, 3—chlorogenic acid, 4—rutin, 5—hyperoside, 6—isoquercitrin, 7—3,3′,4′,5,7-pentahydroxyflavanone, 8—quercitrin, 9—quercetin, 10—I3, II8 bipigenin, 11—pseudohypericin, 12—hypericin, 13—hyperforin analog, 14—hyperforin analog, 15—hyperforin, 16—adhyperforin. (From Brolis M., Gabetta B., Fuzzati N., Pace R., Panzeri F., and Peterlongo F., *J. Chromatogr. A* 825, 9, 1998.)

other chromatographic methods—liquid chromatography (LC), gas chromatography (GC). It is obtained by the combination of planar chromatography or high-performance liquid chromatography (HPLC) with mass spectrometry (MS) (Figure 29.1), tandem mass spectrometry (MS-MS), infrared (IR), Raman and fluorescence spectroscopy (FAB-MS), Fourier transform infrared spectrometry (FT-IR), diffuse reflectance infrared detection (DR-IFT) [12–16]. Fuller and Griffiths [17] showed that following transfer procedures involving scarping off, extraction, centrifugation, filtration, concentration and, good quality DR-IFT spectra can be obtained from small amounts (100 ng) of dyes separated by TLC.

Quinones usually originate from diphenols' enzymatic oxidation. Anthrone derivatives form a large group among polyketides. They coincide by condenzation of one molecule of acetyl-coenzyme A (CoA) and 7 molecules of malonyl-coenzyme A (Figure 29.2). However, in fungi (mainly *Penicillinum* and *Aspargillus*) all rings of this structure are synthesized by polyketide pathway. In higher plants the ring A probably comes from shicimic acid way. Anthraquinones, oxidative forms of anthron mainly occur in plants from Polygonaceae and Rhamnaceae. In reactions catalyzed by phenol oxydase the product composed of more than one molecule of anthraquinone can be built. The dimer compound was found, for example, in *Penicillium islandicum*.

Different chemical or physical signals mediate in intra or extracellular transformations. A lot of them act as effectors discriminating genes expression. Some of

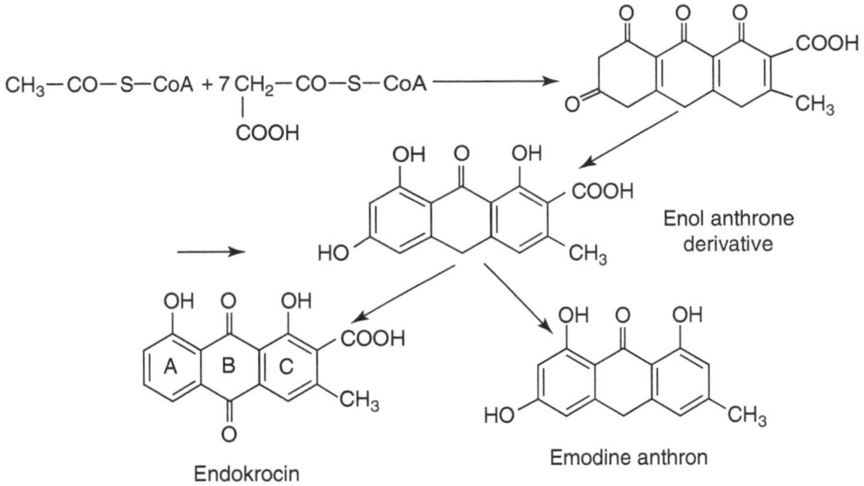

$$CH_3-CO-S-CoA + 7 CH_2-CO-S-CoA \longrightarrow$$
$$\underset{COOH}{|}$$

Enol anthrone derivative

Endokrocin

Emodine anthron

FIGURE 29.2 Synthesis of anthrone in the polyketide pathway.

them belong to phytohormones. It is known that indolylacetic acid (IAA) acts in gene expression in anthraquinone synthesis.

Traditionally, derivatives of quinones can be extracted from plant material by organic solvents such as methanol [18], dichloromethane [19], ethyl acetate containing 2% (v/v) hydrochloric acid, benzene and methanol [20], petrol and chloroform [21], hexane, diisopropyl ether and methanol [22,23], diethyl ether [24,25], water and methanol [26,27]. Isolated quinones are yellow, orange, or red. In reaction with alkali they give characteristic colors. Other reactions of detection are given in Table 29.1. Methods of chromatography of anthraquinones—solvents and adsorbents are listed in Table 29.2.

There are four main groups of quinones: monocyclic benzoquinones, dicyclic naphthoquinones, tricyclic anthraquinones, and phenanthraquinones. All these structures can be found in plants, although the most common are anthraquinones.

29.2 BENZOQUINONES

Benzoquinones appear as pigments in mushrooms and also in herbs. Some of them are antibiotics or cytostatics; however, they are not widely used in medicine so far.

The first compound tested for antitumor activity in 1955 by the National Cancer Institute, USA was a simple quinone, 2-methyl-p-benzoquinone [28]. Only a few of these compounds, for example, irisquinone (Figure 29.3) obtained from plant sources like *Iridaceaelatea pallasii* (Iridaceae) and used in antineoplastic Chinese medicines are effective against transplantable rodent tumors [60,61]. Irisquinone was also isolated from *Iris kumaoensis* in India [31] and used as a sensitizer for radiation therapy of cancer [62].

Benzoquinones were isolated from the aqueous extract of young shoots of Chinese pear, *Pyrus ussuriensis* Maxim, too [63].

TABLE 29.1

Methods of Anthraquinones Detection

Quinone	Reagents for Visualization	Spot Color
Sennosides	HNO$_3$ potassium hydroxide reagent, visible light	Brown
	Sodium metaperiodate reagent, heating for 5 min at 100°C	Green-yellow, dark brownish UV 365 nm
Hypericins	10% Pyridine in ethanol under visible UV light	Red
	NP/PEG in UV 365 nm or visible	Green-brown (visible), red-violet (365 nm)
Aloes anthraquinones	UV 365 nm	Yellow (aloin) dark violet-red (7-hydroxyaloin)
	Bornträger reaction in UV 365 nm	Red (aloe-emodin)
Anthraquinones of Frangulae and Rhamnus cortex	UV 365 nm	Bright orange-red
Rhei radix anthraquinones	UV 365 nm	Yellow (aglycones) Brown-red (8-*O*-monoglucosides) Blue (rhein aglycone)
	Phosphomolybdic acid/H$_2$SO$_4$ reagent, visible	Light yellow
	UV 254 nm	Fluorescence quenching
	UV 365 nm	Yellow, orange-brown
9,10-phenanthroquinone	0.1% HCl + 3% aqueous solution of variamine blue hydrochloride	Yellow-brown
	0.1% HCl + 1% aqueous Ni(NO$_3$)$_2$ + 3% aqueous solution of variamine blue hydrochloride	Green-blue
	1% aqueous Ni(NO$_3$)$_2$ + 1% dithizone in methanol + 1% aqueous Na$_2$CO$_3$	Yellow-brown
	1% Dithizone in methanol + 1% aqueous Na$_2$CO$_3$ + 1% aqueous Ni(NO$_3$)$_2$ + 3% aqueous solution of variamine blue hydrochloride	Brown
	0.1% HCl + 1% aqueous Ni(NO$_3$)$_2$ + 1% dimethylglyoxime solution ethanol	Orange
	0.1% HCl + 1% aqueous Ni(NO$_3$)$_2$ + 1% dimethylglyoxime solution ethanol + 3% aqueous solution of variamine blue hydrochloride	Green-yellow
	1% Aqueous Na$_2$CO$_3$ + 1% aqueous Ni(NO$_3$)$_2$ + 1% 8-hydroxyquinoline in ethanol	Orange
	1% Aqueous Na$_2$CO$_3$ + 1% aqueous Ni(NO$_3$)$_2$ + 1% 8-hydroxyquinoline in ethanol + 3% aqueous solution of variamine blue hydrochloride	Blue

TABLE 29.1 (continued)

Methods of Anthraquinones Detection

Quinone	Reagents for Visualization	Spot Color
	0.1% HCl + 1% aqueous Co(NO$_3$)$_2$ + 3% aqueous solution of variamine blue hydrochloride	Blue-green
	0.1% HCl + 1% 8-hydroxyquinoline in ethanol + 1% aqueous Co(NO$_3$)$_2$	Pink
	0.1% HCl + 1% 8-hydroxyquinoline in ethanol + 1% aqueous Co(NO$_3$)$_2$ + 3% aqueous solution of variamine blue hydrochloride	Orange
1,2-naphtoquinone	0.1% HCl + 3% aqueous solution of variamine blue hydrohloride	Violet
	0.1% HCl + 1% aqueous Ni(NO$_3$)$_2$ + 3% aqueous solution of variamine blue hydrochloride	Blue-violet
	1% Aqueous Ni(NO$_3$)$_2$ + 1% dithizone in methanol + 1% aqueous Na$_2$CO$_3$	Brown
	1% dithizone in methanol + 1% aqueous Na$_2$CO$_3$ + 1% aqueous Ni(NO$_3$)$_2$ + 3% aqueous solution of variamine blue hydrochloride	Violet
	1% Aqueous Na$_2$CO$_3$ + 1% aqueous Ni(NO$_3$)$_2$ + 1% 8-hydroxyquinoline in ethanol	Gray-violet
	1% aqueous Na$_2$CO$_3$ + 1% aqueous Ni(NO$_3$)$_2$ + 1% 8-hydroxyquinoline in ethanol + 3% aqueous solution of variamine blue hydrochloride	Violet
	0.1% HCl + 1% aqueous Co(NO$_3$)$_2$ + 3% aqueous solution of variamine blue hydrochloride	Blue
	0.1% HCl + 1% 8-hydroxyquinoline in ethanol + 1% aqueous Co(NO$_3$)$_2$ + 3% aqueous solution of variamine blue hydrochloride	Gray-pink
1,4-naphtoquinone	0.1% HCl + 3% aqueous solution of variamine blue hydrohloride	Brown-yellow
	0.1% HCl + 1% aqueous Ni(NO$_3$)$_2$ + 3% aqueous solution of variamine blue hydrochloride	Blue-violet
	1% Aqueous Na$_2$CO$_3$ + 1% aqueous Ni(NO$_3$)$_2$ + 1% 8-hydroxyquinoline in ethanol	Flesh
	1% Aqueous Na$_2$CO$_3$ + 1% aqueous Ni(NO$_3$)$_2$ + 1% 8-hydroxyquinoline in ethanol + 3% aqueous solution of variamine blue hydrochloride	Light violet
	0.1% HCl + 1% aqueous Co(NO$_3$)$_2$ + 3% aqueous solution of variamine blue hydrochloride	Blue
	0.1% HCl + 1% 8-hydroxyquinoline in ethanol + 1% aqueous Co(NO$_3$)$_2$	Orange
	0.1% HCl + 1% 8-hydroxyquinoline in ethanol + 1% aqueous Co(NO$_3$)$_2$ + 3% aqueous solution of variamine blue hydrochloride	Gray-brown

TABLE 29.2
Chromatography Sets in Anthraquinones Analyze

Types of Quinons	Extraction Methods	Chromatographical Set	Connection with Other Techniques	References
Benzoquinones	Standards were soluted in chloroform	Chloroform–methanol–water (80:20:2.5, v/v/v) HPTLC plates Si 60 with fluorescent indicator, prewashed in methanol	TLC	[28]
Iryanthera juruensis quinones	Grounded fruits extracted with hexane	Hexane–ethyl acetate (8:2, v/v), TLC preparative plates Hexane–ethyl acetate (1:1, v/v), TLC preparative plates Chloroform–methanol (95:5, v/v), TLC preparative plates	RP-18 HPLC, MS, UV	[29,30]
Naphtoquinones from Drosera sp.	Whole plants were extracted with methanol even for one month	Toluene–formic acid (99:1, v/v), precoated silica gel plates Methanol:water (1:1,v/v), RP-18 Si plates	UV, EI-MS, ^{1}H and ^{13}C NMR	[31,32–35, 36]
Naphthoquinones from Diospyros sylvatica	Extraction with chloroform in room temperature for 24 h	Hexane–chloroform (4:1, v/v) silica preparative gel Detection under UV light and by spraying with 5% sulfuric acid in methanol	EI-HRMS, TLC	[34,37–41, 42,43]
Emodin in Rheum sp.	Extracted with methanol (3 × 100 mL; 15 min for each extraction) in an ultrasonic bath	Toluene:dichloromethane:ethyl acetate (4:4:1, v/v/v), HPTLC Si 60 plates	TLC, densitometry	[6,24,44]
Bianthraquinones from Cassia siamea	Hot methanol	Benzene–acetone (4:1), PTLC Si 60 F$_{254}$ benzene:diethyl ether:ethyl acetate (60:60:4.5, v/v/v) PTLC Si 60 F$_{254}$	CC, NMR, ^{1}H NMR, ^{13}C NMR, TLC	[45]

Compound	Source/Extract	Conditions	Method	Ref.
Cercospora beticola toxins	Crude extract	Chloroform–methanol–acetic acid (100:2:1, v/v/v) and pretreated plates Si 60 F-254 (the pretreatment consisted of dipping plates in an aqueous solution of phosphoric acid and calcium bis-(dihydrogenphosphate) hexane–aethyl acetate (1:1, v/v) and pretreated plates chloroform–methanol–water (100:20:2, v/v/v) regular plates Si 60 F-254	HPLC	[46,47]
2-Methoxy-1,4-naphtoquinone	Dichloromethane extract of *Swerita calycina* (Gentianaceae)	Nova-Pak C_{18} column, MeCN–H_2O gradient: 5:95–65:35	HPLC	[48]
Emodin chrysophanol physcion	Chloroform extracts of: *Vitis vinifera* grape wine leaves, *Rhisoma graminis* couch grass root, *Plantagines lanceolatae* plantain herb, *Fagopyrum esculentum* Buckwheat, *Artemisia annua* worm wood, lettuce, beans, peas	Knauer Eurosher Si-100 C-18 column, 5% methanol–MeCN–0.1% acetic acid (50:50:1), 80% MeCN/20% water	HPLC	[49]
Chrysophanol	*Rheum emodi* (Polygonaceae)	Silica gel 60 F_{254} (Merck), hexane–ethyl acetate (45:5, v/v)	HPTLC	[50]
Emodin	Methanol extracts of: *Rheum undulatum* L., *Rheum rhaponticum* L. (Polygonaceae)	Silica gel 60 F_{254} (Merck), toluene–dichloromethane–ethyl acetate (4:4:1, v/v/v)	HPTLC	[24]
Chrysophanol rhein emodin	Methanol extracts of: *Senna* leaves, *Rhubaeb* roots, *Dock* flowers	Spherisorb-CN column, chloroform–acetic acid (95:5)	HPLC	[51]
Aloin	*Aloe* (Liliaceae)	Micropreparative TLC silica gel 60 F_{254} (Merck), RP-2, RP-8 (Merck), Florisil (Fluka, Buchs, FRG) isopropanol–dichloromethane–acetic acid (30:69:1), methanol–water (6:4); ethyl acetate–MeOH–water (77:13:10)	Micropreparative TLC	[52]

(continued)

TABLE 29.2 (continued)
Chromatography Sets in Anthraquinones Analyze

Types of Quinons	Extraction Methods	Chromatographical Set	Connection with other Techniques	References
Aloin aloeemodin	Cape aloes—the laxative drug	HPTLC plates coated with amino layer–NH$_2$ (Merck), ethyl acetate–2-propanol–water (100:17:13, v/v/v)	HPTLC	[6]
Physcion chrysophanol emodin chrysophanol glycoside	*Rheum emodi* (Polygonaceae)	HPTLC RP-18 (Merck), methanol–water–formic acid (80:19:1, v/v/v)	HPTLC	[53]
2-Azaanthraquinone	Dichloromethane extract of *Mitracarpus scaber* (Rubiaceae)	HPTLC Silica gel 60 F$_{254}$ (Merck), toluene–ethyl acetate–methanol (80:12:2, v/v/v)	HPTLC	[54]
2-Formyl-1-hydroxy-anthraquinone	Dichloromethane extract of *Morinda elliptica* L.	dichloromethane + petrol (2:3) for TLC plates	TLC, preparative TLC	[55]
Cerosporin	Ethyl acetate extract of *Cerospora beticola* fungus	Silica gel 60 F$_{254}$ (Merck), chloroform–methanol–acetic acid (100:2:1, v/v/v) for HPLC-diol phase–hexane–ethyl acetate (50:50) -C18-acetonitrile–water–acetic acid (50:50:1)	TLC HPLC	[46,47]
Bianthraquinones	Diethyl ether, chloroform, ethyl acetate extracts of *Cassia siamea*	Silica gel 60 F$_{254}$, benzene–acetone (4:1, v/v), benzene–diethyl ether–ethyl acetate (60:60:4.5, v/v/v)	Preparative TLC	[45]
Naphtodianthrones	Methanol extract of *Hypericum perforatum*	RP-18 column, water–85% phosphoric acid–acetonitrile–methanol (99.7:0.3:10, v/v/v)	TLC HPLC	[56–59]

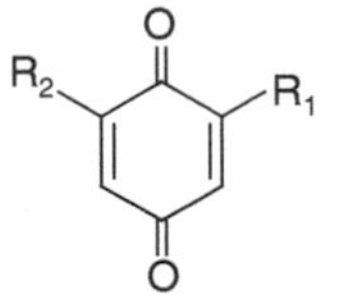

	R_1	R_2
Benzoquinone	H	H
Irisquinone	OCH_3	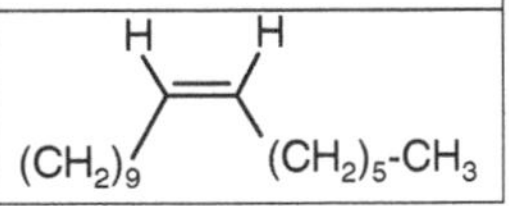

FIGURE 29.3 Chemical structures of benzoquinones.

In the group of benzoquinones can also be mentioned:

- Tymoquinone from *Nigellia sativa* (Ranunculaceae)
- Poliporic acid from *Polyporus nidulans*
- Teleforic acid from fungi genus *Hydnum* and *Sarcodon*

Two acids—sargaquinoic and 3-methyl-sargaquinoic (Figure 29.4a and b) are also included in benzoquinones. They were isolated by Silva et al. [29] from *Iryanthera juruensis* (Myristicaceae).

Amazon Indians were reported to utilize crushed leaves of *I. juruensis* for healing seriously infected wounds. They also used latex from the barks of *I. juruensis* mixed with warm water for treating stomach infections, as *I. juruensis* species have an antioxidant activity [64].

Silva et al. [30] reported the isolation of antioxidant compounds of *I. juruensis* fruits guided by the β-carotene TLC assay, and the evaluation of their oxidation potentials through cyclic voltammetry; β-carotene TLC assay is used to describe antioxidant activity of a substance [65]. The examined compounds were applied on a

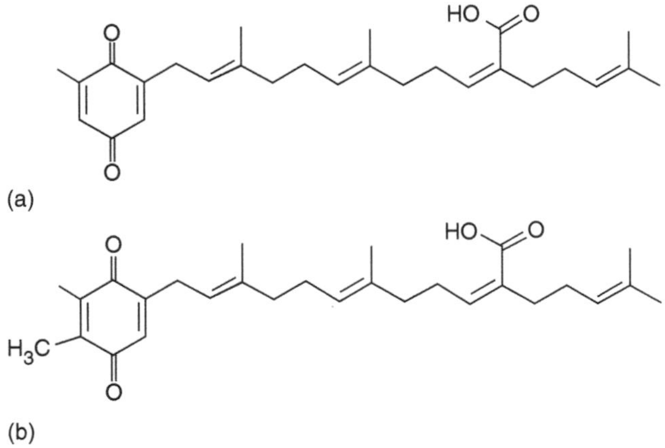

FIGURE 29.4 Chemical structures of (a) sargaquinoic acid and (b) 3-methyl-sargaquinoic acid.

plate with solution of β-carotene and their ability to protect β-carotene from degradation was measured.

29.3 NAPHTHOQUINONES

Naphthoquinones and other related quinone compounds are one of the major natural product classes with varied biological activities; the most interesting is their antiparasitic biological activity.

Quinone named hydrolapachol-2-hydroxy-1,4-naphthoquinone has proved to have an activity against *Plasmodium lophurae*. Plumbagin and other related quinones have cured *Leishmania* subsp. infections, while diospyrin was found to have an action against *Leishmania donovani*. The quinones of *Diospyros*, mainly plumbagin and 7-methyljuglone, were shown to possess good antibacterial activity in various strains of bacteria, while the dimeric benzoquinone, microphyllone was reported to possess antiallergic properties.

There are several substances in naphthoquinones group (Figure 29.5):

- Jouglon from *Juglans regia* (Juglandaceae), with bacteriostatic and fungistatic activity
- Droseron from *Drosera* sp. (Droseraceae), with spasmolitic action in respiratory diseases

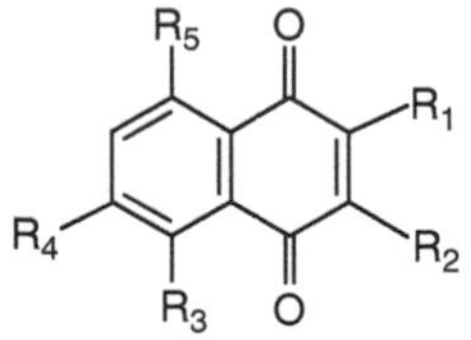

	R₁	R₂	R₃	R₄	R₅
Juglone	H	H	H	H	OH
Plumbagin	H	CH₃	H	H	OH
7-Methyljuglone	H	H	H	CH₃	OH
Lawsone	H	OH	H	H	H
Alkannin	H	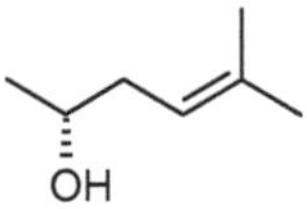 (chain with OH)	OH	H	OH
Shikonin	H	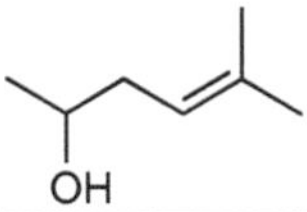 (chain with OH)	OH	H	OH
Phylloquinone	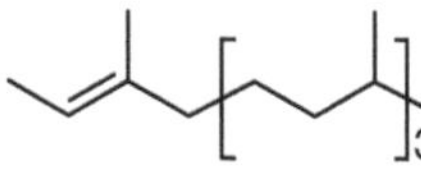 (chain)	CH₃	H	H	OH

FIGURE 29.5 Chemical structures of naphthoquinones.

- Lawsone from *Impatiens glandulifera*
- Plumbagin from *Plumbago* subsp.
- Lapachol from *Tacoma stans* (Bignoniaceae), yellow pigment with cyto-static properties
- Alkannin from *Alkanna tinctoria* (Bignoniaceae), red substance used in food production as pigment
- Shikonin from *Lithospermum erythrorhizon*
- Phylloquinone
- Diospyrin from *Diospyrons melanoxylon*

First from mentioned naphthoquinones—juglone (Figure 29.5) was isolated by Girzu et al. [66] from fresh leaves of walnut tree (*Juglans regia*). Juglone and its 7-methyl and 3-methyl derivatives were found in many tropical plants alone, or in association with other quinones in *Diospyros* and *Drosera* subsp. [32–34]. A more efficient and simple method, for example, HPLC, was developed for separation of the monomeric naphthoquinone derivatives with use of isocratic elution with acet-onitrile–water (40:60, v/v) on a μBondapak C_{18} RP column, followed by UV detection at 255 nm [35].

Methyl derivative of juglone—plumbagin has been isolated from *Plumbago zeylanica* and other *Plumbago* subsp., and is widely used in several systems of medicine in India, China, and Far Eastern countries like Taiwan, Korea, and Malaysia [67].

Hydroplumbagin 4-*O*-glucoside from *Drosera intermedia* and 7-methylhydro-juglone 4-*O*-glucoside (rossoliside) (Figure 29.6) from *Drosera rotundifolia* were isolated by Budzianowski [3].

The above-mentioned naphthoquinones were produced by in vitro micropropa-gation. These glucosides can be detected in plant extracts by reversed-phase TLC (RP-TLC). Appearance of the corresponding free quinones was ascertained after treatment with β-glucosidase. Naphthoquinones were identified by UV, EI-MS, ^{1}H, and ^{13}C NMR [36].

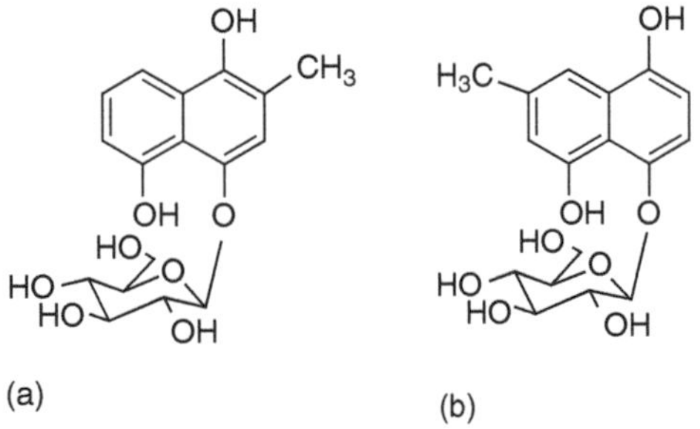

FIGURE 29.6 The structures of (a) hydroplumbagin 4-*O*-glucoside and (b) 7-methylhydro-juglone 4-*O*-glucoside (rossoliside).

Drosera rotundifolia is the most important medicinal *Drosera* species and is used for the treatment of respiratory diseases, for example, whooping cough [68]; the activity is due to plumbagin and 7-methyljuglone.

Shikonin—the enantiomer of alkannin (Figure 29.5) was found in the roots of European dye-plant *Alkanna tinctoria*. This substance was also isolated from the roots of the herb *L. erythrorhizon* Sieb. et Zucc (Boraginaceae) and is found to be an active component present in traditional medicines in the East [69]. Extract of *L. erythrorhizon* roots in Japan was used for wound healing, and also taken internally as an antipyretic and anti-inflammatory agent. Shikonin is currently used in various medicinal preparations in China, Japan, and Korea and also in cosmetics and pigments in Japan, while alkannin is mainly used in cosmetics and for food coloring in Europe and USA. Shikonin and the related compounds, have been reported by Papageorgiou et al. [47]. It accumulates in the roots of *L. erythrorhizon, L. officinale,* and many other Boraginaceae plants like *Arnebia hispidemia*. Presently, shikonin is produced in a large scale by liquid cultures of *L. erythrorhizon* root in Japan. It is the world's first commercial production of a secondary metabolite by plant cell culture [47].

Fujita et al. reported separation of seven shikonin derivatives by HPLC method with mobile phase consisting of acetonitrile–water–triethylamine–acetic acid (70:30:0.3:0.3, v/v/v/v) [70].

Phylloquinone (prenylquinone) is found in plant, animal, and bacterial cells and contains an isoprenoid side-chain, which is bound to a benzo- or naphthoquinone nucleus. Phylloquinone, one of the Vitamin K group of naphthoquinones, occurs in green leafy vegetables and other higher plants. While the therapeutic efficacy of its synthetic analog, i.e., menadione or Vitamin K_3, is well established for treatment of multi-drug resistant leukemia [71], a number of studies have demonstrated the anticancer effects of Vitamin K_1 also against several cell lines [72–75], for example, liver, colon, lung, stomach, etc. Classically, separation of such naturally occurring quinone mixtures was achieved by TLC on polyamide, Ag^+-impregnated silica gel plates, and RP-TLC (RP-18, Merck) [76].

29.3.1 DINAPHTHOQUINONES AND OTHER NAPHTHOQUINONES

Diospyrin is one of the dimeric naphthoquinones abundantly present in *Diospyros* subsp. (Figure 29.7). The antibacterial, antifungal, and termite-resistant properties of *Diospyros* have all been attributed to the presence of naphthoquinones.

The *Diospyros* genus produces a large number of 1,4-naphthoquinone metabolites which include several monomers, dimers, a few trimers, and eleven tetramers. In fact, such quinones are useful as taxonomic markers for this genus as proposed in an extensive review on this topic [34]. The leaves, wood, bark, fruit, seed, and root of *Diospyros* plants have traditional uses in many countries in Africa and Asia, mostly as astringents and chewing stick, for treatment of skin diseases, schistosomiasis, abdominal discomfort, and female diseases.

Diospyrin was isolated by column chromatography and TLC by several groups of researchers in different countries [37–40]. A study on preparative TLC of diospyrin and isodiospyrin along with its monomer (7-methyljuglone) revealed the

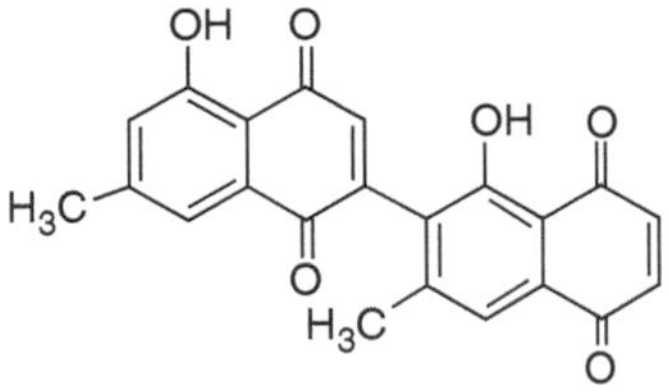

1. Diospyrin

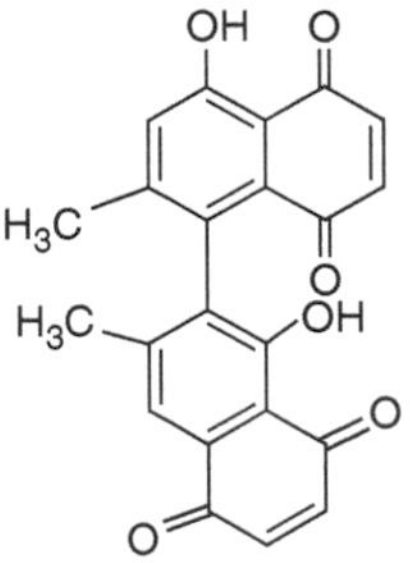

2. Isodiospyrin

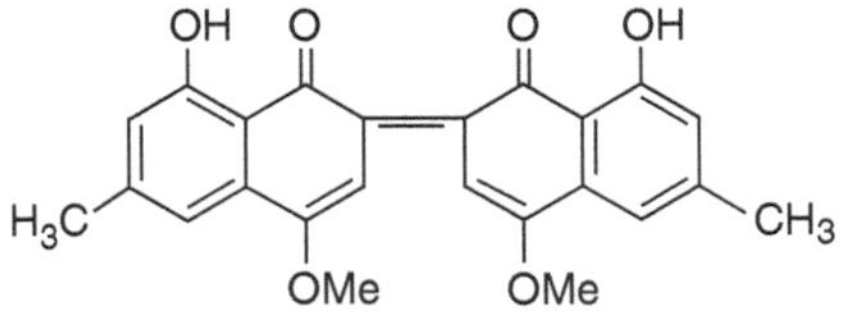

3. Diosindigo

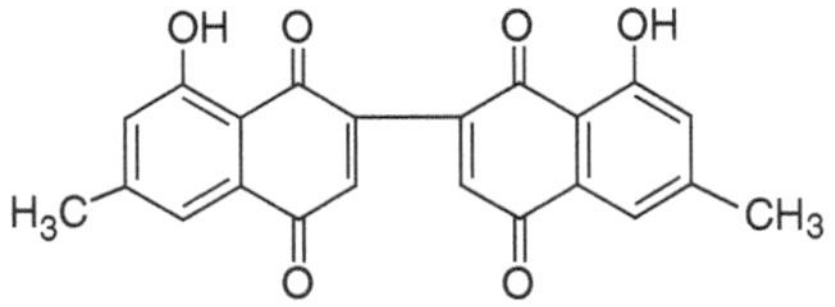

4. Mamegakinone

FIGRUE 29.7 Chemical structures of dimeric naphthoquinones.

sensitivity of these compounds to exposure when adsorbed on silica gel [41]. The estimation of diospyrin in the stem bark of *Diospyros montana* was first reported by Ravishankara et al. [77]. It was achieved with use of reversed-phase high-performance thin layer chromatography (HPTLC) method. The analysis was performed on a precoated silica gel G60 HPTLC plates with the Linomat IV automatic sample spotter application.

Six quinone derivatives: diospyrin, isodiospyrin, diosindigo A, plumbagin, micropyllone, and 2-methyl-anthraquinone were isolated from the chloroform extract of the roots of *Diospyros sylvatica* by TLC on silica gel (Merck) [42]. The extract was chromatographed over the silica gel with portions of chloroform and successively eluted with petrol ether, petrol ether/chloroform, chloroform, and chloroform/methanol. All the quinones were reported from *D. sylvatica* for the first time. The investigated root extract is rich in naturally occurring quinones as the chemicals chiefly responsible for its repellent property and hence the use of these plant-derived compounds can be considered as alternatives to termite control, due to their environmental safety.

The spots on chromatograms were detected under UV light and by spraying with 5% H_2SO_4 in methanol or with 0.1 mol L^{-1} NaOH. All given molecular formulas were determined by EI-HRMS. The HPLC analysis of its isomers, isodiospyrin, and mamegakinone (Figure 29.7) was first achieved by Marston and Hostettmann, with use of a normal phase column [43].

In the North regions of the Russian Far East can be found a typical lichen *Cetraria islandica*. The ethyl acetate extracts of *C. islandica* contain the naphthoquinones and the quinones. By preparative TLC two known quinones: 3-ethyl-2, 7-dihydroxynaphthazarin and 6,6′-bis(3-ethyl-2,7-dihydroxy-naphthazarin) and one

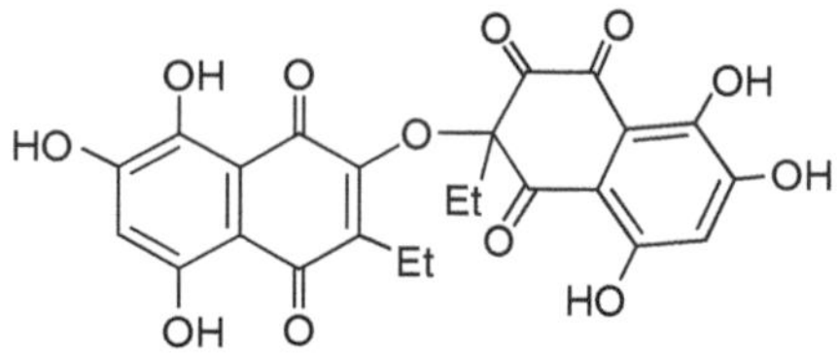

FIGURE 29.8 Chemical structure of islandoquinone.

new quinone: islandoquinone (2-*O*-(3′-ethyl-1′,2′,3′,4′-tetrahydro-5′,7′,8′-trihydroxy-1′,2′,4′-triketo-3′-naphthyl)-3-ethyl-7-hydroxynaphthazarin) (Figure 29.8) were isolated [78].

Rapid detection of biologically active natural products plays a vital role in the phytochemical investigation of crude plant extracts. In order to perform an efficient analysis of the extracts, both biological assay and HPLC analysis with various detection methods may be used.

Such investigations enabled Wolfender et al. [48] to find naphthoquinone for the first time in *Swertia calycina* (Gentianaceae). The authors isolated and identified 2-methoxy-1,4-naphthoquinone (Figure 29.9). The structure of that compound was confirmed by LC-thermospray (TSP)-MS spectrum.

This naphthoquinone was isolated by a combination of column chromatography (with silica gel) and HPLC in RP mode, and was found to be the substance responsible for the strong antifungal activity of the extract of *Swertia calycina* [79].

Diffuse reflectance FT-IR (DR-FT-IR) was used for the quantitative analyses (in situ) of photochemically labile compound—diazonaphthoquinone separated by HPTLC plates [16]. Results were achieved in less than 2 min. Precision of the overall technique (chromatographic and spectroscopic) was 3% relative standard deviation with the use of commercially available plates, linearity of detection was obtained with detection limits of 400 ng for diazonaphthoquinone ester (Figure 29.10).

29.4 ANTHRAQUINONES

Anthraquinones are derivatives of tricyclic aromatic compound—anthracen (Figure 29.11).

These substances are important, naturally occurring pigments which are widely distributed in nature. There are anthraquinones, anthrons, anthranols, and dianthrons in this group. Anthraquinons and anthrons differ in the degree of oxidation of basic structure, while anthranols are tautomeric forms of anthrons (Figure 29.12).

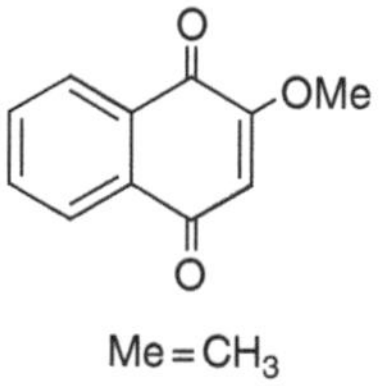

Me = CH$_3$

FIGURE 29.9 Chemical structure of 2-methoxy-1,4-naphthoquinone.

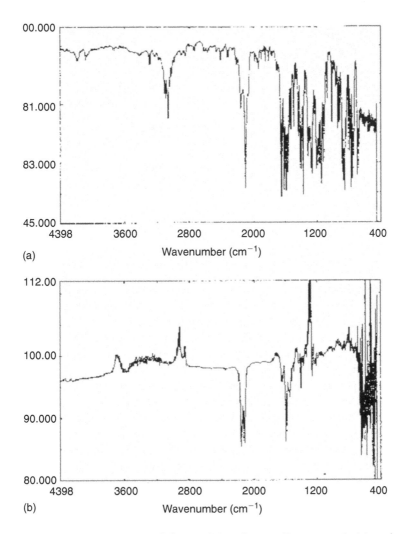

FIGURE 29.10 DR FT-IR spectra of diazonaphthoquinone sulfonate ester in (a) noninteracting KBr matrix and (b) adsorbing RP-TLC matrix. (From Beauchemin B.T. and Brown P.R., *Anal. Chem.* 61, 615, 1989.)

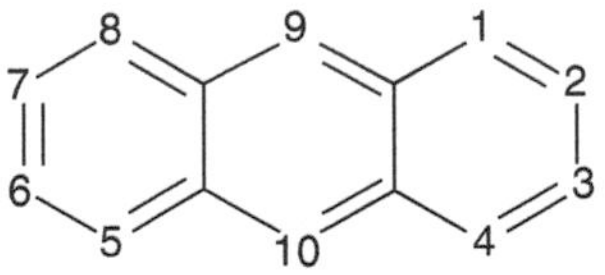

Anthracen

FIGURE 29.11 Chemical structure of anthracen.

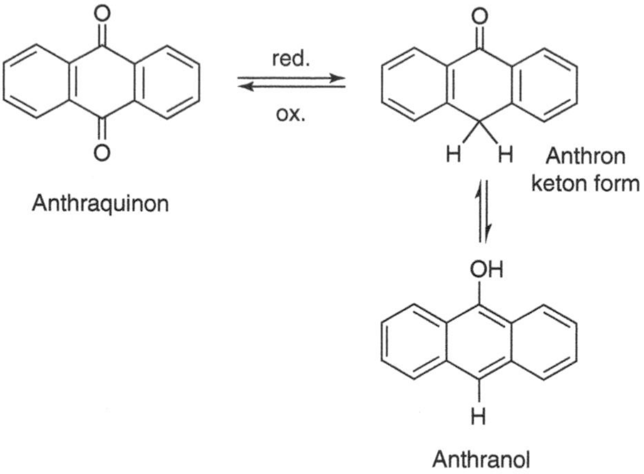

FIGURE 29.12 Chemical structures of anthraquinone, anthron, and anthranol.

The most frequent substitutes of antracen structure are hydroxyl, methoxyl, methyl, and carboxyl groups. In plants it can be found as *O*- and *C*-glucosides with D-glucose, L-rhamnose, or primeverose (Figure 29.13).

Anthranoids are solid, red, orange, or yellow substances. They have a great ability to redox reactions even during drying and storage of plant material. Anthron forms have stronger therapeutic effect.

The difficulty in determination of glycosides and aglycones of anthranoids in whole plant extracts is that the compounds of interest are usually isolated as groups of similar structures. The components are difficult to separate from each other in a group. Also the groups themselves are significantly different in their relative polarity. TLC is well suited to this challenge because it enables crude extracts to be applied to the layers for simultaneous sample preparation (SPE-solid phase extraction) and enables the separation of a sample in parallel for rapid screening of extracts.

Anthranoids occur as glucosides in forms of heterosides or glucozoles where sugar molecule is bonded through a carbon atom. They are located in different organs as aglycones or heterosides (palmidines, reidines).

Materials from Rhamnaceae family undergo changes of quality of active compounds during storage. In fresh plant material there are reduced forms of anthranoid compounds while in material after storage there are oxidized ones. Anthron and anthranol derivatives can be found in fresh material. After long storage and enzymatic transformations, these compounds undergo oxidation changes: anthraquinone derivatives are formed. The same process can be achieved by heating the plant material.

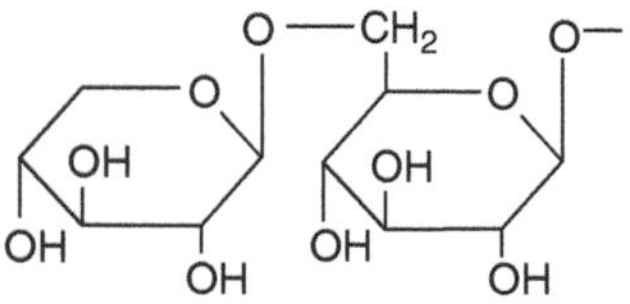

FIGURE 29.13 Chemical structure of primeverose (O-β glu^6→^{1}xyl).

During oxidation process anthranols form anthraquinones which, after reduction form anthranols. Depending on the season, the amount of each form can be changed in plant material. In spring and summer when sun exposition is high, anthranols predominate, while anthaquinones dominate in autumn and winter.

As the anthraquinones content is a quality criterion for plant-derived laxatives, a number of analytical methods have been established for different anthraquinone derivatives. These compounds in plant material have been separated by TLC with use of silica gel and polar eluents, and by classic column chromatography [51]. Mainly, extraction with methanol and water or acetonitrile and HPLC, RP-TLC, RP-HPLC methods were used for analysis [52]. Finally, capillary zone electrophoresis, micellar electrokinetic capillary chromatography and liquid chromatography-mass spectrometry (LC-MS) analysis have been used for detection and, in some cases, for quantification of the different anthraquinone derivatives [80,81].

Anthranoids can be found in fungi (e.g., *Penicillium*), animals, and medical plants from the following families: Polygonaceae, Rhamnaceae, Leguminoseae, Rubiaceae, and Liliaceae. A few raw plant materials are used in medicine. They have large amount of anthranoids that have laxative effect.

In this group we can find *Aloe, Frangulae cortex, Rhei radix, Sennae folium, Rhamni catharticae fructus, Rhamni cortex*.

Hydroxyanthraquinones are the active compounds in a large number of plant-derived laxative drugs. In many fungi imperfecti and some toadstools, anthraquinone derivatives have been discussed as the toxic metabolites and coloring matters (*Dermocybe sanguinea, Penicillum,* and *Aspergillus* species) [67].

A sample transfer approach to TLC/FT-IR was used for the determination of 1,4-dihydroxyanthraquinone (Figure 29.14) and other derivatives. More spectral information is obtained on the analytes and in significantly less time than is required for in situ measurements. The transfer is accomplished with minimal sample loss, decomposition, or contamination compared with previous processes. Spectra are easily identified for samples at a concentration of 40 μg/μL per component [82].

Botz et al. hydrolyzed the extract of the stem bark of *Rhamnus frangula*. Using the optimized TLC mobile phase (tetrahydrofuran–hexane, 1:10, v/v) the two critical compounds: physcion and chrysophanic acid could be separated to baseline within 30 min, 11.5 mg of chrysophanic acid, 9.4 mg of physcion, and 79.6 mg of frangulo-emodin could be isolated from 195 mg extract [10].

Danthron, chrysophanol, emodin, aloe-emodin, physcion, rhein, and chrysarobin (Figure 29.15) are accounted to 1,8-dihydroxyanthraquinones group. The RP-HPLC and RP-LC-MS were used for quantitative analysis. These compounds and their glycosides are the major constituents of *Aloe, Rheum,* and *Polygonum* subsp. They are used in many medicinal preparations. The leaves and pods of *Cassia* (syn. *Senna*) are found to contain dianthrones and their glycosides, represented by the sennosides (see page 843). These species are used as laxative medicine all over the world.

Chrysophanol is the major compound of *Rheum emodi* (Polygonaceae).

Chrysophanol and two other 1,8-dihydroxyanthraquinones (see Figure 29.15)—emodin and physcion were detected by Mueller et al. [49] in a variety of vegetables, some herbs, and herbal-flavored liquors. The vegetables showed a large batch-to-batch variability, from 0.04 to 3.6, 5.9, and 36 mg total anthraquinone per kilogram fresh

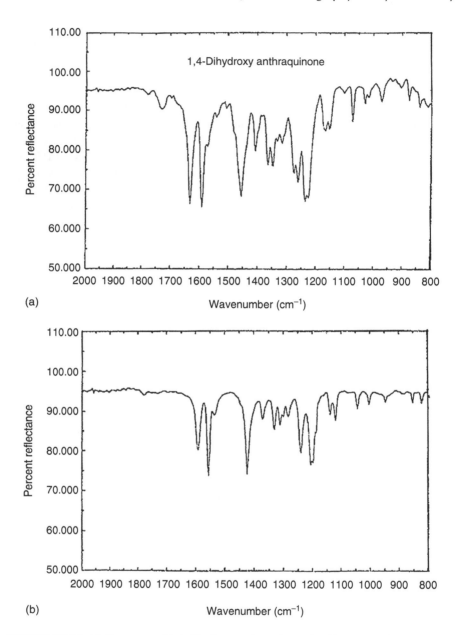

(a)

(b)

FIGURE 29.14 (a) Spectrum of 1,4-dihydroxyanthraquinone for 40 μg quantity spotted on TLC plate and (b) reference spectrum obtained from chloroform solution in a sealed liquid cell. (From Shafer K.H. and Griffiths P.R., *Anal. Chem.* 58, 2708, 1986.)

weight in peas, cabbage lettuce, and beans, respectively. Physcion predominated in all vegetables. In the herbs, grape vine leaves, couch grass root, and plantain herb anthraquinones were above the limit of detection. Contents ranged below 1 mg/kg (dry weight). Emodin, chrysophanol, physcion were also found in seven of eleven

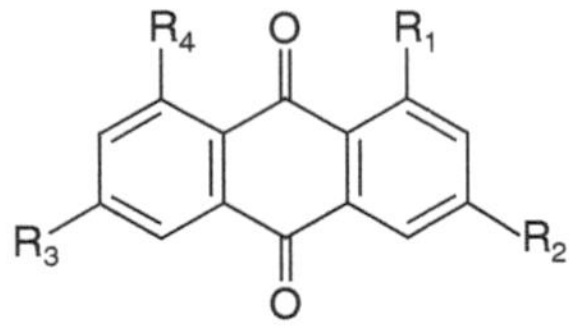

	R₁	R₂	R₃	R₄
Danthron	OH	H	H	OH
Chrysophanol	OH	CH₃	H	OH
Emodin	OH	OH	CH₃	OH
Aloe-emodin	OH	CH₂OH	H	OH
Physcion	OH	CH₃	OCH₃	OH
Rhein	OH	COOH	H	OH

FIGURE 29.15 Chemical structures of emodin and analogs.

herbal-flavored liquors in a range of 0.05–7.6 mg/kg. It should be mentioned that for laxative drugs, the maximum therapeutical dose is 30 mg anthraquinone glycosides. For senna, the maximum dose up to 7 mg anthraquinones per day is used [49].

1,8-Dihydroxyanthraquinones present in laxatives, fungi imperfecti, Chinese herbs and vegetables are in debate as human carcinogens. Some authors stated that emodin was genotoxic, whereas chrysophanol and physcion [83] showed no effect. The complete vegetable extract on its own did not show any effect in the micromuscle test.

HPTLC method has been developed for analysis of chrysophanol, an antioxidant and a major constituent of the rhizomes of *Rheum emodi* (Polygonaceae). In this method chrysophanol was used as the external standard. Detection and quantification were performed densitometrically [50].

Application of densitometric TLC for estimation of anthraquinons has recently been reported by Rai and Turner [84].

Chrysarobin is formed as anthron or anthranol form after chryzophanol reduction. It can be found in the genus: *Rheum, Polygonum, Cassia, Laguminossae*. It is a strong, natural reducing agent with antiseptic properties. It is used in skin infections such as: acne and psoriasis. It was determined alongside emodine and physcion in medical plants (Figure 29.16).

Emodin—(named also franguloemodin, rheum-emodin) is very often an aglycone in glucosides. It is found in many species: *Rheum, Polygonum,* and *Rumex*

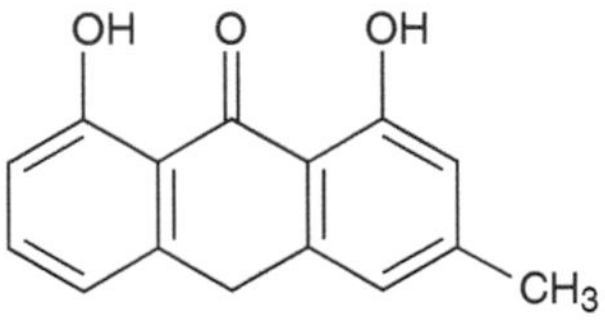

Chrysarobinum

FIGURE 29.16 Chemical structure of chrysarobin.

(Polygonaceae), *Rhamnus* and medical plants: *Frangula* (Rhamnaceae), *Cassica* (Caesalpinaceae), *Polygonatum* (Liliaceae), and in mushroom *Dermocybe sanguinea*.

Rhei radix is gained from *Rheum palmatum* and *Rheum officinale* (Polygonaceae); it contains anthranoids: aloe-emodin, chryzophanol, emodin, physcion, and tannins-gallic acid. The best plant material comes from China. It is collected in 8–10 seasons of vegetation. Therapeutic effect of this material depends on the dose. Dose 0.1–0.3 g gives styptic and obstruent (by tannins) effect while the dose 1.5–3.0 g is laxative (by anthranoides).

TLC method combined with densitometry was applied to quantitative analysis of emodin.

The aim in the work of Smolarz et al. [24] was to examine the emodin content in the petioles of *Rheum undulatum* and *Rheum rhaponticum*. Emodin occurs widely in medicinal herbs prepared from the Polygonaceae family and has been detected in the rhizomes of species belonging to the genera *Rheum* and *Rumex* [85,86]. It has been reported that emodin has a regulatory effect on the proliferation of human primary T lymphocyte [87] and immune responses in human mesangial cells and inhibits the growth of breast cancer [88]. In Smolarz's publications the adsorbent and mobile phase used for HPTLC separation of emodin were optimized experimentally.

The chromatograms were scanned in a zig-zag meander mode at $\lambda = 340$ nm. The calibration curve was determined and the amounts of emodine in the petioles of *Rheum undulatum* and *Rheum rhaponticum* were 4 μg/g and 5.8 μg/g (dry weight), respectively.

Aloe-emodin is an anthraquinone structure with a CH_2OH group. Aloin A and B—the mixture called barbaloin—is C-glucoside of aloe-emodinanthron. Aloe-emodin is found in *Aloe ferox* and has a laxative and antihistaminic effect.

Aloe ferrox or *Aloe berbandense* (Liliaceae) belong to pharmacopeic anthracene plant materials. In therapy they are used as powdered hardened sap, which flows from cutoff parts of the plants. This material is known as Cape-Aloes B.P. or Curacao-Aloes.

The separation of the components: aloin (Figure 29.17) and aloe-emodin of Cape-Aloes by normal phase and RP-TLC has been optimalized by Wawrzynowicz et al. [6]. The system with the best separation efficiency was applied in micropreparative TLC for isolation of aloe-emodin and aloin from *Aloes* (powdered Cape-Aloes B. P.) Normal phase plates were developed with izopropanol–dichloromethane–acetic acid or with ethyl acetate–methanol–water and RP plates with methanol–water.

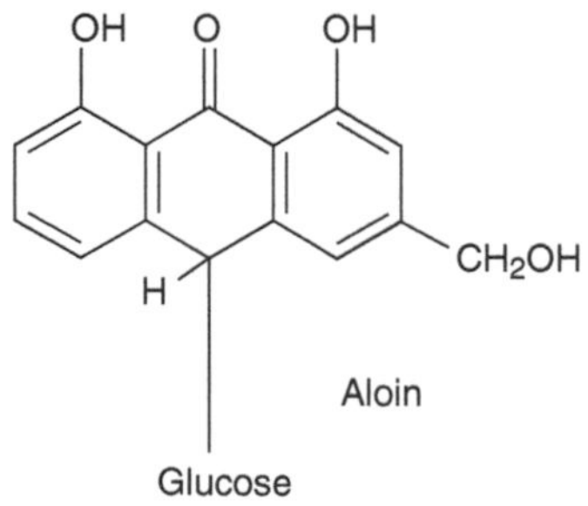

FIGURE 29.17 Chemical structure of aloin.

Plates coated with layers of Florisil, silica gel RP-18 and RP-2 were also used. Pure fractions were obtained with the use of precoated silica plates with preconcentrating zone. Figure 29.18 shows the zonal chromatogram and corresponding densitogram obtained. The bands of *Aloe* components were, owing to the higher efficiency, narrow and well separated from each other.

The laxative action of anthraquinon glucosides basically depends on the cleavage to the corresponding free anthrone [89]. Participation of human bacteria in the cleavage of C-glycosyl derivatives such as aloins has not been clarified [83]. Koch and Müller [90] administered therapeutic doses of aloin to healthy volunteers. It was ascertained that human intestinal flora seems to be capable of metabolizing aloin, and that aloe-emodin anthrone seems to be the sole metabolite in human feces. According to the former observations from in vitro testing [91] outside the human intestine by mixing aloin with human feces was presumed that metabolites such as bianthrones and ten aloine derivatives [92] may be expected to be artificially formed during extraction and storage procedures. The results of investigation imply that

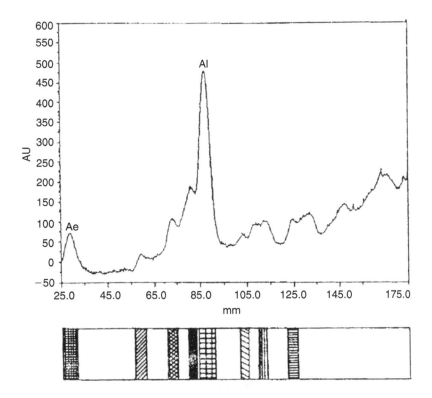

FIGURE 29.18 Micropreparative TLC of Aloe, and densitogram corresponding to the separation; precoated silica plate with preconcentrating zone developed with ethyl acethate–methanol–water, (77:13:10, v/v/v); Al-aloin; Ae-aloeemodin. (From Wawrzynowicz T., Waksmundzka-Hajnos M., and Mulak-Banaszek K., *J. Planar Chromatogr.* 7, 315, 1994.)

nutrition plays an important role in metabolic activity of human intestinal bacteria. It seems that ferro-salts or meat products promote the cleavage of aloin.

Investigation was conducted with the use of HPTLC method. Plates were developed using ethyl acetate–2-propanol–water (100:17:13, v/v/v) as mobile phase, adsorbent—SiO_2 modified with NH_2 groups.

The aglycone of anthranoide—aloe-emodin with glucoside form were subsequently analyzed by Kiridena et al. [93]. The genus *Aloe* (Liliaceae) comprises about three hundred species indigenous of Africa. Leaves of this plant exude a more or less colored juice when cut [94]. From the earliest days *Aloe* juice has been used for therapeutic and cosmetic purposes. Two main products can be recognized: the leaf exudates, used largely as a purgative, and the leaf parenchyma, effective against a range of skin lesions. The USA Food and Drug Administration has expressed concern about the need to monitor the production of substances with *Aloe* juice to ensure they are correctly processed to avoid the undesirable consequences of the biological action of the natural components found in the leaf exudates [95].

In traditional TLC, the plate with applied extracts of *Aloe* was developed in a Camag chamber with ethyl acetate–hexane (18:82, v/v) as the mobile phase. In investigations the AMD (automated multiple development) technique with the use of silica gel HPTLC plates was conducted. The AMD program used for the separation of aloin in the *Aloe* leaf extracts and *Aloe vera* juice consisted of five solvents: 1-propanol, dichloromethane, acetic acid, hexane, and ethyl acetate. There were 12 steps, development time was from 0.2 to 20 min, drying time was from 3 to 10 min.

An anthraquinone derivative present in lichens, fungi, and medical plants is physcion. A HPTLC method of the rapid and simple quantification of the four major anthraquinone derivatives i.e., physcion, chrysophanol, emodin, and chrysophanol glycoside in *Rheum emodi* was described by Singh et al. [53].

A new device is presented which enables a suitable mobile phase flow rate to be used in circular development mode of classical preparative planar chromatography. A solvent reservoir, made of steel, and a rubber sealing ring are placed on the layer and fixed by a magnet located below the chromatoplate. To start the separation, an adsorbent is scratched from the center of the plate and the produced recess is filled with mobile phase. The device can be used for different types of chambers. The entry of sample and mobile phase is regular over the whole cross-section of the preparative layer, regardless of whether the sample is applied in liquid or solid form. The method and device presented ensure rapid, efficient separation with all the advantages of circular development. The resolution is significantly higher than that obtained from linear development.

HPTLC of anthraquinone derivatives was performed on precoated RP-18 F_{254} HPTLC plates. To achieve good separation, the mobile phase of methanol–water–formic acid (80:9:1, v/v/v) was used. The densitometric determination of anthraquinone derivatives was carried out at 445 nm in reflection/absorption mode. The method was found to be reproducible and convenient for quantitative analysis of anthraquinone derivatives in the methanolic extract of rhizomes of *Rheum emodi* collected from three different locations of Western Himalaya, India [96–98].

Rhein is the most popular anthraquinone in *Rheum* subsp. The dried rhizome of *Rheum officinale* Baill, or *R. palmatum* L. commonly known as rhubarb, is a historically popular Chinese preparation (Da-Huang) known to possess stomachic, antifebrile, and carminative properties and has been used for the treatment of bacterial dysentery and menstrual disorders. A RP column was used for analysis of rhein in a crude alcohol extract of methanol–0.5% aqueous phosphoric acid (60:40, v/v). A two-phase solvent system composed of n-hexane–ethyl acetate–methanol–water (3:7:5:5, v/v/v/v) was chosen for purification of rhein from the crude ethanolic extract.

Pharmacokinetic analysis of rhein in human plasma was conducted by feeding *R. undulatum* root extract to healthy volunteers [99]. RP-LC was done by spiking the plasma sample with standard rhein solution with use of a mobile phase composed of acetonitrile–methanol–Mc Ilvaine buffer (35:15:50, v/v/v) at pH 2.2. The presence of rhein in the plasma was also confirmed by a Quattro LC Triple Quadrupole tandem MS and application of positive and negative electrospray conditions.

A normal phase HPLC along with 5 μm Spherisorb-CN column (250 × 4.6 mm i.d.) was used for determination of the monomeric anthraquinones rhein, chrysophanol, emodin, and their glycosides present in rhubarb roots, senna leaves, and dock flower. The effects of various parameters were studied and the conditions for the extraction of the anthraquinones from herbal samples and their chromatographic separation were optimized [100].

Alizarin and analogs—TLC and a liquid chromatographic detection coupled with diode-array and mass spectrometry detection and LC-MS for the separation of both anthraquinone glycosides and aglycones in extracts of *Rubia tinctorum* roots was improved [101]. For on-line MS detection atmospheric pressure, chemical ionization as well as electrospray ionization (ESI) were used. The glucosides were ionized in both positive and negative ionization (NI) mode, the aglycones only in the NI mode. Several LC methods have been described for the characterization of *Rubia tinctorum* extracts [102–104]. Four glycosides (Figure 29.19): alizarin glucoside, lucidin glucoside, lucidin primeveroside, ruberithric acid and aglycones: alizarin, purpurin, pseudopurpurin, and munjistin have been reported from *Rubia tinctorum*. For extraction and sample preparation dried and powdered 3 year old root material of *Rubia tinctorum* was used. Lucidin primeveroside and ruberythric acid were the main anthraquinone components in the extract. These anthraquinones do not have laxative properties. Due to their substituents in 1, 2, 3, and 4 positions they have ability to form complex structures with Ca^{2+} cations. They were used in urolithiasis. The root and rhisomas of *Rubia cordifolia* and *Rubia tinctorum,* i.e., common madder, are rich sources of anthraquinone consisting mainly of alizarin and its analogs. Various uses of this plant have been reported such as remedy for ulcer, inflammation and swelling, poisoning due to snake and scorpion bite, indigestion, and paralysis [105]. Purpurin and lucidin have been found to possess mutagenic properties [106]. Nevertheless, the main interest in madder root is to produce the pigments, such as alizarin and lucidin, along with their respective glycosides

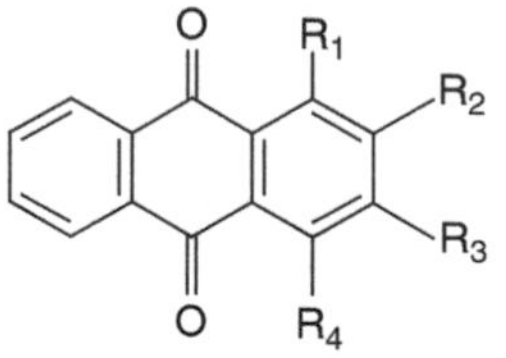

	R_1	R_2	R_3	R_4
Alizarin	OH	OH	H	H
Alizarin glucoside	OH	Glucose	H	H
Quinizarin	OH	H	H	OH
Purpurin	OH	H	OH	OH
Xanthopurpurin	OH	H	OH	H
Pseudopurpurin	OH	COOH	OH	OH
Lucidin	OH	CH_2OH	OH	H
Lucidin glucoside	OH	CH_2OH	Glucose	H
Lucidin primeveroside	OH	CH_2OH	O-β glu$^6 \rightarrow {}^1$xyl	H
Munjistin	OH	COOH	OH	H
Ruberythric acid	OH	O-β glu$^6 \rightarrow {}^1$xyl	H	H

FIGURE 29.19 Chemical structures of alizarin and analogs.

(ruberythric acid and lucidin primeveroside) [107]. Screening and quantitative estimation of these compounds were mainly based on TLC and low pressure CC [108,109].

Phytochemical studies on roots of *Morinda elliptica* (Rubiaceae)* have resulted in isolation of a new anthraquinone, 2-formyl-1-hydroxyanthraquinone (Figure 29.20) and ten known anthraquinones: 1-hydroxy-2-methylanthraquinone, nordamnacanthal, damnacenthal, lucidin-ω-methyl ether, rubiadin, soranjidiol, morindone, rubiadin- 1-methyl ether, alizarin-1-methyl ether, and morindone-5-methyl ether [55].

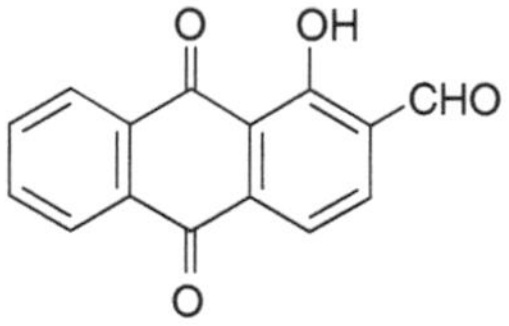

FIGURE 29.20 Chemical structure of 2-formyl-1-hydroxyanthraquinone.

---

* *Morinda elliptica* is a shrub or a small tree. Traditionally, different parts of the plant are used in various ways for a number of health problems and ailments including loss of appetite, headaches, cholera, diarrhea, fever, and hemorrhoids.

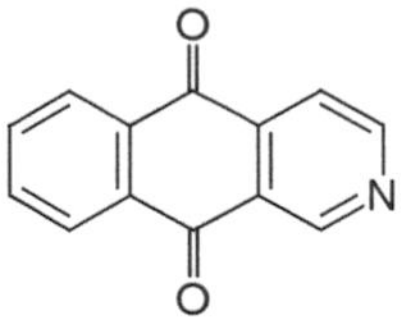

FIGURE 29.21 Chemical structure of 2-azaanthraquinone (AAQ).

All mentioned anthraquinones were also isolated by preparative TLC. Plates were developed with a mixture of CH_2Cl_2–petrol (2:3, v/v) or $CHCl_3$ and Me_2CO (9:1, v/v). The anthraquinones were identified by comparison of their spectroscopic data ^1H NMR and ^{18}C NMR spectra with literature values [46,110–113].

An interesting derivative of anthraquinone is 2-azaanthraquinone (AAQ) (benzo-[g] isoquinoline). This substance may explain the antimicrobial activity of this last extract and part of the effect of the alcoholic extract [54].

This compound (Figure 29.21) was isolated from *Mitracarpus scaber* grown in Benin. It has been found to have *in-vitro* antimicrobial activity against *Dermatophilus congolensis* (MIC 7.5 $\mu g \cdot mL^{-1}$) the consative agent of bovine dermatophilosis* [115–117]. To quantify this compound in *Mitracarpus scaber*, a new simple and rapid HPTLC method was developed and validated for selectivity, recovery, and repeatability. AAQ was separated on silica gel 60 F_{254} plates with mixture of toluene–ethyl acetate–methanol (80:18:2, v/v/v) mobile phase. For AAQ structure mass spectra in positive-ion mode were acquired with an LCQ mass spectrometer, equipped with an APCI source. Data acquisition and processing were performed with Xcalibur software. ^1H and ^{13}C NMR spectra in $CDCl_3$ were obtained on a Bucker 300.

29.5 OTHER QUINONES

From the basal stem of *Cyperus nipponicus* and *Cyperus distans* three quinone derivatives: cyperaquinone, remirol, and scabequinone were isolated (Figure 29.22).

The genus *Cyperus* includes common woods found in upland and paddy fields in temperate to tropical regions. In Asia countries Cyperaceae are used as traditional folk medicines.

Cyperaceae are seldom damaged by phytophatogous insects, because they contain insect antifeedants. In that paper, the identification by TLC and structural determination by ^1H and ^{13}C NMR of these quinones were reported [118].

Emodin, chrysophanol, physcion, knipholone (Figure 29.23), and two new dianthraquinones—10-hydroxy-10-(physcion-7'-yl)-chrysophanol anthrone and 5,10-dihydroxy-2-methyl-9-(physcion-7'-yl) 1,4 anthraquinone—were reported by Alemayehu et al. [44]. Knipholone was isolated from the genus *Senna* for the first time. The compounds were identified (by TLC method) on the basis of their color reactions with authentic samples and spectroscopic data (^1H NMR spectrum in $CDCl_3$).

---

* Bovine dermatophilosis is an enzootic skin infection of cattle in tropical and subtropical countries caused by the Gram-positive bacterium *Dermatophilus congulensis*. Infection may lead to death of the animal in severe acute cases and is difficult to cure, even by parenteral use of antibiotics [114].

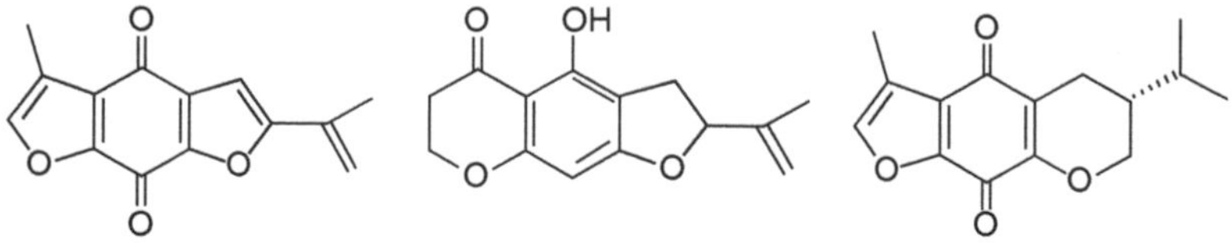

FIGURE 29.22 Chemical structures of cyperaquinone (1), remirol (2), and scabaquinone (3).

29.6 DIANTHRONES

Two molecules of anthranoids can be found in dianthron forms as isodianthrons (symmetric, two identical molecules in the structure) and heterodianthrons (not symmetric, two different anthron or anthranol molecules) (Figure 29.24).

- Sennidin A and B—is an isodianthron (build from two parts of rein) and is found in *Cassia acutifolia* and *Cassia angustifolia* (Caesalpinacea)
- Sennosides A and B—D-glucosides (8 and 8′) of sennidin A and B, main active bodies in *Folium Sennae*
- Sennosides C and D—heterodianthron glucosides of aloeemodin and rein
- Sennosides E and F—are oxalic derivatives of sennosides C and D which occur in roots and rootstocs of *Rheum Palmatum* (Polygonaceae) and in cortex of *Rhamnus purshiana* (Rhamnaceae)
- Palmidin A—heterodianthron of aloeemodin and emodin, occurs in roots and rootstocs of *Rheum Palmatum* (Polygonaceae)
- Palmidin B—heterodianthron of aloeemodin and chryzophanol, occurs in roots and rootstocs of *Rheum Palmatum* (Polygonaceae)
- Palmidin C—heterodianthron of chryzophanol and emodin, occurs in roots and rootstocs of *Rheum Palmatum* (Polygonaceae) and in cortex of *Frangula alnus* (Rhamnaceae)

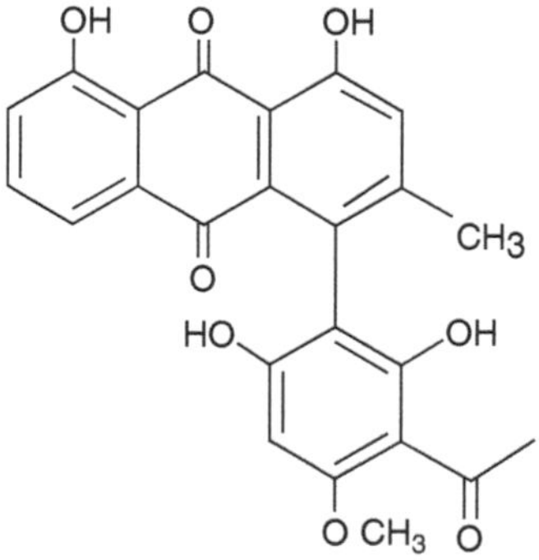

FIGURE 29.23 Chemical structure of knipholone.

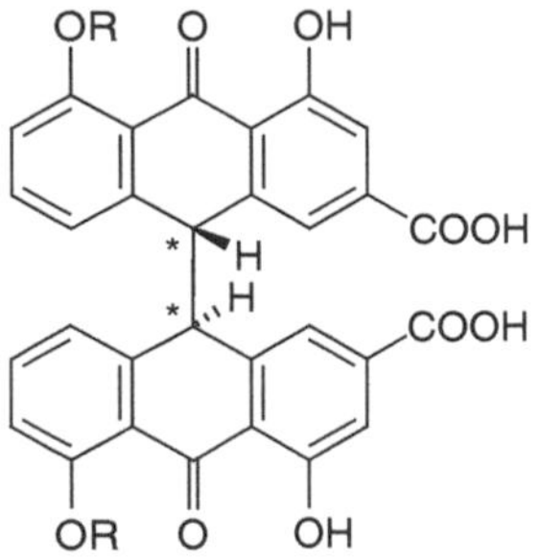

	R
Sennidin A (10,10′-*trans*)	H
Sennidin B (10,10′-*meso*)	H
Sennoside A (10,10′-*trans*)	Glu
Sennoside B (10,10′-*meso*)	Glu

FIGURE 29.24 Chemical structures of sennidins and sennosides.

- Palmidin D—heterodianthron of chryzophanol and fiscion, occurs in roots and rootstocs of *Rheum Palmatum* (Polygonaceae)
- Reidin A (heterodianthron of emodin and rein), reidin B (heterodianthron of chryzophanol and rein) and reidin C (heterodianthron of rein and fiscion)—occur in *Rheum* sp. (Polygonaceae).

Sennidin and sennosides from *Senna* was identified by TLC and HPLC methods [119]. Sennosides A and B present in *Cassia angustifolia* were analyzed on a Symmetry C_{18} column using a mobile phase composed of: methanol–water–acetic acid (20:80:0.1, v/v/v, pH 4) [120].

Sennoside A and B and monoanthraquinones: rhein, emodin, aloe-emodin, aloin, frangulin, 1,8-dihydroxy-3-methyl-anthraquinone formed standard mixture were analyzed by two-dimensional multiple development in HPTLC (Figures 29.25 and 29.26).

Author (for the first time) elaborated that method for separation and identification of plant complex extracts [7]. On the basiss of dependence $R_F = f$ [log % S] the composition of mobile phase to gradient elution was set. This chromatographical set was used to analyze plant extracts and anthraquinone derivatives were identified. Simultaneous separation of both groups of compounds (aglycones and glycoside derivatives) on a single plate can be achived by gradient elution (Figure 29.27) [8].

Dianthraquinones: cassiamin A (1,1′,3,8,8′-penthahydroxy-3′,6-dimethyl [2,2′-dianthracene]-9,9′,10,10′-tetrone) and chlorinate analogs (7-chloro-1,1′,6,8,8′ pentahydroxy-3,3′-dimethyl [2,2′-dianthracene]-9,9′,10,10′-tetrone) (Figure 29.28) were isolated from the root bark of *Cassia siamea* by Koyama et al. [45]. These compounds are minor constituents of *C. siamea*, and were isolated from fractions containing the main constituent, cassiamin A, by using preparative TLC on silica gel (Merck, 60 F_{254}, thickness 0.5 mm) and solvent systems of $CHCl_3$ and $CHCl_3 +$

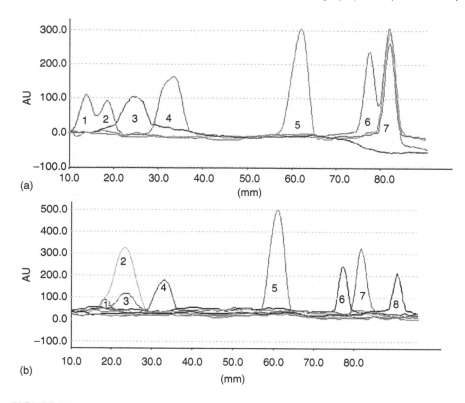

FIGURE 29.25 Densitogram of standard's mixture of quinone derivatives: 1—1,8-dihydroxy-anthraquinone, 2—franguloemodin, 3—aloeemodine, 4—rhein, 5—frangulin, 6—aloin, 7—sennoside B, 8—impurities remained at the start line. Two-dimensional HPTLC; HPTLC Si 60 (Merck, Germany) plate/hexane–dichlorometane–ethyl acetate–methanol. Spatial densitogram (three-dimensional setting) x offset 20%, y offset 70% (a); x offset 10%, y offset 80% (b). (From Matysik G., *J. Planar Chromatogr*. In press (a and b).)

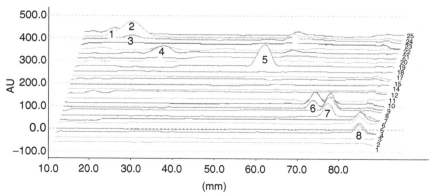

FIGURE 29.26 Densitogram of standard's mixture of quinone derivatives: 1—1,8-dihydroxy-anthraquinone, 2—franguloemodin, 3—aloeemodine, 4—rhein, 5—frangulin, 6—aloin, 7—sennoside B, 8—impurities remained at the start line. Two-dimensional HPTLC; HPTLC Si 60 (Merck, Germany) plate/hexane–dichlorometane–ethyl acetate–methanol. Spatial densitogram (three-dimensional setting) x offset 10%, y offset 80%. Overlay view. (From Matysik G., *J. Planar Chromatogr*. In press (a and b).)

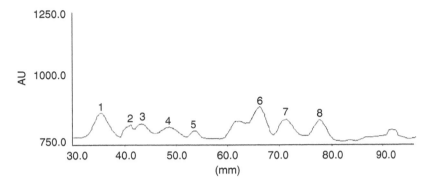

FIGURE 29.27 Densitogram of standard's mixture of quinone derivatives: 1—1,8-dihydroxy-anthraquinone, 2—franguloemodin, 3—aloeemodine, 4—rhein, 5—frangulin, 6—aloin, 7—sennoside B, 8—impurities remained at the start line. Gradient HPTLC on Si 60 (Merck, Germany). Eluent: dichloromethane–ethyl acetate–methanol. (From Matysik G., *J. Planar Chromatogr.* In press (a and b).)

MeOH, respectively. Identified and isolated chlorinated derivatives of bianthraquinone are extremely rare and unique in higher plants. They usually occur in lower plants, lichens, and fungi [120,45]. The structures were established by analysis—NMR:500 MHz for ^{1}H NMR and 125 MHz for ^{13}C NMR, CDCl$_3$, TMS as internal standard.

In the method of extraction, *C. siamea* was put in water and sequentially extracted partitioned with diethyl ether, chloroform, and ethyl acetate.

Hypericin (naphthodianthron) is a dark red dimer compound in *Hypericum perforatum* (Hypericaceae). It is well soluble in alcohol, poorly in water. *Hypericum perforatum* L. is an herbaceous perennial plant, belongs to Hypericaceae family,

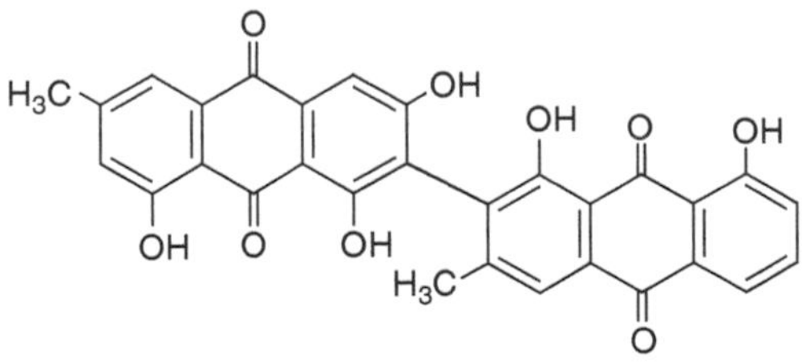

(a) Cassiamin A

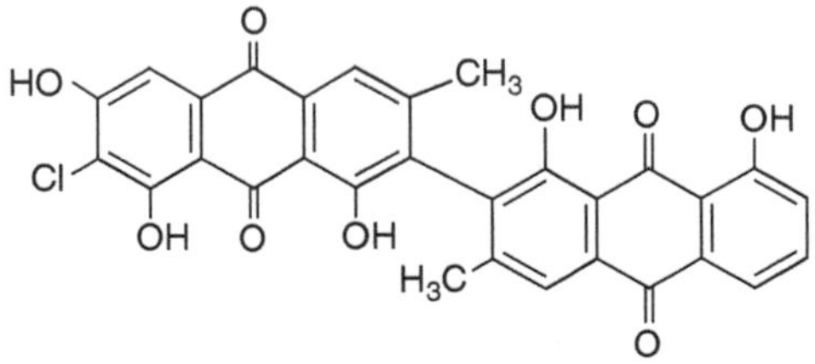

(b) Chlorinate analogue of cassiamin A

FIGURE 29.28 Structures of cassiamin A (a) and chlorinate analog (b).

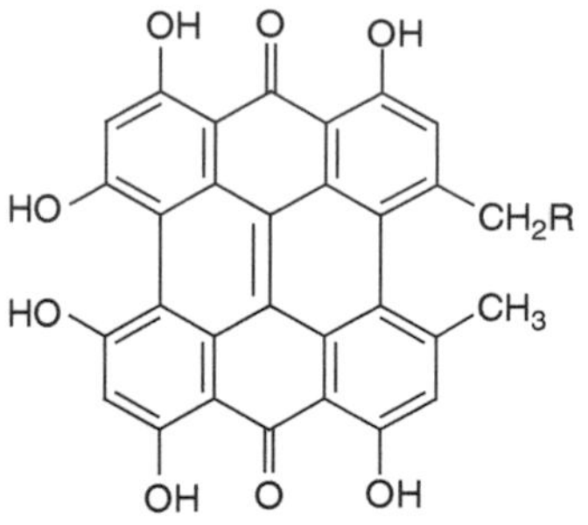

	R
Hypericin	H
Pseudohypericin	OH

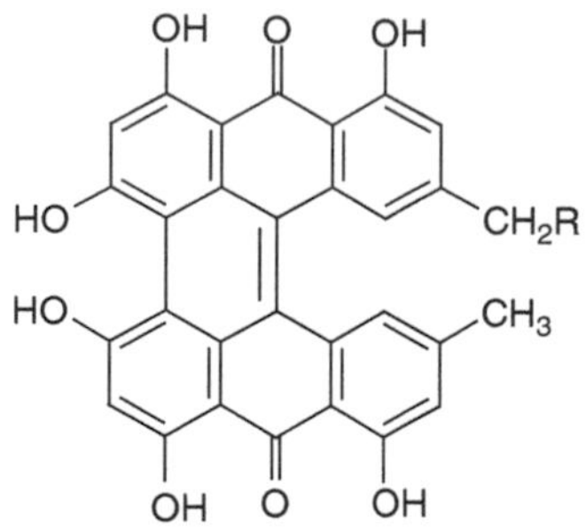

	R
Protohypericin	H
Protopseudohypericin	OH

FIGURE 29.29 Chemical structures of hypericins.

distributed in Europe, Asia, Northern Africa, and naturalized in the USA. It has been a well-known medicinal plant since the antiquity. It was recognized as Erba di S. Giovanni in Italy and St. John's Wort (SJW) in Anglo-Saxon folk medicine. The extract of aerial part in blossom had a high reputation as an anti-inflammatory, antidepressive, and healing agent since the Middle Ages. *H. perforatum* contains a number of substances with documented biological activity [121,122]. Various methods of extraction of SJW were applied for the analysis of hypericins (Figure 29.29) and its cell cultures.

Generally, HPLC analytical methods of *H. perforatum* preparations are concerned with the determination of hypericin and pseudohypericin content [56–59,123].

The US Pharmacopoeia has introduced the method developed by Brolis et al. with use of RP-HPLC elution with aqueous phosphoric acid–acetonitrile–methanol in a linear gradient program [124].

In the extract from *H. perforatum*, both hypericin and pseudohipericin were identified; the other active compounds: chlorogenic acid isomer, 3-*O*-*p*-coumaroylquinic acid, chlorogenic acid, rutin, hyperoside, isoquercetrin, 3,3′,4′,5,7-pentahydroxyflavanone 7-*O*-rhamnopyranoside, quercitrin, quercetin, I3, II8 biapigenin, hyperforin analog, hyperforin analog, hyperforin, adhyperforin were isolated (Figure 29.1).

The eluent system consisted of acidic phosphate buffers with gradients of methanol and acetonitrile. Usually RP columns (RP-18) with a particle size of not more than 5 μm are recommended. As the biological effects of *Hypericum perforatum* are considered arising rather from the whole mixture of the above metabolites, than from the presence of a single constituent the availability of a method allowing the analysis of the entire extract is desirable.

Profile of *Hypericum perforatum* extract with the HPLC-MS attributions of the components was determined (Figure 29.1). For the LS-TSP-MS analysis using a

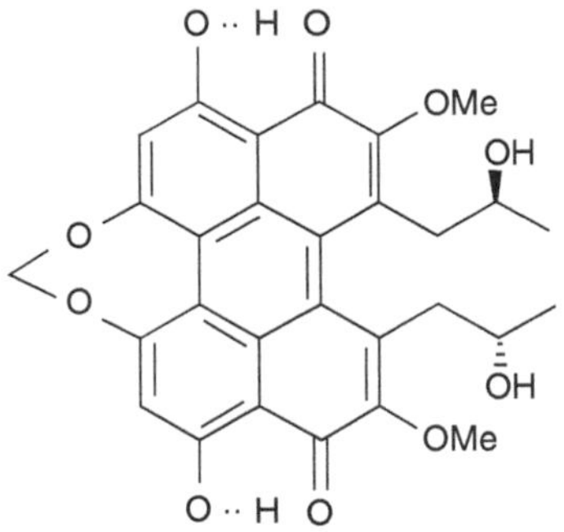

FIGURE 29.30 Cercosporin—a secondary metabolite produced by *C. beticola*.

Finnigan MAT TSQ-700 triple quadrupole mass spectrometer equipped with a 5100 DEC station was used.

The phytopathogenic fungus *Cercospora beticola* is responsible for the main leaf spot disease on sugar beet called cercosporiose. It produces colored secondary metabolites with complex structures. One of them is a red compound named cercosporin; it is a perylenequinone (Figure 29.30). The structure of cercosporin was confirmed using LC-MS method. Cercosporin is a photodynamically active compound which can induce lipid peroxidation and membrane damage. From the extract four bright yellow compounds were isolated. These compounds were identified and named by the authors as beticolins. They represent the leading members of xanthraquinones, a known class of natural toxins which can be divided into two subclasses according to the type of cyclization of the heterocycle. *C. beticola* secondary metabolites were extracted from the mycelium. TLC analyses were useful for monitoring the purity and stability during the different purification steps [46,47].

29.7 PHENANTHRAQUINONES

Denbinobin and moniliformin belong to this group of compounds (Figure 29.31).

Denbinobin was isolated from *Dendrobium moniliforme* Lindl. (Orchidaceae) and related species widely available in Japan, China, and Australia [46]. The plants were historically used for herbal preparations like Shih-hu in China. Denbirobin, also

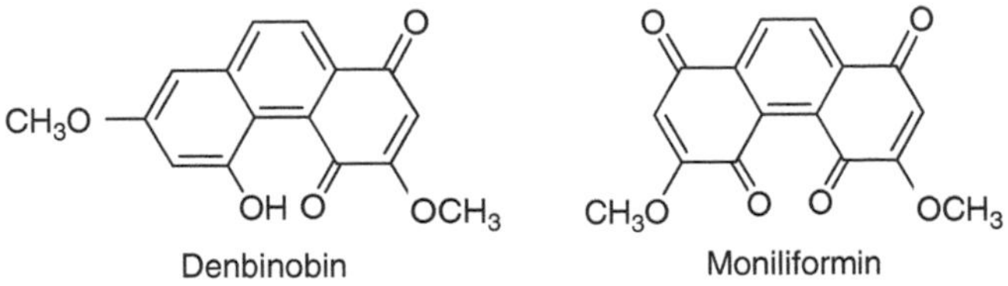

Denbinobin   Moniliformin

FIGURE 29.31 Chemical structures of phenanthraquinones.

isolated from the stems of *Ephamerantha lanchophylla*, induced apoptosis in human colon cancer HCT-116 cells [47].

There is only one report on its HPLC analysis using a Comosil 5C-18 AR column and mobile phase of water–acethonitrile (50:50, v/v) [125].

REFERENCES

1. Bush K.L., in J. Sherma and B. Fried (editors) *Handbook of Thin-Layer Chromatography*, Marcel Dekker, New York, NY, 183–256, 1991
2. Bush K.L., *Trends Analyt.Chem.* 11, 314–319, 2002
3. Budzianowski J., *Phytochemistry* 42, 1145–1147, 1996
4. Budzianowski J., *Phytochemistry* 40, 1145–1148, 1995
5. Scholly T. and Kapetanidis I., *Planta Med.* 55, 611–617, 1989
6. Wawrzynowicz T., Waksmundzka-Hajnos M., and Mulak-Banaszek K., *J. Planar Chromatogr.* 7, 315–318, 1994
7. Matysik G., Application of the new technique—two dimensional multiple development—in TLC anthraquinones analyze, *J. Planar Chromatogr.* In press(a)
8. Matysik G., Application of gradient HPTLC to quinon aglycones and glycoside derivatives analyze. *J. Planar Chromatogr.* In press(b)
9. Wagner H. and Bladt S., *Plant Drug Analysis, A Thin Layer Chromatography Atlas*, 2nd edition, Springer, Heidelberg, pp. 53–73, 1996
10. Botz L., Nyiredy Sz., and Sticher O., *J. Planar Chromatogr.* 3, 401–406, 1990
11. Zogg G.C., Nyiredy Sz., and Sticher O., *J. Planar Chromatogr.* 1, 351–354, 1988
12. Brolis M., Gabetta B., Fuzzati N., Pace R., Panzeri F., and Peterlongo F., *J. Chromatogr. A* 825, 9–16, 1998
13. Zuber G.E., Warren R.J., Begosh P.P., and Donell E.L., *Anal. Chem.* 56, 2935–2939, 1984
14. Otto A., Bode U., Heise H.M., and Fresenius Z., *Anal. Chem.* 331, 376–379, 1988
15. Bush S.G. and Breaux A.J., *Mikrochim. Acta* I, 17–21, 1988
16. Beauchemin B.T. and Brown P.R., *Anal. Chem.* 61, 615–617, 1989
17. Fuller M.P. and Griffiths P.R., *Appl. Spectrosc.* 34, 533, 1980
18. Koyama J., Morita I., Tagahara K., and Aqil M., *Phytochemistry* 56, 849–851, 2001
19. Ali M., Aimi N., Kitajima N., Takayama H., and Lajis N.H., *Phytochemistry* 45, 1723–1725, 1997
20. Stepanenko L.S., Krivoshchekova O.E., Dmitrenok P.S., and Maximov O.B., *Phytochemistry* 46, 565–568, 1997
21. Alemayehu G., Hailu A., and Abegaz B.M., *Phytochemistry* 42, 1423–1425, 1996
22. Morimoto M., Fujii Y., and Komai K., *Phytochemistry* 51, 605–608, 1999
23. Fukui H., Hassan A.F.M.F., and Kyo M., *Phytochemistry* 51, 511–515, 1999
24. Smolarz H.D., Medyńska E., and Matysik G., *J. Planar Chromatogr.* 18, 319–322, 2005
25. Derksen G.C.H., Niederlander, H.A.G., and van Beek T.A., *J. Chromatogr A* 978, 119–127, 2002
26. Milat M.L. and Blein J.M., *J. Chromatogr. A*, 699, 277–283, 2002
27. Kohlmunzer S., Farmakognozja, and Warszawa, *Wydawnictwo Lekarskie PZWL*, 2003, 316–328
28. Driscoll J.S., Hazard G.F., Wood H.B., and Goldin A., *Cancer Chemother. Rep.* 4, 1, 1974
29. Silva D.H., Pereira F.C., Zanoni M.V.B., and Yoshida M., *Phytochemistry* 57, 437–442, 2001

30. Silva D.H., Davino S.C., Barros S.M.B., and Yoshida M., *J. Nat. Prod.* 62, 1475–1478, 1999
31. Mahmood U., Kaul V.K., and Jiorvetz L., *Phytochemistry* 61, 923–928, 2002
32. Caniato R., Filippini R., and Cappelletti E., *Int. J. Crude Drug Res.* 27,129–132, 1989
33. Kamavainen T., Unsitalo J., Jalonen J., Laine K., and Hohtola A., *Phytochemistry* 63,309–312, 2003
34. Mallavadhani U.V., Panda A.K., and Rao Y.R., *Phytochemistry* 49,901–904, 1998
35. Sanyal U., Bhattacharyya S., Patra A., and Hazara B., *J. Chromatogr. A* 1017, 225–231, 2003
36. Budzianowski J., Skrzypczak L., and Kukułczanka K., *Acta Hortic.* 330, 277–279, 1993
37. Lillie T.J., Musgrave O.C., and Skoyles D., *J. Chem. Soc. Perkin Trans.* 1, 2155–2159, 1976
38. Yoshihira K., Tezuka M., and Natori S., *Chem. Pharm. Bull.* 19, 2308–2311, 1971
39. Ferreira M.A., Cruzcosta M.A., Correia Alves A., and Lopes M.H., *Phytochemistry* 13, 1587–1590, 1974
40. Hazra B., Sur P., Roy D.K., Sur B., and Banerjce A., *Planta Med.* 51, 295–299, 1984
41. Vijver L.M. and Gerritsma K.W., *J. Chromatogr.* 114, 443–447, 1975
42. Ganapaty S., Thomas P.S., Fotso S., and Laatsch H., *Phytochemistry* 65, 1265–1271, 2004
43. Marston A. and Hostettmann K., *J. Chromatogr.* 295, 526–530, 1984
44. Alemayehu G., Hailu A., and Abegaz B.M., *Phytochemistry* 42, 1423–1425, 1996
45. Koyama J., Morita I., Tagahara K., and Aqil M., *Phytochemistry* 56, 849–851, 2001
46. Leistner E., *Planta Med. Suppl.* 214, 277–281, 1975
47. Papageorgiou V.P., Assimopoulon A.N., Couladouros E.A., Hepworth D., Nicolaon K.C., *Angew. Chem. Int. Ed. Engl.* 38, 270–276, 1999
48. Wolfender J.L., Rodriguez S., and Hostettmann K., *J. Chromatogr. A* 794, 299–303, 1998
49. Mueller S.O., Schmitt M., Dekant W., Stopper H., Schlatter J., Schreier P., and Lutz W.K., *Food Chem. Toxicol.* 37, 481–491, 1999
50. Kumar S.P., Srinivas P.V., and Madhusudana Rao J., *J. Planar Chromatogr.* 15, 128–131, 2002
51. Lemli I. and Cuveele I., *Pharm. Acta Helv.* 42, 37–40, 1967
52. Metzger W. and Reif K., *J. Chromatogr. A* 740, 133–138, 1996
53. Singh N.P., Gupta A.P., Sinha A.K., and Ahuja P.S., *J. Chromatogr. A* 1077, 202–206, 2005
54. Gbaguidi F., Muccioli G., Accrombessi G., Moudachirou M., and Quetin-Leclercq J., *J. Planar Chromatogr.* 18, 377–379, 2005
55. Ismail N.H., Ali A.M., Aimi N., Kitajima M., Takayama H., and Lajis N.H., *Phytochemistry* 45, 1723–1725, 1997
56. Bombardelli E. and Morazzoni P., *Phytoterapia* 66, 43, 1995
57. Holzl J., Sattler S., and Shutt H., *Pharm. Ztg.* 139, 3959, 1994
58. Brolis M., Gabetta B., Fuzzati N., Pace R., and Peterlongo F., *J. Chromatogr. A* 825, 9–16, 1998
59. Lavie G., Marzur Y., Lawie D., and Merelo D., *Med. Res. Rev.* 15, 111, 1995
60. Xu B., *Mem. Inst. Oswaldo Cruz* 86 (Suppl. 2), 51, 1991
61. Han Y., *J. Ethnopharmacol.* 24, 1–4, 1988
62. Han R., *Stem Cells* 12, 53–58, 1994
63. Jin S. and Sato N., *Phytochemistry* 62, 101–107, 2003
64. Schultes R.E. and Holmstedt B., *Lioydia* 34, 61–78, 1971

65. Pratt D.E. and Miller E.E., *J. Am. Oil Chem. Soc.* 61,104–1067, 1984
66. Girzu M., Fraisse D., Carnat A.P., Carnat A., and Lamaison J.L., *J. Chromatogr. A* 805, 315–321, 1998
67. Thomson R.H., *Naturally Occuring Quinines*, Academic Press, London, 1971
68. Weiss R.F., *Herbal Medicine*, 6th edition, Beaconsfield Publishers, Beaconsfield, p. 210, 1991
69. Shih-Chen L., *Chinese Medicinal Herbs*, Dover Publications, New York, NY, 2003
70. Fujita Y., Maeda Y., Suga C., and Morimoto T., *Plant Cell Rep.* 2,192–196, 1983
71. Nutter L.M., Cheng A.L., and Hung H.L., *Biochem. Pharmacol.* 41, 1283–1287, 1991
72. Wu F.Y., Liao W.C., and Chang H.M., *Life Sci.* 52, 1797–1799, 1993
73. Lamson D.W., and Plaza S.M., *Altern. Med. Rev.* 8, 303–309, 2003
74. Suttie J.W., *FASEB J.* 7,445–447, 1993
75. Shearer M.J., *Proc. Nutr. Soc.*, 56, 915–918, 1997
76. Donnahey P.L., Bart V.T., Rees H.H., and Pennock J.F., *J. Chromatogr.* 170, 272–276, 1979
77. Ravishankara M.N., Shrivastava N., Jayathirtha M.G., Padh H., and Rajani M., *J. Chromatogr. B* 744, 257–261, 2000
78. Stepanenko L.S., Krivoshchekova O.E., Dmitrenok P.S., and Maximov O.B., *Phyto-chemistry* 46, 565–568, 1997
79. Rodriguez S., Wolfender J.-L., Hakizamungu E., and Hostettmann K., *Planta Med.* 61, 362–368, 1995
80. Koyama J., Toyokuni I., and Tagahara K., *Phytochem Anal.*8, 135–139, 1997
81. Stuppner H. and Sturm S., *Chromatographia* 42, 697–707, 1996
82. Shafer K.H. and Griffiths P.R., *Anal. Chem.* 58, 2708–2714, 1986
83. Hattori M., Akao T., Kobashi K., and Namba T., *Pharmacology* 47, 125, 1993
84. Rai P.P. and Turner T.D., *J. Pharm. Pharmacol.* 26, 722, 1974
85. Li Y., Liu H., Ji X., and Li J., *Electrophoresis* 21, 3109–3115, 2000
86. Liu C., Zhu P.L., and Liu M.C., *J. Chromatogr.* 857, 167–174, 1997
87. Kuo Y.C., Tai W.J., and Meng H.C., *Life Sci.* 68, 1271–1286, 2001
88. Zang L., Lau Y.K., and Xia W., *Clin. Cancer Res.* 5, 343–353, 1999
89. Lemli I. and Cuveele I., *Planta Med.* 26, 193–198, 1974
90. Koch A. and Müller M., *J. Planar Chromatogr.* 9, 56–61, 1996
91. Hattori M., Kanda T., Shu Y., Akao T., Kobashi K., and Namba T., *Chem. Pharm. Bull.* 36, 4462, 1988
92. Rauwald H.W., *Pharm Ztg. Wiss.* 135, 169, 1990
93. Kiridena W., Poole S.K., Miller K.G., and Poole C.F., *J. Planar Chromatogr.* 8, 416–419, 1995
94. Reynolds T., *Bot. J. Linn. Soc.* 90, 157–177, 1985
95. Smith W.T., *Laboratory Information Bulletin, Nos. 3865,* US Food and Drug Administration, Washington, DC, USA, 1995
96. Kapoor L.D., *Handbook of Ayurvedic Medicinal Plants*, CRC Press, Boca Raton, FL and London, p. 487, 1990
97. Krenn L., Presser A., Pradhan R., Bahr B., Paper D.H., Mayer K.K., and Kopp B., *J. Nat. Prod.* 66, 1107, 2003
98. Krenn L., Pradhan R., Presser A., Reznicek G., and Kopp B., *Chem. Pharm. Bull.* 52, 391, 2004
99. Lee J.H., Kim J.M., and Kim C., *J. Ethnopharmacol.* 84, 5, 2003
100. Djozan D. and Assadi Y., *Talanta* 42, 861–865, 1995
101. Derksen G.C.H., Niederlander H.A.G., and Beek T.A., *J. Chromatogr. A* 978, 119–127, 2002

102. Brolis M., Gabetta B., Fuzzati N., Peace R., Panzeri F., and Peterlongo F., *J. Chromatogr. A* 825, 9, 1998
103. He X., Lin L., and Lian L., *J. Chromatogr. A* 755, 127, 1996
104. Wolfender J.-L., Roderiguez S., and Hostettmann K., *J.Chromatogr A.* 794, 299, 1998
105. Shih-Chen L., *Chinese Medicinal Herbs*, Dover Publications., New York, NY, 2003
106. Mori H., Ohnishi M., Kawamori T., Sugie S., Tanaka T., Ino N., and Kawai K., *Cancer Lett.* 102, 193–199, 1996
107. Masawaki T., Taya M., and Tone S., *J. Ferment. Bioeng.* 81, 567–571, 1996
108. Rai P.P. and Shok M., *Chromatographia* 14, 599–602, 1981
109. Saronius K., *Farm. Aikak.* 81, 85–88, 1972
110. Burkill I.H., *Gard. Bull. Straits Settlement* 6, 167, 1930
111. *Dictionary of Natural Products on CD-Rom, Release 4.1.* Chapman and Hall, London, 1995
112. Chang P. and Lee K.H., *Phytochemistry* 23, 1733–1736, 1984
113. Adesogan E.K., *Tetrahedron* 29, 4009–4012, 1973
114. Ali N., Moudachirou M., and Akakpo J.A., *Rev. Elev. Med.vet. Pays Trop.* 55, 183–187, 2002
115. Adjanonhoun E.J., *Rev. Med. Pharm. Afri.* 15, 103–111, 2001
116. Ependu T.O., Akah P.A., Adesomoju A.A., and Okogum J.I., *Int. J. Pharmacognosy* 32, 191–196, 1994
117. Okunade L.A., Clark M.A., Hufford D.C., and Oguntimein O.B., *Planta Med.* 65, 447–448, 1999
118. Morimoto M., Fujii I., and Komai K., *Phytochemistry* 51, 605–608, 1999
119. Metzger W. and Reif K., *J. Chromatogr. A* 740, 133–136, 1996
120. Bala S., Uniyal G.C., Dubey T., and Singh S.P., *Phytochem. Anal.* 12, 277–281, 2001
121. Gill M. and Steglich W., *Progress in the Chemistry of Organic Natural Products*. Vol 51, Springer, Vienna, p. 1, 1987
122. Turner W.B. and Aldridge D.C., *Fungal Metabolites II*, Academic Press, London, 1983
123. Holz J. and Ostrowski E., *Dtsch. Apoth. Ztg.* 23, 1227, 1987
124. Haberlein H., Tshiersch K.P., Stock S., and Holzl J., *Pharm. Ztg. Wiss.* 137, 169, 1992
125. Kramer W. and Wiartalla R., *Pharm. Ztg. Wiss.* 137, 202, 1992

Index

A

Abietane, structure, 483
Abietane-type diterpenes, pharmacologic
    role of, 487
Abortifacient isocupressic acid, 62
Abraham's model, of solvent properties, 139
Abrus precatorius, 356
Abscisic acid and derivatives, 71
Accacia prominens, 702
Accelerated solvent extraction (ASE), 378;
    see also Pressurized liquid
    extraction (PLE)
Acetic anhydride reagent, 184
Acetic anhydride sulfuric acid copper-reagent, 184
Acetic anhydride sulfuric acid reagent
    (Liebermann–Burchard reagent), 184
Acetonitrile, 136
Acetylated sterols, 115
Acetylcholinesterase (AChE) inhibitors,
    631–632
Acetylenic carotenoids, 566–567
Acetylsalicylic acid, UV densitogram obtained by
    infusion method, 743
Achillea species, 476, 477
    sesquiterpenes (guaianolides) as lead of
        substances, 477
Acidic cation exchanger, 114
Aconitine alkaloids
    chromatograms of Aconiti tuber extracts, 747
    separation of, 746–747
Aconitum, 45
Active pharmaceutical ingredients (API), 51
Acyclic monoterpenes, 453
Acyclic sesquiterpenes, 454
Adrenaline-like drugs, 72
Adsorption or liquid–solid chromatography, 136
Adulterants and mixtures, detection of, 50–51
Adulteration demonstration, in volatile oils
    application of TLC to, 477
    Anise/Fennel Oil, 477
Aesculus hippocastanum, 17, 376, 396, 418
Aflatoxins, 105, 241
    separation of, 51
Agardiffusion technique, disadvantages of, 534
Agathosma betulina (Berg.) Pillans
    (Rutaceae), 395
Aglycone fingerprints, 411

Alizarin and analogs, 839
    structures of, 840
Alkaloids, 5, 16, 63, 66, 89–90
    classes of, 64
Allelochemicals, 66
Allelopathy, 3
Allethrins, 66
Aloe, 83, 838
    anthraquinones, 820
    micropreparative TLC of, 837
Aloe-emodin, 836
Aloe ferrox or *Aloe berbandense*, 836
Aloin, structure, 836
Alpinia calcarata, 487
Alumina, 29, 113, 136
American Herbal Pharmacopoeia (AHP), 39
Amino acids, 62–63
    chromatographic techniques
        adsorbents and thin layers, 310–311
        detection on thin layer chromatograms,
            312–313
        development of chromatograms, 311
        preparation of thin plates, 311
    coded or primary protein, 300
    of different palms, 320
    noncoded or nonprotein, 300
    occurrence, 300–307
    paper chromatography, 321–324
    of phloem exudates of palm, 318
    in phloem sap from coconut, 319
    separation of
        method for acid hydrolysis, 309
        method for sulfur-containing amino acids,
            309–310
        preparation of test materials, 308
    TLC analysis
        in combination with other techniques,
            318–321
        plant extracts, 313–317
        soybean and kidney bean, 317
        from wine, 321
Amino bonded layers, 111
γ-Aminobutyric acid, 317
Aminoethyl (AE), 114
Aminopropyl, 136
Ammi visnaga L., 376
Ammonium vapor (NH_3), 184

Milton Keynes UK
Ingram Content Group UK Ltd.
UKHW031536071024
449327UK00023B/1841